I0044637

Evolution of the Alabama Agroecosystem

EVOLUTION OF THE ALABAMA AGROECOSYSTEM

ALWAYS KEEPING UP, BUT NEVER CATCHING UP

EDDIE WAYNE SHELL

PROFESSOR EMERITUS
AUBURN UNIVERSITY

NEWSOUTH BOOKS
Montgomery

NewSouth Books
105 S. Court Street
Montgomery, AL 36104

Published in the United States by NewSouth Books, a division of NewSouth, Inc., Montgomery, Alabama.

Library of Congress Cataloging-in-Publication Data

Shell, Eddie Wayne
Evolution of the Alabama Agroecosystem : Always Keeping Up, but Never Catching Up / Eddie Wayne Shell.
p. cm.
Includes bibliographical references and index.

ISBN 978-1-60306-203-9 — ISBN 1-60306-203-3
1. Agricultural ecology—Alabama—History. 2. Agriculture—Alabama--History. I. Title.
S451.A2S54 2013
577.5'509761--dc23

2013026127

Printed in the United States of America

About the Cover Illustration

Local Agriculture–A.A.A 1939 was painted by Aldis Browne in 1939. The mural is one of the hundreds of works of art commissioned during the Depression by the Treasury Section of Fine Arts for US post offices and other government buildings. Browne's mural still graces the former post office in Oneonta, Alabama. The building is now used by the local board of education.

Aldis Birdseye Browne II (1907–1981) was born in Washington, DC, but is most associated with his adopted Connecticut. In 1934, he earned his BFA from Yale School of Fine Arts. His professional career began under the Treasury Art Project and Public Works Art Project, two Depression Era programs to employ out-of-work artists. After the Oneonta TRAP project, he spent seven months at sea painting the International Ice Patrol in action. He also painted murals for the US Coast Guard Academy and so impressed that service, that he received an honorary USCG commission. In 1981, he died in Florida after a long, fruitful career as an artist.

Unlike most Treasury Section post office art works, which were painted on canvas and then mounted on the wall, the Oneonta piece is actually a mural that Browne painted in situ. The local postmaster and citizens of Blount County suggested specific subjects to be included in the painting, and they frequently stopped by to follow Browne's progress. The artist, who was only four feet tall and a Northerner, fascinated the locals, and he developed an affinity for them. In fact, he reported to the Treasury Section that he was "nuts" about Oneonta. Browne not only accommodated the public's suggestion, but he also included such things as terraced field in contrast to an eroded one—a benefit of federal Depression-era agricultural programs.

The photograph of Browne's mural was taken in 2009 by Jimmy S. Emerson, DVM. A native of Hamilton, Alabama, Emerson is a graduate of Auburn University's College of Veterinary Medicine. He lives and works in Dalton, Georgia. Dr. Emerson is dedicated to photographic documentation of aspects of vanishing rural and small town America. Of the approximately 1,200 artworks installed across the country under the Section of Fine Arts program, around 1,000 still exist, and Emerson has managed to photograph 80 percent of those. He continues to travel the country to locate and photograph the remaining post office artworks. He posts pictures at www.flickr.com/groupos/pomurals to bring attention to the Section artworks to people across the country.

To Eugene Butler and Emory Cunningham,

of *Progressive Farmer*

It is unconventional to elaborate on a dedication, but this book is itself unconventional, and an elaboration is called for so that the reader will understand the importance of these two giants in the age-old effort to provide useful information to farmers and their families.

In 1760, Jared Eliot published *Field Husbandry in New England,* the first known book on agriculture in the colonies. In the late nineteenth and early twentieth centuries, literature related to American agriculture literally exploded. The first issue of the *Old Farmers Almanac* came off the press in 1792. This period also saw the appearance of the first periodicals on the subject. The *American Farmer* began publication in Baltimore in 1819, and the *New England Farmer and Horticultural Journal* in 1822.

The *Progressive Farmer* began in 1886 in Winston, North Carolina (Faulkner, 2008). Founded by Leonidas L. Polk, it was initially devoted to North Carolina agricultural issues. In 1899, an 18-year old farm boy, Clarence H. Poe, became editor. In 1908, the publication changed from a weekly newspaper to a magazine with separate editions for the Southeast and Mid-South regions. By 1911, the magazine was growing so rapidly across the South that the main office for the publication was relocated to Birmingham, Alabama.

In 1917, Eugene Butler, with a new master's degree from Iowa State University in his resume, joined his father, Tait, as assistant editor of *Progressive Farmer* at its new office in Memphis (Wright, 1979). In 1922, Butler was appointed editor of the Texas edition of the magazine. Then with his new wife, the former Mary Britt Burns of Memphis, he went to Dallas to try to breathe new life into a magazine that was running a poor second to *Farm and Ranch,* the dominant agricultural periodical in the region at that time.

Under Butler's leadership, the Texas edition of *Progressive Farmer* published the usual material that farmers expect in their farm magazines, but it was his sharply-focused, tightly-written editorials on shady agricultural supply practices, improving soil fertility, boll weevil control, livestock nutrition and health, forage crops, efficient farm management, and rural health that caught their attention. With an obvious concern for the totality of farm life (the agroecosystem), Butler's Texas edition of *Progressive Farmer* finally pulled even with *Farm and Ranch* and then left them eating Texas dust. With his success, it was not surprising that Clarence Poe chose him as president of the *Progressive Farmer* Company when he retired in 1953. Although the company's office was still in Birmingham, Butler continued to live in Dallas. Under his leadership, *Progressive Farmer* absorbed a number of smaller farm papers and magazines, and its circulation quickly reached 1.4 million. In the 1930s, it was emerging as one of the leading farm periodicals in the South.

Butler brought to the presidency the same insight and hard-work that fueled the success of the Texas edition, but the world was changing rapidly. World War II and the Korean Conflict were in the recent past, and American agriculture was downsizing to meet the new realities of supply and demand in food and fiber production. Between 1950 and 1960, the country lost a third of its farms. At the same time, advertising, the life-blood of magazine publishing, began to fall rapidly. Television was becoming the chosen media for advertising investment. Although farm magazines were suffering from the loss of advertising dollars, they were still playing a central role in the rapid modernization of American agriculture. Modernization was even more significant in the South than in the country as a whole. For example, between

1950 and 1959 the number of tractors per 100 American farms increased by 81%, but in Alabama the increase was 171%. Increases in the other Southern states were of similar magnitude. The gospel of farm modernization advocated by *Progressive Farmer* was in the forefront of this major change in Southern farming.

Butler was not only an excellent agricultural writer, but he was also an astute businessman. As the advertising situation deteriorated, he realized that to survive *Progressive Farmer* Company would have to diversify. In 1958, the company bought into the insurance business, and at the same time it purchased Oxmoor House, a commercial publishing firm in Birmingham.

Emory Cunningham joined the *Progressive Farmer* organization as an advertising salesman in 1948–just in time to encounter the beginning of the decline in magazine advertising revenue. Cunningham, son of a Walker County, Alabama, farmer and himself a veteran Navy pilot, had just earned a degree in agriculture from the Alabama Polytechnic Institute.

With his treating-readers-like-neighbors approach, Cunningham was a natural at selling advertising. In 1960, the company soon recognized Cunningham's contributions by making him marketing director. In 1963, likely with significant input from Cunningham, the company decided to discontinue the long-time practice of selling subscriptions to non-farm families. Although these families subscribed to *Progressive Farmer*, they purchased few of the products that the advertisers were trying to sell. This new policy quickly resulted in declining circulation, but it pleased the company's advertisers. Unfortunately, it also meant that the magazine's circulation would be tied to the rapidly declining farm population in the South.

Butler and Cunningham obviously agreed that a farm magazine, to remain effective, should be published for farm families. However, they were also concerned about the increasing number of Southern families, many with farm backgrounds, that *Progressive Farmer* was leaving behind. Consequently, in 1964, Cunningham recommended that the company begin publishing a magazine for urban, suburban, and rural non-farm families. A year later, the company began to publish the highly successful *Southern Living*. Four years later, in 1968, Cunningham became president of the company.

For much of its history, *Progressive Farmer* had been a farm publication of the Southeast, Mid-South, and Southwest, but in the early 1980s its board expanded the magazine's scope by authorizing publication of a regional edition in 12 Midwestern states. With this move, the magazine achieved national prominence and a circulation over 700,000.

In 1984, Eugene Butler, the old Mississippi A & M town boy, retired as one of the country's most distinguished farm journalists and chairman of the board of one of its most prosperous publishing firms. Emory Cunningham was chosen to succeed him. Under Cunningham's leadership, Progressive Farmer Company became Southern Progress Corporation. The company was later purchased by Time, Inc., but Southern Progress continued to operate as a stand-alone unit. Cunningham retired in 1987.

Both Butler (d. 1995) and Cunningham (d. 2000) are gone, their pens forever stilled, but their labor of love lives on. At the beginning of the twentieth-first century, *Progressive Farmer* consists of a cluster of state magazines with the same name in the South and Midwest. With 650,000 subscriptions from throughout the country, it is estimated that each issue reaches 3.5 million readers. In 2007, the American Agricultural Editors Association commented that the magazine "offers a blend of practical lifestyle and agricultural information to meet the needs of people who own land and live in rural America." It is "the largest publication focusing on the information needs of the growing rural lifestyle market."

The cover page of current issues of *Progressive Farmer* includes these words: "Your Farm, Your Land, Your Life." The words accurately embody the philosophies of Eugene Butler and Emory Cunningham regarding the role of their magazine in rural America. They dedicated their careers to improve all aspects of Southern agriculture by providing up-to-date information to farmers and their families. In recognition of their dedication and efforts, it is a privilege to dedicate this book to them.

Contents

Preface

I feel that it would be helpful for those contemplating the investment of a considerable amount of their time in reading this lengthy work to understand my perspective, what I brought to the table in writing it.

I was born in an old, wood-frame house, half-way between the Old-Mill-Hill and Brushy Creek communities in south-central Butler County, Alabama, on June 16, 1930. Although I could not know it at the time, I was making my debut in an especially trying period in the rolling disaster of Alabama agriculture. The Great Depression was just getting underway, and much of the United States was entering one of the hottest-driest periods in American history–the beginning of the Dust Bowl.

By the time of my birth, my paternal grandfather had lost a fine farm to boll weevils, poor management, and the very difficult economics of attempting to produce cotton by using tenant farm labor.

My maternal grandfather had inherited an 80-acre farm in the southern half of the northeast quarter of Section 18, Township 8, Range 14 east, off the St. Stephens Base Line (south central Butler County). Most of the soil on the farm was Luverne sandy loam with 5 to 8% slopes, although there was some Orangeburg in one corner and a substantial amount of the Mantachie, Bibb, and Iuka Complex in two broad creek bottom. The Luverne sandy loam (with 5 to 8% slopes) series has a Capability Class of IVe; consequently, it is poorly suited for cultivated crops. It is characterized by low fertility, low levels of organic matter, poor tilth, and a high erosion hazard.

A substantial share of the total acreage on his farm was so wet during most of the year that it was suitable only for open pasture and woodland-pasture. The woodland-pasture had an excellent stand of mostly hardwood timber on it, but even in the most difficult years, my grandfather steadfastly refused to cut any of it. Considering that the farm had been given to him, the timber represented the only wealth that he had accumulated in his and the family's years of unrelenting hard times. That wealth was to be used only in a true emergency, and he did not consider 9.5-cent cotton to be in that category.

He was the epitome of the yeoman farmer in the Lower South. His family consisted of my grandmother, a son, and three daughters–one of them my mother. My grandmother was a McClain who could trace her family's origin to the Isle of Mull in the Inner Hebrides along the western coastline of Scotland. During the Civil War, my grandfather's father had walked to Tallahassee, Florida, to join the Federal forces. One of his brothers had died in a well that was being dug in the local community. He had gone into the well to save an African American man who had been overcome by methane gas leaking into the well from freshly exposed clay. While he saved the life of his helper, he lost his own.

The family produced virtually everything they ate. They grew and butchered hogs; smoked the meat; made sausage, cracklins, lard, and soap. They had a large flock of chickens, and always had plenty of eggs, but little fried chicken. Two milk cows produced an abundance of milk, kept cool by hanging a metal container in the well, and there was the perennial war with bitter-weeds that caused cow's milk to be bitter. There was always a large garden and an impressive supply of canned fruit and vegetables. An abundance of corn was converted into meal to be cooked into corn bread for the family table or converted into scratch-feed for the chickens. Some of it was used to finish-off hogs before slaughter, and there was always an ample supply of fodder and nubbins for the mule. The family always had a large cane-patch and made plenty of syrup. The skimmings from the syrup-cooking were placed in a wooden barrel and allowed to ferment into a potent sugar cane alcohol. This devil's-brew (my grandmother's description), fueled the neighborhood, "swapped-labor," and "hog-killings." By the middle of the afternoon, after numerous trips to the barn "to kill rats," that was a happy group of hog butchers.

The primary crops produced on the farm were peanuts, corn, and cotton. The peanuts were hogged-off, with only a small amount being harvested for seed. Corn supported the entire operation, and the few bales of cotton were sold. For many years, their only cash income was derived from selling those few bales of cotton, and occasionally there would be a calf to sell. That small amount of money was used to purchase flour, salt, sugar, coffee, and snuff; there was usually a little left over to purchase the girls new dresses and new shoes.

In 1930, Granddaddy Bass harvested crops off about 25 acres, including about four acres of cotton and an acre of peanuts; the remainder was in corn. In that year, he would have harvested around 0.4 bale of cotton an acre, and would have received around 9.5 cents a pound when he sold it to cover all of his production costs including his labor and return on investment.

Their small farm was located about two miles from one of the largest sawmills in the South, but my grandfather never worked a day off the farm. Many of his neighbors worked in the mill during the winter months. Doubtlessly, the family would have been much better off financially if he had done the same.

From the beginning, his farming operation was tied irrevocably to the infamous crop-lien system. In the late winter of each year, without fail, he would enter into agreement with a local merchant whereby he would borrow enough money to purchase seed and fertilizer, plus a little extra for living expenses. For example, family records show that in 1910 he borrowed $50 from W.T. Foster of Georgiana. He borrowed the money on March 19, 1910 with the promise that it would be repaid with interest on or before October 10, 1910. As part of the process, he agreed to the following conditions: "I hereby grant, bargain, sell and convey unto the said W.T. Foster the entire crop raised by or for me or in which I may have an interest during the year 1910 and each succeeding year hereafter until this debt is paid in Butler County, Alabama, or elsewhere, and all my personal property of every description."

Much later, on March 27, 1930, he borrowed $88.50 from Georgiana Hardware Company and had to make the same promise concerning his crops and personal property. In this case, however, the lender required that he specifically name items of personal property that the lender might sell if he could not repay the loan by September 1, 1930: "One mule, color bay, age 10 yrs, name Ella./One, one-horse wagon with harness."

In a later year, he had to use a cow and a calf as collateral for still another short mortgage.

In 1931, my grandfather changed the way he financed his fertilizer and seed requirements. In 1916, President Woodrow Wilson had signed into law the Federal Farm Loan Act, which marked the first involvement of the federal government in agricultural credit. One of the act's provisions was the creation of the Emergency Crop and Seed Loan Program. A family record indicates that Granddaddy Bass applied for a $100 loan under the program in 1931, 15 years after the act went into effect. In his letter of notification, he received his check with the 5% interest charge deducted. The letter stated: "your government loan . . . must be paid from the proceeds of the first crops sold at harvest time." I have no idea why he had not used this source of credit before.

In January 1932, the Federal Farm Loan Act was replaced by an act establishing the Reconstruction Finance Corporation. This new act provided even more options for financing farm credit, and the New Deal legislation of 1933 provided even more. But there are no records showing that my grandfather availed himself of any of these new federal loan options.

There is no indication that he ever failed to repay the loans. They were just part of his farming operation. He never sold enough in a year to get to the point where he could finance his own operation. In this continuing situation, his entire family was just one really bad crop failure away from disaster.

In my early years, my father attempted to make a living producing White Leghorn chicks for sale. This small poultry farm was located on a portion of my maternal grandfather's larger farm. In the process, he and our whole family became deeply involved in selective breeding for egg production. Without any formal training, but with a considerable help from the Alabama Cooperative Extension Service, he produced some really outstanding bloodlines. His hens performed well in several national egg-laying contests. Unfortunately, he had neither the capital nor the knowledge for marketing what he had created. The family suffered through some really tough times before he finally abandoned his dream. Later, he became involved in the production of Red Durocs hogs for sale as breeding stock, and again he produced some of the finest animals in the South, winning numerous regional shows. Unfortunately, on two occasions, he lost virtually everything he owned attempting to time the extremely unpredictable cycles of swine abundance. He probably would have tried a third time, but my mother finally announced that the family was out of the hog business forever. He later managed a feed-and-seed store in a nearby town, and finally established one of his own. He was relatively successful in this venture, but he sold too much to local farmers on credit during agriculture's worst-of-times—the 1980s.

By the time I became a teenager, all the children were gone from my grandparents' farm, and their health was not very good. Consequently, my brother and I were pressed into service as only partially-willing farm laborers. We plowed, hoed corn, chopped cotton, picked velvet beans, pulled fodder, and sawed fire wood. I was never very good at any of this work. I never really learned to operate a turn-plow. On several occasions, my grandfather intimated that I should continue my schooling so I could get a regular day job.

I took my grandfather's advice, and with the turn-plow experience for encouragement, I continued my schooling. After finishing high school in Georgiana, Alabama, I spent six years at the Alabama Polytechnic Institute (today's Auburn University) completing requirements for the BS and MS degrees in fish management. It was during this period that I was first introduced to the concept of the ecosystem. Fish management involves the manipulation of various aspects of the physical and biological components of ecosystems.

Following a stint in the US Army of occupation in defeated Germany and a very rewarding three years at Cornell University, I was invited to take a day job at Auburn. Later, as a faculty member in the International Center for Aquaculture (ICA), I had the opportunity to participate in an effort, funded primarily by the US Agency for International Development, to establish aquaculture in developing countries throughout the world. Because agriculture and aquaculture are so closely allied, working with the ICA provided a unique opportunity to interact with farmers and farming in Asia, Africa, and Central and South America. Similarly, efforts to establish catfish farming in Alabama and the Southeast provided an excellent opportunity to interact with farmers and farming throughout the Southeast United States.

After retiring from Auburn, my wife and I purchased a motorhome, and during the past 15 years, we have taken several trips each year, traveling throughout the United States and Canada. In these travels, I have had the opportunity to visit with farmers and observe farming from the citrus groves and staked tomato farms of central Florida to the grass seed farms and vineyards of the Willamette Valley in Oregon, from the sorghum farms of the Rio Grand Valley of Texas to the dairy farms of Wisconsin, from the table grape farms of the Central Valley of California to tobacco farms of North Carolina, and from the cattle feedlots of New Mexico to the broiler farms of Arkansas and the catfish farms of Mississippi. Along the way, we have been humbled by the majesty of the endless wheat fields in Kansas, the corn and soybeans in Iowa, and potatoes and sugar beets in Idaho. In all of my travels, from the rice paddies of Indonesia to the wheat, potato, and sugar beet fields of the Red River Valley of North Dakota, I have learned that food and fiber production worldwide is an amazing and a very complex process and that farmers everywhere are and have always been truly amazing people. I have also learned that almost without exception, farmers are only meagerly compensated for their contributions to society.

My personal and family experiences have been an inspiration for me to investigate the larger picture of Alabama agriculture and the Alabama farmer and farm family.

Information on Sources

Virtually all of the information included in Chapters 1 to 5 was obtained from copyrighted sources (journal articles, books, etc.). Relatively little was obtained from non-copyrighted or public domain sources. However in Chapters 6 to 17, the situation is reversed. Most of it has been obtained from federal and state publications.

The first US Census was taken in 1790, but information on agriculture was not collected until the 1840 census. From 1840 to 1950, an agricultural census was taken as part of the census. Additionally, a separate, mid-decade agricultural census was taken in 1925, 1935, and 1945. From 1954 to 1974, an agricultural census was taken for the years ending in 4 and 9. After 1976, the schedule of the agricultural census was changed to coincide with the schedule of economic censuses; consequently, one was taken in 1978 and 1982. Afterwards, agricultural censuses were taken at five-year intervals, in years ending in 2 and 7. In 1991, the responsibility for publishing this census was taken from the Bureau of the Census of the Department of Commerce and given to the National Agricultural Statistical Service in the Department of Agriculture.

The US Department of Agriculture (USDA) was created in 1862, and in 1866 it began to publish annual data on various aspects of American agriculture. In 1905, the Crop Reporting Board was created in the department. Later, this organization would become the Agricultural Statistics Board. In 1961, a reorganization within the department resulted in the creation of the Statistical Reporting Service that would later become the National Agricultural Statistical Service (NASS). NASS and its predecessors have accumulated a vast storehouse of data on a broad variety of agricultural subjects, and a large share of it is immediately available in hard copy or on the Internet.

In 1936, the Department of Agriculture published its first annual issue of *Agricultural Statistics*. This publication includes data on the annual production of various crops and livestock, a wide variety of other data related to production, including: prices (support and market), costs, and income; land use; farm workers; food supplies and consumption; fertilizers and pesticides; agricultural wealth; freight rates; and exports and imports of agricultural products and refrigerated warehouse storage. In addition, each issue includes data on commercial fisheries and forest products. Each issue includes current data, plus data for the preceding nine years.

The Economic Research Service (ERS) was created within the Department of Agriculture in 1994. It is the primary source of economic information and research from USDA. ERS periodically issues a wide range of publications on economic and policy matters related to agriculture, food, natural resources, and rural development. Four times each year, the agency issues its flagship publication, *Amber Waves.*

In 1948, the Alabama Department of Agriculture and Industries published the first (Bulletin 1) of its annual reports entitled *Alabama Agricultural Statistics*. The bulletin included data collected as late as 1946, but for many crops, it also included data dating back to 1866. The bulletins are somewhat unusual in that they publish preliminary data for a given year in one issue and in subsequent issues make corrections as new data are collected. Each issue includes detailed current data and summary data for a 10-year period. These bulletins represent the result from a cooperative effort between the agricultural data collecting data agencies of the US Department of Agriculture and the Alabama Agricultural Statistics Service (AASS) of the Alabama Department of Agriculture and Industries.

For most of the citations provided with various sets of data, I have not included a specific year. As noted previously, each of the various USDA data sources include many years of data.

How To Use This Book

It may seem strange to include a section on How To Use This Book; however with a book that is so lengthy, so detailed, and so dry—yes dry, it seems likely that some comments along this line would probably be appreciated. I realize that one should never apologize at such an early stage in a book, so please consider these remarks not an apology, but rather sort of an explanation as to why I have written what I have written and especially why I have written so much. Alabama's agroecosystem is an extremely complex entity. Simply describing some of its more important components, as well as their interaction and the role that they play, requires the presentation of an enormous amount of data and information. Also, the fact that the state's agroecosystem is only a relatively small part of the much larger national one, increases by at least twofold the amount of data and information that must be presented. Further, following their evolution over a period of some 500 million years, expands and complicates the presentation by several orders of magnitude.

To compensate somewhat for dumping this vast amount of material on the hapless readers, I have diligently attempted to develop a logical framework of presentation that will help them navigate it with the least amount of difficulty. For example, I have prepared an introductory section (Chapters 1-4) on some of the more important physical and biological elements (geology, climate, soils) of the agroecosystem and how they have interacted to give us the agriculture that we have in the state today. In the following section (Chapters 5 and 6), I have described some of the historical events that led to its formal establishment in 1819 (statehood). Further, I have divided the presentation on its evolution after statehood on the basis of periods of broad historical importance to food and fiber production in the country and the state. Four of the periods are: the one in which American agriculture was emigrating westward away from the original 13 colonies (1800-1860), the period of the Civil War and the years immediately thereafter (1860-1870, the years of the helter-skelter expansion of agriculture in the period 1870-1895, and the period generally considered to be the Golden Age of American Agriculture (1895-1915).

Beginning with Chapter 6, I have divided each chapter into sections that represent important components of the aqroecosystem (politics, economics, farmers, farms, crop and livestock production, etc.). Further, each chapter also includes a section on the general economic status of agriculture during that period (The Bottom Line). Some of the chapters also include a section on the comparison of some aspects of the agricultural and broader economies of Alabama and three other states that were admitted to the Union around the same time—Illinois, Indiana, and Mississippi. This basic order is followed in all of the following chapters. With it, a reader should be able to follow a subject of interest—such as cattle numbers, farm size, or lumber production—from period to period with only limited difficulty.

Some aspects of the presentation change appreciably in Chapter 12 (From Purgatory to the Inferno: 1930-1940). In 1933, a newly-enfranchised Democratic establishment in Washington—led by a powerful, unified Southern delegation—enacted legislation (Agricultural Adjustment Act of 1933) that would establish a cheap food policy for the country through the use of federal price supports and supply management. This act altered forever the basic nature of the agroecosystems of the country and all of the states. Consequently, in Chapter 12 and beyond, a considerable amount of attention is given to describing the effect of this new gorilla in the room on their continued evolution.

In writing this book, I have fully subscribed to the idea that a picture is worth a thousand words. As a result, I have attempted to simplify the presentation by including an extremely large number of tables and figures (approximately 1,500). Unfortunately, I have included so many that they often interfere with the orderly progress through the narrative. **Instead, the entire complement of both tables and figures is available from NewSouth Books at www.newsouthbooks.com/alagroecosystem/supplement**.

Finally, in an effort to introduce some levity to this enormous and often dry and barren landscape, I have resorted to the occasional use of clichés and colloquial terms. Some of these are sharply constructed, and at first glance may seem to be applied with malicious intent, but please be assured that I never intended for any of them to be interpreted in this way.

Acknowledgments

When someone writes a book that covers a time period of some 500 million years, includes subject matter ranging from geology and meteorology to economic history and human cognition, and when the author takes 15 years to write it, it is patently obvious that many other people must have been involved. Through this long period, many have been asked for help, and all have unfailingly done so–although after so many years, some began to do so with slightly condescending smiles.

First, I must acknowledge the contribution of my precious wife, Jean, who has provided constant encouragement and not a little proding to keep me on task and who has responded to my comment, "no yard work today–have to work on the book," with 15 years of patient equanimity, although there was some indication after 12 years that this comment (excuse) was wearing increasingly thin.

I also wish to acknowledge the role of Emory Cunningham of *Progressive Farmer* and *Southern Living.* In a sense, the book is the culmination of his lifelong interest in and contribution to agriculture. This book is the result of a number of discussions with him, around the time of his retirement, on the status of Alabama agriculture and how it came to be what it is.

The preponderance of the effort to help me assemble the wide range of information required was provided by a large number of faithful, knowledgeable, cooperative, encouraging people at Auburn University's Ralph Brown Draughon Library. At one time or another, virtually everyone there helped, but those in Dr. Dwayne Cox's Department of Special Collections and Archives (The Alabama Room) were called on much more than most.

I also want to acknowledge the contribution of George Martin, soil scientist with the US Department of Agriculture's Natural Resource Conservation Service in Alabama. George helped me immeasurably with the chapter on soils, but he did much more than that. He helped me to understand how Alabama's geology, climate, soils, and vegetation are related. In doing so, he provided me with powerful lesson on the interconnectedness of things. You will encounter the results of this lesson in inter-relationship throughout this book.

Special thanks are in order for Professor Walt Prevatt who, along with professors John Adrian and Gene Simpson of the Department of Agricultural Economics and Rural Sociology in the Auburn University College of Agriculture, reviewed much of the book and steered me around many pitfalls in agricultural economics.

Mike Polioudakis, instructor in the Department of Fisheries and Allied Aquacultures, provided invaluable help to a functionally illiterate computer jockey. He steered me through almost limitless thickets as I attempted to get this mountain of material into and out of this infernal machine. He also made all of the dot-density maps.

Troy Hahn, (Spec V, Info Tech), also in the Department of Fisheries and Allied Aquacultures, helped immeasurably in formatting the innumerable figures used in the text.

Wilmer Rogers, John Jensen, and David Rouse, heads of the Department of Fisheries and Allied Aquacultures, kindly provided me with office space in Swingle Hall for lo these many years, along with considerable secretarial and administrative assistance. They maintained a supportive demeanor long after I had worn out my welcome.

Professor William D. Davies, Butler-Cunningham Eminent Scholar for Agriculture and the Environment, served as a very attentive and supportive listener when plans for the book were being formulated. He also provided substantial funding available to him through the eminent scholar position for use in the early stages of assembling the information.

It is essential that I acknowledge the contribution of thousands of employees of the National Agricultural Statistics Service, the Economic Research Service, the Alabama Agricultural Statistics Service and their progenitors who, for well over a century, have assembled and stored an almost unbelievable amount of data about all aspects of food and fiber production in the country and the states. Writing books such as this would be virtually impossible without the availability of this vast data base and without the ease with which much of it can be accessed.

Acknowledgement is also due to my supportive friends who have typically greeted me for the past 10 years with requests for information as to where they can purchase a copy of the book. These inquiries were really veiled, half-smiling inquiries as to whether or not I was still looking for unicorns.

Finally, in my defense, the inordinate amount of time required to finish the book can be traced directly to my lifelong aversion to touch-typing. Simply put, history happens faster than I can type. In 2002, when I finally realized that I was never going to catch-up, I quit adding new material and have spent the last seven years trying to organize and word-process the data that I had already accumulated.

Introduction

It is obvious that problems have multiple causes and that the real world is not composed of discrete, hermetically-sealed boxes. The real world is completely intertwined. However, because of the knowledge explosion than began with the Scientific Revolution and Enlightenment in the sixteenth through eighteenth centuries, no one person is now capable of being a Renaissance Man. Rather than the generalist, our contemporary ideal is a specialist. (An exception, is a general officer in military service who by definition must be, or at least should be, a generalist, as opposed to junior officers who are specialists.) Consequently, the academic and commercial worlds separate rather than unite knowledge and expertise. To attempt to determine why Alabama's agriculture–and by interpolation, Alabama's socio-economic and political realities–are so disappointing demands a generalist's approach. To even begin to understand why "Alabama is a rich state full of poor people" may not require a polymath, but it certainly requires extensive study and the acquisition of expertise of a broad spectrum of issues and components. The quest itself is worthy of the effort.

In the 1980s, traditional American agriculture went through an extremely difficult period. As a result, there was a widespread effort to encourage farmers to become involved in so-called alternative agriculture. The Alabama Cooperative Extension Service at Auburn University led this effort in Alabama. A team of several Auburn faculty members visited all of the Extension Service districts to lead discussions on this subject with extension agents. Because aquaculture was considered to be a promising alternative, I was asked to join the team and to introduce the presentations, by discussing why alternative agriculture was important. To prepare for this task, I reviewed a considerable amount of detailed information about the nature and characteristics of Alabama agriculture. I was amazed by what I learned. At that time, I had been a student in Auburn's College of Agriculture for six years and a faculty member for 26 years–but no one had ever exposed me to information about the lack of competitiveness of Alabama's agriculture. In my research, I learned that yields of most major crops in the state, and even milk-per-cow, lagged far behind national averages, and that comparisons with specific states were even more dismal. More disturbing still was the fact that these differences had been evident in published data for well over a century. Over this long period, Alabama yields had improved for all of those crops, but while they had generally kept up with increases in national yields, they could never seem to catch up. Yields in other states had increased at about the same or at higher rates.

The published data on the lack of competitiveness of Alabama agriculture is irrefutable, but while this revelation was of considerable interest, the more intriguing question was how this situation had come to be. What was there about Alabama, and I suspected other Southern states as well, that resulted in lower yields of major agricultural crops year after year for well over a century? Agricultural technology (equipment, chemicals, fertilizers, etc.) were available to farmers throughout the country. Further, all of the states had agricultural experiment stations and agricultural extension services, so farmers in all states had essentially the same potential for access to up-to-date technology and information. Then what else could result in such a disparity of outcomes?

At the time, I did not have an opportunity to pursue this matter immediately, but promised myself that I would do so when I retired. This interest was not simply an academic interest. As explained in the Preface, I had long wondered why my grandparents and parents had had such a difficult time making a living in Alabama agriculture, despite their almost superhuman efforts to do so.

I retired as the head of the Department of Fisheries and Allied Aquacultures at the end of January 1994. Then, with the encouragement of Emory Cunningham and with financial support from the endowment that established the Butler-Cunningham position of Eminent Scholar for Agriculture and the Environment at Auburn University, I set up an office in Swingle Hall, bought a computer and began to research, read, and think, and eventually to type. The methodology that I bring to investigating this problem, the methodology that I am most comfortable with, is that of the scientist: to observe, question, hypothesize, experiment with evidence that can be measured, and draw conclusions that are verifiable by repetition. However, an investigation of the magnitude of this book also depends on other kinds of truth, such as reason and historical truth.

By 1995, my mind was abuzz with what I was learning, as well as with what I was remembering that I had learned long ago in high school. I was reminded that it was no accident that civilization first appeared in the Fertile Crescent of the ancient Middle East. This is where agriculture first appeared. In fact, civilization followed agriculture on the world stage, and has played a similar role in all of the succeeding acts. Most evidence suggested that advances in civilization followed advances in agriculture. Advancements that produced an agricultural surplus has historically resulted in increased capital accumulation, the substitution of capital for labor, and the release of excess labor to do all kinds of creative things–the beginning of specialization. In turn, civilization provides inputs that catalyzes agricultural advancement.

Lord Kames (Henry Home), the Scottish philosopher, was among the first to write of this process. In *Sketches of the History of Man*, published in 1776, he described human history as having four distinct stages: hunter-gatherers, herders, farmers (agriculture), and craftsmen and traders (commerce). Other authors have discussed this process. Lenski and Nolan (1998), who discussed it in considerable detail, suggested that the evolution or progression of societies could also be divided into four different stages: hunters and gatherers, simple agriculture or horticulture (without the plow), advanced agriculture, and industry. To these four stages, we can add post-industrial or a service-based economy.

Comments by Robertson (1987) suggest that advances of civilization are the result of changes in their material culture (clothing, domesticated plants and animals, baskets, pottery, cooking pots, plows, wheels, cities, factories, etc.) and their non-material culture (language, ideas, beliefs, customs, skills, political and economic systems, art, etc.) Aspects of the material culture of a society change when some of its members discover or begin to change some characteristic of reality (their environment) that already exists. For example, at some point in the distant past, some human discovered that wet clay was malleable and that it could be formed into various shapes, which would be retained when the clay was air-dried. Later, someone discovered that placing the wet-clay object in a fire would make it hard, but brittle. Tempering reduced fragility. These discoveries provided the basis for the invention of the firing of clay pots to make them non-porous and fire-resistant. Finally, the value of these utilitarian inventions was enhanced by painting pictures on them before firing. These discoveries and inventions would not have had much value to those ancient societies if they had not been widely diffused within and among societies. This process of discovery, invention, and diffusion–largely driven by advances in agriculture–allowed societies to progress from the hunter-gatherer stage to the simple agricultural and advanced agricultural stages, and finally to the industrial and post-industrial stages.

Of course, discovery, invention, and diffusion have not proceeded on the same timescale in all societies. The process moved much more slowly in some societies than in others. For example, when Great Britain began to enter the industrial stage of development in the mid to late eighteenth century, there were still societies in some parts of the world that were still in the hunter-gatherer stage. It likely was no accident that rapid advances in English agriculture (the Agricultural Revolution), accompanied (likely drove) the Industrial Revolution there and certainly played a central role in the dominance of British nineteenth century imperialism. The Agricultural Revolution produced an even greater surplus of food than had previously existed, and agricultural efficiency made many rural workers redundant, and thereby created a workforce for the new factory system. Furthermore, the landowners' economic position improved to the extent that they had the capital to finance industrialization. Two new classes were created: the urban working class and modern middle class.

In my research, it became obvious that some strange things had happened and were continuing to happen in Alabama's agricultural and general economy, but I had no idea as to what had caused them. Further, I hardly knew where to begin to find out. It seemed that the best approach would be to investigate changes in Alabama agriculture from 1819 to the present, looking for clues. I learned that the nature of agriculture would finally develop in Alabama had been largely determined at least two centuries earlier in the English Parliament (the mercantalism system) and in the tobacco fields of Virginia. I also learned that important characteristics of Alabama's geophysical environment had been established between 400 to 500 million years ago and that the genetic characteristics (genomes) of crops and animals found on Alabama farms, including the farmers themselves, were largely determined by Old World environments quite different from those in Alabama. For example, the genome of the soybean was largely determined in East Asia thousands of years ago, and the genomes of many Alabama farmers themselves were partly determined on the steppes of southern Russia, in Celtic Britain, or in the bulge of West Africa. Further, I learned that although geophysical characteristics and genomes were important determinants of agriculture, the ebb and flow of history, politics, and economics were even more important. Finally, I learned that Alabama agriculture was so tightly intertwined with the agricultures of the neighboring states, with those of even more distant states in the country, and with those with other countries around the world that I could not simply separate Alabama's agriculture from the others and study it alone. Alabama agriculture was and is part of a worldwide agroecosystem, and changes would have to be studied and evaluated within that framework.

Early on, I learned that in 1963 the Bureau of Economic Analysis of the US Department of Commerce began to publish data on the Gross State Product for the individual states. The data consisted of estimates of the total market value of each state's finished goods and services for each calendar year. In 1994, these values ranged from $13.7 billion in Vermont to $862.5 billion in California. On a per capita basis, they varied from $19,017 in Mississippi to $38,452 in Alaska. Alabama had the sixth lowest value ($20,926 per person), or 45th of the 50 states. Only values for Mississippi ($19,017), West Virginia ($19,172), Montana ($19,837), Arkansas ($20,473), and Oklahoma ($20,683) were lower. It was difficult, at first, to understand how in 1994 Alabama–with such an abundance of natural resources (limestone, dolomite, granite, marble, clay, iron ore, coal, petroleum, natural gas, timber), with a mild climate, with abundant rainfall, with hundreds of miles of navigable rivers, and with a deep-water port at Mobile–could have had the sixth lowest Gross State Product in the entire country.

In the early 1800s, territory that would eventually become the states of Indiana, Illinois, Alabama, and Mississippi was located on the frontier of the rapidly expanding United States. All four achieved statehood at about the same time: Mississippi (1813), Indiana (1816), Illinois (1818), and Alabama (1819). In 2002, the Gross State Product on a per-person basis of these four states was: Illinois ($38,661). Indiana ($33,287), Alabama

($27,598) and Mississippi ($15,190).

In my attempt to make some sense out of all of this information, it seem likely that–combining the ideas of Kames (1776) and of Lenski and Nolan (1998), as well as Gross State Product data–Alabama society, for some reasons, had moved more slowly through the simple agricultural and advanced agricultural stages than Illinois and Indiana. Those states, probably with fewer natural resources compared to Alabama, apparently moved quickly through the two stages, and by 1994, their GSP values were comparable to those in the older, industrialized states of the Northeast. With this conclusion behind me, I thought I knew what had happened, but I still had to learn why.

The material presented in this book represents a broad agroecosystem approach to understanding the reasons why Alabama agriculture and its economy is like it is–and how it got that way. It includes information on geology, physiography, climate and weather, soils, origins and characteristics of some of the domesticated plants and animals used in Alabama agriculture; origins and cultural characteristics of those peoples who would farm in Alabama; and politics and economics. Finally, most of the book is devoted to the action and interaction of all of these factors, and how they have interacted to determine the history and evolution of agroecosystems, in the world, the United States, and Alabama.

Synthesizing this material has been a daunting task and much more extensive and complicated than I ever imagined. After some 15 years of reading and typing, I am not sure that I yet have it right. I feel fairly confident in the validity of the information that I have included, although I am sure that I have made some errors in transferring it from source to page. I am not, however, nearly so confident in my interpretations and conclusions. Completing this task has forced me to relearn things that I learned in high school and college half a century ago, as well as to learn things that I never learned before. Unfortunately, this exercise had made me a Jack of all trades, but master of none. This unfortunate situation will quickly become obvious to all of the geologists, meteorologists, soil scientists, agronomists, animal husbandrymen, economists, historians, and political scientists who read this material. Many will quickly conclude that I have bitten off more than I could chew by attempting to be a generalist in an age of specialization. I tend to agree with that conclusion; however, by the time that I realized it, I was too emotionally involved to quit. The only solution that I can recommend is that, if you do not accept my interpretations or conclusions, the data is likely to be reliable and verifiable, so accept and use the data to draw your own conclusions.

Evolution of the Alabama Agroecosystem

1

Geology and Physiography

PRECIS

Alabama's geological history is fascinating: the results of eons of creation of today's Alabama have left a topography of amazing diversity. In this chapter, no rock is left unturned in an exhaustive description of Alabama's five physiographic sections and the numerous districts within these sections. The results of the state's geological history is an irregular land surface located in the mid, north latitudes that is brimming with mineral wealth: water, sand, gravel, shale, kaoline, chert, limestone, sandstone, marble, dolomite, coal, iron-ore, oil, and gas. This chapter also describes Alabama's wealth of water resources, including 14 major river systems. A sixth of the surface of the state is water, and a tenth of the water in the United States flows in or across the state. As diverse as Alabama is geologically and as rich as it is in mineral and water resources, the state's agriculture and concomitantly its economy has been adversely affected by hilly and mountainous terrain in much of the northern half of the state. Physiography directed settlement and transportation patterns and fostered subsistence agriculture and, later, extractive industries, whereas landforms and rivers in the southern half of the state fostered one-crop agriculture.

INTRODUCTION

A consideration of Alabama's geology and physiography provides strong support for the old adage that truth is stranger than fiction. A look at the purported events that lead to development of our portion of the planet earth would make good science fiction: a time scale so long that our usual time units—years, decades, centuries, even millennia—are not suitable measures; colliding continents; exploding volcanoes where the very bowels of the earth spew upward and out onto the landscape; cataclysmic earthquakes; new mountains thrust upward thousands of feet; massive avalanches, both large and small that fill valleys with rocks, gravel, sand, and clay; heat that melts the hardest rocks; mechanical pressure so great that simple sand is converted to hard sandstone and soft limestone becomes marble; and changing sea levels that cause the ocean to sweep back and forth across the land time after time. If we have correctly interpreted the story written in the rocks, these are the forces and events that have formed the framework over which hangs the fabric (the soil) where we plant our crops, graze our cattle, and harvest tall trees.

In this incredible story, over time tall mountains composed of the hardest rocks are reduced to low-lying plains by the incessant pressure of elements of the hydrological cycle. Water in the form of rain and ice tears rocks apart, and then transport the pieces into valleys below. In the process, small water-bearing rivulets are joined together to form creeks, then rivers and bays before returning their watery cargo to the oceans from whence it came.

SOURCES OF INFORMATION

Maps produced by the Geological Survey of the United States Department of Interior are invaluable sources of information on Alabama's physiography. The 7.5 minute Series (Topographic) (1:24,000 Scale) is especially useful. Each map covers an area seven miles wide (east and west) and nine miles long (north and south). They provide information on the location and elevation of mountains, valleys, rivers, creeks, reservoirs, highways and roads, cities, towns, villages, and communities. Contour lines (intervals of 10 or 20 feet) provide a real feel for the topography of each area covered. Some of these maps, covering portions of north Alabama, were prepared by the Tennessee Valley Authority. They are almost identical to those prepared by Geological Survey.

The Geological Survey of Alabama is an excellent source of information on the geology and physiography of the state. That organization was established in 1848 for the purpose of gathering data about the mineral, water, and energy resources of the state and to conduct basic and applied research in those areas. Since its establishment, the organization has published hundreds of reports, bulletins, circulars, leaflets, guidebooks, monographs, and maps. Sartwell (1995) has developed a catalog of publications and maps available from the organization. That publication also provides directions for purchasing any of the items listed. *Alabama Stratigraphy* (Raymond, et al., 1988) and *Physiography of Alabama* (Drahovzal, 1968) as well as the maps of *Physiographic Regions of Alabama* (Sapp and Emplaincourt) and *Geologic Map of Alabama* (Osborne, et al., 1989) developed by survey personnel are especially informative.

The publications, also by survey personnel, *Alabama's Water Resources* (Moore and Szabo, 1994) and *Alabama Waterways Guide* (Alabama Geological Survey, 1996) provide a wealth of information about the water geophysical features of the state.

The Natural Resources Conservation Service (formerly the Soil Conservation Service) of the United States Department of Agriculture has been conducting soil surveys in Alabama counties since 1902 (Perry County). To date, reports have been published for 45 of those surveys, and a number of others are awaiting publication. These survey publications contain information on the geology and physiography of the counties.

Summary articles by Hooks (1973), *Physiography and Topography*, and by Hughes (1973), *Geology*, both provide useful and interesting information about Alabama's geophysical environment.

There are a number of excellent books on general geology and physiography. *The Earth Through Time* (Levin, 1994) and *The Tectonics and Geophysics of Mountain Chains* (Hatcher, et al. eds., 1983) have been especially helpful in the preparation of this book. *Physiography of the Eastern United States* (Fenneman, 1938) also contains much useful information on the physiography of Alabama.

Geological Processes

The majestic peaks and deep, sharply incised valleys of mountain ranges are both magnificent and awesome. These physical features represent the current status of an ongoing, eons-long battle between two of the most powerful forces in the universe: the almost unimaginably powerful heat machine of the earth's interior that forces huge masses of rock high into the sky and the unrelenting force of the kinetic energy of falling water, raised high above the tops of the mountains by heat from the sun that tears them down.

This incredibly complex system of geophysical processes gives us our mountains, valleys, hills, rocks, soils, creeks, rivers, bays, and oceans. These competing forces give us much of the substance and shape of our planet. They also determine the location of continents relative to the position of the equator, which largely determines the characteristics of our climate.

It is almost impossible to comprehend the time-scale involved in the operation of these processes. The time units we commonly use—minutes, hours, days, years, decades, centuries, etc.—are certainly unsatisfactory when contemplating the time required for creating a bed of coal under a mountain, a new ocean, or a new micro-continent. Consequently, geologists have devised a new type of scale, one that groups geophysical events together into epochs, periods, eras, and eons. The relationship between these units are presented in Table 1.1.

The Heat-Engine within the Earth

Available scientific evidence indicates that our blue planet earth began to form some 4.5 billion years ago when a giant cloud of interstellar gases and dust swirling in an elliptical orbit about the sun, the center of our disk-shaped solar system, began to condense (Dolt and Batten, 1988 and Levine, 1988). The cloud consisted chiefly of hydrogen and helium with lesser amounts of oxygen, nitrogen, silicon, calcium, aluminum, sodium, potassium, magnesium, carbon, sulfur, and iron and still less of the heavier metallic elements. Gravity within the spinning cloud concentrated the heavier elements at the center of the mass that would become the earth. The resulting compression along with impacting chunks of matter and meteorites and radioactive decay within the mass caused the deep interior to melt (Davidson and Williams, 1996). Later, the exterior surface of the mass would cool down, but the deep interior would remain incredibly hot.

As a result of the nature of its composition, earth is arranged in layers, something like an onion (Figure 1.1). At the center is a spherical core composed primarily of iron (85%) and nickel. The solid inner portion of core has a diameter of 1,500 miles. The molten outer portion, with a thickness of 1,300 miles, is also composed of iron and nickel. The temperature of this outer portion of the core is estimated to be 8,500°F. The next layer is the mantle with a thickness of 1,800 miles. It is stony or rock-like, probably composed of iron-rich rocks. In addition to iron, the rocks probably also contain large amounts of oxygen, silicon, and magnesium. The crust of the earth is the relatively thin (3 to 44 miles thick) veneer that includes the ocean floors and the continents.

The outer, liquid portion of the earth's core is constantly in motion. Heat energy generated in this region by the tremendous pressure—caused by the weight of the mantle above and by the decay of radioactive elements—results in huge, slow-moving convection currents toward the surface in the molten semi-liquid (Moore, 1988). Some of the motion of these density currents is transmitted to the mantle.

The Heat-Engine and the Earth's Crust

The heat in the mantle is finally transmitted, primarily as motion, into the earth's crust. As a results, the crust, floating on the surface of semi-liquid mantle, is shifted about, and in the process is torn apart, forming huge slabs of brittle rocks called tectonic plates (Chronic, 1983). Twelve of these huge tectonic plates, plus an undetermined number of smaller ones, constitute the surface of the earth (Figure 1.2).

The chemical composition of the tectonic plates varies. A few of them are composed primarily of iron compounds; others are composed primarily of lighter compounds. The plates composed primarily of iron compounds are much heavier and float lower on the earth's semi-liquid mantle. These heavier plates carry the oceans, whereas the lighter plates float higher in the semi-liquid and carry the continents (Lambert, et al., 1988). One of these heavier plates carries the Pacific Ocean, and one of the lighter plates carries the North American continent. In some cases, continental plates or portions of the continental crust also extend some distance outward under the ocean. On its eastern margin, the North American Plate extends eastward to the middle of the Atlantic Ocean where it joins the Eurasian and African plates (Figure 1.2). Where the plates meet, molten rock is continuously escaping upward through the fissure, creating a mid-ocean ridge.

Adjacent tectonic plates carrying continents can remain in a relatively passive state for millions of years. Plates in this state are depicted in Figure 1.3, Panel A. As depicted in the figure, the surface of the central cores of the continents (shields), composed of crystalline rock such as granite, have been eroded, and the sediment from that erosion has been carried toward the ocean by stream flow. Large quantities of these sediments have covered a portion of the continents adjacent to the ocean and have spilled into the ocean as well. During periods of low rainfall over the continents, when erosion and sediment flow are reduced, limey deposits formed by the precipitation of carbon dioxide, as carbonate, accumulate in large quantities in the shallow waters of the ocean. Because of the weight involved, the resulting pressure on these deep beds of sediments convert them into sedimentary rocks such as sandstone, mudstone, and limestone. These sedimentary rocks lying on top of portions of the ancient shield form the platforms of the continents.

Because of the heat-induced motion generated in the somewhat plastic mantle, the tectonic plates carrying the continents float (continental drift) about the surface of the earth. The speed at which the plates move can be measured in a few inches a year, but over the millions of years of the geological time scale, they can move great distances. Over time, as the plates float about, they scrape past each other or they collide. As they collide and then separate, the size and shape of both continents and oceans change. In Figure 1.3, Panel B, the two tectonic plates have begun to float toward each other. As they move, the ocean separating them begins to close. Some of the enormous quantity of rock caught between the colliding continents is forced downward, resulting in the development of cracks in the crust that allows molten lava to flow upward from the mantle where it cools at the surface. In this process, volcanoes form along the margins of their platforms in the collision zones.

As the shields come closer together, the ocean between them is completely closed or eliminated. As these massive bodies come closer together, geological debris—such as island arcs, micro-continents, etc.—caught between them are swept along to pile-up at the juncture of the two. Finally, when the continents collide, this rocky debris, plus portions of the margins of the shields, are forced upward as massive, magnificent, mountains thousands of feet high (Figure 1.3, Panel C). The pressure and the resulting heat are so great that the sedimentary rocks that existed on the passive platforms of the two tectonic plates are metamorphosed into other kinds of rock. In this same process, molten magma is forced upward from the mantle where it cools in the crust of the earth far below the surface to form the igneous rock—granite, or at the surface to form the igneous rock—basalt. Finally this mixture of metamorphic rocks and igneous rocks is forced to the surface to form the central core of the mountainous mass. At the same time, rocks hundreds of miles from the so-called suture zone between the colliding continents are also affected, but in these distant regions, the pressure and heat are not great enough to change them physically or chemically. Rather, they are folded like an accordion, creating parallel ridges (anticlines) and valleys (synclines), generally parallel to the suture zone (Figure 1.3, Panel D). From these collisions of tectonic plates, sometimes, supercontinents composed of two or more continents are formed (Figure 1.10, Panel D), only to be rifted apart later when the plates bearing the continents are pulled apart by the relentless heat-induced motion in the semi-liquid mantle.

When a continental plate and an oceanic plate converge, the lighter, higher floating one rides upward and over the heavier one. The heavier plate is carried underneath or is sub-ducted. For the last several million years, the North American Continental Plate, moving slowly westward, has been riding over the Pacific Oceanic Plate, sub-ducting it. The friction of this continuing collision releases enormous quantities of energy that results in the earthquakes and volcanic eruptions occurring periodically from Alaska to California.

Rock Formation and Conversion

When continental plates collide, structural changes are so great that molten rocks deep in the earth's interior may be forced to the surface through volcanic activity. When this molten material cools and solidifies, it becomes igneous rock. Over time, weathering and erosion remove small fragments from these rocks, which collect as sediments. Where enough of these sediments collect, pressure may convert the deeper layers into sedimentary rocks, and if the pressure continues, some of the sedimentary rock may be further changed into metamorphic rock. In this process, extremely small particles of silt, eroding from the igneous rocks, granite or basalt, may be transformed into the sedimentary rock—shale. Then, if the pressure continues to increase, shale may be converted into the metamorphic rock—slate. Further, under the right conditions, when continents collide the slate might be, once again, forced deep into the earth's interior where it will once more become molten rock. Later, it could be, once again, forced to the earth's surface through volcanic activity to become the igneous rock—granite. This process constitutes the so-called Rock Cycle, which is shown in diagrammatic form in Figure 1.4.

Weathering and Erosion

Even while mountains are being thrust up as a result of continental drift and collision, other forces are at work to tear them down. Loose rocks on the peaks, responding to the pull of gravity, fall to the valley below breaking other rocks free as they fall. Simultaneously, the unrelenting force of erosion tears down the highest peak one molecule and one grain at a time, ultimately to form flat plains of soil (sand, silt, and clay). As a result, the landforms that we see about us are products of tectonic events, as well as the Rock Cycle—weathering, erosion, and sedimentation (Figure 1.4).

The processes by which rain, snow, hail, sleet, freezing, thawing, heating, and cooling reduce mountains to level plains is known as weathering, because they are all elements of weather and climate. The kinetic energy of moving water molecules—first as they strike the surface of landforms as rain, then as they collect to run along its surface as rivulets, creeks, and rivers—have a profound effect on the surface of the earth. Because of the density of water, falling raindrops literally blast away the surface of even the hardest rock. Furthermore, because it is the universal solvent, what water does not blast away, it takes away in solution. It seeps and runs into the tiniest cracks, and then expands when it changes state from liquid to solid (ice) to pry larger pieces from the surface. Then when the ice melts, the small pieces are carried along in the liquid phase of water to grind away the rock's surface.

The moving water carries its load of small and large fragments of rock and boulders down the slopes, grinding and tearing into its bed as it moves along. But as it moves down the slope of the mountain to the lowlands below, its kinetic energy is rapidly dissipated. As it loses energy, the size of particle that it can carry is steadily reduced. It leaves the boulders on the side of the mountain to be dealt with later, when the amount of rainfall is greater. Larger fragments are carried into the higher, steeper valleys; they will also be dealt with at a later time, as well. Finally, when most of its kinetic energy is gone, as it reaches the lower valleys it can carry only the larger grains and then not even those, as it finally loses even the silt and clay particles to gravity.

The now slowly-moving water, with most of its kinetic energy dissipated from the bombardment and grinding and prying on the mountains far above, deposits most of its remaining load in the gentle valleys at the mountains' feet. Hour after hour, year after year for millions of years, this process continues. As the mountain is slowly reduced in size, the substance of its grandeur accumulates at its base or in the valleys below. The billions of tons of rocks that had been thrust upward by the collisional dynamics of the tectonic plates have become billions of tons of sediments on a level plain at the mountain's feet. In some cases, the sediments become so heavy that they force the surface of the original valleys downward, creating more space for deposition.

Under the enormous pressure beneath the thousands of feet of accumulated deposits or sediments, the rocks from the ancient peaks are slowly reborn. The mud and tiny fragments (grains) are pressed together and sometimes heated to form sedimentary rocks like sandstone, shale, or limestone (Figure 1.4). These are the stratified rocks that are commonly seen along highway cuts through mountainous areas. Sandstone, probably the most common, is formed when predominantly sand-grain sediments are pressed together by the weight of the overburden. Shale is formed when the sediments are primarily clay and silt particles. Limestone is formed from calcium carbonate when carbon dioxide in solution in lakes or oceans is precipitated and settles to the bottom, or from the remains of tiny plants or animals that die and settle to the bottom in usually warm, clear seas. These plants and animals concentrate carbonate within their structure as a part of their normal cellular metabolism. Chalk is a very pure form of limestone. Coal is another type of sedimentary rock that, like some limestone, is derived from living things.

Coal is formed when vegetation in shallow swamps is inundated with water and mud, and is killed. Then before the vegetative material completely decays, a blanket of mud protects the organic matter in the leaves and trunks from rapid oxidation. Anaerobic bacteria extract the oxygen from the slowly decaying plant tissues leaving a fibrous sludge with a high carbon content. Hundreds and even thousands of feet of sediments accumulate over this carbon-rich sludge pressing it into beds of coal.

Iron ore forms when iron-rich water penetrates softer geological formations such as limestone and forms a precipitate within the rock.

The Heat-Engine and Physiography

It is almost impossible to comprehend the extent of the

changes brought about by the effects of the Heat-Engine on that part of the planet that we now call Alabama. How it came to be what and where it is today, is both a fascinating and important story, especially for Alabama agriculture.

The Curtain Rises

Earth scientists suggest that in the middle on the Silurian Period, of the Paleozoic Era, some 500 million years ago (mya), the shield of the North American continent lay far to the south, along the equator (Levin, 1994). Apparently, the shape of the continent was somewhat different than it is today, but it is likely that the region of the continent that would, much later, become Alabama, lay very near the continent's southwestern margin. Available information also indicates that at that time, most of what is now the Eastern United States was above sea-level. Because of severe erosion from ancient highlands far to the north and west in the central part of the continental land mass, enormous quantities of sediments, primarily quartz sandstone and shale, were carried by swift streams to be deposited along the eastern margin (platform) of the continent. Later, the rising sea slowly covered these deposits, as well as much of the land that would become the United States. As the shallow sea slowly spread inland, additional sand and shale were deposited; however as the sea slowly moved inland, the deposition of carbonates (limey sediments) began to replace the depositions of sand and shale across the floor of the shallow sea. By the end of the Cambrian Period (some 490 mya), enormous quantities of carbonates had accumulated. In some areas, these accumulations were thousands of feet thick. It was this thick layer of limey deposits (later limestone) overlying thick accumulations of sandstone and shale that would constitute Alabama's foundation.

During the Silurian Period of the Paleozoic Era, some 420 mya, the Heat-Engine in the earth's interior began to move the continents together (Figure 1.5), and by the early Mississippian Period of the Paleozoic (320 mya), the tectonic plate bearing Larussia (North America and Europe) finally began to collide with the giant tectonic plate bearing Gondwanaland (South America, Africa, India, Australia, and Antarctica). As the tectonic plates moved closer and closer toward collision, one after another of the micro-continents—island arcs and rock assemblages caught between the platforms of the two super-continents—crashed against the southeastern margin of North America. The collisions resulted in the Talladega platform assemblage, the Piedmont micro-continent, and the Brunswick and Tallahassee-Suwanee terranes being crushed between them. Finally, the section of Gondwanaland corresponding to the modern-day region of Northwestern Africa was tightly sutured to the southeastern margin of North America.

The final collision between Larussia and Gondwanaland literally remade the eastern portion of Alabama's foundation. Volcanos spewed molten lava that would cool into igneous rock. The old sandstone, shale, and carbonates, were converted into a wide variety of extremely hard metamorphic rocks. In the process, large mountains composed of masses of these rocks were thrust thousands of feet into the air. Further, the force of the collision resulted in the folding and thrusting of masses of rocks far inland from the point of contact.

The late Paleozoic collision between Larussia and Gondwanaland did not result in a permanent partnership. During the Triassic Period of the Meszoic Era, some 245 mya, the same forces that brought about the collision of the two super-continents began to separate them. However, the separation between the two was not a clean one. The continents did not separate at the same point where they were joined. As they moved apart, large fragments were left behind. Over time, the continents continued to drift apart, and by around the middle of the Cretaceous Period of the Mesozoic Era, some 70 mya, the North American continent was near its current location (Figure 1.6). This position, with respect to the equator, is primarily responsible for Alabama's subtropical climate.

An Assortment of Rocks

Rock cycles associated with the geological events described in the preceding section, have resulted in formation of a large assortment of different kinds of rocks in Alabama. The collision of the two super-continents resulted in the formation of a large number of igneous and metamorphic rocks in the east-central portion of the state, and sedimentation and metamorphism produced an even larger number in other portions.

Geologists have located, identified, described, named, and mapped the various rocks and assemblages of rocks of the state. One of the maps produced in this process is shown in Figure 1.7. As indicated by data presented in the figure, rocks underlying much of the red soils in the Tennessee Valley have been named the Tuscumbia Limestone Formation. The formation has been described as "light-gray bioclastic or micritic, partly oolitic limestone in thin beds; cross-bedded with massive (up to 10 feet thick), very coarse, bioclastic, crinoidal limestone." Geologists estimate that this rock formation was established during the Upper Mississippian Period of the Paleozoic Era, some 325 mya.

Similarly, geologists have identified a large rock mass in the north-central Alabama, lying principally in Jefferson, Walker, Winston, and Cullman counties, as part of the Pottsville Formation (Figure 1.7). It has been described as consisting of "interbedded sandstone, siltstone, shale, conglomerate, and coal." This same formation is found from north-central Alabama to Pennsylvania. The formation was first described in Pottsville, Schuylkill County, Pennsylvania, 95 miles northwest of Philadelphia. Geologists estimate that this formation was established during the early Pennsylvanian Period of the Paleozoic era, some 300 mya.

Another of these rock masses has been identified in Coosa and Chambers counties in the east-central portion of the state, and named the Dadeville Complex (Figure 1.7). This complex consists of several smaller rock units, including: Agricola Schist, Ropes Creek Amphibolite, Waresville Schist, Waverly Gneiss, Camp Hill Granite Gneiss, and Rock Mills Granite Gneiss. The entire complex has been described by Raymond, et, al. (1988) "as a sequence of volcaniclastic rock including amphibolite, hornblende gneiss, pyroxenite lenses, mafic igneous rocks, and leucocratic granitic rocks, inter-layered with a sequence of biotite-granite-muscovite schist, biotite-muscovite schist, biotite gneiss and thin amphibolite units." The smaller units, such as the Waverly Gneiss, are named for a geographical location in the vicinity where they were identified and described. These highly location-specific rock units are found no other place in the world. Geologists estimate that these formations were established sometime in the Precambrian, possibly as early as 1,900 mya. Rocks this old were likely formed before the Piedmont micro-continent was sutured to North America.

Soils of the so-called Black Prairie (Black Belt), lie diagonally across south-central Alabama from Pickens, Greene, and Sumter counties on the Mississippi line to west-central Russell County (Figure 1.7), was formed from rocks of the Mooresville Chalk and Demopolis Chalk Formations of the Selma Group. The Mooreville Chalk Formation underlies the northern (Lower Selma) portion of the prairie, and Demopolis Chalk (Upper Selma) underlies the southern portion. According to Raymond, et al., (1988), these formations consist primarily of "fossilferous chalk and clayey marl." Geologists estimate that these formations were established during the Cretaceous Period of the Upper Mesoic Era, some 100 mya.

In south Alabama, a group of rocks spanning the state from Henry County in the east to Choctaw County in the west has been identified as the Claiborne Group, Undifferentiated. This group includes the Talahatta and Lisbon formations and the Gosport Sand. The Tallahata Formation is described as consisting of "white to very light greenish-gray thin bedded to massive siliceous clay-stone inter-bedded with thin layers of clay, sandy clay, and glauconitic sand and sandstone." Geologists estimate that these formations were established during the Eocene Epoch of the Tertiary Period of the Cenozoic Era, some 50 mya.

The earth's Heat-Engine has left piles of many different kinds of rocks—now mostly covered by a thick mantle of soil— scattered everywhere across Alabama's landscape, but they are not located in a haphazard manner. Every part of the pile is there for a reason, and every pile has it own peculiar characteristics. After millions of years of mountain-building, weathering, erosion and sedimentation, and after the creation of a hodgepodge of mountains, hills, ridges, valleys, and flats, seemingly thrown helter-skelter across the landscape, there is a magnificent orderliness in the disorder.

Topography

Because of the interplay of the geological forces described in a preceding portion (mountain-building, weathering, erosion, and sedimentation), the land surface of Alabama is highly irregular; however, compared to mountainous land areas in nearby states, such as Georgia and Tennessee, and in the western part of the country such as Colorado and Idaho, Alabama topographically is an area of relatively low relief. Our highest mountains are only 2,400 feet above sea level, and most of the area of the state is much lower. Mountains two times as high are located in the adjoining states. Alabama's highest mountains are located in its northeastern quadrant (Figure 1.8); however, elevations fall away sharply both to the west and south. More than half of the state is at an elevation of 500 feet or lower.

The highest mountains (white areas in Figure 1.8) are located in an area generally running from Jackson County in the northeast to the southwest through Shelby to Clay counties and southward to Cleburne County. Little Coon Mountain (1,699 feet) in northwest Jackson County; Double Oak Mountain (1,480) in northwest Shelby; Talladega Mountain (1,821) in southwest Clay, and Cheaha Mountain (2,405) in southwest Cleburne, are examples. There are also some relatively high mountains in the northwest quadrant of the state, principally in Colbert, Franklin, Lawrence, and Cullman counties—Alabama's so-called Central Mountains. Lagrange Mountain (900 feet) in eastern Colbert County and Brindley Mountain (1,161) in northern Cullman, are examples.

While the state is not mountainous, it is not very level either. With the exception of the area of relatively high relief in northeast Alabama, the entire state can best be described as hilly. Exceptionally hilly areas can be found all over the state. For example, the Lime Hills in Choctaw, Clark, and Monroe counties form one of the more rugged topographic regions in Alabama. Elevations above 400 feet are common across this area, and on an especially high hill in northwest Monroe County, the elevation reaches 570 feet. There are also rugged hills in parts of Pike, Crenshaw, and Coffee counties. For example, the elevation is 650 feet on a high ridge on Mount Carmel in northeast Crenshaw, more than 300 feet above a nearby lowland. The highest elevation on Becks's Mountain in northeast Pike County is 675 feet. Other high hills are located in Bullock and Barbour counties. The top of Mount Andrew in northwest Barbour County is near 630 feet.

Physiography

Geologists have used rock locations, their characteristics, and the processes that created them as a basis for dividing Alabama into five physiographic sections: Highland Rim, Piedmont Upland, Alabama Valley and Ridge, Cumberland Plateau, and East Gulf Coastal Plain (Figure 1.9). For example, the assemblage of highly metamorphosed rocks and groups of rocks, with the northeast to southwest orientation, in east-central Alabama (principally Cleburne, Randolph, Chambers, Lee, Clay, Tallapoosa, Coosa, and Elmore counties) is the Piedmont Upland Section (Figure 1.9). Obviously, the processes that resulted in formation of these sections were not confined to Alabama alone. As a result, the Piedmont Upland Section extends from east-central Alabama to beyond the state of Maine, as part of the Piedmont Province (Figure 1.10). Each state with a portion of this assemblage of rocks has a Piedmont Upland Section. All of these state sections together are given the name Piedmont Province, and in turn, the Piedmont Province is part of the Appalachian Highlands Physiographic Division. The relationship between the sections, provinces, and divisions in the Eastern United States is shown in Table 1.2. The location of other physiographic provinces within the United States is also shown in Figure 1.10.

Further, geologists have also determined that physiographic sections are actually too large to adequately represent the physiographic diversity represented within them. Therefore, they divide the sections into districts, and in some situations, the districts are further divided into sub-districts. The sections, districts, and sub-districts in Alabama are listed in Table 1.3 and are shown in Figure 1.9.

The Highland Rim Section

The Highland Rim Section is located in the extreme north-western corner of Alabama, primarily in Colbert, Lauderdale, Limestone, Madison, Lawrence, and Morgan counties. Small portions also extend into Franklin and

Jackson counties (Figure 1.9). US 43 first encounters the section about two miles north of the village of Liberty Hill in south Franklin County (Figure 1.12). The highway remains in the section until it reaches Tennessee, except for passing-through a small out-crop of the Fall Line Hills District of the East Gulf Coastal Plain Section in the town of Russellville. I-65 enters the section near mm 317, about a mile south of Lacon in south Morgan County and leaves it at the Tennessee state line. US 431 enters the section near the village of Mountain View, southeast of Guntersville, and leaves it at the Tennessee state line (Figure 1.12).

The section generally has a rectangular shape with the long axis running east to west (Figure 1.9). Raymond, et al (1988) generally describe this section as a plateau of moderate relief. The relatively flat, valley-like character of the section is interrupted by a band of narrow mountainous highlands (Little Mountain) that divide it into the larger Tennessee Valley on the north and the smaller Moulton Valley on the south.

The Highland Rim Section is only a small portion of the larger Interior Low Plateaus Province (Figure 1.10 and Table 1.2). The larger portion of the province is located in Tennessee and Kentucky. Only a small portion is located in Alabama. The Interior Low Plateaus Province (Figure 1.10) is the easternmost extension of the enormous area of the Interior Plains Division that includes a third of the entire area of the contiguous United States. The Great Plains Province is the westernmost portion of this division.

The Highland Rim is the only remnant in Alabama of the more or less level foundation of sandstone, shale, and limestone (Figure 1.7) laid down on the southeastern margin of the continent, before the collision of Larussia and Gondwanaland at the beginning of the Paleozoic Era, some 550 mya (Levin, 1994).

Drahovzal (1968) and Sapp and Emplaincourt (1975) divide the Highland Rim Physiographic Section into three districts. From north to south they include: Tennessee Valley, Little Mountain, and Moulton Valley.

Tennessee Valley District

The Tennessee Valley is the larger of the three districts of the Highland Rim Section (Figure 1.9). It is an elongate valley with its long axis running almost the entire width of the state in the northernmost tier of counties. The valley has a maximum width of 35 miles and forms the basin for much of the Tennessee River in Alabama. Florence (Lauderdale County), Athens (Limestone), Decatur (Morgan), and Huntsville (Madison) are located in this district.

The Tennessee Valley District has a rolling surface with a maximum relief of 400 feet, much greater than the 100 feet of relief in Moulton Valley to the south (Figure 1.8).

Little Mountain District

The Little Mountain District consists of a narrow range of rugged hills lying generally along an east-to-west axis, primarily in Lawrence and Morgan counties (Figure 1.9). The district averages about five miles in width, and is around 40 miles in length. It is located between the Tennessee Valley and Moulton Valley Physiographic districts. Russellville (Franklin County), Littleville (Colbert), Hatton and Caddo (Lawrence), and Hartselle (Morgan) are in this district.

In portions of the district, especially in Colbert County, the highlands of Little Mountain rise steeply from the floor of the Tennessee Valley. Many of the higher ridges are between 800 and 900 feet in elevation (Figure 1.8). Harper (1943) called these steep slopes a "bold escarpment."

Moulton Valley District

Moulton Valley is a long, narrow, undulating open lowland with an east-west axis (Figures 1.9 and 1.11). The valley is 85 miles long and from three to eight miles wide. It is located on the south side of the highlands of Little Mountain and north of the Warrior Basin District of the Cumberland Plateau Section, primarily in Franklin, Lawrence, and Morgan counties. Saints Crossroads and Russellville (Franklin County), Moulton (Lawrence), and Falkville (Morgan) are located in the Moulton Valley District.

The surface of the Moulton Valley is relatively flat (Figure 1.8). The elevation of the entire valley ranges from about 575 to slightly less than 700 feet. The higher elevations are generally in the western end. Because it is sandwiched between highlands on the north and south, it is difficult to determine exactly where the valley ends and the mountains begin. While the valley is relatively flat, it generally is tilted to the north. All of the drainage in the valley is into the Tennessee River.

Piedmont Upland Section

The Piedmont Upland Section (Figure 1.9) is generally shaped like an equilateral triangle with its base on the fall line, its right leg on the Georgia line, and its left leg lying against the southeastern boundary of the Alabama Valley and Ridge Section (Figure 1.9). The section is located principally in Cleburne, Clay, Randolph, Coosa, Tallapoosa, Chambers, Elmore, and Lee counties. I-65 enters the Piedmont Upland Section near mm 209, southeast of Clanton (Figure 1.11), and leaves it around mm 227 near Calera. US 431 enters the section around mm 120, near the Lee/Russell line just north of Phenix City. It leaves the section just south of its intersection with I-20, east of De Armanville in southeastern Calhoun County (Figure 1.12).

Williams and Hatcher (1983) concluded that the Piedmont Province of eastern North America (Figure 1.10 and Table 1.2) that includes the Piedmont Upland Section of Alabama (Figure 1.9) was emplaced when the Talladega platform assemblage and the Piedmont micro-continent were sutured to the eastern margin of North America by heat and pressure when Larussia (North America and northern Europe) and Gondwanaland (South America, Africa, India, Australia, and Antarctica) collided, some 300 mya (Figure 1.5) during the Mississippian, Pennsylvanian, and Permian periods of the Paleozoic Era (the Appalachian Orogeny). The Talladega-Cartersville Fault or crustal fracture zone, which runs generally from the Cleburne - Cherokee County line southwestward to east-central Talladega County, generally marks the northwestern boundary of the region where the Talladega platform assemblage rocks were thrust up and over ancient carbonate rocks on the North American continental shield. The Brevard Fault (Hatcher, 1989), which lies diagonally northeast to southwest across southeastern Randolph, northwestern Chambers and central Tallapoosa counties, marks the exposed surface boundary or suture zone (crustal fracture) where the Piedmont fragment was forced against the Talladega fragment.

According to Drahovzal (1968) and Raymond, et al. (1989), the Piedmont Upland Section is a submaturely dissected surface developed upon metamorphosed sedimentary rocks such as quartzite (formed from sandstone) and igneous rocks such as granite (formed from molten material in the earth's mantle). Most of the rocks in the section are metamorphic (Figure 1.7). The degree of metamorphism (metamorphic grade) generally increases across the section, northwest to southeast (Steltenpohl and Moore, 1988). Higher metamorphic grades are generally associated with the exposure of rocks to greater heat and pressure.

Drahovzal (1968) divided the section into two districts: the Ashland Plateau and the Upland Plateau. Sapp and Emplaincourt (1975) also divide it into two parts: the Northern Piedmont Upland and the Southern Piedmont Upland. Raymond, et al. (1989) divided it into three: the Northern Piedmont that corresponds to Drahovzal's Ashland Plateau, the Inner Piedmont, and the Southern Piedmont. The Inner Piedmont and Southern Piedmont correspond to Drahovzal's Upland Plateau. The districts described by Raymond, et al. (1989) will be used in the following discussions. In their districts, the Northern Piedmont contains almost two-thirds of the total land area. The Southern Piedmont probably contains no more than 10%. The districts of Raymond, et al. include: Northern Piedmont, Inner Piedmont, and Southern Piedmont.

Northern Piedmont District

Sapp and Emplaimcourt (1975) describe this district as a "well-dissected upland developed on metamorphosed sedimentary and igneous rocks." It is bounded on the northwest by the Talladega-Cartersville Fault or crustal fracture zone, which lies generally northeast to southwest from a few miles northwest of the Cleburne and Cherokee County line in the northeast to north Chilton County line in the southwest (Figure 1.9). This fault system is also the southeastern boundary of the Alabama Valley and Ridge Section. The district is bounded on the south by the Brevard Fault Zone, which lies diagonally, northeast to southwest, across southeastern Randolph, northwestern Chambers, and central Tallapoosa counties. The district is slightly over 100 miles long and 35 miles wide. It is shaped somewhat like an inverted trapezoid with its base lying along the boundary with the Valley and Ridge Section; consequently, its two parallel sides have a strong northeast to southwest trend. Heflin (Cleburne County), Ashland (Clay), Wadley (Randolph), Rockford (Coosa) and are other towns located in the district.

The Northern Piedmont District is a mountainous area with narrow, steep valleys, and rolling upland surfaces (Raymond, et al., 1988). The central portion of the district in southwestern Cleburne, northeastern Talladega, and northern Clay counties contains some of the highest and most rugged terrain in the state (Figure 1.8). There are ten mountains here with elevations above 1,500 feet.

Inner Piedmont District

This district lies primarily in Chambers County, with portions in southeast Randolph, southeastern Tallapoosa, and northern Lee counties (Figure 1.9). It is 50 miles long and 30 miles wide. Figure 1.11 does not include the Inner Piedmont District. In this figure, it is included within the Southern Piedmont Upland.

The terrain of the Inner Piedmont is not nearly as rugged as that in the Northern Piedmont to the north, but it is hilly (Figure 1.8). Here the high mountainous ridges and deep narrow valleys are replaced by broad low ridges and broad shallow valleys.

Southern Piedmont District

The Southern Piedmont District in Alabama is confined primarily to the southeastern portion of Lee County (Figure 1.9). Its northwestern boundary is the Towaliga Fault Zone, which trends northeast to southwest, passing through the point where Chambers and Lee counties meet at the Georgia line and through Opelika. Its southern boundary is the fall line in south Lee County, which is the northern boundary of the East Gulf Coastal Plain Physiographic Section (Figure 1.9).

The relief in this District is generally equal to or lower than in the Inner Piedmont District , and much lower than in the Northern Piedmont (Figure 1.8).

Alabama Valley and Ridge Section

The Alabama Valley and Ridge Section consists of a series of parallel ridges and valleys, underlain by highly faulted and folded rocks, lying principally in Jefferson, St. Clair, Etowah, Cherokee, Shelby, Talladega, and Calhoun counties (Figure 1.9). It is a portion of the much larger Valley and Ridge Physiographic Province (Figure 1.10 and Table 1.2) covering a sizeable portion of the Eastern United States. The ridges and valleys have a conspicuous northeast-southwest trend, which indicates that most of the force causing the folding came from the southeast. The ridges are underlain by resistant sandstones and chert, and the valleys are developed on less-resistant shale and carbonate rocks (Raymond, et al., 1988) (Figure 1.7). I-65 enters the Alabama Valley and Ridge Section near mm 227, southeast of Calera in Shelby County (Figure 1.12); and leaves it around mm 263, near Sayreton in Jefferson County. US 431 enters the section just south of its intersection with I-20, near De Armanville in southeast Calhoun County. It leaves the section around mm 274 northeast of Attalla in central Etowah County.

As described in a preceding section, the collision of the tectonic plates bearing Gondwanaland (South America, Africa, India, Australia, and Antarctica) and Laurussia (North America and Europe) during the Appalachian Orogeny resulted in the re-melting and violent re-organization of the crystaline rocks of the Talladega Mountains and the Piedmont in Alabama. The intense pressure of the collision also thrust upward the eastern portion of the sandstone, shale, and limestone (Figure 1.7) of the ancient Appalachian Plateau, lying north and west of the Piedmont Upland, to form the Appalachian Mountains. The accumulation and metamorphosis of these sediments was described in a preceding section. This great episode of mountain building resulted in the development of a chain of mountains that extended from central Alabama to southern New York and that rivaled the Alps in Europe for grandeur (Hughes, 1973).

The intense pressure affected to a lesser extent another portion of the ancient sandstones, shales, and limestones of the Appalachian Plateau lying westward of the newly raised mountains, but it was not nearly as great here as it was farther east; consequently, the effect on the old rocks was not as cataclysmic. There was little or no metamorphsim of rocks that far away from the point of impact between the two continental masses. Rather than changing the form of the rocks and thrusting them upward to heights of several thousand feet, they were folded and contorted like the folds of an accordion. This deformational folding, and erosion that followed, resulted in the formation of parallel ridges separated by narrow valleys that make up our Alabama Valley and Ridge Section.

Sapp and Emplaincourt (1975) described seven districts within the Alabama Valley and Ridge Section. These include: Weisner Ridges, Armuchee Ridges, Coosa Valley, Coosa Ridges, Cahaba Valley, Cahaba Ridges, and Birmingham-Big Canoe Valley.

Weisner Ridges District

The Weisner Ridges District consist of a series of parallel mountains, seven miles wide and 40 miles long that trend northeast to southwest in the southeastern portion of the Alabama Valley and Ridge Section. It is located in Cherokee, Cleburne, Calhoun, and Talladega counties (Figure 1.9). Sapp and Emplaincourt describe them as "maturely dissected faulted and folded quartzite mountains of high relief with intervening carbonate valleys."

Some of the highest mountains in Alabama are located in this district. In fact, Dugger (2,140) and Choccolocco (2,063), located in northwest Cleburne and northeast Calhoun counties (Figure 1.8), respectively, are two of the higher elevations in Alabama. In other parts of the district, there are more than 14 mountains with elevations greater than 1,500 feet.

Armuchee Ridges District

The Armuchee Ridges are a series of ridges with northeast to southwest trend, six miles wide and 13 miles long, located in northeast Cherokee County (Figures1.9). This is the smallest physiographic district in Alabama. It is geologically different from Lookout Mountain to the north, the Coosa Valley to the west, or the Weisner Ridges to the southwest. These ridges are not as high as the Weisner Ridges to the southwest.

Coosa Valley District

The long, broad Coosa Valley is 130 miles in length and 20 miles in width. It also has a northeast to southwest trend. It is located primarily in Cherokee, Etowah, St. Clair, Talladega, Jefferson, Shelby and Bibb counties, and includes most of the land area in the Valley and Ridge Section (Figure 1.9). It extends from the Georgia line in eastern Cherokee County southwestward to three miles north of the point where Shelby, Chilton, and Coosa counties join. Childersburg, Pell City, Gadsden, and Centre are located in the district.

Drahovzal (1968) describes the Coosa Valley District as a "relatively broad, flat lowland of low relief." Most of the elevations in the valley are between 500 and 650 feet (Figure 1.8). The upper portion of the valley tends to be much more level than the lower portion. The Coosa River has cut or eroded a channel through mountains near the village of Greensport in northeast St. Clair County. Now the Beaver Creek Mountains lie west of the river, and Greens Creek Mountain lies to the east. This channel was the location of the head of Coosa Shoals before Neely Henry Lake was created. The erosion-resistant carbonate rocks at the head of these shoals served as a low dam that, in effect, slowed the velocity of the river from that point north to the Georgia line. This natural dam reduced the rate of erosion in the upper portion of the valley and encouraged the meandering of the river channel and the more or less level deposition of sediments being carried by the river from the highlands in west-central Georgia. The soils formed from these sediments were the basis of the development of the cotton kingdom in the valley between 1850 and 1860.

Coosa Ridges District

Sapp and Emplaincourt (1975) describe this district as a "series of parallel linear ridges formed by folded sandstone." It is 55 miles long and five miles wide, and lies on the northern shoulder of the southern-western portion of the Coosa Valley and west of the Coosa River in St. Clair and Shelby counties (Figure 1.9). The Cahaba Valley is to the north. The district consists of a series of massive, almost mountainous, long parallel ridges with a northeast to southwest trend separated by narrow valleys (Drahovzal,1968). These accordion-like ridges are timeless indicators of the unimaginable lateral pressure generated at some point southeast of Shelby and St. Clair counties by the collision of tectonic plates hundreds of millions of years ago. There are more than 20 named mountains in the district.

The parallel ridges of the Coosa Ridges District are mountainous and extremely rugged (Figure 1.8).

Cahaba Valley District

The Cahaba Valley District is a long, narrow valley-like lowland "developed on a faulted homocline (monocline)."

This description implies that a vertical movement of rocks resulting from tectonic plate activity was partly responsible for the formation of the valley. The district, trending northeast to southwest, is located principally in Etowah, St Clair, Jefferson, Shelby, and Bibb counties (Figure 1.9). It is less than five miles wide for much of its 100-mile length, but widens somewhat at both the upper and lower ends. It is separated from the Valley of the Coosa by the Coosa Ridges. The Cahaba Ridges form the northern shoulder of the valley (Drahovzal, 1968). The northeastern terminus of the district merges with the Coosa Valley District near the point where Cherokee, Calhoun, and Etowah counties meet, east of Piedmont. Alabama 119, often called the Cahaba Valley Road (Figure 1.12), follows the valley from Montevallo in southwest Shelby County to Leeds in the eastern portion of Jefferson County. Southside (Etowah County), White Chapel (St. Clair), Simsville (Jefferson), Montevallo (Shelby), and Six Mile (Bibb) are in this district.

The Cahaba Valley is not a valley in the same sense as the Tennessee or Coosa valleys in that it does not carry a major river over a long distance. In fact, the Cahaba Valley District carries only a short segment of the Cahaba River. Most of that river is in the Cahaba Ridges District to the west. The Cahaba River finally enters the Cahaba Valley District a few miles northeast of Centerville (Bibb County) just before plunging over the fall line. The district does provide the basin for the Little Cahaba River. It is also interesting that the Coosa River flows across the upper end of the district in Etowah County. Rather than serving as a conduit for a major river, the Cahaba Valley District carries the valleys and channels of several moderate size creeks. At its eastern terminus in Etowah County, it contains Big Cove Valley and Big Cove Creek. A few miles westward, it contains Little Cove Creek and Little Cove Creek Valley, then Greens Creek and Greens Creek Valley. Further west in St Clair County, it contains Beaver Creek Valley and Beaver Creek. In Jefferson County, it carries the Little Cahaba River southwestward into Lake Purdy. Once it leaves Lake Purdy, the river flows west out of the district through a gap separating the northeast end of New Hope Mountain and the southwest end of Pine Ridge. It empties into the Cahaba River just east of US 280 in eastern Jefferson County.

Cahaba Ridges District

The Cahaba Ridges lie between the Cahaba Valley to the southeast and the Birmingham-Big Canoe Valley District to the northwest (Figure 1.9) in St. Clair, Jefferson, Shelby, and Bibb counties. Sapp and Emplaincourt (1975) describe this district as a "series of parallel, northeast striking ridges formed by gently folded sandstone and conglomerate beds with intervening shale valleys." Like the Coosa Ridges to the southeast, the Cahaba Ridges are also the result of the collision of tectonic plates. This series of northeast to southwest trending ridges and valleys is less than five miles in width for most of its almost 60 miles of length, but becomes wider (up to 25 miles) at its southwestern terminus in the northeastern portion of Bibb County. The Cahaba River runs through this district for much of its length. Margaret (St. Clair County), Acton (Shelby), Bluff Park (Jefferson), and Marvel (Bibb) are located in this district.

Birmingham-Big Canoe Valley District

The Birmingham-Big Canoe Valley, a narrow limestone valley, lies northwest of the Cahaba Ridges and the Cahaba Valley, principally in Etowah, St. Clair, Jefferson and Tuscaloosa counties (Figure 1.9). The different districts of the Cumberland Plateau Section lie to the north. The district is 100 miles long and five miles wide, although it widens somewhat southwest of Birmingham. At its northeastern terminus, in the city limits of Gadsden, the Birmingham-Big Canoe Valley District, like the Cahaba Valley District, merges into the Coosa Valley District. In that area, the Birmingham-Big Canoe Valley and the Cahaba Valley Districts are separated by Hines Mountain. The southwestern terminus of the Birmingham-Big Canoe Valley District is east of the village of Vance in the eastern portion of Tuscaloosa County. This district, like the Wills, Murphrees, and Cahaba Valley districts, does not carry a major river. Although erosion certainly played a major role in the formation of this long lowland, rock movement related to tectonic plate activity probably played the major role in determining its length and shape.

As the name implies, the Birmingham-Big Canoe Valley District is divided into two distinct portions. Birmingham Valley is the southwestern portion, and Big

Canoe Creek Valley is the northeastern portion. The dividing line is in northeastern Jefferson County, east of the town of Palmerdale, where Big Canoe Creek originates. A tributary of the Cahaba River also originates in this vicinity. Big Canoe Creek flows eastward to empty into the Coosa River north of Greenport in the northeastern portion of St. Clair County. Birmingham Valley is given that name because much of downtown Birmingham is located in it, but Jones Valley is the older name for the lowland. Jones Valley, as such, generally originates east of the town of Center Point in east-central Jefferson County. Northeast of that point, part of the same lowland, in the vicinity of the origin of Big Canoe Creek, is called Clayton Cove. From Center Point, Jones Valley courses through downtown Birmingham. It provides the basin for Village Creek from near Irondale to near the junction of US 11 and US 31. At that point, the creek leaves the district and flows westward into the Warrior Basin District of the Cumberland Plateau Section where it flows into Bayview Lake. Further downstream, its waters enter the Locust Fork of the Black Warrior River. Southwest of the junction of US 11 and US 31, Jones Valley provides the basin for Valley Creek, which flows southwestward in the district, finally turning westward and leaving it near the town of Bessemer. Much of downtown Birmingham lies within this relatively long, narrow valley. St. Clair Springs (St. Clair County), Trussville, Bessemer, Bucksville, and McCalla (Jefferson) are also located in this district.

Mineral Resources of the Section

Mineral resources being exploited in the Alabama Valley and Ridge Section include limestone, dolomite, shale, kaolin, chert, sand, and gravel. The province contains two of Alabama's coal fields, one in the Coosa Valley and one in the Cahaba Valley (Figure 1.12). The coal beds are found primarily in the Pottsville Formation of inter-bedded sandstone, siltstone, and shale. Although there are quantities of iron ore in the section, they are not being commercially exploited at this time (Raymond, et al., 1988). These deposits, found in the vicinity of Birmingham, are much older on the geological time scale than the coal fields. During the Silurian Period, some 420 mya, a vast shallow sea covered much of the central part of North America, including the area that much later would become the Valley and Ridge Section in Alabama. Enormous quantities of limey, iron-containing sediments accumulated on the floor of this basin. Later, the weight of overlying accumulations would convert these sediments into iron ore.

Cumberland Plateau Section

The Cumberland Plateau Section is located primarily in Winston, Cullman, Marshall, Jackson, Walker, Blount, DeKalb, and Jefferson counties (Figure 1.9). Raymond, et al. (1988) describes this section as a "series of submaturely to maturely dissected sandstone and shale plateaus of moderate relief."

US 43 enters the Cumberland Plateau Section about seven miles north of its junction with US 82 in south-central Tuscaloosa County (1.12). It goes in and out of the Warrior Basin District of the section and the Fall Line Hills District of the East Gulf Coastal Plain Section until it leaves the section about two miles north of Liberty Hill in south Franklin County. I-65 enters the section around mm 263, near Sayreton, north of Birmingham (Figure 1.12). It leaves the section around mm 317, just south of Lacon in southwestern Morgan County. US 431 enters the section around mm 274 just north of Attalla in central Etowah County. It leaves the section around mm 317, just west of Mountain View in southeastern Marshall County (Figure 1.12).

King (1977) comments that the Cumberland Plateau was derived from the vast amounts of sandstone, shale, and silt eroded from the high mountains to the east (Figure 1.7). The sediments were deposited onto the largely carbonate surface of the ancient Interior Lowlands, also described in a preceding section. Over time, they covered vast forests of tropical vegetation, which later would be compressed into thick beds of coal. This immense plateau of sediments were also subjected to severe erosion. Emerson (1920) comments that the plateau has been so maturely eroded that only comparatively small areas of level land remain. In Alabama, only a small area north of US 72 and east of Paint Rock River in Jackson County is identified as the original Cumberland Plateau. The area is generally west of Alabama 79, although the highway runs along the northern portion of the area between Skyline and Hytop. The present day Cumberland Plateau

Physiographic District refers to the general shape of the plateau when it emerged from the sea. The effect of erosion has not been uniform across the plateau resulting in the development of a markedly hilly terrain. In fact, it is so hilly that much of it might more appropriately be called mountainous.

The Appalachian Plateaus Province that includes Alabama's Cumberland Plateau Section (Figure 1.10 and Table 1.4), extends from north-central Alabama northeastward into New York State. It is a much larger, broader area in West Virginia and Pennsylvania than in Alabama. Throughout its length, it lies to the west of the Valley and Ridge Province and the Appalachian Mountains.

Sapp and Emplaincourt (1975) and Raymond, et al. (1988) identified eight physiographic districts within the section. Osborne, et al. (1989) developed an excellent map showing the location of geological formations associated with those districts: Warrior Basin, Jackson County Mountains, Sequatchie Valley, Wills Valley, Sand Mountain, Lookout Mountain, Blount Mountain, and Murphrees Valley.

The Cumberland Plateau Section is roughly divided into three distinct regions (Figure 1.9): a region of high relief in the northeastern portion of the district that includes the Jackson County Mountains; a region of somewhat lesser relief in the southeastern portion that includes Sequatchie Valley, Sand Mountain, Murphrees Valley, Wills Valley, Lookout Mountain, and Blount Mountain; and the western half of the district that includes the Warrior Basin (Figure 1.9).

Warrior Basin District

The Warrior Basin includes most of the area within the Cumberland Plateau Section and all of the western-most two-thirds (Figure 1.9). Much of the basin is located in Blount, Cullman, Winston, Fayette, Walker, Jefferson, and Tuscaloosa counties. Its greatest north-south dimensions are 90 miles in the west and 45 miles in the east. Its greatest east-west dimension is 80 miles. Tuscaloosa, Jasper, Cullman, Double Springs, and Falkville are in this district.

The Warrior Basin District is quite mountainous along its northern boundary (Figure 1.8). These mountains consist primarily of highly erosion resistant limestones and shales. The mountainous highlands in southeastern Franklin, northeastern Marion, southern Lawrence, northern Winston, and western Cullman counties form the so-called Central Mountains of the state. These mountains are at the terminus of an arcuate band of highlands that originates in Jackson County and reaches across the entire state. Several of these mountains have elevations of 1,000 feet or higher.

Jackson County Mountains District

The Jackson County Mountains District is located in northwestern Jackson, eastern Madison, and northeastern Marshall counties (Figure 1.9). Mountains in the district include: Little Cabin, Crow, Putman, Jacobs, and McCoy Mountains in Jackson County and Lewis and McKinney Mountains in Madison and Marshall counties, respectively (Figure 1.8). The greatest north-south and east-west dimensions of the mountainous area are 40 miles and 35 miles, respectively. Skyline, Paint Rock, Trenton, Stevenson, and Princeton in Jackson County are all located in this district.

The Jackson County Mountain District is the most mountainous area in the state (Figure 1.8). Although the mountains here are not quite as high as those in the northwestern corner of the Northern Piedmont District, the mountainous area of this district is more extensive. Also, the entire area is uniformly higher. Several mountains over 1,000 feet in elevation are located in portions of Madison, Marshall, and Jackson counties in this area of northeast Alabama. Relief between the valley floor and some of the higher mountain tops is almost 1,100 feet.

Sequatchie Valley District

Sapp and Emplaincourt (1975) describe the Sequatchie Valley District (Figure 1.9) as an "anticlinal, tripartite valley of moderate relief." This description implies that the valley was once an up-fold (anticline) or broad ridge of rock sandwiched between two down-folds (synclines) and that over hundreds of millions of years, swift streams have eroded most of the ridge (up-fold) away, leaving it lower in elevation than the ancient synclinal valleys, (down-folds) on each side. The Jackson County Mountains on the west and Sand Mountain on the east are high plateaus, remnants of the ancient valleys (down-folds). The Sequatchie Valley consists of three parallel parts: a ridge down the center with a narrow valley on each side. It is long and narrow (four to eight miles wide) with a northeast to southwest trend.

Sequatchie Valley, which originates in east-central Tennessee, enters Alabama at its extreme northeast corner in Jackson County, then trends southwestward through Marshall County into northeast Blount County, a distance of 85 miles (Figure 1.9). From its northeastern terminus to Guntersville, the valley is mostly submerged under the waters of Guntersville Reservoir on the Tennessee River. Some aspects of the physiographic structure of the upper valley can still be seen east of Bridgeport in Jackson County. There the reservoir is still largely confined to the original river channel. For example, Hog Jaw Ridge is a prominent feature of the upper portion of the district. It lies in the valley east of the Tennessee River northeast of Bridgeport. At Guntersville, the river turns westward, and the Sequatchie Valley emerges from Guntersville Reservoir. From Guntersville in Marshall County to its southwestern terminus in northeastern Blount County, the valley is divided into eastern and western portions by the so-called Dividing Ridge. Guntersville is located on that ridge. This low ridge is only a remnant of the ancient anticline (up-fold) that probably, at one time, towered over the Jackson County Mountains and Sand Mountain, which were synclinal valleys at that time. Big Springs Valley lies to the east of the Dividing Ridge, and Brown Creek Valley lies on the west side. Near Brooksville in north Blount County, the Sequatchie Valley crosses the divide between the Tennessee and Black Warrior River Drainage Basins. At that point, the Dividing Ridge becomes the Middle Ridge. Brown Valley, which was west of Dividing Ridge, is replaced by Blountsville Valley west of Middle Ridge that carries the headwaters of the Mulberry Fork of the Black Warrior. East of the Middle Ridge, a valley carries the headwaters of the Locust Fork of the Black Warrior River. The southwestern terminus of the Sequatchie Valley District is located near Blountsville in west-central Blount County where the Middle Ridge merges with the Warrior Basin District. Bridgeport and Scottsboro (Jackson County), Guntersville (Marshall), and Brooksville (Blount) are located in this district.

Sand Mountain District

The Sand Mountain Physiographic District is a broad, flat-topped mountain, actually more like a plateau, located primarily in Jackson, Dekalb, Marshall, and Etowah counties. Sapp and Emplaincourt (1975) describe it as a "synclinal (down-folded) plateau of moderate relief." It is located to the southeast of the Sequatchie Valley and to the northeast of Wills and Murphrees valleys (Figure 1.9). Its long axis with a maximum length of 90 miles is generally oriented southwest to northeast. It has an average width of 15 miles. It is generally much higher in the northeast than in the southwest. The northeastern boundary of Sand Mountain is the Georgia line. The mountain extends northeastward into Georgia for some distance. The southwestern boundary of the district is somewhat indefinite. Generally its north-western shoulder along its boundary with the Sequatchie Valley District merges with the Warrior Basin District in north Blount County east of Blountsville. Most of the mass of the mountain also merges into the Warrior Basin in that general vicinity; however, a narrow ridge of its southeastern shoulder extends all the way to central Jefferson County terminating east of Fultondale. This extension is too narrow to be shown on Figure 1.9. There is an interesting physiographic feature associated with the southeastern shoulder of Sand Mountain. There is a narrow valley at the base of the mountain and running parallel to it from west of Hammondville in northern Dekalb County to Pinson in northeastern Jefferson County. For much of its length, it is called Sand Valley. In northeastern Etowah County, it is called Little Sand Valley, and in Jefferson County it becomes Sand Hollow. For a portion of its length (west of Hammondville to east of Crossville), there is also a narrow ridge associated with Sand Valley, lying parallel to it on the southeast. The ridge is called Little Ridge west of Hammondville; then Shinbone Ridge, Ripshin Ridge, and finally Mt. Vera Ridge as it runs through Dekalb County. Oneonta (Blount County), Albertville (Marshall), Geraldine (DeKalb), Rainsville (DeKalb), and Higdon (Jackson) are located in the district.

The average elevation of Sand Mountain is 1,000 feet, but there are areas on its northeast corner—just east of the Tennessee River and north of Warren in Jackson County—with elevations above 1,600 feet (Figure 1.8).

Wills Valley District

Sapp and Emplaincourt (1975) also describe Wills Valley as an "anticlinal (up-folded) tripartite valley." In this respect, it is similar to Sequatchie Valley described previously. Sand Mountain on the west and Lookout Mountain on the east are plateaus, remnants of ancient synclinal (down-fold) valleys. The Wills Valley District is located southeast of Sand Mountain in Dekalb and Etowah counties (Figure 1.9). It is a long (65 miles), narrow (five miles) valley with its long axis also oriented northeast to southwest (Drahovzal, 1968 and Osborne, et al., 1989). The valley, in Alabama, begins at the Georgia line near Rogers in northeast Dekalb County, and ends in Gadsden in Etowah County where Big Wills Creek empties into the Coosa River. It separates Sand Mountain on the northwest and Lookout Mountain on the southeast. The valley proper is generally divided lengthwise into three parts by two parallel sandstone ridges, remnants of ancient anticlinal (up-folded) ridges. For example, Alabama 40 passes across it at Hammondville in northeast Dekalb County. After leaving Mentone on Lookout Mountain, the highway descends into the Wills Valley District. It passes northwestward over a distance of five miles; across Railroad Valley, Little Ridge, Big Wills Valley, Big Ridge, Dugout Valley, Little Ridge, and Sand Valley; before climbing onto Sand Mountain. Fort Payne, Hammondville, and Valley Head (DeKalb County) and Sardis City and Aurora (Etowah) are located in the district. Given the complexity of the series of valleys and ridges within the Wills Valley District, the topography is quite varied (Figure 1.8).

Lookout Mountain District

Sapp and Emplaincourt (1975) also describe the Lookout Mountain District as a "narrow, synclinal (down-folded), sub-maturely dissected, flat-topped remnant of the Cumberland Plateau." The district is located primarily in DeKalb, Cherokee, and Etowah counties (Figure 1.9). It lies to the southeast of Wills Valley and northeast of the Coosa Valley District of the Valley and Ridge Section. It is similar in many respects to its larger brother, Sand Mountain, to the north across Wills Valley. Its average width is less than ten miles, and it is 45 miles long, about 2/5 as wide and half as long as Sand Mountain. Its long axis is oriented, like Sand Mountain, southwest to northeast. Little River and Little River Canyon divide Lookout Mountain into two parts in the northeastern portion of the district. The mountain is also a plateau, tilted downward to the southwest. The difference in elevations from the northeast near Mentone to the southwest is almost 900 feet (1,840 to 980). The mountain's highest elevation is almost 200 feet greater than the highest elevation on Sand Mountain. A narrow valley and ridge lie just below the southeastern shoulder of Lookout Mountain. Shinbone Valley lies parallel to the mountain for its entire length except where Little River Canyon passes off it onto the Coosa Valley northwest of Cedar Bluff in Cherokee County. Northeast of Gadsden, the name of the Shinbone Valley changes to Owl Valley. Shinbone Ridge lies on the southeast side of Shinbone Valley and parallel to it. The relationship among Lookout Mountain, Shinbone Valley and Shinbone Ridge is similar to the relationship between Sand Mountain, Sand Valley, and Shinbone and Ripshin ridges. Oakdale (Etowah County), Scrougeout (Etowah), and Dog Town (DeKalb) are in the district.

The topography of most of Lookout Mountain is similar to that of Sand Mountain (Figure 1.8); however, it is somewhat higher along its eastern boundary. The highest elevation near the Georgia line is 1,840 feet, some 200 feet higher than the highest elevation on Sand Mountain.

Blount Mountain District

Blount Mountain (Figure 1.9), like Sand Mountain and Lookout Mountain, is a sub-maturely dissected, synclinal (down-folded) sandstone and shale plateau. It is located in northwestern St. Clair, southwestern Etowah, and southeastern Blount counties, and is generally a southwestern extension of Sand Mountain. Except for the northeastern portion of Murphrees Valley and the southwestern portion of Wills Valley, both Blount and Chandler mountains would be part of Sand Mountain. Blount Mountain is 35 miles long and less that ten miles wide. Chandler Mountain is a small arm off the southeastern side of Blount Mountain. Straight Mountain is a narrow ridge off the northwestern shoulder of Blount Mountain. A major tributary of the Locust Fork of the Black Warrior River, begins on the northwestern slope of Blount Mountain and flows southwestward. A dam constructed on it east of Remlap in south Blount County impounds Inland Lake. Drainage off Chandler Mountain flows southward to the Coosa River.

The elevation of Blount Mountain is slightly higher than Sand Mountain (Figure 1.8). Across Murphrees Valley to the north in eastern Blount County, the Sand Mountain District is beginning to merge into the Warrior Basin District. Most elevations there are in the range of 1,000 to 1,200 feet. On Blount Mountain, the elevations generally range between 1,000 to 1,300 feet.

Murphrees Valley District

Murphrees Valley is also an anticlinal (up-folded) valley. It is located on the northwestern side of the Blount Mountain District, primarily in southeastern Blount County with limited extensions into Jefferson and Etowah counties (Figure 1.9). Straight Mountain lies on its southeastern shoulder, and Red Mountain lies on the northwestern side. Murphrees Valley has a northeast to southwest axis, and a length and width of 40 miles and three to four miles, respectively. It originates just south of Walnut Grove, located on US 278 in southwestern Etowah County. Its terminus is located southwest of Remlap in south Blount County. Oneonta in eastern Blount County is near its midpoint and also its highest point. The drainage is northeastward and southwestward from the town.

As noted in a preceding paragraph, Murphrees Valley slopes downward both to the northeast and the southwest from its midpoint in Oneonta (Figure 1.8). To the northeast, the differential is 35 feet (885 to 850). It is much greater to the southwest.

Mineral Resources of the Section

Mineral resources being exploited in the Cumberland Plateau Section include oil and gas from several sandstone and limestone units in the Warrior Basin. Also, coal is being mined from two of Alabama's four coal fields found in this district (Figure 1.12). The Warrior coal field contains 90% of the state's known coal reserves.

East Gulf Coastal Plain Section

The East Gulf Coastal Plain Physiographic Section is considerably larger than the combined areas of all of the other sections (Figure 1.9). All, or part, of 45 of Alabama's 67 counties are in the section. Its northern boundary follows a roughly arcuate line (the fall line) from Lauderdale and Colbert counties in the northwest, through Chilton County in the central portion of the state, to Lee County in the southeast. The fall line generally marks the boundary between the older metamorphic rocks (limestone, shale, sandstone, schist, and quartzite) of north Alabama and the younger sedimentary deposits of gravel, sand, silt, and clay in south Alabama. This line is most apparent where it passes through Chilton, Elmore, Tallapoosa, and Lee counties at the southwestern boundary of the Piedmont Upland Section. Monroe (1941) comments that the "true" fall line does not exist to the east or west of this zone of metamorphic rocks on the southern margin of the Piedmont Upland Section. Many authors divide the coastal plain into two portions. The portion north of the Black Prairie is considered to be the Upper Coastal Plain, and the portion to the south is considered to be the Lower Coastal Plain.

US 43 originates in Mobile in the Coastal Lowlands District of the East Gulf Coastal Plain (Figure 1.11). It leaves the district, temporarily, about seven miles north of its junction with US 82 northwest of Northport in Tuscaloosa County. Northward, it passes in and out of the Warrior Basin District of the Cumberland Plateau Section until it reaches a point two miles north of Liberty Hill in south Franklin County, where it enters the Moulton Valley District of the Highland Rim Section. US 43 finally leaves the last portion of the East Gulf Coastal Plain in Russellville in north Franklin County.

I-65 originates at its junction with I-10 southwest of Mobile in a portion of the Alluvial-Deltaic Plain District (Figure 1.11). It leaves the section around mm 209 near Clanton in Chilton County, where it enters the Northern Piedmont District of the Piedmont Upland Section.

US 431 originates in Dothan in the Doherty Plain District of the East Gulf Coastal Plain (Figure 1.11). It leaves the section around mm 120, near the boundary of Russell and Lee Counties. At that point, it enters the Southern Piedmont District of the Piedmont Upland Section.

The processes of weathering and erosion began on the land that we call Alabama even while the highlands of the Piedmont Upland, the Cumberland Plateau, and the Alabama Valley and Ridge were being formed. Swift streams carried enormous quantities of gravels, sand, and clay southward to the shallow seas covering south Alabama. Intermittingly, the waters of the Gulf of Mexico advanced northward then retreated southward across all of south and parts of west Alabama. As a result, the water-borne sediments were deposited in more or less parallel bands lying northwest to southeast across this area (Figure 1.7). During intervals within this long period of time, carbonate-secreting organisms living in clear, marine seas sank to the bottom after death, covering large areas with thick blankets of carbonate rocks and chalk. These thick blankets of clay, sand, limestone, and chalk sediments deposited from the fall line to the present day Gulf of Mexico comprise our present day coastal plain.

The Coastal Plain Physiographic Province of the Atlantic Plain Physiographic Division, which includes the East Gulf Coastal Plain Section in Alabama (Table 1.2), constitutes a significant portion of the surface area of the Eastern United States (Figure 1.10). The province extends northeastward along the coast to Massachusetts. Cape Cod in that state is considered to be a part of this province. Although the map does not show it, the province reaches southwestward to the Rio Grande River in West Texas and extends another 1,000 miles into Mexico. In the United States, the province is widest in the Mississippi Valley where it reaches as far north as southern Missouri, Illinois, and Indiana.

The Coastal Plain Section in Alabama is a gently dissected plain with nearly level to gently rolling valleys with some uplands with steep slopes and narrow valleys (Clements,1991). This generally flat aspect is interrupted where areas of rugged topography are developed on erosion-resistant geological formations (Rindsberg, 1989). The soil veneer of the Coastal Plain Section is supported on an almost unimaginable quantity of relatively recent sediments—primarily clay, silt, and sand transported onto a relatively-level, marginally marine to shallow marine depositional surface by water from eroding highlands to the north and west of the section. In some areas, of the northwestern part of the section, the sediments are less than 50 feet thick. In the southeastern part, sediments may reach a depth of 8,000 feet; while in the southwestern part, sediments may reach depths of 24,000 feet.

The collision (Figure 1.5) between the tectonic plates bearing Larussia (the ancient North American and European continents) and the super-continent Gondwanaland during the Pennsylvanian Period of the upper part of the Paleozoic Era that resulted in the development of the Appalachian Mountains, also eliminated the sea that lay at the southern margin where the Gulf of Mexico lies today. But some 80 million years later in the Triassic Period of the early part of the Mesoic Era, the same forces that caused the collision began to exert their effect in opposite directions; the tectonic plates and land masses began to separate, however the plates did not separate cleanly. Remnants of the ancient African and South American continents were left behind in south Alabama, south Georgia, and Florida (Figures 1.5 and 1.6) (Patterson, 1993 and Scotese, 1994). Fossils originating in northwest Africa have been found at a depth of 15,000 feet in petroleum explorations in Conecuh County. Apparently the remnants are somewhat shallower (7,000 feet) in southeastern Alabama (Charles Copeland, Alabama Geological Survey, personal communication).

As the separation of the tectonic plates progressed during the Jurassic Period of the mid-Mesosoic Era, some 200 mya, a largely inland sea that later would become the Gulf of Mexico began to develop at the point where the separation was taking place. Over the next 175 million years, until after the end of the Oliogocene Epoch of the Tertiary Period (Cenozoic Era), the ancient ancestor of the Gulf of Mexico would transgress and regress over the Coastal Plain many times. Dr. Charles Savrda, Department of Geology, Auburn University (personal communication), suggested that there might have been as many as 20 distinct incursions of sea water over the Coastal Plan during the Late Mesozoic and the Cenozoic Eras.

The broad sea that alternately covered then uncovered all or part of the Coastal Plain was part of the Mississippi Embayment or Western Seaway that covered, from time to time, much of the Central United States, including the ancestral Mississippi River with all of its tributaries. Rindsberg (1989) notes that the embayment was a major structural trough throughout the Mesozoic and Cenozoic eras. Ancestral streams of the modern-day Alabama, Coosa, Tallapoosa, and Chattahoochee rivers would likely have emptied into the embayment.

During some periods, the transgressions were somewhat limited, but during other periods, the portion that covered the coastal plain was only part of an enormous inland sea. During the Early Jurassic Period of the Mesozoic Era, some 200 mya, virtually all of Alabama, probably with the exception of portions of the Piedmont Upland Section, was covered by an inland sea that reached from the present day Gulf of Mexico through the center of the continent all the way to the Artic Circle (Levin, 1994).

Even at times when the entire section was covered by the sea, rivers flowing from exposed highlands, north and east, continued to carry immense quantities of clay, silt, sand, and gravel southward into near-shore and off-shore waters. Under those conditions, sedimentation near-shore and off-shore was quite different. Heavier materials, such as sand and gravel, were deposited in broad deltas near the mouths of the rivers. Clays and silt remained suspended to be deposited far from the shore; consequently, there were significant differences in the nature of materials deposited north to south across the coastal plain.

Also, a regression cycle required millions of years. As the coastline receded from north to south, vast areas that had been quietly lying under the sea were once again exposed to erosion. Materials that had been deposited millions of years earlier were carried into the sea and re-deposited, only to be exposed later in the cycle and subjected to another round of erosion and deposition. Then, as regression of the coastline came to a halt and the cycle reversed itself so that incursion returned and the coastline moved northward once again, new layers of clays, sands, and gravels were deposited on those newly submerged, eroded surfaces. Again, lighter materials were deposited offshore and the heavier materials inshore, but as the coastline moved further north, deposits of lighter sediments covered the heavier sands and gravels that were deposited in shallow water in an earlier cycle.

Sedimentation in the coastal plain was also dependent on the nature of the climate in the highlands. In periods of high rainfall, erosion and sedimentation increased, and in periods of low rainfall, there was less erosion and sedimentation. During the latter part of the Cretaceous Period of the Mesozoic Era and the beginning of the Tertiary Period of the Cenozoic Era, some 60 to 70 mya, deposition of materials eroded from the highlands seems to have been reduced. Instead of clay and sand deposition, conditions were favorable, probably in relatively calm, clear seas, for the development of vast numbers of single-celled organisms that could incorporate carbonate out of the water column into their cellular structures. Once they died and settled to the bottom, soft cellular structures decayed leaving the tiny pieces of carbonate. Over the millions of years, these tiny pieces accumulated in beds hundreds of feet thick as limestone or chalk. The chalk beds in Sumter County are over 400 feet thick and are composed entirely of tiny calcareous platelets secreted by a species of tiny golden brown algae that lived in that ancient sea in untold numbers. Later, the climatic conditions over the highlands would change, and sand sedimentation would return.

Also there were differences in sedimentation from east to west across the coastal plain because of differences in the geological characteristics of the highlands where the sediments originated. Consequently materials being eroded from the crystaline rocks of the Piedmont Province were different from those from the sedimentary rock highlands in the Valley and Ridge and Cumberland Plateau. Further, in eastern Alabama, the Mississippi Embayment was probably much shallower and more riverine in nature than in the western part of the state. Therefore, the sedimentation on the coastal plain of eastern Alabama was influenced more by river flows than in the west. Western Alabama was much nearer the deeper waters of the Mississippi Embayment, so sedimentation there was less influenced by river flows and more influenced by open sea conditions. For example, the Selma Group of sediments, deposited in the Late Cretaceous Period of the Mesozoic Era some 70 mya (Figure 1.7), consists primarily of chalk formations in western Alabama and sand formations in eastern Alabama, indicating open marine sedimentation conditions in the west and river sedimentation conditions in the east during the same general period of time. Further, the configuration of the ancient rocks underlying the coastal plain dictated that the shoreline of the transgressing and regressing sea generally had a northwest to southeast orientation. Currents that developed parallel to coastlines would have had a major effect on the east-west distribution of sediments entering the sea from the rivers.

As the quantity of sediments from the highlands accumulated over the coastal plain, the weight and pressure increased on the basement rocks to the point that they subsided. This effect increased from north to south across the section. As a result of all of the different factors described in the preceding paragraphs, sediments accumulated in wedge-shaped formations. Generally, the wedges are not completely homogenous, but consist of layers. Geologists have mapped the location of the various groups of formations of sediments at the surface of coastal plain as well as the other sections in the state. Figure 1.7 shows the general location of the surface (outcrop) portion of the wedges.

Fenneman (1938) prepared a map showing the stratigraphy or subsurface location of the various sediment formations (wedges) in the coastal plain (Figure 1.13). As noted in the preceding paragraph, the total accumulation of sediments consists of a series of wedges. The lowermost wedge (the earliest deposited) consisting of varicolored clay, sand, and gravel (Tuscaloosa Group) was deposited first on a platform of ancient rocks during the Late Cretaceous Period of the Mesozoic Era, some 100 mya. Later in the same period, the Eutaw Formation was deposited on top of the Tuscaloosa. One of the uppermost wedges, the Citronelle Formation, was deposited during the Quaternary Period of the Cenozoic (Recent) Era probably less than two mya.

Drahovzal (1968), Hooks (1975), and Sapp and Emplaincourt (1975) describe nine physiographic districts and three sub-districts in the East Gulf Coastal Plain Section. Figure 1.9 shows the general location and relative size of the districts.

East Gulf Coastal Plain Section districts: Fall Line Hills, Black Prairie, Chunnennuggee Hills, Southern Red Hills with Flatwoods and Buhrstone Hills sub-districts, Lime Hills and Hatchetigbee Dome sub-district, Southern Pine Hills, Doughherty Plain, Coastal Lowlands, and Alluvial-deltaic Plain.

Fall Line Hills District

The Fall Line Hills District, which form the northern boundary of the East Gulf Coastal Plain Physiographic Section, is an arcuate shaped area of low hills extending from Colbert and Franklin counties in northwest Alabama through parts of Chilton and Autauga counties in the central portion of the state to parts of Lee and Russell counties in the east (Figure 1.9). The area is 40 miles wide in the northwestern part of the state, but narrows to a width of 16 miles in the east. The towns of Hamilton (Marion County), Vernon (Lamar), Fayette (Fayette), Carrolton (Greene), Tuscaloosa (Tuscaloosa), Greensboro (Hale), Marion (Perry), Clanton (Chilton), Wetumpka (Elmore), Tuskeegee (Macon), and Phenix City (Russell) are towns in the Fall Line Hills District.

The Fall Line Hills Physiographic District is a "dissected upland with broad flat ridges separated by deep valleys (Figure 1.8)." The highest elevations in the district are in the vicinity of Spruce Pine in southeastern Franklin County. A benchmark in the community is located at an elevation of 1,027 feet; however, that high hill is considerably higher than most of the district..

A meterorite impact site is an interesting physiographic feature in this district. The impact left a series of steep, rugged hills just south of Wetumpka (Elmore County) on the east side of US 231. The road cut on the eastern side of the highway rises several hundred feet above the highway. To the west, the flat surface of the Coosa Valley can be seen. Available evidence indicates that these hills are part of a cryptexplosive surface or meterorite impact zone (Neathery, et al., 1976). These authors suggest that a meterorite struck this portion of the fall line, which separates the Piedmont Upland and the Coastal Plain, sometime between the late Cretaceous Period (Mesozoic Era) and the Pleistocene Epoch of the Quaternary Period (Cenozoic Era), or within the last 60 million years. The impact created a deep, circular-shaped crater four miles (6.5 km) in diameter. The structure now stands between 150 and 450 feet above the surface of the Piedmont Upland. Its rim is formed of metamorphic rocks (schist, gneiss, and quartzite). Long after the impact, erosion opened the structure on the southwest part of the rim. Even later, a river flowed into the structure from the southwest covering the crater with alluvial deposits. A popular tourist attraction, Jasmine Hills and an area of homes have been developed within the crater.

Black Prairie District

The Black Prairie District, an area of 4,000 square miles, is a "narrow, elongate, generally flattened crescent-shaped, shallow trough" (Figure 1.9). Its greatest width is 25 miles, near the Mississippi-Alabama line, and its greatest length is 185 miles. It lies across central Alabama from Pickens and Sumter counties in the west through Lowndes and Dallas counties in the central part of the state to Macon and Russell counties in the east.

The district is divided into two distinct portions (Figure 1.7). The Lower Selma portion constitutes the northern half of the district, and the Upper Selma portion constitutes the southern half. The two portions divide the district into two almost equal parts. Lower Selma was apparently deposited first, some 75 mya. This portion is composed principally of Mooreville Chalk that consists primarily of fossiliferous chalk and chalky marl. Because of its composition, this formation has eroded uniformly, leaving its surface relatively level. Upper Selma is composed principally of Demopolis Chalk that consists primarily of fossiliferous chalk. The erosion of this formation has been much less uniform, leaving its surface hilly.

Cochrane (Pickens County), Newbern (Hale), Marion Junction (Dallas), Hope Hull (Montgomery), Fitzpatrick (Bullock), Roba (Macon), and Hurtsboro (Russell) are located on the Lower Selma portion of the district. Geiger (Sumter County), Demopolis (Marengo), Uniontown (Perry), Safford (Dallas), Letohatchee (Lowndes), Sprague (Montgomery), and Hector (Bullock) are located in the Upper Selma portion.

The district is an undulating, deeply weathered and highly eroded plain with low relief (Figure 1.8). Elevations generally increase from west to east across the district. The highest elevations are associated with the cuestas, or narrow, elongate hills. ("Cuesta" is the Spanish word for hill.)

There are several cuestas found in the Black Prairie District (Figure 1.14). They are high hills, generally trending east to west, across a more or less flat surface. They generally are not more than a couple of miles in width, but they may be traceable across the entire state except in alluvial-deltaic deposits associated with large river valleys. They have a sharply-rising front (north) slope, a relatively narrow, level top, and a gently descending back (south) slope. They were apparently formed when a relatively small quantity of sediments, which would later metamorphose into a highly erosion-resistant clay, were deposited across the entire state on the margin of the Gulf of Mexico. Over time, less erosion-resistant clays on either side were eroded away leaving the high hills or cuestas. Monroe (1941) comments that cuestas are common features in regions with gently sloping rocks of variable hardness, such as those found in Alabama's Coastal Plain.

The Arcola, Ripley (High Ridge), and Enon cuestas are found in the Black Prairie District (Monroe, 1941 and Drahovzal, 1968). The northern-most of these, the Arcola Cuesta, is a conspicuous ridge or line of hills lying east to west, generally along the center of the Black Prairie Belt for much of its length (Figure 1.14). The Arcola is composed of a more erosion-resistant clayey chalk and chalky marl (Arcola Limestone member of the Mooreville Chalk Formation of the Lower Selma Group) (Figure 1.7). Less resistant material has eroded away leaving this high, elongate hill (cuesta) standing in the midst of the heavily eroded prairie. The Alabama portion of the Arcola begins at the Mississippi line near Cochrane in southeast Pickens County. It can be traced into northern Sumter County across south Hale through southwestern Perry and into northwestern Dallas. It is an obvious landform feature just west of Uniontown in Perry County at the Perry-Marengo County line. The Arcola, like the other cuestas in the Coastal Plain, is generally not traceable in the alluvial and terraced areas (Alluvial-Deltaic Plain District) of the Tombigbee and Alabama River valleys.

US 80 crosses the Arcola Cuesta just west of Uniontown in Perry County where it is a narrow hilly area in an otherwise generally level landscape. I-65 crosses the cuesta south of Montgomery between mm 156 and 157. It is traceable to the Montgomery-Bullock County line where it appears to join the much higher Enon Cuesta (Figure 1.15). In east-central Montgomery County, US 231 crosses it around mm 106. The exact location of the cuesta in this area is somewhat uncertain because so much of the highway in this vicinity is in the Alluvial-Deltaic Plain.

Higher elevations along the Arcola Cuesta, from west to east, include: north of Dancy, southwest Pickens County (262 feet); north of Uniontown, southwest Perry (292); northwest of Letohatchee, east-central Lowndes (313); and east of Pine Level, southeast Montgomery (571).

The physiographic map prepared by Sapp and Emplaincourt (1975) shows the Ripley Cuesta at the southern boundary of the Black Prairie with the Chunnennuggee Hills in West Alabama, through Sumter County, Marengo, and into western Dallas (Figure 1.14). It is developed on the erosion-resistant clay, sand, and sandstone of the Ripley Formation (Upper Selma Group) (Figure 1.7). This geomorphic feature is traceable across the entire state. It is a prominent landform south of Epes along the Sumter-Greene County line. US 43, south of Demopolis in north Marengo County, crosses the cuesta between mm 135 and 136. It continues into western Lowndes where it terminates in this district. Monroe, in an earlier publication (1941), called this cuesta the High Ridge and traced it all the way across the state, except for a short distance in western Barbour County.

Some of the higher elevations along the Ripley Cuesta from west to east include: south of Epes along the Sumter-Greene County line (346 feet), four miles northeast of Thomaston in eastern Marengo County (348), and three miles west of Braggs in southwest Lowndes County (292).

The Enon Cuesta is located on the southern margin of the Black Prairie District in north-central Bullock County (Figure 1.14). According to Monroe (1941), the portion of this cuesta, from Union Springs to the Enon community, was the Chunnennuggee Ridge of the early pioneers. It almost appears to be an extension of the Arcola Cuesta, described previously. Most of Union Springs is located on this cuesta. The south toe begins on US 29 just north of the Conecuh River near mm 143. The north slope ends just north of mm 144. Ten miles of east Union Springs, this cuesta courses southward into the Chunnennuggee Hills District. The Enon Cuesta is developed on a more erosion-resistant portion of the Cusseta Sand Member of the Ripley Formation (Upper Selma Group) (Figure 1.7)).

Most of the hills of the short section of the Enon Cuesta in the Black Prairie District have elevations between 250 and 500 feet. Five miles west of Union Springs, near the origin of the cuesta, the elevation is 355 feet. Two miles northeast of Union Springs, in north-central Bullock County, the elevation is 580 feet.

Chunnennuggee Hills District

The Chunnennuggee Hills District consists of series of low hills and cuestas south of the Black Prairie District, lying generally west to east from Sumter County through southern Lowndes County in the central portion of the state to Russell and Barbour counties in the east. It is less than ten miles wide in Sumter County, but more than 30 miles wide in Russell and Barbour (Figure 1.9). In the central portion of the state south of Montgomery, this hilly area is eight miles wide. Livingston (Sumter County), Linden (Marengo), Carlowville (Dallas), Braggs (Lowndes), Grady (Montgomery), Inverness (Bullock), Pittsview (Russell), Highland Home (Crenshaw), Saco (Pike), and Clayton (Barbour) are located in the Chunnennuggee Hills District.

Topographically the Chunnennuggee Hills District is quite hilly throughout its entire length (Figure 1.8). In the western part of the state, steep ridges in the district are separated by narrow valleys. The steepness of the hills and the narrowness of the valley give much of the terrain a jumbled appearance. There are deep ravines cut into many of the hills, and in some locations there are steep bluffs adjacent to larger streams (Harper, 1943). There are hills of equal or greater height in the eastern portion of the district, but they are not so steep and the valleys are much broader. Many of the hills in the Chunnennuggee Hills Physiographic District are associated with the cuestas found there; however in the eastern portion in Bullock and Russell counties, there are many high hills that do not seem to be part of these unique land forms.

According to Sapp and Emplaincourt (1975), four cuestas: Ripley, High Ridge, LaPine, and Troy are conspicuous hills within portions of the Chunnennuggee Hills District (Figure 1.14). These authors show the High Ridge Cuesta originating in southwestern Montgomery County near Highland Home. I-65 passes over a high hill where it crosss the boundary between the Chunnennuggee Hills and Black Prairie Districts near mm 147 in southeast Lowndes County (Figure 1.11). Monroe (1941) identified the hill as part of the High Ridge Cuesta; however Sapp and Emplaincourt (1975) did not place any of their cuestas at or near that location. If Monroe is correct, that hill is likely to be part of the cuesta.

Sapp and Emplaincourt (1975) show the Ripley Cuesta of the Chunnennuggee Hills District in east Alabama as occurring only as a short range of hills in east-central Barbour County northeast of Clayton (Figure 1.14). Its western portion had terminated in western Lowndes. One of the higher hills on the eastern portion of the cuesta is located two miles northeast of Clayton in north Barbour County. The elevation there is 657 feet. This cuesta is also developed on the Ripley Formation.

The High Ridge Cuesta is a significant landform in southeast Montgomery County where US 231 crosses it between mm 97 and 98 (Figure 1.11). From that point, it generally courses eastward across southern Bullock County to the Pea River where it merges with the Lapine Cuesta. US 29 crosses the cuesta in south-central Bullock just north of Perote, between mm 131 and 132. This cuesta in southwestern Bullock and eastern Montgomery Counties is part of the Chunnennuggee Ridge that was well known to early Alabama settlers (Monroe,1941).

Higher elevations, west to east, along the High Ridge Cuesta include: east of Pine Level in southeast Montgomery County (571 feet); near the community of High Ridge in west-central Bullock (664), and just west of Perote in southeast Bullock (540).

The High Ridge Cuesta is formed on more erosion-resistant portions of sandstone, sand, and clay of the Ripley Formation of the Upper Selma Group (Figure 1.7).

The Sand Fort Cuesta (Figure 1.14) separates the Chunnennuggee Hills from the Fall Line Hills in southeastern Macon and central Russell counties. It originates in southwest Macon County near the Bullock County line. It can be observed just north of the Macon-Bullock County line where US 29 passes over it between mm 154 and 156 south of Davisville. It is traceable through north-central Russell County to near US 431 (Figure 1.12).

The elevation on a high ridge on the Sand Fort Cuesta in south Macon County northeast of Hardaway is 377 feet. In central Russell County, five miles northwest of Hatchechubbee, it is 564 feet.

This relatively short cuesta is formed from the clay, marl, sand, and sandstone of the Bluffton Formation (Upper Selma Group) (Figure 1.7).The Chunnennuggee Hills portion of the Enon Cuesta (Figure 1.14) enters the district from the north, seven miles east of Union Springs in north Bullock County. It is a very noticeable land-form where US 29 passes over it near the northern town limit of Union Springs (mm 144 to 145) (Figure 1.11). It is traceable across northwestern Barbour into southeastern Russell near Alabama 165.

Higher elevations, west to east, along the Chunnenuggee Hills portion of the Enon Cuesta include: four miles southeast of Union Springs in central Barbour County (563 feet), north of Midway in east-central Bullock County (561 feet), and five miles northwest of Spring Hill in north Barbour County (402 feet).

The Enon Cuesta is supported by the Cusseta Sand member of the Ripley Formation (Upper Selma Group) (Figure 1.7). The western portion of this cuesta is also part of the Chunnennuggee Ridge.

The Lapine Cuesta (Figure 1.14) originates in southwestern Lowndes County between Fort Deposit and Braggs, and generally follows an easterly course through southern Lowndes, northern Crenshaw and Pike, and southern Bullock into central Barbour County. It is joined by the High Ridge Cuesta just east of the Pea River at the Bullock-Barbour line. A portion of this prominent cuesta can be observed in southeastern Lowndes County. Its southern slope is located just north of the intersection of I- 65 and Alabama 187, Exit 142 (Figure 1.11). The north toe of the cuesta ends just south of mm 144. US 331 passes through it just north of Highland Home, north of the Crenshaw-Montgomery County line. It can also be observed just south of the Montgomery-Pike County line where US 231 crosses over it at between mm 88 and 89. The cuesta at this point is not easily recognizable, because it is passing across the swampy, headwaters region of some large tributaries of the Conecuh River and of Patsaliga Creek. North of the Pike-Bullock County line and just south of Perote, US 29 passes over it between mm 127 and 128.

Higher elevations along the Lapine Cuesta include: northeast of Bragg southwest Lowndes County (417 feet), south of Calhoun east-central Lowndes (551), Highland Home along US 331 in north Crenshaw County (635), northeast of Ansley (Beck's Mountain) in northeast Pike (675), and west of Perote in north Pike (540). From northwestern Crenshaw across to Pea River at the Bullock-Barbour County line, many of the hills are in the range of 500 to 750 feet above sea level (Figure 1.8).

The Lapine Cuesta is formed on the Providence Sand and Prarie Bluff Chalk Formations of the Upper Selma Group (Figure 1.7) and is a central member of the most extensive range of high hills in south Alabama.

The physiographic map of Sapp and Emplaincourt (1975) locates the Troy Cuesta just north of the boundary between the Chunnennugee Hills and the Southern Red Hills Districts (Figure 1.9). Monroe (1941) located it on the south side of the boundary. It is the longest cuesta in the Coastal Plain. It appears on US 43 just north of Linden in Marengo County near mm 126 (Figure 1.11). It forms a large jumble of high hills south and east of the junction of Alabama 21 and 263 in southwest Lowndes County. The cuesta appears on I-65 as a high hill with an east-west trend and with a sharp incline on the north side. South of Exit 130 (Alabama 185), the highway passes through a series of fairly steep hills of the Southern Red Hills District, so it is somewhat difficult to identify the south slope of the cuesta. However, between mm 138 and 139, it is easy to identify the top and steep north slope as the highway descends just before crossing a tributary of Pigeon Creek. The Troy Cuesta also appears as a high hill with a sharp incline on the north slope just south of the Conecuh River within the city limits of Troy, between mm 79 and 80, on US 231 (Figures 1.11 and 1.15). East of Banks in northeast Pike County, the cuesta is divided into two portions. The portions are separated by a tributary of the Pea River. US 29 passes over both portions and the creek between mm 115 to 116. A few miles east of Clayton in central Barbour County, the cuesta turns sharply southeastward to terminate near the Barbour-Henry County line and near the Chattahoochee River.

Higher elevations along the Troy Cuesta include: northwest of the town of Linden in west Marengo County (246 feet), north of Thomaston in east Marengo (246), southwest of Braggs in southwest Lowndes (455), south of Ft. Deposit in northeast Butler (548), northeast of Troy in central Pike (567), south of Josie along US 29 in eastern Pike (615), east of Clayton in central Barbour (607), and southwest of Eufaula in southeast Barbour (413) (Figure 1.8).

The Troy Cuesta is formed from sand, marl, chalk, and limestone of the Clayton Formation of the Midway Group (Figure 1.7), a geological group associated with the Southern Red Hills District. All of the other cuestas in the Chunnennuggee Hills District were formed on the older geological formations of the Upper Selma Group.

Southern Red Hills District

The Southern Red Hills Physiographic District is a series of relatively high hills and red levels reaching across the entire width of the south-central portion of the state from Sumter and Choctaw counties in the west through Butler and part of Covington in the central portion to Barbour, Henry, and Houston in the east (Figures1.9). For most of its length, the district is 40 miles wide, but increases somewhat in width in the southeastern part of the state. Sapp and Emplaincourt (1975) include two sub-districts, Flatwoods and Buhrstone Hills, in the western portion of the district. Consequently, the main body of the district, sandwiched between the two sub-districts, is only 10 to 15 miles wide west of the Alabama River. The major portions of the two sub-districts are west of the river, although a narrow portion of the Buhrstone Hills extends into north-central Monroe County. Beginning just east of the Crenshaw-Pike County line northeast of Luverne, portions of the Southern Red Hills and Chunnennuggee Hills Districts are complexly inter-fingered. Generally, the tops of the hills are in the Southern Red Hills District, and the valleys are in the Chunnennuggee Hills District. Ward (Sumter County), Butler (Choctaw), Beatrice (Monroe), Camden (Wilcox), Greenville (Butler), Luverne (Crenshaw), Gantt (Covington), Elba (Coffee), and Ozark (Dale) are located in this district.

There are series of rugged hills located throughout this district (Figure 1.8). The hills just northwest of Greenville in north-central Butler County are almost mountainous. The area around the Cambrian Ridge Golf Course is indicative of the hilliness in that vicinity. The hills in northwestern Coffee Country near the Crenshaw-Coffee line are also especially rugged. On either side of the Crenshaw-Coffee line along Alabama 189 (Brantley to Elba), there are several ridges with elevations between 450 and 520 feet. The hills in this district around Troy in Pike County and north of Ozark in Dale County are equally high, but not nearly so rugged, because the tops of the hills are much broader, the slopes on either side are gentler, and the valleys are somewhat broader.

Flatwoods Sub-District. US 43 (Figures 1.9 and 1.11) passes across the Flatwoods Sub-District, generally south of Linden in central Marengo County. The highway leaves the main body of the Southern Red Hills District and enters the sub-district south of the town, between mm 115 and 116. It leaves the Flatwoods north of town between mm 125 and 126 where it enters the Chunnennuggee Hills District. York (Sumter County), Linden (Marengo), and Flatwood (Dallas) are towns located in these lowlands.

The Flatwoods is an area of relatively low relief compared to the Chunnennuggee Hills on its north side (Figure 1.8). The Flatwoods Sub-District is formed from the massive, dark marine clay of the Porters Creek Formation of the Midway Group (Figure 1.7). The formation was laid down during the Paleocene Epoch of the Tertiary Period (Cenozoic Era) some 60 mya (Raymond, et al., 1988).

Buhrstone Hills Sub-District. The Buhrstone Hills or Cuesta Sub-District (Figure 1.9) is located in the southern part of the Southern Red Hills District in southwest Alabama and adjacent to and north of the Lime Hills District. This area is ten miles wide and 80 miles long. The hills extend from the Mississippi line in Choctaw County across northern Clark and southern Marengo and through southwestern Wilcox into north-central Monroe.

US 43 (Figure 1.11) passes through the Buhrstone Sub-District north of Thomasville in Clarke and Marengo counties. The highway enters it from the Lime Hills District between mm 98 and 99 just north of the junction with Alabama 5. It leaves the sub-district 4.5 miles north of the junction with Alabama 10 in the Dixon's Mill community (between mm 111 and 112) in southeastern Marengo County, where it enters the main body of the Southern Red Hills District. Butler (Choctaw County), Hoboken (Marengo), Bashi (Clarke), and Chestnut (Monroe) are located in the Buhrstone Hills.

The Buhrstone Hills rise sharply above the surrounding streams. Drahovzal (1968) comments that this is the most rugged area in Alabama's Coastal Plain (Figure 1.8). The elevations of these hills are not as great as some of those along the northern boundary of the Southern Red Hills District; however, the tops of the Buhrstone Hills and the intervening valleys are much narrower. Also the slopes of these hills are much steeper. These characteristics give them a rugged, jumbled appearance. There is a series of these sharply jumbled hills located along US 43 at the Clarke-Marengo line, between mm 101 and 102. Elevations of the highest ridges in the sub-district tend to increase from west to east.

The Buhrstone Hills or Cuesta is developed on the highly erosion-resistant, siliceous claystone and sandstone (buhrstone) of the Tuscahoma Sand Formation (Wilcox Group) (Figure 1.7). This formation was laid down during the latter part of the Paleocene Epoch of the Tertiary Period (Cenozoic Era) some 55 mya.

Lime Hills District

The Lime Hills District of the East Gulf Coastal Plain Physiographic Section, in Alabama originates at the Mississippi line in central Choctaw County, and extends across northern Clark, Monroe, and Conecuh counties into northwest Covington (Figure 1.9). It lies on the south side of the Buhrstone Hills Sub-District of the Southern Red Hills District in the southwestern portion of the state. Its width is highly variable, ranging from 30 miles in western Choctaw and eastern Clarke, to less than five miles northeast of Evergreen in Conecuh County. The district is 120 miles long.

Because of the hilly and almost mountainous nature of the terrain at the northern boundary of the Lime Hills District and the southern boundary of the Buhrstone Sub-District of the Southern Red Hills District, it is difficult to know where the actual boundary is located. Thomasville (Clarke County), Monroeville (Monroe), and Loree (Conecuh) are in the Lime Hills District.

The elevations of the higher ridges in the Lime Hills

District are greater than those in the Buhrstone Hills Sub-District of the Southern Red Hills District to the north, but the area does not seem to be quite as rugged (Figure 1.8).

Hatchetigbee Dome Sub-District. The Hatchetigbee Dome (Hooks, 1973), or anticline (up-fold), is a sub-district of the Lime Hills District (Figure 1.9). According to Sapp and Emplaincourt (1975), the dome extends from the Mississippi line in southwest Choctaw County almost to Grove Hill, in central Clarke County, an area with an east-west length of 50 miles and a width of 20 miles. The outcrop of the dome is divided in half, length-wise, by the Tombigbee River and its Alluvial-Deltaic Plain deposits.

US 84 (Figure 1.11) passes along the east-west axis and generally along the northern boundary of the Hatchetigbee Dome Sub-District west of the town of Grove Hill, in Clarke County. From the west, the highway enters the sub-district two miles east of the Mississippi State line and west of Silas in Choctaw County. It leaves the surface of the dome ten miles west of Grove Hill between Satilpa Creek and Zimco.

The Hatchetigbee Dome Sub-District represents the stratigraphic displacement of the sediments in that area (Figure 1.7). The higher elevations are 150 to 250 feet higher than the nearby lowlands (Figure 1.8). Much of the surface of the central east-west axis along US 84 across the center of the dome is relatively flat to gently rolling.

Southern Pine Hills District

The Southern Pine Hills District occupies the southwestern corner of the state from Washington and Mobile counties northeastward through southern Monroe. Conecuh, and Escambia counties and along the Florida line to the southern part of Covington (Figures1.9). The district generally is not recognizable in the Tombigbee, Alabama, and Conecuh River valleys where the Alluvial-Deltaic Plain is extensive. Chatom (Washington County), Bay Minnette (Baldwin), Brewton (Escambia), and Florala (Covington) are located in the Southern Pine Hills District.

The relatively flat surface of the Southern Pine Hills District, on the south side of Monroeville, at the junction of Alabama Highway 21 and US 84 (Figure 1.11), is in sharp contrast to the jumble of high hills of the Lime Hills District, along Alabama Highway 41, north of town. Although the Southern Pine Hills are relatively flat at that point, in general, the topographic relief is substantial throughout the northern portion of the district (Figure 1.8). Drahovzal (1968), describes the Southern Pine Hills "as an elevated, dissected plain, resembling a Cuesta, sloping to the south." The Cuesta-like characteristics are especially prominent along the northern boundary and in the central and eastern portions of the District (Figure 1.8). The northwestern portion west of the Tombigbee River, is relatively low.

Dougherty Plain District

According to Sapp and Emplaincourt (1975), Alabama's Dougherty Plain Physiographic District is the continuation of a limestone upland which originates in Georgia. The district is an irregular area of low hills and relatively flat valleys that extend across the southern tier of counties of southeast Alabama from Houston and Henry in the east to Conecuh and Escambia in the west (Figure 1.9). The district's western terminus is in east-central Monroe County. The northern boundary between the Dougherty Plain and the Lime Hills is only approximate because of the jumble of geological formations in that area. The district is almost 140 miles long, east to west. Its width is highly variable. It is less than five miles wide in western Conecuh County, but almost 40 miles wide in Houston and southwestern Henry. Drahovzal (1968) did not recognize this district in his publication. He included a small portion of it in the Southern Pine Hills District, but most of it was included with the Southern Red Hills District. Dothan (Houston County), Geneva (Geneva), Andalusia (Covington), and Evergreen (Conecuh) are towns in the Dougherty Plain Physiographic District.

A considerable portion of the area along the northern boundary of the Dougherty Plain is above 300 feet in elevation (Figure 1.8). The elevation of the southeastern portion is much lower. Also, the district is much more dissected in the northwestern portion than in the southeast.

Much of the original limestone near the surface in the Dougherty Plain District has been dissolved by water seeping through the formations. Sapp and Emplaincourt (1975) note that most of the minor surface drainage-ways have been transferred to subsurface as a result of the solution of the limestone. Also, as a result of the solution of the limestone and the collapse of the soil above, there are numerous sinkholes in the district. These shallow depressions usually fill with water during the winter, but are dry during the late summer. There are a number along US 29 (Figure 1.11), between mm 32 and 33, near the southwestern boundary of the district south of Andalusia in central Covington County. Blue Lake as well as Open, Buck, and Ditch Ponds located in the Conecuh National Forest along Alabama 137 in southern Covington County are good examples of the larger sinkholes found in that area. These larger sinkholes do not go dry in summer. There is also a relatively large one east of US 29 just north of its junction with Alabama 55 south of Andalusia. There is also a sinkhole just east of US 231 north of the railroad, between mm 30 and 31, in Midland City in southeast Dale County.

Alluvial-Deltaic Plain District

The pebbles, gravel, sand, silt, and clay deposited in and along creeks and rivers are among the most ubiquitous physiographic features in the entire East Gulf Coastal Plain Section in Alabama (Figure 1.9). These are the Alluvial Low Terrace Deposits of Osborne, et al. (1989). The only significant accumulation outside of the Coastal Plain Section is in the upper Coosa River Basin. These deposits in the Coastal Plain are actually much more widespread than shown in the figure. Because of the scale of the map, the accumulations associated with smaller rivers and creeks are not shown. Also, because of the scale of the map both the Alluvial-Deltaic Plain and Coastal Lowlands Districts are shown together.

Of course, the entire East Gulf Coastal Plain Physiographic Section, as noted previously, is an enormous accumulation of relatively old, sand, silt, and clay sediments. Those of the Alluvial-Deltaic Plain are much younger. According to Osborne, et al. (1989), these materials have been deposited during the last 10,000 years or in the Recent or Holocene Epoch of the Quaternary Period (Cenozoic Era). There are thousands of square miles of them south of the fall line (Figure 1.7). Deposits of some combination of pebbles, gravel, sand, silt, clay, mud, and decaying vegetation can be found adjacent to and generally beneath, virtually every stream in the section. These materials were eroded from the terrain upstream, carried downstream, and deposited when the velocity of the flowing water was no longer sufficient to overcome the gravitational force acting on the particles. The large rivers meander on enormous beds of these deposits. The position of the channels on those beds probably has changed fairly often over time.

In some areas, these accumulations are truly enormous. For example, there are probably over 1,000 square miles of these deposits in the Alabama River Basin in Autauga, Dallas, and Lowndes counties alone (Osborne, et al., 1989). I-65 (Figure 1.11) is constructed on an accumulation of alluvial low terrace deposits from near Exit 172 at the southern approach of the I-65 bridge over the Alabama River in Montgomery to just south of Exit 179 in Elmore County, a distance of six miles. I-85 east of Montgomery from near Exit 9 to near Exit 26, a distance of some 16 miles, is constructed on accumulations associated with the Tallapoosa River and one of its tributaries, Calebee Creek. Along I-85, mostly south of it, the width of the bed is six miles. Highway US 80, for almost its entire length, from Montgomery to Selma is constructed on sediments deposited there by an ancient relative of the Alabama River.

The sands, silts, clays, and gravels of the Alluvial-Deltaic Plain accumulated as alluvial and low terrace deposits during the Pleistocene Epoch of the Quaternary Period between 10,000 and two mya. The deposits were left there by ancient ancestors of the current occupants of the stream channels. These deposits are important, because they provide clues to the location of streams in the distant past.

All of these sediments are associated with lowlands. The elevations of most of these accumulations are less than 100 feet. Even as far north as Echola on the Sipsey River in northwest Tuscaloosa County, the elevation of the Alluvial and Low Terrace Deposits is less than 100 feet.

In many areas in the Coastal Plain Section, associated with the Alluvial-Deltaic Plain, there are also Alluvial High Terrace Deposits (Osborne, et al., 1989). These deposits are generally found at higher elevations out of the present floodplain in the stream valleys. The map published by Osborne, et al. (1989) shows the location of many of these accumulations. The High Terrace Deposits are not nearly so extensive as the Low Terrace ones. Most of these older High Terrace deposits have been eroded away, leaving only small outcrops. The most extensive accumulation, in the Coastal Plain Section is located in a discontinuous band from west of the Coosa River in northwest Elmore County, to north of the Alabama River in south-central Autauga County, but there are similar deposits associated with all of the large rivers in the section. Another of large accumulation is located two to three miles north of the Alabama River in south-central Autauga County just west of Autaugaville.

The Alluvial High Terrace Deposits are generally some distance away from current stream channels, often several miles. For example, there is a small area of these deposits located near Hayneville in central Lowndes County, 13 miles from the nearest point on the Alabama River. Generally, elevations of those deposits, located north of the Alabama River are not greatly different from the elevations of the geological formations of the Fall Line Hills District that surround them; however those south of the river have much higher elevations than the surrounding countryside. Deposits in northwestern Lowndes County around the junction of US 80 and Alabama 97 (Figure 1.12) are found on hill tops some 300 feet above the present elevation of the river. In this area, High Terrace Deposits located on a high hill just north of Lowndesboro are at an elevation of 416 feet. The elevation is 100 feet along the Alabama River, some 12 miles north. Even higher accumulations of these deposits are located west of Collirene in western Lowndes County. There are several ridges in that area with elevations between 450 and 500 feet, and the elevation on one of the highest is 520 feet. Elevations of High Terrace Deposits north of the Alabama River west of Autaugaville in south Autauga County are similar to those near Lowndesboro, indicating that an ancient ancestor of the Alabama River once flowed across that part of the Coastal Plain depositing gravel and sand at a much higher elevation than it flows today.

The High Terrace Deposits are generally associated with the channels of the larger rivers, but there is at least one area where these deposits seem to be associated with a large creek. There are several square miles of outcrops of these materials north and west of Persimmon Creek in southeast Butler County south and east of Georgiana. Persimmon Creek is a tributary of the Sepulga River tha flows into the Conecuh River southwest of Andalusia.

Coastal Lowlands District

The Coastal Lowland Physiographic District is located in a narrow strip one to five miles in width along the coastal areas of Baldwin and Mobile counties (Figure 1.9). In Baldwin County, the Coastal Lowlands begin near the point where I-10 crosses the Perdido River and follows the coast line around to the point where I-10 crosses Mobile Bay (Figure 1.11). In Mobile County, these lowlands begin at the Mobile-Washington line and follow the coastline around to the Mississippi line. Geologically, these lowlands consist of sand, shell fragments, silt, clay, peat, mud, and ooze. They generally correspond to the Coastal Deposits of Osborne, et al. (1989).

Traveling along I-65 from its origin in southwest Mobile County to the point where it leaves the state north of Athens in Limestone County provides a good opportunity to view Alabama's portions of all five of the physiographic sections and most of its physiographic districts as well. (Figure 1.11). This interstate highway essentially passes across the heart of the East Gulf Coastal Plain, brushes the western end of the Piedmont Upland, and cuts across the western fourth of the Alabama Valley and Ridge, the western third of the Cumberland Plateau, and the eastern third of the Highland Rim.

The journey provides an opportunity to observe close at hand the physiographic evidence of hundreds of millions of years of continental collisions, mountain building, lithification (rock formation), weathering, erosion, water and wind transport, and sedimentation. Table 1.4 in the online companion resources for this volume lists approximate mile marker numbers along the I-65 route giving the location of boundaries between physiographic sections and districts within those sections. For example, near mm 31, the I-65 leaves the Alluvial-Deltaic Plain District and enters the Southern Pine Hills District (Figure 1.9). The boundary here is quite distinct, but even though it is quite obvious, mile markers do not pin-point exact locations very well. A mile is a considerable distance.

Added to the inexactness of using mile markers is

the indefinite nature of most physiographic boundaries. For example, mm 90 is the approximate location of the boundary between the Southern Pine Hills District of the East Gulf Coastal Plain Section and the Doherty Plain District within that section (Figure 1.9). The location is approximate because the boundary seems to be quite broad. In fact, there appears to be little apparent change in the landscape, either south or north of mm 90. The only indication is that on the Doherty Plain, there are a few more red cedars (Christmas trees) and some Spanish Moss in the trees along the tree line. Apparently, the Doherty Plain was established when limey sediments accumulated in a narrow, shallow body of water across southern Alabama and southwestern Georgia. The presence of the cedar trees and Spanish Moss are indicative of the accumulation of lime in the soil. There are very few of these plant species on the roadside in the Southern Pine Hills south of the Doherty Plain District and in the Southern Red Hills District north of it.

I-65 crosses the Mobile and Tensas rivers around mm 26. The delta formed by these rivers is one of the most interesting geomorphic features on the entire trip from Mobile to Ardmore. As the crow flies this delta, at this point, is only seven miles wide. In terms of streamflow, it is the fourth largest in the country. Each second, an average of 64,000 cubic feet of water flows through this location. With a streamflow of this magnitude, one would expect a much wider delta or floodplain. Geologists suggest that in the fairly recent past there was a large, fast-flowing river passing this location and that the coast line was many miles further south than it is today.

The southwestern boundary of another interesting landform is located around mm 31. Here I-65 enters soils formed on the Citronelle Formation (Figure 1.7). The northeastern boundary is near mm 51. These soils have eroded relatively evenly, leaving a large area with a low relief. Today, the entire area is heavily farmed by members of the Poarch Band of Creeks.

There is an interesting landform between mm 103 and mm 106. At this point, the highway is located in the Lime Hills District. Once across the Sepulga River (the line between Conecuh County on the south and Butler County on the north), it goes up a long steep hill. This hill contains a significant amount of siliceous clay-stone of the Tallahata Formation. This clay is highly resistant to erosion. As a result, softer material on either side have eroded away, leaving the high hill standing. This formation is traceable across the entire state.

Just north of Exit 142, I-65 passes across the LaPine Cuesta. On its north slope, there is an outcrop of the Prairie Bluff Chalk Formation. A sizeable stand of red cedar has developed on the extensive cut adjacent to the south-bound lane. An extensive stand of red bud trees has developed in the tree line on the north-bound lane. These small trees thrive in soils developed on limey deposits. They literally fill the woods in this area with their color in the late winter.

The boundary around mm 147 is also interesting. At this point, I-65 leaves the Chunnennuggee Hills District and enters the Black Prairie District. The relief here is relatively limited, but to the east it is quite different. In Bullock County, the Chunnenuggee Ridge reaches elevations as great as 500 feet. Loblolly pine is quite common in the Chunnennuggee Hills, but much less common on soils derived from the limestone in the Black Prairie. Around mm 137, the highway leaves the relatively thick pine-hardwood forest and enters open pasture land with few pines and mostly hardwoods. At this point, we are entering the southern boundary of what was for many years Alabama's most important cotton-producing region.

There is another interesting change in landforms around mm 156. South of this point, the land is hilly and moderately forested. To the north, it is almost flat and primarily pasture land. The soils to the south were formed on the Demopolis Chalk Formation. Those to the north were formed on Mooreville Chalk, which is a much purer form of limestone that has experienced more even erosion over time.

Near mm 209, the highway leaves the Fall Line Hills District of the East Gulf Coastal and enters the Northern Piedmont District of the Piedmont Upland Section. This is an important boundary within the state. Here the soft rocks of the Coastal Plain are replaced by the ancient hard rocks region. There are few distinct characteristics associated with this boundary. The only obvious one is that the more regularly arranged hills and valleys to the south,are replaced by heavily jumbled ones.

Around mm 227, the highway is at the boundary of the Northern Piedmont District of the Piedmont Upland Section and the Coosa Valley District of the Alabama Valley and Ridge Section. Here the jumbled hills of the Piedmont are replaced with a broadly open valley with low relief. I-65 remains in the Coosa and Cahaba valleys through mm 238.

Around mm 238, the highway is at the southern boundary of the extensive range of rugged mountain-like parallel ridges of the Coosa Ridges District of the Alabama Valley and Ridge Section. At this point, the highway is near the southwestern terminus of Double Mountain, and within in a mile or so it will cross near the southwestern terminus of Double Oak Mountain. The Cahaba Valley and the Cahaba Valley Highway (Alabama 119) lie to the west. I-65 remains in the Coosa Ridges District until near the point where it crosses Alabama 119 at Exit 246 near Indian Springs. At this point, the highway is again in a very narrow portion of the Cahaba Valley District. Beyond this point, it is in the mountainous Cahaba Ridges District that includes Shades Mountain, Shades Valley, and Red Mountain.

The boundary between the Cahaba Ridges District and the Birmingham-Big Canoe District in around mm 257 where I-65 passes over the deeply cut crest of Red Mountain.

The boundary between the Birmingham-Big Canoe District of the Alabama Valley and Ridge Section and the Warrior Basin District of the Cumberland Plateau Section is around mm 263. This boundary is in a heavily urbanized area and has few distinctive characteristics; however within a few miles north of the boundary, cuts along the roadside have exposed extensive areas of the lighter colored sedimentary rocks of the Pottsville Formation. Further north as the district becomes more mountainous, cuts reach heights of 50 feet or greater. Here the rocks are almost black as a result of presence of exposed slate, siltstone, and coal.

There is another interesting geomorphic structure around Exit 287. Here I-65 passes across the southwestern terminus of a narrow finger of Bangor Limestone projected southwestward from the widespread outcrop of this formation in the Highland Rim Section further north. The finger is only about three miles wide at this point. Here the lighter colored limestones and mudstones contrast sharply with the black rocks of this portion of the Warrior Basin on either side of it.

At mm 317, the boundary between the Warrior Basin District of the Cumberland Plateau Section and the Moulton Valley District of the Highland Rim Section, is also quite definite. Near that mile marker, I-65 leaves the extremely hilly, almost mountainous, Warrior Basin, and enters the more or less level surface of Moulton Valley. The soils on the plateau were derived primarily from sandstone, and rocks are mostly just below the surface. In Moulton Valley, soils were derived primarily from the Bangor Limestone Formation, and are much deeper and more erosion prone.

Mineral Resources of the Section

Raw materials for industry and commerce being extracted from the accumulated sediments of the Coastal Plain Section include sand, gravel, calcarerous rocks used in the manufacture of cement, limestone, clay minerals, brine for industrial chemicals, oil, gas, and gas condensate (Raymond, et al. 1988). A large band of lignite, 30 to 60 inches deep, occurs across the entire width of the state from Sumter County in the west to Henry County in the east (Figure 1.12). However, it is not being commercially exploited. Also, a preliminary study indicates that there is a large quantity of commercial grade peat available for exploitation in Mobile and Baldwin counties.

Water, Water Everywhere

Geographical data indicate that the preponderance of Alabama's total area is in land; however, there are few places in Alabama where one is more than 20 to 30 miles from a sizeable river, and usually much closer than that to a large creek. Rivers and creeks are about as common as pine trees. When we throw a rock onto the still surface of a reservoir, watch a river trudging along with its load of silt, or watch a colorful leaf floating jauntily down the riffle of a creek, it is difficult to comprehend that these bodies of water and thousands of their ancestors, were/are responsible for the continual shaping the surface of Alabama as we know it.

Alabama receives a large amount of rainfall, relatively evenly distributed between months during the year, but subject to erratic timing from year to year. If it were all accumulated at one time in one place, it would fill a lake 52,000 square miles in area to a depth of about 4.5 feet. In addition, large quantities of water from rainfall in adjoining states flow across Alabama. In the millions of years that this vast flood has been flowing off the land into the sea, it has created a vast network of large and small rivers and creeks. The larger of these water courses are shown in Figure 1.15. The Tennessee River criss-crosses the state, east to west, along its northern border. The Chattahoochee and Tombigbee rivers traverses much of its eastern and western borders, respectively, and the Black Warrior, Coosa, Conecuh, and Alabama flow diagonally across the state northeast to southwest. Finally, the state is bounded on the south by a large estuarine area, the Inter-coastal Waterway and the Gulf of Mexico. In sum, there are 3,600,000 acres of freshwater wetlands and 27,600 acres of coastal wetlands within the state. Altogether, at least one-sixth of Alabama's surface area is in wetlands, streams, ponds, lakes, reservoirs, rivers, and estuaries (Moore and Szabo, 1994). Alabama is probably unsurpassed in the country with the quantity and variety of its surface-water features.

Some Freshwater Features

The US Geological Survey estimates that 10% of all the freshwater in the continental United States originates or flows through Alabama each year. There are an estimated 77,000 miles of waterways of all sizes in the state. Some 61% flow permanently. An estimated 33.5 trillion gallons of water flow through these streams annually.

As the freshwater features of the state are considered, we should remember that they are constantly changing and that the degree of change, over time, has been almost beyond the limits of our imagination. For much of geological history, the portion of the state that we call the Coastal Plain was covered by the sea, and in more recent times, geologically speaking, the sea level was much lower than it is now. As an example, at one time, Mobile Bay was probably a river valley. What we now call the Mobile River ran through it to the sea coast further south. When the sea rose again to near its present level, the valley was flooded. Also, there probably have been significant changes in the courses of our rivers over recent times. For example, Monroe (1941) suggests that the Alabama River (Figure 1.15) between Montgomery and Selma once flowed south of its present location. High Terrace Deposits consisting of sand, gravely sand, silt, and clay are found as far south as Hayneville in central Lowndes County, 14 miles from the nearest point on the river. These deposits were at one time located in or adjacent to the channel of a flowing river. They were, at that time, similar to the Alluvial Low Terrace deposits found along the river today. Further, weathered gravels are common on flat areas on top of high hills or terraces south of the river in Lowndes County, indicating that the level of the river was as much as 200 to 300 feet higher at one time than it is today. According to geologists, these materials were deposited by some ancient ancestor of the Alabama River, during the Pleistocene Epoch of the Quaternary Period (Cenozoic Era) that began two million years ago (Osborne, et al., 1989). Harper (1943) noted that river gravel could be found on top of the hills around Spruce Pine in Franklin County at an elevation of 1,000 feet. He further commented that a stream gradient of 50 feet per mile would have been required to develop the necessary current velocity to move the gravel.

Mike Szabo of the Alabama Geological Survey (personal communication) suggests that the portion of the Tennessee River in Jackson and Marshall counties was once part of the Black Warrior Basin. He surmises that the Tennesse captured the upper portion of the Black Warrior in the past (Figure 1.15). As the original headwaters of the Tennessee in Morgan and Marshall counties eroded eastward, they eventually cut through the hills dividing the two river basins at that time, capturing the upper reaches of the Black Warrior. Szabo further theorizes that at one time, the Tallapoosa River continued to flow to the southwest to become part of what is now the Conecuh River Basin rather than bending sharply and flowing westward to join the Coosa as it does today. He theorizes that as the Tallapoosa eroded downward in its bed that it encountered geological formations that were resistant to further erosion. At that point, it began to cut through less resistant formations toward the west to join the Coosa.

While Mother Nature has played by far the greatest role in changing the watery environment of Alabama in the long term, humans have effected some significant changes within a relatively short span of time. Because of location, size, topography, and flow characteristics, Alabama's rivers were relatively easy to impound for the production of electric power and for navigation. As a result, there are 16 impoundments in the state constructed primarily for the production of electric power (Figure 1.17). There are 16 impoundments constructed primarily for river navigation. Electric power is also produced at five of these.

In a little over a century, major dams have essentially stilled the free flow of Alabama's largest streams, impounding over 500,000 acres of water (Figure 1.17). As a result of the construction of locks and dams and snagging and dredging, six of the 14 river basins (Figure 1.16) in the states contain navigable rivers (Moore an Szabo, 1994). As a result of the construction to the essential aids to navigation, Alabama leads the nation in the number of miles of navigable streams (Figure 1.18) (Moore and Szabo, 1984).

On a less grand scale, much of the flow of the Coosa River that would normally pass through Jordan Dam in central Elmore County has been diverted to flow through hydroelectric turbines at Bouldin Dam and back to the river below the Devil's Staircase. Many years ago, the source of the Conecuh River, east of Union Springs in Bullock County, was diverted across the divide between its drainage basin and the drainage basin of the Alabama River into Old Town Creek to provide additional water to power a grist mill. Now the Conecuh originates in a group of springs at the southeastern edge of town (Monroe, 1941).

Rivers

The major rivers in the state are listed in Table 1.5 along with information on their annual mean discharge rates (cubic feet per second, cfs) for the calender year 1992, as reported by the United States Geological Survey (Pearman, et al., 1993). Alabama's streams carry enormous quantities of water each year. Data in Table 1.5 show that the annual mean discharge of the Alabama River for calender year 1992 at Claiborne Lock and Dam in southwestern Monroe County was 31,670 cfs. There is considerable variation in the mean annual discharges of Alabama's major rivers. The discharge of the Tennessee (69,210 cfs) is much higher than any of the others. It is almost one-third higher than that of the next highest, the Tombigbee (49,040 cfs) and over twice as high as the Alabama (31,670 cfs). The discharges of the Black Warrior (12,900 cfs), Chatahoochee (10,710 cfs), and Coosa (14,820 cfs) are grouped between those of the larger rivers and those of the smaller rivers such as the Cahaba (2,861 cfs), Conecuh (5,421 cfs), and Tallapoosa (4,298 cfs).

Reservoirs

Reservoirs in Alabama are also significant physiographic features (Figures 1.16 and 1.18) (Moore and Szabo, 1994). These bodies of water (excluding farm ponds) have a total area of about 563,000 acres. Farm ponds would add well over 100,000 acres to the area of impounded waters. Table 1.6 lists the major reservoirs along with their area (Geological Survey of Alabama, 1996). Reservoirs on the Chattahoochee River are not included in the table because a significant portion of each of those is in Georgia. Similarly, Pickwick Lake on the Tennessee River is not included because a significant portion of it is in Mississippi and Tennessee; however Guntersville on the Tennessee River and Aliceville on the Tombigbee River are included because most of the area of these reservoirs are in Alabama.

Data presented in Table 1.6 show that, in general, the impoundments on the Tennessee River are much larger than reservoirs on other rivers in the state. Wheeler and Guntersville are 50% larger than Martin, which is the next largest. Over half of the reservoirs have areas of 10,000 acres or less. The size of reservoirs is generally a function of the height of the dam, the depth of the channel, and the topography of the basin. The R. L. Harris Dam on the upper Tallapoosa River is located in the rugged mountains of the Northern Piedmont. The dam is 150 feet high, and it impounds a reservoir of 10,661 acres. Wheeler Lock and Dam is located in the broad Tennessee Valley of the Highland Rim Section. That dam is only 72 feet high; yet it impounds a reservoir of 67,070 acres. The gentler relief of the land along the Tennessee River channel allows the reservoir to cover a much larger area for each foot of dam height compared to the steep slopes along the upper Tallapoosa. The bluffs along the Alabama and Tombigbee rivers generally are so high that even a high dam will not raise the water level out of the channel. Claiborne Lock and Dam on the Alabama River is 95.5 feet high, but it does not raise the water level out of the old stream channel; consequently Claiborne Lake is over 50 miles long with an area of only 5,930 acres. In many areas of Alabama, the land adjacent to the river is so flat that it is not feasible to construct a dam that will raise the level of the water out of the channel. To do so would create a shallow impoundment with a large surface area but with limited storage capacity. Demopolis Lock and Dam is located in the Black Prairie of the Coastal Plain just below the confluence of the Tombigbee and Black Warrior rivers. The dam, which is 58 feet high, impounds a reservoir with two long narrow arms in the old river channels. A dam 20-30 feet higher would have raised the water level out of the channels onto the flat topography of the Black Prairie, creating a much larger but very shallow reservoir.

Alabama's River Basins

Alabama's total area is 52,423 square-miles. It ranks 30th in the nation in total area. Within its boundaries, there are the usual types of land-forms (flat-lands, valleys, hills, and even mountains), and a large number of examples of each. Scattered among this large number of land features, there is an almost equally large number of water features (rivulets, branches, creeks, rivers, and bays). In sum, these different features constitute our physiography.

Because of the widespread dispersal of waterways, Alabama has been divided into 14 major and three minor river basins (Moore and Szabo, 1994). These river basins are listed below, and their location shown in Figure 1.16: Tennessee, Black Warrior, Upper Tombigbee, Cahaba, Coosa, Tallapoosa, Lower Tombigbee, Alabama, Conecuh, Escatawpa, Chattahoochee, Choctawhatchee, Perdido, and Mobile.

The headwaters of three other river basins (Blackwater, Yellow, and Chipola) are located in the southeasterrn part of the state along the Florida line. None of these headwater streams reach the size to be identified as rivers until after they enter Florida.

Tennessee River Basin

The Tennessee River Basin, with a total area of 6,826 square miles, is the largest river basin in Alabama (Figure 1.16). It is located principally in Lauderdale, Limestone, Madison, Jackson, Colbert, Lawrence, Morgan, Marshall, DeKalb, and Franklin counties. The rivers and the creeks in the basin provide drainage for all of Alabama's Highland Rim Physiographic Section, as well as for the northern third of the Cumberland Plateau Section and a small part of the extreme northwestern portion of the East Gulf Coastal Plain Section (Figure 1.9). From outside of the state, the basin receives water from Tennessee, Georgia, North Carolina, and Virginia. Further, with the construction of the Tennessee-Tombigbee Waterway, the Upper Tombigbee Basin technically has become part of the Tennesse River Basin.

The Tennesse River is the primary waterway in the basin (Figure 1.15). There are three additional smaller rivers: Elk, Flint, and Paint Rock. Bear, Cypress, Big Nance, Town, Flint, and South Sauty are large creeks in the basin. The river is already large when it enters Alabama through a gap (the Sequatchie Valley District—Figure 1.7) three miles northeast of Bridgeport in northeastern Jackson County. On its course to Mississippi, the Tennessee flows through Scottsboro, Guntersville, Decatur, and Florence. The river generally flows in a southwesterly direction until it reaches Guntersville. There it abruptly changes its course to the northwest and flows all the way across Alabama to the Mississippi line about ten miles northwest of Waterloo in Lauderdale County.

The Paint Rock River enters the Tennessee as the line between Marshall and Madison counties. About four miles downriver, west of Meeks Mountain, the Flint River joins it. Then about two miles east of Thornton in southeastern Lauderdale, the Elk River enters.

Throughout the basin, dams and reservoirs have essentially eliminated the riverine characteristics of the Tennessee and the lower reaches of most of its tributaries (Figure 1.18). Only immediately below the dams is there any semblance of free flow. Three major lock and dam structures, which form three large reservoirs, have been constructed in the basin. There is also the upper portion of a fourth large reservoir resulting from the construction of a large lock and dam just outside the state. Guntersville Lock and Dam in Marshall County, three miles northeast of Union Grove; Wheeler Lock and Dam in Lauderdale and Lawerence counties, two miles southwest of Thorntontown; Wilson Lock and Dam just east of Florence; and Pickwick Lock and Dam just north of the Mississippi-Tennessee line form four large reservoirs: Guntersville, Wheeler, Wilson, and Pickwick, respectively, in the Alabama portion of the Tennessee River Basin. Smaller lakes in the basin are Cedar Creek east of White Oak; Little Bear Creek southwest of Russellville; Lower Bear Creek northwest of Hodges in Franklin County; and Bear Creek, northeast of the village of Bear Creek in Lawrence, Franklin, and Marion counties.

The Tennessee River is navigable throughout its entire length in Alabama (Figure 1.18).

Black Warrior River Basin

The Black Warrior River Basin (6,287 square miles) is located principally in Winston, Cullman, Blount, Etowah, Walker, Jefferson, Fayette, Tuscaloosa, and Hale counties (Figure 1.16). The primary drainage trend in the basin is from northeast to southwest. Waterways in the basin provide drainage for more than half of the Cumberland Plateau Section and a small portion of the Coastal Plain Physiographic Section (Figure 1.9).

The Black Warrior River (Figure 1.15) is the primary river in the basin. The main stem of the river is formed from three major tributaries: the Sipsey, Mulberry, and Locust forks, all of which originate within the Warrior Basin District. Strictly speaking, the Black Warrior River originates in western Jefferson County, four miles northwest of Gilmore where the waters from the three forks are joined together. Before the confluence, however, the Sipsey and Mulberry forks had joined just north of the village of Sipsey in northeastern Walker County about 25 miles north of Gilmore. With water from all of the forks combined, the main stem of the Black Warrior flows southwestward through or past Tuscaloosa, Moundville, Eutaw, and on to Demopolis where it joins the Tombigbee and disappears. Once the river passes Oliver Lock and Dam near Tuscaloosa, it begins to meander widely across a relatively level flood plain.

Dams and reservoirs have eliminated a third of the free flow of the main stem of the Black Warrior River and its major tributaries (Figure 1.17). Two large and several smaller reservoirs have been created. Lewis Smith Dam on the Sipsey Fork, six miles southwest of Cold Springs on the Culman County-Walker County line, forms Lake Lewis Smith. John Bankhead Lock and Dam on the main stem of the river in Tuscaloosa County, five miles east of Windham Springs, forms Bankhead Lake. Further downsteam near Holt, the Holt Lock and Dam forms Holt Lake. West of Holt on the North River, a large dam forms Lake Tuscaloosa. Near Northport on the main stem of the Black Warrior, Oliver Lock and Dam forms Oliver Lake. On the Hale-Greene County line, five miles west of Sawyerville, the Warrior Lock and Dam forms Warrior Lake. Finally at Demopolis, the Demopolis Lock and Dam on the Marengo-Sumter County line forms Demopolis Lake. Bayview, north of Mulga in Jefferson County, and Highland and Inland, northeast of Remlap and west of Blount Mountain in Blount County, are smaller lakes in the basin.

The Black Warrior River is navigable past Port Birmingham to northern Jefferson County (Figure 1.18).

Upper Tombigbee River Basin

The Upper Tombigbee Basin (3,650 square miles) is located principally in Marion, Lamar, Fayette, Pickens, Tuscaloosa, Greene, and Sumter counties (Figure 1.16). The basin provides drainage in Alabama for much of the northwestern portion of the Fall Line Hills District of the East Gulf Coastal Plain Physiographic Section (Figure 1.9). In addition, it receives a significant amount from northwestern Mississippi. Also, with the construction of the Tennessee-Tombigbee Waterway, the entire Tennessee Basin is connected to the Upper Tombigbee Basin. The drainage in the basin is almost north to south; although there is a slight northwest to southeast trend. The Tombigbee River is the primary stream in the basin (Figure 1.15). The Sipsey and Noxubee are the only other rivers. Buttahatchee, Luxapallila, Hell's, and Lubbob are large creeks.

The Upper Tombigbee River enters the state in southwest Pickens County within the Black Prairie District

(Figures 1.9 and 1.16) three miles northwest of Pickensville. On its course to Demopolis, it generally flows southeastward through or past Cochran, Gainesville, and Boligee. In Demopolis, it meets and absorbs the waters of the Black Warrior. In all of its travels from the Mississippi line to Demopolis, it is a widely meandering, slow moving, Lower Coastal Plain stream. Its course is especially circuitous once it flows past Gainesville. The Sipsey River joins the Tombigbee near the point where Greene, Pickens, and Sumter counties meet. The Noxubee River flows out of Mississippi to enter Gainesville Reservoir just northwest of Gainesville in Sumter County.

Dams and reservoirs have also eliminated virtually all of the free flow on the main stem of the Upper Tombigbee and the lower reaches of its major tributaries (Figure 1.17). The three major reservoirs in the basin are links in the Tennesse-Tombigbee Waterway. Southeast of Pickensville in Pickens County, the Aliceville Lock and Dam forms Aliceville Lake. Downstream, the Gainesville Lock and Dam on the Sumter-Greene County line forms Gainesville Lake. Here the river has been channeled to eliminate a large bend in the river. Near Demopolis, on the Sumter-Marengo County line, Demopolis Lock and Dam forms Lake Demopolis. There are no smaller lakes in the basin. The Upper Tombigbee River is navigable throughout its length (Figure 1.18).

Cahaba River Basin

The Cahaba River Basin (1,818 square miles), is located principally in Jefferson, St. Clair, Shelby, Bibb, Perry, and Dallas counties (Figure 1.16). The Cahaba Basin provides drainage for a fourth (the northwestern portion) of the Alabama Valley and Ridge Physiographic Section (Figure 1.9). The primary direction of drainage in the basin is also northeast to southwest. The Cahaba is the primary stream in the basin (Figure 1.15). The Little Cahaba is the only river. Shades and Oakmulgee are large creeks in the basin.

The Cahaba River (Figure 1.15) originates in southwestern St. Clair County and Eastern Jefferson County, a few miles southwest of Acmar in the Cahaba Ridges District (Figure 1.9) of the Valley and Ridge Section, where several mountain creeks converge. It originates in the southwestern-most extension of Alabama's Eastern Mountains (Figures 1.8). On its way to join the Alabama, the Cahaba flows past or through: Cahaba Heights, Acton, Helena, Marvel, Centerville, Sprott, and Suttle. About three miles northeast of Big Mountain in east-central Bibb County, the Little Cahaba joins the main stem of the river.

Throughout its journey, until it reaches central Bibb County, the main stem of the Cahaba tumbles down falls, races across shoals, and rests infrequently in shallow pools—always guided by high bluffs and steep banks. Once it reaches Centerville, however, where it flows over the fall line, the velocity of the river slows significantly, and it begins to meander, guided only by low banks of alluvial and low terrace deposits in the Fall Line Hills District of the Coastal Plain Section (Figure 1.9).

There are no reservoirs on the main stem of the Cahaba. It is the only major free flowing river remaining in Alabama. Lake Purdy is the only significant body of impounded water in the basin. Constructed on the Little Cahaba River, the lake is located in eastern Jefferson and northwest Shelby counties between Cahaba Heights and Oak Mountain.

Coosa River Basin

The Coosa River Basin (5,400 square miles) is located primarily in DeKalb, Etowah, Cherokee, St. Clair, Calhoun, Cleburne, Shelby, Talladega, Clay, Chilton, Coosa, and Elmore counties (Figure 1.16). The Coosa Basin provides drainage for more than three-fourths of the Valley and Ridge Section and a third of the Piedmont Upland Section (Figure 1.9). In addition, the basin receives a substantial amount of drainage from northwest Georgia. The Coosa River (Figure 1.30) is the primary stream in the basin, and Chatooga and Little rivers are major tributaries. Tallasseehatchee, Big Wills, Chocolloco, Talladega, Hatchett, Weogufka, Yellowleaf, and Big Canoe are large creeks in the basin.

The Coosa River originates in the northwestern corner of Georgia where the Oostanaula and Etowah rivers join, at Rome (Jackson, 1995). The river flows out of Georgia into the Coosa Valley District of the Valley and Ridge Section (Figure 1.9) through a broad gap in Alabama's Eastern Mountains (Figure 1.8) in east central Cherokee County southeast of Gaylesville. In its course to meet the Tallapoosa south of Wetumpka, the Coosa flows past or through Centre, Gadsden, Pell City, Sylacauga, and Clanton. The Chatooga River enters Weiss Reservoir on the Coosa in Gaylesville in northeast Cherokee County (Figure 1.16). Little River enters the reservoir about two miles southwest of Hurley in the northwestern portion of the county.

Past Centre the river meanders generally southwestward through the upper valley and onto a gap through the Beaver Creek Mountains near the location of the old Greensport Ferry, about four miles north of H. Neely Henry Dam (Figure 1.17). This gap is the natural dam that was the primary cause of the meandering of the river in the upper valley. Before the construction of Logan Martin Reservoir (Figure 1.17) near the point of convergence of St Clair, Talladega, and Coosa counties southeast of Ragland, the Coosa River passed over Peckerwood Shoals and in the next 65 miles fell 275 feet. In its rapid descent, it would pass over the Narrows, the Devil's Race, Hell's Gap, and finally over the Staircase.

There is virtually no free flowing water on the main stem of the Coosa or on the lower reaches of its major tributaries, except immediately below the dams and even there the flow is regulated for power production (Figure 1.17). Six large reservoirs and one small reservoir have been constructed in the basin (Figure 1.17). Weiss Dam in Cherokee County, two miles southwest of Leesburg, forms Weiss Lake; Neely Henry Dam on the line between Calhoun and St. Clair, counties, three miles west of Ohatchee, forms Neely Henry Lake; Logan Martin Dam on the county line between St. Clair and Talladega, three miles northeast of Vincent, forms Logan Martin Lake; Lay Dam on the county line between Coosa and Chilton, four miles southwest of Marble Valley, forms Lay Lake; Mitchell Dam on the county line between Coosa and Chilton, seven miles northeast of Cooper, forms Mitchell Lake; and Jordan Dam, five miles northeast of Speigner, forms Jordan Lake. A few miles west of Jordan Lake, Bouldin Dam forms Bouldin Lake, a much smaller reservoir than the others on the main stem of the river.

Tallapoosa River Basin

The Tallapoosa River Basin (4,023 square miles) is located primarily in Cleburne, Clay, Randolph, Tallapoosa, Chambers, Elmore, Lee, Montgomery, Macon, and Bullock counties (Figure 1.16). The Tallapoosa Basin provides drainage for half of the Piedmont Upland and a small portion of the Coastal Plain Sections (Figure 1.9). The basin also receives a significant amount of drainage from west-central Georgia. The primary drainage in the basin is from northeast to southwest.

The Tallapoosa River is the primary stream in the basin (Figure 1.15). The Little Tallapoosa River is a major tributary. Enitachopoco, Hillabee, Saugahatchee, Uphapee, Emuckfaw, and Sandy are large creeks in the basin. The major stem of the Tallapoosa River originates near Buchanan, Georgia, in the northwestern region of that state. It crosses the Alabama line near Muscadine in east central Cleburn County, in the Northern Piedmont District (Figure 1.9) of the Piedmont Upland Section, through a narrow gap in Alabama's Eastern Mountains (Figure 1.8). From that point on its course to meet the Coosa, the Tallapoosa tends southwest across the Piedmont Upland, through or near: Weedowee, Wadley, Jackson's Gap, Tallassee, and Milstead. At Milstead in Macon County it abruptly turns westward before joining the Coosa southwest of Wetupmka. In central Randolph County, four miles north of Weedowee, the Little Tallapoosa River enters Harris Reservoir.

The elevation of the Tallapoosa is 1,100 feet where it enters the state. At its confluence with the Coosa, some 214 river-miles downstream, the elevation is around 125 feet (Figure 1.8). Because of this rapid change in elevation, the construction of dams has eliminated only a relatively small part of the river's geophysical characteristics. Although much of the river does flow freely, the flow below the dams is closely regulated, depending on power production and flood control requirements. Two major reservoirs and two small ones have been constructed in the basin (Figure 1.17). R. L. Harris Dam, three miles northwest of Malone in Randolph County, forms Harris Lake; 45 miles downstream, Martin Dam on the county line between Tallapoosa and Elmore a mile southeast of Red Hill forms Martin Lake. A few miles downstream, Thurlow Dam forms Thurlow Lake; at Tallassee, Yates Dam forms Yates Lake.

Lower Tombigbee River Basin

The Lower Tombigbee River Basin (4,042 square

miles) is located primarily in Sumter, Marengo, Choctaw, Clark, and Washington counties (Figure 1.16). This basin provides drainage for the southwestern portion of the Coastal Plain Section in southwest Alabama (Figure 1.17). The Lower Tombigbee River (Figure 1.15) originates just north of Demopolis in the Black Prairie District (Figure 1.9) of the Coastal Plain Section after the Black Warrior joins the Upper Tombigbee.

The Lower Tombigbee River is the major stream in the basin (Figure 1.15). Sucarnoochee River is a major tributary. Chickasaw, Kinterbish, Santa Bogue, and Bassett are large creeks. From its point of origin, the Lower Tombigbee River flows generally south through many bends and loops. Along its course from Demopolis to its confluence with the Alabama, the river passes through or near Bellamy, Coffeeville, Saltipa, Jackson. and McIntosh Station The Lower Tombigbee, like the Coosa, is a county line river. Except for a few miles in Pickens County, where it enters the state, it serves as a county line. Approximately five miles southeast of Bellamy in Sumter County, it is joined by the Sucarnoochee River.

The Lower Tombigbee is truly a low-land river. From its beginning at Demopolis to the point where it joins the Alabama, its channel is less than 100 feet above sea level. Higher hills within the basin range up to 250 feet (Figures 1.8).

Although now a part of Coffeeville Lake (Figure 1.17), the waters of the Lower Tombigbee are contained within the original channel, but the already slow flow rate is reduced by the dam. The flow of the river's water in unimpeded below Coffeeville Lock and Dam, but the rate of flow is regulated. The Lower Tombigbee River is navigable throughout its entire length (Figure 1.18).

Alabama River Basin

The Alabama River Basin (6,024 square miles) is located primarily in Perry, Dallas, Chilton, Autauga, Wilcox, Lowndes, Montgomery, Monroe, Escambia, and Baldwin counties (Figure 1.16). It provides drainage for a relatively large portion of the Central Coastal Plain Section (Figure 1.9), as well as the Cahaba, Coosa, and Tallapoosa basins. Drainage is generally northeast to southwest. The Alabama River is the major stream in the basin (Figure 1.15). The Cahaba and Little rivers are large tributaries. There are several large creeks including: Pine Barren, Mulberry, Catoma, Pintlalla, Cedar, and Boguechitto.

On its course from its origin, just south of Wetumpka in Elmore County where the Coosa and Tallapoosa rivers converge, to its confluence with the Tombigbee northeast of Mt Vernon in northeast Mobile County, some 315 miles downstream, the river passes through or near Montgomery, Burnsville, Selma, Boykin, Miller's Ferry, Chrysler. and Perdue Hill. Near the site of Old Cahawba in Dallas County, the Cahaba River becomes part of the Alabama, and near Chrysler in south-western Monroe, Little River joins the river's main stem.

From Montgomery westward to Selma and then south-westward to northern Clarke County, the Alabama meanders widely across a broad expanse of the Alluvial-Deltaic Plain District (Figure 1.9). In its travels, the Alabama seems to want to visit every farm in the valley. Some of its most famous meanders are: Dutch, Durant, Kings, Calhoun, Molette, Gees, Canton, and Packers bends. Before the impoundments were constructed, the elevation of the river fell by less than 0.5 foot per mile for most of its length.

Three major lock and dam installations have virtually eliminated the free flow from the Alabama (Figure 1.17). The Robert Henry Lock and Dam just north of Benton in Lowndes County forms the R. E. "Bob" Woodruff Reservoir. It results in the establishment of reservoir conditions all the way upriver past Montgomery into the lower reaches of the Coosa and Tallapoosa rivers. Millers Ferry Lock and Dam in southwestern Wilcox County, two miles west of Millers Ferry, forms the William "Bill" Dannelly Reservoir. This reservoir extends all the way up-river to the base of Henry Lock and Dam. Upstream, four miles northwest of Claiborne, Claiborne Lock and Dam forms Claiborne Lake. Lake waters generally remain within the old river channel all the way upstream to the next lock and dam. There are no small lakes in the Alabama Basin. The Alabama River is navigable for its entire length (Figure 1.18).

Conecuh River Basin

The Conecuh River Basin (3,849 square mile) is located primarily in Montgomery, Bullock, Butler, Crenshaw, Conecuh, and Covington counties (Figure 1.16). It provides drainage for the south-central portion of the Coastal Plain Section (Figure 1.9). Drainage in the basin is similar to drainage in the other rivers of the Coastal Plain, northeast to southwest. The Conecuh River (Figure 1.15) is the major stream in the basin, but the Sepulga River adds a significant amount of flow. Patsaliga, Pigeon, Murder, Burnt Corn, and Big Escambia are large creeks in the basin.

The Conecuh River (Figure 1.15) originates south and east of Union Springs in north-central Bullock County on the south slope of the Enon Cuesta in the Chunnennuggee Hills Physiographic District (Figure 1.9). It leaves Alabama just east of Flomaton in south Escambia County. From its the origin, the river flows generally southwestward through or past Troy, Glennwood, Brantley, Gantt, River Falls, Roberts, and East Brewton. Just northeast of Dixie in northeast Escambia County, the Sepulga River joins the main stem of the Conecuh.

Generally, the Conecuh is a free-flowing river throughout its length. Two small reservoirs just north of River Falls in central Covington County are the only areas of impounded water in the basin. Point "A" Dam is just north of River Falls, and Gantt Dam is a few miles upstream.

Escatawpa River Basin

The Escatawpa River Basin (767 square miles) is located primarily in southeastern Mississippi, but a small portion extends into Choctaw, Washington, and Mobile counties (Figure 1.16). This basin provides drainage for a small portion of the southwest corner of the Coastal Plain Section (Figure 1.9). The Escatawpa River (1.15) is the primary stream in the basin in Alabama. There are no large creeks.

The Escatawpa originates in west-central Washington County in the Southern Pine Hills District (Figures 1.9 and 1.15). It flows generally southward where it enters northwestern Mobile County five miles west of Citronelle. It exits the state in west-central Mobile County three miles northwest of Wilmer. On its course through the extreme southwestern corner of Alabama, it flows through or past Yellow Pine, Vinegar Bend, Deer Park, and Earlville.

Big Creek Lake in Mobile County is the only impounded water of significance in the basin. The dam is located five miles southwest of Semmes.

Chattahoochee River Basin

The Chattahoochee River Basin (2,570 square miles) occurs primarily in Georgia; however a small portion of the western side of the basin is located in Randolph, Chambers, Lee, Russell, Bullock, Barbour, Henry, and Houston counties (Figure 1.15). The basin provides drainage in Alabama for a small portion of the southeastern corner of the Piedmont Upland Physiographic Section and a narrow north-south section on the extreme eastern side of the East Gulf Coastal Plain Section (Figure 1.9). There are no other rivers in the basin in Alabama. Large creeks include Halawakee, Uchee, Hatchechubbee, and Cowickee.

The Chattahoochee River (Figure 1.15) becomes the state line between Georgia and Alabama at Lanet in southeastern Chambers County in the Northern Piedmont District (Figure 1.9) of the Piedmont Upland Physiographic Section. In most cases where a river is the boundary between two states, the state line is in the middle of the channel; however this is not the case with the Chattahoochee. The line is on the west side of the river. Georgia was an established political entity for well over a century before Alabama became a state, and in 1819, Oglethorpe's great experiment had little interest in sharing the river with a country bumpkin upstart. The river exits Alabama southeast of Crosby in southeastern Houston County. In its journey southward, the river passes through or past Valley, Columbus-Phenix City, Eufaula, Ft Gaines, Columbia, and Gordon.

Several of the cuestas described in preceding sections extend to the Chattahoochee River (Figure 1.14). In fact, most of them extend across the river into Georgia. Generally, however, they are not traceable across the river valley. From Lanett south to southeastern Lee County, the Chattahoochee River has cut into and through the most highly erosion-resistant igneous and metamorphic rocks (granites, gneisses, schists, and amphibolites) in the state (Figure 1.7). Before impoundments were constructed on this stretch, the river was a continuous series of shoals and low falls.

Although dams on the Chattahoochee were constructed primarily in Georgia, significant portions of waters impounded behind them back-up into Alabama (Figure 1.17). Significant portions of West Point Lake formed behind West Point Dam, constructed near West Point, Georgia, are located in Chambers County. Similarly, portions of Lake Harding, formed behind Bartlett's Ferry Dam east of Opelika, extend into Lee County. Walter F. George Reservoir, formed behind Walter F. George Lock and Dam near Ft. Gaines, extends into Barbour and Russell counties. Smaller dams located up-river from Lake Harding (near Langdale and Riverview) and up-river from Columbus, Georgia (Goat Rock, Oliver, North Highlands, City Mills, Eagle, and Phenix) form small impoundments, primarily in the channel of the river. Only small portions of these waters are backed into Alabama. At the lower end of the Chattahoochee on the eastern border of Houston County, Andrews Lock and Dam forms an impoundment, but very little of that reservoir is located in Alabama.

The Chattahoochee River is navigable from the Florida line northward to Columbus, Georgia (Figure 1.18).

Choctawhatchee River Basin

The Choctawahchee River Basin (3,129 square miles) is located principally in Bullock, Pike, Barbour, Coffee, Dale, Henry, Covington, and Geneva counties (Figure 1.16). This basin provides drainage for a significant portion of the Coastal Plain Physiographic Section (Figure 1.9) in southeast Alabama. Primary drainage is also northeast to southwest across the basin. The Choctawhatchee River (Figure 1.15) is considered to be the primary stream in the basin, although it receives a significant volume of water from the West Fork, the Little Choctawhatchee River, and the Pea River. The Pea is almost as large as the Choctawhatcee. Claybank Creek, which flows into the Choctawhatchee and Whitewater Creek which flows into the Pea, are major tributaries in the basin.

The main stem of the Choctawhatchee (Figure 1.15) is formed from the confluence of the East and West forks of the Choctawhatchee, northeast of Midland City in southwest Dale County. The East Fork originates east of Clayton in Barbour County, and the West Fork originates southwest of Clayton. From its origin, the main stem passes through or near Clayhatcheee and Bellwood before it joins the Pea River southeast of Geneva. A short distance southeastward, it crosses into Florida. About three miles south of Daleville in south Dale County, the Little Choctawhatchee joins the main stem of the river.

The Pea River (Figure 1.15) originates in the Chunnennuggee Hills Physiographic District (Figure 1.9) from the confluence of several small creeks in northeast Bullock County east of Union Springs and a few miles south of Chunnennuggee Ridge. From its point of origin, it follows a circuitous course passing through or near Three Notch, Pickett, Jamback, Richland, Ariton, Elba, Alberton, and Royal Crossroads on it way to join the Choctawhatchee near Geneva.

Lake Tholocco on Claybank Creek in western Dale County is the only impounded water of any significance in the basin.

Mobile River Basin

The Mobile River Basin (1,836 square miles) is located principally in Mobile and Baldwin counties below the confluence of the Alabama and Tombigbee rivers (Figure 1.16). Technically, the Mobile River Basin provides drainage for only a small portion of the southwest corner of the Coastal Plain Physiographic Section (Figure 1.9) in Alabama, but in reality waters from six states flow into this basin. In terms of discharge, the Mobile River system is the fourth largest in the United States (Masek, 1996). The Mobile and Tensaw (Figure 1.15) are both large rivers in the basin. The Fowl, Fish, and Magnolia rivers are also found in this drainage. The confluence of the two major rivers result in a maze of ill-defined river channels, creeks, cut-offs, swamps, sloughs, and bays so that it is difficult to determine where one ends and another begins.

Specifically, the Mobile River (Figure 1.15) originates in the Southern Pine Hills District (Figures 1.9 and 1.25) with the confluence of the Alabama and Tombigbee rivers in northeast Mobile and northwest Baldwin counties, a few miles east of Malcom and Calvert. On its course to Mobile Bay, the river passes through or near Mount Vernon, Flat Point, Mary Ida Point, White Horse Point, Twenty-seven Mile Bluff, Twenty-one Mile Bluff, Twelve Mile Island, and Akka.

Four miles southeast of Mt. Vernon, the Mobile provides water for the origin of the Tensaw River (Figure

1.15).

The Tensaw River also receives a substantial volume of water from creeks and sloughs originating in northern Baldwin County. It even receives some flow through a small cut-off creek from the lower Alabama River before it joined the Tombigbee. From its point of origin, the Tensaw flows through or near Devil's Bend, Sandy Hook, Willow Point, Upper Hall Landing, Hurricane, Gravevine Island, and Historic Blakley before passing on to enter Mobile Bay near the USS Alabama Battleship Memorial Park. Just south of Historic Blakely, the Tensaw provides some water for the formation of the Apalachee River.

Fowl River (Figure 1.15) originates northwest of Theodore in south-central Mobile County. It flows generally southeastward to a small bay two miles west of Mon Louis. From there a portion (East Fowl River) flows eastward into Mobile Bay at Mon Lewis just south of Fowl River Point. Another part (West Fowl River) flows southward to Portersville Bay (Figure 1.9).

Fish River originates just south of Stapleton, in south-central Baldwin County. From that point, it flows almost due south, east of Belforest and west of Silverhill, to enter Weeks Bay just south of Marlow (Figure 1.20).

Magnolia River (Figure 1.15) originates northwest of Foley near Magnolia Springs at the south end of Baldwin County. It flows generally westward for only about eight miles before it also enters Weeks Bay (Figure 1.19). The Magnolia is Alabama's shortest river.

There are no impounded waters in the Mobile River Basin. A considerable portion is affected by tidal action._The Mobile River is navigable for its entire length (Figure 1.18).

Perdido River Basin

The Perdido River Basin (831 square miles) is located principally in Escambia and Baldwin counties (Figure 1.16). This basin provides drainage for only a small portion of the southwestern corner of the Coastal Plain Section in Alabama (Figure 1.9). The Perdido River (Figure 1.16) is the primary stream in the basin, but the Styx and Blackwater rivers are also small rivers in the basin.

Perdido Creek originates northwest of Poarch, in northwestern Escambia County in the Southern Pine Hills (Figure 1.9). It flows southward to become the Alabama-Florida line three miles east of Perdido. At that point, Perdido Creek becomes the Perdido River (Figure 1.15). Flowing southward, the river passes through or near Phillipsville, Barrineau Park, Beulah, and Seminole before entering Perdido Bay at Chambers Point. About three miles southeast of Seminole, the Styx River (Figure 1.16) joins the Perdido. The Blackwater joins the Perdido a short distance downriver.

There are no major impounded waters in the Perdido Basin.

Travel along I-65 (Figures 1.11 and 1.15) provides an opportunity to see a number of Alabama's water features. Their locations are shown by mile markers along the highway (Table 1.7). Between mm 24 and 28, the highway crosses the braided flood-plain of the delta of the Alabama River system. The major waterways in this area are: Mobile River (mm 24), Middle River (mm 27), and Tensaw River (mm 28). There are also numerous small creeks and sloughs.

Between mm 64 and 138, I-65 crosses several of the large creeks and one small river (Sepulga). All of these streams flow northwest to southeast into the Conecuh River.

Around mm 145, the highway crosses the dividing line between the Conecuh River and Alabama River Basins (Figure 1.16). It crosses the Alabama River and Woodruff Reservoir around mm172. The floodplain here is seven miles wide. Its expansion on the north side is somewhat limited by the presence of the more erosion-resistant Eutaw Formation. Down river around Selma, the flood-plain is more than twice as wide.

From just north of Montgomery, around mm 236, I-65 passes through the extreme western portion of the Coosa River Basin (Figure 1.9), but at that point, it quickly passes into the narrow Cahaba River Basin; then the highway crosses the Cahaba River near mm 248. Once it crosses Red Mountain in Birmingham and enters the Birmingham-Big Canoe Valley around mm 257, it enters the Basin of the Black Warrior. North of Birmingham near mm 279 and 292, it crosses the Locust and Mulberry Forks respectively of the Black Warrior. River, respectively.

Near mm 313, I-65 crosses the Tennessee Valley Divide, and enters the Basin of the Tennessee. It crosses the Tennessee River and Wheeler Reservoir around mm 337.

Some Marine and Estaurine Features

The number of marine and estaurine features in Alabama's coastal zone is limited because of the state's relatively short coastline (Figure 1.19). A straight line between the Alabama-Florida line at the mouth of Perdido Bay just east of Orange Beach in Baldwin County and the Alabama-Mississippi line on the Mississippi Sound just west of Bayou La Batre in Mobile County would be 52 miles long. Of course the length of actual coastline would be much greater.

Marine and estuarine features in Alabama's Coastal Zone include: the Gulf of Mexico, Mississippi Sound, Mobile Bay, other Mobile County bays, other Baldwin County bays, barrier islands, and other small islands.

The Gulf of Mexico

The Gulf (Figure 1.19) is an important physiographic-economic feature of the state because of the access it provides to worldwide shipping. It is also the source of the state's seafood industry.

Mississippi Sound

The sound (Figure 1.19) extends eastward from coastal Mississippi into Alabama, and is important because of the access that it provides to the National Intracoastal Waterway System.

Barrier Islands

The barrier islands in Alabama's Coastal Zone are technically land features, but they are so closely associated with the coastal and estuarine physiographic features of the state that they are listed here. There are three barrier islands located in Alabama (Figure 1.20). The largest of these, Dauphin Island, lies perpendicular to the north-south axis of Mobile Bay, a short distance to the west of Mobile Point in Baldwin County. Two other small islands, Pelican Island and Sand Island, are located south and southeast, respectively, of the eastern end of Dauphin Island. Generally, barrier islands form off the mouth of large rivers where the current from the river meets coastal currents, so that sands being carried from the river settle to the bottom.

Mobile Bay

Mobile Bay (Figure 1.19) is located in both Mobile and Baldwin counties. The county line separating them generally runs down the center of the bay. It is 30 miles long and averages ten miles in width. It receives a massive amount of fresh water from both the Mobile and Tensas rivers and a relatively small amount from East Fowl, Fish, and Magnolia rivers (Figure 1.16). Mobile Bay is much larger than the others bays in Alabama's coastal zone and much more important to the economy of the state, because of the opportunities for transportation that it provides. Alabama State Docks has a major installation on Mobile Bay.

According to Mike Szabo (Alabama Geological Survey, personal communication), Mobile Bay is unique among major estuaries along the Gulf Coast, because it does not have a well developed delta at its mouth. Its delta is some 30 miles north. This situation has led geologists to conclude that Mobile Bay is really a flooded river rather than a typical bay. During the Pleistocene Epoch of the Quaternary Period (Cenozoic Era) when much of the world's supply of water was tied up in the glaciers of that period, the sea level of the Gulf of Mexico was much lower than it is today; consequently, the mouth of the river and its delta were located much further to the south, and Mobile Bay was part of the Mobile River System. River currents carried their loads of sediments through the Mobile Bay basin to be deposited on a delta to the south of Dauphin Island. Then when the glaciers melted and the sea level rose, the old river was flooded, and part of it became Mobile Bay (Charles Savrda, Department of Geology, Auburn University, personal communication)

Other Mobile County Bays

Other bays in Mobile County include: Grand Bay, Portersville Bay, and Heron Bay (Figure 1.19). Grand Bay, which is a small open bay located between Point of Pines and the Mississippi line, extends southward to Mississippi Sound. Portersville Bay, which is in the same vicinity, extends eastward from Point of Pines to Barron Point and southward to Mississippi Sound. It receives a small amount of fresh water from West Fowl River (Figure 1.16). Heron Bay is located on the extreme southeast corner of the Mobile County mainland west of Cedar Point.

Other Baldwin County Bays

Other bays in Baldwin County include: Weeks, Bon Secour, Oyster, Perdido, and Wolf (Figure 1.19). Weeks Bay is a modest size, almost closed body of water. Contrasted to Grand Bay, its mouth is narrow. Located on the on the southeastern side of Mobile Bay, it receives fresh water from both the Fish and Magnolia rivers (Figure 1.15). Bon Secour is a modest size, open bay in the southeast corner of Mobile Bay just south of Weeks Bay. Oyster Bay is a small closed projection off the southeast corner of Bon Secour Bay. The Intercoastal Waterway passes through the north end of this bay. Perdido and Wolf bays are portions of the same larger bay system. Perdido Bay is the larger of the two. It is on the state line between Alabama and Florida, and receives a moderate amount of fresh water from the Perdido, Blackwater, and Styx rivers (Figure 1.15). Wolf Bay to the west is the smaller portion of the larger system. The Intercoastal Waterway passes through parts of both Perdido and Wolf bays.

Other Islands

Dauphin Island is the largest in Alabama's coastal zone, but there are several others that are not considered to be barrier islands. Virtually all of these are located along the outer boundaries of Grand Bay and Portersville Bay. The largest is Isle Aux Herbes (Figure 1.19). Technically, the largest island in the coastal zone is Mon Louis. It is located on the southeast corner of the Mobile County mainland, but it is separated from it by the narrow channels of East and West Fowl rivers.

Geology, Physiography, and Agriculture

In Alabama, the blanket of soil where we grow our crops, build our homes, and carry on our daily lives is draped over an amazing assortment of rocks. The nature of these rocks, where they are located with reference to the equator, where they are located in the state, and how they are arranged have had a profound effect on the past and present development of agriculture in the state, and will doubtlessly continue to affect it in the future. Natural features and their consequences include, as will be detailed below: Alluvial Deposits and Native American Agriculture; Geology, Physiography, and Emigration; Rivers and Commercial Agriculture; Mountains, Travel, and Subsistence Agriculture; Geology, Stream Gradient, and River Transportation; a Natural Dam on the Coosa; Rivers and Railroads; Chalk and Drinking Water in the Black Prairie; Chalk, Marl, and Boll Weevils; Geology and a Local Market for Cotton; and Use of Excess Agricultural Labor.

Alluvial Deposits and Early Alabama Agriculture

Geological characteristics of the East Gulf Coastal Plain Section determined the meandering patterns and the velocity of the stream flows in the section. Over the last two million years, swift rivers that eroded deeply into the steep hills and mountains of north-central and east Alabama quickly began to drop their loads of gravel, sand, silt, and clay when they reached the more level Coastal Plain. Tremendous quantities of Alluvial Low Terrace Deposits accumulated in and adjacent to their channels to form the Alluvial-Deltaic Plain Section (Figure 1.19). The generally fertile, moist soils developed from these deposits were first use extensively by Native American farmers of the Mississippian Tradition, beginning about AD 700 to 900, to grow their crops of corn, beans, and squash (Hudson, 1976). Those soils, with their relatively high content of sand and silt, were much easier to work with the limited agricultural technology available to those early farmers. As a result of the availability of these easily tillable-soils, the Choctaw Indians were generally the most agriculturally-oriented of all of the Five Civilized Tribes. It is likely that this agricultural orientation played a major role in their earlier relationships with European settlers.

Nine hundred years later, soils formed in these same deposits were the base for the early development, in the late eighteenth and early nineteenth centuries, of the first significant European farming settlement in Alabama—the Bigbee Bottoms in northeast Washington County in the lower Tombigbee River Basin (Figure 1.17). Later, the Cotton Kingdom in Alabama would develop on soils formed in these deposits in both the Alabama and Tombigbee basins.

Physiography and Emigration

The mountainous terrain (Figure 1.8) along the state's eastern border from Randolph County northward had a significant effect on the movement of settlers (primarily farmers) traveling from the east into the state and across it. Generally, they had no choice but to travel around the southern end of this impediment, crossing the Chattahoochee at Fort Mitchell south of Columbus, Georgia.

Similarly, the mountainous terrain associated with Alabama's Central Mountains, stretching from Jackson County in the east to Franklin County in the west (Figures 1.7 and 1.8) impeded north-south travel by settlers. Fortunately, there were several passes through the mountains in Blount, Cullman, and Marshall counties that allowed settlers to travel north-to-south in this region. US 231 generally follows one of the early trails through this mountainous area.

Rivers and Commercial Agriculture

The Tennessee, Alabama, Tombigbee, and Mobile rivers, as well as Mobile Bay and the Gulf of Mexico (Figures 1.16 and 1.19)—which are the result of complex geophysical processes operating over hundreds of millions of years—played a major role in the early establishment of commercial agriculture in the state and its early participation in international commerce. To begin with, these geophysical features were primary determinants of the early discovery and exploration of the state by Europeans. By the 1760s, virtually all of the Alabama River System had been mapped in detail (Jackson, 1995). In the late 1700s and early 1800s, when the soils on the older cotton plantations in Georgia and the Carolinas had been depleted and when the demand for the fiber literally exploded in Europe, the farmers in those states began to look toward the western frontier. Within the first decade of the nineteenth century, commercial production of cotton with its slave labor was literally pouring into the valleys of the Tennessee and Alabama rivers. The new immigrants quickly determined that the deep soils that had developed in those flat, river valleys, were well suited for commercial agriculture and that those waterways provided made-to-order transportation systems. Cotton could be floated downstream to Mobile and New Orleans where it quickly and easily entered international commerce. In turn, supplies originating in the ports of the world could be transported upstream. With plenty of suitable land available and a dependable transportation system assured from the beginning, the plantation agriculture that had been shaped and refined in the Caribbean Islands and in the English enclaves in Georgia and the Carolinas, quickly took root and flourished. This complex fiber and food production system would weigh heavily in the making of Alabama history for the next 200 years.

Mountains, Travel, and Subsistence Agriculture

Geophysical processes in Alabama resulted in the evolution of hard erosion-resistant formations such as sandstone, shale, limestone, schist, and gneiss in the northern half of the state and less resistant formations of sediments of sand, silt, chalk, and clay in the southern half (Figure 1.7). As a result, much of north Alabama, especially the northeastern quadrant, is mountainous (Figure 1.8). Because of the mountainous terrain, travel and transportation were much more difficult in that region, especially in the northeastern and north-central parts of the state. The difficulty of establishing effective transportation systems in this mountainous area was a primary determinant in the early development of subsistence agriculture and the delay of the development of commercial agriculture in those areas.

Geology, Stream Gradient, and River Transportation

Geological processes and the resulting formations also played a major role in the evolution of the gradient of Alabama's river channels, and in turn, stream gradient determined early river transportation patterns and significantly affected the development of Alabama agriculture. The outcrop of Fort Payne Chert (Figure 1.7) on the Tennessee River in Colbert and Lauderdale counties resulted in the development of Mussel and Colbert shoals. Larger boats carrying cargo could not negotiate these shoals except during the winter months (generally February through April). Cotton harvested during the fall would have to be stored until the winter floods came; then there was only a three-month window of opportunity to get it past the shoals down to the Ohio River. Similarly boats bringing goods up the river and on to New Orleans could not negotiate the shoals except during those few months. Florence in Lauderdale County and Sheffield in Colbert County developed as market towns on the river to facilitate the storage and transportation of cotton and supplies during that window of time. The presence of these shoals and their effect on agricultural commerce played a significant role in the early development of railroads in Alabama. The first railroad in the state was completed in 1832 to by-pass a portion of Mussel Shoals (Atkins, 1994).

In the early days of settlement and development ocean-going schooners could not proceed up-river beyond St. Stephens in Washington County on the Tombigbee River or beyond Claiborne in Monroe County, probably because of the shoals on those streams where they had cut through the erosion-resistant Hatchetigbee and Tallahatta formations (Figure 1.7). The width of Alluvial Low Terrace Deposits is significantly reduced in those areas, indicating shallower, more narrow channels and increased current. These restrictions on up-stream travel resulted in many problems trying to get supplies upstream to farmers along the river before to the coming of steamboats in 1821.

River transportation from Rome, Georgia, to Mobile was a dream of many Alabamians from the middle of the nineteenth century onward. In the early 1840s, steamboat traffic was common on the Coosa River from north of the Coosa Shoals located near the village of Greensport in east St. Clair County to Rome. Several boats ran regular schedules for many years (Jackson, 1995). But river transportation over those shoals and southward to Wetumpka was a different matter. Jackson lists 11 major shoals, reefs, and falls between Greensport and Wetumpka. Most of these outcrops of extremely hard rocks are located along the river where it serves as the line between Chilton and Coosa counties; however the most massive one was the Devil's Staircase near Wetumpka in Elmore County. In this area, the river passes through the highly erosion-resistant metamorphic rocks (slate, metasiltstone, phyllite, quartzite, gneiss, and schist) of the Northern Piedmont District of the Piedmont Upland Section (Figures 1.7, 1.8, and 1.9). In the last 60 miles the elevation of the river dropped over 275 feet. In 1867, the Alabama legislature appropriated $3,000 for a survey of the Coosa River and for recommendations on construction that would be required to make the river navigable. Major Thomas Pearsall was employed for the task. No significant obstructions to navigation were found between Rome and Greensport south of Gadsden; however he noted that at least 18 dams and locks would be required from Greensport to north Elmore County and that seven more would be required for the Devil's Staircase. In the 1890s, work began on the structures. Locks # 1, 2, and 3 near Greensport were completed, and plans made to build Lock #4 several miles downriver. In 1905, the engineers responsible for the project recommended that the effort to make the Lower Coosa navigable be discontinued. They had concluded that the projected commercial use of the river did not justify the expense of completing it.

A Natural Dam on the Coosa

the Coosa River in the vicinity where St. Clair, Etowah, and Calhoun counties meet, outcrops of Newala and Longview Limestones and Red Mountain Sandstone (Figure 1.7) created a low, natural dam. Coosa Shoals affected the velocity of the river, to some extent, upstream almost to the Georgia line. The presence of the shoals probably played a direct role in the meandering pattern of the upper river and in the accumulation of large amounts of the Low Terrace Deposits that would play an important role in the development of its cotton economy. In the slack-water just above the shoals, Jacob Green, in 1832, started a ferry service that played an important role in the development of the early agricultural commerce of that region (Jackson, 1995). Coosa Shoals plus other shoals, steep rapids, and low falls in Chilton, Coosa, and Elmore counties made transportation on the river south of that area extremely difficult. As a result, before the coming of the railroads after 1854, most agricultural produce from the Upper Coosa Valley was shipped upriver to Rome, Georgia, and usually on to Savannah, rather than down the Coosa to the Alabama and on to Mobile. This problem with river transportation had a significant impact on the early development of railroads in the central part of the state.

Rivers and Railroads

The geophysical (physiographic) characteristics of Alabama's large rivers had an important impact on the development of transportation, especially railroads in the state. Generally, because of their size and location, these rivers provided the base for an almost ideal river transportation system (Figure 1.15). Unfortunately, there were many areas of the state that were too far away to benefit. In the early years, the system was so effective in meeting the needs of the Cotton Kingdom that there was little incentive to develop roads and railroads for sections that desperately needed them. Even when the development of railroads finally began, they were built to solve mostly local problems related to river transportation. Generally, they were developed to facilitate river transportation, rather than to supplant it (Atkins, 1994; Jackson, 1995).

Chalk and Drinking Water in the Black Prairie

The deep formations of dense chalk underlying the black soils of the Black Prairie District are several hundreds of feet thick (Figures 1.7 and 1.9), and there are virtually no underground water channels or springs and few permanent streams. Consequently, obtaining drinking water on the early farms in that area was a major problem. That problem was not completely solved until equipment was available to drill deep wells that reached water reservoirs below the chalk in the Eutaw and Tuscaloosa formations. For that reason few early towns were established on the Black Prairie. Most of them were established around its edge.

Chalk, Marl, and the Boll Weevil

The chalks and marls of the Selma Group (Figure 1.7) that formed on the bottom of a clear, calm sea during the Upper Cretaceous Period, (Table 1.1), some 100 mya provided the parent material for the development of the unique, deep, black soils of the Black Prairie Belt in west-central Alabama (Figure 1.9). These soils were to become the center of Alabama's cotton economy after about 1830 (Abernathy, 1990). Unfortunately, because of the nature of those same soils, the so-called Blackbelt was the area of the state most severely affected by the boll weevil. Because these heavy clay soils are extremely sticky and difficult to cultivate when wet, it was not practical to plant as early in the year in the Blackbelt as in other areas of the state. Consequently, the cotton crop in that area reached the stage of greatest vulnerability just at the time of the insect population reached its maximum density.

Geology and a Local Market for Cotton

Outcrops of the hard metamorphic rocks (schists and quartzites) of the Jacksons Gap Group in Elmore and Tallapoosa counties (Figure 1.7) at Tallassee resulted in the evolution of Tallassee Falls. Early entrepreneurs built a water-powered textile mill there that would provide Alabama farmers with a local market for their cotton and would provide many farm wives with their first opportunities for off-the-farm employment (Jackson, 1995).

Use of Excess Agricultural Labor

Some 300 million years ago, during the Pennsylvanian Period of the Paleozoic Era, large deposits of coal were formed in parts of the Coosa, Cahaba, and Black Warrior valleys (Figure 1.9). Earlier, large amounts of iron-bearing sediments had been deposited in the same general area, especially in Jefferson County. Beginning in the mid-1800s, entrepreneurs began to exploit these resources for the production of iron. By 1865 there were 16 blast furnaces operating around Birmingham (Rogers and Ward, 1994). The production of iron was one of the first steps in the industrial development of Alabama. Until that time, virtually all of the state's economy was dependent on agriculture and agribusiness. This first halting step toward economic diversification would begin to help absorb some of the vast amount of surplus labor that had been accumulating in agriculture in the state.

2

Climate and Weather

PRECIS

To the overwhelming percentage of the contemporary urban population, climate and weather are synonymous, and a good day is a sunny one. Rain is an un-welcomed inconvenience in their personal lives. The climatic attractions of those drawn to the Sun Belt and for those whose contact with nature is on the golf course or watching football and tailgating are not positives for agriculture. However, to those in tune with the natural world and with the broader environment, the positive and negative impact of both climate and weather on the economy are more readily apparent. The vagaries and erratic weather consequences of Alabama's summer subtropical climate and its winter temperate climate result in major obstacles for agriculture. The last section of this chapter, Agricultural Consequences of Alabama's Weather, delineates the adverse consequences of mild winters, heavy late-winter and early-spring rains, late spring freezes, mini-droughts during the growing season, heavy rainfall during late spring and early summer, excessively high summer temperatures, and tropical storms.

INTRODUCTION

Weather has been defined as the current status of the atmosphere with respect to temperature, moisture content, wind velocity, cloudiness, or any other meteorological phenomenon. Climate is the average condition of weather over a period of years. In effect, climate is what you expect when you get up each day, but weather is what you actually get.

Climate and weather are primary determinants of what food and fiber we produce and how effectively we produce them. Weather and climate effectively set limits on the crops we choose to plant, when we plant them, the nature and quantity of production inputs required (fertilizers, pesticides, and irrigation), the extent to which the genetic potential of the plants will be realized, when we harvest, and to a great extent the quantity, quality, and profitability of crops that we will harvest.

Claude Wikard (1941), secretary of agriculture in the second Franklin D. Roosevelt administration, wrote: "Next to crop prices, nothing is more important to the farmer's business than the weather, and in fact, the weather has a strong influence on prices. So every farmer takes a keen interest in the weather, and in many cases he is a weather prophet of no mean ability, at least for his own local area. One of the first things he is likely to do when he gets up in the morning is to scan the sky, note the direction of the wind, and plan his day's work according to what he thinks the weather is going to be. This remains true even though the radio and newspaper now give him Government forecasts for his general area."

Gove Hambidge (1941), editor of *Climate and Man*, wrote in his "Summary of the Volume:" "Few occupations of man, in fact, involve such varied knowledge and skill as farming. It looks simple to drop some seed into the ground, let it grow under the blessed influence of sun and rain, and then harvest the crops. But at least as far back as the days of the Romans, men realized that the business was not so simple after all. A Roman named Columella wrote a long treatise on agriculture in which he summed up the best knowledge of his time—and it was by no means simple even then. In the 2,000 years since then, it has not become any simpler."

INFORMATION ON CLIMATE AND WEATHER IN ALABAMA

One of the earliest sources of information about Alabama climate and weather is a publication from the Agricultural Experiment Station of the Agricultural and Mechanical College [today's Auburn University] entitled, "Climatology of Alabama" published as Bulletin 18 (New Series) in 1890 (Mell, 1890). The publication was "compiled from Meterological Observations from 1811 to 1890, including General Phenomena from 1711 to 1890." The publication contains a valuable section on the "History of the Weather Work in Alabama." More recently, Emigh (1941) and Long (1978) have published general works on Alabama's climate and weather.

The "Weekly Weather and Crop Bulletin," is a useful source of information on weather and its effects on crops nationwide. It is published as a cooperative effort between the National Weather Service (National Oceanic and Atmospheric Administration, United States Department of Commerce) and the National Agricultural Statistics Service and World Agricultural Outlook Board (United States Department of Agriculture).

The National Oceanic and Atmospheric Administration of the United States Department of Commerce collects information on weather throughout the country continuously, and makes it available through monthly publications. Each month the National Climatic Data Center in Asheville, North Carolina, issues a monthly publication of climatological data for each state. At the end of each year, an annual summary is published. Climatological data, which has been collected and published by this agency and its predecessors for well over a hundred years, provides an invaluable source of information.

In Alabama climatological data are collected at many locations throughout the state, which is divided into eight climatic regions for purposes of reporting (Figure 2.1). The regions generally correspond to major physiographic provinces (Figure 1.9) within the state, except for the Coastal Plain Province that has been divided into four regions because of its size.

A number of observations of rainfall and temperature are made at locations throughout each region. Currently, for example, in the Northern Valley Region (Region 1), daily rainfall totals are recorded and reported from Athens, Belle Mina, Hodges, Huntsville and Moulton (National Climatic Data Center, 1994). Daily maximum and minimum temperatures are recorded at all of the same locations except at Hodges. Generally the number of stations reporting is dependent on the size of the region. For example, in the Gulf Region (Region 8) that includes only Baldwin and Mobile counties, there are only six stations for measuring and reporting rainfall. In contrast, in the Coastal Plains Region (Region 7) that includes 16 counties, there are 31 reporting stations for rainfall. Because temperature is less variable within a region, there are generally fewer stations recording and reporting temperature than rainfall. In the Upper Plains Region (Region 3), there are 26 stations for recording rainfall, but only 12 for recording temperature.

A valuable source of historical information is a book published by the United States Department of Agriculture in the *Yearbook of Agriculture* Series, entitled *Climate and Man* (United States Department of Agriculture, 1941). The volume includes data on rainfall and temperature, dates of last and first frosts, and length of growing season from stations located throughout the state. The data reported generally covers the period from the late 1900s through the late 1930s. The series on average rainfall for the state covers the period from 1886 through 1938. Similar information from the other states is also presented.

Climate and Weather Processes

The geophysical processes and their interactions that give us our climate and weather are exceeding complex. While their "within years" effects tend to remain relatively constant year after year for long periods of time, there is enough variation involved to make it extremely difficult to predict exact day-to-day effects for more than a few days, at most, in advance. Further, there is enough variation "within-years" from year to year to make the production of food and fiber (agriculture) relatively unpredictable. Agriculture involves a rather complex sequence of events that depends heavily on a rather specific sequence of weather events, and departure from this sequence can have extremely unfavorable effects.

Solar Radiation and Planet Earth

Because of the relative positions of the earth and sun and their relative sizes, only a tiny fraction of the energy radiated by the sun has the potential of reaching our planet. Fifty-one percent of the energy that has the potential of reaching earth is absorbed at the earth's surface. The remainder is either reflected back into space or is absorbed by airborne dust and clouds in the atmosphere high above the earth. The 51% that does reach the earth heats the oceans and the surrounding land. This heating of land and water sets in motion the forces that initiate and govern our weather and climate.

While only a fraction of the sun's energy ever reaches the earth, that quantity is sufficient to virtually end life on earth as we know it if it were to accumulate. This continuing accumulation of heat is prevented by the radiation of energy heat back into the atmosphere and finally back into outer space, so that the earth's average annual surface temperature is a relatively cool 59°F. Over a period of a year, the same amount of energy reaching the earth must be balanced by the amount of energy leaving the earth. This balancing of incoming and outgoing energy would seem to be a simple matter, but it is actually an extremely complicated system. So complex, in fact, that weather forecasters have a difficult time accurately forecasting the future status of this dynamic system more than a few hours in advance.

There is relatively little variation in the short term in the amount of radiation leaving the sun; however there is a considerable variation in the amount of energy striking the earth at any one location. The earth makes a complete revolution about its axis of rotation once in 24 hours, the diurnal cycle; consequently, no location receives any radiation while in the shadow. So the amount of radiation received in a 24-hour period varies considerably. Also, one location receives more radiation during part of the year because the earth goes completely around the sun once in slightly more than 365 days, in an elliptical orbit with the earth as one focus of the ellipse. So during part of the year, the earth is closer to the sun than at other times. Further, the axis of rotation of the earth is not vertical relative to the position of the sun. Rather, it is slanted at an angle of 23.5 degrees from the vertical (Figure 2.2). So as it rotates around the sun, different regions receive greater or lesser amounts of radiation depending on the position in the annual cycle. For example, because of the tilt of the earth, the North Polar Region receives very little daily radiation during the winter months when the earth' s axis of rotation is tilted away from the sun. This situation is reversed during the summer when the axis is tilted toward the sun. During the summer, the North Polar Region receives some radiation for 24 hours each day. The tilt of the earth has limited effect at the equator where any location receives 12 hours of solar radiation in each 24-hour cycle throughout the 365-day cycle.

Alabama is located a third of the distance between the equator and the North Pole; consequently there is considerable variation in the amount of direct solar radiation that the state receives from one part of the annual solar cycle to another. For example, day length at Auburn varies from 10 hours and 2 minutes in late December to 14 hours and 15 minutes in mid-June.

Air and Weather

Climate and weather are mostly about the dynamics of the cloak of air surrounding our planet, and its changes hour-by-hour, day-by-day, month-by-month, and year-by-year, as the earth spins on it own axis and as it revolves around the sun.

Air Masses

Air is quite homogeneous from a chemical perspective. The percentage of nitrogen and oxygen does not change appreciably from place to place; however, this does not mean that air is equally homogeneous relative to other characteristics. For example, the temperature of a mass of air at the North Pole at mid-day is quite different from a mass of equal volume at the equator. The temperature of the mass at the equator is much higher, and because of the heat, the molecules are more widely spaced; consequently there are fewer molecules per unit volume, and the mass weighs less than a comparable mass at the North Pole. Air masses also vary considerably in size. A thunderstorm, which is a relatively uniform air mass, may be no more than five miles in diameter; while some cold air masses coming out of Canada across the

United States may cover several states.

The unique character of masses of air may remain remarkably unchanged over a considerable period of time even though they move several hundred miles overland. For example, during the winter, a mass of cold, arctic air moves south across Canada into the United States. Under the right conditions, this mass may move rapidly south into the Gulf States without the temperature increasing appreciably. At this point, it may become stalled and isolated from the cold air source further north. Moreover, the temperature of the mass may remain relatively low for several days. In fact, the temperature in the isolated mass may be considerably lower than air masses that have moved in behind it nearer the Canadian border. Under these conditions, the temperature in Birmingham, Alabama, may actually be lower than in Minneapolis, Minnesota.

Two conditions are required for the formation of air masses with specific characteristics: (1) a large area of the earth's surface where the topography and other physical characteristics are relatively homogenous, and (2) the air mass must remain in place long enough to take on the temperature and moisture characteristics of the associated land area. Examples include the development of a large mass of extremely hot, dry air between mountain ranges in the southwestern United States and northern Mexico during the summer. Similarly, during the winter, the Mackenzie River Valley in northwestern Canada receives cold air from the Arctic Region and from the North Pacific Ocean. As the air fills up the valley, it takes on the characteristics of the extremely cold, snow-covered land mass associated with it.

Air masses are classified generally according to the region where they originate (Figure 2.3). If they originate over oceans they are identified as maritime (m) air masses and continental (c) if they originate over land. They are further differentiated by the latitudinal zone where they originate as either equatorial (E), tropical (T), polar (P) or arctic (A). Note that there is no air mass designation for the middle latitudes, which include the subtropics and the temperate zone. Air moves so rapidly through and around the middle latitudes that masses with specific characteristics do not form or do not have time to take on the characteristics of the land below. Using this scheme of classification, an air mass might be identified as an mT mass, meaning that it originated over water in a tropical region. Similarly, an air mass originating in central Canada would be identified as a continental polar mass, or cP mass. The primary masses of concern in North America are mT, mP, cT and cP. Only two of these, mT air and cP air, are of major importance to the elaboration of weather in Alabama and the Southeast.

Continental polar (cP) air masses that affect the United States usually originate over Alaska or northwestern Canada (Figure 2.3). As the mass grows, it moves southeastward as an airstream through the Rockies and onto the Northern Plains. Under certain conditions, it may move all the way to the southern tip of Florida; however the mass would not maintain unusually low temperatures all the way to the Gulf of Mexico unless much of the intervening path was covered with snow.

The Gulf of Mexico and the Caribbean are the source region for maritime tropical (mT) air masses (Figure 2.3). These masses move northward as moist tropical airstreams into the eastern United States and provide much of the water that falls as annual precipitation east of the Rockies and south of Hudson Bay.

Continental tropical (cT) air masses develop over the deserts of the Southwest and Mexico (Figure 2.3); however, they have relatively less effect on weather outside that region than do cP and mT masses on other regions of North America east of the Rocky Mountains.

If air masses remain stationary, the weather associated with them will generally remain unchanged. They bring change when they move into an area occupied by another air mass. The collision of air masses almost always brings with it a significant change in the local weather. The collision of cP air masses moving south from Canada with mT air masses moving north from the Gulf of Mexico results in cloudy weather, precipitation and under the right conditions, thunderstorms and tornadoes.

Air Movement

The horizontal movement of air around the surface of the earth is of great importance in the elaboration of day-to-day weather phenomena. It is the horizontal movement of air that is primarily responsible for distributing the heat from the sun, which is concentrated near the equator, around the globe and prevents the build-up of heat at any one point to dangerous levels. Air moves horizontally from areas of high pressure to areas of low pressure, in a similar manner in which water runs down hill. High pressure pushes air toward low pressure. Wind is the physical evidence that a mass of air is moving from an area of higher pressure to an area of lower pressure. The moving mass indicated by the presence of wind may vary in size from a few square miles when associated with a thunderstorm to hundreds or even thousands of square miles when associated with the movement of a large body of cold air from Canada into the United States.

The speed of the wind in a moving air mass is indicative of the pressure differential or Pressure Gradient Force between the two areas. For example, neglecting friction, which affects the speed of air movement, wind would accelerate to 80 miles per hour traveling between two areas 500 miles apart, one with a higher atmospheric pressure of 29.79 and the other with a lower pressure of 29.51. Small differences in the Pressure Gradient Force results in large differences in the speed of air movement. In hurricanes, a low barometric pressure of 28.94 can result in wind speeds of between 74 and 95 miles per hour. Pressures below 27.17 can result in wind speeds greater than 155 miles per hour (Williams, 1992).

Air does not move in a straight line as it is forced from an area of higher pressure to an area of lower pressure, but tends to travel in a curved path. As the earth spins from west to east, on its axis of rotation, it provides a moving plane of reference for an air mass moving from north to south, from an area of high pressure to an area of low pressure. Viewed from the North Pole, the air mass would seem to be moving or curving to the left, as it moves southward. This phenomenon is the result of the Coriolis Force. This force is greatest at the poles, and there is little or no effect at the equator. This force operates regardless of the direction of air flow, and it always acts to force the moving air mass or stream to curve to the right of the direction of the flow.

The Coriolis Force is partly responsible for the clockwise flow of wind around areas of high pressure and the counterclockwise flow around an area of low pressure. A description of the mechanism that results in this phenomenon is beyond the scope of this chapter. The reader is encouraged to refer to any good book on climate and weather, for this description.

Because of the angle of incidence of the sun's radiation at the North Pole, and the long periods during winter when no radiation reaches that region, the air is always cold there. Because of the low temperature, air molecules vibrate at a lower rate and over a shorter distance; consequently a given number of molecules occupy a relatively smaller volume, and the air is more dense or heavier. As molecular activity is reduced, the space occupied also decreases, and the volume shrinks. Air from above slowly sinks toward the earth further compressing the lower mass causing it to spread out in all directions at the surface. This compression and sinking combines to form a huge dome of cold air—a huge dome of cold, dense air with high barometric pressure. During the winter months, the production of extremely cold air in this snow-covered region is intense. Because of the size and weight of this mass, extremely cold air spills southward, bringing cold weather deep into the middle latitudes.

While at the North Pole, the air is always cold and dense and the barometric pressure high; the characteristics of the air mass at the equator is the exact opposite. Regardless of the time of year, that region receives 12 hours of sunlight and solar radiation each day. Also the angle of incidence is much higher, so that a lower percentage of incoming radiation is reflected into space and a higher percentage is absorbed at the earth's surface, both land and water. Further, the area of the earth's surface associated with the equator is much larger than that associated with the pole; consequently there is always a large quantity of very warm air present in that region. Because of the heat, the air molecules in the mass vibrate more rapidly, resulting in more space between them. As a result the air mass expands both laterally and vertically, but because there is less resistance to expansion vertically than laterally, the mass expands more rapidly in that direction. Warm air masses extend thousands of feet upward into the troposphere above the equator. As the mass expands upward, it encounters cooler temperatures and the molecules lose some of the heat energy that they received near the earth's surface; consequently, they vibrate less rapidly.

The atmosphere cools at a rate of 6.5°C (3.5°F per thousand feet) per kilometer of altitude. At the higher, cooler altitudes, the air mass no longer expands upward, but because air continues to be forced upward from below, the air at the top of the mass begins to move laterally both north and south, toward the poles. Because of the heat in the mass and the upward expansion, there are fewer molecules per volume of air and relatively fewer molecules of air in a vertical column reaching from the surface of the earth up into the troposphere. As a result, the barometric pressure is lower. Relative to a cooler air mass further north, the atmospheric pressure over the equator is lower. This continuous low pressure belt at the equator and for several hundreds of miles on either side, both north and south, is one of the most consistent characteristics of the world's climate. It remains in place, essentially unchanged, throughout the year.

As this great column of air at the equator is heated at the surface and rises into the troposphere, cooler air masses both north and south with higher barometric pressure are pushed along the surface from the area of high pressure toward the low pressure area at the equator in much the same way that cool air is forced into an open fireplace (Figure 2.4). This inflowing air is then also heated and expands upward to the top of the troposphere where it flows laterally high above the earth, both north and south, toward the poles. This flow of warm air aloft toward the poles is accompanied by a flow of cold air at the surface from the huge domes of frigid air at the poles toward the equator, thus setting up a huge circular system of air flow. Around the earth at the equator, there is an enormous mass of air rising almost continuously. There is some difference in the intensity of heating between day and night, and this difference results in some variation in the speed with which air rises and of the absolute amount of heat going into the air, but the earth is always hot and hotter at the equator. Consequently the process of heating, expansion, ascension, and lateral flow pole-ward in the upper troposphere goes on continuously.

As the radiant energy from the sun strikes the earth at the equator where the heated air begins to rise, tremendous quantities of water evaporates at the same time and rises with the air mass toward the mid-troposphere. As moist air rises, it encounters the cooler temperatures aloft where the moisture condenses as clouds and falls back to the earth as rain. This continuing cycle of evaporation, condensation, and rain is responsible for the daily thunderstorm cycle so characteristic of the equatorial region. Consequently, by the time the rising air mass has reached its maximum altitude and begins to spread horizontally north and south toward the poles, it has lost virtually all of its moisture and is extremely dry.

The flow of heated air upward and away from the equator toward the poles and the concurrent flow of cold dense air from the poles (Figure 2.4) along the surface back toward the equator describes the global circulation of air and heat as it generally occurs. If it were not for this general flow of heat away from the equator, this area of the world would be unbearably hot and if it were not for the countercurrent flow of cold air away from the polar zones, those areas would become even more cold than they are currently. The system, however, does not function as simply as that. Because of friction and the Corolis Force described previously, air does not flow in a straight north-south or east-west direction either toward or away from the poles. For example, air flowing at the surface from the North Pole toward the equator, actually flows from east-northeast to west-southwest.

The warm, now dry, equatorial air flowing horizontally toward the poles, in the cold upper troposphere, loses heat fairly rapidly so that the molecular activity is reduced and there is less distance between molecules. As a result, the density of this huge, dry air mass increases, and it begins to subside and settle toward the earth at about 30 degrees north latitude and south of the equator. In the Northern Hemisphere, the center of the subsidence is generally on a line that passes through New Orleans, Louisiana; south of Jacksonville, Florida; south of Casablanca, Morocco; through Cairo, Egypt; south of Basrah, Iraq; south of New Delhi, India; through Chunking, China; north of the Midway Islands; and through Rosario on the Baja peninsula of northern Mexico. Of course, this huge mass of dense, dry air settles in a band around the earth, several hundred miles on either side of this "line."

There is another important factor involved in the subsidence of this huge mass of air. The air mass began its journey at the equator where the circumference of the earth is greatest; however, it is returning to earth where the circumference is much less, so a large mass of air is being forced into a smaller area. This phenomenon also leads to increased pressure. At this latitude, therefore, the earth is surrounded with an enormous mound of high pressure air. Meteorologists call this phenomenon the subtropical highs as contrasted to the previously described equatorial lows found at 0 degrees latitude.

When this enormous mass of dry air settles to the earth, it is forced to spread out over the surface of the earth in all directions. This circulation of air aloft from the equator to 30° north latitude, then along the surface back to the equator is called the Hadley Cell after the meteorologist who first described it (Figure 2.4).

Some of the air falling back to earth as part of the subtropical high at 30° north latitude also flows northward at the surface toward the North Pole. At the same time there is a counter-flow of air aloft from the polar region toward the equator where it becomes part of the Hadley Cell circulation. This mid-latitudinal circulation system is called a Ferrel Cell (Figure 2.4). The Ferrel Cell extends from 30° latitude to 60°. The flow of air is not as well defined in the Ferrel Cell as in the Hadley Cell, but the general effect is the further movement of warm air heat away from the equator toward the poles and the movement of cold air toward the equator. There is relatively little vertical circulation similar to that seen in the Hadley Cell (Lydolph, 1985). For that reason the circulation pattern is depicted with broken lines in the figure. There is a third cell, the Polar Cell (Figure 2.4), that extends from 60° north latitude to the poles (Henderson-Sellers and Robinson, 1986). So rather than the single circulation system moving heat between the equator and the poles as suggested previously, there are three interconnected general circulations. Henderson-Sellers and Robinson (1986) commented that this three-cell model is an oversimplification of this complex process, but that it is a useful conceptual tool. Of the three, the Hadley Cell is the most clearly defined throughout the year. All three are heat driven.

These three interconnected cells of circulating air are the mechanism by which the tremendous heat-load at the equator is dissipated. They provide for the flow of heat in the form of heated air away from the equator. They also provide the mechanism for the flow of cold polar to replace the heated air as it flows northward.

The huge mound of dry, dense, high pressure air settling around the entire circumference of the earth at about 30 degrees north and south latitude has a profound effect on the weather and climate of the world. For example, this huge mound of dry, high pressure air results in the establishment of the huge deserts (Mojave, Colorado, and Sonoran) of the southwestern region of the United States and of North Africa (Sahara). Similar deserts are also found at about the same latitude in the Southern Hemisphere. The heating and drying potential of this mass of air was evident in the extremely hot, dry weather throughout Alabama during July and August, 1995.

The huge mass of air settling to the earth over the ocean at 30 degrees latitude north and south, results in the formation of great subtropical high pressure regions that influence much of the world's weather and climate. For example, the great subtropical high of the Atlantic Ocean, the Bermuda High or Azores High (Figure 2.5), has a profound effect on the weather and climate of eastern North America and Western Europe and is especially important to Alabama's agricultural weather.

As the great high pressure air mass that originated at the equator as a low pressure mass settles to the earth around its circumference at near 30 degree north latitude, it spreads out horizontally just as water being poured from a pail onto a flat surface, but the Coriolis force causes the air to circulate clockwise. The effect of this huge mass of air being poured onto the earth behaves somewhat differently depending on whether it is settling on land or on the oceans. When it settles on a large land mass, it significantly reduces cloud formation and precipitation over an area of hundreds of square miles and is a primary cause of desertification around the world especially on the western side of continental land masses.

Over the oceans, the high pressure air mass settling to the surface establishes an enormous clockwise circulation system that affects hundreds of thousands of square miles of open ocean and adjacent land areas. These enormous, slowly-spinning systems result in the development of the great so-called trade wind circulations in world's oceans. These are the most persistent and predictable of the planetary wind systems (Akin, 1990). They are called trade winds because they provided much of the energy for sailing ships for hundreds of years. In the North Atlantic, the trade winds blow clockwise around the Bermuda or Azores High (Figure 2.5), along the coast of Western Europe toward the south and southwest, down to the coast of Africa where they turn westward across the Atlantic, then northward and northeastward over the eastern Gulf of Mexico, then along the eastern coast of North America and then eastward across the North Atlantic to Western Europe. Until early in the nineteenth century, European merchants used the power of the "trades" to transport manufactured goods south to Africa. There they loaded slaves to transport westward across the Atlantic to South America, the Caribbean, and to the United States. In the Americas, they exchanged their human cargo for rum, sugar, rice, indigo, cotton, tobacco, timber, naval stores, and fish. From the Caribbean, they sailed on the trade winds northeastward along the coast of North America then eastward to Europe.

A similar subtropical high pressure center is located in the North Pacific, with the center located near the Hawaiian Islands (Figure 2.6). A clockwise circulation also develops around this center similar to the one in the North Atlantic. Sailors in the Pacific also used those trade winds to transport people and cargo aboard sailing ships. Similar high pressure regions are also found over the oceans of the Southern Hemisphere.

The high pressure cells do not remain in the same position throughout the year. In the summer, in the Northern Hemisphere, when the earth's axis of rotation is inclined toward the sun, the Bermuda High is located in the western North Atlantic, much closer to North America than to Europe (Figure 2.5). In fact, during long periods in the summer, a ridge from the center of the high pressure cell often extends into the Southeastern United States, as it did in the summer of 1995, which resulted in hot, dry weather in the region.

During the winter when the earth's axis of rotation is inclined away from the sun and the Northern Hemisphere is cooler, the center of high pressure moves southeastward away from North America until it is located off the coast of North Africa (Figure 2.7). The cell is also much reduced in size. The other high pressure zones described previously also move and expand and contract with the seasons.

During the winter in the Northern Hemisphere when the subtropical highs have moved generally southward and have been reduced in size, they have considerably less effect on the weather of the continental United States than in the summer. As the cold air from the North Pole moves further south with the onset of winter, the land masses lose their heat much more rapidly than do the oceans. Consequently, the land becomes much colder relative to the oceans, and the air over the water is warmer than air over the land. Because the air is warmer, it is also lighter, and it begins to rise into the troposphere in the same way that air at the equator rises. Obviously, the air over the oceans is colder during the winter than rising air at the equator, but it is much lighter and much more buoyant than the relatively colder air over the adjacent land masses. As the oceanic air mass rises, it is replaced by more dense air from over land. The result of these vertical and horizontal air movements is the creation of very large low pressure cells (Figure 2.7). But in contrast to the subtropical highs where the descending air flows downward and away from the center of high pressure, in these large low pressure systems, air flows inward toward the center of low pressure and then upward above it. As the air flows into the center of low pressure, the Corolius Force causes it to move counterclockwise around the center. This process results in the development, in the North Atlantic, of the enormous Icelandic Low (Figure 2.7). This enormous low pressure system fills the North Atlantic and the Norwegian Sea basins between North America and Europe north of 60 degrees latitude, as well as covering Iceland. This large mass of low pressure air has relatively little effect on the weather of the United States, but is extremely important to the development of weather in Europe.

Another massive low pressure center develops in the North Pacific during the winter months. The Aleutian Low, is located generally between the Aleutian Islands and the Kamchatk Peninsula (Figure 2.7). This low pressure region is not as massive as the Icelandic Low, and it is located a little further south. This low pressure system has a major impact on the weather of the entire North American continent. The counterclockwise flow of air around this center of low pressure generates smaller low pressure storms that spin-off and that are literally thrown against the northwestern coast of the United States and the southwestern coast of Canada. These storms come ashore at three- to five-day intervals throughout the winter. These low pressure cells travel west to east across the northern United States and southern Canada where they are finally pulled into the Icelandic Low, described previously.

Fronts are zones where moving air masses collide or meet. This name was given to these zones by meteorologists during World War I, because the collision of air masses reminded them of the fronts established between opposing armies that developed during battles in that war (Lydolph, 1985). For example, when a mass of continental polar air (cP) (Figure 2.3), advances southeastward across the United States, it encounters a mass of maritime tropical (mT) air advancing northwestward from the Gulf of Mexico. Where these two masses meet, a so-called front is established. This particular point of collision has been given the name: polar front. Generally when these two air masses collide over the central United States, the heavier, cooler mass prevails, and pushes the lighter, warmer air eastward.

Weather associated with individual air masses is usually relatively stable and clear; however where different masses collide (fronts), the weather is usually turbulent and stormy. There are several primary frontal boundaries in the world (Lydolph, 1985), but only the so-called polar front is of concern in the southeastern United States and Alabama. Figure 2.8 shows the position of a cold polar front advancing across the US in June, 1993. The collision of these two large contrasting air masses at this front results in the development of some of the most variable and turbulent weather on earth, especially during the spring and early summer months.

The polar front changes position between seasons and within seasons, as air masses advance and recede. During the summer, when North America is receiving its largest annual influx of solar radiation, mT air covers much of the United States east of the Rockies, and the polar front is moved far to the north along the Canadian Border. It usually does not move very far from that location during the warm months; although it does remain more or less active throughout the year, and on occasion it may move southward as a weak front, even during the summer months. In 1993, the polar front stalled over the upper Midwest. The abnormal location of the front was a major contributing factor in the severe flooding that occurred in that region (Scott,1993).

With the return of the fall and winter seasons, the supply of mT air is diminished and the supply of cP air begins to increase; consequently, the polar front moves southward. In some situations, when the production of cP air is quite high, the front may move all the way to the Gulf Coast only to be pushed back north with the influx of a greater supply of mT air. On average, during the winter months, the western terminus of the front is in the lower Mississippi Valley and it extends over New England.

In moving fronts, if cold air is displacing warm air, it is called a cold front. If warm air is displacing cold air, it is called a warm front. If the front is not moving, and neither the cold nor the warm air is being displaced, it is called a stationary front. As fronts move, they bring changes in weather. For example, when a cold front is advancing at the expense of warm, humid air, the lighter moist air is forced upward where it cools, resulting in the condensation of its water vapor into clouds and finally into precipitation. Depending on the degree of contrast in the temperatures and water vapor content of the colliding air masses, strong storms with high winds, heavy rainfall, and even tornados, may develop along the front.

As a cold front moves from northwest to southeast in the Eastern United States, it tends to rotate counterclockwise around a low pressure center. In the beginning, a cold front may be moving essentially parallel to the Canadian border, but as it moves south and east, it rotates so by the time it has moved into the Southeastern United States, it may be almost perpendicular to the Canadian Border. Also, as the cold front rotates, the warm air portion is pushed far to the north as a warm front displacing cold air.

The polar front, normally extending eastward from the Rocky Mountains across the central United States, provides a sort of track for storms that move from west to east across the country during the winter months. During the winter, low pressure disturbances spin off the huge Aleutian Low (Figure 2.7.) over the North Pacific and strike the western coast of North America. Usually the point of impact is along the Canadian border, but these storms may reach land anywhere along the coast. Once the storms reach land, they move across the continent following the frontal boundary until they reach the huge Icelandic Low in the North Atlantic.

Storms also may move southeastward once they reach the coasts of Washington and Oregon. Then they break-up crossing the Cascades and Rockies and are regenerated in the Denver area. From there, the low pressure storms move southeastward into Texas and Oklahoma then eastward across the South and finally northeastward into New England. These low pressure disturbances that originated far out in the Pacific Ocean are the primary source of winter storms in Alabama and the Southeast, during the cooler months of the year. Some of the strongest storms that affect Alabama during the winter and spring, however, originate as low pressure cells in Texas or over the western Gulf of Mexico before they follow the polar front eastward. The Blizzard of '93, which brought snow and extremely high winds from Alabama to eastern Canada, originated as an area of low pressure in the western Gulf of Mexico, on March 12, before it moved generally northeastward on March 13 along the polar front drapedover the South and into New England (LeComte, 1993).

Water, Air and Weather

Solar heat and the perpetual motion of the daily rotation of the earth with its annual circling of the sun are two of the three primary determinants of Alabama's weather (Lydolph, 1985). Water is the third. Water is essential for life, but it also performs important functions in redistribution of heat over the earth's surface. For example, the Gulf Stream transports enormous quantities of heat from the equatorial region along the coast of North America to the coast of Western Europe. This transported heat maintains a much higher winter temperature in the British Isles and parts of Northern Europe than would be possible without that translocated energy. Also, water stores an enormous amount of energy as latent (hidden) heat in water vapor that has evaporated from the surface of the earth. Water or sweat transported from our bodies as vapor also carries heat with it, giving us the benefit of evaporative cooling.

Clouds

One of the most important result of the interaction of water and air is the formation of clouds. At the surface of the earth, only a narrow layer of air is cooled sufficiently to condense and reach the dew point, so that dew, fog, or frost are formed. The formation of clouds through condensation involves the cooling of much larger volumes of air. In the summer, when radiant energy from the sun strikes the surface of the earth, it is converted into heat energy. As a result, the land and water over a large area are heated. These warm surfaces then heat the air. The added energy also increases the molecular activity of water molecules in the air at the surface of the soil and of streams, ponds, lakes, and oceans, so that they are evaporating into the vapor state and mixing with molecules of oxygen and nitrogen and other gaseous components of air, at a high rate. The heated molecules in the mixture, vibrate more rapidly resulting in the development of a mass of warm, moist air at the surface. Because heat-activated molecules occupy more space, a given volume of air contains less molecules, so it is lighter. This lighter mass of air is more buoyant; consequently, it is forced upward into the troposphere (Figure 2.9-A). As these activated molecules from the surface are forced higher and higher into the troposphere, they encounter fewer and fewer other molecules. With fewer molecules per volume of air, the pressure is lower. Following the General Gas Laws, when there is a reduction in pressure, there is a proportional change in the temperature of the gas. So as the air rises, the pressure is lower so that the volume of the mass increases. Under these conditions, the temperature of the mass decreases or is cooled at a rate of 10°C per kilometer (5.5°F per 1000 feet) as it rises. Descending air would be warmed at the same rate. This change of temperature of an air mass as it rises or descends is called an adiabatic process. If the mass rises high enough and the temperature decreases so that the dew point of the mass is reached, clouds form as the water molecules condense into tiny droplets.

In Alabama, in the summer, clouds usually form by mid-morning through this process, and although they are billowy in appearance and extend for hundreds of feet into the troposphere, they have sharply flattened bases. The flat bases correspond to the height above the earth's surface where the temperature of the air mass decreases just to the "dew point"—the temperature at which air becomes saturated with water and condensation begins (Figure 2.9-B). Below this rather sharply defined line, the water in the air mass is in the vapor state. Above the line, it is in the liquid state. The air mass containing the clouds extends downward below the so-called "cloud line," but below it the water as vapor is invisible.

All clouds form when an air mass containing water vapor rises to a height where its temperature reaches the dew point, but there are three different processes by which the mass ascends to that level: convection, orographic lifting, and cyclonic lifting. The process described in the preceding paragraph that resulted in the air mass ascending as a result of heating at the surface of the earth until its temperature reaches the dew point is called convection. Orographic lifting is the result of an air mass moving at the surface that encounters a mountain slope. As the mass is forced up the slope, its temperature decreases until the dew point is reached and clouds form. Cyclonic lifting occurs when a mass of cool, heavier air moving across the earth's surface encounters a mass of warmer, lighter air, and because it is more dense and heavier, it forces the lighter air upward. When the temperature of the ascending air mass reaches the dew point, clouds form.

Precipitation

Condensation is virtually an automatic process. When the temperature and/or the quantity of water in a volume of air high above the earth changes so that the saturation point (dew point) in the mass is reached, water vapor becomes liquid, condensation occurs, and clouds form. However precipitation of the water from those clouds is much less predictable. The precipitation process begins when water molecules in the liquid state are attracted to a hygroscopic condensation nucleus, such as a tiny particle of sea salt or a dust particle suspended in the atmosphere, to form a water droplet. As the water molecules attach to the particle, the tiny mass becomes heavier and may begin to fall toward the surface of the earth. However, once they begin to fall, the temperature of the tiny mass begins to increase, and the molecular activity of the attached water molecules begins to increase. Quickly, the activity becomes so great that the attachment of the molecules to the condensation nucleus is overcome, and the liquid water goes back to the vapor state. Consequently, the water evaporates before it reaches the ground (Figure 2.9-C). In clouds, however, physical characteristics are conducive for the coalescing of many tiny droplets into a larger water drop. Its mass is great enough that it does not heat rapidly enough during its fall, so that evaporation takes place before it reaches earth as rainfall (Figure 2.9-D) (Lydolph, 1985).

As warm, moist air rises and cools, and if condensation occurs (the point of saturation is reached) at a temperature above the freezing point of water, the water drops fall to the surface as rain. Under some weather situations, the water drops must pass through a layer of much colder air before they reach the surface of the earth. If the temperature of the lower mass of air is below freezing, the tiny falling masses fall as sleet. In some situations, the surface of the earth is so cold that the rain drops freeze into glaze ice as they reach the ground. When condensation occurs in rising warm, moist air, and clouds are formed where the temperature is below freezing, water droplets coalesce into snowflakes rather than rain drops. Unless they melt back to rain drops as they fall toward earth, they fall as snow.

Hail occurs when rain drops falling to earth are carried aloft again by strong updrafts or rising currents of air usually associated with strong thunderstorms. The violent updrafts carry the rain drops above the freezing level in the atmosphere where they become ice or hail stones. As the stones are forced upward, they pass through layers of clouds containing varying concentrations of water droplets. These droplets freeze to the surface of the stone increasing its size and weight until it finally falls to earth.

Thunderstorms

Thunderstorms represent one of the most complex forms of the interaction between water and air. Williams (1992) provides an excellent description of the formation and development of thunderstorms, and according to Lydolph (1985), they are the most violent and turbulent storms. He estimated that 40,000 take place on earth each day. Thunderstorms are extra-tropical storms. They do not need the warm waters of the tropical ocean for their development. Rather, their development depends on the presence of warm, moist air at the surface of the earth and cold, dry air aloft. As already noted, when air is heated by the sun it begins to rise because, as it is heated, it becomes lighter. As it rises, the pressure of the mass decreases and it begins to cool, but it is still warmer than the surrounding air. Usually, air reaches its dew point between 10,000 and 20,000 feet; then condensation of the water molecules begins, and clouds form. As the water molecules condense, they release the heat that they acquired when they were evaporated far below. This released, latent heat slows the cooling process, and the air mass continues to rise. At this point, there is enough heat being released from the condensing water molecules to keep the air rising. Meteorologists call this the free convection zone, because at this altitude, convection is free of the heating at the earth's surface. Thunderstorms cannot develop unless the cloud system reaches the point of free convection.

When the air mass reaches 40,000 to 50,000 feet, water drops and/or ice crystals grow heavy enough to overcome the updraft, and they begin falling (Figure 2.9-E). During the summer, the ice crystals usually melt on the way down, and they reach the earth as rain. As the drops of water or ice crystals fall, they drag some of the air molecules with them, creating a downdraft. At the same time, the updraft of warm moist air is continuing. With updrafts and downdrafts occurring simultaneously, the storm reaches its most violent stage. Thunderstorms that develop as a result of the process described in the preceding paragraph generally do not last very long. As the rate of precipitation increases, the strength of the downdrafts also increase. Finally the downdrafts become so strong that they choke off the updrafts that provide the warm, humid air to the storm. Without the infusion of this warm, moist, unstable air, the storm ends.

The thunderstorm developed through the process described in a preceding paragraph, is called a convection storm because the warm air is carried aloft through convection. During the summer months, these are the most common thunderstorms in Alabama; however on an annual basis, most of the thunderstorms in the state develop when warm, moist air is lifted when a cold, dry air mass associated with the polar front (Figure 2.8), pushes underneath. Most often, several separate thunderstorms will develop in the frontal boundary and move as a squall line along the front. In Alabama, front-related storms occur more commonly during the late winter and spring months when the polar front is increasingly being challenged by the flow of mT air from the Gulf of Mexico. The front-related storms generally are more violent than those that result from convection.

The formation of clouds as a result of convection is a common phenomenon in Alabama throughout the summer. The sky may be essentially cloudless at dawn, but by mid-morning the sky is filled with white, fluffy cumulus clouds. All of these clouds form as a result of the convection of parcels of warm, moist air from the earth's surface, and rise until they reaches the dew point where the water vapor in the parcel condenses to form the cloud. Obviously, there are many clouds in the summer sky and only a very few continue to grow to the point that a thunderstorm develops. With information currently available, meteorologists are unable to predict which cumulus clouds will become thunderstorms. For some reason, a specific cloud seems to accumulate more warm, moist air from the surface of the earth than those around it, so that when condensation takes place, there is sufficient latent heat released to keep the parcel rising until it reaches the free convection zone.

As a thunderstorm develops, different areas of the cloud acquire different electrical charges, and as a result, electrical currents move rapidly between these areas, causing lightning that, in turn, heats the air in its path to as much as 40,000°F—hotter than the surface of the sun. After the current passes, the air cools quickly. The rapid heating and cooling and the rapid movement of air molecules back and forth create sound waves that we hear as thunder. Most of the electrical currents move within and between clouds, and pose no real danger to people below; however, under certain circumstances, the attraction between positive and negative charges becomes so strong that the air's high resistance to electrical conductivity is overcome and a lightning bolt reaches the ground, splintering trees, killing animals including people, and interrupting power transmission. Lighting is the second largest weather-related people killer in the United States after floods (Williams, 1992).

Thunderstorms may also be accompanied by damaging downdraft winds as the falling rain drags air molecules down with them. Also, it is not uncommon for these storms to involve extremely heavy local rainfall and flash flooding; however most of these storms are relatively small (three to five miles in diameter) and non-violent, and they dissipate within a short period of time, usually an hour or so. Under special circumstances, ordinary thunderstorms develop into super-cells that are 10 to 15 miles in diameter and that may reach a height of over 60,000 feet. The super thunderstorms may last for several hours and travel several hundred miles before dissipating. These extremely large and potent storms may result in hail, extremely heavy rainfall, straight-line winds, and tornadoes.

Tornadoes

Tornadoes are the most violent of all weather phenomena (Lydolph, 1985). They occur more frequently in the United States east of the Rocky Mountains than anywhere in the world. These violent storms build downward out of low-hanging clouds associated with multi-thunderstorm systems or supercells. The violent storm takes the form of a large irregular funnel or inverted cone with the apex moving along the earth's surface. Winds in the vortex of the funnel near the ground may reach speeds of 300 miles per hour. Within the vortex, air pressure is so low that buildings literally explode as a tornado moves over them. Typically the apex, where most of the property damage occurs, is less than 400 yards in diameter. Tornadoes generally move northeastward at 45 mph. As the tornado moves along its path, it typically recedes upward into its parent cloud, only to drop to earth again a short time later.

Monsoons

Most everyone who has spent time in coastal areas is familiar with the sea breeze phenomenon. During the day, radiation from the sun heats the land adjacent to the coast more rapidly than the ocean, and, in turn, the air at the surface begins to rise. As it rises, air from over the ocean rushes to replace it so that a steady breeze blows from the ocean inland carrying with it a large amount of water vapor. Then at night, conditions are reversed. The land cools more rapidly than the ocean, and as air rises off the warm water, wind from onshore rushes to replace it. So the direction of wind flow changes 180° twice during each 24-hour period.

The ocean-to-shore-sea-breeze phenomenon is active on a much larger scale in monsoons. In monsoons, there is an annual change in wind direction. In parts of tropical and subtropical Asia, heating of the land during the warmest months is so extreme and so extensive that a massive transfer of warm, moist air from the ocean to the land takes place. As the moist air moves inland, it continues to rise until it condenses and finally falls back to earth as rain. The monsoon-related rains transfer enormous quantities of water from the ocean to the land. For example during June in eastern India, the winds shift and blow steadily inland off the Bay of Bengal and the Indian Ocean, carrying with them enormous quantities of water. These moist winds are pushed upward as they ascend the Khasi Hills of Assam and the Arakan ranges of Burma, where they release torrents of water as rain. At Cherrapunji near the Burmese border the average annual rainfall is 436 inches—more than 36 feet (Thompson and O'Brien, 1965).

The monsoon phenomenon is most active over the Indian subcontinent and adjacent water bodies—the Arabian Sea to the west and the Bay of Bengal to the east (Lydolph, 1985); however, there are other areas of the tropical and subtropical world where monsoons play a major role in weather and climate. Akin (1991) notes that there are monsoon air flow patterns on the southeastern coast of the United States and north and west of the Gulf of Mexico. Fein and Stephens (1987) note that the Southeastern United States is included in the North American Summer Monsoon System. The moderate increase in rainfall in Alabama during July and part of August is probably the result of some monsoon activity.

Storms That Originate in the Tropics

Storms that originate in the tropics are giant weather machines that convert the heat energy of the warm water of tropical oceans to strong winds (Williams, 1992). Meteorologists divide these storms into several groups depending on the average speed of the winds in the storm (Williams, 1992). Storms in which average wind speeds are less than 39 miles per hour are called tropical depressions. When average wind speeds are between 39 and 74 mph, they are called tropical storms, and when average wind speeds of the storm are 74 mph or greater, they are called hurricanes. (In the Pacific, storms with winds that exceed 74 mph are called typhoons.)

Tropical storms are most common and most intense in the western North Pacific. According to Lydolph (1985), it is not uncommon to have two or three of these storms in different stages of development, following each other across the western North Pacific several hundred miles apart. The second most important area for tropical storm formation is the western part of the North Atlantic, the Caribbean Sea, and the Gulf of Mexico.

Tropical storms that finally reach the coast of the Southeastern United States begin as simple thunderstorms over the tropical waters as far east as the west coast of Africa. For some reason, several of these tropical thunderstorms come together in a counterclockwise swirl to form a tropical depression. If the right ingredients are present, the depression intensifies to become a tropical storm, and with further intensification, a hurricane.

Tropical storms that form in the tropical Atlantic, generally move westward with the trade winds around the Bermuda or Azores high pressure zone (Figure 2.5). Finally, the storms turn northerly with the trade winds until they reach the southeastern coast of the United States. When the Bermuda Subtropical High is relatively weak, storms may turn northward and go into the North Atlantic without striking the coast; however if the subtropical high is strong and ridging onto the southeastern coast, the storms may be guided onto the coast of Georgia or the Carolinas or further west into the Gulf of Mexico before they reach landfall on the Gulf Coast. Tropical storms that form in the Caribbean or Gulf of Mexico move northward until they enter the trade winds circulation. Once they enter this circulation, they respond essentially the same way to the forces guiding the Atlantic storms. However, once a storm from the tropics reaches land, it no longer has the warm ocean to provide the necessary energy to maintain its strength, and it weakens. Consequently, it is not uncommon for hurricanes to be downgraded to tropical storms or even tropical depressions after moving only 20 to 30 miles inland; however, this is not always the case.

Even after a hurricane has moved inland in Alabama, Mississippi, or Louisiana, and has weakened to become a tropical storm or tropical depression, it can still cause tremendous damage as it turns northward up the Mississippi Valley, then northeastward up the Ohio Valley, guided by the subtropical high or trade winds circulation until it moves out over the Eastern Seaboard into the Atlantic. In August 1969, after causing widespread destruction and loss of life in Mississippi, Hurricane Camille became a tropical storm as it moved northeastward across western Tennessee, then eastward across Kentucky, West Virginia, and Virginia, until it reached the ocean near the mouth of Chesapeake Bay. As the storm passed through the mountains of western Virginia, it deposited 28 inches of rain in Nelson County resulting in the death of 200 people (Sullivan,1993).

Climate Classification

Because the primary determinants of weather and climate—constant solar radiation, the annual cycle of the seasons as the earth circles the sun, the daily rotation of the earth about its axis, the earth's size and spherical shape, the earth's distance from the sun, and the physical and chemical characteristics of the earth's atmosphere—are relatively constant, the weather and climate patterns on earth are relatively predictable and recognizable from year to year and within years (Trewartha, 1981). The climate is always cold at the poles and hot at the equator, and always at some intermediate level between those two constant extremes. Consequently, it is possible to define or classify climates on earth at least to a general level. The first published efforts to classify climate were made in 1918 by Waldimir Köppenan, Austrian botanist. Later, in 1954, Glenn Trewartha, an American geographer, modified the Köppen system, particularly as it applied to the United States (Akin, 1990). The primary classification of climates is shown in the following table:

Type of Climate	Climate Symbol
Tropical humid	A
Dry	B
Subtropical	C
Temperate	D
Boreal	E
Polar	F
Highland	H

In general these different types of climate, with the exception of the Dry and Highland climates, reflect latitudinal zones on the planet; because temperature generally decreases as the latitude or distance from the equator increases (Figure 2.10). In the Northern Hemisphere, Type A (Tropical humid) climates are generally found between latitude 0°N and 30°N. C (Subtropical) climates are located between 30°N and 40°N. D (Temperate) climates are generally located between 40°N and 60°N. The boundaries of the E and F (Boreal and Polar) climates are even less clearly defined, but they generally occur between 60°N and the North Pole at 90°N. Both the B (Dry) and F (Highland) climates are special climates that are located within the other major latitudinal zones; however they are so important that they are given climate type recognition. For example, the huge dry land area east of the Rocky Mountains, in the Western United States, extending from Mexico to Canada is a B (Dry) climate, but it is located within both the C and D zones.

The C (Subtropical) climate is most important to Alabama, because—except for the southern tip of Florida, which has a true A (Tropical) climate—it is the climate type found in the Southeastern United States (Figure 2.11). Trewartha and Horn (1980) commented that this type of climate is transitional between the climates of the tropics and the mid-latitudes. In the C climate zone, the mean temperature of the warmest month is above 10°C (50°F), and the mean temperature for the coolest month is above 0°F (32°F), but below 18°C (60°F) (Akin, 1990). In the United States, the eastern boundary of the C zone extends generally northeastward between Fort Myers, Florida, and New Brunswick, New Jersey (Figure 2.11). Its southern boundary extends westward to Corpus Christi, Texas. The western boundary extends generally northward from southwest Texas to the Oklahoma Panhandle. The northern boundary of the C zone is a relatively straight line between Oklahoma and New Jersey, except for a small, narrow area with D (Temperate) climate that extends down through the mountains from northern Virginia and West Virginia into northern Georgia. The C zone is divided into sub-zones depending on the temperature of the warmest month and the annual precipitation regime (Figure 2.11). The climate in Alabama is classified as Cfa (Subtropical humid), because the mean temperature of the warmest month is above 22°C (71.6°F), and there is no dry season or periods with little or no rainfall, during the year.

Because of the characteristics of global air circulation, the Cfa climates occupy the southeastern portions of continents in both the Southern and Northern Hemispheres, at latitudes between 25° and 37°. Rainfall in the Cfa sub-zone in the United States, ranges from 47 to 55 inches annually in the eastern part, and decreases to 35 inches on the western boundary. Similarly, there is a temperature gradient from south to the north. Average January temperature in the Cfa sub-zone varies from 38°F in the north to 62°F in the south. Summer temperatures are more similar, varying from 77°F north to 80°F south.

Except for a relatively small area in the northeastern portion of the C zone (North Carolina, Virginia, Maryland, and Delaware), all of the climate in the Southeastern United States is classified as Cfa. The climate in that small mid-Atlantic region is classified Cfb because the summer temperatures are a little cooler than in the Cfa sub-zone, as a result of the marine influence.

The southeastern coast of China and part of Japan are the only other large land areas in the world with the same combination of temperature and precipitation regimes. Both the C zones in China and Japan and in the Southeastern United States are located on the extreme western sides of subtropical high pressure cells. At this position in the circulation, the air is moving generally northwestward around the center of high pressure; consequently, during the warm months both receive mT air that has passed over thousands of miles of warm oceans before it reaches land. As a result, the climate in both areas is extremely sultry during the summer months. This sultry weather is probably the most distinct characteristic of Cfa climates. Lydolph (1985) comments that the combination of temperature and humidity in the Cfa sub-zone result in weather that is "truly oppressive."

The warm, moist Maritime Tropical (mT) air passing inland from the oceans into Cfa sub- zones is usually highly unstable, which leads to the development of some of the most violent weather on earth in the form of hurricanes (typhoons). These storms develop thousands of miles away from the Cfa sub-zones, but the global circulation system guides them ashore in the Southeastern United States and in Southeastern China. In the hurricanes' fully mature stages, they often cause massive destruction and the loss of many lives. Also, in the spring months in the Southeastern United States, highly unstable mT air streaming inland from the Gulf of Mexico leads to the development of especially strong and violent thunderstorms and tornadoes when it collides with cooler Continental Polar (cP) air (Lydolph, 1985).

Climate and Weather in Alabama

As detailed in a preceding section of this chapter, Alabama's so-called subtropical climate is actually a hybrid. It tends to be Dc (Temperate Continental) in the winter and Ar (Tropical Humid) in the summer, but there are times when the climate becomes confused, especially in the spring, when it often becomes tropical earlier than expected, only to shortly afterwards become temperate again. And then there are times when it remains temperate far longer than expected in the spring. Further, there are even times when it resembles a Bs (semi-arid) climate. This unpredictable changeableness has extremely important implications for crop production.

Temperature

As detailed in a preceding section, temperatures in Alabama tend to be a mixture of those found in the Dc (Temperate Continental) area (Figure 2.10) and those found in the Ar (Tropical Humid) area. As a result, temperature ranges can be extreme. For example, data published by Kincer (1941), indicated that in the period 1902–1938, the lowest temperature recorded by a weather station in Alabama was -18°F. (Valley Head, DeKalb County). During the same period, the highest temperature recorded was 111°F (Madison, Madison County). These temperatures are obviously extreme examples, but it is not uncommon to encounter winter readings in the lower teens and summer readings above 100°F .

Data on mean annual temperatures in Alabama, during the period 1902-1938, were published by Kincer (1941), and some of it is presented in Figure 2.12. The data indicate that these temperatures ranged between 62.2°F (1903 and 1917) and 66.4°F (1927). The average for the 37-year period was 64.21°F. The standard deviation (actually standard error of the mean, because the data are means) was 1.14. The data in the figure show that the year-to-year estimates followed a somewhat sinuous pattern throughout the period. Average temperatures generally increased for a short period, only to decline for a few years.

Temperature Gradients

The National Weather Service of the US Department of Commerce has accumulated data on various aspects of Alabama's weather over a considerable time period. The Cartographic Research Lab of the Department of Geography at the University of Alabama has used some of this data to construct Temperature Gradient Maps for January (Figure 2.13) and July (Figure 2.14) for the state.

During January, there was a clear gradient of average temperature of about 10°F (44 to 54°F) from north to south (Figure 2.13). Mean January temperatures in the northern tier of counties were in the low 40s, while in the southern tier, they were in the low 50s. The temperature gradient reflects not only a latitudinal difference from north to south, but also a difference in elevation. Elevations in Alabama generally trend upward in a series of undulating plains from sea level at the coast to an elevation of 600 to 700 feet in the north central counties and even higher in the eastern counties (Figure 1.8). West to east differences in temperature were relatively minor, generally averaging a degree or less, except where there are relatively large differences in elevation such as where the Appalachian Mountains (Figure 2.1, Region 2), and the Eastern Valley (Figure 2.1, Region 4) are located in the state, in the counties in the northeast (See also Figure 1.8). The July gradient from north to south was much less dramatic than the January gradient.

These data indicate that in January the average temperatures in Alabama are generally under the control of weather systems originating in the Temperate Continental (Dc), climate area (Figure 2.10), but that in July the Tropical Humid (Ar) area is in control. From October through early June, the polar front (Figure 2.8) is quite active over the state, but during the remainder of the year it remains far to the north and only occasionally affects our weather.

Length of the Growing Season

Gallup (1980) defines growing season as "the climatic interval during which meteorological conditions permit plant growth." He further states that there is no clear standard for measuring the exact growing season, but that shelter temperatures that have reached freezing are more applicable, in general, for describing growing season than subjective observations of frost damage to vegetation. He suggests that growing season boundaries be determined by temperatures of 32°F, or lower, measured five feet above the ground.

Length of the growing season is a function of two temperature variables: date of the last 32°F temperature in the spring and date of the first 32°F temperature in the fall. Alabama has a long growing season. It is well over 200 days in most areas of the state.

Data published by Kincer (1941) provide estimates on the average length of the growing season in Decatur, Tuscaloosa, and Citronelle, during the period 1899-1938, are presented in Table 2.4. Data for Columbus, Ohio, are presented for comparison. These data show, for example, that during the period that the average length of the growing season at Columbus, Ohio, was 186 days, that the range in the length of the growing season was 71 days, that the shortest growing season was 150 days, and that the longest growing season was 221 days. The length of the average growing season was almost three months longer in Citronelle than in Columbus (260 versus 186 days).

Figure 2.17, developed by Gallup (1980), shows lines of equal freeze-free periods (growing season length) for the average year in Alabama. For example, the growing season in the northwestern portion of Lauderdale County is 190 days or less, while in the southern portions of Baldwin and Mobile counties it is 270 days or more. For most of the State, the growing season ranges between 200 and 220 days. Below the northern portion of Washington County, it increases rapidly, and from northern to southern Mobile County it increases 30 days, from 240 to 270.

Figure 2.17 also includes information on areas where the length of the growing season is somewhat different, both longer and shorter, than expected. For example, there is a small area in eastern Colbert County where the growing season is just 10 days (210 days) less than in a small area in eastern Escambia County (220 days). Further, there is a small area in eastern Cleburne County where the length of the growing season is the same as in northwestern Lauderdale (190 days).

Areas associated with shorter than expected growing seasons are usually lower in elevation than surrounding areas and serve as collecting basins for heavier, colder air. Similarly, areas with longer than expected growing seasons are usually higher than surrounding areas and are locations where warmer, lighter air accumulates (Gallup, 1980). For example, the night-time, winter temperatures recorded in the valleys that form the eastern watershed of the Tombigbee River in Fayette, Lamar, and Marion counties, in northwest Alabama are among the lowest in Alabama, when cold air flows down off the surrounding hills and mountains and collect in the lowlands. As a result, the length of the growing season is considerably shorter than expected (Figure 2.17). In fact, this accumulation of colder air affects the growing season southward all the way into southern Washington County.

A similar situation results in relatively colder temperatures and a shorter growing season in portions of Russell, Barbour, and Henry counties. This anomaly is a result of local terrain favorable to the down-slope drainage and accumulation of cold air and of the low heat capacity of the light-textured, sandy-loam soil in those areas. Another relatively colder, freeze-prone, area is found along the valleys of the Conecuh and Pea Rivers in Crenshaw, Coffee, Covington, Geneva, and Escambia counties of southeast Alabama. Other areas with shorter than expected growing seasons are found in Escambia, DeKalb, and Cleburne counties (Figure 2.17).

A relatively warmer zone is associated with broken highlands beginning in Houston County in southeast Alabama and running generally northwestward through Dale, Pike, Butler, Montgomery, Lowndes, and Dallas counties and then northeastward through Hale, Tuscaloosa, and Jefferson counties (Figure 2.18). So-called warm centers are also found in parts of Butler, Lowndes, Dallas, Wilcox, Monroe, and Conecuh counties; in parts of Clay, Talladega, and Calhoun; in parts of Marshall and Madison and in parts of Colbert and Lawrence.

Rainfall

There are two primary characteristics of Alabama's climate and weather: warm and moist. As noted in a preceding section, enough precipitation falls on the state in most years to fill a 52,400-square-mile lake to a depth of between four and five feet. Data reported by Kincer (1941) indicated that average annual rainfall in Alabama was second in the nation in the period 1886 through 1938. The top five states, in descending order, were: Louisiana (55.22 inches), Alabama (53.00), Mississippi (52.86), Georgia (50.16), and Arkansas (48.06). Almost all of the precipitation falls as rain. Only a small fraction is the result of snow fall and an even smaller fraction falls as hail.

Most of our precipitation occurs as a result of the collision between cold or cool, dry cP air north and west of the polar front (Figure 2.8) and warm, moist, mT air south and east of that front. When the mass of heavy, cold, high pressure air flows southeastward, it collides with the lighter, warm, moist air moving northward from the Gulf. The heavier air flows underneath the lighter air forcing it upward, and as it rises, it cools. When it cools sufficiently, the water vapor in the mass condenses, and under the right conditions rain falls.

Annual Statewide Rainfall

Total annual statewide rainfall averages for the period 1886 through 1938 were reported by Kincer (1941). These averages are presented in Figure 2.19, which graphs the data year by year for the entire period. Figure 2.20 graphs a frequency distribution of the data. The average annual rainfall for the 53-year period was 53.00 inches, and ranged between 39.21 and 76.48 inches. The standard deviation (really the standard error of the mean, since the data are means) was 7.20. The frequency distribution indicates only a limited degree of central tendency in the data. The average only broadly defines the expected rainfall. Rainfall amounts between 44 and 55.9 inches occurred with essentially equal frequency. Two-thirds of the observations were within this range.

The rainfall data presented in Figures 2.19 and 2.20 reflect a long-term rainfall pattern that is relatively unpredictable, especially during the period 1920 through 1938. Comparatively, the period from 1890 through 1899 was less variable. Between these two extremes there was a general trend of increasing rainfall during the period 1910 through 1920.

For comparative purposes, the rainfall data for Illinois for the same period (1886-1938) are presented in Figures 2.21 and 2.22. In Illinois, the average annual rainfall for the 53-year period was 35.80 inches with a standard deviation of 4.58. The range was from 27.75 through 49.39 inches. The frequency distribution (Figure 2.22) for the Illinois data indicates a relatively higher degree of central tendency compared to that characteristic in the Alabama data (Figure 2.20) during the same period. This higher degree of central tendency is reflected in a lower standard deviation, 4.58 compared to 7.20. This data indicate that while average rainfall was considerably less (35.80 inches versus 53.08 inches) in Illinois during that period, it was considerably more predictable from year to year.

More recent annual statewide rainfall data are presented in Figure 2.23. Data published by the National Climatic Data Center (1972-1991) were used for the figure. Annual averages for the eight weather reporting regions (Figure 2.1) in the state were averaged to obtain annual statewide averages for the period 1972 through 1991. The average annual rainfall for the 20-year period was 59.07 inches, or about six inches greater than in the 1886-1936 period. The average ranged between 46.53 and 75.19 inches. The standard deviation was 8.20. Thus the variability noted in the period 1886 through 1938 continued during the period 1972 through 1991. In fact, comparing the standard deviations (7.20 versus 8.20), it appears that the latter period was somewhat more variable.

Annual Totals

Data on the average annual rainfall over a 20-year period (1972-1991) in Alabama's weather reporting regions (Figure 2.1) are shown graphically in Figure 2.24. These data indicate that average annual rainfall for that period was generally similar in all of the regions except for the Upper Plains (Region 3) and the Gulf (Region 8). The Gulf Region, with a 20-year annual average of 67.06 inches, was considerably higher than the others. Except for the Appalachian Mountains (Region 2), averages for the other regions were clustered around 55 inches.

The Department of Geography at the University of Alabama has published a map showing the distribution of rainfall within the state. Areas with the same average rainfall are connected by lines (Figure 2.25). The data presented in the figure show that rainfall in the state ranged from 50 inches in a small area in Lee, Barbour, and Macon counties and in Pickens County to 66 inches in a small area in south Baldwin County. Rainfall was 54 inches or less in three-fourths of the state. It is above 60 inches only in extreme southwestern Alabama. The relatively high rate of precipitation in the southwestern corner of the state persists in a narrow band along the Gulf Coast from the Texas-Louisiana border eastward to the Apalachicola River in Florida.

Monthly Distribution

Data presented in Figure 2.26 indicate that during the period 1972-1991 that total annual precipitation varied from around 3 to 6.7 inches a month. This is another unique characteristic of our climate and weather. There are no pronounced or unusually wet or dry periods. Generally, rainfall trended downward from January through October; then increased through January. It was highest in March and lowest in October. It was also relatively higher in May. July and September.

Monthly distribution of rainfall in the different regions for the 20-year period is shown in Figure 2.27. Generally, the annual rainfall pattern in the regions was the same as the statewide patterns shown in Figure 2.26; except that the annual downward trend was less obvious in the Coastal Plain and Gulf Regions.

Colliding air masses cause virtually all of the rain during the cool and cold months (November through March) when the polar front is most active over the state. It is during this period that Alabama receive the highest percentage of its rainfall. In April, the polar front begins to shift northward. From that time until November, colliding air masses produce a smaller percentage of our total rainfall, and local convective storms or thunderstorms produce more. However even in mid-summer, colliding air masses continue to provide most of the state's rainfall.

During the warmer period (May-August), a large portion of Alabama's rainfall is a result of convective activity when a small parcel of mT air, heated at the surface, rises (Figure 2.9). The parcel continues to rise until the water vapor condenses forming a cloud, and when the right conditions are present, the cloud will continue to rise and to grow until a thunderstorm develops and rain falls. Thunderstorms are a locally important source of rainfall, but they usually do not provide a significant amount of rainfall over large portions of the state at the same time. Thundershower activity begins to abate when the surface temperatures begin to decrease reducing convective activity. October is our driest month, because there is virtually no convective activity as the surface cools and because the polar front has not yet become well established over the state.

Although convective storms provide some rainfall during the warmer months, even then there is a continuing presence of weak polar front activity. During the warmer months, the front is generally located far to the north along the Canadian border, but it can move weakly and temporarily far to the south. In 1994, the polar front remained active throughout the summer, bringing the state one of the highest rainfall totals on record. Lydolph (1985), notes that North America is the only continent where the polar front remains somehat active throughout the continent throughout the year.

Lydolph (1985) comments that in areas with Cf (Subtropical, humid) climates that rainfall in the warmer months typically is greater than in colder months. He further comments that this general rule does not seem to apply in the southeastern United States. There is an area centered on Tennessee where the rainfall in the colder months is generally greater than in the warmer months. Information on this phenomenon in Alabama is presented in Table 2.5. This table presents data on the number of years in the 20-year period in which the total rainfall during January, February, and March was greater than the total for June, July, and August. These data indicate that Lydolph's comments on the relative wetness of winters versus summers in the southeastern United States applies generally to all regions in Alabama with the exception of the Gulf Region. In that region, only in eight of 20 years did the general rule apply. In the other years, the summer months were wetter.

Daily Totals

The amount of rain falling in a 24-hour period is of major importance to agriculture. Information on this phenomenon was obtained by averaging the daily rainfall amount at representative weather stations in all eight regions for May, June, July, and August in the even years (1972, 1974, 1976…1990) during the period 1972-1991. Average recorded daily rainfall amounts varied between 0.01 inch and 4.98 inches. This range was divided into intervals of 0.25 inch. Then the average daily rainfall amounts were assigned to the various intervals. Finally the numbers of events in the intervals were converted to percentages. A graph of these data is presented in Figure 2.28.

The figure represents the frequency distribution of 2700 individual daily rain events recorded in even years over the 20-year period. In the graph, the percentage occurrence of a rainfall amount is plotted against the upper limit of the related interval. For example, 45.33% of the 2700 rainfall events recorded, yielded 0.24 inch of rain, or less. Another 20% of the events yielded between 0.25 and 0.49 inch. Thus, almost 65% of the May through August, daily rainfall amounts recorded in all even years and all regions were less than 0.50 inch. Only 16.25% of the rainfall events were greater than 1.00 inch, and less than 1.00% were greater than 2.99 inches.

Rainfall Probabilities

Rainfall probabilities provide an interesting perspective on rainfall in Alabama. Getz and Owsley (1979) used daily rainfall records for a number of Alabama locations to calculate the probabilities of given locations receiving more than 0.00, 0.10, 0.25, 0.50, 1.00, 2.00, 4.00, or 10.00 inches of rain in either one-, two- or three-week periods, throughout the year. Data has been taken from their publication for presentation here on the probabilities of receiving more than 1.00 inch of rain, in one-week periods, in the 20 weeks from May 3 through September 19. Probability data for one location in each region has been combined for presentation in Figure 2.29. In Figure 2.30, probabilities for all 20 weeks have been combined to show the probability of each region receiving more than one inch of rainfall, per week, in the 20-week period.

Data presented in Figure 2.29 shows that the probabilities of receiving more than one inch of rain in a one-week period generally increased from slightly less than 30% (approximately one in three) in week 1 (May 3-9) to 47% in week 11 (July 12-18); then decreased to slightly above 30% in week 20 (September 13-19). It is important to remember that these are probabilities of receiving more than one inch in a week. Subtracting these percentages from 100 gives the percentages of not receiving more than one inch of rain in a particular week. The probability of receiving more than one inch of rain in one week during the period, May 3 through May 9, was slightly less than 30%. Based on this probability, a farmer could expect to receive more than one inch of rain during that particular week, three years in ten, and would not expect to receive that amount during that week in seven of ten years.

Figure 2.30 shows the probabilities of receiving more than one inch of rain in one week for eight region locations, with the data averaged across the 20 weekly periods. Generally the probabilities increase from left to right in the figure. This increase is likely the result of the general location of the regions. Generally the regions on the left of the figure are located in the northern half of the state and those on the right side are located in the southern half. Remember that the southern half of the state (especially the Gulf Region) receives more rainfall during this period of the year (Figure 2.25). But even in the Gulf Region, on the average, a farmer would not expect to receive more than one-inch of rain in a week during the 20-week period five years in ten.

Storms

Storms of one kind or another are common events in Alabama, and their effect on Alabama agriculture is a mixed-bag. Convectional thunderstorms in summer bring badly needed rain to the state's crops, and in the dry fall months, tropical depressions can also bring welcome rain. Unfortunately, when the conditions are right, these weather phenomena can bring too much rain, resulting in serious flooding. Tornadoes are devastating storms, but generally do not result in widespread damage to Alabama agriculture. Hurricanes are also devastating storms, although their wind velocity generally is never as high as in a tornado. These storms can destroy crops and infrastructure, but not usually of a large portion of the state at any one time.

Thunderstorms

As noted in a preceding section, much of the rainfall that Alabama receives during the summer months comes from thunderstorms. Data from Carter and Seaquist (1984) shown in Table 2.6 give the average number of days with thunderstorms at representative locations in six regions in the different months of the year.

The data presented in Table 2.6 show that the number of days with thunderstorms generally increased from January through July; then decreased through the end of the year. The lowest number of days was in October; although there was little difference in October and December. In most of the months, there was little difference in the number of days with thunderstorms from north to south; however during June, July, August, and September, there were generally more days with thunderstorms in the regions in the south than in those in the north.

Tornadoes

According to Chermock (1976), Alabama ranked ninth among all states in the number of tornadoes that occurred within its boundaries between 1916 and 1964. During the period 1950 through 1982, the annual number of tornadoes occurring in the state ranged from two in 1950 to 45 in both 1957 and 1973 (Carter and Seaquist, 1984), with an average of 18 a year. Table 2.7 gives the number of tornadoes and number of deaths caused by these storms in Alabama and surrounding states during the period 1953 through 1971. Alabama ranked second among the surrounding states in the number of tornadoes during the period, and second in the number of deaths resulting from those powerful storms. Mississippi ranked first in the number of deaths per tornado (ten deaths in each 12 tornadoes). Alabama ranked second with ten deaths in each 37 of the storms. Florida ranked last with ten deaths in each 143 of the storms.

Tornadoes occur in all Alabama counties, but they tend to occur more commonly in the so-called tornado alley through a band of counties from Sumter and Pickens in the west to Jackson, DeKalb, Cherokee, and Cleburne counties in the east (Chermock, 1976). They also occur in all months, but are most common in March (3.64 a month) and April (3.70 a month) and least common in the period June through October, with the lowest number in August (0.21 a month). Approximately 82% of all tornadoes occur between noon and midnight with 23% of them occurring between 4 and 6 pm.

Storms Originating in the Tropics

Storms originating in the tropics that finally reach Alabama form as tropical disturbances in warm, moist tropical air, often near the equator. These disturbances consist of discreet systems of clouds, rain showers, and thunderstorms. Under the right set of conditions, these disturbances become increasingly better organized, and are carried along by the trade winds circulation (Figure 2.5), northwestward toward the south-eastern coast of the United States. Most of these storms originate in the Western Atlantic ocean, in the Caribbean Sea, or the Gulf of Mexico, between the equator and 10° North Latitude. Of the ten storms entering Alabama in the period 1981-

2000, five originated in the Gulf, one originated in the Caribbean, and four originated in the western Atlantic.

In his book on hurricanes, Pielke (1990) included annual North Atlantic Hurricane Tracking Charts for the years 1871 through 1989. These charts were published first by the National Weather Service (1871-1986) then by the National Hurricane Center (1987). They track the paths of individual hurricanes, as well as all other forms of tropical storms in the North Atlantic Ocean, the Caribbean Sea, and the Gulf of Mexico and over adjacent land areas. By tracing individual storm tracks, it was possible to determine the number of storms entering Alabama each year and their paths across the state. Between 1871 and 2002, some 92 storms of tropical origin entered Alabama directly from the Gulf or from adjoining states. This total includes hurricanes, tropical storms, or some form of tropical depression. Table 2.8 gives the number of these storms by decade that entered Alabama.

Most storms of tropical origin that reach Alabama develop during the late summer months, but some of our storms develop as early as May and some as late as October. The distribution of storms by month of development is given in Table 2.9. Generally, but not always, the month in which they developed is the same month that they entered the state.

Because of the small amount of Alabama coast line, relatively fewer of the 82 storms entering the state from 1871 through 1989 did so directly from the Gulf of Mexico. A total of 17 storms (14.6%) entered directly from the Gulf, 29 (35.4%) entered from Florida, 24 (29.3%) entered from Mississippi, and 12 (20.7%) entered from Georgia.

As detailed in a preceding secion, it is not uncommon fo a hurricane to be downgraded to a tropical storm after traveling overland for 20 to 30 miles, but it is not always the case. For example, in September 1975, Hurricane Eloise reached landfall on the Florida Panhandle near Panama City. It moved into Alabama near Geneva County. The storm continued to move northward with hurricane force winds intact until it became a tropical storm in the vicinity of Chambers and Randolph counties, some 200 miles away from landfall. As Eloise moved northward in an area with virtually no history with sustained winds of this magnitude, it left a tremendous amount of property damage in its wake.

Alabama's Weather Effect on Agriculture

In general, Alabama would seem to have an ideal climate and weather for agriculture. On average, we have mild winters, long growing seasons, and an ample supply of rainfall that is generally evenly distributed across the state and from month to month through the growing season. Unfortunately, we do not produce food and fiber on the average, and the day-to-day, month-to-month, and the year-to-year characteristics of our weather present a somewhat different picture.

Specific problems resulting from the interaction of weather and agriculture in Alabama, include the following as will be detailed below: Mild winters; Heavy, late-winter and early-spring rains; Late spring freezes; Mini-droughts during the growing season; Heavy rainfall during late spring and early summer; Excessively high summer temperatures; and Storms from the Tropics entering the state.

Mild Winters

Genomes of harmful insects and plant pathogens are adapted to the specific environments in which they naturally occur, and it is only when they encounter extremes in those environments, such as extremely low or extremely high temperatures, that their population numbers decrease significantly (Humphrey, 1941 and Hyslop, 1941). All winters in Alabama are relatively mild; however some are so mild that there is little natural control of pests or diseases that limit the quantity and/or quality of plants and animals produced for food or fiber. For example, the relatively mild winters in recent years may have contributed to the potency of viruses that have devastated commercial tomato farming on Straight Mountain in Blount County (Sikora, 1995a). The winter of 1994-1995 was so mild that damage from Southern pine beetle infestations continued throughout the colder months; then increased rapidly in several of the central counties in the early spring (Alabama Farmers Federation, 1995).

Information was obtained on the frequency of relatively mild winters in Alabama from data published by the National Climatic Control Center (National Climatic Control Center, 1972-1991). Data in Table 2.10 show the number of winters (December, January, and February) in the period —winter 1972-1973 through winter 1990-1991 (a total of 19 winters)—in which the monthly average of the daily minimum temperatures was above 32.0°F at representative stations in each of the regions.

Data in the Table show that, for example, in the Northern Valley (Nova) Region, in the 19 winters there were only two in which the monthly average daily minimum temperature was above 32.0°F. In contrast, in the Gulf Region, in 18 of 19 winters, the monthly minimum daily temperature was above freezing.

Heavy Late Winter and Early Spring Rains

Excessive rainfall during the late winter and early spring has always created problems for Alabama farmers. Mell (1890) reported that in the late winter of 1850 the weather was so wet that the land could not be properly prepared for planting. Cultivation resulted in the formation of wet clods that had not pulverized when the time for planting arrived. Then in 1867, late winter and early spring rains were so heavy and continued so long that all crops had to be replanted. More recently, during the period February 17-24, 1961, quasi-stationary polar fronts and a maritime tropical (mT) air mass interacted over the Gulf and southeastern states, to produce heavy rains and a flood of historic proportions in Georgia and Alabama. Rainfall was especially heavy in the central counties of Alabama. During the 24-hour period ending on the morning of February 25, 9.27 inches of rain fell at Milstead in Macon County, 9.13 inches at the Yates Hydropower Plant in Tallapoosa County, and 8.32 inches at Greenville in Butler County (Weather Bureau, 1961). The heavy rainfall resulted in flooding of varying magnitudes throughout the entire Mobile River system including its major tributaries. Approximately 7,800 Alabama families sustained flood losses. Many low-lying areas remained under water for several days after rainfall diminished. It was estimated that 2,500 head of cattle were lost in Elmore and Montgomery counties alone.

A flood in late March 1990, in the central and southern counties, resulted in significant top soil erosion over thousands of acres of farmland. Some early planted corn and vegetables had to be replanted. There was also extensive damage to farm equipment, buildings, fences, rural homes, roads, and bridges (Alabama Agricultural Statistics Service, 1990).

Excessive rainfall during the late winter and early spring usually is accompanied by unusually cool weather. In 1997, wet, cool weather during this period in the Tennessee Valley resulted in the stunting of young cotton plants to such a serious extent that the entire crop was lost (Sikora, 1998).

Late winter and early spring flooding is a persistent problem for Alabama agriculture. Even when the flooding is not excessive, as it was in 1961 and 1990, virtually any amount of rainfall above the average during February or March will be detrimental to agriculture. Excessive rainfall erodes recently plowed fields, delays field work, washes away recently applied fertilizers and pesticides, and encourages development of weeds and plant diseases. Further, wet fields and access roads are detrimental to the operation of farm equipment.

Data on the incidence of unusually heavy rainfall during February and March are presented in Table 2.11. Included are the number of years in a 20-year period in which average rainfall in a region was 8.00 inches or greater in either February or March. Rainfall levels of this magnitude (2.00 inches a week) would be detrimental to all phases of agriculture. Data in the table indicate that, for example, in the Northern Valley Region (Nova) that in two years out of 20, average monthly rainfall was 8.00 inches or greater during February and five out of 20 years in March. Similarly, in the Gulf Region rainfall was 8.00 inches or greater six years out of 20 in February and five years out of 20 in March.

The frequency of excessive rainfall was relatively consistent from region to region across the state; although the frequencies tended to be somewhat higher in the southern portion. A comparison of the totals for the two months show that excessive rainfall was almost two times as frequent in March as in February.

Late Spring Freezes

Late spring freezes have plagued Alabama agriculture from the earliest days. Mell (1890) described a killing frost that occurred in the northern half of Alabama on June 8, 1816, and a freeze that killed everything green on April 16, 1849. Findley (1982) commented on the high degree of irregularity in the last spring freezes in his Alabama Crop-Weather Bulletin. It is not uncommon in the state to have several days or even two to three weeks of relatively mild weather during March, only to have two to three days of weather with killing morning frosts late in the month or in early April. Conversely, it is not uncommon to have mild weather with no killing frosts after March 15. During the week ending April 5, 1987, after several weeks of mild weather, freezing temperatures across the state caused extensive damage to fruit and vegetable crops and to wheat and corn (Alabama Agricultural Statistics Service, 1987). Three years later, in 1990, freezing temperatures resulted in light to heavy losses in early producing peach varieties (Alabama Agricultural Statistics Service, 1990). Altogether, in the 33-year period from 1970 to 2002, late freezes reduced peach production in ten of those years.

In the period 1899 through 1938, the last spring frost at Tuscaloosa, Upper Plains Region, ranged—counting from January 1—between day 55 (February 24) and day 115 (April 25)—a range of 60 days. The average day of last-frost was March 23 (Day 82) (Kincer, 1941). In 13 of those 40 years, it occurred after March 31. The standard deviation was 14.34 (Table 2.12). The year-to-year data are presented in graphical form in Figure 2.31.

A relatively high degree of variability is evident in the saw-tooth distribution of the data in Figure 2.31. Also, there is a general downward trend over the 39-year period indicating earlier last-killing frosts. These same data are graphed as a frequency distribution in Figure 2.32. The distribution is relatively flat and elongated, suggesting only a limited degree of central tendency in the data; consequently, the mean was not a very accurate predictor of the date of the last killing-frost during that 39-year period.

During the same time period (1899-1938), at Citronelle in north Mobile County (Gulf Region), the date of the last killing-frost ranged from January 10 (Day10) to April 12 (Day 102)—a range of 92 days. These data are graphed in Figure 2.33. The saw-toothed distribution seen in the Tuscaloosa data (Figure 2.31) is also evident in the Citronelle data; however there was no downward trend. In fact, there seems to be a slight upward trend.

The average last killing-frost date in Citronelle was March 8 (Day 67), 15 days earlier than at Tuscaloosa. The standard deviation of 18.42 (Table 2.12) was somewhat greater than the value of 14.34 for the Tuscaloosa data. The average date of the last killing-frost at Columbus, Ohio, during the same period, was April 17 (Day 107). The standard deviation there was slightly more than one-half as large as the standard deviation of the Citronelle data (8.3 compared to 14.34). The year-to-year data for Columbus are graphed in Figure 2.34. The higher degree of variability in the Tuscaloosa and Citronelle versus the Columbus data is obvious by comparing Figures 2.31, 2.33 and 2.34. A frequency distribution of the Columbus data is shown in Figure 2.35. Note that the central tendency in the frequency distribution of the Columbus data is greater than for the Tuscaloosa data (Figure 2.33); consequently predictions for agribusiness use based on these data would have been much more reliable using the Columbus mean than either those of Tuscaloosa or Citronelle.

Generally, the ranges of the earliest and latest dates of the last killing frosts tend to increase as the distance to the equator decreased. Variances of the data sets show this relationship even more clearly. Table 2.12 presents variances and standard deviations from the Columbus, Ohio; Tuscaloosa and Citronelle data, plus similar data for Decatur, Alabama. The relatively high degree of variation in the date (Table 2.12) of the last killing frost at three locations is indicative of the highly variable nature of Alabama's subtropical climate, during the late winter and early spring. The control of our weather swings rapidly from the cool, dry, temperate (Dc) climate zone to the north to the warm, humid (Aw) zone to the south. Polar fronts from the north bring cold, dry high pressure (cP) air into the state at approximately weekly intervals. As the front passes, clockwise circulation brings cold Canadian air down along the leading edge of the cell into the state, but after 24-48 hours, the high pressure cell generally moves beyond the state, and this cold air is replaced by mild air—as the clockwise circulation around the cell of high pressure brings warm mT air northward along its trailing edge into the state from the Gulf of Mexico. This battle moves back and forth across the state from late February until mid-April with no clear winner. Although there is no clear winner until mid-April, the increasing variability of the dates of the last frost from north to south (Table 2.12), demonstrates the increasing influence of the humid tropics zone compared to the influence of the dry temperate zone.

Mini-Droughts During the Growing Season

Mini-droughts during the primary plant growing months of May, June, July, and August, probably have always been the single most important problem limiting crop production in Alabama. For example, Mell (1890) noted that in 1851 that the lack of rain from May to August resulted in the poorest corn and cotton crops ever made on the sand and clay soils of Alabama, and in 1860, rainfall in east Alabama was so limited in June and July that the cotton plants were "bare of fruit." Forest shrubs died for lack of moisture. Findley (1982) noted that severe local droughts occur nearly every year someplace in Alabama. Ward et al. (1959) commented that agricultural drought probably occurs every year in Alabama and that these water deficits are the principal cause of failure not to obtain maximum yields.

Periods of two to three weeks with little or no rainfall, coupled with the high air temperatures during the growing season, often slow plant development and growth. Data in Table 2.13 show the number of days in each of ten even years (1972,1974,1976 . . . 1990) that the different regions experienced periods of either 7 through 14 days or 15 through 35 days without rainfall during the growing season (May, June, July, and August). Data for the different months are combined. The data show, for example, that in the Northern Valley (Nova) Region, in 1972, there were two periods of 7 to 14 days within the growing season when there was no rainfall, but there were no periods when the region received no rain for 15 through 35 days. Data in Table 2.13 also indicate that 1980 would have been a particularly poor year for crop production. In that year all regions except Gulf had at least one period when they received no rain for 15 through 35 days.

Another source of information on the frequency of mini-droughts is monthly rainfall totals. Table 2.14 contains data on the number of years over a 20-year period (1972-1991) that average rainfall in May, June, July, or August was 4.00 inches or less in the different regions. Data in the Table show, for example, that in the Northern Valley Region (Nova) in 8 of 20 years during the 20-year period that average rainfall in May was 4.00 inches or less. For August, in the same region, in 15 of 20 years rainfall totals were 4.00 inches or less. Summed across months, the total for the Northern Valley Region in the right hand column was 45, meaning that in 80 possible months (4 months times 20 years) the average monthly rainfall in that region was 4.00 inches or less in 45 of those months. Totals for the other regions, with the exception of the Coastal Plain and Gulf Regions, were generally similar to the pattern in the Northern Valley Region. In those two regions, in south Alabama, the totals were significantly lower, indicating significantly fewer months with 4.00 inches or less of rainfall.

The rainfall data in Table 2.14, summed by months, across regions, are shown across the bottom of the table. The total for May was 62. These data indicate that during May in the 20-year period that monthly rainfall totals of 4.00 inches or less occurred 62 times in the 160 possible time-location (20 years times 8 regions) combinations. These totals indicate that monthly rainfall events of 4 inches or less, was more likely to occur in June (89) and August (105). The total for August would have been much more striking except for the low numbers for the Coastal Plain (8) and Gulf (5) regions.

Monthly rainfall of 3.00 inches or less would indicate a somewhat increased tendency for drought. In Table 2.15, data are presented on the number of years in the 20-year period that average monthly rainfall in each region was 3.00 inches or less during May, June, July, or August. These data show, for example, that average monthly rainfall in the Northern Valley Region, was 3.00 inches or less, 3 years in 20, during May and 4 in 20, 3 in 20 and 11 in 20, during June, July, and August, respectively. Again, this lower level of rainfall was much less likely in the Coastal Plain (13 of 80 months) and Gulf (11 of 80 months) in May, June, and July.

Table 2.16, contains data on the number of times in the 20-year period that average rainfall in the regions, for May, June, July, or August was 2.00 inches or less. As expected, rainfall amounts of 2.00 inches or less per month were less common. For example, the Northern Valley Region received 2.00 inches or less in May only once in 20 years. A comparison of the totals at the right hand side of the table shows that in 7 of the 80 months (20 years times 4 months), farms in the Eastern Valley Region (Eava), for example, would have received rainfall in the amount of 2.00 inches or less. In all of the other regions, rainfall amounts at this level were generally similar; although those in Coastal Plain and Gulf continued to be somewhat lower. With rainfall levels this low and rainfall likely not evenly distributed within the months, water needs of plants in cultivation would be significantly greater than water availability for much of the month. In those years in which rainfall totals were 2.00 inches or less in any month—during the critical period of growth, flowering, and filling—crop yields would have been severely affected.

By considering data from Tables 2.14, 2.15, and 2.16 as a whole, it is obvious that environmental resistance in the form of reduced rainfall during the growing season likely resulted in a heavy burden on the genetic potential of crops, and in turn on the livelihoods of most Alabama farmers during the 20-year period.

Ward et al. (1959) contributed valuable information on the subject of agricultural drought in Alabama. They used data on rainfall and estimates of evapotranspiration and soil moisture storage capacity to compute the probability of the minimum number of drought days during the period March through November that would occur in different areas of Alabama, in soils with either 1 or 2 inches of soil moisture storage capacity for either 1 out of 10, 2 out of 10, 3 out of 10, or 5 out of 10 years. Some of their data is shown in Figures 2.36 and 2.37. Figure 2.36, presents data on the minimum number of drought days expected in different areas of the state during the period March through November (275 days) in 5 out of 10 years if the soil moisture storage capacity is 1 inch.

Figure 2.36 shows that in extreme northeast Alabama that if there is 1 inch of soil moisture storage available that in 5 of 10 years, the area can expect a minimum of 90 drought days in the 275-day period. Under the same set of conditions, farmers in south Baldwin County could expect a minimum of 100 drought days in the 275-day period. Figure 2.37 provides similar data, but with the assumption that there would be 2 inches of soil moisture storage available. This figure shows that with the increased soil moisture storage capacity available that the minimum number of drought days expected is reduced from 90 to 60 in northeast Alabama and from 100 to 60 in south Baldwin County in the 275-day period.

These authors concluded that "this study has clearly shown that drought occurrence is to be expected during the summer and fall in Alabama. As a result, there are even odds that optimum crop yields cannot be attained because of this drought hazard."

Heavy Rainfall During Late Spring and Early Summer

Heavy, prolonged rainfall and flooding in February and March can have serious repercussions to Alabama agriculture. During those months when the amount of vegetative cover is limited, the primary effect of flooding is related to soil erosion. Under those conditions, large quantities of soil may be washed from recently plowed fields, carrying with it seed, fertilizer, pesticides, and young plants. Heavy rainfall later in the year, during the growing season can also result in serious problems, but because there is so much vegetation present, the adverse effects are generally less related to soil erosion. Heavy rainfall during the growing season interferes with most farm activities that should be taking place. It is difficult to work in the fields during and immediately following heavy rains. Weeds grow rapidly, and it is difficult to control them either by cultivation or with herbicides. Undesirable insect populations and plant disease organisms numbers can increase rapidly with the added moisture and the high temperatures (Findley, 1982).

Heavy rains in late spring and early summer have long plagued Alabama agriculture. Mell (1890) reported that excessive rainfall in July and August, 1852, resulted in the rapid development of weeds in cotton fields and the rapid multiplication of injurious insects that reduced yields significantly.

More recently, from mid-June through mid-July, 1989, heavy rains fell throughout the state, resulting in flooding of low-lying fields and pastures. Significant flooding occurred along the Tallapoosa, Coosa, and Alabama rivers. Record rainfall filled Lake Martin, so that flood gates had to be opened. As a result of the long period of heavy rainfall, field work was severely limited for almost a month; hay on the ground was damaged; spraying and cultivation was delayed, resulting in increased weed and plant disease problems; fruit and vegetable quality suffered from water damage; and wheat that had not been harvested previously deteriorated badly (Alabama Agricultural Statistics Service (1989a, 1989b, 1989c, 1989d, 1989e). In other years, flooding damaged stands of corn, cotton, and soybeans so severely that replanting was necessary, cutoff dates for planting peanuts and cotton were almost missed, and field rot badly damaged Irish potato crops in south Alabama (Alabama Agricultural Statistics Service, 1983,1991a,1991b).

Data on the frequency of excessive rainfall during May, June, July, and August are presented in Table 2.17. Data in the table indicate the number of years in 20, during the period 1972-1991, in which average monthly rainfall for each of the eight regions was 8.00 inches or greater during those months. Data in the table indicate that in 5 of 20 years average rainfall in the Northern Valley Region (Nova) was 8.00 inches or greater during May, but that rain did not fall at that rate in either July or August and only once in 20 years in June. In contrast, excessive rainfall was generally much more common in the Gulf Region. During the 20-year period, that region received 8.00 inches of rain or more in May (6 years of 20) , June (3 in 20), July (6 in 20), and August (5 in 20).

The totals at the right-hand margin of the table show that the excessive rainfall was much more common in the Coastal Plains (Copl) and Gulf regions. The totals at the bottom of the table show that there was a considerable difference between the months. Excessive rainfall was much more common in May and July than in June and August.

Excessively High Summer Temperatures

Excessively high temperatures over a several-day period during the summer months always have a negative effect on crops and livestock. Hildreth, et al. (1941) commented that above the "optimum growth temperature" that rate of growth drops rapidly and plants become dwarfed. Rhoad (1941) noted that high air temperatures reduced milk production in dairy cows and negatively affected animal reproduction, especially the breeding efficiency of males. Unusually high temperatures also cause stress in animals making them more susceptible to attacks by microorganisms and parasites and physiological disorders. Extreme heat in August, 1990, resulted in higher than normal death loss in poultry in Alabama (Alabama Agricultural Statistical Service, 1990); more than 300,000 were lost in July, 1995, and more than a million were lost in early August when temperatures were extremely high for more than four weeks (Sikora, 1995b).

Summer temperatures in Alabama are relatively mild; however there are periods during June, July, and August and occasionally in September when maximum daily temperatures can be extremely high. Often these periods will last for two to three weeks. Data showing the number of times in a 20-year period in which at least one station in a region recorded a monthly average maximum temperature for June, July, August, or September of 93.0°F or higher during the period 1972-1991 is presented in Table 2.18. The maximum temperature values were taken from monthly climate data reports published by the National Climatic Data Center.

Based on the analysis of a number of years of maximum temperature data, when the average maximum temperature for a month was 93.0°F or higher, there were at least 12-15 days during that month with maximum temperatures 95°F or higher. Data in the table show that in the Northern Valley Region (Nova) during the 20-year period (1972-1991) that there were 7 of the 20 months of July, in which the average monthly temperature was 93.0°F or greater. There were 4 of 20 years in which these high temperatures were recorded in the region during August.

Variation in the totals at the side of the table indicate that there were some differences among regions. The months with the higher temperatures were a little more frequent in the Appalachian Mountains (16 of 80 months), Upper Plains (19), Prairie (16) and Coastal Plain (26) regions. During the 20-year period, there were generally fewer really hot months in the Gulf Region and more in the Coastal Plain Region. The relatively cooler temperatures in the Gulf Region are a reflection of the moderating influence of the Gulf waters on the climate of that region during the warm months.

The totals at the bottom of the table also show that more of the high temperatures were recorded in the regions during July and August, and that significantly more were recorded in July than in August.

Tropical Storms Entering the State

Only after 1900 were storms classified as hurricanes, tropical storms, or some form of tropical depression. Of the 70 of these reaching Alabama in the period 1900–2000, 13 were classified as hurricanes (wind speeds 74 mph or more), 28 were classified as tropical storms (wind speeds between 39 and 74 mph), and 29 were classified as a form of tropical depression (wind speed less than 39 mph).

Because of the geography and climate of the region, hurricanes have probably ravaged the Gulf Coast and what is now Alabama for hundreds of millions of years. Mell (1890) described hurricanes that caused great damage on the Gulf Coast in 1740, 1746, 1772, 1819, and 1852. Later, during the period 1900 through 2002, hurricane force winds reached Alabama in 1911, 1912, 1916, 1926, 1932, 1950 (Hurricane Baker), 1975 (Eloise), 1979 (Frederick), 1985 (Danny), 1992 (Andrew), 1995 (Erin), 1997 (Danny) and 1998 (Georges). Of these, Hurricane Frederick was probably the most destructive. It caused widespread destruction on Dauphin Island and in south Mobile County.

Tropical storms can bring tremendous rainfall. During July 4–11,1994, Tropical Storm Alberto moved north across the Florida Panhandle and stalled over the eastern Coastal Plain Region (Region 7) in southeast Alabama and southwest Georgia. Rainfall totaled 14.55 inches at Dothan in Houston County, 12.56 inches at Headland in Henry County, and 10.51 inches at Geneva in Geneva County. The storm increased rainfall as far west as Brewton in Escambia County (8.54 inches) and as far north as Milstead in Macon County (9.12 inches) (Alabama Agricultural Statistics Service, 1994). Rainfall was even greater in Georgia. Hurricane Frederick brought heavy rain to Baldwin and Mobile counties on September 12–13, 1979. Mobile received 8.23 inches on the day of landfall. On September 13, Dauphin Island received 8.45 inches (National Climatic Data Center, 1979).

From earliest recorded times, high winds and heavy rains associated with tropical storms periodically result in major losses to Alabama crops, livestock, timber, and agricultural infrastructure. Mell (1890) provides some of the earliest records of tropical storm activity that adversely affected agriculture. He recorded that a hurricane on September 11, 1740, washed away half of Dauphin Island and drowned 300 head of cattle. On September 18, another violent storm struck resulting in destruction of Indian plantations in the area. In 1846, a destructive storm along the Gulf Coast severely damaged plantations and destroyed the rice crop. In late August and early September, 1772, a hurricane of almost unprecedented force drove ships into the heart of the town of Mobile and pushed salt water inland, destroying vegetation and blowing the leaves off trees. In the modern era, hurricanes still cause serious losses to Alabama agriculture. On September 12, 1979, Hurricane Frederick entered southwest Alabama and moved generally north, just west of the Alabama-Mississippi line. In northeast Mississippi, winds diminished to below hurricane strength. Then the storm re-entered northwest Alabama before moving into Tennessee. In the 24 hours that the storm was active in the state, it caused an estimated $334 million in damage to agriculture (Pendergast and Freeman, 1979). In 1994, Tropical Storm Alberto caused extreme damage to crops and agricultural infrastructure in southeast Alabama (Alabama Agricultural Statistics Service, 1994), and in 1995 Hurricane Erin destroyed between 25 and 50% of the south Alabama pecan crop. Pecan growers in Baldwin County experienced the greatest losses.

3

Soils

PRECIS

A cursory look at Alabama the Beautiful is deceptive. The qualities of the state's underlying soils have not and do not bode well for highly productive, competitive agriculture. Whereas only 13% of the soils in the United States are old or ultimate soils (Ultisols), 65% of those in Alabama are. In fact, the entire Southeast is dominated by old soils that are deeply weathered because of millions of years of exposure to high rainfall and high temperatures while lying in a heavily forested landscape covered with mixtures of coniferous and deciduous trees.

As a result of the conditions under which Alabama soils were formed, many of them have problems with water, perhaps surprisingly, sometimes too much and sometimes too little. About 20% have restricted drainage problems; they are too wet. In others, water flows through the soil so rapidly that little remains for maintaining the integrity of cultivated plants such as corn, cotton, and soybeans. As a result, they are drought-prone. This problem is exacerbated by the state's often erratic rainfall. Other soil-water problems associated with the state's soils are permeability, water holding capacity, and tendency to erosion. The latter is especially true in highly cultivated crops, such as corn, cotton, and soybeans. Alabama's worn-out soils are highly acidic and low in nutrients. Consequently, they require liming and expensive fertilizers. However, even with these additives—which, of course, make them less competitive with newer, more productive soils—there are inherent obstacles. Low pH or high acidity reduces the amount of nitrogen and phosphorus available, but causes high amounts of aluminum hydroxide to be retained in the soil. Soils that have low capacities to absorb, hold, and exchange plant nutrients are not very fertile. Another negative associated with high acidity—partially the results of low organic matter and hence low nitrogen—is the adverse effects on plant's root zones caused by the prevalence of hard pans. Alabama's worn-out soils increase the amount of work and fertilizers required to farm, resulting in higher production costs for the farmer, with still uncompetitive yields.

INTRODUCTION

> The soil is the great connector of lives,
> the source and destination of all.
> It is the healer and restorer and resurrection;
> by which disease passed into
> health, age into youth, death into life.
>
> —Wendell Berry, *The Unsettling of America*

Ezra Taft Benson (1957), secretary of Agriculture in both of Eisenhower's administrations, wrote in the "Foreword" of Soil, the 1957 Yearbook of Agriculture: "All my life I have had direct experience of the importance of soil. As a boy and young man I tilled it, worked with it, and got from its bounty or, in bad years wrested from it its reluctant yield. Then later I learned to love it, respect it, and appreciate its values and limitations. I learned what every farmer knows—that each of thousands of different kinds of soils requires its own care and skillful use, which changes from season to season as conditions of moisture, temperature and crops change."

Charles E. Kellog wrote in the same volume: "Two hundred generations of men and women have given us what is in our minds about soils and soil fertility—the arts and skills and the organized body of knowledge that we now call science. Men in ancient times used many practices that we use—manuring, liming and crop rotations with legumes. In the Odyssey, Homer told how Odysseus the far-wandering was recognized in his homecoming first by his old dog 'lying on a heap of dung with which the thralls were wont to manure the land'."

To most people, soil is nothing more than dirt, and frankly, it is difficult to give a more enlightened definition, because like beauty, the definition is in the eye of the beholder. Farmers would have a definition that is different from an engineer designing a highway construction project or a parent working to remove it from a child's clothing. Soil, however, can be generally defined as the collection of living and non-living materials that form the exposed surface of the crust of the earth. Akin (1990) defined soil as "the link between the rock core of the earth and the living things at the surface." Fitzpatrick (1983) noted that soil is a space-time continuum, because it is constantly changing. Kellogg (1938) commented that soil is the natural medium for the growth of plants, and that society has its roots in soil. Regardless of how the term is defined, Alabama has lots of it, and how it got where it is and the effects that it has on our lives make a fascinating story.

INFORMATION ON ALABAMA'S SOILS

There is a wealth of information available on the soils of Alabama. Extensive and intensive work on the description and properties of Alabama's soils have been ongoing since 1902 when a soil-survey in Perry County was completed. These county surveys have been a cooperative effort by the United States Department of Agriculture and the Agricultural Experiment Station and the Cooperative Extension Service of Auburn University. From the mid-1930s, when the Soil Conservation Service was established as an agency of the Department of Agriculture, it has played a leadership role in conducting the surveys and publishing the reports. These reports provide an enormous amount of descriptive as well as quantitative information. Most of the more recent reports are available from the United States Government Printing Office in Washington. The Scientists and Extension Specialists of the Department of Agronomy and Soils of the College of Agriculture at Auburn University have also published a wealth of material on Alabama's soils. Without question, there is more information available on Alabama's soils than on any of its other natural resources.

Some Aspects of Soil Science

The blanket of dirt that hangs over the rock base of our land forms, is a truly unique structure, and the story of how it came to be there is even more remarkable. The most remarkable story of all, however, is how important this dirty blanket has been to our progress from hunting-gathering to industrial societies. Further, it holds the key to what we will become tomorrow.

Soil Formation

Weathering of rocks provides the parent material for soil formation. Continued weathering of the parent materials results in the formation of the soils (Byers, 1938). Rocks are simple to complex chemical compounds that contain greater or lesser quantities of all of the known elements (Byers, et al., 1938). For example, the mineral quartz, a product of the weathering of granite, is a common parent material of sand in soil; it is a simple oxide of silicon or silicon dioxide (SiO_2). However minerals such as feldspar, hornblende, pyroxene, and mica that are also the products of the weathering of granite, are complex compounds of silicon plus aluminum, oxygen, iron, calcium, magnesium, potassium, sodium, and trace amounts of other elements. Feldspar, which is the parent material of many clay minerals in soils, is a complex compound of potassium, aluminum, silicon, and oxygen.

Soil formation is a matter of geological stability. A rock is a less stable geologic unit in a given physical and chemical environment than a soil; consequently, the weathering of the rock progresses until a more stable form is reached. For example, a large piece of granite located in the humid tropics of Africa would be geologically unstable, but an Oxisol (Ferrasol)—a common group of soils of the tropics—formed from that rock in that environment would have a higher degree of stability. Therefore when weathering reached that stage relatively little further development would be expected. However, if the environment changed appreciably, a relatively stable soil could again become unstable and further weathering could take place.

There are five general factors that determine the rate in which the physical and chemical weathering processes proceed and soil is formed. These are: parent material, climate and weather, biological activity, relief, and time.

Parent Material

The nature of the rocks that give rise to soil is the least important of the factors that determine development. Byers, et al.(1938) note that a rock (parent material) that contains all of the chemicals required for optimum plant growth may in the process of development become a poor soil for agriculture. Similarly, under the right conditions of development, parent material poor in the chemicals required for plants may develop into a good soil.

Parent material can have direct as well as indirect effects on the kind of soil produced and the rate at which it is produced. Soils being developed from quartz rocks are affected much more by the long-term effect of weathering by water than soils being developed from shales and slates. Water does not pass readily through these rocks metamorphosed from silt. Also rocks containing more lime are more resistant to the effects of weathering by water. Generally, soils developed from rocks containing limestone will develop into better soils for agricultural crops and grasses than those developed from the more acid shales, sandstone, and other more acidic rocks. The effect of parent material on the soil formation process is much more important on younger, less completely developed soils. In the longer term, soil development processes overcome most beneficial or detrimental effects of the nature of the parent material.

Climate and Weather

The most important direct effect of climate and weather in the formation of soil is on the weathering of rocks and the alteration of parent material. Daily heating and cooling causes rocks to expand and contract, and because rocks are not good conductors of heat, the exterior becomes hotter than the interior resulting in considerable tension within the structure. Further, different minerals in the rock have different expansion and contraction characteristics. Over time, the daily changes in tension results in the flaking off of tiny pieces of rock and finally the entire structure slowly crumbles. Alternate wetting and drying associated with rainfall also result in the weathering of rocks that lead to soil formation. Carbon dioxide, dissolved in rainfall forms a weak acid that assists in the chemical weathering of rocks. Soluble materials are dissolved (hydrolysis), and new chemical compounds are formed that occupy more space in the structure, resulting in increased tension and ultimately, in the breaking apart of the rock. Freezing and thawing exert tremendous pressure on rocks over time. Water entering cracks in a rock and then freezing can quickly split even the largest rocks.

Weather also affects the weathering of rock and the formation of soil indirectly. Rainfall collected in creeks and rivers, carry sand that grinds away rocks along the stream channel. Wind driven waves along the shore also carry sand to grind away rock surfaces. Wind carrying sand across the surface of the earth blasts the surface of rocks in its path. Broken rock and sand carried in slow moving glaciers are ground into smaller and smaller fragments as the enormous mass moves slowly along. Rocks in the path of the glacier are also broken and carried along to be further reduced in size. Floods carry partially weathered minerals to new locations, often far away from the location where primary weathering occurred, where soil formation continues under different environmental conditions. Avalanches associated with snowfall can move large quantities of partially weathered materials to new locations while contributing to the weathering or breaking apart those materials in the process. Erosion by water and wind carries partially weathered materials away from the surface of a weathering rock, exposing it anew to the full effects of weather.

Biological Activity

Organic matter resulting from the activities of living organisms is an essential component of soils. Byers, et al. (1938) suggest that there really is no soil without organic matter. Living things contribute organic matter to soil, and they also contribute significantly to the formation of soil both directly and indirectly. The roots of plants grow into cracks in rocks, and as they grow they increase the mechanical tension on the crack, ultimately breaking it apart. Burrowing mammals bring partially weathered material to the surface exposing it to the full force of climate. Water enters the burrows of animals undermining rocks causing them to topple. Earthworms contribute significantly to the final stages of soil development by mixing different weathered components into a more homogeneous mass. Humans have also had some effect on soil formation since they have been on the scene. Plowing of soil has contributed to its continued development. In many cases, changing or removing the vegetative cover over developing soils has also contributed to rapid changes.

Living things probably exert their most important effects on soil development indirectly. The death and decay of living things contributes enormous quantities of organic matter to the weathered rock that makes up most of the total soil mass and is a major factor contributing to soil fertility. Also, as organic matter decays, organic acids are formed that speed up the weathering process.

Relief

On steep slopes, geological stability is extremely low. Products of rock weathering are quickly eroded away, either by water or wind, continually exposing the remaining rock to the full forces of the weathering process. On flatter surfaces, these products tend to remain in place, insulating the rock to a degree from the weathering forces. Also on steep slopes, the potential energy associated with a rock perched there is extremely high. Often, relatively little force will send it careening downhill to be broken into smaller pieces as it bounces along down-slope. As it falls, it also contributes to the cracking and breaking of rocks in its path. A steep slope, or high relief, maximizes the effect of gravity as a soil development agent, but at the same time it minimizes the effect of water. Water from rainfall passes quickly down-slope, minimizing infiltration and solution. On relatively flat surfaces where water tends to accumulate and the water table remains high for much of the year, soil development is profoundly affected. Under those conditions, the effects of hydrolysis and solution are unrelenting.

Time

An enormous amount of time must pass before a large rock is completely weathered into an agricultural soil. FitzPatrick (1983) suggests that the complete physical and chemical weathering of a rock to form a Ferrasol (Oxisol), a common group of soils found in the tropics and subtropics, would require more than a million years. Ferrasols are some of our most completely weathered soils. They have a high degree of geological stability; very little further development is likely to occur. He further notes that most of the soils of the world began their development within the last 100 million years.

Soil development is a slow process, but there are differences in the rate at which the process proceeds. Soil development is much more rapid in subtropical, humid areas where there is heavy vegetative cover than in areas with a cool, dry climate and grassland vegetative cover. All of the forces of physical and chemical weathering operate at a maximum rate in the subtropical region.

Soil Profile

Physical and chemical weathering of a large rock, exposed on a relatively level area at the surface of the earth, would proceed slowly downward into its structure over time. As weathering proceeded, geological stability would increase at the surface, but would remain high in the subsurface. After a half-million years of weathering, the material at the surface would be quite different from that several feet below the surface. If the weathering rock were located in an area with forest vegetation, there would most likely be a layer of fresh and decaying leaves and twigs at the surface. Under this top layer, there would be an accumulation of insoluble sand and silt, plus some completely decomposed organic matter, remnants of the weathering process. Smaller particles such as clay minerals and aluminum and iron compounds would have washed out of this layer downward into the layer below. These smaller particles would accumulate in a third layer, just above an accumulation of relatively well weathered, but generally stable parent material. All of these layers would be resting on a base of solid rock. A trench with vertical sides dug into a soil from the surface down to the un-weathered rock would reveal the series of layers of the products of the weathering process. Soil scientists call this the soil profile, and the layers are called horizons. All of the horizons above the bedrock, or un-weathered rock, form the solum or true soil.

A solum is a relatively specific manifestation of the weathering process of a particular parent material under a rather specific set of environmental conditions. Weathering of the same type of parent material under different environmental conditions would result in the development of a different solum.

The Soil Survey Staff (1975) of the United States Department of Agriculture has developed a system for describing the different horizons of the solum, so that identification and comparison of soils can be facilitated. The different horizons are identified by capital letters: O, E, A, B, and C. The capital letter R is used to designate that portion of the profile that is not part of the solum, or true soil, but is bedrock. Horizon properties and characteristics related to the letters are as follows:

O Horizon

Surface layer of the profile, containing 20%, or more, of fresh or decomposing organic matter (Figure 3.1).

A Horizon

Horizon containing primarily mineral soil and some decomposed organic matter that has filtered downward from the O horizon above. Plant roots, insects, small animals, bacteria, and fungi are most concentrated in this zone. Sand and silt may accumulate following the loss through leaching of clay minerals and aluminum and iron compounds formed in the weathering process to the B Horizon below. Decomposed organic matter coats the mineral particles, giving this horizon a darker color than the horizon below.

E Horizon

The primary feature of this mineral soil horizon is the loss of silicate clay particles such a kaolin, and iron or aluminum compounds, leaving an accumulation of larger, resistant sand and silt particles.

B Horizon

This horizon is usually called the subsoil, and it contains only products of completely weathered parent material. Clay minerals, compounds of aluminum and iron, and humus predominate. The concentration of organic matter is limited. No trace of the parent material remains.

C Horizon

Weathered parent material, but does not include solid, un-weathered bedrock. This horizon is not considered to be part of the true soil.

R Horizon

Underlying, non-weathered bedrock, such as granite, sandstone, or limestone. Presumed to be the parent rock from which the overlying material, with the exception of the organic matter, was formed.

In the USDA system, descriptions of the horizons may be expanded by the use of Arabic numeral subscripts attached to the capital letters. For example, O_1 would be used to describe a situation in which the original form of the organic matter in the O horizon would be visible to the naked eye, and O_2 would describe a situation in which the original form of the organic matter could not be identified in that manner.

Small letters are also attached to the capital letters to provide more specific information about a horizon. For example, A_p indicates that portion of the A horizon affected by plowing. B_t indicates a situation where there is an accumulation of trans-located silicate clay in the B horizon (subsoil).

Table 3.1. gives a description of an actual soil found on a farm in Marshall County in the northeastern area of the state. This particular soil belongs to the Hartsells Series of the Ultisol Order. This soil is widely distributed in the Appalachian Plateau of Alabama, as well as in Georgia, Kentucky, and Tennessee. The name of the series is taken from the name Hartselle, a town in Morgan County. Because of space limitations, only a limited amount of information about each horizon is included.

In the table, the term A_p indicates that this horizon reaches to plow depth. The term B_A idicates that this horizon has some characteristics usually found in both A and B horizons. The terms B_{t1} and B_{t2} indicate that the B horizon contains translocated silicate clay, and that because of some other characteristic, it should be divided into two parts. The term BC indicates that this horizon has characteristics normally associated with both B and C horizons. The term R indicates that this horizon is bed-rock, in this case, an acid sandstone. This soil was weathered from the sandstone of the Pottsville Formation found in the Warrior Basin District of the Cumberland Plateau Physiographic Section (Figure 1.11).

The concept of the soil profile presents a somewhat misleading picture of the nature of soils. The profile represents a soil as a two-dimensional entity, when in fact it is three-dimensional. Theoretically, a large rock of more or less homogeneous composition weathering in place over a long period of time would be developed into a large, three-dimensional mass of soil. Vertical sections taken anywhere in the mass should reveal relatively similar soil profiles.

Actually soils have a fourth dimension—time. Soils are constantly changing as they weather toward geological stability. There are always changes taking place that affect the structure and function. Every day there is heating and cooling. Frequently there is wetting and drying, freezing and thawing. Erosion is a constant fact of life for a soil. In many situations, erosion removes the O and A horizons almost as quickly as they form.

The Aging of Soils

As weathering of a rock mass proceeds, the weathering of primary by-products also proceeds. For example, feldspar is one of the primary by-products of the weathering of granite (an igneous rock). Then with the weathering of feldspar, secondary minerals such as kaolinite, appear in the soil. Then as the process continues even the secondary minerals are further broken down into more resistant forms. In general terms, as a soil ages, there is a loss of more unstable secondary by-products that results in the reduction in natural fertility and potential productivity.

Akin (1991) divided the sequence of events of soil aging into four stages: young soils, soils in early maturity, soils in late maturity, and old soils. Some of the characteristics of these different stages are as follows:

Young Soils

In young soils, many primary minerals inherited from the parent material are still present.

Soils in Early Maturity

Clay-sized, primary mineral particles, such as hornblende, biotite, and plagioclase feldspar, containing calcium, magnesium, and iron, have disappeared from the A horizon of the soil. Silicate clays are forming and accumulating in the B (subsoil) horizon.

Soils in Late Maturity

Most primary minerals have disappeared from the clay and silt fraction of the soil. More complex silicate clays are changing into kaolinite. Iron oxide and aluminum oxide compounds are increasing.

Old Soils

The clay and silt of the A horizon have almost all leached away, leaving mostly quartz sand. Kaolinite and oxide clays are accumulating in the B (subsoil) horizon. Concentrations of the oxides of aluminum and iron continue to increase.

Soil Taxonomy

Taxonomy is the systematic assignment of things to classes, according to natural relationships. The objective of soil taxonomy is to have hierarchies of classes that help us to understand the relationship between the classes and the factors responsible for their character (Soil Survey Staff, 1975). Scientists have been interested in the classification of soils for many years, and there are numerous systems currently in use around the world. The system described in the following section was developed by the Soil Survey Staff (1975) of the United States Department of Agriculture.

The USDA system divides the world's soils using the following scheme: orders, suborders, great groups, families, soil series, and soil phase. Only three of these classifications (order, soil series, and soil phase) will be of interest to us.

Orders

Soils are grouped into eleven different orders based on those aspects of morphology thought to be related to the way in which they were formed (soil genesis). The distribution of these orders within the continental United States, in Alaska, and in the Hawaiian Islands is shown in Figure 3.2. The figure shows that the Order Ultisol is widely distributed in the Southeastern United States and especially in Alabama. The soils in this particular order are very old in terms of the weathering process. Essentially, they have reached the ultimate or last stages of soil formation or geological stability. These soils have reached Akin's old soils stage of development, described in a preceding section. The Soil Survey Staff (1975) uses the Latin term *sol* in the names of all of the orders. The formative element *Ult* in the name is from the Latin term *ultimus* for last. Some characteristics of the different orders are as follow:

Alfisols

(Alf from Al and Fe = chemical symbols for Aluminum and Iron)—soils in which silicate clays have been trans-located to form an argillic (clay) layer in the Bt (subsoil) horizon, but leaching has not removed all of the bases. Water is generally available for plant growth for most of the season, but these soils can be hard and massive when dry. There are often problems with tilth (physical condition of the soil). Thirteen percent of the soils in the United States and 4% in Alabama are in this order.

Entisols

(Ent from recent = new)—soils are very young or have not weathered to the point that horizons have developed. Occasionally there is a marginal A horizon. Entisols often form in recent river floodplain deposits and in cover sands. Eight percent of the soils in the United States and 7% in Alabama are in this order.

Histosols

(Histo from tissue = soil developing from decaying plant tissue)—soils composed primarily of accumulated decaying organic matter generally saturated with water. Less than 1% of the soils in the United States and in Alabama are in this order.

Inceptisols

(Incept from beginning = young soils)—young soils in which only the more rapidly formed horizons are developed. Horizon development usually has been retarded because of water-logging or lack of time. In some cases, the soil is so young that weatherable minerals are still present. Often there are rocks at shallow depths. Eighteen percent of the soils in the United States and 7% in Alabama are in this order.

Mollisols

(Moll from soft = high surface accumulation of organic matter)—soils with relatively thick, dark colored, humus-rich, surface horizons. Bivalent cations (Calcium and Magnesium) are dominant. Mollisols are the primary soils in the Upper Midwest where so much of the nation's corn and soybeans are produced. They are among the most productive soils in the world. Twenty-five percent of the soils in the United States but only 7% in Alabama are Mollisols.

Oxisols

(Oxi from oxidized = highly oxidized soil)—soils that are excessively weathered. Few of the original minerals remain un-weathered. Iron and aluminum oxide clays predominate. Less than 1% of the soils in the United States are Oxisols. There is only a very small area of Oxisols in Alabama.

Spodosols

(Spod from wood ashes = refers to light gray color of the surface or A2-horizon)—soils with a sandy, coarse-loamy or coarse-silty (Spodic) horizon of trans-located humus combined with aluminum. Only 5% of the soils in the United States but only 1% of the soils in Alabama are in this order.

Ultisols

(Ult from ultimate = long, intensive weathering)—deeply weathered, strongly acidic, primarily red and yellow soils indicating the accumulation of iron oxide clay minerals. Considerable amounts of trans-located silicate clays have accumulated in an argillic horizon in the subsoil (Bt horizon). Kaolin and gibbsite (silicon poor clay minerals) are common in the argillic horizon. Thirteen percent of the soils in the United States and 65% in Alabama are in this order.

Vertisols

Vert from turn = soil at surface falls into large cracks)—these are the so-called self-swallowing soils. Because of the high content of clay that swells when wet, large cracks develop in the soil when it dries. Surface materials fall into these cracks, so that over a period of time, the soil undergoes a churning process. Montmorillionite is the dominant clay mineral in these soils. Only 1% of the soils in the United States are Vertisols. In Alabama, 4% of the soils are in this order.

Series

Series names tell nothing about the soil. They are generally place names associated with the location where that specific soil was first identified. For example, the Pacolet Soil, or Pacolet Series, is an important soil in the Piedmont Upland Section of east-central Alabama, but it was first identified and described near the town of Pacolet in eastern Spartanburg County in South Carolina's Piedmont Upland.

George Martin of the National Resource Conservation Service of the United States Department of Agriculture described a soil located near the Alaga Community in Houston County as follows: "Grayish brown loamy sand surface layer, overlying a brown and yellowish brown loamy sand substratum to a depth of 80 inches or more." Soils with these same characteristics are found in several locations in southeast Alabama. Because this particular set of characteristics was first described near Alaga, Alabama, this soil and others like it were given the name Alaga. Similar soils in other locations are said to belong to the Alaga Series.

Nearby, in Covington County, a different soil was identified with the following characteristics: "Brown sandy loam layer overlying a red to dark red sandy subsoil." These characteristics had already been described in a soil near Orangeburg, South Carolina; consequently, the Covington County soil was given that name. A portion of the soil blanket near the village of Dewey in northeastern Cherokee County, Alabama, was described as a "Reddish brown loamy sand surface layer overlying a yellowish red silty clay loam and dark red clay subsoil." This soil was given the name Dewey,

Phases

A soil series often occurs over a wide geographical area within a county or even over several states. Within this relatively large area, certain characteristics of the series vary. This range can have significant management implications. When this situation exists the series is subdivided into soil phases. For example, a soil series can occur over a range of slopes, and the slope of the soil at a given location is often an important consideration in its management; consequently, the series is subdivided into phases representing different segments of its total range. For example, the Pacolet Soil Series is widely distributed in Lee County, and the series occurs there in three phases: Pacolet Sandy Loam, 1 to 6% slopes; Pacolet Sandy Loam, 6 to 10% slopes; and Pacolet Sandy Loam, 10 to 15% slopes.

Soils of Alabama

A Scottsboro mule trader once remarked to a prospective buyer that the animal that they were inspecting, "It really is not as old as it looks. It just looks that way because it has been tied out in lots of bad weather." His statement is an appropriate characterization of Alabama's soils. Our soils appear to be old because they have been exposed to lots of bad weather. As noted in a preceding section, the weathering of rock to form soil is primarily a physical-chemical process, and the rate at which it proceeds is dependent to a great extent on moisture and temperature. This process occurs much more rapidly in the warm, humid subtropical climate in Alabama than, for example, in the cool, drier, temperate climate of the Upper Midwest. Soils that are essentially the same chronological age would appear to be much older or much more weathered in Alabama than in South Dakota. Also, the larger quantity of organic acids formed during the decay of the leaves from deciduous forest vegetation would play a much larger role in soil formation in Alabama than the acids formed in the natural grasslands, such as those found in the Upper Midwest.

Nine of the 11 soil orders found throughout the world are represented in the soils of Alabama. The estimated percentages of each of the nine orders in seven soil areas of the state (Figure 3.3) are presented in Table 3.2. These are the soil areas described by Mitchell and Meetze (1994). The percentages were provided by George Martin, Soil Data Quality Specialist, USDA-NRCS (personal communication).

From Table 3.2, it is apparent that from the perspective of the percentages of the soil areas involved that the Ultisols are by far the most prevalent in the Alabama (Figure 3.4). The Inceptisols and Entisols are much less common, but generally much more widespread than the Alfisols. While the Vertisols and Histosols are not widespread across the state, they, never-the-less, are important in those soil areas where they are found. The Vertisols are common in the Black Prairie, and the Histisols are common in the Coastal Marshes and Beaches area. The Mollisols and Oxisols have only a very limited distribution in Alabama. Oxisols have been mapped in only a very small area in Lee County (Piedmont Upland), and Mollisols have been mapped in only a small area in Greene County (Black Prairie Physiographic District) (George Martin, personal communication).

Most of Alabama's soils (65%) are included in the Order Ultisol, or Ultimate soils, indicating a stage of advanced maturity in the soil genesis or development process. Ultisols have been described as soils that are strongly weathered and leached, and therefore nutrient-poor and acidic. Only the Oxisols are more heavily weathered. Ultisols are found in heavily forested areas, and in areas with warm climates that are moist all year. They are found throughout the Southeastern United States, as far west as eastern Texas and into the Middle-Atlantic States as far east as Virginia and Maryland (Soil Survey Staff, 1975). Ultisols are also found in parts of South America, Africa, and Asia, where there are similar climates (Akin, 1990). The Ultisols worldwide were derived from a variety of parent materials, but because of the overriding effect of a common climate operating over a long period of time, the soils have similar characteristics. This distribution of the order in Alabama (Figure 3.4) is not as homogenous as it appears. There are small areas of soils representing other orders located in parts of the state, but they are too restricted to be shown on this map.

Based on the soil surveys conducted by the Soil Conservation Service of the United States Department of Agriculture and cooperating state agencies, Alabama's soils are divided into seven major areas (Mitchell and Meetze, 1994). These include the following: limestone valleys and uplands, Appalachian Plateau, Piedmont Plateau, prairies, coastal plain, major flood plains and terraces, and coastal marshes and beaches.

These general locations of these soil areas are shown in Figure 3.3 A-G. Obviously, this classification is based on the location of Alabama's major physiographic sections and districts that were described in a Chapter 2. This classification is based on the premise that soil parent materials are likely to be different in the different physiographic areas and that, as a consequence, the soils in those areas are likely to be more similar to each other than they are to soils in other areas with different parent materials. A consideration of some of the characteristics of the soils in these different areas follows:

Limestone Valleys and Uplands

The Limestone Valleys and Uplands Soils area (Figure 3.3 A) includes two different physiographic sections—the Alabama Valley and Ridge and Highland Rim (Figure 1.9). The soils in the river valleys (Tennessee and Coosa) of this area were derived from weathered limestone; while those of the uplands were derived primarily from cherty limestone (Mitchell and Meetze, 1994). The parent material of this limestone was deposited out of warm, calm, clear seas when small organisms whose bodies contained calcium carbonate died and settled to the bottom (Levin, 1988). Later, the weight and pressure of the thick accumulations pressed the lower portion into limestone. The limestones in the Tennessee Valley was deposited during the Mississippian Period of the Paleozoic era some 350 million years ago (mya). The limestones in the valleys of the Valley and Ridge Section, such as the Coosa Valley, are much older. They were laid down during the Upper Cambrian and Lower Ordovician Periods of the Paleozoic era, some 525 to 470 mya (Table 1.1). These limestone valleys are level to gently rolling with some flat-topped hills. Decatur and Dewey Series Soils are extensive in the valleys, but Bodine and Fullerton are more common in the uplands.

The surface layers of these soils are a mixture primarily of silt and sand and are predominantly red in color. Subsoils are mostly clay. Ponding of water and internal drainage following moderate rainfall are problems as a result of the generally flatness of the fields in this area. Gully erosion is not a major problem. The soils have become moderate to strongly acidic as a result of intensive weathering and cropping. There is little accumulation of organic matter and natural fertility is highly variable depending on the parent material. (Soil Survey Division, 1938 and Pearson and Ensminger, 1957).

Appalachian Plateau

The soils of the Appalachian Plateau area (Figure 3.3 B) are primarily found in the Appalachian (Cumberland) Plateau Physiographic Section (Figure 1.9). These soils were derived from inter-bedded sandstone or shale (Mitchell and Meetze, 1994). The sand and mud accumulated there in shallow seas some 300 mya when highlands far to the west were being weathered and eroded and carried to the continental margin by swift streams (Levin, 1988). This ancient continental margin is now the Appalachian Plateau. In most areas, weathered sandstone lies only 2.5 to 3.5 feet below the surface. The soils occur primarily on a series of flat-topped ridges or plateaus. The soils are mostly fine sandy loams and silt loams. Montevallo and Townley Series Soils are extensive in this region.

Those soils developed from the shales tend to be finer, more susceptible to erosion and cause problems related to water infiltration than those developed from sandstone. Crust formation following rainfall and drying winds is a problem in soils developed from either sandstone or shale. Regardless of the parent material, they are generally acidic in reaction and low in mineral plant nutrients, and with a mixture of sand and silt at the surface and in the subsurface. (Pearson and Ensminger, 1957).

Piedmont Plateau

The Piedmont Plateau Soils area (Figure 3.3 C) essentially corresponds to the Piedmont Upland Physiographic Section (Figure 1.9) described in a preceding section. The soils in this area were derived from crystaline rocks such as the igneous rock (granite) and the metamorphic rock (micaschist). Consequently, they contain a considerable amount of quartz. These rocks are thought to have been put in place during the Carboniferous-Permian Periods (Paleozoic Era) some 365 to 240 mya (Table 1.1) as a result of tectonic plate activity when the supercontinent Gondwanaland was approaching its collision with Laurussia, the supercontinent containing the tectonic plate carrying ancient North America (Williams and Hatcher, 1983). Scattered over the surface locally are fragments of quartz and outcrops of granite. The soils are a mixture of sand, silt, and clay at the surface, but have a red clayey subsoil. Soils of the Madison, Pacolet, and Cecil Series are extensive in this area (Mitchell and Meetze, 1994).

Because of the generally hilly topography of the entire section (Figure 1.8), erosion has removed the sandy surface from the soil over most of the area. In fact, erosion has been so severe over much of the area that the original surface layer of the soil has eroded away, and the subsoil is being cultivated where farming continues. Both sheet and gully erosion have been active. Because of the accumulation of clay in the subsoil, the workability of the soil is impaired and the infiltration of water from rainfall is limited (Pearson and Ensminger, 1957).

Coastal Plain

The soils of the Coastal Plain area (Figure 3.3 D) are found in the Coastal Plain Physiographic Section (Figure 1.9). They are widely distributed from northwest Alabama to the Georgia line in the east-central part of the state and southward to the Florida line and the Gulf of Mexico. The soils occur on a widely varying topography ranging from the high hills with steep slopes found in the Fall Line Hills District in Franklin County to the relatively flat terrain of the lower portion of the Doherty Plain District in Houston County (Figure 1.8). The soils in the upper part of this large area, or Upper Coastal Plain (north of the Black Prairie District), generally were derived from sediments deposited there by water flowing from highlands to the north and east and probably the west as well (Savrda, personal communication). These have mixtures of mostly sand and silt at the surface. The subsurface also contains sand and silt, but tends to include a higher amount of clay. Smithdale, Luverne, and Savannah Series Soils are extensive in this area.

The soils in the eastern part of Coastal Plain Soil area, or those in the eastern portion of the Lower Coastal Plain (south of the Black Prairie), in east-central and southeast Alabama were derived primarily from materials settling out of brackish-water and shallow marine bays that once covered that area. Sand and silt are predominant in the surface layer of these soils as well as the subsurface, although there tends to be an increase in the amount of clay in the subsoil. Dothan and Orangeburg Soils are common here.

Soils of the western part of the Lower Coastal Plain in southwest Alabama were also derived from shallow marine and brackish-water sediments. Because of excessive leaching, these soils developed from marine sands and clays and under forest vegetation are relatively sandy both at the surface and and the subsurface. Also, they are acid in reaction and low in natural fertility and organic matter. Soils in the southern part of this area tend to contain more sand at the surface, whereas soils in the north contain somewhat more clay. Generally the soils with more clay in the north contain more organic matter and more mineral nutrients than the sandy soils in the south. At least 20% of the soils in this area have moderate to severe erosion hazards, because of composition and the rolling, hilly topography. (Pearson and Ensminger, 1957). Smithdale and Troup Soils are extensive here.

Blackland Prairie

The soils of the Prairies area (Figure 3.3 E) are found in the Black Prairie Physiographic District of the East Gulf Physiographic Section (Figure 1.9). They were derived from chalk and clay that settled to the bottom of a warm, clear, calm sea that once covered that area during the Upper Cretaceous Period of the Mesozoic Era, some 70 mya (Levin, 1988). These soils generally have a predominance of clay both at the surface and in the subsurface. Sumter Soil is common in the western, or alkaline soils portion, of this area, but Oktibbeha is more common in the eastern, acid soils portion. Wilcox, Mayhew, and Eutaw Soils are dominant on the rolling hills along the southern boundary of this area with the Coastal Plain. The land is gently rolling (Mitchell and Meetze, 1994).

Because of the physical characteristics of the clay in the Prairies soils, their natural fertility was originally relatively high compared to other Coastal Plain Soils, but deficiencies have developed as a result of cropping over a relatively long period of time.

In soils of the Prairies Soils area, erosion is not as important as in some of the other areas. These soils (Vertisols) become extremely sticky when wet and large cracks form when they are dry; consequently, field preparation in the early spring can be a problem (Pearson and Ensminger, 1957).

Major Flood Plains and Terraces

Soils of the Major Flood Plains and Terraces (Figure 3.3 F) were derived from sediments deposited during the last two million years (Quaternary Period, Cenozoic Era) from rivers flowing generally northeast to southwest across much of the state (Mitchell and Meetze, 1994). Most of these materials had been weathered extensively before being transported there by the water. The surface layers of these soils are generally a mixture of sand and silt with a small amount of clay. Subsurface layers contain mostly clay with smaller amounts of sand and silt. The Cahaba, Myatt, and Urbo Soil Series are dominant soils in this areas.

Level terraces on the higher ground in these areas can be cultivated, but the lower areas are generally in the flood plains of streams and are in hardwood forests. The higher terraces were once used extensively by Native Americans for crop production.

Coastal Marshes and Beaches

The marsh soils found in the Coastal Marshes and Beaches area (Figure 3.3 G) were derived primarily from decaying marine or brackish-water plants and generally consist of muck at the surface and in the subsurface. The beach soils were derived from sandy marine sediments and consist primarily of sand at the surface and in the subsurface. Dorovan, Lafitte, Axis, and Osier are important soil series in this area. These soils are deep and poorly drained (Mitchell and Meetze, 1994).

Properties of Some Alabama Soils

Properties of soils (drainage, permeability, available water capacity, etc.) are the result of the formation process and are extremely important considerations in their use and management. Alabama farmers, first Native American and then Euro- and Afro-Americans, have been working with and around soil properties on a daily basis for at least a thousand years.These properties, along with climate and weather, have played a major role in the evolution of Alabama's agriculture to this point, and they will continue to determine how successful it will be in the future. There are a large number of different kinds of soils (soil series) in Alabama. It is beyond the scope of this book to consider the properties of all of them, but it is worthwhile to consider in detail some of the properties of at least a few of them.

As a point of reference, I have also included some data on the Nicollet Soil Series found in north-central Iowa (Humboldt County). It is one of the more productive soils in Iowa. Fenton and Miller (1977) comment that "Iowa's soils ('Black Gold') comprise the largest concentration of prime farmland in the world."

Expected yields of corn and soybeans on the Nicollet Soil are around 153 and 49 bushels an acre, respectively. In another soil series, Tama, found in eastern Iowa, expected yields for corn and soybeans are 170 and 57 bushels an acre, respectively. The Nicollet Series was first described in Nicollet County, Minnesota, in 1949. The series is extensive in south-central Minnesota and north-central Iowa.

Alabama Soil Surveys provide a wealth of data and information on the characteristics and properties of all of the recognized soils in each county. However, for our purposes and because of space limitations, it is not possible to consider more than a small portion of what is available in each of those publications. Information on the general characteristics of the soils chosen for consideration is presented in the following sections:

Names and Locations

The series names of the 14 soils (soil series) chosen for comparison are shown in Table 3.3—along with the county where the soil being considered is located, the acreage of that soil in the county, the percentage of the total acreage of soils in the county represented by that particular soil, and the reference to the soil survey publication describing it.

Figure 3.5 shows the general location within Alabama where 13 of the soils are found. Information on the location of the soils was provided by George Martin (personal communication). The maps generally indicate the counties where the largest acreage of each of the soils is found. Because a particular county is completely shaded on a map does not mean that that soil is found in all parts of that county. Rather, the shading means that some of that soil is found there. Because of space limitations, a map for the Leon soil is not included; this soil is found in only a small area along the coast in Baldwin and Mobile counties.

Taxonomy

Table 3.4 gives part of the taxonomy (subgroup, great group, suborder, and order) for each of the 14 soils. As noted in a preceding section, discussions will be primarily limited to information about the orders, soil series, and soil phases. However, some additional taxonomic information is shown in the table. Names of the respective subgroups, great groups, and suborders are also included. These names provide valuable additional data about each of the soils.

The 14 Soil Series represent eight of the 11 orders (Alfisols, Entisols, Histosols, Inceptisols, Mollisols, Spodosols, Ultisols, and Vertisols). Because Oxisols are so limited in Alabama, no soils from that order are included. Ten of the most common of the 47 suborders (Aqualfs, Aquepts, Aquods, Aquults, Ochrepts, Psamments, Saprists, Udalfs, Uderts, and Udults) found in Alabama are also represented. Because the Ultisol Order is so common in Alabama, five of the soils from this group are included, one from each of the five physiographic sections. Similarly, because the Udult Suborder is so common, four of the soils are from this group.

Several of the soil series to be considered are found on a range of different slopes; consequently they have been subdivided into soil phases. For example, the Pacolet Soil Series as mapped in Lee County is divided into three soil phases based on the slopes where they are found. These are: Pacolet sandy loam on 1 to 6% slopes, Pacolet sandy loam on 6 to 10% slopes, and Pacolet sandy loam on 10 to 15% slopes. If a specific soil series has been subdivided into different soil phases, only one of them will be considered. The soil series and soil phases to be considered are listed in Table 3.5.

The general location of each of the 13 soils in a representative county is given in Table 3.6. The table includes the general location of each soil in only one county in the state; however most of the soils have a much wider distribution. For example, the Orangeburg Soil is widely distributed throughout Alabama's Coastal Plain, and in the Coastal Plain Provinces along the Gulf Coast and Eastern Seaboard. The characteristics of the soil would be generally the same at all locations. Marine sediments that provided the parent material for the Orangeburg Series are widely distributed in the Southeast and in the Atlantic Coastal States.

The Pottsville Formation—consisting of inter-bedded sandstone, siltstone, shale, conglomerate, and coal—is widely distributed throughout the Appalachian Plateau Province in north-central Alabama (Osborne, et al.,1989). This formation provided the parent material for the Montevallo Series; consequently, this soil is widely distributed through that entire area. At the other extreme with respect to location and distribution, the Lafitte Series is restricted primarily to the marshes along Mobile Bay. Its parent material, decaying herbaceous plant material of brackish water bays, is confined to that limited area of the state.

Parent Materials of the Soils

After climate, parent material probably has the greatest effect on the formation of soils. Alabama's soils, as shown in Table 3.7, were formed and continue to be formed from a wide variety of materials. For example, the Auburn Gneiss (Opelika Complex) in Lee County has been in place since at least the Paleozoic Era some 500 mya (Osborne, et al., 1989). This very old material is extremely resistant to weathering and to soil formation. It is one of the parent materials of the Pacolet Soil. At the other end of the spectrum, the Una Soil found in depressions and sloughs along the Alabama River in Monroe County (Alluvial-deltaic Plain Physiographic District) was formed and continues to be formed in alluvium or water transported sediments that began to accumulate during Pleistocene Epoch (Quaternary Period, Cenozoic Era) some two mya (Table 1.1). These sediments had already been weathered to a considerable degree before being carried by water to the flood plain of the lower Alabama River. The names of the geological formations are those of Raymond, et al. (1988).

Descriptive Characteristics

Some descriptive characteristics of the 14 soils follow. It is beyond the scope of this book to present more than a minimal amount of information about each of them. If more information is needed, the references cited in Table 3.3 should be consulted. These soil survey publications contain a wealth of information about each soil.

Soil Profiles

Soil profiles provide important information about properties used in the classification of soils and characteristics that play a number of roles in the dynamics of soils and their use. Information on the profiles of the 14 soils in question are given in Table 3.8.

Mineralogy Class

Mineralogy class is an important property of soils. It relates to the approximate mineralogical composition of selected size fractions and provides important information on the natural fertility of soils. For example, the natural fertility of soils with high levels of kaolintic minerals is much lower than those soils with high levels of montmorillionitic minerals. In the soils under consideration, there are four mineralogy classes: siliceous, kaolinitic, smectitic, and mixed (Table 3.9). The definitions of these various classes are as follows:

1. Siliceous—contains more than 90% (by weight) of silica minerals (quartz, chalcedony, or opal) and other durable minerals in the 0.02- to 2.00-mm fraction that are resistant to weathering.

2. Kaolintic—contains more than half (by weight) of kaolinite or other non-expanding clays. Contains less than 10% (by weight) of smectite.

3. Smectitic—has more (by weight) smectite (montmorillionite, beidellite, and nontronite) than any other clay mineral.

4. Mixed—have properties outside the limits of the other classes.

Quantitative Properties

Some descriptive characteristics of the soils were presented in the preceding section. The nature of these descriptive characteristics provide the basis for a number of important quantitative properties:

Soil Depth

Soil depth is an important quantitative characteristic. The depth of a soil or the distance from the surface to bedrock, influences the soil's potential to support root, and ultimately, plant growth. Broderson (1991) lists five depth classes for soils. These are given in Table 3.10. The depth classes of the 14 soils are given in Table 3.12.

Montevallo Soils are shallow. Fractured siltstone and sandstone occur below a depth of 12 inches. Hartsells, Pacolet, and Sipsey are moderately deep; while Alaga, Dewey, Kipling, Lafitte, Lenoir, Leon, Nicollet, Orangeburg, Una, and Vaiden are very deep. There are no very shallow soils included in this group.

Soil Drainage

Soil drainage is the another important quantitative soil property. Singer and Munns (1987) list seven drainage classes used in soil survey work. These are given in Table 3.11. The soil drainage classes of the 14 soils used in the comparison were given in Table 3.12.

On the drainage class scale, the water table is near or at the surface during most of the year in the very poorly drained soils; while at the other end of the scale, there is virtually no long-term accumulation of water in the excessively drained soils. In this latter group, water flows through the soil rapidly or very rapidly. Six of the soils (Dewey, Hartsells, Montevallo, Orangeburg, Pacolet, and Sipsey) are considered to be well drained (Table 3.11). Alaga is the only soil in the somewhat excessively-drained class.

In a well-drained soil, there is no standing water or water in the profile. Water is removed from the soil readily, but not rapidly. Wetness does not inhibit growth of roots for significant periods during the growing season. The soils in this group contain loamy sand, sandy loam, or silt loam and relatively less clay (See Table 3.9).

Kipling, Lafitte, Leon, Lenoir, Una, and Vaiden are loam, silty clay loam, silty clay soils, and clay. These soils contain relatively more clay and are very poorly drained to somewhat poorly drained. Lafitte is very poorly drained, and Leon and Una are poorly drained. In a poorly drained soil, water is removed so slowly that the soil is saturated periodically during the growing season. Free water is commonly at or near the surface long enough that crops cannot be grown without artificial drainage.

As indicated by the data presented in the preceding paragraphs, soil drainage is correlated to some extent with the amount of clay that it contains. Generally, the higher the clay content, the more poorly the soil drains; however, there are also other factors that affect drainage, including the structure of the soil and slope.

Soil Permeability

Soil permeability is still another important quantitative property of soil. It is a measure of how rapidly (inches per hour) water moves through the soil. Permeability determines how rapidly water from the surface will flow downward to fill the root-zone reservoir when there is a deficiency, or how rapidly excess water in the soil will flow downward to be drained away. Broderson (1991) lists seven soil permeability classes (Table 3.13.)

Data on the permeability of the 14 soils is given in Table 3.15. These data indicates that permeability is moderate to rapid in Alaga, Dewey, Hartsells, Lafitte, Leon, Montevallo, Nicollet, Orangeburg, Pacolet, and Sipsey. It is moderately rapid in Lafitte and Leon and rapid in Alaga. Permeability is very slow in Kipling, Una, and Vaiden and slow in Lenoir.

Particle size and soil texture (whether or not the structure is granular, prismatic, blocky, or platy) are primary determinants of the movement of water through saturated soil; however as Scofield (1938) notes, there are also other factors involved. The nature of the subsoil plays a role, as does the presence of a hardpan or an impermeable layer, which formed when soil particles become cemented together. Also, under certain conditions, the colloidal material in the soil may expand to form an impervious gelatinous mass. All of these conditions are largely independent of soil texture.

Available Water Capacity

Available water capacity is one of the most important properties of soils. This characteristic is defined as the difference between the amount of soil water at field moisture capacity and the amount at wilting point of the crop. McNutt (1981) listed five classes of available water capacity of soils (Table 3.14). Data for this characteristic, for the 14 different soils, is given in Table 3.15.

Available water capacity is considered to be low to very low (6 inches or less of water in 60 inches of soil, or to the first soil barrier) in Alaga, Leon, Montevallo, Pacolet, and Sipsey and moderate to high (6 to 12 inches) in Dewey, Hartsells, Kipling, Lafitte, Lenoir, Nicollet,

Orangeburg, Una, and Vaiden.

The amount of clay and, in turn, the amount of internal surface area in a soil, is an important determinant in its available water capacity. With all other factors being equal, the more clay in a soil, the greater is its internal surface area and the greater its available water capacity. Obviously, Leon (3 to 6 inches of capacity) with its generally sandy texture would have significantly less surface area than Una (9 to 12 inches of capacity) with its silty clay loam texture. This difference in texture is reflected in the difference in available water capacity of the two soils.

Montevallo also has a very low available water capacity (0 to 3 inches). Part of the problem is the amount of clay in the soil, but the primary determinant is the shallowness of the soil. The bedrock barrier begins at 12 inches.

A map published by the Southern Global Change Program of the US Forest Service showed that available water storage capacity is generally similar throughout Alabama, except where the soils are derived from sandstone. As a result values are generally considerably lower in the Cumberland Plateau Physiographic Section and in a small portion of the Valley and Ridge Section. This is the Limestone Valleys and Uplands Area of Mitchell and Meetze (1994) (Figure 3.3). The map also showed that values for Alabama and Iowa are generally the same, except that values are somewhat higher in the western portion of Iowa, along the Mississippi River, and in the southeast.

Soil Erosiveness

Soil erosiveness is another extremely important quantitative property of soils. Susceptibility of a soil to erosion is dependent on a number of factors, including the amount of silt and sand in the soil, organic matter content, soil structure, amount and intensity of rainfall and the angle of the soil slope. These factors plus others are brought together in an equation, named the Universal Soil Equation (USLE), to predict the long-term average soil loss for a location (Singer and Munns, 1987). Two elements of that equation, the Soil Erodibility Index ("K") and the angle of the soil slope, are shown for each of the Soils, in Table 3.16.

"K" is a single number (index) that combines several of the properties that determine the effects that rainfall and runoff will have on a soil. It includes measures of the amount of silt and very fine sand in a soil, the amount of other sand, organic matter content, soil structure, and permeability. The index ranges from 0.02 to 0.64 or more (Broderson, 1991). The magnitude of the index is directly (positively) proportional to the susceptibility of a soil to erosion. The higher numbers indicate a higher susceptibility. The "K" values are for the surface layer of the soil.

Another erosion factor, "T," is also included in the table. The "T" factor is an estimate of the maximum amount of soil that can be lost from an acre of that soil each year without reducing its productivity. The "T" factor ranges from one for very shallow soils to five for very deep soils.

The "K" values shown for most of the soils are generally near the middle of the range (0.02 to 0.64), indicating only a moderate susceptibility to erosion. The values for Alaga, Leon, Nicollet, and Sipsey are relatively low. In Alaga and Leon, their permeability is so high (6 to 20 inches per hour in Alaga and 2 to 6 inches per hour in Leon) that rain water runs through them rather than over the surface. Lafitte soil contains so much organic matter in its surface layer that it is not given a "K" value.

The "T" factor for the soils ranges from two to five (Table 3.15,) and indicate that the soils could sustain erosion losses of from two to five tons an acre per year without reduction in productivity. Eight of the 14 soils (Alaga, Dewey, Kipling, Lenoir, Leon, Nicollet, Orangeburg, and Una) can sustain losses of five tons an acre per year without a reduction in productivity. At the other end of the scale, Hartsells and Montevallo can sustain losses of only two tons an acre without losing productivity. Vaiden (four tons) and Lafitte Pacolet and Sipsey (three tons) have intermediate soil loss "T" factors.

Of the three factors listed in Table 3.16, the angle of the soil slope is probably the more important one affecting erosion, because soil loss increases exponentially with the steepness of the slope (Singer and Munns, 1987). The soils can be divided into three general groups with respect to the angle of the slope. Lafitte, Lenoir, Leon, Orangeburg, Una, and Vaiden are on slopes of 0 to 2%. Alaga, Dewey, Hartsells, Montevallo, Pacolet, and Sipsey generally have slopes greater than 2% and Hartsells (6 to 10%), Montevallo (30 to 60%) and Sipsey (4 to 18%), are on critically steep slopes.

Published soil surveys indicate that the Dewey, Hartsells, Montevallo, Pacolet, and Sipsey soils are subject to excessive erosion, and according to the "T" values in Table 3.15, productivity can be lost in all except Dewey with relativey low annual soil losses, two to three tons an acre (Sherard, 1977; Bowen et al., 1979; McNutt, 1981; Stevens, 1984; Harris,1989 and Stevens, 1992).

Soil pH

Soil pH is another important property of soils. It is a measure of whether a soil is acid or alkaline. It is extremely important to Alabama farmers. Table 3.1, includes pH values for the 14 Soils. Note that a range of values is given. Liming of soils has become so common in Alabama that it is difficult to obtain soil pH data that are unbiased (George Martin, personal communication). The low values in the ranges are probably relatively unbiased; although certain agricultural practices also tend to increase soil acidity. Given the nature of most Alabama soils, the high values probably reflect the effect of liming.

The lower portions of the ranges of pH values for all of these soils, with the exception of Lafitte and Nicollet, are in either the extremely acid (Below 4.5) or very strongly acid (4.5 to 5.0) categories as defined by soil scientists (Harris, 1989). The upper portions of the ranges of Dewey, Hartsells, Lenoir, and Una are in the strongly acid (5.1 to 5.5) category. The upper values for Alaga, Kipling, Leon, Montevallo, Orangeburg, Pacolet, Sipsey, and Vaiden are catergorized as either medium acid (5.6 to 6.0) or slightly acid (6.1 to 6.5), probably indicating that the particular location where the sample was taken for analysis had been limed. The range of the Lafitte soil (6.1 to 8.4). lightly acid to mildly alkaline probably represents the influence of brackish water on the soil.

According to Broderson (1991), soils with pH values less than 5 usually have lower levels of nitrogen, phosphorus, calcium, and magnesium available for plant feeding; while the amounts of aluminum, iron, and boron are increased.

Cation-Exchange Capacity (CEC)

The cation-exchange capacity of a soil is a measure of its capacity to adsorb, hold, and exchange plant nutrients (Broderson, 1991). In general, it is dependent on the amount and kind of clay and the amount of organic matter present. Coleman and Mehlich (1957) commented that soils containing organic matter and/or montmorillinite clay have substantial CEC values. Soils with the lowest CEC values are those containing substantial amounts of sand or kaolinite clay (Table 3.7).

The cation-exchange capacity values of the 14 soils are presented in Table 3.18. The multiple values for each soil in the table indicate the CECs for different parts of the soil profile. The data in the table indicate that the CEC for most of these soils are relatively low. Except for Lafitte, Nicolett, and Vaiden, which have CEC values from 25 to 100, the values for the other soils are generally below 10. This situation is more or less to be expected; because only Vaiden, which contains a relatively large amount of montmorillionitic clay, and Lafitte, which contains a significant amount of organic matter, have characteristics conducive to attracting, holding, and exchanging relatively large quantities of plant nutrients. Four of the soils (Alaga, Hartsells, Leon, and Sipsey) contain significant amounts of sand (silica minerals), and three of the soils (Dewey, Orangeburg, and Pacolet) belong to the kaolinitic soil mineralogy class (Table 3.9). All of these have very little capacity to attract, hold, or exchange nutrients.

Data published by Sawyer (2003) indicated that the CEC values of 12 representative soils in Iowa ranged between 14 and 43.9. The average was 25.6. Values for these Iowa soils were higher than those for most of those in Alabama, but none of the Iowa values were as high as those for Lafitte or Vaiden.

Cumulative Soil Effects

In the preceding sections, several characteristics and properties of 13 soils in Alabama and one in Iowa were considered, but no attempt has been made to evaluate their cumulative effects in terms of the advantages and disadvantages that they convey in determining any limitations on soil use or in the return on investment that their use might potentially provide. Two measures of this cumulative effect are Land Capability Class and Woodland Ordination Symbols.

Land Capability Class

Based on the soil surveys conducted by the Natural Resources Conservation Service (NRCS), recommendations on land uses and limitations to those uses can be made for each soil in a county. These recommendations are made in a general sense as Land Capability Classes developed by the US Department of Agriculture and indicate the limitations of soils for the sustained production of agricultural crops (Singer and Munns 1987). The level of limitation associated with each class is described by Birdwell (1987). The list is reproduced in Table 3.25. The Roman numerals indicate the degree or level of limitation on the general use of a soil. In general, the higher the number, the greater the degree of limitation.

Generally, the Soil Capability Class classification system presented in Table 3.19 indicates that only those soils in the first four classes are suitable for cultivation or are arable soils, and with the restrictions on Class IV soils, even some of them are not really suitable for this purpose. Classes V through VIII are considered to be non-arable soils. Limitations for soils in Class VIII, for example, are so severe that they have virtually no value in commercial agriculture.

There are also Land Capability Subclasses, as shown in Table 3.20, which more clearly define the nature of the limitations. The table includes three subclasses. In areas at more northern latitudes, where soil temperature can be a limiting factor, a fourth Subclass *c* for climate is added to the list.

The Soil Capability Class and Subclass designations for each of the 14 soils are given in Table 3.21. In reviewing these data, it is important to remember that the class designation of a specific soil series is dependent on the particular soil phase involved. For example, the Pacolet Soil in Lee County is mapped or identified in soil surveys in three phases: Pacolet on 1 to 6% slopes, Pacolet on 6 to 10% slopes, and Pacolet on 10 to 15% slopes. The class designations for these three soil phases are: II, III, and IV, respectively. The degree of limitation increases with the slope. All three of the phases have the same subclass designation *e*.

All of the 14 soils in our list have some severe to very severe limitations except the Nicollet in Iowa and the Orangeburg in Covington County, Alabama, which have no limitations (Class I). However, while both of these soils have the same classification, this is not the end of the story. For example, the expected yield of corn grown on Orangeburg soil in Covington County is not the same as the expected yield of corn grown on Nicollet soil in Humboldt County, Iowa—although they have the same Soil Capability Class (I). The expected yield in Iowa is 53% higher (153 versus 100 bushels an acre).This difference reminds us that the quality of the soil is but one of the factors involved in the production of a crop. In this situation, there are apparently other factors that severely limit the production of corn in Covington County.

Two of the soils, Dewey and Pacolet, can be cultivated with moderate conservation practices and with only some limitations on the kind of plants that can be used (Class II, Subclass *e*).

The Alaga, Kipling, and Vaiden soils can also be cultivated, but they have severe limitations that reduce the choice of plants or that require special conservation practices (Class III, Subclasses *s, e*, and *w*, respectively).

The Leon and Sipsey soils have very severe limitations. They require very special management and special considerations in choosing plants for cultivation (Class IV, Subclasses *w* and *e*, respectively).

Lenoir is the only soil among the fourteen in Class V (Subclass *w*). Its limitations are so severe that it should be used for pasture, woodland, or wildlife habitat rather than for cultivation.

Montevallo and Una soils have such very severe limitations that their use should be restricted to recreation, wildlife, water supply, or to esthetic purposes (Class VII, Subclasses *e* and *w*, respectively).

Lafitte is the only soil in Class VIII (Subclass *w*). Soils in this class have such serious limitations that they should be used only for recreation, wildlife, water, or for esthetic purposes.

The Dewey, Kipling, Pacolet, and Sipsey soils have erosion and runoff limitations (subclass *e*); consequently, moderate soil conservation practices such as conservation tillage, terraces, contour farming, and cover crops are recommended (Fox, 1989).

The subclass designation for the Lafitte, Lenoir, Leon,

Una, and Vaiden soil, is *w*. This limitation is related to wetness and drainage. Wetness is much more of a limiting factor in soil use than erosion and runoff, because it is so much more difficult to control. Except for Una and Vaiden soils, it is not recommended that any of the soils with this subclass designation be used for cultivation. Even with Vaiden, special precautions such as cultivating only within a narrow window of soil moisture content and using proper row arrangements, field ditches, and vegetated water outlets are necessary. There are even limitations to using the soil for pasture. Grazing should be limited to periods of high soil moisture to reduce soil compaction and to protect the sod (Harris, 1989).

Alaga soil is the only one of the 14 with an *s* subclass designation. Because of its sandy (loamy sand) nature, it is extremely droughty and applied plant nutrients (fertilizer) leach rapidly from it. There are no conservation practices that are useful in reducing the effect of this limitation.

Woodland Ordination Symbol

Woodland Ordination Symbols are another measure of the cumulative effects of soil characteristics and properties on their use. They are measures of the potential production of a specific indicator tree on a particular soil; consequently, the symbol is a summary statement about a number of tree-growth supporting characteristics and properties in a specific soil.

Woodland Ordination Symbols are somewhat similar to the Land Capability Classes described above. Both are general indicators of soil usefulness. Numbers (symbols) are assigned to each soil, but Woodland Ordination Symbols as contrasted to Soil Capability Classes, indicate the level of potential production rather than the level of limitation. These symbols are assigned as part of the soil survey procedures described previously.

The Woodland Ordination Symbols also include a letter of the alphabet that indicates the nature of the limitations, if any, related to soil characteristics that would affect forest management (Johnston,1992). The letters used to denote limitations are given in Table 3.22. The letters are listed in descending order of the importance of the indicated limitation to tree management. In the Table, the letter A indicates that there are no significant limitations. The letter R indicates that the steepness of the slope of that particular phase of the soil can affect management because of the potential for soil erosion and because of restrictions on the use of equipment on steeper slopes. The letter W indicates that general wetness of the soil could result in limitations on the use of equipment, on seedling mortality, and on competing plants. The letter D indicates that wind-throw or the loss of trees to high winds can be a limitation. The letter C indicates that seedling mortality may be a problem because of the amount and kind of clay in the surface of the profile.

Table 3.23 gives Woodland Ordination Symbols for 12 of the 14 soils. No symbol has been assigned to Lafitte and Nicollet, because there is little likelihood of commercial production of trees in marshes or on prairies. Note that a number has also been linked with each symbol. The numbers indicate the annual potential production in cubic meters per hectare (one cubic meter per hectare is equivalent to 171.4 board feet an acre) of an indicator tree species in a well-stocked natural stand. In the table, the tree name in parentheses indicates the specific tree or indicator species (the loblolly pine, in most cases) to which the ordination symbol refers.

Data in Table 3.23 indicate that the potential production loblolly pine for most of the soils is similar. Potential production of this indicator species is highest (9 m^3/ha) on Kipling, Lenoir, and Orangeburg and only slightly lower (8m^3/ha) on Alaga, Dewey, Pacolet, Sipsey, and Vaiden. Potential production of loblolly pine is slightly lower on Leon (7 m^3/ha) and Hartsells (6 m^3/ha), and lowest on Montevallo (5 m^3/ha). The highest potential production for any indicator species, on any of the soils, is for the slash pine on Leon (10 m^3/ha), and the lowest is for bald cypress on Una (3 m^3/ha).

There are no limitations (Symbol A) to tree management on five of the twelve soils. These include: Dewey, Hartsells, Orangeburg, Pacolet, and Sipsey. There are limitations on five of the twelve. Slope (R) is a limitation on Montevallo; wetness (W) on Lenoir, Leon, and Una; heavy clay (C) in the top soil of Vaiden and Kipling; and deep sand (S) in Alaga.

There is little, if any, relationship between potential production and limitations to management. Potential production on Kipling and Lenoir (Loblolly pine, 9 m^3/ha) and Leon (Slash pine, 10 m^3/ha) are among the highest of any of the soils, and all three have severe limitations to management. Kipling is limited by the amount and kind of clay in the subsoil (C), and Lenoir and Leon are limited by general soil wetness (W).

Soils and Alabama Agriculture

In a preceding section, some of the characteristics of 13 of Alabama's soils and one soil in Iowa are considered in detail. In this section, some of the specific effects that some of these characteristics have on the production of food and fiber will be considered. These characteristics include: age of the soils, soil pH, soil erosion, root zone problems, soil wetness, droughtiness, soil fertility, and quantity of arable farmland.

Age

Pearson and Ensminger (1957) noted that Alabama's soils developed in a warm, humid climate and under forest vegetation—conditions that favored a high degree of weathering. In our undisturbed, highly-weathered soils virtually all of the primary minerals have been converted to secondary ones; leaching has removed most of the soluble materials, and as a result of downward movement or illuviation, silt and clay (a secondary mineral) accumulate in the subsurface leaving a layer of sand at the surface. Akin (1991) suggests that soils that have been subjected to weathering of this intensity for this duration are in a state of old age, and they are destined to remain in that state until geological processes such as glaciation or new surges of mountain building set the stage to begin new cycles.

In a preceding section, it was noted that 65% of the soils are in the Order Ultisol or Ultimate soils, and they are by far the most common order in every soil area of the state except in the Blackland Prairie Belt and the Coastal Marshes and Beaches areas (Table 3.2). Only 13% of all of the soils in the United States have reached this stage of maturity. Most of the soils in old age are found in the Southeast United States. Of the eleven soil orders, only the order Oxisol (Oxidized soils) have been weathered more severely. Fortunately, there is only a trace of this order in the state.

Ultisols by definition are deeply weathered, strongly acidic, primarily red and yellow soils indicating the accumulation of iron oxide clay minerals. Considerable amounts of silicate clays, such as kaolin and gibbsite, trans-located from the surface layers of the soil, have accumulated in an argillic or clayey layer in the subsoil.

The presence of the iron oxide clay minerals indicates further intense weathering. These minerals, which give the soils their bright colors, are generally formed from the weathering of secondary minerals such a kaolin. The weathering of kaolin also releases alluminum hydroxide. These by-products of the intensive weathering of soil are highly resistant to further weathering and leaching. As a result they accumulate in the soil.

While most of the soils in Alabama have been aging for millions of years, the situation in Iowa is completely different. Virtually all of the soils there were developed on parent materials (loess and glacial till) deposited by glaciation relatively recently. As a result, the Nicollet Soil is relatively young. Further, it was developed under prairie grass rather than a deciduous forest and in a region with limited rainfall and much lower annual temperatures. As a result, it is high in organic matter and nitrogen. Also, relatively less of the other plant nutrients have leached out of its profile. It also has good structure and holds water well (Pierre and Riecken, 1957).

pH

Soil pH, which is a measure of whether the soil is acid or alkaline, has always been important to farmers, especially to those in Alabama. Our soils have been subjected to millions of years of those conditions (acid-forming parent materials; high rainfall with its prolonged, recurrent, vigorous leaching; and a warm climate that encourages a maximum level of biological activity) that favor the development of soil acidity. The intense weathering that produced our Ultisol soils also caused their inherent acidity. Pearson and Ensminger (1957) noted that most of Alabama's soils are predominantly acid in reaction. The data in Table 3.17 are indicative of this phenomenon. In contrast, the pH of Nicollet, in the top 25 inches of soil, was 6.1 to 7.3. However, the soils in Iowa have been heavily limed over the years.

Soil acidity is especially important to Alabama farmers, because some of our most important crops are sensitive to low pH. For example, cotton, alfalfa, common beans, peas, and some clovers are highly sensitive (Singer and Munns, 1987). Wheat, soybeans, and sorghum are sensitive. Fortunately, peanuts, potatoes, oats, and corn are mildly tolerant. Truog (1938) comments that the broad range of common agricultural plants function most efficiently when the pH of the soil is in the range of 6 to 8. He also lists six effects that soil pH outside of that range, especially the acid part, has on plant growth: physical condition of soil, availability of essential elements, activity of soil microorganisms, solubility and potency of toxic agents, prevalence of plant diseases, and competition with other plants.

Three of the effects are particularly important to Alabama farmers. First, the solubility and potency of aluminum in soils at low pH levels can be particularly damaging to the roots of cotton, soybeans, corn, and small grains (Miller, 1986). High levels of free aluminum in the soil inhibits cell division in the tips of roots, resulting in stubbiness or stunting (Singer and Munns, 1987). Nutrient intake by these damaged roots is impaired. Also, aluminum toxicity reduces rooting depth and the degree of root branching in the subsoil. These effects are especially apparent during periods of stress. As discussed in a preceding section, aluminum is a significant component of silicate clays, such as kaolin and aluminum oxide clays. As the soil becomes more acid, these minerals become more soluble, releasing aluminum into the soil-water solution. Toxic levels of this free aluminum are common in soils with pH values 5 or lower (very strongly acid) and toxic levels can occur with soil pH values as high as 5.5 (strongly acid), especially in kaolintic soils and soils low in calcium. Part of the problem of aluminum toxicity in soils is easily solved by liming, which raises the pH, resulting in the reduction of the solubility of the clay minerals. Unfortunately, common agricultural practices generally only gets the lime into the plow layer (Ap horizon) of the soil and almost none of it gets into the deeper parts of the surface horizon and in the subsoil. As a result, roots do not grow into those deeper layers, and nutrients stored there, especially water, are not available to the plant. Plants are unable to obtain stored water in soils even with a high storage capacity if the free aluminum concentration is at toxic levels. It is possible to get the lime down into the subsoil with special plowing equipment, but the process is expensive.

As noted in a preceding section, phosphorus is an essential plant nutrient. It was also noted that there is relatively less phosphorus in soil than either nitrogen or potassium. The amount of phosphorus in virgin soils of the coastal areas of Alabama is especially low (Olsen and Fried, 1957). There is generally less in virgin soil, and the amount present is quickly depleted by cropping and erosion. However, there is another factor that significantly affects the availability of phosphorus to plants, especially in Alabama's acid soils. Free phosphate in the soil solution quickly forms sparingly soluble compounds with calcium, iron, and aluminum. Also, these compounds fit tightly into the structure of clay minerals such as kaolin. With pH values of 6.5 (slightly acid) or lower the phosphorus compounds become increasingly insoluble. The compounds are least soluble at pH 4. Consequently, under acid soil conditions, phosphate bound in the soil is generally unavailable to agricultural plants (Loomis and Connor, 1992). In this situation, although there is a considerable amount of phosphorus in a soil, it is mostly unavailable. Liming reduces the acidity of the soil, which increases the solubility of the compounds. Solubility increases as the pH increases from 4 to 8.5, but begins to decrease again beyond that point. Phosphorus availability is greatest when pH is in a range of 6.5 to 7.0.

Several important processes mediated by microorganisms are negatively affected by soil acidity. For example, the rate of break-down of soil organic matter by bacteria is reduced in highly acid soils, thus reducing the amount of nitrogen and phosphorus potentially available from that source. Root nodule bacteria of alfalfa do not persist or function very well when the soil pH is below 6.5, so that nitrogen fixation is reduced. This is a primary reason why the soybean plant is so sensitive to acid conditions in soil.

Erosion

The loss of soil from erosion is another important

problem in Alabama agriculture. Pearson and Ensminger (1957) noted that many of our soils are naturally erosive. In the comparison of the properties of the 14 soils presented in a preceding section, it was noted that five of them (Dewey, Hartsells, Montevallo, Pacolet, and Sipsey) were subject to excessive erosion. Factors indicating susceptibility to erosion (K and slope) suggest that several of the others are also candidates for severe erosion. A generalized erosion map published by Blakely, et al. (1957) (Figure 3.6) suggests that 25 to 75% of the soil in two-thirds of the state may have been lost to erosion and that more than 75% may have been lost in the remainder. According to Pearson and Ensminger (1957), erosion has been particularly severe in Piedmont Upland soils. In much of the area, erosion has removed all of the top soil, and the subsoil is being cultivated.

As noted in a preceding section, intense weathering over a long period of time has led to the accumulation of a high percentage of soils of old age or ultimate soils (Ultisols) in the state. As a result of this weathering process, some of the most erosion-resistant components of the soils, the clay minerals, have filtered downward into the subsoil, leaving a thin layer of organic matter plus a somewhat thicker layer of sand and silt at the surface. These surface materials are especially prone to erosion given the frequency, intensity, and duration of rainfall events. These combined effect of these two factors, the erosive nature of the surface layer and rainfall patterns, are further compounded by the inherent hilliness of the state's landscape.

The natural erosiveness of Alabama's soils has been magnified over the years by farming practices. Moore (1986) notes that early upcountry farmers in the Lower South had been described as the "worlds most destructive agriculturalists." On those early upland cotton farms, especially in the Piedmont, it was customary to use shovel plows for cultivation, and rows were almost always arranged up and downhill, rather than on the contour. Each winter new ground had to be cleared to replace the washed-out fields. Finally, when there was no more land to be cleared on the farm, the family closed the door on their log cabin and moved westward. These destructive agricultural practices persisted well into the 1840s, over much of the upland, erosion-prone lands of the region (Moore, 1986). After that time, practices that would reduce erosion were widely implemented only to be discarded after the Civil War, and it was not until about 1930 that concern for soil erosion surfaced again.

Blakely, et al. (1957) note that corn, cotton, and soybeans are considered to be the most erosion-prone crops because of the excessive amount of cultivation (plowing-under cover crops, seedbed preparation, planting, and plowing for weed control) associated with their culture. All three of these crops have been widely used in Alabama.

A review of information on Alabama's Land Capability Classes and Subclasses provides a good picture of the magnitude of the erosion problem in the state. A discussion of the Land Capability Classification System used by the US Department of Agriculture and a comparison of the classes and subclasses of 13 representative Alabama soils and one Iowa soil were presented in a preceding section (Table 3.21). George Martin (Soil Data Quality Specialist, USDA-NRCS, personal communication) provided data on the estimated share of the total in each of the classes and subclasses of 31,135,000 acres of Alabama land. These data have been converted into percentages and presented in Table 3.24. They indicate that the usefulness of 62% of Alabama's soil is limited to some degree by susceptibility to soil erosion (Subclass e).

Root Zone Problems

Some 16% of Alabama's soils have use limitations related to water storage capacity in the root zone (Subclass *s* in Table 3.24). In contrast, only 4% of the soils in Iowa, carry Subclass s designations. For example, the Alaga Soil described as a representative Alabama soil has class and subclass designations of III and *s*, respectively (Table 3.21). These classifications mean that this specific soil phase has severe use limitations related to root-zone problems. The root zone of the soil is deep, and there is no hindrance to root penetration. However, the soil consists of 80 inches or more of loamy sand (Table 3.8), and because of the sandy nature of the soil, water rapidly moves through it and out of the root zone (Table 3.15).

In some of the Montevallo soils in north-central Alabama, the depth to bedrock is 10 to 20 inches (Table

3.8). So the depth of the root zone is also 10 to 20 inches; consequently, the size of the soil water reservoir is extremely small (Table 3.15).

Loose rocks in a so-called channery soil can also limit its water-holding capacity. The Montevallo channery silt loam (30 to 60% slopes) has, by volume, more than 15% coarse, flattened fragments of sandstone embedded in it (Table 3.8). Some of the fragments are as much as six inches in length. The presence of these rock fragments also restricts water-holding capacity (Table 3.15) and results in problems in the root zone for plants.

Under certain conditions, a loamy, brittle horizon consisting primarily of silt, very fine sand, and some clay will form fragipans. These horizons often become cemented and restrict root penetration so that the effective size of the water reservoir is somewhat reduced.

Excessive cultivation of Ultisol soils, with their accumulation of clay minerals in the subsoil, can result in the development of an induced tillage pan or hard pan, a thin, dense layer of soil just below the plow layer (Ap horizion). These layers that restrict root growth and penetration into the subsoil also restrict the effective size of the soil water reservoir for plants. Under these conditions, crop yields can be reduced significantly (Blakely, et al., 1957). Pearson and Ensminger note that these plow pans or compacted layers,are fairly common on Coastal Plain soils.

Aluminum toxicity was discussed in a preceding section on soil acidity. One of the practical problems resulting from the accumulation of toxic levels of aluminum in soil horizons can be the inhibition of root penetration and a reduction in the quantity of water available to plants. Generally, root zone problems resulting from aluminum toxicity are not included in assigning soils to capability subclasses. This is apparently not a major problem outside of the Southeastern United States where there are no Ultisols. Perhaps if aluminum toxicity were taken into consideration, a considerably higher percentage of Alabama's soils would be in Subclass *s*.

Wetness

Edminster and Reeve (1957) commented that excess water in agricultural soils can affect crop production and interfere with the scheduling of planting, cultivation, and harvesting. Excessive soil moisture can also affect nutrient uptake and root development. Roots will not penetrate saturated soil, and if saturation persists and oxygen is eliminated from the soil's pore spaces, they will die. Elimination of oxygen from the pore spaces of soils also reduces the rate of decomposition of organic matter, thus reducing the availability of nutrients normally released in that process. Although excessive wetness can cause severe problems in the root zone, soils with this limitation are assigned to Subclass *w* rather than Subclass *s*. The use of six million acres (20%) of land in Alabama is limited by wetness (Subclass *w*) (Table 3.24). Five of the representative soils (Lafitte, Lenoir, Leon, Una, and Vaiden) have wetness-related limitations (Table 3.21). Wetness is apparently even more of a problem in Iowa soils. Data presented in Table 3.25 indicate that 28% of soils in that state carries the *w* classification.

Drougthiness

Available-water capacity was defined in a preceding section as the difference in the field moisture capacity and the amount of moisture in the soil at the plant wilting point. It is also a measure of the potential drougthiness of a soil. Plants growing in soils with a limited available-water capacity quickly reach the wilting point, unless the amount of water in the soil is regularly replenished through rainfall. Data on the available-water holding capacity of the 14 soils in inches of water in 24 inches of soil profile is presented in Table 3.15. The water-holding capacity in only one of the soils (Alaga) is below 2 inches in the top 24 inches of the soil profile. In five of the soils (Leon, Montevallo, Orangeburg, Pacolet, and Sipsey), it is between 2 and 2.9 inches. It is between 3 and 3.9 inches in Dewey, Hartsells, Lenoir, and Vaiden, and above 4 inches in Kipling, Lafitte Nicollet, and Una. Probably all of these soils with capacities of less than 2.9 inches would be considered droughty.

Pearson and Ensminger (1957) suggested that in the Southeastern Uplands Region, an area that includes Alabama, that the frequent shortage of rainfall during the growing season, coupled with the limited water-holding capacity of all but a few soils, means that moisture is deficient for crop growth during some periods in most years. As an example, they noted that the water-holding capacity of a typical Orangeburg Soil, which is widely distributed in the Coastal Plain, is 2.11 inches in the top 24 inches of the profile, and that a typical corn crop during the period of maximum growth will remove 0.4 inch of water from the soil each day. At this rate of removal, there would be sufficient moisture stored in the top 24 inches of soil to last for less than six days.

Fertility

Pearson and Ensminger (1957) noted that soils of the Southeastern Uplands, which includes Alabama, are naturally low in plant nutrients, and that crop yields have been limited as a result. Information presented in Table 3.26 on the natural fertility of the 14 soils described in preceding section is indicative of the severity of this problem in Alabama.

Data in Table 3.26 show that all of the representative soils, except Vaiden and in some instances Una, are low in natural fertility. Only the Nicolett Soil of Iowa is high. These generalized statements of natural fertility usually parallel the information on cation-exchange capacity presented in a preceding section (Table 3.18). Apparently, one of the reasons that Alabama's soils in their natural state generally have such low fertility is because they have only a limited capacity to store plant nutrients. Also, 65% of the soils in the state have weathered to the point (Ultisols) that they contain virtually no other minerals required for the release of additional nutrients.

Obviously, soil fertility is a problem for Alabama farmers, and it can be partially solved through the use of commercial fertilizers. Field tests have demonstrated significant responses to the addition of nitrogen, phosphorus, potassium, and lime plus minor nutrients, in some cases, for most crops, on most soils. Unfortunately, soil fertility is just one of many factors affecting crop yields. In most years, even with adequate fertilization, crop yields are relatively low as a result of other limiting factors (Environmental Resistance); consequently the cost of fertilizer per unit of crop produced is relatively high (Pearson and Ensminger, 1957). These complex relationships between inputs and outputs will be discussed in considerable detail in subsequent sections.

Quantity of Arable Land

The quantity of arable soils in Alabama is also a concern. Arable soils are those that can be cultivated indefinitely for the production of crops. Generally, the soils in Land Capability Classes I through IV are considered to be arable (Table 3.19). Those in Classes V through VIII are not. Soils in Class VIII generally are not suitable for any form of commercial agriculture, even pastures or woodlots (Birdwell, 1987). Table 3.27 includes data on the percentage of land in each of the classes of soil in the United States. These percentages, from the 1982 Land Resources Inventory, are based on data on non-federal land in the contiguous 48 states. The survey indicated that there were slightly less than 1.5 billion acres of land in the country. Approximately 52% of it is arable. Approximately 2% has no use limitation (Class I). Comparable percentages for Alabama and Iowa soils are also given in the table.

Approximately 59% of Alabama's land is arable. Five of the 14 soils, included in the study of representative soils, are considered to be non-arable. These include Hartsells (VI), Lafitte (VIII), Lenoir (V), Montevallo (VII), and Una (VII). The percentage of arable land in Alabama is similar to the country (59 and 52%, respectively), but far below Iowa's 92%. Iowa has four times as much land in Class I, and twice as much in Class II, as either Alabama or the nation. In general terms, the overall productivity of land decreases and the cost of making it productive increases as the degree of use limitation increases. A comparison of the percentage of land in classes I and II, in Iowa, with the percentages in Alabama, indicate the potential degree of comparative advantage that Iowa farmers enjoy in crop production.

4

Some Plants and Animals of Alabama Agriculture

PRECIS

Because of the nature of some of its geophysical characteristics (geology and climate), Alabama is home to an unusually large number of species of plants and animals. Surprisingly, the first representatives of our species (Native Americans) to arrive domesticated almost none of them to meet their food and fiber needs, choosing instead to continue their practice of hunting and gathering. When the next wave of our species (Europeans) arrived, they brought their domesticated plants and animals, including human slaves, with them. Except for loblolly pines, pecans, channel catfish, and white-tailed deer, virtually every plant and animal that we depend on for food and fiber evolved in another part of the world.

Without question, the most important species of plant or animal to be found in Alabama's agroecosystem, at least for the last 40,000 to 50,000 years, is the hairy vertebrate animal with opposing thumbs, *Homo sapiens.* After coming out of Africa some 1.7 mya, his ancient ancestors proceeded to spread out in every direction until he became firmly established on virtually every land mass on the planet. However, they would wait until very late in this long period to complete their journey to the Americas and Alabama.

For whatever reason, long before the exodus from Africa, ancient ancestors of *Homo sapiens*, had begun an altogether different kind of journey, an evolutionary journey through the complex world of anatomy and physiology. Along the way, they began to walk upright, freeing their arms to carry things or throw rocks; they developed forward focusing eyes that allowed for depth perception; their reproductive systems evolved to allow for continuous mating; and their evolving nervous system reached a completely new realm, the realm of cognition— of knowing that you know and of knowing that you do not know. With cognition, they learned to make and use tools; to plan; to maintain and manage fire; to make clothing and shelter; and with the combination of cognition, stereoscopic vision, bipedialism, and a highly flexible shoulder-thorax connection, they became formidable hunters. Soon they held dominion over all living things. Likely, many of these attributes were the result of millions of years of Darwin's selection of the fittest, but through the use of cognition, the gaps in what evolution had provided could be bridged with an entirely new, highly flexible building material—culture.

The factors (the bear went over the mountain, the grass is always greener, fight or flight) that led or pushed our early ancestors out of Africa, continued to push him around the face of the earth for thousands of years. This sloshing back and forth was especially obvious in early Europe. Well before the time of Christ, wave after wave of our species began to flow westward out of the cold steppe region of Southwestern Russia and neighboring countries in Central Asia. As the surge continued, native peoples in its path were displaced, their cultures either assimilated, altered, or replaced. The wave continued to flow across Western Europe until it temporarily lost its momentum when reaching the shores of the vast Atlantic. Not to worry—soon their cognition and cultural adaptations (ocean-going ships, the compass, astrolabe, water and sand clocks, primitive nautical charts) allowed them once again to be on their way, westward to the Americas. When they arrived, they were likely somewhat surprised to learn that another wave of Asian people (Native Americans) that had also emigrated from Central Asia thousands of years earlier had gotten here first by taking the overland route.

INTRODUCTION

Alabama's subtropical climate (Caf), described in Chapter 2, could be considered to be a hybrid. It has characteristics of both the humid tropical climate (Af) to the south and of the temperate climate (Daf) to the north (Figure 2.10). Alabama can be extremely hot and humid in mid-July, but cold and dry in mid-February. This broadly variable, split-personality climate has resulted in the establishment of an unusually large number of ecological niches for plant and animal species. Surprisingly, however, even though there was a high level of diversity of living things within the state's borders, there were few suitable candidates for domestication. Our native plants did not include the wealth of wild fruits associated with the tropics or the wealth of wild grasses (grains) associated with more dry, temperate regions. Consequently, only a few species of pines and pecans from our stock of native plants have been domesticated and used widely. Our native ecosystem produces unusually large quantities of cellulose annually, but because of its chemical composition and characteristics, its usefulness is limited.

The situation is similar with respect to animals. Only the channel catfish has been domesticated and is used widely. At one time in the distant past, Alabama likely had horses as part of our ecosystem, but by the time mankind arrived on the scene, the prehistoric horses were extinct. Alabama had no native pigs or goats, and there was apparently little interest in the domestication of our native herbivores, the whitetail deer or the elk.

Plants

Plants, especially those with green leaves containing chlorophyll are essential for life on earth as we know it. In their life processes, they use the energy in sunlight to split molecules of water into oxygen and hydrogen. The oxygen is released into the atmosphere, and the hydrogen is combined with carbon dioxide to form carbohydrate. Oxygen is an essential reactant of most of the processes of life, and carbohydrate is a basic building block of all organic matter on earth. All food used by animals has its beginning within the tiny cells in the leaves of green plants. Carbohydrate fills the seeds of wheat, rice, and corn and the tubers of potatoes and cassava that are primary food sources for most of the peoples of the world. In many countries, these same carbohydrate-filled seeds are fed to domesticated animals.

While the production of oxygen and carbohydrate are of primary importance to living organisms, plants give us many other things that make our lives easier and more predictable. Lumber made from the trunks of trees plays a central role in providing shelter for a large portion of the world's population. Fibers (lint) taken from the coat of seeds of the cotton plant are used to manufacture clothing. From the time that man domesticated fire, plant materials provided him with a source of energy for heating his shelter. After the Industrial Revolution transitioned from water power, the remains of ancient plant accumulations (coal, oil, and gas) provided the energy to run factories. Later, these fossil fuels powered transportation systems. In earlier times, wood and other plant materials played an important role in the production of spears, arrows, and traps needed by our ancient hunting and gathering ancestors. Wood, with other materials such as bones and antlers, was used to make digging sticks and other tools that were important in the development of agriculture. Our early ancestors also learned of the treasure trove of materials found in plants that could be used to heal the sick. Today, plants continue to provide a source of an ever-growing array of medicines and pharmaceuticals. Last, but not least, there are few if any people on earth who do not respond favorably to the beauty of vegetation and especially the blooms of plants. World-wide, untold amounts of time and money are expended to bring these beautiful and interesting living thing into our lives.

Sources of Information on Alabama Plants

There has never been a major effort to describe the native flora of Alabama, and the information that is available is relatively old. Work on the herbaceous plants is especially limited. Much more attention has been accorded the woody plants (trees, shrubs, and vines). The most complete work ever completed on the plant life of the state was Dr. Charles Mohr's (1901) *Plant Life of Alabama*. This book of over 900 pages provides information on the history, general and local distribution, and comments on economic value (if any) for all of the 2,400 species and varieties of plants known in Alabama over a century ago. There are no descriptions of the species, or keys or aides for their identification. Unfortunately, there are relatively few copies of this publication remaining. Mohr's work has not been updated, and there have been many changes in plant taxonomy during the intervening century.

Harper's (1928) *Economic Botany of Alabama, Part 2*, provides a catalog of the trees, shrubs, and vines (mostly woody plants) of the state along with notes on their description, local distribution, and economic value. He also provides a useful description of Alabama's physiography and climate and related plant distribution to those environmental characteristics. However, he does not provide keys or other aides to identification, and he provides essentially no information on the herbaceous plants. As with Mohr's book, it is also difficult to find copies of Harper's books.

Small's (1933) *Manual of the Southeastern Flora* provides descriptions of the seed-bearing plants known to occur in the Southeastern United States at that time. Along with the taxonomy and descriptions of individual species, he provides information on their habitats and ranges. The book also contains extensive keys to the identification of the species, including many line drawings of flowers and flower parts. This publication is also out of print and difficult to locate, except in larger libraries.

Harper's (1943) *Forests of Alabama* lists the large and small trees occurring in each of Alabama's physiographic sections and districts. It also contains useful descriptions of the topography in those areas. Davis and Davis (1975) provides a useful and descriptive guide and key to the trees of Alabama. Aquatic and wetland plants of the Southeastern United States are described by Godfrey and Wooten (1979 and 1981). These books provide good descriptions of the botanical characteristics of these plants along many excellent line drawings. There are extensive keys for their identification.

Clewell's (1985) publication includes extensive keys for the identification of the vascular plants of the Florida Panhandle, and includes many of the plants found in the lower part of the southern tier of Alabama counties. He also provides limited information on habitat and distribution, but there are no illustrations.

Godfrey's (1988) *Trees, Shrubs, and Woody Vines of Northern Florida and Adjacent Georgia and Alabama* provides keys, detailed species descriptions, many excellent illustrations, habitats and ranges of the woody plants in those parts of the Southeast, but like Clewell's book, it does not provide coverage for all of Alabama.

Alabama is blessed with an abundance of wildflowers. Dean, et al. (1973) published a beautifully illustrated book, featuring many of the wildflowers found in Alabama and adjoining states. Later, Tondera, et al. (1987) published a similar book featuring the wildflowers of North Alabama.

Information on the forest resources of Alabama is available from the Alabama Forestry Commission in Montgomery and the Southern Forest Experiment Station in New Orleans. The latter is a division of the US Forest Service of the US Department of Agriculture. The Southern Forest Experiment Station publishes an inventory of forest resources for the state at 10-year intervals (see McWilliams, 1992).

The Alabama Department of Conservation and Natural Resources has information on the endangered species of plants in the state.

Information on the cultivated plants of Alabama is available from the Alabama Agricultural Statistics Service of the Alabama Department of Agriculture and Industries, as well as the Department of Agronomy and Soils and the Department of Horticulture in the College of Agriculture at Auburn University. Information on aquatic plants is available from the Department of Fisheries and Allied Aquacultures in the same university. The county offices of the Alabama Cooperative Extension Service have more localized information on plants.

County offices of the Natural Resources Conservation

Service, which is a division of the US Department of Agriculture, also have information on plants. These county offices, plus the state office in Auburn, have available a substantial amount of information on wetland plants.

Botanists who serve as faculty members at Alabama's universities are excellent sources of general information on the plant life of the state. For example, faculty members at the University of South Alabama are especially good sources of information on marsh plants. Both the University of Alabama at Tuscaloosa and Auburn University have collections of Alabama plant specimens.

Plants in Alabama—General Considerations

Plants have a much different relationship with their environment than do animals. If, for some reason, the environment becomes inhospitable, many kinds of animals can move. In some case, they move hundreds of miles to find more suitable environments. Plants are not so endowed. Fortunately, evolution has provided them with special characteristics that allow them to survive and even thrive in changing environments. They can become dormant, and remain in that state for long periods of time. Many of them have an annual lifestyle where the connection between generations is maintained by the production of seeds. In some plants, seeds are extremely resistant to adverse environmental conditions and can remain dormant for many years.

Plants and Their Environments

As was discussed in Chapter 3, the geophysical world is in a state of constant change. Over long periods of geological time, continents are continuously changing position on the face of the earth. New oceans are formed and new mountain chains are thrust upward then eroded away. As continents change positions with respect to the equator and the poles, climates are also constantly changing. There are some shorter term climate changes, such as the periodic ice ages that are generally independent of tectonic plate activity. These shorter term changes probably reflect changes also occurring in the solar system. The overall result of these geophysical changes is that the environments for plants are constantly changing. Because they are so dependent on the geophysical environments in which they live (especially temperature and precipitation), they must also change if they are to survive over the long term.

Variability In Plant Environments

Temperature, light, carbon dioxide, oxygen, water, mineral nutrients, soil, etc. are all vitally important in the life processes of plants. These environmental factors are ubiquitous; that is, they are found literally all over the surface of the earth, from high mountains to the floor of the sea. Although they are present everywhere, they are not distributed uniformly. The amount of oxygen in water is much lower than in air. There is very little light at the bottom of a lake compared to that present on a mountain top. There is very little water in the A Horizon of the soil in the desert, while the soil at the bottom of a shallow pond is saturated. The temperature of air and soil decreases with distance north and south from the equator, and at a specific location it changes both diurnally and annually. The nature of soil changes with depth and often from one field to another nearby. As a result of these differences, related to time and space, there are an almost infinite number of environments where plants can live. Because of their specific characteristics, some of these environments are extremely poor sites for virtually all plants. These would include the tops of high mountains where the air temperature is always below freezing, the bottom of lakes where there is no light, and in hot springs. However, except for the most inhospitable of all of these environments some form of seed-bearing plant, lives, grows, flowers, and fills seed there. In the 300 to 400 million years that seed-bearing plants have been on earth, evolution has shaped and re-shaped them countless times, so that they fit a large portion of an almost-infinite number of environments. This shaping and re-shaping of biological characteristics by evolution has resulted in the development of plants that function relatively well even in extremely harsh environments and others that thrive in more hospitable locations.

Climate and Plant Distribution

In general terms, climate (primarily temperature and precipitation) is the major determinant of the environments in which plants live. Whitaker (1975) and Ricklefs (2001) published a diagram showing how temperature and rainfall affect the distribution of forests (biomes) worldwide (Figure 4.1). The diagram provides only a general guide, because the boundaries between the different zones are indistinct. Shantz (1938) notes that soils are also extremely important in determining the distribution of plants. He comments that an evaluation of the kinds of plants growing in a given area can be used as an indicator of the climatic conditions under which they developed and of the soils on which they grow. But then, as noted in Chapter 3, the distribution of soils is also largely dependent on climate. Climatic regions and the environments related to them have been, within general limits, distributed relatively stably around the world for thousands of years (Lydolph, 1985). As a result, patterns of vegetative distribution have evolved to fit them. Ritter (2007) developed a forest distribution map for North America (Figure 4.2). There is considerable similarity between the distribution of forests and the distribution of climate types and subtypes proposed by Lydolph (Figure 2.10).

Plant Evolution and Environmental "Fit"

Evolution has resulted in the development of thousands of different species of plants that fit a vast number of different environments. However, in most cases, a given species of plant does not precisely fit a broad spectrum of different environments. As an example, the slash pine is well adapted to grow and reproduce in the seasonably wet flatwoods, inter-dune hollows, and near-coastal sands in south Alabama, generally below a line running from northern Washington County across to the extreme south-central portion of Butler and across to northern Houston (Harper, 1928). It is not found in self-sustaining populations north of this line. Similarly, the eastern hemlock grows and reproduces in cool ravines and gorges in Franklin, Marion, Jefferson, and Jackson counties (Harper, 1928). It is not found south of these counties, or at lower elevations or more open areas, even in adjoining counties. The distribution of slash pine and eastern hemlock seems to be determined by temperature of the environment, but there are many other environmental factors that limit the distribution of other species. For example, the red cedar is found in most areas of Alabama, but it is common only where the soil in which it lives was derived from limestone. The button-bush is also found throughout the state, but only in ponds, sloughs, stream banks, and other wet areas. In contrast, the longleaf pine grows best in somewhat dry, deep sandy soil.

Many species that fit a relatively limited number of environments will grow relatively well outside of those environments if planted and if protected from competition from species that are better fits. For example, the southern magnolia is found in self-sustaining populations only in the southern portion of the state, generally below a line drawn from southern Lee County in the east to central Sumter County in the west. If planted in a garden or park environment, however, the plant will survive and grow as far north as eastern Pennsylvania (Harper, 1928). Nevertheless, a naturally sustaining population would not be established that far north.

Introducing Plants To New Environments

The success of agriculture in meeting the food and fiber needs of a growing human population has been largely based on moving plants from one environment in one part of the world to a generally similar environment in another part. Moving the potato from South America to Europe brought about enormous benefits to the food supply situation there. Similarly, corn from the Americas has become a primary food crop in parts of Africa. The soybean from China, as well as wheat and oats from Europe, are primary crops in the highly productive system of American agriculture. Generally, movement of a plant between widely separated geographical locations will be successful if the climates and soils of the two environments are similar. In most cases, the plant would fit equally well into both; however, the exactness of the fit is extremely important in agriculture.

Loomis and Conner (1992) note that when crops are grown outside their optimal environment even small environmental differences can result in large variations in yield. Under conditions where the fit is not optimal, not only will average yields be lower, but the year-to-year variability will be larger. Soybean yields in different areas of the United States provide a good example of these phenomena. Table 4.1 contains data on the annual average yields of soybeans (bushels an acre) in five states during the period 1987-1998 (National Agricultural Service, US Department of Agriculture). The data in the table show that soybean yields in Alabama, Georgia, Illinois, Indiana, and Iowa averaged 24.29, 23.67, 39.92, 40.21, and 41.58 bushels, respectively during the period. Yields were 69% greater in the three Midwestern states than in the two Southern states.

The states included in the table generally represent five somewhat different environments in which soybeans were produced commercially in the country during that period. It is obvious from these data that soybean yield varied considerably in the states of the Southeast as compared to those in the Midwest. At the same time, there was relatively little difference in production between the states in the same region. Following the reasoning of Loomis and Connor (1992), the significantly higher yield in the Midwestern states, especially in Iowa, suggest that the environmental fit of the soybean is much better in that region than in the Southeast.

Of course, the estimates of yields include a genetic response factor as well as a technology factor. Even if the genome fit the environment extremely well, the use of poor agricultural technology would likely mask its effect to some extent. However, it is not likely that differences in agricultural technology and practices could explain more than a small fraction of the differences shown in the table.

Soybeans have been grown in the United States for a relatively short time compared to corn. Some part of the corn genome has been used in Alabama for centuries, but even in that long period of time, the fit in the South remains much poorer than in the Midwest. Data presented in Table 4.2 presents the annual average yields of corn in Alabama, Georgia, South Carolina, Illinois, Indiana, and Iowa for the period 1987-1998. The data show that yields are 60% greater in the three Midwestern states than in the three Southern states.

Data presented by the 1850 US Census show that annual average corn yields in Alabama, Georgia, South Carolina, Illinois, Indiana, and Iowa for the period 1849-1850 were: 15, 16, 11, 33, 33, and 32 bushels an acre. Even at that time, average yields were 132% greater in the Midwest than in the South. At that time, it is highly likely that a considerable portion of these difference was the result of a poorly fitted genome. Between 1849-1850 and 1987-1998, average yields in the three Southern states increased by 461% (14 to 78.5 bushels an acre). In the Midwest, they increased by only 284% (32.6 to 125.1 bushels an acre). The relatively large increase in the South is almost certainly the result of a vast improvement in technology and practices. It is obvious from all of these data that the corn genome does not fit the South much better than the soybean genome.

The Seed-Bearing Plants

Mohr's *Plant Life of Alabama* (1901) is probably the most comprehensive work on the state's plants. Mohr estimated that a complete list would contain between 2,500 and 2,550 species and varieties of seed-bearing plants. He did not include any of the non-seed plants, such as algae and mosses, and they are not included in the considerations in this chapter. Mohr also commented that while the flora of Alabama and adjoining states is similar, Alabama surpasses all of them with respect to the number of species and varieties and the diversity of their associations.

General Characteristics

Most Alabama seed plants are Angiosperms, plants whose seeds are covered or develop within a flower as a fruit or nut (oaks, persimmon, blackberry). Mohr's list included only 12 species of Gymnosperms, plants with naked seeds such as pines, hemlocks, and cypresses. Within the Angiosperms, most of the forms (over 1,700) belong to a group of plants known as dicotyledons (class *Magnoliopsida*), because there are two seed-leaves in each seed. This group includes oaks, poison-ivy, wisteria, and black-eyed Susan. Most of the plants in this group have broad leaves. Another group, the monocotyledon (class *Liliopsida*) have only a single seed-leaf in each seed. Most of these plants have grass-like or strap-like leaves. This group—which includes corn and other grasses, lilies and cattails—is represented by 700 forms.

Thickets, Forests, and Fields

Information presented in the previous three chapters summarize some of the characteristics of Alabama's geography, physiography, climate, and soils. It was noted that the state is located on the southeast corner of a large continent between the tropics to the south and the temperate zone to the north. Alabama's topography ranges from 2,200 feet at the tops of mountains in the northeast to sea level in the southwest. Throughout the state, a myriad of large and small streams have carved-out thousands of acres of lowlands. In the winter, Alabama's climate varies from cold and moist in the Tennessee Valley, to cool and wet in south Mobile County. In the summer, it varies from warm and moist in the north to hot and wet in the southwest. Its soils varies from rocky on the Cumberland Plateau, to heavy clay in the Black Prairie, to deep sand and muck in the Coastal Lowlands. With this relatively wide range of geophysical characteristics and their interactions, it would be expected that there would be a large number of different environments for plants within the state.

According to Stauffer (1960), the name Alabama is derived from two words in the Choctaw Indian language that together mean thicket clearers. For much of the state, this is an appropriate name. Alabama truly is a land of thickets. Associations of larger trees with smaller trees, saplings, seedlings, and shrubs all tied together by a number of different kinds of vines and briars form millions of acres of thickets throughout the state, especially in the southern half. Many are so thick that they are virtually impenetrable. The impenetrability of the titi thickets in the creek swamps of the Southern Pine Hills (Figure 1.11) and of the pine-sweetgum-yellow jessamine-catbrier-blackberry thickets found on cut-over lands of the upper portion of the Coastal Plain are legendary among those who have occasion to pass through them.

Thickets are really forests, and these unique assemblages of plants dominate the landscape of Alabama. Before the coming of the Native Americans, some 10,000 to 20,000 years ago, virtually all of the state's 32.5 million acres must have been covered by forests. There is, however, some question as to whether the land of the Black Prairie Physiographic District was ever completely forested. Early inhabitants reduced the percentage of land in forests somewhat; although it is difficult to determine to what extent. Hudson (1976) suggested that the native peoples reduced the amount of forest cover in some areas of the state far out of proportion of their numbers. They routinely burned large areas of land to improve grazing for the whitetail deer. They started fires in the forest to surround and concentrate wild animals to make it easier to harvest them with their spears and with their bows and arrows. They also routinely burned their farming areas before planting. The result of this widespread use of fire was to create large expanses of grassland with only a few large trees, especially in the uplands. William Bartram, the naturalist, traveled widely in the region from 1773 to 1778. He wrote that the fields adjacent to the ancient town of Apalachula in what is now Russell County stretched beyond the scope of sight (Stauffer,1960). Stauffer concluded that there might have been more forestland in Alabama in 1850 than in 1650.

From the early 1800s, as more and more settlers came into Alabama and established farms, an increasing acreage of forestland was converted to cropland; however the percentage of land in forest probably never fell below 59% (Stauffer, 1960). From about 1960 onward, the percentage has slowly increased. It was 65.6% in the early 1970s (Murphy, 1973), 66.8% in the early 1980s (Rudis, et al.,1984), and 67.6% in the early 1990s (McWilliams, 1992). The percentage is probably slightly higher today, as the number of farms continues to decrease in the state. In the 10-year period between 1987 and 1997, the amount of Alabama farmland decreased by a million acres (10.7 to 9.7 million) (Vanderberry, 1998). A high percentage of this land will ultimately go into forest. The approximate percentages of total land area in forests in each of Alabama's 67 counties in 1990 are shown in Figure 4.3.

The term forest is a simple collective term for a very complex assemblages of different species of plants, each of which has relatively specific environmental requirements. As a result of these different requirements, forests are composed of different patches of species located on different parts of the landscape (Figure 4.4). Patches of loblolly pine are found on dry hillsides. Cypress trees are found in wet swamps. Live oak trees are found in the southern part of the state. Virginia pine are found in the north. This patchy nature of Alabama's forests has been described by Thomas (1973), Akin (1991), and McWilliams (1992).

While fields, pasture, and open places may not be as common today as they were in past years, enough of them remain to provide environments for those plants that require open areas. Because of the nature of plant ecology in Alabama, fields and open places do not remain fields and open places very long without almost continuous attention. Without annual plowing, mowing, or the use of fire, these areas quickly revert to thickets. As a result of the constant attention they receive, plants that live in open areas must grow and reproduce quickly. Seed germination, root-and-shoot development, flowering, and seed development must begin in the late winter or very early spring, and the entire process must be completed in a short time. Plants that live in temporary waste areas around fields and pastures usually have a little more time to complete their life cycles, but it is also to their long-term advantage to complete the process early and quickly. Some plants inhabit permanent waste areas along fence-rows or in wet places. They can go through the life cycle more leisurely, but even there, competition from other plants also places restrictions on how much time they have.

Herbaceous Plants

While the woody plants constitute the majority of the biomass of plants in Alabama, there are many more species of herbaceous plants, such as clovers, violets, and goldenrods. About 85% of all of the approximately 2,500 different kinds of plants in Alabama are herbaceous, rather than woody (Mohr,1901). Half of those herbaceous species belong to just four families. Members of the sunflower family (*Compositae* = *Asteraceae*) are most common, and there are some 300 species of plants in the state that belong to this family. The ubiquitous black-eyed Susan seen blooming profusely along Alabama roadways in late summer is a *Compositae.* Most of the members of this family have large, showy blooms, which attract insects that carry pollen from flower to flower.

Members of the grass family (*Gramineae* = *Poaceae*) are the second most common of the herbaceous plants. There are 280 species in Alabama that belong to this family. Grasses are found throughout the state, but it is the corn, wheat, and oats in the fields, as well as the grasses in the pastures and on lawns that are the most important. Except for the large swatches of green, there is not much about these plants to command our attention, because grasses do not have showy flower petals to attract insects.

The sedges (*Cyperaceae*) are the third most common family of herbaceous plants. There are 140 species of this family in Alabama. Unfortunately, although there are lots of species, they are not very obvious on the landscape. Most of them are found in marshy areas. The tall wool grass stems with their yellowish mop-like heads of tiny flowers seen blooming in wet areas in the late summer are probably the most noticeable species of this family. Farmers and gardeners generally have an unfavorable familiarity with another member of this family: nut grass. However, wildlife managers and hunters plant large quantities of a closely related species (*Chufas*) to provide food for wild turkeys.

The fourth most common family of herbaceous plants, in terms of the number of species, is the *Legumiosae* = *Fabaceae*. This family includes both small and large trees (redbud and honey-locust), shrubs (bi-color lespedeza), woody vines (wild-wisteria), and a large number of herbaceous species. Many of the herbaceous species of the family are welcomed, because they accompany spring as it sweeps across Alabama. Least-hop and low-hop clovers and some of the wild vetches are among the earliest blooming herbaceous plants on Alabama's landscape. A little later, crimson clover seems to set on fire the roadsides and medians of the interstate highways.

While the overpowering presence of our massive woody forests captures our imagination, it is the little herbaceous wildflowers that grow along their margins, along the roadsides, around the edges of fields, on wasteland, and even within the forests themselves that color our lives and lift our spirits. Dean, et al. (1973) commented that Alabama is truly blessed with the diversity and beauty of the wild flowering plants that we have within our borders. Alabamians make wonderful use of the wealth of cultivated flowering plant material commercially available to them to enhance the space around and within their homes, but it is the beauty of the flowers of the wild herbaceous plants that fills in the background.

The beautiful wildflowers of the herbaceous plants are literally everywhere. The small greenish-yellow flowers on the long stalks of the sea oats nodding in the sea breeze along the upper sand dunes in Gulf State Park in

Gulf Shores break the monotony of sea and sand. Not far away, along the edges of the brackish-water marsh, west of the outlet for Lake Shelby, the large, showy, pink to purple flowers of the wild cotton plant color a tangle of vegetation. In the northern part of the state, the bright yellow flowers of the golden ragwort seem to be on fire in moist meadows. Between these geographical extremes, the broad pastures of the Black Prairie lying across the central part of the state often appear to be solid blankets of white or pink blooms in late spring, furnished free, courtesy of the evening primrose. Fortunately, this species is not particular about where it displays its color for us. It is also found along roadsides, in ditches, and in fields in virtually every part of the state. Tondera, et al. (1987) commented that this species can be found showing-off brightly even along city sidewalks and in pavement cracks.

Hundreds of species of herbaceous flowering plants —such as passion flower (the may-pop), morning glory, stiff verbena, butterfly weed, sensitive brier, and wild geranium—color the roadsides, byways, fields, and wasteland. Around the ponds and in the wet ditches along the way, arrowhead, pickerel-weed, primrose willow, and cardinal flower provide color. In those areas, the tiny, brown, compressed flowers in the long cylindrical heads of the cattail do not provide much color, but they do provide wonderful perching sites for another colorful inhabitant of wet places, the red-winged blackbird. The forests are dominated by trees, shrubs, and vines, but even there some colorful herbaceous flowering plants brightens the shadows. These include: Jack-in-the-pulpit, Atamasco lily, wood sorrel, partridge berry, trillium, and mayapple.

Wild, herbaceous flowering plants are found everywhere throughout the state, and fortunately there is something blooming someplace almost throughout the year. Henbit, bluets, toadflax, bird's foot, and common blue violets color the warming landscape of late winter and early spring. In late spring and early summer, crimson clover, black-eyed Susan, Queen Anne's lace, and common mullein appear. Then, in late summer and early fall, rattlebox, partridge pea, common goldenrod, and Sweet-Joe-Pye weed cheer up the drying, cooling landscape, getting us ready for leaf-fall.

The wildflowers above are only a few of the more common of the wealth of herbaceous wildflowers found in Alabama. Dean, et al. (1973) and Tondera, et al. (1987) describe many more, and provide excellent color illustrations of all of them. Tondera's illustrations are especially attractive.

Woody Plants

As noted in a preceding section, most of the species and varieties of plants in Alabama are herbaceous. Slightly more than 13% of all species are woody (oaks, huckleberries, wild grapes). Woody plants are divided into three general groups: trees, shrubs, and vines. Both large and small trees are those woody plants that have a relatively long, straight stem. When mature, their crowns occupy the upper-story of the forest. Shrubs generally have crooked stems with several crooked branches. When mature, they occupy the under-story of the forest where they form the undergrowth. Vines are unusual plants. As they grow, they generally creep and climb among the shrubs in the undergrowth, but some of the species are able to grow upward along the trunks of trees until they reach the very top canopy of the forest. Mohr listed 172 trees, 171 shrubs, and 32 vines in his book. In a later work, Harper (1928) listed 136 trees, 150 shrubs, and 32 vines in his *Economic Botany of Alabama*, Part 2. Harper also listed the various families of woody plants (Table 4.3). In a later publication, Davis and Davis (1975) listed 174 species of trees in their *Guide and Keys to Trees of Alabama.*

Trees

In terms of total volume, trees, vines, and shrubs (forests), especially trees, comprise most of the plant material found in Alabama. Data presented in Figure 4.3 show the extent and distribution of these plants in 1990. McWilliams (1992) estimated that in that year the total marketable woody biomass (stems) of 109 of the most important tree species in Alabama was 840 million tons (dry weight) (Table 4.4). The biomass of loblolly pine alone was estimated as 173 million tons.

Harper (1928) identified 169 species of large Alabama trees such as loblolly pine and small trees such as the flowering dogwood. There were more different kinds (24) of oaks (Family *Fagaceae*) than for any other family of trees. There were also 13 different cone-bearing trees

(Family Pinaceae) and 10 different hickories and walnuts (Family Juglandaceae) on his list.

The Alabama Forestry Commission's map (Figure 4.4) shows the distribution of different types of forests in Alabama. It is generally similar to one published by Shantz (1938), who used it to show the relationship between the distribution of soils and vegetation in the United States. The commission's map shows five different forest-types: oak-hickory, oak-pine, loblolly-shortleaf pine, longleaf-slash pine, and oak-gum-cypress.

Some of the characteristics of these different plant associations are described in the following sections:

The Oak-Hickory Forest

This plant association in Alabama (Figure 4.4) is located primarily on the northern portion of the Limestone Valleys and Uplands Soil Area in extreme north and northwest Alabama (Figure 3.3). This particular forest-type extends far beyond the state from southern Louisiana north to southern Michigan and northeastward along the Appalachian Mountains to eastern New York. The most common trees found in the Alabama portion of this large region include four species of oak, three species of hickory, sweet-gum, yellow-poplar, black walnut, winged elm, and basswood. There are relatively few pines. Usually less than 5% of the trees in this forest-type are conifers.

The Oak-Pine Forest

This plant association (Figure 4.4) is primarily restricted to the Cumberland Plateau, Piedmont Upland, and Coastal Plain Soil areas (Figure 3.3). A small portion of it is also located on the southern portion of the Limestone Valleys and Uplands Soil Area. Most of this forest-type is found on the Cumberland Plateau and on the northwestern portion the Coastal Plain Soil Area. The most common trees include four species of oak, three species of hickory, sweet-gum, yellow-poplar, plus loblolly, shortleaf, and longleaf pine. The loblolly pine is the most common species in the association. The pines represent a significant part of the forest, but generally not a plurality.

The Loblolly-Shortleaf Pine Forest

This plant association (Figure 4.4) generally covers the southern two-thirds of Alabama, including the southern portion of the Cumberland Plateau, the Piedmont Upland, and all but a small, narrow portion of the Coastal Plain Soil Areas (Figure 3.3). However, it is primarily restricted to the Piedmont Upland and Coastal Plain areas. This particular forest type is also part of a much larger one. The entire region lies along the Coastal Plain Physiographic Province (Figure 1.10) that reaches from eastern Louisiana to southern New Jersey. Loblolly pine, shortleaf pine, Virginia pine, sand pine, and spruce pine are all found in this region. Sweet-gum, oaks, hickories, yellow-poplar, red maple, and black-gum are common associates. The pines are a plurality in the association. One of the reasons for the predominance of pines in the southern portion of the state has been the shift of hundreds of thousands of acres of land from cultivation into woodland over the last 50 years. Because of the broad wing attached to their naked seeds, the pines, especially the loblolly (old field pine), are extremely effective in colonizing abandoned fields. Wind carries the seed over such relatively long distances that they regularly establish almost pure stands in abandoned fields. Akin (1991) suggests that over a long period of time that deciduous broadleaf species such as the oaks, hickories, and sweet-gum will slowly move into those stands of the loblolly pine-shortleaf pine forest-type converting them back into the same oak-pine forest-type now found on Cumberland Plateau Soil Area. Obviously, where forest managers regularly use controlled burning to kill hardwood seedlings and saplings and use clear cutting, chemicals, and pine seedling planting to establish new stands, this gradual return to a more diverse forest will likely not occur.

The Longleaf-Slash Pine Forest

This forest-type, as described by Harper (1943) and Akin (1991), is a narrow band of woody vegetation along the Gulf of Mexico from east Texas to coastal South Carolina. In Alabama, it is restricted to the higher elevations in the southern tier of counties in the Southern Pine Hills District (Figures 1.25 and 4.4). This portion of Alabama lies within the south-westernmost part of the Coastal Plain Soil Area (Figure 3.3). This is a unique forest association that can survive only in areas where winters are essentially frost-free or nearly so. The uniqueness of the forest rests in the fact that a number of species of both large trees (live oak, water oak, and southern magnolia) and small trees (wild olive, red bay, and yaupon) are evergreens. While they do shed their leaves annually, shedding is not generally related to shorter day length that triggers leaf fall in deciduous trees. Akin suggests that this evergreen forest-type was, at one time, probably much more widely distributed in the state, and that its northernmost margin was pushed southward by the cold accompanying the movement of glaciers into southern Illinois and Indiana some ten thousand years ago. Longleaf pine and slash pine constitute a majority of the association, but there is also a substantial amount of loblolly pine present. Deciduous trees in the association include sweet-gum and white oak.

The Oak-Gum-Cypress Forest

This forest-type (Figure 4.4) is confined primarily to the western portion of the Coastal Plain Physiographic Section (Figure 1.7) and is closely associated with the Alluvial, Coastal, and Low Terrace Deposits (Figure 1.27) within that section. According to Burgess (1943), sweet gum, black gum, and tupelo gum are common in this forest-type, along with several species of oaks. In the southern portion, the white bay is the most common species. Cypress is also prevalent in the larger swamps.

The Prairie Grassland Forest

There is at least one additional forest-type not included on the Forest Commission map (Figure 4.4). Shantz named it the Prairie Grassland Forest. It has never been firmly determined that Alabama's Black Prairie was ever a true grassland. When the first European settlers arrived, it was primarily forested, but it has many of the characteristics of the Tall Grass Prairie shown in Figure 4.2. Shantz concluded that because of its soils it should be included in this vegetative association. This forest-type would be restricted to the Black Prairie District of the Coastal Plain Section (Figure 1.11) and to the Black Prairie Soil Area (Figure 3.3) reaching southeastward across central Alabama. The most common trees found within this region include five species of oak, one hickory, plus red cedar. Ironwood, hackberry, cottonwood, spruce pine, and magnolia are also found in some areas of the forest. Further, it is the only forest-type in the state that includes the Osage orange (*bois d'arc*/bodark). There are few pines, except where they have been planted.

Thomas (1973) also prepared a publication on the distribution of trees in Alabama. He used a map developed by Kuchler (1970) as a basis for his presentation. Although Thomas's distribution is generally similar to that of the Forestry Commission, he does describe several additional associations (forest-types). The most important one is the Mixed Mesophytic Forest. It is found on the moist slopes and in the cool ravines associated with the Tennessee River and its tributaries northeast of Guntersville in extreme northeastern Alabama. Thomas comments that this association contains more different kinds of vegetation than any other forested area of Alabama. Several species of oak and hickory are found there, along with sweet-gum, maple, beech, yellow-poplar, and basswood. There are almost no pines in the association, but this area is one of the few places in Alabama where the eastern hemlock is found. Apparently, this vegetative association is a small remnant of a vast forest that once covered a large portion of the Northern Hemisphere.

Shrubs

Harper also identified 150 kinds of shrubs in Alabama. Table 4.5 lists some of the more common ones. Among the shrubs, members of the azalea and honeysuckle family (Family *Ericaceae*), the St. John's Wort family (*Guttiferae* = *Hypericaceae*), and the blackberry and rose family (*Rosaceae*) were the most abundant. No picture of the woody plants of Alabama would be complete without considering these plants. Shrubs are the second-class citizens of the forest. They generally live in the shadow or shade of both the larger and smaller trees. Many never get taller than four to five feet, and they receive only a limited amount of the sunlight, except where they grow along the edges of clearings. Furthermore, if always living in the shadows of the trees is not enough of a cross to bear, they must constantly compete with up-start seedlings and saplings of their large neighbors, for space, water, nutrients, and sunlight. They receive little of our direct attention; although when we walk in the forest, we are constantly stepping around them or pushing aside their limbs and leaves as we pass through. There are exceptions. An encounter with three of our common shrubs (poison ivy, eastern poison oak, and poison sumac) may not merit our immediate attention, but for many, it will be a most memorable one.

Fortunately, shrubs are survivors, and they live quite well on the floor of the forest. Their presence is of great importance to wildlife there. Many produce fruits (southern gooseberry, huckleberry, American snowbell, and common elder). Others produce fleshy drupes (arrow-wood, grandsie-gray-beard, and swamp-privit) that serve as food for a variety of birds and mammals for much of the year.

Vines

Among the 32 different species of vines that Harper identified, members of the wild grape family (*Vitaceae*) and smilax family (*Smilacaceae*) were more common. The most common ones are listed in Table 4.6. Vines in our forests often get our attention, and hold it very effectively compared to the lowly shrubs. Kudzu, introduced to the United States in 1876, appears well on its way to covering the entire state, unless it is eaten by another Asian import, bean plataspids, introduced about 2009. Another introduced species, Japanese honeysuckle, will likely cover those areas left uncovered by kudzu. Some species such as the rattan-vine and cross-vine, climb to the tops of our highest trees. A bodily encounter with the prickles or spines of the various species of the vines of the Genus *Smilax*, in a tangle of shrubs, on the forest floor definitely grabs one's immediate attention and creates a lasting impression. Feasting on wild grapes and muscadines, fruits of the vines of the genus *Vitis* in the early fall provide a worthwhile encounter. Several of the vines are also important to wildlife. The foliage of the Japanese honeysuckle is a favorite food of the whitetail deer, and many species use wild grapes and the berries produced by the various species of *Smilax*.

Flowers of Woody Plants

The colorful flowers of some of the herbaceous plants found in Alabama were discussed in the preceding section. Woody plants also have flowers, and while the vast majority do not flaunt themselves, a few species put on spectacular displays. The flowering dogwood is a smaller tree in the forest, but it puts on a display of color in the early spring that is far out of proportion to its size. Its profusion of white bracts surrounding its flowers actually seems to light-up the dark forest just awakening from winter. The redbud, also a smaller tree, adds a contrasting bright pink to magenta color to the creamy white of the dogwoods. These beauties literally hang like banners of color on the edge of the forests of some sections of our interstate highways. The display of redbud is especially striking in the early spring along Interstate 65 in the Chunnennuggee Hills and Southern Red Hills Physiographic districts (Figure 1.11) south of Montgomery. The display is especially striking on the north slope of the LaPine Cuesta (Figure 1.15) just north of Exit 142. This floral display is located on a narrow outcrop of the Prairie Bluff Chalk Formation of the Upper Selma Group. The red maple is also a smaller tree, although it does grow to a relatively large size in some locations. Its deep red to pinkish flowers are not large and showy like the bracts of the flowering dogwood, but they always capture our attention because they bloom so early (February and March) in the year, sometimes six to eight weeks before the dogwood.

The flowers of most of our large trees are small and add relatively little color to our forests. The yellow poplar, or tulip tree, is an exception. Although they are not numerous, the large, green to yellowish green flowers with a conspicuous blotch of deep orange at their base, add splashes of color to the highest part of the forest canopy. The southern magnolia is the real show-off among the trees. Its large, cream-white flowers are also very fragrant. The flowers appear a few at a time over a long period, so we have them with us for much of the summer. Unfortunately, the tree is not common in the wild outside the Coastal Plain.

Our shrubs give us some of the most beautiful flowers to be found in Alabama. In the northern portion of the state along the open mountain slopes and bluffs, usually at least 1,000 feet above sea level, beautiful clusters of rose to purple blooms of rhododendron, a moderate size shrub, hang like a pastel blanket over the gray rocks. Harper (1928) comments that this is one of the most gorgeous of American shrubs. Two other members of this same family, the Alabama and Piedmont azaleas, provide vivid splotches of color in the greening-up of the early spring forest. The oak-leaf hydrangea is a handsome shrub with beautiful, elongate clusters of white flowers found on shady ledges, wooded bluffs, and stream banks in rich forests throughout the state. The buttonbush, with its spherical heads of white flowers, is a common shrub found in swamps, around lakes, and along streams. Another shrub of wet places is the hazel alder. Its reddish brown catkins, which appear long before the leaves, are not very showy, but they begin the annual parade of plant flowering in Alabama, as early as January and February. Harper comments that this species is probably the most common shrub in the state.

The flowers of a number of species of vines also color or landscape. The cow-itch-vine celebrates its conquest of the heights in our forests by producing showy, large orange and yellow flowers in the late summer. An encounter with the fragrance of a profusion of the beautiful yellow blooms of the yellow jessamine vine, twining along old fence rows and in low trees in the late winter and early spring, is also a memorable experience, one that is anticipated year after year. The Japanese honeysuckle seems to want to atone for its boisterous pestiness by providing masses of fragrant, cream-colored, sometimes pinkish, blooms from May to November. Few of our plants bloom continuously for such a long period of time. There is no way that the kudzu vine could atone for all of the harm that it causes in our forests and around our homes, but it tries to atone by sporadically producing beautiful, elongate masses of sweet-smelling, lavender flowers that resemble bunches of grapes.

Cultivated Plants

Cultivated plants add a different dimension to Alabama's plant landscape. The neatly arranged stands of loblolly pine in the state's extensive forest plantations, the orderliness of row upon row of cotton plants, and the manicured surface of a well-maintained pasture are in direct contrast to the seemingly disordered thickets and natural assemblages of trees and shrubs. The thousands of home lawns with their plantings of azaleas, camellias, hollies, and shade trees add an even different dimension. Cultivated plants have been important in Alabama for hundreds of years, from the time when Native Americans began to plant and harvest rather than hunt, fish, and gather; although hunting, fishing, and gathering would continue to be important for many centuries. For much of the first century of statehood, Alabama's economy was almost entirely dependent on the cultivation of plants. Land in farms and in cultivated crops increased gradually from the end of the eighteenth century until about 1950 when the total reached 22 million acres, 68% of Alabama's total land area. Since that time, the acreage in farms and in cultivated crops has decreased fairly rapidly. In 1997, the total acreage in farms was 9.7 million acres.

As noted in a preceding section, Dean, et al. (1973) estimated that the flora of Alabama includes 3,000 species of native or naturalized seed plants. Of course, only a handful of this number are actively cultivated. Bulletin 45, Alabama Agricultural Statistics (Vanderberry and Placke, 2003) lists the cultivated plants and groups of cultivated plants contributing the greatest value to the Alabama economy in 2002 (Table 4.7).

It is fascinating that few of the cultivated plants listed in Table 4.7 are native to North America. The genomes of these plant were largely determined in parts of the world far removed from Alabama (Table 4.8), and in most cases, in environments different from those where they are being grown today. In some of the introduced plants, the environmentally-determined genomes that they brought with them were poorly suited to function effectively in the poor soils and subtropical climate in which they found themselves. As a result, using them for crop production has been akin to fitting round pegs in square holes. For example, cotton, a foreigner, fits the Alabama environment rather poorly. The crop would probably see limited use without the protection that government subsidies provide and have provided for almost a century. Further, as detailed in a preceding section, soybeans are so poorly suited for Alabama's soils and weather that even subsidies have not been sufficient to make it a suitable crop for our agroecosystem. Even corn, which was important to farmers in Alabama centuries before the Europeans arrived, cannot be produced competitively.

The evolution of vegetation in Alabama gave us vast forests containing almost untold quantities of fiber, but few grasses that could be used for the production of grain, few native pulses with large seeds, and few small trees whose fruits were suitable for commercial production. It is likely that our soils and our warm, humid climate are primarily responsible for the evolution of the heavily forested landscape. Trees, once established, are able to wait out the bad months and years of bad weather, and then flourish in the good times. Their genomes were determined in this environment, and they perform exceedingly well in it.

Our most valuable plant, the loblolly pine, is a native son. However, although it evolved here, its genome is not the primary reason for its importance. Its importance is guaranteed only by the more or less continuous care given to it by foresters (silviculture). The genomes of the oaks and hickories, likely fit Alabama's overall environment better. Pecans and some nursery, sod, and greenhouse plants are also native sons, yet, individually their commercial value is far less than that of the loblolly pine.

In 2002, Alabama cash receipts from the sale of livestock and poultry and all crops, including farm forest products and non-farm commercial timber, totaled $3,697.8 million (Table 4.7). The sale of crops and all forest products, accounted for 35.7% of the total ($1,319.5 million). Trees from all sources (farm, commercial, and government) contributed the highest percentage of receipts from all plant sources (55.7%). A variety of greenhouse, sod, and nursery plants contributed the next highest percentage (18.4%). Poinsettias, geraniums, petunias, impatiens, Easter lilies, and garden mums were the most important plants in this group. Cotton was the next most important (9.8%). Peanuts, corn, soybeans, and hay contributed 4.8, 2.4, 1.4 and 1.3%, respectively, to the total.

Table 4.5 does not include any vegetables other than Irish potatoes and sweet potatoes. All others were included in the other vegetables category. This group accounted for 2.7% of total receipts. The group included sweet corn, tomatoes, and watermelons. Of course, a number of other vegetables were harvested for sale in 2002, but the quantity involved did not warrant collecting data on them.

Most of the species used in plant agriculture are annuals, which means that the crop is produced in a single annual cycle. Within a single year, as a result of series of highly complex, integrated chemical, and physical processes, a seed germinates, roots and shoots develop, reproductive structures (cones or flowers) are formed, and finally seeds mature within these structures. These processes constitute the biology of the plant under cultural conditions. This is a highly simplistic description of a process that provides food and fiber, directly or indirectly, for all of the earth's animals, including people. Unfortunately, these four seemingly simple stages of the growth and development of a seed plant that are programmed in the genetic complement of its cells must take place in an inhospitable world (drought, floods, late frosts, diseases, insects, limited nutrient supply, etc.). Fortunately, however, evolution and selective breeding have provided the plants with genomes that make it possible for them to be highly successful under average environmental conditions, but not when they appear or occur very far outside of the average range. Because the demand for food is continuous, the success of individuals and populations of individuals, depend on how successful the plants are over time. Through agriculture, humans attempt to manage the biology or life processes of individual plants in large populations (crops), so that they can be as successful as possible within the limitations of their genetic constitution, so that they provide a positive return on the investment required for their management, and so that the level of success can be predicted with a relatively high degree of certainty year after year.

Characteristics

It is impractical to describe the biological characteristics of all of the hundred or so kinds of plants that are cultivated in Alabama. Consequently, I have chosen to provide limited information on five of the most economically significant, based on cash receipts: the most important tree crop (loblolly pine), as well as the most important field crops (cotton, peanuts, corn, and soybeans). I have chosen to provide the same information on the Irish potato, pecan, and azalea; because including them provides a broader perspective of the wide range of plants used in Alabama agriculture.

Loblolly Pine

The loblolly pine (Figure 4.5) is one of the so-called Southern pines or yellow pines. Other species in this group include the Virginia pine, shortleaf pine, spruce pine, longleaf pine, and slash pine. In Alabama, in terms of dry biomass present, the loblolly is more abundant than all of the others combined (McWilliams, 1992). In descending order of importance, the biomasses of the various species were as follows: loblolly pine (173.7 million tons), shortleaf pine (36.6), longleaf pine (28.1), slash pine (25.6), Virginia pine (16.2), and spruce pine (5.9). The second most abundant tree species was the sweet-gum with 65 million tons.

Taxonomy

The taxonomy of the loblolly pine (Figure 4.5) is as follows (After Wahlenberg, 1960 and Mauseth, 1991): Family - Pinaceae; Genus - *Pinus*; Species - *taeda L.*

According to Mauseth (1991), there are five divisions of living seed plants. Included are the divisions Coniferophyta (the cone bearing plants) and Magnoliophyta (the flowering plants). The loblolly pine is in the Coniferophyta Division. It is also a gymnosperm. Seeds of pine trees, as well as other gymnosperm plants, are naked as contrasted to seeds of the angiosperm plants of the Magnoliophyta Division, which are enclosed in a fruit formed from a flower. The loblolly pine also belongs to the family Pinaceae. Other families within this division are the junipers (*Cupressaceae*) and redwoods (*Taxodiaceae*). Within the family Pinaceae, the loblolly pine belongs to the genus *Pinus*. Other genera within this family include the true firs (*Abies*), spruces (*Picea*), true cedars (*Cedrus*) and hemlocks (*Tsuga*). The genus *Pinus* has evolved over the last 200 million years to include 100 different species and varieties and to spread throughout the Northern Hemisphere (Farjon, 1984).

The species name of the loblolly pine is *taeda*, which comes from the Latin word for torch. It is also an ancient name for resinous pines.

Origin

When the early settlers arrived on the southeastern Atlantic coast of the United States, they were confronted by a broad expanse of swamps, swales, lowlands, moist depressions, and mud holes. They applied the name loblolly to some of these areas, and they gave this name to one of the pines they found there (Wahlenberg, 1960).

The ancestors of the loblolly pine first appeared 100 million years ago. Its relatives are distributed throughout the Northern Hemisphere, from hot tropical beaches to cold mountaintops. These plants are well adapted to the environments in which they live. In Alabama, the loblolly pine literally evolved along with the environment.

Although the loblolly pine is well adapted to the Alabama environment, scientists suggest that it was not nearly so abundant historically as it is today (Schultz, 1997). They suggest that before fire began to be used widely by Native Americans, the species was not very abundant. The widespread use of fire killed the competing hardwoods, thus opening up the forest while burning the detritus down to mineral soil. These burnt-over conditions created an optimum environment for colonization by the loblolly pine with its winged seed.

The use of fire by Native Americans provided some opportunity for an increase in the abundance of loblolly pine. However, the most important factor in its expansion was the more recent widespread, regional abandonment of cultivated fields as row-crop agriculture became increasingly unprofitable (Wahlenberg, 1960). The loblolly pine is also often called the old-field pine, because it so readily colonizes abandoned fields.

Economic Importance

In terms of acres planted, pine trees are by far the most important plant crop in Alabama. In 1990, there were 3.4 million acres of planted pines in the state, most of them planted with the loblolly. There were an additional four million acres of unplanted woodland in which pine trees were the predominant species. According to Hartsell and Brown (2000), between 1990 and 2000, Alabama woodlandowners planted 255,000 acres of pines each year. Most of these newly established plantations were planted with loblolly pine.

The Alabama Forestry Commission (2003), reported that in 2002 the total estimated value of stumpage harvested in Alabama was $735.8 million. Pine saw timber, primarily loblolly pine, accounted for 72.8% ($536 million) of this total, and the sale of pine pulpwood accounted for another 9.5% ($69.8 million). It is likely that most of this harvested pine was loblolly (Table 4.4).

Critical Stages of Development

Schultz (1997) comments that it is practical to manage loblolly pines using rotations of 25 to 120 years, depending on economic objectives. Compared to field crops where one-year rotations are the rule, a 25-year or longer rotation puts this crop at risk for an exceedingly long time. Compared to field crops, most of the phases of the development of the loblolly pine require extended periods of time. While the germination time of the loblolly pine seed is comparable to that of the field crops, the seedling of the pine is dependent on food supplies in the seed for a much longer time than in most other crops.

In its earliest years, the young pine seedling and sapling in a mixed forest environment does not compete very well for sunlight, space, water, and nutrients. Young loblolly pine seedlings do not capture enough energy through photosynthesis to allow them to develop a root system that is adequate to provide the young plant with sufficient moisture during periods of drought. By comparison, at six months of age, seedlings of the flowering dogwood grown in the same environment will have 3.5 times as many roots as the young pine seedling of the same age. The loblolly pine is extremely susceptible to competition until it has secured a space in the upper canopy of the forest. In natural, mixed species forests, the species has only limited opportunity of obtaining that critical space.

Because the loblolly pine grows to maturity over such a long period of time, individual trees are subject to attack or damage from a wide variety of insects and mammals, from diseases, and from adverse weather. Schlutz (1997) reported on research by other scientists that indicated in Alabama in 1983 that 23.9% of the loblolly pine saplings, 25.8% of pole timber, and 26.4% of the saw timber were damaged by one or some of those factors. The overall effect of all of these factors have increased in the past half century. Before the early 1950s, most loblolly pines grew in mixed-age, mixed-species forests. Since that time, however, more and more of them grow in even-aged, pure stands. Conversion to pine monoculture has reduced or eliminated the overall impact of several of these adverse environmental factors.

Cotton

According to Lewis and Richardson (1968), when sixteenth century Spanish explorers visited the West Indies and the Aztec, Maya, and Inca civilizations of Mexico and Peru, they found the spinning of cotton fibers to form threads, and the weaving of those threads into cloth, flourishing over this broad geographical area. These Native Americans were using woven cotton for bedding, clothes, armor, awnings, carpets, and tapestries. Cotton was seen commonly in the local markets, and it was an important item of trade. Seeds from some of those cotton plants were introduced into the United States in Virginia early in the seventeenth century.

In British North America, cotton culture was of limited importance. It could not compete favorably with tobacco, rice, and indigo in the marketplace, especially for the export market. There was some spinning and weaving of homespun clothes, using cotton fiber, but most cotton textiles were imported from the West Indies. In England, during the latter part of the eighteenth century, the steam engine plus several pieces of machinery were invented to spin cotton fibers and weave cotton thread. A few years later, the cotton gin was invented in the United States. These inventions effectively industrialized the making of cotton cloth and forcefully thrust cotton culture into a new era, especially in the Southeastern United States.

Perkins (1990) noted that cotton was first cultivated in Alabama in 1772, in the general area where the Alabama and Tombigbee rivers flow together in modern-day Clarke County, and that by 1808 it had become the leading agricultural product of the southwest Alabama settlements. Over the next 150 years, the culture of this crop, along with the spinning and weaving of its lint, would play a major role in the economic and social development of the state.

Taxonomy

The taxonomy of the cotton plant (Figure 4.6) is as follows (After Munro, 1987 and Mauseth, 1991): Family - Malvaceae; Genus - *Gossypium*; Species - *hirsutum*.

Cotton belongs to the family Malvaceae, which also includes the hibiscus, hollyhocks, okra, and mallows. It belongs to the genus *Gossypium*. There are many species of this genus growing wild throughout the tropical and subtropical regions of Africa, Asia, Australia, and the Americas. Lewis and Richmond (1968) noted that worldwide, there are 31 recognized species of cotton. Of these, 27 are wild and four cultivated. According to Brown and Ware (1958), in 1536 the Spanish explorer de Vaca found wild cotton plants growing in the area that was to become Louisiana and Texas. Today, semi-cultivated, mostly wild, native forms of cotton grow in Arizona. All cottons are annual shrubs, perennial shrubs, or small trees.

According to Munro (1987), all of the world's cultivated cottons were derived from four species of the genus *Gossypium*. These include:

Old World diploid cotton:

G. arboreum

G. herbaceum

New World tetraploid cotton:

G. hirsutum (Upland cotton)

G. barbadense (Sea Island cotton)

Lewis and Richmond (1968) suggest that it is possible that both of the New World species developed from a common ancestor.

Origin

According to Brown and Ware (1958), the English word cotton comes from the Arabic words *qutun* or *kutun*. Munro (1987) notes that in Peru and India, between 4000 and 3000 B.C., that lint from cotton seed was being twisted into threads and the threads woven into textiles. Tetraploid cotton provided the lint in Peru and diploid cotton in India.

The Greek historian Herodotus (484-425 B.C.) wrote about cotton in India: "There are trees which grow wild here, the fruit of which is a wool exceeding in beauty and goodness that of sheep. Indians make their clothes out of it."

Because of the nature of their chromosome complements that carry their genes, Old World cottons are referred to as diploid cottons. The original cottons found in the New World were also diploid cottons; however, possibly as long as 100 mya, cotton seed (probably *G. herbaceum*) from the Old World somehow reached the New World and developed into plants. They probably were carried from Africa to South America by ocean currents. Later, the Old World form crossed with a New World form, creating tetraploid cotton in which the chromosomes of the two parents are combined and retained. All the original, or diploid cottons, found in the Americas are lintless; although they do have short hairs or fuzz attached to the seed. This means that they do not have the long fibers needed for making thread and cloth. The Old World diploid cottons are all linted, or have the fibers attached to the seed. Fortunately, the tetraploid cottons inherited the trait for producing lint.

Lewis and Richmond (1968) comment that over a long period, stable populations or a center of origin of *G. hirsutum* developed in southern Mexico and Guatemala, and that when it was domesticated, the range of this species generally expanded to the north and west. At the same time, a center of origin developed for *G. barbadense* in northern Peru. Later, its range expanded northward and eastward toward the islands of the Caribbean.

In the sixteenth century, the Spanish took seed of the New World tetraploid cottons to Europe. Over the next century, these cottons were introduced to many parts of the Old World.

Native Americans were not growing cotton in the Southeastern United States when the first European settlers reached this region. However, cotton was fairly widely used among Indians in the Southwest at that time. Europeans had become familiar with New World cotton, at least a half century earlier through their colonization efforts in the tropics. The first seeds of *G. hirsutum* were probably brought to Virginia by travelers, traders, or officials very early in the seventeenth century. Brown and Ware (1958) comment that there is evidence that cotton was being grown in the colony in 1607. It was soon determined that it could only be grown successfully in the uplands, away from the Tidewater region. From that time *G. hirsutum* was known as Upland cotton.

Sea Island cotton, *G. barbadense,* was introduced into the United States from the West Indies after the colonial period, about 1785. Because of its physiological characteristics, it could be grown on the islands off the coasts of South Carolina and Georgia and on a narrow strip of land along the coast. The lint of this species was unusually long and silky. Consequently, it was much sought after for use in weaving high quality textiles. Over time, high production costs and heavy damage by insects and diseases reduced the profitability of this species.

Economic Importance

Data on the acres of cotton planted annually in Alabama during the period 1950-2000 are presented in Figure 4.7. The data indicate that in 1950 Alabama farmers planted 1.3 million acres. Acres planted increased slightly through 1953 (1.6 million), but then began to trend downward. In 1983, it reached the lowest level in the twentieth century (219,000 acres). Since that time, government subsidies have made it more profitable to grow the crop, and as a result acres planted has trended slowly upward, reaching 590,000 acres in 2002. The downward trend from the early 1950s through the mid-1980s generally reflects the loss of farms during that period.

In 1992, cotton acres harvested accounted for 19.4% of total acres of cropland harvested in Alabama. In 1997 and 2002, the percentages were 21.8 and 25.6%, respectively. In 2002 (Table 4.7), cash receipts for cotton accounted for 4.1% of total cash receipts for all commodities (livestock plus crops), including farm forest products. In 1980, the comparable percentage was 6.8%.

Critical Stages of Development

The cotton plant evolved in relatively warm, dry areas of the tropics and subtropics. Under those conditions, it was a perennial plant. The same plant would live and produce flowers and fruit for several years. Considerable progress has been made in adapting it to temperate climate areas around the world, but it has not been possible to change many of its basic characteristics. In subtropical and temperate areas, cotton is an annual plant. As it dies when exposed to freezing temperatures, populations must be re-established each year by planting seed. Some five to six months, depending on the variety, are required to complete the life cycle from seed germination, to the opening of the mature fruit, and to the harvest of the lint. As a result, the plant is at risk from a variety of adverse environmental factors for a long time. The level of risk is increased because of its indeterminate flowering and fruiting habit. There are flowers being formed, fertilization taking place, and fruit growing continuously for several months.

To fit the life cycle into the available growing season in most temperate areas, seed must be planted early in the spring while there is still considerable likelihood of several days of relatively low temperatures, and farmers are usually racing against time to get the crop harvested before the first freezing-day in the fall. In the intervening months, cool weather, floods, drought, wind storms, hail, weeds, insect pests, diseases, and nutrient balance in the soil can, and often do, significantly affect the physiology of the plant to the extent that the yield of lint is reduced. Unfortunately, even after fruit development is complete and the lint erupts, the crop is not necessarily safe. In the Coastal Plain portion of the Old Cotton Belt, which stretched from Virginia to Texas, the wind and rain associated with fall storms originating in the tropics (tropical storms and hurricanes) destroy or significantly damage entire crops with some regularity.

Corn

About AD 800, an eastern variety of tropical flint corn became the primary plant in the farming system that supported the development of Native American cultures in the Southeastern United States (Hudson, 1976). When the Spanish explorer, Narvaez, entered the territory of the Apalachee Indians in the north Florida area, he found large supplies of corn in the fields. Eastern flint corn was extremely important to the development of the Mississippian Culture (AD 900-1600) throughout the Southeast, and to the very survival of the early European settlers in the area. Without the availability of the corn production technology that the Europeans adopted from the Native Americans, colonization and the expansion of those colonies inland would have been much more difficult. Well into the nineteenth century, and to some extent into the early part of the twentieth, corn was the most important food in the diets of a majority of Alabamians.

Taxonomy

The taxonomy of corn (Figure 4.8) follows that of Aldrich, et al., 1986 and Mauseth, 1991: Family - Poaceae (= Graminae); Genus - *Zea*; Species - *mays.*

The corn plant (Figure 4.8), *Zea mays*, is a member of the grass family Poaceae (= Graminae) that includes some 10,000 species and most of the important food plants grown throughout the world including barley, buckwheat, grain sorghum, oats, rice, and wheat. In a larger sense, corn is an angiosperm, because its seeds develop in a covered structure, and it is a monocot, because as the seed germinates only a single seed leaf, or dicotyledon, will be produced. Corn is used as the name of the plant throughout North America, but in the remainder of the world, the name maize seems to be preferred.

Origin

There is still some disagreement about the origin of corn. A mural on the wall of the Corn Palace in Mitchell, South Dakota, depicts a Native American myth concerning its origin. In the mural, the first corn plant is seen rising from a drop of milk that has fallen from the udder of the sacred buffalo. The likely origins are almost as mythical. Some scientists suggest that it was developed from an ancient corn ancestor that has become extinct. Solbrig and Solbrig (1994) suggest that the most likely explanation is that Indian corn was developed through selective breeding of a domesticated, corn-like grass called teosinte. They further suggest that much of this early selection took place in Mexico, and that an early form of corn was introduced into Peru at least 5,500 years ago. From 4000 BC to AD 1000, Native American farmers, without any understanding of genetics or plant breeding, accomplished the remarkable feat of developing varieties of corn that could be used in different environments throughout the Americas. They apparently paid special attention to the development of varieties that grew and matured at different times in the growing season, so that by growing a mixture of these, they could expect some production every year. As a result of these and later breeding programs undertaken by other plant breeders, corn is cultivated in more widely diverse climates than any other cereal crop (Jenkins, 1941).

Economic Importance

Data on the acres of corn planted annually in Alabama during the period 1950-2000 are presented in Figure 4.7. The data show that Alabama farmers planted considerably more corn than cotton in 1950 (2.678 versus 1.335 million acres). After 1950, acres of corn planted trended downward until 1972, when it reached 631,000 acres. It increased slightly during the mid-1900s; only to begin to decline again in 1985, and by 2000 reached 230,000 acres. The 1999 and 2000 acres planted were somewhat lower than expected, as a result of poor crop weather at the time of planting. A portion of the decline in corn acreage in the state can be ascribed to the loss of farms during the period, but the more important reason is that Alabama has little comparative advantage in corn production.

In 1992, acres of corn harvested accounted for 14% of total acres of cropland harvested. In 1997 and 2002, the percentages were 12 and 9%, respectively. In 2002, cash receipts for corn harvested in Alabama, accounted for only 1% of total cash receipts for all farm commodities (livestock and crops), including farm forest products. In 1980, the comparable percentage was 0.9%.

The corn genome is not a very good fit for the agroecosystem of Alabama and the South. Average yields of corn in the state are typically at least 40 bushels an acre below the national average. The differential is considerably greater when the comparison is made with states in the Corn Belt. Data on average yields of the crop, during the period 1987-1998, in Alabama, Georgia, South Carolina, Illinois, Indiana, and Iowa are presented in Table 4.2. Over this 12-year period, the average yield of the three states in the South was 60.2% lower than in the states in the Midwest.

Critical Stages of Development

The growing corn plant in Alabama is always at risk from negative environmental factors, and it is probable that no plant ever reaches its full genetic potential. However, there are some stages of development that are more critical or where the degree of risk is greater. The first of these is at germination. The seed is highly resistant; however once the seed coat is broken in the first stages of germination, its susceptibility increases rapidly. At this stage, the food-rich endosperm is subject to attack by a host of soil-borne bacteria and fungi. The starch and protein found there provide an excellent medium for their development. If the pace of development is slowed by cool, damp weather, populations of these organisms grow so large that they destroy the food supply for the embryo. Under the worst conditions, they can grow on the tissues of the embryo itself, leading to its death. If a hard crust develops at the surface of the soil, as a result of rapid drying caused by high wind velocity and dry air, the tiny developing stem will be unable to push through the soil to the surface.

The second critical stage in development comes when all of the leaves have formed and the internodes of the plant have begun to grow rapidly in length. This spurt of growth requires immense quantities of water and nutrients from the soil, as well as a metabolic system functioning at high efficiency to provide the necessary carbohydrate substrate or food to energize the entire process. Drought is especially damaging at this stage as is a shortage of nitrogen and phosphorus.

The third critical stage occurs when the metabolism of the plant is shifting from vegetative growth to the production of tassels and pollen, and ears and kernels. If growing conditions are unfavorable at this point, especially if nitrogen metabolism and protein production is negatively affected, the size of the developing ears will be adversely affected. Ear development and silking are more sensitive to these shortages, because more of the plant's resources are going to tasseling and pollen formation. The shortage of nutrients, drought, insect damage, and overcrowding of plants in the field will have their greatest negative affect here.

The first few days after pollination are also critical in the life cycle of the plant. Aldrich, et al. (1986) comment that if proteins and sugars are in short supply as a result of drought, disease, or nutrient deficiencies because of cloudy weather, competition for light among heavily crowded plants, or other adverse conditions, kernel size and weight will be reduced and kernels near the tip of the cob will not develop or be filled even though they were fertilized.

Soybeans

According to Liu (1997), soybeans were introduced into the United States in the latter part of the eighteenth century, but during the following century the plant attracted little attention. During that period, it was used primarily as a hay crop. It was not until the early twentieth century that the value of the plant for its oil was recognized. During the early part of the last century, many of new varieties were being brought to the United States from China. In the 1920s, there were significant developments in harvesting and processing equipment. Although interest in the use of the soybean seed for its oil and as an animal feed grew rapidly, it was not until 1941 that production for seed exceeded production for hay, silage, green manure, and pasture (Caldwell, 1973). In 1940, approximately 4.8 million acres of soybeans were planted for hay in the United States (Probst and Judd, 1973). Production increased rapidly in the following half century. It had reached 500 million bushels by 1960, over a billion bushels in 1970, and two billion bushels by 1990. In a little over a half century, the soybean has attained a vital significance in the economic welfare of the United States.

Taxonomy

The taxonomy of the soybean (Figure 4.9) is as follows (after Mauseth, 1991 and Liu, 1997): Family - Fabaceae (= Leguminosae); Subfamily - Papilionoideae; Genus - *Glycine*; Species - *max*.

In the scientific name of the soybean plant (Figure 4.9), the Greek word *Glycine* means sweet and the word max refers to the large nodules on the roots of the plant (Liu,1997). The plant belongs to the family Fabaceae (= Leguminosae). The family includes a large number of plants found in Alabama, including herbaceous ones such as the clovers and lespedezas, shrubs such as false- or bastard-indigo, vines such as wisteria and kudzu, and small trees such as the redbud, mimosa, and honey-locust. Mohr (1901) listed 116 species of the family in Alabama. Several of those listed above, those with flowers shaped somewhat like a butterfly (clover, wisteria, and kudzu) belong to the Papilionoideae subfamily. There are more than 1,000 species worldwide.

Origin

Liu (1997) states that the soybean was domesticated in the northern and central regions of China 4,000 to 5,000 years ago, and that it is one of the oldest crops in East Asia. About 1711, it was introduced into Europe as a botanical curiosity. Recent research has suggested that one Samuel Bowen, after living in China for five years, brought the first soybeans with him to Savannah, Georgia, in 1765. Seeds were planted on a nearby plantation. A letter from Benjamin Franklin written in 1800 stated that he had sent soybean seeds from England to America.

Economic Importance

Data on the acres of soybeans planted annually in Alabama during the period 1950-2000 are presented in Figure 4.7. The data show that acres of soybeans planted remained around 150,000 from 1950 through 1960. Afterwards, it began to increase exponentially, finally reaching 2.2 million acres in 1980. After 1982, acres planted declined sharply, reaching 290,000 acres in 1992. Since that time, it has changed relatively little. Alabama also has little comparative advantage in producing soybeans. Farmers in the state typically harvest 18 to 20 bushels an acre fewer than the national average (See Table 4.1).

In 1992, acres of soybeans harvested accounted for 12.8% of total acres of cropland harvested in Alabama. In 1997 and 2002, the comparable percentages were 16.4 and 7.8%, respectively. Data presented in Table 4.6, indicates that cash receipts for soybeans in 2002 accounted for only 0.6% of all farm commodities (livestock plus crops, including farm forest products). In 1980, the comparable percentage was 13.6%.

Critical Stages of Development

Scott and Aldrich (1983) comment that because the soybean plant produces more flowers and pods than it can sustain and that it does so over such a long period of time that it is much less susceptible than corn to adverse environmental conditions, especially drought. In the corn plant, flowering, pollination, and fertilization occur in the entire crop over a relatively few days. Even a minor drought at that time can be extremely detrimental. Although the soybean plant is not especially susceptible to adverse weather conditions, they still can affect the overall yield.

Maderski, et al. (1973) state that water is the dominant environmental factor in seed germination. Seed water content must increase from 15 to 50% before germination will begin. If there is not enough water in the soil to raise the level in the seed that much, it will often be attacked by fungi and decay. Abnormally high soil moisture can also reduce seed germination. Under these conditions, populations of microorganisms may develop on the seed and the developing root to the point that growth will stop.

Crusts on the soil surface can also reduce germination significantly. Because of the unusual way that the hypocotyl grows upward through the soil, it is unable to penetrate the crust. Often it will break off in the process.

Anything that interferes with normal plant function, especially while the ovules are developing into seeds, can reduce yield. While the maximum number and size of seed that will develop in a fruit or pod is genetically controlled, the actual number and size are largely determined during the seed-filling period. Dry weather at this time will not only affect seed size, but may also affect the number of seeds. Some ovules may not develop at all. If the stress become severe enough, immature pods will abort or fall off.

Mederski, et al. (1973) comment that although soybean plants seldom lack sufficient water, at least in the major production areas, growth-limiting water stresses within the soil and plant are common. Under most field conditions, an optimum water environment is seldom realized, and some growth-limiting water stress is the rule rather than the exception.

Soybeans are relatively tolerant of a wide range of temperatures. For example, temperatures during the night must exceed 85°F over a period of time before any noticeable reduction in yield occurs. On the other end of the scale, sustained temperatures below 75°F will delay the beginning of flowering.

A combination of high temperature and high humidity during seed development and even after maturity can result in poor seed quality if harvest is delayed. Populations of seed-attacking organisms can increase rapidly under this combination of environmental factors.

Irish Potatoes

The lowly (Irish) potato has come a long way from its use as a subsistence food on the high plateaus of South America's central Andes Mountains 5,000 years ago to become the essential companion for Big Macs at the Golden Arches. The potato has been carried to and planted for food virtually throughout the world. It has become a major staple crop and an important food commodity in international commerce. After corn (maize), the potato is the most widely distributed crop in the world. It is especially important in the northern latitudes where the relatively cool temperatures during the growing season limit the production of cereals. Northern Europe and Russia are major world centers of production.

The United States is not one of the major potato producing areas of the world. It accounts for well under 10% of the world's production. Considerably less than half of the potatoes produced in this country are prepared for consumption in the home. However, with the growth of the fast-food and snack food industries, the consumption of French fried potatoes and a variety of chips has skyrocketed. Agricultural statistics reported in 1997 (National Agricultural Statistics Service, 1997) that the production of frozen French fries in the United States was 129 million pounds, while the production of chips and shoestrings was 48 million pounds. Production for table use was 124 million pounds.

Taxonomy

The taxonomy of the Irish potato (Figure 4.10) is as follows (after Hawkes, 1978 and Mauseth, 1991): Family - Solanaceae (nightshade family); Genus - *Solanum*; Species - *tuberosum L.*

The scientific name for the potato plant that is cultivated in Alabama is derived from two Latin words. The name of the genus *Solanum* is the Latin word for nightshade, which is the name given to a large group of plants that have medicinal properties. The genus is an extremely large one. It includes 2,000 species distributed throughout the world, except for the far north and south. There is a strong concentration of diverse species in Australia and in Central and South America. The specific name, *tuberosum*, is derived from the Latin word for hump or knob. In addition to *S. tuberosum*, there are some seven other species of potatoes that are cultivated. There are also 154 other species of so-called wild potatoes. Tuber-bearing members of the genus are found only in the Americas.

The family Solanaceae is large and complex. It includes many members that are used extensively for food, medicinal, and other purposes. It includes, in addition to the potato, the tomato, eggplant, and red pepper. Tobacco and the petunia, a common annual flower, are other widely-used family members. Atropine, a drug used widely for the dilation of the pupil of the eye, is derived from one of the plants in this family.

As early as 1538, the Spanish were referring to potatoes as *papas*. This name is said to be derived from the name given to the tubers by the Incas. Columbus encountered the sweet potato, *Ipomoea batatas*, in his first voyage to the Americas in 1492. They soon began to refer to them as *patatas*. Because of the similarity of the sweet potato and the Irish potato, *S. tuberoseum* were also called *patatas*. Later *patatas* would become potatoes, which would become the name used throughout the English-speaking world.

Potatoes are also classified according maturity group. This classification indicates the time of year when the tubers of the different varieties are mature and ready for harvest. Howard (1978) used the following system: early (first early), mid-season (second early), early maincrop, and late maincrop.

Origin

Several species of tuber-bearing *Solanum* grow wild in Central and South America, Mexico, and as far north as Colorado in the United States (Smith, 1977). The parent stock of our cultivated potato grew wild in the highlands of the Andes Mountains between Colombia in the north and Chile in the south. Possibly as early as 4,000 years ago, native people in South America domesticated the plant. It soon became a primary food item, in the diets of people, living in those semi-arid highlands. When the Spanish entered the highland valleys of the Andes, around 1530, several varieties of the potato were being grown by native farmers. At that time, potatoes were not known in the lowlands and along the coast.

The potato seems to have first entered Europe in the last quarter of the sixteenth century (1570) when Spanish explorers took it home to Spain from Peru. From Spain, it appears to have been carried to Italy. It was carried to England sometime between 1588 and 1593, and from there, on to Ireland, Scotland, and Wales and finally to Northern Europe. The English also took the plant to their overseas colonies. Bermuda had received it about 1613. From there, it was brought to British North America about 1621. Still later, in 1719, a group of Scotch-Irish immigrants, who would establish a settlement in Londonderry, New Hampshire, brought potatoes with them from Ireland.

As noted in a preceding paragraph, potatoes probably were first domesticated in the high mountains of South and Central America, not far south (15° to 20° south latitude) of the equator. In this area, day length is relatively short. At 20° south latitude, day length varies from 12 to 13 1/2 hours. Also, the growing season in that region is a relatively long six to eight months. Because tuberization or potato formation is a photoperiodic response, tuber initiation in the plants that had evolved there began under relatively short-day conditions (12 hours), and they matured over a long period of time. In Europe, the day-length and growing-season situation is different. For example, Ireland lies at 50° north latitude. Day length there varies from 8 1/2 to 16 1/2 hours, and the length of the growing season is much shorter. There short-day potatoes would not set tubers and mature them before frost destroyed the above-ground portion of the plant; consequently, potato culture was not successful in Europe and especially Northern Europe until varieties could be developed that set tubers under long-day conditions and that matured in a much shorter period of time.

Economic Importance

Data on the acres of Irish potatoes planted annually in Alabama during the period 1950-2000 are presented in Figure 4.11. In the early 1950s, Alabama farmers planted 30,000 acres of potatoes. Afterwards, however, acres planted began to trend downward to 20,000 acres in the early 1960s. Afterwards, it changed relatively little until the late 1970s, when it began to decline rapidly once more. By 2000, Alabama farmers were only planting 5,000 acres. Alabama has never been a major player in the production of potatoes, but because of the soils of Baldwin County became warm enough for planting in the late winter, farmers there could often get a crop of new potatoes harvested early enough to beat producers further north, to the market. Over the years, however, this comparative advantage has almost disappeared.

Data presented in Table 4.7, indicated that in 2002 cash receipts for Irish potatoes accounted for 0.2% of total cash receipts for all farm commodities (livestock plus crops), including farm forest products. In 1980, the comparable percentage was 0.4%.

Critical Stages of Development

Because potatoes cultivated for the market are grown by planting seed pieces, problems associated with the production of seed (pollination, fertilization, and seed formation) and the germination of individual seed are bypassed. Also, a seed piece has multiple buds or growth points compared to the single one in the embryo of a seed. Of course, the seed piece is much larger; consequently, the probability of establishing a mature plant is much greater than the planting of a seed. Because there are much greater food reserves in the seed piece compared to the seed, stems originating from the eye (bud) of a seed piece can grow for a much longer period before it must begin to produce its own food. However, a major disadvantage of growing potatoes from seed pieces is that developing stems spend much more time underground than when a single seed germinates. As result, apical meristematic tissue is at the mercy of plant pathogens and pests for a much longer period of time. When a seed germinates, the apical meristem in the epicotyl is fairly quickly lifted from the soil by the growth of the hypocotyl.

In cereal crops such as corn, wheat, and rice, the part of the plant that is marketed is the seed, and these are produced in the open air. While these seed are susceptible to attack by a variety of plant pathogens and pests, radiation in sunlight, the drying effect of moving air, and diurnal fluctuation in temperature work to the farmer's advantage in controlling plant diseases. Also when a problem develops, it is easily observed and can be dealt with relatively easily. The marketable product of the potato plant is the tuber, and it remains underground for its entire period of development. In this dark, moist environment where diurnal temperature fluctuation is rather limited, plant pathogens literally thrive. Further, when a problem develops, it is difficult to observe it and even more difficult to deal with it.

According to Moorby (1978), tissue water relationships, appear to be generally similar to other crops; although reduced water content did seem to a have a somewhat greater effect on photosynthesis and food production. Apparently the stomates of the leaves of the potato close at a relatively low level of drought stress, thus restricting the entry of carbon dioxide needed for photosynthesis. Smith (1977) commented that water needs of the potato plant is greater at the time when stolons and tubers are developing than at any other time. For example, variation in water supply can sometimes result in deformed tubers. Growth is slowed when water becomes limiting. Then when it again becomes available, knobs develop when growth and development of secondary tubers takes place at the eyes of the tuber.

Potatoes have different temperature requirements for

different stages of growth: (1) In the period between the emergence of the stem to tuber initiation, low temperatures are detrimental because the emergence and growth of the vegetative shoots of the plant are slowed. When the temperature is relatively low, susceptibility to diseases is increased. (2) During the period of tuber initiation and early tuber development, high temperatures are detrimental to high yields. High temperatures during this period also seem to increase the incidence of some diseases. (3) During the period of tuber enlargement and maturation, high temperatures increase yields.

Peanuts

According to Rosengarten (1984), in the Civil War, Confederate troops were sometimes issued peanuts as part of their daily rations allowance. This practice led to writing of a song that became popular in that day, "Eating Goober Peas."

Sitting by the roadside, on a summer day
Chatting with my messmates, passing time away
Lying in the shadow underneath the trees
Goodness how delicious, eating goober peas!
Peas! Peas! Peas! Peas! Eating goober peas!
Goodness how delicious, eating goober peas!

Peanuts were introduced into the Southeastern colonies in the mid-eighteenth century. Thomas Jefferson commented that peanuts were grown in Virginia, but implied that the crop was of little commercial importance (Woodroof, 1966). Peanuts did not become really important in the agriculture of the Lower South until after the Civil War when production increased rapidly. After Appomattox in 1865, soldiers returning home to the Lower South, carried peanuts into areas where they had never been known before. Production increased significantly in Alabama and Mississippi in the early 1920s, after the introduction of the boll weevil. Now, in Alabama, among the cultivated plants, it trails only loblolly pine and cotton in terms of value as a farmed commodity (Table 4.7).

Taxonomy

The taxonomy of the peanut (Figure 4.12) is as follows (after Mauseth, 1991 and Singh and Simpson, 1994): Family - Fabaceae (= Leguminosae); Subfamily - Papilionoideae; Genus - *Arachias*; Species - *hypogaea L.*

Subspecies - *hypogaea*
    Variety - *hypogaea* (Virginia)
    Variety - *hirsuta* (Peruvian)
Subspecies - *fastigiata*
    Variety - *fastigiata* (Valencia)
    Variety - *vulgaris* (Spanish)

The peanut plant is one of the Papilonoid legumes because the petals of its flowers somewhat resemble a butterfly. There are 25 species in the genus *Arachis*. The species name, *hypogea*, is derived from the Greek word meaning growing beneath the ground. The two subspecies generally have different growth habits: The growth habit of the Virginia and Peruvian is spreading or running, while that of the Spanish and Valencia is erect or bunched.

Origin

Hammons (1994) comments that the peanut, or groundnut, is a native of the New World. Fruit hulls were found in archeological sites in Peru dating some 3,750 to 3,900 years ago. Early European explorers of the sixteenth century observed the peanut in cultivation in the Antilles, on the northeast and east coast of Brazil, and in the warmer regions of Paraguay, Bolivia, Argentina, and Peru. Cultivation of the plant was somewhat limited in Mexico. According to Singh and Simpson (1994), the center of origin of the peanut (*Arachis sp.*) was probably central Brazil, but that its cultivation originated in Southern Bolivia and Northern Argentina.

Different varieties of the peanut were widely distributed by the Europeans in the early sixteenth century. The Portuguese took the peanut to Africa, then on to India and the East Asia. The Spanish took it to the Western Pacific and China. Although the peanut was well known throughout the tropical world from the latter part of the sixteenth century onward, it apparently did not reach Europe until almost two hundred years later (Hammons, 1994). There is no record of the introduction of the peanut to North America. Apparently, it was introduced in British North America about the same time that the European explorers were distributing it around the world. Some evidence indicates that it was introduced from South or Central America or possibly from some of the islands in those regions. Other evidence indicates that the seed may have come on ships carrying slaves directly from Africa.

The first records of peanut cultivation in what is now the United States are from the colonial period.

Economic Importance

Data on the acres of peanuts planted annually in Alabama during the period 1950-2000 are presented in Figure 4.13. The data show that Alabama farmers planted 405,000 acres in 1950, and that acres planted declined sharply in the following two years to reach 260,000 in 1952. Thereafter, acres planted trended slowly downward, finally reaching 183,000 in 1967. Subsequently, acres planted generally trended upward through 1991 (278,000 acres), but declined afterwards, finally reaching 190,000 acres in 2000. Throughout this entire period (1950-2000), acres planted has largely been determined by farmer response to government regulation (acreage allotments, quotas, and price supports).

Data published in 2002 (Table 4.7) indicate that cash receipts for peanuts accounted for 2% of total cash receipts for all farm commodities (livestock plus crops), including farm forest products. In 1980, the comparable percentage was 3.5%.

Critical Stages of Development

The peanut plant and its fruit are at risk for a relatively long period of time. The pods of Spanish-type varieties are ready for lifting after seven weeks of underground development, or 120 days after planting. Virginia-type varieties require significantly longer periods of time to reach maturity, some as long as 180 days.

Peanuts have a substantial tolerance for dehydration (Wright and Nageswara, 1994). Nevertheless, drought is recognized as one of the most important constraints to production of peanuts in areas where all crop water is provided by rainfall. Lack of water has a deleterious effect on all stages of plant growth and seed production, because it strongly affects virtually all of the physiological processes within the plant. Water deficits restricts photosynthesis and nitrogen fixation, as well as root and shoot development, including leaf expansion. Water deficit can significantly reduce the number of flowers and a delay to time of flowering; however, peanuts are able to compensate somewhat by producing a large number of flowers when water again becomes available. Peg entry into the soil is extremely sensitive to soil hardness that is related to soil water content. Water deficit during pod set and pod development reduces overall yield by causing the reduction in the number of pods, as well as reductions in pod and seed weight. There is little question that water stress reduces peanut yield, but Coolbear (1994) suggests that it also reduces seed quality.

Temperature also affects the production of peanuts. Research has shown that low temperatures can depress seed yield by reducing flower numbers, and that high temperatures may inhibit fertilization and embryo growth.

The final critical stage, especially from an agricultural perspective, comes during the drying phase. The space around the drying kernels within the shuck provides an optimum environment for the development of fungal and bacterial populations. Unless the kernels dry quickly, there is a danger that these organisms will use the nutrients in them to form microbial toxins that will negatively affect their value.

Pecans

Thomas Jefferson so enjoyed eating pecans that while he was minister to France he requested a friend send him two to three hundred from the Western Country, preferably packed in a box of sand to keep them fresh (Manaster, 1994). A Frenchman serving in General Washington's army commented that the general was forever munching on pecans and that he always had some in his pocket.

The pecan is one of the few native species of North American plants to be developed into a major agricultural crop. In 1997, over 338 million pounds of pecans were produced in the United States. Georgia, Texas, and New Mexico together produced well over half of all the pecans produced in the country (Figure 4.14). Alabama produced almost 4% (13 million pounds) of this total.

Taxonomy

The taxonomy of the pecan (Figures 4.15 and 4.16), is that of Mauseth, 1991 and Manaster, 1994: Family - Juglandaceae (walnut family); Genus - *Carya*; Species -*illinoinensis* (Wangenh.) *K. Koch.*

Since the pecan was first given the name Juglans pecan about 1785, the scientific name of the pecan plant has been changed several times over the years. It has been named *Juglans illinoensis*, *Carya olivaeformis*, and *Hicoria pecan.* Finally, in 1969, it was given its current name in deference to the traditional belief that the pecan evolved in present-day Illinois.

The word for the name of the genus, *Carya*, comes from the ancient Greek word for walnut. The species name, *illinoensis*, as noted above, is derived from the name of the state. The common name, pecan, apparently comes from a word in the language used by the Algonquian Indians meaning a nut too hard to crack by hand. It is similar to the Ojibwa Indian word *pagin*, meaning a hard nut. Over the years, the name has been spelled and pronounced differently in many languages, including: pacane, paconos, pecane, and poccon, before finally becoming the pecan we use today. Although we have generally agreed on the spelling, there are two different pronunciations in use. In the Southeast and Southwest where most of the nuts are produced, we eat pek·hns while those uninformed citizens farther east eat pÈcans.

Origin

Pollen from an ancient ancestor (family Juglandaceae) of the pecan, has been identified in sediments that were deposited during the early part of the Cretaceous Period of the Mesoic Era, some 135 mya. It was during this period that the flowering plants first made their appearance. More recent and closer relatives of the pecan (genus *Hicoreae*) were identified in sediments deposited 70 mya in the latter portion of the Cretaceous.

Native (unplanted) pecan trees were originally found primarily on alluvial flood plains in an egg-shaped area reaching from southwestern Illinois to northeastern Mexico. According to traditional belief, the pecan evolved someplace in present-day Illinois, and river and ocean currents distributed them along the lower Ohio and Mississippi rivers and along the coasts of Louisiana and Texas. Some evidence, however, indicates that they originated at the other end of this area, or along the Rio Grande River in Texas (Figure 4.14).

Native Americans certainly found enormous numbers of these magnificent nut-bearing trees along the river and stream banks when they arrived in that area of North America (Figure 4.14), and they apparently transported and planted the nuts widely throughout the area. Spanish explorers, who visited the Central Gulf Coast early in the sixteenth century, made note of the pecans and Native American use of them. Early European settlers used pecans for food when they were available in the fall. They planted them around their farms for shade and for the nuts, and they used wood from the trees for furniture and for handles for farm implements. The handsome pecan tree had so many good characteristics that they were soon dispersed far and wide. However, success was limited where the trees were established outside of the southeastern quadrant of the country.

It was soon noted by early settlers who planted pecans around their homes that when seed from good trees were planted that most often they did not develop into good trees themselves. While this problem was recognized early, very little was done to correct it for many years. There were just too many good native pecan trees scattered throughout the river bottoms to worry too much about the propagation of improved trees. In the mid-nineteenth century, this problem was finally solved when a slave gardener on a plantation in southern Louisiana grafted small pieces of the limbs from a good tree onto the stem or rootstock of another tree. The limbs that developed from those grafts retained and displayed the good characteristics of the donor tree. Later, he succeeded in grafting a number of additional trees and establishing the first orchard to produce pecans for sale. From that inauspicious beginning, pecan culture spread both eastward and westward across the southern tier of states.

Economic Importance

Data on the pounds of pecans harvested annually in Alabama during the period 1950-2000 are presented in Figure 4.17. Because of the alternate-bearing (masting) characteristic in pecans, it is difficult to determine whether-or-not there have been trends in pounds harvested. It does appear from the data, however, that it might have trended upward from the 1950s through the early 1970s, only to trend downward afterwards. In 2002, Alabama pecan farmers harvested 5,000 pounds of the nuts. In that year, the state ranked seventh in the country in production (Vanderberry and Placke, 2003); however, not many states grow and harvest pecans.

Critical Stages of Development

The length of time from flowering to maturity for the pecan is about two hundred days; consequently, it is at risk for a relatively long period of time. In this long period, it is often subjected to at least some unfavorable growing conditions. Brison (1974) suggests that the effect of unfavorable growing conditions on the development of the pecan is dependent on when the conditions occur. If they occur while the pecan is increasing in size, the nuts will be smaller than average. If the conditions occur after the shell has reached full size, the nuts will be poorly filled.

Continued high relative humidity (85% or higher) during pollination can be detrimental to pollen shed. Without sufficient drying, the anthers do not split open to release the pollen grains. The tissues of the stigmatic surfaces of female flowers are highly sensitive to cold. They are destroyed by temperatures slightly below freezing.

Tender new shoots and leaves of the pecan are killed by temperatures that are only slightly below freezing. When the meristematic cells of the vascular cambium begin to divide in the early spring, they are highly susceptible to extremes of both heat and cold. Either of these extremes can also result in the complete failure of early grafts when the cambium tissue of both the growing stock and the grafted twigs (scions) are killed. Roots of the pecan tree are more susceptible to low temperatures than the shoots. Young pecan trees are often girdled at the ground line by extreme heat at the soil surface.

Maturing pecan fruits are at particular risk from high winds from storms originating in the tropics. Much of the pecan growing area is in the region of the country visited relatively frequently by these storms. Nuts reach their maximum size around mid-summer; although they are still some three months from maturity. When they reach maximum size, they are at their greatest susceptibility to being blown off the tree. Even if the nuts are not blown off, high winds may defoliate the tree to the point that total photosynthesis and food production is reduced significantly. Without an adequate food supply during this critical period of nut filling, its development may be seriously affected. When the storm is severe enough so that both nuts and leaves are blown off, not only is the crop for the current year lost, but the crop for the following year will most likely be lost as well. With the loss of leaves in early fall, the reduction in photosynthesis and food production may be so great that there are insufficient food reserves for fruiting, as well as for development the following spring. Hail that often accompanies the powerful thunderstorms, which develop over the pecan-growing region of the country, can strip-off leaves and fruits as effectively as the high winds of tropical storms and hurricanes. Fortunately, hail seldom causes so much widespread damage.

Strong storms also do considerable damage to the main stem and branches of mature trees. Because of the rigidity of the wood, pecan limbs often break rather than bend under the pressure of high winds. Most pecan trees grow in light sandy loam soils with sandy clay subsoils. The heavy rains over several days that usually accompany storms quickly saturate these soils. The softened soils plus the heavy weight of the large canopy of the tree with its growing nut crop, make the mature pecan tree a prime candidate for uprooting. Also pecan trees are usually planted in open areas where there is little protection from the full force of the wind. Major hurricanes often uproot large numbers of these trees. Lightning accompanying thunderstorms also destroys substantial numbers of trees each year, across the growing region. The height of pecans, which are usually located in relatively open areas, makes them prime targets for lightning strikes.

Azaleas

From a world-wide perspective, plant agriculture is primarily undertaken to provide food for people and to a lesser extent for their animals. The production of plants for fiber is extremely important, but the human population requires a much smaller quantity of fiber than food. Cultivated plants are also important sources of industrial chemicals and pharmaceuticals. All of these uses are generally related to the useful or material requirements of our species. There is, however, still another important reason why plants are cultivated that is not related to our material needs. We seem to have an innate need for the beautiful in our lives, and we humans seem to be unique within all the species of animals in this respect. This need for the aesthetic is realized through the creation of and interaction with architecture and with the visual, performing, and literary arts; through interaction with the natural world; and through the cultivation and interaction with living plants. The cultivation of plants for their aesthetic value is widespread in the world, but without exception, the need for food, shelter, and clothing takes precedence over the need for aesthetics. Consequently, their production is concentrated among those people who experience higher standards of living.

Taxonomy

The taxonomy of the azaleas is that included in the work by Galle(1987). The Family Ericaceae includes 50 genera of plants and 1,400 species. The Genus *Rhododendron* is the largest in the family, comprising 800 species. Other large genera are *Erica* (heaths) with 500 species and *Vaccinium* (blueberries, cranberries, and huckleberries) with 400 species. When the rhodendron-like plants were first being described, the azalea-like plants were placed in a separate genus, *Azalea*. Later, these plants were placed within the genus *Rhodendron* as a subgenus. More recently, *Rhodendron* has been divided into eight subgenera. The azaleas are included in two of these: *Pentathera* and *Tsutsutsi*. In turn, the two subgenera are further subdivided into four and three sections respectively. This taxonomic system is based primarily on the characteristics and relationships of plants found in the wild and has been accepted by botanists for many years; however, in 1930, another system (the *Rhododendron* Series) was proposed for use by horticulturists. It is based primarily on the characteristics and relationships of cultivated plants. In this system, the Genus *Rhodendron* is divided into 45 series, one of which is *Azalea*. The *Azalea* Series is further divided into seven subseries that generally equate to the sections of the botanical system. Below the sub-series level the taxonomy of the azaleas becomes extremely complicated, and well beyond the scope of this book.

Origin

Apparently the rhododendron-like plants began to evolve from magnolia-like ancestors some 75 mya (Mauseth, 1991). This ancient differentiation apparently began in the eastern Himalayan Mountains where they extend through Tibet into South-central China and on into Northern Burma (Berrisford, 1964 and Reiley, 1992). Over time, they spread westward into Europe; southward into Indonesia, the Philippines, and Northern Australia; and eastward into Northeastern Asia. From Northeastern Asia, they spread, probably on ocean currents, into the northwestern portion of North America and finally southward all the way to the Gulf Coast.

Although the home range of the azalea is probably centered in Eastern China, the Eastern United States is an equally hospitable habitat. Both areas lie on the eastern margin of large continents, and have similar climates and about an equal number of species.

Economic Importance

Plant production to meet aesthetic needs is an important agricultural industry in the United States. In 1998, Alabama farmgate sales of greenhouse, sod, and nursery products reached $210 million (Vanderberry, 1998). Included within this larger group, were the so-called greenhouse plants such as geraniums, impatiens, petunias, and poinsettias; sod plants such as Bermuda and zoysia grasses; and nursery crops such as azaleas, camellias, and hollies. Among the various species produced in the state to beautify our landscapes, the azaleas are the most important to Alabama producers. They account for from 20 to 25% of the total value of all greenhouse, sod, and nursery crops. Further, the production of azaleas, which is centered in southwest Mobile County in the vicinity of Semmes, is increasing rapidly, probably faster than any other plant produced in Alabama.

Critical Stages of Development

Virtually all of the azaleas coloring the landscapes around our homes, in parks, and around public buildings were obtained for planting well-established plants with numerous branches, a good distribution of healthy leaves, and a dense root system. This process is in contrast to the establishment of most of our field crops where seed are placed in the soil in a large field and left, essentially, to nurture themselves. With azaleas, most of the critical stages of development are in the past by the time they are finally planted in a landscape. They occur while under the careful protection of nursery personnel. Hundreds of thousands of cuttings are established under the most optimum conditions for the formation of adventitious roots. After roots are formed, plants may be transplanted several times into larger and larger containers until they reach market size.

If the plants are suitable for the area where they are

being established, and if planted correctly and given even minimal care (watering, feeding, insect control, etc.) they will thrive. If there is a critical period, it comes immediately after they are planted. Ample watering is essential while the roots are growing into the surrounding soil.

Vertebrate Animals

Vertebrate animals (animals with backbones) have internal skeletons, paired appendages, closed circulatory systems, highly organized nervous systems, and separate sexes. They also tend to be mobile, some highly so, and they have a proclivity to form associations (schools of fishes, flocks of birds, prairie-dog towns, packs of wolves, prides of lions, parties of chimpanzees, and villages of humans). All of these characteristics, in one way or another, contributed to the establishment of agriculture and ultimately to its arrival in Alabama.

Vertebrates in Alabama's agroecosystem serve as a sources of food (meat, milk, and eggs); power (horses and mules); industrial materials (hides and bones); companionship (cats and dogs); and recreation (largemouth bass, wild turkeys, and whitetail deer) with one of their kind, humans, orchestrating all of their myriad roles. How this complex situation evolved is a long and fascinating story.

Sometime in the middle Pre-Cambrian Era, 900 mya, two groups of living things were set on widely-divergent paths. One group, the plants, would be able to manufacture their own food from basic elements (sun light, water, carbon dioxide, and minerals) found in their environments, and the second group, the animals, would capitalize on this unique characteristic of plants by using them for food.

By the early Paleozoic Era, which began 543 mya, there were both small green plants (algae) and invertebrate animals in the warm seas of the earth. Then around 543 mya (Ordovician Period), the animal line itself divided into two distinct branches: invertebrates and vertebrates. Both branches thrived in the evolving environment of the earth; however, evolution steered the two groups onto quite different pathways. In invertebrates, the emphasis generally was on large numbers of individuals with small size. In the vertebrates, the emphasis was on larger body size, but smaller numbers. Today, in terms of numbers, the invertebrates are the most numerous animals on earth, probably by many orders of magnitude. But in terms of body mass, the vertebrates enjoy the advantage. In fact, the combined weight of one species of vertebrate alone (humankind) is greater than that of all other animals (invertebrate and vertebrate) combined.

Even after these hundreds of millions of years, both invertebrates and vertebrates continue to be totally dependent on plants for food, and as is always the case when the same creatures must have the same resource to survive, the competition is fierce. As an example, humans spend billions of dollars each year combating invertebrate animals, primarily insects, who seek to use food and fiber crops in meeting their own needs for food. The competition is not limited to the use of the same plant food resources. Invertebrates and vertebrates use each other for food. For example, the primary food items in the diets of fish and birds are insects. In turn, parasitic mites, worms, and protozoans, feed more or less continuously on birds. Similarly, virtually all fish are infested with invertebrate parasites.

Competition between invertebrates and vertebrates are both omnipresent and important, but competition and similar relationships within these primary divisions of animal life are probably even more fierce. Predation by one group of invertebrates on another is widespread in the natural world. Similar relationships exist in the vertebrates as well. For example, animals and animal products are used to meet many different kinds of our life needs, and they have served this purpose since the appearance of our species. A primary use is for food. This use is doubtlessly growing throughout the world, as humans—who once ate plants and plant products primarily—are now adding more animals and animal products to their diets. The use of animals as beasts of burden in still widespread in the world, although this use tends to diminish whenever agricultural income grows. The use of animals as companions has also grown rapidly in some countries, especially the more affluent ones.

One of the most fascinating (and important) of these intra-divisional relationships is the domestication of animals. The first domestication likely took place relatively recently in human history (10,000 to 12,000 years ago), but since that time, it has come to play an extremely important role in our culture and as a foundation of societies in general. A fascinating aspect of the domestication of animals is the age-old practice of "domesticating" members of our own species through the practice of slavery. It is also fascinating that for such a long period of time slaves, along with domesticated plants and animals, would have been thrown together in the production of food and fiber.

Generally speaking, humankind is the only species with the necessary mental characteristics to domesticate other animals. Although this statement is generally true, it is not completely so. Wenke (1990) cites work with the African cultivator ant carried out by David Rindos that clearly demonstrates that these social insects also practice a primitive form of domestication. The ants carry cut-up plant debris into their underground nests where it is deposited in specially prepared beds. Numerous ventilation passages are excavated into the beds that can be opened and closed to provide some control of ventilation. To develop the beds, the ants chew the plant material into a pulpy mass that is deposited in layers. The bed is then planted with propagules from a previous bed that contains spores of an edible fungus. The tiny farmers are constantly tending to their beds, removing unwanted fungal growth, and adding anal and salivary secretions to it. Over time, the cultivated fungi produce small, whitish round structures that are the principal food of the ant colony. While the evolution of agriculture in this other species of animals is of considerable philosophical interest, it is only our species, which over some 10,000 years, has reached this unique co-dependent relationship (domestication) with other animal and plants on such a widespread and colossal scale.

Sources of Information on Alabama Vertebrates

As was the case of Alabama's plants, there is relatively little recent specific published information available on the vertebrates of Alabama as a group. The publication edited by Mount (1984), *Vertebrate Animals of Alabama*, is the most extensive one available. It lists all of the vertebrates in the state and includes a small amount of information on the distribution of each one. Another publication, also edited by Mount (1986), *Vertebrate Animals of Alabama in Need of Special Attention*, covers only a small portion of the vertebrates of the state, but contains a considerable amount of information about each one, including a description, range, habitat, life history, and ecology.

Information on all the of mammals of the state, other than those listed above, is limited. There are only two comprehensive publications. One was written by Howell (1921), *Mammals of Alabama*, and the second is an unpublished PhD dissertation by Holliman (1963) also entitled, *Mammals of Alabama*. There are several other publications about mammals that are of a regional nature, but their coverage of Alabama species is somewhat limited. Also, there are numerous scientific papers about specific species. Additionally, there are several commercially available nature and field guides that describe many of the mammal species found here.

The only comprehensive publication on Alabama birds is one written by Imhof (1976), *Alabama Birds*. As was the case with mammals, however, there are regional publications and several commercially available nature and field guides that describe many of the bird species found here.

Mount's (1975) publication, *Reptiles and Amphibians of Alabama*, is the only comprehensive work dedicated to Alabama amphibians and reptiles.

There is generally more information available on the fishes of Alabama than any of the other groups of vertebrates. Boshung's PhD dissertation (1957), *The Fishes of Mobile Bay and the Gulf Coast of Alabama*, was the first publication dealing specifically with a major segment of Alabama's fish fauna. Later, Smith-Vaniz (1968) wrote the first comprehensive publication on the state's freshwater fishes, *Freshwater Fishes of Alabama*. By far the most comprehensive treatment of our fishes is contained in *Fishes of Alabama and the Mobile Basin*. This publication of Mettee, O'Neil and Pierson (1996) contains a wealth of information, including distribution maps, identification keys, characteristics, and excellent color pictures of all of the species with their biology.

Although the amount of recent published material on the Alabama's vertebrates is somewhat limited, there is a wealth of current information available from

Alabama's major universities. Several of the institutions have extensive collections of preserved specimens. The Alabama Museum of Natural History at the University of Alabama in Tuscaloosa has a particularly good collection of all of the groups. There are also good collections at Auburn University. Generally, all of the faculties include specialists in the major groups of vertebrates who welcome the opportunity to provide information about Alabama's vertebrates. The Alabama Department of Conservation and Natural Resources also has a number of specialists knowledgeable about Alabama game fish, birds, and mammals. They probably also have the most up-to-date information on endangered species of vertebrates in the state.

Scales, Feathers, and Hair

Vertebrate animals with either scales, feathers, and hair have been on earth a long time. One of the earliest vertebrate ancestors, a Chordate (animal with a notochord), Pikaia, appeared 535 mya in the Cambrian Period of the Paleozoic Era (Table 1.1). About 30 million years later (505 mya) at the beginning of the Ordovician Period of the Paleozoic, the first vertebrate, a jawless fish (Ostracoderm), appeared. Then 480 mya, Placoderms (jawed fishes) appeared. Strangely, the first vertebrate animal with four limbs, the amphibian (Acanthostega), did not appear until 165 million years later (315 mya). Hylonomus, the first true reptile, probably appeared around 300 mya in the Carboniferous Period (Pennsylvanian) of the late Paleozoic Era. True birds apparently evolved from the reptiles, first appearing on earth between 200 and 145 mya. At about the same time (220 mya), the earliest mammal-like ancestor (Eucynodont) appeared. Mammals roamed the earth for millions of years before the first primate (Primatomorpha) appeared, some 65 to 85 mya. The earliest ancestor of our human family, the Hominidae, was a relatively late-comer on the scene when this family evolved from the lesser apes or gibbons around 15 mya. Then our Genus *Homo* did not appear until around 2.5 mya.

Scales on fish and reptiles, feathers on birds, and hair on mammals are important characteristics of a large groups of living things (vertebrate animals) that are found throughout the state. A number of them play important roles in Alabama's agroecosystem. Altogether, there are slightly over 850 species of these animals occurring naturally in Alabama (Mount, 1984). Few other states have more species. A little over a third of the species belong to the Class Aves (birds), and a little less than a third belong to the Class Osteichthys (bony fishes). Amphibians (frogs, toads, and salamanders); reptiles (lizards, snakes, and turtles); and mammals (humans, whitetail deer, dogs, cats) together comprise the other third.

Vertebrates with Scales

The scaled vertebrate animals found in Alabama belong to two quite different groups. One of the groups primarily extracts the oxygen that it needs in its life processes directly from the air. The other group extracts oxygen from water. The air breathers (Class Reptilia) are represented by 80 species (Table 4.9). Half of the scaled species are snakes (Order Squamata, Suborder Serpentes). Less than one-fourth of the species are lizards (Order Squamata, Suborder Lacertilia). The remainder are species of turtles (Order Testudines).

As Alabama becomes more urbanized and fewer people live and work in rural areas, contacts between humans and snakes are occurring less frequently. A high percentage of younger Alabamians have probably never seen a live snake in the wild. Except for exhibits in zoos and similar displays, many people would never see one. One small town in southeast Alabama is apparently determined to reverse this trend. If people will not go to the snakes, they will bring the snakes to the people. In late winter, thousands of visitors travel to Opp, in east-central Covington County, where a few foolhardy individuals go into the forests to round-up substantial numbers of one species of these scaled animals, the one with a forked tongue, its own personal set of rattles, and a fearful reputation: the eastern diamondback rattlesnake. In earlier times, when a high percentage of Alabamians lived and worked on farms, contacts were common and often highly detrimental to snakes. Unfortunately, for most people in those days, there were no good snakes.

Approximately 14 species of lizards (Order *Squamata* - Suborder Lacertilia) are found naturally in Alabama (Mount, 1984). The skinks (Eumeces sp. - Family Scincidae) comprise almost half of this total. Lizards, like the snakes, are seldom encountered by the average Alabamian, and few zoos provide exhibits for species commonly found in the state. In earlier times, every farm homestead had its complement of gray, formidable-looking, fence runners (Fence Lizard), and skinks were common around the open farmhouse yards and along the sandy portions of unpaved county roads.

Approximately 26 species of turtles (Order Testudines) are found naturally in Alabama. Half of these species belong to a single family (Emydidae), and more than half of these belong to two genera (*Graptemys* and *Pseudemys*). Because the vast majority of turtles are found in and around ponds and streams, contacts with humans have always been limited. This situation certainly has not changed with the increased urbanization of the state. Except for the box turtle that seems always to be looking for a public road to cross, relatively few people ever encounter turtles in the wild.

The fishes are the second major group of scaled vertebrates. They extract their oxygen from water with the use of specialized structures called gills. There are three different classes of fishes found in Alabama: Agnatha (jawless fishes, lampreys), Chrondrichthyes (cartilaginous fishes, sharks), and Osteichthyes (bony fishes). Altogether there are some 27 orders of fishes represented in Alabama's inland and marine aquatic environments. One of these orders, the Pertromyzontiformes (lampreys) belongs to the Class Agnatha (jawless fishes). Two of the orders, Squaliformes (sharks) and Rajiformes (sawfishes, rays, and skates) belong to the Class Chrondrichthys (cartilaginous fishes) (Table 4.10). The remaining 24 orders belong to the Class Osteichthys (bony fishes).

The fish habitat of Alabama can be divided into three major parts: freshwater, brackish water, and marine. Mettee, et al. (1996) further divide the freshwater habitat into three regions: Tennessee River, Mobile River Basin, and coastal rivers. This division has been compromised somewhat by the completion of the Tennessee-Tombigbee Waterway that potentially links the Tennessee River with streams throughout the Mobile River Basin. The basin includes all of those rivers (Tombigbee, Black Warrior, Cahaba, Coosa, and Tallapoosa) whose waters eventually enter the Gulf of Mexico through Mobile Bay (Figure 1.17). The coastal rivers include the Escatawpa that flows through Washington and Mobile counties, the Perdido in Baldwin County, plus those four larger rivers in the southeastern corner of the state (Conecuh, Blackwater, Yellow, and Choctawhatchee), and the Chattahoochee near the eastern boundary with Georgia. The location of the brackish and marine waters of the state are shown in Figure 1.32.

Around 318 species of native, breeding species of bony fishes (Class Osteichthys) are found in the fresh water and estuarine environments in Alabama (Mettee, et al., 1996) (Table 4.11). Walls (1975) listed 500 species of fish found in the Northern Gulf of Mexico (Table 4.12). There is some duplication in these totals, because both include species of estuarine and marine fishes found in Mobile Bay and the interconnected estuaries. Half of the species of bony fishes belong to a single order (Perciformes). A hundred of these species (freshwater basses, sunfish, and darters) are found in freshwater, and 250 are marine species (groupers, amberjacks, cobia, pompano, snappers, mackerels, seatrout, and mullet). Other important orders with both freshwater and marine species include *Clupeiformes* (shads, herrings, and anchovies); *Siluriformes* (catfishes); and *Atheriniformes* (topminnows, silversides, and killifish).

Because of their habitat requirements, the distribution of fishes is discontinuous. Encounters with them are seldom incidental. However, for thousands of years, humans in what is today Alabama made encounters a matter of priority. Fish played an important role in the food supply of the earliest human inhabitants. Obviously, most of the attention was devoted to a limited number of the larger species living in freshwater or estuarine areas. The vast majority of species are so small or lived so far from shore that they were not harvested. When Europeans displaced the aboriginal inhabitants, they continued to harvest significant quantities of fish, and over time extended their harvest far out into the Gulf of Mexico.

In recent years, the harvest of freshwater fishes from natural waters, specifically for food, has diminished. There is almost no commercial fishing on the rivers and reservoirs of the state. However, the harvest of marine species for food has increased. In fact, the demand for marine fish has increased to such high levels that species once considered to be inedible (sharks, various jacks, sheepshead, various drums, and triggerfish) are regularly harvested and marketed.

The most dramatic change that has taken place in encounters between humans and fish has been the increase in recreational fishing. Currently Alabamians purchase 330,000 licenses to fish in fresh water and 25,000 licenses to fish in marine waters, annually. Obviously, some individuals purchase licenses to fish in both fresh- and saltwater. Hundreds of thousands of hours of fishing effort are expended each year in both environments. Fishermen interests in fresh water center on a relatively few species such as largemouth bass, bluegill sunfish, crappie, and channel catfish. Marine recreational fishermen interests include a somewhat larger number of species; although in recent years the speckled trout, red snapper, and grouper probably continue to attract the largest number of anglers. Interest has increased in catching jacks, mackerel, cobia, billfish, tuna, and even sharks. Each year, fishing contests bring thousands of anglers to our rivers and reservoirs where prizes are awarded to anglers catching the most and/or the largest fish. Most of the contests center on the largemouth bass; however there are a number of contests, especially in the northern part of the state, where crappie are the center of attention. Contests are generally less common in saltwater fishing.

Recreational fishing in Alabama is big business. The total value of all sales and services associated with angling in Alabama is estimated to be well over a billion dollars annually. Recreational fishing is a major element in tourism in the state. Half century ago, virtually all of the fish taken by anglers were used as food. More recently, catch and release has increased in popularity. Even so, a significant portion of fish taken in recreational fishing continue to be used as food.

The growth of fish farming in the state has changed the relationship between people and, primarily, one species of fish, the channel catfish. Also, many thousands of Alabama homes have aquaria containing small numbers of valuable fish. Purchase and care of these animals also involve the expenditure of several millions of dollars each year. Few of the fish involved are native.

Vertebrates with Feathers

Approximately 360 species (Mount, 1984) of those vertebrates with feathers (Class Aves birds) fill our airways and brighten our treetops and yards with color and song. This is a large and complex group of animals. There are 17 orders. Most of the species belong to two of them. The Order Passeriformes includes 165 species of perching birds, such as purple martins, blue jays, chickadees, wrens, and mockingbirds. The Order Charadriiformes includes 64 species, such as plovers, sandpipers, snipe, gulls, and terns. Other orders and representative species are presented in Table 4.13.

We are more likely to encounter one of these kinds of vertebrates on a daily basis than any other, although less than half of them are permanent, native breeding residents. A majority are found in the state only during migration or during the winter. Mocking birds, brown thrashers, blue jays, mourning doves, woodpeckers, hummingbirds, chickadees, pigeons, house wrens, starlings, sparrows, and many others are thriving urban and/or suburban residents. They seem to be as well adjusted to urban living as are humans. Our species seems to be especially appreciative of the presence of these indicators of wildness in our midst, considering the millions of dollars spent each year to attract these small colorful animals to backyard feeders. Another indication of our love affair with feathered vertebrates is the thousands kept as pets, usually in small cages within our homes. The cost of their upkeep totals millions of dollars each year. Many different species are used for this purpose, virtually all of them imported from countries in the tropics.

Our largest feathered vertebrate, the one in which the male of the species gobbles and struts (wild turkey) about in our woodlands during the spring, has essentially the same effect on the human species as the antlered mammal described in a following paragraph. Thousands of the supposedly more intelligent species rise early in the morning, drive deep into the forest, stumble in the dark for hundreds of yards through brush and briars, over fallen logs and rocks and through various kinds of water courses to be in position to hear the supposedly less intelligent species issue its mating call.

Another of the feathered forms (pigeons) has taken a special liking to courthouses and other public buildings. This form of public housing program is generally less acceptable than others currently in vogue; consequently, considerable effort and expense is devoted each year in an effort to evict these unwanted tenants, usually with limited success. Unfortunately, the proclivity of humans to feed these feathered grifters in the parks and on the grounds surrounding public buildings adds to the problem.

Vertebrates with Hair

The vertebrate hairy animals (mammals) of Alabama are a varied lot, and they play multiple roles in the daily affairs of this small portion of planet earth. Approximately 60 species of mammals are found naturally in the state (Mount, 1984). Almost half of that total belong to one of nine orders (Rodentia rodents). Other orders of the Class Mammalia along with the numbers of families and approximate number of species in each are listed in Table 4.14.

As more and more people moved from rural to urban areas, ecological relationships between people and the other mammals in Alabama have changed significantly in the last half century. At the same time, even many of those people continuing to live in rural areas no longer farm. In this period, the total number of farms has decreased from 200 thousand to 50 thousand. The acres in farms has also decreased, but not so dramatically. As a result, the nature of encounters between ourselves and other wild mammals has changed. Generally speaking, Alabamians are much less likely to encounter as many species of wild animals—such as a mouse, rat, bobcat, fox, rabbit, or bat—as they would have 50 years ago. This relationship is not the same for all species, however. For example, squirrels and raccoons moved to town with the people; consequently, one is much more likely to see one of these mammals today than a half century ago. As a result of the increase in the amount of woodland succumbing to urbanization, the number of white-tail deer has literally exploded in the state. As a result, the average Alabamian is much more likely to encounter a wild deer today than 50 years ago. Hunting also continues to bring a sizable number of Alabamians in contact with mammals. Currently, 200 thousand Alabamians buy hunting licenses annually. While only a small percentage of the 60 species of mammals are actually hunted, the hunting process itself increases the number of encounters with many more species.

In another of the hairy species (whitetail deer), the males sport a pair of antlers on their heads, and they attract hundreds of thousands of humans to the state's woodlands each winter. These antlered animals elicit a range of strange behavioral responses, primarily among the males of the human species who seem to be driven to buy millions of dollars worth of vehicles, clothing and footwear, guns, and ammunition. In 1996, hunters in Alabama bought an estimated $610 million of goods and services. Then with all of this paraphernalia in tow, thousands of them are led to take to the woods to endure cold and often rainy weather for hours, many perched on small platforms on the sides of trees 10 to 20 feet above the ground.

Hundreds of thousands of two other hairy species (cats and dogs) live out their lives in the homes of the humans as members of their extended families. Millions of dollars are spent each year for their care and feeding.

One of the hairy species, the one with the scaly tail and embarrassed grin (opossum) seems to have evolved primarily to be killed by speeding vehicles on Alabama's highways. They are generally distributed throughout the state in substantial numbers. Unfortunately, they are most commonly encountered as road-kills. An immigrant from the American Southwest, the so-called possum-on-a-half-shell (armadillo) seems determined to challenge Pogo's relatives for the title of most common road-kill, especially in south Alabama.

Two fish-like species of haired vertebrate (dolphins) delight young and old as they swim in small groups (pods) along our beaches. The pleasure is intensified when we finally realize that they are not really sharks after all.

Many of the contacts between humans and their mammalian kin are detrimental to their relatives; however, in at least three situations, the tables are turned. Foraging deer cause millions of dollars worth of damage each year to agricultural crops, home gardens, and ornamental plantings. Similarly, the beaver has been responsible for destroying hundreds of thousands of acres of valuable bottomland hardwoods and a considerable acreage of pine plantations with their dam building and its resultant flooding. Feral hogs cause widespread damage as they root in pastures and fields.

Vertebrates without Scales, Feathers, or Hair

The use of scales (reptiles and fish), feathers (birds), and hair (mammals) as characteristics to delineate the different groups of vertebrates, omits a most interesting group frogs, toads, and salamanders (Class Amphibia). They have a smooth moist skin that contains many glands, but no scales, feathers, or hair. This is an interesting group because some of the species extract their oxygen from water, while the remainder are air breathers. Approximately 65 species (Table 4.15) of these vertebrates are found naturally in the state (Mount 1984). Some 30 species are frogs and toads (Order Salienta), and 35 species are salamanders (Order Caudata). Most species of the amphibians are never encountered by people. This situation existed even before the exodus to suburbia. Most of these species have limited ranges; they do not move around very much. Also, they tend to remain hidden for much of the time. We do occasionally encounter species of tree frogs and toads, even in suburban settings. Some of these species persisted even after their woodland habitats were converted into subdivisions

Domesticated Vertebrates

Almost all of the animal agriculture in Alabama involves vertebrates. Only the honey bee, among the invertebrates, is produced in large numbers. There is some crayfish production for food, some production of worms and crickets that are used for fish bait, and there are a few marine shrimp farms operating in west-central Alabama. In a different way, invertebrates are extremely important in the Alabama agriculture. Many millions of dollars are spent each year in an attempt to reduce the negative effects of insects and nematodes and other invertebrates on food and fiber production.

As was the case with plants grown by Alabama farmers, most of the domesticated animals used in Alabama agriculture are introduced from Africa, Asia, or Europe. With the exception of the channel catfish, all were introduced by European explorers or colonists when they first came to America. All of these introduced species evolved in substantially different environments from those where we grow them today. Both cattle and hogs had been in Europe for hundreds of thousands of years before they were brought to the New World. The chicken likely evolved in the tropical environments of Southeast Asia, and probably is better adapted for life in subtropical Alabama than any of the other introduced species.

Vertebrate animal agriculture (livestock production) is an important business in Alabama. In 2002, cash receipts for all livestock, poultry, and crops (including farm forest products) totaled $3.1 billion. Marketing of livestock and poultry and their products, principally milk and eggs, accounted for 76.3% of the total (Vanderberry and Placke, 2003). The sale of broilers alone accounted for 51.6% of the total (Table 4.16). Receipts for cattle, calves, and eggs accounted for almost equal shares (9.8 and 9.5%, respectively). While receipts for catfish were relatively low on a statewide basis (2.4%), they represent a highly significant contribution to the economies of several counties of west-central Alabama. The state's once proud dairy and hog industries have fallen on hard times. They only accounted for 1.2 and 0.6% of the total, respectively.

Chickens

In many ways, the chicken is an integral part of American culture. In 1992, American poultry farmers produced 12 billion pounds of poultry meat and six billion dozen eggs. In that year, the average American consumed 60 pounds of poultry meat. These edible products represent significant contributions to our material culture (machines, buildings, food, and clothing), but it is surprising how much the chicken has also contributed to many aspects of our nonmaterial culture (language, ideas, beliefs, myths, and customs). For example, there is the Little Red Hen who is the paragon of diligence and industriousness, and Henny Penny who almost lost her head to Foxy Loxey when she was trying to reach the king to tell him that the "sky is falling." "Which came first the chicken or the egg" has provided endless grist for the philosopher's mill. The crowing of the rooster played a significant role in the events leading to the establishment of the early Christian Church (Matthew 26:34). How many jokes have their origin in "why did the chicken cross the road?" Every young person knows that a friend who will not go out alone into the dark is a "chicken." Finally, the slogan "a chicken for every pot" was the promised result of voting correctly in an earlier national election.

The production of broiler chickens is an interesting phenomenon in the development of Alabama's agriculture. Chickens certainly must have accompanied some of the first European settlers who came to the state. For 150 years, chickens were kept primarily for egg production. Consumption of the meat was done only on very specially occasions ("We'll kill the old red rooster when she comes"). Chicken and eggs also played an important role in the barter economy, especially in the South. Commercial production was extremely limited until after World War II. Since that time, however, production has increased rapidly. Although broilers are by far the most important of our livestock, commercial production is limited to a relatively small number of farms.

Taxonomy

The taxonomy of the domestic chicken (Figure 4.18), as described by Weichert (1958), Nesheim, et al. (1979), Moreng and Avens (1985), Hickman, et al. (1990), and Ensminger (1992) is as follows: Family Phasianidae; Genus *Gallus*; Species *domesticus*.

There are at least four other species of the genus *Gallus*. These include:

G. gallus - red jungle fowl
G. lafayetti - Ceylonese jungle fowl
G. sonnerati - grey jungle fowl
G. varius - black or green jungle fowl

Over the centuries, different names have been given to the different sexes and to different ages of animals within each sex. A young chicken is called a chick from the time that it is hatched until it reaches an age of five weeks. A female through her first laying year is referred to a pullet. When she is more than a year old, she is called a hen. A male chicken less than a year in age is called a cockerel. When he is older than a year, he is a cock or a rooster.

Origin

The ancestor of the domestic chicken (*G. domesticus*) is probably the red jungle fowl (*G. gallus*) found in the jungles of South and Southeast Asia. Although it has the widest distribution of the wild species, it is likely that the other species listed in the preceding section also contributed genes to the genome of the present breeds. Moreng and Avens (1985) note that wild chickens of the genus *Gallus* can still be found in the jungles of Southeast Asia. The earliest record of poultry dates to about 3200 BC in India. Chickens were being bred in captivity, the eggs artificially incubated, and the chicks grown for meat and egg production in Egypt as early as 1400 BC. Domestication of chickens in China began about 1400 BC. In the following centuries, the red jungle fowl was dispersed by man through much of the Middle East and Europe. When the Romans invaded Gaul and Britain in the first century AD, the domestic chicken was already there. By the year AD 1, the domestic chicken was distributed throughout Western Asia and Eastern Europe. From there, early explorers took them to South Africa, Australia, Japan, Russia, Siberia, and Scandinavia. The English brought them to what is now the United States in 1607. They probably were introduced to South and Central America even earlier.

It is important to note that much of the evolutionary history of the domestic fowl took place in the humid tropics, or under climatic conditions not greatly different from those found in Alabama for much of the year. However, cattle, swine, and sheep evolved primarily in the temperate climate zone. The channel catfish is the only one of our domestic animals that evolved here.

Economic Importance

Data on the number of Alabama chickens (excepting broilers) on December 1 of each year for the period 1950-2000 are presented in Figure 4.19. These data show that there were eight million chickens in the state on December 1, 1950. Afterwards, the numbers generally declined through 1956, but as the numbers of broilers produced began to increase, the demand for broiler hatching eggs also began to increase. Chicken (layers) numbers followed. Numbers increased sharply until the late 1960s when they reached 20 million. Numbers remained generally unchanged until the early 1980s, when they began to trend downward, reaching 16 million in the mid-1980s and remaining at near that level through 2000. Data on the annual production (number) of broilers in Alabama during the same period is presented in Figure 4.20. Generally, numbers increased from 13 million in 1950 to a billion in 2000.

As is evident for Table 4.16, poultry and poultry products are extremely important to Alabama's agricultural economy. In 2002, the sale of broilers accounted for 51.6% of the total value of all products, including farm forest products, sold from Alabama farms (Vanderberry and Placke, 2003).

Cattle and Calves

For several millions of years, ancient ancestors of humans and cattle may have coexisted in Eurasia and Africa. In the earliest times, they probably simply shared the same environment, but with the passage of time and with evolutionary change, the relationship changed significantly. By at least a million years ago, the pattern of co-existence had changed to a predator-prey relationship. Then 9,000 years ago, the relationship slowly changed again in an extremely important way when humans assumed . . . dominion over the . . . cattle . . . . (Genesis 1:26), and cattle were domesticated. This relationship has taken some fascinating turns, as human culture and the relationship has evolved. Ensminger and Perry (1997) suggest that cattle were first domesticated because of their projected use as beasts of burden. The first book on agriculture written by the Greek epic poet Hesiod, between 600 and 700 BC, discusses the value of using older, less frolicsome oxen for plowing. In Roman times, personal wealth was based on the number of cattle owned. The words cattle and chattel, which means "possession," come from the same Latin word. In the American colonies, cattle were kept mainly for their use as draft animals, for milk and butter production, and for hides. There was little interest in the animals as a source of meat. In the film, *The Wild Breed*, actor Jimmy Stewart characterized the longhorn Spanish cattle of the Old West as being "meatless, milkless, and murderous."

Over the last 9,000 years, the relationship between people and cattle has changed, but the more it has changed, the more it has remained the same. In many parts of the world, cattle are still used primarily as beasts of burden. In the United States, in the latter part of the twentieth century, they are no longer used in this manner, but there are few areas of our lives where they do not play some role. Most importantly, in 1992 some 260 million Americans consumed an average of 63 pounds of beef each. In the same year, 9.8 million dairy cows produced 151 billion pounds of milk and 5.6 billion pounds of milk-fat. In the process, these animals received billions of pounds of high quality feed and produced billions of pounds of manure, which had to be disposed of within our environment. While edible meat (steak, roasts, hearts, kidneys, tongue, and tripe) is the most important product obtained from cattle, numerous other non-edible byproducts play important roles in in our lives. These include hides, bone products (bone china), products made from the hair of the animal, animal feeds, glue, fertilizer, lanolin, lubricants, and soap. In addition to the edible and inedible portions, parts of the animal are used to make over 130 pharmaceuticals including cholesterol, renin, estrogen, insulin, heparin, thrombin, and thyroid extracts.

Taxonomy

The taxonomy of domestic cattle (Figure 4.21), as proposed by Fowler (1969) and Ensminger and Perry (1997), is as follows: Family Bovida; Subfamily Bovinae; Genus *Bos*; Subgenus *Taurine.*

Species *taurus*

Species *indicus*

Most of the beef cattle in Alabama are domesticated descendants of *B. taurus* that are primarily domesticated cattle of the temperate zone; however, there are a few descendants of *B. Indicus*, or humped (zebu) cattle that are common to tropical areas of the world. There are also a small number of crossbreeds of these two species.

Bos taurus is thought to be a domesticated animal derived from a cross between the giant ox or Aurocs *(B. primigenius)* and the Celtic shorthorn ox (*B. longifrons*).

Other important members of the family Bovidae include:

Bibos gaurus - gaur

Bibos sondaicus - banting

Bison bison - bison or buffalo

Bos grunniens - yak

Capra species - goats

Ovibos moschatus - musk-ox

Ovis species - sheep

Both sexes of cattle are called calves until they reach one year of age. After they are one year old, females are called heifers until they begin to develop the prominent hips and large middle associated with sexual maturity. When these secondary sexual characteristics appear, females are called cows. Until the male reaches two years of age, it is called a bullock whether or not it has been castrated. Over the age of two years, the un-castrated male is called a bull. A male castrated when young is called a steer.

Origin

Fossil forms of ancestors of the genus *Bos* are known from Eurasia and are associated with the Pliocene Epoch of the Tertiary Period (Cenozoic Era), three mya (Table 1.1). Much later, during the Pleistocene Epoch of the Quaternary Period, two mya, one species of the genus reached North America (Alaska), but became extinct there. There were no members of the genus in North America when the Native Americans arrived (Darlington,1957).

Fowler (1969) suggests that cattle might have been the first of the animals domesticated specifically for agricultural purposes. The giant ox or Aurocs, *Bos primigenius*, was probably first domesticated some place in Asia Minor or Europe south of the Alps or in the Balkans during the New Stone Age (Neolithic), some 9,000 to 10,000 years ago, after the glaciers receded from Europe. The Celtic ox is thought to have been domesticated north of the Alps, or in Northwest Asia, at about the same time. Both of these ancestral forms are extinct. The wild Aurocs became extinct in the early part of the seventeenth century. The Celtic ox was never known from the wild.

Several authors have suggested that the economic system of the ancient Indo-European peoples were based primarily on the possession of cattle (Lincoln, 1981). They further suggest that when some of them emigrated to India, Iran, and Europe they took their cattle-based culture with them. Some of their descendants in Central Europe, the Celts, are known to have provided Julius Caesar with cattle for his army. Further, cattle were the primary base for the agricultural economy of the Celtic people in the Scottish Highlands.

The role of cattle in agriculture is well documented in all of our recorded history. Some of the most ancient stories recorded in the Bible mention herds of cattle. Hesiod, the Greek poet (776 BC), who wrote the first known treatise on agriculture, mentioned cattle. As was noted in a preceding paragraph, there were no cattle in the Americas when Columbus made his first voyage in 1492. On his second voyage, he brought some with him. Cattle were brought to Mexico around 1520, and from there they were carried by missionaries to California, Central America, and later over the whole of South America. Ponce de Leon brought them to Florida at about the same time. In 1540, Deigo Maldonado, transported cattle to the Pensacola area, as did Tristian de Luna in 1559. Blevins (1998) suggests that in the latter part of the sixteenth century, Spanish missionaries traveling up the rivers into Georgia and Alabama to establish missions probably took cattle with them and gave them to the Native Americans that they encountered. The first documented introduction of cattle to Alabama was in 1701 when the French brought them to the Mobile area. Like the Spanish before them, the French helped spread the herding of cattle to Native Americans. Missionaries and traders traveling up the Alabama and Tombigbee rivers carried their merchandise and agricultural methods to the Choctaw, Chickasaw, and Creek peoples. At about the same time, cattle were being taken into the backcountry of the Carolinas from Charleston. By 1712, Cherokees who lived in western North Carolina, eastern Tennessee, northwestern Georgia, and northeastern Alabama had large stocks of these animals.

Abernethy (1965) commented that in the 1820s settlers in southeast Alabama owned large herds of cattle that were allowed to roam wild throughout the forests of the region, feeding on whatever plants they could find. It is likely that a substantial share of these backcountry settlers were direct descendants of people who had emigrated to the colonies from the Scottish Highlands in the preceding century.

Economic Importance

Figure 4.22 presents data on the inventory (thousands) of "all" cattle (excepting milk cows) in Alabama taken on

January 1 of each year during the period 1950-2000. The data clearly show the presence of Cycles of Abundance. They also show that numbers increased from 800,000 in 1950 to 2.8 million in the mid-1970s. Afterwards, however, numbers began to decline, and by 2000, they were down to 1.4 million. The decline in numbers in the mid-1970s coincides with the period in which American consumers were beginning to prefer beef (grain-fed) with a heavy, white fat marbling, rather that the more fibrous (grass finished) meat with yellowish fat.

In 2002, Alabama ranked 14th in the country in the number of cattle inventoried. There were 47,000 farms in Alabama in 2002, and 25,000 of those operations included beef cattle (Vanderberry and Placke, 2003). There was an average of 29.3 animals on each farm. The Alabama Cattlemen's Association is the largest in the country. Apparently, the ancient Indo-European and Celtic love for cattle still thrives in the state at the beginning of the twenty-first century.

Alabamians love their cattle and their beef. Unfortunately, the only practical way to raise cattle in the state is to grow cows only to produce calves that are grazed on prepared and maintained pasture grasslands. Then when they reach a certain size, they are sold to feedlot operators in the several Western states where they are finished on a grain-based feed before slaughter. Unfortunately, grass does not produce efficiently in poor soils or in Alabama's subtropical climate. Grass pastures are expensive to establish and equally expensive to maintain. Further, they do not produce very well in dry weather or when it is unusually cold.

Data presented in Figure 4.23 indicate that in the early 1950s there were 370,000 milk cows in Alabama; however, beginning in 1953, numbers began to trend downward exponentially. By 2000, there were only 25,000 remaining. Through the years, the federal government made a determined effort to make milk production in the state competitive by intervening directly in the marketplace. Unfortunately, none of their efforts made much difference. The weather is simply too hot in the summer, and the cow's genome, determined in temperate climates, simply could not accommodate the Alabama heat.

In 2002, Alabama ranked 43rd in the country in milk production, producing only 0.2% of the national milk supply. In the same year, only 200 of the state's farming operations included milk cows, down from 900 a decade earlier.

In 2002, cash receipts for cattle and calves sold from Alabama farms totaled $304.7 million (Table 4.16) and accounted for 9.8% of the total receipts from the sale of all farm products, including farm forest products (Vanderberry and Placke, 2003).

Swine

Ensminger and Parker (1984) have provided a fitting introduction to the life and ways of the humble pig:

> Who, from the remote day of his domestication, has artfully mirrored the world around him;
> Who has been maligned as unclean, regarded with contempt, fed by the prodigal son and considered abominable to the Lord;
> Who has been people-downgraded as hog wild, wallowing in it, as a pig in a poke and as fat as a pig;
> Who has been extolled for being the little pig that went to market, for bringing home the bacon, for being the mortgage lifter and for being in pig heaven;
> Whose living legacy, to a hungry world is: Root hog or die.

The humble pig is literally all things to all people: We eat ham, bacon, pork chops, pork tenderloin, hogs-head cheese, fried pork skins, and pickled pig's feet; we use the pig's thick hair or bristles to manufacture paint brushes; in earlier times, its small intestines were used as sausage casings and are still used today in chitterling-eating contests; its skin is tanned into a fine quality leather used in manufacturing gloves, shoes, billfolds, and footballs; extracts from its adrenal glands have been used to treat Addison's disease; and insulin extracted from its pancreas is used to treat diabetes.

Taxonomy

The taxonomy of the domestic pig (Figure 4.24), as described by Weichert (1958), Ensminger and Parker (1984) and Hickman, et al. (1990), is as follows: Family Suidae; Genus *Sus*; Species *scrofa*.

The species is divided into five subspecies. These

include:

Sus scrofa scrofa—Central European wild boar
Sus scrofa leucomystax—Japanese wild boar
Sus scrofa vittatus—Southeast Asian pig
Sus scrofa christatus—Southeast Asia pig
Sus scrofa domestica—domestic pig

In addition to *Sus*, the family Suidae includes five other genera. The more important of these are: *Phacochoerus* the genus of the warthog; *Hylochorerus* the genus of the giant forest hog, and *Potamochoerus* the genus of the bush pigs. Altogether there are 31 species in the family. In addition to *S. scrofa*, the more important species include: *S. salvanis* the pygmy hog; *S. barbatus* the bearded pig; and *S. verrucosus* the Javan pig.

The domestic pig, S. *scrofa domestica*, apparently was created by crossing domesticated Southeast Asian pigs (S. *scrofa vittatus* and S. *scrofa christatus*) with Central European wild boars (S. *scrofa scrofa*). It is a highly plastic species. Different breeds and varieties are easily produced by the application of the principles of animal breeding. Ensminger and Parker (1984) commented that there were over 400 different breeds of hogs, worldwide. There are probably over 100 of these in Europe and the United States. The most important breeds in this country include: American Landrace, Berkshire, Chester White, Duroc, Hampshire, Hereford, Poland China, Spotted, Tamworth, and Yorkshire.

In the practical taxonomy of swine, a pig is a young animal of either sex, weighing less than 120 pounds and less than four months old. A hog is a large mature animal of either sex, generally weighing more than 120 pounds. A gilt is a female under one year old that has not had her first litter. A female that shows evidence of having produced pigs or that is an obvious state of pregnancy is called a sow. A boar is a male being used for breeding purposes. A male castrated before reaching sexual maturity and before developing the physical characteristics of a boar is called a barrow. If a male is castrated after reaching sexual maturity, he is called a stag.

Origin

Archeological evidence indicates that swine were first domesticated in the East Indies and Southwest Asia, some 11,000 years ago, during the Neolithic Period or New Stone Age (Ensminger and Parker,1984). Another center of domestication seems to have been near the city of Jerico, Jordan.

Hoyt and Chodorow (1976) commented that in the agricultural economy of Early Medieval Europe that meat in the diets of peasants was supplied primarily from the half-wild pigs that roamed in woodlands near villages where they eked-out a tough-and-stringy existence on acorns and beechnuts. The meat of half-wild swine was especially important to the inhabitants of Central Europe at the end of the Roman period. The forests were so thick and the soils so heavy that traditional primitive European plant agriculture of the period was virtually impossible.

Sir Walter Scott's novel, *Ivanhoe*, provides an insight into the importance of swine to the old Anglo-Saxon nobility in twelfth century England. In fact, Gurth, one of the principal characters was a swineherd.

There were no members of the family Suidae or true pigs in the Americas before the coming of the Spanish in the latter part of the fifteenth century; however, a related family (Tayassuidae) or peccaries had been here for millions of years (Darlington, 1957). Columbus brought the first hogs to this hemisphere when he made his second voyage in 1493. The first of these animals to reach North America were probably brought here by the Spanish explorer de Soto when he landed near the present day Tampa, Florida, in 1539. He landed only 13 of these animals, and they were taken along on his journey through the Southeast. Three years later, the herd had grown to some 700 animals. No doubt some of this herd escaped during this long journey and reverted to the feral or semi-wild form, as the so-called razorback. Abernethy (1990) suggests that the descendants of de Soto's traveling herd probably were roaming the woods when the first European settlers reached the region that would later become Alabama. The French carried swine to Canada in 1604, and in 1608 the London Company transported 600 of the animals to Virginia. They were also taken to the Massachusetts colony in 1624 (Shepard, 1886). With the vast quantities of acorns, beechnuts, chestnuts, and other native forage available to them, the imported swine flourished in their new environments and within a few years were becoming pests because of their rooting habits.

In 1630, the inhabitants of Jamestown were compelled to palisade the town to keep the wandering pigs out of their gardens.

Locher and Cox (2004) commented that hog mania was decidedly a Southern phenomenon. They noted that an 1848 study by the US Patent Office (there was no US Department of Agriculture at that time) revealed that the average Southerner ate 173 pounds of pork that year, as compared to the 51 pounds consumed by Northerners. Southerners of every walk-of-life loved their pork. According to the authors, Dr. John S. Wilson labeled the region as the "Hog-eating Confederacy" or the "Republic of Porkdom."

Economic Importance

Figure 4.25 presents data on the number of hogs and pigs in Alabama on December 1 in each year from 1950-2000. The data show that in 1950 there were 1.25 million of the animals in the state. Afterwards, although there was considerable year-to-year variation, numbers began to trend downward, and in 2000 only 165,000 were counted in the inventory. The data also show that the Cycles of Abundance described in a preceding section for cattle were also quite common in hogs, although the cycles were more abbreviated.

In 2002, Alabama ranked 26th in the country in the number of hogs and pigs inventoried that year, only 0.3% of the total. In the same year, only 300 of Alabama's 47,000 farming operations included hogs, down from 2,500 a decade earlier.

In 2002, cash receipt from the sale of hogs produced on Alabama farms accounted for only 0.6% of total cash receipts for all farm products, including farm forest products (Vanderberry and Placke, 2003) (Table 4.16).

For maximum performance, hogs require a unusually high quality diet, a diet not very different from that of humans. Unfortunately, ingredients for their diets are not readily available in Alabama, or the South, but must be imported from the Corn Belt. Further, they have a relatively poor rate of conversion of feed to meat. Also, their genome was largely determined in temperate climates, and because they do not have sweat glands (that is why they love mud holes), they are at a disadvantage in the extreme heat of Southern summers. For these reasons, Alabama has relatively little comparative advantage in pork production. While the state has little comparative advantage in pork production, the old Anglo-Saxon love of hogs and pork (especially bacon) kept farmers trying year after year, long after it became economically unreasonable to do.

Channel Catfish

With the exception of the channel catfish (Figure 4.26), the animals that we produce for food in Alabama were domesticated thousands of years ago far away from the state. In this respect, the channel catfish is unique. Its domestication is no more than a century old, and its development as a major agricultural crop has occurred in the past half-century.

The value of the channel catfish as a food source was likely recognized by the first hunting and gathering people who entered what is now the Southeastern United States. For example, Jackson (1995) comments that early-on Native Americans built weirs in the major streams of Alabama by piling tons of rocks in an inverted V shape with the point upstream so that most of the flowing water was diverted around the ends. In this way, migrating fish were guided into nets and traps, placed in the point of the V. Traps were constructed of reeds found along the banks of the streams. In this manner, many different species of fishes including the channel catfish could be collected. The weirs also created pools and eddies where schools of fish would concentrate; they could be easily stunned using poisons made from local plants such as the buckeye and black walnut. Later, European American settlers used those same weirs. For decades, fish taken around those structures provided a highly nutritious supplement to foods available to local farmers who were fishermen in their spare time. After family needs were satisfied, the excess were sold in local communities.

Although the channel catfish was first sought for their food value, it was not this characteristic that led to their domestication. The channel catfish is also an excellent game fish. It reaches a relatively large size, and when taken on hook and line fights powerfully. Smitherman and Dunham (1985) commented that as early as 1910 Kansas State Fish Hatcheries began to produce the species in ponds for stocking as a sport fish. Later, personnel at the Arkansas

State Fish Hatcheries refined procedures for producing large quantities of fingerling fish (Nelson, 1956). In the mid-1950s, H.S. Swingle at Auburn University began a highly successful series of studies on the commercial production of the channel catfish in man-made ponds (Swingle, 1954, 1956, 1958). Farmers in west-central Alabama, using information developed by Swingle, began the commercial farming of this species in 1959. Later, in 1963, commercial production began in Mississippi. In 1997, Alabama fish farmers sold $53.7 million worth of catfish from 21,000 acres of ponds. Sales for the United States were more than $400 million.

Taxonomy

The taxonomy of the channel catfish (Figure 4.26), as described by Weichert (1958), Hickman, et al (1990) and Bond (1996), is as follows: Family Ictaluridae; Genus *Ictalurus*; Species *punctatus*.

The family Ictaluridae is known as bullhead catfishes, because of the relatively large blunt head on individuals of many of the species. The name of the family is derived from two words that together mean fish cat. The name, of course, is related to the presence of barbels or cat-like whiskers around the mouth. There are six genera in the family. The genus *Noturus* (madtoms) contains 25 species: *Amerius* (bullhead catfishes), 7; *Ictalurus* (forktail catfishes), 4; *Pylodictus* (flathead catfish), 1; *Satan*, 1, and *Trogloglanis*, 1. The latter two genera are blind subterranean catfishes found in Texas and Mexico. The species *punctatus* was described by Rafinesque in 1818. The specific name is related to the presence of numerous black spots on the body.

Origin

Darlington (1957) suggests that the ancestors of the freshwater catfishes of North America may have originated in the Old World tropics, probably tropical Asia. From there they may have moved to temperate Asia and then to North America. Bond (1996) suggests that the catfishes of the family Ictaluridae were once native to all of North America, but in recent times (after the Pliocene Epoch, some two mya), they disappeared from the area west of the Rocky Mountains. Now they are native only in the area east of the mountains; however they have been introduced to virtually all areas of the continent where the climate is suitably warm. The channel catfish is native to all of the river systems in the Southeastern United States. Domestication involved simply taking them from local streams and developing procedures for reproducing and culturing them in man-made ponds. In this respect, it is important to remember that domestication of the channel catfish is just beginning. All of the fish currently taken to market each year are only a relatively few generations away from the wild. The fish that will be cultured in the middle of the twenty-first century will likely be quite different from the one in use now.

Economic Importance

Figure 4.27 presents data on the number of surface acres in catfish ponds on January 1 in each year during the period 1989-2000. The Alabama Agricultural Statistics Service did not begin to keep these data until 1989. The data in the figure show that acres increased sharply from 13,500 in 1989 to 17,700 in 1990, and remained near that level through 1995. Afterwards, they generally trended upward, reaching 22,0000 acres in 2000. Half of the surface acres were in Hale County. In 2002, Alabama catfish sales ranked second in the country, and Alabama farmers harvested 18.6% by number of all the fish harvested.

It is likely that the genome of the channel catfish was largely determined in moderate-sized streams in Southeastern North America, so the species is generally at home in Alabama. Of course, ponds are a new environment for the genome, but the species has adapted well. Consequently, production is extremely efficient. In fact, the species is so efficient in the use of feed that farmers half a world away are able to import feed ingredients necessary for its diet and still make a profit.

Cash receipts from the sale of channel catfish accounted for 2.4% of total cash receipts for all farm products, including farm forest products in Alabama in 2002 (Table 4.16). Cash receipts for catfish were greater than for either hogs (0.6%) or dairy (1.2%) (Vanderberry and Placke, 2003).

While channel catfish perform extremely well when stocked in man-made ponds and fed large quanties of high quality feed, the biological degradation of the large quantity of manure produced sometimes results in the production of complex chemicals by blue-green algae in the water, which under some conditions accumulate in the muscle tissue of the fish giving it a distinct and unpleasant off-flavor.

Catfish is primarily a food of the South, and it is unlikely that large numbers of people out of the region will ever consume substantial quantities of this local delicacy. Per capita consumption of catfish increased from 0.91 pound in 1992 to 1.02 pounds in 1997, and has changed very little since. In fact, the consumption of all fish and shellfish in the United States has changed relatively little since the mid-1980s. In 2002, Americans consumed 15.6 pounds per capita. In the same period, consumption of chicken increased from 45.5 to 70.7 pounds. In 2002, beef consumption was 64.5 pounds per capita.

White-Tailed Deer

It might seem strange to include the white-tailed deer in the list of Alabama's domesticated vertebrates, but when one considers how much money and effort are expended each year on seed, fertilizer, labor, and equipment for food-plots and on harvest, it would be unreasonable not to include them, especially since so many of them are produced and harvested on Alabama farms.

The relationship between the white-tailed deer and Alabamians has been a fascinating one. No other vertebrate has played such a central role in Alabama's history and economy over such a long period of time. When the Amerindians arrived in this area, they found large numbers of white-tailed deer in the woodlands, and they immediately began to use them for many things. First and foremost, deer provided a steady and reliable source of food, and they used almost every part of the animal including the bone marrow for that purpose. Hudson (1976) estimated that deer provided early hunters from 50 to 70% of the animal protein eaten. Antlers were used to make several kinds of primitive tools, including the one used in flaking flint and other hard rock fragments to shape them into projectile points, knives, scrapers, and other tools. Bone shards were used as needles and sinews (tendons) were used for bow strings and for sewing. Skins were used for clothing, footwear, and housing. Later, when agriculture finally emerged among the hunting and gathering people, scapulae or shoulder blades were attached to wooden handles and used as hoes or shovels for planting and cultivating crops. Still later, European traders exchanged rifles, ammunition, pots, axes, cloth, hoes, and beads for literally tons of deer skins that were shipped to Europe to be used in making men's trousers, coats, and gloves.

In the latter half of the nineteenth century, improved firearms and widespread habitat destruction reduced Alabama's deer population to a small, highly localized remnant. Then, beginning around the middle of the twentieth century, once-cultivated land began to return to forest. With increased habitat, protection, and enlightened regulations, plus some judicious restocking, the deer population literally exploded. From a low point of less than 10,000 animals in the early 1900s, the population has grown to in excess of one million in the intervening century.

Taxonomy

The taxonomy of the white-tailed deer (Figure 4.28) is that of Putman (1988) and Hickman, et al. (1990): Family Cervidae; Subfamily Odocoilinae; Genus *Odocoileus;* Species *virginianus.*

There are four subfamilies of deer: Moschinae (musk deer), Hydropotinae (water deer), Muntiacinae (tufted deer), Cervinae (Eurasian deer), and Odocoilinae. Only two of these, Cervinae and Odocoilinea are found in North America. The genus *Cervus* of the Cervinae and *Alces, Odocoilinus*, and *Rangifer* of the Odocoilinae are the only genera found here. The genus *Cervus* contains the wapiti or elk (*C. canadensis*), the genus *Alces* includes only a single species, the moose (*A. alces*), and the genus *Odocoilinus* includes both the white-tailed deer (*O. virginianus*) and the mule deer (*O. hemionus*). The genus *Rangifer* also contains only a single specie, the reindeer or caribou (*R. tarandus*). The latter species is found only in Alaska and Canada.

Putman (1988) comments that there are as many as 38 sub-species of the white-tail. The most common subspecies in Alabama is *O. v. virginanus*, the prototype of all white-tailed deer. The Florida coastal subspecies, *O. v. osceola*, is found in extreme south Alabama, adjacent to the Florida Panhandle. In recent years, some northern woodland white-tails, *O. v. borealis*, have been stocked in some of the northern counties.

Both sexes of the newly born deer are called fawns. They retain this name until they shed their spotted coat, and grow their first winter coat. Afterwards, males are called bucks and females are does.

Origin

The first true deer, all relatively small forms, appeared in Eurasia during the Miocene and early Pliocene epochs of the Tertiary Period (Cenoic Era) some 20 mya. By the end of the Pliocene, the fossil record is replete with a great variety of types, many of which have survived to modern times. The Cervinae and Odocoilinae probably invaded North America during the Pleistocene Epoch of the Quaternary Period some two mya (Darlington, 1957).

Ecoomic Importance

The white-tail deer continues to be important to today's Alabamians just as they were to the people who lived here thousands of years ago. Alabamians as well as visitors from other states spend hundreds of millions of dollars a year for firearms, licenses, food, lodging, transportation, leases, insurance, clothing, seed, fertilizer, guide services, and other related services and equipment associated with hunting the white-tail.

Bolton (2003) reported that in the 2001-2002 hunting Alabama's 422,800 resident and non-resident hunters spent $799 million on hunting trips. Money derived from hunting accounted for 16,900 jobs, $364 million in wages, $31 million in taxes paid on purchases of fuel, $37 million in federal income taxes, and $13 million in state income taxes. He further reported that of 422,800 resident hunters, 378,800 or about 90% were deer hunters. Deer hunting is big business in Alabama.

The white-tailed deer is a wild animal in every sense of the word. As part of the wild portion of the ecosystem, it is constantly beset by a host of predators. Rue (1978) comments that constant vigilance is the payment for life in the species. In many habitats, it is a primary food item for several species of carnivores. This constant danger has guided the evolution of the five senses of the animal—especially smelling, hearing, and seeing—to a high level of development. The animal must always be alert, always monitoring the air for (odor) particles, listening for unnatural sounds, and watching for any movement that might indicate danger.

If the deer's brain receives any signal from any of its senses that is outside the normal range, the animal immediately responds. The lower levels of response may involve raising the head with the muzzle thrust forward and nostrils flared as it searches the air for additional information concerning a suspicious smell. If the suspicious signal is received by its ears, the animal will pivot its head in the direction from which the sound is coming with its ears also pointed in that direction. If the signal is a visual one, the animal will stare intently in that direction, and it may move its head up and down to slightly change the perspective of the object. The animal may also snort. This explosive, blasting sound is produced by expelling air forcefully through the nostrils with the mouth closed. The snort indicates that the animal is startled or surprised.

If the level of the stimulus is sufficiently intense, the animal may begin to stamp first one forefoot and then the other, as it takes a few cautious steps forward. In this process, the animal may stand with its body tensed and rigid with the head bobbing up and down while being held below the level of the back. At this level of response, the animal may emit a higher-pitched whistling snort. This sound, made with the mouth open and combined with a vocal sound, means danger, and in seconds that animal and all others in the vicinity will flee. As the animal quickly bounds away from the area, it lifts its white tail to the vertical while flaring the white hairs on its fringe. The up-lifted tail is almost always present in the fleeing doe, but apparently in some cases, the buck chooses to run with its tail clamped against it rump. All of these characteristics of wildness endear the white-tailed deer to hunters. Hunters are captivated and at times spellbound (buck fever) in matching their senses with those of this magnificent wild animal.

The human hunter is the only predator of any consequence for the mature white-tailed deer. This relationship, which is 10,000 to 20,000 years old, dates to the arrival of the first hunter-gatherers into the range of the animal. Over time, this relationship has been altered as laws have been enacted to prevent the extirpation of the species. As a result, the animals generally may be hunted for only a portion of the year. Night hunting with the aid of spotlights is forbidden, as is killing them after they have been attracted to a specific location with bait. Seemingly, the animals have responded to this set of rules by modifying their behavior. Throughout most of their range, if hunted heavily, both sexes, especially the older ones, remain hidden in thick vegetation during daylight, emerging to feed only after dark. Younger animals that survive their first hunting season also adopt this strategy. Unfortunately for the bucks, those behavioral drives related to procreation over-ride those related to self-preservation, for a short period each year. During this period, they throw caution to the wind as they attempt to locate and breed with as many does as possible (the rut). Bucks move about with little concern for concealment almost continuously. Without this annual change in behavior, once an animal reached three or four years of age, it would most likely die from old age.

tebrates with Opposing Thumbs (Humans)

One species of the hairy vertebrates, the one with opposing thumbs (humans), dominates the landscape of the state, in fact the entire world. Individuals of this species can be found virtually everywhere in Alabama. Thousands of the species move about and through the state each day. Most of the species lives in urban areas, but a sizable minority lives in rural areas. They are involved in a myriad of activities. A dwindling number till the soil, producing food and fiber.

As noted in the preceding paragraph, humans are unique among vertebrates. The Subspecies (*Homo sapiens sapiens*) is the only extant member of the Family Hominidae. The first representatives of the Subspecies, the Amerindians, emigrated here as long as 10,000 years ago and possibly much earlier. Then some 400 years ago, Europeans began to arrive. Soon thereafter, the Europeans began to introduce various groups of West Africans as slaves.

Concerning humankind, William Shakespeare has Hamlet say in Act II, scene 2: "What a piece of work is man, how noble in reason, how infinite in faculty, in form and moving how express and admirable, in action, how like an angel, in apprehension how like a god: the beauty of the world, the paragon of animals; and yet to me, what is this quintessence of dust?"

As Hamlet comments, man is first among the animals; yet they share many of the general characteristics of all vertebrate animals. However, there are some additional ones that place humans in a more restricted group—the primates. We along with lemurs, tarsiers, monkeys, and apes have an unusually well-developed nervous system, particularly the cerebral section of our brains; long pentadactyl limbs; eyes directed forward and completely encircled by bony orbits; thumb and big toe usually opposable; plantigrade foot posture; and usually only one young per birth event.

Several additional characteristics place humans in an even more restricted group the Family Hominidae and the Genus *Homo*. One of these characteristics is the potential for year-round mating. There is no breeding season in humans. As a result, mates tend to form stable long-lasting bonds that are important in the rearing of young. Of course, there are other reasons for human pair-bonding, but year-round mating plays an important role. Another important characteristic is the human infant's long dependence on adults. This lengthy period allows the young human to acquire important information (learning) needed to function as an adult before becoming one. This characteristic is also important in the formation of extended families. Further, their proclivity to form complex associations (communities) has provided them with unique opportunities as well as a wealth of associated problems. Also, while most vertebrates are mobile, the mobility in the Genus *Homo* is exceptional. From our early appearance in Africa, we have made the entire planet and now part of outer space our oyster. Probably, however, the most important characteristic of *Homo* is our huge complex brain that allows us behavioral flexibility and an extensive capability for creativity, thinking, knowing, and for accumulating the products of thinking and knowing for planning.

Taxonomy

The taxonomy of the human species is that of Weichert

(1965), Robertson (1987), Hickman, et al.(1990), and Stein and Rowe (1996): Family Hominidae; Genus *Homo*; Species *sapiens* (archaic man); Subspecies *sapiens* (modern man).

For many years, anthropologists attempted to subdivide the human subspecies into a number of different races; however, there was little agreement on the number. Some suggested that there were not more than three. Others identified as many as 100. Stein and Rowe (1996) reported on work conducted by Cavalli-Sforza in which he analyzed gene frequencies of 42 aboriginal populations from Africa, Asia, Oceania, Europe, and North and South America. Analysis of the data indicated that the gene frequencies fell into three major, more or less discrete groups of populations: African, North Eurasian, and Southeastern Asian. Different populations included within these great groups are shown in Figure 4.29.

Origin

Origins are important because every animate or inanimate object is the result of the accumulation of a vast number of prior events. For example, farmers and the plants and animals of Alabama agriculture are the result of a vast number of physical and biological characteristics that have been accumulating for 3.5 billion years. Their, and our, entire complement of genes have been determined by the trial-and-error process that is the result of millions of years of the application of the natural selection for beneficial mutations. Each of those mutations was evaluated by natural selection, and only those that conveyed increased odds of surviving, in a specific environment, were kept. In this sense, we have been shaped by a vast number of environmental challenges, over an extremely long period of time. Understanding something of the nature of those changes helps us to understand how agriculture came to this place and what happened once it arrived.

Origins also have to do with the evolution of cognition, behavior, and culture. Culture also accumulates over time and is significantly affected by environment. Tools (knives, wheels, hoes, plows) are cultural artifacts that allowed humans to do things we could not easily do with our physical characteristics (hands, fingers, finger nails) that evolution had provided us. Over time, the nature of these tools changed in significant ways. In fact, over the last 10,000 years, tools have changed much more dramatically than the genotypes of humans. Similarly, statecraft, which is also a cultural artifact, has changed significantly from the early hunter-gatherer organizations to modern democratic states.

Further, origins also have to do with the establishment of human associations and their movement. The three groups of human beings who would farm in Alabama (Asians, Europeans, and Africans) came from three widely separated areas of the world; however, it is fairly clear that their oldest common direct ancestors originated in Africa (Wenke, 1990). Apparently, as early as 1.5 mya, members of the Genus *Homo* had colonized much of East, Central, and South Africa, and had begun to move outward from that area. One branch of this genus ultimately evolved into our own species; however there is considerable debate as to where this evolution actually took place. According to one model, the individuals of that genus spread from Africa outward into Asia and Europe, and later all three branches (African, Asian, and European) evolved independently into our species, *Homo sapiens*. This model accounts for some of the variability in the species between geographic regions. In a second model, *Homo sapiens* evolved in Africa and later spread into Asia and Europe. In the process of evolution, at least two other species of the Genus *Homo* appeared (*H. habilis* and *H. erectus*). There is considerable debate as to how these two are related to our species. There also is considerable uncertainty as to when *H. sapiens sapiens*, or modern man, first appeared. Stein and Rowe (1996) suggest that they may have been present in Africa as early as 130,000 years ago, in Asia as early as 115,000 years ago, and in Europe 60,000 to 38,000 years ago. The first modern men appeared in Europe at about the time their distant relatives the Neanderthals disappeared. It is more certain, however, that by about 35,000 years ago, modern man was distributed widely over the entire globe with the possible exception of the Americas. Suggested relationships among the various groups of modern man are shown in Figure 4.29.

Cognition, Behavior, and Culture

Without question the most important characteristic, which differentiates humans from other vertebrates, is our more complex brain and some of the unique products of its function. Of these products, cognition is certainly the most significant. So far as we can determine, it is likely that no other vertebrate species has it. Many other important characteristics, in turn, are derived from our cognitive ability. Behavior is another product of our brain. It explains what we do and why and how we do it. Then both cognition and behavior serve as the bases for the elaboration of culture. It is through culture that products of this unique mental ability are organized, communicated, manipulated, stored, and passed on to succeeding generations. Cognition, behavior, and culture were essential for the domestication of plants and animals and for the evolution of agriculture.

Cognition

Cognition refers to all those mental processes associated with remembering, knowing, and thinking (Myers, 1989). It is an extremely important part, probably the defining part, of our biology. While many animals are able to remember, at least to a limited degree, we seem to be the only species that is aware that we can remember. We are able to remember or know many things, and we also are able to be aware that there are things we do not know. We also are able to know in the abstract. We are able to know what the advantages or disadvantages might be if we knew more about a particular subject.

When vertebrate animals first appeared 450 mya, they already had a functional brain and nervous system. Two mya, one group of these vertebrates (*Homo*) had reached the point where they could make tools. Tool-making represents a significant milestone in the evolution of the vertebrate brain and in its use. Over the following two million years, continued improvement of their tools and their use indicate continued evolution of mental capacity. Much of the early tool technology probably came about as a result of learning from observation and from trial and error, and remembering and repeating what had been learned. Over time, however, people began to know or to understand that there were things about the production of tools and their use that they did not know. Then using the information they did have, and using the thinking process, they were able to project the likely outcomes of trying new ways of doing things. These new ways were then evaluated through trial and error. These processes are self-accelerating. Each thing that was learned and remembered increased the pool of things that they knew, and as the knowledge base grew, the thinking process became ever more efficient and effective.

Behavior

Virtually all animals do things that can be classified as behavior, which concerns the things that animals do and why and how they do them (Keeton, 1967). For example, virtually all animals exhibit some behavioral reaction to irritation and some rather specific behavior patterns associated with reproduction. Many animals exhibit behavior patterns related to survival, such as fight or flight. Further, many kinds of animals are able to communicate with each other. Humans have the same basic behavior patterns as all other animals, but in almost all cases the behavioral responses of humans to specific situations are much more complex than is generally the case in the other vertebrates. In reproduction, for example, our behavior related to the care of our young is much more complex than even in our closest animal relatives. Similarly, our responses to threats to our survival are much more involved. However, the greatest departure between our behavior and that of our animal relatives is related to obtaining food and providing for our clothing and shelter. Lower vertebrates do not herd; choose crops for culture; or plant, weed, or preserve food—although many species do harvest and store food. Certainly none would be involved in the even more complex situations such as making decisions concerning agricultural research, protecting the environment, fertilizing their crop land, or caring for their animals during the winter. They do not know the kinds of plows to use or understand general versus commercial farming. Our history books are replete with descriptions of the kinds of behavioral responses we exhibited in many different situations. However, there is relatively little information concerning why we did those things or why we behaved as we did.

Innate And Inherited Behavior

Much of what we know about behavior is based on

our observation of lower animals. For example, when the embryo of the domestic fowl is ready to hatch, it uses a tiny, temporary spur on the tip of its beak, as it pecks to break through the shell of the egg. This action is essential if it is to emerge. This same behavior at hatching takes place if the egg is incubated in a nest underneath the hen or in a mechanical incubator far removed from her. What chicks do and how they do it is well known, but why all of them peck their way out is not nearly as well understood. Apparently, this entire process is somehow passed down from the parents who engaged in exactly this same activity when they were hatching. This pecking behavior is pre-programmed in the brain of each chick to be activated only when it is ready to hatch. There must be a code written in the genes that prepares the spur and initiates and directs this activity. Apparently, the same type of genetic mechanisms that determine form and function in cells of the animal body also determines at least some of the what, how, and why of animal behavior. It follows then that the gene-based activities of animal behavior must also be subject to evolution or the selection of the fittest and that over time there generally would be an increase in those behavioral activities and characteristics that allow the individual to function more effectively within its environment and a decrease in those that allow the individual to be less well adapted.

The pecking of the chick to effect its release from the eggshell does not end its pecking activities. As soon as the chick hatches, without any instructions from its mother, it begins to use its beak to peck to obtain its food. Later, as it grows to maturity, it will peck at an oil gland at the base of its tail. Oil adhering to the surface of the beak will be rubbed onto the feathers to waterproof them. The beak and pecking also are used in social interactions with other chickens (pecking-order).

Nature Versus Nurture

Keeton (1967) notes that for much of the first half of the twentieth century, a battle raged between biologists and psychologists over the relative importance of inheritance (nature) and learning (nurture) in human behavior. As a result of the early work in animal behavior (ethology), which was done primarily by biologists observing animals in natural settings, it was concluded that all behavior was the result of inheritance. Later, as the new field of psychology began to develop, the pendulum of opinion began to swing until many scientists believed that all behavior was learned. Since about the middle of the twentieth century, however, scientists have increasingly come to the position that both are involved. Now, there seems to be an increasingly broad consensus on the question as a result of exhaustive studies on the lives of identical twins reared separately in different environments, as well as advances in molecular biology and molecular genetics. These studies have uncovered a more significant genetic (nature) component to human behavior and personality than was previously imagined. The anthropologist Melvin Konner suggested that everything from risk taking to aggression is strongly influenced by the power of heredity (Sanoff, 1991). He further suggested that the relative importance of biology and culture (nurture) in determining behavior might be near fifty-fifty. Kuehn, et al. (1998) summarized the results of the studies with twins and stated that such attributes as alienation, extroversion, traditionalism, leadership, career choice, risk aversion, attention deficit disorder, religious conviction, and vulnerability to stress seem to owe as much to heredity as to environment (nurture).

Keeton (1967) suggests that nature and nurture interact so that the genes, or inheritance, determine the limits within which a particular type of behavior can be modified and that learning, acting within those limits, determines its precise nature. Obviously, limits set by the genes vary dramatically with different types of behavior. For example, the genes are ultimately responsible for the development of the behavioral trait by which we quickly move our finger when it touches a hot object. In this case there is no learning involved. It is all nature; however, in making the sounds by which we communicate with each other when we talk, the limits that the genes place on the full development of the trait are extremely broad.

The presence of vocal cords, plus the ability to use them to make sounds, is inherited. It is apparently encoded in the genes. Apparently some aspects of the ability to learn to make and use some of these sounds as language is also encoded in the genes; however, most of it is learned, and most of it is learned by observation. The ability to convert sounds into symbols through writing is also learned. Although learning plays an important role in the writing and reading of books of history, in the development of computer programs, in the development of architectural plans, and in the development of operating budgets, genes also play an important role. All of the higher vertebrates have some form of the physical tools (vocal cords, appendages, nervous systems) required to vocalize however, even the most closely related species (the great apes) cannot be taught to perform more than restricted portions of the very simplest elements required to sing opera. Obviously, genes control the capacity to learn or set the limits at different levels in the different groups of animals.

Learned Behavior

Obviously, not all behavior is inherited or instinctive. Some of it is learned through experience or by trial and error (Hickman, et al.,1990). For example, by using its instinct to peck to eat, a chicken can learn or be taught to peck a lever to receive food. In this situation, the chicken has learned to repeat an action that results in a favorable outcome, and as a result, its behavior is changed. Similarly, animals, especially the higher animals, readily learn to avoid those actions that bring bad results. This is the way children learn to avoid touching a hot stove. These are behavioral actions that are obviously learned. Animals also learn by observing the behavioral activities of other individuals. This is an important means of learning in the primate animals, especially our species. In this process, animals observe and repeat the behavior of others.

Insight learning is another type of learning that is important in the higher animals and especially so to our species. In this process, learned behavior is somehow stored in our brain, and used later in such a manner that the animal is able to respond correctly the first time to a situation that it has never before encountered. It is able to respond correctly without going through the process of trial and error. This process is also called reasoning.

Behavioral Change

The combined and interactive effects of the inherited and learned aspects of social behavior, in virtually all species except our own, remain generally the same or fixed, generation after generation, except as they are very slowly changed through the evolutionary process. In humans, however, changes in the learned aspects of both individual and social behavior can change much more rapidly because of the evolved and inherited capacity effectively to pass learned behavior from one generation to another through the use of language. Many species of vertebrates have vocal cords and can produce sound, and in virtually all species of these animals, the sounds convey information from one individual to another, but only in humans are these sounds accumulated (myths, beliefs, rules) and passed on to succeeding generations to help them function more effectively in their environments. Certainly no other species is able to convert sounds into symbols (alphabets and numerals), so that information can be disseminated and passed on in the absence of communication by vocalization. The development of writing (and reading) expanded the capacity to accumulate the effects of learned behavior to a degree almost beyond our imagination.

Although behavior in humans can be changed more readily than in the lower animals, specific behavior patterns can persist for long periods of time. Fischer (1989) discussed in detail the differences in the behavior patterns of discrete groups or people in the British Isles who came to America before the American Revolution. He comments on the persistence of some of these patterns in the establishment of the different colonies, in the American Revolution, in westward migration, in the Civil War, and in Reconstruction. He also presents evidence that many of these patterns persisted well into the last half of the twentieth century, some 400 years after these British people left their homeland for the New World.

Behavior and Making Decisions

Most animal behavior rests on a deep, complicated bed of continuous decisions: deciding when and where to seek shelter, deciding where to search for food, deciding whether or not another species of animal is a friend or foe, deciding where to build a nest, deciding whether or not to wear a coat, deciding what to prepare for dinner, deciding how much corn to plant, deciding whether or not to ask for a raise, and deciding which stock to purchase are but a few examples of the myriad decisions that different species of animals must make each day.

These are examples of simpler decisions that animals have to make, and the effects of making a wrong one is generally limited. In our species, there are much more complex ones: whether or not to have a complete or partial mastectomy when dealing with breast cancer, how aggressively to treat prostate cancer, or when to place a parent in a nursing home. These are much more complex and have far-reaching consequences.

Myers (1989) discusses two decision making process: systematic reasoning and intuition (rule of thumb or seat of the pants). In the use of systematic reasoning to make decisions, the desired outcome is clearly defined, along with alternative choices available to realize that outcome. Then the advantages and disadvantages of each choice are carefully and clearly evaluated. Once this evaluation is completed, only then is a decision made. This is the preferred process decisions, but it is extremely time consuming. The rate of return on time invested is often extremely low, but the likelihood of realizing the desired outcome is relatively high. In cases where success or failure is extremely critical this process is much preferred.

The process of using intuition, also called heuristics, to make decisions proceeds much more rapidly, but the success of the outcome is usually, but not always, much more uncertain. Intuitive decisions are made on the basis of whether or not the desired outcome along with possible alternatives represent a prototype or a problem with similar outcomes and alternatives encountered before. This approach is termed the representativeness heuristic. A slightly different process involves information stored in our memories, the availability heuristic.

In human societies, decisions made by farmers are unique. The quantity and quality of our food supply is dependent on the aggregated decisions made by an extremely large number of individuals. Decisions made by a single farmer have little to do with it. At the same time, decisions made by individuals can determine the well-being of the family unit. Decisions by the individual farm family on whether to continue to farm, on what commodities to produce, and on how much to invest in production have little effect on the food economy of the country, but a wrong decision can have a disastrous effect on the family's finances.

During the period 1942-1948, prices that farmers in Alabama received for their commodities were sufficient to cover costs of production and to allow for a small amount of profit. In other words, the ratio of prices received to prices paid remained above 1 for the entire period. During this period, each year some 223,000 Alabama farm families made the decision to commit to farming for another year; however, in 1949, the ratio fell from 1.1 to 1.0. At that level, they could pay for most of the costs of production, but there was little left over as profit. The situation improved very little in 1950, and sometime late that year or early 1951, some 14,000 farm families (6.4% of the total) made decisions to stop farming. One can only imagine the serious, often tearful discussions that took place around those dinner tables and at the local feed and seed stores in the rural areas of the state. Obviously, this was not an easy decision to make. This ratio was the average for the entire country, and by 1949, it is likely that it was considerably below 1 for some of these Alabama families that decided to quit. They had few alternatives. Afterwards the ratio continued to decline, and individual decisions to leave farming became a flood. Between 1950 and 1970, Alabama had lost 63% of its farms. There had been 138,000 individual decisions to quit. By 1970, all except Alabama's very best farms were in serious trouble.

These decisions were agonizing affairs. Some of these farms had been in continuous operation for generations. Their owners had continued to hope and pray that things would get better. Further, they knew what they could expect in farming, but were extremely concerned about what would happen to them away from their farms. They had good reason to be concerned. Off-farms jobs were scarce. Faced with this uncertainty, many of them had wrestled with this decision for years.

Farm families in Illinois also faced similar decisions as to whether or not to leave farming after 1945, but the environment within which they made their decisions was somewhat different than in Alabama. Between 1950 and 1951. only 2.5% of Illinois farm families made the decision to quit. Between 1950 and 1970, that state lost only 36.9% of its farms. Yearly changes in the ratio of prices received to rices paid were not as punishing to

Illinois farm families.

When farm families decide to continue to farm another year, decisions on what crops to plant and how many acres to plant still present farmers with complex choices. Decisions regarding how many acres to plant are extremely complex, because farmers can never know how the weather will be after they plant their crops or what the price will be when they harvest them. In general, they depend on their experiences to make decisions. Generally, they expect that crop weather the following year will approximate some long-term norm and that they will receive the same price in the coming year that they received the past year. This would seem to be a good strategy; however, wishing to maximize their returns, a good price the preceding year invariably results in decisions to increase acres planted the following year. Unfortunately, all farmers seem to read the tea leaves the same way, and all make the same decision: plant more acres. Just as invariably with large increases in acres planted, supply exceeds demand, and price declines.

For example, in 1966 American corn farmers planted 66.3 million acres of the crop, and when they harvested it, they received an average of $1.24 a bushel, the highest in 11 years. With that experience behind them, they made the decision to increase acres planted to 71.2 million in 1967. When they harvested that crop, the price had fallen to $1.03, the lowest in seven years. With that sharp decline in price in 1967 firmly fixed in their decision-making process, they reduced acres planted to 65.1 million acres in 1968. As a result of this aggregated decision process, acres planted and prices teeter-totter across the years. Very seldom are increased acres planted associated with increased prices.

It seems likely that these examples of the teeter-totter relationship between acres planted and price were the result of the widely held assumption that more is better; that is, if a single squirt of a pesticide gives good results, two squirts would provide twice the effect, or if a single pill is good for you, two would be even better. Using the more-is-better assumption as a basis for making decisions was probably important in the evolution of our species, in those ancient times when there was seldom enough of anything. At the same time, they had also learned that when you find yourself in a hole, the worst thing that you can do is to keep digging. This experience also became extremely important in the decision-making process.

In some cases, increased acres planted and increased prices rceived can occur in the same year when bushels harvested actually declined as a result of a reduced yield. For example, in 1979 corn farmers planted 81.4 million acres of the crop and received $2.52 for it. In 1980, they planted 84 million acres and received $3.11. However the yield in 1980 was considerably lower than in 1979 (91 versus 109.5 bushels an acre). As a result, bushels harvested were also much lower (6.6 versus 7.9 billion). This outcome probably occurs just often enough to keep farmers believing that more is better.

Culture

Culture represents another special milestone in the evolution of vertebrate animals. It is a unique product of our highly-specialized brain and its function. More specifically, it is another product of our cognitive ability. As a result, genetic evolution is not the only determinant of the nature of our everyday lives. Culture and cultural evolution also play an important role. Culture allows us to chink the cracks in our basic gene-determined foundation.

For example, for a number of very basic reasons, humans are unable to fly; however, through the use of cognition and culture they have created airplanes, air terminals, jet fuels, strip searches and lost luggage, and for all practical purposes they can now fly anywhere they want.

Cultural evolution allows us to respond more quickly and effectively to environmental challenge than natural selection. For example, clothing that is an artifact of our material culture allows us to extend our geographical range far beyond what our genetic makeup allows. As clothing evolved as a result of environmental challenge and selection, geographical ranges are extended even further. Similarly, myths and legends that are artifacts of our nonmaterial culture allow us to pass on important information to following generations that would not be passed on through the reproduction of our genetically determined structures.

Nature of Culture

Stein and Rowe (1996) define culture as "learned, nonrandom systematic behavior and knowledge that is transferred from person to person and from generation to generation." They further note that culture changes through time and is a primary contributor to human adaptability within the environment. Our species appears to be the only one fully capable of using culture in development of ever-increasingly complex systems.

As a result of cognition or consciousness and the ability to think, humans are able to plan, to act on those plans, to evaluate the effectiveness of what they have created, and to make corrections or changes where necessary. If the products of these processes were lost with each generation, human progress would be virtually impossible. Fortunately, we are able to accumulate and store them and to share them with generations through culture. Culture is a product of this complex process. It consists of both nonmaterial (plans, poems) and material (knives, wheels) products of this creative effort that we share in our society. However, the most important aspect of culture is not any of these products specifically, but that they represent steps in an ongoing process whereby our thinking process seeks to create those cultural elements that allow us to fit our lives more effectively into our environments and in some cases allow us to actually alter those environments. Alabama agriculture with all of its component parts is a cultural artifact created as a result of a broad application of this cognitive process.

The evolutionary process (natural selection of the fittest) provides a mechanism by which all animals including humans, over time, were adapted to their environments. However in our species, this process seems to have provided us with a higher level of adaptive capability. Evolution has provided us with an increased capacity to learn and to use what we have learned, to adapt to our environment more quickly and effectively, and to be able to pass this accumulation of learned behavior on to our offspring. On this genetic base of increased capacity, we have learned how to make an almost unbelievable variety of physical or material objects (clothes, buildings, fires, wagons, ploughs, improved kinds of plants, medicines, pesticides, books, schools, computers) that allow us to effectively use a broad range of physical environments and to be able to move from one of those environments to another with relative ease. At the same time, our increased capacity has also provided us with the ability to develop or produce a wide variety of nonmaterial objects (languages, computer programs, myths, religious practices, rules, customs, legal systems, economic systems, political systems) that allow us to function more or less effectively in our social environments or in the realm of our intrapersonal and interpersonal or social relationships. To sociologists, the sum of these material and nonmaterial products of our shared endeavors constitute our culture. In their view, culture consists of the shared products of society, and society consists of the interacting people who share a culture (Robertson, 1987).

Evolution of Culture

Historically, material objects and abstract creations or culture represented the evolution of all aspects of the behavior of a specific group of people to a specific set of environmental conditions. For example, the cultures of Native Americans that developed in the cold woodlands of Alaska were very different from those that developed among the tribes of Native Americans on the Great Plains east of the Rocky Mountains, and these were different still from those that developed in the eastern woodlands of the United States, the jungles of Central America, and the mountainous highlands of South America. In these different environments, highly different cultures evolved among the descendants of the relatively small group or groups of hunters who entered North America from Siberia 10,000 to 20,000 years ago. Different cultures also evolved among this same broad group of Native Americans, even when physical environments were not so widely divergent. For example, the cultures of the Native American Choctaw in southwest Alabama, the Chickasaw in northwest Alabama, the Cherokee in northeast Alabama, the Upper Creeks in east-central Alabama, and the Lower Creeks in southeast Alabama were all different in many important respects (Hudson, 1976).

Many characteristics of the environments of the eastern woodlands of the Eastern United States and those found in Western Europe are similar. Both regions share the same temperate climate (Group D of Lydolf, 1985) (Figure 2.10). The major difference is that there is more oceanic influence on the climate in the western portion of Western Europe than in the eastern portion of the

United States. Both regions have large areas of forested soils (Alfisols and Inceptisols) characteristic of temperate climates, and both have similar broadleaf and mixed broadleaf-coniferous forests (Akin, 1991). Yet with the strong similarities of the environments in the two regions, the cultures developed there were strikingly different. By the middle of the seventeenth century in Western Europe, agriculture was advancing rapidly. For the first time in history, farmers were producing enough food so that they had a surplus to sell, and part of the surplus was being fed to animals. Brick homes were being constructed, and factories were making wooden furniture for consumers. Large quantities of metal utensils (knives, spoons, pots, pans) were being manufactured. The publication of books was widespread. The magnificent Shakespearean literature had been available for almost a half century. The characteristics of the cultures of Western Europe of this period contrasted sharply with those of the cultures that the first people from that region found when they reached Eastern North America.

Cultures are continually changing as people learn by trial and error and as they accumulate more experiences on which they can base reasoning. Although change is continuous, historically it occurred very slowly, especially when peoples were isolated from each other. Some cultures discovered earlier in the twentieth century, in geographically isolated areas of the world, had changed very little for thousands of years. While most cultures might have evolved relatively slowly from a historical perspective, the rate seems to have increased rapidly as communications improved in the last half of the twentieth century. The changes that have occurred in both the material and nonmaterial elements of culture in virtually all environments, and among all people throughout the world following World War II, has been nothing short of astounding.

When different cultures are brought together over a short period of time in the same location, changes can take place rapidly. When the Spanish colonized Central and South America in the early sixteenth century, many aspects of the cultures of most of the Native Americans living there changed drastically, suddenly, and permanently, and the results were essentially the same when the English colonized the Eastern Seaboard of North America. In these examples, both the Spanish and English brought large parts of their own local cultures with them when they came to the Americas, and the pressure of those combined cultural elements was so great that the cultures of the Native Americans were unable to resist. However, rapid and major cultural changes are not always the result of massive pressure from competing cultural systems. When the Spanish first passed through portions of the Great Plains, the domesticated horses that they left there changed the culture of the Native Americans rapidly and permanently. Similarly, the invention of the cotton gin had a significant and permanent effect on the cultures that would evolve in the Southeastern United States.

According to Robertson (1987), when culture changes, its two basic elements (material and nonmaterial) change at different rates. Changes in the material things that we create (wheels, plows, wagons, houses) take place much more rapidly than abstract, nonmaterial creations (ideas, beliefs, rules, customs, myths). Material things are much more easily and successfully altered by trial and error and by reasoning than nonmaterial things. In Europe in the millennium from the fifth century until the beginning of the sixteenth, world-changing advances were made in the material aspects of culture (ships, navigation, metallurgy, energy production, manufacturing, agriculture). Yet in this same period, many aspects of the religious beliefs and practices changed relatively little.

Migration and Persistence of Culture

When people move from one environment to another, they usually take much of their culture with them. Fisher (1989) commented on both the material and nonmaterial elements of the culture that the first shipload of Puritans had with them when they left East Anglia in the east of England for North America in 1630. Material elements included: horses, cattle, sheep, goats, pigs, fowl, dogs, cats, provisions, and personal treasures. Nonmaterial elements included: a Christian creed of extreme austerity (Puritanism), congregational churches, the so-called Puritan work ethic, unique patterns of speech, a specialized form of architecture, distinctive ideas about marriage and the family, nucleated settlements, town meetings, and a tradition of ordered liberty. These cultural elements provided the base on which the Massachusetts Bay Colony would evolve.

Many elements of the culture that the Puritans brought with them did not fit the specific characteristics of the coastal environment of New England very well, and as a result, the early immigrants suffered extreme hardships, and for many years it appeared that the colony might not survive at all. Fortunately, enough changes were made in their culture that it began to grow. Over time, many of its elements including the adaptations were carried into much of southern New England, New Jersey, northern New York, then northward into Maine and Canada and then westward to Ohio and beyond. According to Fischer, many elements of this culture would have a lasting effect on the development of the United States as a whole.

Fischer (1989) also commented on the elements of culture that the Scotch-Irish brought with them when they immigrated first to Ulster in Northern Ireland and then to North America, a movement that began in the early part of the seventeenth century and continued well into the latter part of the eighteenth. Compared to the Puritans of New England, these people carried relatively few items of material culture with them because they had none to carry. They were primarily very poor. Most of them were tenant farmers and farm laborers. A large minority were semiskilled craftsmen. While these people brought with them relatively few items of a material nature, they brought a wealth of abstract (nonmaterial) culture including a powerful, militant Christianity with its field meeting, prayer societies, and free grace. They also brought with them a fierce and stubborn pride, ideas about the construction of housing, unique patterns of speech, family ways, and child-rearing customs. They also brought with them ideas about public life that were dominated by a distinctive ideal of natural liberty. These people with their distinctive cultural elements did not fit well at all into the environments (culture) of the English settlements along the Atlantic; consequently, a high proportion of them immigrated westward until they reached the foothills and highlands of the Appalachian Mountains. This environment matched very closely the one they had left in Scotland, especially the Highlands. These immigrants and their culture came to dominate this mountainous region from Virginia to Georgia. Over time, they moved westward from the southern terminus of the Appalachians into the mountainous regions of northern Alabama and on into northern Mississippi. They carried much of their culture with them. Fisher provided evidence that many of the cultural elements that had characterized the Scotch-Irish for hundreds of years persisted well into the twentieth century. Bolt (1983) found that a considerable amount of it is still alive and well in the Chattooga River country of northeast Georgia, northwest South Carolina, and southwest North Carolina.

Gladwell (2008) wrote powerfully of the history of the endemic violence of the Southern Highlands. He commented that the simple truth is—if you want to understand what happened in those small towns in Kentucky in the nineteenth century—you have to go back into the past and not just one or two generations. You have to go back two or three or four hundred years, to a country on the other side of the ocean, and look closely at how the people in a very specific geographical area of that country actually lived. That is a strange and powerful fact. It is just the beginning, though, because upon closer examination, cultural legacies turn out to be even stranger and more powerful than that.

Another example of the consequences of the movement of cultures can be observed in the forced migration of large numbers of African peoples to the New World as slaves (Conniff and Davis, 1994). In this process, these people were able to bring essentially none of the elements of their material culture with them. Also, as a result of the way in which families were separated, tribes mingled and ethnic groups mixed indiscriminately; consequently, much of their nonmaterial culture was also lost. Only a cursory evaluation of history, as well as current events, attests to the consequences of the losses of these cultural elements and the difficulty of replacing them.

Coevolution of Genes, Behavior, and Culture

Lumsden and Wilson (1981) have published a book in which they propose that genes, behavior, and culture evolve together. For example, virtually all human infants are tolerant of the milk sugar (lactose) in their mother's milk while they are nursing; however, in most populations, this tolerance disappears shortly after nursing is discontinued. Surprisingly, in many of those persons whose ancestry can be traced either to Western Europe or to certain parts of Africa, this tolerance persists after nursing ends. This continued tolerance seems to be linked to descendants of those populations in which cattle herding was an important part of their cultures. It is suggested that lactose intolerance after nursing was at one time widespread in the human genome; however, in those populations constantly exposed to lactose in cow's milk, those genomes were altered over time. In turn, it is likely that this increased tolerance resulted in cattle herding becoming a more important part of their culture, because they were better able to use the extremely nutritious cow's milk throughout their lives.

Cognition, Behavior, Culture, and Domestication of Plants and Animals

We are surrounded by the products of our cognition, behavior, and culture (families, friends, clerks, mechanics, teachers, schools, prisons, books, newspapers, buildings, highways, computers, automobiles, dishwashers, hydrogen bombs, trash dumps), and all of these are important in our lives. However, the domestication of plants and animals and the agriculture that followed is probably the most significant one.

The Nature of Domestication

The root of the word domesticate is the Latin "domus," which means house. In general terms, domesticate means "to bring to the house." In domestication, some aspects of the biology of a plant or animal species are brought under the purview or control of humans. When people become involved in the management of any aspect of the biology of a species, it changes the environment of the species, which in turn changes the selection pressure operating on its genetic makeup. Domestication changes the evolutionary pathway of the captive portion of the species. Further, when people began knowingly or unknowingly to select certain individuals for more attention, for example selecting larger seeds for planting, the rate of evolutionary change increases significantly.

Domestication is a coevolutionary process (Wenke, 1990). The process changes the evolutionary pathway of both the plant and animal and the people managing their biology. Domestication of plants and animals has also had a significant effect on the biology of humans. In effect, the domestication of plants and animals has resulted in the domestication of humans. Before the development of farming, humans were not greatly different from other small bands of roving animal omnivores. Along with the domestication of their food supply, permanent villages for relatively large numbers of people appeared, and within these primitive collections of people, civilization began (Solbrig and Solbrig, 1994). In a sense, domestication of both humans and plants proceeded simultaneously.

The Process of Domestication

The early domestication of plants and animals almost certainly was not the result of a planned or deliberate effort. It simply grew out of the age-old foraging and hunting relationship between animals and plants and the predator-prey relationship between animal and animal. Long before domestication was accomplished, people certainly must have been aware that seeds shed from harvested wild plants would germinate under some conditions after being accidently sown around their camp sites. Also, evidence for the domestication of animals (sheep and goats) appears in the archeological record a millennium before evidence for the domestication of plants (wheat). Ponting (1991) suggests that the domestication of plants may have resulted from collecting them for animal feed and fodder. Under these circumstances, seeds from the collected plants would be sown around the areas, rich in animal manure, where animals were being held. Beginning 10,000 to 12,000 years ago, the relationship slowly began to change from one of foraging to primitive husbandry or farming. Then 10,000 years ago, plant culture became important and widespread enough that its presence can be detected through recent archeological research. It probably began in primitive or very restricted form thousands of years earlier.

Unfortunately, we will never know with certainty how the process of domestication began or where it began. Domestication was almost certainly well advanced several thousand years before recorded history begins. Wente (1990) discusses in some detail how the process might have begun and the factors that may have contributed to it. He suggested that while there are many proposed models, most of them include three factors: (1) Post-Pleistocene

(Ice Age) climate change, (2) human population growth and dispersion, and (3) growing human intellect and improved technology. I have added a fourth, geophysical and ecological relationships.

Geophysical and Ecological Relationships

The hydrologic cycle and erosion over time have created valleys, large and small, across the earth's landscapes. In this process, steep, narrow valleys with swift streams in the uplands are replaced by broad valleys with meandering, slow-moving streams in the lowlands. At the same time, these same processes were continually filling those valleys with rock fragments removed from adjacent mountains and highlands. These fragments are graded by weight and size as they were carried downstream. In the grading process, the smallest fragments collected far down the valleys where stream flow was much reduced. In this process, large quantities of very small fragments (sediments) accumulated in the broad valleys. In this geophysical relationship, environments in the highlands were extremely severe. Changes in the seasons brought accumulations of snow and ice. Spring thaws brought catastrophic floods. In contrast, environments far down the valleys were much more benign. While seasonal changes were still active, the effects were much reduced. In these broad, well-watered valleys with their vast accumulation of fine (small) fragments, there were an exceedingly large number of ecological niches (homes) for different kinds of plants and animals. Diversity was several orders of magnitude greater here than in the steep valleys in the highlands. For hundreds of thousands of years, mankind was but one of a large number of animals at home in these benign lowland environments. They were a paradise for hunters and gatherers. In the wet valleys and the gently sloping, drier, heavily timbered, higher lands adjacent to them, they found virtually everything that they required in their lives: food, shelter, and clothing. Under these ecological conditions, the establishment and maintenance of more or less sedentary, extended families and communities was to be expected. This highly diverse mix of plants and animals provided an ideal environment for the beginning of domestication.

Geophysical and ecological processes played an even broader role in domestication than indicated in the preceding paragraph. In the distant past, the movement of tectonic plates brought together the land masses of Eurasia (Europe/Asia and North America) and Gondwanaland (Africa and South America) during the Carboniferous Period of the late Paleozoic Era, some 354 to 290 mya (See Figure 1.5) (Scotese, 1994). The Fertile Crescent lies in the general region (Middle East) where these two supercontinents were temporarily fused together. This region generally found in present-day Turkey, Syria, Jordan, Iraq, and Iran served as a land bridge that allowed for the mixing of plants and animals from both supercontinents. This does not imply that this bridge allowed the exchange of plants and animals involved in early domestication. In the Carboniferous, there were no mammals on earth, and all of the plants were seedless ones. However, this fusion of the supercontinents did serve to enlarge the gene pool in that particular location, which would have significantly increased the possibilities for future evolution of plants and animals that would be domesticated.

Further, geological processes associated with these events resulted in the formation of a mild-climate zone that protected these ecosystems from the wild swings associated with the various ice ages. As a result of this mixing effect, 10,000 to 20,000 years ago, progenitors of several domesticated crops (emmer wheat, einkorn, barley, flax, chick peas, peas, lentils, and bitter vetch) would have been encountered growing wild on this land bridge in the valleys associated with the Tigris and Euphrates rivers (Solbrig and Solbrig, 1994). Further, cows, goats, sheep, and pigs would have also been found in the same region, and wild horses could be found on its northern margin. This geophysical and ecological mixing pot would have also provided an ideal environment for the establishment of a relatively dense and largely sedimentary population of humans. Given the growing intellectual capacity of humans at that time, domestication would have been almost an expected outcome.

Climate Change

As noted in Chapter 2, the climate of the world began to become warmer about 14,000 years ago. Akin (1991) comments that the last ice advance (the Wisconsinan) reached its greatest advance around 12,000 BC when it covered a quarter of the earth's surface. However by 6500 BC, all of the continental glaciers had disappeared from most of Europe leaving in place only those in the high mountains. Likely, most of the land mass in the Fertile Crescent (Figure 4.51) of Southwest Asia had been ice-free since 10,000 BC. With the end of the ice age that corresponds generally to the end of the Pleistocene Epoch of the Quaternary Period (Cenozoic Era) and the beginning of the Holocene Epoch of the Quatenary (Table 1.1), the earth's climate changed dramatically from cold and wet to warm and dry. For 5,000 years, the average temperature of the world increased gradually until it reached its highest level between 5000 and 3000 BC. Akin (1991) suggested that it was probably warmer then than at any other time in the 10,000-year period since the end of the Pleistocene. There is also evidence that climate in some parts of the world, especially Southwest Asia, became much drier during this time. Climatologists have named this period the Altithermal Climatic Epoch (also sometimes referred to as the Climatic Optimum). It was during this period of ever increasing temperatures that domestication and agriculture began.

During the cold, wet period associated with the Wisconsinan Ice Age (60,000 to 10,000 years ago), humans in the Fertile Crescent likely subsisted primarily by hunting the herds of reindeer, horses, and other large mammals, concentrated around the southern margins of the glaciers. However, even though the large animals provided most of their food, they still must have foraged for plant materials, especially wild cereals. As the glaciers retreated, the herds of foraging mammals also moved northward, following the vegetation belts as they expanded. Some of the hunting bands followed, but apparently many of them remained in the areas where they had been living for several thousand years. Because they could no longer depend on the large grazing mammals for food, however, they had to turn to other sources. At about this time, they must have begun to forage widely and intensively for smaller mammals, such as gazelles, wild cows, sheep and goats, as well as for waterfowl and fish. They also must have begun to make much greater use of wild cereals that were increasingly available as a result of the warming, drier climate. These changing conditions suggest that shifting their food base may have required a significant increase in the effort expended to capture or collect sufficient quantities of these smaller animals or wild cereals to feed themselves. This intensification of effort was probably also accompanied by a closer and more continuous contact with their living food sources and a growing awareness of characteristics of their cycles. Domestication and farming seems to have evolved out of this increased contact and awareness associated with the slow warming of the earth.

Population Growth

The gradual increase in human populations following the retreat of the glaciers 8,000 to 15,000 years ago must have also played a role in plant and animal domestication. As the number of humans increased in a given environment, there was a significant increase in the number of hunting bands that resulted, in turn, to restrictions on the sizes of foraging and hunting areas available to each band. These restrictions may have lead to a gradual reduction in the movements of groups of people and a gradual shift toward the establishment of more permanent settlements (associations). Also, there is some evidence that the warming climate, especially in Southwest Asia, was becoming much drier, so that hunting and foraging in the highlands became more difficult, concentrating a more or less static population of people into the river valleys.

These changes would have required changes in the relationships between the forager and plant populations and the predators (hunters) and their prey (wild animals). With restrictions in the size of foraging and hunting areas, planting and herding may have evolved. As part of this overall process, more attention probably would be given to those plants and animals in their environments whose biological characteristics made them more amenable to cultivation and herding. The obvious choices were those plants like barley and emmer and einkorn wheat that grew in thick stands all around them and animals like wild cows, sheep, and goats that were generally found in herds nearby. These kinds of plants would have been good candidates because they grew in thick stands and produced relatively large and relatively visible seeds. Further, grazing animals would have been good candidates for domestication, because they require so much time to secure the large amount of low quality, fibrous grasses and herbs that they use as food. They tend to confine themselves within a relatively small area for long periods of time while they are feeding. Simple herding is the most primitive form of domestication. In its most primitive form, it changed the relationship between predator and prey relatively little, but certainly this primitive relationship led almost inevitably to more intensive forms.

Growth of Human Intellect

Anthropologists suggest that the development of domestication and, in turn, agriculture, could not have begun until our species reached a certain level of cultural, social, and technological sophistication (Solbrig and Solbrig, 1994). They suggest that without advanced stone tools such as lightweight stone hoes, sharp scythes, and knives, as well as containers such as baskets for holding cereals at harvest and earthen vessels for storage that humans could not have become agriculturalists. The ever-increasing sophistication in the kinds of tools available and in tool use in the making of clothing, in the development of structures, in the use of fire, and in the use of language as a vehicle for planning and securing cooperation were extremely important. Humans seem to have reached this stage of development at about the same time throughout the world—around 10,000 years ago, usually referred to as the Upper Paleolithic (Old Stone Age) and Neolithic (New Stone Age) periods. The appearance of these improved tools generally coincides with the retreat of the glaciers and the beginnings of agriculture.

Plant Domestication

The domestication of plants involves more than simply learning to systematically collect and sow plant seeds. After all, the accidental collection and dispersal of seeds by animals had likely been occurring for hundreds of thousands of years before agriculture began. Domestication could only begin when the genetic makeup of the cultivated plant began to change either through the process of evolution or as a result of human manipulation (artificial selection), so that it became more useful and valuable to people than its wild ancestor. In this scenario, the domestication of wheat did not occur when the hunters and gatherers in the Fertile Crescent in Southwest Asia learned to collect and sow seeds. It began when the existing genotype of the wild plants changed so that the seed heads could be threshed with limited loss of individual kernels, so that the glumes (husks) surrounding the kernels became thinner and softer, and so that the two rows of kernels in the seed heads of the wild forms was replaced by seed heads with six rows.

For an extremely long period of time, the relationship between humans and plants was likely not greatly different from the relationship between plants and any other foraging animal. Humans simply harvested what was available growing wild in their environment. Then 10,000 to 12,000 years ago, the human-plant relationship began to change, at least for a few species. Humans began to be involved in at least some aspects of the production of plants through artificial selection and cultivation. Both domestication and farming began when humans started to be deliberately involved in the actual production of plant materials.

Fortunately, the process of plant domestication did not end with the culture of lentils, peas, wheat, and barley in the Middle East. The technology spread relatively rapidly from family to family and then from settlement to settlement throughout that area and much later into Europe. Also, the technology of domestication involving other plants probably evolved independently in other parts of the world shortly thereafter. For the next 11,000 years, until the fifteenth century of our era, plant domestication developed relatively slowly. The next great wave of plant domestication took place in the fifteenth century. New plants were domesticated, and the genomes of most of these cultivated plants were slowly altered through primitive selective breeding. Then during the Age of Exploration, the third great wave of domestication began when cultivated plants were widely redistributed throughout the world. As these plants were moved from continent to continent, both their geophysical and biological environments were altered, in some cases dramatically. These changes resulted in a new burst of change in the genomes of the plants. These changes progressed slowly until the latter part of the nineteenth century when the science of plant breeding began to evolve. The fourth great wave of plant domestication followed. This wave resulted in the development of thousands of new varieties of some plants. In the latter quarter of the twentieth century, a fifth wave began emerging. For 12,000 years, the genomes of plants were changed through selective breeding that depended on the rather rigorous laws of inheritance. Changes could be effected only by rearranging by selection of naturally occurring variation. However, with the new technology (wave) of genetic engineering, it is possible to manipulate individual genes directly and to put completely new genes into cells. In this way, changes in plant genomes can be effected much more rapidly, and in many situations can now be made that would have been almost impossible to make with selective breeding. For 12,000 years, sexual reproduction has been the primary tool for advancing plant domestication. Unfortunately, it is not very precise. Each new generation includes individuals with a range of genomes. Now, the evolving technology of cloning, where exact copies of a plant can be made by culturing small pieces of it, essentially eliminates the range of genomes among offspring. At this point, it is difficult to even imagine or visualize the nature of the next wave of domestication, but if past is prologue, it is safe to say that the relationship between ourselves and the plant kingdom will continue to evolve. There seems to be almost no limit to the changes that may occur as we continue to bring plants into the house.

Animal Domestication

The process of the domestication of vertebrate animals was somewhat different from that of the domestication of plants, because the relationship between humans and other animals is different than our relationship with plants. Over the millions of years of the human-animal association, the relationship changed dramatically. Initially, the ancestral forms of the human species were just wild animals eking out an existence along with all of the other species in a particular environment. With time, however, these pre-humans began to change in astounding ways. As a result of these changes, where pre-humans had co-existed probably more or less peacefully with the other animals in an environment, they became extremely proficient predators or hunters, and virtually all of the other animals came to fear them and to actively avoid them to the extent possible.

Compared to domesticated plants, domesticated animals were somewhat unusual in that while many of them were likely first used for food, over time they were used for other purposes. For thousands of years, humans hunted horses as a source of food. The horse was not domesticated until around 3000 BC, several thousand years after the sheep, goat and cow. However, once it was domesticated, its use as food probably diminished, because of its overriding value as a beast of burden and its importance in warfare. The domesticated cow provided meat, milk, and power. Strangely, cattle met another need of these primitive people. They came to believe, over the long course of history, that the slaughter (sacrifice) of cattle, under certain conditions, would result in fewer bad things happening to them. In South America, the Native Amerindians hunted the guanaco, a relative of the camel, as a food source. Then, around 3500 BC, this species was domesticated as a beast of burden. Later, through selective breeding, it evolved into the llama, which remains an important beast of burden in mountainous areas today. In Southeast Asia, the red jungle fowl provided both eggs and meat, although eggs were probably more important for a long period. It was likely at this point that ancient humans learned the important lesson of not killing the goose that laid the golden egg.

The dog was probably the first animal domesticated, and from early times these animals became related to humans in a novel way. They served as companions, participated in hunts, protected their masters and their property, and warned them when strangers approached. It was well known in that role as early as 12,000 BC. They probably were eaten only as a starvation food. Available archeological evidence indicates that the cat was domesticated relatively recently, probably as late as 1600 BC. It was found to be an effective means of ridding the household of rats and mice. There is no indication that this animal was ever used as food.

Human Associations—Establishment of Early Old World Communities

As noted in a preceding section, one of the unique characteristics of vertebrate animals is their natural propensity to establish and maintain associations (schools, flocks, herds, prides). This propensity is especially strong in our species. The establishment of early communities was a key element in our eventual success in the peopling of the earth. For example, all of the people who would finally come to farm in Alabama could trace their ancestors to a relatively few, loose human associations (communities) established thousands of years ago in the Old World (Africa, Asia, and Europe). The Old World is an enormous land mass encompassing a large part of planet earth. As our species dispersed from Africa to fill it, they were separated by great distances (Figure 4.30), and with time they evolved into different groups of people. It is truly a fascinating story to trace the path of these ancient peoples to farms in this "land of the thicket clearers."

Africa

Africa was almost certainly the continent where the most ancient form of our species originated, but there is little agreement concerning the evolution and movements of the different groups that the Arabs would encounter in the ninth century when they began to visit that continent. Stavrianos (1971) suggests that the modern African peoples originated in the general area of Lake Victoria in central East Africa, partly in the present-day countries of Kenya, Uganda, and Tanzania. There the species differentiated into the four major groups of African peoples: Bushmen, Pygmies, Negroes, and Caucasoids. Later, the four groups migrated to different regions of the continent: the Bushmen to Southern Africa, the Pygmies to the Congo River Basin and into the coastal rain forests of West Africa, the Negroes to West Africa and on northwest into the fertile (at that time) Sahara region, and the Caucasoids northwest to Egypt and North Africa as well as northwest into Arabia and even to West Africa (Figure 4.31).

Asia

Asia probably received its first *Homo sapiens* 100,000 to 115,000 years ago. In the succeeding millennia, variation in the basic genotype of the group increased until it included several different types of people: Uralic, Mongols, Tibetian, Japanese, Chinese, Indonesian, and Melanesian (Figure 4.29). Generally, these types evolved in the major geographic and widely separated areas of this huge continent. For thousands of years, the Mongols, or Mongoloid people, inhabited Central and Eastern Asia, which includes modern-day Eastern Siberia (Figure 4.32). Present evidence indicates that before 30,000 years ago, Siberia had extremely low population densities or was unsettled altogether. The great frigid interior swamps and forests of Siberia may have barred human colonization until huts were developed. In any case, the earliest known human habitation sites in Northeastern Siberia date to about 23,000 years ago. The sites reflect a hunting economy focused on mammoths and other large mammals (Wenke, 1990). Unfortunately, the archaeology of this area has been poorly studied, so it is possible that modern man has been there much longer.

Europe

Europeans did not appear on the world's stage until relatively recently. Wenke (1990) comments that until 9,000 years ago, the great expanse of forests and grasslands of the British Isles, Central Europe, Scandinavia, and Russia was a wilderness inhabited only by roving bands of hunters and foragers whose major cultural achievements included great skill at deer hunting and a theology based on the worship of oak trees. According to Langer (1952), by 2500 BC, there were three major groups of native people established in their respective areas in Europe: the Mediterraneans (Iberians) in Southern Europe, the Alpines in Central and Western Europe, and the Nordics in Northern Europe (Figure 4.32).

There were also other communities of people living on the fringes of Europe that would ultimately determine much of the course of its history (Childe, 1950 and 1958) (Mallory, 1989). One of these groups was the Aryans, or Indo-Europeans peoples, who lived far to the east, on the western side of the southern end of the Ural Mountains in the modern-day South Central Russia (Figure 4.32) . These Neolithic (New Stone Age) Aryans seem to have been a cow people with few horses. Their livelihood depended on the herding of cattle. They ate beef, and only after many centuries did they begin to use these animals for pulling sleds, wagons, and plows. They were nomadic, wandering over the meadows and forests of their homeland following pasturage. Neolithic Mongols lived east of the Neolithic Aryans (east of the southern end of the Urals). They were a horse people. Initially using horses for food, over time the Neolithic Mongols learned to use them as beasts of burden. However, their learning to use the horse in warfare would have the most far-reaching effect on the world: Mongol use of horse cavalry would change the face of Europe.

Moving About—An Important Verterbrate Behavioral Trait

One of the most important elements in the process of the peopling of the earth has been the natural propensity for our species for moving about, to move from place to place. Actually, moving about seems to be a basic vertebrate characteristic. Several species of fish follow increasing water temperature northward in the Gulf of Mexico each spring, only to follow declining temperatures back south in the fall. Elk herds migrate out of the higher mountains into more favorable feeding areas as winter intensifies. Purple martins migrate thousands of miles north and south each year. Even the common box turtle and the possum are forever moving about, looking for just one more busy highway to cross. Most commonly, movements in the lower vertebrates tend to be migrations. They move in one direction at one time of the day or year, only to return later to the same general locality. There are many examples of migration in our species. The economy of Florida is heavily dependent on the annual migration of "snow-birds" from the Northern states each fall and winter. However, emigration, permanent moves from region to region, has been much more important than migrations in the peopling of the earth.

One of the most important examples of emigration or moving about was the exodus of our ancient ancestors out of Africa some 1.7 mya. We cannot know with certainty why this move was made. Of course there are several possibilities such as: the bear went over the mountain to see what he could see, or the grass is always is always greener, or even fight or flight. Instead the most likely one was localized over-population or exceeding the carrying capacity of the environment.

Moving About—Migration of the Indo-Euroeans

Another, extremely important historical emigration began around 1900 BC when groups of the fierce, pastoral (herding) Aryan peoples began to move out of the steppes of Russia into both Asia and Central Europe (Childe 1925, 1950, and 1958) (Figure 4.33). In the process, they spread their Indo-European languages throughout that part of the world. One of these groups, the Celtic-speaking peoples, moved west through Northern and Western Europe as far as modern-day France and the British Isles. Later, a second group, Greek-speaking peoples, moved down through the modern-day Balkans into Greece. Another group, the Italic-speaking group, moved into Italy (Figure 5.5). Later, a Teutonic-speaking group, followed much the same route as the Celts, until they reached Northern Europe. By 1700 BC, these warrior peoples with their Indo-European languages had reached the westernmost limits of Europe. They intermingled with and married with Europeans (Iberians) already living there, but they imposed significant and permanent changes on those local cultures. Probably the most significant change was the warrior ethos. This warrior-herdsmen lifestyle brought from the western steppes would become a permanent part of European culture into our own time.

According to McWhiney (1988), Celts (sometimes called Gauls) as a distinct group of people first appear in written history between the eighth and sixth centuries BC living in Central Europe (Figure 4.34). They likely were descendants of those Indo-Europeans who had emigrated from South Central Russia (Figure 4.33). From earliest times, the Celts were a nomadic, warrior people who herded animals for food and had little inclination for planting and tilling crops. In Central Europe, they lived by demanding tribute in food and other goods from the peoples they had conquered. They were the first of the Central European peoples to master the art of cavalry tactics in warfare. They had learned to do so from their neighbors, the Mongols, on the steppes of Russia. They also developed the art of the use of the two-handed broad sword in battle (McNeill, 1963). For the next four centuries, this warrior race terrorized Europe from the Balkans in the east, to the Iberian peninsula in the southwest, and to the British Isles in the far west. Beginning about 225 BC, however, the Romans, along with the Germanic peoples (descendants of those early Teutonic nomads), began to drive the Celts northward and westward into Western Europe. In 50 BC, Julius Caesar with his legions finally conquered them in Gaul (mostly modern-day France). Caesar describes these years of intense warfare in his *Commentaries on the Gallic Wars*.

Germanic peoples (Teutons) first appeared in history where they were established in Scandinavia (Denmark) and between the Elbe and Oder rivers in the vicinity of modern-day Northern Germany as early as 2000 BC. According to Langner (1952), the Germans had followed the Celts into Western Europe from the east. The Balts were located to their east beyond the Oder, and the Celts to their west beyond the Elbe. After 500 BC, the Germans would continue to expand their territory by moving southward up the Elbe and westward to the Rhine (Figure 4.35) while at the same time pushing the Celts westward. By 200 BC, the Celts had moved up the Rhine Valley as far as the mouth of the Main River, near present-day Mainz.

From around 3000 BC when the first city states appeared in the Middle East, until around 350 BC, the central power of the known world resided in the empires of the Egyptians, Hittites, Assyrians, Chaldeans, and Persians; however around 336 BC when Alexander the Great became leader of the Macedonian Greeks, the center of power began to shift away from Western Asia to Southern Europe, first to the empire of the Greeks and then to the empire of the Romans. As Spielvogel (1994) comments, by the beginning the fifth century AD, the Western Roman Empire was beginning to disintegrate under the onslaught of various groups of Germanic peoples from the north.

As problems for the Roman Empire mounted, they were forced to withdraw their legions from Britain. Faced with the loss of their protection and concerned that some other army might take their territory, the native Britons invited fierce warriors (Angles and Saxons), living west of the Rhine in Northern Germany, to establish colonies in their midst. These ferocious Germans became a much-feared occupying force, and by the early seventh century much of the old Romanized Britain was Anglo-Saxon. The old Celtic-Britons had been driven into the marginal lands of Wales, Ireland, and Scotland.

Moving About—The Indo-Europeans and Their Neighbors Settle Down

Meanwhile, what remained of the Western Roman Empire was being replaced by a series of kingdoms, such as the Ostrogothic kingdom in Italy and the Visogothic kingdom in Spain that were ruled by Germanic kings. Unfortunately, they spent most of their time fighting among themselves, trying to decide who would be kings. A number of these kingdoms were also established in the old Celtic region of Central and Western Europe (Figure 4.35). None lasted very long, except the Frankish kingdom (510-751), established by Clovis (the Merovingian dynasty) in old territory of the Gauls, in what is much of modern-day France. Finally, in 768 Charlemagne became the king of the Frankish Empire, and by 814 had unified all of the Germanic kingdoms in Western Europe, west of the Elbe and Danube rivers, and in northern Italy.

Charlemagne was succeeded by his son, Louis the Pious. When Louis died, Charlemagne's Empire (the Carolingnian) was divided among his four sons, but they were ineffectual rulers constantly warring among themselves. Ultimately, out of this division came the territorial outlines that would eventually become modern-day France and Germany. Unfortunately, none of Charlemagne's grandsons was strong enough to protect their subjects from almost constant raids by Vikings from the north and Magyars from the east. Consequently, the common people had to turn to local, landed aristocrats for protection. These local political arrangements that developed in the period of the Early Middle Ages (750-1000) lead to the emergence of feudalism (Spielvogel, 1994).

The forces, which had led to the coming of feudalism began to abate in the latter part of the Early Middle Ages, and a new type of civilization slowly emerged. Spielvogel (1994) notes that the High Middle Ages (1000-1300) became a period of recovery and growth for Western Civilization. There was a greater sense of security among the people that resulted in a burst of energy and enthusiasm. Improved agricultural practices resulted in a more dependable food supply. Life in the towns and cities improved. Trade and commerce expanded steadily. The static life of feudalism was soon forgotten. Unfortunately, this revival of European civilization did not proceed smoothly. Life in the new world of trade and commerce and growing cities was still unstable. In this period of transition, there were wars and rumors of wars (Crusades and the Hundred Years War), famines and pestilence. But, eventually, the lives of people did improve.

European civilization entered the so-called Renaissance, the rebirth of classical learning and arts, around the middle of the fifteenth century or near the end of the Late Middle Ages (1300-1500). It first appeared in the city-states of northern Italy (Milan, Venice, Florence), but because of the sharp increase in trade and commerce associated with it, the Renaissance quickly spread to much of the Continent. Hoyt and Chodrow (1976) commented that the Renaissance was a rebirth of interest in learning, science, art, literature, music and statecraft, as well as both the Protestant and Roman Catholic reformations.

The rebirth of learning resulted in far-reaching changes throughout Europe, and it re-kindled an interest in the world beyond European shores. Between 1488 and 1522, Diaz sailed around the Cape of Good Hope, Columbus discovered the Americas, and Magellan's expedition circumnavigated the earth. These and similar voyages contributed to a completely new world perspective for Old Europe and ushered in the Age of Discovery and Expansion. Unfortunately, Old Europe's institutions and people were not yet ready for such a change. This brave new world of the Renaissance had come too quickly, leaving the Continent, once again, in a growing state of instability.

The Reformation, which grew out of the period of rebirth, brought about a series of wars of religion that would dominate European history from 1560-1650 (Spielvogel, 1994). These highly destructive wars exacerbated the instability resulting in constitutional crises, economic upheaval (inflation and depression), and social disintegration.

Fortunately, during the seventeenth century, order and harmony were slowly restored throughout the Continent; as a result of the declining power of the Church and the rising power of monarchies and the beginning of nation-states. This new stability encouraged the revival of those interests in learning that had been unleashed in the Renaissance. The age-old interest in the natural world was rekindled, resulting in the Scientific Revolution. Actually this revolution had its beginning in the fertile mind of Aristotle, the Greek philosopher (384-322 BC), but it lay largely dormant until St. Thomas Aquinas (1225-1274), in the middle of the thirteenth century, was able to convince the church that it was acceptable to study the nature of man's soul (theology) and the nature of man (science) at the same time. Later Leonardo de Vinci (1452-1519) and Nicolaus Copernicus (1473-1543) put legs under the revolution with their practical studies in astronomy, human anatomy, engineering and geo-science.

Still later, an Englishman, Roger Bacon (1561-1626), contributed mightily to this revolution when he promoted a scientific method of discovering truth, using inductive principles and experimentation. Later, in 1637, a Frenchman, Rene Descartes published his *Discourse on Method* in which he described a systematic approach to discovering truth, one emphasizing deductive reasoning and mathematical logic.

The outburst of learning associated with the Scientific Revolution had widespread implications for the people of Europe and far beyond in both space and time. Intellectuals (philosophs) began to use the methods of Bacon and Descartes for discovering truth by undertaking a dramatic re-examination of all aspects of life. In these studies, the philosophs came to believe that learning could create better human beings and better human societies. These beliefs laid a foundation for the a new worldview, one based on rationalism and secularism, rather than one based on religion. Their far-reaching studies had such an impact that historians began to call the period the Age of Reason, or the Enlightenment. Their studies also resulted in rising tide of humanitarian sentiments. They believed passionately that there were natural laws that determined the course of civilization, as natural laws determined the course of the material world. They believed just as passionately that there were natural rights that ought not to be withheld from any person. They further believed that Europe's Old Order—its long-standing social order (clergy, nobility, inhabitants of towns and cities, and peasants)— was hostile to the progress of society (Spielvogel, 1994).

Agriculture was one of the primary beneficiaries of the Scientific Revolution. New crops, fertilizers, and new crop rotations were introduced. The understanding of crop nutrition was advanced, and new methods of meeting crop needs were introduced. As a result of experimentation, selective breeding of livestock was begun. Increased production of crops and livestock lead to increased commercialization of food production. By the mid-eighteenth century an Agricultural Revolution was underway in Western Europe. Although it began in the Netherlands, it quickly spread to England. It resulted in a much larger, more varied, and more stable food supply. Unfortunately, the new agriculture was also much more efficient in the use of labor, and there was no longer a need for hundreds of thousands of peasants and agricultural laborers.

Despite all of the changes that had taken place in Europe from the beginning of the Renaissance through the mid-eighteenth century, much of the Old Order remained well entrenched (Spielvogel, 1994). The nobility still determined the course of human affairs on the Continent. Some 85% of the total population were still peasants with the restrictions and obligations placed on them by the seigneurial system. Poverty was still a highly visible rural and urban problem. Even as late as the end of the century, in both France and Great Britain, 10% of the population depended on charity or begging for their food. But things were slowly changing. While monarchs were enlarging their bureaucracies and armies and increasing the collection of taxes to support them, it was clearly evident that the Old Order was no longer capable of dealing with changes that were taking place.

As detailed in a preceding paragraph, during the High Middle Ages (1000-1300) with the growth of trade, commerce, and urbanization, there was increased opportunity for merchants, craftsmen, artisans, government officials, and professionals (academics, lawyers, and physicians, as well as clerics). Slowly, a more affluent middle class emerged. Over time it diverged sharply from the lower class of common laborers from whence it had come. However, with the Industrial Revolution a completely new class emerged: the urban working class. The middle class was transformed into the bourgeois, which included some of the lower nobles, businessmen, industrial managers, bankers, merchants, clergy, lawyers, teachers, scientists, physicians, and government officials. These groups were relatively much more affluent, much better educated, and much more concerned with the world around them than the rural peasantry and the new urban working class.

The Enlightenment, with its increased humanitarian concerns born of a rapidly expanding middle class, also saw an increase in the growth of representative institutions in which an increasing number of the governed participated in government by electing persons to represent them in the governing process. Participation of the governed can be traced at least to the fourth century BC, if not earlier to the city-states of Greece. Later, citizen participation appeared in the Roman Republic, but only a small portion of the population were citizens and thus able to participate. Progress in the development of these institutions was extremely limited during the Early Middle Ages. It began to increase again, especially in England in the thirteenth century when the barons forced King John (1199-1216) to sign the Magna Carta. Unfortunately, as was the situation in the earlier Roman Republic, participation was limited to only a small portion of the population, and this portion did not increase very rapidly. As late as 1782, the English Parliament was elected by only 3% of all of the people in the realm. Fortunately, the franchise of the governed was slowly expanded at the expense of the monarch and the nobility, but really significant changes would not come until the nineteenth century when Parliament enacted the Reform Act of 1832. Under this act, the percentage of the population that could participate in the selection of representatives was increased by 60%, but still only a share of the population could participate. The act required that participants had to own property valued at ten pounds sterling. This requirement excluded hundreds of thousands who could not meet this requirement, especially rural farm laborers and industrial workers.

In the latter part of the eighteenth century, the Industrial Revolution began in England. It developed on a broad base established earlier by the Scientific and Agricultural revolutions. By 1850, this new revolution had made the country the wealthiest on earth. It also resulted in the emergence of a completely new class of people–the urban working class.

The Agricultural and Industrial revolutions resulted in important socioeconomic changes in society in eighteenth century Western Europe. Changes in agriculture, especially enclosure of common lands and an end of the medieval, open-field system of England, forced hundreds of thousands of poverty-ridden rural workers, without access to land, to flee into the cities where they sought jobs as domestic servants, common laborers, and industrial workers (Kishlansky, 2005). It was estimated that unemployment during this period reached as high as 40% of the population. As a result, despite increased availability of food, additional housing, and more generous poor relief, the share of the population in dire poverty soared, especially in the cities, and unfortunately an ever-increasing population exacerbated the growing problem. As the birth rate increased and more children survived early childhood, the ranks of the impoverished swelled. Population growth was unchecked by the sorts of periodic demographic crises (wars, famines, and pestilence) that had limited the population in previous centuries.

The cascading flood of human misery of the late eighteenth century provided excellent grist for the mills of early bourgeois social reformers born of the Enlightenment. Soon they were demanding increased government involvement in this ever increasing human tragedy. One of these early reformers, the French politician Louis Blanc (1811-1882), wrote in his 1840 book *The Organization of Labor* (Spielvogel, 1994):

1. Competition was the main source of the economic ills of the day. 2. Social problems could be solved by governmental assistance. 3. The path to a more just society was government responsibility for the welfare of its people. 4. Government ought to be considered the supreme regulator of production and endowed for this duty with great power.

Earlier, the French aristocrat, Henri de Saint-Simon (1760-1825), founder of French socialism, had voiced his deep concern for the impoverished masses, but he felt that they had only a limited innate capacity to change their situation. He was convinced that their salvation lay in the establishment of a strong central government, planned and controlled by a small elite of engineers, scientists and technocrats (Booth, 1871 and Hewett (2008). The early American patriot and the first American secretary of the treasury, Alexander Hamilton (1755-1804), held similar views (Morison, 1965).

This new European movement to increase government intervention in the everyday welfare of ordinary people was given the name socialism. It did not immediately have an effect on the human misery in cities. Conservative governments in power opposed it. Over time, however, it became part of the ever-increasing stream of national liberalism that would finally change the face of the Continent socially, economically, and politically. It doubtlessly contributed to the revolutions of the 1820s, 1830, and especially the French revolution of 1848. Although these revolutions failed to destroy the Old Order of national conservative leadership, they forced it to appreciate the fact that the next revolution might be successful. As a result, the Old Order moved decisively to make concessions that would result in better life for their peoples (Spielvogel, 1994). Unfortunately, while there were some positive changes, the more things changed the more they remained the same. This slow progress in improving the lot of the vast majority of the working class (the proletariat) lead Marx and Engels to publish their work, *The Communist Manifesto* in 1848. It included these words: "The proletariat have nothing to lose but their chains. They have a world to win. WORKING MEN OF ALL CONTINENTS UNITE." While these stirring words did lead to an increase in the pace of labor organization, around the middle of the nineteenth century, it was not until the latter part of the century that the movement really had a lasting significant effect on government intervention in the welfare of the working classes throughout the entire industrialized world.

Human Associations—Establishment of Early New World Communities

Until relatively recently in the long history of our species, there were no humans in the Western Hemisphere.

Apparently, when our species was emigrating from Africa, conditions were not suitable for them to reach this New World. But when conditions changed, they came with a rush. Modern man has been around for 100,000 years, and it took us over 75,000 years to find our way to the Americas; however, once we found it, within 20,000 years, we have peopled it from one end to the other. First the Asians came. They were followed 10,000 to 20,000 years later by the Europeans. The Europeans had barely unpacked when they forced Africans to join them. As we shall see later, the immigrants from all three of the continents of the Old World would play important roles in the development of agriculture in Alabama.

Asian Communities

Asians who emigrated to the New World must have done so as part of wandering bands of hunters. Available evidence suggests that sometime within the last 50,000 years bands of Mongol peoples, living as hunters and foragers on the Chuckchi Peninsula of far eastern Siberia (Figure 4.36), crossed a land bridge into what today is western Alaska. This land bridge (named Beringia) would have become exposed when a significant amount of the waters of the oceans were frozen during periods of extensive glaciation. It has probably been uncovered at least four times in the last 60,000 years—as the sea level went down when great amounts of the earth's water was frozen. Unfortunately, there is little agreement regarding the exact time of these migrations. Estimates vary from as long ago as 30,000 years, and possibly even longer, to as recently as 12,000. It is likely that were several migrations over time. Human anatomical evidence indicates that there have been three. The Inuit (Eskimos) of Alaska certainly seem to have arrived at a much later date than the others.

Asians who crossed the Beringia land bridge into Alaska were most likely the ancestors of the Native Americans or Amerindians. Probably after leaving their Mongol relatives and traveling across Beringia, the emigrants evolved into a distinct subgroup (American or Amerind) with somewhat different characteristics from their Mongol ancestors (Stein and Rowe, 1996). Within a few thousand years after entering the Americas, they had spread throughout the length and breadth of the American continents (Figures 4.37 and 4.38). They were wandering hunters in the southeastern woodlands of America during the latter part of the of the Pleistocene Age, as early as 9,000 years ago when the mammoth, horse, camel, and other large mammals were present.

People of the Paleo-Indian tradition were the earliest Amerindian inhabitants of the Americas. Much of their livelihood came from specialized hunting of large mammals. It is also likely that they used some smaller animals, as well as a considerable quantity of plant materials, such as nuts and berries.The Paleo-Indian tradition ended around the end of the last ice age, some 10,000 years ago, as the Southeast became warmer and the large mammals began to disappear (Hudson, 1976).

The Archaic tradition replaced the Paleo-Indian tradition as it began to decline. The people of this new tradition became much more sedentary, making their living in smaller and smaller territories. They became increasingly efficient exploiting all of the food resources found in their environments. They fished, hunted deer, and trapped many kinds of small woodland animals, and they continued to gather a variety of seeds, fruits, and nuts. Late in the period related to the Archaic tradition, the first evidence of plant cultivation appeared when the cultivation of bottle gourds and squash evolved in Northern Mexico and diffused northward into the Southeast of what is today's United States.

The Woodland tradition began to replace the Archaic tradition around 3,000 years ago. It would last until about AD 700. In this period, the first evidence of permanent housing appears. The Indians continued to hunt and gather, but they became more dependent on the gathering and storage of seeds and nuts. They began to show a decided preference for living near the flood plains of large creeks and rivers. It was here that many seed-bearing plants thrived. They continued to cultivate bottle gourds and squash, and then around 2,000 years ago, the technology for the cultivation of tropical flint corn reached them from Indians in the Southwest.

The Mississippian tradition that replaced the Woodland tradition appeared in areas along the middle course of the Mississippi River around AD 800. From there, it spread in all directions, but especially toward the Southeast. This was the tradition associated with mound building. During this period, in about AD 400, the cultivation of tropical flint corn disappears. When corn cultivation reappears, some 800 years later, it is based on an improved variety—eastern flint corn, a variety better adapted to the moist conditions of the Southeast. At about the same time, bean cultivation appears. When de Soto explored the Southeast between 1539 and 1543, he observed all aspects of the Mississippian tradition; however in about 150 years, when the next Europeans traveled through the area, they found Creeks, Cherokees, Choctaws, and Chickasaw peoples and relatively few traces of the Mississippian tradition. De Soto had encountered large numbers of these people in his travels through the Southeast in the middle of the sixteenth century; however, he made little effort to describe the various groups or to delineate their territorial boundaries.

European Communities

Until near the end of the fifteenth century, Europeans had been confined to a relatively small geographical area for almost a millennium, and for a time during the westward expansion of the Muslims in the eighth century, it appeared that Christian Europe might become even smaller. The one attempt to expand beyond those frontiers, through the Crusades, had largely failed. By the middle of the fifteenth century, the monarchs of Spain, Portugal, France, and England, had generally unified their positions within their respective countries and had accumulated sufficient wealth to begin to look beyond their borders for additional resources, new opportunities for trade, and new souls to save. These interests sent explorers and missionaries from the shores of Europe into the unknown for God, Gold, and Glory. As a result, the Europeans came to spread Christianity and a considerable amount of other aspects of their culture to a large part of the world. From the end of the fifteenth century to the beginning of the seventeenth, Portugal, Spain, France, and England would strive mightily to put their national imprint on the Americas. In the end, it would be the descendants of the emigrants from the southern half of the British Isles who would be the most successful in North America.

Spanish

Portugal and then Spain took the lead in European expansion to the New Worlds in the fifteenth century. Initially, the driving force behind these efforts was the extermination of Moslems in Africa ("to quinch the fire of the sect of Mahamede") (Figure 4.39), and the deliverance of heathen from idolatry (Stavrianos, 1971). The Roman Catholics of the Iberian Peninsula had lived under Moslem rule for over seven centuries (715-1492). Now it was pay back time. Quickly, however, the Portuguese realized that there was little profit to be had in destroying Islamic strongholds, killing Moslems, and converting the heathen. They quickly found far greater returns in taking pepper, gold, ivory, sugar, and slaves from Africa home with them to sell in Europe.

Following the successes of the Portuguese under Prince Henry the Navigator, in their explorations along the African coast after 1419, Christopher Columbus finally persuaded the Spanish Queen Isabella, in 1492, to finance his effort to reach Asia by sailing west across the Atlantic. After his return home, the Spanish, concerned that the Portuguese might move into their new-found world, petitioned Pope Alexander VI for recognition of their sovereignty in the New World. With the exception of what is now Brazil, the Spanish petition was approved in 1493.

By 1550, the Spanish conquistadors had completed the conquest of the great Aztec and Inca empires, and had established control over Central and South America and the Atlantic and Caribbean islands that would provide the basis for the establishment of the first European empire in the New World. Soon, Spaniards were flooding into the vast new territory to establish plantations and mines and to convert the heathen. In the process, they destroyed indigenous cultures, seized countless treasure objects, and enslaved hundreds of thousands of Amerindians. Later, large quantities of gold and silver began finding their way to Spain with 20% of it going into the royal treasury.

Gray (1958) noted that as early as 1526, the Spanish established a settlement in the Chesapeake Bay region, near the site where the English would establish Jamestown, some seven decades later; however, it was abandoned after a year. The Spanish attempted to establish a colony in the area of Mobile Bay and upstream on the Alabama River, during the period of 1559 to 1561, but they had to abandon the effort because of the shortage of food. Later, in 1565, they established St. Augustine on the northeast coast of Florida (Figure 4.40), after destroying the nearby French settlement of Fort Caroline. In the following years, they established a series of trading posts and missions northward to St. Catharine's Island, just south of the site of the present-day Savannah and westward across the Panhandle of Florida to Pensacola. Spain maintained control of East Florida continuously for 2 centuries until it was brought under the control of the English, at the end of the French and Indian War (Seven Years War) in 1763. The Spanish, however, regained control of the Floridas, including all of the territory along the coast to and including New Orleans (West Florida) at the end of the American Revolution in 1783. In 1819, Spain finally ceded all lands east of the Mississippi River to the United States.

In the early years of settlement, European traders of various nationalities (Scottish, German, Dutch and Spaniards) established the trading posts, with Spanish approval, at strategic points along trade routes in the Southeast. Here they traded with the Indians and lived among them, usually taking Indian wives. These so-called Indian Countrymen were present in South Alabama long before the first English pioneers entered the area.

Spanish monarchs apparently were never interested in establishing colonies of their citizens in their vast territory in North America. They established large numbers of military installations, but few of these included any families. Instead, they encouraged families from other countries to settle in the backcountry as farmers and to pledge allegiance to the Spanish crown. For example, in 1789, they established Fort San Esteban de Tombeche (St. Stephens) on the western side of the Tombigbee River in what would later become the northeastern portion of Washington County, Alabama. The fort was established to protect settlers and their families on farms nearby from marauding Indians. This policy led to the establishment of the first farms in the Bigbee Bottoms on the land that would later become a portion of Alabama (Atkins, 1994).

French

In France, the growth of Calvinism in the early part of the sixteenth century led Francis I (1515-1547) and then Henry II (1547-1559) to persecute the Protestant converts in an attempt to stop its spread. Catholic converts to the Protestant faith (Huguenots) came from all parts of French society, but the lower nobility was especially receptive. Because of its rapid growth, especially among the nobles and their families, the Catholic monarchy felt that the Huguenots could be a dangerous threat to royal power. Outright warfare erupted in 1562 when a powerful Catholic nobleman massacred a peaceful congregation of Huguenots in Vassy. This massacre led to a series of wars (French Wars of Religion) that ended in 1598 with the issuing of the Edict of Nantes by Henry IV (1589-1610). This edict gave Huguenots equal political rights with Catholics, but by no means gave them complete freedom to worship and to practice their religion as they chose.

In 1564, the French landed four shiploads of Huguenots north of St. Augustine at the mouth of the St. Johns River (Figure 4.40) near present-day Jacksonville where they established Fort Caroline. The French enclave lasted for only about a year when the Spanish attacked, killing all of the fort's inhabitants. Morison (1965) comments that the only significant result of this ill-fated French effort was the introduction of tobacco to both England and France. A European visitor to the fort before it was destroyed obtained a parcel of tobacco from the local Indians. When he returned home, he took some of the herb with him. He described the plant as follows: "The Floridians have a kind of herb dried, who with a cane and an earthen cup on the end, with fire doe suck through the cane the smoke thereof, which smoke satisfieth their hunger."

In 1685, Louis XIV, determined to destroy Protestantism in the country, revoked the Edict of Nantes. As a result, more than 50,000 Huguenot families made their way to foreign countries. Included in the emigrants was a large proportion of France's artisans. These highly skilled people took with them not only their skills, but also their Puritan work ethic. Some of these exiles would help fuel the fires of the Industrial Revolution in England. Many of them would also be welcomed with open arms in the English colonies along the Atlantic coast of North America (Spielvogel, 1994).

Hostility increased between France and England after William and Mary succeeded to the English throne in 1688. This change, which was detrimental to French interests in Europe, encouraged France to try to blunt the expansion of British settlements in America. There were already established forts in Canada and along the Mississippi River; however, France had no presence in the Gulf of Mexico. As a result, in late 1698, Louis XIV (1643-1715) dispatched Pierre Le Moyne de Iberville along with a detachment of soldiers and 200 colonists, including some women and children, to the Gulf of Mexico to establish a fort at the mouth of the Mississippi River (Figure 4.41). Later they would establish military outposts as far up the Mississippi as the mouth of the Arkansas River, as far up the Red River as Natchitoches, and as far up the Alabama River as Fort Toulouse.

British

In the final analysis, it was peoples from the British Isles (English, Welsh, Scots, and Irish) who colonized North America and established its institutions and the outlines of its culture. The English and the Scottish bore the brunt of settlement and western expansion in this New World.

English

The English joined Spain and Portugal in searching for the so-called Northwest Passage, a northwest route to the riches of the Orient when Henry VII (1485-1509), the first Tudor king, dispatched John Cabot westward in 1497. Cabot reached what is thought to have been Cape Breton Island at the northeastern end of Nova Scotia. From there, he sailed along the southern coat of Newfoundland before returning to England. In 1498, he and his brother Sebastian returned to the supposed land of the Great Khan where they actually explored the region from Greenland in the north to (possibly) Delaware in the south. However, they found no spices on their journey, so both they and the king lost interest in the New World.

In the last half of the sixteenth century, commercial interests began to lobby Queen Elizabeth for a charter to establish an English presence in the New World, with the goal of providing new markets for English goods and new sources of raw materials for their growing industrialization. They insisted that English fortunes depended on finding buyers for their wool cloth, hats, caps, shoes, and cutlery—an ever increasing quantity of manufactured goods. The crown had already become deeply involved in international trade before the reign of Elizabeth when English merchants began to seek royal charters to establish companies specifically for developing overseas markets. The charter for the Russia Company was granted in 1554, and Elizabeth had issued letters patent for the establishment of the Levant (Turkey) Company in 1581. However, the queen was reluctant to heed their wishes to grant a charter for North America, because she did not want to antagonize the powerful Spanish Empire that had established St. Augustine in 1565 and had just destroyed the nearby French settlement at Fort Caroline and killed all of its inhabitants. Over time, Elizabeth reluctantly agreed to the wishes of the merchants, but did not provide the support of her government in doing so. It would have to be done with private capital.

The first serious attempt to establish an English colony in North America came in 1583 when businessmen in London, financed an exploratory voyage led by Sir Humphrey Gilbert, to the present-day site of St. John's Harbour, Newfoundland. They found a number of fishermen of various nationalities already living in a small settlement there. Gilbert took possession of the site and soon set sail for England to obtain settlers for a permanent colony. Unfortunately, on the voyage, his ship was lost, and no further effort was made to colonize Newfoundland.

Beginning in 1585, Sir Walter Raleigh and some of his friends began an attempt to establish a colony on Roanoke Island in Pamlico Sound in territory that would later become North Carolina. After a miserable year, some of the settlers decided to return to England. A second effort was made in 1587, but in 1591 no trace of the original settlers could be found.

With the accession of James I (1603-1625), a new age of English colonization began. The concern for Spain's objections had been virtually eliminated after the defeat of the Spanish Armada in 1588. In this age, colonies would be established as a result of men willing to invest their own funds in an overseas wilderness in search of profits. In 1606, a group of London merchants organized the Virginia Company of London. The Virginia Company of Plymouth, created at the same time, founded a colony at Jamestown, Virginia, in 1607. In 1612, English settlers were sent to Bermuda, which provided a base for colonization of the entire Caribbean. In 1627, settlers financed by a London promoter, Sir William Courteen, were sent to Barbados. English colonization of the Atlantic coast of North America continued sporadically for another century. English settlers went to Massachusetts in 1620, the Carolinas in 1663, Delaware Valley in 1675, and Georgia in 1732.

The rise in prices in England (the Price Revolution) at the end of the sixteenth and the beginning of the seventeenth century acted as a powerful catalyst for all traditional economic, social, and political relationships. Prices doubled, trebled, and even quadrupled within a century. It has been suggested that the Price Revolution was a result of the enormous influx of precious metals from the Americas to Spain, and eventually to most of Europe. As a result of the rapid decline in the value of currency, renters, wage earners, and even governments suffered a serious erosion of income. This revolution put a serious strain on the expectations of nearly every rank of society. The consequent distress and uncertainty probably accounted for the peculiar violence of economic, political, and religious controversy of the period. Early in the seventeenth century, the period of steep price inflation gave way to a depression that would continue for almost a century.

Within a period of 11 years from 1629 to 1640, some 80,000 English men, women, and children swarmed outward from their island home; this was called the Great Migration. Most of them were not attracted to foreign lands. Rather, they were fleeing the intolerable conditions that were developing at home. During this period, Charles I attempted to run the country without Parliament, and Archbishop Laud, with the king's blessing, purged the Anglican church of its Puritan members. These 11 years were also a time of economic depression, epidemic disease, and so many hardships and sufferings that John Winthrop commented that "it seemed that the land itself had grown weary of her inhabitants." During this time of troubles, some emigrants went to Ireland and others to the Netherlands and the Rhineland. Another group (20,000) emigrated to Barbados and a fourth group, mostly Puritans, chose to settle in Massachusetts.

The intolerable conditions persisted well into the latter part of the seventeenth century, and waves of migrants continued to flee their homeland. Fischer (1989) noted that between 1645 and 1670 that 40,000 to 50,000 additional Englishmen arrived in the Chesapeake colony alone. Migrations to other colonies and to other lands were correspondingly large. A significant element of the emigrants were the so-called Cavaliers. They were the higher nobles who had supported the royalists cause throughout the unsettled period of the English Civil Wars. Many had fought for Charles I, then for Charles II. They suffered severely in this struggle, and they were badly treated and even persecuted for their efforts. As a result, many of them "did fly from their native country, as from a place infected with the plague." Most of these political refugees went to Europe, but many were recruited by Sir William Berkley, himself a royalist, who was governor of the Virginia colony from 1642-1676. Later, some of these royalist emigrants to Virginia would play a role in the governance of the colony, and the development of other colonies on the Atlantic coast. Some of those who fled to Europe would later play an important role in the Agricultural Revolution that would have a lasting effect on the British Isles.

Jamestown. Jamestown became the first permanent English settlement in America (Figure 4.40). It was established in 1607 by the Virginia Company of Plymouth, England. The specific purpose was to establish a permanent English presence in the region between 34° and 41° North Latitude (Figure 4.40). It was a business enterprise operated as a stock company for profit. There was every expectation that the owners of the company would receive huge returns on their investment. Few of these first emigrants looked on Virginia as a place of permanent abode. Most expected quick riches from gold and precious stones before they returned home. The majority were adventurers. Hardly more than a fourth of them were artisans and laborers.

Unfortunately, many of Jamestown's first founders either died in their new home or speedily returned to England, but in spite of all of the problems, the colony survived; however, even with almost superhuman effort, they could not achieve prosperity. The 1622 Indian massacre of a third (347) of all of the white settlers in the colony contributed significantly to their problems. Finally, after several years of quarrels within the company, disagreements over its governance and mismanagement, plus the problems with the Indians, James I annulled the Charter of the Virginia Company in 1624. Henceforth, Virginia became a royal province with a governor and council appointed by the crown.

The Virginia Colony made steady progress under royal governors. In 1629, the population stood at 2,500, and by 1640, it increased a little over threefold to 8,000. In 1641, Charles I appointed Sir William Berkley, a young Royalist or Cavalier, as royal governor. He would remain in that position for over 30 years. As soon as he arrived, he began to recruit Royalists from England for the colony, and when they arrived, he promoted them to high office and granted them large estates. In this manner, he created a ruling oligarchy that would govern the colony for many generations (Fischer, 1989). Cavaliers would continue to move to Virginia throughout Berkley's tenure, but the flow increased substantially after Royalists were defeated in the English Civil War during the 1640s.

One the most influential things that Governor Berkley and his Royalists brought to Virginia from the mother country was the belief in hegemonic liberty. Fischer (1989), describes this Royalist concept of liberty in detail. Liberty was thought to be a special birthright of free-born Englishmen. It set them apart from other mortals, and gave them the right to rule the less fortunate in the world. But it also had a hierarchal element. Men of high estate were thought to have more liberties than others of lesser rank. Servants possessed few liberties and slaves none at all. This idea of hegemonic liberty was entirely consistent with the idea of involuntary servitude and Indian removal. It saw no inconsistency, in the latter part of the eighteenth century, in railing at the English Parliament about no taxation without representation while maintaining a firm hold on their slaves. This concept of liberty eventually became well established in all of the English colonies in the South, but never in the Puritan strongholds in the North. It was finally transported into the Old Southwest (Alabama and Mississippi) as the descendants of the first Virginia planters moved there.

By 1660, the population of Virginia had increased to 30,000. Royalists comprised only a small portion of the total flow of people to the colony. Most were people of lower social rank. More than 75% of them came as indentured servants. Their travel costs had been paid for them, and they were required to work as servants, usually for seven years until they repaid their benefactors. Most of these servant-migrants were half-grown boys and young men between the ages of 15 and 24. The immigrants came from every county in England, but the vast majority of both Royalists and indentured servants came from 16 counties in the south and west of the country. Very few of them came from East Anglia or Cornwall.

West Indies. The West Indies colonies (Figure 4.42), established by the English in the early part of the seventeenth century, became important centers for trade and for the development of plantation economies. Later these off-shore settlements would have a tremendous influence on the development of other English colonies on the South Atlantic coast of British North America. In 1612, a colony established in the Bermudas, served as base of operations for further expansion southward into the Caribbean. The first colony established in that region was on the island of St. Christopher in 1624. Then in 1627, settlers funded by a London promoter, Sir William Courteen, occupied Barbados. The English population of the island was augmented in the 1640s when large numbers of Irish and Scotch-Irish, either fled there or were deported there during the war following the Irish Rebellion of 1641.

Plymouth Colony. The Plymouth Colony was founded by the Mayflower Pilgrims who brought Puritanism in its purest form to America (Figure 4.40). In December 1620, a hundred men, women and children arrived in Plymouth Bay, and began to build Plymouth Plantation. In those first years, the colony grew slowly, but in the 1630s it exploded as thousands of Puritans emigrated to escape religious practices being forced on them by Archbishop Laud on behalf of Charles I. By 1640, some 21,000 Puritans had emigrated to the Plymouth colony, but after that time the movement of these Englishmen to New England virtually ceased, and even began to be reversed, as some of them returned to England to become involved in the events leading up to the English Civil Wars (1642-1646 and 1648-1649).

These Englishmen, who left the security of their homes to come to the New World, came for one purpose—to build a New Zion in America. These Puritans intended to build a Bible commonwealth that might serve as a model for mankind. Religion was not merely the leading purpose for their immigration, it was their only reason. They were rich in spirit but poor in worldly goods. The only way they could finance their undertaking was to borrow money from a group of loan sharks in London who required seven years of virtual servitude in return for the loan. They survived by fishing and hunting, and by setting up a trading post where they traded corn they had grown, for the beaver pelts of the Indians. Through this fur trade, they eventually got out of debt.

Lest the inhabitants settle in a haphazard fashion, carefully laid-out villages were planned. They were patterned after the manorial farms of the Middle Ages. Each one consisted of lots in the village with space for a house and a kitchen garden. The main farms were to be located nearby. This same village plan would be followed in the early development of the Georgia colony in the next century.

Most of those emigrants who settled in Massachusetts came from the far eastern portion of England or East Anglia. These people with their beliefs and work ethic provided the base on which the whole of New England developed, and over time, as the country spread westward, some of those Puritans and the influence of their religious and economic philosophy also moved westward. They had relatively little direct effect on the colonies developing south of New England along the Atlantic coast; although later, New England ships would be very active bringing slaves to the Southern colonies.

Delaware Valley. The Delaware Valley began to receive English settlers in 1675 when the first group of Quakers arrived (Figure 4.40). After William Penn received his charter in 1681, the migration continued sporadically until about 1715. Religious persecution also played a part in the migration of these Englishmen and their families, but, like the Puritans before them, the primary reason was their desire to seek a place where they could practice their religion, a place where they could show Quakerism at work, freed from hampering conditions. With this desire firmly in place, the English Friends founded West Jersey. Penn's colony in Pennsylvania resulted from this effort. During the period 1675 to 1715, some 23,000 colonists moved to Delaware Bay. Most of them came from the North Midlands of England and from portions of eastern Wales.

From the beginning, these settlements also attracted many emigrants from other countries, all attracted to the tolerant, liberal political, religious and social environment that the Friends established there. Penn made a special effort to recruit German and Dutch Quakers. The lower Rhineland was well represented in the area near the present-day Germantown.

The Carolinas. The Carolinas became the site of the first permanent English colony south of Virginia when it was established in 1670 (Figure 4.40). After the middle of the seventeenth century, the vast wealth being accumulated by English planters on the sugar plantations of Barbados began to attract the attention of businessmen in England. Because that island was already badly overcrowded, they began to look for other lands where colonies might be established that would make them equally wealthy. An obvious choice was the vast landholdings claimed by the crown between Virginia and Spanish Florida. One attempt to establish a colony in the Carolinas was made in 1630, but the ship carrying those settlers landed its passengers in Virginia by mistake. In 1663, Charles II granted a charter to eight wealthy merchants and traders, to establish a colony on land between the southern boundary of Virginia and the present-day location of Datona Beach, Florida. The list of proprietors included Sir William Berkley who was still royal governor of Virginia. For administrative purposes, the proprietors separated the Carolina colony into northern and southern portions (North and South Carolina).

1. South Carolina. In 1699, three ships provisioned in England set sail for Barbados to pick up settlers, primarily small farmers who were being squeezed out as plantations replaced farms on the island. Unfortunately, a severe storm wrecked one of the ships and damaged the second one so severely that it was no longer seaworthy. The remaining vessel picked up two hundred emigrants from that severely overcrowded island before continuing to the mainland. In the spring of 1670, after visiting several locations along the coast, the settlers disembarked at a site upstream on the western bank of the Ashley River. A few years later, many of them would move downstream to the present site of Charleston (Figure 4.40).

Once the colony was established, the proprietors began a massive advertising campaign to attract more settlers. A steady stream of Barbadians responded. In 1680, a small group of French Huguenots arrived, and a band of rugged Scottish emigrants arrived in 1683. Also, in the 1680s, the colony received a large number of Protestant dissenters who were fleeing religious persecution in Britain brought on by the Act of Conformity of 1663 and the Test Act of 1673. Five hundred of them arrived in a single month. From the very beginning, African slaves were being regularly imported. By 1718, the colony included 6,000 Europeans and 10,000 African slaves. A substantial group of Scotch-Irish arrived in 1732 to settle along Black River north of Winyah Bay, 50 miles northeast of Charleston. These were just the forerunners of thousands of Scottish peoples who would arrive in the next 40 years.

In 1719, the South Carolinians, who were already extremely unhappy with the management of the southern colony, used a rumor of a threat of an imminent Spanish invasion as a reason to declare themselves independent of the proprietors, and they invited the English government to take control. Consequently, in 1721, the first royal governor reached Charleston.

A succession of royal governors provided inducements in the form of land grants to encourage emigrants to settle in the backcountry. Germans were among the first to respond when, in 1735, a large number arrived, and agreed to move up the valley of the Edisto River to establish the settlement of Orangeburg (Wallace, 1951). Later, German settlements would continue to expand northward, finally reaching well up into the Congaree and Saluda river valleys. These settlements resulted in the development of a relatively large German population in the western half of the colony. By 1750, the population of the future state of South Carolina had increased to 25,000 Europeans and 39,000 Africans.

From the earliest years of South Carolina, colonists were interested in trading with the Indians. It began in a small way as plantation trade in which Indians would bring deerskins to the nearest plantation to trade for English goods. Soon, however, British (usually Scottish) traders began to move inland. By 1690, they were moving up the Savannah River to trade with the Cherokees, and in 1698 a Charleston trader traveled overland from the coastal town to the mouth of the Arkansas River, just west of the present-day Cleveland, Mississippi. In the succeeding years, Charleston traders would transport, on packhorses, literally tons of ammunition, knives, steel tomahawks, brass kettles, hoes, axes, cloth, glass beads, and rum, as far west as the lands of the Creeks and Chickasaws in what is now Georgia, Alabama, and Mississippi. In the early days, Indian slaves were the most important items of trade. Labor was scarce in the colony, and native slaves brought a high price. Later, when African slaves became plentiful, deerskins became the most sought after item of exchange. During the period 1699-1715, South Carolina exported annually on the average of 54,000 deer skins, and in 1707 exports reached the highest level with 121,000 skins.

2. North Carolina. The North Carolina colony, first established on Roanoke Island in 1585 and 1587 through the efforts of Sir Walter Raleigh, failed. A formal effort would not be made to establish another until well over two decades later. The early days of the North Carolina colony were quite different from those of South Carolina. North Carolina continued under control of the proprietors for several years, after the crown assumed control of South Carolina. However, in 1729, the charter that had provided for the development of the Carolinas was annulled, and North Carolina also came under control of the English crown.

Although the Carolina charter included the entire

region from Virginia to Spanish Florida, no formal efforts were made to settle the northern portion of the region until well into the eighteenth century. However, disgruntled farmers and poor whites from southern Virginia had begun to move southward into North Carolina before the end of the seventeenth century. They established small farms along the shore of Albemarle Sound, east of the present-day Rocky Mount, west of the port of Edenton (Figure 4.40). As early as 1691, a few French Huguenots moved from Virginia to North Carolina, and in the succeeding years more would follow to establish several settlements of these French Protestants. The would-be colony received its first formal settlement in 1710, when a Swiss nobleman arrived with a thousand settlers from Switzerland and Germany. Unfortunately, soon after arriving, they were attacked by Tuscarora Indians, and their settlement, New Bern, was almost wiped out.

Georgia. The Georgia colony (Figure 4.43) was established in 1733. South Carolina had grown prosperous by that time, but they continued to have grave concerns about the possibility of invasion by the Spanish at St. Augustine. Consequently, they were interested in the establishment of a colony to provide a buffer on their southern flank. Fortunately for them, in 1729 General James Oglethorpe was a member of Parliament and chairman of a committee to look into the status of English jails. As a result of his investigation, he developed a keen interest in finding some solution to the problem of the large number of people accumulating in debtors' prisons in England. Discussions between Oglethorpe and his friends resulted in a plan to send some of the distressed debtors to a new colony in America. In 1732, a charter was granted to establish a colony between the Savannah and Altamaha rivers. The charter established a board of trustees to conduct the affairs of the colony for 21 years, at which time its management would revert to the crown. Parliament even appropriated money for the enterprise.

The trustees determined that the colony would remain free of slaves. They had no humanitarian objections to slavery, but they were afraid that the availability of slave labor might interfere with the economic rehabilitation of the debtors. They also decided that no spirits would be imported; this rule, however, was disregarded almost from the beginning. Its importation was legalized in 1742, and the prohibition against slavery was lifted in 1749. It was also the intent of the trustees that there would be no Catholics or Jews in the colony, although the trustees were determined that no one in the colony would be persecuted because of their religion. Soon after it was established, there were a few people belonging to both groups in the colony.

Large numbers of people were interested in moving to the new colony. As a result, the trustees found it necessary to establish a committee to screen prospective settlers. When the selection process had been completed, they had actually accepted very few debtors. In fact, the first group moving to Georgia were small businessmen, tradesmen, and unemployed laborers from London. Initially, the colony's make-up was very much like that of London without the elite or the large debtor element (Spalding, 1991).

The first settlers accompanied by General Oglethorpe arrived in Charleston early in 1733, before going on to Georgia. They finally disembarked at Yamacraw Bluff, 18 miles up the Savannah River, to establish a town that would bear the name of the river. Lest the inhabitants settle in a haphazard fashion, the trustees planned carefully laid-out villages, patterned after the manorial farms of the Middle Ages. Villages consisted of lots in town with space for a house and a kitchen garden. Main farms were to be located nearby. This village plan had also been followed by the Massachusetts colony in its early years.

Within a few years, the colony became a haven of distressed people. Among the early arrivals were a group of German Lutherans from Salzburg, Austria, who were fleeing religious persecution there. These were added to the 1,500 original settlers. The colony did not develop as it was intended, and the settlers became unhappy with the trustees. Consequently, they relinquished their charter in 1752, a year earlier than planned. However, the government did not send a royal governor until 1755. The farming economy based on the production of rice was not very productive. As a result, most of the original settlers departed for other colonies or returned to England. When Georgia became a royal colony there were only two thousand Europeans and a thousand

African slaves living there. Although the colony grew slowly, it did survive to become important to the British Empire in the conflicts of the mid-eighteenth century. Most of Georgia's development came after the middle of the eighteenth century, and the colony slowly stabilized. Peace had come to its southern frontier after Oglethorpe attacked St. Augustine in 1743, and the city of Savannah was moved down near the mouth of the river, where it became a flourishing seaport and trading center. After 1760, the colony exploded as thousands of Scotch-Irish, having crossed the Carolina Piedmont from the north (Figure 4.45), then crossed the upper valley of the Savannah River, and began to settle in the backcountry. The flow was especially heavy in the early years of the 1770s.

At the beginning of the American Revolution, there were a few more Europeans than Africans in Georgia. Most of the slaves were found in the low country between the Savannah and Altamaha rivers where most of them worked on plantations near the coast. The rich customs and speech habits of the slaves mingled with those of the English and helped produce a coastal society far different from that found on the Georgia frontier and especially in the backcountry.

Scots

Richards (1991) suggests that an emigration ideology evolved in the culture of the Scots long before the Act of Union in 1707. Even before the first English settlement in the New World, Scots in significant numbers left their homeland for Europe. They were not wanted in the English colonies in America. In fact, before 1707, they were officially barred from immigrating to any of them. As a result, the few Scots who reached America, entered by un-official and untoward avenues. After 1707, however, Scottish subjects, then British subjects, were free to enter any of the settlements, and over the next century hundreds of thousands of them would make the journey. Those arriving first came directly from Scotland; however, by far the larger number would come later after first going to Ulster in Northern Ireland, a journey that would not be completed until over a century later.

Lowland Scots fled to Ulster (Northern Ireland) from their homeland after 1610 because of changes in the land tenure system taking place in Scotland. For centuries, Scottish farmers had been part of the ancient manorial system that existed throughout Europe; however, in the latter part of the sixteenth century, changes in the old system began to appear, and they caused great hardship among those who gained their livelihood from tilling the soil. The most far-reaching change resulted in a change in land tenure. Under the old system, farmers living on a manor, generally had access to land as long as they met their obligations to the lord. Under the new system, farmers were forced to pay a fixed cash rent each year, with no obligations for service. They could continue to farm a specific piece of land as long as they paid their rents. Unfortunately, many of the poorer farmers could not pay rent, and were forced to become agricultural laborers or in the worst case—beggars. Consequently, in 1610, when the offer was made for farms on the plantations in Ireland, thousands of the suffering Lowland Scots responded. They responded in such large numbers to the opportunity to go to Ulster that whole neighborhoods of them were established in the plantations there. Most clearly felt that their prospects in Ireland were much better than in Scotland. The early opportunities to immigrate were limited to Lowlanders only. James I had required of his Scottish friends that in arranging for settlers for Ulster that no Highlanders be included.

Economic problems and food shortages in the Scottish Lowlands coupled with good prospects in Ulster were the primary causes for the migration of the Lowland Scots to Ireland after 1610; however, after 1615 religious difficulties became increasingly important. James I precipitated the difficulties when he undertook to correct Knox's hallowed Book of Common Order to bring it closer to the English Book of Common Prayer. The Scots refused to accept the changes, and the matter remained in stalemate until James's death in 1625. Later, in 1637, James's son, Charles I (1625-1649), required that Knox's order of service be supplanted by a new form (Archbishop Laud's liturgy), which included many Anglican High Church elements. The Scots responded to this threat by joining the Puritans in England who were also experiencing religious difficulties. Together, they drew up the Solemn League and Covenant in 1643 that stated their intent to oppose all religious practices that contained even a whiff of prelacy or popery. Unfortunately, this effort to gain allies in their religious battle with the king did not solve their problems. Later, even their Puritan friends turned on them when, under Cromwell, their persecution was intensified. With the Restoration of the Stuart monarchy, Charles II (1660-1685) was crowned king in 1660. Shortly thereafter, Parliament, which was sympathetic to the king's religious interests, began to pass a series of repressive acts designed to bring all religious practices in the British Isles in line with those of the Church of England. In 1666, the Scots (Covenanters) rebelled, and were defeated in the so-called killing times at the battle of Pentland Hills. With this defeat, thousands of these religious non-conformists left Scotland for Ulster. Later, in 1679, the covenanters would try one last time to establish their religious independence. They were soundly defeated at the Battle of Bothwell Bridge, and a new wave of emigration began. Finally, in 1689, the Bloodless Revolution, which brought William (of Orange) and Mary to the throne gave the Scots freedom in religious matters.

In the last decades of the seventeenth century, and the beginning decades of the eighteenth, thousands of additional Scots fled the destitution visited on their entire country by the so-called Seven Hungry Years. The agricultural economy of the country had never been very good because of the generally wet, cold climate and poor soils; however, beginning around 1670, the climate of Scotland became even worse than usual for agriculture. In 1674, snow and frosts were so severe that large numbers of sheep were lost. Then, in seven of the years during the eight-year period (1693-1700), there was essentially no harvest of oats, the primary food source of the Scots in all of the upland parishes of Scotland. In parishes all over the country and especially in those in the Highlands, from one-third to two-thirds of the population died. Some 20% of those who survived were utterly destitute. By this time, there was no longer any restriction on the immigration of Highlanders; consequently, because they suffered most from these famines, thousands left their homeland for Ireland.

The Emigration of Ulster Scots (Scotch-Irish) to America, in large numbers, did not begin until almost a century after the English had arrived in America. However in the first quarter of the eighteenth century, rack-renting, years of drought, and the religious Test Act began to have dire consequences in Ulster. In 1717, after the fourth successive year of drought had ruined their crops, the Scotch-Irish who had already left one homeland in Scotland, prepared to do so again. Whole villages lost their Protestant element when entire families departed for America. Altogether, during 1717 and 1718, some 5,000 Ulstermen made the journey to the American colonies. Reports coming back to Ulster, especially those from tolerant Pennsylvania, were uniformly and highly favorable. These reports plus the continuing problems of rack-rents, high taxes, and poor harvest worked together to send another wave of Scotch-Irish to America during the period 1725-1729. This second wave was so large that it attracted the attention of the English Parliament. The emigration portended the loss of the entire Protestant element in Ulster.

Severe famine struck Ulster again in 1740. An estimated 400,000 people died from causes associated with poor nutrition, and the third wave of Scotch-Irish sailed for America. A fourth wave left Ulster during the period 1754-1755. Continuing problems with calamitous drought and an effective propaganda campaign by royal governors of North Carolina regarding the advantages of removing themselves to that colony worked together to fuel this fourth wave. During the period 1771-1775, the final wave of Scotch-Irish moved to America. By the 1770s, the situation in Ulster had grown increasingly worse. The people were living chiefly on potatoes, oat-bread, and milk. To pay the exorbitant rents, they were trying to subsist on smaller and smaller farms. They really had no other choice but to seek better opportunities elsewhere. The emigration in 1775 was essentially the last from Ulster. Within a year, the English colonies of North America would begin their revolution for independence. This action effectively curtailed further emigration. Between 1717 and the American Revolution, at least a quarter of a million Scotch-Irish immigrated to America. This was a momentous loss of human capital for Northern Ireland. It deprived the old Plantation of a large proportion of its hardest-working and most productive citizens.

Scottish emigration directly to America began soon

after the Act of Union was accepted (1707), but by far the highest proportion of the total number of Scottish people reaching the New World came by way of Ulster. Considerably less than half came directly from Scotland. This so-called direct emigration involved a far larger number of Highlanders than Lowlanders. Also, direct emigration from the Highlands began earlier than from the Lowlands. Emigration from the two areas was largely driven by different motives. Lowlanders also suffered somewhat from the same famines that ravaged the Highlands, and these events played a role in their decisions to emigrate, but they were also seeking better economic opportunities. Numbers of mobile and individualistic Lowland laborers emigrated in response to the growing opportunities for employment and advancement in America compared to those in Scotland. In some cases, cooperative associations of Lowland farmers funded well-articulated emigrations to cheap land in the colonies. In contrast, large numbers of poor Highlanders were literally expelled from their homelands by poverty and famine, exacerbated by rapid population growth. Their emigration was a desperate response to the social and economic conditions of that region. They had little choice but to leave, and leave they did, in large numbers. Most went to Ulster, but those who could afford to do so, sailed to America.

These Scottish emigrants were a different people from the English. McWhiney (1988) described them as clannish, herding, leisure-loving Celts who relished gambling and combat and who despised hard work, anything English, most government, fences, and any other restraint on their free-ranging livestock. They were also different with respect to their ideas of liberty and freedom. According to Fischer (1989), they brought a concept of natural liberty with them from the Celtic borderlands of Britain. A traveler to the backcountry commented that "they shun everything that appears to demand of them law and order and anything that preaches restraint." Another traveler commented that they hate the name of justice, and yet they are not transgressors. Their object is merely wild. Altogether, natural freedom is what pleases them. Also included in their concept of natural liberty, was the idea of elbow room, another trait from the North of Britain. One writer from the region commented that every man at nature's table has the right to elbow room. Apparently, however, natural liberty was not a reciprocal concept. It did not recognize the right of dissent or disagreement. When there was disagreement, it was to be suppressed by force. The Scots seemed to regard the terms opponent and enemy as being synonymous. At the time that these Celts began arriving in the New World in significant numbers, the English were making good progress in putting an end to the idea of natural liberty in all of Britain. However, there were no restraints on natural liberty in the backcountry of the English colonies in the Upper South; consequently, it flourished there and moved with these pioneer-settlers as they moved southward through the Carolinas and Georgia toward the Old Southwest.

Scots in the Delaware Valley. A high percentage of Scotch-Irish emigrants from Ulster to British North America landed in the Delaware Valley at Philadelphia, New Castle, and Chester in southeastern Pennsylvania (Figure 4.40). Therefore, Pennsylvania played an important role in the development of the backcountry of all the colonies from Pennsylvania south to Georgia. Between 1717 and the American Revolution, a large number of Ulstermen (Scotch-Irish) immigrated to America. Others came from other areas bordering the Irish Sea—the Scottish Lowlands and the North of England. These people were members of a traditionally Gaelic (Norse-Gaelic) society who had been moving back and forth between Ulster, the Scottish Highlands and Lowlands, and the Isles for nearly a thousand years (McWhinney, 1988). Peak periods of the wave-like emigration came in 1718, 1729, 1741, 1755, 1767, and 1774. Two-thirds came in the decade from 1765 to 1775. Most immigrants were under-tenants, farmers, and farm laborers. A large minority were semiskilled craftsmen (Fischer, 1989).

Scots in the Susquehanna and Shenandoah Valleys. When the Scotch-Irish arrived in Pennsylvania, they found much of Pennsylvania already full of their age-old enemies—Englishmen. So they soon moved westward across the Great Valley of the Susquehanna, along the Great Philadelphia Wagon Road (Figure 4.44) toward the Blue Ridge Mountains. They crossed the Potomac River at Williamsport and turned southward into the Shenandoah Valley of Virginia. As each new wave of immigrants arrived, until about 1775, they would leapfrog over established settlements of Celts, and move further westward and southward until by the beginning of the nineteenth century they had literally filled-up the backcountry of the Piedmont (Figure 4.45) along the eastern slope of the Appalachian Mountains from western Pennsylvania to North Georgia. Soon, they would move through Alabama and on to Mississippi. These succeeding waves of Scotch-Irish immigrants generally moved to find suitable farm sites; however, this was not the only reason they moved. Leyburn (1962) commented that the Scotch-Irish never seemed satisfied with the location of their homes or farms. Many seemed to have the compulsion to move again and again. McWhinney (1988) suggested that many of them moved frequently because they were restless and because they were programmed to do so by their culture. A Scotch-Irish autobiographer admitted, "I had been reared to a belief and faith in the pleasure of a frequent change of country."

The great Shenandoah Valley of backcountry Virginia received much of its population from Scotch-Irish emigrants from Ulster and Germans from the Rhineland. (Leyburn, 1962). A high proportion of the Scotch-Irish settlers were either Highlanders themselves or descendants of Highlanders. Richards (1991) quotes W.R. Brock when he comments that the Highlanders in colonial America were shock troops who threw themselves at the frontier. Rendered desperate in their homeland, they were the ideal colonists to endure the hardships required to settle a wilderness. In the longer term, their endurance provided a pathway for waves of others of their kin to join them in the Highlander districts in America.

Scots in South Carolina. The filling of the Upcountry, or the mountainous portion of the South Carolina colony around present-day Greenville and Spartanburg, was largely accomplished by large numbers of Scotch-Irish families moving southward through the Shenandoah Valley, across the foothills of southeast Virginia, onto and across the broad Piedmont of western North Carolina (Figure 4.45), and into northwestern South Carolina. Leyburn (1962) comments that this movement of people from 1740 to 1750 was part of the largest migration in colonial times. So great was the tide of Ulstermen pouring in from the North that by the beginning of the American Revolution, there were 83,000 of them living in the Upcountry or Piedmont. That group of people constituted 80% of the population of the entire colony at that time.

A number of Lowland Scots traveled directly from Scotland to South Carolina. They were a mean and hungry mercantile corps. Ultimately, they became links between the wilds of America and the domestic source of industry in Scotland, England, and the Continent. They were especially important links between the Highlanders in the backcountry and coastal merchants.

Scots in North Carolina. A few Highland Scots came to North Carolina directly from Scotland after the abortive Jacobite rebellion of 1715; after the second rebellion of 1745, a great flood of the clansmen poured into the country at Wilmington (Figure 4.40), from which entire counties were populated by these Celtic peoples. Many of the names of towns in the area around Fayetteville reflect the country of origin of the people who founded them.

Around the middle of the eighteenth century, a succession of North Carolina governors made an effort to attract Celtic emigrants from Ulster and Scotland to the colony. The Scotch-Irish responded with a mighty rush. After 1740, thousands of them flooded into the Piedmont (Figure 4.45). To get there, they traveled over 400 miles from southeastern Pennsylvania and then southward through the Shenandoah Valley. From Roanoke, they crossed the hilly country of southeast Virginia into the Piedmont of North Carolina. By the outbreak of the American Revolution, there were 65,000 settlers there. As with South Carolina, they constituted 80% of the entire population of the colony. Leyburn (1962) comments that although there were a few Swiss and Welsh and a considerable number of Germans in the backcountry that there were more Scotch-Irish than all the others combined, and that most of them had gotten there by beginning their travels over the Great Philadelphia Wagon Road.

Scots in Georgia. A number of Lowland Scots traveled directly from Scotland to Georgia early in the colonial period. As in South Carolina, they formed a mercantile corps linking the American frontier, especially Highlanders in the backcountry, with coastal merchants and with Scotland, England, and the Continent. They also became a highly vocal group opposed to early policies of the Georgia trustees.

A sizable number of Highland Scots entered the Georgia backcountry from the coast. They landed there after abandoning a failed Scottish effort to establish a colony at Darien, near the present-day, western entrance of the dug portion of the Panama Canal. In 1733, they established a colony that they named Darien, near the mouth of the Altamaha River (Figure 4.43).

Germans

The Germans played an important role, although a subsidiary one, in the peopling of the New World. As a result of continuing internal conflicts and divisions, there was no central German government that could mount a colonization effort in the New World—in competition with the Spanish, French, and English. Consequently, relatively few Germans entered the stream crossing the Atlantic in the early seventeenth century. While few made the voyage during that period, however, conditions that would change that situation were beginning to emerge.

In Germany, in the early part of the seventeenth century, Pietist religious sects began to grow in the Rhineland. Pietism placed the emphasis in religion on the inner or spiritual life as opposed to the formal ecclesiasticism and ritual of orthodox Lutheranism. They were much more sympathetic with Quaker beliefs. Over time, the Lutheran clergy in Germany became concerned with the religious practices of the Pietists, and began to push for conformity in worship. When changes were not forthcoming, they requested that the civil authorities require that all Christians follow the same practices. When the Pietists refused, they began to be persecuted.

In the latter half of the seventeenth century, reports of the success of the settlements in the Delaware Valley of Pennsylvania began to reach the persecuted and economically depressed Pietists of the Rhineland. These reports provided them with visions of a land where they could worship as they pleased, among the Quakers, where the fruits of their labor would not be snatched away by greedy landlords or marauding armies and where their sons would not be conscripted as soldiers. The first group of Pietists sailed for Philadelphia in 1683. Their emigration signaled the beginning of a significant movement of Germans to the Atlantic coast of British North America that would continue until the middle of the eighteenth century.

The great Susquehanna and Shenandoah valleys of backcountry Pennsylvania and Virginia, (Figure 4.40) received much of its population from Scotch-Irish emigrants from Ulster and Germans from the Rhineland. Over time, this valley was populated with Germans and Scotch-Irish (Leyburn, 1962). Many German emigrants also first arrived in America at Philadelphia; over time, they would become the most numerous nationality in the 13 colonies, other than the British. The first significant movement of Germanic people to Philadelphia began as early as 1683. Initially, religious persecution played a significant role in causing them to leave their homeland; in time, however, economic hardship became the primary reason. Wright (1959) commented that by the beginning of the eighteenth century 20,000 Germans, primarily from the distressed areas of the Rhineland and Palatinate, were happily farming the valleys and uplands of Pennsylvania. Germans made up 23% of the population of Pennsylvania in 1726, and it had grown to 42% by 1755. The German immigration continued in waves until it became especially heavy in the decade following 1727. After 1750, the flow began to diminish.

From about 1715 until about 1750, arrivals of large numbers of Germans coincided with the arrival of large numbers of Scotch-Irish. In some years, the number of Rhinelanders and Palatines arriving was equal to the number of Celts. Both groups established farms in the Pennsylvania backcountry, outside the English settlements. As a result, by about 1725, Pennsylvania was filled with small farmers representing these two groups.

After the Pennsylvania backcountry was filled, succeeding waves of Germans, like their Scotch-Irish counterparts, also leap-frogged established settlements to move westward to the mountains and then southward along the foothills and valleys as far south as the Upcountry Piedmont of South Carolina. Although the two peoples tended to move through the countryside more or less together, they did not become close neighbors. Generally, the Germans would settle in one part of a valley and the Scotch-Irish in another. Once settled, with a home and farm established, Germans tended to remain there.

African

African slaves were being taken to the countries of the Near East from the time that the Muslims invaded Africa in the eighth century, and it is likely that a few of them were, in turn, taken to Europe. However, it was not until the end of the fourteenth century that the Europeans themselves began to bring African slaves to Europe. (Franklin and Moss, 1988). Then, when European exploration of the New World began, at the end of the next century, they brought African slaves along. In 1502, the Spanish government authorized the introduction of slaves to their new colonies, and within the next three decades, thousands of slaves entered the Spanish islands (Cuba, Puerto Rico, Jamaica) annually (Figure 4.42) (Conniff and Davis, 1994). Slaves were introduced to Guatemala by the Spaniards as early as 1524, and to Mexico about 1528. The Portuguese took African slaves to Brazil about 1538.

The English did not become involved in the slave trade until long after the other European powers. Around the middle of the seventeenth century, however, they decided that they would try to monopolize the industry to the extent possible. By 1780, the English were transporting twice as many slaves from West Africa as their competitors.

The first African slaves arrived in an English colony in North America in 1619 when a few were offloaded in Virginia from a Dutch ship. By 1620, the Dutch were also supplying enslaved Africans to their colony in New Netherlands. Slaves were present in Maryland shortly after the first settlements were made in 1634. In 1638, a Salem ship unloaded several slaves in Boston, and within a decade slaves were being used throughout New England. Negroes were in the Carolina colony virtually from its beginning, about 1663. The French introduced slaves at Mobile soon after 1704. Georgia did not adopt slavery until the 1749.

The American colonies received 45 of every 1,000 slaves landed in the Americas. In the early years, most of the slaves reaching the colonies were first landed in the Caribbean islands (St. Christopher, Antigua, Barbados, Jamaica) where they were seasoned. There they were taught the rudiments of either the French or English language, taught to work in gangs, and introduced to the regime of slavery before being transported (re-exported) onward to the colonies (Figure 4.49). Over time, however, because these seasoned slaves were more expensive and generally more troublesome, mainland planters began to prefer slaves brought directly from Africa.

Initially, the settlements on the Chesapeake were the most lucrative markets for slaves. During the 40-year period from 1680 to 1720, thousands were imported each year. In 1708, the population of Virginia included 18,000 Europeans and 12,000 Africans. By the 1730s, however, Charleston supplanted the Chesapeake as the most important slave market. Soon after, importations to South Carolina reached 2,000 annually. After the decision was made to import slaves to Georgia, the merchants there made up for lost time. By the late 1750s, Georgia was importing shiploads of slaves from the West Indies. Later, most of the slaves arriving in Savannah were brought directly from Africa.

By 1720, more slaves were being born in America than were being imported, and by 1740 most of the blacks in the thirteen colonies were American-born (creoles). At its peak in 1770, blacks made up 22% of the population in the 13 colonies. Figure 4.50 shows the distribution of slaves in the Southern states in 1790.

There is only limited information available on the ethnic origins of the slaves shipped to the New World; apparently, most of them came from an area bounded on the north by the Senegal River near Cape Verde and by modern-day Angola and Zaire in the south (Figures 4.46 and 4.47). However, the slaving frontier shifted continuously as the supply changed, but the principal areas included: the Sene-Gambia, Sierra Leone, the Gold Coast, the Bight of Benin, the Bight of Biafra and Central Africa, and the vast Congo Basin.

Fischer and Kelly (2000) provided information on the African origins of slaves in Virginia. A majority, 37.7%, were brought from the Bight of Biafra, which was in the vicinity of the present-day Cameroon (Figures 4.46). Other important areas were the Gold Coast (16%) in the vicinity of present-day Ghana, Angola (15.7%), and Sene-Gambia (14.9%), which was in the vicinity of modern-day Senegal and Gambia.

The United States National Park Service has researched

the source of slaves coming to South Carolina (Figure 4.48). They suggested that 25% were Mande peoples taken from the general area of present-day Senegal, Gambia, Guinea Bissau, and Guinea. A total 7% came from the Mano River area along the border between modern-day Ivory Coast and Liberia. Akan peoples from the present-day areas of Ghana, Togo, and Benin (Dahomey) comprised 11% of the total. Surprisingly, only 1% came from the Niger Delta in present-day Nigeria even though for many years, as many as 100,000 slaves a year were funneled out of the Niger Delta to other parts of the New World. The Bantu people along the Lower and Upper Congo River provided more slaves than any other area of Africa for South Carolina. Approximately 11% (Bakongo/Bantu) came from an area north of the Congo River in modern-day Gabon, Congo, and western Zaire. Another 28%, the Ovimbundu/Bantu, came from an inland area south of the Upper Congo River, generally in the area of present-day Zambia and southern Zaire. The origins of a sizable portion could not be determined.

5

Antecedents of Alabama Agriculture in the Old World and the British Colonies of North America

10,000 BC–AD 1800

PRECIS

French settlers in the early eighteenth century likely planted seeds in the sandy loam soils north of Fort St. Stevens, in the area that would later become the northeastern portion Washington County, Alabama. The Choctaw Tribe of Native Americans had been farming in this same area for several hundred years. At the same time, farmers in the south central Andes had been growing potatoes and guinea pigs for food for well over 4,000 years, and farmers in the Jordan River Valley had been growing einkorn wheat and herding goats for well over 8,000 years. How English settlers finally came to grow food and fiber in the land we call Alabama is a long and fascinating story.

It all began a long time ago. Apparently, around 9,000 years ago, gathering wild grass seeds and hunting wild goats in the Fertile Crescent of the Middle East slowly evolved into domestication and agriculture. In the next few thousand years this technology slowly spread throughout the Middle East, but it took some 3,000 years before it reached Greece and another 3,000 before it reached Britain and Scandinavia. The production of wheat, barley, and oats was generally well established, but highly unpredictable at the time of the Roman-Gallic Wars (first century BC). Julius Caesar in his *Commentaries on the Gallic Wars* underscores the importance of the production of all cereal grains in Western Europe at that time.

Although agriculture was widely practiced throughout Europe in the early centuries of the Common Era, it was not well systematized until the coming of the manorial system in the early Middle Ages. Even then it was not very productive and predictable. Agricultural technology improved slowly throughout this period, but was not really effective in meeting the food needs of the Continent's population until the middle of the fifteenth century.

When the English began to come to America in the early seventeenth century, they brought their agriculture with them, and almost starved before they could modify it to fit the environmental characteristics they encountered in Virginia and Massachusetts. Agriculture as practiced in Virginia was not adequate to meet the food needs of the population for almost three decades after the colony was established.

The colonists in both Virginia and Massachusetts Bay muddled along for years, working to change their old English agriculture to more effectively meet their needs. In the meantime, the Virginians added a completely new dimension to their farming system—the production of tobacco to meet the growing addiction to nicotine in their mother country. Initially they grew the crop as part of their general food and fiber production enterprises, but it eventually morphed into a replica of the sugar plantations operated by slave labor in South and Central America and the Caribbean.

So at the beginning of the American Revolution, there were two quite different farming systems operating in the colonies—one in the North producing wheat, cabbages, turnips, onions, radishes, carrots, apples, plums, quince, cherries, beef, pork, chickens, eggs, milk, and cheese for consumers in the rapidly growing urban centers in the region, as well as one in the South producing tobacco, rice, indigo, and eventually cotton for the nicotine addicts and machines in the industrial centers of Europe. After the Revolution when the gates were finally opened, the westward flood of settlers took these two systems with them—the general farming system into the Old Northwest and the plantation-slave system into the Old Southwest.

INTRODUCTION

Very little of the agriculture practiced in Alabama at the time it became a state actually evolved here. Virtually all of it evolved someplace else and was brought here as a result of technology transfer from tribe to tribe and village to village, and as a result of emigrants (primarily Europeans and Africans) bringing their technology with them. Consequently, to fully understand the development of agriculture in Alabama, it is necessary to understand the evolution of agricultural traditions (culture) of the peoples who came to live and farm here.

Information, especially written information on early agriculture, accumulated slowly. One of the first known works was a treatise on agriculture written by the Greek poet Hesiod around 776 BC. Another Greek author, Theophrastus of Ereos (371-287 BC), wrote a treatise on botany and horticulture. Among the Romans—Marcus Porcius Cato (Cato the Younger) (234-149 BC), Marcus Terentius Varro (116-22 BC), Publius Vergilius Maro (Virgil) (70-19 BC), Caius Plinius Secundus (Pliny the Elder) (AD 29-79), and Lucius Junius Moderatus Columella (first century AD)—all wrote books about the technical aspects of agriculture as they knew them. But information about its broader characteristics and practices did not begin to accumulate until the Early Middle Ages, or around AD 700.

Ponting (1991) comments that life under the constant threat of malnutrition and even starvation has been the common lot of mankind since the beginning of agriculture. Only slowly in a few areas of the world did some cultures emerge from this ages-long struggle to survive. Over the centuries, numerous small-scale technological improvements slowly raised agricultural output and productivity. Unfortunately, improved technology is just one aspect of hunger. Natural increases in population, ignorance, adverse weather, diseases of domesticated plants and animals, and political upheaval are other equally important, and generally negative, aspects. As the result of the constant battle between the positive and negative, hunger continues to plague humanity even at the beginning of the twenty-first century.

The Appearance of Agriculture

According to Smith (1995), plant and animal domestication and agriculture evolved independently in seven different locations around the world (Figure 5.1) between 4,000 and 10,000 years ago (Figure 5.2). During this long period, humans began to cultivate and herd some of those plants and animals that they were encountering in their daily hunting and gathering activities. Over time, they began to develop very specific processes and procedures for herding animals and producing crops. These processes and procedures were passed from generation to generation, as they became part of their cultures. They became, in a broader sense, a part of their biology. Some aspects of this accumulated experience and understanding (domestication of specific animals and plants, fences, tools, seed, ropes, containers) became part of their material culture. Other aspects (general attitude toward agriculture, the role of the sexes in food production, food likes and dislikes, food taboos, conservatism in outlook, the role of the supernatural in their practices) became part of their nonmaterial culture. Elements of material culture tend to change much more readily than nonmaterial elements. The interplay of these two basic elements gave rise to and helped to maintain the agricultural traditions of people as they lived from day to day within their villages or even when they moved from village to village or over long distances to other parts of the world.

Agriculture apparently first appeared in communities of hunting-gathering peoples settled around the margins of lakes, marshes, or rivers (Smith, 1995). Generally, these environments were so rich in natural food resources that the people were able to establish relatively permanent communities (associations) using the wealth of foods from local plant and animal communities that they found there. This environment provided the optimum set of conditions for the operation of cognitive processes. Under these conditions, the elements of cognition (observation, thinking, communication, planning) functioned in tandem to add a significant new dimension to food security. These environmental conditions allowed the people the luxury of performing primitive experiments without jeopardizing their food supply.

It is unlikely that we will ever know with certainty how, where, why, or when plants and animals first became domesticated and how agriculture actually began. However, it seems likely that the process occurred over a long period of time, and that for much this period, humans continued to depend primarily on hunting and gathering as means of meeting their food and fiber needs. Only very gradually did domesticated plants replace wild plants in their diets. The first agriculture appeared in Southwest Asia 10,000 years ago; however, a society and culture totally dependent on herding and cultivation did not appear there until 4,000 years later, and agriculturalists did not come to dominate the entire planet until 2,000 years ago.

In the Beginning—The Fertile Crescent

The Fertile Crescent of the Middle East (Figure 5.3) was apparently the site of the earliest domestications of plants and animals (Smith, 1995). Supposedly, about 10,000 years ago (Figure 5.2), people living in permanent communities in the Levantine corridor began to experiment (trial and error learning) with ways to increase the production of wild grass seeds that were an important part of their diets. This area is located in the Jordan River Valley, generally from the present-day Jericho (Ariha) in Jordan, northward to Damascus in Syria. Out of this experimentation came the domestication of emmer wheat, einkorn wheat, and barley. By 9,000 years ago, the cultivation of these cereal crops had spread northeastward, around the arc of the Crescent. Somewhat later, additional plants (lentils, peas, chickpea, bitter vetch, and flax) were domesticated and included in the mix of plants supporting the development of agriculture in this region.

The domestication of animals also played a central role in the development of agriculture in the Fertile Crescent. About 9,000 years ago, people in valleys in the Zagros Mountains (southwestern Iran) in the eastern end of the Crescent began to manage captive goat herds. They had hunted herds of these animals for food for thousands of years before they began to manage them. Later, goat herding spread westward through the Crescent at the same time that cereal culture was spreading eastward. Sheep were domesticated nearer to the center of the Crescent and somewhat later than goats. Most evidence indicates that they were first domesticated in the area of Northern Syria and Southeastern Turkey around 8,500 years ago. Apparently pigs were also domesticated near the center of the Crescent in the area of present-day Northwestern Syria and Southeastern Turkey around 8,000 years ago. Cattle were the last of the four primary animal species to be domesticated in the region, and while they would become extremely important later, they had a minimal role in the emergence of agriculture until after about 8,000 years ago. Although they arrived late on the scene, domestic cattle made possible an extremely important addition to agriculture—the use of animal power in pulling the plow.

After around 8,500 years ago, the cultivation of plants and animals converged around the center of the Fertile Crescent, in the upper valley of the Euphrates River in the northwestern portion of present-day Syria and, modern balanced agriculture began. It was there and then that agricultural economies took shape that fueled a long and complex developmental sequence leading to statecraft and the development of the first city-states.

East Asia

Apparently, domestication of plants in East Asia began some 1,500 years after it first appeared in the Fertile Crescent (Figures 5.2). Hunter-gatherers in the Yangtze River corridor of Eastern China (Figure 5.1) first domesticated rice and provided the impetus for the development of agriculture based on the culture of this species throughout Asia and beyond (Smith, 1995). This valley with its rich subtropical wetlands, mild winters, and reliable summer rainfall provided the ideal environment for the development of rice-centered agricultural societies that would ultimately dominate much of Asia. This climate, on the eastern side of the continent, is similar to that of the Southeastern United States. Domesticated pigs, chickens, and water buffalo were also important to these early farmers, but archaeologists are not sure where and when their actual domestication occurred. Pigs had been domesticated in the Fertile Crescent at least a thousand years earlier, but it is not likely that these animals in Eastern China originated there. It is likely that they were domesticated again someplace in Asia. Chicken bones have been found in sites of ancient villages that are over 7,000 years old and represent the oldest remains of that animal yet found; however, at this point, archaeologists are not certain that they were from domesticated animals.

The Ch'in Ling Mountain range, which extends westward from the Yellow Sea, forms the northern shoulder of the Yangtze River Valley. North of this range of mountains lies the Yellow River Valley. In this valley, the climate is much cooler and drier than in the valley south of the mountains. Here lies a second center of domestication, and the center for the development of East Asian agriculture (Figure 5.1). Hunter-gatherers in this great valley, between 7,000 and 8,000 years ago (Figure 5.2), domesticated and began to cultivate two more important food crops: broomcorn millet and foxtail millet. Archaeological studies indicate that these two plants were important food sources long before they were domesticated. Domesticated chickens and pigs were also important to these early agricultural communities.

Sub-Saharan Africa

In the Fertile Crescent first and in Europe later, wheats and barley became the major crop plants; however, Africans south of the Sahara (Figure 5.1) domesticated additional plants that formed the basis of their agricultural way of life (Smith, 1995). This region generally stretches eastward from present-day Senegal on the South Atlantic to Northern Ethiopia on the Red Sea.

Plant domestication and cultivation began much later in Sub-Saharan Africa than in the Fertile Crescent (Figure 5.2) when hunter-gatherers began to grow sorghum some 4,000 years ago. Pearl millet and African rice were domesticated later. Other important crops of African origin include: cowpeas, coffee, sesame, watermelons, oil palm, and castor oil. These plants, even today, are important crops across Africa and Asia.

Although Sub-Saharan agriculture developed later, it apparently developed independently even from that in the Nile Valley north of the Sahara, which like Europe, mirrored developments in the Fertile Crescent. Animal agriculture south of the Sahara began somewhat earlier than the farming of plants. The herding of domesticated cattle began about 5,000 years ago, probably using animals introduced into the region from the Nile Valley (Smith, 1995).

The South Central Andes

This region encompasses a relative small area near the shared boundaries of present-day Peru, Bolivia, and Chile (Figure 5.1). The region represents still another independent site of plant and animal domestication (Smith, 1995). Research carried out in the high valleys and basins of this region indicate that distinctive agricultural economies developed there independently beginning around 4,700 years ago (Figure 5.2). These hunter-gatherers domesticated five species (two plants and three animals) from their environment that they used for food: quinoa, potato, llama, alpaca, and the guinea pig (Smith, 1995). These five species would define the high-altitude food production economies of that region, although two other Andean species (common bean and lima bean) were domesticated and cultivated to some degree. There is relatively little information regarding when, where, or under what conditions these species were domesticated and included in the farming economies. Recent information suggests that quinoa was probably domesticated between 4,000 and 5,000 years ago in the Southern Andes. Wild potatoes were being used as food by hunter-gatherers as early as 12,000 years ago and were probably domesticated at about the same time as quinoa. Camelids (llama and alpaca) were hunted for their meat and skins at least 9,000 years ago, but they were not being widely herded in the region until some 4,000 years ago. The domestication and early history of guinea pigs seems in general to parallel that of the camelids. They were clearly an important food source long before they were domesticated. Archaeological sites indicate a significant increase in the quantity of guinea pig bones around human settlements inhabited around 4,500 years ago, which suggests that domestication was well under way at that time.

Central Mexico

Central Mexico (Figure 5.1) was a second site in the Americas where early domestications occurred independently (Smith, 1995). The long history of human occupation of the Americas began sometime between 15,000 to 20,000 years ago; although there is some evidence that Asians may have arrived earlier. Regardless of the time of immigration, they apparently did not bring domesticated plants and animals or agriculture with them. They were hunters and gatherers and would be for several thousand years. Agricultural villages slowly emerged independently out of hunter-gatherer settlements in Central Mexico some 4,700 years ago (Figure 5.2). These early farmers contributed three additional important plant crops to the world's food supply: maize, beans, and squash (Smith, 1995). The early domestication and cultivation of these plants was a logical extension of the hunter-gatherers' wide-ranging experimental (trial and error) efforts to increase the predictability and abundance of food resources and thereby reducing the effects of good and bad years. Of course, it is not likely that they understood what they were doing. They first began to grow maize (around 4,700 years ago). It probably was not domesticated there, but had spread out of the southwestern part of the country where it was domesticated somewhat earlier. Domesticated beans and squash also played and important roles in the evolution of agriculture in Central Mexico. It has not been firmly established when these two plants were domesticated.

Eastern Woodlands of North America

Domestication of some native plants also apparently evolved independently in the eastern woodlands of North America (Figure 5.1). Some 5,000 years ago, dense deciduous forests extended from the Atlantic Ocean westward to the eastern margin of the prairies of the Great Plains and northward from the Gulf of Mexico to Canada. Like the great forests of Central Europe, it was an inauspicious place for the independent development of agriculture. Amerindians in this broad area received several domesticated crops, including maize and squash, from Mexico and the Central Andes that were incorporated into local farming economies. However, three species of floodplain plants (goosefoot, marsh elder, and gourds) were domesticated independently in the Eastern United States at an early date. Smith (1995) suggests that these plants were being deliberately planted as early as 5,000 years ago and that they were being cultivated around 4,500 years ago (Figure 5.2). Later, knotweed, may-grass, and little barley would be added to the list of North American domesticates.

Agriculture Comes to Europe

Independent domestication of plants and animals was limited in Europe, although the Europeans did apparently domesticate oats. While there was limited independent domestication, the agricultural economies that later developed around domesticated plants and animals carried distinctive European stamps. The differences increased as agricultural technology migrated westward and northward from the Fertile Crescent. The climate in most parts of Southern Europe was generally so similar to that of the Middle East that agriculture from that region could be adopted with little or no change. Such was not the case in Central Europe. There, the winters were much cooler and longer. Also, there were significant differences in vegetation. Much of Central Europe at that time was covered by dense forests of large deciduous trees, especially oaks. There was virtually no open grassland.

Most of the technology related to the planting of domesticated species reached Europe from the Middle East or Western Asia (Solbrig and Solbrig, 1994 and Smith, 1995). This transfer required a long period of time. For thousands of years, farming, hunting, and gathering existed in the same villages. Wheat and barley agriculture did not appear in Greece until some 3,000 years after it was developed in the Fertile Crescent, and another 3,000 years passed before those plants were widely used in Scandinavia and Britain. The practice of fallowing, or not planting a field for one or more seasons to allow the fertility to recover, was also transferred from Western Asia, but thousands of years passed before it was widely used.

European Agriculture from the Eighth through the Seventeenth Centuries

The period from around AD 700 through 1600 was an extremely tumultuous one for European agriculture. Good economic times were quickly followed by a period of severe contraction; then, just as quickly, by another period of rampant inflation. In this almost 1,000-year period, farmers were whip-sawed unmercifully. Unfortunately, as we will see, time- after time, as we trace farming down through the ages, farmers and whip-sawing seemed to be joined at the hip. A summary of these changes, for the period from the eighth through the seventeenth centuries, are shown in a simplified tabular format in Table 5.1; those changes will be discussed in some detail in the following sections.

Early Middle Ages

As detailed in Chapter 4, beginning in the Early Middle Ages (AD 750 to 1,000), in much of Europe, depredations by roving bands of robbers became so severe that in parts of Europe free peasants gave up their freedom to the lords of large landed estates (manors) in return for protection and for sustenance in those years of poor harvests (Spielvogel, 1994). In return for providing these serfs with protection, the lords demanded their piece of flesh as compensation. Under this arrangement, serfs—who technically were not slaves, because they could not be bought or sold—were bound to the land and to the lord (Table 5.1). They could not leave his estate without his permission. Serfs also had servile obligations that included the payment of a fixed rent in food and/or money and the duty to work the lord's land for a few days a week. In turn, the lord had certain obligation to his serfs. He was expected to provide them with protection and to provide them with food when harvests had been poor. He also granted them use of the so-called common land on his estate. Generally, he did all that he could reasonably do to keep them healthy and ready and able to work. The overall prosperity of the entire estate depended on their labor. It has been estimated that by the end of the ninth century that some 60% of the peasant population of Europe had been reduced to serfdom.

For hundreds of years in some parts of Europe, agriculture would be organized around serfdom and these large estates, or manors (Solbrig and Solbrig, 1994). Serfs lived in cottages provided by the lord, in small villages. Generally, there were several of these villages on each estate. The primary goal of these manorial farms was to produce enough food and fiber to sustain the lord, his extended family, his servants, and his serfs. Custom established everyone's social position, and regulated the pattern of daily life. Agriculture was designed to reduce the risk of crop failure. Loss of a crop meant near starvation for the whole estate for at least a year. There was no experimentation. Time-tested methods became inviolable traditions. Cereals (wheat, oats, barley) made up the greater portion of the diets of most of the people on a given estate; consequently, most of the farming effort was given to producing them. Some animals (cattle, pigs, sheep, goats) were produced to provide meat, primarily for the elite. Because few cattle were grown on the estates during this period, the lord and his extended family ate mostly sheep and goat meat, supplemented by venison and bear meat from animals taken from the lord's hunting preserve. When peasants had red meat to eat, they had pork. The estates also produced a considerable number of horses.

Various kinds of cereals provided most of the calories consumed on the estates. These included: ordinary wheat, dwarf wheat, barley, oats, rye, horse beans, peas, knotgrass for porridge, spurge, and flax.

Open fields on the estate were divided into a number of plots. There were no hedges or fences separating them. Each plot was further subdivided into a series of strips. Generally, an individual family would have strips assigned in several different plots and fields. The lord also had strips distributed in the same manner. Serfs were responsible for producing their own food on their strips, but they were also required to provide a specified amount of labor to produce food for the estate. The men worked on the lord's strips, while the women helped in the lord's house. Each serf was also provided a cottage by the manor. Each cottage had a small kitchen garden located near it.

Strips were further divided into two groups and managed under the two-field or two-strip system. The strips in one group were planted with crops while the strips in the second group were allowed to lie fallow for a year to rebuild their fertility. This practice was generally followed throughout Europe, and although it was extremely wasteful of land resources, it was used for hundreds of years. In the early days, there was no chemical fertilizer and only a limited amount of animal manure, so fallowing was the only way to maintain the fertility of strips.

Management of the strips generally was not a matter for the individual serf farmers to decide. These decisions were made by councils of elders in a village or group of villages. The succession of crops; plowing, planting and harvest schedules; sowing rates; the fallow year; the raising of stock; irrigation of meadows and forestry practices, were set by the councils for the entire estate, and everyone adhered to those schedules. Strips were reassigned periodically; consequently, serfs were reluctant to make improvements on their assigned land, because someone else would soon reap the benefit of their efforts.

On all of the estates, there was a substantial common area. It consisted of meadowland, land poorly suited for cultivation, woodland, ponds, and streams. All of the people working or living on the estate could use these areas. In some cases, however, the lord required that any products taken from these areas be shared with him. Access to these areas was extremely important to the serfs.

High Middle Ages

Beginning around 1,000 (the beginning of the High Middle Ages), the population of Europe began to increase dramatically. This increase, fortunately, was accompanied by a significant increase in food production (Table 5.1). Much of the increase was likely the result of the expansion of the amount of land in cultivation and a general improvement in climate throughout the region. There were also a gradual improvement in agricultural technology during this period. A new type of horse collar had appeared in the tenth century, and the use of horseshoes spread during the eleventh and twelfth centuries. There were also improvements in the use of iron in making scythes, ax heads, hoes, saws, and heavy wheeled plows with iron ploughshares.

The use of new management techniques also played a role in the increased output of agricultural products. One of the major changes in management was the development and use of the three-field system whereby the strips were divided into three groups instead of two. Strips in one group were planted in the fall with winter grains, such as rye and wheat. A second group was planted in the spring with grains such as oats and barley. The third group of strips was fallowed for the year. This new management strategy apparently originated between the Seine (France) and Rhine rivers (Germany). It slowly spread both to the east and west from that region. It resulted in an increase in food production, because more of the estate's land was in production at any one time. Also, it spread the year's work more evenly.

With the increased production of food and the increase in population, there was a re-birth of towns and cities that had languished throughout Europe during the Early Middle Ages. Slowly, there emerged in these urban areas artisans who would sell or barter their wares for food. The growing number of people in these urban centers lead to an increased demand for food that, in turn, lead to an increase in prices. This increased demand precipitated an agricultural boom throughout Europe in the twelfth and thirteenth centuries, and there emerged, for the first time in Europe, the opportunity to produce food for profit. The old subsistence or direct agricultural consumption system was beginning to be replaced by a market-driven or indirect agricultural consumption system. This change in the economics of food production had a serious impact on feudalism, the manorial system, and the life of the serf. In most cases, the lord of the manor reacted to the new economic reality by leasing his lands to the serfs for money payments or fixed rents. Servile labor was reduced somewhat. In some parts of Europe, landlords allowed their serfs to become share-croppers. Under this system, the serfs worked the lord's land and shared the crops they grew with their lords. Over time the serfs would be transformed into peasants.

The increased prices associated with agricultural boom

also begin to bring about changes in the old open field system of farm management. To take advantage of the developing market economy for agricultural products, estate owners (lords) needed more control of their land; consequently, they began to enclose their common lands or to restrict their peasants use of them. By taking control of these lands, they were able to increase the production from them. They also consolidated their scattered strips and plots into more manageable units. The losers in this enclosure process was the landless or near landless peasants who lost their ancient rights to the commons and its resources. Gras (1946) noted that this movement actually began around the middle of the thirteenth century. It would bring even greater changes in European agriculture in the fourteenth and fifteenth centuries.

In the twelfth century, the forerunner of modern industrial agriculture first appeared in the Mediterranean basin in the form of sugar plantations (Solbrig and Solbrig, 1994). The Arabs brought sugarcane and the technology for making sugar to that region from India in the seventh century. Sugar plantations were operational in Cyprus and Crete in the twelfth century, but Sicily is where they became a model for the future of the industry. Later this model would be used in the establishment of sugar plantations, first on the Atlantic Islands (Azores, Maderia, Canaries) and then on the islands and mainland of the New World.

As sugarcane and sugar production increased, the enormous requirement for labor quickly became obvious. It also seemed apparent that the only way to meet those requirements was through the use of forced labor. Because of the nature of the process by which sugarcane is grown, harvested, and converted into sugar, it was realized that large amounts of labor and capital would have to be concentrated on a relatively small area of land; thus the concept of the narrowly specialized farming operation or plantation evolved. Further, because of the high cost of establishing and running a plantation, it was necessary to have markets in large centers of population for the product.

From the beginning, slaves were a central element of the plantation system of sugar production. At first, slaves from the region were used, but after the Portuguese began to trade in Africa in the fifteenth century, that continent furnished virtually all of the forced labor required for sugarcane and sugar production.

Late Middle Ages

Well into the thirteenth century, Europe experienced good harvests and an expanding population (van Bath, 1963). By the beginning of the fourteenth, however, a long period of disastrous change began (Table 5.1). The population had essentially exceeded the level that the so-called agricultural boom conditions could support. Also, about this time, the climate of Europe began to deteriorate somewhat. Food shortages and actual famines began to appear early in the fourteenth century, and they came with startling suddenness. From 1308 to 1311, harvests were somewhat below average, which led to some price increases (Parry, 1978). Then in the period 1315 to 1317, there was an extraordinary series of wet summers and mostly wet springs and autumns. In 1315, the grain failed to ripen all across Europe. According to Lamb (1982), 1315 was probably the worse year of the miserable sequence of years that followed. The poor harvests resulting from especially wet summers led to almost unimaginable human misery. The poor died in large numbers, and many of those living turned to robbery in an attempt to obtain food. Huge bands of starving peasants swarmed across the countryside. Famine was widespread. In the first half of the fourteenth century, that famine resulted in an estimated 10% reduction in Europe's population.

Farm animals also suffered during the miserable years. Many more animals than normal had to be slaughtered for food. Many more had to be sold because there was no feed for them. Finally, the bad weather, plus the poor nutritional condition of those remaining, resulted in widespread outbreaks of animal diseases. The loss of draught animals was so severe that it affected tilling of the soil in many areas.

It has been suggested that the combination of crop and animal losses during the period 1315-1317 generated the worse agricultural crisis in England since the Norman invasion in 1066. The crisis was equally bad, and in some cases worse, all over Europe. Everywhere farmland and even agricultural villages were abandoned, and in the same period, wheat prices rose to three times their normal level, and in some areas with acute shortages, they were eight times higher.

About 1320, harvests began to improve somewhat. Unfortunately, because such a high proportion of the loss of human life had occurred in the towns, the farmers found that they had significantly fewer customers to purchase their crops. As a consequence, an agricultural depression developed. Then beginning about the middle 1340s, the bubonic plague or Black Death, arrived in Europe with a vengeance. When it had run its course, around 1350, at least half of Europe's entire population had died. Again, the loss of life was greatest in the towns, which reduced even further the markets for the farmers' crops. As a result, the agricultural depression that had begun about 1320, deepened (van Bath, 1963). Cereal prices fell rapidly, while money wages and real wages increased, leading to rampant inflation. The depression would continue until the middle of the fifteenth century.

Farmers coped with the plunging cereal prices by converting arable land to grass and pasture, because as a result of the heavy loss of animals earlier, the price of meat remained relatively stable. Other farmers changed from producing cereals to producing industrial crops. Farmers in England shifted to sheep production in response to the increased price for wool. In Germany, they shifted to hops, oil-producing seeds, flax, and hemp. Many farmers left agriculture altogether and sought employment in the towns.

The loss of population from famine and plagues in the first half of the fourteenth century had an important effect on the structure of European agriculture. In most areas, there was a severe labor shortage. Medieval agriculture was labor intensive, and with the shortage of labor, there simply were not enough people available in the old servile-labor or serf system to meet the needs of the estate farms. As a result, estate owners had to employ farm laborers and pay them relatively high wages. The shortage of labor and the increased use of wage labor further reduced the importance of the old servile-labor system. The system had begun to break down during the agricultural boom days of the latter half of the twelfth and all of thirteenth centuries. Now, the labor shortage and the depression almost finished it in much of Europe. However, it would remain in place for another 400 years in the northeastern portions of the continent.

The agricultural depression of the fourteenth and first half of the fifteenth centuries also resulted in the intensification of the enclosure movement, especially in England (Solbrig and Solbrig, 1994). Large consolidated or enclosed holdings were converted from cereal to sheep production when the demand for wool began to grow in response to the rapidly expanding Flemish woolen cloth industry. Many marginal tenant farmers were pushed off the land. Whole villages were wiped out to make room for the pastures where sheep could graze. Sheep farming required considerably less labor. These changes led to the use of the expression, "sheep eat people."

The great agricultural depression of the fourteenth and fifteenth centuries placed severe restrictions on the profitability of the large estates; consequently, the owners began to look for ways to accommodate this new economic reality. The move toward the production of industrial crops and the intensification of the enclosure movement, described in preceding paragraphs, resulted from efforts to improve returns on their holdings. However, probably the most important response was a general improvement in agricultural technology, more specifically in animal production. As noted previously, during the depression, meat prices did not fall nearly as much as cereal prices; consequently, more land was put into the production of cattle and especially into the production of fodder. With more animals, there was a significant increase in the production of manure. The added manure stimulated an increase in the yields of both cereals and industrial crops. Also, with the increased availability of manure, there was no need to fallow land. Manure could be used to maintain and even increase fertility.

The enclosure movement and the improved agricultural technology associated with it also had another far-reaching effect. It changed forever the relationship between the landowner and the people who worked on his hand. The old relationship of the manor in which the lord had a responsibility to look after the needs (protection and sustenance) of his serfs was replaced by a more formal one governed by the marketplace and profitability. This change doubtlessly contributed to the improved efficiency of agriculture that was desperately needed by an entire continent of people who had suffered through some four centuries of one famine after another. Unfortunately, it created a extremely large number of landless poor who were inadequately prepared to take care of themselves. Many of these would ultimately seek new opportunities far to the west in the Americas. Many of these people and their descendents ultimately would determine the nature of agriculture in Alabama.

From a historical perspective, the events described in the preceding paragraphs concerning the development of agriculture in the Late Middle Ages represent a true watershed in man's ageless efforts to guarantee his food and fiber supply. From this time onward, food and fiber production would increasingly become a commercial rather than a subsistence venture. These events would reverberate down through history, and in doing so, would change the course of world events. As we shall see, these halting moves toward commercial agriculture would finally have a long term, far-reaching effect on the development of the economic, social, and political way of life for the peoples of Alabama.

Early Renaissance

Around 1450, European agriculture, in general, began to slowly emerge from the deep depression of the preceding 150 years (Table 5.1). Farmers throughout the continent were successfully growing a variety of crops, including the cereals: wheat, rye, barley, and oats, as well as two pulses: peas and beans. They also kept cattle, sheep, pigs, and poultry. The farming year began after the harvest of summer crops when the fields would be plowed in preparation for planting fall crops. In the process, a mould-board plow was pulled by a team of up to eight oxen or from one to four horses. Winter cereals, wheat and rye, were sown in September or October. From late February to April, the spring crops, barley, oats, peas, and beans, were sown when the soil was warm enough to get germination and when it was dry enough so it could be plowed effectively. Seeds were covered by harrowing. Once the crops were growing, the labor-intensive job of protecting the plants from weed competition began. Harvest usually began in August or September with winter grains being harvested before spring-sown crops.

Also during this period of slow agricultural recovery, deaths from plagues were beginning to be reduced as epidemics gave way to pandemics. As a result, the population began to increase, and because there were more customers, cereal prices increased, and because there were more laborers, real wages decreased. This slow recovery lasted for a hundred years or until about 1560.

Beginning at that time, the slow recovery of agriculture from the deep depression of the Late Middle Ages gave way to a period of steep inflation (price revolution) as food prices, especially the price of wheat, increased rapidly. While real wages increased, they did not keep pace. Within a relatively short time, wage earners, especially agricultural laborers and workers in urban areas, began to experience a lower standard of living. At the same time, this inflationary period fueled prosperity for the estate owners and the commercial class. While there is considerable debate on the cause of rising prices, the growing European population, which increased the demand for food and land, and the influx of precious metals from the Spanish mines in the New World, played major roles.

This inflation-fueled prosperity in Europe, at least for some, coincided with the development of a flourishing trade involving much of the Continent. Later, trading patterns would be expanded to include the Americas. Trade began to make all of Europe into a more integrated market that became increasingly vulnerable to localized price shifts. Statesmen on the Continent began to understand that a flourishing commerce was a direct asset to the state, and as a result, they began to openly encourage trade and manufacture.

Agriculture also began to reflect the expansion and intensification of the commercial economy associated with the price revolution. Calculations of cost and profit began to enter into decisions regarding crop rotations, methods of tillage, and the balance between animal and crop production. Even land and rents were sometimes treated as commercially negotiable commodities. The manner of producing and exchanging agricultural commodities that evolved during this period essentially set the stage for the market economy for agricultural products in use today more or less throughout the world. A largely local economy based on self-sufficiency and barter evolved into a regional, market-oriented, monetary economy. The cooperative manorial system of production—based on conservative customs and collective decisions by a lord and his serfs—was replaced by specialized farming units with each owner or tenant making individual decisions regarding what, when, and how to plant to maximize profits. Higher productivity became an important aspect of the whole process, as important, if not more so than simply avoiding failure. These changes were far-reaching, but they did not take place immediately. Several centuries passed before they were firmly established throughout Europe.

Unfortunately, agricultural technology in much of Europe in the latter part of the sixteenth century and the beginning in the seventeenth was little unchanged from the mid-1500s. Rye, rather than wheat, continued to be the predominant grain produced in many countries. Oats and oatmeal cakes were still the mainstay of Scottish diets. There had been little progress in livestock husbandry for over a hundred years. While many aspects of the old manorial system had disappeared as trade and commercialism had expanded, the large estates remained, and the great majority of all agricultural production still took place on those large plantation-like farms. Serfdom had been replaced by tenants, subtenants, and day laborers, but most of the profits still accrued to the landowner. Despite expanding markets and rising prices, increased rents imposed by the landlord, the rising cost of agricultural inputs, and higher fees, plus increased taxes levied by the state, meant little or no improvement in the life of the peasant.

While the new emphasis on trade had not had much effect on agricultural technology, one change was in the making that would literally transform food production in Europe and throughout the world in the next century. This change involved exchange of crops and animals between continents. Over time, the Columbian, or Grand, Exchange, would result in an almost worldwide increase and stabilization of food production. In this worldwide exchange, Southeast Asia provided sugar cane, rice, and the banana. Europe provided wheat, barley, oats, sheep, cattle, horses, pigs, and bees. The Americas provided maize, tobacco, potato, tomato, manioc, cocoa, rubber, upland cotton, pineapple, peppers, squash, and pumpkin.

Late Renaissance

The inflationary period in Europe began to come to an end in the early part of the sixteenth century, only to be replaced by a time of severe economic contraction, if not depression, that would last until the end of the century (Table 5.1). Early in the century, the importation of precious metals from the New World began to decline. In Germany, the Thirty Years' War (1618-1648) had a devastating effect on the economy, and it did not help the economies of primary combatants: Spain, France, Bohemia, Denmark, and Sweden.

Also, poor weather of the so-called Little Ice Age affected harvests and gave rise to famines all over the Continent during the seventeenth century. For over a century, average temperatures were lower, winters were colder, summers were wetter, and devastating storms seemed more frequent. The first famine associated with this period of poor weather actually occurred during the latter part of the sixteenth century (1594-1597), and it affected virtually all of Europe. It was especially bad because food reserves had slowly dwindled for some three decades before this time as a result of generally poor summers. Crop failures were especially bad in 1649, 1660-1661, and the 1690s. The failures of the 1690s were especially devastating in Northern Europe. The effects of the famine were especially extreme in Scotland. As an example, on an estate in Northeast Scotland that earlier had accommodated 169 individuals, only three families, the proprietor's included, survived the ordeal. Between 1693 and 1700, harvest (largely oats) failed in seven out of eight years, in all of the upland parishes of the country. Lamb (1982) comments that these famines caused many Scots to abandon their farms and migrate to the Ulster Plantation in Northern Ireland. In Finland, as a result of both food shortages and epidemic disease in the malnourished population, at least a quarter and possibly as much as a third of the people died in the so-called Great Famine of 1696-1697.

As if war and famine were not enough, pestilence returned to Europe. The last great visitation of bubonic plague spread across the Continent in the middle to late years of the century. It struck London in 1665, killing some 20% of the city's population.

Trade, industry, and agriculture were deeply affected by the economic contraction, but agriculture, especially the peasants working the land, was devastated. At that time, 80 to 90% of all Europeans still worked on the land. As the depression deepened, cereal prices plummeted. In fact, all foodstuffs fell to two-thirds to a quarter of their former value. Real wages were relatively high. The wages of farm laborers became excessive. Landholders converted their arable land into pastures. Animal husbandry was expanded, as was the cultivation of industrial crops. Throughout the period, there was little innovation in farming techniques on most of the Continent, and everywhere there were lost agricultural villages, empty farms, and neglected fields.

Throughout much of the seventeenth century, lowered agricultural production, widespread food shortages, malnutrition, famines, disease outbreaks, and exorbitant mortality rates especially among the very young and elderly had always resulted in increased internal social and political instability. This period of excessive social instability coincides with the widespread emigration of Western Europeans to the Americas. It was during this century that all of the British North American, colonies with the exception of Georgia, were established.

Surprisingly, the Scientific Revolution appeared amid the calamitous economic and social conditions of the late sixteenth and seventeenth centuries. Roger Bacon (1561-1626) had set it in motion with his insistence that the only way to understand the natural world was through the use of the scientific method or by the use of observation and experimentation. This was the age of Galileo (astronomy), Harvey (anatomy and physiology), Cavendish (natural philosophy), and Newton (physics). While this revolution expanded rapidly during this period, without a doubt many of its elements began to take form much earlier, in the Renaissance. The Scientific Revolution was born amid calamity, but its influence would grow steadily down through the ages. In the eighteenth century, it doubtlessly played an important role in the Agricultural Revolution.

Agricultural production began to stabilize and then increase somewhat around the middle of the seventeenth century (Table 5.1). At about the same time in much of Europe, the harsh, highly variable, severe climatic conditions that had characterized the past 300 years began to ameliorate, even though the poor climate returned for a decade at the end of the century. Generally after about 1650, agricultural production improved. Coincident with the improved climatic conditions, there were also improvements in agricultural management and technology. For example, more land was brought into production as a result of the continued reduction in the amount in fallows and with planting of legumes, such as alfalfa and clover to improve soil fertility. Land was planted in legumes rather than being fallowed. These crops while improving the land also provided an excellent source of animal feed. Additional land was also brought into production by the continuation and even intensification of the enclosure movement. During this period, farmers began to plant turnips (a native plant of Northern Europe) to provide winter fodder for their animals. The value of turnips rested in the fact that they could be used in two ways: the leafy portion could be used directly by grazing, or the roots could be harvested and stored to feed to cattle held in stalls. The improved animal fodder situation allowed the landowner to maintain even larger numbers of animals, which resulted in increased meat and milk production. More animals also meant more manure that could be applied to the fields to increase soil fertility. Also, during the latter part of the seventeenth and beginning of the eighteenth centuries, the potato and maize began to add to the food supply, especially of peasants. Both crops had been introduced to Europe in the sixteenth century, but contributed very little to food production for over a century. At first, the potato was used primarily as a fodder crop, but after some selective breeding by farmers resulted in the development of varieties that produced good tuber crops, their use for human food increased dramatically. Maize had very little impact until the weather improved after about 1650.

This Agricultural Revolution would finally have far-reaching consequences throughout Europe. As a result of its application, more people could be fed, food prices were lower, and less labor was required. These positive changes also resulted in the increased accumulation of capital that would help fuel the Industrial Revolution in the eighteenth and nineteenth centuries. With the improved availability of food, economies of the various countries improved significantly. This improving economic situation resulted in the rebirth of a highly specialized, intellectual group, the *philophes*, who had both time and money to examining "all aspects of life." It was the Age of Reason, or the Enlightenment—the age of Voltaire, Hume, Diderot, Rousseau, Smith, Montesquieu, and Gibbon. It was the age of philosophy, art, music, poetry, economics, law, the writing of history and an age of secularization. This new age of liberalism changed Europeans and their societies in important ways, but surprisingly it had little effect on their extremely conservative views of the plight of the poor and disadvantaged among them. However, while it effected little immediate change in the plight of this largest segment of the population in all European countries, it did begin to create an environment in which concern for their welfare would increase among future generations.

Development of Agriculture in Specific European States

The general development of European-wide agricultural traditions was discussed in some detail in the preceding paragraphs. Although these comments present a broad overview, they do not provide a picture of important traditions developing in the individual states within the region. Often these traditions were quite different, for example, from England to Russia within the same time period. Obviously, it is beyond the scope of this book to follow all of those changes for very many countries. Instead, I have chosen to look at the development of traditions in the Netherlands, Great Britain, and Germany. Traditions that developed in these specific states would most affect the course of the evolution of agriculture in the English colonies of North America and finally in Alabama.

Netherlands

The Netherlands was the source of many of the improvements in agricultural technology in the latter part of the seventeenth century. Farming in the Netherlands during the latter part of the great European recession of that century became, in many respects, the school of agricultural improvers. One of the most important reasons for this must surely have been the density of the population in Flanders, Brabant, Zealand, and Holland in the sixteenth and seventeenth centuries. Agriculture became more intensive because the peasants of this densely populated area were trying to wrest a living from the land during a depression that had devastated the price of cereals. To do so, they began serious efforts to increase the productivity and profitability of their agriculture. They began to give special attention to industrial crops that provided better returns on their investments. In turn, improvements made in the production of these crops were quickly transferred to the production of food.

Woolen textiles had first become important in Flanders after the twelfth century, but when the population throughout Europe began to increase rapidly in the seventeenth century, the demand for textiles also increased rapidly. Quickly, much of the farmland was converted to the production of crops needed to support that industry. Large quantities of flax were produced for the weaving of linen cloth. Also, many farmers produced crops that were used in the making of dyes for coloring cloth. As a result of the conversion of lands to the production of industrial crops, the Netherlands lost their self-sufficiency in the production of cereals. Fortunately, as a result of the development of ship building there and their long history as international traders, they were able to import the requisite quantities of their grain from the Baltic states. In effect, they were using their lands to produce high-valued crops while importing crops of lesser value. Intensive dairying, which had developed in Friesland in the sixteenth century, became increasingly important in the seventeenth, so that the production of fodder crops increased significantly. Van Bath (1963) suggested that the most important contribution of the Netherlands to the improvement of European agriculture was their developments in the production and use of fodder (vetches, peas, spurrey, turnips, rapes, clover). Farmers also began to produce tobacco, as well as barley and hops for the brewing industry. During the sixteenth century, the Dutch developed sophisticated techniques for reclaiming wetlands thereby increasing the acreage of farmland.

England

As noted in a preceding section, agriculture was essentially the same throughout most of Europe until the middle of the seventeenth century; however, near the end of the sixteenth century, some changes began to take place in English agriculture that would have worldwide implications.

The importance of trade and commercialism and the growing world economy in the development of agriculture was noted in a preceding section. Nowhere did this increasing emphasis on commercialism and trade affect agriculture more than in England. Elements of commercialism had appeared in English agriculture with the beginning of the enclosure movement in the thirteenth century, and farming for the market became increasingly important in the succeeding years. However, the English never abandoned their concern for self-sufficiency, as the Netherlands had done. It was also apparent in the large estate or plantation-like agriculture that was common throughout the country. Most of the agricultural production throughout England took place on large plantation-like farms or estates. Of course, this situation was centuries old, but what was new about it was that the owners were beginning to be concerned about profit and loss and about maximizing returns on investment. Instead of serfs or peasants, the farms were operated by tenants, subtenants, and wage laborers. There were relatively few independent farmers operating small holdings.

Another indication of the effect of commercialization on English agriculture was the effect it had on year-to-year farm management. Farmers there practiced a mixed agriculture where they produced primarily wheat, grass, and livestock. Several combinations of these crops were grown depending on the condition of their land, the place of livestock in their management systems, and the state of the market. For example, they might grow wheat for the town market and keep animals chiefly for the sake of maintaining fertility in the fields through their manuring. As an alternative, they might grow much the same acreage in crops, but use most of the produce for feeding animals for the butcher. They also might fatten for the butcher by feeding the animal mostly grass, while growing a smaller amount of wheat. All of these combinations, required orderly and usually extensive tillage as well as close control over animals. Farm animals were either penned or restrained in various ways. All animals were to be kept secure at night, so they would not damage crops in the fields. Village bylaws also required that pigs must have rings in their snouts to retard them from excessive rooting.

The increasing amount of land devoted to sheep growing for the production of wool is another indication of the growing commercialization of English farms. English wool had been a commodity of international trade since before the time of Chaucer (1340-1400), but production expanded rapidly as world trade expanded in the latter part of the sixteenth century while Elizabeth was queen. Also, the severe economic depression of the latter part of the sixteenth and the first half of the seventeenth centuries, forced landowners to shift from the production of crops that had fewer buyers, to the production of wool for export. Unfortunately, the ever increasing number of hungry looms of the cloth weavers demanded more and more raw wool. As a result, sheep-raisers began to encroach on land that had long been used in crop production. More and more cropland was lost to enclosure. Overton (1996) notes that some 24% of the enclosure of English land took place in the seventeenth century—almost twice the percentage enclosed in either the sixteenth or eighteenth centuries. The loss of cropland threw many small farmers and farm laborers out of work. Many of these joined the throngs suffering through this period of economic hardship and who wandered about the countryside as vagabonds and sturdy beggars. Later, many of these dispossessed farm laborers would allow themselves to become indentured servants in order to go to America to acquire farms of their own (Wright, 1947). Several of these characteristics of commercial agriculture would accompany these farmers when they emigrated to America around the middle of the seventeenth century.

All of English agriculture would finally have a major impact on world food production. It would be the farmers in the South and West of England, however, who are of special interest to Alabama, because this region furnished a major proportion of the emigrants who established and maintained the colony of Virginia beginning in 1607. Over the next century and a half, they would furnish many farmers for the Carolinas, Georgia, and finally for the Old Southwest in Alabama and Mississippi.

In the early seventeenth century, the landscape of South and West England was shaggy and heavily wooded. A large amount of the land was in royal forests that were governed by forest law, a judicial system of exceptional rigor. Most of this royal forestland was kept as parks and chases for the sport of country gentlemen. A majority of the population in this region lived in scattered manorial settlements, each of which consisted of a large manor house. These estates provided a home for the lord and his family, plus some servants. The manor house was surrounded by clusters of small cottages, inhabited by tenants and subtenants. The classic manorial system of medieval Europe died very slowly in the South and West of England. By the time of the American colonization, both slavery and serfdom were long gone from this region, but other forms of social obligation remained very strong. A smaller part of the population were freeholders here than in East Anglia, which furnished most of the settlers for the Massachusetts Colony.

The region of the South and West of England was marked by deep and pervasive inequalities, a staple agriculture, and rural settlements. It was ruled by powerful oligarchies of landowners; its rural economy, like most of the country, was deeply depressed in the 1640s and 1650s. From 1642 to 1666, the Four Horsemen of the Apocalypse rode freely through this troubled countryside. It suffered much from the English wars of the mid-seventeenth century. The woolen trade was disrupted, and once again, major epidemics of the plague appeared. Social anarchy became a serious problem. These grave conditions gave rise to the mass migration of people from the region to Virginia, many as indentured servants. Most of them had belonged to families of poor tenant farmers and country laborers—a degraded rural proletariat.

The period of 1650 to 1700 has been called the period of agricultural incubation in England. In the English Civil War during the mid-seventeenth century, many of the wealthy supporters of the Royalists cause took refuge in the Netherlands when the Parlimentarians triumphed. While there, they had a good opportunity to observe the new developments taking place in agriculture. When they returned home, they carried with them the details of the high intensity crop production systems that they had observed. As a result, England was the first country to completely follow the Flemish model for growing fodder crops. However, the new system was not sufficiently developed to spread widely at home or abroad until after the mid-eighteenth century. This Agricultural Revolution saw the decline of legume rotations and probably field-grass husbandry. At the same time, the ancient medieval fallowing-out system completely disappeared from the country. All of these changes were incorporated into the Norfolk, four-course rotation, which was at the center of the revolution. Although it was thoroughly English in its basic character, it was constructed out of Continental materials: Spanish clover, Burgundian and French grasses, the Dutch plow, the horse-hoe of Languedoc, and the Flemish method of cultivating turnips in fields. After the middle of the eighteenth century, English agriculture advanced rapidly and by the middle of the following century, it led the world. From England the New Husbandry conquered the world by books and writings produced by English farmers.

The English Agricultural Revolution had far reaching consequences. The world would never be the same. The new farming used less land, fewer laborers, produced more and cheaper food, and was extremely profitable. With this vastly improved, advanced agricultural system in place, England quickly moved into the next stage of Lenski and Parson's (1998), societal evolution—the industrial stage. By the end of the eighteenth century, English factories were producing immense quantities of a wide variety of consumer and industrial goods. The English Industrial Revolution produced a large number of extremely wealthy factory owners, and an even larger number of persons in the industrial middle class, (bankers, lawyers, teachers, physicians, government officials, master craftsmen, and shopkeepers) (Spielvogle, 1994), but an equally important byproduct of this and the earlier Agricultural Revolution were the thousands upon thousands of displaced agricultural workers turned industrial workers, the new working class. These industrial workers were chewed up and spit out with little regard for their humanity. They were as expendable as the coal that powered the factories. Their pathetic living conditions were matched by their equally wretched working conditions. This deplorable collection of human misery would ultimately result in the development of an opposing force (socialism) that would have world-wide implications ("to every action there is an equal and opposite reaction," Isaac Newton's Laws of Motion).

The Celtic Fringe

Celts had been open-range pastoralists since antiquity. In the Celtic Fringe of the British Isles (Wales, Scotland, and Ireland), crop tillage was minimal, and open-range herding was customary. In early Ireland, for example, forests abounded, and animals were turned out to feed on mast or anything else they could find. An early report to the English government commented, "The Irish do not dwell together in any orderly form, but wander with their cattle all the summer in the mountains and all winter in the woods." Every corn (wheat) field had to be provided with fencing to keep the free-roaming animals from entering and destroying the crop. Scottish Highlanders opposed enclosures for their livestock. One observer commented that the whole pasturage of the country became a cattle pen in the winter and spring. A visitor to the Highlands commented that local farmers viewed enclosing their animals as a much more expensive kind of husbandry than simply letting them graze free in the hills.

As soon as the grain was harvested, cattle were allowed to roam through the fields. Then in the winter when there was no forage left in the pasturage, both horses and cattle were fed straw and boiled chaff. The herdsmen generally made no other provision for them. As a result of the use of these poor quality feeds, the animals were so starved at the end of the winter that they had difficulty standing. Many of them were mere skeletons that had to be lifted to their feet to get them to spring pasture. In many instances, the animals were bled copiously because their owners thought that it would improve their condition. Also, cattle blood cooked together with oat meal was an important part of the diet of common people during the hard winter months.

Cattle were an important source of income for the people of the Celtic Fringe. They drove them long distances to market. Overton (1996) comments that cattle from Wales were driven to England as early as the fifteenth century where they were dispersed through the Southern Counties. From the mid-seventeenth century, cattle were being driven to England from Scotland, mainly Galloway in the southern part of the Scottish Lowlands; although some were driven all the way from the Highlands. Considerable numbers of cattle were also imported from Ireland until 1667 when the trade was banned.

As late as the beginning of the seventeenth century, agricultural methods in the Celtic Fringe were extremely primitive. The people were utterly ignorant of developments elsewhere that had come to agriculture since the Early Middle Ages. One of the chief reasons, and possibly the primary cause for the continued backwardness of agriculture, was the insecurity of life and property from 1400 to 1600. For that entire period, there were very few years when there were not wars either with enemies from without or among themselves. Banditry and cattle stealing were widespread.

Throughout the North of England, the Scottish Lowlands, and Northern Ireland a large part of the best land was owned by a small number of people—many of whom were absentee landlords. The land and farms were essentially managed under the ancient manorial system that had been used throughout much of Europe for a thousand years. Fields were divided into strips and were assigned by lot. Assignments of strips were changed periodically. As a result of changing leases, people did not build good houses although there was ample stone available. Neither did thy plant hedges for their orchards or manure their land. The strip system of land assignment not only discouraged permanent improvements on the land, but it also resulted in constant bickering about management.

Strips nearest to their house was called the "infield," and most of the care was given to it. All manuring, when they bothered to do it, was done there. When they used manure, it often resulted in a rank growth of weeds that necessitated the constant cultivation of the oats and barley usually grown there. The grain sown as seed was the poorest and least prolific kind. It had long since been abandoned throughout the remainder of Europe. A crop, once planted, often received little or no care, so that it might easily degenerate into empty chaff or straw. Crop cultivation, when it was done, was primitive. It was mainly a hoe-husbandry with some use of the light plow. Crop yields were low. It was not uncommon for men renting from 40 to 100 acres to purchase grain for their families.

Strips located away from the house were called "outfield." A portion was sown in oats for three years in succession or until production diminished so that yields were reduced to three seeds for each one planted. Then the strip was fallowed for up to six years. Strips generally ran vertically down the slopes, which contributed to severe erosion.

Oaten cakes or oatmeal, greens or pot herbs from the yard, and beer or ale were the steady diet year in and year out. Milk could be used only sparingly because the ill-fed cows produced so little. Pork was not eaten. Visitors to Scotland commented on the complete absence of wheat bread.

In Scotland, food and fiber production were generally poorer than in the other Celtic countries. Mountainous terrain, poor soils, and severe agricultural climate made agriculture a very tenuous undertaking even in the best years. Some 60% of the entire country is covered by mountains, moors, and hills. The Highlands alone in the north and west of the country comprise some one-half of the total area. This mountainous area is underlain by schist and granite. These erosion-resistant, igneous rocks are covered by a thin soil. The average elevation of the whole region as 1,500 feet above sea level.

Handley (1953) commented that in all of Europe, it would be difficult to find a territory of equal area that offered so much evidence of poverty, misery, and barrenness as did Scotland in the early eighteenth century. Of its 19 million acres of land, only some five million were considered to be suitable for cultivation. The remainder consisted of "mountain, moor, shaggy heath and watery waste."

Great stretches of Scotland (the Highlands) were regions of natural poverty, and food production over much of the remainder of the country was acceptable only when the very best agricultural practices were used. The food famines of the seven hungry years during the period, 1693-1702 have already been mentioned, but serious scarcities were common as far back as the middle of the sixteenth century, and they continued until near the end of the eighteenth. Food shortages in the following years were especially noteworthy:1568, 1574, 1577, 1586, 1595-1596, 1598, 1600, 1612, 1616, 1622-1623, 1634-1635, 1642-1643, 1649, 1651, 1655, 1680, 1688, 1740, 1756, 1766, 1778, and 1782-1783 (Handley, 1953).

The agriculture of the Scottish settlers in Ulster varied somewhat from what they had practiced in their Scottish homeland. The most important change resulted from the fact that the old manorial system of agriculture, which had largely controlled their lives in Scotland, did not follow them to Ireland. In Ulster, they secured long-term leases to the land that they farmed. Also, the soil was somewhat better in Northern Ireland than it had been in the hilly terrain of Scotland. The differences in soils between Ulster and the Scottish Highlands were especially great. Also, the climate was somewhat more conducive to crop production. Virtually all of Scotland lies north of 55° North Latitude. Most of Ulster lies south of that line. Further, Ulster lies somewhat closer to the Gulf Stream and its mediating effect on the harsh Northern European climate.

Two discoveries made by the Scots greatly assisted their enterprise and made farming in Ulster much more satisfactory than it had been in Scotland. First, they learned from English tenants the art of draining bogs and swamps so that these areas could be planted. They also began to plant the potato, which gave them a new and healthy staple of diet.

The Scots also learned that sheep would flourish on the Ulster meadows. From the beginning, the settlers made woolen cloth from the wool they produced. Later, they began to market cloth produced in their homes. This cottage industry grew steadily until these woolen goods were in demand throughout the European world.

In 1689, King William III (1689-1702) invited a colony of French Huguenots to settle in Ulster. There they stimulated the widespread cultivation of flax and initiated the large-scale manufacture of linen. They encouraged the development of the industry on a domestic basis, with piecework distributed widely through the homes of the region, providing employment for farm women.

Germany

In contrast to the Celtic peoples who never seemed to develop a strong agricultural tradition, McNeill (1963) comments that German tribesman began to settle down to a more intensive agriculture in the Rhine River Valley as early as AD 100. He further notes that the success of Clovis (481-511), the Merovingian king, in unifying the various tribes of Germanic peoples was a result of the strong base of support that he had in the well-established agriculture of the Rhineland.

Throughout the Middle Ages, German agriculture was essentially a mirror image of manorial agriculture widely found throughout Europe, and the general evolution that took place there during that period was similar to that which took place throughout the Continent. One striking difference was the emphasis of the Germans on forestry and the production of forest products. It has been suggested that Germans of the Middle Ages inherited from ancient ancestors a mystical concern for trees, and that this intense feeling was translated into what was probably the first efforts in forest management, certainly in Europe and probably over much of the rest of the world. Forest culture was important and was regulated according to well established rules. Dating from the mid-fifteenth century, there were many forest laws in place in parts of the country. Special attention was given to regulations for thinning the forests. Janssen (1966) comments that the cultivation of trees in Germany in the sixteenth century left little for modern times to improve. By this time, wood production was an area of agricultural specialization. Wise management of the forests was a unique concern of the Germans. Elsewhere in Europe, there was widespread destruction of woodlands, as human settlements spread far and wide. Ponting (1991) comments that forests originally covered about 95% of the land area of Western and Central Europe. By the end of medieval colonization, however, this large expanse of woodlands had been reduced to some 20%.

Throughout the Middle Ages, agriculture and forestry were overwhelmingly dominant sectors of the German economy and of sources of revenue for governments, the church, and nobility. A fundamental problem was the inflexibility of agriculture. Yields of agriculture were very low, and it was very difficult to convert from subsistence farming to the production for the market. In the latter part of the sixteenth century, the inevitable crisis caused by population growth combined with the built-in limits of agricultural production had arrived. Germany, especially in the western part, began to experience the problem of overpopulation and food shortage. Pauperism and vagabondage were increasing before 1618, the beginning of the Thirty Years' War.

Parts of Western Germany, especially the Rhineland and Palatinate, had benefited directly from agricultural developments taking place in the latter part of the sixteenth and early part of the seventeenth centuries in the Netherlands because of the close proximity of the two areas. Any technology developed in the Netherlands would quickly be adopted in the nearby German states. As a consequence, farm productivity advanced rapidly in the Rhineland and the Palatinate during this period; however, although they did improve their agriculture significantly, the primary emphasis remained on the production of grain.

Unfortunately, any progress that had been made in German agriculture over the years came to a halt and was even reversed in the Thirty Years' War (1618-1648). The Rhineland and Palitinate were especially hard hit. Many of the battles in which Spain fought France and her agents took place in the valley of the Rhine. Armies spoiled the fields and damaged growing crops beyond recovery. They burned houses and took stored grain and livestock for their own use. When the war was finally ended with the Treaty of Westphalia, the Rhine Valley including the Rhineland and Palatinate lay devastated and ruined. Only the poorest and most remote mountain villages had been spared. Before the war, much of German agriculture was experiencing serious troubles because of falling prices. The war simply intensified those problems. Landlords pressured tenants until they were near starvation. The several wars of succession that followed and that continued for over a century made recovery difficult. In 1658, the prices for cereal were so low that farmers had to use it for barter or give half of it away. The loss of animals that occurred continued to plague German agriculture into the beginning of the eighteenth century. During the War of Spanish Succession (1701-1714), armies fought throughout the Valley of the Rhine. There was a serious subsistence crisis from 1708 to 1712.

Because of the lack of a German central government and almost continuous warfare, Germany did not play a very important role in the expansion of world trade that was taking place in the seventeenth century. As late as 1770, German agriculture as a whole was still very inefficient, because the great majority of those engaged in it were subsistence farmers and were little concerned with farming for profit or for the creation of capital. The backwardness of agriculture was a major barrier to modernization. Rural economic individualism was very rare, as there were few incentives or opportunities to improve cultivation even when markets were available.

German agriculture was inefficient only in terms of profitability or capital formation. The returns on investment were relatively low, but the Germans, in the absence of war, were very good general or subsistence farmers, even if they did not specialize in the production of crops that brought high prices in international markets. Janssen (1965) notes that in the summer, cattle were put out to graze, and the herdsmen were instructed to be diligent in preventing them from damaging crops. In winter, the cattle were housed where the herdsmen gave them fodder to eat and straw to sleep on. Owing to the winter housing and feeding of cattle, there was always an abundance of manure available. Particular attention was given to the care of meadows and to the production of hay. There were large expanses of vineyards, and the vines were tended with skill and care. McWhinney (1988), reviewing notes of an English visitor to Ireland in the eighteenth century regarding German settlers who had migrated to Ireland, suggested that they were excellent farmers who were very industrious. They were better fed, clothed, and lodged than the Irish peasants. They tilled more land than the Irish. They drilled their potatoes, and housed their cattle, feeding them with hay. They were cleaner and neater and lived much better. All of them had neat little kitchen gardens. These same characteristics would be apparent in the German farmers who migrated to America in the first half of the eighteenth century.

Development of Amerindian Agriculture

Agriculture probably first originated in the eastern woodlands of North America around 4500 BC when Indians there began to actively domesticate plants that they encountered in their hunting and gathering activities. Agriculture evolved very slowly on the continent over the next 6,000 years. However, when the Europeans began to establish permanent colonies in the early part of the seventeenth century, the pace of evolution increased considerably.

Amerindian plant domestication and rudimentary agriculture began in the eastern woodlands within the time period (Figures 5.1 and 5.2) covered by the Archaic Tradition that began to take shape around 8000 BC and lasted until around 1000 BC (Terrell, 1971 and Hudson, 1976). Sometime around the middle of this long period of time, the Indians in Eastern North America apparently began domestication by cultivation of some of the native seed plants that they been harvesting from their environment for thousands of years. Toward the end of the period of the Archaic Tradition, around 2,000 BC, technology for the culture of the bottle gourd and squash slowly filtered into Eastern North America from northern Mexico where the plants had been domesticated much earlier.

Around 1000 BC, the Archaic Tradition began to slowly give way to the Woodland Tradition that was to last until around AD 700. Around 200 BC, technology for the culture of tropical flint corn also arrived in eastern North America from the Southwest. Apparently, the culture of this corn variety, in turn, disappeared here around AD 400, and when corn was cultured again (around AD 800), it was with a different variety (eastern flint)—a variety better adapted to the cooler, more moist environment of the woodlands of eastern North America. Technology required for the cultivation of both eastern flint corn and the common bean, arrived at about the same time.

It was during the Woodland Tradition that the Indians began to show a preference for building their permanent houses, villages, and even towns on higher ground near the flood plains of large creeks and rivers. Flood plains were the primary habitats for many of the seed plants that served as sources of food. The soils in flood plains were generally more sandy than the soils on the uplands;

consequently, they were much easier to work with their primitive tools. Also, drought was generally less of a problem there, although this advantage was offset to a degree by the danger from flooding. Generally, they established smaller, private home gardens around their houses and around the village, but their larger fields were established in the flood plains of nearby streams. The large fields consisted of a number of private plots, separated by a narrow strip of untilled soil.

Fields were cleared by burning the underbrush and girdling the larger trees. Their tools were poorly suited to cutting trees down or cutting them into pieces once they were on the ground. Before the arrival of the Europeans with their metal tools, the Indians had to use stone axes for cutting away the bark in the girdling process. The deadened trees were burned once they fell, and the ashes used as fertilizer. They worked the soil with digging sticks and with hoes made by attaching a large sea shell, a flat piece of flint, or the shoulder blade of a deer to a short wooden handle or haft.

They planted their crops on small mounds of soil among the standing stumps. Each mound was a foot or more in diameter and six to eight inches high. Individual mounds were three to four feet apart. They were generally laid out in straight lines, in both directions. This regular pattern facilitated weeding. In each mound, farmers made a cluster of small holes. Seeds, which had been soaked for a day to hasten germination, were placed in them. Quite often they planted corn and beans together so that the vines of the bean plant could climb up the corn stalk. Between the mounds of corn and beans and around the edges of the field, they planted gourds, squash, pumpkins and sunflowers. Over time, varieties of corn, which required different amounts of time to mature, became available. They planted the early-maturing variety in their home gardens. When mature, it would be eaten green. A second crop of a later-maturing variety of corn or other vegetables was planted in the home gardens after the first crop of corn was harvested. Because of the danger of flooding in the large floodplain fields, corn was planted there later, using seed of a late-maturing variety. It would be allowed to air dry in the field before harvest. This dried corn was then stored for winter use.

As the corn grew during the summer, periodically the farmers would remove, by hand, any weeds growing on the mounds and would use their hoes to cut the weeds around them. Weeds would have been extremely abundant in those moist flood plain soils. Also, the Indian farmers would periodically strew loose soil on the individual mounds. These added layers of soil helped to control weeds germinating there.

The private home gardens were planted and cultivated by the women in the family. The larger fields were planted and cultivated by both sexes, and the labor was communal. Everyone worked on all of the individual plots; however, harvesting was the responsibility of the plot owner.

Agriculture Among the Five Civilized Tribes

The Creek, Cherokee, Choctaw, and Chickasaw Indians shared the agricultural technology used by all of the tribes of the Southeast. In some cases, however, there were subtle differences related to the cultures of these four specific groups.

Creeks

Creek Indian food and fiber production was typical of the flood-plain agriculture being practiced by most of the Southeastern Indians when the first Europeans arrived in Alabama. Although agriculture played a significant role in the lives of the Creeks, they continued to spend a considerable amount of effort hunting and gathering. Hunting white-tail deer and wild turkey were important, as was the trapping fish in streams. Hudson (1976) commented that the Creeks burned the leaves on the hillsides to make it easier to harvest chestnuts. Among the Creeks, there was no such thing as private ownership of land. However, individual families did own the food and fiber that they produced, but they were required to contribute a certain amount of their produce to a reserve for public use.

Cherokees

The Cherokees shared the agricultural technology of most of the Indians of the Southeast. Cultivation of corn, beans, and squash was important to them. Agriculture, however, only mildly influenced the fundamental nature of their cultural and social organizations. Hunting shaped them.

Choctaws

Choctaws hunted and fished, but were also known as a farming people. They raised beans, squash, melons, and pumpkins, as well as corn. Griffith (1972) commented that the Choctaws were the best farmers among those Indians who inhabited Alabama. Once they were exposed to European farming methods, they quickly adopted many of them. They planted European fruits and vegetables, cultivated cotton for spinning and weaving, and raised chickens. Of all of the Indians of the Southeast, the Choctaws were the only ones who produced a surplus of food that could be sold or traded.

Chickasaw

The Chickasaws were the fierce independents of the Southeastern Indians (Terrell, 1971). They disdained agriculture. They were hunters and warriors, not hoe farmers. They demanded complete freedom to roam and hunt at will through the trackless woodlands and prairies of the Mid-South. Although they disdained farming, they did have to produce a minimal quantity of beans, corn, and squash, using the same technology employed by the other Southeastern Indians.

Development of Agriculture in Africa

Harlan (1992 and 1995) has provided a considerable amount of information on the development of early African agricultural traditions. Nevertheless, he comments that studies on the evolution of indigenous agriculture in Africa have been seriously neglected, and that they are inadequate to provide more than a tenuous outline of events. Endemic farming systems that evolved there supported extensive village populations with local markets and important early regional trading centers in Benin, Ghana, and Mali. These agricultural economies were based on a wide variety of crops, including cereals, grain legumes, vegetables, oil seed crops, fibers, drugs, narcotics, fish poisons, as well as those used for magic and rituals.

Early African hunter-gatherer societies established settlements on the edges of lakes and around marshes where they fished, hunted savannah herd animals including wild cattle, and harvested seeds of wild grasses and other plants. The harvesting of wild grass seeds was important to early African peoples, and the practice continues there today. Apparently, there was no center where crop production evolved. Activities related to plant domestication and the development of agriculture seem to have been widely distributed all across Sub-Saharan Africa from the Atlantic to the Indian Ocean. Indigenous African agriculture evolved around the use of the hoe, rather than the plow.

Native African crops of the Sub-Sahara were typically savannah plants.The most important were pearl millet, sorghum, and African rice, all drought-resistant plants that are native to the savannahs. These same crops are still important sources of food for millions of people across Africa and beyond. Cowpeas were another ancient African crop. These highly variable plants were important sources of protein for early subsistence farmers. Other important crops of African origin include: coffee, sesame, watermelons, oil palm, and castor oil. During the latter part of the fifteenth century when the Portuguese began trading, both in Africa and South America, maize was introduced into the Sub-Saharan region. Later, wheat and Asian rice were also introduced. These two importations became major agricultural crops.

For several thousand years, animal agriculture has been extremely important to people in the Sub-Sahara; however, there is considerable disagreement on whether or not animals used in the agricultural economies were domesticated there or were introduced from the Nile Valley in the northeast. Domesticated sheep, goats, and cattle had been introduced into the great valley from the Fertile Crescent some 6,500 years ago. Apparently, the pig was introduced later. The domestic chicken was introduced much later. One bird, the guinea fowl, was domesticated in Sub-Saharan Africa and included in household farming economies, but when the domesticated chicken was introduced to the region, it became much more popular than the guinea fowl.

Early Agriculture in the European Colonies of North America

As detailed in a preceding section, the English Agricultural Revolution did not begin until around 1750. As a result, the agricultural traditions that the first settlers in both the Southern and Northern colonies brought with them to the New World in the early seventeenth century were little different from the agricultural practices that their ancestors in England had been using for some 200 years, at least. Farmers in the Northern colonies continued to follow those traditions until after the American Revolution, and even after the methods of English Agricultural Revolution reached them in the latter part of the eighteenth century, most of them were reluctant to abandon their traditional agricultural practices. Some farmers in the Southern colonies also continued to follow the old traditions, but for most of them, tradition was quickly abandoned when they learned how to grow and cure tobacco. After 1613, most of the farmers in Virginia began to march to the tune of a different drummer: staple or export agriculture.

Early Problems

Several problems affected the development of agriculture during the early years for all of the European colonies in North America. These include: inadequate food supplies, profit-oriented development, lack of experience in colonization, differences in climate, inexperienced farmers, massive forests, Indian hostility, land policy, and mercantilism (Gray, 1958). Each of these problems will be described in the following paragraphs.

Inadequate Food Supplies

The first Spanish colony established in Florida in the 1560s had to be maintained on food supplies obtained from the Spanish colony in Cuba. Unfortunately those supplies were not always adequate. For long periods of time, half-starved settlers had to subsist on fruits of the palmetto plant, fish, and other supplies precariously obtained from the Indians. The early English colonies fared no better, at least initially. Bidwell and Falconer (1941) comment that in the early years the food supply situation in the Northern colonies was pitiful. For at least three years after its founded, the Plymouth colony was very near starvation. They survived by rationing grain, purchasing grain and other foods from the Indians, and hunting and gathering fish, game, and other foods from their surroundings. In the Massachusetts Bay colony, as well as in Providence and Hartford, the scarcity of grain is referred to repeatedly in documents from that period, until around 1640. But scarcity was not completely eliminated even at that late date. As late as 1648, Massachusetts Bay had enough grain to meet the needs of it inhabitants for only two months. The Swedish colony in Delaware, founded in 1638, had to depend on outside sources of food until around 1658. By around 1680, the food supply situation had improved considerably. Most farmers were producing all of the food that their families needed, with some extra that could be bartered or sold. Penn's colony in Pennsylvania, established in the early 1680s, had few problems with food availability. In the early years, they were able to purchase all that they needed from other colonies and from pioneer settlers already living in the vicinity.

The food supply situation had been even worse in the Virginia colony (Gray, 1958). From 1607 to 1611, literally hundreds of people in Jamestown starved to death, and poor nutrition contributed to death from diseases for many more before local food production and shipments from England began to stabilize the situation. Only after 1627 was the food supply (production and storage) in the colony considered to be adequate.

The first South Carolina settlers arrived with enough provisions to last them for only about three months. Subsequently, food became so scarce that they were forced to purchase it from the Indians. Fortunately, they were able to obtain enough food that starvation, like that encountered by the Jamestown settlers, was averted. Settlers in this colony were not able to produce enough food to meet their own needs until the harvests of 1674. There was virtually no agricultural development of consequence in the Lower Mississippi Valley for more than two decades after the French established their first colony at Biloxi. For this entire period, there were chronic food shortages, and the settlers were largely dependent of supplies from abroad. However, this source was generally precarious during much of this period because warfare between France and England (1702-1713) made food shipments in the region unreliable. At intervals, food supplies became so limited that settlers had to rely on hunting and fishing, eating acorns, or maize obtained from the Indians.

Although food production and food supplies were problems in the early days of all the colonies in the South, the level of human misery was reduced considerably with the establishment of each new one. There was no widespread starvation after the early days at Jamestown, and by the time the Georgia colony was established, it was recognized that the uncertainties of food supply could be averted only by maintaining public storehouses. At this later time, there was also more food available in the region. Local shortages in Georgia, for example, could relatively quickly be alleviated by bringing supplies from the West Indies, Virginia, the Carolinas, Maryland, or New England. Also, throughout this 175-year period, there was continuing improvement in agricultural technology, as people learned what crops to plant and when and how to plant them. There was also continual improvement in the implements used in agriculture. Early food production in Jamestown was based on the hoe, but by the time the Georgia colony was established, plows and the necessary animals to pull them were generally available. However, for many years, the hoe remained an essential farm implement among the limited-resource farmers in the backcountry.

Profit-Oriented Development

Profit-oriented development dominated economic thought and practice in seventeenth and eighteenth century Europe, especially England and its empire. Subscribing to an economic theory now called mercantilism, major European governments attempted to maintain a balance of trade in which the value of goods exported fom the mother country were supposed to exceed the value of goods imported. A corollary was that wealth was defined as precious metals, so the goal was to import more gold than was exported. The practical application of mercantilism required governments to take an active roll in stimulating and protecting its export trade by fostering monopolies, founding colonies and establishing empires, and building a merchant marine and a powerful navy to protect them and their shipping lanes. Early European colonization paid a central role in this system. The English purchased raw materials cheaply from their colonies, and shipped them high-priced finished goods in return. The mother countries did not allow their colonies to be involved in manufacturing. Ideally, colonies would not trade outside the empire of their mother country. Under this arrangement, agriculture in the colonies was forced to develop in an artificial environment—in which the colonies provided the mother country with raw materials and the mother country provided the colonies with finished goods—where the normal progression from the simple to the advanced economic stage proceeded very slowly.

The English colonies in Maryland, Virginia, the Carolinas, and Georgia were financed primarily as commercial ventures by well-connected English investors who expected good returns on their investments. Establishing colonies was expensive, and those financing the venture expected that the profits would quickly accrue as the settlers began to send home cargoes of furs, fish, precious metals, timber, naval stores, medicines, dyewoods, and spices. Mistakes were made that resulted in much human deprivation and suffering. For example, in every colony, costly and generally futile efforts were made to produce commodities such as hemp, flax, silk, and wine that were supposed to bring good prices in England and that would increase the profits of the company stockholders. Emphasis on producing commercial, industrial, staple crops for the market economy detracted from the production of food.

Lack of Experience in Colonization

Lack of experience in colonization was a problem in the establishment of all of the colonies. For example, those who planned their establishment had little concern for agriculture. They expected the settlers to provide most of their own food from the very beginning. After all, food was generally plentiful in England, and they expected that food production would come easy in the new land as well. Further, they expected that only a relatively few farmers would be required to supply all of the settlements' needs. The remainder would quickly go about the business of making the investors rich. These basic problems were compounded, especially in the South, by the practice of sending new settlers to the colonies without adequate advance notice and without sending adequate food supplies along with them. The existing food supply was barely adequate for those already there, and certainly not adequate to meet the needs of a hundred or so additional mouths.

Another significant problem related to the inexperience of the Virginia Company arose from the fact that the settlers owned no land. They had no economic stake in the colonization process. They worked for the Virginia Company in an almost quasi-military environment. Twice each day they were marched to the fields or woods to the beat of a drum and twice marched back to church (Morison, 1965). They led an almost hopeless existence. There seemed to be no future except death by starvation, disease, or Indian attack.

The fact that the expectations of the planners were unrealistic was recognized very slowly and at great expense. This short-sightedness carried with it an enormous price in human suffering and death (Sauer,1941). In the first few years of Virginia, the survival of the colony was continually in doubt. In the first winter, 67 of the original 105 settlers perished of disease and malnutrition. In October of 1609, after the arrival of more settlers, there were an estimated 500 people in the colony. The following May, only about 60 remained, and they were subsisting on roots, herbs, acorns, walnuts, snakes, with an occasional "little fish." Their food situation remained extremely precarious until 1610.

Differences in Climate

Differences in climate between England and the colonies initially had a significant effect on efforts to produce food. Traditional English food production ways were tried in the colonies and usually found wanting because of differences in climate. Massachusetts winters were much longer and much colder than English winters, there were long period with snow on the ground. and the summer growing season was short. In Virginia, the summers were much more humid than in England. The sun was brighter and hotter, and the rain was more tempestuous. Winters were colder, but shorter, than in England. These differences were even greater southward along the Atlantic coast. Consequently, food production from the beginning was highly unpredictable, so new crops and new ways of growing had to be discovered and tried quickly—but the colonists had little time for experimentation.

The colder winters, at first, resulted in considerable misery for the settlers. They had significant difficulty in providing for changes in their shelter and clothing needs.

Inexperienced Farmers

Inexperienced farmers compounded the early problems of food production in the colonies. The men who first farmed in the New World were most likely to be amateurs: "all sorts of people turn farmers . . . no mechanic or artisan-sailor-soldier-servant, etc., but what if they get money, take land and turn farmer" (Boorstin, 1958). Gray (1958) commented that few of the settlers were experienced farmers. Some were gentlemen unaccustomed to hard labor, while others were adventurers seeking an easy road to wealth. Also included were religious dissenters, tradesmen seeking new opportunities, prisoners taken from English jails, and poor children picked off the streets—all indentured as servants. Morison (1965) commented that the first settlers were poorly prepared for the daunting task before them. Many of them were simply not capable of hard work, and the remainder were unwilling to do so.

Massive Forests

The massive forests of large, mature trees were almost overwhelming to the first settlers who needed to clear fields to plant their European crops. They were awed by the wonderful wealth of trees along the Atlantic coast, and the trade potential that they contained. They were completely unprepared, however, for the task of converting these vast forests into enough fields to feed even their small numbers. By that time in history, virtually all of the forests in the South of England, the homeland of most of the settlers, had been cleared; consequently, they had no experience with land clearing. They also lacked the labor required to fell the trees and remove the stumps. It did not take long to adopt the methods used by the Indians. They simply girdled the trees to kill them, and left them standing until they rotted and fell. Unfortunately, even this process was too slow to meet their needs for fields in the early days of settlement. As early as 1610, the Virginia settlers began to seize fields of the Indians, and in Massachusetts settlers found numerous open areas that had apparently once been cultivated by the Indians.

Indian Hostility

Indian hostility was a constant worry and often a

serious threat to the early colonists. Initially, contacts between the Europeans and the Native Americans in the colonies along the Atlantic Seaboard were beneficial to the settlers. The Indians taught the settlers many important lessons about survival in a wilderness, and on many occasions, the food they provided helped to avert certain starvation. However, over time, when it became apparent that the English had come to stay and that they expected to expand their settlements at Indian expense, the Native Americans reacted violently. In 1622, Indian attacks on farms and plantations all along the James River nearly destroyed the Virginia colony. More that three hundred men, women, and children were slain, and hundreds of others were forced to leave their homes and move to the fortifications at Jamestown. In 1715, the Yemassee Indians in South Carolina attacked and slaughtered traders and settlers all along the outer fringes of the colony. The onslaught was so ferocious that its very existence was severely threatened, as hundreds of settlers abandoned their farms in the interior and flocked to Charleston. In 1729, the Natchez tribe massacred some 250 French settlers at Fort Rosalie on the Mississippi River. This tragedy sent shock waves all the way to Paris and discouraged immigration to the Lower Mississippi Valley.

As destructive as these major Indian attacks were, most came relatively late in the colonization process, after the central core of settlements had been established; consequently, there was no long-lasting damage. Further, by the time these attacks took place, there were sufficient military forces available in the region to counter them, and usually retaliation against the Indians was relatively swift and far-reaching. In each case, long-term Indian suffering and hardship resulting from the retribution exacted by the colonists far exceeded the damage the Indians had inflicted. The most widespread effect that the presence of the Indians had on food production was largely an indirect one. The psychological effect of the continuing threat of Indian attack, in the early days of each of the Southern colonies, seriously reduced the enthusiasm and willingness of settlers to venture into the fields to plant, weed, and harvest their crops. Further, the building and maintenance of fortifications and the maintenance of military readiness exacted a high price in labor and other resources that could have been used to produce food adn infrastructure.

In the Northern colonies, relations with the Native Americans were positive in the years immediately after colonization; however, not many years passed before the situation changed. By 1637, deteriorating relations resulted in open warfare. In that year, members of the Pequot tribe murdered an English trader in Connecticut. Quickly, colonists organized a military force to punish the Indians. They attacked and burned the villages of the Pequots, and some 500 men, women, and children were massacred. Some were taken as captives and made slaves by the colonists themselves or sold as slaves in the West Indies.

Sporadic warfare regularly erupted in outlying areas of the expanding colonies when Indians would provoke the settlers with attacks on isolated farms or by stealing livestock. In 1675, colonists hanged several members of the Wampanoag tribe of Massachusetts. Although colonists generally had good relations with this tribe, some of their members had murdered a member of another tribe who had been extremely helpful to the colonists. In retaliation, colonists hanged the Wampanoag. As a result, the Indians went on a rampage in what has been called King Philip's War. King Phillip (Metacon) was the chief of the tribe. The war began with an attack on Swansea, a frontier settlement on Naragansett Bay, Massachusetts. Soon attacks were leveled across the frontier. By the fall of that year, the Indians had wiped out all of the westernmost settlements of the Massachusetts Bay and Plymouth settlements. At around the same time, the Abneki tribe forced the evacuation of every English settlement in Maine. By 1678, these wars had run their course, and the colonists had restored peace throughout the region. However, the destroyed and abandoned settlements were not completely restored until about 20 years later, and the frontier did not begin to expand westward again for some 40 years (Morison, 1972).

As settlement in the Northern colonies extended westward, pioneers began to encounter French soldiers and their powerful Indian allies. At this juncture, Indian hostility became much better organized and more serious. In 1701, the French organized a conference (the

Great Peace), involving some 30 Indian tribes. During the conference, the Indians agreed to join the French in fighting further British expansion westward. In 1704, a combined force of French and Indians attacked the small outlying English settlement of Deerfield, Massachusetts. In the battle, 56 men, women, and children were killed, and another 109 were taken as captives. While this massacre slowed westward expansion for a time, it did not stop it. Northern colonists continued to move westward toward the Ohio Valley.

In 1752, Louis XV sent the Marquis Duquesne to New France as governor general with specific instructions to take possession of the Ohio Valley and to remove all British presence from the region. Following a series of minor battles over a period of years, the French established a fort (Fort Duquesne), near present-day Pittsburg, at the point where the Allegheny and Monongahela rivers join to form the Ohio. In 1755, a large British force was dispatched to capture the fort. The force was badly mauled. After this decisive victory, the French sent large numbers of Indians as raiding parties deep into British territory, creating great panic on the frontier as far south as the Shenandoah Valley. These attacks on outlying farms and settlements were generally brutal affairs resulting in horrible mutilation of men, women, and children. This effort to drive the British settlers back to the Atlantic continued until 1758 when a large British force was again sent to dislodge the French from Fort Duquesne. This attack was also unsuccessful, but because the size and effectiveness of the British force made a strong impression on the Indian allies of the French forces, they changed their allegiance from the French to the British. Subsequently, their attacks on the frontier were reduced considerably. Later, the French were forced to abandon Fort Duquesne, and by 1763 they had lost all of their North American territory. Battles between settlers and Indians would continue in the North for many years, but the attacks would never again threaten to dislodge them from their western settlements.

Land Policies

Policies used for the disposition of land probably also played a negative role on the production of food and fiber and the orderly development of the new colonies, at least in the beginning. Land acquisition and ownership finally became very flexible in all of the colonies, but it was more difficult and complicated in the beginning. When preparations were being made to establish each of the colonies, no consideration was given to the fact that Native Americans had been occupying all of the territory in North America for thousands of years. English plans for colonization were based on the assumption that all of the land belonged to the king of England and that only he could dispose of it. After all, this was the basis for the land policy that had been in place in England for hundreds of years. Over time, English kings gave large quantities of land to relatives and supporters. These gifts would become the large hereditary estates that provided the land base for the manorial system described in a preceding section. Under this system, land was owned by an important family, and by the period of colonization, the land was worked by either employees or tenants of the landowner. Landownership in England had been the key to economic, social, and political privilege (Gray, 1958), and in the early planning for colonization, it was intended to establish a similar system in the New World. At least some of this old system was put in place in the colonies.

Although the first Southern colony was established over a decade before the Massachusetts colonies, the extension of old English traditions to Plymouth and its environs will be detailed first. This approach will provide a useful point of reference for considering the amazing events unfolding along the James River in Virginia. In New England, colonies were established by trading companies that had received large grants of land from the English Crown. Very few grants were made directly to individuals. The trading companies, in turn, provided for the establishment of local governments that were given the responsibility of disposing of the land in their grants. The primary characteristics of land disposition in this region were based on community action and the absence of a profit motive (Edwards, 1940).

Under the disposition system that evolved in New England when an existing settlement became crowded, individuals wishing to move to a new area applied as a group to the local governing body for a grant of land in an adjacent uninhabited area. The governing body then appointed a committee of citizens in the existing settlement to evaluate the request. The committee evaluated the suitability of the location of the new settlement and the qualifications of the leaders of the group to establish and govern it. If the report of the committee was favorable, the grant of land was made without charge.

The group moving to the new land grant, established a "quasi" corporation that had the responsibility of subdividing it. They had it roughly surveyed. Before land was assigned to individuals, the corporation or local government fixed the location of streets and roadways, the village green, the community meeting house, the minister's house, the school, the marketplace, the burial ground, a community common pasture, and house lots for individual families. Each house lot was provided with enough space for the dwelling, a door yard, out-buildings, a garden, and an enclosure for feeding livestock.

After the locations of the public areas were established, the remaining land was divided into strips for the use of individual families for food and fiber production. These strips were similar to those used by individual families for centuries on the large manors in England. Strips were assigned by lot with the provision being made that each family would receive both good and poor land.

Edwards (1940) comments that this system of land disposition, in which local people were involved in all decisions regarding land disposition, played an important role in the evolution of the unique form of democracy put in place in America after the American Revolution. Further, the system provided for secure land titles for individual families and a protection against inaccurate surveys. It also discouraged any form of land speculation. There was almost no land speculation in the New England colonies for almost 100 years after the establishment of the first settlements. Not until about 1725 did both Massachusetts and Connecticut began to sell large tracts of generally uninhabited land to individuals for resale. This new policy resulted in a period of wild land speculation.

The system of land disposition used in the New England colonies was not widely applied in the Middle Atlantic colonies. The system used there was more similar to the one being used in the Southern colonies.

In contrast to New England, as late as the end of the colonial period, virtually all of the land in the Southern colonies beyond the coastal settlement belonged to the English king and about a half dozen English noblemen. However, the manorial system never worked in the colonies, as there was simply too much land to manage effectively under that system, and there were too few tenants available, at least initially, to work the estates. Therefore, the old English manorial system quickly gave-way to the plantation system in all of the Southern colonies, and slaves replaced the tenants of the manors. Although plantations were numerous in all of the colonies, they never occupied more than a tiny fraction of the land available.

In establishing the Virginia colony, the trustees had decreed that there would be no private property; all the colonists would work for the Virginia Company. It became obvious quickly that this system would not work. Without landownership, the colonists had no economic future and hence had little incentive to work (Morison, 1965). This system was changed around 1615, and the company's trustees decreed that laborers who would remain in the colony would henceforth become tenants, and even though they did not own the land that they farmed, they would own what they produced. Later, this policy was changed again so that the land was given to them outright. This change provided the basis for the development of the so-called head-right system of distribution of the king's domain. Under this system, individuals who paid for their own passage from England to Virginia would be given a tract of 50 acres of land, their head-right. Families would be given 50 acres for each member if they paid for their passage. Further, a landowner would be given 50 acres for each immigrant for which they paid passage. This system later became the primary means of land distribution in all of the Southern colonies.

The head-right system worked extremely well from the beginning and was responsible for attracting thousands of settlers to the colonies. There was, however, no similar provision for land acquisition when succeeding generations moved into the backcountry and away from the head-right grants of their parents. Under those conditions, land had to be purchased from some owner; in many instances, individual owners operated their own head-right system to encourage individuals to settle on their usually large holdings. Many of the settlers going into the backcountry had little money to purchase land, so they simply squatted on someone else's property until forced to move. Squatting on vacant public land was a popular, and in frontier communities, an accepted way of acquiring a farm in the latter part of the colonial period.

The Northern Colonies

The agricultural traditions of their English ancestors were of little use to the first settlers who arrived in New England in 1620. They did not bring very much food with them, and they certainly were in no position, in the short-term, to produce very much of what they needed. Their only hope was to follow the example of their Native American neighbors: to combine limited agriculture with hunting and gathering natural foods from their environment. This was a difficult choice. Their ancestors had spent thousands of years trying to escape hunting and gathering, and they were reluctant to revert to that primitive level of subsistence. They almost starved before they finally accepted the obvious. The Indians knew what they were doing, and the settlers were soon following the Native American methods of producing Indian corn, pumpkins, squash, and beans, and harvesting wild strawberries, a variety of berries, chestnuts, wild turkeys, ducks, pigeons, clams, and oysters (Bidwell and Falconer, 1941).

It was probably the production of Indian corn (maize) that saved the early settlers. They had known little or nothing about this crop before they reached Massachusetts, but it quickly became their primary grain crop. Initially, they had poor success trying to grow English wheat. They quickly learned that maize produced much more food an acre than wheat. Further, the yield of grain per planted corn seed was much greater than for wheat, barley, or oats. The planting of corn was much simpler. With wheat, a seedbed had to be prepared to accept the large number of broadcast seeds. Corn was planted in discrete mounds, and with mound-culture it was much easier to control the weeds. The corn was also much more tolerant of the vagaries of the New England climate. Because the length of the growing season for corn was also relatively short, it matured much earlier than the European grains. Corn continued to be an important source of bread in New England throughout most of the seventeenth century, but later, farmers there found that they could not compete with producers in the Middle and Southern colonies. If there was a disadvantage to the production of corn, it was that it was not a very good crop for export. Europeans did not like the bread that was baked with it.

While corn (maize) remained the primary grain crop of the New England colonies for many years, the settlers finally learned how to grow the European grains. In the early years, wheat production was especially good in the Connecticut Valley. Later, however, farmers in the Middle Atlantic colonies, especially New York, New Jersey, and Pennsylvania, became much more proficient in the production of the crop. No doubt, the longer growing season and the generally warmer climate there played a central role in the final outcome of the competition. As a result, by the beginning of the eighteenth century, the urban centers in New England became almost totally dependent on the importation of bread-stuffs from the Middle and Southern colonies.

The first vegetables grown by the New England settlers were those also used by the Native Americans, including: pumpkins, squash, and beans. Later, they began to experiment with the production of European vegetables, including: cabbages, turnips, onions, radishes, and carrots. The production of these vegetables became increasingly important as the urbanization of the towns along the coast proceeded. The urban population provided a ready market for vegetables produced in close proximity of the towns. While this market was not large in the beginning, it continued to grow steadily, and it was one market that farmers in the Middle and Southern colonies could not profitably enter. Unfortunately, because of transportation problems, farmers in the backcountry could not easily participate in the marketing of fresh vegetables in the towns either.

In the beginning, settlers harvested large quantities of wild strawberries, blackberries, and gooseberries from the waste lands around their farms. Later, they began to produce a variety of European fruits, including: apples, pears, plums, quince, and cherries. Like the vegetables, the fruits were important to all the farm families in the region, but became especially important to those living near the growing population centers.

As late as 1623, there were no cattle in the New England colonies and no sheep or horses until 1627. Settlers there quickly learned that the native grasses in that region were inferior forages for all grazing animals. As a result, they quickly imported and established English grasses and clovers, and in a few years, they had spread throughout the region. From the beginning, most settlers, following the practice throughout England, did not provide their animals with winter shelter. For many years, they did not really appreciate the fact that winters in New England were much more severe than in the mother country. Before 1650, the production of all kinds of livestock, except hogs, was inadequate to meet the needs of the population. For example, before that time, they imported much of their beef from Europe or Virginia. After 1650, production increased to a point that they began to have a surplus for sale, and these commodities found a ready market in the sugar-producing colonies of the West Indies.

Dairying and raising cattle for beef were not differentiated in the early days. After the establishment of better forage, milk production began to increase in the colonies, and individual families were soon able to produce a surplus. Obviously, it could not be stored for very long, and marketing it in local towns was difficult because of inadequate transportation. In fact, obtaining milk was such a problem for the growing towns that many of the people, even in the largest towns, kept cows on their lots. Surplus milk was generally converted into butter and cheese, but for many years, the quality was extremely poor. Much of what was consumed in the towns was imported.

Swine adapted themselves to the harsh New England environment better than any of the other livestock. They could be fed a wide variety of readily available native foodstuffs, such as clams and other shellfish. However, even without supplemental food, they were able to fend for themselves by foraging in the forests.

Agricultural labor in the New England colonies was extremely scarce, and there were few indentured servants available. As a result, wages were extremely high. Because land was readily available to virtually anyone, unless laborers were paid extremely well, they would quit and establish their own farms. In the beginning, some farmers attempted to use slaves as agricultural labor; however, given the small size of most of their farms, and the high cost of obtaining and supporting slaves, they quickly determined that using them was not practical. Given this set of conditions, most of the labor required on the individual farms had to be provided by the families themselves or through community cooperation. This labor situation tended to place severe limitations on farm size.

There were no staple crops that could be produced in the New England colonies similar to the tobacco, rice, indigo, and later long-staple cotton produced in the Southern colonies. Farmers in the Plymouth and Massachusetts Bay colonies could produce only wheat, oats, barley, sheep, and cattle, and there was only a very limited market in Europe for these commodities. Only in time of crop failure and regional wars, could they be sold there. Fortunately, as specialized agriculture advanced in the South, markets developed there for the foodstuffs that were finally being produced in surplus in New England. Also, over time, sizeable quantities of foodstuffs were marketed in the West Indies whose farmers produced almost nothing other than sugar cane. Trade with this region finally became extremely favorable. In fact, it became so favorable that it lead to a spate of shipping, ship-building, commercialism, and urbanization in the Northern colonies. Trading in agricultural commodities, and the increased interest in international commercialism, lead local merchants there to begin the importation of English goods for resale in both the Northern and Southern colonies. Later, local tradesmen would learn to manufacture these goods using the water power in the numerous streams in coastal New England. As a result of all of these factors, commercialization and urbanization developed much more rapidly in the North. These developments also brought significant changes in agriculture in New England. Increased commercialization and urbanization along the coast resulted in an increasing demand for all kinds of foodstuffs. Local farmers began to specialize in the production of commodities (meat, milk and milk products, eggs, fresh fruit, and vegetables, etc.) that could not be practically transported from the Southern colonies or from Europe. Finally, this agriculture became almost as specialized as that of the Southern colonies.

The Virginia Colony

The development of the food supply in Virginia worked through the set of problems to arrive at a general level of self-sufficiency and acceptable dependability by about 1627 (Gray, 1958). Nevertheless, there were some other developments in agriculture in Virginia that would have even more far-reaching consequences than their development of a dependable food supply.

The Indians were growing an inferior strain of tobacco when the first colonists landed at Jamestown. Then around 1612, some of the settlers began to experiment with the cultivation of seed obtained from Trinidad in the British West Indies. In 1612, John Rolfe began his efforts to cultivate tobacco. In 1613, the first sample of Rolfe's tobacco was taken to England where it was well received. Tobacco had been grown in England from about 1570, and had become extremely popular by 1585. Europeans apparently first encountered the herb in the 1560s when they received some of it from the ill-fated French Fort Caroline in what would later become northeast Florida (Morison, 1965). The extremely addictive nature of nicotine was apparent even in those early days. The quality of the Virginia tobacco was superior to that generally produced on the Continent. As a result, the demand for the Jamestown product escalated rapidly. In 1620, some 100,000 pounds was exported to the mother country (Cochrane, 1993). It is difficult to fully comprehend the overall effect that the early and continued production of this crop, with its highly addictive nicotine component, would have on the future of the Southern colonies and on the country itself. Certainly, the long-term effects were not readily apparent at that time.

The profitability of tobacco production soon became apparent, and the industry literally exploded throughout the colony. It was soon being planted in virtually every open field, even in the streets of the towns. Tobacco production became so widespread that food production began to be reduced. In 1616, Governor Dale found it necessary to require that "each person responsible for his own maintenance had to plant 2 acres of maize for himself and for each manservant as a condition for raising any tobacco, on pain of forfeiture of his tobacco crop." But nothing was going to slow the production of the "leaf." As the tobacco habit swept through Europe and as the colony's population grew, tobacco production in Virginia kept pace. During the years 1637-1640, annual exports of tobacco to England averaged almost 1.4 million pounds. A century later, in 1736, they reached 19.2 million. Thus John Rolfe and his friends began the trade of a commercial crop that would dominate the agriculture of the Chesapeake region for more than a century, and in the colony as a whole for a much longer period.

Tobacco production not only affected food production in the colony. It also had a far-reaching effect on the use of land and labor in agriculture. The production of the enormous quantities the plant required to meet export demands demanded large amounts of land. Further, tobacco production in the absence of fertilization quickly reduced the fertility of the land to the point that production of the crop was no longer practical. Acreage began to be reduced, and wheat was substituted for tobacco. Consequently, new lands had to be constantly brought into production just to maintain existing export levels. To solve this problem, successful tobacco planters began to develop large land-holdings (plantations) along the rivers (James, York, Potomac, and Rappahannock) above Chesapeake Bay. Production simply shifted away from the Tidewater to the southwest and northwest, to and above the fall line, and into the Shenandoah Valley. By the middle of the eighteenth century, large quantities of tobacco were being hauled to Richmond, at the great falls on the James River, in rolling hogsheads from as far away as Culpepper in northern Virginia almost at the foot of the Blue Ridge. Wealthy Virginia tidewater tobacco planters, seeking better lands, were responsible for establishing the town of Petersburg at the head of navigation up the Appomattox River. By 1730, there was so much of the leaf being produced there that the government authorized the construction of a tobacco inspection warehouse in Petersburg.

Tobacco production, like the production of sugar, required large amounts of labor. In the early days, tobacco planters used family members and indentured servants to produce the crop, but there was never enough of either to meet their ever-increasing needs, so they turned to a third source—black slaves purchased from slave traders. Without the use of large amounts of slave labor, tobacco production would never have advanced very far in the colony. After 1650, the number of slaves in Virginia increased rapidly. The actual statutory recognition of slavery in Virginia came in 1661 (Franklin and Moss, 1988). There were around 2,000 there by 1670, some 6,000 by 1700, and 12,000 in 1708 when slaves comprised about 40% of the colony's 30,000 residents. A half century later, there would be over 120,000 blacks there. Hurst (1994) suggested that John Rolfe and the other Jamestown tobacco planters—in beginning the production of a commercial crop that would ultimately "require" the use of slave labor—provided a basis for the development of an economic and social relationship between whites and blacks that would greatly influence the course of American history.

Thus, the plantation system of agricultural production that would later manifest itself throughout the other Southern colonies had its beginnings in the tobacco industry of the Virginia colony. The tobacco planters, however, did not invent plantations. At the time they first appeared in Virginia, plantations had already existed for some 600 years; they had first appeared in the Mediterranean islands of Cyprus and Crete as early as twelfth century where sugar was produced with slave labor. However, it was plantations on Sicily that provided the model for the sugar plantations that developed on the Atlantic Islands of Maderia, the Canaries, and São Tomé in the middle of the fifteenth century. A century later, sugar was being produced on plantations in Brazil.

While plantations were not invented in colonial Virginia, once the sugar-production model was put in place for the production of tobacco, they drove the economy of the colony, shaped social structure, and determined political power. In the process, plantations slowed the development of a merchant and artisan class, discouraged investment except in land and slaves, fostered rural isolation, and slowed the growth of urban areas (Hurt, 1994). Following the Virginia model, these large, slave-driven commercial farms soon dominated all aspects of Southern colonial agriculture.

Corn (maize) quickly became the principal bread grain of Virginia, as it eventually did for the South as a whole, and it retained that status for many years. Compared to wheat, land preparation for planting corn was simpler. Further, corn was simpler to harvest, the yield per amount of seed planted was greater, preparation for consumption was simpler, and fodder from the corn plant was superior to wheat straw as forage for animals. In the early years, Corn, both that obtained from the Indians and the small amounts produced in their own fields, helped save the Jamestown colony from starvation and failure. Fortunately, it did not take long for the settlers to learn of the merits of this "foreign grain" and to learn how to grow it. In 1614, settlers had 500 acres of land planted in corn, and by 1616 they were growing an abundance of the grain in the new settlements spreading inland along the James River. With the expansion of the tobacco plantations, the role of corn changed somewhat in the lower Tidewater. Well-to-do planters imported wheat flour for their bread, and grew corn to feed their slaves and indentured servants. Also, some of it was used to fatten livestock before they were slaughtered or taken to market. Corn was so important to the colony until the last quarter of the seventeenth century that laws were occasionally passed to prevent its export. In fact, there were some restrictions as late as the early years of the eighteenth century.

While domestic consumption of corn greatly exceeded the amount exported throughout the colonial period, Virginia farmers were able to export a considerable amount of the grain. In 1768, the colony exported 76,000 bushels to England; 4,800 to Ireland; 113,000 to Southern Europe and Africa; 365,000 to the West Indies; and 129,00 to other colonies along the Atlantic coast (Gray, 1958). Corn was also extremely important to the settlers on the frontier. To those backcountry farmers, it was the staff of life. Corn not only was an essential part of their food supply, but it also provided them with one of their favorite drinks—distilled spirits. This corn product was especially important to the Scotch-Irish settlers who by-passed the Tidewater plantations to settle in the backcountry.

English wheat had been planted in the first year in the Virginia colony, but apparently it was given little attention until after 1617. Wheat did not fit well with traditional Indian, hoe-cultivation methods that the settlers were forced to use in the early years. However, wheat began to gain some importance once plows began to be used in the colony; although hoes were used to some extent to cultivate the crop until near the end of the colonial period. Production of the grain increased slowly, and it was not until about 1640 that tzhe Virginia was producing enough to export a quantity to the West Indies. In 1667, the colonial government offered a premium to be paid in tobacco to anyone exporting large quantities of wheat, and in 1688 some tax exemptions were allowed for producers of "English grain." During the first half of the eighteenth century, planters began to substitute wheat for tobacco on their plantations. Soil exhaustion and unsatisfactory marketing conditions both contributed to the change. Nevertheless, tobacco was not extensively abandoned in the Tidewater before the Revolution. Also, wheat production was more important in the mixed agriculture operations of those farmers who began to settle further away from the Tidewater. Wheat production was much more important to the German farmers in the Shenandoah Valley than to the English planters in the Tidewater. As a result of these factors, Virginia wheat production began to increase. In 1768, Virginia shipped over 55,000 bushels of wheat and 119 tons of bread and flour to England, and over 11,000 bushels of wheat to Ireland (Gray, 1958). During the same period, large quantities were also being exported to the West Indies and to Southern Europe.

The commercial success of the cultivation and export of tobacco in the Virginia colony was in direct contrast to the generally poor results from attempts to establish the commercial production of other agricultural crops. Significant efforts were made to establish wine and silk production in the colony, but with generally poor results. There were also experiments conducted on the production of hemp and flax. The early experiments with these crops were mostly discouraging, but later those two crops became fairly important in colonial agriculture. In the early years, production trials were also conducted on cotton, indigo, and rice. None of these crops became important commercial successes in colonial Virginia.

Many garden vegetables, including turnips, radishes, onions, peas, potatoes, cabbage, cauliflower, carrots, parsnips, pumpkins, lettuce, watermelons, muskmelons, cucumbers, and chickory were introduced during the early years of the colony. Most of these came to play important roles in the food production in the later years. As early as 1611, experiments were conducted on the production of oranges, potatoes, and pineapples, and after 1621, figs, pomergranates, lemons, plantains, sugarcane, cassava, paw-paws, and red peppers were introduced. Gray (1958) suggests that it is probable that most of the English fruits were introduced in the early years. In 1629, it was reported that apples, pears, apricots, vines, figs, and other fruits had "prospered exceedingly."

Some livestock were brought to the Jamestown colony with the first settlers, and additional shipments were in the ensuing years. In the fall of 1609, there were six mares and a horse, some 500 swine, and an equal number of chickens, plus some goats and sheep. In the following year, the Starving Time, colonists and Indians had slaughtered all of the livestock, including the horses. Only a single sow survived. Beginning in 1610, a determined effort was made to rebuild the colony's livestock base. Numbers of all types of animals increased slowly through 1618. Progress in increasing cattle numbers came so slowly that the colony found it necessary to pass an act forbidding the slaughter of any cattle without the consent of the governor. Attempts to improve the livestock situation suffered a serious setback in 1622 during the Indian massacre of so many settlers. Animal thefts by the Indians had been a problem from the beginning, but in 1622, with the destruction of so many farms, an inordinate number were either slaughtered or taken away during the widespread attacks. Vigorous efforts to punish and subdue the Indians followed, and apparently improved the prospects for livestock production significantly. In 1629, it was estimated there were more than 2,000 cattle in the colony along an abundance of swine, goats, sheep, and poultry. Although the production of livestock increased slowly, it was not the result of an enlightened animal husbandry. Cochrane (1993) commented that as late as the latter part of the seventeenth century that "the state of the arts with regards to animal husbandry was appallingly bad." Good pastureland was scarce. Fenced enclosures and protective sheds were virtually non-existent. All types of livestock were forced to run wild in the forests and to find what food and shelter they could. Almost no efforts were made to produce and store feed and hay that could be used to get the animals through the winter. Although the number of animals gradually increased under these conditions, the general quality of all kinds of livestock deteriorated. Horses became small in size, and cattle produced little meat and almost no milk.

Agriculture in the Virginia backcountry was quite different from that of the Tidewater. Although traders and fur trappers had penetrated the backcountry all the way to the Appalachian Mountains and beyond by the end of the seventeenth century, the settled population of the Jamestown colony was largely confined to the coastal plain up to the fall line (Figure 4.45). In fact, there was no important migration beyond until about 1720. Between 1726 and 1730, Germans from eastern Virginia and Pennsylvania established several settlements in the Shenandoah Valley. During the next 20 years, the great valley was rapidly filled, primarily with Germans and Scotch-Irish (Gray, 1958). This stream of settlers, consisting largely of relatively recent European emigrants, brought with them the instincts and habits engrained in them by peasant life in Germany and Northern Britain, but modified to a degree by conditions and people that they had encountered on their travels overland from eastern Pennsylvania. Once settled, their largely European peasant culture was further modified by the exigencies of pioneer life, far from the nearest town or city. As a result of these factors, their lives and culture evolved into a completely different pattern from that of the Englishmen in the Tidewater.

Those peasants coming from Germany had intensively farmed small landholdings there, and they continued this practice in the Shenandoah even though land was plentiful. They tended to settle in compact communities and to move infrequently. Both their German peasant background and their remoteness from markets pushed them toward a system of multi-crop farming involving a large measure of self-sufficiency. Their agriculture centered on the production cereal grains. They used wheat, oats, barley, and corn to make their bread, fatten their hogs, brew their beer, and distill their whiskey. They produced butter and cheese. They were also accustomed to the production and use of potatoes. They developed abundant orchards. They produced their own clothing from flax and hemp they had raised. They were compelled for a time to let their livestock fend for themselves in the forests, but when possible, they constructed fenced enclosures and shelters (usually substantial barns) to contain them. They also produced and stored hay and grain to provide them with winter feed. Usually, these compact settlements included a number of farmer-artisans. The presence of these skilled individuals, plus the remoteness from markets, were favorable to the gradual development of community industries (Gray, 1958).

By the middle of the eighteenth century, Scotch-Irish, German, Welsh, and Dutch farmers along the western reaches of the Shenandoah Valley had developed an important cattle-raising industry. The mountain valleys and hillsides were ideal for grazing cattle. All of these farmers were engaged in mixed agriculture that included herding, but some of them began the practice of feeding surplus corn to their animals to fatten them for markets in Philadelphia. Some of these farmers were among the first importers of European purebred cattle in the colonies.

The North Carolina Colony

The first permanent settlements in the region that would become North Carolina were created by people migrating southward out of the Virginia colony as early as 1650. These settlements were much more isolated than those in the Virginia Tidewater or those that would later develop on the Ashley and Cooper rivers in South Carolina. In both of those colonies, large navigable rivers allowed ready access far into the interior. However, in North Carolina extensive sandbars and capes along the coast prevented all but the smallest vessels from reaching the mainland. For most of the colonial period there was little direct trans-Atlantic trade. Most of the trade that was conducted with Europe had to pass through either Norfolk or Charleston. Some of the planters owned sloops or small ships that could pass through the shallow waters. These were used to conduct trade with the other Atlantic Seaboard colonies and with colonies in the West Indies. Also, small ships owned by New England merchants were able to navigate North Carolina's treacherous coastal waters. These merchants provided still another outlet for the products of the colony. As the backcountry developed, a large portion of the products from western North Carolina was shipped to Charleston down the Yadkin and Catawba rivers. Thus the colony was not without trade outlets, but they were expensive to use. Transportation remained a problem throughout the colonial period and effectively discouraged the development of plantation agriculture. Instead, an agricultural economy evolved that was primarily self-sufficient and essentially local and isolated (Gray, 1958). The colony's isolation encouraged the development of considerable diversity in community economies and household industries. Farmers raised sheep, hemp, flax, and cotton; their wives made substantial amounts of cloth from these various fibers. Under those conditions, men, including some slaves, had to be involved in many trades such as carpenters, joiners, blacksmiths, wheelwrights, coopers, butchers, tanners, and shoemakers.

The emigrants from Virginia brought with them all of the provision crops (peas, beans, potatoes); garden crops (carrots, turnips, lettuce, spinach, squash, cucumbers); and fruits (peaches, apricots, plums, figs) that they had grown there. All of these crops fitted well into their largely self-sufficient agriculture.

Throughout the colonial period, North Carolina farmers produced a quantity of tobacco in the region around Albemarle Sound. Although it was their most important trade product, it was difficult to get it to Europe where it would bring the best price. In 1679, Virginia prohibited the importation of North Carolina tobacco and its re-exportation, apparently because of its poor quality. Subsequently, the Albemarle farmers had to first ship their tobacco to New England. From there, it was taken to Newfoundland where it was finally shipped to the Continent. Later, as settlers began to move into the backcountry, they took tobacco culture with them. For many of these, largely self-sufficient and limited-resource farmers, this was their primary cash crop. They produced relatively small amounts of the crop, and it was difficult to get to market. Most of it had to be shipped down the Yadkin and Catawba rivers to Charleston.

Tobacco was the primary plant crop exported from the colony, but small quantities of other products were also exported, including beans and peas, some corn, and a little wheat. North Carolina also produced some upland rice, but apparently only for home consumption. Livestock herding was a large industry in the colony, and produced substantial amounts of beef, pork, hides, tallow, butter, and cheese for export. They apparently exported a small amount of indigo. The immense pine forests in the colony, provided tar, pitch, turpentine, clapboards, staves, barrel headings, masts, and several kinds of lumber. Substantial quantities of these forest products were exported. Officials in Charleston, however, made it difficult to do so by levying almost prohibitive duties on the export of naval stores as well as indigo through that port.

Limited-resource, self-sufficient farmers began to trickle into the North Carolina Piedmont (Figure 4.45) between 1740 and 1750. Leyburne (1962) comments that these families were harbingers of one of the most massive immigrations in colonial times. By the end of the colonial period, there were 60,000 settlers in the backcountry above the fall line. These frontier settlements contained some 80% of the entire white population of the colony. A few of them had moved westward from the Tidewater, but the great majority had come from Pennsylvania and Virginia, down the Great Philadelphia Wagon Road. A few of these settlers were Swiss and Welsh, and a large number were German, but there were more Scotch-Irish than all others combined. Many of the Scotch-Irish families were new arrival from Ulster. Many of these farmers, especially the Scotch-Irish, in the beginning, lived off the bounty of the surrounding forests. They hunted and fished and gathered nuts and berries. They also depended heavily on the Indian corn that they planted among the girded trees. Squash, pumpkins, and an assortment of beans were produce in kitchen gardens around their primitive cabins. Virtually all of them had a few cattle and swine, which ran wild in the surrounding forests. Over time, if they did not move further south, they began to produce some tobacco and probably a little wheat that they could sell for cash.

The South Carolina Colony

General problems described in a preceding section that plagued the Southern colonies in their attempts to provide themselves with adequate supplies of food were apparent, at least to a degree, in the early days of the colony on the Ashley River. The most important of these was the practice of sending large numbers of new settlers to the colony without adequate food to support them until they could produce for themselves. This practice placed a heavy burden on those people already there, and made it necessary for the colony to import large quantities of food until 1674. Fortunately, some lessons about food production had been learned in the Jamestown colony, so that the hardship and suffering experienced there was averted in South Carolina. At times, there was some hunger and considerable anxiety about the future, but there were no starving times. In the early years, the mortality rate was extremely high in the colony, but it apparently was not the result of a shortage of food. In Jamestown, food production did not reach a level of adequacy until some 20 years (1607-1627) after the colony was established. In South Carolina, adequate food supplies were being produced locally after five years (1669-1674).

From the beginning, the proprietors had urged the settlers to plant maize, beans, peas, turnips, carrots, and potatoes as their primary provision crops. Within a relatively short time, they were producing all of these basic food crops, plus a wide variety of garden vegetables including cucumbers, radishes, lettuce, parsnips, turnips, pumpkins, squashes, muskmelons, and watermelons. They also produced a wealth of all kinds of berries and fruits. While the proprietors were concerned about domestic food supplies, they still expected to realize some profit from their investment, so they encouraged the settlers to experiment on the cultivation of money crops: ginger, sugar, indigo, and cotton. In the short term, all of these trade crops failed to meet expectations. As in the other colonies, there were perennial experiments to produce wine and silk, but these also failed to produce the desired results. The early efforts to produce indigo and rice were not promising either. It was expected that tobacco would be a good export crop, just as it had been in Virginia and Maryland. Good quality tobacco could be produced in coastal South Carolina; however, by the time the colonists there were ready to enter the market, so much of the leaf was already being produced in the older colonies to the north, the South Carolinians were discouraged from investing the necessary capital to be competitive. In fact, in the early years they could not seem to find any cultivated plant crop that could be exported, so they turned to the abundant forests all around them. Soon they were exporting large quantities of deer skins, which they obtained through trade with the Indians, as well as timber, clapboards, barrel headings, pipestaves, pitch, tar, and turpentine (Gray, 1958). Oddly, during the latter part of the colonial period, South Carolina exported a small amount of orange juice.

The proprietors had made certain that there would be adequate livestock available for food in the new colony on the Ashley, but they had no intention of providing more than would be required for domestic consumption. It was their intent to build a colony of planters and not graziers. However, as a result of the mild winters and abundant natural range, the population of horses, cattle, hogs, and sheep literally exploded after their introduction. Within ten years, South Carolina was well on its way of becoming an important herding region. By 1682, it was reported that settlers, primarily those with larger landholdings, owned many thousands of cows, hogs, and sheep. Virtually all of these animals were produced on the open range. Generally, they were provided neither food nor shelter during any part of the year. Over time, the colony would become the center of the British-American cattle industry. Beginning about 1682, South Carolina traders began exchanging barrels of salted beef for sugar, slaves, and cash in other English plantation colonies—the Bahamas, Jamaica, and Barbados. By-products of cattle production (tallow, candles, leather, and hides) were also exported in large quantities. The trade in beef became so lucrative that the owners of the larger herds purchased slaves from Gambia to serve as cattle herders. These slaves were especially sought after for that purpose, because many of them had experience in herding cattle in their

African homeland (Hurt, 1994).

Early attempts to cultivate rice in South Carolina had been generally discouraging; although a small amount was produced during the early years. Then in 1685, a new variety was introduced, probably from Madagascar. By 1690, rice had become an important cash crop. At the end of the century, planters exported some 394,000 pounds of the grain. Until around 1720, rice was grown on dry upland fields, and production was heavily dependent on rainfall. After that time, increasing amounts of rice were produced using irrigation in inland swamps along the coast. This change provided a new impetus for the industry, and in 1726 the colony exported nearly ten million pounds. Rice production and exports continued to grow until around 1740 when the War for the Austrian Succession (1740-1748) began. This war involved England, France, Austria, and the German states of Prussia and Bavaria. The European war soon became a worldwide conflagration, and it severely reduced international trade. As a result, rice prices fell sharply. In 1745, prices dropped to 20% of their former level. By 1752, however, they had returned to their prewar level. After 1720, rice became South Carolina's most important staple crop, and it provided the economic basis for the development of the most extensive plantation system in colonial America. Rice plantations were large, and rice production was labor intensive. Therefore, slave labor became the foundation of the plantation production system. Planters purchased thousands of slaves for that purpose; each rice plantation required between 50 and 100 slaves. Those from Ghana and Sierra Leone, the Rice Coast of Africa, were preferred because of their experience in cultivating the crop in their homelands (Hurt, 1994).

Although a small amount of indigo was produced in the early years, these efforts were also discouraging. In the war years (1740-1748) when rice prices became so depressed, planters searched for a substitute for the grain, and they rediscovered indigo. Also, during this period, the growing textile industry in England was demanding more and more of the rich blue dye produced from the plant. As a result of these factors, cultivation increased rapidly, and in 1746 South Carolina planters exported 5,000 pounds and 134,000 pounds in 1748. In that year, England established a bounty on the crop that added significantly to the value of exports. Wallace (1951) commented that in 1754, indigo production "had people in high spirits and the slave markets booming." Indigo production flourished until the Revolution. Afterwards, of course, England no longer imported the dye from any of the North American colonies.

Early experiments with the production of upland cotton in colonial South Carolina were also generally discouraging. For many years, the small amount being produced was for home use. It was just too difficult to separate the seed from the fibers. Beginning in 1744 when rice prices were depressed as a result of the war, the colony established a bounty on the fiber to encourage its production. Under this provision three pence would be paid for every pound of "neat well cleared and merchantable cotton" exported. The bounty may have helped increase production somewhat. Gray (1958) comments that 3,000 pounds was exported to England in 1768.

Corn played essentially the same role in South Carolina as it did in Virginia. Early on, a substantial amount had to be purchased from the Indians while the settlers were establishing their own fields. There were periodic shortages until 1674. With the development of the extensive plantations associated with the production of rice and indigo, there was little interest in producing more corn than the amount needed for domestic consumption. Although there was some export of corn periodically from the port of Charleston, the amount was significantly less than that exported from Virginia. The grain was just as important to the South Carolina Upcountry settlers as it had been to those settling the Shenandoah Valley in Virginia.

There was little early interest in the production of wheat in the colony. Apparently, it did not produce well in the soils and under the weather conditions of the coastal plain; however, with the settling of the Upcountry on the soils of the Piedmont, production increased significantly. Toward the end of the colonial period, large quantities of flour and bread were being exported from Camden, South Carolina.

The migration of limited-resource farmers down the Great Philadelphia Wagon Road from Pennsylvania,

Virginia, and North Carolina between 1750 and the end of the colonial period quickly spilled over into South Carolina, literally filling up the Upcountry (Piedmont). Leyburn (1962) comments that by the beginning of the American Revolution that there were 83,000 settlers in this region, 80% of the total white population of the entire colony. Wallace comments that this immigration consisted of largely Scotch-Irish (1951) people. The agriculture of these immigrants followed closely those of friends and relatives that they had left behind in the uplands of Virginia and North Carolina.

The Georgia Colony

The Georgia colony was intended largely as a philanthropic enterprise. In its planning, it was intended to avoid some of the problems that had developed in the other colonies. Its trustees especially wanted to prevent the development of large plantations worked by black slaves. Instead, they wanted to establish a colony of agricultural peasants farming small landholdings. Further, they wanted to avoid competing with the other colonies in the production of tobacco. Although they were interested in rice and sugar, they concluded that farmers with limited resources could not effectively produce those crops. Indigo was mentioned as a possible crop, but the primary emphasis was placed on silk, wine, cotton, hemp, flax, potash, medicinal plants, dyestuffs, and tropical fruits and nuts. Unfortunately, despite considerable investment by the trustees, none of these crops or products became commercially important. Silk exports, however, did reach almost a ton (the highpoint) in 1767, and eight bagsful of cotton was shipped to Liverpool in 1764.

Agricultural production developed slowly in the colony, probably because the trustees provided the settlers with a well-stocked public store, which met almost all of their food needs for several years. With adequate food being provided by the trustees, many of the settlers put forth little effort to help themselves. The trustees continued the operation of the public store until 1739 when it was closed; nevertheless on several occasions in the next decade, they had to re-open it to prevent widespread hunger.

By 1740, colonists were meeting most of their food needs. They were producing adequate quantities of the same provision, garden and fruit crops being grown in the other colonies. However, much of the rice and virtually all of the wheat and flour were imported from the Carolinas. They also imported apples, onions, butter, and cheese.

Although the Georgia colonists were largely self-sufficient by the 1740s, there was considerable discontent with the management provided by the trustees. There was widespread disagreement over their policy of prohibiting the importation of slaves; consequently, in 1749, they petitioned the Crown to remove the restriction. It was lifted by an act of Parliament in 1750. In the next few years, a great flood of planters with large numbers of their slaves entered the colony, and the lowland districts were soon transformed into a plantation region similar to that in the South Carolina Lowcountry (Gray, 1958). By 1771, there were 13,000 slaves laboring on the large number of great rice and indigo plantations in the colony.

After 1750, the amounts of rice and indigo, to a lesser degree, available for export increased rapidly. These products were added to the corn, hogs, pork, beef, horses, tallow, leather, deerskins, timber, staves, shingles, pitch, and tar that were already being exported. In the latter part of the colonial period, exports also included small quantities of chickens, geese, ducks, and turkeys.

The export of deerskins was especially noteworthy. In 1755, almost 50,000 skins were shipped to Europe, primarily England, and by 1772, the number had increased to well over 200,000. The level of exports indicates the importance of Indian trade in the colony. Further, the lists of the livestock, livestock products, and forest products exported are indicative of the importance of animal agriculture and timber harvesting in the young colony.

Georgia's backcountry, primarily the area above the fall line (Figure 4.45), developed in much the same way as the Virginia and North Carolina Piedmont upland sections and the South Carolina Upcountry. It was first thinly settled by Indian traders and cattle herdsmen. Later, after the Indian land cessions of 1773, north of Augusta waves of limited-resource Swiss, Welsh, German, and especially Scotch-Irish farmers moved out of the Piedmont uplands of Virginia and the Carolinas into this area. By the end of the colonial period, the majority of the colony's population lived in this region. While the agricultural economies they developed was based largely on self-sufficiency, they did export small amounts of tobacco, hemp, livestock, and other miscellaneous products through the port at Savannah (Coleman, 1977).

Other significant population changes were also taking place in the backcountry above the fall line. Virginia's tobacco planters who had moved from the Tidewater to the Piedmont in that colony during the first half of the eighteenth century, because of declining yields, began to experience the same problems in the Piedmont as well. The land was becoming exhausted, and as soil fertility declined as did tobacco yields. As a result, a number of planters purchased land on recent Indian cessions (1773) in the Broad River Valley of Georgia, some 25 miles east of Athens, and moved there with their slaves. Once there, they changed the name of the old colonial town of Dartmouth to Petersburg and began to produce tobacco. At one time, the town was an important center for the production of the leaf, and a number of warehouses were erected. However, with the coming of the cotton revolution, the tobacco warehouses were no longer needed, and the town built on tobacco soon was abandoned.

The Lower Mississippi Valley

There was little agricultural development in the French settlements of the Lower Mississippi Valley for almost two decades after the establishment of the first Biloxi garrison in 1699 (Gray, 1958). In fact, very little progress was made until 1717 when they established a colony at the present site of New Orleans. From that time until the French surrendered Louisiana to the English in 1763, the land along the Mississippi and its tributaries became the focus of economic activity, primarily agriculture and lumbering. Various governmental regulations and inadequate facilities for marketing retarded efforts to establish a viable commercial agriculture throughout the colonial period. In the first 50 years, commerce was uncertain, irregular, and subject to many restrictions (Gray, 1958). The settlers in Louisiana never enjoyed the freedom of trade with France that the English colonies did with their mother country. The French colonists were generally prevented from producing any crop or engaging in any commerce that would provide undue competition to any industry in France. Also, from the beginning, commercial agriculture and the establishment of the plantation system in the colony suffered from an unstable currency, the result of generally chaotic economic conditions in France throughout the colonial period. France's periodic wars with England, coupled with their generally inferior naval forces, made their efforts to ship products from the colony problematic. The 1744-1748 war between the two countries virtually stopped all trade between the colonies and the mother countries. Later, the Seven Years' War (1756-1763) again interrupted trade, resulting in the accumulation of large quantities of commodities in storehouses along the lower Mississippi.

Peas and beans were cultivated from the beginning of the Louisiana colony, and they quickly became important elements in their domestic food supply. As early as 1731, they were also being exported to the West Indies. For many years, these crops continued to be regular items of export. The French colonists also produced the traditional garden vegetables and fruits that were being grown in the English colonies on the Atlantic Seaboard at that time. Probably, the most unusual crop which they produced was the pecan, which grew wild in the area. Although fruits did not produce well around New Orleans, orchards were more successful on the higher elevations around Natchez.

From its beginning, corn was also grown in the colony. It was extensively produced for domestic use of rich and poor alike. Corn was an especially important source of food for the many slaves there. Because of the warm, humid climate in the region, the grain deteriorated rapidly, so relatively little of it was exported.

As a result, no doubt, of the commercial success of tobacco production in the English colonies, the French intended to make production of the plant an important commercial enterprise in their colony. From the beginning, tobacco production efforts were centered around Natchez, although there was some production west of the Mississippi, around the present-site day site of Natchitoches, Louisiana. The Natchez area was an especially good location for the production of the crop, because of the wind-deposited soils (loess) found there. Unfortunately, production of the leaf had a checkered history in the colony. Periodic changes in laws, regulations, subsidies, product quality, availability of slaves, shipping, marketing, and prices slowed production from time to time. Nevertheless, the amount of tobacco exported generally increased over time. In 1720, Louisiana exported 400,000 pounds, but in the following year they fell to 300,000. In the early years of the 1750s, exports varied between 170,000 and 450,000 pounds. In 1763, the Peace of Paris ended the Seven Years' War (French and Indian War in North America). France ceded her lands east of the Mississippi to England, and the remainder of the Louisiana colony and all French territory west of the Mississippi was ceded to Spain. Production continued in the old French colony throughout the remainder of the colonial period. In fact, Gray (1958) comments that the industry seems to have made its greatest advances after the English took control of lands east of the Mississippi.

Indigo became another important commercial crop of the French colony. The dye was being exported as early as1725, and by 1728 considerable quantities were being shipped to Europe. By 1728, there were 14 or 15 indigo plantations in the vicinity of New Orleans with a total production of 70,000 pounds annually. In 1754, there were 47 planters in the colony. Although the number of planters had increased significantly, production only reached 82,000 pounds. Apparently, a serious limitation to increased production was the difficulty in obtaining the necessary slaves. Production was seriously affected when Spain took over the colony after 1763, because the crop exported from Louisiana competed with a superior dye being exported from the Spanish colony in Guatemala. Even with these problems, overall production in the old French colony seems to have increased through the end of the colonial period. One of the reasons was that the English began to develop commercial transportation facilities on the east side of the Mississippi.

By 1720, rice was being produced in considerable quantities using slave labor. Annual rice production was generally good and dependable. In some years, production was adequate to meet the needs of the colony and to export small amounts. However, rice never became an important export crop.

Forest products (timber, pitch, tar) were also exported from the colony in fairly large quantities. The industry became quite important in the winter on plantations, because they could keep their slaves fully employed during the harvesting and processing these products. It is interesting that one of the more important of Louisiana's exports was roots of the sassafras tree, which had been second only to tobacco as the largest Virginia export for a short time early in the seventeenth century.

The potential for the production of silk was relatively good because of the abundance of several kinds wild mulberry trees in the forests. The industry never really developed, however, because of the shortage of labor to take care of the silkworms. A small amount of silk continued to be produced throughout the colonial period, and a small amount of the product was exported periodically.

There were early experiments with the production of sugarcane and sugar. Early efforts to grow the crop commercially were generally unsuccessful. In fact, there was very limited production of the crop until after the end of the Revolutionary War when the United States gained control of the region.

By 1733, colonists had tried unsuccessfully to grow Sea Island cotton, but in these efforts, they discovered that the short-fiber, Siam cotton (a perennial) could be produced. At the middle of the eighteenth century, large quantities of the fiber was being produced for domestic use, and a small amount was being exported. About this time, one of the local planters, M. Dubreuil, invented a gin of the roller type that effectively removed the fibers from the seed.

From the beginning, the French attempted to establish a viable livestock industry in the Louisiana colony, but the number of animals increased slowly, in part because the French were generally unsuccessful in obtaining suitable breeder stock from any of the Spanish or English colonies in the region. However, by 1746, there were 10,000 cattle, many flocks of sheep, and large droves of swine in the colony. Unfortunately, in 1748 an epidemic resulted in the death of so many of the cattle that a shortage of beef resulted. Fortunately, herd numbers were rebuilt quickly, and within a few years an extensive herding industry emerged in the central and southern portions of the colony.

Colonial Agriculture at the End of the Colonial Period

As detailed in the preceding sections, agriculture in the Europe's North American colonies developed sometimes slowly and sometimes rapidly before the American Revolution. Fortunately, the experience gained from even failed trials made it easier in subsequent efforts and provided a solid base for the successful development and expansion that eventually came (Cochrane, 1993).

By the end of the colonial period, the outlines of commercial agriculture were firmly established in the thirteen English colonies of North America. In the northern colonies, highly productive, highly successful commercial farms were clustered around the major urban centers that had developed in that region. These farms produced the meat, milk, grain, fruits, and vegetables that inhabitants of those cities could no longer produce for themselves. At the same time in the southern colonies, large commercial farms (plantations) were producing tobacco, rice, and indigo for international markets, principally in Europe. In both regions, however, a substantial majority of farmers operated farms where the primary goal was to produce enough food and fiber just to meet the needs of their families; hopefully, with enough left over to be bartered or sold to obtain things (axes, plows, sugar, coffee, etc.) that they could not produce themselves. Also during this perod, farmers firmly established in the New World the commercial mentality that involved producing commodities for local, regional, and internaional markets.

Unfortunately, in their haste to increase production and to develop commercial agriculture, early colonial farmers had paid relatively little attention to the conditions of their farms. As late as the 1740s, Peter Kalm, a Swedish botanist, traveling in the Middle Atlantic Colonies commented that "the grain fields, the meadows, the forests and the cattle, etc., are treated with equal carelessness . . . [The farmers'] eyes are fixed upon the present gain and they are blind to the future." Many other observers noticed the broken fences and the stunted cattle, running at large and unprotected. Their manure was put to no use. They would farm their fields until the fertility was exhausted then move to clear new ones. Established pasture was a rarity, and few farmers stored feed for their animals in the winter (Boorstin, 1958). As late as 1791, George Washington wrote a friend in England: "An English farmer must entertain a contemptible opinion of our husbandry or horrid idea of our lands . . . the aim of farmers in this country, if they can be called farmers, is, not to make the most that they can from the land . . . the consequence of which has been, much ground has been scratched over and none cultivated or improved as it ought to have been."

Fortunately, not all farmers in the New World were equally disdainful of their farms and its resources. In contrast, German immigrants in Pennsylvania gained a good reputation for careful husbandry. They practiced crop rotation, used manure for fertilizer, and regularly fallowed their fields. Near the end of the eighteenth century, an observer, Benjamin Rush of Philadelphia commented that "A German farm may be distinguished from the farm of other citizens of the state, by the superior size of their barns; the compact form of their houses; the height of their enclosures; the extent of their orchards; the fertility of their fields; the luxuriance of their meadows, and the general appearance and neatness in everything that belongs to them." In their efficient farming methods, the Germans were using specialized skills that they brought with them to the New World.

At the time of the American Revolution, the Agricultural Revolution was well under way in England. It involved the Norfolk Rotation in which the production of the lowly turnip and other root crops were alternated with the production of cereals. Normally, the complete rotation was completed in four years, at which time, the cycle would begin again. This system helped to maintain the fertility of the soil. While this system played a major role in agricultural development in England and finally much of Europe, it played essentially no role in colonial agriculture or, for that matter, in the early decades of agricultural development in the young United States (Cochrane, 1993).

Colonial Agriculture and the American Revolution

The American Revolution generally benefited the farmers of the colonies, especially those with the most resources, because there was a ready market for all that they could produce. However, not all farmers benefited. Many had their crops destroyed, as well as their homes and buildings burned. Others paid the supreme sacrifice in the loss of their lives. The fearsome local struggles between English loyalists and American patriots resulted in terrible actions against persons and property on both sides. In some rural areas of the South, these conflicts assumed monstrous proportions, resulting in widespread human misery and deprivation. In 1775, Congress authorized the confiscation of Loyalist property in Massachusetts. This policy was quickly adopted in other states and was continued throughout the war. Confiscation had only a limited effect on agriculture, because relatively few people who had remained loyal to the crown were farmers (Hurt, 1994).

The war had an important effect on the institution of slavery in the South. During those periods of intense military conflict in the South, large numbers of British soldiers were located in the coastal areas where many of the large plantations were found. Many slaves used the general turmoil of the war to escape either to the more lenient North or to British protection. This flight was so extensive in some areas that production on plantations was seriously reduced. Approximately 30,000 slaves escaped to British protection in Virginia and around 25,000 in South Carolina. The British used this opportunity to offer freedom to any slave who escaped to British lines, and who would take up arms against their former masters. These escapes on the part of slaves served to harden the attitude of planters regarding slavery. Plantation owners found it difficult to accept the loss of so much of their investment.

During the American Revolution, colonial farmers and merchants lost their markets in Britain, but they continued to trade with both British and French colonies in the West Indies. This trade in flour, wheat, tobacco, indigo, cattle, and barreled pork was important to both the sugar planters in the islands as well as the colonists. Neither could afford to lose it.

6

American Agriculture Migrates Westward

1800–1860

PRECIS

After the French and Indian War, the English king issued a proclamation in 1763 that forbad establishment of permanent settlements west of the Appalachian Mountains and south of the 31st parallel. With the end of the American Revolution, however, the long closed gates were flung open, and an enormous flood of immigrants immediately began to pass through. They quickly flowed around or through the massive Appalachian Mountain barrier and brutally brushed aside the Native Americans who had lived on the land for centuries. The new national government encouraged this vast westward flood by making arrangements for the Louisiana Purchase with France in 1803 and by completing the Federal Road and National Roads shortly thereafter.

In 1787, the Congress of the Confederation of the United States enacted legislation establishing the Northwest Ordinance. This legislation provided the blueprint for the orderly development of the new country as its people moved westward. Later, legislation was enacted establishing specific territories to provide for the development of individual states. The Northwest Territory legislation provided for the creation of Ohio, Indiana, Illinois, Michigan, Wisconsin, and part of Minnesota. In 1798, Mississippi Territory was established under the provisions of the Northwest Ordinance. Later Mississippi and Alabama would be established in this territory. By 1860, the country's land base had been divided into 34 states and three territories. Only portions of the Great Plains, Rocky Mountains, and Great Basin remained undivided.

When the citizens of the new country left for the West, they intended to go and stay. They took everything they owned with them—all of the elements of their material and non-material cultures, including their farming systems. General farming went into the Old Northwest Territory, and plantations and slaves went to the Old Southwest. The two systems were vastly different from the beginning, and wherever they went different societies and agricultural economies emerged. By 1860, the agricultural economies developing the North Central and South Central regions were quite different, and the societies developing in the two were even more widely divergent. Further, other census data would show that significant portions of the wealth of the North Central Region were concentrated in homes, streets, roads, bridges, railroads, and factories. In contrast, in the South Central Region much of he region's wealth was concentrated in the value of slaves. At that time, it was not clear if two such different societies and economies could remain yoked together in the same union.

INTRODUCTION

For almost two centuries following the establishment of the first English colonies in North America, agriculture remained tied securely to the Atlantic margin. That was where the people were. Most of the population remained, figuratively speaking, within a stone's throw of the Atlantic. Slowly over the years, a few hardy souls moved up the major river valleys into the interior, but they were never very far from the old settlements. By the beginning of the American Revolution, there were sizeable farming settlements as far west as the base of the Appalachian Mountains; however, most of the people remained along the coast. The English government strongly discouraged the establishment of settlements west of the mountains. Title to those lands had been hotly contested for 2 centuries, and the English had no desire to do battle with the French, Spanish, and Native Americans, unless absolutely necessary. After the United States was established, the old restrictions no longer applied, and settlers quickly began to look for passes westward through the highlands.

In 1803, the United States purchased Louisiana from Napoleon, opening millions of acres of new land for settlement. After 1803, while there were still some land title issues to be resolved with the Native Americans, for all practical purposes the gate to the West was wide-open, and American farmers with their families quickly began to flood through it. It would take a few years, but finally that flood would reach the land to be called Alabama.

Thus began the largest migration in recorded history. The westward movement of so many people was an awesome story; the movement of American agriculture was even more awesome. With the movement of those families west of the mountains, American agriculture moved onto some of the most productive farming areas in the world. Over time, this westward movement of agriculture would have important implications for food and fiber production throughout the country and the wider world.

NATIONAL POLITICS

Basic differences between the commerce and manufacturing of the North and the agriculture in the South, began to take on a political life of their own in the first Washington administration (1789–1793). These differences were expressed in the political philosophies of Alexander Hamilton of New York and Thomas Jefferson of Virginia. Hamilton believed that republican government could succeed only if directed by a governing elite. He wanted to concentrate power in the federal government, create a strong national bank, and use legislative enactment to encourage shipping and create manufactures. In contrast, Jefferson felt strongly that a republic must be based on an agrarian democracy. He wanted to diffuse power within the republic, and he was inalterably opposed to the creation of a national bank. Over time, various opposing groups in the new country coalesced around the opposing philosophies of Hamilton and Jefferson, and as a result, two political parties emerged—the Federalist Party of Adams and Hamilton and the Democratic-Republican Party of Jefferson, Madison, and Monroe.

The Presidency

The Federalist Party controlled the office of president from 1797-1801. Democrat-Republicans controlled the office from 1801-1829. Democrats, under Andrew Jackson, were in control from 1829-1841. The Democrats, after Jackson, held the office in the periods 1845-1849 and 1853-1861. Whigs controlled the office from 1841-1845 and from 1849-1853.

John Adams, a Federalist from Massachusetts, was the first to hold the office in the period 1800–1860. He served from 1797–1801. Other persons serving as president were: Thomas Jefferson (Democratic-Republican, VA) (1801–1809); James Madison (DR, VA) (1809–1817); James Monroe (DR, VA) (1817–1825); John Quincy Adams (DR, MA) (1825–1829); Andrew Jackson (Democrat, TN) (1829–1837); Martin Van Buren (D, NY) (1837–1841); William Henry Harrison (Whig , IN) (1841); John Tyler (W, VA) (1841–1845); James Knox Polk (D, TN) (1845–1849); Zachary Taylor (W, LA) (1849–1850); Millard Fillmore (W, NY) (1850–1853); Franklin Pierce (D, NH) (1853–1857); and James Buchanan (D, PA) (1857–1861).

Secretaries of Agriculture

The Department of Agriculture was not established until 1862, and the secretary of agriculture was not given cabinet status until 1889.

The Senate

During the period 1797–1860, Federalists controlled the Senate in three of the 32 Congresses (4th–6th, 1795-1801). Democratic-Republicans were in control in 12 (7th–18th); John Quincy Adams, Democratic- Republicans, in one (19th); Andrew Jackson, Democratic-Republicans, in one (20th); and Democrats (former Andrew Jackson Democratic-Republicans) in 14 (21st–26th and in 29th–36th). The Whigs controlled the body only in the 27th and 28th Congresses.

Senate Majority Leaders

Senate majority leaders were not designated by any party at this time.

Chairmen of the Senate Committee on Agriculture

The Senate Agriculture Committee was established in 1825 (19th Congress). Warren Findlay, a Democratic Republican from Pennsylvania served as the first chairman. Other persons serving in that capacity were: Calvin Willey (Adams Party, CT) (last part of 19th Congress); John Branch (DR, NC) (20th); Charles D. J. Bouligny (Adams-Clay Republican, LA) (last part of 20th); William Marks (Adams Party, PA) (21st); Horatio Semour (W, VT) (22nd); Bedford Brown (Jacksonian Democrat, NC) (23rd and 24th); Perry Smith (D, CT) (last part of 24th and the 25th); Alexander Mouton (D, LA) (26th); Lewis Linn (Jacksonian, KY) (27th); William Upham (W, MA) (28th); Daniel Sturgeon (D, PA) (29th-31st); Pierre Soule (D, LA) (32nd); and Phillip Allen (D, RI) (33nd and 34th). There was no Senate Agriculture Committee in the 35th and 36th Congresses, 1857-1861.

The House of Representatives

Federalists controlled the House of Representatives in three of the 32 Congresses (4th–6th) during the period 1797-1860. Democratic Republicans were in control in 12 (7th–18th); John Quincy Adams Democratic Republicans in one (19th); Andrew Jackson followers, Jacksonians, controlled the 20th, Democrats under Andrew Jackson, controlled 12, 21st–26th, 28th–30th, 32nd–33rd and 35th. The Whigs controlled the House in the 27th and 30th Congresses. Republicans were in control for the very first time in 34th. They also had a majority in the 36th.

Speakers of the House

In the period 1800–1860, Federalist Theodore Sedgwick of Massachusetts served as the first speaker of the House. He was elected to the office in 1799 (6th Congress) and served until 1801. Other persons serving in that capacity subsequently were: Nathaniael Macon (DR, NC) (last half of 6th-9th); Joseph B. Varnum (DR, MA) (10th and 11th); Henry Clay (DR, KY) (12th-18th); Langdon Cheves (DR, SC) (last half of 13th); John W. Taylor (DR and American, NY) (last half of 18th and the 19th); Andrew Stevenson (DR and Jacksonian, VA) (20th, 21st-22nd); James K. Polk (D, TN) (24th and 25th); Robert M.T. Hunter (W, VA) (26th): John White (W, KY) (27th); John W. Jones (D, VA) (28th); John W. Davis (Independent, IN) (29th); Robert C. Winthrop (W, MA) (30th); Howell Cobb (D, GA) (31st); Linn Boyd (D, KY) (32nd and 33rd); Nathaniel P. Banks (D, MA) (34th); James Orr (Republican (R), SC) (35th); and William Pennington (R, NJ) (36th).

Chairmen of the House Committee on Agriculture

The House Committee on Agriculture was established in 1820 (16th Congress). Federalist Thomas Forrest of Pennsylvania was the first person to serve in that office. He was elected to the position in 1820 and served until 1821. Other persons serving in that capacity during the period 1820–1860 were: Josiah Butler (DR, NH) (17th); Stephen Van Rensselaer (Adams-Clay Federalist, NY) (18th-20th); Ambrose Spencer (Anti-Jackson, Democratic Republican, NY) (21st); Erastus Root (Jacksonian, NY) (22nd); Abraham Bockee (Jacksonian Democrat, NY) (23rd and 24th); Edmund Deberry (W, NC) (25th-28th); Joseph H. Anderson (D, NY) (29th); Hugh White (R/W, NY) (30th); Nathaniel Littlefield (D, ME) (31st);

John G. Floyd (D, NY) (32nd); John L. Dawson (D, PA) (33rd); David P. Holloway (Opposition People's Party, IN) (34th); William G. Whitely (D, DE) (35th); and Martin Butterfield (R, NY) (36th).

The Political Situation

In 1801, Thomas Jefferson, a Democratic-Republican, was finally elected president of the United States in an extremely close election. The defeated Federalists never again mounted a serious challenge to lead the government. Farmers everywhere had finally reached their promised land, or had they? (Morison, 1965). Jefferson served two terms as president (1801–1809). His purchase of the Louisiana Territory from Napoleonic France in 1803 was a clear signal that he fully expected the westward expansion of agriculture. He also proposed to use funds from the federal treasury for public improvements (roads, bridges, canals, etc.). If enacted into law, these acts would have been of immeasurable benefit to farmers everywhere who needed access to markets for their surplus crops.

Jefferson had also wrestled with the problem of how to get Indians to give up their lands east of the Mississippi without having to resort to force. He unofficially proposed an overly simplistic and unethical solution, but he left office without making any progress on this very pressing matter.

Jefferson had wanted to spend his time as president promoting a more decentralized federal government while at the same time encouraging the development of an agrarian democracy. Unfortunately, in his second term, the British and French presented him and the country a very complex problem in international relations. In the late 1790s, Great Britain and France went to war once again. At that time, the United States was exporting large quantities of fiber and some foodstuffs to both countries. Although Great Britain and France needed American imports, as hostilities intensified these two world powers began to view American trade as providing direct support to the enemy. As a result, Great Britain began to take steps to prevent American exports from reaching France, and the French reciprocated. Both used their navies to intercept American merchant ships and cargoes on the high seas. With the more powerful navy, Great Britain carried the interdiction to extreme lengths. British actions finally became so onerous that the young United States was ready for war. Jefferson was able to persuade Congress that instead of war, the country should to try to stop all American exporting to Great Britain in hopes that a shortage of tobacco, cotton, rice, and indigo would force the British to stop interfering with American commerce. Consequently, Jefferson got Congress to enact an Act of Embargo. The act was to prevent American ships from sailing to foreign ports for the shipment of goods and to prevent foreign vessels from loading cargoes in American ports for shipment overseas. The embargo finally did result in some economic problems in Great Britain, but it was much more damaging to the American economy. Finally in 1809, the embargo was lifted. Jefferson had tried to prevent another war with England, and he was successful during his term in office. Unfortunately, the problems with the British continued, and Jefferson's successor finally had no other choice but to take Great Britain to task in the War of 1812.

The war officially began on June 18, 1812, and lasted only 30 months. It was fought in three principal theaters: the American-Canadian frontier, at sea, and in the American South. Except for the burning of Washington, the United States suffered little physical damage. Hostilities ended in a dramatic victory at New Orleans, a battle that led to the presidency for General Andrew Jackson. The Treaty of Ghent, which officially ending the war, was signed on December 24, 1814.

Madison, another Democratic-Republican and Virginia planter, was inaugurated in 1809. He also served two terms (1809–1817). He and his administration did several things that rather quickly altered the future of American agriculture. In 1806, his administration arranged with the Creek Indians for the development of a horsepath from central Georgia to southwestern Alabama (Figure 6.1). Although designed to provide a primitive road for federal mail delivery from Washington to New Orleans, it soon became a heavily used road for settlers traveling from the Carolinas and Georgia to federal land in southwest Alabama. Fearing war with Great Britain, in 1811 Madison's administration ordered that the horsepath be improved for wagon traffic. Also in 1811, work was begun on the National Pike. When completed, it would link Cumberland, Maryland, with Columbus, Ohio, in the Old Northwest, and it too, like the federal horse-path, would have long-lasting effects on the evolution of American agriculture. Both the Federal Road across central Alabama and the National Pike from Maryland to Ohio directly resulted in opening millions of acres of virgin land to agricultural settlement.

The Federal Road was soon flooded with hundreds of settlers infected with Alabama Fever who believed that they could get rich quick by planting cotton in this new territory. This sudden influx of settlers alarmed the Creeks, and in 1812 they began to attack outlying settlements in the area. Finally, in August 1813, they attacked and massacred all of the settlers gathered at Fort Mims in southwest Alabama. A year later in 1814, General Andrew Jackson ended the military resistance of the Red Stick Creeks at the Battle of Horseshoe Bend. In the Treaty of Fort Jackson, which officially ended the Indian uprising, the Creeks ceded to the United States almost the entire southern half of Alabama.

After the end of the War of 1812, the great manufacturing centers in Great Britain began to flood the United States with inexpensive manufactured goods of all kinds. American manufacturers could not compete with these dumped imports. Soon, hundreds of factories either reduced their production or closed. Thousands of mill workers in the North and East were out of work. Quickly, the old Hamilton and Federalist principle of using the power of the federal government to advance the welfare of industry was resurrected. As a result, Congress passed and Madison signed the Tariff Act of 1816. This overtly protective tariff was designed primarily to keep coarse British cotton goods off the American market. The passage of the act was strongly supported by Henry Clay and John C. Calhoun, two Southerners who saw the tariff as a source of almost unlimited funds for more internal improvements. Madison supported the tariff, but not public improvements. He signed the tariff bill and vetoed the other. Southern cotton farmers were not at all pleased with the tariff. With the inflow of British goods limited by the tariff, Northern manufacturers were able to increase their prices. At the same time, cotton farmers had to sell their crop on an open world market. As a result, they were selling cotton low and purchasing Northern goods high (Coit, et al., 1963).

James Monroe, still another Democratic-Republican and Virginia planter, was inaugurated as president in 1817. He also served two terms (1817–1825). He had hardly taken office when faulty management by officials of the Second National Bank lead first to over-expanding credit and then to quickly reducing it. In 1819, over-expansion of protected American industries, loss of foreign markets after the enactment of the protective tariff, and collapse of the easy-credit-fed Western land boom resulted in a national financial panic. Monroe, like Madison was not a supporter of publicly financed internal improvements, and in 1822, he vetoed a bill that would have raised funds to repair the rapidly deteriorating National Pike. The bill had proposed charging a toll of turnpike users to pay for the repairs. People in the Old Northwest were furious with the president (Coit, et al., 1963). Monroe, near the end of his second term, succumbed to pressure from the growing manufacturing power in the North Atlantic Region to increase the rates on the 1816 tariff. It was also important to the future of the Democratic-Republican Party that a large share of the American population (44% in 1820), and a large share of votes, lived in that region. In 1824, Congress enacted and Monroe signed a bill further increasing rates. Southern cotton farmers were forced into a position of even greater disadvantage.

John Quincy Adams, another Democratic-Republican and a man with strong Federalist antecedents, was inaugurated in 1825. He had defeated Andrew Jackson in an extremely close contest that had to be decided in the House of Representatives. He served only a single term (1825–1829). Adams had a broad and progressive vision of the American future. In his inaugural address, he urged Congress to undertake a broadly-based program of public works and to enact legislation to encourage both culture and commerce. He held a Federalist view of the development and use of the country's vast resources. He strongly believed that the federal government should carefully shepherd these resources for the benefit of the public good. Public lands should be released slowly for development, and the proceeds from sales should be used for public improvements. His view was diametrically opposed to those pioneering settlers and supporters of Andrew Jackson who firmly believed that all those resources were gifts to them from God, and that they should be seized and exploited as quickly as possible. A large share of the electorate viewed Adams as a New England aristocrat who had lived his entire life on the public dole. Adams had just taken office when the Democratic-Republican Party became irrevocably divided. During the dispute, followers of Adams and Clay became the National Republicans (later Whigs), and the followers of Jackson retained the old name—Democratic-Republicans. Later, they would simply be called Democrats.

From the beginning, Adams's administration was so unpopular that he had hardly taken office before the campaign began to elect the next president. A major battle in that bitter campaign was fought by agricultural interests who wanted the lowest possible tariffs, and industrial interests who wanted tariffs set as high as possible. In 1828, the industrial interests won that battle with Congress enacting legislation to increase the tariff rates—thereby providing additional protection for New England mills. Southern agrarian interests were outraged; Southern planters called the act the Tariff of Abominations.

Andrew Jackson was inaugurated in 1829 and served two terms (1829–1837). He was the first person to be elected president who was not from one of the original thirteen colonies/states. Immensely popular, Jackson was elected because of his strong support for political and financial reform and because of his devotion to protecting states rights against the increasing power of the federal government. By 1828, most of the states had eliminated many of the restrictions that had limited the adult, white male franchise. As a result, the list of eligible voters grew rapidly. Jackson was swept into office on the shoulders of these newly enfranchised voters. Some commented that his election was a victory for the common man. Jackson had only been in office for little more than a year when Congress enacted the Indian Removal Act of 1830. In 1825, President Monroe had placed before Congress a master plan for the removal of all Eastern tribes to permanent territory west of the Mississippi. Nothing was done with this proposal until Jackson (an advocate of forced Indian removal) was in office. The Cherokee resisted their removal by appealing to the Supreme Court, which ruled in their favor. Jackson, however, refused to use the forces available to him to enforce the Supreme Court's decision. In fact, he abetted forced Indian removal, which became the Trail of Tears. Soon the old Indian barrier to westward expansion of agriculture was swept away.

Many Southern farmers and politicians, especially those in South Carolina, had reacted sharply to the enactment of the Tariff of Abominations in 1828. Politicians in that state had voted to nullify that federal legislation within the state of South Carolina. The resistance became so powerful and such a danger to the stability of the United States that the Jacksonians in Congress enacted new tariff legislation in 1832. The act, which returned the tariff rate to the level set by the 1824 act, did not provide nearly as much relief as the Southerners felt that they needed, and they intensified their resistance. A year later, Congress enacted the Clay Compromise. This act provided for the gradual reduction of tariff rates to the level set by the 1816 legislation. The New England mill interests strongly opposed the provisions of the legislation.

Jackson was never a strong supporter of the widespread use of federal funds for internal improvements. His general position on the matter was underscored in his veto of the Maysville Road Bill. The passage of this bill would have resulted in the use of some of those funds to build a road inside the state of Kentucky. In his veto message, in 1830, he voiced his general support for federally-supported, internal improvements only when they would result in benefit to a larger segment of the population than a single state.

Jackson had always opposed the idea of a national bank. In 1832, he had the opportunity to bring his opposition into play. The charter for the bank was up for renewal, and Congress had approved it. Jackson quickly vetoed it, and was sustained. The veto started a series of events that directly and indirectly lead to the disastrous financial Panic of 1837.

Martin Van Buren had been hand-picked by Jackson as his successor. He was elected by a wide majority and inaugurated in 1837, but he served only a single term.

From the beginning of his term, his administration was largely occupied with attempting to deal with all of the widespread effects of the Panic of 1837.

Jackson's earlier fight against a national bank was continued into the administration of Martin Van Buren (1837–1841) when the president vetoed a bill, sponsored by Henry Clay, to establish a new one.

By 1840, the Cotton Kingdom and slavery were growing rapidly in the Southern States. With this growth, the South was rapidly becoming a region wholly unlike the North with its growing urbanization, commercialization, industrialization, and diversified agriculture.

James K. Polk (1845–1849) was elected president on a platform of westward expansion (Morison, 1965). Soon after his inauguration, he began to put wheels under that platform by wresting the so-called Republic of Texas from Mexico. In February 1845, he sponsored a joint resolution in Congress to officially annex the Republic of Texas. Then in May 1846, the United States officially began the war with Mexico. The annexation of Texas did not end the quest of the American government for more Western land. In June 1846, Congress approved a treaty that established the boundary between the US and Canada at the 49th parallel. This treaty gave the US sole possession of territory that would become the states of Oregon and Washington. Finally, in 1848, the Treaty of Guadalupe Hidalgo with Mexico ceded Texas, New Mexico (including Arizona), and upper California, including San Diego, to the United States.

Westward expansion was strongly supported by most Americans. There was, however, widespread disagreement as to whether slavery should be allowed into those newly acquired territories. This question largely determined the politics at both local and national levels for the remainder of this period (1840–1860). An attempt at compromise was passed by Congress in 1850, but it did not really satisfy any of the protagonists in the rapidly-escalating battle.

In July 1854, a group of discouraged members of the Free-Soil Party, Conscience Whigs, and Anti-Nebraska Democrats met in Jackson, Michigan, to establish the Republican Party. Its members were committed to two primary issues: free land and preventing the spread of slavery to the Western territories. Although extremely small to begin with, the new party grew rapidly, and by 1860, it was having a significant impact on national politics.

ALABAMA POLITICS

The first four decades of Alabama's history provided a real roller coaster ride for the state's politics. Establishing a functioning government in a time of severe turbulence in American politics was a major task. Even the seemingly simple task of deciding where to locate the state capital became almost a life-and-death affair for many of the politicians. Then, before the government was firmly settled in, the question of secession reared its hydra-like head.

Governors

Winthrop Sargent, a Federalist from Massachusetts, served (1799–1801) as the first governor of Mississippi Territory. Other men serving in that capacity were: William C. C. Claiborne (Democratic-Republican, TN) (1801–1805); Robert Williams (DR, NC) (1805–1809); and David Holmes (DR, VA) (1809–1817) (Rogers, et al.,1994). William Wyatt Bibb, a Virginian by birth and a Democratic-Republican, served (1817–1819) as the first and only governor of Alabama Territory.

William Wyatt Bibb, who had been governor of Alabama Territory, was also the first governor of the state of Alabama. He served from 1819 until 1820 when he died in office. His brother, Thomas Bibb, also a Virginian by birth (DR, Limestone County), served the remainder of his brother's term (1820–1821). Other men serving as governors of the state were: Israel Pickens, a North Carolinian by birth (DR, Greene County) (1821–1825); John Murphy (NC) (DR, Monroe) (1825–1829); Gabriel Moore (NC) (DR, Madison) (1829–1831); Samuel B. Moore (TN) (DR, Jackson) (1831); John Gayle (SC) (DR, Greene) (1831-1835); Clement Comer Clay (VA) (DR, Madison) (1835–1837); Hugh McVay (SC) (DR, Lauderdale) (1837); Arthur P. Bagby (VA) (DR, Monroe) (1837–1841); Benjamin Fitzpatrick (GA) (D, Autauga) (1841–1845); Joshua L. Martin (TN) (D, Limestone) (1845–1847); Reuben Chapman (VA) (D, Madison) (1847–1849); John A. Winston (AL) (D, Sumter) (1853–1857); and Andrew B. Moore (SC) (D, Perry) (1857–1863).

Commissioners of Agriculture

The office of commissioner of agriculture did not exist in Alabama before 1883.

The Political Situation

The land that would later become the state of Alabama was included within Mississippi Territory, organized in 1798, under provisions of the Northwest Ordinance of 1787 (Figure 6.2). Within the territory, there were enough people residing in settlements along the Tombigbee River to organize a county there; consequently, Washington County was established in 1800 with St. Stephens as its county seat. In 1805, the Choctaw Indians ceded additional land in the Tensas basin to the US Government (Figure 6.3). Also in 1805, the Chickasaws in north Alabama ceded land in the western portion of the Tennessee Valley to the government. The following year (1806), the Cherokees ceded additional land in the eastern portion of the valley. These cessions would have important implications for politics in the eastern (Alabama) portion of Mississippi Territory for some time.

With the Southern cotton boom gaining steam in the original states of the South Atlantic Region in the early 1800s, well-to-do planters and politicians in the Broad River region northeast of Athens, Georgia, watched with great interest the developments taking place in the eastern portion of Mississippi Territory. When the lands from the Chickasaw and Cherokee cessions in the Tennessee Valley were auctioned, their representatives were first in line to make large purchases. They also benefited from the establishment of the federal horsepath and later (1810) the wagon road from central Georgia to the Choctaw cessions in Washington County (Figure 6.1). [This, of course, was part of the Federal Road from Washington to New Orleans.] After the 1814 Treaty of Fort Jackson, these Broad River Georgians were also first in line to purchase land ceded by the Creeks in south Alabama. Some of these same individuals were also extremely active in gaining statehood for Mississippi in 1817 and the establishment of Alabama Territory at the same time. As a result of their efforts, one of their group, William Wyatt Bibb, was appointed by President Monroe to be the Alabama territorial governor. Later, after the territory became a state (1819), Bibb would be elected its first governor. These old Virginia planters and their descendents were so active in all of these events that they became known in Alabama politics as the Georgia Faction and later as the Royal Faction.

The Georgia Faction and many of their supporters were generally well-to-do with large land-holdings, large plantations, and large numbers of slaves. It did not take long after Alabama was admitted to the Union for resentment of their wealth and influence to surface among the overwhelming majority of poor yeomen farmers in Alabama. When Governor Bibb died in 1820, resentment began to turn into open resistance. His brother Thomas, a planter-merchant from Limestone County, was chosen to complete his term. The depth of resentment became apparent in 1821 when the Royal Faction candidate, Dr. Henry Chambers, a physician from Madison County, was easily defeated by the non-Royal Faction candidate Israel Pickens in the election for governor. Faction influence finally disappeared completely in the campaign for governor in 1823. Pickens again defeated Chambers. In the legislative elections, Pickens's supporters won control of both houses. The Royal Faction had played an extremely important role in the early history of the state. Much of their effort was, no doubt, largely self-serving, but they did provide much needed leadership when there was none available locally.

The demise of the Royal Faction left a vacuum in Alabama politics. The yeomen knew what they were "agin," but they had great difficulty in deciding what they were "fer." The only thing that came close to uniting them was their undying dislike and suspicion of wealth and position (natural liberty). The lack of political consensus was apparent in the legislature as it wrestled with banking problems, location of a permanent capital, and reapportionment. Intertwined in all these political problems was the slowly declining price of cotton. At that time, it was the sole engine pulling the Alabama's economy along. Unfortunately, each year resistance to forward movement was getting stronger and stronger.

In the early 1820s, Alabama's electorate was casting about looking for a hero to lead them and the new state into the future. They did not have to wait very long or look very far to find one. Andrew Jackson, from neighboring Tennessee, began his rise as a military hero in the War of 1812, enhanced his fame in 1814 when his ragtag army ended the power of the Red Stick Creeks at Horseshoe Bend and in the Treaty of Fort Jackson forced the Creeks to cede 20 million acres in Georgia and Alabama. He achieved national, or even international, fame with his defeat of British forces in the Battle of New Orleans in 1815, just after the peace treaty had been concluded between the United States and Great Britain. His invasion and capture of Spanish Florida in the First Seminole War (1817–1818) further enhanced his fame. Jackson was the ideal hero for a substantial majority of Alabamians. After all, he was an outspoken supporter of the common man in their battle against oppression from wealth and its power (natural liberty). Furthermore, it probably was of no little importance that he and most of them had thousands of years' worth of ancestors buried in the Highlands of Scotland. He was one of them in many ways. When Jackson ran for president in 1824, he received 69.3% of the popular vote in the state. Alabamians' long-term infatuation with Democratic Party views had begun. Although Jackson received more electoral votes than any other candidate, he did not receive a majority. He was finally defeated when the US House of Representatives chose the New England aristocrat John Quincy Adams. Wealth and its power had, once again, defeated the common man. The results of the 1824 election only served to inflame and unify the growing electorate in the West, including Alabama. In the election of 1828, Alabama's yeomen swarmed to Jackson's banner. He won 89.9% of the popular vote in the state. Although there was some opposition to Jackson's position on the National Bank, Alabama voters gave him virtually all of their popular votes in 1832. Unfortunately, his popularity quickly began to dissipate, resulting in another short-term vacuum in Alabama politics.

Jackson, the old Indian fighter,was a strong proponent of Indian removal and of the Indian Removal Act of 1830. When in 1831, the Supreme Court held that the United States should honor its treaties with the Cherokees, Jackson initially refused to enforce the judgement. Then in 1832, federal troops were sent to an area near the present-day town of Eufaula to remove settlers from land guaranteed to the Creeks by treaty. In fact, it was a treaty that Jackson himself had worked out with the Indians. The matter created a huge uproar in Alabama over the rights of states to make and enforce laws within their boundaries versus the power of the federal government to enforce its treaties within those same boundaries. [Legally, the Creek Nation was not part of the state of Alabama.] The matter was settled in a manner that allowed the president to save face, but his popularity was somewhat diminished. Even then, however, he continued to enjoy widespread support in Alabama for the remainder of his administration.

As the uproar over federal efforts to remove settlers from Creek lands began to subside, opposition to Jackson began in the state. At the national level, anti-Jackson people had adopted the name Whigs for their party after he was elected president. By 1834, Alabama opponents of Jackson had adopted the same name for their organization. This party, which primarily represented cotton-plantation owners, was active in Alabama politics for many years, and although they won a number of local races, they were never successful at the state level. There were just too many Jacksonian Democracy advocates in the state, especially outside of the cotton counties.

Alabama politics, after the Creek affair at Eufaula in 1832 and through the end of the decade, was primarily concerned with the vicious, ongoing battle between Whigs and Jacksonians, the rising tide of states' rights concerns, Indian removal, disposition of Creek lands, the Creek War of 1836, and the financial Panic of 1837.

Although banking problems held much of the attention of Alabama's politicians throughout the 1840s, concern over the possible abolition of slavery by the federal government also began to play a role in Alabama politics in this period (Atkins, 1994). It was difficult to consider or pass legislation needed in the state, because of the continuing struggle for supremacy between the Democrats and Whigs. The Whig Party invited Alabama's citizens to adopt a broad national philosophy that could lead to the diversification of productive power and to the enhancement of commerce and manufacturing in the state. Unfortunately, the yeoman farmers who controlled elections in the state, suspected the motives of the Whigs. After the US Congress passed the so-called Compromises of 1850, Alabama politics quickly became even more consumed by bitter fights between Alabama Unionists and Secessionists. By far, the most ferocious of all the political battles taking place in the state during this period (1840-1860) involved elections for the state's congressional delegation. The ferocity of these disagreements increased in intensity through 1860, and virtually precluded getting the people's business done.

NATIONAL ECONOMY

Morison (1965) intimates that there was no real American economy until well after the American Revolution ended. Before the Revolution, the colonial economy was an integral part of the much larger economy of the British Empire. For example, there was not a single bank in the 13 colonies, and the Revolution was well underway (1781) before the first one was established. When peace was finally realized, the new country had three complex economic problems to deal with: first, it had to recover from a long and expensive war; second, it changed from occupying a special place in the most powerful economy (English) on earth to fending for itself in a highly competitive, cut-throat world; and lastly, it had to develop every aspect of its own economy, so as to be efficient and competitive, within a very short period of time.

The Economic Situation

One of the most significant events in the development of the economy of the new country was the writing of the Constitution, so that it established an economic charter for the whole country. For example, it forbade the levying of taxes or tariffs on interstate commerce. It directed that the federal government alone could regulate commerce among the individual states and with foreign powers. It directed the federal government to establish uniform bankruptcy laws, to create money and regulate its value, to establish standards of weights and measures, to establish post offices and build roads, and to fix rules governing patents and copyrights (Conte and Karr, 2001).

The American Revolution (1776–1783) resulted in an economic boom for the newly independent United States, as merchants responded to the need for all kinds of materials needed in the war effort. Unfortunately, this expanded capacity was mostly unnecessary after Yorktown, resulting in a postwar depression with rapid deflation of prices. Fortunately, while trade with England did not return to prewar levels for many years, especially in the North, the shipping economy in the Northeast recovered rapidly as a result of trade established with East Asia and, later, with the West Indies.

The South recovered from the economic problems resulting from the Revolution much more rapidly than the North. The English really had no suitable alternative as a source of tobacco, indigo, rice, and naval stores. For example, by 1786, the value of Virginia's exports had returned to prewar levels. Unfortunately, in the same year, the value of exports from Massachusetts were only 25% of the 1774 level. This disparity would persist for many years, and would ultimately lead to serious political differences between the two sections (Morison, 1965).

The young country's first tariff legislation, enacted in 1789 during the first Washington Administration, was primarily to provide much-needed revenue, but it was also intended to provide protection for infant industries just getting established.

The Embargo of 1807 severely damaged the American economy. American exports fell from $108 million in 1807 to $22 million in 1808. Warehouses were filled with goods with no markets. Merchants, manufacturers, and farmers everywhere suffered heavy financial loss (Coit, et al., 1963). In the South, prices for tobacco and cotton fell 40 to 50%. The embargo was lifted in 1809, but several years passed before the economy fully recovered.

Probably the most significant change, ever, in the American economy began to take place when manufacturing began to move from homes and farms to shops and factories. In the early colonial period, virtually all manufacturing was done in homes and on farms, and it was done primarily to meet the needs of individual families. For much of this period, the policy of British mercantilism had strongly opposed the development of manufacturing in the colonies. Over time, however, some of the homes and farms began to produce goods for their neighbors. The logical extension of these changes was to enlarge the houses into shops and to employ local people as workers. The shops soon produced enough goods to meet the demands for larger and larger areas. Finally, the shops were replaced by factories employing large numbers of people. Associated with these changes, was the replacement of human labor skills with machines. This change began in the eighteenth century in England when machines for spinning cotton fiber into thread began to replace spinning wheels, and water- and steam-powered looms began to replace hand looms. Shortly after the end of the American Revolution, all of these changes came to America. The machines were so productive that goods had to be marketed over even wider areas. Changes in the kind of power used in manufacturing ran parallel to changes in the organization of labor and hand in hand with the increasing use of machines.

The War of 1812 resulted in an economic boom for some segments of the economy, but at the end of hostilities, the young country itself was almost bankrupt. As a result, worthless paper money was soon being issued everywhere by state banks. In 1811, the Madison Administration and Congress had not moved to recharter the National Bank, so there was nothing to prevent states or individuals to issue as much paper money as they wanted. This worthless money was used to finance widespread speculation in Western land purchases. Finally, in 1816, the Second National Bank was chartered, but rather than moving to control the speculation, officials of the bank encouraged it by making large sums of money available on easy credit terms.

Inexpensive English manufactured goods had begun to flood into the United States in 1814, following the end of the War of 1812. The importation of these goods had had a severe adverse effect on the fledgling manufacturing industries in the Northeast. In 1816, Congress reacted by enacting a protective tariff. It quickly provided some relief for the new industries, but just as quickly caused severe problems for Southern cotton producers. The Tariff of 1816 (Madison) was followed by additional ones in 1824 (Monroe) and 1828 (Adams); both increased rates on imports. These higher rates provided additional protection for industries in the Northeast, but additional problems for Southern cotton farmers. This trend toward increasing rates on imports was reversed in 1832 (Jackson) when Congress voted to reduce them. In 1832, Congress voted to gradually reduce rates to the 1816 level. Obviously, these later changes were widely popular in the South, but equally unpopular in the Northeast.

In late 1818, officials of the National Bank became anxious about the speculation and the inflation in land prices, and they began to curtail credit. Branch banks were ordered to accept no currency not printed by the National Bank, to present all state bank notes for payment at once, and to renew no personal notes or mortgages (Morison, 1972). As a result, many state banks failed, and the Western land boom collapsed. In 1819, these financial problems, plus unemployment in New England, resulted in a national panic and deep depression. Prices fell sharply, and property values contracted. Many people were soon deeply in debt with no hope of repaying them. The economy began to recover in 1823, but it did not get completely back to normal for almost a decade.

The charter of the Second National Bank was scheduled to expire in 1836, but for political reasons, Congress passed a bill rechartering it in 1832. President Jackson vetoed it, and Congress was unable to override. The bank's director responded by reducing the loans it made to state banks by $10 million over a six-month period. This action resulted in a banking recession in 1833 and 1834. When factories began to close, construction slowed abruptly, and grain prices began to decline, the bank reversed its policy. It began to make substantial amounts of money available for credit on easy terms. Worthless paper money appeared everywhere. As a result, a wild boom ensued. Land sales doubled and redoubled as speculation in Western lands exploded. Inflation ran wild. Jackson was dismayed by this turn of events, and he issued a regulation requiring that all deposit banks and receivers of public money accept nothing but "coin" for the sale of public lands.

In March, 1837, a financial panic in England also forced them to reduce credit being used to purchase American cotton. These events severely reduced the amount of gold held by the national government, and a severe depression set in. Later in 1837, three great cotton marketing businesses in New Orleans failed. Cotton prices fell to catastrophically low levels. Some 126 companies in New York closed. The stock market crashed, and in May a run on the country's banks force them to suspend payments in coin. Soon thousands of unemployed people were on the streets. The depression was especially bad for some of the newly-founded towns in Ohio, Indiana, and Illinois. Many of them lost substantial numbers of their residents. When they could not find work, they had to return to the larger towns and cities in the East. The depression did not end until 1843. Jackson's battle with the National Bank was not the sole reason for the depression. Worldwide trade in silver and gold, financial schemes being pursued by other countries, and even the Chinese opium trade were involved, but none of these other causes alone probably would have tipped the balance if there had been no battle. Fortunately the Panic of 1837 had spent itself by1841 (Morison,1965). The panic had temporarily halted the Western movement, but after 1841, it quickly picked up speed again. Soon settlers from New York and Ohio were moving onto the prairies. A few of the more hardy were beginning to hew farms out of the forests of Wisconsin.

As late as 1850, most freight in the North was hauled on a rapidly growing canal system, and in the South on the many rivers in the region. Farmers on the prairies had to transport their grain on wagons and livestock on the hoof to the canals; however, by 1860, a network of rails crisscrossed the entire region from the East Coast to Michigan. Unfortunately, because of the low cost of transporting cotton downriver, railroad development in the South lagged far behind.

Construction of canals and railroads demanded the availability of a large amount of cheap labor that was not available in the settler population. As a consequence, thousands of immigrants from Europe entered the United States.

Expanded transportation led to the creation of large urban trading centers. This movement to urban centers involved even more people than the move westward. These centers concentrated the demand for food, shelter, and clothing that provided a powerful stimulus for increased commercialization and industrialization in the North. Except for a few river towns, there was little urbanization in the South during the period 1840-1860. Throughout this period, cotton production and slavery tied to a largely rural landscape increased in the South.

In the 1830s, cotton spinning and weaving began to move rapidly from England to North America. In 1840, there were 1,200 cotton factories in the United States. Two-thirds of them were in New England. Almost none of them was in the South, which produced cotton and shipped it to the North, while purchasing cloth and other manufactured products from the urban centers in that region.

In New England, most of the new factories were powered by falling water. In Connecticut, almost every waterfall was linked to small factories producing machine tools, firearms, furniture, tinware, and a wide variety of other manufactured products. Unfortunately, while small-scale manufacturing thrived, it would be some time before the United States produced as much iron as it needed.

Widespread speculation in land, banking, and transportation in the late 1830s, resulted in the Panic of 1837 that resulted in a depression that lasted from 1838 through 1843. In 1844, the economy began to recover; then expanded through 1857. In 1857, a Democrat-led Congress passed a new tariff that seemed to set off a short-lived panic. A period of declining prices followed.

The Gross National Product

Gross National Product (GNP) has been defined as a measure of the productivity of a country. It is the sum of all purchases of goods and services by individuals and government and the value of gross domestic investment, including changes in business inventories and exports minus imports (Darnay, 1992). A slightly different version of the same measure is called the Gross Domestic Product (GDP). While the GNP includes the value of goods manufactured in foreign countries by American companies, the GDP includes only the value of goods produced within the geographical boundaries of the United States.

The federal government did not begin to collect data that could be used for the estimation of the GNP until near the end of the 1860s, and it did not begin to publish annual data on the measure until 1889. Recently, however, Johnston and Williamson (2007) published data based on US censuses that provided estimates of GDP, beginning in 1790. Some of these data, for the period 1790-1860, are presented in Figure 6.4. The data in the figure are based on the value of the dollar in 2000. In other words, the original data are corrected for inflation and deflation of the currency that might have occurred during the period.

The data presented in Figure 6.4 show that the GDP of the new country was $3.6 billion (constant 2000 dollars) in 1790. Because of the scale of the graph, the effects of the panics of 1817 and 1837 on GDP are not very obvious; however, the recovery of the economy after 1840 is readily apparent. Between 1840 and 1855, the GDP increased from $27.86 billion to $61.71 billion. The downturn in the economy after 1855 is also fairly obvious.

The Consumer Price Index

The Consumer Price Index (CPI) is defined as changes in the weighted average price of a basket of consumer goods and services, compared to some base period. It is also sometimes called the Cost of Living Index.

The federal government did not begin to publish an official CPI until well into the twentieth century; however, enough economic data has been pieced together to obtain estimates beginning in 1800 (*Historical Statistics of the United States to 1970,* Tables E135-166). These estimates were computed using 1967 (Index=100) as the base year. These estimates, for the period 1800-1860, are presented in Figure 6.5. They reflect several of the disruptions in the American economy during the period 1795-1860. The sharp decline (down to 43 in 1802) in the CPI after 1800 was the result of the post-American Revolution depression and deflation. The effect of the 1807–1809 Embargo also resulted in minor changes. Price inflation, associated with War of 1812, raised the CPI to 63 in 1814. Afterwards, it generally trended downward relatively rapidly through 1824. The decline was especially sharp (40 to 33) during the 1819–1822 panic and depression. The inflationary period associated with Jackson's banking problems and fluctuating national gold reserves resulted in the increase in the CPI (29 to 34) from 1834 through 1837. In the depression that followed the Panic of 1837, it began to decline again.

The estimated US CPI in 1840 was 30. It had been 42 in 1820. With the country in a depression in 1840, the CPI was in the midst of a declining trend. It had been declining since 1838. It generally continued to decline through 1853 to 25. The CPI had never been that low. Afterwards, business began a slow expansion that continued into the mid-1850s. As a result, the CPI began to increase slightly, reaching 28 and 29 in 1854 and 1855, respectively. It remained near that level until it fell-back to 26 after the 1857 panic. It returned to 27 in 1859 and remained at that level through 1860.

Inflation Rate

As might be expected, considering the CPI data presented in Figure 6.5, the national inflation rate remained relatively benign throughout the period 1801-1840 (Figure 6.6). In 18 of the 41 years in the period, it was negative. It was above 10% (12.1%) in only one year (1813). This was during the inflationary period associated with the War of 1812. Even in the inflationary period leading up to the Panic of 1837, the inflation rate reached only 6.1%. The negative rates of inflation (deflation) were much more prominent in the data. They were lower than 10% in 1802 (-16.3), 1815 (-14.5), and 1824 (-11.1). They were just slightly less than 10% in two years (1820 and 1824).

During the period 1840-1860, the annual inflation rate varied between -7 to slightly more than 6.0. It was greater than 0 in five of the years, equal to 0 in nine years, and less than 0 in seven of the years. There was no indication of trends during the period.

Interest Rates

The Federal Reserve Board did not publish data on interest rates before 1934.

THE ALABAMA ECONOMY

As might be expected, there is relatively little information available on economic conditions in Alabama during the period 1819-1860. Some of the data that is available is reported in the 1840 Census and are presented in Table 6.1. These data show the level of involvement of the residents of Alabama and several other states in the trades, manufacturing, commerce, and agriculture. The states were chosen to show the relative levels of commerce and industrialization in three states created from the original 13 colonies (Georgia, Massachusetts, and Pennsylvania), two Midwestern states about the same age as Alabama (Illinois and Indiana), and two states created out of the territory of the Old Southwest (Alabama and Mississippi).

The data in Table 6.1 show that as early as 1840 the states in the South were already beginning to fall behind in commerce and industrialization. Although the difference is not great, it is obvious that industrialization has increased at a greater rate in Illinois and Indiana than in Alabama and Mississippi, even though the four states are essentially the same age. It is also obvious that agriculture involved a much larger share of the populations in Alabama, Georgia, and Mississippi than in any of the other states. It is also interesting that although Georgia was created from one of the original 13 colonies, its economy was at essentially the same level of industrialization, as measured by citizen involvement, as the economies of Alabama and Mississippi.

Some data concerning Alabama's economy and that of six other states from the 1860 Census are presented in Table 6.2. These data show that the level of industrialization of the economies of Alabama and Mississippi—as measured by manufacturing establishments, per million persons, and annual value of products of manufacturing, in dollars per person—were considerably lower than those of its sister states of Illinois and Indiana. In fact, the levels of those two variables were lower in the three Cotton States than in any of the other states.

THE AGRICULTURAL ECONOMY

Until the beginning of the nineteenth century, American agriculture essentially remained both spatially and functionally where it had been throughout its colonial history (Edwards, 1940). It was still generally confined to the same North Atlantic, Middle Atlantic, and South Atlantic regions where it had been during the lifetime of the original 13 colonies. Most of the farming practices in New England were those used by farmers in England before the Agricultural Revolution. In fact, they were still following the general crop rotation scheme (grain-grass-fallow-grain) practices of medieval England. It was a time of growing food and fiber for home needs with any surplus to be exchanged for salt and sugar and other necessities at a local store. They made their own clothes and most of their primitive farm implements. They used very little fertilizer of any kind, and neglected their livestock, orchards, and woodlands. In New England, most farmers continued to produce grain and livestock for home consumption and for barter; however, after the end of the American Revolution, growing urbanization around the commercial centers of Boston, New York, and Philadelphia was beginning to provide them with a rapidly growing cash market for their products, and many of them quickly began to manage their farming operations to supply those markets.

While early nineteenth century, New England farmers were plodding slowly along in the footprints of their English ancestors, their cousins in the South Atlantic Region were beginning to follow the footprints of a completely different farming system—footprints made by the slave-operated, sugar plantations in Brazil and the Atlantic islands (Azores, Canaries, Madeira, etc.) As detailed in a preceding chapter, in 1614 John Rolfe had devised a process for curing tobacco, a Native-American crop. Soon, this cured product was being exported with great success. In fact, it was such a success that farmers were unable to find enough labor to produce the quantity of tobacco that they could readily sell locally or export. The obvious solution was the same that was used on the sugar plantations—slaves. Slave-operated plantations would also soon be established for the production of rice and indigo in the Southern colonies, and by the beginning of the nineteenth century, they would be producing cotton. It should be noted, however, that from the beginning, in the South Atlantic Region there were far more small, family-operated, general farms than slave-operated plantations. The social, economic, and political agendas were set and controlled by the plantations and the production of staple crops, primarily for the export market.

From the beginning of the colonial period, in the South Atlantic Region, tobacco was the most important crop produced on the plantations and the farms of the yeomen; however following the end of the American Revolution, the crop went into a period of rapid decline. By the last decade of the eighteenth century, virtually all of the lands suitable for producing tobacco in the coastal plain had been cleared and was being used for that purpose. During the period of 1790 to 1815, only those farmers on newly cleared lands in the Piedmont of Virginia, North Carolina, and Georgia were able to make a profit on tobacco. This industry would not begin to grow again until much later when production was shifted beyond the Appalachian Mountains into the backcountry of Kentucky and Tennessee.

Upland, short-staple cotton had been produced in small amounts throughout the backcountry of the South Atlantic Region for many years before the American Revolution, but because of the difficult and time-consuming task of removing the seeds from fibers, it was grown primarily for home use. However, in one year in the decade before the Revolution, after roller gins came into regular use for removing the seed from the lint, the Southern colonies had exported 46,000 pounds of lint to England. Around 1786, Sea Island, long-staple cotton was introduced to coastal South Carolina and Georgia from the Bahamas. It could be produced in substantial quantities on the Sea Islands of those two colonies, and its long fibers were easy to separate from the seeds with the use of the newly invented roller gins. Unfortunately, it could not be grown further inland than about 50 miles; consequently, it had serious limitations as a commercial crop.

The roller gin was not very effective in separating lint and seed in upland cotton, and in the late eighteenth century, this problem limited the production of cotton away from the sea coast, but when the Eli Whitney machine was introduced, this situation changed rapidly. The Whitney gin removed lint quickly and effectively; although it did a considerable amount of damage to the individual fibers. It was estimated that United States cotton production totaled 3,135 and 16,719 bales (500 pounds each) in 1790 and 1795, respectively—most of it the Sea Island variety (Gray, 1958).

Whitney invented or perfected the cotton gin in 1793 and patented it in 1794. By 1800, US cotton production had increased to 73,145 bales. Most of that increase was the result of increased production of upland cotton. Then, in 1800, it increased to 146,290 bales, and in 1810 production reached 177,635 bales. With the availability of the new gin, upland cotton could be produced virtually any place in the South, but it grew extremely well in the Piedmont of the South Atlantic Region (Figure 4.45), an area containing large numbers of yeoman farmers involved largely in subsistence agriculture.

The explosion of upland cotton production in the South Atlantic Region at the end of the eighteenth and the beginning of the nineteenth centuries was only partly the result of the availability of the Whitney gin. Demand for cotton fiber had exploded in England some two decades earlier—a result of new technical developments in the textile production industry. Spielvogel (1994) commented that the English textile industry provided the first major step in the evolution of England's Industrial Revolution when the flying shuttle, Hargraves's spinning jenny, Arkwright's water frame spinning machine, Crompton's so-called mule, and Cartwright's power loom provided the ingredients for the establishment of the first modern factory. With this new machinery, it was possible to significantly increase the production of cheaper cotton cloth. Soon, it began to replace wool and linen textiles for the production of clothing. Consequently, the demand of cotton soared. Soon, English cotton goods were selling all over the world. There were other sources of cotton in the world, but America was much closer to England; so, as the worldwide demand for cotton material soared, the demand for American cotton soared along with it.

Cotton production brought about many changes in the agriculture of South Carolina and Georgia. As Wallace (1951) notes, the price of cotton would become "the barometer of southern prosperity," and at the same time, of "southern discontent." Just as the tobacco craze had commandeered agriculture, including food production, in the Virginia colony two centuries earlier, cotton production changed the agriculture of South Carolina and Georgia. In 1820, farmers had to be warned of the danger of placing so many of their resources in a single crop, but they paid no heed. Wallace (1991) further commented that South Carolina exported much cotton and rice and imported virtually every necessity.

Soon cotton production also expanded into the Lower Mississippi Valley. At the end of the French and Indian War in 1763, the English had taken over the French colonies east of the Mississippi, including the one at Natchez. In the following years, plantations with large slave holdings were producing tobacco and indigo for export. Following the invention of the Whitney gin, the production of cotton increased rapidly on the wind-blown soils (loess) in the Old Natchez District. Production of the crop became extremely profitable there, and over the succeeding decades, it spread onto the alluvial plains on either side of the great Mississippi. At Natchez, the planters soon abandoned the upland cotton being used in the Piedmont in South Carolina and Georgia and replaced it with a variety obtained from Siam (Thailand). This new cotton produced a higher quality lint, and yields were about twice as great. The planters were not completely satisfied with this new cotton, however, and they soon began breeding experiments to improve it. Within a few years, they had developed a new variety that would become the standard throughout the South, and later, throughout the world. Sales of lint from this new variety, plus sales of seed to farmers wishing to use it, added significantly to their already fabulous incomes (Moore, 1986). This improved variety of cotton, plus their excellent soils, gave the Mississippi River planters an enormous financial advantage over the Piedmont farmers to the east.

The changing nature of Southern commercial agriculture also affected the institution of slavery. Immediately before the American Revolution, slaves were primarily tied to the tobacco, rice, and indigo plantations on the coast; however, when upland cotton production exploded into the Piedmont, plantation owners moved there with their slaves. Also, some of the local yeoman farmers were able to grow enough of the crop to buy additional land, which necessitated the purchase of their first slaves. In 1787, South Carolina had stopped the foreign slave trade, and from 1792 to 1802, importation from other states was forbidden as well. However, the Napoleonic Wars, which began in Europe in 1803, resulted in a rapid increase in prices for cotton and rice. At the same time, production of those two crops, especially cotton, was expanding rapidly, creating a growing need for more farm labor. In that same year, South Carolina reopened its slave trade. Coleman (1991) commented that the rapid and extensive increase in cotton cultivation served to "fasten" the "yoke" of slavery much more securely on the South than in the past.

Export agriculture added another dimension to problems of plant production in the South Atlantic Region. With subsistence agriculture, such as that practiced by the Native Americans, exploitation of the limited natural fertility of the soils (Ultisols) was at a relatively low level, and generally governed by local population density. However, with the coming of tobacco production and later cotton production, those poor soils quickly became the base for satisfying the demands for an extremely large, almost insatiable, population base far away. Where native soil fertility had been shared only by a relatively small number of local Indians, by the mid-seventeenth century it was being shared by much of the population of England and some of Europe. This situation lead to the evolution of an agricultural system based on shifting cultivation. Under this system, once the limited, natural soil fertility in a field was exhausted, it was abandoned to fallow, and a new field was cleared from the farm's woodland. This system required a large reservoir of land, as well as a large amount of labor.

The mixed-crop farming system practiced in the North Atlantic Region and the plantation-slave labor system practiced in the South Atlantic Region provided the basic foundation of American agriculture at the time of the American Revolution. However, there was a third system, and although it did not involve many farmers, it was nevertheless extremely important because it lead agriculture westward in the beginning of the nineteenth century. This was the system used by the farmer-pioneers. These were the people who followed hunters, trappers, and Indian traders into the Western wilderness. They lived and farmed on the edge. They owned no land, few tools, and little livestock. Their hand-to-mouth agriculture never provided them with enough food and fiber to meet their needs. Without hunting, fishing, and gathering to supplement their meager farm production, they could not have survived. By 1800, large numbers of these farmer-pioneers were already pressed against the Appalachian Mountains from New England to Georgia. They produced almost no surplus of any crop that could be traded. However, these were the people who would lead the way westward. They would find the passes through the mountains and establish the trails that would become the roads to be followed by the general farmers and plantation owners as they began to move their farming operations to those new lands.

Although agriculture in America is at least 10,000 year old, there was never any systematic attempt to collect data on its characteristics until 1840. A considerable amount of data had been collected before 1840, but it generally included only local data collected over a relatively short period. Books by Bidwell and Falconer (1941), Gray (1958), and Otto (1994) summarize much of the available data obtained from the beginning of the sixteenth century to 1860. These books were used liberally in the preparation of this chapter.

The 1840 Census published data on the major and some minor plant and animal crops harvested in the nation, the individual states, and their counties. Although these data were not very comprehensive in scope, they allow us to begin to systematically document a very important milestone in the history of American agriculture—its early movement away from the Atlantic Seaboard.

Agricultural Legislation

In the period 1800-1860, the US Congress involved itself in matters related to agriculture only to a very limited extent. Most of the legislation was related to land policy or to getting a substantial portion of the enormous amount of federal lands into the hands of private individuals.

Land policy decisions faced the young country from the time it declared its independence in 1776. To begin with, much of the land in the original 13 colonies already belonged to either the states created from that land base or to people already living there who had titles to what they owned. Several of those states also claimed lands far beyond their frontiers. For example, Virginia, North Carolina and Georgia claimed territory lying far beyond their western-most settlements. In 1780, a year before the Battle of Yorktown, Virginia decided to cede all of her Western lands north of the Ohio River to the newly established Confederation of States. As a result, Congress passed an ordinance the same year with the provision that any land ceded to the Confederation would "be settled and formed into republican states which shall become members of the Federal Union." At the same time, the Confederation requested that all of the other states also agree to cede their unsettled Western lands.

After the final peace agreement with England was signed (Peace of Paris) in 1783, the Confederation was faced with the problem of how millions of acres of public domain land east of the Mississippi, south of Canada, and north of the Floridas should be distributed to its citizens. As a result, Congress passed the Territorial Ordinance in 1784. It provided that all of the public domain within the United States would be divided into ten rectangular territories. Each territory would be given a governor when its population reached 20,000 inhabitants and would become a state when its population equaled that of the smallest of the original 13 states. From the beginning, two contrasting points of view emerged regarding land disposition. The more liberal view preferred by New Englanders wanted the land sold in small tracts, at low prices, and on easy credit terms. The more conservative view favored by the Southerners wanted the land sold in large tracts for cash. Generally speaking, the large-tract-for cash view prevailed; consequently, the Ordinance of 1785 provided for the sale of federal lands at public auctions to private persons in minimum lots of 640 acres at a price of not less than $1 an acre. Unfortunately, very few of the people who so desperately wished to purchase land were able to do so for the simple fact that almost none of them had $640. As a result, much of the land was purchased by land speculators who would then attempt to sell it to the pioneers and squatters in small tracts for inflated prices and on credit. In many instances, these schemes failed because many of the prospective buyers could not even afford to purchase the smaller tracts, and if purchases were made on credit, many defaulted on their loans. The ordinance also provided for the surveying of federal lands. Under its provisions, all land that had not been surveyed would be described according to the so-called New England System whereby it would be divided into six-mile square townships consisting of 36, 640-acre sections. It also provided for the sale of surveyed land at public auction.

In 1787, the Congress of the Confederation passed the Northwest Ordinance that was expected to apply primarily to those lands north of the Ohio and east of the Mississippi. It further defined the fundamental principles of federal land and territorial policy including provisions by which territories could be formed, requirements for the admission of states to the Confederation, provisions for the establishment of territorial governments, and a provision that "There shall be neither slavery nor involuntary servitude" in a territory established under the Northwest Ordinance.

In 1790, North Carolina ceded its western lands to the federal government. In turn, these lands that included settlements west of the Appalachian Mountains were incorporated into the Southwest Territory. In 1796, some of the territory became the sixteenth state, Tennessee. In 1796, Congress passed the Land Act that essentially continued the policies as the 1785 Ordinance, except that the price for land was increased to $2 an acre. The 1796 Land Act did include a provision for credit, but stipulated that the full price had to be paid in one year. As a result, squatting on vacant public lands continued to be the popular, and in frontier communities, an acceptable way of acquiring a farm.

Land sales continued at a slow pace following the passage of the 1796 Land Act, and agitation increased in the agrarian South and from Western representatives in Congress to liberalize the land-disposal program. Consequently, the Frontier Land Act was passed in 1800. It kept the $2 an acre price, reduced the minimum purchase to 320 acres, and provided for more liberal credit. Under this act, purchasers had up to four years to pay for the land. Other important provisions included the establishment of land offices in those areas where land was being sold. It also established the office of surveyor general and a corps of surveyors to conduct the system of rectangular survey on a systematic and permanent basis (Cochrane, 1993).

Although Georgia had not yet ceded her Western lands, the federal government organized the Mississippi Territory in 1798, under the provisions of the Northwest Ordinance of 1787. The no slavery provision of that ordinance was not enforced in the establishment of the new Mississippi Territory.

In 1804, Congress further liberalized land sales policies. They reduced the minimum bidding price to $1.64 an acre and the minimum purchase to 160 acres. These changes coupled with the liberal credit provision from the 1800 Frontier Land Act were much more acceptable to the average person who wanted to buy land. Unfortunately, the more liberal terms served to encourage a greater degree of land speculation. Even with these liberal provisions, however, there were still thousands of farmers and many speculators who could not make their payments. Cochrane (1993) commented that even as late as 1819, half of the land that had been bought from the public domain had not been paid for.

In all of the early efforts to develop an effective and equitable land policy, no provision was made to protect the "rights" of squatters on the public domain. Even though many of them had provided a significant service for the country by pushing the frontier westward and by making improvements on their holdings, they could not have title to their land without paying for it. Many of these families had settled on their land before it was surveyed. To protect their "property rights," squatters often formed protective associations to force concessions when land offices were opened to sell land in their areas. They used various means, including open intimidation, to prevent speculators and others seeking to purchase land from bidding on land they felt that they "owned."

Again and again, in the years before 1840, Congress had passed laws sanctioning the action of particular groups for the purchase of public lands. Finally, in 1841, Congress passed the Preemption Act, whereby heads of families and widows were allowed to settle on 160 acres of un-surveyed land with the right to purchase their holdings at the minimum price when they were actually put on sale (Edwards, 1940). In effect, the Preemption Act sanctioned the rights of squatters on un-surveyed federal land.

Details regarding the enactment of early tariff legislation were discussed in a preceding section. Here I will remind you only of the different tariff acts that were enacted before 1860. Tariff acts were enacted in 1789, 1816, 1824, 1828 (Tariff of Abominations), 1832, 1833, 1842, 1846, and 1857. These acts were generally detrimental to the agricultural economies of the South, especially to the cotton producers who exported most of their crop.

In 1839, Congress decided to enact legislation to involve the federal government directly in American agriculture. In that year, they appropriated a small amount of money to the commissioner of patents in the Department of State for the purpose of collecting agricultural statistics, conducting agricultural investigations, and distributing seeds. Beginning in 1847, they began annual appropriations for these activities. In 1849, they transferred the Patent Office and its agricultural activities into the newly created Department of the Interior.

In 1840, the National Reform Association began to support the doctrine that public land should be distributed in equal amounts to actual settlers. After several years of political maneuvering, Andrew Johnson introduced a homestead bill in 1852. It easily passed the House, but it could not pass in the Senate because it became part of the slavery controversy. It would not receive additional consideration until Southern congressmen withdrew from the body after secession.

In 1854, Congress passed the so-called Graduation Act. Under the provisions of this act, land that had been on sale for ten years would be sold for a minimum of $1 an acre, and land that had been on sale for 30 years would be sold for 12.5 cents an acre.

Agricultural Exports

Data on agricultural exports were not reported on an annual basis during the period 1800-1860. Only averages for the different decades are available. These data are presented in Figure 6.7. The data indicate that the annual average value of exports for the decade 1800-1809 was $23 million. In that decade, agricultural exports accounted for 75% of all exports (Figure 6.8). Without annual data, it is not possible to determine the effect of the Embargo Act of 1807. It reduced total exports from $108 million in 1807 to $22 million in 1808, so it must have had a similar effect on agricultural exports. Certainly, the average for the decade ($23 million) must have been reduced significantly. The Napoleonic Wars in Europe also raged through the first 15 years of the nineteenth century, and must have had some negative effect on American exports.

Average annual exports almost doubled in the next decade (1810-1819) to $40 million. Agricultural exports accounted for 82% of all exports. The War of 1812-1814 must have reduced agricultural exports somewhat, but it is difficult to know how much. Also, in 1815, England enacted the Corn Laws that placed an extremely high tariff on the importation of grain. While the Corn Laws might have negatively affected the export of grain to Great Britain, the growing demand for cotton obviously had a positive effect. During the period 1816-1820, cotton accounted for 40% of all American exports.

The annual average for the decade of the 1820s ($42 million) was only slightly higher than it had been in the 1810s. The share of total exports from agriculture declined to 61%. One of the contributing factors to the limited increase was the sharp decline in export prices for cotton. Prices had been as high as 34 cents a pound in the preceding decade, but in 1820 it was near 17 cents. Then they generally declined for the remainder of the decade, reaching nine cents in 1830. Because cotton accounted for such a large share of agricultural exports, the decline in prices would have had a strong negative effect on the total value of those exports. However, while low cotton prices were exerting a negative effect on exports, the continued expansion in the quantity exported had a positive effect. During the decade of the 1820s, cotton exports accounted for one-half to two-thirds of the value of all exports. The export of wheat and wheat products was considerably lower in the period 1821-1830 (4.1 million bushels) than in the period 1815-1820 (4.7 million bushels). By that time, Europe had sufficiently recovered from the Napoleonic Wars to be able to meet much of their need for grain. The severe financial Panic of 1819-1822 and the depression accompanying it probably also helped to restrain exports somewhat.

In the decade of the 1830s, the value of agricultural exports increased to $74 million (73% of the total). Cotton exports continued to increase, and prices rebounded from the lows of the preceding decade. Export prices were near 17 cents in 1835. Although they declined again by the end of the decade, the average price for the 1830s decade,was substantially higher than in the 1820s. The value of wheat exports also increased during this decade.

The data presented in Figure 6.7 also indicate that average annual exports of farm products increased to $90 million in the decade of the 1840s. Agricultural exports comprised some 65% of all US exports. In 1846, Great Britain repealed the Corn Laws that had kept the importation of grains to very low levels. After the repeal of the laws, grain exports to the British Isles began to increase.

In the 1850s, annual agricultural exports averaged $189 million (Figure 6.7) and accounted for 81% of all exports. The Crimean War (1854-1856) provided a boom for American agricultural exports during the 1850s. Throughout the period 1840-1860, cotton remained the primary agricultural export of the United States.

Agricultural Prices

There is little data available on prices that farmers received for the crops that they sold or for the prices they paid for farm inputs during the period 1800-1860. Bidwell and Falconer (1941) published data on prices paid for all products purchased and for prices of all farm commodities. Rather than publishing the actual data, they published indices of the data where the value of these products purchased in 1825 was equated to 100. These index numbers are presented in Figure 6.9. The data indicate that for most years the relative prices paid for agricultural products were higher than for prices in general. The data also show the effects of economic booms and busts on prices. All prices declined as a result of the Embargo of 1807-1809; then prices increased sharply with the inflation associated with the War of 1812. They declined sharply around the time of the Panic of 1819-1822, but were up sharply during the panic and depression of 1837-1843. Then they began to increase again during the period of recovery and business expansion from 1844-1856.

Farmers

The essential role of farmers in the advancement of civilization and society has been detailed in a preceding section. As producers of food and fiber and as citizen soldiers, farmers had played a major role in the development of America before and during American Revolution. Yet the Revolution and its aftermath began to push farmers into the background. The war had demonstrated that the new country could not survive without increased emphasis on government service and bureaucracy, commerce, trade, diplomacy, manufacturing, and urbanization. The importance of the farmer would remain unchanged, but his relative position in society began to diminish. This situation is reflected in the general lack of information collected on farmers in the censuses. There was little data collected on farms in the country until 1840, and specific data on the farmers themselves would not be included until much later.

The Total Population

To fully appreciate the place of the farmer in the early evolution of American agriculture, it is necessary to place them alongside the larger population of the country. Data on the distribution of the total population in 1790, 1800, 1810, 1820, and 1830 are presented in Figures 6.10, 6.11, and 6.12. Data on the population in the Southeast (South Atlantic and South Central Regions) in the period 1810-1860 are presented in Figures 6.13, 6.14,6.15, 6.16, 6.17, and 6.18. Numerical estimates of the total population of the United States, its major regions, and Alabama are presented in Table 6.3. Numerical estimates of the populations of the states in the South Central Region are presented in Table 6.5. All of the numerical data, with the exception of that for the Western Region, were obtained from census data published by the University of Virginia, Geospatial and Statistical Data Center. There were no census data for states in the Western Region until 1850, but, of course, there were large numbers of people living in the region long before 1860. Coulson and Joyce (1999) have provided estimates of numbers of persons living in the various areas that would later become states. These estimates were used to compute the totals for the regions and are included in Table 6.3 for 1820, 1930, 1840, 1850, and 1860.

Both the 1790 and the 1800 data indicate that at the

beginning of the nineteenth century almost all of the population remained in the territory of the original 13 colonies (49% of the total in the North Atlantic Region, Maine south to Pennsylvania; 43.4% in the South Atlantic Region, Virginia south to Georgia). By that time, 327,000 people had moved into the states of Tennessee and Kentucky in the South Central Region. This region would finally include all states from Kentucky south to Alabama and west to Texas and Oklahoma.

The 1810 Census counted slightly more than 7 million people in the country and included, for the first time, a count in the North Central Region—231,000. This region would finally include all states from Ohio south to Missouri and west to Nebraska and the Dakotas.

The 1820 Census counted 9.9 million people and included, for the first time, a count in the Western Region—52,000. This region would finally include all states from Montana south to New Mexico and west to the Pacific.

The population of the country totaled 13 million in 1830 and 17.3 million in 1840. In 1840, for the first time, the population in the North Central Region exceeded that of the South Central Region with its thousands of slaves (3,353 thousand versus 3,098 thousand). In 1840, the third largest concentration (3.4 of 17.3 million) was in the North Central Region. The fourth largest concentration (3.1 of 17.3 million) was in the South Central Region. Very few Americans had settled in the Western Region; by 1840, some 67,000 of 17.3 million US citizens lived in that region.

By 1850, vast new areas of land were available for purchase and settlement, and farmers, merchants, etc. were increasingly moving west. As a result, the population in the inland areas was beginning to increase faster than along the coast. By 1850, slightly less than 57% lived in the original 13 states, and by 1860, the percentage had declined to 50%. The largest increase during the period 1840-1860 was in the North Central Region. From 1850 to 1860, it almost doubled (5.4 to 10.6 million). The increase in the South Central Region was not nearly as great (4.3 to 5.6 million), as the number of people moving west from the region was only slightly smaller than the number moving into it from the East.

The Farm Population

In 2000, the Economic Research Service of the United States Department of Agriculture published a history of American agriculture from 1776 through 2000, in which they estimated the number of people in the farm population in each decade beginning in 1840. They estimated that the total was 9.0, 11.7, and 15.1 million in 1840, 1850, and 1860, respectively.

These estimates can be used, in turn, to estimate the percentage of the farm population as a percentage of the total population. In 1840, farmers accounted for 52% of the total population, then declined to 50.1 and 48% in 1850 and 1860, respectively. There are no data for earlier decades, but in 1800 it was probably 90%. These data indicate the rapid growth of commercial agriculture in the early nineteenth century. By the middle of the century, farmers were producing enough food and fiber to meet their own families' needs plus a surplus large enough to meet the needs of an equal number of other people.

A more useful way of expressing the relation of farmers to the entire population is to compute the ratio: total population/number of farmers. This ratio provides a rough estimate of the number of persons that an individual farmer can feed. In 1840, this ratio was only 1.9 (17.3 million persons per 9 million farmers). In 1850 was also 1.9, but in 1860 it increased to 2.0. Even with the level of technology available to farmers during that period, they still could do little more than provide food and fiber for their families.

The Nature of Early American Farmers

It has always been extremely difficult to arrive at a satisfactory definition of a farmer. Of course, all farmers are involved in some way in farming, but having arrived at that point, it is difficult to go much further. This difficulty was apparent from the earliest colonial days. For example, virtually all of the early colonists were directly involved in farming. However, many of the farms were owned by absentee owners in England who never once in their entire lives, visited their farms. Those English landlords were obviously involved in farming, but it was difficult to count them as farmers. This problem has been extremely persistent, even into modern times.

Bidwell and Falconer (1941) characterized the early farmer (including the entire family) in the North Atlantic Region as a jack of all trades. They were strongly self-sufficient, producing their food and fiber, clothing, furniture, and farm implements. Virtually all of them were yeomen farmers. These authors commented that this versatility was not a result of choice, but rather the result of persistent efforts to adjust to their geophysical and economic environments. They really had no alternative to being versatile. Fortunately, commerce developed rapidly in the coastal towns and provided many of the farmers with some opportunity for off-farm work. Although the opportunity for working away from the farm increased with time, lives of the farmers in that region did not change very much for a long time.

Comments made by Gray (1958) suggest that there was a much broader range of farmers in the South. He identified five groups that were present in the region during the period 1800-1860. These included: poor whites, Highlanders, commercial farmers, aristocratic planters, and plutocratic planters.

Of course, as the author notes, these were not completely discrete groups. They tended to merge at the margins. Generally, there was an increasing level of resources (wealth) at their command in their farming operations from the poor whites through the plutocratic planters. Further, there was a continuum across the five groups in the roles they played in shaping the political system that would govern their affairs. This characteristic would become increasingly important over time, as plutocratic planters wielded political power far in excess of their numbers. Although there was no continuum involved, farmers in the first three groups were involved in general farming and produced a variety of crops. Both groups of planters produced a much more limited variety of crops, most of them destined for the market economy and export.

Farmers in the Work Force

The Economic Research Service (2000) included estimates of the percentage of the labor force made up by farmers. They estimated that in 1790 farmers comprised 90% of the entire American labor force. There are no estimates for other decades until 1840. By that decade, the percentage had declined to 69%, and it continued to decline in 1850 and 1860 to 64% and 58%, respectively.

Census data on the number of persons employed in agriculture in the different US regions in 1840 are presented in Table 6.4. No comparable data are available earlier than 1840 or for 1850 and 1860. Data collected in the 1840 Census, indicated that the largest number of persons employed in agriculture in the country that year was in the North Atlantic Region (1.1 million), and that there were more persons employed in agriculture in New York (456,954) than in any other state. Although the total number of employed was high, it represented only 16.8% of the estimated population of that region.

In 1840, a million persons were employed in agriculture in the South Atlantic Region. Almost a third of these (318,771) worked on farms in Virginia. Slightly more than 693,000 persons were employed in agriculture in the North Central Region, most of them in Ohio, Illinois, and Indiana.

The highest percentage of the total national population employed in agriculture (including slaves) was in the South Central Region (27.4%). In 1840, there were 317,200 persons employed in agriculture in the states of the Old Southwest (Alabama and Mississippi). Approximately 177,400 were employed on farms in Alabama.

As would be expected with slaves included, a large share of the persons employed in agriculture in Alabama lived and worked in the cotton counties of the Black Prairie, Coosa Valley, and Highland Rim (Figure 1.11). Employment, ranging from 5,800 to 10,200 persons, was concentrated in Madison County and on the Black Prairie from Montgomery County westward. A total of ten counties in these two areas accounted for 42.6% of all persons employed in agriculture. Employment was especially high in Dallas (10,906), Montgomery (10,120), Greene (9,729), and Madison (8,528) counties.

Slaves

Slaves were generally not considered to be farmers, because few of them owned farms. However, they played important roles in the daily lives and fortunes of many of the farmers in the country, especially in the South.

Almost from its beginnings, African Americans had played a role in the development of colonial agriculture. It quickly became obvious, however, that slave labor was not economical on the relatively small, multi-crop farms in the North Atlantic Region. Conversely, farmers producing export crops in the South Atlantic Region quickly found out that they could not operate without them. While white farm laborers might have been more economical in the long term, they were simply not available in the quantity required to produce the amount of tobacco needed to meet the rapidly growing addiction to nicotine in Europe. While slaves played a vital part in the evolution of agriculture in the country, few of them actually owned farms until after the Civil War. For this reason, changes in their number and distribution are considered separately from those people who did own farms.

Data on the number of slaves in the different regions and in Alabama are presented in Table 6.5. Data on their distribution in the South Central Region in 1830, 1840, and 1850 are presented in Figure 6.19. By 1800, the number of slaves in the North Atlantic Region was already declining. The census that year counted 36,100 there. The count in 1790 had been 40,100, and the number would decline steadily in the next three decades. From 1800 through 1830, more slaves lived in New York than in any state in that region. There were 21,000 in New York in 1810, but less than half that number (10,100) in 1830. It is likely that only a very few of these slaves would have been involved in farming.

The number of slaves in the South had been increasing slowly since the coming of the tobacco plantations in Virginia in the early part of the seventeenth century. The 1790 Census indicated that there were 654,100 in the South Atlantic and South Central regions (the South). The invention of the cotton gin in 1793 revolutionized cotton production and ignited a vast explosion in commercial agriculture and the slave trade in the two Southern regions. By 1800, there were 851,000 slaves in the two regions combined (Table 6.5), and by 1810, 1820, and 1830, this total had grown to 1.1 million, 1.5 million, and 2 million, respectively. At the same time, the center of the distribution of slaves was rapidly shifting southwestward. In 1800, 40.7% of slaves in the South were in Virginia. In 1830, only 23.9% lived there. In that year, there were over a half million in the territory stretching from Georgia to Louisiana. It was not that the number was decreasing in Virginia; rather it was that the number was increasing rapidly in the other states to the southwest

Table 6.5 indicates that there were no slaves in the North Central Region in 1800 and 1810; the Northwest Ordinance of 1787 had decreed that all land west of New York and Pennsylvania, north of the Ohio River, and east of the Mississippi River would be Free Soil. Later the Missouri Compromise (1820) provided that slavery would be prohibited in all lands north of N 36° 30'. The act excluded Missouri, which had been admitted to the union as a slave state, as part of the Missouri Compromise. The Census of 1820 counted 11,300 slaves in that region, all of them in Missouri Territory. By 1830, the count had more than doubled (25,900).

By 1840, many of the states in the North Atlantic Region had freed their slaves, and there were only 800 remaining in that region that year (Table 6.5.), and most of them were in New Jersey. In 1860, there was less than 100 in the entire region.

Since colonial times, most of the slaves in the country had been in the South Atlantic Region; however, by 1840, the South Central Region was rapidly catching-up, as cotton production exploded westward. In 1840, there had been a million in the region, and by 1860 there were slightly less than two million. Sometime in the decade of the 1850s, the number of slaves in the western portion of the South had by-passed the number in the eastern portion. Figure 6.19, clearly shows the rapid westward expansion of slaves numbers in the region during the period.

Gray (1958) discussed the question of slave prices in the United States in considerable detail. He commented that prices were affected by age, sex, physical condition, temperament, skill, and experience; by the price of cotton; and by general economic conditions. He also commented that newly-imported slaves were sold for less than those that were reared here. From the time of the invention of the cotton gin (1793) until the beginning of the War of 1812, prices of slaves in Virginia increased from $300 to $500. In South Carolina and Georgia, the increase was from $400 to $600.

By about 1835, the price of prime field hands ranged from $1,000 to $1,200, but then prices declined sharply after the Panic of 1837. Around 1855, prices began to increase rapidly again, and soon reached levels between $1,500 and $2,000. In 1860, the average was $1,800. It is not known how many of Alabama's 435,080 slaves were prime field hands, but it is obvious from these data that the total investment in slaves must have been in the hundreds of millions of dollars in the state.

Tenancy

For thousands of years, relatively few farmers had owned the land where they grew their crops. They were forced to farm on land owned by the elite. This ancient situation prevailed in the early days of the colonial period; however, after the American Revolution there was so much land available in the new country that tenancy became relatively rare. Unfortunately, with the passage of time, as the country began to fill up, the old order began to be re-established, especially in the South (the South Atlantic and South Central regions), and tenancy began to increase.

Amid the growing wealth of the Southern planters, the growing population of the region (both whites and blacks), the increase in land in cultivation, and rapidly escalating land values, the numbers of farmers forced to become tenants and to establish and operate their farms on land belonging to someone else increased. The origins of tenancy in the South are shrouded in obscurity, but they certainly had their roots in the antebellum years. For decades before the Civil War, in fact, from the earliest days of the country, many yeoman farmers, specially in the backcountry were barely surviving. From the beginning, many of them with their small farms were simply not economically viable units. And there were literally thousands of such units in the region on the eve of the Civil War. For example, in 1860, there were just over 50,000 farms in Alabama. Some 43.6% of them were less than 50 acres, and on virtually all of these farms only a fraction of the total acreage was actually cultivated.

The high reproductive rates among Alabama's people was almost certainly a factor in the origin of tenancy there. Flynt (1989) commented on the large number of children in many of the early farm families. Odum (1936), writing over a century after Alabama's statehood, commented that the white population of the Southeast tends to reproduce at a higher rate and have a larger ratio of children and young people per mother than the remainder of the country. As children of these early families grew to adulthood and formed families of their own, only a few of them could share their parent's resources and farm on family land. Few of their families had sufficient wealth to provide their offspring with separate farms with the required farming equipment and livestock. With the large number of children involved in each family, the resources of the parents could be divided only so many times. Further, in those early days there were no lending sources where a young family might borrow money to start their own farm. Ultimately, most of the children would have to find their own way in the world, and tenancy was likely the best option they had.

Even in the best years, many yeomen throughout the South did not own enough land, livestock, or equipment to produce crops sufficient to meet the needs of their families. As a result, more and more of them had to turn to wealthy landowners for help. Some of them had sufficient livestock and equipment, but they just did not have enough land, so they moved onto a larger farm and rented some land from the owner. Those who paid cash for the rent were called cash tenants. Others, who paid their rent in crops usually gave the landowner a third of their grain or a fourth of their cotton. As a result, they were said to be farming on thirds or fourths.

Before the Civil War, tenants were overwhelmingly white, and they tended to be more like their landowning, yeoman neighbors. In those early years, when there were few renting tenants, they were allowed considerable control over their farming operations. They were allowed to choose the crops they produced, and most of them usually chose food crops over cotton. Unfortunately, some yeomen who needed help had virtually nothing but their labor to pay the rent on the landowner's property. In this situation, the landowner furnished land, housing, fuel, working stock, livestock feed, and seed. If fertilizer was used, the landowner determined the kind to use, then included its cost as part of the inputs that he was providing. The tenant provided the labor, and fed and clothed his family. When the crop was harvested, the tenant and owner shared it equally. This arrangement gave rise to the term, sharecropping or farming-on-halves; the farmer was called a share tenant. Sharecropping came to be the most pervasive form of tenancy in the South, and it was the least rewarding form for the tenant and his family. Under this arrangement, in many years, the crop would be so small that the landowner could not recover the cost of his inputs. In that situation, he would carry the shortfall over into the next year. When this occurred, the sharecropper would begin the following year in debt.

Generally, yeoman farmers reacted negatively to becoming tenants, even though they might improve their incomes in the process. To them, it represented a decline in status. In the beginning, when they were forced to become tenant farmers, most of them expected that it would be a temporary expedient to get them back on their feet; however, over time, it became more and more apparent that this would not be the case. Under the economic conditions developing in the South, it was highly unlikely that a tenant farmer would ever be anything but a tenant farmer.

Immigrants in US Agriculture

This period (1800-1860) also saw the dramatic addition of thousands of immigrants to the American farm population. Initially, most of them found employment in the construction of canals and railroads, linking the East and West. Morison (1965) notes that only around 129,000 immigrants entered the US in the decade of the 1820s. In the 1830s, this number increased to 540,000 (44% Irish, 30% German, and 15% English). The number almost tripled in the 1850s, and reached 2.8 million in the 1860s. A large share of these newcomers remained in towns near where they entered the country, and entered the work force there. A significant number took jobs building canals and railroads in the West; however, when those construction projects were completed, many of them and their children entered the labor force of the growing farm economy in the new states of the Old Northwest.

Edwards (1940) comments that over a half million Germans came to America between 1830 and 1850, and another million before the beginning of the Civil War. These immigrants generally settled around Cincinnati, in the lake counties of Wisconsin, and in Indiana, Illinois, and Michigan. Only a relatively small number of them settled in the South Central Region.

Farmer Migration within the Old Colonial Boundaries

Migration of farmers westward literally exploded beyond the old colonial boundaries, following the end of the Revolutionary War, but in reality migration was commonplace within those boundaries almost from the beginning. As detailed in a preceding chapter, by the mid-1600s tobacco farming and tobacco farmers had begun to migrate up the river valleys west of Jamestown in search of land to replace that worn out around the original settlement. Similarly, in the latter part of the 1600s, disgruntled farmers in the southern part of the Virginia colony had begun to move southward into the territory that would later become the North Carolina Colony. At the same time, farmers in the Plymouth Colony began to move northward and westward soon after the first settlements were established. From earliest times, for farmers within the old boundaries the grass was always greener.

Migration within the North Atlantic Region

By the 1800s, there were substantial numbers of farmers settled throughout the North Atlantic Region and beyond the boundaries of the original seven colonies of the North Atlantic Region (Figure 6.11). Vermont had become a state in 1791. There were already enough settlers in Maine that it had reached territory status; the 1800 Census indicated that there was already 151,719 persons there. Also, large numbers of settlers had already reached western New York by 1788. The density of the population there was still relatively low, but there were settlements throughout the region, and by 1800, it was beginning to fill up. Soon, many of the early settlers were looking for more elbow room, and as a result they began to pack-up and move again—this time into the Old Northwest (Figures 6.11 and 6.12).

Migration within the South Atlantic Region

By 1800, virtually all of the non-mountainous territory in the original six colonies of the South Atlantic Region had filled up. Farmers in every state except Georgia were pressing against the Appalachians. In Georgia, western migration was effectively blocked by the large number of members of the Creek Confederacy living there. In fact, the region was already filled to the point that substantial numbers of its population were migrating beyond the mountains into Tennessee and Kentucky (Figure 6.11).

Migration Beyond the Old Colonial Boundaries

As detailed in a preceding section, as late as the end of the 1700s, most of the citizens of the new country continued to live along the Atlantic (Figure 6.10), but this situation was about to change dramatically. Within the next half century, the greatest emigration in history would take place. In 1860, only slightly more than 50% (15.8 of 31.5 million) of the population was still living along the Atlantic Seaboard. Not only had they leaped the Mississippi, but also a substantial number were living along the Pacific.

There were several reasons for the exodus:

1. Virtually all of the soils in the old settlements, along the coast, were worn out. Generally, crop yields were much lower that they had been earlier.
2. With the general increase in population (Table 6.3), land prices were increasing.
3. After the American Revolution, millions of acres of land became available in the West at very low prices.
4. After the Treaty of Paris (1768), danger from Indian attack began to diminish. By the early 1800s, there was very little physical resistance from Native Americans to westward emigration.

For at least 6,000 years, Caucasians have been slowly moving westward. Westward migration of Indo-European people into Eastern Europe probably began as early as 4,000 BC (Figure 4.33). By 500 BC, they had occupied much of the territory in Eastern and Western Europe. The wide Atlantic stopped their westward movement for a time while ocean-going technology evolved, but around 1400 they resumed their westward migration. They clung precariously to the eastern edge of the New World for some two centuries, until they were sufficiently strong enough to overcome their mother-country and the indigenous peoples of the Americas. Then they began to move westward again.

By the late eighteenth century, many farmers in the original 13 colonies were pressing hard against the eastern slopes of the Appalachians from New England to Georgia (Figure 6.10). A few of them had found gaps in the mountains and had passed through to establish farming communities in regions that would later become Vermont, Ohio, Kentucky, and Tennessee.

As early as 1763, after the end of the French and Indian War, English settlers from Connecticut and Massachusetts began to flow northward into the Connecticut River Valley, into the region that would later become Vermont. Six years later, a group of Virginians passed through Cumberland Gap in extreme southwestern Virginia into territory that would later become Tennessee to establish a settlement on the Watauga River. In 1774, Virginian James Harrod established the first permanent settlement (Harrodsburg) in territory that would later become Kentucky (Figure 6.10).

Soon after the American Revolution, settlers from New Jersey and New York had followed the Susquehanna and Tioga river valleys into western New York. The first settlers arrived in the Ohio territory in 1788 when a group from Massachusetts founded the village of Marietta north of the Ohio River. Later in the same year, settlers from New Jersey established a community near where the Little Miami River enters the Ohio. By 1800, these early settlers had blazed the trails through the once impregnable Appalachians. Now what had been just a trickle was about to become a flood. The speed with which the early emigration of farmers moved westward is evident in comparing Figures 6.10, 6.11, and 6.12.

The Romance of Pioneering

There seemed to be something romantic about a family with all of its belongings—on a wagon, on the back of a horse, being rolled along in a large barrel (a hogshead) with handles on it, or even being carried on their own backs and in their hands—going into a wilderness, a virgin forest full of Indians, determined to establish a new life and make their fortunes. The truth is that many of them never had a chance to realize their dreams. They had been economic casualties where they came from, and there was no magic in the wilds of the West that would reverse that situation. Even the most determined application of sweat equity could carry them only so far. When they arrived, they did not have time to enjoy the scenery. They immediately faced three very difficult problems: preparing a permanent shelter, clearing the forest, and making a crop.

Preparing a Permanent Shelter

The pioneers' immediate problem was to provide some kind of a permanent shelter for their families. There was plenty of raw material available for constructing houses, but it was not an easy task to convert those huge trees standing in the forest to walls and roofs. They usually had some sort of temporary shelter with them that they used on the trail, but this was not sufficient to protect the family from the elements for very long, especially during the winter months.

Clearing the Forest

There were some natural clearings in the forests, but much of the Old Northwest and Old Southwest was covered by thick stands of virgin timber. Crops could not be grown under the closed canopies of those thick stands. Somehow these impediments to sunlight reaching the soil had to be removed. This could be accomplished by killing the tree in place and waiting for the leaves to fall off, or cutting and removing them to create an open clearing. For most of the settlers, there was only one option. While they did have axes for cutting the trees, they did not have the necessary means of moving them out of the way once they were cut; consequently, they had to use the old Indian method of girdling, or cutting away the bark and conductive tissue in a narrow band around the circumference of the tree. Deprived of water and nutrients supplied by the roots, the leaves would begin to fall as the tree slowly died. Of course, this process left the tree standing, but without its leaves, it no longer shaded the crops planted around its slowing decaying skeleton. This was the same solution to the problem that the first settlers in Virginia had use 2 centuries earlier.

Making a Crop

The vast majority of early settlers brought little food with them, and there were no supermarkets; consequently, they had very little time to eliminate the forest canopy, and get a food crop planted, matured, and harvested before the small amount of provisions they had brought with them was exhausted. Fortunately, most of these pioneers had not completely given up the age-old method for providing food for their families—hunting and gathering, and there was plenty of wild game, fish, nuts, and fruits for the taking.

These immediate difficulties were not equally troubling to all settlers. Settlers with older children, especially older boys, had extra labor to help with the difficult tasks associated with creating a life in a wilderness. Also, the common practice of extended families, neighbors, and church congregations traveling together and settling near each other provided a large pool of labor that was available through the sharing of difficult tasks. Further, in case of calamity, these groups functioned as a sort of safety net for the individual families.

Wealthy planters, who emigrated into the wilderness of the Old Southwest with slaves, had to deal with essentially the same problems as the small family traveling alone. However, with the resources available to them, they were usually able to get their plantations to an operational state in a relatively shorter period of time. To begin with, they were usually able to bring more provisions with them; although, of course, they needed a great deal more. Also, with the large amount of labor available, and usually some animal power, they were able to clear away the forest to create large open fields. Further, clearing and planting could be carried on simultaneously. They did not have to wait for leaves to fall off the trees. In fact, with the planters' labor supply, they could clear, plant, build buildings, and hunt and gather at the same time.

Development of Transportation

One of the important steps in the movement of farmers westward and the commercialization of American agriculture during the period 1800-1860 was the changes that took place in transportation (Edwards, 1940). In 1800, virtually all of the transport of goods related in any way to the evolution agriculture had to be moved along the waterways of the new Western territories in boats of various kinds or on poorly defined, overland trails. Earlier, the Connecticut, Hudson, Susquehanna, York, James, Cooper, Savannah, Alabama, and Mississippi rivers had been used by settlers to get away from the coast. As they moved further westward, they used some of these same waterways and their tributaries to move their families and belongings. Unfortunately, waterways did not always go where the settlers wanted to go. When this happened, they had no choice but to strike out overland.

In 1800, overland transportation westward was limited to a few Indian trails leading into the wilderness from settled areas along the coast. As late as the American Revolution, there were only three roads leading west or east out of New York City, and only one leading west out of Philadelphia. In the South, a single rude trail passed from the coast through the Harpers-Ferry area, and another through the Cumberland Gap. Traders traveling overland out of Charleston or Savannah had only Indian trails to follow.

From this very humble beginning in 1800, within the next 60 years, transportation east of the Mississippi river literally exploded across the countryside in a series of discrete steps: privately-financed toll roads, government-financed regional roads, government-financed canals, and government-financed railroads.

Construction of Roads and Turnpikes

After the American Revolution, as westward immigration began to grow, businessmen looking to make money from tolls began to form stock companies to provide funds for the construction of turnpikes. One of the first of these connected Philadelphia to Lancaster, Pennsylvania, a distance of 62 miles. This private venture constructed between 1792-1794 was so successful that other stock companies were established throughout the Northeast to finance the construction of similar roads. By 1832, 86 of these companies in Pennsylvania had constructed 2,200 miles of roads. In 1810, there were some 180 similar companies active in New England. Between 1800 and 1807, 88 companies had constructed 3,000 miles of roads in New York.

While the use of private funds for road construction quickly began to meet the needs for intrastate transportation, it was soon obvious that they would not provide for the transportation that was needed for interstate use. To that end, after 1806, Congress began to show an interest in road construction. The first, and one of the most important, of these regional roads constructed with federal funds was the National Pike (Cumberland Road). Begun in 1811, it connected Cumberland in western Maryland with Columbus in central Ohio. Later, it would be extended to Vandalia in south-central Illinois. State funds authorized by the New York legislature in 1794 were used to construct the Genesee Road from near Utica westward, just south of Syracuse, to Canandaigua southeast of Rochester. When completed, as the Seneca Turnpike, it reached into the wilderness a total of 157 miles.

steamboats

In 1807, when Robert Fulton piloted his steamboat up the Hudson River from New York City to Albany, a new era in the evolution of American agriculture and commerce began. Until the appearance of the railroad, steamboats were primarily responsible for the increase of commercial agriculture west of the Appalachians. In 1811, the first steamboat traveled down the Ohio from Pittsburg to the Mississippi. Afterwards, most of the commerce in the Old Northwest involved steamboat transportation on the Mississippi, Missouri, and Ohio rivers to and from the port of New Orleans; while in the Old Southwest commercial agriculture still depended on the Mississippi, Alabama, Tombigbee, and Tennessee rivers. By 1851, there were nearly 600 steamboats on those major rivers.

Canals

River transportation did play an important role in establishing commercial agriculture, but it was limited somewhat by the fact that most of the large rivers did not go where people wanted to go. There was no suitable river, for example, that connected the growing commercial centers in the Northeast with the growing farming communities in western New York or northern Ohio. To remedy this situation, in 1817, New York began to construct the 364-mile Erie Canal, connecting the Great Lakes at Buffalo to the Hudson River at Albany. It was completed in 1825. Soon all kinds of farm produce from the Old Northwest were flowing to and through the international shipping port of New York City. Farm produce from that region that had once gone down the Ohio and the Mississippi to New Orleans could now go through New York to Europe and the Caribbean much quicker.

In 1828, the federal government began the construction of the Chesapeake and Ohio Canal. It was to begin at Georgetown in northwest Washington, DC, and follow the Potomac River northwestward to terminate in Cumberland, Maryland. At one time, there were plans to extend it beyond Cumberland to Pittsburg, so it could be linked to the Ohio River. This would have connected the Chesapeake Bay with New Orleans. But the construction of the early sections of the canal proved so difficult that the engineers soon gave up the idea of going on to the Ohio. Construction reached Harpers Ferry in 1833, but that was the easy part. It did not reach Cumberland until 1850.

Railroads

The success of the Erie Canal resulted in a sharp increase in canal building. Unfortunately, by the time that many of the projects were being completed, they were made obsolete by the coming of railroads. The first railway was constructed in Massachusetts, with horses being used to pull the cars. In 1828, the Baltimore and Ohio Railroad ushered in the Railroad Age. Until 1840, most of the railroad construction in the country was limited to states on the Eastern Seaboard, and as late as 1850 there were only 9,000 miles of railroads in use in the whole country. However, in the next decade that total increased to 30,000 miles. New York was connected to Chicago by rail, as was Charleston, Savannah, and Chattanooga. Gray (1958) notes that by 1860 all of the major commercial centers in the South were connected by railways.

Migration into the South Central Region

The westward expansion of American agriculture beyond the Appalachians began in the South Central Region before it did in the North Central Region. As noted in a preceding paragraph, there were settlers from Virginia and the Carolinas in territory that would become Tennessee and Kentucky as early as 1769 and 1774, respectively. But it would be many years before many settlers would move into other parts of the region. The French were in New Orleans, and they controlled the lower Mississippi River until 1803. Spain controlled Mobile until 1813, and the powerful Creek Confederacy controlled most of western Georgia and three-fourths of Alabama until 1814. Consequently, it was not possible for many settlers to move into the lower portion of the region until after the Battle of Horseshoe Bend in 1814 (Figure 6.13).

Kentucky and Tennessee became states in 1792 and 1796, respectively. These were the first states established in the South Central Region. The federal government established the Mississippi Territory in 1798. It included much of the land that would later become the states of Mississippi and Alabama, but the Census of 1800 did not include counts of any people in the territory, although there certainly were several hundred living in the far western portion along the Mississippi River. Also, there were substantial numbers of people already living along the Mississippi from New Orleans northward to the general location of Natchez. Further, after the end of the French and Indian War (1755-1763), exiled French Acadians began to settle in the southwestern portion of territory that would become Louisiana. By 1800, some 5,000 of the so-called Cajuns were settled there. However, because none of these settlers were located in an American state, they were not included in the 1800 Census. As a result, the 1800 Census indicated that there were 326,577 persons in the South Central Region, all in Kentucky (220,955) and Tennessee (105,602).

In 1804, following the Louisiana Purchase, the federal government established the Orleans Territory. Later, it would become the state of Louisiana. Between 1800 and 1810, a substantial number of settlers moved onto Chickasaw and Cherokee land cessions (Figure 6.3) in north Alabama and onto Choctaw cessions in southwest Alabama and southeast Mississippi. Also, with New Orleans and the Mississippi River in American hands, settlers were moving into western portions of the region as well. However, because there were still no states established south of Tennessee, those settlers do not appear in the 1810 Census. However, census data indicated that there were 406,511 and 261,227 persons in Kentucky and Tennessee, respectively (Figure 6.11). At the same time, settlers from Pennsylvania, Virginia, and the Carolinas were literally flooding through the Cumberland Gap into the Nashville Basin in Tennessee and into the entire state of Kentucky. However, Indian resistance to settlement was still preventing migration into western Tennessee (Figure 6.13).

The 1820 Census estimated that there were 1,360,000 persons in the South Central Region (Table 6.5 and Figure 6.14). After 1800, Louisiana (1812), Mississippi (1817), and Alabama (1819) had become states. Numbers of persons in those new states, in 1820 were: 153,407; 75,448; and 144,317, respectively. Between 1820 and 1830, the number of persons in the region increased from 1,360,000 to 2,062,000 (Table 6.5 and Figure 6.15). The 1830 total included settlers from Arkansas Territory (30,400) for the first time. Most of the regional increase (52.5%) occurred in Tennessee and Kentucky. The increase in Tennessee (422,800 to 681,900) was especially large. The Indian problem in Tennessee was largely solved during this decade, and settlers took advantage of it by flooding into the western portion of the state. Most of the new settlers immigrating to the region were coming from Virginia, the Carolinas, Georgia, and eastern Tennessee, and they probably had a strong Celtic (Scotch-Irish) flavor. McWhiney and McDonald (1989) noted that people of Celtic ancestry "dominated the interior from Pennsylvania southward, ranging in various areas from three-fourths to nearly a hundred percent of the population."

The population in the South Central Region increased from 2,061,800 in 1830 to 3,025,400 in 1840, an increase of 963,600 persons, or 46.7% (Table 6.5 and Figure 6.16). The larger share of the total population in the region in 1840 was still in Kentucky and Tennessee, 25.8 and 30.7%, respectively. An increase in the number of slaves accounted for much of the overall increase in the population during the decade (Table 6.7 and Figure 6.19). The slave population increased by 398,000, or 41.3%, of the total population increase. Alabama lead the region in the number of slaves (253,500). In 1840, slaves accounted for 42.9% of Alabama's total population.

During this decade (1830-1840), most of the remaining Native Americans in Tennessee, Alabama, and Mississippi were moved out of those states, making millions of acres of additional land available for settlement. By 1840, virtually all of this land had been occupied (Compare Figures 6.14 and 6.16).

As early as 1823, a wedge of Americans thrust across Mexico's northern border into territory that would eventually become the state of Texas. By 1834, a colony established on the Brazos River by Stephen F. Austin with the approval of the Mexican government, included 20,000 whites and 2,000 slaves (Morison, 1965). However, even though these were primarily American settlers, they were not included in the 1840 US Census, because Texas was not then a state.

Between 1840 and 1850, the population in the South Central Region increased from 3,025,400 to 4,303,400, an increase of 1,278,000 persons, or 42.2% (Table 6.5 and Figure 6.17). However, the larger share continued to be in Kentucky and Tennessee, 22.8 and 23.3%, respectively. In this decade, the increase in the slave population again accounted for a significant share to the total population increase (451,000 of 1,278,000) (Table 6.7 and Figure 6.19). In 1850, Alabama continued to lead the region in the number of slaves, with 342,800, or 44.4% of its total population. For the first time, the census (Table 6.7) included the number of persons in Texas (212,600). That state had been admitted to the Union in 1845. With the Texas situation settled after 1845, settlers literally swarmed into the Lone Star State (Figures 6.18 and 6.19). Between 1850 and 1860, the number of persons counted there increased from 212,600 to 604,200.

During the decade between the 1850 and 1860 censuses, the number of persons in the South Central Region increased by 1,465,200 (4,303,400 to 5,768,600), or by 34% (Table 6.5 and Figure 6.18). Settlers were continuing to come into the region from the East, with a substantial share of the total ending up in Texas. Even though the flow of settlers into Kentucky (20%) and Tennessee (19.2%) had slowed somewhat, as late as 1860, 39.2% of the total population in the region lived in those two states.

Between 1850 and 1860, the slave population of the region increased by 545,000, or by 37.5% (Table 6.7). This increase accounted for 37.2% of the total increase of the population in the region. In 1860, there were slightly more slaves in Mississippi than in Alabama (436,600 versus 435,100). In that year, slaves in Mississippi and Alabama accounted for 55 and 45.3%, respectively, of the total population.

Going to Alabama

Immigration into Alabama represented an important stage in the westward movement of American agriculture. The Appalachians, stretching from northern Maine southwestward to northeastern Alabama, were a formidable barrier to the movement of farmers and their families westward from the Atlantic states. But once they passed around the southern flank of this mountainous barrier and entered territory that would later become the state of Alabama, there was little more to impede them than relatively level coastal plain terrain reaching westward for a thousand miles or more. There were, however, thousands of Cherokees, Creeks, Chickasaws, and Choctaws astride this pathway west (Figures 6.20 and 6.21). The new American government had already forcefully removed a similar barrier to western immigration in the North Central Region, and it was prepared to repeat the process in Alabama.

Native American Barrier. As detailed in a preceding paragraph, in the early 1800s, a growing wave of thousands of pioneers were impatiently biding their time in Tennessee, the South Carolina Upcountry, and western Georgia (See Figure 6.13)—waiting for someone or something to break the barrier of Native Americans who were preventing them from taking possession of the lands in the Old Southwest that their Creator had made just for them. The unfortunate process by which the barrier was breached achieved the same result with all of the major groups of Indians (Choctaws, Creeks, Chickasaws, and Cherokees), but there were some differences in the way it progressed in each case.

Choctaws. The Choctaws were the major Indian nation of the lower Mississippi region. When the French arrived at the beginning of the eighteenth century, the Choctaw occupied a narrow strip of land in present-day Alabama in the Tombigbee River Valley, from central Mobile County northward to the northern portion of Sumter County (Figure 6.20). The principal Choctaw landholdings were in southeastern Mississippi.

In October 1540, the first contacts between the Choctaws and Europeans, resulted in a six-hour losing battle between the Indians under their chief, Tuskalusa, and the forces of de Soto. Almost a century and a half later, in 1702, the French moved their garrison from Biloxi to a site upriver from Mobile and immediately concluded an alliance with the Choctaws. The Choctaws, located for the most part in French-claimed territory in southwest Alabama, benefited from trade with the French; consequently, they sided with them in their conflicts with the British.

The end of the French and Indian War in 1763 left England in control of West Florida, from the Apalachicola River westward to New Orleans and all of French territory west of the Mississippi. Then in 1765, the Choctaws ceded their lands between the Tombigbee River on the east to the Chickasaha River (southeastern Mississippi) on the west and from the coast on the south to Bladon Springs (southeast Choctaw County) on the north to the English. This was the first tract of land ceded by any Indians to white settlers in what would later become the State of Alabama (Moore, 1934). This ceded land included the Bigbee Bottoms, which later would become the site of the first agricultural settlement of whites in Alabama.

At the end of the American Revolution, Spain controlled East and West Florida, including New Orleans and the Mississippi River as far north as the present site of Vicksburg (Walnut Hills). The treaty that ended that war gave Spain control of all of West Florida north to the 31st parallel (Figure 6.2); however, the Spanish actually claimed lands above the 32nd, or as far north as the confluence of the Tombigbee and Black Warrior rivers, and they constructed a fort even further up the Tombigbee. These Spanish claims included the early Choctaw cessions (1765) to the British. As a result of the Spanish claims to so much of south Alabama, the Creeks, Choctaws, and Chickasaws, or at least some of them, made treaties putting themselves under Spanish protection (Atkins, 1994).

In 1795, the United States and Spain, signed the Treaty of San Lorenzo (the Pinckney Treaty), whereby the Spanish agreed to accept the 31st parallel boundary, and to evacuate all forts north of that line. Unfortunately, that territory had never been surveyed, so no one knew exactly where that line was. The Spanish were confident that the old French fort at St. Stephens was below that line, so they established Fort San Esteban de Tombecbé on that site. The Spanish garrisoned the fort with the expectation of providing protection for settlers from Indian attacks in that region. The Spanish also promised generous land grants to Americans on the old Choctaw lands and on some of the Creek lands there. In addition, they promised freedom of religion and no taxes. In return for these provisions, the American settlers had to take oaths of loyalty to Spain. During this period, Americans were arriving in goodly numbers to settle there.

In October, 1802, following the signing of the Treaty of San Lorenzo, the Choctaws confirmed to the United States the land cession they had made to the British in 1765. Three years later (1805), they ceded another tract of land in southwest Alabama to the United States (Figure 6.3). Two years after the defeat of the Creeks at Horseshoe Bend (1814), the Choctaws ceded the bulk of their remaining lands in Alabama (Figure 6.3).

In 1813, the great Shawnee Chief Tecumseh,attempted to enlist the Choctaws in resisting the encroachment of the white settlers from the East. The great chief actually visited the Choctaws before going on to Creek territory. Despite his great oratory, he was not able to convince them to join the resistance. Each of his arguments for resisting the whites was countered by the Chief Pushmataha of the Choctaws who insisted that they were friends of the whites and that such a war would ruin their tribe (Atkins, 1994).

Removal of the Choctaws from their land stirred more public opposition to their removal than any of the other Indian nations. Opposition was especially strong in Mississippi; however, it was not strong enough to stem the greed of the Americans for more and more Indian land. As a result, the Choctaws became the first of the so-called Five Civilized Nations/Tribes to begin forced removal.

In 1829, the Choctaws, under severe pressure from Andrew Jackson, signed the first treaty providing for their removal (Treaty of Doak's Stand). However, they became apprehensive about the land promised to them in the West, and refused to leave. Then after Jackson became president, the Choctaws were tricked into accepting the Treaty of Dancing Rabbit Creek. The first group departed for the West in December, 1830. Most of the removal was completed by 1833; although a sizeable number again refused to leave. However, by about 1854, the remnants finally moved westward, away from the unrelenting pressure of the settlers and their governments.

Creeks. Before the middle of the sixteenth century, the Creeks apparently controlled most of the territory that would eventually become the State of Georgia, but about that time the Cherokees in northwestern Georgia and Tennessee began to pressure the Creeks to move generally southward toward the western part of Georgia. When the Georgia colony was established in 1733, the location of the Creeks was generally as shown in Figures 6.20 and 6.21. At the end of the American Revolution in 1783 (the last battle was in 1782), the Creeks were the most powerful of the Indian nations in the Southeast. Because of their location in the middle of the Federal Road that led from western Georgia to Alabama and Mississippi, they probably suffered the most at the hands of American settlers from the East. In Alabama, they finally lost the most territory of all of the Indian groups.

The relationship between the Creeks and the British was good in the beginning. At the end of the French and Indian War in 1763, the British controlled West Florida from the Apalachicola River to New Orleans and all French territory west of the Mississippi. Once they gained control, they began to establish good relationships with the Indians in the area, primarily the Creeks. As a result, the Creeks signed the Treaty of Pensacola in 1768; it provide for the cession to the British of Indian lands from the coast to 35 miles inland.

In 1805, the Creeks allowed the Americans to develop a horsepath (the Federal Road) through their territory to facilitate the delivery of mail from the East to southwest Alabama (Figure 6.1); actually from Milledgeville, Georgia and points east to New Orleans. Later, in 1810, the road was widened. In 1811, the Indian agent, Benjamin Hawkins, reported that the Federal Road was crowded with settlers moving westward through the Creek Nation to the Choctaw land cessions (Bigbee Bottoms) in southwest Alabama.

By the end of the first decade of the nineteenth century, the Alabama Creeks were being squeezed between the Georgians moving relentlessly westward and the growing American settlements along the lower Tombigbee. Every day there was less living space and hunting ground. By this time, decades of intermarriage between the Indians and European traders had produced a bicultural and bilingual people. Many of these people lived as Americans. They herded cattle, used African slaves to grow cotton, and produced food in their barnyards, orchards, and gardens. More and more, they came under the influence of the old British, now American, market economy, and they enjoyed the status that came with their new wealth.

At the beginning of the second decade of the nineteenth century, conservative elements within the Creek Nation began to rebel under the onslaughts of the Americans against their living space and their culture. Encouraged by the great Shawnee Chief Tecumseh and led by a group of religious zealots called prophets, the Indians began to increase their depredations against the white interlopers. In July 1813, the Indians won a decisive victory over a small group of American militiamen at the Battle of Burnt Corn Creek in Escambia County. This victory encouraged the Red Stick Creeks, while at the same time, increasing the power of the prophets. The flush of that victory encouraged them to strike a lethal blow at the Americans; and within a few weeks (August 30, 1813), they attacked a large gathering of settler families at Fort Mims on the Alabama River in northern Baldwin County. They killed 250 men, women, and children, and took a few away with them as slaves. The Creeks could not know it, but this was a senseless, as well as a hopeless move. They had struck a fearful blow at a small number of terrified American settlers. Unfortunately, the end of their story had already been written hundreds of years earlier in the steady westward march of Europeans through history. As Atkins (1994) notes when European thrusts of discovery and exploration evolved into occupation and settlement, all Indians in the New World, including the Creeks, lacked not only the technology, but also the vitality and the strength of numbers, to defend themselves. In less than a year, on March 27, 1814, Andrew Jackson—with Indian allies, including White Stick Creeks—attacked a large group of Red Stick Creeks with their religious leaders at Horseshoe Bend on the Tallapoosa River. In the ensuing battle, more than a thousand Red Stick Creek warriors were killed. In less than two years, the Red Stick dream of redeeming their culture and reclaiming their living space was lost. Creek power in Alabama was destroyed. A few months later (August) at Fort Jackson (old Fort Toulouse), the Creeks signed a treaty that ceded to the Americans almost half of the territory that would become the state of Alabama some six years later. Under terms of the Treaty of Fort Jackson, the Indians were able to keep some land in the eastern portion of the state(Figure 6.3). Unfortunately, when news of the American victory at Horseshoe Bend spread to the cities and towns of the the old Southeastern states, settlers determined to squat and claim Creek land began to rush westward.

Jackson's victory over the Red Stick Creeks at Horseshoe Bend did not completely end Indian efforts to slow the rush of settlers. Attacks along the Federal Road continued after the Treaty of Fort Jackson with a randomness that defied defense (Southerland and Brown, 1989). On March 13, 1818, the Ogly and Stroud families were killed at Fort Bibb near Greenville (Butler Springs) in northwestern Butler County. More settlers were killed a week later in the same vicinity. Later, the Stokes family was wiped out near Claiborne in Monroe County. After these murders occurred, a sweep of the region was conducted by a military unit stationed at Fort Crawford, located on the Conecuh River south of Brewton in Escambia County. Later, Governor Bibb wrote to a Creek chief informing him that as a result of the murders and the difficulty of differentiating between innocent hunters and killers in that region that all Indians would be required to leave those lands included in the 1814 cessions.

By the beginning of the 1820s, the Creeks along with other Indians in the Southeast realized that they could not win a war with the Americans; consequently, they began trying to become "civilized" and to meet the Americans on their own terms (Hudson, 1976). More than ever, they began to adopt settler farming practices, to build mills, and to acquire titles to their property. At that time, the Creeks still controlled a large amount of territory in the northeastern quarter of Alabama, under provisions of the Treaty of Fort Jackson. Unfortunately, the waves of migrating settlers wanted every single acre of it, and with the help of federal and state authorities, planters, politicians, and land speculators, proceeded to get what they wanted. Where legal means were insufficient to accomplish their goals, they resorted to illegal ones. Hudson (1976) comments that seldom in modern history has one people's aggression against another been so unforgiving and relentless and marked by such terrible results.

The Creeks strenuously resisted removal. Alabama brought pressure to bear on the Indians to do so by refusing to give them any legal protection. Settlers and renegades moved in illegally to stake claims on their lands, and the state of Alabama afforded Creeks no legal redress. By 1831, many of the Creeks were starving, and to make matters worse, an epidemic of smallpox broke out among them. In 1832, the Creeks finally gave up and signed the Treaty of Washington (Treaty of Cusseta). Under the treaty's provisions, the Creeks ceded all five million acres of their territory to the United States. Out of this total, the US Government reserved a little over two million acres to be allotted to the Creeks for plantations and farms. Parcels of this land were to belong to individual Creeks who were to be given deeds after they had occupied the land for five years. Alternatively, they could sell their parcels and move west. The treaty also specified that the United States would guarantee their right to property and that they would remove white intruders and troublemakers. The US government further agreed to pay their expenses for their removal to the West and to finance their subsistence the first year there. The Creeks were explicitly assured under the terms of the treaty that they were free to stay or immigrate. A census taken in May 1883 revealed a population in the Upper Towns (Upper Creeks) of 14,000 people, including 450 black slaves. The population in the Lower Towns (Lower Creeks) was somewhat lower: 8,500 people including 457 black slaves (Hudson, 1976).

The United States immediately broke the terms of the treaty when it allowed white settlers to invade Creek lands where they looted, burned, and killed without federal interference. The newly-formed State of Alabama continued its policy of allowing living conditions of the Creeks to become so miserable that they would much prefer removal to the West. Creek removal began in December, 1834, and was essentially completed by 1838. In the process, many Creeks perished. In the first group of immigrants, only 469 of 630 made it to Indian Territory in the West. Many of them froze to death or died of hunger and disease. Later, 311 of 611 drowned when the steamboat *Monmouth* carrying Indians to Mobile for shipment to the West collided with another boat and sank.

So by the end of the third decade of the nineteenth century, 18 years after the Alabama Territory became a state, the hegemonic liberty of the English and the natural liberty of the Scotch-Irish (Fischer, 1989) had prevailed, at least as far as the Creeks were concerned. God's good green earth, in much of Alabama, was now, in the minds and hearts of the settlers, finally and completely in the hands of those who could make best use of it (Hudson, 1976).

Cherokees. The Cherokees claimed a relatively small amount of land in the territory that came to be the state of Alabama, and a considerable amount of that which they claimed in the western half of the Tennessee Valley was also claimed by the Chickasaws (Figures 6.20 and 6.21). Although the Cherokees had relatively little land to lose, when they finally relinquished it to the United States, it would have an important impact on the establishment of early agriculture in the state.

Early efforts by traders from Virginia to reach the Cherokees in the mountains of the Southeast were largely unsuccessful, but by 1690, traders from Charleston were traveling up the Savannah River to trade with them. However, even though they were much closer geographically, their trade with the Cherokees did not assume a great volume until later.

For almost a quarter century after its founding in 1760, the primary business of the South Carolina colony was raising livestock and cutting timber and barrel staves, for export to the West Indies; trading for furs and deerskins with the nearby Catawba Indians or the more distant Cherokees; or trading for captives taken by these Native Americans in their tribal wars and selling them as slaves to the other colonies. Some of the trading with the Cherokees would have involved those in north Alabama.

At the close of the French and Indian War in 1763, George III issued a proclamation to try to reduce the squatting of British subjects on Cherokee lands by separating them. The king decreed that neither Indians nor colonists would cross the Proclamation Line of 1763 running down the Appalachian Mountains from Maine to Georgia. Unfortunately, the British did not have the military manpower to police the decree. As a result, neither group took it seriously, and colonists poured onto Cherokee lands. As described in a preceding section, as early as 1779, settlers moving from northeast Tennessee by water to the site of Nashville passed through Cherokee Territory in north Alabama.

In the 1791 Treaty of Holston, the United States guaranteed the independence of the Cherokees, but the Georgians, who at that time had been chopping away at their territory for some 30 years, regarded the treaty as obsolete (Morison, 1965).

During the American Revolution, most of the Southeastern Indians remained neutral. The Cherokees, however, who were angry that British colonists had encroached into their territory, decided to settle the score. In July, 1776, they attacked loyalist and rebel alike, over a wide area. The colonists struck back with a terrible vengeance. An army of several thousand frontiersmen stormed through Cherokee territory, burning towns, destroying crops, and killing men, women, and children. After this attack, the Cherokees were never the same.

In 1806, the Cherokees ceded their first land in Alabama. It consisted of a small triangular-shaped tract near Huntsville (Figure 6.3). This cession provided the land for the establishment of the second major agricultural settlement (after the Bigbee Bottoms) in Alabama. Other cessions followed in 1816, 1818, and 1819. Then finally in 1835, all remaining Cherokee territory within the state was ceded to the United States

Of all the Southeastern Indians, the Cherokees made the most determined efforts to acquire the trappings of American civilization (Hudson, 1976); however as early as 1820, some of them began to understand that the loss of their lands was inevitable, and they began to migrate voluntarily to Arkansas. Most of them, however, refused to leave their homeland. Under the leadership of Sequoyah, they learned to write and read the Cherokee language. In 1828, they began publishing a bilingual newspaper, *The Cherokee Phoenix*. The Cherokees also set up a centralized government modeled after governments in the surrounding states. In 1827, they adopted a formal constitution that chartered a government with separate legislative, executive, and judicial branches. Their constitution disenfranchised women and blacks.

The Cherokees also made a determined effort to remain in their homeland. Unfortunately, their strongly held desire to remain in the land of their ancestors had little effect on those who coveted it. In 1832, Georgia established a land lottery for the disposal of Cherokee lands in that state. In December, 1835, the Cherokees were told that they would have to meet with representatives of the federal government at New Echota, the Cherokee capital in northeast Georgia, to sign a treaty for their removal. When the meeting was held, only 300 to 500 Cherokees out of a total population of 17,000 appeared. Although, the meeting represented a small minority of the total, they were forced to sign the Treaty of New Echota. Under the provisions of the treaty, all of the Cherokees agreed to move to the West in exchange for a payment of $5 million. The treaty was considered to be fraudulent by the majority of Cherokees, and they refused to leave their homes.

When the Indians signed the treaty, fraudulent or not, whites began to move onto Cherokee lands. Moreover, whites began lawsuits to take what little money the Indians had, and because Indians could not testify against whites in court, they automatically lost any suit brought against them. When the Indians refused to leave, whites resorted to less civil means. They invaded the lands where the Cherokees lived, and beat both men and women with "cowhides, hickories and clubs." In north Alabama, a federal military officer, General Wood, attempted to protect the Cherokees. For his efforts, he was accused of disturbing the peace and trampling on the rights of Alabama citizens, and he was promptly court-martialed.

By 1838, only 2,000 of the 17,000 Cherokee had moved west, leaving 15,000 in their homeland. This lack of response was unacceptable to those states in which the Indians lived; consequently, General Winfield Scott was ordered to remove them, which he did with a force of 7,000 soldiers. In June 1838, some 5,000 Cherokees departed for Indian territory in the West. In the fall, an even larger party departed, on what has come to be called the Trail of Tears. Altogether, 4,000 Cherokees out of a total population of 17,000 died during the process of being arrested, confined in stockades, and finally being marched to Indian territory.

Chickasaws. Chickasaw territory included portions of Tennessee and Kentucky and extended southward into northeast Mississippi and northwest Alabama (Figure 6.20). Apparently, they also claimed territory in the central Tennessee Valley in north Alabama that was also claimed by the Cherokees (Figure 6.3).

The first contact between the Chickasaws and Europeans resulted from de Soto's exploration of the Southeast during the years 1539 to 1542. The Chickasaws were the most implacable and effective warriors that de Soto encountered. The Indians harassed the Spaniards continuously while they were in Chickasaw territory, attacking mainly at night. There were no contacts for the next century and a half, until the English traders from Charleston arrived in the latter part of the seventeenth century; although it is likely that the Chickasaws may have received some European trade goods from other Indians who had contacts with traders representing the Spanish in Florida. From the 1690s onward, Carolina traders exchanged English goods with the Chickasaws. In fact, the bulk of Carolina trade after that time was with the Creeks and Chickasaws.

Early in the eighteenth century, the Chickasaws became embroiled in the war between France and England for control of the interior of North America. Probably because Carolina traders were active in Chickasaw territory before the French entered their area, the Indians sided with the British. French attempts to make a meaningful peace with the Chickasaws proved futile. In 1707, envoys from Charleston were able to get the Chickasaws to agree to help them dislodge the Spanish from Pensacola. In that year, they attacked and burned the town, but were unable to capture Pensacola's presidio or the central military post.

Also, during the early part of the eighteenth century, the Chickasaws joined forces with the Cherokees to drive the Shawnees from the Cumberland and Tennessee river basins in Kentucky and Tennessee. About 1765, the Chickasaws began to build villages along Big Bear Creek in today's Franklin County. Then later, they moved further eastward and established a settlement on the great bend of the Tennessee River south of Huntsville. The Cherokees considered the central Tennessee Valley to be their territory, and they promptly went to war with the Chickasaws, their former allies. In 1769, the Chickasaws won a decisive battle over the Cherokees, but it came at such a high price in lost warriors that they decided to abandon that area; although they continued to claim title to it. At the same time, the Cherokees, although defeated in battle, never relinquished their claim either. These lands remained contested until they were taken by the United States early in the nineteenth century. The early land cessions by the Cherokee (1806), which became the site of the second American agricultural settlement in Alabama (Figure 6.3), were part of these contested lands.

Although the Chickasaws were formidable warriors and paid little heed to the boundaries of their neighbors, they never openly confronted the Anglo-American settlers. From the beginning, the relationship between the Chickasaws and the settlers was generally excellent, at least from the settler's perspective. The Chickasaws claimed that in all of the contacts between the two groups that they had never spilled a drop of American blood.

Terrell (1971) suggests that the decline of the Chickasaws can be traced to the conflict (the French and Indian War) between Britain and France over the control of the Eastern North American interior. Unfortunately, most of the Indian nations around them were allied with France. After the 1763 British victory in the French and Indian War, the Chickasaws suffered severely in their relations with their neighbors.

In 1813, Shawnee Chief Tecumseh visited the Chickasaws in an attempt to get them to join a general Indian uprising to counter the continued encroachment of the white settlers. He was foiled by members of the Colbert family, especially James, who argued forcefully against their participation.

The Chickasaws ceded most of their territory in Alabama to the United States in 1816 (Figure 6.3). They signed the Treaty of Pontotoc Creek in October 1832. Under its provisions, they ceded all of their lands, including the remainder in Alabama, to the United States. The lands were to be sold by the federal government with the proceeds being held for the use of the Indians in the removal process. Also, the treaty's provisions allowed the Indians to remain in their homeland until suitable land had been purchased in the West. Finally, in 1837, a suitable tract of land was purchased in the western portion of territory held by their relatives, the Choctaws. A roll prepared at that time included 4,900 Chickasaws who owned 1,100 black slaves.

In the summer of 1838, a group of 500 Chickasaws began their westward journey. Later, in the same year, most of the remainder followed. In general, they were better outfitted and better organized than groups of other Indian nations that had emigrated earlier. They used funds that had accumulated from the sale of their lands to purchase supplies and to make travel arrangements. Most of them traveled up the Arkansas River on steamboats after boarding at Memphis. Small numbers of the Chickasaws continued to emigrate as late as 1850. Overall, the removal of the Chickasaw Nation was much less traumatic than it had been for the Creeks or Cherokees. Relatively few of the Chickasaws died during removal.

The worst hardships for the Chickasaws occurred after they reached their new homes. They were soon attacked by Shawnees, Kiowas, and Comanches, who resented their settlement in their territory. In making arrangements for the removal of Indians from the East, the US Government had not adequately arranged with the Indians of the West for the Chickasaw arrival and settlement . They also suffered grievously from outbreaks of smallpox in which hundreds died. They suffered as well from hunger and malnutrition when the federal government failed to supply them with food that had been promised under the terms of the treaty.

Origins of Settlers. Hilllard's (1984) map showing the distribution of the population in the South in 1810 (Figure 6.13) provides a basis for determining origins of the early American settlers in Alabama. Concentrations of people closest to the land that would later become Alabama, were located in eastern Georgia between Athens and the Savannah River, especially in the Broad River Valley; in the South Carolina Upcountry; along the upper Tennessee River in northeastern Tennessee; and in the Nashville Basin around the town of Nashville. Hillard's maps, presenting the 1810 and 1820 census data, shows that in the period concentrations of people in those four source areas had swelled considerably (Figures 6.13 and 6.14). One can easily imagine several thousands of people in those communities champing at the bit waiting for someone to do something about all those Indians and their lands in the Old Southwest.

Abernethy (1990) published a map (Figure 6.22) showing the general origin of settlers in Alabama soon after the first Creek land cession in 1814. The map shows the general location of settlements of emigrants from Virginia, the Carolinas, Tennessee, and Georgia. According to the map, settlers from Tennessee generally established homes and farms in the Tennessee Valley. Settlers from Virginia and the Carolinas mostly found homes in the Black Warrior and Tombigbee river valleys, and most of the immigrants from Georgia settled in the valleys of the Alabama, Conecuh, Pea, and Choctawhatchee rivers.

D'Andrea (1973) provides some later information on the origin of settlers who came to Alabama. He determined that the state could be divided into four general regions (Figure 6.23).

Region I. Region I includes the Tennessee Valley, the central portion of the Warrior Valley, the western portion of the Alabama Valley and Ridge, and the Fall Line Hills portion of the East Gulf Coastal Plain (Figure 1.11). According to D'Andrea, this region was settled primarily by emigrants from interior Virginia, North Carolina, South Carolina, and east Tennessee (Figure 6.14). Many Scotch-Irish and Germans were included among the emigrants who settled here:

Region II. Region II includes the eastern portion of the Cumberland Plateau, the Piedmont Upland, and the eastern and southwestern portions of the East Gulf Coastal Plain (Figure 1.11). This region included many relatively large landholdings owned by people who had emigrated from the coastal regions of Virginia, North Carolina, and Georgia. By 1860, many of the emigrants from the Lowcountry of South Carolina were settled in the Black Prairie.

Region III. Region III is essentially the Black Prairie District of the East Gulf Coastal Plain (Figure 1.11).

Here was the largest concentration of plantation agriculture in Alabama. In 1860, most of its settlers were from the Tidewater of Virginia and North Carolina, and the Lowcountry of South Carolina and Georgia.

Region IV. This region includes the extreme south-central portion, plus a long, narrow strip of the East Gulf Coastal Plain (Figure 1.11). It was settled primarily by emigrants from eastern Tennessee and western South Carolina and Georgia.

McWhiney and McDonald (1989) commented that 250,000, so-called Scotch-Irish had migrated to America before the American Revolution, and that from 1776 to the end of the nineteenth century they were the largest single ethnic group in the Southern interior. They further commented that Celtic peoples (Scottish, Irish, Scotch-Irish, Welsh, and Cornish) dominated the interior from Pennsylvania southward, ranging in various areas from three-fourths to nearly one hundred percent of the population. Given this apparent concentration of Celtic peoples in the interior of the Lower South, it is likely that a substantial share of the early settlers in Alabama were of this ethnic group.

Substantial numbers of settlers were Englishmen. Large numbers of Englishmen from around Petersburg, Virginia, had earlier emigrated to the Broad River Valley in northeastern Georgia. In turn, many of them were among the earliest settlers in the Tennessee Valley and the upper Alabama River Valley.

Travel Routes to Alabama. Settlers entered Alabama both by water and overland. Most of the very earliest ones to arrive came by water, entering the area through Mobile (Figure 1.20). The largest number, however, entered by three overland routes, but they had a variety of options in reaching those three points of entry (Figure 6.24). The actual route they chose was largely determined by where they were when they decided to migrate, where they wanted to go, and the quickest and easiest route for getting there.

The Water Route. Early Spanish and French entered Alabama through Mobile Bay and by traveling up the Mobile, Tombigbee, and Alabama rivers (Figure 1.20). Unfortunately, entry by water was not always dependable. The mouth of Mobile Bay, which was the primary point of entry to the interior, often became so filled with silt and so shallow following floods that larger sea-going ships could not enter. Under these circumstances, both people and materials had to be shifted to smaller boats before proceeding upriver. North of the bay, near the location of St. Stephens on the Tombigbee and Claiborne on the Alabama, both rivers passed through erosion-resistant geological formations (Figure 1.7). As a result, currents were so swift that ocean-going schooners from Mobile could go no further upstream. Both the villages of St. Stephens and Claiborne were established at the upper limits of pre-steamboat navigation.

Overland Routes. There were three primary overland routes to Alabama (Figure 6.24). These were not new routes to the Old Southwest. Rather, they were trails established by traders earlier in the settlement of the old wildernesses of backcountry Virginia, the Carolinas, Georgia, and Tennessee.

1. The Federal Road. In terms of use, the most important of the three routes that led to Alabama was the Federal Road (Figures 6.1 and 6.24), which entered Alabama in today's Russell County near Fort Mitchell and Columbus, Georgia. It originated in the region of the headwaters of the Oconee River, near the site of Milledgeville, Georgia. At that point, it connected to an older road. linking Fredericksburg, Richmond, Raleigh, Cheraw, Camden, Milledgeville, and Augusta. This was the so-called Fall Line Road, because it generally followed the fall line, or the eastern margin of the Piedmont Physiographic Province (Figure 6.25), from Virginia to Georgia. Another interstate road, the Piedmont Road—which linked Baltimore, Greenville, and Athens—joined the Federal Road before it reached the Chattahoochee at Columbus.

2. The Great Philadelphia Wagon Road. The second most important route to Alabama, was an extension of the Great Philadelphia Wagon Road or Great Valley Road (Interstate-81) that linked Philadelphia, Lancaster, York, Winchester, Staunton, Roanoke, Knoxville, and Nashville. A branch off the Knoxville-Nashville segment entered Alabama just north of Huntsville (Figure 6.24).

3. The Natchez Trace. The Natchez Road, or Trace, was the oldest route to Alabama. It also was an extension of an older road, linking Pittsburg, Lexington, Nashville, and New Orleans. The trace linked Nashville and Natchez, passing through Florence, Alabama and Jackson, Mississippi. There also was a branch off the Trace in east-central Mississippi that led into Alabama near the Choctaw-Washington County line (Figure 6.24).

Routes of Choice. Virginians coming from the Shenandoah Valley generally traveled to Alabama by following the southern part of the Great Philadelphia Wagon Road from Staunton to Roanoke to Knoxville to Huntsville. This was the route followed by a majority of the Scotch-Irish and German immigrants who had settled the backcountry, the eastern foothills of the Appalachians. Virginians coming from the eastern portion of the Piedmont (Figure 6.25) reached Alabama by more eastern roads: the Fall Line Road and the Piedmont Road, which connected to the Federal Road.

People in western North Carolina were relatively close to the Great Valley Road (Staunton-Roanoke-Knoxville-Huntsville). However, to gain access to it, they first had to cross the rugged Appalachian Mountains, so most of them chose to come to Alabama by the Fall Line Road and the Piedmont Road, and then to the Federal Road.

South Carolinians could follow the Fall Line Road to the eastern terminus of the Federal Road, or they could take the more difficult route across the mountains through Saluda Gap from Greenville to Asheville, then to Knoxville and on to Huntsville.

Settlers from the central Piedmont region of Georgia had easy access to the eastern terminus of the Federal Road by traveling along the Augusta-Milledgeville-Fort Mitchell or Athens-Fort Mitchell roads. However, people in the northwestern portion of Georgia probably had easier access to Alabama by the Athens to Chattanooga, then to Knoxville-Huntsville roads.

Travel Routes within Alabama. Two factors drove the evolution of travel within Alabama in the early nineteenth century: migration and cotton production. After the early Indian land cessions (Figure 6.3), there were large amounts of new land available to a growing population of unsettled people in the older states, but no suitable way for them to reach it except by Indian trails.

Indian Trails. In the beginning, Indian trails were important. In the eighteenth century, Scottish and English traders from Charleston followed Indian trails into Alabama where they settled at strategic locations and established trading posts. Farms of those traders who married Indian women were often established near one of these locations. One of the most important of these trails, was the Great Pensacola Trading Path (Wolf Path) that connected Cherokee territory in the Tennessee Valley with Pensacola, a major Spanish trading post from about 1696 onward (Figure 6.26). Southerland and Brown (1989) describe how the naturalist, William Bartram, followed an old Indian trading path from Milledgeville, Georgia to Mobile, generally along the path that later would become the Federal Road.

Dedicated Roads. Indian trails were suitable for foot traffic, horses and pack animals, and could be used only in a very low level of commerce; however, they were almost useless for the passage of carts and wagons, and they were even more useless for the development of the export of cotton to England. Years before Alabama became a state, officials in Washington realized that if they were to adequately administer the affairs of the southwestern portion of the new country (the Old Southwest) that it was essential that permanent, all-purpose roads must be established.

1. Natchez Trace. The Natchez Trace was the first dedicated road in Alabama (Figure 6.27). As noted in a preceding section, it connected Nashville and Natchez (and New Orleans by the Mississippi). It represented the last leg of the route between Washington and the mouth of the Mississippi. Initially, it followed a series of old Indian trails. It was generally in service after 1795. Mail service over the trace was begun in early 1800. The following year, the US government received permission from the Chickasaws and Choctaws to improve the road. The trace barely touched Alabama, passing through the western portions of Lauderdale and Colbert counties. However, it did provide a route by which settlers from the Nashville Basin could reach the western end of the Tennessee Valley. Later, a branch off the road, in Mississippi (Figure 6.24), connected it with the settlements on the lower Tombigbee (Bigbee Bottoms). The trace through Alabama was largely abandoned about 1819 and played essentially no role in the immigration of settlers after that time. Dedicated roads in the state included the following:

2. Federal Road. Some of the details regarding the construction of this road are presented in an earlier chapter..

The Federal Post Road was generally ineffective for the purpose for which it was intended, but in the minds of the waiting pioneers in Virginia, the Carolinas, and Georgia, it was an American road leading to public lands in the Bigbee Bottoms. Once established, as inadequate as it was, they began to flow over it. The 1810 Census showed that there were 2,000 people living in southwest Alabama (Figure 6.14). In 1811, the Indian agent, Benjamin Hawkins, commented that "the road is now crowded with travelers moving westward, for the safety of whose property at times I have some anxiety" (Atkins, 1994).

In 1809, anticipating war with the British and expecting an attack along the Central Gulf Coast, the United States decided that it needed a more dependable road across the Mississippi Territory, which included both present-day Mississippi and Alabama. The Creeks objected to this proposal, but were told that the military road would be constructed over their objections. In July 1811, General Wade Hampton, military commander in New Orleans, was given the order to build it, generally along the route of the original post road horsepath. The Military Road was opened in November 1811. Southerland and Brown (1989) commented that the construction of this road, over the objections of the Creeks, was a primary cause leading to the wars of 1813-1814 that generally lead to the end of Creek power in central Alabama. With the Military Road in place, and the Creek land cessions of 1814 (Figure 6.3), the flow of settlers became a flood. By the time of the 1820 Census, a year after Alabama became a state, there were 50,000 people living around the southern terminus of the Military Road (Figure 6.15). Atkins (1994) commented that most of this population growth came after 1815.

3. Huntsville Road. Chickasaw land cessions in 1805 and Cherokee cessions in 1806 in the Tennessee River Valley (Figure 6.3) opened for settlement the area in present-day Madison County (Atkins,1994). About the same time that preparations were being made to sell land there at public auction, a road connecting the area with the Knoxville-Nashville Road was completed (Figure 6.24). By 1820, some 50,000 people had crowded into the area around Huntsville (Figure 6.15).

4. Georgia State Road. This road was opened in 1812. It extended from Dallas, Georgia (northeast of Atlanta), westward into the Coosa River Basin, near the present location of Centre in Cherokee County (Figure 6.27). From that point, it passed through Collinsville, and on across Sand Mountain to Fort Deposit on the Tennessee River.

5. Bear Meat Cabin Road. After the construction of the road from the Nashville Basin to Huntsville was completed in 1807 and the Military Road in 1811, travel from the old Southeastern states to the Tennessee Valley and to central Alabama was relatively easy; however, there were no roads through the mountainous portion between. This problem was partially solved with the 1818 construction of the Bear Meat Cabin Road (Figure 6.27). It extended southward from Huntsville, crossed the Tennessee River at Ditto's Ferry, thence on to Blountsville (this was the site of the Bear Meat Cabin), and through Jones Valley in Jefferson County to the head of navigation on the Black Warrior River, at Tuscaloosa. The Bear Meat Cabin Road was especially important because it opened the heartland of north-central Alabama to the large population in the Nashville Basin (Figure 6.15).

6. Byler Road. The Byler Road from Courtland (also in the Tennessee Valley) to Tuscaloosa (Figure 6.27) was opened about a year after the Bear Meat Cabin Road (1819); however, the Byler Road was of lesser importance. This road passed through some of the most rugged terrain in the state, and at that early date there were not very many people who had a reason to use it.

7. Fort Crawford to Fort Gaines. As a result of Indian depredations that continued in southeast Alabama for some time following the Battle of Horseshoe Bend, a number of forts were constructed in the region to protect settlers. One of these, Fort Crawford, was established in 1817 near today's Brewton in Escambia County. In that year, military personnel constructed a road from Fort Crawford eastward along the west side of the Conecuh River to Brooklyn in Covington County, then through today's Montezuma, Daleville, Block, and Abbeville to Fort Gaines on the Chattahoochee River (Figure 6.27).

8. Three Notch Road. In 1824, Three-Notch Road was constructed from the Federal Road, at a point near the location of the line between Russell and Macon counties (Old Fort Bainbridge), southwestward, passing near Enon and down the Conecuh River Valley, through the sites of Andalusia and Brewton (Figure 6.27). This road provided access to southeast Alabama and the Wiregrass region and to portions of the Creek land cessions of 1814 (Figure 6.3). This road passed on beyond Brewton, down the Conecuh and Escambia rivers to Pensacola.

By the late 1820s, the roads needed to get people into Alabama and into the major regions of the state were in place (Figure 6.27). The Federal Road with its Three-Notch branch, the Bear Meat Cabin Road from Huntsville to Tuscaloosa, and the Fort Gaines to Fort Crawford Road opened the entire state to those thousands of people in the original Southeastern states who were seeking land of their own or more elbow room. Soon, even more roads would be constructed. Atkins (1994) comments that by the late 1850s a system of connecting roads blanketed Alabama.

Rivers. River transportation was essential to establish commercial agriculture in Alabama. In the early 1800s, demand for cotton began to increase sharply in England, and by 1820, substantial quantities of the fiber were being produced on the Chickasaw and Cherokee land cessions in the Tennessee Valley and on the Creek cession along the Alabama River in the central portion of the state. The only way to get the cotton to Europe was first to get it to either Mobile or New Orleans. Fortunately, three large rivers (Tennessee, Alabama, and Tombigbee) and several smaller ones (Black Warrior and Cahaba) (Figure 1.19) passed through some of the best cotton-growing land in the state. In fact, river transportation for cotton was so efficient that there was little interest in either roads and railroads in the state until well after they widely blanketed the North. Cotton went downriver, and supplies from the markets of the world came upriver.

In the beginning, cotton was loaded onto flatboats and floated down to Mobile or New Orleans. When the cotton was off-loaded, the boats were dismantled, and the lumber sold. Afterwards, the boat operators walked back upriver to construct other boats before making another journey.

Atkins (1994) notes that in 1819 a steamboat plied the Tombigbee River as far north as Demopolis and that steamboat shipping between St. Stephens and Mobile was common as early as 1820. The stream currents in the Alabama River were much stronger than in the Tombigbee; consequently, it was not until 1820 that the *Tensas* made the trip from Mobile to Cahaba. According to Abernethy (1990), there were 11 steamboats on the rivers in 1823, and by 1826 the number had increased to 18. He also commented that the first steamboat reached Florence on the Tennessee River in 1823. Cotton planters above Musssel Shoals still had to ship their cotton to Florence by keelboat, but from there steamboats towed them to New Orleans.

Steamboats did not appear on the Coosa until much later than on the other major rivers of the state. Jackson (1995) comments that the steaamboat *Coosa* was launched near the site of the ferry at Greensport in southwestern Etowah County in 1844. The Coosa River above the first shoals near Greenport was much shallower than the Alabama and Tennessee. As a result, steamboats in use there were much smaller, and their use much more seasonal. During the high-water months, however, steamboat traffic on the Coosa was heavy. In 1851, the railroad line between Chattanooga and Atlanta was completed. Afterwards, cotton shipped upriver from the Coosa Valley to Rome, Georgia, could then be transported by rail to Charleston and Savannah. By the mid-1850s, there were some 30 steamboats operating on the Coosa. A rail connecting the Coosa and Alabama rivers was not completed until around 1861. With the coming of suitable rail transportation for cotton, steamboat traffic on the Coosa waned. However, the shipping of forest products kept some of the boats operating until the beginning of the twentieth century.

Railroads. River transportation of cotton within Alabama was so efficient that there was limited early interest in railroads. Actually, Alabama was among the first states in the country to become involved in the construction of railroads (Atkins, 1994), but its first effort was to expedite river transportation, not replace it.

1. Shoals Bypass. In 1832, a two-mile railway was

completed along the Tennessee River (Figure 6.28). It was constructed to allow the shipment of goods up and down the river to by-pass some of the shoals east of Florence. The railroad was not well constructed or equipped and provided few of the expected benefits. Later, it was improved and incorporated (1834) into a much longer line connecting Tuscumbia and Decatur. This line allowed shipping to bypass all of the shoals. It was later incorporated into the Memphis-Charleston Line that was completed in 1861.

2. Montgomery-West Point. In 1832, the state chartered its second railway—the Montgomery to West Point, Georgia Line. It was planned to provide transportation between Montgomery and areas of rapidly growing cotton production in east-central Alabama and west-central Georgia. The last section of the line was not completed until 1858 (Figure 6.28).

3. Selma and Tennessee. The state chartered the Selma and Tennessee Railway in 1836. It was planned to connect the Alabama River at Selma with the Tennessee at Gunter's Landing (Guntersville) via Montevallo, Talladega, and Jacksonville. As planned, it would have connected Alabama's three major cotton-producing regions with Mobile. The segment to Montevallo was completed in 1853, and by 1861, it was extended through Talladega to Anniston. It could not be extended to the Tennessee because of the Civil War (Figure 6.28).

4. Alabama and City of Montgomery. In 1836, the state also chartered the Alabama and City of Montgomery Railroad. It was planned to connect Montgomery and the upper Alabama River with the seaports at Pensacola and Mobile. This railroad experienced considerable difficulty in getting started, and by 1860, it had been completed only as far as Pollard in Escambia County (Figure 6.28).

5. Mobile and Ohio. Plans to develop a rail connection between Mobile and the Ohio River (the Mobile and Ohio Line) were first developed in 1847. It was chartered by Alabama, Mississippi, Tennessee, and Kentucky in 1848, but construction did not begin until 1852. It connected Mobile with Meridian, Mississippi; Jackson, Tennessee, and the Ohio River near Paducah, Kentucky. The last section of the line was completed at Corinth, Mississippi, in 1861 (Figure 6.28).

Where They Settled. Humans began settling in the land now called Alabama at least 10,000 years ago; however, in this section focuses on the coming of the Europeans.

Before the American Revolution. The French established the first permanent European settlement in Alabama in 1702, just north of Mobile Bay. At the beginning of the American Revolution in 1776, the only settlements were essentially in the same location where they had been three-quarters of a century earlier. By 1704, French subjects who had accompanied the Le Moyne brothers in their move from Fort Maurepas (Biloxi) were herding animals and managing fields near La Mobile. Although the French were nominally involved in farming, most of the actual field work was done by Indian slaves. Moore (1934) comments that over the years, it is likely that some of these French settlers established farms up the Tombigbee River as far north as Fort St. Stephens. This location was in the lower portion of the Bigbee Bottoms, where the first farming communities of British pioneers in Alabama would begin to develop later in the century.

The French ceded West Florida, including the settlements around Mobile, to Great Britain in 1763; the British would rule there for less than two decades. However, in that relatively short period, new agricultural settlements were established and arable land was expanded. A number of large plantations were established along the Mobile and Tensaw rivers. Some of the planters resided in Mobile, allowing overseers and large numbers of slaves to operate the farms. Early in the British regime, a number of families settled in the Lake Tensas region. Then during the American Revolution, a number of Tory families from the Carolinas and Georgia joined them there. Moore (1934) comments that these settlements gave rise to Alabama's first planter "aristocrats."

Under Spanish Rule. In 1779, near the end of the American Revolution, Spain entered the conflict against Great Britain. In early 1780, Bernado de Gálvez led a Spanish expeditionary force from New Orleans into Mobile Bay. After a 14-day bombardment, the British at Fort Charlotte surrendered. With the surrender of the British, Spain gained control of all of the territory in south Alabama, as far north as the confluence of the Tombigbee and Black Warrior rivers (32°28' North) (Figure 6.2).

The Spanish only established a military presence in the area. Except for military personnel, few Spaniards moved very far into the newly acquired territory. However, under the Spanish extensive ranches and plantations were established along the coast. Large tracts of marshland in the Mobile Delta were converted into ranches. The French and British had determined that these lands were too wet for effective crop production. A number of cow pens were constructed on the ranches to accommodate the large herds of cattle. A large variety of crops were produced on the various plantations. Even rice culture was attempted on some of the wetter areas, but before Spanish rule ended, the emphasis had shifted to the production of corn and cotton. Most of the management and labor required was supplied by French and British settlers and their slaves. who had come to the area during the preceding political regimes.

The Spanish encouraged agricultural settlements on the Lower Tombigbee in the area where the Choctaws had ceded land to the British earlier (Figure 6.3). To provide support for the settlements, they located a military garrison at the site of the old French fort at St. Stephens, which they renamed Fort San Esteban de Tombechbé. This fort provided protection for British and American farmers who had already begun to move into the Bigbee Bottoms from the east, in the latter part of the eighteenth century.

In 1795, Spain agreed to vacate the territory north of 31° North (Figure 6.2); although they did not complete the move until 1798. With this move, the old Choctaw cession on the Tombigbee now belonged to the United States, and settlers began to flood into the area. The US Government moved to provide protection for these growing settlements by constructing Fort Stoddert in 1799, near the site of Mount Vernon in northeastern Mobile County.

In 1803, the itinerant Methodist preacher, Lorenzo Dow, visited the Lower Tombigbee Valley. He recorded that at that time that there was a thick settlement at McIntosh Bluff, and that there were straggling settlements up the Tombigbee beyond St. Stephens, even to Wood's Bluff, which is near the extreme northern portion of Clarke County.

In Mississippi (and Alabama) Territory. The federal government established Mississippi Territory in 1798, under the provisions of the Northwest Ordinance of 1787. This territory included land that would eventually become the states of Mississippi and Alabama; however, long before the eastern portion became a state in 1819, Mississippi Territory was filling up with settlers from the older Southeastern states.

1. Tennessee Valley. In 1806, Chickasaw and Cherokee land cessions (Figure 6.3) in north Alabama made land in the Tennessee Valley available for settlement. Some of the first settlers to take advantage of these Native American cessions in the Tennessee and Moulton valleys (Figure 1.11) were wealthy planters from the Broad River Valley of Georgia. By 1819, the Cherokees had ceded virtually all of their remaining land in the two valleys. When the federal government put these new lands on sale, settlers from Virginia, Georgia, Tennessee, the Carolinas, and other states in the east, streamed into them. Within little more than a year after the lands were made available, there were sufficient people present to organize the counties of Cotaco (now Morgan County), Jackson, Limestone, Lawrence, Lauderdale, and Franklin (Figure 6.29). Moore (1934) comments that these people came into this "wilderness . . . to deny with the help of nature, the power of economic law."

Census data for 1820 (Figure 6.15) indicated that 12.1% of all persons counted in Alabama were in Madison County, and 35% were in the Tennessee Valley Physiographic District (Figures 1.11 and 6.29). The early concentration of people there is not unexpected given the large number clustered near that area in 1810 (Figure 6.13).

2. Alabama and Conecuh River Valleys. As shown in Figures 6.14 and 6.29, a goodly number of people were counted in the counties of the Alabama River Valley in the 1820 Census. Monroe County (8,838 persons) had the third largest population in the state, and the largest outside the Tennessee Valley. Populations of Montgomery and Dallas counties were 6,604 and 6,003, respectively.

The development of the Federal Road and later the Federal Military Road in 1806 and 1811 (Figure 6.27), respectively, provided the first access to the lands in this region. It was the Creek cession of 1814 (Figure 6.3), however, that really made it available to settlers. Wealthy planters from the Broad River Valley of Georgia were instrumental in the establishment of the first plantations in the Tennessee Valley. Later, they also played a major role in establishing many of the large slave-operated cotton farms in the area that would become Montgomery County. With these first settlers along the upper Alabama River as a nucleus, large numbers of Georgians moved into the region east of the river. Many of these early emigrant families were of Scotch-Irish descent.

South of Montgomery County, the Federal Road (Figures 6.1 and 6.27) followed a high ridge separating the Alabama and Conecuh river valleys (Figure 1.17). As a result, settlers traveling southwestward along the road had equal access to either valley. For example, settlers on the Federal Road near Fort Dale in the northern part of Butler County could go about a mile either east or west and establish farms in either valley. However, as shown in Figures 6.15 and 6.29, by 1820, relatively few people had opted to settle in the valley of the Conecuh. Until 1832 some of the eastern portion of the valley of the Conecuh still belonged to the Creeks.

3. Black Warrior and Tombigbee River Valleys. By 1820, the lands of the Tennessee Valley District were beginning to fill up (Figures 1.11, 6.15, and 6.29). When they did, the new arrivals leap-frogged over the established settlements and over the mountains south of the settled valleys into the isolated pockets of the mountainous region of the northern Warrior Basin (Figures 1.8 and 1.11). Some of these settlers in the more isolated areas lived under pioneer conditions long after other settlements in the state had advanced well beyond that stage of development.

The new arrivals were not the only ones to move south into the mountains. As more and more people crowded into the Tennessee and Moulton valleys, some of the established settlers began to be concerned about elbow room, and they soon joined the stream flowing southward. Some of them settled again, further south in the Warrior Basin; while others, along with some of the newcomers, passed southeastward into Jones' Valley and Roupe's Valley, located in the area that would become Jefferson County. Still others continued down the Bear Meat Cabin Road (Figure 6.27) to present-day Tuscaloosa County. Some 5.7% of all persons counted in Alabama in the 1820 Census (Figures 6.15 and 6.29) were living in Tuscaloosa County. It had the second highest population (8,229 persons) of any county outside the Tennessee Valley.

Settlement Patterns from 1830 to 1860. Land cessions by the Chickasaws and Creeks in 1832 and the Cherokees in 1835 (Figure 6.3) ended all claims of the Native Americans in Alabama. These cessions opened the remainder of land for settlement, although land surveys would not be completed for some time. But the way was now clear for the filling out of the new state, and it grew rapidly as Alabama Fever raged in Virginia, the Carolinas, Georgia, and Tennessee.

Maps prepared by Hilliard (1984) using federal census data tell the story of how Alabama grew from 1830 to the eve of the Civil War.

1. The 1830 Census. By 1830, Alabama's population had more than doubled (144,000 to 309,500) since 1820 (Table 6.5). The distribution of these settlers in 1830 is shown in Figure 6.15. Although the numbers were higher, the distribution was not greatly different than a decade earlier. Most of the people continued to reside in the Tennessee Valley or the Upper Alabama River Valley. The increase in numbers in those two areas are partly a reflection of the rapid increase in the number of slaves (Figure 6.30). Note that there were still large amounts of Indian lands remaining in the state; the larger share being the Creek lands in the eastern portion (Figure 6.29).

2. The 1840 Census. By 1840, all the Native Americans had been removed, and in this short period between 1830 and 1840, their lands had been ceded to the United States and most of it sold to settlers. Also within the decade, all of the ceded lands had been subdivided into counties. Within the decade, the number of persons counted in the state increased from 309,500 to 590,800 (Table 6.6). Their distribution within the state is shown in Figure 6.16. During the same period, the number of slaves in the state increased from 117,500 to 253,500, an increase of almost 16%.

The distribution indicates that by 1840 the Choctaw

cession in southwest Alabama was quickly filling up. A considerable portion of this land had already been in limited cultivation by the Choctaws. There was probably less clearing required than in the Creek and Chickasaw cessions.

Data presented in Figure 6.16 show that most Alabamians continued to live in belts of counties across north and central Alabama. The heavy concentration of people in those two regions was primarily the result of the growth of cotton production with slaves (Figures 6.30 and 6.31). The population continued to be more limited in the north central and extreme southern portions where cotton production was limited.

3. The 1850 Census. By 1850, the number of persons enumerated in the census had increased to 771,600. The number had been 591,000 in 1840 (Table 6.5). The distribution of people in 1850 is shown in Figure 6.17. It is obvious from the figure that the 1832 Creek cession in east Alabama had quickly filled-up. It is also apparent that cotton production with the use of slaves was increasing rapidly in the upper Coosa Valley (Figure 6.30). Settlers were continuing to avoid the more mountainous areas of the Warrior Basin, as well as several of the counties in extreme south Alabama.

While the number of persons counted in the census continued to increase, the rate of increase diminished. From 1820 to 1830, the number counted increased 115%. In the period 1830 to 1840, it increased 91%, but from 1840 to 1850, the rate of increase had declined to 31%. However, it is important to remember that by 1850, many of the yeomen had already worn-out their small farms in Alabama and had succumbed to so-called Texas Fever. After the 1836 Battle of San Jacinto in which the American Texans defeated the Mexican forces commanded by General (and Mexican President) Santa Anna, hundreds of disgruntled Alabama farmers began to cross the Mississippi bound for Texas.

The number of slaves increased from 253,500 in 1840 to 342,800 in 1850, an increase of 35% (Table 6.7). During the decade, the rate of increase for slaves was considerably greater than for whites.

4. The 1860 Census. In 1860, the census counted 964,000 persons in Alabama. The 1850 count was 772,000 (Table 6.5). Their distribution within the state in 1860 is shown in Figure 6.18). Most of the people continued to live in the Tennessee Valley and the Cotton Arc of central Alabama. The population continued to increase in most counties, but losses in a few provided a clear warning of future problems in Alabama's Cotton Kingdom. During the decade, the population of white persons declined in Chambers, Greene, Lauderdale, Lawrence, Limestone, Macon, Madison, Marengo, Pickens, and Sumter counties. These counties had been extremely important in the establishment of the Cotton Kingdom, but relentless increases in cotton production with virtually no efforts to control soil erosion or to maintain soil fertility had worn out the land. At that time, the only practical solution was to buy more land or to move. Note that in relative terms, farmers were continuing to avoid the sandy soils of the Southern Pine Hills (Figure 1.25), as well as the central, hilly spine of the state. During the decade, the population of Mobile County increased from 9,479 to 15,730. Most of this increase represented the growth of the city of Mobile. With the completion of the several railroad projects and the construction of roads, Mobile had become an extremely important international seaport.

The number of slaves continued to increase until the eve of the Civil War. Between 1850 and 1860, slaves increased from 342,800 to 435,100, an increase of 30%. In 1820, slaves constituted 32.9% of the total population. By 1860, that percentage had increased to 45.1.

The Settlement of Slaves. With the vast amount of cheap government land made available in Alabama after the battle of Horseshoe Bend, thousands of white settlers optimistically flooded in, but there were thousands of others who came with them who had considerably less optimism. The first significant settling of slaves in Alabama began after 1702. At that time, the French above Mobile Bay were clearing land and beginning to start farms, and they were using Indian slaves for labor (Atkins, 1994). They soon concluded that the Indians were not suitable for that purpose, and requested that African slaves be sent to them. As a result, in 1721 a French ship of war, *Africane*, arrived in Mobile with a cargo of 120 native peoples from Guinea in West Africa (Conniff and Davis 47). Shortly thereafter the *Marie* arrived with 338 and the *Neride* with 238, both from Angola.

After about a century, the largest concentration of slaves was still located in the southwestern portion of Alabama. Hilliard's map (1984) for 1810, which is based on the census for that year, indicates that there were at least 2,000 slaves in the Bigbee Bottoms of Washington County. It does not indicate that there were any in the Huntsville area of the Tennessee Valley. The map does not include African Americans that might have been present in Mobile; because the city was still under Spanish control. By that time, there were a few slaves located throughout the central portion of Alabama. Even some Indian farmers owned a few; however there were no concentrations large enough to be included on Hillard's map.

The early settlement patterns of African American immigrants were an important aspect of the settlement of Alabama. As slaves came to Alabama with their masters, their settlement patterns indicate that they settled where men with enough wealth to purchase and support slaves were settled. Although it was not uniformly the case, a considerable amount of the wealth invested in slaves had originated in the coastal areas of Virginia, the Carolinas, and Georgia. Some had obviously come with yeoman farmers who had gotten wealthy growing cotton, but certainly the majority must have originally come from the tobacco, rice, and indigo plantations on the coast.

In the first census taken after Alabama became a state (1820), there were 47,400 slaves here (Tables 6.6 and 6.7). The larger share was on cotton plantations on the Chickasaw and Cherokee land cessions in the Tennessee Valley (Figure 6.30). There was a much smaller number scattered in the 1814 Creek cession in the central and southern portion of the state. By 1830, the number had more than doubled to 117,500 (Table 6.7). Although the number was greater, the distribution was about the same as it had been in 1820 (Figure 6.30).

The distribution of slaves within Alabama in 1840 is also shown in Figure 6.30. The census that year indicated that there were 253,500 here (Table 6.7). In that respect, Alabama lead the South Central Region in slave holding. By this time, agricultural technology was available to fully exploit the soils of the Black Prairie (Figure 1.20), and cotton production exploded in that district. Except for a lesser concentration of slaves in the Tennessee Valley, most of them were distributed along Alabama's Cotton Arc that included 19 counties stretching from Chambers in the east to Greene and Pickens in the west (Figure 6.31). Those 19 counties accounted for 67.8% of all Alabama slaves.

Approximately 6.8% of all slaves in Alabama in 1840 were living in Dallas County. Other leading counties were: Greene (6.5%), Sumter (6.3), and Montgomery (6.1). Some 14.7% were in the five cotton counties of the Highland Rim, and 4.4% were in the four cotton counties of the Coosa Valley.

By 1860, the number of slaves in Alabama had increased 71.6% (253,532 to 435,080) over the 1840 population (Table 6.7). In Mississippi, the number increased almost 124% during the same period. The distribution of slaves within Alabama in 1850 and 1860 is shown in Figure 6.30. Although the number of slaves was larger in both years, they were still generally concentrated in those same areas where they had been in 1840. The largest percentage was still in Dallas County (5.9%). Other leading counties included: Marengo (5.6%), Montgomery (5.4), Greene (5.4), and Lowndes (4.4). Together, these five counties accounted for 26.8% of the total. In 1860, there were 10,000 or more slaves, in 16 of Alabama's counties.

A considerable portion of the increase in the number of slaves in Alabama from 1820 through 1860 was due to natural increase, or birth of children on the plantations. Where natural increase was not sufficient to meet their needs, planters had to purchase slaves from professional traders who obtained them from markets on the Atlantic Seaboard, usually Virginia and Maryland. In the early days, they were shipped to Mobile and then upriver to the plantations, or they were walked in coffles overland. Later, when railroad transportation became available, they were moved by train.

In 1860, there were 33,730 slave owners in Alabama. There are no comparable data for earlier years. The estimated population of the state in that year, including slaves, was 964,000 persons, so only a relatively small fraction of the total population actually owned any slaves. Some 16.6% of all owners had only a single slave. An estimated 64.6%, owned nine or less, and 82.1% owned 19 or less.

A total of ten owners had between 300 and 500 slaves. None owned more than 500.

Slavery was an enormous burden on planters in Alabama. Because so much of their wealth was tied up in slaves (Table 6.5 and Figure 6.32), there was little money to do anything else. Further, once a man established himself as a slaveholder and invested most of his money in them, he found it extremely difficult to alter his farming practices, especially the amount of cotton planted. As a result of the fact that they owned their labor supply, planters were almost helpless in adjusting to fluctuations in cotton prices. Even when prices declined because of oversupply, enough cotton had to be planted to keep the slaves occupied, even if it meant losing money on the crop.

Management of slaves was a major concern for planters. The work of the slave revolved around the production of cotton. Between the demands of the production of the crop (planting, chopping, picking, ginning, and baling); slaves mended fences, cut wood, cleared new ground, built and maintained barns and outbuildings, and constructed new slave quarters. Labor tasks were assigned according to the skills and health of the individual slaves. During the winter months, however, when the requirement for labor on the plantation was limited, planters would hire out their slaves to other plantations or to commercial or government interests who were involved in projects requiring large amounts of manpower.

The Alabama Immigrants Settle-In. With Settlements established either by yeoman farmers or by planters were generally different in character. These differences were quite obvious even when the first ones were established in the later part of the eighteenth century on land in Mississippi Territory that would later become Alabama; the differences persisted well into the nineteenth century. In those early days, yeomen followed their route of choice into the territory, continued to follow it until they reached an area to their liking, squatted there, and established their largely self-sufficient homesteads. Soon, they were joined by other pioneer families to establish communities, and in time, enough communities were established that counties had to be created to provide government, especially court systems, for the communities.

When they had a choice, yeomen farmers tended to establish their settlements apart from areas where most of the plantations were located. When it was not practical, however, they simply established their small farms in the nooks and crannies among the large landholdings and tried to ignore the planters and their slaves.

Corn was the basic stabilizing element in the farming systems of the yeomen farmers and their settlements. To meet the family's food needs, corn could be prepared in several ways to provide some variety to their meals—ash cakes, corn pone, bread, and grits (Locher and Cox, 2004). Corn could also be used to make whiskey that was almost a staple with many of the yeoman families. It was so important and so common in some areas that, for a period, it served as a medium of exchange. As has too often been the case, excessive consumption of this intoxicant was accompanied by family and community violence. Corn was supplemented with game, fish, or a slaughtered hog from the droves running wild in the forest in the vicinity of the family dwelling.

Life on the Alabama frontier, especially for those self-sufficient yeoman farmers, generally consisted of grinding, unremitting labor. However, Little (1971) in his *History of Butler County, Alabama,* suggested that many of the settlers there did not spend as much time as they should have providing for their families, preferring instead, to spend a considerable amount of their time hunting deer and turkey. These early settler families also had to make do with few of the amenities of life. Their meager, self-sufficient farms provided them with very little money to buy anything, had anything been available in the Alabama wilderness.

The self-sufficient yeomen farmers, at least initially, were not dependent on the price of cotton. They had come to Alabama in search of economic freedom rather than a fortune. They built their cabins of logs, cleared land for their corn patches, and let their hogs and cattle roam free in the forest. Probably the majority of these farmers had come to Alabama because they had not fared very well where they had been. Their lands to the east had badly eroded and were unproductive, and for some of them, there were simply too many neighbors around. They had great expectations that the grass would be greener west of the Chattahoochee.

The piney woods region of much of southeast Alabama was quickly considered to be too infertile for crop production, especially cotton (See Figure 6.31). From the beginning, the region was inhabited primarily by yeomen. Land in this backwoods section was not offered for sale until about 1824; consequently, yeomen settlers who began to enter the area after about 1817 were able to squat on their land and make improvements on it for several years before it was placed on the market. Also, by 1824, much of the land speculation that had plagued squatters earlier, in other sections, never became a problem in the piny woods of southeast Alabama. Consequently, the settlers there were able to buy their farms with little competition and at reasonable prices.

The early piney woods pioneers were heavily dependent on livestock grazing (herding). They were first and foremost, subsistence farmers, like their north Alabama counterparts, and heavily Anglo-American (probably Scotch-Irish). The wild pasturage in this region was excellent. It allowed cattle to sustain themselves throughout the year on the open range; consequently, many of the settlers kept large herds. Occasionally, the herders would round up their livestock and drive them to market in Tallahassee, Pensacola, or Mobile.

Many of the early settlers (primarily yeomen) from the backcountry of Virginia, North Carolina, and Tennessee moved through the thickly-settled communities in the Tennessee and Moulton valleys, across the Central Mountains (Figure 1.9), and down into the secluded creek valleys of the Warrior Basin where they re-established the same rigorous self-sufficient life styles that they had left. Many of the loosely organized communities that they established would, in this isolated wilderness, retain their pioneer characteristics for decades.

Interspersed among the yeomen farmers was another group—the so-called poor whites or poor white trash. Flynt (1989) in his *Poor But Proud* describes them as "Refusing to pursue whatever meager opportunities came their way, they were satisfied with a subsistence consisting of a bit of corn and whiskey, freedom to hunt and fish whenever they chose, the most casual kind of living arrangements with the opposite sex, little effort at child rearing . . . [They were legendary for their fecundity] . . . and no institutional involvement in churches, political parties, farmer's organizations or schools. Illiterate and transient, they moved through Alabama's history like a shadow leaving little if any impression."

From early statehood, the philosophy of yeomen farmers regarding taxes was to levy the highest rates on slaves and slave holders with lower rates on land. Also, they believed that secondary taxes should be levied on ostentatious symbols of wealth, such as gold watches, private libraries, race horses, etc. As a result of the yeomen's ability to get their tax interests enacted into law in the early years, the wealthiest one-third of Alabama's population paid two-thirds of the taxes.

Planters, in contrast to yeomen farmers, faced a completely different future when they emigrated to Alabama. After the Cherokee land cession of 1806 in the Tennessee Valley and the Creek land cession in the Upper Alabama River Valley in 1814 (Figure 6.3), wealthy Georgia planters quickly purchased large tracts of land in both areas and established their cotton plantations with their large numbers of slaves. Following the end of the War of 1812, cotton prices increased rapidly. In the period 1817 to 1818, the price reached over 30 cents a pound. As a result, the fortunes of the Georgians (descendents the old Petersburg, Virginia, tobacco planters) who had settled in the Tennessee Valley soared. Because of the growing wealth of those planters, other men of extensive wealth moved to that area and began to purchase land. They were able to pay whatever price was necessary to purchase the land they wanted; consequently, cotton planters soon filled the central portion of each valley (Figures 6.14 and 6.31).

Some planters owned several plantations in the same areas. They lived on one that they supervised, and let their son(s) or hired overseers manage the others. Owners tried to operate their plantations in such a manner that they would be as self-sufficient as practical; however, some supplies such as cloth, tools, and specialty foods such as coffee, tea, flour, and sugar were purchased wherever the cotton crop was sold. Often, the planter would send a list of supplies needed down the river with the cotton. Planters' agents or factors who arranged the sale of the crop would purchase the supplies and get them shipped back upriver. These arrangements provided planters and their families a wider choice of goods than would be available locally; however, this system slowed the development of general stores in the small towns along the rivers. New farm animals might be purchased from other plantations or from drovers from Tennessee or Missouri who came south driving herds of animals for sale.

Immigration into the North-Central Region

In 1790, there were no states in the vast territory west of New York and Pennsylvania, north of the Ohio River, south of the Great Lakes, and east of the Missisippi River (Figure 6.10). In 1787, Congress had established the Northwest Territory that included land that would later be included in the states of Ohio, Indiana, Illinois, Michigan, and eastern Minnesota. The 1800 Census did not include data on the Northwest Territory, because there were no states there until 1803 when Ohio became a state.

By 1810, the census estimated that there were 231,000 people there, principally in Ohio. Most of them were in the southwestern corner of the state, pressed against the northern side of the Ohio River. Foreign immigrants began to appear in the settler stream during this period. Between 1816 and 1817, some 20,000 Germans, emigrating to escape a famine, entered the country. A large number of them were farmers, and many of them joined the stream to the Old Northwest. In 1816 and 1818, Indiana and Illinois, respectively, became states (sister states of Alabama and Mississippi). Many of the settlers in the northern part of both Indiana and Illinois were New Englanders, but those in the southern part were primarily yeoman farmers from the Virginia and North Carolina Piedmont. Later substantial numbers of yeomen settlers would be pushed northward into Indiana and Illinois out of Kentucky and Tennessee by wealthy planters moving from the South Atlantic Region (Edwards, 1940). The 1820 Census indicated that there were 867,000 persons in the region, whereas there had been only 231,000 in 1810 (Figure 6.12).

Between 1820 and 1830, large numbers of settlers moved into the North Central Region, primarily from New York and New Jersey. By 1812, there were sufficient numbers in Missouri that it was declared a territory. Most of the settlers in that territory had come, along with their slaves, from Tennessee, Kentucky, Virginia, and the Carolinas. Between 1820 and 1830, the North Central Region increased from 867,000 to 1,615,000 (Table 6.3). Missouri had become a state in 1821, and the Census of 1830 included data from Michigan Territory for the first time. During this period, the number of foreign immigrants settling in the region increased. Some 151,800 aliens, many of them Germans, entered the country during the decade, and large numbers of those finally settled in the North Central Region, especially in Ohio around Cincinnati, as well as in Indiana, Illinois, Michigan, and the lake counties of Wisconsin (Edwards, 1940).

By 1840, enough settlers had arrived in the portions of the North Central Region that would later become the states of Iowa and Wisconsin that census data was reported from them as the Iowa Territory and Wisconsin Territory. Michigan had become a state in 1837. The Old Northwest was rapidly filling up. In 1840, the population in the region reached 3.3 million (Table 6.3). In the 1830s, a large number of European immigrants arrived in the United States. Some 44% of them were Irish, and 30% Germans. Morison (1965) comments that the Irish were tired of farming, so they remained in the port cities where they landed. Many of the Germans were also farmers, but they soon followed their countrymen into the Old Northwest.

Between 1840 and 1850, the North Central Region's population increased from less than 3.5 million to more than 5.4 million (Table 6.3). During the period, Iowa (1846) and Wisconsin (1848) became states, and in 1850 census data was first reported from Minnesota Territory. During the period 1.6 million immigrants arrived from Europe. Some 60,000 of them were Germans, fleeing the failed 1848 revolutions. Only a small proportion of these Germans were farmers, but a substantial portion still followed their countrymen westward.

From 1850 to 1860, the number of persons in the North Central Region increased from 5.4 million to 10.6 million (Table 6.3). By 1860, the population of this region, was slightly greater than that of the North Atlantic Region (10.64 versus 10.59 million), and is very likely that there was a much larger share of farmers in the population of the North Central Region. A significant share of the population in the North Atlantic Region was located in Boston, New York, and Philadelphia. During the period, Minnesota became a state, and Kansas and Nebraska were designated as territories. Also during the period, 2.5 million Europeans emigrated to the United States.

By 1860, the populations of both the North Atlantic and North Central regions were almost two times as large as those of the South Atlantic and South Central regions (Table 6.3). Further, the populations of Alabama's two sister states, Indiana and Illinois, were considerably larger. than that of Alabama. Apparently, the new lands in the North Central Region were much more attractive to immigrants than those in the South Central Region.

Immigration into the Western Region

There were no states in the Western Region (Montana south to New Mexico and west to the Pacific) until the end of the war with Mexico in 1848 (Treaty of Guadalupe Hidalgo). California was admitted to the Union in 1850. Utah, which included land that would later become Nevada, became a territory in 1851. Nevada Territory was established in 1861. Censuses did not include any data on the region's population until 1850. Nevertheless, long before 1820, there were people living in all of those areas that would later become states. Coulson and Joyce (1999) have provides estimates of some of those numbers. These estimates are included in Table 6.3

Gold had been discovered in California in 1848, and the 1850 Census reported that there were already 92,600 persons in California and 188,000 in the Western Region. San Francisco grew from a squalid village to a city with 20,000 to 25,000 inhabitants within a few months. Relatively few of these settlers, at least initially, were farmers.

Morison (1965) comments that in 1842 so-called Oregon Fever struck folks on the frontier in Iowa, Illinois, Missouri, and Kentucky, and in 1843, waves of settlers, primarily from those states, began to follow the Oregon Trail out of Independence, Missouri, across prairies, mountains, and deserts into the Willamette Valley, an area that became part of Oregon Territory in 1848. The northern boundary of the Western Region at 49° North had been settled by treaty with the British in 1846, and Oregon Territory was granted statehood in 1859. Even though there were several thousand settlers (mostly farmers) there in 1850, the census that year did not include them in its count. Further, by 1848, some 5,000 Mormon settlers had arrived in the future state of Utah, and Congress organized the territory in 1851, but none of those people were counted in the 1850 Census either.

By 1860, there were 380,000 persons in California. There had been 92,600 a decade earlier. An additional 52,500 persons were counted in Oregon, and although the future state of Nevada was not given territorial status until 1861, the Census of 1860, reported that there were 6,875 persons there that year. Altogether, the estimated population of the Western Region was 628,000 in 1860 (Table 6.3).

The Final Filling-In of the Country

Although there was still some filling-in remaining to be done, most of the country between the original 13 colonies and the Pacific was at least thinly settled by 1860. In 1800, some 7.8% of the total population lived beyond the original English colonies. Just 60 years later, 51% lived beyond those old boundaries.

Even though the earliest westward immigration had been to the South Central Region (Tennessee and Kentucky), settlement in that region did not keep pace with settlement in the Old Northwest. One of the reasons for this disparity was the much larger population base in the states in the North Atlantic Region. Also, to a prospective settler in New York and New Jersey, it was much closer to good farmland in Ohio than to good farmland in the Old Southwest. This was especially true for most of the immigrants from Europe who arrived in the United States at the ports of Boston, New York, and even Philadelphia. Further, it is likely that many of those prospective settlers in the North were much more attracted to the opportunities in general farming in the Old Northwest and less to opportunities in cotton farming in the Old Southwest..

Farms

Decennial censuses before 1850 did not collect quantitative data on farms in the United States; however, a considerable amount of information on the general characteristics of farms in the colonies and the young country before 1850 is available. Two especially detailed sources were prepared by Bidwell and Falconer (1941) and by Gray (1958). Bidwell and Falconer prepared a detailed descriptions of early farms and agriculture in the Northern United States for the period 1620 to 1860, and Gray prepared a two-volume description of agriculture in the Southern United States from before the coming of the English to 1860. Additionally, material prepared by Edwards (1940) on the first 300 years of American agriculture presents a general overview of farms during that period.

The Census of 1850 contained a considerable amount of information on characteristics of farms in the country, states, and counties. In 1839, Congress appropriated a small amount of money to the commissioner of patents in the Department of State for the purpose of collecting agricultural statistics, conducting agricultural investigations, and distributing seeds. Beginning in 1847, they began annual appropriations for these activities. In 1849, they transferred the Patent Office and its agricultural activities to the newly created Department of the Interior. The 1850 Census began a trend in which each succeeding census would include more information on various aspects of American agriculture. Some of the information for 1850 and 1860 is summarized in the following sections.

Geophysical Determinants of Farm Characteristics

Climate and physical geography (soils, topography, geography, etc.) have always played an important role in determining where humans produced their food and fiber, many of the characteristics of the processes they used to do so, and many important characteristics of the farmers themselves. For example, the generally rocky, rough topography found in much of the North Atlantic Region required that most of the farming be done in relatively small fields. Further, the climate was cool and humid, and the growing season short. These climatic characteristics placed severe restrictions on the crops which could be produced there economically at the beginning of the nineteenth century.

A considerable amount of the land in the South Atlantic Region is hilly, and a substantial amount of it is mountainous. In that respect, its topography, especially in its western portion, is little different from most of the land in the North Atlantic Region; consequently most of its farms had to be relatively small. However, much of the eastern portion of the region lies in the Coastal Plain (Figure 1.10) where the relief is much lower. Here large farms were much more practical.

While the climate in the South Atlantic Region is much more amenable to the production of a wide range of crops, its soils (mostly Ultisols) are uniformly poor; consequently it was poorly suited for the long-term production of staple crops such as tobacco, rice, and indigo. This combination of geophysical characteristics, in turn, also determined some of the characteristics of farms that developed there.

As farmers moved west of the mountains into the North Central Region, they encountered a different set of geophysical characteristics. Here the relief is generally much lower than in either of the other regions, and the soils (mostly Mollisols) are uniformly good. While the climate is extremely cold in winter, the growing season is long enough to allow for the production of a number of valuable food crops. Here the combination of geophysical characteristics were ideal for the development of farming systems for the production of grains on extremely large farms. Initially, farmers there struggled with the heavy native grass thatch, but with the coming on larger, more powerful work horses and heavier plows, they were able to overcome this obstacle.

Geophysical characteristics of the South Central Region are not greatly different from those in the South Atlantic Region. A significant amount of the land, especially in its western portion is of high relief—extremely hilly or mountainous, and yet a large portion lies in the Coastal Plain where the relief is much lower. The annual range of temperatures is generally somewhat higher, and annual rainfall totals are also generally higher. This higher level of rainfall has resulted in the development of wide, level floodplains along the major rivers in the region. This geophysical anomaly made it possible to develop large farms in areas where the relief of the surrounding area was much higher.

Geophysical Characteristics and Farms in Alabama

Land in Alabama can be roughly divided into two major regions (Figure 1.8) with respect to relief: a High Hills region (Figure 1.8) north of the Fall Line (Figure 1.11) and the Low Hills region of the Coastal Plain south of it. Soils to the north were largely derived from so-called hard rocks (Figure 1.7), and soils to the south were derived largely from ancient sediments (Figure 3.3) eroded from the hard rock region north of the Fall Line and deposited in marine waters south of it.

Included in the High Hills region are two areas of unusually low relief—the flood plains of the Tennessee and Coosa rivers. Both of these river basins are associated with soils derived from ancient deposits of limestone. Areas of unusally low relief, associated with the basins of the Alabama-Tombigbee rivers and the Conecuh River are found among the Low Hills. A third and larger area of unusually low relief in this region, the Black Prairie, lies just south of the Fall Line. Historically, most of the cotton produced in the state was produced on these areas of especially low relief.

With this geophysical perspective as a basis, Alabama's counties have been divided into five groups:

1. High Hills counties
2. Low Hills counties
3. Cotton Arc (Black Prairie) counties
4. Tennessee Valley counties
5. Coosa Valley counties

Dividing Alabama's counties into these five groups, generally based on physiographic differences, will facilitate discussion of the characteristics of farms found here. (In 1840, there were 18 fewer counties than there are today.) Three new counties (Coffee-1848, Choctaw-1847, and Winston-1850) were created out of older, larger ones in the period 1840-1860. Consequently, several of the 1840 counties were much larger than they are today. For example, in 1840 Franklin County in northwest Alabama extended northward to the Tennessee River. Later, Colbert was created out of the northern portion of Franklin. Macon and Russell were also much larger than they are today. Lee was created out of the northern portions of those two counties.

The **High Hills** counties are generally located north of the Black Prairie and south of the Tennessee Valley and include the Fall Line Hills District of the Eastern Gulf Coastal Plain Section, the Inland Piedmont and Northern Piedmont districts of the Piedmont Upland Section, portions of the Alabama Valley and Ridge Section, several districts of the Cumberland Plateau, and portions of the Highland Rim Section (Figure 1.11). Large portions of the High Hills Section are mountainous with elevations to 2,000 feet; elevations above 1,000 feet are common. Much of the entire region was too steep for farming; however, in the 1830s and 1840s there were substantial numbers of small, limited-resource, inefficient farms established in the narrow, relatively-small, flat creek valleys.

The **Low Hill**s counties generally lie south of the Black Prairie (Lower Coastal Plain) and include the Chunnennuggee Hills, the Southern Red Hills, the Lime Hills, the Southern Pine Hills, and the Dougherty Plain—all districts of the Eastern Gulf Coastal Plain Section (Figure 1.11). Elevations of hills in these districts reach a maximum of 500 feet, and most are much lower. Geophysical characteristics of this region also dictated the establishment of mostly small farms. However, there are some large areas where the land was level to gently rolling, where larger fields and larger farms were much easier to establish. Soils were generally poor in all of these counties and were uneconomical to farm with the technology available at the time. Consequently, farms were established at a much slower rate there than in some of the other parts of the state.

The **Cotton Arc** counties are primarily those located within the Black Prairie District and extend from Russell, Macon, and Barbour counties in the east, through Montgomery, Lowndes, Wilcox, and Dallas to Marengo, Sumter, and Greene in the west, plus a few counties lying around its margin (Chambers, Tallapoosa, Autauga, Pike, Monroe, Clarke, Tuscaloosa, and Pickens) (Figure 6.31).

Coosa Valley counties include Cherokee, Calhoun, Talladega, and Coosa, and Tennessee Valley counties include Madison, Limestone, Lauderdale, Lawrence, Franklin, and Colbert—all located in the western portion of the Highland Rim Physiographic Section (Figure 1.11).

Farm Numbers

The 1840 Census did not include data on the number of farms in the United States or any of the states. Even the 1850 Census does not provide a complete picture of the numbers of farms. Census data were collected only in areas that had achieved statehood and those areas that had been officially designated as territories. Unfortunately, there were substantial numbers of farms in other areas that were not counted. Furthermore, portions of the North Central and Western regions were still on the frontier. Settlers were still moving into those areas to establish farms. South Dakota in the North Central Region and Montana in the Western Region did not become states until 1889.

The 1850 Census

The 1850 Census data indicated that there were slightly more than 1.4 million farms in the country when these data were obtained (Table 6.8 and Figure 6.33). The largest share, 489,754 farms (33.7% of the total), was in the North Atlantic Region, but the share in that region was not much greater than the 30.2% (437,597 farms) in the North Central Region. Approximately 18% of all of the farms in the country (260,814) were located in the South Central Region. Some 41,964 of that total (15.7%) were located in Alabama.

The distribution of farms in the different counties of Alabama in 1850 is shown in Figure 6.34. The figure indicates that a large share of Alabama farms was located in the eastern counties from Barbour in the south to Cherokee in the north. Most of these counties were formed from land ceded to the United States by the Creeks in 1832. There was a smaller concentration in the valleys of the upper Tombigbee and the lower Black Warrior rivers, as well as a small concentration in the Tennessee Valley (Figure 1.16).

In terms of the farming systems used within the state, the largest number of farms were located in those counties where large amounts of cotton were being produced. There was an average of 1,000 farms per county in the Tennessee Valley, the Coosa Valley, and the Cotton Arc. In the High Hills, the average was only 673 farms per county. In the Low Hills, it was 412. The average was especially low in the western portion of this section. There were 138, 121, 249, and 141 farms in Covington, Baldwin, Mobile, and Washington counties, respectively.

The 1860 Census

The 1860 Census indicated that the total number of farms in the United States was two million (Table 6.8 and Figure 6.35). Data presented in the table show that the largest number of farms was in the North Central Region. The total there was 200,000 farms, greater than in the North Atlantic Region (772,000 versus 565,000). Numbers in the South Atlantic and South Central regions were relatively similar (302,000 and 370,000, respectively).

Between 1850 and 1860, farm numbers in the country increased 595,000, or 41%. During the same period, the population of the country (Table 6.3) increased from 23.2 to 32.9 million (36.6%). Farm numbers increased faster than the population. More than half the total increase (334,600) in numbers was added in the North Central Region. In the 1850s, the North Central Region was still in the process of filling up. Between 1850 and 1860, the number of farms in Ohio only increased by 25.1%; however in states west of Ohio, the rate of increase was much higher. Further, the rate of increase was higher as the distance westward from Ohio increased. For example, there were 40.4% more farms in Indiana in 1860 than in 1850, 104.1% more in Illinois, 243.3% more in Wisconsin, and 11-fold more in Minnesota. South Dakota would not even become a state until 1869. No one would be aware of it at that time, but these new farms being established in the 1850s in the North Central Region were the foundation of what would soon become the most productive grain-producing region in the world.

Edwards (1940) commented that the movement of farmers onto the prairie lands of western Indiana and Illinois was a slow process. The wooden and cast-iron plows that they had used east of the Alleghenies were not suitable in the thick grass sod of the prairie soils. Also, pulling those plows took much more animal power. From three to seven yoke of oxen were required to break the land, and when the sod was broken-up, the roots of the grass decayed so slowly that fields could not be effectively cultivated for two or three years.

The South (South Atlantic and South Central regions

combined) added 157,300 farms from 1850-1860, some 65% of those were in the South Central Region. During that period, the numbers of farms in Tennessee, Kentucky, and Alabama increased 13.3, 21.4, and 31.4%, respectively. In Arkansas (119.6%) and Texas (251.6%), the percentage of relative increase was much greater than in states to the east. These data indicate that the westward movement of farmers had all but stopped in the eastern portion of the region, but was increasing steadily in the western portion. However, the magnitude of the westward movement in the South was not nearly as massive as in the western portion of the North Central Region.

Alabama farmers established 13,200 new farms (41,964 to 55,128) between 1850 and 1860, an increase of 31%. The distribution of farms within the state in 1860 remained about as it was in 1850 (Figures 6.34 and 6.36). In 1860, farms continued to be more concentrated in the Cotton Counties, but numbers were increasing at a rapid rate in Low Hills and High Hills sections. Numbers continued to be relatively lower in the western portion of the Low Hills section.

Surprisingly, many of the counties in the Cotton Arc lost farms from 1850 to 1860. Macon lost 427 farms (1,203 to 776), Sumter lost 170 (668 to 498), and Montgomery lost 170 (962 to 800). Altogether, in 11 of the 19 counties in this section, numbers declined. Atkins (1995) notes that in the period 1830-1860, hundreds of Alabamians moved to Texas. Most of them moved westward to get away from the increasingly wornout cotton lands in Alabama.

In the Cotton Counties of the Highland Rim numbers declined in two of five counties. Both Madison and Lauderdale were major cotton-producing counties, and both lost farms during the period, 184 (1,180 to 996) in Lauderdale and 165 (1,080 to 915) in Madison.

There was little loss in numbers of farms in the Cotton Counties of the Coosa Valley during the decade. The total in the five counties of the section increased from 4,481 to 5,828. Cotton production, while important, did not dominate agriculture in that section. There were large numbers of yeoman farmers in the valley in 1850, and their number increased throughout the following decade.

There were 2,879 more farms (4,117 to 6,996) in the ten counties of Low Hills section south of the Black Prairie in 1860 than there had been in 1850. The largest increases were in Dale (697 to 1,066), Covington (138 to 517), Butler (553 to 1,027), and Washington (141 to 725). Baldwin remained at the bottom of the list. There were only 67 more farms (121 to 189) there in 1860 than in 1850.

Blount County in the High Hills section was the only one of 14 counties in the section in which the number of new farms created was less than the number abandoned (753 to 711). Altogether, the number of farms in that section increased by 4,438 (9,427 to 13,865). The largest increases were in Jackson (856 to 1,214), St Clair (573 to 1,074), Marion (573 to 1,127), and Randolph (969 to 1,233). Winston had the smallest number of farms in the section in 1850 (144) and in 1860 (339).

Farm Size

Data on average farm size in the United States, its major regions, and Alabama in 1850 and 1860 are presented in Table 6.9. These averages include quantities of both improved and unimproved land. Improved farmland generally includes both cropland and pasture (meadowland).

The 1850 Census

In 1850, the largest farms in the country (695 acres) were in the Western Region. There were not that many of them (Table 6.8), and they were primarily extremely large, cattle grazing operations. At that time, there was very little crop production in that region.

As might be expected, farm size in the North Atlantic Region (113 acres) was lower than in any of the other regions in 1850. These were relatively small farms where the owner-operator and his immediate family provided most of the labor required. With a limited supply of labor, and with machinery and equipment suitable for use on small farms, it was not practical to operate larger ones. Further, the topography of that region played a significant role in determining farm size. Farms were somewhat larger in the North Central Region (143 acres), but the availability of labor and the suitability of machinery and equipment still mandated relatively small ones, even though topography was generally more suitable for larger ones.

Farm size (Table 6.9) in the South Atlantic (376 acres) and the South Central (291 acres) regions demonstrates the effect of the availability of land and labor (slaves)on the size of farm that a single owner could operate. Many of the farms in the South Atlantic Region had been in use for almost 2 centuries. Much of the land was largely unsuitable for crop production as a result of prolonged use without good conservation practices. Acceptable production could be obtained only by using land that had not been in cultivation recently. All of these factors and their interaction required larger farms.

The average size of farms in Alabama in 1850 was 289 acres (Table 6.9), very nearly the same as in the region as a whole. Data on average size in Alabama's counties were not available in the 1850 Census. Those estimates were computed from census data on total number of farms and total number of acres (improved and unimproved) in each county. As might be expected, farms in the 19 counties in the Cotton Arc were considerably larger (401 acres) than the average for the state. In Dallas County, the average was 749 acres. In Wilcox, Montgomery, and Greene counties it was 565, 507, and 402 acres, respectively. These averages probably do not adequately represent the distribution of farm sizes within those counties. Each average represents a relatively small number of very large cotton plantations, each with a large number of slaves, as well as a much larger number of much smaller farms operated by yeomen farmers and their families.

Farms were somewhat smaller in the five Cotton Counties of the Tennessee Valley (384 acres) and much smaller in the four counties in the Coosa Valley (240 acres). In both of these cases, and especially in the Coosa Valley, these smaller averages probably represent higher ratios of yeoman farms to cotton plantations than in the Cotton Arc counties.

The average size of farms in the ten counties of the Low Hills section was 257 acres. While this section was largely the bastion of yeomen farmers, the number of slaves (Figure 6.30) indicates that there were also substantial numbers of large cotton farms, especially in Butler, Conecuh, and Washington counties. Coffee and Dale counties had the smallest farms in the section, 101 and 146 acres, respectively. In 1850, there were few fences in this section, and livestock ranged widely without regard to property lines. While individual farmers might own relatively small farms, their livestock used a much larger area.

In the High Hills section, the percentage of yeoman farms was obviously quite high. The average farm in the 14 counties in that section was only 174 acres. In Winston and Walker counties, the average was 84 and 94 acres, respectively.

The 1860 Census

In 1860, the largest farms in the country continued to be in the Western Region (367 acres), but they were not much larger than those in the South Atlantic (353 acres) and South Central (321 acres) regions. Average size in the North Atlantic (108 acres) and North Central (140 acres) regions was lower than in the other regions.

Between 1850 and 1860, factors determining average farm size had not changed much in the North Atlantic and North Central regions; it remained at very near the 1850 level in both regions, although there was a small decline (less than 5%) in both. Family farms were still dominant in both regions. Individual families could not operate large farms, and there were few, if any, slaves.

There was a much larger decline in size in the South Atlantic Region (376 to 353 acres) between 1850 and 1860. It is likely that in this region size was beginning to reach a point of diminishing returns. With mostly wornout land available, it was no longer economically feasible to continue to operate larger farms.

Between 1850 and 1860, cotton production and slave ownership (Table 6.7) exploded in west-central Georgia, central Alabama, and the Delta of Mississippi and Arkansas. The topography and physiography, especially in the Black Prairie of Alabama and the Mississippi River Delta, were ideal for establishing large plantations, and for expanding their size. As a result, average size increased from 291 to 321 acres (10.3%) between 1850 and 1860 in the South Central Region (Figure 6.42). Economic returns resulting from expanding size in those areas were apparently favorable, even though a much greater investment in slave labor was required.

Average farm size in Alabama increased from 289 to 347 acres (Table 6.9) during the 1850s, a much higher rate (20.1 versus 10.3%) than in the region as a whole. The relative increase in farm size in Mississippi compared to the region was similar to that in Alabama (19.7%), but was considerably lower in both states than in Arkansas (67.8%). The large increase in Arkansas probably was the result of the rapid growth of plantation-type, cotton production in the state between 1850 and 1860.

The increase in average farm size in Alabama between 1850 and 1860 was the result of increases in all sections, and was probably the result of the sharp increase in the production of cotton in the state during the decade. The largest increases in the state were in the 19 counties of the Cotton Arc. Average size in those counties increased from 401 to 577 acres. Some of the larger increases were in Montgomery (507 to 691 acres), Autauga (396 to 601), Marengo (514 to 744), and Sumter (513 to 799).

Increases in average farm size in the Cotton Counties of the Coosa Valley and the Tennessee Valley were generally lower than in the Cotton Arc, and it actually declined in Calhoun County (316 to 227 acres), in the Coosa Valley, and Madison County in the Highland Rim (609 to 445 acres). The largest increase in either section was in Lauderdale County in the Highland Rim (217 to 428 acres).

In 1850, average size in ten Low Hills counties south of the Black Prairie, excepting Monroe and Clark, was 257 acres. A decade later, it had increased to 361 acres. Some of the larger increases in the section were in Covington (196 to 332 acres), Dale (146 to 329), Henry (206 to 396), and Mobile (180 to 383) counties.

In 1850, average farm size in the 14 counties of the High Hills section was 174 acres, and by 1860 it had increased to 277 acres. Some of the larger increases in that section were in Marion (199 to 322 acres), Winston (84 to 251), Blount (94 to 271), and Fayette (137 to 324) counties. It decreased in Jackson County from 279 to 269 acres.

Table 6.10 presents data on the percentage of farms throughout Alabama in several different size-classes. Note that only farms three acres or larger were included in the 1860 Census data. The data in the table indicate that there were many large farms (100-499 acres) in the state in 1860, but the size-class with the largest number (16,049) was 20 and 49 acres. When the percentage of total farmland that was actually being cultivated is considered, it is obvious that there were a large number of farms in the state, even at that early date, that simply were not large enough to be economically viable. This situation has continued to plague Alabama agriculture even into the twenty-first century.

The decline in farm size in the Western Region (695 to 367 acres) (Table 6.9) probably reflects that region's growing number of crop producing farms in relationship to the number of cattle-grazing operations. For example, wheat production increased more than tenfold in the Western Region from 1850 to 1860.

Total Farmland

Data on the total quantity of farmland (improved plus unimproved) in the United States, its major regions, and Alabama, in 1850 and 1860, are presented in Table 6.11. Total quantity of farmland is a product of farm number and farm size.

The 1850 Census

Data in Table 6.11 show that total farmland in the United States was 294 million acres in 1850, and that there was more farmland in the South Atlantic Region (93.40 million acres) than in any of the other regions. The second highest total was in the South Central Region. In 1850, the Cotton South (the combined South Atlantic and South Central regions) was the country's primary agricultural area, at least in terms of total acreage of farmland. Slightly more than 58% (171.04 of 293.56 million acres) of all acres in the United States was in these two regions combined. Some 15.6% of all the farmland in the South Central Region, and 4.1% in the county was on Alabama farms.

In considering the distribution of farmland within Alabama in 1850, it is important to remember that many of the counties were substantially larger than they are today. There were only 52 Alabama counties at that time. A total of 15 counties were created after 1850. These included: Bullock, Clay, Cleburne, Crenshaw, Elmore, Etowah, and Lee—all created in 1866. Three more new counties (Colbert, Hale, and Lamar) were created in 1867, and three (Chilton, Escambia, and Geneva) in 1868. Cullman County was created in 1877, and Houston in 1903.

In 1850, some of the leading counties in the state with respect to total farmland included: Dallas (577,600 acres); Greene (526,700); Montgomery (487,800); Macon (478,700); and Wilcox (376,600). All of these counties were in the Cotton Arc that stretched across central Alabama at that time (Figure 6.31).

There were also relatively large amounts of farmland in the Cotton Counties of the Coosa Valley (388,029 acres in Calhoun and 279,644, in Talladega) and the Tennessee Valley (401,831 acres in Franklin and 329,006,in Madison). The total in Franklin was somewhat higher than the average for that section. In 1850, that county reached northward all the way to the Tennessee River. As noted in the preceding paragraph, Colbert County was created out of the northern portion of Franklin in 1867.

There were also many counties on the other end of the scale with respect to quantity of farmland. These counties were mostly the domain of yeoman farmers. They had few slaves and produced relatively little cotton. In 1850, a number of these counties were in the Low Hills section south of the Black Prairie, excepting Monroe and Clarke counties. There were only 27,102; 44,695; 46,347; 57,111; and 61,253 acres of farmland in Covington, Mobile, Baldwin, Washington, and Coffee counties, respectively.

There were also a number of counties in the High Hills section north of the Black Prairie with relatively small farmland totals. There were only 12,087; 62,215; 64,736; 71,170; and 79,006 acres of farmland in Winston, Marshall, Blount, Walker, and DeKalb counties, respectively.

The 1860 Census

Data on the quantity of total farmland (improved plus unimproved) in the United States, its regions, and Alabama in 1860 are also presented in Table 6.11. These data show that the quantity was nearly the same (around 110 million acres) in the South Atlantic, North Central, and South Central regions and that the quantity in each of those three regions was significantly greater than in the North Atlantic (61 million) and Western (13 million) regions.

With farm numbers increasing substantially in the country between 1850 and 1860 (Table 6.8), total farmland increased by almost 114 million acres (Table 6.11). It increased, in all regions. Even where average farm size decreased, the increase in farm number was large enough to keep the quantity of farmland growing.

The largest real increase in total farmland was in the North Central Region where more than 45 million acres were added (62.69 to 107.90 million). The largest relative increase was in the Western Region where quantity increased more than threefold (4.66 to 12.72 million acres).

By 1860, the South Central Region took the lead in total farmland with 119 million acres. The South Atlantic Region was in second place with 107 million. Together, these two regions (the Cotton South) accounted for 55% of the total for the entire country. Between 1850 and 1860, cotton ginned in the United States increased almost 2.9 million bales.

In both the North Central and Western regions, the large increase in the total quantity of new farmland (Table 6.11) was the result of agriculture expanding into vast new areas. For example, the quantity in Minnesota increased from 28,881 to 2.71 million acres between 1850 and 1860, and from 290,600 to 1.41 million acres in New Mexico.

In 1860 Census data indicated that there was 19 million acres of total farmland in Alabama. This acreage represented 16% of that in the region (19.1 of 119 million acres) and 4.7% of that in the country (19 of 407 million acres).

In 1860, the distribution of total farmland was similar to that in 1850. Most of it was still concentrated in the Cotton Counties. In the 19 counties of the Cotton Årc, average total farmland per county was slightly over 493,000 acres. Averages in the Coosa Valley (397,000) and the Tennessee Valley (372,000) were relatively lower. Averages in the Low Hills (245,000) and High Hills (269,000) were considerably lower.

Total farmland in Alabama increased 57% (12.14 to 19.1 million acres) between 1850 and 1860 with both farm numbers and farm size increasing around 20% from 1850 to 1860. Consequently, both factors contributed equally to the increase in total farmland. In the same period, cotton production increased by 426,000 bales

(75.4%). As a prerequisite to the sharp increase in cotton production, farmers in the 19 counties of the Cotton Arc increased total farmland by an average of 118,070 acres per county. Some of the larger than average increases during the decade were in the counties around the margin of the Black Prairie itself. Total farmland increased by 157,342 acres in Autauga County; 158,270 in Marengo; 184,034 in Tallapoosa; 254,242 in Pike; and 334,545 in Clarke.

Surprisingly, total farmland actually declined in Dallas County (577,589 to 547,473 acres) and Macon (478,743 to 445,492). Further, in the other counties in the heart of the Black Prairie cotton-production area, Greene, Montgomery, and Sumter added only 34,417; 55,211; and 65,264 acres, respectively, to their farmsteads during the decade.

The amount of farmland in the Cotton Counties of the Coosa Valley also increased significantly in the period 1850-1860. The relative increase was even greater than in the Cotton Arc, increasing 127,145 acres in the decade. Farmers in Cherokee, Talladega, and Coosa counties added 143,718; 170,262; and 210,704 acres of new farmland, respectively. Total farmland in Calhoun declined during the period (388,029 to 371,925 acres).

Increases in total farmland in Cotton Counties of the Tennessee Valley were much lower than in the Cotton Arc. In the four counties in that section, average total farmland per county increased only 75,861 acre. Farmers in Lauderdale added 186,431 new acres. There were 78,237 new acres added in Madison, but only 12,287 in Franklin.

New farmland exploded in the Low Hills section between 1850 and 1860. In 1850, there were 94,700 acres per county in the ten counties of that section. In 1860, the average per county had increased to 245,287 acres, an increase of 159%. Much of that new land was used for cotton production. The yeoman farmers of the Low Hills had joined the Cotton Kingdom. Farmers in Choctaw, Henry, Dale, and Butler counties added 179,490; 230,652; 248,468; and 257,305 acres of farmland, respectively..

The magnitude of the increase in total farmland in the High Hills section was similar to that in the Low Hills. Farmers increased their acreage significantly. Average acreage per county in the ten counties increased from 120,330 to 268,689 acres. Total farmland increased 239,077; 248,614; and 280,776 acres,in Fayette, Marion and Randolph counties, respectively. However, the largest relative increase in total farmland in Alabama during this period was in Winston County. There were only 12,087 acres in that county in 1850. A decade later, there were 84,992, an increase of 603.2%. "Cotton, cotton everywhere, but very little to eat."

Improved Farmland

The quantity of total farmland, as estimated in the censuses of 1850 and 1860 consisted of both improved and unimproved land. Improved farmland generally included land that had been cleared of forest and that was being used for the production of crops or as pasture (meadowland). Estimates of improved farmland are much more variable than total farmland. As improved farmland, used for the production of cotton wore out, it might be abandoned to become unimproved farmland again. Conversely, when improved farmland was abandoned, an equal amount of unimproved farmland had to be added, if the farm was to retain its productive capacity.

Data on the quantity of improved farmland and the percentage of total farmland considered to be improved on farms in the United States, its regions, and Alabama, in 1850 and 1860, are also presented in Table 6.11.

The 1850 Census

In 1850, the quantity of improved farmland in the United States ranged between less than 100 thousand acres in the Western Region to 34 million in the North Atlantic Region (Table 6.11). In the regions (North Atlantic, South Atlantic and South Central) where the political boundaries (statehood) had been determined, quantity was relatively stable; compared to those regions (North Central and Western) that were continuing to expand westward.

The quantity and percentage of improved farmland varied considerably across the country in 1850. Both were generally inversely related to farm size. For example, farms were generally smaller in the North Atlantic Region (Table 6.9), and farmers in that region improved a larger share (61.6%) of the total amount of land on their farms

(Table 6.11). Conversely, farms were almost three-times larger in the South Atlantic Region, and a significantly smaller share of their total acreage was considered to be improved (32.1%).

Improved farmland in Alabama totaled 4.4 million acres in 1850. Approximately 20% of the improved farmland in the South Central Region and 3.9% in the country was on farms in Alabama.

The quantity of improved farmland in Alabama counties in 1850 was generally related to the amount of total farmland in each county. The average quantity per county in the 19 counties of the Cotton Arc was 142,000 acres. In the Cotton Counties of the Coosa and Tennessee valleys, averages were 68,000 and 120,000 acres ,respectively. Where there was less cotton produced, the quantity was much lower: 29,000 acres per county in the Low Hills and 40,000 in the High Hills.

In 1850, the percentage of farmland in Alabama that was considered to be improved was 36.6%. In the South Central Region and the country as a whole, it was 28.4 and 38.3%, respectively.

In 1850, farms in the Cotton Arc across central Alabama, generally had percentages of improved farmland higher than the state average. For example, in Russell, Marengo, Montgomery, Sumter, and Greene counties, 40.0, 41.4, 41.6, 45.2, and 45.4%, respectively, of total farmland was improved.

Farms in some Cotton Counties of the Coosa Valley (Cherokee, Calhoun, Talladega, and Coosa) generally had lower percentages of improved land than the state as a whole. For example, only 19.4% of the farmland in Calhoun County and 29.3% in Coosa County was considered to be improved.

Farms in some of the Cotton Counties in the Tennessee Valley (Madison, Limestone, and Lauderdale) had percentages of improved farmland well above the average for the state. Approximately 50.1% of all the farmland in Madison County was in the improved category. In fact, Madison was the only county in the state in that year with more improved than unimproved farmland.

Several counties south of the Black Prairie in the Low Hills section also had percentages of improved farmland well below the state average. There was relatively less commercial agriculture in those counties. Counties in the southwest corner of the state generally had the lowest percentages of improved farmland in that section and among the lowest in the entire state. Percentages were 11.5, 13.1, 21.0, and 25.2% in Mobile, Baldwin Washington, and Choctaw counties, respectively.

Percentages of improved farmland in the 14 High Hill counties ranged from 29.2% in Bibb County to 56.5% in Winston; although none of the others were nearly as high as Winston. The average for the section was 35.7%, very close to the average for the state (36.5%). At the same time, none of the counties in this section had percentages as low as some of those in the Low Hills section.

The 1860 Census

In 1860, there were 163 million acres of improved farmland in the United States. There were generally equal amounts (around 35 million acres) in the North Atlantic, South Atlantic, and South Central regions. Quantities in the North Central and Western regions were 52.3 and 3.7 million, respectively, but in both of those regions quantities were increasing as settlers move westward. The share of total farmland considered to be improved continued to be highest in the North Atlantic Region (63.8%). It was 48.5% in the North Central Region, where family farms still dominated. The percentage was near the same level in the South Atlantic (34.9) and the South Central (27.9) regions, where a large share of the land was in plantations.

Farmers in the South Atlantic Region added 13 million acres (93 to 106 million) of land (improved plus unimproved) to their farms from 1850 to 1860 (Table 6.11), but the quantity of improved land increased by only five million acres. As a result, the percentage of improved land increased only from 32.1 to 32.8%. The percentage remaining nearly the same from 1850 to 1860 means that they either did not improve 8.23 million (13.12 - 4.89 million acres) of their newly acquired acres, or that they abandoned about that amount of land that had been improved previously. The latter explanation is more likely to be correct. In 1850 and 1860, much of the suitable land in this region had been in farms for more than 200 years, and it is likely that a substantial amount of it was re-cycled from improved to unimproved, and from unimproved back to improved, several times in that period. Also, with the cultural practices in use for the production of cotton in the South at that time, land would not support the production of the crop for very many years before yields became so low that it could no longer be used for that purpose. With few other crop options, it was usually abandoned, at least for a number of years, before being re-cycled.

The data on the percentage of improved farmland in the North Central Region is also likely to be somewhat misleading. In 1850 and 1860, many new farms were still being established on the frontier, and many farmers were still struggling with the conversion of forestland and prairie into improved farmland. The increase in the percentage of improved land from 1850 to 1860 (42.1 to 48.5%) probably supports this conclusion.

In the South Central Region, farmers added 41 million acres of new farmland, but the quantity of improved land only increased by 12 million (Table 6.11). The percentage of improved land actually decreased 28.4 to 27.9%. They either did not improve 29 million acres of their newly acquired land, or they abandoned that amount that had been improved earlier. In Alabama, farmers added almost seven million of farmland, but the percentage of improved land declined from 36.5 to 33.4%.

Census data indicated that in 1860 there were six million acres of improved farmland in Alabama. This total represented 19.2% of all of the improved farmland in the South Central Region and about 3.9% of all of the improved farmland in the entire country. The share of total farmland in Alabama considered to be improved was slightly higher than in the region (33.4 versus 27.9%), but both were lower that the percentages in the North Atlantic and North Central regions, 63.8 and 48.5%, respectively.

As noted in a preceding paragraph (Table 6.11), total farmland in Alabama increased by 57% from 1850 to 1860; however, the percentage of improved land declined. It was 36.6% in 1850, but down to 33.4% in 1860. The rate of decline in the state was greater than in the region (8.5% versus 1.8%). Alabama farmers added about seven million acres of new farmland during the 1850s, but in 1860 there were only an additional two million acres of improved land.

The percentage of improved farmland decreased in Alabama from 1850 to 1860 (36.6 versus 33.4%); although surprisingly, the percentages of improved land in some of the counties in the Cotton Arc increased substantially during the decade. Percentages in Macon, Montgomery, Dallas, Sumter, and Greene counties were all greater than 46.6% in 1860. None had been that high in 1850.

Percentages of improved land generally increased in the Cotton Counties of both the Coosa and Tennessee valleys, although in both areas, the percentage increased in some of the counties, but declined in others. In Cherokee County, the percentage increased from 30.4 to 32.9%. In contrast, it declined from 34.7 to 31.1% in Talladega. In Madison, the percentage increased from 50.1 to 52.6%, but in Lawrence the percentage declined from 45 to 32.2%.

Farmers in counties in the Low Hills and High Hills sections generally had relatively smaller quantities of total farmland in 1850, and they added significantly to their acreage by 1860; however, they apparently improved very little of it when they purchased it. For example, in Covington County, farmers added 144,824 acres of total farmland to their farmsteads between 1850 and 1860, but in 1860 only 17% of the total was improved. In 1850, they had a much smaller amount of land, but 33.9% of it had been improved. Farms in many of the Low Hills counties had lower percentages of improved land in 1860 than in 1850. For example, in Baldwin County, it declined from 13.1 to 12.2%, 11.5 to 7.4% in Mobile, 21 to 13.6% in Washington, and 33.9 to 17 in Covington.

As detailed in a preceding paragraph, total farmland had also increased significantly in the High Hills section, but in almost all of them, percentages of improved farmland declined. For example, it declined from 26.3 to 10.7% in Marion County, 30.6 to 13.7 in Walker, 31.8 to 14.7 in Fayette, and 56.5 to 14.5 in Winston.

The yeoman farmers in both the Low Hills and High Hills sections were able to obtain substantial quantities of additional land between 1850 and 1860, but they did not have the resources to improve very much of it.

With limited labor at their disposal, they could farm only so much land each year. There was little incentive or opportunity to improve very much above that level.

Value of Land and Buildings

Trivelli (1997) suggested that there are many factors that act and interact to determine the value of land on farms at any one time. Its value for the production of food and fiber is only one of them. For example, when Alabama was only a portion of the Mississippi Territory, the value of land that had been obtained from Native American cessions was determined largely by political fiat. Its value for the production of food and fiber had little to do with it. At about the same time, the 1800 Census indicated that there were 422,800 persons in Massachusetts. This population was equivalent to about one person for each 11.9 acres of land. So, at the time when there was virtually no competition for land in Alabama, Massachusetts was well along the path to urbanization. As a result, there was rapidly increasing competition for land for many uses.

Data on the total amount of money invested (value) in land, fences, and buildings on farms in the United States, its major regions, and Alabama are presented in Table 6.12. Although these data reflect the costs of all buildings, fences, and other improvements, most of the total is related to costs for land. Regional difference are primarily the result of differences of the cost or value of farmland, although there are also some differences in the characteristics of improvements from region to region.

The 1850 Census

Data in Table 6.12 reveal that in 1850 slightly more than 44% ($1,455 million of $3,271.6 million) of the total value of all farm real estate (land, buildings, fences, and other improvements) in the United States was on farms in the North Atlantic Region, although farms in that region represented only 18.8% of total farmland acreage. Farms in that region were generally located near good, growing markets and had access to the best transportation facilities in the country. As a result, farmland values were higher, and farmers could afford to invest more money in improvements. Further, because of the extremely low winter temperatures, farmers had to invest a considerable sum to provide suitable housing for their families and their livestock.

Total value of land and farm improvements was lowest in the Western Region ($8.6 million), primarily as a result of the small amount of total farmland there (1.5% of the national total), and extremely low land prices. It was next lowest in the North Central Region. Only some 21.3% of total farmland was located in that region, and land prices were moderate, especially in the frontier areas of Wisconsin, Minnesota, and the Dakotas. Only 1.6% of the total value was in that region.

Values were generally similar in the South Atlantic and South Central regions ($576.6 and $479.6 million), respectively). Cotton production with slave labor dominated both regions, and the economic condition of both were heavily dependent of the export of the fiber either to the Northeast or to Europe. It is also likely that land prices were relatively similar in both. Further, the same kinds of farm infrastructure would have been required in both.

The total value of land and improvements on Alabama farms in 1850 was $64.3 million, or 13.4% ($64.3 of $479.6 million) of the total for the region (Table 6.12). Approximately 15.6% of total farmland of the region was located in Alabama. A substantial portion of total value was concentrated in the counties of the Cotton Arc. Some 59.8% of the total for the state was in those 19 counties. In contrast, only 5.6% was in the ten counties of the Low Hills section, south of the Black Prairie. The largest investment (value) in a single county was in Dallas ($3.9 million). The lowest investment in a single county was in Winston ($40,139).

Table 6.12 also includes data on the value of land, buildings, and other improvements on a per-acre basis. This conversion adjusts the data to some degree for differences in the number and size of farms from region to region. It also increases the emphasis on the cost or value of land. With per-acre estimates of value, the cost of the land is the primary element.

The converted data do not change the general conclusions regarding the level of total value in the different regions, but the differences are somewhat more pronounced. These converted data also demonstrate, even more strikingly, the relatively low level of investment in farm infrastructure in the Cotton South ($6 an acre in both the South Atlantic and South Central regions versus $26 an acre in the North Atlantic Region). Relatively lower land costs probably account for most of the difference, yet some of it was certainly the result of less investment in fencing and in winter shelter for livestock. The average per-acre investment in Alabama was even lower than the average for the North Central Region ($5 versus $6 an acre).

Investment in farm infrastructure (buildings, fences, and other improvements) on a per-acre basis in Alabama in 1850 varied substantially. In the Cotton Arc, the average investment per county was $5 an acre. In the Coosa Valley, Tennessee Valley, Low Hills, and High Hills, it was $6, $7, $3, and $4, respectively. Throughout the state, per-acre value was highest in Madison County ($10) and lowest in Marion ($2). These estimates likely reflect differences in land prices in the different sections, which in turn primarily reflect differences in the availability and demand for land for the production of cotton.

The 1860 Census

Value of land, buildings, and other improvements in the United States totaled $6.6 billion in 1860. The larger shares of the total were in the North Atlantic ($2.1 billion) and North Central ($2.1 billion) regions. Total value was relatively equal in the South Atlantic ($1 billion) and South Central ($1.3 billion) regions.

Total value in the country increased 103% ($3,271.6 million to $6,645 million) from 1850 to 1860 (Figure 6.12), but during the decade the quantity of total farmland had increased by 38% (293 to 407 million acres) and improved farmland had increased by 44% (113 to 163 million) (Table 6.11). This means that almost half of the total increase in value was likely the result of the increase in the amount of farmland in the country.

The increase in total value was especially large in the North Central ($51.7 to $2,130 million) and Western ($8.6 to $70.5 million) regions where many new farms were still being established on the frontier, and the quantity of land in farms was also increasing rapidly. The increase was not so great in the North Atlantic ($1,455 million to 2,121.9 million), the South Atlantic ($576.6 million to 1,008.6 million), or the South Central ($479.6 million to $1,314 million) regions where the number of farms and the quantity of farmland did not increase very much. For example, between 1850 and 1860, the quantity of total farmland in the North Atlantic Region only increased by 10.6% (55.2 to 61.1 million acres); whereas, in the North Central Region, the increase was 72.1% (62.7 to 107.9 million acres). Total value was significantly lower in the Western Region where land prices were extremely low. Farmers there also tended to invest much less money in fences and other improvements.

Over time, inflation of the currency can play a significant role in changes in the total value of land, buildings, and other improvements on farms, but it seemed to play a relatively minor role in the increase in value in the United States between 1850 and 1860. The Consumer Price Index (Figure 6.5) was at a level of 25 in 1850 and 27 in 1860. The increase of the CPI from 25 to 27 was equivalent to an inflation rate of 8% for the decade, while the increase in total value was 103%.

Data on per-acre value of land, buildings, and improvements on American farms in 1860 are also shown in Table 6.12 These data show that the value in the North Atlantic Region ($35) was almost twice as high as those in any of the other regions. This higher value almost certainly reflects the competition for land in that region, plus some additional costs for infrastructure there. Values in the South Atlantic ($9) and the South Central ($11) regions were generally similar. The slightly higher value in the South Central Region might indicate some comparative advantage in the relative quality of farmland for the production of cotton. The per-acre value in the North Central Region ($20) was considerably higher than either of the Cotton States, but not nearly as high as in the North Atlantic Region. There might have been some competition for land in Ohio and Illinois by 1860, but not likely in any of the other states of that region. Economic returns from farming in that region were certainly higher than in either of the Cotton States, but not as high for the specialized agriculture around the growing urban centers in the Northeast.

The total value of farmland and infrastructure in the United States increased from $3,271.6 million in 1850 to $6,645 million in 1860, an increase of 103% (Table 6.12). Per-acre value did not increase to the same extent, increasing only 46% ($11 to $16 an acre). While per-acre value continued to be highest ($35) in the North Atlantic Region, its relative increase between 1850 and 1860 was less than for the country as a whole (32 versus 46%). In 1860, it was second highest in the North Central Region ($20 an acre) and had increased at a much higher rate than the country (65 versus 46%). Per-acre value in the South Atlantic and South Central regions was generally similar, $9 and $11, respectively, although it had grown faster in the western portion of the Cotton South during the decade (79 versus 54%). Land prices were probably growing more rapidly in the South Central Region as cotton production surged.

Investment in farm infrastructure on a per-acre basis in the Western Region increased by 200% between 1850 and 1860 ($1.86 to $5.54), but it was still much lower than in any of the other regions in 1860.

The total value of Alabama farm real estate in 1860 was $175.8 million and represented 13.4% of the total value in the region ($175.8 of $1,314 million), the same relative level as in 1850. In 1860, 16% of total farmland in the South Central Region was in Alabama. As was the situation in 1850, in 1860 a large share of total value of land and improvements in the state ($112.1 million of $175.8 million) was concentrated in the 19 counties in the Cotton Arc. In fact, that share increased during the decade from 59.8 to 63.8%. In 1860, Marengo County had the largest share of total value ($10.3 million). Farms in Winston continued to have lowest value ($213,261).

In 1860, per-acre value continued to be lower in Alabama than in any of the regions, except the Western. Per-acre value in 1860 varied widely across the state, from $2 in Fayette County to $18 in Montgomery County, which had the highest per-acre value ($11.80) in counties in the Cotton Arc. Average investment in six of those counties (Macon, Montgomery, Dallas, Perry, Sumter, and Greene) was $16. It was almost equally as high ($11.53) in counties in the Tennessee Valley, but considerably lower ($6.79) in the Coosa Valley. Per-acre values in the counties of the Low Hills and High Hills sections ($5.75 and $4.78, respectively) were the lowest in the nine counties in these two sections (Baldwin, Blount, Covington, Fayette, Marion, St Clair, Walker, Washington, and Winston).

Between 1850 and 1860, per-acre value of Alabama farmland, buildings, and improvements (Table 6.12) increased from $5.30 to $9.20, an increase of 74%. Much of this increase was in the 19 counties of the Cotton Arc, where it increased from $5.15 to $11.80 (129.1%). The increase in the Tennessee Valley was also substantial (57.5%). The increases in both of these sections are probably indicative of the growing cotton economies in both. The increase in the Low Hills (76.4%) was surprising high and probably is the result of the movement of cotton production onto yeoman farms in that section. Increases in the Coosa Valley (22.5%) and Low Hills (17.7%) were the lowest in the state.

Value of Farm Implements and Machinery

Edwards (1945) comments that for thousands of years agriculture was dependent on hand labor with the use of relatively few rudimentary tools. Later in history, with the evolution of metallurgy and metal working, there was a steady improvement in the quality and quantity of hand tools and implements available to farmers. Later, farmers began to use animal power to a limited degree, primarily in breaking-up the soil. Initially, however, animals were of little use in planting, cultivating, and harvesting crops. There were no suitable implements or machines available for the conversion of animal power into crop work.

Toward the end of the eighteenth century, inventions began to appear in England and the United States that would change the world of agricultural work forever. In the 1790s, the cradle and scythe were introduced. In 1793, Eli Whitney invented the cotton gin. In 1797, Charles Newbold patented the first cast iron plow. In 1819, Jethro Wood patented an iron plow with interchangeable parts. Then in the 1830s, three inventions appeared that were extremely important. In 1834 John McCormick patented the mechanical reaper, in 1837 John Deere and Leonard Andrus began to manufacture steel plows, and also in 1837 the first practical threshing machine was patented..

In 1841, the first practical grain drill was patented. In

1842, the first grain elevator was put into use in Buffalo, New York. In 1844, the first practical mowing machine was invented. Irrigation was first used in Utah in 1847. Mixed chemical fertilizers began to be available for limited purchase after 1849. A rudimentary corn picker was invented by Edmund Quincy in 1850. After 1850, most of the grain in the prairie regions was being threshed by machine. By 1851, McCormick was producing 1,000 of his reapers each year. The first self-governing windmill was perfected in 1854, and in 1856 the two-horse, straddle-row cultivator was patented. By 1857, John Deere was producing 10,000 steel plows a year. As early as 1860, the wheat drill and the horse-drawn corn drill were in general use in the Middle Atlantic States.

The growing use of those labor-saving devices quickly revolutionized American agriculture. Production and productivity increased rapidly while the amount of farm labor required on individual farms began to decline. However, the widespread adoption of those new contraptions came at a high price. They increased the farmer's need for large amounts of ready cash. The machines could not be purchased with the proceeds of subsistence farming. Agriculture in America was quickly forced to become commercialized. Yeomen farmers everywhere had to produce something to sell for cash or perish.

Unfortunately, the use of these new developments were not equally distributed nationwide. Virtually all of them were used in the production of grain. It would be many years before new machines would be widely used in cotton production. Certainly, before 1860, with so many slaves involved in cotton production, there was little incentive to seek labor-saving devices. There was little else for slaves to do, but "plant, hoe and pick."

The 1850 Census

The Census of 1840 gathered a great deal of information on many aspects of American agriculture, but nothing on the value of farm implements and machinery. The first census to obtain this kind of data was taken in 1850, and it estimated the total value of farm implements and machinery to be $151.6 million (Table 6.13). As would be expected, the value of implements and machinery is lower than the value of land and improvements (buildings, fences, etc.) on those farms. For example, the 1850 Census estimated that the total value of farmland and improvements in the United States was $3.3 billion (Table 6.12). For each dollar invested in land and buildings, American farmers invested 4.6 cents in implements and machinery.

Census data (Table 6.13) also show that the larger share of the total investment in implements and machinery was in the North Atlantic Region ($54.2 million). Levels of investment in the South Central and North Central regions were similar, $36.7 and $35.6 million, respectively. It was substantially lower in the South Atlantic Region ($24.7 million). With a mostly cattle-grazing agriculture in the Western Region, investment was only a small fraction of the national total ($400,000).

When investment in implements and machinery is computed on a value per-farm basis, the South Central Region ($138), replaces the North Atlantic Region ($111) in first place (Table 6.13). Those two regions are followed, in descending order, by the South Atlantic, North Central, and Western regions with investments of $99, $81 and $67 per farm, respectively. These changes clearly show the increased investment in South Central Region resulting from the rapidly increasing level of cotton production and on-farm processing (ginning) in the region, especially in Alabama, Mississippi, Louisiana, and Arkansas.

The 1850 Census data indicate that total value of implements and machinery on Alabama farms totaled $5.1 million and represented 13.9% of the total in the South Central Region. In 1850, some 16.1% (41,964 of 260,814) of the farms in the region were in Alabama. Investment (value) varied widely across the state. It was $369,856 on all farms in Dallas County and only $4,745 on all farms in Winston County. The largest average per-county investment was in the 19 counties in the Cotton Arc across central Alabama ($159,574). Average per-county investments in other sections were, in descending order: Tennessee Valley ($136,697); Coosa Valley ($98,005); High Hills ($45,803); and Low Hills ($37,701).

Total value of implements and machinery was highest in the counties in the Cotton Arc and in counties around its margin. In this section, average investment per county in five of the major cotton producing counties (Dallas, Greene, Montgomery, Perry, and Marengo) was

$246,179. Cotton production and on-farm processing (ginning) generally required a relatively large quantity of specialized machinery and implements. In contrast, in 1850 there were a number of Alabama counties (principally in the Low Hills and High Hills sections) where cotton production was much less important. There yeomen farmers still produced crops primarily for their own use and for barter. Cash agriculture was less advanced, and implements and machinery requirements were much lower. In five counties in the High Hills (Blount, DeKalb, Marion, Walker, and Winston), the average value per county was $26,487. In five similar counties in the Low Hills (Baldwin, Butler, Covington, Mobile, and Washington), it was $23,068 per county.

The data discussed in the preceding paragraph is biased somewhat by differences in the number of farms in the different counties. Some of this bias is eliminated by computing investment in implements and machinery on a per-farm basis. This data does not change the primary conclusions about the distribution of investment within the state: the largest investment was in those counties that produced the largest amounts of cotton. However, it does provide a more realistic picture of the relative levels of investment on individual farms, in plantation and yeoman agriculture. On a per-farm basis, the average value in those major cotton-producing counties listed in the preceding paragraph was $268 per farm, and in the High Hills and Low Hills counties, it was $67 and $98, respectively. Note that the average investment on a per- farm basis in the High Hills counties was much lower than in the Low Hills counties. It had been higher on a per-county basis. This is a result of the generally larger number of farms in those counties north of the Black Prairie. Also, there was likely a little more cotton production on farms in the Low Hills. These are astounding numbers. When considering the value in implements and machinery required to operate a modern farm in Alabama, it is difficult to imagine that farmers in Winston County, in 1850, operated their farms with an average investment of only $32.

The 1860 Census

Data presented in Table 6.13 indicate that the total value of implements and machinery on American farms in 1860 totaled $246.1 million. Value was still highest in the North Atlantic Region ($73.8 million), but it was not much higher than in the North Central Region ($72.8 million). It was somewhat lower in the South Central Region ($61.3 million), and much lower in the South Atlantic Region ($34 million). It was only $4.1 million in the Western Region.

From 1850 to 1860, value of implements and machinery in the United States increased 62.3% ($151.6 to $246.1 million) (Table 6.13) partially as a result of the increase (41%) in the number of farms in the country (Table 6.8) during that period. The increase, in total value was greatest in the North Central Region ($37.2 million) where farm numbers increased 76.4% during the decade. On a relative basis, however, the increase was greatest in the Western Region (tenfold). Increases (percentages) in the other regions were: 67 in the South Central, 37.6 in the South Atlantic, and 36.1 in the North Atlantic regions.

Obviously, increases (percentages) in value were related to the increases in farm numbers, but in all cases increase in value was greater than increase in number, so increase in numbers was not the only factor involved. As detailed above, a world-changing evolution of farm implements and machinery was taking place during this period. By the 1850s, all kinds of new, better, and more efficient machines were being manufactured. These new machines were quickly revolutionizing farming, but they invariably cost more money than traditional tools. Because most of the improvements was of greatest benefit to grain production, farmers in the North Central Region likely purchased more of them than farmers in the other regions.

On a per-farm basis, value in the United States increased by 15.1% ($104.61 to $129.40) between 1850 and 1860. Obviously, when corrected for the increase in farm numbers, increases in total value shown in Table 6.13 were much more limited. There was also inflation of the currency involved in the increase in total value. Between 1850 and 1860, the Consumer Price Index increased from 25 to 27, which is equivalent to an inflation rate of some 8%. When taking inflation into consideration, it is clear that American farmers did not purchase as many of the new machines as it first appeared.

On a per-farm basis, the increase (78.8%) in value in the Western Region between 1850 and 1860 far exceeded increases in the other regions (South Central (20.2%), North Central (16.0), North Atlantic (18.2), and South Atlantic (13.5). Agriculture in the Western Region was changing rapidly during this period. As detailed in a preceding section, it had originally been based primarily on grazing, but in the decade of the 1850s grain farmers with their higher requirements for implements and machinery were flooding in.

The increase (20.2%) in value per farm in the South Central Region was somewhat higher than in the other regions; however, cotton production increased by 156% (1.6 to 4.1 million bales) during the decade. Much of the increase in the value was likely the result of the added requirement for implements and machinery in the increased production of cotton.

The value of implements and machinery on Alabama farms in 1860 totaled $7.4 million. This total represented 12% and 3%, respectively, of the totals in the region and in the country. The largest share ($4.4 million) of the state total was in the 19 counties of the Cotton Arc. In that section, average per county was slightly more than $229,300. Average value per county in the other sections was: Coosa Valley ($161,800), Tennessee Valley ($137,000), Low Hill ($63,582), and High Hills ($79,100).

Total value of farm machinery and equipment in Alabama increased from $5.1 to $7.4 million (10.4%) in the 1850s (Table 6.14). Much of the absolute increase was again concentrated in the Cotton Arc where the average per county went from $159,600 to $229,300 (43.67%). However, the relative increase in investment per county during the period was substantially greater in counties in the Low Hills (68.7%) and High Hills (72.7%) sections, as a result of the unusually high rate of establishment of new farms.

Average investment per farm in the Cotton Arc was $277 in 1860. In the other sections, it was, in descending order: Coosa Valley ($114), Highland Rim ($164), Low Hills ($93), and High Hills ($83). Two of the highest levels of investment were in the Cotton Arc counties of Wilcox and Sumter: $818 and $500, respectively; the two lowest counties were: $33 per farm in Washington in the Low Hills and $48 in Randolph in the High Hills.

Value per farm in Alabama increased from $122.14 in 1850 to $148.47 in 1860, an increase of only 21.6%. Alabama farmers were apparently not taking part in the better implement and machinery revolution. While value per farm increased little in the state between 1850 and 1860, it increased by 64% ($169 to $277) in the Cotton Arc. In contrast, it declined by 4% ($98 to $94) in the Low Hills.

Animal Power on Farms

Human muscle provided virtually all of the power required for the production of food and fiber for thousands of years, and only very slowly did the species begin to find ways to share that burden with domesticated animals. From the early days of the colonial period, horses, mules, and oxen, played increasingly important roles as sources of power in agriculture.

Horses are more versatile work animals than either mules or oxen. The nature of their gait makes them much more useful for riding. They are also very adaptable in pulling light loads over long distances at a relatively rapid rate. Working oxen are powerful draft animals; they generally cost about half as much as horses and required about half the feed. However oxen could not be easily ridden. They were not very useful in crop cultivation. They worked at a much slower pace, and because of the nature of their hooves, they were virtually useless on frozen ground. The mule was a good compromise between the horse and ox. They had several useful characteristics of both. In the young United States, George Washington became convinced of the superiority of mules as draft animals compared to horses, and he worked diligently to get them into general use.

In the early pioneering days, yeomen farmers in the Upper South generally chose the horse as their primary work animal; however, in those areas where slaves performed much of the field work, the mule was more common. As pioneers moved inland from the coasts of Virginia and the Carolinas, the working ox was indispensable.

With the growing mechanization of agriculture, especially in the North Central Region and the growing size and weight of machines used in the fields of the open prairie, none of the three work animals really met the needs of the farmers. As a result, farmers began to import specially bred draft horses from Europe. Over time, the heavier animals became the principal work animal in America, not only on farms, but in cities and towns as well.

The 1840 Census

The 1840 Census collected data on the value of horses and mules in the various states and counties. Unfortunately, the data are not as useful as they might have been because the numbers of both animals were reported as a single number. Also, that census did not report the number of working oxen. Further, Hilliard (1984) commented that the census underestimated numbers of horses and mules in the country. Those animals kept in non-farm, urban areas were not included.

The census data indicated that there were 4.3 million horses and mules on American farms in 1840. Surprisingly, the numbers were fairly evenly divided among the four regions. The highest percentage was in the North Atlantic Region (27.3%). Percentages in the other regions were: South Atlantic (20.7%), North Atlantic (25.7), and South Central (26.3). No numbers were given for the Western Region. Because there were no estimates of the number of farms or the number of acres of farmland made in the 1840 Census, it is not possible to explain why the numbers in the different regions were so similar.

In 1840, there were more horses and mules in New York than in any other state (10.9% of the total). Other states with large numbers of the animals, included, in descending order: Ohio (9.9%), Kentucky (9.1), Tennessee (7.9), and Virginia (7.5). Some 45.4% of all the horses and mules in the country were in these five states. There were 143,147 horses and mules in Alabama in 1840 (3.3% of the total).

In 1840, some 4.7% of all the horses and mules in Alabama were in Greene County. Other counties with large numbers, in descending numbers were: Montgomery (4.4%), Perry (3.9), Marengo (3.7), and Wilcox (2.8). Only 19.6% of the total were in those five counties, indicating their widespread distribution in the state.

The 1850 Census

Data from the Census of 1850 on the number of horses, mules, asses, and working oxen in the United States, its regions, and Alabama in those years are presented in Table 6.14. The data show that there were 4,336,700 horses; 559,300 mules and asses; and 1,700,700 working oxen in the country at that time. Horses were generally equally distributed, around one million, between the North Atlantic, North Central, and South Central regions, although there was a slightly larger number in the North Central Region. The share in the South Atlantic Region was somewhat lower.

There were relatively few mules and asses in the North Atlantic Region (7,700). In contrast, there were 324,000 of the animals on farms in the South Central Region. More than half of all the mules and asses in the country were in that region. The mule was the animal of choice in cotton production.

In the 1850 data, working oxen were more numerous than mules and asses, an indication that American agriculture in general was still firmly in the pioneering stage. These animals were more common (around 400,000) in the North Atlantic, North Central, and South Central regions, although there were a few more in the North Atlantic Region than in either of the others. There were substantially fewer in the South Atlantic Region (270,400).

In 1850, horses were also more common (128,000) in Alabama than either mules and asses (59,900) or working oxen (30,400).

Data in Table 6.14 showing actual numbers of work animals in the different regions have been corrected for differences in farm numbers. These data are shown in Table 6.15, as the number of animals per 100 farms. The use of 100 farms solves the problem of dealing with a small fraction of an animal. These data present a somewhat different picture of the distribution of work animals on farms in the different regions in 1850. Using the corrected data, all of the animals were relatively much more numerous on individual farms in the South Central and Western regions than they appeared to be with the uncorrected data.

The corrected data for Alabama, like the data for the South Central Region, show that the numbers of animals on individual Alabama farms were relatively much higher than they appeared to be using the uncorrected numbers.

The 1860 Census

Data on the numbers of horses, mules and asses, and working oxen in the United States in 1860 are presented in Table 6.14. Horses were considerably more common (6,249,200) than either mules and asses (1,151,100) or working oxen (2,254,900). They comprised well more than half of all animals.

There were considerably more of the horses in the North Central Region (2,541,800) than in either of the other regions. By 1860, the use of horse-drawn machines was becoming much more common there.

Well over half of the mules and asses (678,000 of 1,151,100) in the country were on farms in the South Central Region.

Working oxen were more common (2,254,900) than mules and asses (1,151,100) in the United States in 1860. There were about equal numbers in the North Central (712,100) and South Central regions (716,500). Almost two-thirds of all the work animals in the country were in these two regions.

Data on the number of animals per 100 American farms in 1860 are also presented in Table 6.15. As was the case with the 1850 data, these 1860 corrected data show that all the animals were relatively more numerous on individual farms in the South Central and Western regions than the data showing the uncorrected numbers.

Numbers of all of the work animals increased substantially from 1850 to 1860 (46.4%). This was a slightly greater rate of increase than in farm numbers (41.1%). The largest increase in all work animals was in the North Central Region. The 1860 Census counted 1.5 million more (82.2%) there in 1860 than in 1850. This increase reflects the large increases in both farms and farmland in that region (Tables 6.8 and 6.11) between those two censuses.

The number of mules and asses increased at the highest rate (105.8%) between the two censuses, and probably reflects their increased use in cotton production in the South Atlantic and South Central regions. Numbers of horses and working oxen increased by 44.1 and 32.6%, respectively.

The almost sixfold rate of increase (37,300 to 217,300) in horse numbers in the Western Region between 1850 and 1860 was phenomenal and considerably greater than the fivefold increase in farm numbers.

The corrected data indicate that in 1860 all work animals were more numerous on farms in the South Central and Western regions than in any of the other regions. They also show that most of the increase in animal numbers between 1850 and 1860 was the result of increases in the number of farms. In fact, the corrected data show that horses were actually less numerous in the South Atlantic and South Central regions in 1860 than in 1850.

Data from the 1860 Census indicated that there were 2,254,900 work animals in Alabama. Horses still represented the largest share (38.8%). Shares of mules and asses (34.1%) and working oxen (27%) were smaller.

The total number of all working animals increased by 28.3% in Alabama between 1850 and 1860 (254,900 to 327,100), whereas farm numbers increased by 31.4%. Most of the increase in animal numbers was the result of a sharply higher number of mules and asses (86.5%). There were actually slightly fewer horses in 1860 than in 1850 (127,000 versus 128,000).

The corrected data for Alabama indicate that mules and asses were almost as numerous as horses on farms in 1860, and that working oxen were also extremely important. Between 1850 and 1860, horse numbers declined sharply in the state, while mule and ass numbers increased sharply. Numbers of working oxen remained at about the same level. Alabama farmers were quickly discovering that mules were more suitable than horses in cotton production.

Use of Fertilizer

Some American farmers had understood for many years that, under the severe pressure of annual cropping, the native fertility of the soils that they tilled would not last indefinitely. In the early years, however, land was cheap, and it was easy to move to new fields when the old ones were no longer productive. Under those conditions, there was limited interest in adding fertilizers to maintain fertility. There are no estimates of the use of fertilizers in the country before 1850. Lerner (1975) supervised the publication of the estimates of the quantity of commercial fertilizer used in the United States in 1850 and 1860. His data indicate that in those two years consumption was 116 and 328 million pounds, respectively. There are no estimates of the quantity of cropland in use in those early years. The only estimate of the amount of land used for crops was acres of improved land (Table 6.11). Using those estimates for 1850 and 1860, farmers used 1 and 2 pounds, respectively, of commercial fertilizers an acre of improved land in those two years. Obviously, use on an acre of cropland was higher in 1860 than in 1850, but still extremely low considering the condition of many of the soils in those years.

Plant Agriculture in the New, Rapidly Expanding Country

The year 1840 was a watershed in our knowledge of some of early characteristics of agriculture in the country, the states, and the counties. For the first time, the census began to systematically collect data on a wide range of subjects related to American agriculture. It collected production statistics on all of the major plant and animal crops (and a few of the minor ones). For many of the early years, the value of all plant crops produced in the country far exceeded the value of animal crops.

The data collected in the 1840 Census was not nearly as comprehensive as it would be later. For example, the census did not collect data on acres harvested for any of the plant crops, and data on several of them (for example, orchard crops) were combined, making it impossible to determine the production of individual crops within the group. However, the 1840 data did provide a valuable opportunity to establish a baseline of the what-where - how for much of American agriculture.

Although that 1840 data is somewhat limited in scope, there is enough of it to make it a daunting task to adequately document and discuss it, while at the same time not becoming hopelessly tedious. That problem will be compounded as the amount of data collected increases exponentially over the following 160-years. I must confess that I probably lean toward reporting too much detail, because to me, the statistics are not just numbers. They are vital human history tightly compressed into tiny numeric characters. Virtually every number reported in these pages represent untold amounts of dawn-to-dusk, mostly back-breaking, un-ending labor by large numbers of ordinary people just trying desperately to make ends meet. Many of the numbers represent untold numbers of days of anxiously watching for a gathering of rain clouds and for a little rain, or alternatively, for enough blue sky to signal the end of too much rain. Numbers also represent the anxiety of watching, sometime hopelessly, while a crop-pest infestation marches relentlessly across planted fields. The numbers mean, in some cases, purchasing more human lives to plant more tobacco, rice, indigo, or cotton to live higher on the hog, or to try to compensate for declining prices. The numbers also represent the annual trip to the landlord to try to arrange for enough supplies to put in a new crop and enough credit to feed and clothe the family until the crop could finally be sold. The numbers represent the soul-rending desperation following the downward spiral of prices resulting from events (financial panics caused by banking scandals or stock-market crashes) over which the farmers had no control. They also represent the queasy feeling that one gets when the contract is being signed to purchase an expensive new piece of farm machinery with the realization of all of the things that have to go right to be able to pay for it. The numbers also represent untold numbers of wise, unwise, and just plain stupid decisions along the way. Finally, they represent those few good years when everything comes-up roses—when exceptionally good crops go to market at good prices, and new overalls, new shoes, a new stove, new dresses, a new wagon or combine, a new piece of choice farmland, and new toys appear on a magic carpet. Statistics, as ordinary as they appear, are not just numbers when they relate to farming.

Many plant crops have played important roles in the evolution of agriculture in the United States and in Alabama. Vital information on their origins and some of their biological characteristics were considered in a preceding chapter. In this section of Chapter 6, primary consideration will be devoted to the data available on the production of some of them.

The order of the discussions is determined, generally, by the relative importance of each of the crops in Alabama in the late twentieth century. Each chapter's section on plant agriculture will begin with a discussion of the data on cotton and corn production, because they have always been such important crops in Alabama. In the early days, there were significant acreages of wheat and oats in the state, so they will be discussed next, followed by a consideration of the production of hay, sorghum, and peanuts. In later chapters, a considerable amount of attention will be given to soybeans, because of their importance for the last 20 to 30 years. Sweet potatoes, Irish potatoes, peaches and pecans have long been important to the state, although there were never large acreages here. After World War II, Alabama farmers began to produce commercial quantities of truck-garden crops (tomatoes, sweet corn, various kinds of beans, watermelons, blueberries). For several years, farmers in the state also produced a small quantity of tung nuts for oil. Finally, a considerable amount of attention will be given to Alabama's timber production. This was probably the first plant crop exported from America. It has always been an important crop in the state, and it has tended to increase in importance with the passage of time. For a number of years, cash receipts from the sale of timber from farms has been greater than for any other plant crop. A standard format will be followed in the discussion of the more important crops in each of the time periods (chapters). When the data are available, the discussion will begin with exports of the crop, followed by prices, acres planted and harvested, yields, and quantity harvested. Each of these characteristics of production will be discussed first for the United States as a whole and then for Alabama.

When I began work on this book, it was my intention to place primary emphasis on the production of plant crops in Alabama; however, I quickly learned what I should have known all along. There is no such thing as Alabama agriculture as a separate entity. There is only American agriculture of which agriculture in Alabama is a relatively small part. Trying to follow the evolution of Alabama agriculture separately really makes little sense. Its evolution derives from the evolution of American agriculture, and even of world agriculture in most cases. For this reason, for most of the crops, considerably more discussion is given on the data for the country than for the state.

The Census for 1840 was actually published several months before plant crop production for that year was available; consequently, crop data published in that census was actually obtained a year earlier, in 1839. The same situation prevailed in both the 1850 and 1860 censuses.

It is also important to remember in the discussions to follow that census data provide a very limited picture of the status of production of any of these plant crops. For example, data on yields were not collected and published for any of the major crops on a year-to-year basis, and variation in yields between years could be quite high as a result of inclement weather, pests, or both. A sharp decline in harvest from one census to the next most likely was the result of an unusually poor yield in that particular year, rather than a major change in agricultural practices. Further, no data was widely collected on acres planted and harvested for any of the crops in the early censuses.

Of course, accuracy of the data on the various aspects of crop production is a concern. Census designers, data collectors, and recorders and those summarizing and analyzing the vast amount of data were well aware of the problems of making good estimates within a given census year; and of the comparability of data and estimates between censuses. In every census, the writers go to great lengths to explain problems related to estimates. Unfortunately, in most cases, there was little that could be done after the fact. Throughout these chapters, I have mostly reported the data as published without any effort to comment on the reliability of any of it.

Cotton

The evolution of cotton production in the United States was described in a preceding chapter. Until eighteenth century English inventions of improved machinery for the manufacture of cotton textiles and the American invention of the cotton gin to separate seed from lint, cotton production played only a minor role in American agriculture. As late as 1790, farmers harvested only about two million pounds of the fiber, and much of that was used for home manufacture of textiles. However, with the improvements in manufacturing, the demand for cotton soared, especially in England, and farmers in the South Atlantic Region responded quickly. In 1800, they harvested 40 million pounds.

Cotton could be grown north of 37° N (north of a line

from Norfolk, Virginia westward to Cairo, Illinois). While it was possible to do so, however, it was not economically feasible north of that line, because the climate was not conducive to its production. Cotton production began in Virginia, but because of climatic limitations there, moved southward with the establishment of colonies in the Carolinas and Georgia.

Cotton was never a very good crop for the South, but it was the only choice available for the plantation (slave labor) system of agriculture that had been embraced by the region early in the eighteenth century. Doyle (1941) commented that "the most favorable conditions for cotton production are a mild spring with light but frequent showers; a moderately moist summer, warm both day and night so as to maintain even and continuous growth and fruiting; and a dry, cool and prolonged autumn." Over the long-term, these are characteristics of the weather in the South with its "*Caf*" (humid, subtropical with no dry season) climate (Figure 2.11). Unfortunately, cotton is not produced on the long-term. The cotton crop is at risk for an extended period of time, and given the high degree of year-to-year weather variability throughout the region, ideal conditions for the production of the crop are not frequently encountered. Doyle further comments that while all of the properties of the plant that determine yield and fiber quality are under the control of the plant's genome, weather interactions are the final determinants in the process. It is the frequency of the occurrence of the deleterious effects of these interactions that are particularly troublesome.

Genome-weather interactions can be especially troublesome in the production of cotton, but the addition of a third variable—soil—further exacerbates the overall process. Most of the soils in the South belong to the Order *Ultisol* (Figure 3.2). Even under the best of conditions, *Ultisols* are not well suited for crop production without costly emendation (lime and fertilizer). These soils are heavily weathered, deficient in major plant nutrients, low in organic matter, quite acid, and with limited capacity to store applied plant nutrients.

When properly managed, *Ultisols* can produce quite good yields of cotton, but good management comes at a high price, especially when late floods destroy the young stand, when all of the fruit falls off during a mid-summer drought, when a tropical storm defoliates the crop before it matures, or when extended fall rains result in discolored or even rotted lint.

Exports

As late as 1791, the United States provided less than one-sixth of 1% of all the cotton imported by England (Census of 1900). In 1794, the United States exported only about two million pounds of cotton and 60 million pounds of tobacco. As a result of rapidly growing demand, exports of cotton to England increased steadily through the early years of the nineteenth century and reached 128 million pounds (485 thousand 264-pound bales) in 1820 (Figure 6.37). At that time, most of the crop produced in the country was exported. Total harvest that year was 180 million pounds.

The War of 1812 with England helped establish a textile industry in the United States itself, and by 1820, it was growing rapidly. However, this increase in local demand did not restrict exports of raw cotton. By 1840, cotton exports had increased substantially to 744 million pounds (1.9 million 385-pound bales). Although there was considerable year-to-year variation, exports continued to trend upward throughout the period 1840-1860. They reached 1,386 million pounds in 1859. Then in 1860, on the eve of the Civil War, exports increased by some 27.5% to reach 1,768 million pounds. It would not reach that level again for another 20 years. The 1860 exports were equivalent to 74% of the 2,397 million-pound crop harvested in 1859. As these data indicate, exports provided the primary market for cotton throughout this period.

Prices

In the early part of the nineteenth century, demand for raw cotton increased rapidly in England and some other European countries. As a result, prices were extremely good during that period (Figure 6.38). However, the South's slave-based plantation production system quickly responded to the increased demand and higher prices, and production soared. Rather quickly, supply overtook demand, and prices declined rapidly. The financial Panic of 1837 that resulted in a severe depression from 1838 through 1843 also put increased downward pressure on cotton prices.

In 1839, the average price for cotton was 13 cents a pound. It had been 17 cents in 1820. Even though exports and internal demand continued to grow, there was apparently just too much cotton on the market. In 1849, the census recorded that the average price reached eight cents. It had never been that low. Anecdotal reports indicate that the price actually reached five cents during that period. Prices were so low that a large number of farm foreclosures occurred throughout the South. The burden was especially severe on the planters with large numbers of slaves to support.

US Cotton

The systematic collection of production data on crops in the United States did not begin until 1839 (1840 Census); however cotton was such an important crop in the country that some data was reported on aspects of its production as early as 1790 (Decennial Census of 1800). Unfortunately, interpretation of data on cotton harvested is somewhat complicated by the fact that there was no agreement on a standard weight of a bale of cotton until the beginning of the twentieth century. In 1839, 1849, and 1859, average weights in use were: 385 pounds, 400 pounds, and 445 pounds, respectively (Table 6.15). The 1840 Census reported production in pounds only. Bales produced were obtained by dividing total pounds by the bale weight generally in use at that time. In the 1850 Census and afterwards, cotton harvested is reported in bales only.

Pounds of cotton harvested had soared after the beginning of the nineteenth century (Figure 6.39). In 1800, American farmers, primarily in the South, were harvesting around 40 million pounds of the fiber. In 1859, they harvested 2,397.2 million pounds.

Data on the quantity of cotton (thousands of bales) harvested in the United States, its regions, and Alabama in 1839, 1849, and 1859 are presented in Table 6.16. These data were taken from the 1890 Census. The 1839 data represent the first reports on county- and state-level harvest of cotton. In that year, American farmers harvested 2,053.2 thousand, 385-pound bales. Approximately 37% of that crop was harvested in the South Atlantic Region, and 63% in the South Central Region. Cotton could be grown profitably only below the 37th parallel; consequently, there was none produced in the North Atlantic Region and relatively little in the North Central Region, except in southern Illinois, southern Indiana, and Missouri. Most of the crop harvested in that region came from Missouri.

In 1839, farmers in Mississippi harvested the largest share (24.5%) of the country's cotton crop. Other states with large harvests included: Georgia (20.7%), Louisiana (19.3), Alabama (14.8), and South Carolina (7.8). These five states accounted for 87% of the total crop. At this time, 80% of all the cotton harvested in the South came from states east of the Mississippi River.

Data in Table 6.16 and Figure 6.39 indicate that bales harvested in the United States increased from around 2 to 5.4 million between 1839 and 1859; however, most of that increase came between 1849 and 1859,when it more than doubled (2,469.1 to 5,387 thousand bales). The percentage harvested in the South Central Region declined slightly in 1849 to 62.2%, but then increased sharply in 1859 to 75.5%. Cotton production was also growing rapidly in the South Atlantic Region during the period, but in the South Central Region, it exploded. The growth came from throughout the region. Bales harvested between 1849 and 1859 increased by 302,000 in Arkansas, 373,000 in Texas, 426,000 in Alabama, 595,000 in Louisiana, and 718,000 in Mississippi. Cotton production was growing rapidly and also moving westward. In 1859, only 65% was harvested east of the Mississippi River.

Farmers in Mississippi harvested the largest share (22.3%) of the 1859 crop. Other states harvesting large shares included: Alabama (18.4%), Louisiana (14.4), Georgia (13.0), and Texas (8.0). These five states accounted for slightly over 76.2% of the total crop.

Alabama Cotton

In the years immediately following the end of the War of 1812, there seemed to be no limit to how much wealth the cotton planters in Alabama Territory might amass. Average annual cotton prices were 17 cents a pound in 1820 (Figure 6.38), and in 1821 Alabama farmers harvested 20 million pounds of cotton (Table 6.17). Six years (1826) later, they harvested 45 million, and in 1859 they harvested 440 million. According to Griffith

(1972), in 1821 Alabama farmers harvested 11.3% of all of the cotton in the country. By 1859, that percentage had grown to 21.7%.

Hilliard (1984) published a dot-density map showing the distribution of cotton harvested (bales) in the state in 1819 (Figure 6.40). [Note that the date that I have used here (1819) and the date on Hilliard's map (1820) are not identical.] Hilliard used the date of publication of the census in the heading for his map: however, data for the harvest of cotton was actually for 1819. Data for the 1820 harvest was not available when the census was prepared and published in 1820. Hilliard follows this same practice for all of his crop harvest maps.

As the data in Figure 6.40 indicate, virtually all of the 1819 crop was harvested either in north Alabama in those counties created from the Choctaw and Cherokee cessions of 1806, or in Central Alabama in those counties created from the Creek land cessions of 1814. The larger share of the total crop was harvested in the central portion of the state.

In general, the distribution of bales harvested in Alabama in 1830 was similar to that of 1820, but was increasing rapidly. However, bales harvested seemed to be growing faster in central Alabama. Griffith (1972) reported that production in 1833 was 65 million pounds. A visitor to Alabama in 1834 remarked that "To sell cotton in order to buy negroes—to make more cotton to buy more negroes, 'ad infinitum,' is the aim and direct tendency of all the operations of the thorough going cotton planter: his whole soul is wrapped up in the pursuit."

Decennial censuses provide information on the harvest of cotton in Alabama in 1839, 1849, and 1859 (Table 6.16). Bales harvested were 304,000; 564,000; and 990,000, in 1839, 1849, and 1859, respectively. Approximately 47.8% of the 1839 crop was harvested in just five counties: Montgomery (12.7%), Perry (10.8), Franklin (9.2), Madison (8.8), and Morgan (6.3). These five counties accounted for 48% of the state's total. The large share in Franklin County is somewhat misleading because the county had not been divided into two counties at that time.

Data on the distribution of cotton harvested in the state in 1839 was shown on a dot-density map (Figure 6.31). These data show that production was still generally concentrated in the same areas as in 1820 and 1830—the Black Prairie (Figure 1.11) from Montgomery County westward through Perry and in the Tennessee Valley from Madison County westward through Franklin and Lauderdale counties. Although, most of the cotton was harvested in a relatively few counties, production was increasing rapidly across central Alabama, from Chambers County in the east to Sumter, Greene, and Pickens counties in the west. This line of counties formed the Cotton Arc. For many years, farmers in these central counties would harvest a large share of the cotton produced in the state. The map also shows that by 1839 cotton production was beginning to move into the Coosa Valley. Lands in this valley were not formally ceded to the United State by the Creeks and Cherokees until 1832 and 1835, respectively.

In 1849, Alabama farmers harvested more cotton (564 million bales) than any other state, and almost 23% of the total US crop (Table 6.18). By 1859, production had increased to 990,000 bales. Between 1849 and 1859, cotton harvested increased throughout the state from 564,400 to 990,000 bales, and the increase was especially large in western portion of the Black Prairie. Figure 6.41 presents data depicting the distribution of Alabama cotton harvested in 1859. Harvest was increasing in the Tennessee Valley, and it was slowly intensifying in the Coosa Valley, but it was exploding in the Black Prairie. Farmers in Dallas (6.4%), Marengo (6.3), Montgomery (5.9), Greene (5.8), and Lowndes (5.4) counties were the five leading counties in cotton harvested. They harvested almost 33% of the state crop that year.

In terms of total harvest, the increase was much greater in those counties of the Highland Rim and Black Prairie; but in relative terms, some of the largest increases were realized in the counties around the margin of the Black Prairie, especially in the east-central portion of the state (Barbour, Russell, Chambers, Macon, and Tallapoosa counties). Harvest was also growing rapidly in the Low Hills section, south of the Black Prairie in Pike, Butler, Monroe, and Clark counties.

Even though bales harvested in Alabama increased substantially between 1849 and 1859, its share of the total American crop declined to slightly over 18%. Cotton production had been moving westward for years. By 1859, the center of production in the country was in Mississippi, and soon it would move across the Mississippi River. But in Alabama, cotton was still king. In 1858, a British visitor to Mobile commented that the city was a place "where the people live in cotton houses and ride in cotton carriages. They buy cotton, sell cotton, think cotton, drink cotton, and dream cotton. They marry cotton wives and unto them are born cotton children." The Cotton Kingdom had truly established itself in the state.

Yeomen and the Cotton Kingdom

Until the late 1840s, plantations produced most of the cotton harvested in Alabama. Yeomen farmers produced a small amount along with the other crops in their general farming operations; however, the times were changing for those highly independent individuals. Merchants and craftsmen were beginning to move into their communities. Soon, some of their most pressing needs (flour, sugar, salt, medicine, clothes, shoes, tools) could be met by purchases, rather than making things themselves or making do. This growing division of labor was altogether practical. Complete self-sufficiency is extremely inefficient and wasteful of scarce resources, especially time. The only problem with purchasing instead of making necessities is that they require some money. Unfortunately, self-sufficient general farming did not generate much legal tender in the 1840s. As a result, yeoman farmers slowly reduced their level of self-sufficiency; they had to change their farm economy in some way to generate cash for the market economy. Unfortunately, the only game in town was cotton. There was almost no cash market for anything else that they could produce.

There is no specific information on the number and distribution of Alabama yeomen farmers in any period. Any definition of the category is highly subjective, but their farms were relatively small—limited by the amount of labor available in their immediate families. Generally, most of them did not have enough resources to own slaves. Data on the average number of improved acres of farmland per farm in each county in 1850 provides some insight into the rough distribution of these farmers in the state. Improved acres generally included areas where crops were being grown plus pastures (meadowland).

In the Cotton Arc counties (See Figure 6.41)—where most of the cotton was produced and where most of the slaves in the state lived—the average number of improved acres per farm was 148. While there were obviously some yeomen farmers in these counties, most of those farms required considerably more labor than a single family could provide. In the Cotton Counties of the Tennessee Valley, the average number of improved acres per farm was 162. In contrast, in the High Hills counties north of the Black Prairie, the average number of improved acres was only 59. It is likely that in these counties most of the farms were operated by yeomen. The number of acres of improved land per farm (68 acres) was similar in the Low Hills south of the Black Prairie.

As noted in a preceding paragraph, in 1850 cotton production was just beginning to expand into the Coosa Valley, but apparently most of the farms were still operated as general farms by individual farmers and their families. The average number of acres of improved land per farm was 61. From these data, it appears that in 1850 much of the entire land area of the state was still being farmed by limited resource farmers.

Alabama cotton production almost doubled between 1849 and 1859 (Table 6.16). Approximately half of the growth was a result of increased production in the counties of the Cotton Arc. However in relative terms, production was increasing at a faster rate in the Low Hills section. Between 1849 and 1859, bales harvested increased by an average of 112% in the Cotton Arc counties of Dallas, Lowndes, and Montgomery. During the same period, it increased by an average of 225% in Butler, Henry, and Choctaw, three counties in the Low Hills. A comparison of Figures 6.41 and 6.42 suggests the same influx of yeomen farmers in the hills north, south, and southeast of the Cotton Arc into the Cotton Kingdom. In those sections, yeomen farmers were entering the Cotton Kingdom with a vengeance. The movement of these farmers into the export market economy with the production of cotton represents a watershed event in the evolution of the Alabama agroecosystem. These limited-resource farmers produced only a few bales most years, but there were so many of them.

Seeking Salvation

After 1820, the cotton bubble burst, and in 1823 prices declined to 11 cents a pound (Abernethy, 1990). The average price rebounded to 17 cents in 1825, but by 1827 it was 10 cents (Figure 6.38). This uncertainty in cotton prices might have given rise to some concern in the minds of the cotton farmers, but not enough to deter them from buying more land and more slaves and planting more of the crop. They could still see the Cotton Kingdom coming over the horizon, and in it they would all be kings, carried along forever on the backs of their slaves.

After 1840, declining prices, plus decreasing production on the exhausted lands in the Carolinas and Georgia, created an atmosphere of growing concern among planters across the South. Underlying this concern for the future of the Cotton Kingdom's economy was the realization that the South was rapidly falling behind the North in population, education, transportation, general agricultural production, industrialization, economic growth, and wealth.

The North's growing industrial and commercial power was of great concern to Southern planters. Increasingly, goods used in the South were flowing southward across the Mason-Dixon Line, instead of from Europe. Several efforts were made to remedy this situation, but none of them worked. Out of these efforts, came the realization that the South's economic backwardness was a result of fundamental weaknesses that could not be easily or quickly rectified. Gray (1958) comments that "it was the misfortune of the South that the great mass of its labor was of exceeding low quality for non-agricultural activities." He further comments that it was not easy to transform the plentiful supply of "poor whites and mountain whites into efficient factory employees. Accustomed to rude independence, they were frequently undisciplined, slovenly, careless and intractable."

Accompanying the rapidly growing economic power of the North was the increasing clamor for the abolition of slavery. As a result, there was widespread agreement among Southern planters that some drastic changes were needed in the Cotton Kingdom. There were far-ranging discussions of changes that might be made, but there was one reservation—slavery, the essential stackpole of the entire South, could not be abandoned.

Concerns for the future and general discontent within the South led to a series of planters' conventions. The first of these was held in Montgomery in 1845. The second was held in Memphis, a few months later. With all of the possible changes that might be made to improve conditions in the economy of the South, the planters in convention determined that everything would be satisfactory if they could find some way to increase cotton prices.

At Memphis, the delegates to the convention voted to reduce production by a fixed percentage to reduce supply that they predicted would result in increased prices. They further recommended that the labor released in the reduction of production be put to work in regional agricultural diversification. At later conventions, it was decided that while the control of supply might be helpful, the control of marketing would be more effective; consequently, they set to work designing a system whereby the planters would establish a marketing monopoly on the trade of cotton.

Finally, amid all of the discussions on reducing production and controlling the market, some interest began to grow in increasing the efficiency of production and to prevent the further deterioration of the soil of the Cotton Kingdom. Consequently, at the convention held at Montgomery in 1853, the delegates approved the formation of Southern Planters Association. Among other objectives, the association was asked to initiate studies of the relationships among soil chemistry and crop yields, the control of insect pests, means of improving exhausted soils, and the mechanic arts as they related to agriculture. Unfortunately, the convention movement as a means of dealing with problems within the Cotton South quickly lost its agrarian fervor when cotton prices began to increase toward the end of the 1850s . The price climbed back to 12 cents in 1859 (Figure 6.38), still not a very high price, but cotton production was marginally profitable at that level.

While the various conventions generated little interest in reducing production, they did result in growing interest in regarding the efficiency of production and in crop diversification. Near the end of the antebellum period, a planter, Dr. Noah B. Cloud, with plantations in Macon and Montgomery counties, began to encourage the application of science to agriculture. He had conducted experiments on his farms and had kept extensive records on yields. As a result of his studies, he concluded that Southern farmers committed too much of their resources to acquiring more land and labor (slaves) and not enough to maintaining the productivity of their soils or farms. He advocated that farmers begin to apply large amounts of fertilizer (guano, cottonseed, barnyard manure, etc.) to the soil to increase yields. He further suggested that better seeds should be planted and that greater care should be exercised in picking, ginning, and forming the raw cotton into bales. Dr. Cloud also urged farmers to produce more livestock, to plant vineyards and orchards, and to grow melons, peas, and beans. He promoted his ideas through several publications and by supporting agricultural societies, planters' conventions, exhibitions, and cattle shows. There is little indication that Cloud's efforts had any effect on the basic problems in the Cotton Kingdom in Alabama or elsewhere. Cotton planters throughout the South had generally dealt with declining prices by purchasing more slaves and planting more acres of the crop. Cloud's efforts were not going to change that mindset. Further, the influx of thousands of limited-resource, yeomen farmers into the cotton economy made salvation even more difficult.

Atkins (1994) suggests that Cloud's efforts might have resulted in some increase in food production in the state, and that without them, food shortages during the Civil War years might have been much worse.

In the decade before the Civil War, with the cost of slaves and land rising in Alabama, many non-slaveholders saw no possibility of moving into the planter class or the Cotton Kingdom. As a result, large numbers of them immigrated westward, seeking a better future (Atkins, 1994).

Corn

Early in the colonial period, Indian corn or maize became the most important grain in all of the colonies (Bidwell and Falconer, 1941 and Gray, 1958), and it retained its importance through the American Revolution. It was called Indian corn to establish the fact that this was the grain that the Native Americans were growing when immigrants first arrived in America. It had probably been in use for at least two thousand years. The British term corn was essentially synonymous with any cereal grain: wheat, rye, oats, barley, and maize. The term Indian corn was used in censuses through 1890, but not generally afterward, because by that time common American usage had substituted the term corn for maize or Indian corn.

Corn quickly became the principal crop produced on the new farms established west of the Appalachians at the end of the eighteenth and beginning of the nineteenth centuries. In the antebellum South, the acreage devoted to corn production was greater than for cotton. Unfortunately, reliable quantitative data are generally lacking before 1839, as the censuses before that time did not collect any data on corn production.

Locher and Cox (2004) comment that "European settlers came to the South with no taste for corn. However, they soon developed a liking for it out of necessity. Celtic immigrants had been fond of oats in the old country, but soon replaced the old grain with the new. Those immigrating from France, England or the Northern states brought with them a fondness for wheat bread, but the South had few mills capable of processing the sticky, glutinous grain, particularly in Alabama or Mississippi. But almost any small burg that sprang up along a stream in Alabama soon had a working grist mill capable of milling the non-glutinous corn.

Exports

In the early part of the nineteenth century, agricultural prices in Great Britain began to trend downward. In response to the financial interests of large British landowners, in 1815 Parliament enacted the Corn Laws to stem the flow of inexpensive grain from abroad. These laws placed extremely high tariffs on foreign grain, thereby keeping the price of British-grown grain high to satisfy the large landowners. These laws had the effect of reducing, to some degree, the export of corn to Great Britain. The Corn Laws were not repealed until 1846.

Although there is little quantitative information on corn production in the United States before 1840, the young country began to collect data on foreign commerce, including exports, much earlier. Data on corn exports are available for the period 1791-1846 (Bidwell and

Falconer, 1941) and 1849-1860. Data show that in the year ending September 30, 1791 that the United States exported 1.7 million bushels of corn and 70,339 barrels of corn meal. Generally, around four bushels of the grain were used to make a barrel of meal. Conversion of the meal to bushels of corn indicates that total exports of the grain were two million bushels that year. Total exports (bushels of grain and the grain equivalent of meal) for the period 1791-1846 are presented in Figure 6.42. Similar data for the period 1849-1860 are presented in Figure 6.43. In most of the years between 1791 and 1840, corn exports varied between one and three million bushels. Year-to-year variation was apparently related to the level of harvest and economic cycles. Exports varied between 1.1 and 2.8 million bushels during the period 1791-1802. They began to decline sharply after 1804, as international shipping problems began to increase; they reached 0.4 million in 1808. Exports were up sharply just before the beginning of the War of 1812 with Great Britain, but declined during the period of hostilities. After the end of the war in 1814, exports increased rapidly, but began to decline during the period of panic and depression (1819-1822). From 1820 until around 1830, exports remained relatively stable, but afterwards they trended downward. During the banking recession of 1833-1834 and the panic and depression of 1837-1839, they generally trended downward, reaching a level of slightly less than 0.7 million bushels. After 1840, exports increased sharply, and reached three million bushels in 1846. They continued to increase rapidly through 1849 during a period of business expansion, reaching 15 million bushels in 1849 (Figure 6.43). Exports declined sharply in the early 1850s, increased again in the middle of the decade, only to decline again after 1856. These gyrations were apparently the result of changes in US corn prices during the period (Figure 6.44).

According to the 1850 Census, American farmers harvested 592.1 million bushels of corn in 1849 (Table 6.19). Exports in 1850 totaled eight million bushels, equivalent to 1.3% of the 1849 harvest. In 1860, exports were equivalent to an even smaller share of the 1859 crop—0.5% (4.2 million of 838.8 million bushels). These small percentages, in contrast to cotton exports, indicate the importance of corn as a food and feed in the United States at that time. Corn was simply too important in the country, especially in the South, to allow it to be exported in significant quantities.

In 1791, 14.1% of total corn exports was in the form of meal. Generally, for the remainder of the period, meal exports accounted for a larger share of the total. In 1815, it reached 68.9%. Annual meal exports were not as variable as exports of the grain. Annual average meal exports for the entire 1791-1817 period accounted for 27.2% of total exports. Most of the grain and meal exported was probably used as animal feed. Corn meal was never widely used for human food outside of the American South. The United States continued to export substantial quantities of corn meal in the years before the Civil War. In 1849, some 405,168 barrels (equivalent to 1.6 million bushels of grain) were exported; this represented 10.9% of total exports. In 1860, 233,709 barrels (0.9 million bushels of grain) were exported and accounted for 22% of total corn exports.

Prices

Data on corn prices in the United States for the period 1800-1860 are limited. Data published by Gray (1958) gives the average annual price for shelled corn in Virginia for the period 1801-1860, but there are no comparable data for the country as a whole or for any of the other states. Gray's data are presented in Figure 6.44. Generally, annual average prices varied between 40 and 80 cents a bushel during the period. They apparently were affected by the economic cycles of the times. For example, prices declined sharply from 1806 to 1807, as a result of the shipping embargo described in a preceding section. They were much higher in the period 1814-1816, as a result of the speculative boom, but declined sharply during the panic and depression of 1819-1822. Prices trended upward again during the speculative boom of 1834-1837, but declined again during the panic and depression of 1837-1843. As detailed in a preceding section, the country entered a period of business expansion after 1843 that continued through 1856. During this period, prices generally trended upward.

Gray also published data on the wholesale price of shelled corn at New Orleans during the period 1837-

1860. Generally, wholesale prices in that port city were 5 to 15 cents a bushel lower, in most years, than in Virginia during the same period. It is not surprising that prices there were lower. By 1840, substantial quantities of corn from the North Central Region were being transported down the Mississippi to New Orleans for trans-shipment to the Cotton States and to international markets. This ever-increasing supply of the grain certainly must have put a lid on prices.

Bidwell and Falconer (1941) published data of the average price of corn at New York on January 1, each year, for the period 1840 to 1860. These data are presented in Figure 6.45. In most years, prices at the port of New York were substantially higher that those in Virginia or New Orleans. These authors suggest that the nature of markets in New York resulted in higher prices. Growing industrialization, urbanization, and international trade put considerable upward pressure on prices, even though the quantity of grain reaching that area from the North Central Region was increasing rapidly. Demand was increasing as rapidly as supply.

US Corn

Data on the bushels of corn harvested in the United States, its major regions, and Alabama in 1839, 1849, and 1859 are presented in Table 6.19. Those data show that in 1839 (the earliest data available) harvest of Indian corn in the United States was 376.4 million bushels. Farmers in the South (South Atlantic and South Central regions combined) harvested well over half (62.4%) of the country's crop that year (28 and 34.5% in the South Atlantic and South Central regions, respectively). Corn was more important as a human food in the South than in the other regions. Farmers in the South Central Region, with its enormous number of slaves, had little choice but to produce large quantities of the grain. These data also show that while Indian corn had been produced for some 200 years along the Atlantic Seaboard, within about a half-century after European farmers began to farm west of the Appalachians 34.5% of the 1839 crop was being harvested there. Tennessee was the leading state in the region for the harvest of corn in 1839. Farmers in that state harvested 11.9% of the total 1839 national crop. Other leading corn-harvesting states were: Kentucky (10.6%), Virginia (9.2), Ohio (8.9), and Indiana (7.5). These five states accounted for 48% of the total.

Data from the 1850 and 1860 censuses on the bushels of corn harvested in the United States, its regions, and Alabama are also presented in Table 6.19. While these data were reported in the censuses for those years, it was collected a year earlier, in 1849 and 1859, respectively. The US harvest increased by 215.7 million bushels (57.3%) from 1839 to 1849 and 246.7 million (41.7%) between 1849 and 1859. Bushels harvested increased substantially in all of the regions over the 20-year period; however, harvests were much larger in the North Central Region than in any of the other regions. In 1839, bushels harvest in the South Central Region were the highest in the country (34.5% of the total crop). Tennessee and Kentucky were still at the heart of America's Corn Belt; however, just two decades later, the situation had changed drastically. By 1859, bushels harvested in the South Central Region trailed far behind the North Central Region. In those two decades, the Corn Belt moved northward and westward. In 1859, farmers in the North Central Region harvested 48.4% of the national crop. Much of the corn in that region was being marketed in the form of whiskey and livestock, especially hogs. Even after transportation to the East improved, corn was used primarily as an animal feed. Also, during this period, cattle feeding that earlier had been centered in the Connecticut Valley had followed corn production into the Old Northwest. In 1839, farmers in Ohio and Indiana, harvested almost half of the total crop in the entire North Central Region. In the following two decades, farmers were still rapidly moving into the western portions of that region, and the Corn Belt was moving westward with them. Illinois farmers harvested only 22.6 million bushels in 1839. Twenty years later, they harvested the largest share (115.2 million). Farmers in Iowa only harvested 1.4 million bushels in 1839, but 42.4 million in 1859.

In 1859, Illinois farmers harvested the largest share (13.7%) of the national crop. Other states with large shares of the crop that year were: Ohio (8.8%), Missouri (8.7), Indiana (8.5), and Kentucky (7.6). These five states accounted for some 47.4% of the total national corn crop.

Alabama Corn

In 1839, Alabama, with 3.3% of the country's population, harvested 5.6% (20.9 million bushels) of the total US corn crop (Table 6.19). Harvest was generally higher (Figure 6.46) in the central portion of the Tennessee Valley section. However, as data presented in the figure show, substantial quantities were harvested throughout the state. Farmers in Madison County harvested the largest share of the Alabama total (6.7%). Other counties harvesting large shares were: Montgomery (6.7%), Lawrence (6.5), Jackson (5.2), and Limestone (4.8). Together, these five counties accounted for only 29.9% of the corn produced in the state. That the five leading corn-producing counties harvested such a small percentage of the total Alabama crop underscores the fact that corn was produced in relatively large quantities all over the state. It was the most important food crop everywhere in Alabama.

In 1839, a large share of the corn in the state was harvested in those same counties that harvested the most cotton (Figures 6.31 and 6.46). Foods prepared from corn were extremely important in meeting the nutritional needs of slaves, so farmers in those counties either had to produce large quantities of corn or purchase it. However, there were also substantial quantities of the corn harvested in areas beyond those Cotton Counties.

Corn production in 1839 in the Tennessee Valley was much more broadly based than cotton production (Figures 6.31 and 6.46). The same was true in the Coosa Valley and in the northern portion of the Low Hills. Of course, all of those counties were inhabited by large numbers of yeomen whose lives tended to revolve around the annual corn crop. Production was still somewhat limited in the southern portion of the Warrior Basin in the High Hills section, and the counties of the lower portion of the Low Hills section.

From 1839 to 1859, corn harvest in Alabama increased 58.8% (20.9 to 33.2 million bushels) (Table 6.19). During the same period, this rate of increase was substantially lower than in the South Central Region as a whole (77%). Limited data indicate that Alabama farmers, especially on the larger farms, were beginning to increase their acreages of cotton at the expense of corn. For example, bushels harvested per capita declined from 37.3 to 34.4 in the period 1849 to 1859. Farmers compensated for this change by purchasing corn from Tennessee and Kentucky.

Data presented in Figure 6.47 depict the distribution of bushels of corn harvested in Alabama in 1859. Farmers in Montgomery County harvested the largest share (4.8%) that year. Other counties with large shares of the total crop were: Marengo (4.2%), Dallas (4.1), Greene (3.9), and Lowndes (3.9). All of these counties were in the heart of Alabama's Cotton Arc. Of these, only Montgomery County had been on the list of leading corn-producing counties in 1839.

Table 6.19 also contains data on corn harvested in Alabama's sister states (Illinois, Indiana, and Mississippi), states admitted to the Union at about the same time. In 1839, bushels harvested were similar in Alabama, Illinois, and Indiana, at around 25 million, but were much higher than in Mississippi (13 million). However, by 1859, harvests in Alabama and Mississippi had not increased very much, while harvests in Illinois and Indiana had increased from three- to fivefold. Many farmers in Alabama and Mississippi grew corn to fuel their cash crop (cotton) production. In Illinois and Indiana, corn was their cash crop.

Peanuts and Soybeans

Some peanuts were being produced in some of the states of the South Atlantic Region as early as 1840, but no data on the production of the crop was published in censuses earlier than 1900. Data on soybean production were not available until at least a decade later.

Hay

The physiographic characteristics of the Northern colonies were not conducive to the development of an agricultural economy based on staple or export crops such as tobacco, rice, indigo, and cotton. Instead, the North Atlantic Region turned its attention to shipbuilding and international commerce. This path led to a rapid increase in urbanization, which in turn resulted in a sharp increase in the demand for livestock products (beef, milk, cheese, pork, eggs) that the city dwellers could no longer produce themselves. In the development of the production of beef, milk, and cheese, hay production and harvest was extremely important. Because of the Northern climate, there was very little forage available in the fields during the winter and early spring, and it was not economical to feed the animals grain during that time of the year. Consequently, the production and harvest of hay was well advanced in that region by the early nineteenth century.

US Hay

There are little data available on the harvest of hay in the country until 1839, when the Census of 1840 provided estimates of harvest in the counties and states of the different regions. These data, plus data for 1849 and 1859, are presented in Table 6.20. Total hay harvest in the United States in 1839 was 10.2 million tons. Almost 78% (7.9 million tons) of that total was harvested in the North Atlantic Region where beef, milk, and cheese production was growing rapidly. Only 801,000 tons were harvested in the South (South Atlantic plus South Central regions). The difference in harvests in the two regions clearly demonstrate the difference in the general farming practices in the North and the staple production practices in the South. Also, there was much more natural forage available during the cold months in the South; consequently, the production and harvest of hay was not as important. Although cattle and hay production was still centered in the North Atlantic Region in 1839, they were beginning to increase rapidly in the North Central Region. That year, farmers in the region harvested 1,594,000 tons, 15.5% of the total national crop.

Farmers in New York harvested 30.5% of the 1839 US crop of hay. Other states harvesting large shares were: Pennsylvania (12.8%), Ohio (10.0), Vermont (8.2), and Maine (6.7). These five states accounted for 68.2% of the total national hay crop.

Total national harvest of hay increased to 13.8 million tons in 1849 and to 19.1 million in 1859 (Table 6.20). Much of this increase was a result of the rapid growth of beef production in the North Central Region. As a result, by 1859 harvest in that region was rapidly approaching that in the North Atlantic Region. Harvest also increased substantially in the other regions, but in 1859 over 90% of the total for the country was in the North Atlantic and North Central regions.

Farmers in New York still harvested the largest share (20.6%) of the 19 million tons of hay harvested in the United States in 1859. Other states harvesting large shares of the total were: Ohio (10.6%), Pennsylvania (10.4), Illinois (10.0), and Iowa (6.5). These five states accounted for 58% of all the hay harvested in the country.

Alabama Hay

Grass has probably always grown extremely well in Alabama, and from the early days of the Native Americans, farmers have used it to feed animals. When the first Europeans arrived in Alabama, they found that the native people were burning the forest to encourage grass to grow to feed the whitetail deer. This practice was followed for many years after the first European settlers began to herd cattle, and over the years hay became an increasingly important crop. However, the importance of the crop was not recognized by the persons in the state who were responsible for submitting data for the 1840 Census. Only 11 of 49 counties reported any harvest of hay. Harvest that year ranged from one ton in Lauderdale County to 4,478 tons in Montgomery. The results were somewhat better in 1850. Thirty of 52 counties reported the harvest of the crop. It ranged from one ton in Bibb, Jefferson, and Marengo counties to 6,467 tons in Dallas. Either hay production was improving by 1859, or farmers were more responsive to the request for census data when farmers in 43 of 52 counties reported some harvest of hay. Harvests ranged from one ton in Walker to 7,545 tons in Russell.

Alabama farmers harvested slightly less than 62,200 tons of hay in 1859 (Table 6.20), whereas they had harvested some 32,700 tons in 1849. Farmers in Russell County harvested the largest share (12.1%) of the total in 1859. Other counties with large shares were: Madison (9.1%), Lowndes (8.9), Dallas (7.8), and Tuscaloosa (7.7). These five counties accounted for some 46% of the total hay harvested.

Table 6.20 also contained data on hay harvested in Alabama's three sister states (Illinois, Indiana, and Mississippi) in 1839, 1849, and 1859. In 1839, harvests were tenfold greater in Illinois and Indiana than in Alabama and Mississippi. By 1859, the difference was even greater.

Wheat

Corn is a more versatile crop than wheat, and generally easier to grow over a wide range of environmental conditions. Further, the return from a single, planted kernel of seed corn is much higher than for wheat. Finally, it is much easier to process the non-glutinous corn into meal. In the final analysis, however, bread made from the two grains is quite different. Bread made from corn was tolerated in the South because Southerners had little choice in the matter, but where wheat could be grown and processed reliably, only the so-called white bread was readily acceptable.

Late in the colonial period, wheat production was centered in New York and Pennsylvania, in the southern portion of the North Atlantic Region. Smaller amounts of the grain were produced both north and south of that area. Physiographic conditions (climate and soils) there were more conducive to the production of the crop than in the northern portion of that region or in the South Atlantic Region. As will be quickly evident, however, there were other areas of the United States with even more friendly physiographic characteristics for raising wheat.

Exports

For a considerable part of the period 1815-1846, the Corn Laws were in effect in Great Britain. These laws likely had a limiting effect on the export of the grain to that country for a considerable part of this period. Fortunately, the market for American wheat was not wholly dependent on Great Britain. Edwards (1940) comments that America had been a wheat exporter since the colonial period. Data on wheat exports during the period 1790-1813 are presented in Figure 6.48. These data were obtained along with the data for corn and other grains (Bidwell and Falconer, 1941). The data include the export of the grain and wheat flour. The wheat flour data that were reported in barrels have been converted to bushels by multiplying by five. About five bushels of the grain were required to produce a barrel of flour. In 1790 exports of grain and flour totaled 1.1 million bushels, or 724,623 barrels. With the conversion of flour to grain equivalents, total exports were 4.7 million bushels. Almost three-fourths of that total was flour. Average annual exports of wheat for the period 1790-1813 was slightly more than 4.7 million bushels. There was considerable year-to-year variation, but there was no indication of a general increase or decrease during the period. While there seemed to be no trend in total exports, the quantity exported as flour did tend to increase.

Data on the total exports of wheat (grain plus flour) for the period 1850-1860 are presented in Figure 6.49. In 1850, total exports were considerably higher than in 1813 (Figure 6.48). Total exports generally trended upward through 1857. In that year, the country exported 14.6 million bushels of grain and 18.6 million bushels as flour. Exports of grain and flour generally trended downward after 1857 and through 1860. During this period (1849-1860), the average annual percentage of wheat exported as flour was 78.3%. It had been somewhat higher during the earlier period.

The 1850 Census reported that American farmers harvested 100.5 million bushels of wheat in 1849. Total (grain plus flour) 1850 exports (Figure 6.49) were equivalent to 7.5% (7.5 of 100.8 million bushels) of the 1849 harvest. A decade later, the percentage of the US crop being exported had increased to 9.9% (17.2 of 173.1 million bushels). Percentages of harvests exported for wheat were much higher than for corn, but significantly lower than for cotton.

Prices

There are no national data on wheat prices available for years before 1866; however, Gray (1958) published a data set on prices in Virginia for the period 1801-1860. It is likely that they are generally representative of prices throughout the country for that period. These data are presented in Figure 6.50. During this period, wheat prices varied between 72 cents a bushel in 1820 and $2.03 in 1816; however, in most years, they ranged around $1. Prices seemed to be highly responsive to changes in the national economy. Peaks in prices (1804, 1810, 1816, 1836, 1847, and 1854) were associated with times of national prosperity, often resulting for short periods of speculative booms. Periods with unusually low prices were also associated with business recessions. For example, prices in 1807 and 1808 were 87 and 85 cents a bushel, respectively. These low prices were the result of the American embargo on exports to Great Britain, which was begun to try to force Britain to refrain from interfering with American shipping on the high seas. Lower prices during the periods 1819-1834 and 1838-1843 were the result of a series of economic panics and recessions. There was a severe panic during the period 1857-1860, and prices were affected, but they did not plunge to the extent that they had earlier.

US Wheat

The Census of 1840 was the first to include data on the harvest of plant crops in the United States. That census included data that showed American farmers harvested 84.8 million bushels of wheat in 1839 (Table 6.21). Farmers in the North Atlantic Region harvested the largest share (33.3%), but only slightly more than in the North Central Region (32.6%). Bushels of wheat harvested in the South Atlantic and South Central regions trailed far behind the two other regions. In 1839, wheat production was just not important enough or reliable enough to use good cotton land for its production in the South.

In 1839, Ohio farmers harvested the largest share (19.5%) of the total national wheat crop. Other states harvesting large shares of the total crop were: Pennsylvania (15.6%), New York (14.5), Kentucky (5.7), and Tennessee (5.4). These five states accounted for 60.7% of the total national wheat crop. At the end of the colonial period, wheat production was centered in New York and Pennsylvania; however, after the American Revolution, as these data show, wheat production was rapidly moving westward beyond the Appalachians.

The 1850 Census

Wheat harvested increased in all regions, except the South Central from 1839 to 1849 (Table 6.21) where it declined sharply (10.5 to 4.4 million bushels). Harvest in Tennessee and Kentucky (the major wheat-producing states in the South Central Region) declined by 64.6 and 55.4%, respectively. Mell (1890) reported that the weather in the winter and spring of 1849 was extremely wet in Alabama. This wet weather probably was responsible for the reduction in the state (838,000 to 294,000 bushels), and most probably, inclement weather was responsible for the poor harvest across the region. Harvest in the North Central Region increased sharply (27.5 to 43.8 million bushels). Much of this increase was the result of the rapid growth of the number of farms in the western portion of that region. For example, harvests in Iowa increased from 155,000 to 1.5 million bushels during the decade. In Wisconsin, they increased from 212,000 to 4.3 million.

The 1860 Census

Wheat harvest in the United States increased from 100.5 to 173.1 million bushels from 1849 to 1859 (Table 6.21). It increased in all regions except the North Atlantic. Wheat farmers in that region were beginning to discover that they could not compete with the farmers in the expanding agriculture of the Old Northwest in wheat production. Further, by this time, rail transportation from the North Central Region eastward was so efficient and transportation costs so low that it was much less expensive to purchase the grain that the North Atlantic Region needed.

Harvest was much lower than expected in the North Central Region (43.8 to 45 million bushels). Probably, inclement weather was also responsible for the poor harvest in that region in 1859. Of special interest in these data, is the rapid growth of wheat production in the Western Region. In 1850, farmers there harvested 533,000 bushels, and a decade later they harvested 7.7 million. Agriculture, especially grain production, was continuing to move westward.

The largest share of the 1859 crop (15.3%) was harvested by farmers in Pennsylvania. Other leading wheat-producing states were: Illinois (13.8%), Indiana (9.7), Wisconsin (9.0), Ohio (8.7), and Virginia (7.6). Farmers in these five states harvested 48.8% of the total national wheat crop in 1859.

Alabama Wheat

Some wheat had been grown in Alabama,during the formative years of its agriculture. However, the environmental characteristics of the state were really not conducive to wheat production. The relatively high levels of precipitation during the period when the plant is producing grain usually encourages development of large numbers of insect pests and diseases that attack the maturing plants. Data obtained in the 1840 Census, for the 1839 crop (Table 6.21) indicated that Alabama farmers harvested only 838,000 bushels of wheat that year, or slightly less than 1% of the national total.

Distribution of wheat harvested in Alabama in 1839

is shown in Figure 6.51. These data indicate that a large share of the grain was harvested in the Tennessee Valley and the Coosa Valley; while relatively little was harvested south of the Black Prairie. The largest share of the 1839 crop was harvested in Madison County (10.3%). Other counties with large shares of the total Alabama wheat crop were: Calhoun (9.4%), Lauderdale (5.6), Cherokee (5.2), and Talladega (5.0). Farmers in these five counties harvested 36% of the state total.

The 1850 and 1860 CensUses

The 1849 Alabama harvest was smaller than expected (300,000 bushels). As detailed in a preceding paragraph, a small harvest was likely the result of adverse crop weather. The 1859 crop was considerably better (1.2 million bushels). While the Alabama harvest had increased, , the state's share of the national total declined to 0.70%. The distribution of the harvest of the 1859 wheat crop was generally similar to the distribution of the crop in 1839 (Figures 6.51 and 6.52), except that harvest in the Tennessee Valley appeared to be somewhat lower.

In 1859, farmers in Calhoun County harvested 8.5% of the state's crop. Other counties reporting large harvests were, in descending order: Cherokee (7.5%), Chambers (6.5), Randolph (5), and Talladega (4.8). All of these counties are located in the Coosa and Tallapoosa river valleys of east-central Alabama. These five counties accounted for only 32% of the total Alabama wheat harvest.

Table 6.21 also includes data on wheat harvested in Alabama's three sister states (Illinois, Indiana, and Mississippi). These data show that as early as 1839 harvests in Illinois and Indiana were at least fivefold greater than in Alabama and Mississippi. By 1859, the differences were even greater.

Oats

The early English settlers generally did not use oats as food. Virtually all of the oats produced in the country was used for the feeding of livestock, principally horses and mules. For people in the Scottish Highlands, however, oats had long been a principal food, so when they immigrated to this country, they brought this custom with them. As their numbers began to increase in the American population, the use of oats as a food for people also began to increase.

Exports

American oat exports for the period 1790-1817 are presented in Figure 6.54. During this period, oat exports were much lower than those for either corn (Figures 6.42 and 6.43) or wheat (Figure 6.48 and 6.49). They generally ranged between 40,000 and 80,000 bushels annually. Oats were one of the principal crops produced throughout Europe, so except in time of war or when the crop there failed, Europeans really did not need very much of the grain produced in the United States,

During the period, exports of oats tended to be somewhat cyclic, but the overall trend was downward. Exports did seem to recover somewhat after the War of 1812. The estimates for both 1810 (448 bushels) and 1811 (211,894) appear to be erroneous. Unlike data for both wheat and corn, there are none available for the period 1849-1860. In those years, reports for the exports of oats were combined with rye and other small grains.

Prices

Apparently very little data was collected and published on the price of oats in the country during the period 1800-1860.

US Oats

As was the situation with both corn and wheat, censuses before 1840 did not report data on the harvest of oats. According to the 1840 Census, American farmers harvested 123.1 million bushels of oats (Table 6.22) in 1839. The North Atlantic Region led the country in harvests with 51.9 million bushels (42.2% of the total crop). In 1839, the largest percentage (27.5) of all [farm] horses and mules in the country were in that region (Table 6.14). The numbers of those animals in urban areas were generally not counted in the census; consequently, it is likely that the number of horses and mules in that region, especially horses, requiring oats as feed was much larger than is indicated by the census numbers. Farmers in the North Central Region harvested 25% of the 1839 oat crop.

Oat harvests in 1839 in the South Atlantic and South Central regions were lower (Table 6.22) than in the other two Northern regions, although there were large numbers of horses and mules in both (Table 6.14). By 1839, it is likely that Southern farmers had already realized that the production of oats was not dependable enough to justify the expense, especially when they considered that an acre of land not planted in oats could be planted in cotton. As a result, they probably planted cotton and bought oats. As stated above, they probably did the same with corn.

The largest share of the 1839 crop was harvested in New York (16.8%). Other states harvesting large shares were: Pennsylvania (16.8%), Ohio (11.7), Virginia (10.9), and Kentucky (5.8). These five states accounted for 62% of the total national crop of oats.

The 1850 Census

The number of horses and mules in the United States increased between 1840 and 1850, and oat harvest followed (123.1 to 146.6 million bushels). Harvests increased in all regions except the South Atlantic (Table 6.22). It was up sharply in the North Central Region by 12 million bushels (30.3 to 42.3 million). While the number of working farm animals in that region increased during the period, it is likely that the increasing number of horses and mules used in the growing urban centers there also contributed to an increased demand for oats.

The 1860 Census

Oat harvest increased sharply in the North Atlantic and North Central regions (30.9 and 48.9%, respectively) from 1849 to 1859 (Table 6.22). During this period, the number of horses, mules, and asses (mostly horses) on farms in the North Atlantic Region increased only 20%. Therefore, it is likely that the large increase in oat production was driven by the growing numbers of these [uncounted] animals in the rapidly expanding urban areas of the region.

The number of farm horses, mules, and asses (still mostly horses) in the North Central Region increased by 85% from 1850 to 1860 (Table 6.14), so the sharp increase in oat harvest is not surprising. In fact, it is surprising that farmers there did not produce even more, assuming that they were capable of doing so. In fact, the estimated bushels of oats available for each animal declined from 28.9 to 23.3 in the decade.

Oat harvest declined in both the South Atlantic and South Central regions (13.3 and 56.3%, respectively) between 1849 and 1859 (Table 6.22). It is likely that these declines are the result of poorer than normal harvests in both regions in 1859. The number of horses, mules, and asses (especially mules) increased substantially during the period. Whatever the reason for the declines, that year Southern farmers probably had to purchase a considerable quantity of oats from the other regions.

In 1859, New York continued to lead the country in the harvest of oats. Farmers there harvested 20.4% of the 172.6 million-bushel national crop. Other states harvesting substantial shares of the total were: Pennsylvania (15.9%), Ohio (8.9), Illinois (8.8), and Wisconsin (6.4). These five states accounted for 60.4% of the total crop.

Alabama Oats

Oats have always been a minor crop in Alabama. The environmental characteristics of the state are not really suitable for the reliable production of the crop. There has been no data published on the production of the crop in the state for several years, but it was such an important crop at one time that it must be included in any discussion of the evolution of Alabama agriculture. The crop was extremely important when a substantial share of the power required on Alabama farms was provided by horses, mules, asses, and working oxen.

The 1840 Census

In 1839, Alabama farmers harvested 1,406,000 bushels of oats, 1% of the national crop (Table 6.23). Most of the crop was harvested in 17 counties north of a line between Randolph County in the east-central part of the state and Fayette County in the northwest. Within that general area, harvest was highest in counties in the Tennessee Valley. Unfortunately, it is likely that the total harvest in the state is overstated to some degree. The 1840 Census indicates that farmers in Wilcox County harvested 256,500 bushels of oats in 1839. It is likely that this estimate was erroneous. If not, it would have been the largest harvest—by over 50,000 bushels—of any Alabama county in either 1839, 1849 or 1859. If that estimate is eliminated from the data, farmers in Madison County harvested the largest share of the total crop (13.3%) in 1839. Other counties harvesting large shares of the crop were: Limestone (9.9%), Lauderdale (9.3), Montgomery (6.3), and Jackson (6.3). These five counties accounted for 45.1% of the total crop that year.

The 1850 Census

Harvest of oats in the state increased by some 114% from 1839 to 1849 (Table 6.22). Much of that increase was in the counties of the Cotton Arc. For example, it increased in Macon, Montgomery, and Greene counties by 181,000, 119,000 and 107,000 bushels, respectively. Because cotton prices were eight cents a pound in 1849 (Figure 6.38) farmers in the Cotton Arc probably turned to oat production in desperation.

The 1860 Census

Alabama farmers harvested 2.3 million fewer bushels of oats in 1859 than they had in 1849 (0.7 versus 3 million bushels). Cotton prices were up to 12 cents a pound in 1859, but inclement weather was probably the primary cause of the much lower oat harvest. The number of horses, mules, and asses had increased 51,000 animals (27%) from 1850 to 1860 (Table 6.14). In 1849, farmers harvested enough oats to provide each animal with 16 bushels each. In 1859, each animal could have received three bushels each. Whatever the reason for the sharp decline in harvest, farmers must have had a difficult time trying to find feed for their work animals that year.

Figure 6.54 presents data showing the distribution of the much-reduced, 1859 crop of oats (1860 Census). Most of the harvest was in counties in the eastern half of Alabama, in the Cotton Arc and around its margins, and in the Cotton Counties of the Coosa Valley and the Tennessee Valley. Very little of the crop was harvested in either the Low Hills in south Alabama or the High Hills in the north central part of the state. Farmers in Talladega County harvested the largest share of the 1859 crop (9.4%). Other counties with large shares of the crop were: Lowndes (6.6%), Chambers (6.6), Madison (6.5), and Montgomery (4.9). These counties accounted for only 34% of the total Alabama oat crop.

Table 6.22 also included data on oats harvested in Alabama's sister states (Illinois, Indiana, and Mississippi) in 1839,1849, and 1859. These data show that harvests were some fourfold higher in Illinois and Indiana than in Alabama and Mississippi in 1839. In 1859, the difference in harvests in Alabama and Mississippi and in Indiana was about where it was in 1839, but the difference was considerably greater in Illinois.

Sweet Potatoes

The sweet potato is most productive when grown in hot, moist climates. It produces especially well in those areas where cotton is grown. In the United States, in 1849 and 1859 most of the crop was produced in the Coastal Plain from Virginia southward. It was first cultivated in that state as early as 1648 and possibly earlier. Because the crop is easy to grow and generally dependable, from the earliest days it was an important food crop for pioneering families and yeoman farm families, and it was one of the most important foods used by slave families.

US Sweet Potatoes

The 1840 Census combined estimates of Irish and sweet potatoes harvested in the country in 1839, so it is not possible to determine the distribution of the national harvest of this crop in that year. In the 1850 and 1860 censuses, data on the two crops are reported separately. These data on the harvest in 1849 and 1859 are presented in Table 6.23.

According to these data, American farmers harvested 38.3 million bushels of sweet potatoes in 1849. Using the generally accepted conversion of 54 pounds a bushel, farmers harvested 2,066.2 million pounds or slightly less than 20.7 million hundred-weight (Cwt). The hundred-weight measure is used in all of the more recent data on sweet potato production. As expected, most of the 1849 crop (96.2%) was harvested in the South. Farmers in the South Atlantic Region harvested slightly more than those in the South Central Region (50.4 versus 45.8%). Georgia farmers harvested the largest share of the 1850 national crop (18.2%). Other states with significant harvests were: Alabama (14.3%), North Carolina (13.3), Mississippi (12.4), and South Carolina (11.3). These five states accounted for 69.5% of the national sweet potato crop.

The US harvest of sweet potatoes increased only by 10% from 1849 to 1859 (20.7 to 22.7 million Cwt). The same five states with the largest harvests in 1849 also had the largest in 1859; North Carolina, however, replaced Alabama in second place.

Alabama Sweet Potatoes

The 1850 Census also published separate data on the harvest of both the Irish potato and sweet potato in Alabama in 1849. Those data (Table 6.24) indicate that

Alabama farmers harvested slightly less than 3 million Cwt that year. Alabama counties with large numbers of slaves (Figure 6.30) generally harvested large quantities of sweet potatoes. Farmers in five counties (Montgomery, Macon, Greene, Marengo, and Dallas) harvested between 120,000 and 160,000 Cwt each. However, combined harvests in these counties only accounted for 28.5% of the total Alabama crop. Sweet potato production was important in every county in the state, and especially in counties in the Coastal Plain. Only 43,500 Cwt of the crop (1.5%) was harvested in Madison County, although they had one of the largest populations of slaves in the state. Sweet potatoes were not a very good crop out of the sandy Coastal Plain.

The harvest of sweet potatoes declined slightly (19,000 Cwt) between 1849 and 1859 (Table 6.23). As a result, the Alabama share of the national harvest was slightly lower (14.3 versus 12.8%) than in 1849. In general, the same counties in the Cotton Arc were responsible for a large share of that crop in the state in 1859. Figure 6.55 shows the heavy concentration of harvest in the Cotton Arc and around its margins. It also shows the somewhat lower concentration in the Tennessee Valley and in the High Hills.

Irish Potatoes

It was noted in the preceding section, sweet potatoes were an important food crop in the South, but of considerably less importance in the North. In the case of Irish potatoes, the situation is reversed. This situation probably has less to do with food preferences than with the fact that Irish potatoes are much easier to produce in the cooler climate in the North.

US Irish Potatoes

Data on the harvest of Irish potatoes and sweet potatoes was combined in the 1840 Census; consequently, it is not possible to obtain any specific data on the harvest of either. The 1850 Census, which separated the two, also reported the harvest of Irish potatoes in bushels. These estimates have been converted into hundred-weights (Cwt) by using the conversion (1 bushel = 60 pounds).

The 1849 data (Tables 6.24 and 6.23) show that the harvest of Irish potatoes in the United States was almost twice that of sweet potatoes (39.5 versus 20.7 million Cwt), In contrast with the concentration of sweet potato production in the South, Irish potato production was concentrated in the North, primarily in the North Atlantic Region (67.2% of the national crop). Irish potatoes were an extremely important food in the urban centers of that region. Only 10.3% of the crop was harvested in the South, and it was divided almost evenly between the South Atlantic and South Central regions (5.1 versus 5.2%). In 1849, farmers in New York harvested the largest share (23.4%) of the national crop of Irish potatoes. Other states harvesting large shares of the crop were: Pennsylvania (9.1%), Ohio (7.7), Vermont (7.5), and New Hampshire (6.5). These five states harvested 54.2% of the total national Irish potato crop.

The harvest of Irish potatoes increased 69% (39.5 to 66.7 million Cwt) from 1849 to 1859. It increased in all regions (Table 6.24), but the most significant increase came in the North Central Region where farmers harvested 139% more potatoes (21 versus 8.8 million Cwt) in 1859 than in 1849. However, harvest there continued to be lower (21 versus 38.2 million Cwt) than in the North Atlantic Region. New York (23.4%), Pennsylvania (10.5%), and Ohio (7.8%) continued to lead the country in the harvest of the crop in 1859, but Maine (5.7) and Illinois (5.0) replaced Vermont and New Hampshire in the top five.

Alabama Irish Potatoes

Farmers in Alabama only harvested 147,600 Cwt (0.37% of the national crop) of Irish potatoes in 1849 (Table 6.25). Some were harvested in all of Alabama's counties, except in the Low Hills. Generally, harvests in these counties were among the lowest in the state (13 Cwt in Butler, 32 in Covington, and 53 in Coffee). Surprisingly, farmers in Mobile County harvested 8,100 Cwt. Harvests greater than 6,000 Cwt were recorded in only eight of Alabama's 52 counties. Five of these were in the Tennessee Valley and one in the northeastern portion of the High Hills. Madison County led the state with 12,500 Cwt.

The harvest of Irish potatoes in Alabama almost doubled between 1849 and 1859 (147,600 to 295,000 Cwt) (Table 6.24), but the percentage of the national crop remained less than 0.5%. Most of the 1859 crop was harvested north of Montgomery (Figure 6.56). Harvest increased substantially in counties in the Low Hills, but they still were among the lowest in the state. For example, over the decade they increased from 13 to 347 Cwt in Butler County, 32 to 156 in Covington, and 53 to 535 in Coffee. The level of harvest in Mobile County had been unusually high in 1849, and it remained relatively high in 1860 (8,100 versus 9,079 Cwt). Harvest in that county was third highest in the state. Farmers in 24 of 52 counties harvested 5,000 Cwt, or more, in 1859. Counties in the northern part of the state continued to harvest a large share. Of the 24 counties where the harvest was 5,000 Cwt or more, 12 of them were north of Jefferson. Madison continued to lead the state with 12,676 Cwt.

Sorghum and Sugarcane

Sorghum and sugarcane were grown in the colonies as a source of sweetener (molasses) from the time the first agricultural settlements were established here. In early pioneer homes, sugar was a scarce commodity. Instead, most of families used sorghum molasses, cane syrup, or honey. The censuses of 1840, 1850, and 1860 did not include any data on the harvest of sorghum or sugarcane in the country. The 1850 Census reported that American farmers produced 12.7 million gallons of molasses, but there was no indication of the source, whether sorghum or sugarcane. There has always been some uncertainty about the difference between molasses and syrup. In this, and the remaining chapters, molasses refers to the syrup-like liquid obtained by eliminating most of the water from the juice of sorghum cane by boiling. Syrup (or sirup) is the syrup-like liquid obtained by eliminating most of the water from the juice of sugarcane stalks.

The 1860 Census published data indicating that American farmers produced 15 million gallons of syrup from sugarcane in 1859 (Table 6.25). Some 93% of it was produced in the South Central Region. The grasslike, sugarcane plant is not cold tolerant. It generally cannot be grown economically north of 32° North. Even South Carolina did not report any production in 1859. Only nine states in the entire country reported any production. Louisiana led the country with 89% of the total. Other states with large shares were: Georgia (3.6%), Florida (2.9), Texas (2.7), and Alabama (0.6).

The production of molasses from sorghum cane was more widespread than the production of syrup from sugarcane (Table 6.25); although total production was only about half as great (6.7 versus 15 million gallons). Farmers in the North Central Region produced 70% of the 6.7 million gallons of sorghum molasses produced in the United States that year. Producers in five North Central Region states (Iowa, Indiana, Illinois, Missouri, and Ohio) made more than 775,000 gallons each. Iowa farmers alone made 1.2 million gallons. The South Central Region was in a distant second place with 1.4 million gallons. Tennessee (707,000 gallons) and Kentucky (357,000 gallons) accounted for most of the product made in the region. Together, these two regions accounted for 90% of the total. Surprisingly, farmers in the South Atlantic Region produced only 642,000 gallons of molasses. They would have had much better access to sugar from ports along the Atlantic coast.

Alabama farmers made 55,653 gallons of sorghum molasses, or 4.1% of the total of 1.4 million gallons made in the South Central Region that year, or 0.8% of the national total (Table 6.25). Data published later in the nineteenth century indicate that a single acre of sorghum cane would yield 58 gallons of molasses. Using a more conservative 50 gallons an acre, farmers in Alabama must have harvested 1,100 acres of the cane in 1859.

Produce of Market Gardens

There are no early data on the production of vegetables, as such, in the United States. Rather, beginning in 1849, censuses reported the "value of produce of market gardens." Census data indicated that total value in the country in 1849 was $5.3 million (Table 6.26). According to that census, farmers in the North Atlantic Region accounted for 59.7% of the total value for the entire country. Home gardens played an essential role in the early development of the colonies. Over time, in the North Atlantic Region, as urbanization began to advance more rapidly, farmers quickly seized the opportunity to expand their home gardens to provide fruits and vegetables for the ever-growing number of city dwellers. In 1849,

New York, the state with the largest urban population, accounted for 17.2% of the total value. Pennsylvania with the second largest urban population accounted for 13%. Massachusetts, New Jersey, and Kentucky accounted for 11.3, 9.0, and 5.7%, respectively. Farmers in the South Central, South Atlantic, North Central and Western regions accounted for 13.6, 11.7, 11.2, and 3.7% of the total, respectively.

These data do not mean that farms in the South did not have gardens. Home gardens were extremely important to Southerners, as well. Most slave families also tended gardens. In the 1850s, the entire region was essentially rural, and all the towns were relatively small. There simply was limited opportunity to sell the products of market gardens.

The 1860 Census reported that in 1859 the total "value of produce of market gardens," in the United States was $16 million, an increase of 201% above 1849 (Table 6.27). During the same period, the American population only increased by 42% (Table 6.3). Obviously, people in urban areas were purchasing ever-increasing amounts of products of home gardens.

In 1859, the North Atlantic Region's share of total value declined to 52.9%, but the New York share increased to 21.1%. Other states with large shares were: New Jersey (9.6%), Massachusetts (8.7), Pennsylvania (8.4), and California (7.2). It is likely that these percentages are good indicators of the degree of urbanization in each of the states. Shares of the other regions were: North Central (18.3%), South Central (11.6), South Atlantic (9.1), and Western (8.1).

The 1850 Census reported that the total value of the produce of market gardens in Alabama in 1849 was $84,800 (1.6% of the national total) (Table 6.26). Only 18 of the state's 52 counties reported any value. Mobile County with the largest urban population accounted for 79.7% of the state total. Montgomery (10.2%) and Baldwin (3.9%) accounted for most of the remaining value.

Total value in Alabama increased from $84.8 million in 1849 to $163 million in 1859 (Table 6.26). Alabama's share of the national total had been 1.6% in 1849, but by 1859 it had declined to 1%. Only 32 of 52 of the counties reported any value. Again, most of it was reported from the Mobile area: Mobile County (54.7%) and Baldwin (6.9%). Montgomery County and adjacent Autauga County accounted for 10.1 and 5.3%, respectively. Surprisingly, only 1.2% of the state's total was reported from Madison, another large population center. Instead, nearby Lauderdale reported 10.2%.

Orchard Products

There are no early data on the production of specific products of farm orchards. Beginning in 1840, the censuses included a category of data on value of orchard products without any indication of the specific products involved. Census data for that year indicated that the total value in the United States was $7.2 million (Table 6.27). More than half ($4.1 of $7.2 million) was reported by farmers in the North Atlantic Region. The largest share (23.5%) of the total value for 1839 was reported by farmers in New York. Other states reporting large shares were: Virginia (9.7%), Pennsylvania (8.5), Ohio (6.6), and New Jersey (6.4). These five states accounted for 54% of the total.

These data on orchard products are similar to the data on market gardens. The largest share was produced in the regions and states with the largest urban population. The large number of people living in the growing urban areas of the Northeast apparently had enough surplus income to allow them to purchase products of the orchard from nearby farms. Alabama's share of the total was only 0.8%. No data was collected on values in the counties.

By 1849, US value of orchard products increased to $7.7 million (Table 6.27). The North Atlantic Region further solidified its position as the primary location for the production of orchard crops. In 1849, states in that region accounted for 61% of the national total. Four of the five top states (New York, Pennsylvania, Massachusetts, and New Jersey) in the production of these crops were located there. Alabama's share of the total was 0.2% ($15,408). Only 27 of the 52 counties reported any production. Coffee County reported the largest share (27%). Other counties with large shares were: Mobile (21.8%), Walker (17.4), Autauga (4.8), and Greene (4.0). These five counties accounted for 75% of the Alabama total in that year.

By 1859, the value of orchard products in the country had increased to $20 million (Table 6.27). Much of it continued to be in the North Atlantic Region, but its share, which had been 61% in 1849, declined sharply to 42.2%. New York continued to lead the country with 18.6% of the total, but three states in the North Central Region were included in the top five. Apparently, the rapidly growing urban population in that region was also beginning to demand more orchard products. After New York, other states with large shares were: Ohio (9.6%), Pennsylvania (7.4), Indiana (6.3), and Illinois (5.6).

Data presented in Table 6.27 indicate that value of orchard products harvested from Alabama farms totaled $55,200 in 1839. Only 26 of Alabama's 49 counties reported any value. Values reported ranged from $7 in Marengo County to $26,800 in Chambers. Most of the values reported were below $5,000.

Values declined sharply between 1839 and 1849 ($55,200 to $15,400). None of the counties reported values above $5,000.

Alabama's share of total national value increased from 0.2 to 1.1% ($15,400 to $223,300) between 1849 and 1859 (Table 6.27). In 1859, 47 of 52 counties reported values. Tuscaloosa County accounted for the largest share (11.8%). Other counties with large shares were: Coosa (8.8%), Chambers (7.5), Autauga (7.4), and Wilcox (5.6). These five counties accounted for only 41.1% of the state's total. Some 14 of 52 counties reported values of $5,000, or more.

Nursery Products

The 1850 Census reported that the total value of nursery products in the United States in 1839 was $593,000 (Table 6.28). The North Atlantic Region accounted for 50.8% of the total. Shares of other regions were: South Atlantic (17.4%), North Central (13.1), and South Central (12.0). Value was highest in Massachusetts (18.9% of the total). Other states with large shares were: New York (12.8%), Tennessee (12.0), Pennsylvania (8.4), and North Carolina (8.2). The value of nursery products in Alabama accounted for 0.1% ($370) of the national total. Apparently, no county data were collected. Further, no data were collected from any of the states in 1849 and 1859.

Alabama Timber

Timber production and utilization has always been highly variable across the country. As a result, it did not seem worthwhile to include very much national data in this section. Consequently, national data will be included only when it is required to provide a point of reference for the Alabama data.

Long before the first permanent European settlements were established in North America, timber was an important crop in the part of the world that would become the United States. Native Americans used timber to build shelters, for fuel, and for making many of their utensils and tools. Timber was one of the first products to be exported from North America. Although it had long been an important crop, very little data on timber production or its use was included in the censuses published during the period 1800 through 1860.

The Census of 1840 reported the number of cords of wood sold and the value of lumber produced in counties in the different states. Unfortunately, these data are most likely very incomplete. For example, in Alabama only 31 of the state's 52 counties provided data on cords of wood sold, and only 16 reported data on the value of lumber produced. These incomplete data indicate that farmers in those 31 counties sold a total of 60,955 cords of wood in 1839, and that farmers in Mobile County sold the largest share of the total (18.3%). Other counties with large shares were: Montgomery (13.2%), Baldwin (8.2), Monroe (7.7), and Washington (6.8). These five counties accounted for 54.2% of the state's total.

Total value of lumber produced in the 16 reporting counties was $169,008. The limited amount of data available indicated that mills in Baldwin County produced by far the largest share of the total (61.4%). Other counties producing significant shares were: Mobile (9.8%), Conecuh (8.9), Dallas (3.1), and Tallapoosa (2.9). These five counties, accounted for 86.1% of the total reported. Although the data on wood sold and value of timber produced were limited, they clearly indicate, even at that early date, the importance of timber production in the southwestern region of Alabama.

Animal Agriculture

Livestock were important on American farms from earliest times. Generally, however, all farm animals were largely forced to fend for themselves. Crops planted for food required attention (planting, cultivation, and harvest). Animals would generally survive with very little attention until they were needed for the table. This benign neglect of livestock was commonplace in the country until industrialization, commercialization, and urbanization reached a level where there was a significant off-farm demand for livestock and livestock products. This point was reached around the beginning of the nineteenth century. From that time onward, livestock and livestock products became a readily marketable product and a crop demanding attention.

Although there were many changes taking place in animal agriculture in the early nineteenth century, there is little real data to document them. The 1840 Census published the first limited data on the status of livestock and livestock products in the United States, the individual states, and the counties. Apparently, the data were obtained by requesting it from people who were operating farms in that year. This would seem to be a logical approach; however as Hilliard (1984) observes, this approach probably underestimates the status of livestock in the country. At that time, a significant number of milk cows, hogs, and chickens were kept by people living in towns. Further, in the herding areas of the South, it is likely that no one really knew with any degree of accuracy how many cattle and hogs there were out there. Fortunately, or unfortunately, these are the only data that we have.

Cattle

The first cattle reached Jamestown in 1611 and Plymouth in 1624. In the beginning of the colonial period, there was little interest in beef production. Beef generally was produced as a byproduct of milk production and the production steers for use as draft animals. Most beef was produced as a part of the general farming enterprise. Cattle were given little winter shelter and very little feed or fodder of any kind. However, with the development of Boston, New York City, and Philadelphia as centers of national and international commerce, demand for beef began to increase (Bidwell and Falconer, 1941). These rapidly growing markets forced the specialization of cattle production on farms around those cities. After 1800, large centers of beef production developed in the Connecticut Valley and in southeastern Pennsylvania. By the 1830s, cattle from Kentucky, Ohio, and Illinois were beginning to appear in the important Brighton, Connecticut, cattle market. Their appearance marked the beginning of competition for the beef cattle market between farmers west of the mountains and those in New England. With the availability of an abundance of cheap corn for feeding cattle in the Old Northwest once the railroads reached those areas, the competition was essentially ended.

In the early days of the Southern colonies, concerns for the plantation production of staple crops such as tobacco, rice, and indigo precluded their owners from giving but little attention to beef production. The situation was somewhat different, however, in the backcountry away from the coastal plantations. In these areas, a few decades after settlement, open forests around yeoman homes were swarming with wild cattle (Gray, 1958). In fact, there were so many of them wandering about that they were a detriment to crop production. The presence and abundance of these animals, which no one really owned, gave rise to the widespread re-awakening of the ancient Celtic practice of herding. Periodically, these animals would be rounded up and given cuts on an ear for purposes of identifying them as property of a specific herdsman. The pioneer herdsman would follow them around on the open range, but would provide them with no shelter or feed. Often herds of several herdsmen would be merged, but with the ear marks they could be identified when it came time to sell them. Every year or so, the animals would be driven to the nearest seaport, mostly for export. For many years, urbanization was extremely limited in the South Atlantic and South Central regions, and as a result there was no widespread demand for beef. Most of it was shipped out of those regions.

Over time, as the pace of settlement in the South proceeded, the amount of open range decreased and herding began to be replaced by systematic husbandry of cattle. However, even with these altered circumstances, the practices of providing shelter and supplemental feed and fodder developed very slowly.

The 1840 Census reported data on the total number of "neat" cattle in each county in the country. Unfortunately, it is not completely clear what those totals mean. The word cattle was apparently derived from the ancient word chattel, which was a general word for property. Therefore, in the early nineteenth century, the word cattle included all kinds of livestock. When one wanted to limit the discussion to animals of the genus *Bos,* the word cattle had to be modified to indicate that it meant only those animals of that genus. Apparently, the word "neat" was chosen, because it was derived from a very ancient Indo-European term that specifically identified animals of that genus. Later, this modifier was eliminated when the use of cattle became widely accepted as a name for animals of that specific genus.

In the 1850 and 1860 censuses, cattle are divided into three sub-groups: milk cows, working oxen, and "other" cattle. It is not clear whether neat cattle in the 1840 Census includes the three sub-groups or just "other" cattle; consequently, it is not clear to what degree the 1840 and 1850 data are comparable. I will assume that they are not.

Unlike crop data that was collected the year before the census was published, data on numbers of cattle could be determined in the same year of publication. As a result, the years on Hilliard's maps are the years in which the cattle were actually inventoried.

US Cattle

The 1840 Census indicated that there were 15 million, neat cattle in the United States in that year. All the regions had at least three million animals; however, the North Atlantic Region led the country with 4.8 million. It is likely that the larger share in that region was the result of relatively larger numbers of milk cows there.

Census data on the number of "other" cattle (excludes milk cows and working oxen) in the United States, its regions, and Alabama in 1850 and 1860 are presented in Table 6.29. The 1850 data indicate that in the country there were 9.7 million "other" cattle, or cattle that might be used for beef. The states of New York (767,000 animals), Ohio (749,000), Georgia (690,000), Virginia (669,000), and Texas (661,000) led the country in numbers of "other" cattle that year.

In 1850, the South Atlantic Region had the most "other" cattle (2.7 million) and cattle per farm (10.9). With the exception of the Western Region, the North Atlantic Region had the lowest total number (2 million) and also the lowest number per farm (4.1). The relatively large numbers in the South Atlantic (2.7 million) and South Central (2.4 million) regions likely reflects the level of open range herding in the backcountry of the Southern states. Further, it is likely that a large share of the owners of cattle in the South, especially the herders, were descendents of immigrants from Celtic Britain with their ancient cattle herding tradition.

The number of "other" cattle in the United States increased from 9.7 to 14.8 million between 1850 and 1860. Numbers increased substantially in the North Central Region (2.4 to 4 million), primarily as a result of the rapidly growing number of farms there, as well as the increase in cattle feeding. Numbers also increased in the South Central Region (2.4 to 5.1 million). Most of that increase was the result of the explosion of numbers in Texas. According to the 1850 Census, there were only 661,000 animals in that state. In 1860, there were 2.8 million. In one decade, Texas became the overwhelming leader in cattle production in the United States. Numbers also increased at a relatively high rate in the Western Region (0.3 to 1.1 million). In contrast to increases in all the other regions, numbers declined in the South Atlantic Region (2.7 to 2.4 million) as a result of lower numbers in South Carolina, Georgia, and Virginia.

In 1850, there were 42 head of "other" cattle for each 100 persons in the American population. A decade later, there were 47. Apparently, the number of these animals was increasing only slightly faster than the population.

Alabama Cattle

Figure 6.57 presents the 1840 Alabama "neat" cattle inventory data in the form of a dot-density map. Note that no data were obtained from Pickens and Sumter counties. Generally, there were relatively large numbers of neat (all) cattle in most Alabama counties, except in several of those in the High Hills section. At that time, there were also fewer farms there. There were many more cattle in the Low Hills section south of the Black

Prairie than might have been expected considering the smaller number of farms. These larger-than-expected numbers probably also reflect the degree to which open range herding was practiced there at that time. The larger concentrations in and around Montgomery, Randolph, Perry, and Madison counties probably reflect the larger number of farms concentrated in those areas.

In 1850, there were 433,300 "other" cattle in Alabama (Table 6.29). Except for the somewhat larger concentrations in and around Russell (14,500), Macon (14,300), Greene (12,800), and Madison (15,700) counties, and the somewhat lower concentration in the High Hills. cattle numbers were relatively evenly distributed across the state. Surprisingly, three counties—Conecuh (12,400 animals), Butler (12,000), and Coffee (11,900)—in the Low Hills section in the Lower Coastal Plain had numbers greater than 11,500. Average number per farm was 25, 22, and 20 animals, respectively, in those counties, which was far higher than the 8.8 per farm in the state. These counties were well known for their open range herding.

According to the 1850 and 1860 censuses, during the decade neat cattle numbers increased only 4.9% (433,300 to 454,500) in Alabama (Table 6.29). At the same time, the distribution within the state remained essentially the same as in 1850 (Figure 6.58). The distribution of cattle in the state generally mirrored bales of cotton harvested (Figure 6.42). During the decade, some counties lost cattle, while others gained some. For example, Madison lost 8,000 animals; while Cherokee gained 5,200.

In 1850, there were 56 head of "other" cattle per 100 people in the population of the state. A decade later there were only 47, about the same as the country as a whole. By 1860, with the rapidly growing production of cotton, it is likely that farmers were devoting less attention to herding and cattle production, in general.

Hogs

Within a few decades of settlement in both the North Atlantic and South Atlantic regions, enormous herds of wild hogs were ranging through the forests. Hogs were the ideal pioneer animal. They mature early and produce several large broods a year. The sows in the wild were ferocious defenders of their young against all predators regardless of size. They were also voracious, efficient foragers and found plentiful food in the hardwood forests of the Atlantic Seaboard. It is no surprise that their numbers increased so rapidly. In the early years, they became so numerous near Jamestown that the inhabitants had to palisade the town to protect their gardens from the hordes of rooters. Hogs were especially well adapted for the almost endless forests of the South with their wet swamps, abundance of acorns, chestnuts and other wild forage, and the relatively mild winter weather.

US Hogs

The 1840 Census reported that there were 26.3 million hogs (including pigs) in the United States (Table 6.30). The largest share (8.4 million) was in the South Central Region. Within that region, most of the animals (5.2 million) were in Kentucky (2.3 million) and Tennessee (2.9 million). Almost 20% of all of the hogs in the country were in those two states. In those two states, commercial hog production, associated with the growing production of corn, was already developing rapidly. In that year, the two states led the country in the production of hogs, as well as the production of corn. They marketed their hogs throughout the South and into the Northeast as well. In 1840, there were no railroad connections with the population centers in the East. As a result, hogs from these states had to be driven, sometimes hundreds of miles, to market. Ohio was the country's third leading hog-producing state in 1840 (2.1 million) with Virginia (2 million) and New York (1.9 million) rounding out the top five.

Hog numbers in the country increased from 26.3 to 30.3 million head between 1840 and 1850 (Table 6.30). This increase, however, did not keep pace with the increase in population. In 1840, American farmers produced 150 hogs for each 100 persons. By 1850, this number had declined to 130. A significant share of the increase in the American population in that decade took place in the North Atlantic Region, but people in that region much preferred beef to pork. Locher and Cox (2004) reported that in 1848 Northerners consumed less than a third of the amount of pork consumed by Southerners (51 versus 173 pounds).

Numbers of hogs increased in four of the five regions during the 1840s, declining only in the North Atlantic

Region (Table 6.30). As noted above, people in that region much preferred beef to pork. With beef production increasing in the region (Table 6.30), and with quantities of both beef and pork coming from the Old Northwest increasing each year, there was no need to increase hog production.

The increase in hog production was especially significant in the South Central Region (8.4 to 11.7 million head). Farmers in Tennessee and Kentucky still led the country in production of the animals with 3.1 million (10.2% of the national total) and 2.9 million (9.5%), respectively. Other states with large shares of total numbers were: Indiana (7.5%), Georgia (7.1), and Ohio (6.5).

Hog numbers continued to increase in the country (Table 6.30) between 1850 and 1860 (30.3 to 33.5 million), but continued to fall behind the increasing national population. In 1860, total hog production was equivalent to only 106 animals per 100 people. Further, numbers continued to decline in the North Atlantic Region, and began to decline in the South Atlantic and South Central regions as well. Increasing corn (Table 6.20) and hog production (Table 6.30) in the North Central Region, as well as the ever improving rail transportation between that region and markets in the East and South, meant that the other regions were rapidly losing whatever comparative advantage they might have had in the past. Both Tennessee (2.4 million) and Kentucky (2.3 million) were still among the top five hog-producing states, but they were rapidly losing ground to states in the North Central Region.

Alabama Hogs

Inventory data presented in Table 6.30 indicate that there were 1.4 million hogs in Alabama in 1840, 5.3% of the national total. Primary concentrations were in the Cotton Counties of central Alabama, the Coosa Valley and the Tennessee Valley (Figure 6.59). Generally, the distribution of hog numbers paralleled the distribution of the population (Figure 6.16). Lowndes County had the largest share of the state total, 6.9%. Other counties with large shares were: Cherokee (5.9%), Madison (5.7), Jackson (4.2), and Montgomery (4.0). These five counties only accounted for 26.7% of the total, which indicates that many of the other counties also had relatively large numbers of hogs.

By 1850, hog numbers in Alabama had increased to 1.9 million (Table 6.30). Numbers generally increased throughout the state, but the increase was especially large in the western portion of the Cotton Arc. In 1840, none of the counties west of Lowndes were included in the top five in hog numbers. In 1850, both Dallas and Greene counties were included in that group, and there were almost as many in Marengo County as in Dallas County. Again, the distribution of hogs was generally similar to the distribution of people (Figure 6.17). The increase in numbers in Alabama between 1840 and 1850 kept pace with the increase in population. In 1840, there were 24 hogs per 100 persons in the state. In 1850, there were 25.

Between 1850 and 1860, hog numbers decreased by 200,000 animals in the state. There were 1.7 million of the animals here in 1860 (Table 6.30). While there are no data to support it, the decrease probably represents a reduction in the open range production of hogs. About this time, many farmers were beginning to try to improve the quality of their livestock by introducing new breeds. However, to do so they had to pay more attention to the confinement of their animals, and they likely confined fewer animals than they had owned under open range conditions.

By 1860, hog production was firmly established in the Cotton Arc (Figure 6.60). All of the top five counties (Montgomery, Marengo, Lowndes, Barbour, and Pike) were in that section, and Dallas had only a few less animals than Pike. Numbers in all of these six counties were greater than 55,000. None of the other counties in the state had as many as 50,000. The five Cotton Arc counties listed above also produced a large share of the corn in the state (Figure 6.47).

Milk Cows

The first cows arrived in Jamestown as early as 1611, and Plymouth in 1624. In the early years, there was no differentiation between cows used for milk production and those used for beef (Bidwell and Falconer, 1941).

US Milk Cows

The 1840 Census did not include any data on the number of milk cows in the country. It reported only the number the number of "neat" cattle. This category included cattle used for milk production and beef and those used as working animals.

Beginning in 1850, data on the number of milk cows in all of the counties in all states were included in the census. Table 6.31 includes data on the number of milk cows in the United States, its regions, and in Alabama in 1850 and 1860. The data indicate that there were 6.4 million of the animals in the country in 1850. Slightly over a third (2.2 million) of the total herd was in the North Atlantic Region. Numbers in the other regions were: South Atlantic (1.2 million), North Central (1.6), South Central (1.4), and Western (less than 100,000). The largest share of the national total was in New York (14.6%). Other states with large shares were: Ohio (8.5%), Pennsylvania (8.3), Georgia (5.2), and Virginia (5.0). These five states accounted for 41.6% of the national total.

The large number of milk cows in the North Atlantic Region reflected the growing urbanization in and around Boston, New York City, and Philadelphia. The North Central Region had the second highest total of milk cows (1.6 million), and a third of those were in Ohio. In fact, in 1850 some 8.5% of all the milk cows in the country were in Ohio. In the same year, 6.2% of the US population lived there. Surprisingly, Georgia, with only 2.9% of the American population, accounted for 5.2% of the milk cows. Approximately 21.9% of all milk cows (1.4 million) were in the South Central Region. These were fairly evenly distributed at levels of around 220,000 in all of the states except Arkansas (93,500) and Louisiana (105,600), which had substantially lower numbers of people.

By 1860, the number of milk cows in the United States had increased to 8.6 million (Table 6.31). The North Atlantic Region continued to lead the country with 30% (2.6 million) of the total. Totals for the other regions were: South Atlantic (1.2 million), North Central (2.6), South Central (1.9), and Western (300,000). New York State continued to have the largest share with 13.1% of the national total. Other leading states were: Ohio (7.9%), Pennsylvania (7.8), Texas (7.0), and Illinois (6.1). In 1850, neither Texas nor Illinois had been in the top five states. This change reflects the continued westward movement of farm families with their milk cows.

The largest increase (62%) in milk cow herd size in the decade of the 1850s took place in the North Central Region (1.6 to 2.6 million animals). In 1842, the first whole milk was shipped by rail in the United States. After that time, it became increasingly easy to ship milk from Ohio, Indiana, and Illinois and other states in the North Central Region, with their growing surplus of grain, eastward to the growing urban markets.

In 1850, there had been 27 milk cows per 100 persons in the United States. In 1860, the ratio remained at essentially the same level. Nationally, herd size had kept pace with population growth during the decade. Although the ratio remained about the same nationwide, it probably increased substantially in the North Central Region with its large increase in herd size.

Alabama Milk Cows

The 1850 Census reported that there were 228,000 milk cows in Alabama (Table 6.31). The largest number of animals was in the counties of the Cotton Arc across central Alabama where the largest number of people were located (Figure 6.17). The average number per county in these 19 counties was 5,608 animals. Leading counties were Macon (7,206 milk cows), Greene (7,070), and Montgomery (6,863). The lowest concentration of milk cows was in the High Hills counties north of the Black Prairie. The average per county in those 14 counties was 3,258 animals. Counties with the smallest number were: Walker (2,019), Marshall (2,293), and Blount (2,375). The difference in the number of milk cows in the two sections seems to be primarily the result of differences in the number of farms. The average number of farms per county in the Cotton Arc and the High Hills was 5,608 and 3,258, respectively.

Between 1850 and 1860, the number of milk cows in the state increased by only 1.3% (228,000 to 231,000) (Table 6.31). The primary reason seems to have been the result of the fact that there was almost no increase in the number of farms in the Cotton Arc during the decade. The number of milk cows there was lower in 1860 than in 1850 (102,000 versus 107,000). Some 11 of 19 Cotton Arc counties actually lost farms during the 1850s.

In 1860, the largest share of Alabama's milk cow herd

was in Tuscaloosa County (3%). Other counties with large shares were: Pickens (2.8%), Pike (2.8), Chambers (2.6), and Coosa (2.6). These five counties accounted for only 13.8% of the state's total. This low percentage indicates that milk cows were widely distributed across the state. The average per county was 4,400 animals. A total of 33 counties reported herds of 4,000, or more, animals.

Milk and Dairy Products

Milk and milk products began to play an important role in the nutrition of colonists when the first cows arrived. In the beginning, almost all of the colonists were involved in general agriculture, and most of them produced small amounts of raw milk, butter, and cheese for family use. These products were much more important to those families than beef. Over time, however, with growing urbanization, especially in the Northeast, farm families began to produce a surplus of milk, butter, and cheese that could be sold in the growing towns. Dairying and the production of butter and cheese became a large and very specialized enterprise in the North Atlantic Region.

In the Southern colonies, there was only limited interest in dairying on the tobacco plantations. The planters preferred to devote their attention to the production of staples and to purchase their dairy products from general farmers in the backcountry (Gray, 1958). With limited urbanization in the South Atlantic Region, a large market for dairy products never developed as it did in the North Atlantic Region. Although development of dairying was limited in the South, milk cows were extremely important to the individual families. Locher and Cox (2004) commented that most Southern families in the early years considered a milk cow to be one of their most prized possessions and a vital source of nutrition.

Value of US Dairy Products

Censuses before 1870 did not include data on the quantity of milk produced in the United States. The 1840 Census, however, did include data on the value of dairy products ($33.8 million). Farmers in the North Atlantic Region were responsible for 71.4% of the total. Values in the other regions were South Atlantic (11.7%), North Central (10.3), and South Central (6.6). The census did not include any data on value of production in the Western Region.

Value of dairy products of New York farmers in 1840 accounted for 31.1% of the US total. Other states with significant shares were: Pennsylvania (9.4%), Massachusetts (7.0), Vermont (5.9), and Maine (4.4). Note that all five of the leading states were in the North Atlantic Region.

Almost from their beginnings, the colonies/states of the North Atlantic Region followed a path toward international commerce, industrialization, and urbanization. Colonies/states in the South Atlantic Region chose the path of staple-production agriculture using slave labor. Consequently, by the middle of the nineteenth century, many of the people in the North Atlantic Region were able to afford more dairy products in their diets, and local farmers responded by providing them with what they wanted and could afford. Part of the reason for the disparity of value in the two regions, however, was differences in the number of people. In 1840, there were 6.8 million people in the North Atlantic Region and 3.9 million in the South Atlantic Region. However, putting value data on a per-person basis shows that in the North Atlantic Region each person had dairy products available with a value of $3.57, whereas in the South Atlantic Region the value available was only $1.01 a person. In Alabama, it was only $0.45.

Production of US Butter and Cheese

The 1850 Census included data on the quantity of butter and cheese produced in the individual states. In that year, total estimated butter production in the country was 313.3 million pounds (Table 6.32). Farmers in the North Atlantic Region produced the largest share of the total (55.2%). Shares of the other regions were: South Atlantic (9%), North Central (25.7), South Central (10.0), and Western (less than 0.1).

As expected, dairy farmers in New York State produced the largest share (25.4%). Other states with large percentages were: Pennsylvania (12.7%), Ohio (11.0), Indiana (4.1), and Vermont (3.9). These five states accounted for 57.1% of the total. Alabama's share was 1.3%.

In 1850, national cheese production totaled 105.5 million pounds (Table 6.32). The cheese share of the North Atlantic Region was even greater than its butter share (75.6 versus 55.2%). Shares of the other regions were: North Central (23.3), South Atlantic (0.6), South Central (0.5), and Western (less than 0.1).

Farmers in New York also produced the largest share of this dairy product (47.1%). Other states with large percentages were: Ohio (19.7%), Vermont (8.3), Massachusetts (6.7), and Connecticut (5.1). These five states accounted for 86.9% of the national total. Alabama's share was less than 0.1%.

The 1860 Census included data that showed national butter production totaled 459.7 million pounds (Table 6.32), an increase of 47% over the 1850 total. The North Atlantic Region percentage was 48.7%, considerably less than the 55.2% it had been in 1850. Farmers in the North Central Region were catching up; their percentage, which had been 25.7% in 1850, had increased to 33.3%. Percentages in the other regions were: South Central (9.6%), South Atlantic (7.4), and Western (less than 0.1).

In 1860, farmers in New York State continued to lead in butter production with 22.4% of the national total. Other states with large percentages were: Pennsylvania (12.8%), Ohio (10.6), Illinois (6.1), and Indiana (4.0). These five states accounted for 55.9% of the national total. The Alabama share was 1.3%. Note that the North Central Region had three states in the top five. In 1850, there had been only one top producing state in the North Central Region.

National cheese production totaled 103.7 million pounds in 1860 (Table 6.32). The share of the North Atlantic Region was slightly less in 1860 than in 1850 (70.3 versus 75.6%). Shares of the other regions were: South Atlantic (0.4%), North Central (27.2), South Central (0.6), and Western (1.5). In the Western Region, California had entered the fray. Between 1850 and 1860, cheese production there had increased from 150 to 1.3 million pounds. This was a sign of things to come."

In 1860, cheese producers in New York continued to lead in production (46.8%). Other states with large shares, included: Ohio (20.8%), Vermont (7.9), Massachusetts (5.1), and Connecticut (3.8). These five states accounted for 84.4% of the national total. The Alabama share was less than 0.1%.

Production of Alabama Butter and Cheese

In 1840, farms in Alabama accounted for 0.8% ($265,200) of the total national value of dairy products. Farmers in five counties across the central part of the state accounted for a large share of Alabama's total. Perry was the leading county with 13.6%. Other counties with large percentages were: Russell (12.3%), Chambers (10.1), Montgomery (7.4), and Bibb (5.8). These five counties accounted for 49.2% of the state total.

There was no data on value of Alabama dairy products in the 1850 or 1860 censuses.

According to the 1850 Census, butter production in Alabama totaled 4 million pounds (Table 6.32), or 1.3% of the national total. Farms in Lawrence County reported the largest share (4.2%). Other counties reporting large percentages were: Macon (3.6%), Madison (3.5), Greene (3.4), and Marengo (3.3). These five counties only accounted for 18%, which is indicative of the widespread distribution of butter production in the state. In fact, 19 of 52 counties reported productions of 1,000 pounds or more.

In 1850, Alabama cheese production totaled only 31,400 pounds (Table 6.33), 0.1% of the national total. Only 40 counties reported any production, and 33 reported productions of 1,000 pounds or less. Cherokee (17.8%) and DeKalb (10.3) counties accounted for more than a third of the total.

In 1860, Alabama butter production totaled 6 million pounds (Table 6.32), or about 1.3% of the national total. Farms in Tuscaloosa County reported the largest share (4.7%). Other counties with large percentages were: Chambers (3.7%), Randolph (3.7), Talladega (3.1), and Calhoun (3.1). However, these five counties accounted for only 18.3% of the state total. As was the case in 1850, this relatively low percentage is indicative of the widespread distribution of butter production. A total of 33 of Alabama's 52 counties reported productions of 100,000 pounds or more. It is interesting that all of the top five counties were in the High Hills section, three of them in the eastern, more mountainous region.

In 1860, Alabama cheese production was 15,900 pounds (Table 6.32) and accounted for 0.1% of the national total. Only 34 of the counties reported the production of any cheese, and 29 of the counties reported productions of 1,000 pounds or less. Only Coosa County reported a production greater than 2,000 pounds.

Chickens

Bidwell and Falconer (1941) and Gray (1958) commented that chickens were included in the livestock accompanying the first settlers to both the Massachusetts and Virginia colonies. Indians in the Lower Mississippi Valley obtained chickens from a wrecked European ship even before there were European settlements on the Gulf Coast.

In all the regions, chickens were given very little attention once they were introduced, but because of their high biological potential, their numbers increased rapidly. By 1609, there were an estimated 500 chickens in Virginia, and by 1627 they were considered to be abundant. By the middle of the eighteenth century, all families, Indian and European, in all of the Southern colonies had a flock of chickens. In every situation, chickens were left to shift for themselves. To prevent their becoming too wild, however, they were given a little feed from time to time. Production of chickens became fairly widespread throughout the Northeast, especially in Rhode Island, as urbanization of the region proceeded.

Value of US Poultry

In the period 1800-1860, only the 1840 Census included any data about poultry of any kind. That census published estimates of the value of poultry. In that year, the value of poultry on American farms totaled $9.4 million. It was highest in the North Atlantic Region ($3 million); however, it was not much lower in either the South Atlantic ($2.5 million) or South Central ($2.3 million) regions. Total value in the North Central Region was considerably lower ($1.6 million). Value was highest in New York with 12.3% of the national total. Other states with large percentages were: Virginia (8.1%), Pennsylvania (7.3), Tennessee (6.5), and Ohio (5.9). These five states accounted for 40.1% of the national total. The geographical distribution of these five states are generally indicative of the market for poultry near the major urban areas of the Northeast. Slightly more than 4% of total value in the country was on Alabama farms.

The 1850 Census only reported total national value ($12 million) and provided no additional data. It did comment that the 1850 estimate was made on the same basis as one published the 1840. The 1860 Census did not report any data on poultry.

Value of Alabama Poultry

In 1840, the total value of poultry on Alabama farms was $405,000 (4% of the national total). Based on data, reported to census takers, 5.6% of the value in the state was on farms in Montgomery County. Other counties with large percentages were: Perry (4.8%), Madison (4.2), Chambers (3.8), and Marion (3.6). These five counties accounted for only 22% of the state total. Value was above $10,000 in 17 of Alabama's 52 counties. These data indicate that poultry were widely distributed in Alabama and generally distributed in the same manner as the population. Where there were more people, there was more poultry.

Eggs

The production of eggs did not receive much attention in the early colonial period. They were a by-product of the production of chickens for their meat. By the middle of the 1800s, however, with urbanization proceeding rapidly throughout the North Atlantic and North Central regions, egg production grew rapidly. During this period, Cincinnati became a national center for the production and shipment of eggs. Producers there shipped barrels of eggs throughout the country and to some foreign countries.

Decennial censuses before 1880 did not include data on national egg production. The 1850 Census reported that the value of eggs produced in the country totaled $5 million, but provided no additional data.

Agricultural Production

The purpose of this section is to provide data on the overall contribution of agriculture,to the economy of the United States and Alabama. Unfortunately, at this early date there was very little data available. Virtually all of the data reported in the censuses of 1840, 1850, and 1860 are estimates of only the quantities of the various products of agriculture in the states and the country. However, those persons who prepared the 1850 Census made an effort to provide estimates of the total value of products. They made no effort to provide similar data for the individual states. It is obvious in considering the data that those were rough estimates at best; but they are all that is available, and even though they are rough, it is likely that most of the individual estimates are, at least, in the ballpark. Probably the most important aspect of these data is that they provided an early model for the collection and publication of data that is extremely important in describing and quantifying American agriculture in the various states and in the country.

Data on the estimates of value for the various products are presented in Table 6.33. They indicate that the estimated total value for all crops was $869.2 million, and that among these crops the value for feed grains ($392.5 million) was much greater than those for food grains ($132.2 million) and for cotton ($111.8 million). In these estimates, the value for feed crops combines values for corn and oats. The value for food grains combines values for wheat, buckwheat, rice, and barley. The estimate for the value of feed crops indicate the importance of livestock feeding in the country even at this early date.

Data on the total value of livestock (Table 6.33) indicate that the value for "meat animals" (cattle, hogs, and sheep) was fourfold greater than the values for either dairy products or poultry and eggs ($280 versus $62.4 and $17 million, respectively).

Data in Table 6.33 also show that the value of all products of agriculture in the country in 1850 was $1,248.8 million. This estimate does not include the value of any products from the towns and cities, although in 1850 a considerable amount of the production of some agricultural products took place around the individual houses in urban areas. It is likely that including these data would have increased the values for vegetables, fruits and tree nuts, dairy products, poultry and eggs, but would likely have had little effect on the values for food grains, feed crops, cotton, tobacco or "meat animals."

The Bottom Line

As detailed in preceding sections, American agriculture remained tightly bound to the Atlantic margin throughout most of the seventeenth and eighteenth centuries. At the beginning of the American Revolution, there were many agricultural settlements along the eastern edge of the Appalachians, and a few venturesome souls with their families had passed through Cumberland Gap in southwestern Virginia into territory beyond the mountains. Nevertheless, most American farmers still lived, grew their crops, and raised their livestock within a relatively short distance of where their ancestors had first landed in the New World. Then, with the end of the American Revolution, farmers became a tidal wave flowing westward.

First, the Americans removed the Native American barrier; then within little more than a generation, they had purchased the Louisiana Territory from the French, wrested the remainder of the land west to the Pacific from Mexico, and established the northern boundary in the Northwest with England (Canada) at the 49th parallel. By the 1848 Treaty of Gadalupe Hidalgo, just 63 years after the signing of the Treaty of Paris (1783), American agriculture was firmly entrenched from the Atlantic to the Pacific and from roughly the 30° North in the south to roughly 49° North in the north.

By 1860, most of the essential characteristics of American agriculture had been established. In the North Atlantic Region, farmers on relatively small farms produced wheat, vegetables, fruits, milk, cheese, butter, poultry, and eggs for a rapidly growing, prosperous, commercial, urban population. In the North Central Region, farmers on larger farms were beginning to produce enormous quantities of food grains, feed crops, hogs, and cattle that would be shipped throughout the country, as well as to Europe. In the South (South Atlantic and South Central regions), farmers were bound with an ever growing disaster. There were 3.8 million slaves in the two regions, and cotton was selling for around 12 cents a pound. Given the highly variable weather and poor soils, the South never had very much comparative advantage in agriculture, and the situation was about to get much worse. In fact, the South's poorly competitive, slave-labor, plantation agriculture was a rolling plague on the country for the foreseeable future and beyond. In the 1976 Clint Eastwood movie, *The Outlaw Josey Wales,* Chief Dan George remarked at the beginning of a horrorific showdown that "hell had come to breakfast." In 1860, in Southern agriculture, "hell had come to breakfast."

With such widely divergent models of agricultural and economic development in the North and the South, it was almost inevitable that these differences would lead to equally widely divergent social, as well as political differences. Eventually, these differences became so pronounced that a majority of white Southerners decided that they could no longer remain yoked with the people in the North. Subsequently, around 1860, they began to effect a permanent separation and to establish a confederation of like-minded people.

In 1860, some 2.8% (997,000) of the American population lived in Alabama; 43.6% of them had virtually no economic goods, and a host of others (the yeoman farmers) were little better off. Alabama farmers harvested 18% and 4%, respectively, of the country's cotton and corn crops. They sold their cotton for an average of 12 cents a pound and their corn for about 60 cents a bushel. There was almost no industry in antebellum Alabama. For all practical purposes, Alabama's economy was bankrupt. It teetered perilously on the backs of slaves and bales of cotton. With this seriously unbalanced and defective economy, Alabama's leaders and elites were chomping at the bit to lead this blighted land—of mostly poor farmers—into a horrible war that they could not win.

ALABAMA AND HER SISTER STATES IN 1860

In the Introduction to this book, I noted that in 1994 Alabama's Gross State Product (on a per-person basis) was 45th in the United States. Further, using Lenski's (1998) concept of Societal Evolution, I posited that the state's poorly competitive agriculture had slowed the advancement of its society through the Simple Agriculture and Advanced Agriculture stages, and that as a result its advancement into the Industrial stage had also been slowed, resulting in lagging economic development.

I proposed to test the hypothesis by following the changes in the some of the characteristics of Alabama's agricultural and general economies along with those of its sister states (Mississippi, Illinois, and Indiana). These states, of course, were located on the frontier when they were admitted to the Union around the same time (1816-1819). Some of the geophysical characteristics of these four states are presented in Table 6.34, and some of their economic characteristics, as reported in the 1860 Census, are presented in Table 6.36. Comments on some of these characteristics follow:

Population. Only about 40 years after becoming states, the populations of Illinois and Indiana were almost two times greater than those of Alabama and Mississippi. Of course, the size of their primary source populations (North Atlantic versus South Atlantic regions, respectively) was much larger. Most European immigrants entered the United States through ports in the North Atlantic Region.

Percentage of Slaves in the Population. Of course, this is a glaring difference. In Alabama and Mississippi, these unfortunate persons were tools of agricultural technology, and given their initial cost and the cost of their maintenance, it is not likely that slave owners in those two states ever realized an adequate return on investment.

Cash Value of Farms (Dollars Per Person). By 1860, the contribution of Alabama's agriculture to its people's welfare was lower than its sister states. The Mississippi value was probably greater than the Alabama value, because the large cotton crop (1.2 million bales) harvested in that state in 1859 exceeded the 990,000 bales harvested by Alabama farmers.

Annual Average Yields of Field Crops (Bushels Per Acre). Note that these are data from the 1850 Census. There were no yield data reported in the 1860 Census. As early as 1849-1850, the average yields of corn and oats in Alabama and Mississippi were already only about half as large as those in Illinois and Indiana. Yields of oats in the two Southern states were much closer to those in the two Midwest states, but they were still lower. These differences, at this early date, did not bode well for the future.

Value of Animals Slaughtered or Sold for Slaughter. In 1860, Illinois and Indiana still trailed far behind Alabama and Mississippi in livestock production. Even at this early date, it was apparent that Alabama and Mississippi needed to produce more field crops, and it was also apparent that Illinois and Indiana should produce more livestock.

Percentage of Males Employed in Manufacturing.

These data indicate that even as early as 1860, Illinois and Indiana were already well into the Industrial stage of development, while Alabama and Mississippi were largely stuck in the Simple Agriculture stage. The percentage of males employed in manufacturing was twice as great in Illinois and Indiana as in Alabama and Mississippi.

Capital Invested in Manufacturing. These data show that the amount of money (dollars per person) invested in manufacturing was also about twice as great in Illinois and Indiana as in Alabama and Mississippi.

Annual Value of Products of Manufacturing. This is the most telling characteristic of all. In just 40 years, return on investment per person in manufacturing was about three times greater in Illinois and Indiana than in Alabama and Mississippi.

True Value of Personal Property. This is also an important characteristic. These date include the value of slaves as personal property, and they are about four-times as great in Alabama and Mississippi as in Illinois and Indiana. Assuming that the true value of personal property, excluding the value of slaves, was really no greater in Alabama and Mississippi (it was probably much smaller), these data clearly show the magnitude of the investment in these "tools "of farming in Alabama and Mississippi.

It is clearly obvious that as early as 1860 Illinois and Indiana were clearly superior in grain production, but Alabama and Mississippi were clearly superior in livestock production. It is also obvious that Illinois and Indiana were clearly much deeper into the Industrial stage of societal development than Alabama and Mississippi.

At this time (1859), with the agricultural technology available in the South and Midwest generally equal, it is likely that the differences in yields shown in Table 6.35 are the result of the action and interaction of poorly adapted genomes: poor soils and unpredictable weather in Alabama and Mississippi. As detailed in Part I, *Ultisol* soils are poorly suited for field crop production. Poor soils coupled with the highly variable weather in the humid-subtropics presented farmers with almost insurmountable problems year after year. Further, genomes of domesticated plants—determined in environments far removed from Alabama and Mississippi—compounded the problems. Time will tell whether better agricultural technology will allow Southern farmers to overcome these obvious geophysical disadvantages.

7

The Civil War and Its Aftermath

1860–1870

PRECIS

Southern agriculture changed drastically during the 1860s. The system established during the colonial period was a spectrum that extended from the agricultural market-economy at one extreme to subsistence-economy farming at the other. This Southern system continued right up to the Civil War. Simplified, it was characterized by market-economy, large-scale, plantations growing staples (tobacco, rice, indigo, and sugar) with slave labor and by subsistence, small-scale, yeoman farming.

The Civil War crippled this system by rendering the international, market-economy agriculture impossible (because of the Union naval blockade) and disrupting labor resources, especially for yeomen farmers who were serving in Confederate grey. The war changed the South's agricultural system in other important ways: the quantity of improved farmland fell sharply when a large share of farmers became soldiers; the quality of farm real estate deteriorated; machinery and equipment were not maintained or repaired; severe hunger developed in some areas as a result of reduced production, hostility limited transportation, and food requirements of the army; cotton production declined sharply as a result of the loss of international markets and requests by the Confederacy to produce more foodstuff for the war effort. With the defeat of the Confederacy, of course, came the end of the use of slaves as the labor source for market-economy agriculture.

Northern agriculture—and industry thrived during the war, so the relative wealth of the two regions diverged, especially with the loss of the Southern slaveholders' greatest financial asset, their slaves.

The North might have helped the South through this trying period if they could have found a way to invest in the region. Unfortunately, they seemed more interested in revenge and exploitation, and frankly there was very little to invest in. The land was almost worthless, and there was almost no industry. After the war, they had hung the South out to dry, and they did little more than watch them twist slowly in the wind. The magnanimous spirit that would lead the American people to rebuild Germany and Japan after World War II was nowhere to be found after the end of the Civil War.

Consequently, during the last few years of the decade, a new Southern agriculture system began to evolve. In essence, various forms of tenancy replaced slavery and, to a large degree replaced yeoman and subsistence farming as well. The 1860s saw the beginnings of the crop-lien system, commonly called share cropping. This new system of peonage affected poor whites as well as freedmen.

Southern agriculture emerged from the Civil War badly bruised and beaten. Of course it had been severely ill for generations, but the forced surgery of Emancipation nearly killed it. Then reconstructionists almost administered last rites when they refused to help it deal effectively with the freedmen problem. As it was, it finally emerged half dead and half alive to become an invalid that would linger in that condition until it would finally be placed on life support in the twentieth century.

INTRODUCTION

In the early days of colonization, the settlers in both Jamestown and Plymouth had known extreme hunger and in some cases literal starvation. Slightly over 250 years later, in 1860, the young United States, having occupied the east central third of the entire continent, was now in a position not only to feed, clothe, and shelter its rapidly growing population extremely well, but also to feed and clothe a large share of the world as well. Unfortunately, the American people would not have the opportunity, at least for a time, to enjoy their fortuitous situation. From the early days, there had been two separate countries east of the Appalachians: a mixed farming, industrialized, commercialized, urbanized one in the North and a rural, slavery-driven, export crop agricultural one in the South.

By the middle of the nineteenth century, a number of issues arising from the differences in the economies and societies of these two countries were becoming increasingly divisive. In 1860, the situation was about to reach a flash point. It seemed that the United States could not continue to move forward until the question was settled as to whether the obvious differences between the North and South would finally be legitimized by establishing two separate, independent, sovereign countries.

NATIONAL POLITICS

In 1860, the old political realities that had served the country relatively well for almost a century were about to be thrown into a cocked hat. There was a new kid on the block. The newly-established Republican Party had come to town, and politics would never be the same. Politics as usual had ended, and this new reality resulted in a great deal of uncertainty in the South.

The Presidency

Republicans controlled the Office of President for most of the period, 1860-1870. Even Andrew Johnston, a Democrat who replaced Lincoln, followed the assassinated president's established policies as closely as possible.

James Buchanan, a Democrat from Pennsylvania, held the presidency in 1860. He served from 1857 until 1861. Persons serving as president during the period 1860-1870 were: Abraham Lincoln (R, IL) (1861-1865); Andrew Johnson (D, TN) (1865-1869); and Ulysses Grant (R, IL) (1869-1877).

Secretaries of Agriculture

The US Department of Agriculture was established in 1862, but it was not given cabinet status until 1889.

The Senate

The Democrats held the balance of power in the US Senate in the 36th Congress (1859-1863). The newly established Republican Party first gained control of that house in the 37th Congress (1861-1863), and they retained it in the 39th, 40th, and 41st (1869-1871).

Senate Majority Leaders

There was no office of Senate majority leader before 1920.

Chairmen of the Committee on Agriculture

There was no Senate Agriculture Committee in the 36th (1859-1861) and 37th congresses. John Sherman (R, OH) was chairman of the committee at the beginning of the 38th Congress. He only served until the middle of this term, but he was returned to the office in the 39th. Other persons serving in that capacity were: James Lane (R, KS) (last half of 38th) and Simon Cameron (R, PA) (40th and first half of 41st).

The House of Representatives

The Republicans controlled the House of Representatives for most of the period, 1860-1870, (36th, 37th, 38th, 40th and 41st congresses). The Union Party controlled it in the 39th.

Speakers of the House

William Pennington (R, NJ) was speaker of the House in the 36th Congress (1859-1861). Other persons serving in that capacity during the period were: Galuska Grow (R, PA) (37th); Schuyler Colfax (R, IN) (38th, 39th, and 40th); and James Blaine (R, MA) (41st).

Chairmen of the Committee on Agriculture

Martin Butterfield (R, NY) served as chairman of the House Committee on Agriculture in the 36th Congress. Others serving in that capacity were: Owen Lovejoy (R, IL) (37th); Brutus Clay (Unionist, KY) (38th); John Bidwell (R, CA) (39th); Rowland Trowbridge (R, MI) (40th); and John Wilson (R, OH) (41st).

The Political Situation

In the early colonial years, the presence of large numbers of slaves in the South was of limited concern to people in the North who also owned sizeable numbers. After the American Revolution, however, opposition to slavery began to build in the North. Great Britain abolished slavery in 1808, and many people in the North, and a few in the South, wanted Congress to abolish it throughout the United States as well. Quickly, the welcomed Western expansion of settlement beyond the mountains became conflicted with the concern for the expansion of slavery westward. These concerns just as quickly became a part of the politics of the new country. Southerners wanted to expand slavery into as many new territories and states as possible, to insure that the balance of power in Congress would be held by pro-slavery interests. Northerners were just as determined that the balance of power be held by anti-slavery interests. Slavery was not the only issue related to agriculture that divided the two sections. The increasingly industrialized North wanted Congress to continue to protect growing industries there through the use of high tariffs. Unfortunately, high tariffs worked against the interests of the agricultural South. High tariffs resulted in a decrease in the inflow of foreign goods, which allowed Northern merchants to increase the prices of goods that they sold in the South. Further, Southerners had to sell a large share of their cotton, tobacco, and rice on the world market where they faced serious tariff reprisals from countries responding to high American tariffs. For a number of years, national politics had been increasingly divided between the commercial, industrial, anti-slavery North and the agricultural, slave-holding South. By 1860, the differences had become essentially irreconcilable. Despite herculean efforts on the part of a relatively small number of compromisers in Congress, tensions between the two interests increased almost continuously from around 1840 onward, until they finally boiled over into open warfare in April 1861.

The period 1860-1870 was a true watershed period in the history of American politics. Within this decade, the decision would be made whether two countries would emerge from one. It was a time of enormous need for great statesmen. Unfortunately, by 1860 the politicians and their supporters had become so polarized that about the only thing they could do was to scream past one other.

The Republican Party, which appeared on the national scene in the 1850s, was born and bred in the North central Region among those hardy pioneers who earlier in the century had moved beyond the Appalachians, primarily out of the North Atlantic Region. With a religious zeal, they were determined to stop the spread of slavery (Atkins, 1964). They also supported federal aid to railroads, enactment of a homestead law, land grants to agricultural and mechanical colleges, and a high protective tariff. Most Southerners were strongly opposed to all of these proposals, especially any effort to limit slavery. Through the late 1850s, the new party grew rapidly in the North and Northeast, and in 1860 the Republican candidate, Abraham Lincoln, running on a platform that would prohibit slavery in the new territories, was elected president in a four-party race. The Republican Party had taken took control of the House of Representatives in the 36th Congress (1859-1861) and the Senate in the 37th Congress (1861-1863).

Rightly or wrongly, a majority of Southern politicians viewed the election of Lincoln as a sure indication that their concerns regarding economic matters and slavery would not be addressed by the dominant Republican Party and the new administration. Consequently, on December 20, 1860, South Carolina seceded from the Union. By the end of February 1861, Mississippi, Florida, Alabama, Georgia, Louisiana, and Texas had also seceded. By the middle of April, Virginia, North Carolina, Tennessee, and Arkansas had joined them. The border states of Missouri, Maryland, and Kentucky decided to remain with the Union. On February 8, 1861, delegates from those states that had seceded at that time met in Montgomery, and formally established the Confederate States of America. The following day, they elected Jefferson Davis as provisional president.

The military prosecution of the war was a relatively simple matter for the North. It simply tied a noose of soldiers and ships around the entire South and slowly pulled it ever tighter. The war began with the firing on Fort Sumter on April 12, 1861. The last Confederate flag was lowered on November 6, 1865, when the CSS *Shenandoah* surrendered to British authorities. Assuming that the North had the stomach for a war to preserve the Union, and that none of the European states would actively intervene, there was never any real question concerning the war's outcome.

Abolitionism had a long and bitter history in the North, and when it finally reached its goal of freeing the slaves, there was too much pent-up desire for vengeance to stop there. Its most devoted supporters in Congress seemingly wanted to pursue the rebels onto their homesteads, punishing them every step of the way.

The most significant political developments of this period were the sometimes divided efforts to prosecute the war, and the efforts to officially return the South and its people to the Union. Lincoln had determined the outlines of the program to do so before he was assassinated. In March 1865, Congress created the Freedmen's Bureau within the War Department with general power of relief and guardianship over all refugees. Johnson put Lincoln's program into effect by appointing provisional civil governors in each formerly Confederate state where Lincoln had not already done so. By January 1866, civil administrations were functioning in every state of the defeated Confederacy except Texas. Unfortunately, some members of Congress felt that Lincoln and Johnson's plans were inadequate to protect the freed slaves. In April 1866, a joint Congressional committee reported a plan that, in essence, denied statehood for all Southern states until the guarantee of Negro equality was incorporated into their laws.

In March 1867, by virtue of Congressional action, US military rule replaced the civil administrations. The first Reconstruction Act divided the South into five military districts under the supervision of general officers who initially took their orders directly from General Grant, instead of the president. Alabama was under Presidential and then Congressional Reconstruction control from 1865–1874. (Reconstruction continued until 1877 in Louisiana and South Carolina, the last two formerly Confederate states to be governed under the Reconstruction Act.)

After the end of the Civil War, the Radical Republicans in Congress were determined to establish party ascendancy in the South through reconstruction, but to do so they had to reduce the executive branch of government to a position of merely rubber stamping congressional actions. Of course, Johnson would not accept this role, so the Congress began impeachment proceedings. In February 1868, the House moved to impeach the president before the Senate, citing twelve articles of high crimes and misdemeanors. On one after another of the articles, Johnson was found not guilty. Finally on May 26, 1868, the Senate fell a single vote short of the two-thirds majority required to impeach him on the last two articles. At that point, Justice Chase adjourned the proceedings, effectively bringing to an end efforts to remove Johnson from office. This failure also seemed to reduce efforts to radically reconstruct the South.

ALABAMA POLITICS

In 1833, South Carolina had begun the process of leading the South out of the Union, when the South Carolina legislature had enacted a law declaring that the Tariff of 1832 was null and void. So it is not altogether surprising that South Carolina would finally lead the South in the exodus from the Union. In December 1860, that state's legislature declared "that the Union between South Carolina and the other States . . . is hereby dissolved," but South Carolina was only the stalking horse in this unfolding tragedy. Quickly, several other Southern states rushed to form a new confederacy.

Governors

Andrew Moore (D) served as governor of Alabama from 1857 through 1861. He was replaced by John Shorter (D) who served as governor from 1861 through 1863. Thomas Watts (D) served from 1863 until he was replaced by the provisional Reconstruction governor, Louis Parsons (D) in April 1865. Parsons served in that position through December 1865. He was replaced by Robert Patton (R) in December 1865. Patton served through July 1867 when he was replaced by the US military governor, Wager Swane. General Swane remained governor until July 1868, when he was replaced by William Smith (R), who served until November 1870.

Commissioners of Agriculture

The Alabama Department of Agriculture and Industries was not established until 1883.

THE POLITICAL SITUATION

By 1860, it had become obvious to a substantial number of Southern political leaders and commercial men that they could not compete with the North economically, and they had so much money invested in slaves that they could not give them up either. To a large percentage of Southerners, there could be no acceptable compromise. They felt strongly that their only course was to leave the Union and form their own country. Alabama support for secession generally increased throughout the 1850s, paralleling the national growth of the newly formed Republican Party. After Lincoln's election, Alabama's legislature passed a resolution calling for a convention to vote on an ordinance of secession. Delegates to the convention were to be chosen in statewide elections. After months of acrimonious debate, a slate of elected delegates met in Montgomery, and on January 11, 1861, they voted by a margin of 61 to 31 to secede immediately. On that day, Alabama became a free, sovereign, and independent state. The matter of secession was never voted on by the general electorate. The decision was made that the election of delegates to the convention was all that was necessary to dissolve Alabama's tie with the United States.

In retrospect, when one looks at these on-rushing events in Alabama and the South, one is struck by the parallels of these events and those associated with the efforts of ancestors of many of these same Southerners to overthrow the Hanoverians and restore the Stuarts to the English throne. This so-called Lost Cause began in 1745, when Highland and Lowland Scots followed Charles Edward Stuart (Bonnie Prince Charlie) southward to invade England. After some initial success, this insurrection ended in April 1746, with the terrible slaughter of the remnants of Charles's ragtag army at Culloden at the hands of Lord Cumberland and his forces.

The American Civil War ended in April 1865. In Alabama, it ended four years of death and destruction (Rogers and Ward, 1994). There has been considerable disagreement on the number of Alabamians who were killed or wounded, but there was widespread agreement that Alabamians lost heavily. For a state with a relatively small population, the loss of life and the number of wounded who had lost arms and legs exacted a terrible price. Unfortunately, along with the high human price, the loss of property added significantly to the total. As battles surged across the state, both sides were guilty of planned destruction. Soldiers, especially the Confederates, often were forced to live off the land, further exacerbating the already dismal situation in local communities. Ironically, north Alabama, which had generally opposed secession, suffered the greatest damage to property—in large part because an internal civil war that raged among Alabamians in the northern part of the state. The Black Belt, where support for secession had been greatest, suffered relatively little. There was almost no damage to property in most of the Lower Coastal Plain.

The turmoil of war quickly gave way to political turmoil. Federal military occupation, the Ku Klux Klan, unrepentant rebels, newly enfranchised freedmen, and Radical Republicans acted, reacted, and interacted to make a political witches' brew of unparalleled proportions. There was martial law, intimidation, and political chicanery of all flavors. There was a new constitution, new acts of Congress to punish the rebels, blacks voting for the first time, and rapid changes in political fortunes. In 1868, with large numbers of blacks voting, with most ex-Confederate leaders disenfranchised, and with many Democrats sitting out the election, a Republican governor and a Republican legislature was elected in Alabama. The state's Senate was composed of 32 Republicans, including a single black and only one Democrat. The House consisted of 94 Republicans, including 26 blacks and three Democrats. Following the 1868 election, driven by Klan intimidation and the resurgence of Bourbon Democrats, the political world began to change quickly. Between 1868 and 1870, the Klan intimidated elected officials and potential voters, especially freedmen throughout the state (Rogers and Ward, 1994). In 1870, a Democrat was elected governor, and the Democrats controlled the legislature. The turmoil did not end there, but the old political power structure was well on its way to re-establishing itself. Phoenix was rising from the ashes.

THE NATIONAL ECONOMY

As late as 1840, 69% of the entire American labor force worked on farms, and it is likely that the percentage was even higher in the South. The country's economy was still based largely on agriculture. It would be close to the end of the nineteenth century before income from manufacturing would exceed income from agriculture. Things were beginning to change, however, around the middle of the century.

The Economic Situation

Throughout the period 1835-1860, the United States and especially the South was essentially an agrarian country with most of the total national labor force working on farms. However, beginning in 1860 and for the next decade, this agrarian economy, both North and South, would quickly be changed into military-industrial complexes.

Morison (1965) commented that Northern industry became "fat and saucy" during the war. Wartime demands stimulated production throughout the industrial economy. Initially, the war and the shortage of cotton resulted in the closing of many Northern textile mills. However, by 1862 enough of the fiber was available from states that did not leave the Union and from Union-occupied areas of the Confederacy to allow many textile mills to resume production. The loss of men from Northern factories and farms was quickly replaced by large numbers of European immigrants. The printing of large amounts of paper money and the imposition of a high protective tariff resulted in an increase in prices in the North of around 117% during the period 1860-1865. Generally, increases in wages did not keep pace with increases in prices. The rapid increase in prices during the war was quickly followed by the recession of 1866-1867. A nationwide boom in the expansion of railroads began in 1868, and in 1869 the first transcontinental link was completed.

The Tariff of 1857 is said to have caused the Panic of 1857. This panic resulted in a short-lived recession that lasted until 1858. The following year, the economy began to recover slowly. The slow recovery continued into 1861. Then in the period 1861-1865, wartime prosperity resulted in a high rate of inflation (Figure 7.3). After the war ended in 1865, the economy entered a postwar recession that generally continued through the end of the decade.

According to Morison (1865), the Civil War had a completely different effect in the South where it absorbed literally everything. By 1863, people throughout the South were already feeling the deprivations of poverty. As the war progressed, many people, especially the families of soldiers, became increasingly poverty stricken. State aid to the indigent families of men in uniform began in 1861, not long after the war began. By 1863, a quarter of the entire white population of Alabama was dependent on this aid. After 1865, the South was left without capital or currency. No Southern bank was solvent, and few stores had anything to sell. Where Sherman and Sheridan had passed, almost the entire apparatus of civilized life had been destroyed. Fortunately, almost as soon as hostilities had ceased, normal trading relations with the North and the world were quickly restored (Morison, 1865).

Otto (1994) comments that from an economic perspective, the South was poorly prepared to go to war with the North. When the war began, the Confederate States had a combined population of only one million white males of military age (18 to 45 years). In contrast, the North had 4.5 million. In 1861, there were 20,600 Southern manufacturing establishments, with a value of $96 million. There were 118,900 Northern establishments, with a value of $907 million. Further, Southern manufacturing establishments were primarily non-industrial (sawmills, tobacco factories, tanneries, textile mills, etc.). Almost none of these establishments could manufacture locomotives, steamboats, firearms, ammunition, or even wagons.

At the beginning of the Civil War, the North was able to provide ample food for its armies and for its civilian population. In contrast, farms in the Confederate States produced relatively small amounts of foodstuffs, choosing instead to use most farmland and labor for the production of export (staple) crops, primarily cotton. Farmers in Kentucky, Missouri, and Maryland produced relatively more corn, wheat, vegetables, and livestock–but with those states remaining in the Union, those commodities were not readilly available to their Confederate neighbors.

To survive a long war, the Confederacy had to continue producing and exporting cotton, tobacco, and rice. That was the only way that the Confederacy could purchase industrial items needed for the war effort and consumer goods for the Confederacy's civilian population. With its vastly superior navy, however, the North quickly blockaded all the major Southern rivers and ports. When the blockade was completed, from a military perspective the war was over. The question remained as to how much punishment the Confederate leadership was willing to inflict on its soldiers and civilians. The North, however, did not stop when they had completed the blockade. Union troops quickly moved to destroy the meager railroads in the South. Furthermore, those railroads were so poor that they played only a minor role throughout the conflict. By early 1865, less the four years after the beginning of the war, Union forces had essentially destroyed the entire transportation system of the Confederacy. After that time, there was no possibility of even continuing the war, much less winning it.

At the beginning of the Civil War, the South had no currency and few banks. Most of its wealth was tied up in slaves, land, and baled cotton. Its primary cash crop (cotton) would not be sold to Northern (enemy) factories, and as a result of the Union blockade could not be sold to its usual European markets. Further, the Confederate government initially made the decision to embargo the shipment of cotton overseas in an attempt to force foreign governments to recognize the Confederacy as a sovereign country. This strategy failed, and it resulted not only in a serious loss of income, but also in a reduction of large quantities of foreign goods needed for the war effort and the homefront. (Otto, 1994).

The Civil War destroyed most of the South's capital. Some $3.1 billion in the value of slaves was lost. Billions in bonds, notes, and currency issued during the war had been virtually worthless when they were issued, but not even a trace of value remained when hostilities ceased. About the only things left of value in the South were its farms, but they had been steeply devalued. Further, it appeared for a time that they might even lose those assets when the federal government began to try to find a solution to the problem of what to do with hundreds of thousands of freedmen who were left with nothing at all. During the war, the federal government had enacted the Confiscation Acts of 1863 and 1864 that allowed the US Treasury Department to confiscate the real and personal property of rebels. As a result, in 1865 the US Treasury transferred title of some 800,000 acres in the former Confederacy to the newly created Freedmen's Bureau. Later, President Johnson had the bureau return half of this total to the original owners.The end result of these efforts was that Southerners retained most of their private property, but the individual states lost most of their remaining public lands.

Some 1.8 million bales of cotton survived the war. After the cessation of hostilities, it was sold for extremely good prices in markets both at home and abroad, markets that were beginning desperately to need additional supplies of the fiber. A considerable amount of these cash receipts were invested in restoration of individual farms, but some of it was invested in the restoration and expansion of transportation, especially railroads. Not all Southern farmers were fortunate enough to have cotton that survived the war. Those without hoards had only one option to raise capital for the restoration of their cotton farms–sell some of their land. After Reconstruction began, there were many Northerners who wanted to become cotton farmers, especially with the profits that could be made from the sale of the crop shortly after the war ended. As a result, Northern businessmen did purchase considerable amounts of land in the South. Unfortunately, Southern farmland had depreciated so much during the war that the sale of land did not raise much capital for Southern landowners. The only other source of money was to borrow it, but there was little available for lending in the South. Much of it was provided by Northern banks and businessmen. As might be expected, however, borrowers had to pay a dear price for borrowing. For several years after the war, Southerners had to pay exorbitant interest rates on borrowed money.

The Gross National Product

Before 1889, there were no annual estimates of the US Gross National Product (GNP). The earliest estimate is a yearly average for the period 1869-1878. During that period, it was estimated that the GNP remained at a level of $7.7 billion a year in current dollars or $84.6 billion in 1982 dollars. However, in 2007, Johnston and Williamson published estimates of the Gross Domestic Product (GDP) dating all the way back to 1790. The GNP and the GDP are very similar. GPD data for the period 1790-1870 are presented in Figure 7.1. These data show that the GDP increased rapidly from $72.7 billion in 1860 to $99.4 billion in 1870.

Consumer Price Index

Although specific data required for the computation of a national Consumer Price Index (CPI) was not collected before 1913, enough economic data has been pieced together to obtain estimates beginning in 1800 (*Historical Statistics of the United States to 1970*, Tables E135-166). These estimates were computed using 1967 (1967 = 100) as the base year. Estimates for the period 1800-1860 are presented in Figure 7.2. Data presented in the figure indicate that the CPI was 27 in 1860. It remained near that level through 1861, and then began to increase in the first full year of the war, and by 1864 it was 47. After the battle for Atlanta and Sherman's March to the Sea in 1864, the steam seemed to escape from the inflationary spiral in the economy, and in 1865 the CPI declined to 44. In 1866, the postwar recession began, and the CPI began to trend downward rapidly. By 1870, it had reached 38.

The Inflation Rate

Annual CPI shown in Figure 7.2 was used to calculate annual rates of inflation for the period 1850-1870. These data are presented in Figure 7.3. The rate was equal to zero in seven of the 20 years, and negative in seven of the years. It reached levels of 23.3 and 27 in 1863 and 1864, respectively. The year the Civil War ended (1865), the economy entered a period of severe deflation and postwar recession that continued through 1870. Between 1865 and 1870, the rate was negative in five of the years and zero in one year.

THE AGRICULTURAL ECONOMY

American agriculture during the period 1840-1860 was busting out all over. In the Northeast, farmers increasingly re-oriented their farming practices to respond to the growing incomes and growing needs of an expanding urban, commercialized, and industrialized population. On the other hand, the Southern Cotton Kingdom with its slave labor was becoming more deeply entrenched year by year. Cotton production was growing at an exponential rate. However, the most significant changes in American agriculture were taking place in the Old Northwest (northwest of the Ohio River) where wheat, corn, and hog production were moving onto the prairies of western Indiana, Illinois, and beyond. This movement lead to the establishment of the wheat, corn, and hog belts. On April 12, 1861, the second revolution on American soil in less than a century began. On that fateful day, the three farming systems (Northeast, Old Southwest, and Old Northwest) were thrown into the white hot crucible of a conflict between brothers. The conflict would forever change American agriculture, especially in the South.

Federal Agricultural Legislation

The newly-created (1850) Republican Party understood that their margin of victory in the national election of 1860 came primarily from the rapidly expanding farming population in the Old Northwest (North Central

Region). Recognizing this important contribution of farmers to the party, the Republicans committed themselves to agrarian reform. As a result, in his first address to Congress Lincoln called for the establishment of an agricultural and statistical bureau within the federal government. In 1862, he established the independent Bureau of Agriculture to be headed by a commissioner without cabinet status. Further, during the period 1860-1870, the Republican Congress enacted several pieces of extremely important legislation pertaining to agriculture:

The Department of Agriculture Act of 1862. Without question the most important legislation of this period was the act establishing the US Department of Agriculture. Although the act did not provide the department with cabinet status, it did firmly underscore the importance of agriculture to the national economy. Soon after signing the bill establishing the department, Lincoln named Issac Newton as its first commissioner. In the beginning, the department consisted of nine employees plus the commissioner. Fortunately, one of the employees was a statistician, and he immediately began to set in motion the process of collecting data on many aspects of agriculture. By 1863, the collection of statistics had become so important that a division of statistics was established within the department. Data were collected through the war years within the states of the Union, but statistics on a national-level could not be obtained until after mid-1865. In 1866, the department published its first national data.

Homestead Act of 1862. The Democratic Party had been working tirelessly since the 1830s to encourage the federal government to provide free land to settlers. In 1862, the Republican Congress enacted the Homestead Act, and the Republican President Lincoln signed it into law. The act provided 160 acres of public domain free to any person who was the head of a family or who had reached his majority, and who was an American citizen or who had filed intentions of becoming one. After living on the land, cultivating any portion of it for five years, and paying a small registration fee, the homesteader received title to the land. The homesteader might also purchase the land for $1.25 an acre after living on the land for six months. For many years, the government continued to sell land for cash, and even though land could be obtained free as a result of the act, more land was obtained by direct purchase than by homesteading (Morison, 1965).

Morrill Land Grant of 1862. This act provided grants of land from the public domain to each state. Proceeds from its sale were to be used for the endowment, support, and maintenance of at least one agricultural and mechanical college in each state. These colleges were to be for the teaching of science rather than for the teaching of the classics. They were expressly for the teaching of children of farmers and other working people.

Confiscation Acts of 1861 and 1862. These acts were passed primarily to "suppress insurrection and rebellion" and to seize property of "Confederates and their abettors." Very little property was actually seized under the provisions of these acts, and virtually all of it was returned shortly after the war ended.

Southern Homestead Act of 1866. In 1866, Congress approved an act expressly designed to provide free land from the public domain to the thousands of freedmen. Unfortunately, at that time there was very little public land with commercial agricultural potential remaining in the South. In Alabama, some 16,284 homestead applications were made by freedmen for land under the provisions of the act. However, only 6,293 titles were approved. Altogether, 721,000 acres were transferred, most of it completely unsuitable for farming (Rogers and Ward, 1994). Congress repealed the act in 1876.

Alabama Agricultural Legislation

Probably the most far-reaching legislation enacted by the Alabama legislature during this period dealt with the question of the crop lien. Share tenants before the Civil War had repaid their loans by sharing their crops with the lender. Unfortunately, in many cases, the shared crop was not sufficient to repay the entire loan. As a result, over time, lenders began to look for a way to increase their chances of recovering the entire amount. The crop lien system evolved out of this unfortunate situation. Under this system, the lender(s) took a lien on the farmer's crop. If, at settling-up time, he was unable to repay the loan, the lender could take his livestock or his entire crop, or anything he owned, if necessary, to meet the borrower's obligations. In this process, the lender might obtain three-thirds of the farmer's crops, instead of the quarter, third, or half share that he could take under older share-tenant agreements. In 1866, the crop lien system became even more formalized when the legislature passed a law making the system legal. The legislation did not completely satisfy any of the parties involved, and as a result it would be revisited by the legislature many times over the next three decades (Atkins, 1994).

Free-roaming livestock had been of some concern to the legislature before the Civil War. Some forward-looking farmers were beginning to try to improve the quality of their livestock (especially cattle) by purchasing purebred stock. Unfortunately, there was little hope of making much progress as long as the half-wild cattle of the herders roamed at will across the state. As a result, in 1866 the legislature took the first step in what would be a long and bitter battle to close the range (Blevins, 1998). In that year, the Canebreak Agricultural District was created. It consisted of the Black Prairie counties of Dallas, Greene, Marengo, and Perry. Running at-large livestock was forbidden in those counties.

Agricultural Exports

Before 1900, there are no annual data on the value of US agricultural exports. Previously, only annual decade averages are available. For example, in the decade 1800-1809, average annual US exports were valued at $23 million. Averages for the other decades through the 1860s are presented in Figure 7.4. Data presented in the figure indicate that the value of average annual agricultural exports during the years of the Civil War and immediately thereafter was $182 million. It had been at a level of $189 million during the decade of the 1850s. The decline was certainly the result of the reduction in the exports of cotton during the early years of the period. Immediately after the end of the war, the export of cotton was resumed. Also, beginning in the early 1860s, Europe, which suffered through a series of poor harvests, began to export large quantities of American grain. In 1862 alone, Europe imported 40 million bushels of wheat (grain and flour), whereas in 1859 they had imported only 100,000 bushels (Morison, 1965). These data show the importance of the export of US cotton. From 1815 through 1860, cotton was the country's most important agricultural export, as well as probably the most important single item of all exports.

In the 1860s, the export of the products of American agriculture comprised 75% of total exports. In the 1850s, the percentage was higher (81%). Of course, the reduction of cotton exports from 1861 through 1865 was responsible for this decline.

Agricultural Prices

There are no published annual data on general agricultural prices before 1910. As early as 1866, however, there are annual data on "prices received" for some crops, such as corn, wheat, and oats. The first annual data on prices received by farmers for cotton was not published earlier than 1876. From 1866 onward, annual data on the market price for cotton in New York is available. There are no annual data on prices received for any livestock in the US or Alabama before 1924. There are annual data on farm value per head for cattle and hogs in the US and Alabama from 1866 onward.

Farmers

At the end of the Civil War, agriculture in the North was even stronger than it had been at the beginning of the war. The loss of manpower to military service had little effect on production, and the war effort enhanced the potential for distributing it where it was needed. It was a far different story, however, in the South. Rogers and Ward (1994) commented that following the war Alabama's agriculture was left in a "shambles of loss and confusion." On those farms that had been operated with slave labor there was "vast uncertainty." Many of the plantation owners had exhausted their personal wealth in the vain attempt to finance the war effort, and virtually all of them had lost a large percentage of their wealth as a result of the emancipation of their slaves. Many of those in the yeoman farming class, diminished in number by the war, had lost what few assets they had. Fields had been abandoned. Improvements that they had struggled to make were lost as a result of inattention. Much of their equipment was broken or missing, and losses of livestock were especially severe. Another important consequence of the war was that it lowered the economic scale of the South's agriculture, so that many farming units that had been only marginally viable before the war now fell below the level of viability.

Decennial censuses began to include data on farms in 1840, and the Division of Statistics of the US Department of Agriculture began to publish annual data on plant and animal agriculture in 1866; however relatively little data was published on the farmers themselves.

The Total Population

In 1850 and 1860, the US population totaled 22.3 and 31.4 million, respectively (Table 7.1). From 1860 to 1870, it continued to grow at a rapid rate, reaching 38.2 million persons at the end of the decade. Increases in the North Atlantic and North Central regions were responsible for most of the growth. In the North Atlantic Region, most of the growth occurred in just two states, New York (4.4 million) and Pennsylvania (3.5 million). In the North Central Region, growth was more widely distributed. There, a total of seven states had populations above one million, and there were more than 2.5 million in Ohio and Illinois. In the war-torn South the population changed relatively little during the period. In the South Atlantic Region where hostilities were more concentrated, the population declined slightly (5.29 to 5.28 million). It increased by about 665,000 (11.5%) in the South Central Region.

In Alabama, the total population increased from 964,000 in 1860 to 997,000 in 1870. The Census of 1860 indicated that there were around 435,000 slaves and slightly less than 2,700 free blacks in the state. The 1870 Census counted 475,500 blacks. The 1860 data indicated that the slaves were concentrated in the Cotton Arc in and around the Black Prairie (Figure 6.30). The 1870 data indicated that the spatial distribution of Alabama blacks remained essentially the same.

The Farm Population

The USDA estimated that in 1850, 1860, and 1870 the farm population in the country totaled 11.7, 15.7, and 18.4 million, respectively. Data published by USDA also indicate that in those years the percentage of farmers in the population was 50.4, 48.2, and 47.6%, respectively. The number of farmers was increasing, but not as rapidly as the total population. The ratio of total persons:number of farmers was 2 in 1850, 2.1 in 1860, and also 2.1 in 1870. Farmer efficiency was still very limited. They were not able to provide food and fiber for their families and very many additional people.

There are no estimates available on the farm population; however, given the significant role that agriculture played in the state's economy during that period, it is likely that farm families comprised a larger percentage of the total population than in the country as a whole.

The number of men on farms in the South declined rapidly after the war began. For example, Atkins (1994) comments that by 1862 some 17% of all white Alabama males were in military service, and two years later the number exceeded the state's voting population. In Shelby County by 1862, only 300 men remained of the 1,600 who had been there in 1860. Atkins also commented that while records do not exist to correctly estimate the number of white Alabama men who served in the Confederacy, it can be cautiously estimated at between 90,000 and 100,000. Because such a high percentage of the state's population was involved in production agriculture, most of the loss of manpower to military service resulted in a significant reduction in the production of food and fiber. Atkins noted that in Coosa County, it was estimated that the loss of 800 men could be translated into the loss of 15,000 acres of cultivated land.

Farmers in the Work Force

USDA estimated that farmers comprised 69, 64, 58, and 53% of the labor force in the country in 1840, 1850, 1860, and 1870, respectively. These data, showing the decline in the percentage of farmers in the labor force, demonstrate the effect of rapid growth of manufacturing and urbanization on non-farm employment during this period. Even with the number of farms increasing rapidly, non-farm employment had to be growing at an even faster rate. Most of this increase in non-farm employment had to take place mostly in the East. During this period, there was only a very limited amount of growth in Southern manufacturing.

Tenancy

Before the Civil War, it had become increasingly obvious to yeoman farmers in the South, with their small acreages of subsistence food and fiber crops, that they had no choice but to enter the market economy of staple (cotton) agriculture. Unfortunately, many of them did not have the necessary resources to make the move. With all of the problems that cascaded down on them after the war, the need to make a change from subsistence to market agriculture became more pressing. This need seemed obvious to many of them before the war; however, economic conditions resulting from the war accelerated it. Some of these farmers had been giving up yeoman-status for tenancy before the war, but conditions after the war made the need even more pressing.

While after the Civil War significant numbers of white yeomen became tenants by choice; the hundreds of thousands of black freedmen had no other option. Work-for-wages had not worked very well for them. They appreciated receiving a direct monetary reward for their labor, but wages did not provide them with a house, a garden plot, or access to livestock. Wages did not provide them with a homeplace. Also, after it became obvious that the federal government was not going to provide many of them with 40 acres and a mule, thousands entered into tenancy arrangements with landowners or landowner-merchants. There was really nothing new about this system. As discussed in some detail in Chapter 5, similar arrangements had been the norm throughout Europe for centuries.

Some of the tenants were able to pay cash for the land and buildings that they used. These were the so-called cash tenants. However, by far the largest portion of tenants never accumulated enough cash to pay rent. The large majority (especially freedmen) paid their rent by sharing the crops they produced with the landowners. These were the share-croppers or croppers as they were most often called. The amount of a particular crop (almost always cotton), shared depended on what the owner supplied for the tenant and his family. If he supplied land, buildings, equipment, and mules, he might demand as much as two-thirds of a crop. Specific arrangements on crop-sharing between owner and tenant varied widely across the South.

Tenancy was not the best solution to the problems of either landowners or their tenants, but the arrangement did work to bring land and labor together and to restore food and fiber production on larger Southern farms and plantations (Fite, 1984). With all of its inherent problems, tenancy grew rapidly, and by 1880 some 26.5% of all farms in the United States and 46.8% in Alabama were being operated by tenants. There are no earlier data available. In the country, 17% were cash tenants and 30% were share tenants.

Slaves

In 1860, there were 3.9 million slaves in the United States (Table 6.6). The total had been 3.2 million in 1850. Some 46.5% and 50.6% of the total was in the South Atlantic and South Central regions, respectively. There was also a substantial number in the North Central Region, but almost all of those were in Missouri. In the South, the slave population was between 400,000 and 500,000 in Virginia, South Carolina, Georgia, Alabama, and Mississippi.

In 1860, there were 437,000 slaves in Alabama, ranking the state fourth in the country for slaveholding. Some 5.9% of Alabama's slave population lived and worked in Dallas County. Other leading counties included: Marengo (5.6%), Montgomery (5.4), Greene (5.4), and Lowndes (4.4). These five counties, were home to 26.8% of Alabama's total slave population. That the top five counties would account for only 26.8% of the total indicates how widespread slave ownership was in the state at the end of the antebellum period. A total of 16 Alabama counties had more than 10,000 slaves each. Most of these counties were clustered in and around the Cotton Arc (Figure 6.30).

Freedmen

It is difficult to fully comprehend the catastrophe visited upon hundreds of thousands of slaves on Southern farms who were "freed" in principle by Lincoln's Emancipation Proclamation of 1862 and by actual emancipation at the end of the war in 1865. Slavery itself was catastrophic, but the sudden release from this relationship, especially into the hostile environment of the war-ravaged South, was likely even more catastrophic. According to the Census of 1860, there were almost 437,000 slaves in Alabama. There were similar numbers in Virginia, South Carolina, Georgia, and Mississippi. Almost overnight, these unfortunate people were suddenly cast adrift from the only life and society (as unfortunate as it was) that they had ever known.

Fite (1984) comments that the great majority of freedmen had to quickly move from slavery to freedom with no money or capital of any kind. They had nothing marketable except their labor and considerable experience in just surviving. They owned no land, no housing, no household goods, no transportation, no equipment, no livestock, no seed, little clothing, and almost no food. The only life that most of them knew was producing cotton, and the only future they had was farming, but to become farmers they had to somehow obtain land and the wherewithall to farm it and to survive until they could harvest their first crop.

The Union victory in the Civil War and the enactment of the 13th Amendment, afterwards, technically made freedmen out of slaves, but neither of the events made any provision for their livelihood in their new world. The Union victory, in effect, made the freedmen wards of the federal government with undetermined status and a dubious future (Morison, 1965). Congress, anticipating ultimate victory by Union forces, recognized the federal responsibility for the former slaves by creating the Freedmen's Bureau within the War Department in March 1865. The bureau was given general powers and guardianship over all refugees. Morison (1965) comments that it performed wonders in relief, but little in social adjustment. It issued emergency rations, established hospitals, encouraged former slaves to return to the farms where they had lived and to work for agreed wages, restored thousands of white refugees to their homes, established courts to adjudicate disputes between employers and employees, and established the first schools that African Americans had ever had in the South.

Emergency rations, hospitals, and schools were extremely important measures, but they did not deal with the problem of the freedmen's long-term future. In recognition of this inadequacy, in 1866 Congress enacted the Southern Homestead Act for the express purpose of providing land for freedmen. Despite good intentions, few former slaves ever applied for these land grants. In late 1869, only about 4,000 applications had been filed. Most of the land available was part of the public domain, and most of it had remained in that status because of its poor quality. Further, most of it was quite remote. Very little of these public lands was anywhere near where the freedmen lived. Coupled with the remoteness and poor quality of the available land was the fact that most freedmen did not have the capital necessary to establish even a modest farm and live there until they could harvest and market their first crop. Later Congress repealed the act (Fite, 1984).

Despite the uncertainty of the times, most former slaves remained on their home farms. However, substantial numbers initially moved to the nearest small town where they sought non-farm employment. Unfortunately, in the shattered economy of the rural South, there was little non-farm work available. Further, to compound the problem, there was no place for them to live. Most ultimately returned to their homes. During the same period, thousands of freedmen moved westward into Mississippi, Arkansas, and Texas, into areas that had been somewhat less adversely affected by the war.

Landowners first attempted to employ freedmen for wages. Unfortunately, the farmers had very little cash available, and the workers wanted to be paid when they worked, rather than waiting for the crop to be sold. On the urging of the Freedmen's Bureau, beginning in 1866 many former slaves agreed to work in the fields under contract. Some contracts were based on the payment of wages, and other contracts on a combination of wages and a share of the crop. Over time, many former slaves moved away, and as a result landowners had to offer better contracts to attract the labor they required. Contracts involving wages never worked very well for either the employer or employee. Consequently, over time, many of them also moved inexorably toward tenancy.

Farms

Characteristics of American farms were changing rapidly in the mid-nineteenth century. In the North and West, the fast pace of change continued right through the Civil War. Characteristics of farms in the South also were changed by the war, but the nature of the changes were quite different.

Farm Numbers

Data obtained by the censuses on the total number of farms in the United States, its regions, and Alabama in 1850, 1860, and 1870 are presented in Table 7.2. The total number increased from two million in 1860 to 2.7 million in 1870, an increase of 30.1%. The westward movement of farms in the North had halted at the edge of the Great Plains until 1850—as a result of the federal government's policy on settlement on lands held by Native Americans. However, from 1850 onward there was a steady flow of settlers across the Missouri and Ohio rivers. This westward movement was not slowed appreciably by the Civil War. Sometime between 1850 and 1860, the number of farms in the North Central Region had grown larger than in the North Atlantic Region, and by 1870 there were twice as many in the Old Northwest. In fact, by 1870 slightly more than 42% of all farms in the country were west of the Ohio River.

Although total numbers were growing rapidly west of the Ohio, there were more farms in New York in 1860 and 1870 than in any other state. Farm numbers increased in the South Atlantic and South Central Regions between 1860 and 1870, by 72,162 and 140,625, respectively. It is likely, however, that much of this increase came as a result of subdividing larger farms into smaller tenant farms. For example, in the North Atlantic Region, there was a substantial increase in numbers from 1860 to 1870 (565,000 to 602,000). Numbers also increased in the South Central Region (370,000 to 602,000)..

Census data showed that farm numbers also increased sharply in Alabama (Table 7.2) during the war years and immediately thereafter. In 1860, there were 55,100 farms in the state. By 1870, the number had grown to 67,400, an increase of 22.3%. There had been 42,000 here in 1850. The increase in Alabama had to be the result of either establishing new farms on poorer lands that had not been farmed before, or by subdividing existing farms so that there would be two or more farms where there had only been one before. As will be obvious, in a following section, the increase in Alabama, came primarily as a result of the subdivision of existing farms through an increase in tenancy.

In 1860, a substantial share of Alabama's farms were clustered in the east-central region, from Cherokee County in the north to Chambers in the south (Figure 6.36). Some 4.4% of all farms were located in Randolph County. Other counties with substantial numbers of farms included: Tallapoosa (3.9%), Pike (3.4), Calhoun (3.3), and Coosa (3.2). In 1870, as a result of the subdivision of larger farms (plantations), the distribution of farms numbers had changed significantly. That year, the largest share of farms was in Wilcox County (4.8%). Other counties with substantial shares of farms included: Madison (4.1%), Lawrence (3.0), Lauderdale (2.7), and Coosa (2.7).

Farm Size

Census data on the average size of American farms in 1850, 1860, and 1870 are presented in Figure 7.3. The data indicate that average size in the country remained essentially unchanged between 1850 and 1860 (202.6 versus 199.2 acres). However, by 1870 acreage had declined sharply to 153.3 acres. There was some decrease in all regions, but these were relatively slight, except in the South. During the 1860s, farm size declined 46.5% in the South Atlantic Region (353 to 241 acres) and 91.2% in the South Central Region (371 to 194 acres).

Census data on average size of Alabama's farms in in 1850, 1860, and 1870 are presented in Table 7.3. The data show that average size increased sharply from 1850 to 1860 (289.2 versus 346.5 acres). However, it fell even more sharply from 1860 to 1870 (346.5 versus 222 acres). This decline was the result of the combination of the rapid increase in the number of farms (Table 7.2) and the abandonment of four million acres of farmland (Table 7.4).

Improved Farmland

In 1860, some 40% of all America farmland was considered to be improved. In 1850, the percentage had been 38.3%. Throughout the war years, farmers in much of the United States continued to improve more and more of their farmland. By 1870, a total of 200 million acres was considered to be improved, an increase of 22% over 1860. This general increase occurred nationally, even though quantities in the South Atlantic and South Central regions declined (Table 7.4). Given the problems of attempting to operate farms, especially Southern farms, during the war, it is surprising that improved farmland did not decline even more. There were significant increases in all of the other regions, indicating general positive benefits of the war. Improved land in the North Central Region actually increased by 25 million acres, an increase of 49%.

Census data on the total acres of improved Alabama farmland in 1850, 1860, and 1870 are presented in Table 7.4. These data indicate that there was 6.4 million acres in the state in 1860. There had been 4.4 million in 1850. In 1870, there was 5.1 million. Alabama farmers had abandoned (improved to unimproved) some 20% (1.3 million acres) of their improved farmland between 1860 and 1870, even though the number of farms was increasing (Table 7.2). Farmers in the South Central Region, which includes Alabama, abandoned only about 6%.

Value of Land and Buildings

Census data on the total value of land and buildings on American farms in 1850, 1860 and 1870 are presented in Table 7.5. Value was $6.6 billion in 1860. It had been slightly less than $3.3 billion a decade earlier (1850). The larger share of these values is the value of land itself. The value of buildings, fences, and other improvements represents a smaller share. Total value increased from $6.6 to $7.4 billion (11.5%) in the United States between 1860 and 1870. During the period, value increased substantially for farms in the North Atlantic and North Central regions, although much of this increase was the result of wartime inflation. The Consumer Price Index had increased sharply during those years (Figure 7.2). Between 1860 and 1870, it increased from 27 to 38, an increase of 40.7%. Of course, this 1870 total value included those severe declines in the South Atlantic and South Central regions of almost 50%. These two regions included all of the Confederate States. This decline in total value show clearly horrendous effects of the Civil War on agriculture in the South. The war resulted in the loss of all of the money that Southern farmers had invested in slaves; plus the loss of about half the value of their farm real estate. In purely economic terms, these losses were catastrophic.

In the North Central Region, total value increased from $2.1 billion to $3.4 billion, or by 62%. The rate of increase in the North Atlantic Region ($2.1 to $2.5 billion) was less than the increase in the CPI during the period (19 versus 41%).

With changes in the number and average size of farms, raw data on the value of land and buildings do not accurately reflect changes in the investment in those farm elements. The only practical way to evaluate these changes is to relate value to acreage. For example, in 1860 the value of land and buildings in the United States was $6.6 billion on 407.2 million acres of total farmland, or $16 an acre. This conversion is based on total farmland rather than improved farmland. Improved farmland would likely be the better variable, but data on this characteristic is not available after 1920.

By 1870, value of land and buildings per acre for the country had increased to $18. This national total includes values from farms in the North that benefited economically from the war and farms in the South that were devastated by it. According to census data, between 1860 and 1870 in the North Atlantic and North Central regions, value per acre increased from $35 to $40 and from $20 to $25, respectively. In contrast, in the South Atlantic and South Central regions, value per acre declined from $9 to $7 and from $11 to $7, respectively.

Census data on the value of Alabama's farmland and buildings in 1850, 1860, and 1870 are presented in Table 7.5. In 1860, census data indicated that this total value for Alabama farms was $175.8 million. It had been $64.3 million in 1850. Then, as a result of the war and its aftermath, this value declined to $54.2 million in 1870. This decline came at a time when the annual rate of inflation was rampant (Figure 7.3).

As in the case with the national data, a much clearer

picture of changes in the total value of farm real estate in the period 1860 to 1870 is available by comparing value per acre. In 1860, total value on a per-acre basis totaled $9 on Alabama farms (Table 7.5). A decade later, it was down to $4. It had been $5 in 1850. As will be seen in a following chapter, the bleeding of the value of land and buildings did not end in 1870. The loss plagued Southern and Alabama farmers for decades.

Value of Implements and Machinery

Census data on the total value of implements and machinery on American farms in 1850, 1860, and 1870 are presented in Table 7.6. Before the Civil War began, total value was $246.1 million. It had been near $152.6 million in 1850. The 1870, data indicated that the total value declined to $101.3 million during the decade. However, as was the situation with the total value of land and buildings, the total value for the country as a whole was biased downward by the steep decline in the South during and after the war. In the North Atlantic and North Central regions, total value in the same period increased from $73.8 to $89.7 million (21.5%) and from $72.8 to $123.6 million (69.8%), respectively. In the North Central Region, the rate of increase was greater than the rate of increase in the Consumer Price Index (69.8 versus 40.7%). It was lower in the North Atlantic Region (21.5 versus 40.7%). In contrast, in the South Atlantic and South Central regions, total values actually declined from $34 to $20 million and from $61.3 to $29.8 million, respectively.

A somewhat clearer picture emerges from these data if total value is related to number of farms (Table 7.6). For example, in 1860 all of the implements and machinery on two million American farms (Table 7.2) was valued at $246.1 million, or $120 per farm. By 1870, however, value per farm in the country had actually declined to $101. This overall decline was primarily the result of sharp decreases in the South Atlantic ($113 to $54) and South Central ($165 to $58) regions. These decreases offset increases in the North Atlantic ($132 to $149) and North Central ($94 to $110) regions.

Census data presented in Table 7.6 indicate that the total value of implements and machinery on Alabama farms in 1860 was $7.4 million. It had been $5.1 million in 1850. However, between 1860 and 1870 total value had declined by 65% to $2.6 million.

Value per farm on Alabama farms declined from $135 in 1860 to $39 in 1870. It is difficult to imagine how a farmer could operate a farm on an average of $39 dollars worth of implements and machinery. The $39 per farm represents an average for all farms. A substantial number apparently were operated with considerably less than that amount. Unfortunately, this was the result of the conditions afflicting a majority of Southern farmers as a result of starting a war and then losing it.

Animal Power on Farms

During the period 1860-1870, the power required in farming–for plowing, planting, cultivating, and harvesting crops and for providing transportation–was supplied primarily by draft animals, either mules and asses, horses, or working oxen. In this regard, little had changed since 1800. Of course on smaller farms, especially in the South, a substantial amount of muscle power was still being supplied by the farmers and their families.

In 1860, there were 9.7 million draft animals of all kinds in the United States (Table 7.7). There had been 6.2 million in 1850. The Census for 1860 reported that there were 6.2 million horses (64.7% of all draft animals), 2.2 million working oxen (23.3%), and 1.2 million mules and asses (11.9%) in the country. Horses were much more abundant than either mules and asses or working oxen, and surprisingly, there were more working oxen than mules and asses.

In 1860, the largest share of all draft animals was on farms in the North Central Region (35.4%). Shares in other regions were: South Central (29.3%), North Atlantic (18.2), South Atlantic (13.9), and Western (3.2). There were also large differences in the percentages of the different kinds of animals in the different regions. For example, the percentage of horses was highest in the North Central Region (74.3%) and lowest in the South Central Region (50.7%).

As a result of the war, the total number of draft animals in the country declined slightly from 9.7 million in 1860 to 9.6 million in 1870. Total number increased substantially in the North Central (3.4 to 4.2 million) and Western (305,000 to 373,000) regions, but declined slightly in the North Atlantic Region (1.8 to 1.7 million). Losses were much greater in the South Atlantic (1.3 to 1 million) and the South Central regions (2.8 to 2.3 million). The total number of horses in the country increased substantially from 6.2 to 7.1 million. The total number of mules and asses remained at near the same level (1.2 versus 1.1 million), while the total number of working oxen declined sharply from 2.2 to 1.3 million.

By 1870, the North Central Region increased its share of all draft animals from 35.4 to 44.1%. Shares in the other regions were: South Central (23.8%), North Atlantic (17.3), South Atlantic (10.8), and Western (3.9). Horses also became more important in all of the regions except the South Atlantic. In the North Central Region, the percentage of horses increased from 74.3 to 85.5, and in the South Central Region it increased from 50.7 to 57.3.

In 1860, there were 322,000 draft animals in Alabama. There had been 254,900 in the state in 1850. Horses were also more abundant in Alabama, as they were in the country; however, they were not as important here as in the country as a whole. There were 127,100 in the state (39.4% of all draft animals). Mules and asses were much more abundant than working oxen, 111,700 (34.7%) versus 88,300 (25.9%), respectively.

The war resulted in a much greater loss of all draft animals in Alabama than in the United States as a whole. In the country, the reduction was equal to only 0.6% of the total. In Alabama, however, the total declined from 322,000 in 1860 to 217,000 in 1870, a decline of 32.7%. In the state, the reduction was essentially the same in all three kinds of animals. In 1870, there were 81,000 horses (37.3% of all draft animals); 77,000 mules and asses (35.4%); and 59,000 working oxen (27.3%).

Use of Fertilizer

Data published by Lerner (1975) indicated that commercial fertilizer consumption was 106 million and 328 million pounds in 1850 and 1860, respectively. The same publication indicated that in 1870, American farmers applied 642 million pounds to the land on their farms. These totals would translate into rates of application an acre of improved land of: 1 pound, 2 pounds, and 3.2 pounds in 1850, 1860, and 1870, respectively. It is somewhat surprising that the rate of use did not decline between 1860 and 1870; however, it is likely the South used such a relatively small percentage of the national total that there could not have been much reduction during the war.

PLANT AGRICULTURE

Plant (crop) agriculture in the United States had increased rapidly in the period 1800-1860. According to census data, American farmers added 50 million acres to their improved farmland (primarily cropland) base between 1850 and 1860. During the same period, 11 million acres of improved farmland were added in the South Central Region, and two million acres were added in Alabama.

All agricultural crop data published in the censuses from the 1840 Census onward were obtained the year before the data were actually published. As a result, all crop data are reported for the years 1859 and 1869. Between 1860-1870, the amount of data collected on crop production for the census was very limited.

The first series of national, annual crop data collected by the newly-established Division of Statistics within the US Department of Agriculture was published in 1866. As a result, from 1870 onward, at the beginning of each decade there are two collections of published statistics on agriculture in the country. During this period, data were collected by two different organizations using different sampling procedures. As a result, the two estimates are seldom the same. For example, the 1870 Census (1869 data) reported that cotton harvested in the United States totaled 3,012,000 bales, while the Division of Statistics reported that it was 3,013,000. Here the difference was relatively small; only 1,000 bales out of a three-million bale crop. However, for some of the crops the difference is much greater, although generally never great enough to affect major conclusions about the status of American agriculture. Even though the division's data are available from 1866 onward, throughout this section the census data will continue to be used. The early data collected by the Division of Statistics are now included in the data base maintained by the National Agricultural Statistics Service (NASS).

In discussing the evolution of the production of the different crops in the following sections, data on exports and prices will be considered first. In most cases, the number of acres that a farmer planted of a given crop in a year was closely related to the price that he received for his previous year's crop. Further, the price he received for the previous crop was strongly influenced by demand, and with many of the crops, exports were a major element of demand.

Cotton

As the Civil War progressed, the Confederate government made a determined effort to get cotton producers to forego the production of that crop and to plant provision crops (corn, wheat, sorghum, peas, beans, potatoes, melons, sugar cane, etc.). Although some plantation owners secretly continued to grow a little cotton, the production of that crop declined sharply. When the war ended in April 1865, cotton was selling at 87 cents a pound. Regardless of the unusually high price, only a few fortunate farmers were able to get a crop planted in that year. Many fields had already been planted in food crops, and there was a shortage of viable cotton seed. Those that were able to plant cotton had considerable difficulty in finding the necessary labor to chop, hoe, and pick it. The 1865 crop in the states of the former Confederacy amounted to 500,000 bales. In 1859, those same states had harvested 5.3 million bales.

In 1866 and 1867, Southern cotton farmers were beginning to recover from the war and to get larger acreages planted with the crop. Unfortunately, in both years the crops were decimated by bad weather and insect depredation. The 1866 crop amounted to two million bales, and in 1867 farmers harvested 2.4 million bales. By 1869, bales harvested increased to three million. Fortunately, farmers received 29 cents a pound for the 1869 crop. They had received only 12 cents in 1859. As a result, although the 1869 crop was not as large as in 1859, they received considerably more money for it ($395.4 versus $285.4 million).

Exports

Cotton exports had trended slowly upward from 1820 through 1859, then in 1860 they increased sharply to between 1.7 and 1.8 billion pounds (Figure 6.37). After the beginning of the war, no cotton could be shipped from Southern ports. Although there are no data available, exports from the Confederacy were certainly reduced significantly by the Union blockade. Further, the Confederate government attempted to halt the flow of cotton to European markets in an effort to force those countries to support the Confederacy during the war. As a result, there was very little cotton was exported; in fact, it is surprising that any was exported. Yet, throughout the war years, some Southern cotton was exported by running the Union blockade, but the amount declined as the interdiction intensified. After mid-1865, national data collection resumed, and in 1866 the United States exported 651 million pounds. A significant share of these exports were likely cotton that had been accumulated on Southern farms during the war.

From 1866 through 1870, exports were severely limited by production, as the machinery of cotton production was heavily damaged by the war. It would be a number of years before it could be repaired so that production could return to its prewar level. Because poor environmental conditions (weather and insects) severely limited production in most of the latter part of the 1860s, cotton exports only reached 959 million pounds in 1870, far below where it had been in 1860.

Prices

The average price of cotton generally trended downward throughout the first half of the nineteenth century to reach eight cents a pound in 1849 (Figure 6.38); however, just before the beginning of the Civil War, prices had begun to increase. It had risen to 12 cents a pound by 1859. Then, as a result of growing worldwide demand and reduced production of the crop during the war, prices increased sharply to an average of 29 cents in 1869 (Figure 7.5). Had it not been for this price increase, cotton farmers would have been in an even worse financial condition than they were immediately after the war.

US Cotton

Census data on bales of cotton harvested in the United

States in 1839, 1849, 1859, and 1869 are presented in Table 7.8. The data for 1839 and 1849 are presented for comparative purposes. In 1859, American cotton farmers harvested 5.4 million 445-pound bales. In 1849, they had harvested 2.5 million 400-pound bales. Farmers in Mississippi harvested the largest share of the 1859 crop (22.3%). Other states harvesting large amounts of the total crop included: Alabama (18.4%), Louisiana (14.4), Georgia (13.0), and Texas (8.0). These five states harvested 76.2% of the total American crop that year. These data show that the South Atlantic and South Central regions produced almost all of the cotton in the country in 1859. Farmers in the North Central Region harvested only some 42,731 bales, most of it (96%) in Missouri. Farmers in Illinois and Indiana harvested very small shares of the total.

The data in Table 7.8 also show that bales harvested declined from 5.4 million in 1859 to 3 million in 1869, a decrease of 44%. In some states, the decline in bales harvested was much greater than the national average. For example, Mississippi farmers harvested 53.5% fewer bales in 1869 than they had before the war began (1859), and the decline was even greater in Louisiana (55.4%).

In 1869, the same states that had produced most of the country's cotton before the war continued to do so afterwards. Mississippi still lead with 18.2% of the total; however Georgia (15.3%) replaced Alabama (13.9%) in second place. Louisiana and Texas rounded-out the top five, each with 13.3%.

The decline of Southern cotton production during the war years was actually worse than the data indicate. By the time the 1869 crop was tallied, production was beginning to recover. Otto (1994) commented that in 1864 it was estimated that farmers in the Confederacy harvested 299,000 bales, or 6% of the quantity they had harvested in 1860. Some of this reduction was the result of severely limited marketing opportunities because of the Union blockade of Southern ports and of the Confederate government's effort to halt the flow of cotton to European markets. However, the primary reason for the reduction was probably the effort to get farmers to produce food crops (corn, beans, peas, potatoes, fruit, melons, etc.) rather than cotton. Although the decline in bales harvested was substantial, the increase in the average price during the same period more than offset that loss. The average price increased from 12 cents a pound in 1859 to 29 cents in 1869 (Figure 7.5), an increase of 142%.

The Division of Statistics of the US Department of Agriculture reported that cotton production reached 4.4 million bales in 1870, some 80.8% of the 1859 level. In just five full crop years, Southern cotton production recovered from its state of devastation to reach a level of slightly less than where it had been before the war. Given the difficulties farmers had to contend with, this was a phenomenal accomplishment. Farmers did not realize it at the time, but this ability to increase production so rapidly would soon become the bane of their very existence. It was evident at this juncture that Southern cotton farmers likely had the capability to increase production much more rapidly than the increase in demand.

Alabama Cotton

Data on the number of bales of cotton harvested in the United States, its regions, and in Alabama in 1839, 1849, 1859, and 1869 are presented in Table 7.8. The data for the different years are not exactly comparable as a result of differences in the weights of bales in the different periods (Table 6.15).

Alabama cotton farmers ginned 989,955 bales of cotton in 1859. Data presented in Figure 6.41 show that cotton production in the state at that time was concentrated in the Cotton Arc, most of it west of Macon Country. The largest share of the total, 63,410 bales (6.4%) was ginned in Dallas County. Other counties with large shares of the total included: Marengo (6.3%), Montgomery (5.9), Greene (5.8), and Lowndes (5.4). Together, these five counties accounted for 32.9% of the state's total. Eleven counties ginned more than 30,000 bales. Farmers in the western portion of the Tennessee Valley section also harvested a considerable amount of cotton in 1859, but production there was not nearly as high as in the Cotton Arc.

The war years also exacted a severe toll on cotton harvest in Alabama. In 1869, Alabama farmers harvested 429,500 bales, a decline of 57.1% from 1859. The decline of the state's cotton production in relative terms was much greater than in the country as a whole and slightly greater than the other major Southern cotton-producing states. Further, the reduction in harvest was generally greater in Alabama's primary cotton-producing counties as a whole. The decline (1859 to 1869) in Dallas, Marengo, Montgomery, Greene, and Lowndes counties was 60.9, 62.2, 56.7, 82.8, and 65.8%, respectively.

Corn

Maize or Indian corn quickly became the most important food crop produced in England's North American colonies, and it accompanied the westward migration of farmers beyond the Appalachians. Almost as soon as the pioneers reached the western side of the mountains, it became obvious that corn had found its true home. (Figure 1.10).

Exports

During the period 1849-1859, around 17% of corn exported from the United States was in the form of meal. In the early years of the Civil War, this percentage had declined to around 7%. Much of this exported corn meal seems to have originated in the South. No export data was collected in Southern ports during the war.

Data on the bushels of corn exported, in several years, during the period 1860-1870 are presented in Figure 7.6. One data set on corn exports in the early nineteenth century ended in 1863, and the US Department of Agriculture did not begin to publish its data set until 1866.

Before the war, a large share of the corn produced in the North Central Region was shipped to the South. With the beginning of hostilities, this market was no longer available; the US government redoubled its efforts to export more of the grain. The data show that corn exports trended upward sharply in the Civil War years from 4.2 billion bushels in 1860 to 17.2 bushels in 1863. By 1866, exports were beginning to decline, and by 1869 they had returned to the prewar level.

During the 11-year period (1860-1870), the United States simply did not export much of its corn. Except for slightly higher exports in 1863 (17.2 million bushels), the other quantities of corn exported annually during this period were not much different from the levels exported during the preceding decade (Figure 6.42). The 17.2 million bushels exported in 1863 was the highest ever recorded to that time. However, that relatively high level of exports was equivalent to 2% of the 1859 harvest (838.8 million bushels). The 1870 corn crop was the largest harvested in the United States to that date–slightly over one billion bushels. Apparently exports responded to this enormous crop, as they increased from 2.1 million bushels in 1869 to 10.7 million bushels in 1870.

Prices

There is relatively little data available on corn prices in the United States before the Civil War. Gray (1958) published a set of data on prices in Virginia from 1801 to 1860. These were presented in Figure 6.44. During this entire period, prices varied around 60 cents a bushel. During the war, no data was collected on prices in the United States. USDA began to publish data in 1866. According to these data, the average price for the period 1866-1870 ranged from 52 to 78 cents, and averaged 66 cent a bushel, or near the same level in Virginia in the preceding decade.

US Corn

Data on the bushels of corn harvested in the United States, its regions, and in Alabama in 1839, 1849 1859, and 1869 are presented in Table 7.9. Data from 1839 and 1849 are included for comparison. American farmers harvested 838.8 million bushels of corn in 1859. They had harvested 592.1 million bushels in 1849. Farmers in Illinois harvested the largest share of the 1859 crop (13.7%). Other leading corn-producing states included: Ohio (8.8%), Missouri (8.7), Indiana (8.5), and Kentucky (7.6). These five states harvested 47.4% of the total 1859 crop. As these data show, by 1859 the North Central Region was well on its way of establishing its pre-eminence in American corn production. In 1839, only 28.1% of the country's corn crop was harvested there. By 1849, this percentage had increased to 37.5, and then in 1859 it reached 48.4%. In 1839, Tennessee (12%) and Kentucky (10.6%) led the country in corn production. Some 20 years later (1859), only Kentucky (7.6%) remained on the top-five list.

As a result of the war, bushels harvested fell to 760.9 million in 1869, a decline of 9.2%. During the same period, bales of American cotton harvested had declined 44.4%. Obviously, a much higher percentage of the corn crop was produced outside the South during and immediately after the war. As a result, the hostilities did not affect corn production nearly as much as they did cotton production. Corn harvested in the North Central Region increased from 406.2 to 439.2 million bushels (8.1%) during that period, but declined by 27.9% (229.6 to 165.6 million bushels) in the South Central Region. Harvest increased only slightly in the North Atlantic Region.

According to data published in the 1870 Census, the leading corn-producing state in 1869 was Illinois with 17.8% of the total national crop. Other important corn-producing states that year were: Iowa (9%), Ohio (8.9), Missouri (8.7), and Indiana (6.7). These five states harvested 50.3% of the total crop. All of these states were in the North Central Region. By 1859, Indian corn had finally found its true home.

The magnitude of the decline of corn production in the South Atlantic and South Central regions combined from 1859 to 1869 (30.4%) (364.1 to 252.1 million bushels) is somewhat surprising. This is especially so considering that the Confederate government had made a determined effort to get farmers to increase the production of food crops. Unfortunately, there is little data available on the production of corn in the South Atlantic and South Central regions during the war years. It is likely that data collected in 1864 might have shown that corn production was much higher. By 1869, farmers in the former Confederacy were increasing their acreage in cotton as rapidly as possible, taking much of it out of corn production. Further, by 1869 the South had to acknowledge that it could not compete economically with the North Central Region in corn production. It was quickly becoming obvious that it was cheaper to purchase corn from that region than to grow it themselves.

Alabama Corn

Census data on bushels of corn harvested in Alabama in 1839, 1849, 1859, and 1869 are presented in Figure 7.9. As noted in a preceding paragraph, Alabama farmers harvested 28.8 million bushels of corn in 1849 and 33.2 million in 1859; Figure 6.47 presented data on the distribution of corn harvested in Alabama in 1859. The data presented in the figure show that the distribution of the harvest of corn continued to be closely related to the distributions of the harvest of cotton and the population in the state (Figure 6.41). Montgomery (4.8%), Marengo (4.2), Dallas (4.1), Greene (3.9), and Lowndes (3.9) counties led the state in corn harvested in 1859. These same five counties lead the state in cotton harvested as well.

By 1869, bushels of corn harvested had fallen to 17 million, a decline of 48.8% from 1859. During the same period, bales of cotton harvested declined by 57%. As noted in the previous paragraph, it is likely that if data on the harvest of corn were available for the war years, they would have shown that it was much higher than in 1859. However, it is likely that by 1869 cotton was rapidly replacing corn on Alabama farms. By 1869, the pattern of corn harvest was beginning to change in the state. Only one of the top-five counties (Montgomery) in 1859 remained on the list in 1869. Perry (5%) replaced Montgomery (3.5%) as the leading corn-producing county. Also, the Cotton Arc was beginning to lose its dominance. In 1869, Madison County (3.9%) in the Highland Rim section was the second leading county in the harvest of corn. It had not been near the top of the list in 1859. Further, Lawrence County entered the top of the list as the fifth leading county. Corn production in the country had been moving northward (Kentucky and Tennessee to Illinois and Ohio) for decades. Now, it seemed to be moving northward in Alabama as well. Most of the corn in the state was still being harvested in the Cotton Arc, but at the same time, it was increasing in several counties in north Alabama.

Peanuts and Soybeanss

Censuses did not included data on peanuts until 1900. USDA did not publish data on the crop until 1909. The department first began to publish data on soybean prices in 1913, but nothing on production statistics until 1924.

Hay

During the period 1839-1859, the quantity of hay harvested in the United States increased at a rate of around 3.6% a year. This rate of increase was generally lower than the increase in numbers of working animals (Figure 6.14), beef cattle, and milk cows on American farms during this period. Farmers were increasingly interested in providing feed for their livestock during the winter and during periods when there was little forage available, but they were barely increasing hay production as fast as their animal numbers were growing.

US Hay

Data on the harvest of hay in the United States, its regions, and in Alabama in 1839, 1849, 1959, and 1869, are presented in Table 7.10. The data indicate that between 1859 and 1869 that hay harvested increased in the country by slightly more than 43% (19.1 to 27.3 million tons). The relative increase in hay production was much greater than in total animal numbers. From 1860 to 1870, the number of horses, mules, and asses increased by 12% (7.4 to 8.3 million), but the number of "all" cattle declined by 7% (25.6 to 23.8 million). American farmers were finally getting serious about providing feed for their livestock year-round. This overall increase in hay harvested was achieved even with sizeable declines in the South Atlantic (21%) and South Central regions (20%). The largest increase was recorded in the North Central Region (76.4%) where beef production was increasing rapidly.

Of all the crops produced in the South during the war, hay would have required the most labor, an input in extremely limited supply; consequently, harvest of the crop declined by about one-fifth. The numbers of animals in the South that would have been somewhat dependent on the availability of hay, also declined, but at a slightly lower rate.

Alabama Hay

From 1859 to 1869 (Table 7.10), tons of hay harvested in Alabama declined 83% (62,200 to 10,600 tons). The harvest of no other plant crop in the state declined so sharply during the war years. During the period 1860-1870, the total number of "all" cattle declined 37% (773,400 to 487,200), and the total number of all draft animals declined 34% (238,800 to 157,400), in either case not nearly as sharply as the harvest of hay. These changes are in sharp contrast to those taking place in other parts of the United States, especially in the North Central Region. It is likely that the decrease in harvest in the state was even more severe during the mid-war years, and by 1869 production was probably increasing again.

Wheat

In 1839, the estimated per capita availability of wheat in the United States was 260 pounds. It increased to 363 pounds in 1849. Between 1849 and 1859, the population increased at a much more rapid rate than wheat harvest (35.6 versus 7.9%). During this period, almost two million immigrants entered the United States. With this disparity, per capita availability declined to 291 pounds. These data indicate only the amount of wheat available per capita. Of this amount, a large fraction would have been consumed by humans in the United States. A substantial amount would have been exported, as both flour and grain. A small amount would have been used for animal feed.

In 1859, Illinois farmers harvested 13.7%, of the 173.1 million-bushel, national wheat crop. Other states reporting large harvests included: Indiana (9.7%), Wisconsin (9.0), Ohio (8.7), and Virginia (7.6). These five states accounted for 49% of the total. Alabama farmers harvested only 0.7% of the total crop.

Exports

Wheat had been a well-established crop in Europe for thousands of years before the nineteenth century. As a result, opportunities for export from the United States were limited. In the early years, substantial quantities of the grain could be exported only during those years when wars and/or weather reduced European production. With this limitation, at the beginning of the eighteenth century, the United States was exporting around five million bushels of wheat a year (Figure 6.48). By the middle of the century, although there was considerable year-to-year variation, the annual level of exports had increased to around 20 million bushels (Figure 6.49). In 1860, the country exported 17.2 million bushels 76% of it in the form of flour. The 1860 exports were equivalent to 10% of the 1859 crop of 173.1 million bushels.

Data on exports in several years, during the period 1860-1870, are presented in Figure 7.7. Before the beginning of the war, substantial quantities of wheat produced in the North were shipped to the South. When hostilities began, this market was no longer available. Also, during this period, wheat production was increasing rapidly, especially in the North Central Region (Figure 7.8). As a result of the market constriction and increased production, efforts were apparently intensified to increase exports from the United States. In 1863, wheat exports reached 58.1 million bushels; however, in 1866, with the war ended and at least some of the Southern market restored, exports totaled only 12.6 million bushels. Harvests continued to increase sharply. By 1870, it was 71% higher than in 1866. As a result, prices declined rapidly, making American wheat extremely attractive to European importers. By 1869, the country was exporting 53.9 million bushels of wheat. Harvest declined slightly in 1870, and exports also declined. Even with the large increase in exports, a relatively small amount of the American wheat crop was being exported. The 52.6 million bushels exported in 1870 were equivalent to only 18% of the 1869 crop.

Prices

There are no data available on American wheat prices for the period 1860 to 1865. The Division of Statistics of USDA began to publish data on prices in 1866 (Figure 7.8). With markets recovering quickly in the food-limited South, the annual average price of wheat was at a level of $2.06 a bushel in 1866; however, as production increased after the end of hostilities, prices began to decline. By 1869, the grain had lost almost half of its value, to reach $1.04. Bushels harvested declined by 12% between 1869 and 1870, and as a result the average price increased to $1.25 a bushel.

US Wheat

Data on the quantity of wheat harvested in the United States, its regions and in Alabama in 1839, 1849, 1859, and 1869 are presented in Table 7.11. These data indicate that the American wheat harvest increased by about 66% (173.1 to 287.7 million bushels) during the period 1859-1869, although harvest in the South Atlantic and South Central regions declined by 22.3 and 15.8%, respectively. The increase in the quantity harvested in the North Central Region was rather astounding. Between 1859 and 1869, it increased from 95 to 194.9 million bushels. Farmers in that region harvested more of the grain in 1869 than farmers in the entire United States harvested in 1859. It was apparent that farmers in the North Central Region could produce enough wheat to feed much of the world.

The decline in the Southern wheat harvest is somewhat surprising given the emphasis to increase food production during the war. In the South, returns in the amount of food produced, considering the amount of labor, seed, etc., required, were always lower for wheat than for corn. Further, production of the Southern wheat crop was much more unpredictable because of diseases and pests than for corn. Given these disparities, it is likely that Southern farmers elected to place more emphasis on corn during the war.

Alabama Wheat

Alabama farmers harvested 1.2 million bushels of wheat in 1859 (Table 7.11). They had harvested only 249,000 bushels in 1849. In 1869, they harvested one million bushels, 13% less after the Civil War than before it. With no data on either acres harvested or yield an acre, it is impossible to know which factor contributed more to the reduced harvest. However, with a reduction of that magnitude, it is likely that most of it resulted from planting and harvesting fewer acres. This is not really surprising considering that the total amount of improved farmland in Alabama declined by 21% between 1860 and 1870 (Table 7.4).

Figure 6.52 showed the distribution of wheat harvested in Alabama in 1859. In that year, a large share of it was harvested in the east-central counties, from Chambers northward to Cherokee. Data published in the 1870 Census indicated that most of the crop was harvested in the same region in 1869.

In 1859, the estimated population in Alabama was 944,943 persons, and the quantity of wheat harvested per person was 1.29 bushels. It had been 0.39 bushels in 1849. With the reduction in harvest in 1869, bushels per person declined to 1.06. During the period 1860 to 1870, the quantity of the grain harvested per person in the United States increased from 5.6 to 7.6 bushels. This overall national increase includes the lower levels in the South during that period.

Oats

The increase in the harvest of oats in the United States during the period 1849 through 1859 had not kept pace with the increase in the number of horses in the country. The number of horses increased 44% (4.3 to 6.2 million), while the quantity of oats only increased by 17.7% (146.6 to 172.6 million bushels).

American farmers harvested 172.6 million bushels of oats in 1859. They had harvested 146.6 million bushels in 1849. Farmers in New York harvested the largest share of the 1859 crop (20.4%). Other leading oat-producing states included: Pennsylvania (15.9%), Ohio (8.9), Illinois (8.8), and Wisconsin (6.4). These five states harvested 60.4% of the entire 1859 crop.

Exports

The export of oats continued to be limited during the period 1860-1870. In most of the years, exports ranged around 130 million bushels, and generally accounted for only a small share of the total oat harvest. For example, in 1866 American farmers harvested 232 million bushels of the crop, but in 1867 when much of it would have been exported, exports totaled only 123,000 bushels, or considerably less than 0.1% of production. For some reason, exports were substantially higher in 1866 (826,000 bushels) and 1868 (426,000). However, even with those larger quantities, exports still accounted for only a small share of production. There never was much of a market for American oats in Europe.

Prices

Throughout the period 1866-1870, oat prices remained relatively steady at between 25 and 29 cents a bushel. There are no other data available on prices for this period.

US Oats

Census data of bushels of oats harvested in 1839, 1849, 1859, and 1869 are presented in Table 7.12. In 1859, American farmers harvested 172.6 million bushels of oats. Ten years later (1869), bushels harvested had increased 63.4% to 282.1 million. By 1869, oat production was also shifting westward. Three of the top-five oat-producing states (Illinois, Ohio, and Iowa), were west of the Ohio River, and Illinois had replaced New York as the leading state. Further, the North Central Region accounted for a large share of the national increase between 1859 and 1869. Nationally, harvest increased by 109.5 million bushels. Some 96.8 million bushels of the total increase (88.4%), were harvested in that region.

The large increase of oats harvested in the North Central Region was likely related to the large increase in the number of horses there. By 1869, farmers in the region were using large numbers of the animals to power the equipment they were using in grain production. Between 1860 and 1870, the number of horses in the region increased from 2.5 million to 3.6 million (Table 7.7), an increase of 42%. From 1859 to 1869, oats harvested increased by more than 154%. It is likely that the region had a surplus of oats in 1869 and that large quantities were being shipped to other regions.

Alabama Oats

Alabama farmers harvested 682,200 bushels in 1859, or 0.4% of the national crop. Farmers in Talladega County harvested the largest share (9.4%) of the total state crop. Other counties with large shares included: Lowndes (6.6%), Chambers (6.6), Madison (6.5), and Montgomery (4.9). These five counties accounted for 34% of the total crop in Alabama.

Figure 6.55 shows the distribution of the oat harvest in Alabama in 1859. It was generally distributed in the same way as cotton harvested (Figure 6.41). Strangely, the distribution of harvests seemed to bear little resemblance to the distribution of horses. It was much more similar to the distribution of mules and asses.

By 1869, oats harvested had increased to 770,900 bushels, or less than 0.3% of the total national crop. Alabama farmers harvested enough oats to provide their 80,800 horses (Table 7.7) and 76,700 mules and asses with 6.7 bushels of oats per animal for the entire year. At the same time, Illinois farmers harvested enough to provide each of their animals with 43.2 bushels. Alabama farmers either had to purchase oats from other states or feed their animals something else.

By 1869, Lee County, with 5.7% of the state's oat crop, had replaced Talladega County (5.6%) as the leading oats-producing county in the state. Other counties with large shares included: Chambers (4.7%), Tallapoosa (4.3), and Marion (4.1). Four of the counties are located on the Piedmont Upland (Figure 1.11) of east Alabama. These five counties accounted for only 24.4% of the total harvest. Average harvest per county in Alabama's 65 counties was 11,900 bushels. Some 17 counties reported harvests of at least 15,000 bushels.

Alabama Sweet Potatoes

Sweet potatoes are primarily a Southern crop (Table 6.23). As a consequence, national data will be given less attention. The national harvest of the crop totaled 22.7 million Cwt in 1859 (Table 7.13). Alabama farmers harvested 2.9 million Cwt, or 12.8% of the national crop. Georgia lead the country with 15.5%.

In 1869, the national harvest of sweet potatoes totaled 11.7 million Cwt (Table 7.13). Alabama farmers harvested only 1.10 million Cwt, or 8.6% of the national total. Alabama's 1869 crop represented a decrease of 66% from the 1859 total. This sharp reduction was probably related to some environmental problem (weather, disease, or pests). In 1868 and 1870, farmers harvested 1.5 and 1.4 million Cwt, respectively. Even if the quantity harvested in 1869 had been closer to the 1868 or 1870 levels, it would still have been about half of the quantity harvested before the Civil War. The leading five counties (Wilcox, Mobile, Pike, Butler, and Lee) only accounted for 18.2% of the total crop harvested in the state in 1869, indicating that substantial quantities of sweet potatoes were produced in all the counties.

Alabama Irish Potatoes

In the mid-nineteenth century, the South produced a relatively small share of the country's Irish potato crop. In 1849 and 1859, Alabama farmers harvested less than 0.5% of the national total. For this reason, national production of the crop will be given much less attention from this point onward. However, Alabama farmers have consistently produced small quantities of the crop for well over a century-and-a-half, and in some small areas, they have made an important contribution to the local agricultural economy.

American farmers harvested 66.7 million Cwt of Irish potatoes in 1859 (Table 7.13). The Alabama share of that crop was 295,000 Cwt (0.4%). In 1869, the national harvest increased to 86 million Cwt, and the Alabama share of that crop was 97,500 Cwt (0.1%). The 1870 Census data indicate that in 1869 Irish potatoes were being produced in relatively sizeable quantities all over the state. Leading counties were: Limestone (11%), Jackson (7.6), Mobile (6.4), Madison (6.4), and De Kalb (5.1). These five counties accounted for only 36.5% of the total crop in the state. The average per county in Alabama's 65 counties was 1,500 Cwt. Some 16 of the counties reported harvests of 2,400 or more Cwt.

Alabama Sorghum, Molasses, and Sugarcane Syrup

Alabama has produced only relatively small quantities of sorghum molasses and sugarcane syrup from the early days of statehood. These products contributed to such a limited extent to national and state economies that national data will be given less attention. In 1860, production of sorghum molasses in the country totaled 6.7 million gallons (Table 7.14). Production in Alabama represented 0.8% (55,700 gallons) of national production. National production of sugarcane syrup totaled 15 million gallons in 1860. The Alabama share was 0.6% (85,100 gallons).

In 1870, national production of sorghum molasses totaled 16 million gallons (Table 7.14), more than double the production in 1860 (6.7 million gallons). Production in Alabama also increased sharply (55,700 to 26,7300 gallons). Production in the state accounted for 1.7% of the national total. Apparently no county data on molasses production were collected.

The 1870 Census reported that the production of sugarcane syrup in the United States totaled 6.6 million gallons (Table 7.14), or less than half of the production of 1860 (15 million). Alabama production, however, more than doubled, increasing from 85,100 to 166,000 gallons. Apparently no county data were collected.

Products of Market Gardens

Censuses of 1850, 1860, and 1870, included information on products of market gardens in the United States and Alabama, but no information on specific crops. According to these reports, the total value of these products on American farms in 1859 was $16 million (Table 7.15). The Alabama share was $163,000, or 1% of the national total. The North Atlantic Region accounted for 50.8%.

In 1869 (Table 7.15), four states in the North Atlantic Region—New York (16.6%), New Jersey (14.), Massachusetts (9.6), and Pennsylvania (8.7) accounted for half of the national value of $20.7 million. With the large urban centers continuing to grow in those states, the produce from those market gardens would have played an important role in meeting their food requirements. It is interesting that even at this early date, California was included in the top five states with 7.2% of the national total. In 1870, New York City (including Brooklyn) was the most highly populated urban area in the country. Philadelphia, Boston, and San Francisco ranked 2nd, 7th, and 10th, respectively.

By 1869, the value of market gardens in Alabama had declined from $163,000 to $139,636 (Table 7.15). Although some of this decline was the result of a general decrease in prices for all agricultural commodities in the South after the war, some of the decrease was likely the result of a real decline in abundance. In 1869, the value in Alabama represented only 0.7% of the national total of $20.7 million. Some 25 of Alabama's 65 counties did not report any value for market garden produce. Mobile County reported the largest share of total value in the state (44%). Other counties with large shares included: Montgomery (32.8%), Madison (7.6), Bibb (3.3), and Russell (2.2). These five counties accounted for 89.5% of the total, and all of them were large centers of population or near such centers.

Orchard Products

Censuses in 1860 and 1870 included information on the value of orchard products, but no information on specific crops. The census data indicated that value increased (Table 7.16) in the United Sates by $27.2 million between 1859 and 1869 ($20.1 to $47.3 million). Value increased in all of the regions, but much of the total increase was a result of the gain of $14.9 million in the North Central Region ($6.4 to $21.3 million). With this increase, it became the leading producer of orchard products in the country. Value increased by $9.3 million in the North Atlantic Region ($8.4 to $17.7 million). In 1869, New York continued to lead the nation in the value of orchard products (17.6%). Other states with large shares included: Ohio (12.3%), Pennsylvania (8.9), Illinois (7.5), and Michigan (7.3). These five states accounted for 53.6% of the national total.

Census data showed that value of orchard products in Alabama declined by some $66,100 ($223,300 to $157,200) between 1859 and 1869 (Table 7.16). This is an unusually large decline considering that value increased substantially in the South Central Region. Some local environmental factor, such as a severe late spring freeze, might have been the cause of the reduction. In 1859, orchards in Alabama accounted for 1.1% of total national value; however, with the sharp increase in national value from 1859 to 1869, coupled with the decline in Alabama, the state's share declined to 0.8%.

In 1869, 13 of Alabama's 65 counties did not report any value. The state's average value was only $578 per county. Mobile lead the state with 17.3% of the total. Other counties with large shares included: Blount (8.6%), Cherokee (5.9), Butler (5.7), and Shelby (5.5). These five counties accounted for about 43% of the state's total.

Nursery Products

The censuses of 1860 and 1870 did not include any data on the production or value of nursery crops in the country or the states.

Alabama Timber

The Census of 1860 did not include any data on the production or sale of timber in the United States or any of the states. The 1870 census included data on the value of forest products. In that year, the estimated value in the United States was $37.5 million (Table 7.17). Almost half ($17.3 million) was reported from the North Atlantic Region. During the war years, the growth of the populations (Table 7.1) and economies of the states of that region was requiring the use of tremendous quantities of forest products. This growing demand and use (drain) of their forest resources was well above the level of sustainability.

The Civil War also fueled raging economic growth in the northern portion of the North Central Region. As a result, the value of forest products there was just a little lower (Table 7.17) than in the North Atlantic Region ($13 versus $17.3 million). These two regions accounted for

80% of the national total. Of course, that is where the people were, and that was where economic growth was taking place. As might be expected from the status of the size of their populations and economies, value was much lower in the South Atlantic and South Central regions ($5 and $1.2 million, respectively). Given their status in 1870, they could not imagine such an eventuality, but their time would come.

In 1870, Alabama's share of the total value of forest products was $85,900, or less than 0.1% of the national total (Table 7.17). Only 42 of Alabama's 65 counties reported any value. The average per county was $1,322. The largest share of the state's total was reported from Jackson County (19%). Other counties reporting significant shares included: Barbour (14%), Dallas (10.4), Mobile (5.2), and Russell (4.2). These five counties accounted for 52.8% of the total.

Data published in *Statistical History of the United States from Colonial Times to the Present* (Wattenberg, 1976) showed that lumber production in the country totaled 12.8 billion board feet (BBF) in 1869. In contrast to the data on value of forest products (Table 7.17), the North Central Region led the country in lumber production with 49.4% of the national total. Lumber production in the North Atlantic Region was somewhat lower with 35.7% of the total. The South Central and South Atlantic regions trailed far behind with 7.2 and 2.9% , respectively. Even at that early date, lumber production in the Western Region (4.9%) had already surpassed production in the South Central Region.

Animal Agriculture

The value of livestock on American farms increased dramatically from 1850 to 1860. From a value of $544.2 million in 1850, it increased to $1,089.3 million in 1860, a few years before the Civil War. This increase was a welcome addition to the American economy. The increase in value in Alabama was equally dramatic. It was at a level of $21.7 million in 1850, and increased to $43.4 million in 1860. Unlike data on plant crops, which were collected the year before the census was taken, data on animal numbers were collected in the same year.

Cattle

Inventory data presented in Table 7.18 indicate that the number of "other" cattle ("all" cattle excepting milk cows and working oxen) increased 52.5% (9.7 to 14.8 million) from 1850 to 1860. During the same period, the American population increased by 41.4% (22.3 to 31.5 million). The supply of beef was apparently increasing faster than the population and rapidly becoming available to more people. The sharp increase in numbers in the North Central Region (2.4 to 4 million) also indicated that beef production in the United States was rapidly moving westward.

Number of "Other" US Cattle

Inventory data on the number of "other" cattle in the United States in 1850, 1860, and 1870 are presented in Table 7.18. When the cattle inventory was taken in 1860, there were 14.8 million of the animals in the country. The largest share (5.1 million) of the total herd was on farms in the South Central Region, which includes Texas. This is the primary reason why such a large share of the total was located in this region.

After the Civil War, by 1870 the total number in the country had declined to 13.6 million (Table 7.17). As might be expected, numbers declined in the South Atlantic and South Central regions by a combined 1.8 million, although the number in Texas actually increased by some 172,800 head. Agriculture in Texas was much less severely affected by the war than in other Southern states.

Except for Texas, numbers declined significantly from 1860 to 1870 in most of the Southern states. It declined between 40 to 55% in Virginia, South Carolina, and Alabama, and 35% in Georgia, Mississippi, and Louisiana. In Kentucky, which was largely unaffected by the war, numbers declined by only 16%.

As noted in a preceding paragraph, in 1870 the largest share (21.6%) was on farms in Texas. Other states with large shares included: Illinois (7.8%), Ohio (5.6), Missouri (5.1), and New York (4.6).

It is somewhat surprising that "other" cattle numbers would decline to such a large degree in the Lower South during the war. Unlike plant crops that had to be planted and cultivated, during this period cattle in the Lower South generally took care of themselves. A large percentage were simply allowed to graze on open land until time to market or slaughter them. With the data available, it is not possible to determine whether the decline was the result of reduced production or increased use.

The decline of "other" cattle in the South during the war likely had a different effect on the economy and life in the region than did the decline in some of the important food crops. Southerners apparently were not beefeaters. By all accounts, they consumed large quantities of pork and relatively little beef (Locher and Cox, 2004). This is to be expected from those early settlers of Scottish ancestry. In the Scottish Highlands in the eighteenth century, cattle were a money crop, and very little beef was eaten by the herders themselves.

Surprisingly, numbers also declined in the North Atlantic Region (2.1 to 1.8 million). By 1870, cattle farmers in that region were finding it increasingly difficult to compete with those in the North Central Region in the production of beef. With railroad traffic between the Midwest and the East Coast growing almost daily, competition increased rapidly.

Number of "Other" Alabama Cattle

Inventory data on the number of "other" cattle in Alabama in 1850, 1860 and 1870, are also presented in Table 7.18. There were 454,500 animals on farms in Alabama when the inventory was taken in 1860. There had been 433,300 in 1850. By 1870, numbers had declined to 257,300. As a result of the war and its aftermath, and possibly cattle cycles, herd size in Alabama declined by 43% from 1860 to 1870. In 1860, 3.1% of all "other" cattle in the country were on Alabama farms. By 1870, that percentage had declined to 1.8

Working Oxen

Working oxen were not included in the numbers of "other" cattle discussed in the preceding paragraphs; yet they were an extremely important part of the total cattle population in the war years. In 1860, there were 2.2 million of these animals in the United States. A decade later, the number had declined to 1.3 million. Some of this decline was obviously related to the deleterious effects of the war; however, a major share of the change resulted from a shift from working oxen to horses for farm power. For example, in the North Central Region, from 1860 to 1870, the number of working oxen declined from 712,000 to 312,500. However, during the same period, the number of horses increased from 2.5 million to 3.6 million. Horses were much more suitable to furnish the power required for the new machines and implements being used on the farms in that region. In the South Central Region, the numbers of working oxen also declined during the period 1860-1870. In Alabama, the number declined from 88,300 to 59,200.

Hogs

The number of swine (hogs and pigs) in the United States increased 10.6% (30.3 to 33.5 million, head) from 1850 to 1860. During the same period, the American population increased by 41.3%. Obviously, during that period, hog numbers were not increasing as fast as population. During the same period, cattle numbers increased 52.6%.

US Hog Numbers

Inventory data presented in Table 7.19 indicate that the number of hogs in the United States declined from 33.5 million in 1860 to 25.1 million in 1870. While it was expected that hog numbers would decline in the South as a result of the war, the decline in the United States as a whole is somewhat surprising. Of course, in 1860 well over half of all hogs in the country were in the South, so it is likely that the decline (6.8 million animals) in that region was so great that it had a significant effect on the total for the country. However, even the number in the North Central Region declined. Generally speaking, this region was less negatively affected by the war; yet numbers declined. It is likely that with the increasing quantity of beef available and with increasing rail transportation, the American population was beginning to eat more beef and less pork.

In 1870, as in 1860, the same five states—Illinois (10.8%), Missouri (9.2), Indiana (7.4), Kentucky (7.3), and Tennessee (6.9)—lead the country in numbers. However, after the war 41.8% of the country's herd was in those states–up from 37.7% from a decade earlier.

Alabama Hog Numbers

As Locher and Cox (2004) noted, pork has always been important to Alabamians. In 1840, 1850, and 1860 hogs per person in the state were 2.4, 2.5, and 1.8 animals, respectively. In contrast, in the North Atlantic Region hogs per person were 0.6, 0.3, and 0.2 , respectively, for those same years.

The inventory data in Table 7.19 indicate that the number of hogs in Alabama declined 59% (1,748,300 to 719,800) from 1860 to 1870, and that the percentage of the total national herd declined from 5.2 to 2.9%. The average per Alabama farm declined to 10.7 animals on each of 67,400 farms in the state. The average had been 31.7 in 1860. There was less than one (0.7) animal per person in 1870. The decline (37.2%) in numbers was widespread in the South, but not as great as in Alabama (59%).

Jackson County led the state in numbers in 1870, with 34.6% of the total. Other counties with large totals were: Pike (3.4%), Henry (3.1), Lawrence (2.6), and Madison (2.5). Only 15.2% of the total were found in these counties. This low percentage indicates the widespread distribution of hogs in the state. There were 15,000 hogs in 11 of Alabama's 65 counties.

Milk

There are no published data on the annual production of milk in the United States before 1924. The Census of 1890 provides the only data on milk production during the period 1870-1895. That census provided estimates of the quantity of milk sold in 1870, 1880, and 1890, the data apparently are not comparable for any of those years.

Milk Cows

Inventory data on milk cows in the United States were not published in the censuses before 1850. Between 1850 and 1860, the number of milk cows in the country increased by 34.4% (6.4 to 8.6 million) (Table 7.20). During the same period, the American population increased by 41.5% (22.3 to 31.5 million) (Table 7.1). The first shipment of milk, by rail, came in 1841; by 1860 it could be shipped in large quantities from the Midwest to the rapidly growing urban centers in the North Atlantic Region.

US Milk Cows

Table 7.20 presents data on the number of milk cows in the United States, its regions, and Alabama in 1850, 1860, and 1870. In 1850, there had been many more milk cows in the North Atlantic Regions (2.2 million) than in any of the others; however, by 1860 there were essentially the same number in both and the North Atlantic and North Central regions (2.6 million).

The number of milk cows in the country increased from 8.6 million in 1860 to 8.9 million in 1870. Even though there was a substantial reduction in the South (19.4%) from 1860 to 1870, increases in the other regions were large enough to offset those losses. The increase in the North Central Region was especially large (2.6 to 3.3 million), and for the first time there were considerably more animals in that region than in the North Atlantic Region (3.3 versus 2.8 million). With the production of feed grains growing rapidly in the North Central Region (Tables 7.9 and 7.12) and with rail traffic increasing regularly between that region and the urban centers in the Northeast, the North Atlantic Region had lost its edge in the production of milk.

In 1850 and 1860, there were smaller shares of milk cows in the South Atlantic and South Central regions, and between 1860 and 1870 numbers of both declined (1.2 to 1 and 1.9 to 1.5 million, respectively).

Even though by 1870 the North Atlantic Region was falling behind the North Central Region in the number of milk cows, numbers were continuing to increase in New York and Pennsylvania. The rail shipment of milk into the region was growing, but the number of milk cows around the urban centers apparently continued to increase. Local production and marketing of milk seemed to have some real advantages.

In 1870, the largest share of milk cows in the nation was in New York (15.1%). Others states with large shares included: Pennsylvania (7.9%), Ohio (7.3), Illinois (7.2), and Texas (4.8). Alabama's share was 1.9%. This was the first time since colonial days that a Southern state (Texas) would have been included in the top-five in the country.

Alabama Milk Cows

Inventory data on the number of milk cows on Alabama farms in 1850, 1860, and 1870 are also presented in Table 7.20. Data in the censuses show that as a result of the Civil War and its aftermath that the number of milk cows in the state declined from 230,500 in 1860 to

170,600 in 1870, a decline of 26%. In 1860, there was an average of 5.2 milk cows on each of Alabama's 55,100 farms; however, by 1870 with milk cow numbers declining and farm numbers increasing, the average number on each of 67,400 farms was down to 2.5.

In both 1860 and 1870, the largest share of milk cows was found in the Cotton Arc (Figure 6.41) and in counties on its margin, but large numbers of the animals were distributed throughout the state. In 1860, the largest share of the total herd was in Tuscaloosa County (7,046 animals), but there were 18 counties with more than 5,000. In 1870, the largest share of milk cows was still in the Cotton Arc. Wilcox County led the state with 4,722 animals, but the herd was still widely distributed among all counties. There were 18 counties with at least 3,000 animals.

Milk Sold

The 1870 Census included data on the gallons of milk sold in the country and states. Unfortunately, there are no similar data available in preceding censuses. These data provide little more than a limited insight into the status of the fledgling commercial dairy industry. Certainly it reflects only a small fraction of the total quantity produced in the country, especially milk consumed by farm families who produced it.

US Milk Sold

A summary of the 1870 Census data is presented in Table 7.21. The data show that American farmers sold 235.5 million gallons of milk in 1870. Farmers in the North Atlantic Region sold 79.2% of the total. The North Central, Western, South Atlantic, and South Central regions sold 16.5, 1.7, 1.3, and 1.2%, respectively. These data reflect the number of people financially able to purchase milk, and the capability for milk storage and transportation in the different regions.

Data on gallons sold per person in the different regions are also presented in Table 7.21. It is obvious from these data that there was virtually no commercial dairy industry in the South at that time.

Farmers in New York sold the largest share of the national total (57.6%). Other states with large quantities sold included: Ohio (9.4%), Massachusetts (6.5), Pennsylvania (6.1), and Illinois (3.9). These five states accounted for 83.5% of the total sold. Sales in Alabama only accounted for 0.1%. The percentages sold in the five states are a direct reflection of the levels of industrialization, commercialization, and urbanization in New York City, Boston, Philadelphia, Cleveland, Chicago, and San Francisco.

Alabama Milk Sold

The situation regarding data on milk sold in Alabama during the period 1860-1870 is the same as described regarding national data. The censuses included only data for 1870 (Table 7.21). In that year, Alabama farmers sold 104,600 gallons, or 3.7% of the total quantity sold in the South Central Region and 0.1% of the national total. While a considerable amount of milk must have been produced in the state, only a small fraction of it was sold.

Dairy Products

The 1840 Census included data only on the value of dairy products produced in the country and in individual states. The 1890 Census contained data on the production (pounds) of butter and cheese in the country and the states in 1850, 1860, and 1870.

Butter

American dairy farmers produced an estimated 313.3 million pounds of butter in 1850 and 459.7 million in 1860 (Table 7.22). Census data for those years indicate that the larger share was produced in the North Atlantic Region in both years. Data published in the 1890 Census showed that the national total had increased to 514 million pounds in 1870, and that farmers in the North Central Region produced only a slightly smaller share than those in the North Atlantic Region (43.9 versus 41.7%). Production in the other regions trailed far behind. The effects of the war on butter production is quite obvious when the 1860 and 1870 data are compared for both the South Atlantic and South Central regions. Pounds declined by 15.8 and 22.8%, respectively, in the two regions, while it increased by 40% in the North Central Region. In 1870, Alabama butter production accounted for 0.6% of the national total.

Alabama butter production declined from 6,000 pounds in 1860 to 3,200 pounds in 1870, a decline of 46.7%. This rate of decline was considerably greater than in the region (46.7 versus 22.8%). Some 61 of the state's 65 counties reported data on butter production in 1870. Average production was 49,400 pounds per county. Farmers in Lawrence County reported the highest production, 5.4% of the state total. Other counties reporting high levels included: Marengo (5.1%), Randolph (3.9), Clay (3.8), and Jackson (3.8). These five counties, which accounted for 22% of the total, indicate the widespread distribution of butter production in the state. Some 44 of the state's 65 counties reported production of at least 25,000 pounds.

Cheese

American dairy farmers produced 105.5 million pounds of cheese in 1850 and 459.7 million pounds in 1860 (Table 7.22). Census data for those years indicate that the much larger share was produced in the North Atlantic Region in both years. Data published in the 1890 Census showed that in 1870 the national total had declined to 53.5 million pounds and that the decline was shared by all the regions, except the Western. The North Atlantic Region continued to produce well over half of the national total. Production in the other regions trailed far behind. The effects of the war on cheese production is quite obvious when the 1860 and 1870 data are compared for both the South Atlantic and South Central regions. Production declined by 48.3 and 83%, respectively, in the two regions.

In 1870, farmers in New York led cheese production in the country, with 42.6% of the national total. Other states with large totals included: Ohio (15.3%), Vermont (9.0), California (6.3), and Massachusetts (4.2). These five states accounted for 77.4% of the national total. Cheese production in Alabama accounted for 0.1% of the national total.

The 1890 Census reported that dairy product manufacturers in Alabama produced only 2,700 pounds of cheese in 1870, a decline of 83% from 1860.

Poultry and Eggs

Data on poultry and eggs were not included in censuses of 1860 and 1870.

Agricultural Production

In 1870, American agriculture had just come through its most trying period in its history. In the Civil War, there were both winners and losers. It has been fairly evident that Southern farmers were the losers, but it has been exceedingly difficult to determine just how bad their losses were. Census data are available for both 1860 and 1870, but not for the intervening years. The Statistics Division of the US Department of Agriculture began to collect annual data in 1866, but there is no annual data for the war years themselves. Further, in 1866 some recovery was already occurring. For example, a large quantity of cotton that had been accumulating on plantations during the war was quickly put on the world market at extremely high prices. So all that we have are census data, and with that data there are only two measures of what happened to agriculture during this period of rural disaster. These include: valuations of livestock and value of farm products.

Value of Livestock on Hand in the United States and Alabama

There are no census data on the valuations of crops produced in the United States in either 1860 or 1870, but the 1890 Census included data on valuations for livestock on hand in the country and the states as of June 1, 1850, 1860, and 1870. A summary of these data are presented in Table 7.23. The actual estimates for 1870 as reported in the Census of 1890 have been reduced by 20%, as recommended, to make those data comparable with other years. The 1890 Census provides the rationale for this conversion.

Data presented in Table 7.23 show that total valuations were $544.2 million in 1850, but then increased to 1,089.3 million in 1860. Valuations increased in all of the regions and in Alabama. The increase in the North Central Region was especially large; it almost doubled in this period.

Valuations for the country as a whole also continued to increase (Table 7.23) between 1860 and 1870 ($1,089.3 to $1,220.1 million). The estimate for 1870 has been corrected as recommended. The valuations in 1870 were 12% higher than in 1860. However, the Consumer Price Index for the United States had increased by about 41% during the same period (Figure 7.2). Taking the inflation of the currency into consideration, real valuations apparently declined during the period. This national decline was largely the result of declines in both the South Atlantic (33%) and South Central (33%) regions. It declined by 50.8% in Alabama. These declines in the South were probably even more devastating than they appear, as a result of the currency inflation involved.

Valuations increased substantially in both the North Atlantic (27.9%) and North Central (66.4%) regions during the decade. However, only the rate of increase in the North Central Region was greater than the rate of currency inflation. In 1870, 70.8% of total valuations in the country were reported from those two regions.

The largest share of total livestock valuations in 1870 was reported by New York (11.5%). Other states reporting large shares included: Illinois (9.8%), Ohio (7.9), Pennsylvania (7.6), and Missouri (5.5). These five states accounted for 42.3% of the total. Four of these same five states were among the top-five in numbers of milk cows.

Livestock valuations in Alabama declined from $43.4 million in 1860 to $23.4 million in 1870. In 1870, the largest share of the valuation in the state was reported by Wilcox County (3.6%). Other counties reporting large shares included: Montgomery (3.1%), Marengo (2.9), Dallas (2.8), and Bullock (2.7). These five counties accounted for only 15.1% of the total. Livestock were widely distributed in Alabama. The war had savagely diminished Alabama's livestock industry, but the descendents of those old Anglo-Saxon immigrants still loved their hogs, and the descendents of those old Scotch-Irish immigrants still loved their cattle.

Value of Farm Products

The 1870 Census also published data on the estimated total value of all farm products for the United States and the states for 1870. These data are presented in Table 7.24. These data generally parallel those of Table 7.23. The largest shares of total value were reported from the North Central and North Atlantic regions (40 and 25.9%, respectively). Shares were considerably lower in the South Central and South Atlantic regions (18.6 and 12.6% respectively). The Western Region trailed far behind, with only 2.8%.

The total value data for both the South Atlantic and South Central regions arc somewhat misleading. In 1869, the price of cotton reached 29 cents a pound, and as a result, farmers harvested a crop worth $384 million. It would be many years before they would harvest another crop of equal value. Without such an unusually high-valued cotton crop, the values for those regions, reported in Table 7.24, would have been considerably smaller.

In 1869, Alabama reported a value of $67.5 million, or 3% of the total for the country (Figure 7.24). In the same year, farmers in the state harvested 429,000 bales of cotton with a value of $50 million. The total for all products was only $67.5 million. As might be expected, the larger shares of the values were reported from those Cotton Counties in the Cotton Arc and in the Coosa and Tennessee valleys.

THE BOTTOM LINE

The Civil War had diametrically opposing effects on agriculture in the North and South. Agricultural production in the North surged as a result of the war. During the same period, a series of poor harvests in England and Europe required them to import large quantities of agricultural commodities, especially wheat from the Northern states. During the war, food and fiber of all kinds was generally readily available in the North, although there was a temporary shortage of cotton in New England. While there was no shortage of food in the North, some wage laborers had difficulty purchasing it because of the rapid increase in prices.

There was very little of the devastation of the Northern agricultural resource base (land, buildings, equipment, machinery, livestock, etc.). It remained essentially unchanged throughout the hostilities. In relative terms, the wartime emergency had actually improved the economic position of Northern farmers. In fact, the Civil War resulted in the evolution of a transportation, commercial, and manufacturing base for the unparalleled technological revolution of agriculture that would, in the twentieth century, transform much of American agriculture into the envy of the world. In the short term, however, the Civil War resulted in a real windfall for Northern farmers, but some of their good fortune quickly became the source of their undoing.

In sharp contrast, the war left agriculture in Alabama and the South in shambles and confusion. All parts of Southern agriculture that had been based in any way on slave labor were destroyed, leaving only vast uncertainty. Southern farmers lost $2 billion in capital investment from the loss of slaves. The loss to Alabama farmers was around $200 million (Rogers and Ward, 1994). Further, many Southern planters had purchased large amounts of Confederate bonds that became essentially worthless long before the war ended. Much of the land in the South had gone untended and was growing weeds and bushes when defeated soldiers trudged back to their farms. Acreage of improved cropland declined 10% between 1860 and 1870. Farm buildings had not been adequately maintained. The average price per acre of Southern farms had not returned to the 1860-level even as late as 1900. Farm machinery and equipment had not been maintained or replaced; their value also had not returned to the 1860-level by 1900. Cotton production was severely impacted. Production was 45% lower in 1869 than it had been in 1859. It did, however, return to the prewar level by 1879. Livestock losses were severe, and the total value of all livestock did not return to the 1860-level until 1900. The war had devastated Southern agriculture to the extent that it has never recover its antebellum position as the center of commercial farming in the United States. Further, it was so severely damaged that it could not benefit from the technological revolution taking place in American agriculture until well into the twentieth century.

By 1870, the Civil War was slowly receding into history. The slaves had been freed, the Union had been preserved, and the rebels had been taught a lesson. No one, however, seems to have given any real thought to what was going to happen to 3.9 million impoverished freedmen suddenly freed from slavery and then expected to live on their own recognizance. If President Lincoln had any such plans, they apparently died with him. The Freedmen's Bureau helped prevent outright starvation, but accomplished little else. The Radical Republicans temporarily found a use for the freedmen: as pawns in their game of expanding their political reach. There was very little sympathy for the freedmen in the South where hundreds of thousands of poor whites and yeomen who had not owned any slave and who were little better off in any way than the recently freed slaves. In the North where the war had been primarily about preserving the Union, there had never been much sympathy for the slaves, and subsequently there was little for the freedmen.

In the final analysis, there was only one place for freedmen to go. They had to squeeze themselves in among the hordes of poor whites and failed yeomanry–those who had been trying to eke out a living for generations on hundreds of thousands of small, worn-out, poorly capitalized, unproductive, inefficient farms strewn across a Southern landscape that tobacco and cotton farming had destroyed long ago. Their coming simply added another straw to the camel's back that had been slowly breaking for almost 250 years. In many ways, the lives of these freedmen, poor whites, and failed yeomen farmers were little better off than serfs on feudal estates in the Middle Ages. Agriculture never had much future in the South because of its poor soils and unpredictable climate. The coming of black slaves in the seventeenth century did not change that reality, and their emancipation 250 years later did not change it either. For generation after generation, most Southerners had just hunkered-down, for a spell before moving on while hoping and praying that something better would come along. Much of this ebb and flow of Southern humanity can be traced directly to Celtic Britain three centuries earlier. The freedmen hardly made a splash when they squeezed into this deperately poor mass of people.

8

Seriously Seeking Salvation

1870–1895

PRECIS

At the end of the Civil War, agriculture in the North was actually more prosperous than it had been before the war. The quantity of land in farms was growing rapidly in the North Central Region. At the same time, new machines and larger, more powerful horses were being translated into rapidly-increasing quantities of wheat, corn, and oats. Quickly, production began to outstrip demand, and prices began to decline. In 1895, wheat prices were 50 cents a bushel. They had been $1.60 in 1873. There were similar declines in the prices for corn and oats. With their larger investments in land and buildings and farm machinery and equipment, and with declining prices, the region was rapidly becoming an agricultural disaster.

Southern agriculture was already a disaster before the war, and the hostilities simply degraded it even further. To climb out of the deep hole that they were in, they did the only thing they thought that they could do under the circumstances—plant more cotton. Acres of cotton harvested in the United States (virtually all of it in the South) increased from 9 million in 1870 to 22 million in 1894. Not surprisingly, cotton prices went in the other direction. They were around 10 cents a pound in the late 1870s. Then in 1894, they reached 4.6 cents. Agriculture in both the North and South were in serious trouble as the twentieth century approached. They both desperately needed salvation.

Various strategies were employed in attempts to rectify the plight of farmers. One was the attempt to make farming more scientific through research and Extension Services. Another was the attempt to organize farmers, such as in the Grange, the Wheel, and the Farmers Alliance. Diversification was a third major strategy to improve the financial lot of American farmers. Although the South and Alabama produced more corn, peanuts, etc. as well as timber, cattle, hogs, and chickens, cotton was still king, but not a benevolent king. Reluctantly, fiercely independent farmers attempted political action, but the Populist Movement also failed. Then just as it appeared that there was no hope for deliverance, the Age of Progress emerged from the telephone booth, and agriculture in the North began to recover. Unfortunately, it would take much more than a caped crusader to deliver agriculture in the South.

INTRODUCTION

In 1870, memories of the Civil War were fading from the hearts and minds of Southerners, and they were turning attention to seek ways to improve the lives of their families. Unfortunately, in the South the full cost of the war was not yet paid. Installment payments on the debts they had incurred would have to be made regularly for generations to come.

Until the end of the Civil War, the American economy was largely an agrarian one. In 1860, 48% of the country's population lived on farms, and farmers made up 58% of the labor force. In the South, the economy was much more closely tied to agriculture, especially the production of cotton. For a number of reasons, generally unrelated to the war, the United States and the world were beginning to change. In the United States, commercialism, industrialization, and urbanization were begining to replace agrarianism as the driving force in the national economy. These elements had become increasingly important in the North Atlantic Region following the end of the American Revolution, but by the 1870s they were spreading rapidly across the Northern portion of the country.

There was also a major change taking place in agriculture itself. It was becoming increasingly efficient, and as a result, much more productive. After the American Revolution, the Industrial Revolution had spread rapidly from Britain to the United States, and by 1870, it was well established throughout the North. There were steam engines everywhere, and factories were being established to manufacture all kinds of equipment and machinery that would revolutionize food and fiber production. Quickly, American agriculture came to the point where it could produce much more food and fiber than its population could use. Unfortunately, neither the country nor the world had much experience with this phenomenon. For much of human history, there had never been enough.

This chapter begins in 1870, more or less, and ends in 1895. The Civil War has ended, and Radical Congressional Reconstruction had about run its course. The Alabama legislature had ratified the 14th Amendment (Due Process and Equal Protection), and would soon ratify the 15th (Voting Rights). Its Congressional delegation, including its two senators, had been sworn-in. Politically and officially, Alabama was once again part of the United States. In 1866, the Division of Statistics of the US Department of Agriculture began to collect data annually on different aspects of agriculture in the country and the individual states. Until 1866, collections were made at 10-year intervals, and published as part of the censuses.

By 1870, the general pattern of American agriculture of the future was well established, so in this and subsequent chapters, most of the emphasis will be on changes in agriculture in Alabama and in the United Sates as a whole. With some crops, emphasis on national data, will be limited even further. There is also a practical reason for making this change. It was difficult maintaining continuity and coherence, even with the limited amount of data available from the censuses. With annual data, it would be virtually impossible.

NATIONAL POLITICS

Much of the period 1870-1895 was taken up with the Reconstruction of the South and with bringing the former Confederate States, their people and their institutions, back into the Union. The period also saw the development of a Republican political dynasty in Washington, especially in the presidency and the Senate. The Democrats held the balance of power in the House for more than half of the period. In the United States, the period also witnessed the rapid rise of uncontrolled, unregulated commercialism. This situation provided numerous ways for elected officials to profit from their positions. As a result, it was also a period of unprecedented political corruption (Morrison, 1965).

The Presidency

Republicans controlled the the presidency for almost all of the period 1870-1895. Ulysses S. Grant was elected president in 1869 and re-elected in 1873. Rutherford B. Hayes (OH) was elected to the office in 1877. James A. Garfield (OH) became president in 1881, and was succeeded by Chester A. Arthur (VT) in 1881. The first Democrat since Buchanan (1857-1861) became president when Grover Cleveland (NJ) was elected in 1885. Cleveland was succeeded by the Republican Benjamin Harrison (OH) in 1889, only to be re-elected in 1893.

Secretaries of Agriculture

The Department of Agriculture was established by Congress in 1862, but there was no position of secretary of agriculture in the president's cabinet until 1889. Norman Coleman (D, MO) (1889) was the first individual to serve as secretary. He was replaced by Jeremiah Rusk (R,WI) in 1889. When Cleveland was re-elected president in 1893, he appointed Sterling Morton (D, NE). He served in that capacity until 1897.

The Senate

Republicans also held the balance of power in the Senate for most of this period 1870-1895. Republicans were in control in 11 of the 13 Congresses, and Democrats were in control during only two. The Republicans controlled the body from the 41st Congress (1869-1871) through the 45th (1877-1879) and from the 47th (1881-1883) through the 52nd (1891-1893). Democrats were in control in the 46th (1879-1881) and 53rd (1893-1895).

Senate Majority Leaders

Neither party designated Senate majority leaders until 1920.

Chairmen of the Senate Committee of Agriculture

Chairmen of the Senate Agriculture Committee during the period 1870-1895 were as follows: Simon Cameron (R, PA) (1867-1871); Oliver H.P.T. Morton (R, IN) (1871-1872); Frederick Frelinguysen (R, NJ) (1872-1877); Algeron Paddock (R, NE); John Johnston (D, VA) (1879-1881); William Mahone (Readjuster, VA) (1881-1883); Warren Miller (R, NY) (1883-1887); Thomas Palmer (R, MI) (1887-1889); Algernon Paddock (R, NE) (1889-1893); and James George (D,MS) (1893-1895).

The House of Representatives

Democrats controlled the House of Representatives in eight of the 13 Congresses (44th-46th, 48th-50th, and 52nd-53rd), whereas Republicans controlled the body in 41st-43rd, the 47th, and the 51st Congresses.

Speakers of the House

James Blaine (R, MA) served as speaker of the House of Representatives from 1869 through 1875. Other individuals holding the office during the period 1870-1895 included: Michael Kerr (R, IN) (1875-1877); Samuel Randall (R, PA) (1875-1881); William Keifer (R, OH) (1881-1885); John Carlisle (D, KY) (1885-1889); Thomas Reed (R, MA) (1889-1893); Charles Crisp (D, GA) (1893-1895); and Thomas Reed (R, MA) (1895-1899).

Chairmen of the House Committee on Agriculture

Chairmen of the House Agricultural Committee during the period 1870-1895 were as follows: John Wilson (R, OH) (1869-1973); Charles Hays (R, AL) (1873-1875); John Caldwell (D, AL) (1875-1877); Augustus Cutler (D, NJ) (1877-1879); James Covert (D, NY) (1879-1881); Edward Valentine (R, NE) (1881-1883); William Hatch (D, MO) (1881-1889); Edward Funston (D, KS) (1889-1891); and William Hatch (D, MO) (1891-1895).

The Political Situation

In May 1868, the Radical Republicans finally gave up their efforts to impeach President Johnson, and in the same month they nominated Ulysses S. Grant as their candidate for president. With few former Confederate Democrats in the South able to vote, Grant was easily elected. Although radical efforts to remake the South were waning in 1868, much of Grant's administration was taken up with Reconstruction concerns. With efforts to guarantee a better life for the freedmen through direct local action failing, Congress attempted to legislate that outcome by enacting the 15th Amendment, which declared that it was unlawful to deny anyone the right to vote on account of race, color or previous condition of servitude. While the amendment was being debated and ratified, Southern politicians were actively seeking to limit or counter its effect through election chicanery and outright terrorism. The amendment was ratified in 1870, but never enforced until well into the twentieth century.

When the Republican Party was being established just before the beginning of the Civil War, one of the central planks of their platform reflected their abhorrence of slavery. As a result, before, during and after the Civil War, there were very few Republicans in the South. However, during Reconstruction the few Republicans took the reins of state governments in all the former Confederate states when most of the Democrats either were disenfranchised or refused to vote. Later, when Reconstruction efforts slowed, the Democrats quickly re-established one-party majorities in all the Southern states. Consequently, the South became and remained the Solid South in the hands of Democrats until near the end of the twentieth century.

Grant was not well suited for political life, and especially for the important office of president of the United States. In the first place, he had a very low opinion of the presidency as an institution. He felt that the job of the president was to see that the laws passed by Congress were enforced. Not being an effective politician, he made many poor decisions in his choice of people to assist him in his administration. As a result of his ineptitude and the corruption of many of his appointees, very little positive was accomplished in his two terms. Because of the ineffectiveness of the Grant administrations, and the corruption within the federal government during those years, it was widely expected that the Republicans could not maintain their hold on the presidency in 1876. They nominated Rutherford B. Hays (OH), and he was elected in an extremely close race that required the appointment of a special commission to look into the validity of electoral votes originating in some of the Southern states.

Hays was a conscientious and better than average president, but he accomplished very little in his administration because the professional politicians of his party disliked him. Even with limited support from the Republican Party, he was able to regain some of the power of the presidency that had been lost under Johnson and Grant. He was able to remove from federal service some of the worst of Grant's appointees, and in removing some of those ineffective public servants, Hays became interested in the reform of the federal civil service. He became so disenchanted with politics that he refused to be nominated in the 1880 election. In his place, the Republicans nominated General James A. Garfield (OH). He narrowly defeated the Democratic nominee, General Winfield Scott, but was assassinated in 1881, several months after being inaugurated. On his death, Vice President Chester A. Arthur (NY) became president.

President Arthur continued Hay's efforts to reform the civil service. In 1883, he signed into law the Pendleton Act, which established the US Civil Service Commission. The act established a new set of rules for the operation of the civil service, including the requirement that appointments be made as the result of competitive examinations and forbidding federal officials taking kick-backs from individuals that they had appointed. Arthur, like Hays, received little support from the professionals in his party, so other than getting the Pendleton Act passed, he was able to accomplish very little. He was successful in thwarting Congress in some of its most outrageous efforts to reward their more important constituents. In the process, he made enough of the politicians angry that he had little chance of being nominated for a second term. To replace him, the Republicans nominated James G. Blaine (PA) for the 1884 election. The Democrats nominated Grover Cleveland (NY). Partly as a result of charges of political corruption against Blaine, Cleveland won a narrow victory.

Cleveland was an austere and unbending president. Once he had decided on a position or made a decision, he stuck with it regardless of the political consequences. Cleveland was the first Democratic president in 25 years, and members of his party wasted no time in demanding the spoils of victory. Early in the Cleveland administration, he replaced most of the Republican-appointed postmasters and a large percentage of other federal officials. He soon antagonized just about every special interest group in the country. He deeply offended the Union veterans. He vetoed hundreds of bills passed by Congress that awarded federal pensions to constituents. He vetoed a bill authorizing the expenditure of $10,000 for providing feed grain to farmers in Texas who were suffering through a severe drought. In his veto message, he commented that federal aid in such cases encourages the expectation of parental care on the part of the government and weakens the sturdiness of our national character. He was able to get one piece of important legislation approved by Congress. In 1887, Congress enacted the Interstate Commerce Act, which was to prove especially important in the regulation of interstate freight shipping rates. Cleveland was renominated by the Democrats for the 1888 election, but he had angered so many politicians and special interest groups that he was defeated by the Republican candidate, Benjamin Harrison of Indiana.

Morison (1965) comments that Harrison was an ineffective president, although he had sizeable majorities in both houses. Unfortunately, the Republican Congress had little interest in government reform or progressive legislation. Instead it was primarily interested in raids on the US Treasury on behalf of powerful constituents and in direct attacks on the welfare of consumers.

The consolidation of companies to form powerful monopolies and trusts had been growing since the Civil War. This consolidation had the effect of sharply increasing business profits, while driving up the cost of living. Over time, the public began to understand that unrestrained consolidation was not in the public's best interest. Finally, public opposition became so powerful that both Republicans and Democrats had to support efforts to pass legislation to regulate it. As a result, in 1890, Congress enacted the Sherman Antitrust Act. This was probably the single most important legislation enacted in the Harrison administration.

The Republican Party, which was founded in the Midwest, was initially composed of farmers and small-town businessmen. Over time, however, control of the party moved northeastward toward the industrial, commercial, and financial centers of the North Atlantic Region. With this shift, came an increase in Republican support for protection of American commercial interests through high tariffs. In 1890, the Republican Congress enacted the McKinley Tariff, the highest protective tariff ever passed in the United States. Soon after it was enacted, prices on many products began to increase rapidly. As a result, many of the Congressmen who had supported it were defeated at the polls. Republicans lost control of the House, and their majority in the Senate was seriously eroded.

Harrison was renominated by the Republicans for the 1892 election. The Democrats nominated Grover Cleveland again. Harrison lost the election, likely as a result of the price increases following the enactment of the McKinley Tariff. At the same time, however, there was a growing concern among the electorate that the Republicans had lost their way in efforts to provide a square deal for the common man. This concern was increasingly voiced by a broad coalition of farmers, laborers, and owners of small businesses.

In the early 1890s, the worsening plight of American farmers moved them to take direct political action. The Farmer's Alliance had been formed to provide farmers with fraternal and economic support. But with their efforts resulting in few long-term benefits, its members were politicized. Out of these efforts came the Populist Movement, whose political agenda included the following:

1. Free and unlimited coinage of silver.
2. The establishment of sub-treasuries to force the federal government into direct financial support of agriculture.
3. Government ownership of railroads, telegraphs, and telephones.
4. A graduated income tax.
5. Creation of a parcel post.
6. Restriction of immigration.
7. An eight-hour day, for wage earners.
8. Popular election of US senators.
9. The Australian (secret) ballot.

Eastern newspapers greeted these proposals with cries of creeping socialism; however, within a generation, most of the Populist agenda had been adopted through actions by both Democratic and Republican Congresses. In 1892, the Progressive Party nominated General James Weaver of Iowa as its candidate. However, a large share of the unhappy farmers voted for Cleveland. They felt that the Democrats were more likely to have the political power to enact their agenda than the Progressive Party.

The organized labor movement and the Populist Movement, which appeared in American politics in the latter part of the nineteenth century, actually had their beginnings at least a century earlier in Europe. The Agricultural and Industrial revolutions were extremely important products of the eighteenth century in Europe, especially in England and France. The eighteenth century Enlightenment, or Age of Reason, was equally important. Among its more important aspects was a rising tide of humanitarian sentiments. This movement gave rise to a number of reformers who had grave concerns for the plight of displaced agricultural workers and for the new industrial workers—the so-called working class. One of these idealistic reformers, the French aristocrat Henri de Saint-Simon (1760-1825), the founder of French Socialism, had a deep concern for the impoverished masses, but he felt that they had only a limited capacity to change their situation. He was convinced that their salvation lay in the establishment of a strong central government, planned and controlled by a small elite of scientists, engineers, and technocrats (Booth, 1871 and Hewett, 2008). European aristocrats, in general, had long ago developed an overwhelming disdain for the value and capability of peasants, a group that had exchanged its freedom for protection in the Early Middle Ages (Spielvogel, 1994). Oddly, the early American patriot and the first secretary of the treasury, Alexander Hamilton (1755-1804), held similar views (Morison, 1965).

Later, another Frenchman, Louis Blanc (1813-1882) wrote that the treatment of the industrial workers was unjust and that the only path to a more just society was governmental responsibility for the welfare of its people (Spielvogel, 1994). In his book, *The Organization of Work* (1840), he commented: "The government ought to be considered as the supreme regulator of production and endowed for this duty with great power." His work was instrumental in the development of the movement that we call Socialism and of one of its branches, Progressivism. Less than a century later, some of his philosophy would play an extremely important role in the evolution of the American economy and its agriculture.

Early in the Cleveland administration, a series of bank failures and industrial collapses led to the Panic of 1893. The suffering of laborers led to a period of severe unrest and bewildering complexity in the unions (Morison, 1965). Out of this unrest came a series of violent acts by both labor and business. Although both sides were guilty of violence, the courts generally sided with the owners by allowing them to obtain injunctions against striking laborers.

ALABAMA POLITICS

Someone has remarked that politics is warfare fought with different weapons. The South and Alabama had their opportunity to separate themselves from the United States in the Civil War, and they lost. In fact, they had never had a real chance to win, but for a majority of the political leaders, they just replaced their Confederate uniforms with the garb of politics and continued to fight the war ferociously.

Governors

Governors of Alabama during the period 1870-1895 were as follows: 1868-1870, William Smith (R, Randolph County); 1870-1872, Robert Lindsey (D, Colbert); 1872-1874, David Lewis (R, Madison); 1874-1878, George Houston (D, Limestone); 1878-1882, Rufus Cobb (D, St. Clair); 1882-1886, Edward O Neal (D, Lauderdale); 1886-1890, Thomas Seay (D, Hale); 1890-1894, Thomas Jones (D, Montgomery); and 1894-1896, William Oates (D, Henry).

Commissioners of Agriculture

The Alabama legislature established the Department of Agriculture and Industries in 1883. Governor O'Neal appointed Edward Betts of Morgan County as the first commissioner. He served until 1887. In 1887, Reuben Kolb was appointed. He served until 1891. In that year, legislation was enacted to have the commissioner elected. Hector Lane was the first elected commissioner. He held the office until 1896.

The Political Situation

Ward and Rogers (1994) commented on politics in Alabama during Reconstruction:

> Reconstruction took the existing ingredients of Alabama life and shuffled them and dealt them out again to match a vision that was itself incomplete and unfinished. The old planter class, weakened materially for awhile, had a flawed and challenged claim to leadership. The huge majority of small farmers survived, but now were sunk from merely being poor to being utterly impoverished and often they hated the planter architects of their misfortune. And there were the blacks, cast adrift as pawns and sacrifices to the interests of others, and struggling to find some substance to the promises and the laws of freedom. . . . The war changed Alabama life; it added industry

> and the entrepreneur to the social mix. But as externals changed, the old verities and the old arrangements stayed the same.

The political situation in Alabama, during the late 1880s and early 1890s was especially bitter and divisive. In the early part of the period, Southern agricultural interests suffered terribly; however, over time farmers throughout the country began to suffer as well. Cotton prices were below 10 cents for most of the years after 1876, and wheat prices fell below one dollar in 1881. After trying several kinds of self-help efforts (the Grange, the Agricultural Wheel, Farmers Alliance), farmers finally decided that the only solution was to take over all levels of state and federal government. In Alabama, this meant taking over Alabama's Democratic Party; unfortunatley this proved to be an extremely difficult task. Operating within a loosely organized group of so-called Populists, agricultural interests offered candidates for the legislature in most Alabama counties. However, to elect a governor, they had to first get their man chosen as the candidate of the Democratic Party. They fought hard to get Commissioner of Agriculture Reuben Kolb chosen as the Democratic candidate for the elections of 1890, 1892, and 1894. The contests were unbelievably bitter and fraught with election fraud. Agricultural interests had never been well organized in their efforts, and with the passage of time they became even more disorganized. The high point of their efforts came in the election of 1894. However, they had little or no impact, except at the county level in 1896; afterwards, they more or less disappeared. Results at the national level were essentially the same. The Populists had tried to change government to make it more concerned about the troubles of the common man—troubles made by government—but the Populists failed, at least for a time.

THE NATIONAL ECONOMY

After the Civil War, the American economy began to change in many ways. The country was rapidly becoming more urban. Industrialization was also growing rapidly. Entrepreneurship was literally exploding everywhere. The country was becoming an important player in world affairs. All of these factors seemed to place more of a burden on our economic institutions than they could bear.

The Economic Situation

In September 1869, an effort to corner the US gold market (Black Friday) led to a national economic and financial contraction that lasted for 18 months. In May 1873, a worldwide stock market crash began in Germany. By September, the US stock market collapsed. The collapse created a financial panic in the country and brought about the longest economic contraction in US history (65 months). Other economic contractions (1882-1885, 38 months; 1887-1888, 13 months; and 1890-1891, 10 months) occurred during this period.

One of the most serious of the economic contractions began in 1893, although its foundation was laid earlier. Around 1877, the Kansas farm boom ended as a result of a disastrous drought. Thousands of farmers there abandoned their farms. The resulting economic collapse quickly spread to other regions. Then, beginning in 1890, currency problems began to plague the federal government. Also, overexpansion in industrial development and railroads began to weigh on the economy. During the same period, exports began to decrease. All of these factors led to the beginning of severe economic contraction (depression) in 1893. It was the most serious depression since the administration of George Washington. It lasted for 17 months, finally ending in 1894.

Government extravagance played an important role in the onset of this depression. Federal pensions for Union veterans of the Civil War had virtually eliminated the treasury surplus. Further, for years American industrialists had been investing in new plants and new, more productive machines. Then as productivity increased, fewer laborers were required. Unemployment increased rapidly, along with diminishing consumer purchasing power. Many railroads became insolvent.

In 1894, there were between two and three million unemployed; more than 12,000 businesses failed. The amount of gold owned by the US Treasury reached such a low level that the government had to borrow large amounts of the precious metal from private banks, to maintain the country on the gold standard. Citizens began calling on the government to do something to relieve the rapidly worsening situation.

The Gross National Product

There are no annual government data on the US Gross National Product (GNP) before 1889; however Johnston and Williamson (2007) have published data on estimates of the Gross Domestic Product (GDP) for the period 1790-1900. There is relatively little difference between GNP and GDP. Their data for the period 1860-1895 is presented in Figure 8.1. The estimates are in 2000 dollars. They show that the GDP was $73 billion in 1860. It increased slowly in the years after the Civil War, but in the late 1870s, it began to grow more rapidly, finally reaching $300 billion in 1892. Throughout the period, it clearly reflects the dynamism of the growing American industrial and commercial economy. It also indicates those periods of economic recessions and depressions.

The Consumer Price Index

Data on the Consumer Price Index (Lerner, 1975) for the period 1860-1895 are presented in Figure 8.2. The CPI was at a level of 27 (1967=100) in 1860. It almost doubled in the first three years of the Civil War, but began to decline sharply in 1865. It returned nearly to its antebellum level by 1878 and remained relatively constant during the business expansion of 1878-1893, but began to decline again in the Depression of 1893-1894.

The Inflation Rate

Data on US inflation rates during the period 1860-1895 are shown in Figure 8.3. From 1870 through 1895, the inflation rate was above zero for only one year (1880). That rate (3.6%) was the result of the beginning of a long period (1879-1893) of business expansion. The lowest rate was recorded near the end of the Depression of 1873-1878.

THE AGRICULTURAL ECONOMY

Following the Civil War, American agriculture (especially in the North) began to be pulled into the vortex of a sweeping urban and industrial revolution. In the process, it was transformed from a simple, pioneer and largely self-sufficient enterprise, into a modern business, organized on a scientific, capitalistic and commercial base (Edwards, 1940). The components of this far-reaching revolution were:

1. Passing of the public domain into the private domain as a result of liberal land policies.
2. Initiation and completion of the westward movement of agricultural settlement.
3. Invention and popularization of improved farm implements and machinery.
4. Widespread development and extension of transportation facilities.
5. Migration of industries from farms to factories.
6. Expansion of domestic and foreign markets.
7. Establishment of agencies and organizations for the promotion of agriculture.
8. Use of organized political activities by farmers.

All of these sweeping changes, in effect, seemed to steer agriculture toward something that it could never be—a machine. The Civil War had clearly shown that machines (factories, railroads, ships, telegraph, gun, etc.), were much more effective than valor. So, why could not agriculture be a machine as well? After all, agriculture was a process. Unfortunately, at that time no one knew anything about genes and genomes. They knew that bad weather was not good for agriculture, but they knew nothing about genome-environment interactions.

Unfortunately, while few people seemed to realize it at the time, agriculture could not be made into a machine. After the Civil War ended, the federal government quickly became deeply involved in efforts to take the uncertainty out of farming. This was a commendable goal, and considerable progress was made, but human efforts were doomed because man would never be able to control the farming environment completely. To make matters worse, not only could they not control some critical aspects of the environment, but they also could not predict what nature might do next.

Scientists would make significant progress in controlling some aspects of the agricultural environment. For example, they would learn some amazing things about agricultural soils and plant-soil interactions. They would learn that chemical fertilizers could be used to compensate for deficiencies. Unfortunately, they never seemed to realize that compensation was a costly process and that such costs were the deadly enemy of comparative advantage.

Nowhere was this false sense of promise more obvious, than in the South. With the amazing advances being made in agricultural science, it was widely assumed that the sow's ear of Southern agriculture would quickly become a silk purse. Unfortunately, there is no region in the country where weather is more unpredictable. Further, there is no region in the country where deficiencies in soil quality are so difficult to deal with in a cost-effective manner.

For thousands of years, farmers have known about the unpredictable nature of raising crops and animals, and they generally compensated for this uncertainty by planning for an excess of production. The story of Joseph and the seven good years and the seven bad years, related in Genesis 41, reminds us of the wisdom of this strategy. Agricultural surplus was not only valuable for the farmers themselves, but it was also valuable in that it enabled a portion of the population to engage in activities other than farming. Agricultural surplus made it possible for division of labor, labor specialization, and cities to develop; agricultural surplus is the essential foundation of civilization.

This age-old problem of feast and famine in food and fiber production and consumption accompanied the first settlers to Jamestown, and over the mountains to Tennessee and Kentucky, across the Ohio, and into the Old Southwest, but it would become even more intractable, especially for the farmer after the Civil War when science, engineering, and industrialization began to remove some of the rough edges from genome-environment interactions. As a result, it became increasingly difficult to balance the two. In fact, the problem became so serious in the latter part of the nineteenth century that farmers began to insist that the government adopt, as a national policy, the purchasing and storage of surpluses.

Federal Agricultural Legislation

In the 1870s, there were relatively few people in the country who thought that the Congress had any role to play in agriculture. As the period (1870-1895) drew to a close, however, there was a sizable majority of farmers who were beginning to feel very strongly that there was something the federal government could and should do, but the national Republican dynasty certainly did not see it that way; consequently, they enacted very little legislation during the entire period for the direct benefit of farmers. However, they did do some things that were beneficial to farming in general:

Hatch Experiment Station Act of 1887. This was probably the most important legislation enacted during the period, although it did not immediately help farmers with their growing problems. The Morrill Land Grant College Act had been approved by Congress in 1862, but it had not provided adequate funding. The 1887 act corrected that problem to a limited extent by providing grants to the states for conducting research in the agricultural sciences.

Department of Agriculture Act of 1889. Proposals for the establishment of US Department of Agriculture were made as early as 1776, and President Washington in his final message to Congress encouraged them to establish a national board of agriculture. In 1839, Commissioner of Patents Henry Ellsworth established the Agricultural Division within that agency. In 1849, the Patent Office, along with its agricultural activities, was transferred to the newly created Department of the Interior. After repeated attempts to do so, in early 1889 Congress passed and President Cleveland signed a bill establishing the cabinet level Department of Agriculture.

Meat Inspection Acts of 1890 and 1891. In the years following the Civil War, Europeans became increasingly concerned with the cleanliness of meat that they were importing from the United States. As these concerns grew, meat exporters here became concerned about their markets. These concerns, coupled with a rising tide of outrage in national newspapers over filthy meatpacking practices, forced Congress to act. These acts required that USDA become directly involved in inspecting meatpacking plants. While these acts did not directly involve farmers, they served as the initial step in the direct involvement of government in the food supply system of the country. The acts resulted from the inability of the private sector to police itself in the matter of food safety.

It is problematical to try to guess whether the meatpackers could have ever been pressured into developing their own inspection system. The fact is that they did not do so in a timely fashion. This foot-dragging provided a good opportunity for those who supported an active role for government in the regulation in private business to become involved.

Sherman Antitrust Act of 1890. Congress passed this act to limit the power of monopolies in the rapidly growing commercial economy. While it was not designed to help farmers directly its provisions were used effectively by the judiciary to prohibit farmers from creating cooperatives to market their crops.

Alabama Agricultural Legislation

Fertilizer quality problems in Alabama became so important in the years following the Civil War that in 1871 the state's Reconstruction legislature established an office for the inspection of fertilizers. Unfortunately, the appointment of the inspector became embroiled in politics, and the office was abolished in 1874 (Kerr, 1985). In 1880-1882, President Isaac Tichenor of the Agricultural and Mechanical College of Alabama (Auburn) asked the legislature to enact new legislation that would require fertilizer manufacturers expecting to sell their products in Alabama to submit samples to the state chemist at Auburn for analysis. The bill was modeled after one that had been approved in Georgia in 1874. The legislature approved the bill, but unfortunately Governor Cobb vetoed it, and the legislature was unable to over-ride his veto. In 1883, the legislature approved another fertilizer inspection bill, and this time Governor Cobb approved it.

In 1883, the Alabama legislature established the Department of Agriculture and Industries along with the position of commissioner of agriculture and industries. In the beginning, the commissioner was appointed by the governor; however, in 1891, the legislature enacted legislation providing for the election of the commissioner. One of the primary responsibilities of the commissioner was to supervise the analysis of fertilizers sold in the state.

The growing concern for free range livestock and the first legislative response to the problem was detailed in the preceding chapter. In 1879, this concern forced the legislature to reconsider the problem. During the 1870-1895 period, the initial Act of 1866 that prohibited the open range in certain districts was amended to include nine additional districts. In 1881, an act was passed authorizing the Board of County Revenues or court commissioners in 12 counties to close the range in a particular section, or beat, of a given county on the petition of ten landowners. The same act also authorized local officials to call for an election on free range in the entire county, if requested by a certain number of landowners.

Agricultural Exports

No annual data was published on total American agricultural exports before 1901. Before that year, average annual estimates for each decade were published. For example, it was estimated that in each year during the period 1870-1879 that the country exported agricultural commodities valued at $453 million. It was also estimated that during the period agricultural exports comprised, on an average, 79% of all exports each year.

Agricultural export data for the decades from the 1870s through the 1890s are presented in Figure 8.4. Data for the 1860s, are included for comparison. Data show that agricultural exports were increasing fairly rapidly throughout the period. The average annual estimate for the 1880s was 27.6% higher than for the 1870s ($574 million versus $453 million). The increase in the average was slightly lower from the 1880s to the 1890s (22.5%) ($703 million versus $574 million). In the decade of the 1890s, agricultural exports comprised only 71% of all exports. The exports of the products of our industrialization were slowly overtaking the export of agricultural commodities, even though the export of agricultural products were continuing to increase at a fairly high rate.

The increases in the value of exports shown in Figure 8.4 are generally real ones. There was virtually no inflation of the currency during this period. The Consumer Price Index in 1870 was 38 (1967 = 100). In 1895, it was down to 25 (Figure 8.2).

Exports of cereals, cotton, meat, and meat products all increased after the Civil War. Most of these commodities went to Europe, with Great Britain being our major trading partner. However, there was no way that

Europe could continue to absorb such a large volume of agricultural products without adversely affecting their own agricultural economies.

While there was no annual data on total exports of all agricultural products before 1901, annual data on the export of some individual crops are available for years as early as 1866. Generally these data are reported on a fiscal year basis (July 1 to June 30). For example, exports of corn for 1870 represent foreign sales between July 1, 1869 and June 30, 1870. Generally speaking, most of the corn exported in fiscal year 1870 would have been harvested in 1869. Estimates of exports of cotton are an exception to the rule. The fiscal year for that crop began August 1 and ended July 31 of the following year.

Agricultural Prices

There are no annual data on prices for agricultural commodities, in general, in the United States before 1910. However, after 1865 the US Department of Agriculture began to publish annual data on average prices received by producers for the major plant crops such as cotton, wheat, corn, and oats; for livestock; and for some livestock products. At the same time, USDA began to publish annual data on acres harvested for the major crops and annual inventories on livestock numbers. With these data, for the first time it was possible to see something of the relationship between the two. Ekelund and Tollison (1988) state that "in a market economy, prices are the essential signals that tell producers and resource suppliers what and how to produce." Obviously prices serve the same purpose in agriculture as in industry, but in the production of agricultural commodities, the process is much more complex. In industry, producers can respond to changes in prices relatively quickly, but in agriculture, once crops have been planted it is difficult to make many changes until the following year. Further, even when a decision has been made to either increase or decrease planting, once the crop is planted there is no way to know in advance what the yield for that year will be. In livestock production, especially in the production of cattle, increasing numbers requires at least two years once the decision has been made to do so. Of course, reducing numbers can be accomplished more quickly.

Not surprisingly, the data available after 1865 showed that prices and acres harvested of the major crops were inseparably interrelated. When one goes down, the other, almost without fail, goes up. When the price increases in a year, the following year farmers tend to increase acres planted, which in most cases results in increased production, which in turn, results in a decrease in prices. The following year with prices lower, the preceding year's acres planted are reduced. This game continues year after year, and almost always yields a loss to the farmer. As a result of this relationship, harvesting a large crop in a year with good prices, almost never happens.

Generally, prices for the major grain crops (wheat, corn, and oats) trended downward during the period 1870-1895. For example, farmers received $1.04 a bushel for their wheat in 1870, but slightly less than half that amount (50.5 cents) in 1895. Prices of corn and oats were also much lower in 1895 than they had been in 1870. There were no data published on the average annual price received by farmer for their cotton before 1876. They received an average of 9.7 cents a pound that year. Seemingly, cotton could not go any lower, but it did. In 1894, farmers received an average of 4.6 cents a pound for their crop. The price rebounded to 7.6 cents in 1895.

Unfortunately, the decline in prices represented only one of two disasters that befell American farmers in the period 1870-1895. The other was the rising prices of things that farmers had to purchase. In the last quarter of the nineteenth century, the Northern standard of living and prices of manufactured goods increased. Further, the industrialists there, wishing to protect their industries against foreign competition, had been successful in getting Congress to place highly restrictive tariffs on imports. Without competition, there was nothing to slow the inflation in prices of manufactured goods. As a result, Southerners were selling their cotton at the lowest prices in history, while buying those things they needed and which could not be produced in the South at prices being driven upward by the North's highest standard of living in the world.

As a result of the general decline in prices received for agricultural commodities and the increase in prices paid for farm inputs during the period 1870-1895, the ratio of the two (prices received/prices paid) was well below 1 for most of the period. This ratio is a measure of the general health of the agricultural economy. If it remains below 1 for many years, farmers cannot continue to produce crops or livestock without finding some way to subsidize their operations. After 1910, USDA published annual estimates of the ratio and its components, and it was used as an indicator for the need for government intervention.

Farmers

When the 1860 Census was taken, the total population of the United States was 31.4 million, and it was estimated that the American farm population was 15.1 million. Approximately 58% of the country's total labor force were farmers. In the early 1870s, freedmen continued to establish farms in the South, and the explosive movement of agriculture onto new lands in the West was just getting underway. Further, immigration from Europe was increasing rapidly, and many of these newcomers were establishing farms in the hinterlands. The number of farmers in the country was about to explode.

The Total Population

From 1870 to 1895, the American population increased from 38.6 million to 69.7 million (Figure 8.5). In the same period, the population of Alabama increased from 997,000 to 1.7 million (Figure 8.6). In 1870, 32.2% of the total population lived in the North Atlantic Region and 15% in the South Atlantic Region.

Even as late as 1895 after the westward migration had been underway for almost a century, 27.6% of the population continued to live in the North Atlantic Region. Approximately 21.4% of the total lived in just three states: Massachusetts (2.5 million), Pennsylvania (5.8 million), and New York (6.6 million).

Included in this rather large increase in the population between 1870 and 1895 was a large number of immigrants. Between 1871 and 1890, some 8 million came to this country. Most of them were from Europe, yet 800,000 were from the Americas.

The Farm Population

In 1870, the American farm population was 18.4 million of a total US population of 38.6 million. A decade later (1880), it had increased to 23 million. The rush to become farmers among the country's citizens continued unabated in the 1870s and 1880s. From the 1890 Census, it was estimated that in the 1880s the number of farmers had increased to 29.4 million. Although the number of farmers was increasing at a substantial rate, their numbers were not increasing as rapidly as the population as a whole. In 1860, farmers constituted 48.2% of the total population; however, by 1870 it was down to 47.6%. The percentage continued to decline in 1880 and 1890 to 45.8% and 39.4%, respectively.

Farmers in the Work Force

As detailed in the preceding paragraphs, the American population continued to increase rapidly during the period 1870-1895. At the same time, the percentage of farmers in the population was decreasing. While these changes were occurring, the percentage of farmers in the labor force was also declining. In 1870, it had been 53%. In fact, it had been declining since the first estimates were made in 1840 (69%), and it continued to fall to 49% in 1880 and to 43% in 1890.

Tenancy

Apparently, little annual data were collected on tenancy before 1880. In that year, USDA estimated 25.6% of all farms in the country were operated by tenants (all classes). By 1990, the percentage had increased to 28.4, and it was even higher in 1900 (35.3%). Without a doubt, this increase in national tenancy was largely the result of what was happening in the South. In 1880, tenancy in Alabama was 46.8%, 48.6% in 1890, and 57.7% in 1900. It was probably equally high, if not higher, in all the Southern states. These high rates resulted in a biased estimate on tenancy outside the region.

Farms

During the period 1850-1870, the number of farms in the United States increased at a rate 57,500 a year. Much of this increase occurred in the North Central Region where an average of 30,400 farms were added each year. However, while numbers were increasing, average size was decreasing.

Farm Numbers

There are no annual data on the number of farms in the censuses; USDA did not begin to publish that data until 1910. In the period 1870-1990, the enumeration of farms was limited to those of three acres or greater in size, unless a smaller farm had cash receipts of $500 or more in the year of the census. Census data on the estimated number of farms in 1870, 1880, 1890, and 1900 are presented in Figure 8.7. Data for 1860 are shown for comparison, and because there are no census data for 1895, the number for 1900 is included as well.

In the period 1870-1895, the truly amazing increases in farm productivity were years in the future. To meet the food and fiber needs of a rapidly growing American population in the latter half of the nineteenth century, the number of farms and/or the amount of farmland had to be increased. In 1870, there were 2.7 million farms in the United States. There had been 2 million in 1860. In 1890, there were 5.7 million. A large share of this increase was a result of the large number of new farms being established in the South (South Atlantic and South Central Regions, combined), Between 1870 and 1880, the number of Southern farms increased from 885,100 to 1,531,100, an increase of 73%. Further, by the 1880s, many thousands of settlers were establishing farms in the western portions of the Mississippi and Missouri river valleys. However, by 1890 this westward exodus had largely run its course. In fact, in parts of this enormous region, severe drought was forcing some of the farmers to abandon their land and return to the East.

Trends in farm numbers in Alabama in the period 1870-1890 were similar to those in the country as a whole (Figures 8.7 and 8.8), except that the relative increase in farm numbers between 1870 and 1880 was considerably greater in Alabama than in the country (101.6 versus 50.7%). In 1870, there were 67,400 farms in Alabama. There had been 55,100 in 1860. These large increases were doubtlessly the result of freedmen establishing small farms of their own.

The number almost doubled by 1880 (135,900). Then it changed relatively little in the 1880s, to a total of 157,800 in 1890. As was the case in the country as a whole, the number increased sharply between 1890 and 1900.

While the trends in farm numbers were generally similar in Alabama and the country as a whole, another primary cause for the increases of farm numbers was quite different. Industrialization came exceedingly slow to the state after the war. Alabama's population increased, and it increased rapidly (997,000 in 1870 to 1,262,000 in 1880), but there was simply nothing else for people to do but farm. For example, in 1880 there were 1,640 manufacturing establishments in Alabama per one million persons. In its sister state, Illinois, there were 4,728 manufacturing establishments per one million persons.

The average number of farms per county in Alabama in 1890 was 2,390, ranging from 330 in Baldwin to 5,580 in Dallas. The five leading counties, in terms of percentage of the total farms were: Dallas (3.5%), Lowndes (3.3), Montgomery (3.2), Wilcox (2.8), and Madison (2.4). These five only accounted for 15.2% of the state's total. The counties with the smallest number of farms lay along the Florida and Mississippi state lines, including: Geneva (1,420); Covington (1,222); Escambia (604); Baldwin (330); Mobile (525); and Washington (543). Farm numbers in a group of counties in the southwestern portion of the Warrior Basin (Figure 1.16) were also well below the average for the state. These included: Winston (1,050); Marion (1,713); Walker (1,809); Fayette (1,904); and Cullman (1,978).

Total Farmland

Data on the total acres of farmland in the United States in 1870, 1880 ,1890, and 1900 are presented in Figure 8.9. Data for 1860 are shown for comparison. There were 407.7 million acres of land in farms in the country in 1870. There had been 407.2 million in 1860. From 1870 to 1890, acres in farms followed farm numbers upward to 536.1 million in 1880, then to 623.2 million in 1890. Sometime in the 1890s, it began to grow even more rapidly, and increased over two million acres (to a total of 838.6 million) by 1900. With the enormous quantity of land available in the Western Region, the upper limit on the amount of available farmland in the

United States was extremely high.

After 1860, the total farmland situation in Alabama was in sharp contrast to the country as a whole. As a result of the effects of the Civil War, land in farms was much lower in 1870 than it had been in 1860 (15 million acres versus 19.1 million), and it increased relatively little in the next two decades. In 1890, it was only about 700,000 acres above where it had been in 1860 (19.8 million). Furthermore, it did not increase much in the 1890s (another 900,000 acres). There was a finite amount of land suitable for farming in Alabama, and by 1860 most of that was in use. Increasing it beyond that level meant that farms were being established on generally more unsuitable land.

Farm Size

Data on average farm size for the United States for 1860, 1870, 1880, 1890, and 1900 are presented in Figure 8.11. In 1870, average farm size in the country was 153.3 acres, sharply smaller than in 1860 (190.2 acres). This sharp national decline is probably the result of the significant reduction in the size of Southern farms after the Civil War. For example, in Alabama the average declined from 346.5 acres to 222 acres in the decade following the war, and it is likely that the other Southern states experienced equal or greater reductions. Further, in that period Southern farms represented a large share of all American farms. By 1880, however, the average size of Southern farms was near that of the country as a whole, so there would be little bias involved.

Farm size in the country continued to decline in 1880 (134 acres), but then remained almost unchanged in 1890 (137 acres). By 1880, the really large declines in the South were in the past.

As noted in a preceding paragraph, with the amount of Alabama farmland most likely fully in use in 1870 as the population of the state and the number of farms continued to increase, the only thing that could happen was to subdivide existing farms to make five where there had been three. Of course, the result of this subdivision was a rapid decline in average size. It was 227 acres in 1870 (Figure 8.11), 138.8 acres in 1880, and 125.8 acres in 1890, and the end was nowhere in sight.

Value of Land and Buildings

Data on the total value of land and buildings on American farms for the period 1850-1900 are presented in Figure 8.12. Value amounted to $7.4 billion in 1870. In 1860, it had been $6.6 billion. Because the price of land is the primary component in value of land and buildings, as the number of acres of farmland increased (Figure 8.9), the value of land and buildings also increased. Value of land and buildings was $10.2 billion in 1880, $13.3 billion in 1890, and $16.6 billion in 1900.

By relating value of land and buildings in a given year to the number of acres of total farmland that year, a somewhat different picture emerges (Figure 8.13). For example, in 1870, value per acre was $18 ($7.4 billion/407.7 million acres). In 1860, it had been $16. Value per acre increased to $19 in 1880 and to $21 in 1890; however, it was somewhat lower in 1900 ($20). It appears from these data that value per acre did not increase at the same rate as total value did.

The value of land and buildings in Alabama (Figure 8.14) also increased from 1870 ($54.2 million), to 1890 ($111 million), and to 1900 ($134.6 million). However, between 1860 and 1870, it had fallen sharply ($175.8 million to $54.2 million), and even as late as 1890, it was still substantially lower than it had been in 1860 ($175.8 million versus $111 million). Apparently, the price of land declined sharply during and immediately after the Civil War. Also, acres per farm (Figure 8.11) also declined during this period.

On a per-acre basis, value was much lower in Alabama even before the war than in the country as a whole (Figure 8.13). In 1860, it had been $9 per acre ($175.8 million/19.1 million acres of land). In the same year, it had been $16 nationally. Then in the state in 1870, it was about half ($4) of what it had been in 1860. It remained at about the same level in 1880 ($4), but increased a little in 1890 ($6), but even in 1900, it remained well below the 1860 level.

Average value of farm real estate per Alabama county in 1890 was $1,682,600. Values ranged from $265,600 in Washington County to $4,330,400 in Madison. The five leading counties were: Madison (4%), Jackson (2.9), Montgomery (2.9), Marengo (2.8), and Jefferson (2.7) accounted for only 12.4% of the state total.

As a result of the Civil War, many Alabama farmers lost an extremely large percentage of their wealth with emancipation, and all of them lost an even larger percentage when the value of their farm real estate (collateral for loans) declined so sharply. It is not surprising that Alabama and Southern farmers were still playing catch-up well into the twentieth century. In fact, it could be argued that they never fully recovered from that double whammy.

Value of Farm Implements and Machinery

Data on the value of implements and machinery on American farms in 1860, 1870, 1880, 1890, and 1900 are presented in Figure 8.15. Value increased only slightly ($246 to $271 million) from 1860 to 1870; then increased sharply until 1900. The 1870 Census estimated that total value was $271 million. It increased to $406 million in 1880, to $492 million in 1890, and to $750 million in 1900. The increase between 1890 and 1900 was especially large (51%). The upward trend in these values generally parallels that of number of farms (Figure 8.7).

Between 1870 and 1890, the total value of implements and machinery increased by 82% ($271 to $494 million). During this period, the Consumer Price Index was actually declining (38 to 27), so it is apparent that American farmers added significantly to their stock of implements and machinery during the period. The rate of increase in the country is even more important when the much slower rate of increase in the South is considered.

As expected, total value of implements and machinery on American farms generally increased as the number of acres of farmland increased (Figures 8.9 and 8.15). It generally follows that the more acres a farmer has, the more machinery and equipment is required to farm it; however, it certainly is not a one-to-one relationship. A single piece of equipment can be used on a wide range of acres. Only when acres under cultivation reach a certain level will a farmer have to purchase an additional piece of equipment.

When the total value of implements and machinery in the country is converted to value per farm, the conversion corrects for changes in farm numbers during the period. In 1860, American farmers owned implements and machinery worth an average of $120 per farm (Figure 8.16). In 1870, it was down to $102 as a result of the Civil War. A sharply lower value in the South immediately after the war certainly forced the national value downward, Afterwards, the national value increased slowly to reach $101, $108, and $131 in 1880, 1890, and 1900, respectively. During this period, a wide variety of new implements and machines (increased use of draft horses, better plows, more efficient harvesters, improved machines for planting and cultivating, the spring-tooth harrow, etc.) were being made available to farmers. These implements and machines improved farm efficiency and productivity, but they also required greater investment.

Data on the value of implements and machinery in Alabama during the period 1850-1900 are presented in Figure 8.17. Value declined sharply between 1860 and 1870 ($7.4 to $ 3.3 million); then increased slowly to $4.5 million in 1890. It did not return to the 1860-level until 1900 ($8.7 million).

After the war, value per farm was considerably lower on Alabama farms than on farms in the country as a whole (Figure 8.16). It had been higher in 1860 ($135 versus $120), because of the high value associated with the production and on-farm processing of cotton. However, after the war values in the state were much lower. It was $102 versus $39 in 1870; $101 versus $28 in 1880, and $108 versus $29 in 1890. Some of the difference was obviously related to the average size of farms (Figure 8.11), but differences in the value of implements and machinery were considerably greater than differences in farm size. In 1890, Alabama and the South had a long way to go to catch up with the rapidly increasing level of mechanization on American farms.

Animal Power on Farms

In the latter half of the nineteenth century, the age of the farm tractor in the United States was still a half century away. At the end of the Civil War, many American farmers were still in the process of replacing human labor with power provided by horses and mules. All of the new plowing, cultivating, and harvesting machinery was designed to replace human labor. However, this change in

American agriculture required a significant increase in the amount of animal power. The 1870 Census estimated that there were 7.1 million horses and 1.8 million mules and asses, in the country. In that census, the relatively small number of asses was combined with mules. According to the data, by 1900 those totals had increased to 18.3 million horses and 3.4 million mules and asses. Considering horses and mules together, the total increased 177% (8.9 to 24.7 million) over the period 1870-1890. As might be expected, with so many new machines available that required animal power, the total number of horses and mules increased at a faster rate than the increase in the amount of farmland (53%) (Figure 8.10).

After the war, Alabama and Southern farmers did not increase the total number of horses and mules as rapidly as elsewhere in the country. For example, in Alabama from 1870-1890 the total increased from 157,400 to 256,000 (62%). However, this rate of increase in farm power animals was considerably greater than the increase in total farmland (32%).

In 1860, just before the war, the average number of horses was somewhat higher on American farms than those in Alabama (Figure 8.18). Numbers here are expressed on a per-100-farms basis. This was done to correct for differences in farm numbers and to eliminate those situations where the average number was less than a single animal per farm. The national number declined somewhat after the war (269 per 100 farms in 1870), likely reflecting losses in the South, and it continued to decline through 1880 (258). However, with the rapid growth of horse-related mechanization in the North Central Region, after around 1880, horse numbers in the country began to increase, and by 1890 reached 328 animals per 100 farms. The decline between 1890 and 1900 was likely the result of the national economic situation in the middle of the decade.

The effect of smaller numbers of horses in the South on the national data in Figure 8.18 is obvious in comparing numbers in the North Central and South Central regions. In 1880, there were 322 horses per 100 farms in the North Central Region, but only 217 in the South Central Region. The difference is even more obvious when comparing numbers in Alabama and Illinois in 1880 (45 versus 400).

Following the war, horse numbers per 100 farms declined in Alabama as farm size declined (Figures 8.11 and 8.18). Horses were more expensive and not as effective on those small cotton farms.

Throughout the period 1870-1900, mules per 100 farms were much more popular on Alabama farms than nationally (Figure 8.19). In the state, there were 202 animals per 100 farms in 1860, but afterwards as farm size declined, numbers also declined. The data indicate that in 1870, 1880, and 1890, with numbers below 100 per 100 farms, that many Alabama farms were operated with no mules.

Nationally, mule numbers declined somewhat after the war, but then began to increase slowly thereafter. Throughout the period 1870-1900, however, there were substantially more mules on Alabama farms. In 1870, the comparative numbers were 114 versus 42 per 100 farms. In 1890, they were 85 versus 50. In 1880, numbers in Alabama and Illinois were 89 and 48 per 100 farms, respectively.

Use of Fertilizer

In an 1889 USDA publication, Atwater commented that plant nutrients in the long-cultivated soils of the North Atlantic, South Atlantic, and South Central regions were so depleted that the use of fertilizers was a necessity in crop production. He also added that farmers faced two difficulties in purchasing and using fertilizers. First, the quality and price of available fertilizers were widely variable, and farmers had no way to know which were good and which were useless. Second, even if good fertilizers were available, they did not know which they needed for different crops and soils, or how much to apply, or when and how to apply them.

Fortunately some work had already begun to deal with both of these difficulties. Beginning in the 1870s, state governments throughout the country began to be concerned with fertilizer quality. In 1871, Alabama began to act to rectify this problem. By 1889, agricultural experiment stations in 24 states were analyzing fertilizers, following requirements established by legal authority (Atwater, 1889). Fertilizer inspection in the several states finally made it possible for farmers to know what they were purchasing, but it did not immediately lead to the availability of better fertilizers.

Concern for the technical aspects of crop fertilization (what fertilizer for what crop? how much to use? when and how to apply it? etc.) also began to receive attention in the 1870s. For example, the Alabama Agricultural Experiment Station, became part of the national system after the passage of the Hatch Act in 1887, but the Alabama-financed station was conducting experiments on the use of fertilizer in the production of cotton and corn as early as 1873. The University of Illinois began field trial work on fertilizer material evaluation in 1879, and similar work began at the Pennsylvania Agricultural Experiment Station in 1881. Soon experiment stations throughout much of the country were working on the use of chemical fertilizers in crop production. After the establishment of the national system of experiment stations in 1887, it quickly became obvious that the research would be much more efficient and effective if done cooperatively. Consequently, in 1889 directors of agricultural experiment stations from a number of states met in Washington to plan cooperative work on the use of field trials to determine fertilizer requirements for production of different crops in different cultural environments.

According to Lerner (1975), in 1870 American farmers applied 642 million pounds of commercial fertilizer to their land. They had applied 116 million in 1850. In 1880, 1890, and 1895, usage levels reached 1,506 million; 2,780 million; and 3,156 million pounds, respectively. These usage levels were equivalent to 5.2 and 7.8 pounds an acre of improved farmland in 1880 and 1890. There were no estimates made of improved farmland in 1895.

Plant Agriculture

Annual data on various aspects of agriculture in the United States and the individual states were not collected until 1866. These early collections were made by the Division of Statistics of the US Department of Agriculture. Fortunately, all of this early data is now readily available through the National Agricultural Statistics Service (NASS). While annual data for the individual states have been available from the beginning, it was not included in a separate report for Alabama until 1948 when Volume 1 of Alabama Agricultural Statistics was published by the Alabama Agricultural Statistics Service (AASS). This report included data from 1866 through 1946.

As suggested in a preceding section, plant agriculture in much of the Northern United States emerged essentially unscathed from the Civil War. Unfortunately, while plant agriculture was improving in the North (North Atlantic and North Central regions), Southern and Alabama farmers found themselves dealing with a completely different situation. Plant agriculture in the South had almost been destroyed by the war. Except for the production of cotton, Southern farmers had years ahead of them simply trying to get back where they had been in 1861.

Cotton

Census data indicated that American farmers had harvested 5.4 million 445-pound bales of cotton in 1859. USDA/NASS data indicated that they harvested 2.1 million in 1866. For the remainder of the 1860s, Southern cotton farmers desperately attempted to increase the production of their primary cash crop. Unfortunately, poor crop weather, crop diseases, and pests, hurt them severely during part of the period. By 1870, however, USDA data indicate that they had gotten their cotton harvest to 4.4 million bales. Unfortunately, it took almost a decade to get back to where it had been in 1860.

US Cotton

Census data on cotton production has been limited just to bales harvested; however beginning in 1866, in addition to bales harvested, USDA began to publish data on acres harvested and on yields. Then in 1876, they began to publish annual data on prices. The presentations in the following sections reflects the availability of that expanded data base. In 1870, American farmers harvested 4.4 million bales of cotton (Figure 8.24). Mississippi reported the largest share (19.1%). Other states with large shares included: Alabama (13.4%), Georgia (13.0), Louisiana (13.0), and Texas (12.2). These five states accounted for 70.8% of total American production.

Exports

The United States exported 2.9 million bales of cotton in 1870 (Figure 8.20). Data are based on a foreign trade year that began in August each year. Exports that year were equivalent to 96% of the entire 1869 crop of three million bales. They declined sharply in 1871 (1.8 million bales). In July of 1870, the French declared war on Prussia. The war continued into early 1871. Apparently this conflict, in the heart of Western Europe was sufficient to result in a reduction of imports. Generally, after 1871 exports trended upward through about 1890 (5.9 million bales). They were down slightly in 1874 following the collapse of stock markets throughout Europe, and were higher than expected in 1880 and 1882. These unexpected increases probably were the result of sharp increases in the production of the crop that were beginning to take place in the early 1870s (Figure 8.24). They were also somewhat above trend in 1890 and 1891. Bales harvested were increasing rapidly during this period. Exports were below trend in 1892, 1893, and 1895, likely fueled by the widespread bankruptcies associated with the Depression of 1893-1894. The high level of exports in 1894 was probably the result of the extremely low price of cotton that year, 4.6 cents a pound. The price had never been that low. This low price made cotton extremely attractive to importers everywhere. Then another economic contraction began in 1895, and exports plunged.

Prices

There are no annual data on the annual average prices received by producers for their cotton before 1876. Prices generally ranged between 9.5 and 10.7 cents a pound from 1876 through 1881 (Figure 8.21). It was somewhat lower in 1877 (8.5 cents) and 1878 (8.2 cents). Bales harvested were substantially higher in those two years than they had been previously. With exports increasing smartly in the period 1878-1880 (Figure 8.20), prices also began to increase, although bales harvested were increasing simultaneously. Apparently, exports were sufficiently large to clear the market and allow for better prices. After 1881, however, prices generally began to trend downward even though exports continued to grow. Production was generally increasing rapidly throughout this period. Farmers harvested their first seven-million bale crop in 1887, their first eight-million bale crop in 1890, and their first nine-million bale crop in 1891 (Figure 8.24). In 1894, farmers harvested almost ten million bales (9.9 million). There was no way that they could market that much cotton at any reasonable price either at home or abroad; consequently, in 1894, the price fell to 4.6 cents. It had never been anywhere close to that low in history. The 1895 crop was substantially smaller (7.2 million bales), and as a result, average price recovered somewhat, to 7.6 cents.

Eklelund and Tollison (1988) commented that "in a market system, demand and supply determine prices, and prices are the essential pieces of information on which consumers, households, businesses and resource suppliers make decisions." At least as early as 1833 when cotton prices fell below 15 cents per pound, farmers producing the fiber should have known that there was something seriously wrong with their supply-demand-price relationship. They either did not recognize the problem, or they did recognize it, but could not figure out what they could do about it. In fact, there was virtually nothing that they could do. Unfortunately, in the late nineteenth century South cotton-and-more-cotton was the only game in town. It seemed that the South would never be able to get beyond the pall cast by the adoption of plantation-slave agriculture in Virginia almost three centuries earlier.

Acres Harvested

In 1870, American farmers harvested cotton from 9.2 million acres (Figure 8.22), but it declined slightly in 1871. Generally, acres harvested trended upward through 1891, where it remained relatively unchanged through 1895. Acreage was slightly lower than expected in 1874, probably as a result of the economic depression (Figure 8.1). It was also lower than expected in 1882. Exports had fallen sharply the year before (Figure 8.20). There was considerable year-to-year variation in acres harvested from 1892 to 1895. The nine-million-bale crop (Figure 8.24) and the 7.2-cents per-pound-price in 1891 (Figure 8.21) finally got the attention of American cotton farmers, at least in 1892. However, with bales harvested down and the average price up in 1892, cotton farmers were quickly off to the races again, planting and harvesting more cotton. They harvested more acres in 1894 than ever before (21.9 million), and average price responded quickly by falling to 4.6 cents. With the 1894 disaster, they did plant and harvest less cotton in 1895, but the reduction was not nearly as much as the magnitude of the disaster called for.

These data provide a chilling insight into the degree of desperation of Southern and Alabama cotton farmers. They had only one money crop—cotton. If they reduced their production of cotton, they reduced their already pitifully low annual income. The only response available to them when prices began to fall was to plant and harvest more cotton. As detailed in a preceding section, cotton prices began to trend downward after 1881, but with prices steadily declining they increased acres harvested some 5.9 million (37.5%) acres from 1882 through 1891. It was a desperate replay of the Uncle Remus story about Brer Rabbit and the Tar Baby: the more and the harder he hit, the stucker he got. Unfortunately, in the short term, there was no other choice in Alabama, but to keep hitting.

Readers should write the description of this unfortunate human tragedy indelibly into their memory banks. The history of the tragedy is long and complex, but it will ultimately be converted, through political action, into a unique, but not always helpful, relationship between farmers and government that will last far into the future.

Yields

Data on average annual cotton yields, primarily in the states of the former Confederacy, for the period 1870-1895 are presented in Figure 8.23. Obviously, there was considerable year-to-year variation. This variability is a continuing reminder of the poor fit of the genome of the cotton plant in the agroenvironment of Alabama and the South.

There appeared to be no general trend in the data. In 12 of the 26 years, the average was slightly above or slightly below 170 pounds an acre. There did seem to be an upward trend from around 1884 through 1891; however in 1892, 1893, and 1895, it was again close to 170 pounds. The average for the years 1870 (208 pounds an acre), 1882 (209 pounds), and 1894 (294 pounds) were generally indicative of the potential for cotton production in the South when all factors (primarily related to weather and pests) are ideal for the cultivation and harvest of the cotton plant. Unfortunately, these ideal conditions are an infrequent occurrence in the South.

Bales Harvested

With average yield generally ranging around 170 pounds an acre for the entire period 1870-1895 (Figure 8.23) and acres harvested generally trending upward for the period (Figure 8.22), the trend in bales harvested was similar to that of acres harvested. Bales harvested generally trended upward from around four million in the early 1870s to nine million in 1891. The country got its second nine-million-bale crop in 1894.

Although the trend in bales harvested generally closely followed the trend in acres harvested, large departures from expected yields were quite evident. The effects of the unusually high yields in 1870, 1880, 1881, 1890, 1891, and 1894 on bales harvested are obvious. The effect of the lower than expected yields in 1871, 1881, 1892, and 1895 are also evident.

Alabama Cotton

As detailed in a preceding chapter, the quantity of cotton harvested in Alabama declined from 999,000 bales in 1859 to 429,500 bales in 1869. Fortunately, the price increased from 12 cents a pound to 29 cents during the same period. As a result, Alabama farmers realized slightly more from their crop in 1869 than in 1859.

According to census data, farmers in Montgomery County harvested the largest share of cotton in the state in 1869 (5.9%). Other counties with substantial shares included: Dallas (5.8%), Marengo (5.5), Russell (4.8), and Wilcox (4.7). These five counties accounted for 26.7% of the state's total, which indicates the widespread production of the crop. Some 14 counties reported harvests of 10,000 or more bales

Prices

There are no annual prices received data published for Alabama cotton before 1882, and the data that are available for the period are essentially the same as those shown in Figure 8.21.

Acres Harvested

Alabama farmers also generally increased acres harvested for at least a portion of the period 1870-1895. They harvested the crop from 1.5 million acres in 1870, and 2.8 million in 1885; however after that year, acres harvested changed very little. Alabama farmers apparently were much more concerned about the downward trend in prices (Figure 8.21) than were farmers in the other cotton states. The beginning of the Franco-Prussian War in 1870 seemed to give pause to Alabama cotton farmers. Acres harvested declined in 1881 (1.4 million acres). They also declined in 1874 (1.6 million acres) during a period of national depression and inflation. Poor crop weather in 1881 (Mell, 1890) apparently interrupted the general upward trend in acres harvested. It was also lower than expected in 1882 (2.5 million) at the beginning of another economic contraction. Farmers apparently responded to generally declining prices after 1890. Acres harvested were generally lower than expected from 1892 through 1895.

Yields

Figure 8.23 also includes data on the average annual yield of cotton from Alabama farms during the period 1870-1895. The year-to-year changes in Alabama yields were surprisingly similar to yields in the Cotton Belt as a whole. As was the situation with the entire region, there was no general trend in the data except for the period 1884-1890. Although there was considerable year-to-year variation, the average yield in 1893 was essentially where it had been in 1871—near 140 pounds an acre. Throughout the period, the average yield in Alabama was 40 pounds an acre lower than in the country. If anything should have given pause to cotton farmers in Alabama, it should have been these lower yields. The market for cotton was global almost from the very beginning. As a result, with yields some 40 pounds below the national average, the only way that Alabama cotton farmers could compete was to accept a lower return on investment.

Bales Harvested

Bales harvested in a given year is the product of acres harvested and yield. Although both affect harvest, acres harvested is generally much more variable than yield within short periods of time. As a consequence, trends in the quantity of the crop harvested tend to follow trends in acres harvested more closely. Of course, unusually high or low yields can have a significant short-term effect.

There was a large amount of year-to-year variation in bales harvested from Alabama farms during the period 1870-1895 (Figure 8.26). It generally ranged around 500,000 bales from 1870 through 1878, but then began to trend upward, reaching 1,058,000 bales in 1890. An abnormally wet crop year in 1892 reduced bales harvested to 2,663,000, and it was almost that low in 1894 (2,840,000 bales) as a result of an excessively dry crop year. It is apparent from these data that cotton farming in Alabama during this period was a highly unpredictable business. Farmers never could be certain about what kind of crop they might get from their efforts; however, as was the situation with all of their fellow cotton farmers throughout the Cotton Belt, they did not have many choices.

There are no data available on the harvest of cotton in Alabama's counties in 1895. Data in 1890 Census indicates that farmers in Montgomery County led the state in cotton harvested with 5% of the total. The same five counties that led the state in 1869 continued to do so in 1889, except that Barbour replaced Russell. Farmers in the five leading counties harvested 21.3% of the total crop that year. A total of 16 counties reported harvests of 20,000 or greater bales in 1889.

Corn

Corn production in the United States grew faster than the population from 1839 through 1859. During the period, the number of bushels available per capita increased from 22.1 to 26.7. Unfortunately, by 1869 it had declined to 19.7 bushels. Obviously much of this decline resulted from lower production in the South. Farmers in this region had lost 36 million acres of farmland between 1860 and 1870. Although there are no data available on acres planted, it is likely that in their zeal to get cotton production back to normal, they simply planted less corn.

US Corn

In 1870, American farmers harvested 1.2 billion bushels of corn. Farmers in Illinois harvested the largest share (20.8%) of the total national crop. Other states reporting large shares included: Iowa (8.3%), Ohio (8.3), Missouri (7.2), and Kentucky (5.6). These five states accounted for 54.9% of the total crop that year. These data reflect the fact that the immigration of corn production out of Tennessee and Kentucky to the North Central Region was almost complete.

Exports

The United States exported only 10.7 million bushels

of corn (Figure 8.27) in 1870 (foreign trade year beginning July). The exports that year were equivalent to only 1.4% of the 1869 crop (782.1 million bushels). Exports more than tripled in 1871 (35.7 million bushels). They fell in 1873 and 1874 as a result of the period of depression and deflation. Also, two poor crops in 1873 and 1874 likely played a role. When the economic recovery began after 1874, corn exports began to trend rapidly upward, reaching 99.6 million bushels in 1879. Exports in 1879 were equivalent to 6.4% of the 1878 crop (1.5 billion bushels). They were down sharply in 1881 (41.7 million bushels) as the result of an extremely poor crop. They remained near that relatively low level through 1883. Probably the beginning of the 1882-1885 economic contraction also had an effect on exports during this period. Exports during the period 1884-1895 were highly variable, ranging from 28.6 million bushels in 1894 to 103.4 million in 1889. Throughout this period, the corn crops in the country were also highly variable (Figure 8.31). For example, in 1890 American farmers only harvested 1.6 billion bushels, and exports were down to 32 million bushels. In contrast, in 1889 they had harvest 2.3 billion bushels, and exports were up to 103.4 million. Widespread depression and bankruptcies from 1893 through 1894 probably also contributed to the high degree of variability in exports during this period.

Prices

Data on the annual average price received for corn by American farmers during the period 1870-1895 are presented in Figure 8.28. Given the low level of exports relative to bushels harvested, it is unlikely that exports had very much effect on prices received most years. Although there was considerable year-to-year variation, as supply increased, prices generally trended downward slightly throughout the entire period (Figure 8.31). It was 80 cents in 1870 and 50 cents in 1894. The larger swings in price during the period were generally related to poor crops in those years. For example, between 1880 and 1881 average price increased from 39 to 63 cents a bushel. In the same period, bushels harvested declined from 1.7 to 1.2 billion.

Acres Harvested

Data on acres harvested for corn during the period 1870-1895 are presented in Figure 8.29. Generally, acres harvested trended upward throughout the entire period, but it was more apparent during the early years. Farmers harvested the grain from 38.4 million acres in 1870 and 80 million acres in 1894 (an increase of 108%). Prices seemed to have a significant effect on acres harvested in some years. For example, lower prices in the early 1870s (Figure 8.28) seemed to reduce the increase in acres harvested in 1873. Similarly, the low price in 1889 (28 cents) apparently resulted in farmers planting and harvesting less corn in 1890. The price was up modestly in 1894 compared to the preceding three years, and in 1895 farmers harvested corn from 90.5 million acres, an increase of 10 million acres over 1894.

Yields

Data on annual average corn yields in the United States during the period 1870-1895 are presented in Figure 8.30. Year-to-year variation in yield was extremely high during the entire period, and there did not seem to be a trend. The average generally varied between 20 and 30 bushels an acre, with most of the values falling around 25 bushels an acre. Yield was especially low in 1874, 1881, 1887, 1890, and 1894. Plots of Palmer Drought Severity Indices (Cook, et al., 1999) indicate that crop weather in the Corn Belt was excessively dry to severely dry in three of those years (1874, 1887, and 1894) and abnormally moist in 1881. Surprisingly, crop weather in 1890 was normal in that region. The average yield for that year was one of the lower ones (22.1 bushels an acre) of the entire period. Possibly a corn disease outbreak or a problem with plant pests caused the low yield, but it did not seem to be related to weather. Oddly, average yield was also low in Alabama that year.

Bushels Harvested

Data on bushels of corn harvested in the United States during the period 1870-1895 are presented in Figure 8.31. With no specific trend in yield (Figure 8.30) bushels harvested generally trended upward along with acres harvested (Figure 8.29) throughout the entire period. Farmers harvested 1.1 billion bushels in 1870 and 2.5 billion in 1895, an increase of 127%. While the general upward trend was primarily the result of the same upward trend in acres harvested, the effects of high levels of year-to-year variation in yield is also evident. The unusually low harvests (1874, 1881, 1887, 1890, and 1894) are associated with those years when yield was also unusually low.

In 1895, Illinois farmers harvested the largest share of the national corn crop (16.2%). Other states reporting large shares included: Iowa (14%), Missouri (10.5), Kansas (8.1), and Indiana (6.1). These five states accounted for 54.9% of the total crop. Kentucky was no longer in the top five. Immigration of corn production to its natural home was essentially complete. The game of national corn production was mostly over. Unfortunately, not all of the contestants knew it.

Alabama Corn

In 1859, Alabamians had available to them 35.1 bushels of corn per capita, or 10 bushels more than the national average. Corn was extremely important in the Alabamians' diet. Unfortunately, in 1869 it was 17.1 bushels, slightly less than half what it had been in 1859.

In 1869, Alabama farmers harvested 17 million bushels of corn. The largest share, 5%, was reported from Perry County. Other counties reporting large shares included: Madison (4%), Wilcox (3.9), Montgomery (3.5), and Lawrence (3.1). These five counties accounted for only 19.5% of the total. The state's average was near 261,200 bushels per county. Some 44 of the Alabama's 65 counties reported harvests of 200,000 or more bushels.

Prices

Figure 8.28 also includes data for prices for corn for Alabama during the period 1870-1895. Although there was considerable year-to-year variation, it appears that prices in Alabama also generally trended downward during the period. They were between 70 and 80 cents a bushel in the early 1870s, and between 60 and 70 cents during the 1890s. Over the period, the relative increase in bushels harvested was somewhat larger in the country as a whole than in Alabama.

The general similarity in the year-to-year changes in corn prices in the United States and Alabama is somewhat surprising. Of course, prices received in Alabama was 10 to 30 cents a bushel higher than the national average, but the similarity of the changes suggest that the same factors (supply and demand) were operating with the same general intensity in both cases. However, national events that resulted in major change in prices usually trumped local changes. For example, bushels harvested in the United States (Figure 8.31) increased slightly (1.9%) from 1875 to 1886, putting slightly downward pressure on prices that year. However, the corn crop in Alabama was 8.9% higher in 1876 than in 1875. As a result, the degree of decline in price in the state was much greater than in the country as a whole. In contrast, a major shortfall in bushels harvested nationally in 1881 resulted in a sharp increase in price, and there was similar increase in the Alabama price, although bushels harvested in the state actually increased slightly that year.

In general, by the end of the nineteenth century, corn prices were primarily determined far away from Alabama. Of course, corn production in Alabama represented only a small fraction of national production. As noted in a preceding section, it was 1.5% of the total in 1870. It was just slightly higher (1.7%) in 1895. Also, just five states harvested 50% of the total crop in 1870. But at that time, Alabama farmers were able to get a higher price for their crop by selling locally.

Acres Harvested

Acres of corn harvested in Alabama generally trended upward throughout the period 1870-1895, just as it did nationally (Figures 8.29 and 8.32). Alabama farmers harvested corn from 1.3 million acres in 1870 and 2.8 million in 1895, an increase of 115%. Acres harvested were much lower than expected during two three-year periods, 1879-1881 and 1892-1894. Crop weather was apparently normal over most of the state during the growing season in those periods. Also, yields were not abnormal. The problem must have been the result of poor weather at the time of planting.

Yields

Figure 8.30 included data on annual average yields for corn, for both the country and state. Although there was considerable year-to-year variation, annual yields in Alabama seemed to trend up just slightly during the period 1870-1895. In the early 1870s, they remained generally unchanged at around 12.5 bushels an acre; however from 1884 through 1895 they generally increased from around 13 to around 15 bushels. In any case, corn yields in Alabama were substantially lower than in the national average yields. In Alabama, they were around 10 bushels lower than in the country for most of the period; however, when compared with the states producing most of the corn, the difference is much greater. For example, in 1895 the average yield in Alabama was 15.5 bushels. In the same year, the average for Illinois, Iowa, Missouri, Kansas, and Indiana (the five leading corn-producing states), as a group, was 36.6 bushels.

As detailed in the preceding chapter, Alabama agriculture suffered terribly during the Civil War, and the data for yields of both cotton and corn in the years immediately following the conflict indicate that the state was still years away from being competitive with the remainder of the country.

Bushels Harvested

Alabama farmers harvested 18 million bushels of corn (1.5% of the national total) in 1870. They had harvested 33.2 million in 1860. With yields (Figure 8.30) changing very little during the period, bushels harvested generally followed acres harvested (Figure 8.32), upward from 18 million in 1870 to 38 million in 1894 (Figure 8.33). The effects of lower than expected acres harvested during the periods 1879-1881 and 1892-1894 are quite noticeable in the bushels harvested data. Bushels harvested was also lower than expected in 1890, a result of a lower-than-expected average yield.

There are no data available on the harvest of corn in Alabama's counties in 1895. The 1890 Census included data on the harvest of the crop in the counties in 1889. Those data indicated that farmers in Jackson County harvested the largest share of that crop (3.9%). Other counties with large shares of the total included; Madison (3.7%), Lowndes (3.5), Marengo (2.6), and Sumter (2.6). These data indicate that by 1889 corn production was growing more widespread in the state. The five leading counties harvested only 16.3% of the total crop. That year, a total of 25 counties reported harvests of 500,000 bushels or more.

Peanuts

The first data on the production of peanuts in the United States in the censuses was published in 1890. According to that data, 203,900 acres of the crop were harvested in the country in 1889. Virginia farmers harvested 28.9% of the total. Other states harvesting significant acreages included: Georgia (25.6%), Florida (12.8), Alabama (11.7), and North Carolina (8.7). These five states harvested 87.7% of the total. Even at that time, it was obvious that the peanut was primarily a crop of the Southern coastal plain. The total harvest amounted to 3.6 million bushels, and the average yield was 17.6 bushels an acre.

Soybeans

As noted in a preceding chapter, USDA only began to publish data on soybean prices in 1913, but no production data were published until 1924.

Alabama Hay

In 1870, there were 257,300 "other" (primarily beef animals) cattle on Alabama farms. In 1869, farmers in the state had harvested 10,613 tons of all kinds of hay. From these data, it can be roughly estimated that there would not have been enough hay harvested to provide more than 80 pounds (0.04 ton) per animal for the year. Obviously, most of the animals were obtaining most of their feed from another source.

Prices

Data on the annual average price per ton for hay in Alabama during the period 1870-1895 are presented in Figure 8.34. Generally, prices trended downward throughout the entire period. It was $18 a ton in 1870, but declined to $10 a ton in 1895. Data presented in a following section (Figure 8.37) will support the conclusion that prices declined sharply as tons harvested increased. The relatively large amount of year-to-year variation in prices seemed to be related to variation in annual yields (Figure 8.36). For example, the sharp decline in 1877 seemed to follow two years of relatively high yields. Similarly, the spike in prices in 1890 came in the same year when the yield was relatively low.

Acres Harvested

Data on acres of hay harvested annually in Alabama during the period 1870-1895 are presented in Figure 8.35. Data to be presented in a following section will show that the number of cattle (all) in the country and the state increased rapidly during this period. As a result, farmers likely increased acres harvested of the crop to meet the needs of the growing cattle industry. Alabama farmers harvested hay from 10,000 acres of land in 1870 and 85,000 in 1895. Acres harvested generally increased at an exponential rate during the period. The rate of increase was relatively low from 1870 through 1880, higher from 1881 through 1889, and even higher from 1890 through 1895. It is interesting that in determining how much hay to harvest annually, farmers seemed to be paying more attention to the increase in cattle numbers rather than to rising or declining hay prices.

Yield

Data on the annual average yield of hay on Alabama farms during the period 1870-1895 are presented in Figure 8.36. Generally, yields during the period ranged between 0.8 and 1 ton an acre. Although there was considerable year-to-year variation, they seemed to trend very slightly upward, especially in the latter part of the period. It is likely that those yields that were much lower than expected (1873, 1879, 1885, and 1890) were the result of poor years for the production and harvest of hay. Even short periods of drought can seriously reduce the amount of hay produced on most Alabama soils.

Tons Harvested

Data on tons of hay harvested from Alabama farms during the period 1870-1895 are presented in Figure 8.37. With average yield remaining relatively unchanged during the period, tons harvested generally followed acres harvested upward from 10,000 tons in 1870 to 86,000 tons in 1895. Lower than expected harvests (1880, 1885, and 1890) were generally associated with lower than expected yields.

Tons of hay harvested increased by 760% between 1870 and 1895. During the same period, cattle (all) numbers only increased 39% (620,000 to 863,000). In 1895, farmers harvested only 0.1 ton per animal. They had harvested considerably less in 1870. By 1895, farmers were feeding their cattle more hay, but grazing was obviously the primary feed source for their animals.

There are no data available on the quantity of hay harvested in Alabama's counties in 1895. The 1890 Census contains data on the quantity harvested in 1889. In that year, farmers in Madison County harvested the largest share of the state's total crop (15.1%). Other counties reporting large shares of the total included: Jackson (7.6%), Tuscaloosa (7.2), Mobile (5.2), and Dallas (4.7). These five counties accounted for 39.8% of the total crop. A total of 19 counties reported harvests of 1,000 tons or greater.

Alabama Wheat

There were two major changes in wheat production in the United States between 1839 and 1869: Production had moved rapidly westward, and it was increasing much faster than the population. In 1839, 32.4% of all wheat was harvested in the North Central Region. In 1869, farmers in that region harvested 67.8% of the total crop. Between 1839 and 1869, the American population increased by 55% (16.8 to 37.9 million), but wheat harvested increased by more than 239%. The rate of increase in Alabama during the period was much lower (25%).

American wheat farmers harvested 287.7 bushels of wheat in 1869. They had harvested 173.1 million in 1859. Iowa farmers produced the largest share of the 1869 crop (10.4%). Other leading wheat-producing states included: Illinois (9.7%), Wisconsin (8.9), Ohio (8.8), and California (8.1). At the time, USDA was not differentiating between the types of wheat in its reports. The data are for "all" wheat, but from earliest times Alabama farmers planted only the winter variety. In 1869, they harvested 0.4% of the total national crop. With Alabama wheat production playing such a minor role, little of the national data will be used in the following discussion.

Although relatively little US data will be reported on wheat, it is necessary to consider some of it to understand the Alabama data. Data on US exports and prices are essential to understanding some of the changes seen in Alabama prices. Further, data on US yields are necessary to fully understand changes in US prices. Also, comparison of data on yields helps to understand the long-term attitude of Alabama farmers about wheat production.

US Exports

American wheat (all) exports were 52.6 million bushels in 1870 (Figure 8.38). They were equivalent to 18.1% of the entire 1869 crop, and the proportion exported generally increased during the period 1870-1895. Exports were substantial enough that small changes could have influenced prices. They were around 70 million bushels throughout the early 1870s; then increased rapidly to 188.3 million bushels in 1880. During the next decade, they drifted down to around 100 million bushels. They were up sharply in 1891, but then trended downward through the end of the period.

Year-to-year changes in exports were the result of several factors, including American production and prices, as well as production of the crop in Europe and international events. For example, the Franco-Prussian War was probably responsible for the decline in 1871. Lower exports in 1885 were likely the result of the harvest of the smallest American crop in a decade. In contrast, the unusually high level in 1891 was probably the result of the fact that the largest American crop in history was harvested in that year.

US and Alabama Prices

Although there was considerable year-to-year variation, American "all" wheat prices mostly trended downward throughout the period 1870-1895 (Figure 8.39). The annual average was around $1.10 a bushel at the beginning of the period and near 50 cents a bushel at the end. The downward trend in prices was probably a direct result of the upward trend in bushels harvested. In the early 1870s, American farmers were harvesting around 27 billion bushels annually, and in the early 1890s they were harvesting around 55 billion.

Cotton, corn, and hay are of little or no direct importance as human food. Wheat is another matter. This crop is primarily used as food for people. Never before in history had farmers anywhere been able to produce a surplus of the crop over an extended period of time. In the 1880s, American farmers had every reason to believe that it would be only a matter of time before demand would overtake supply, but they failed to fully appreciate the long-term effect that the revolution in agricultural technology would have on production of the crop.

Data on Alabama's average annual prices for winter wheat are also presented in Figure 8.39. The data show that in both Alabama and the United States prices trended downward during the period at about the same rate; however, farmers in the state received substantially more than their counterparts throughout the country. In 1881, the differential was almost 40 cents a bushel; generally, it was between 10 and 20 cents. Apparently by selling relatively small quantities of the grain locally, they were able to receive higher prices.

Acres Harvested

Data on acres of winter wheat harvested in Alabama during the period 1870-1895 are presented in Figure 8.40. With local prices increasing sharply in the early 1870s (Figure 8.40), Alabama farmers apparently decided to increase the amount of land devoted to wheat production. At the time, cotton prices were hovering around 10 cents a pound. The farmers had considerable incentive to increase wheat production. Consequently, acres harvested grew sharply from 140,000 acres in 1870 to 270,000 acres in 1877. Unfortunately, the average price fell rapidly after 1873; however, farmers decided to stick with the cereal grain through 1879. Then, when it became obvious that prices were not going to increase in the near future, they got out of the wheat business as fast as they had gotten into it. By 1890, they were harvesting the crop from less than 50,000 acres.

The decline in acres of winter wheat harvested was not continuous. Prices increased sharply in 1881, to near $1.20 a bushel, and farmers could not stand it. In 1882, they increased acres harvested substantially (30,000 acres). Again, they were a day late and a dollar short. When they harvested those 185,000 acres, the price was down to $1.10. As detailed in a preceding section, Alabama farmers did not, or could not, respond to negative price signals in the case of cotton, but they were probably over-responsive in the case of wheat. However, as will be demonstrated in a subsequent section, the demand-supply-price relationship was not the only problem that wheat farmers in Alabama had to deal with. Yields were even more troubling,

US and Alabama Yields

Data on annual average yield for both the country and the state for the period 1870-1895 are presented in Figure 8.41. Although there was considerable year-to-year variation, American yields seemed to trend up slightly during the period from around 12 to 13 bushels an acre.

Year-to-year variation was generally lower in Alabama than in the country as a whole; however, the coefficients of variation (standard deviation/mean) would probably have been about the same. There is little indication that yields changed at all in the state during the period. It was between six and seven bushels an acre in the early 1870s and near the same level in the early 1890s. Throughout the period, they were consistently about half of those in the country. Wheat yields provided absolutely no incentive for Alabama farmers to continue to plant the crop. Some years, they hardly got more than their seed back.

Bushels Harvested

Data on bushels of wheat harvested in Alabama during the period 1870-1895 are presented in Figure 8.42. With average yield (Figure 8.41) changing relatively little during the period, year-to-year changes in bushels harvested generally followed changes in acres harvested (Figure 8.40). Farmers harvested around one million bushels in 1870 and 1871. Then, as acres harvested trended upward, bushels harvested followed. It reached 1.9 million bushels in 1877. Then, with acres devoted to the crop declining rapidly, bushels harvested also fell rapidly. In the early 1890s, Alabama farmers harvested less than 400,000 bushels, considerably below the level of the 1870s. For the time, at least, Alabama s flirtation with the cereal was at an extremely low ebb.

There are no data available on the wheat harvest in Alabama's counties in 1895. The 1890 Census includes data on the harvest in 1889. Farmers in Cherokee County harvested the largest share of the 1889 Alabama crop (11.3%). Other counties reporting large shares included: Calhoun (8.5%), Cleburne (7.0), Madison (6.2), and Talladega (6.3). All of these counties are located in the northern half of the state. A total of 21 counties did not report any wheat harvest that year.

Alabama Oats

During the period 1870-1895, the primary use of oats in the United States was as feed for livestock, principally horses and mules. The number of horses and mules combined increased from 8.9 to 20.6 million animals nationally; in Alabama, the number increased from 180,000 to 282,000. These increases provided a rapidly growing market for oats.

In 1869, Illinois farmers harvested the largest share (15.2%) of the 282.1-million-bushel crop of oats. Other states reporting large shares of the total crop included: Pennsylvania (12.9%), New York (12.5), Ohio (9.0), and Iowa (7.4). These five states accounted for 57% of the total national crop. That year, Alabama farmers harvested only 0.3% of the national crop. With the state's oat farmers playing such a minor role, relatively little of the national data will be used in the following discussions.

US Exports

Data on the quantity of oats exported from the United States during the period 1870-1895 are presented in Figure 8.43. As the data indicate, exports were highly variable during the period, especially after 1877. Throughout the period, export of oats seemed to be closely tied to harvest. Apparently, only when bushels harvested was sufficiently high to meet the national feed requirements of the growing number of horses and mules were any oats exported. In 1875 and 1881, when harvests were especially low the preceding year, large quantities of oats were imported to meet the needs for animal feed.

In 1870, the United States exported 148,000 bushels that was equivalent to less than 0.1% of the 1869 crop. Throughout the early 1870s, the country exported considerably less than one million bushels per year; however, after 1874 production reached a level where there was a surplus. Bushels harvested increased very little from 1877 through 1881, and exports fell sharply. In 1882, harvests began to increase again, and exports followed them upward to 7.3 million bushels in 1885. After 1885, exports were extremely variable. In 1888, farmers harvested 773.1 million bushels and only exported 1.2 million bushels. The following year (1889), they harvested 831 million bushels and exported 15.1 million. The following year (1890), the harvest was down again (609.1 million bushels), and exports declined sharply to 1.4 million bushels. This seesaw relationship between bushels harvested and exports continued throughout the remainder of the period. In 1895, a total of 15.2 million bushels of oats was exported. This total represented 1.6% of the crop (924.9 million bushels) that year. With exports this low, it is not likely that they had much effect on prices.

US and Alabama Prices

Data on annual average prices of oats received by American farmers during the period 1870-1895 are presented in Figure 8.44. Although there was considerable year-to-year variation, in response to the supply situation, national prices generally declined from around 75 cents a bushel in the early 1870s to near 50 cents in the early 1900s. Even though prices declined, year-to-year variation was extremely high. A substantial share of this variation was the result of the variation in yields (Figure 8.46), and consequently, the variation in harvests. For example, yield declined sharply from 1873 to 1874, and in 1874 price increased sharply. In contrast, yields trended slightly downward during the period 1885-1887. In response, prices trended slightly upward.

Year-to-year changes in oat prices in Alabama during the period 1870-1895 were generally similar to those in the country. In fact, it is surprising how similar they were. This comparison indicates that prices in the state were largely determined by factors outside of the state. This is not unexpected. Alabama was such a small player in the national oat economy that it was a price taker rather than a price maker. Although the year-to-year changes were similar, Alabama farmers were able to command a higher local price for their oats, but always within bounds set outside the state. In most years, Alabama prices were 20 to 30 cents higher than the average for the country. Even with all of the year-to-year variation, prices in Alabama trended downward during the period 1870-1895 from around 70 cents a bushel to near 60 cents.

Acres Harvested

Data on acres of oats harvested in Alabama during the period 1870-1885 are presented in Figure 8.45. Farmers harvested 85,000 acres in 1870. Acreage harvested increased rapidly through 1879 (325,000 acres), as it did elsewhere in the country. At the same time, the number of horses and mules was also increasing rapidly in Alabama. However, it is surprising that acres harvested increased at all. There certainly was no indication in prices (Figure 8.44) that supply was not adequate to meet demand. It is even more surprising that acres harvested increased substantially when prices were declining. They were justified in staying the course, however, when prices increased in 1879, 1880, and 1881. Unfortunately, prices did not respond when supply declined sharply between 1881 and 1882. In 1883, they harvested the crop from 400,000 acres, the highest level in history. When they harvested that crop, the average price was 60 cents. That 1883 mismatch between acres harvested and prices finally got the farmers' attention, and acres harvested generally trended downward for the remainder of the period.

US and Alabama Yields

Data on annual average yields of oats in the United States and Alabama during the period 1870-1895 are presented in Figure 8.46. Nationally, yields generally ranged between 21 and 30 bushels an acre during the period, but most were clustered around 28 bushels. There did not seem to be a trend. The extremely low yields, in 1874, 1876, and 1890 are likely the result of poor crop weather, insect depredation, or outbreaks of diseases. There was some year-to-year variation in yields in Alabama, but it was generally much less than elsewhere in the country, although it is likely that coefficients of variation (standard deviation/mean) would have been similar in both. In Alabama, it generally ranged between 11 and 12 bushels, which was 15 to 16 bushels less than in the United States. There is no apparent trend involved in the Alabama data. With these differences in yields in the state and the country, Alabama farmers had to give strong consideration to purchasing their oats rather than attempting to grow them themselves.

Bushels Harvested

Data on bushels of oats harvested in Alabama during the period 1870-1895 are presented in Figure 8.47. With no trend in average yields during the period, year-to-year changes in bushels harvested generally followed changes in acres harvested. From 1.1 million bushels in 1870, bushels harvested increased to 5.3 million in 1883. Lower yields (Figure 8.46) resulted from the slight decline in acres harvested in 1880 and 1881. Bushels harvested also followed acres harvested downward after 1882, but slightly improved yields after 1890 boosted harvests somewhat. Alabama farmers harvested 3.9 million bushels of oats in 1895. In 1870, farmers had harvested 5.9 bushels of oats per work animal. In 1895, they harvested 13.9 bushels. If farmers were feeding their animals oats, they were purchasing them from some other source. They certainly were not producing enough in Alabama to meet their needs.

There are no data available on bushels of oats harvested in Alabama's counties in 1895, but 1900 Census data indicate that Alabama farmers harvested 3.2 million bushels of the crop in 1899 and that Elmore County led the state in harvest with 4.3% of the total. Other counties reporting significant harvests included: Tallapoosa (4.2%), Talladega (3.9), Montgomery (3.3), and Lee (3.1). These five counties only accounted for 18.8% of the total. The state average was 50,000 bushels per county. Some 33 counties reported harvests of 40,000 or more bushels.

Alabama Sweet Potatoes

Sweet potatoes are likely to have originated in Africa and are now widely cultivated there, in Latin America and Southeast Asia.

During the period 1870-1895, sweet potatoes were becoming increasingly important to Alabama farmers. In 1870, they harvested 140 pounds of sweet potatoes for each Alabamian. By 1895, this ratio had increased to 165 pounds.

From 1870 through 1876, farmers received $1.30 to $1.48 per Cwt for their potatoes, but in 1879 prices declined to 85 cents and remained near that level through 1895.

Data on acres of sweet potatoes harvested in Alabama during the period 1870-1895 are presented in Figure 8.48. With the growth of the population during this period, the demand for sweet potatoes also increased. Alabama farmers responded by increasing their acres harvested each year. They harvested the crop from 28,000 acres in 1870 and from 61,000 in 1895. Even declining prices after 1879 did not deter farmers from planting and harvesting more acres of the crop. It is likely that many farmers were growing sweet potatoes to feed their families, and prices did not concern them very much.

Year-to-year variation in average yield was relatively high. Excepting an extremely low value in 1883, yield varied between 34 and 52 Cwt an acre. The average for the 26-year period was 44 Cwt. Yields tended to be slightly higher toward the end of the period. With no trend in yields, Cwt harvested generally followed acres harvested upward from 1.2 million Cwt in the early 1870s to 3.1 million in the early 1890s.

There are no data on the quantity of sweet potatoes harvested in the counties in 1895. The 1890 Census includes data on the harvest in 1889. In that year, farmers in Dallas County harvested the largest share of that crop (4.9%). Other counties reporting large shares of the total included: Wilcox (3.7%), Lowndes (3.5), Henry (3.0), and Clarke (2.7). These five counties accounted for only 17.8% of the total crop. The average Cwt per county was 35,400 Cwt. A total of 12 counties reported harvests of 27,000 Cwt or greater.

Alabama Irish Potatoes

Irish potatoes would eventually become an important commercial crop in some parts of Alabama, but in the period 1870-1895 farmers did not spend much effort on the crop. They only harvested potatoes from around 2,000 acres in 1870 and 6,000 acres in 1895. It is surprising that records were kept on production, yields, and prices for a crop with such a low level of acres harvested. Prices were actually better for Irish potatoes than for sweet potatoes. Also yields for the so-called white potato were slightly better, but apparently there was only a very limited market for the crop during this period.

Alabama Sorghum and Sugarcane

In the latter part of the nineteenth century, farms in the South Central Region continued to be the primary source of sorghum molasses and sugarcane syrup in the country. In 1889, farmers in the region produced 43.8% of the country's molasses and 80% of its sugarcane syrup. Farms in the North Central Region produced 38.2% of the molasses.

Census data published in 1890 included data on the production of sorghum molasses and sugarcane syrup in the country and the states in 1869, 1879, and 1889 (Table 8.2). These data indicated that in those years American farmers produced 16.0, 28.4, and 24.3 million gallons of sorghum molasses, respectively. These levels of production are roughly equivalent to 0.41, 0.56 and 0.38 gallon per capita, respectively. The data also show that they produced 6.6, 16.6, and 25.4 million gallons of sugarcane syrup, respectively. These production levels are equivalent to 0.17, 0.33, and 0.40 gallon of sugarcane syrup per capita, respectively.

Data in Table 8.2 also showed that Alabama farmers produced 0.27, 1.16, and 1.24 million gallons of sorghum molasses in 1869, 1879, and 1889, respectively. These production levels were equivalent to 0.27, 0.92, and 0.82 gallon per capita, respectively. Census data also show Alabama farmers produced 0.17, 0.80, and 2.3 million gallons of sugarcane syrup, respectively, in those same years. These levels were equivalent to 0.17, 0.63 and 1.54 gallon(s) per capita, respectively.

Alabama Market Gardens

As was the situation in the preceding period (1860-1870), there is no specific information on vegetable production in Alabama during the period 1870-1895. The 1890 Census only included data on cash receipts for products of market gardens, including small fruits. These values for the United States for the years 1869, 1879, and 1889 were: $20.7, $21.8, and $29 million, respectively. In those same years, cash receipts for Alabama were: $139,600; $135,100; and $431,800, respectively. In those three years, the Alabama share of the national total was 0.7, 0.6, and 1.5%, respectively.

In 1889, the largest share (34.8%) of total cash receipts of products of market gardens in Alabama was reported by Mobile County. Other counties with large shares included: Cherokee (17.3%), Escambia (14.1), Dallas (6.8), and Jefferson (6.4). In that year, 10 of Alabama's 66 counties did not report any value.

Other Crops

There are several crops that Alabama farmers have produced over the years that do not fit the other categories (major field crops, orchard crops, vegetables, etc.). Some of these will be considered in this section. In the period 1870-1895, there was only a single crop being grown in sufficient quantity in the state to be included in this other-crop category.

The so-called cowpea (black-eyed pea, crowder pea, etc.), *Vigna unguiculata unguiculata*, is likely to have originated in Africa, and is now widely cultivated there. It is also widely cultivated in Latin America and Southeast Asia as well. It was probably brought to North America as early as the 1500s.

Farmers learned through experience that when cowpeas were grown along with corn that the production of the grain was enhanced. Later, they would learn that this plant was able to fix elemental nitrogen out of the atmosphere. The plant requires a hot climate; consequently, it was grown primarily in the South. Although, it was produced on Southern farms from early pioneering days, no specific data on the crop appeared in the censuses until 1880. Before that time, it was probably reported as part of a larger group of crops, the pulses. Most commonly, the seeds of the plant were planted alongside stalks of corn at the time of last cultivation. Some of the peas were picked for consumption by the farm family or for sale, but probably the larger share was left in the field when the corn was harvested. Then farm animals were allowed to harvest the beans and vines by foraging.

Data recorded in the 1890 Census indicate that Alabama farmers harvested 414,400 bushels of cowpeas in 1879 and 321,000 bushels in 1889. Farmers in Russell County harvested the largest share (7.9%) of the 1889 crop. Other counties reporting large shares of the total include: Autauga (7.2%), Barbour (6.2), Elmore (4.5), and Lee (4.2).

Alabama Orchard Crops

In the twentieth century, the production of peaches and other orchard crops would be relatively important in some parts of the state, but from 1870-1890 there was very little data published on their production. The 1890 Census published the first detailed information on the production of fruits in the states and counties. Table 8.3 includes data on the production of apples, peaches, pears, cherries, apricots, and plums and prunes in Alabama in 1889. The data indicate that peaches and apples were the most important fruits harvested in Alabama at that time. Farmers harvested 2.4 million bushels of peaches. Farms in Jackson County led the state in the number of bearing peach trees, with 47,700. Other counties with large numbers included: Chambers (46,400); Lee (45,800); Blount (41,800); and Jefferson (37,800).

The Alabama apple harvest in 1889 totaled 1.2 million bushels. DeKalb County led the state in the number (53,100) of bearing apple trees. Other counties with large numbers included: Jackson (39,800); Blount (36,200); Cleburne (32,600); and Cullman (32,500). All of these counties are located in north Alabama.

Alabama Greenhouse and Nursery

Near the end of the twentieth century, the floriculture and nursery industries were among the fastest growing in Alabama; however, there is very little information available on either before 1900. The 1890 Census published the results of the first national survey of the nursery industry and its characteristics. Some of the results of the study are included in Table 8.4. These data show that there were 4,510 nurseries in the country. More than half of these were in the North Central Region (52.3%). The other large share (26.6%) was counted in the North Atlantic Region. New York State reported the largest share of nurseries in the country (11.8%). Other states reporting large shares included: Illinois (9.6%), Ohio (8.7), Kansas (7.5), and Pennsylvania (7.0). These five states accounted for 44.6% of all nurseries in the country. Alabama reported only 0.3% of the total.

The authors of the report on nurseries commented that the value data seemed to be underestimated. Apparently, the nursery owners did not price their stock on hand at fair market value. All of their estimates appeared to be low. The total estimated value of all nurseries was $42 million. The North Atlantic Region reported the second highest number (26.6%) of nurseries (Table 8.4), but the largest share of total value (40.9%).

Alabama Timber

Data on lumber production in the different regions of the United States in the period 1869-1899 are presented in Table 8.5. These data provide a useful indication of the status and use of timber, over time, in the different geographical regions. The data show that in 1869 lumber production was centered in the North Central and North Atlantic regions with 49.4% and 35.7%, respectively, of total production. Until about 1815, lumber production was centered in the North Atlantic Region where the population increase plus urbanization would have required large amounts of lumber; however, by 1869 the population and urbanization was growing so rapidly in the North Central Region that lumber production was rapidly shifting to that region. Further, by that time much of the saw timber had been cut in the North Atlantic Region, and lumber production was rapidly shifting to the enormous, virgin white pine forests of the North Central Region. By 1899, lumber production was falling rapidly in the North Atlantic Region, as timber resources were being rapidly depleted. Also, it was beginning to decline in the North Central Region where timber resources were also being depleted. As a result of the rapid changes taking place in the North, lumber production was beginning to shift southward and westward. By 1899, almost half of the country s lumber was being manufactured in the South and the West.

By the beginning of the twenty-first century, considering all products marketed from Alabama's land base, cash receipts from all forest products would be exceeded only by receipts from the sale of broilers. However, there is almost no information on the sale of or value of timber in the state before 1900, and not very much even then. The 1870 Census recorded that the value of forest production in Alabama that year was $85,933, or about 0.2% of the national total.

The 1880 Census reported that there were 10.4 million acres of woodland and forests in the state and that in 1879 slightly less than 1.9 million cords of wood with a value of $2.2 million had been harvested. Baldwin, Madison, Pickens, Talladega, and Hale counties reported the largest shares of total wood cut. There were no data in the 1890 Census on forests or forest products in the United States or the individual states.

Animal Agriculture

In June 1870, the value of all livestock in the United States totaled $1.5 billion. The value of livestock in the North Central Region was the highest in the country, representing 43.7% of the total. Values in the North Atlantic, South Central, South Atlantic, and Western regions accounted for 27.1, 16.3, 9.0, and 3.8% of the total value, respectively. Value in Alabama accounted for only 1.7%.

The South's livestock industry was ravaged by the Civil War. The data presented in the preceding paragraph are only partly indicative of its status just five years after the hostilities ended. It would be many years before it would return to its antebellum condition, and one can only imagine how the industry might have developed had it not been subjected to such severe devastation.

Cattle

According to census data, in 1850 there were 43 "other" cattle (excluding milk cows and working oxen) in the United States for every 100 persons. In 1860, the ratio increased to 47 per 100. It declined to 36 in 1870, primarily as a result of the loss of animals in the South during the war, and would not return to the antebellum level for many years. In 1850, the ratio of "other" cattle to Alabamians was larger than in the country as a whole (53 versus 43 per 100 people). However, the ratio declined in 1860 (47 per 100) and was even lower in 1870 (26 per 100).

US Cattle

In 1867, the Statistics Division of USDA began to publish annual data on "all" cattle and calves. Apparently, "all" cattle data included milk cows. For discussion, in this section, I have chosen to remove the milk cow numbers from the "all" cattle totals. The USDA estimate for the number of all cattle (excepting milk cows) in the United States on January 1, 1870 was 21.4 million. Using the USDA data, the largest share of the 1870 total (21.8%) was on Texas ranches and farms. Other states with large inventories included: Ohio (5.9%), Illinois (5.3), New York (4.9), and California (4.6). Some 42.5% of the national herd was in these five states.

Value Per Head

In 1867, USDA began to publish data on value per head for "all" cattle. Data for the period 1870 to 1895 are presented in Figure 8.49. They indicate that during the period, value of "all" cattle in the United States appeared to be somewhat cyclic, but that there was a general downward trend for the entire period. Value was $23 in 1870 and $17 in 1895.

It appears, in this particular period that if there was a relationship between values and numbers of cattle (Figure 8.50), it was limited. Values did generally decline throughout the period, while numbers generally increased. At that time, the cattle market in the country was not well organized, In fact, cattle drives were still originating in Texas as late as 1900.

Cycles of Abundance

Annual data on the numbers of "all" cattle (excepting cows that had calved-milk) on January 1 of each year during the period 1870-1895 are presented in Figure 8.50. Although there were changes in the rate of increase from year to year, numbers trended upward from 1867 (20.4 million) through 1890 (45 million), and downward afterwards, finally reaching 34.3 million in 1895.

Mathews, et al. (2001) have written about cyclic changes that occur in the abundance of cattle over a period of years. They attribute these cycles to the nature of the reproductive cycle in the cow, economic factors, weather, grain exports, government programs, etc. They note that a typical cattle cycle lasts 10 to 12 years, and consists of an Expansion Phase of six to seven years when numbers increase, a Consolidation Phase of one to two years when numbers change relatively little, and a Declining Phase of three to four years when numbers decline. At that point, a new cycle begins. Changes in the relative importance of the factors controlling the cycle can result in lengthening or shortening the amount of time in each of its phases.

In the Expansion Phase of a cattle cycle, a large majority of the farmers who produce calves for the market have to add brood cows to their herds continuously. With cows producing only a single calf each year, a large number of brood cows must be added to individual herds to make a difference in the total herd size of the country. The surprising thing, however, is that at some place in the cycle a large number of calf producers have to decide at about the same time based on some market signal to begin to sell off their brood cows. Obviously, it would not take a large number of farmers to make the initial decision to sell, but enough would have to do so to catch the attention of the remainder who quickly join the stampede to sell brood cows.

In the data presented in Figure 8.50, there appears to be portions of three cycles in the 1870-1895 period, although two of them were not very robust. The first one apparently began sometime before 1870, but there was no clear indication of the end of its Declining Phase. Numbers began to increase rapidly again in 1877 in what seemed to be an Expansion Phase of a second cycle. Again, there was no definite end to its Declining Phase. In 1883, once more numbers began to increase rapidly in what was the beginning of Expansion Phase of a more typical cycle. The Consolidation Phase of this third cycle continued from 1889 through 1892 when the Declining Phase began. The data in Figure 8.50 are not sufficient to determine when this phase ended.

It is likely that economic conditions played some role in the elaboration of the cycles in numbers shown in Figure 8.50. This situation is much more complicated than it appears. In the latter part of the nineteenth century when farmers became aware of an economic contraction, and they began to sell all their excess brood cows, some period of time elapsed before their decisions to sell were translated into a reduction in national numbers. Similarly, once the economic contraction ended, an even longer period elapsed before they could rebuild their brood herds and begin to produce significantly larger numbers of calves.

Data published by NASS indicated that in 1895 farms in Texas held the largest share of the nation's cattle (all) with 16.2% of the national total. Other states with large shares included: Iowa (6.5%), Kansas (4.7), Missouri (4.4), and Nebraska (4.1). These five states accounted for 35.8% of the total national herd.

Alabama Cattle

According to census data, by 1860 the number of "other" cattle ("all" cattle, excepting working oxen and milk cows) had reached 1,748,000 in Alabama. There are no estimates of the number remaining at the end of the war, but by 1870 it had declined by 58.8%, to less than 719,700. The decrease in Alabama was over five times greater than in the South Central Region as a whole (58.8% versus 11.2%). Unfortunately, the 1870 Census did not include county data on "other" cattle, so it is not possible to determine where most of this loss occurred within the state.

Census data for 1880 indicates that there were 404,200 cattle (excepting working oxen and milk cows) in Alabama. Large shares of the total were still located in the Cotton Counties of the Tennessee and Coosa valleys and in the Cotton Arc across central Alabama. The largest concentration on a per-county basis, however, was in the Low Hills section south of the Black Prairie. In that section, there were an average 7,740 animals per county. The second highest concentration was in the Cotton Arc with an average of 6,850 per county. By 1880, some legislation had been enacted to restrict the herding of cattle on the open range, but at that time none of the legislation applied to the Low Hills. Cattle production there was little changed from the early days of statehood.

Value Per Head

USDA also published data on the value of "all" cattle for all of the states beginning in 1867. The data for Alabama for the period 1870-1895 are included in Figure 8.49, along with the national data. The Alabama data indicate that there was considerable variation during the period, but that value generally declined from $13 in 1870 to $8 in 1892. Generally, year-to-year trends in numbers were similar in both the state and the country. The data also show that value in Alabama was from $10 to $15 per head lower than in the country in each year throughout the period.

Cycles of Abundance

Annual data from the inventory of "all" cattle and calves (excepting cows that had calved-milk) in Alabama on January 1st of each year in the period 1870-1895 are presented in Figure 8.51. These data also indicate that there were apparently all or portions of three cycles affecting the abundance of cattle in the state during that period. It appears from these data that a cattle cycle, which had begun sometime earlier, was at the end of its Declining Phase in 1870 and that the Expansion Phase of a new cycle began in 1872 with 400,000 animals. The Consolidation and Declining phases of this cycle are not well defined, but apparently it ended in 1882. A second cycle began in 1883 with 575,000 animals and ended in 1888. It also appears that a third cycle began in 1889. With the data in Figure 8.51, it is not possible to determine when it ended. If these are indeed separate cycles, they are of a much shorter duration than expected.

Data on cattle numbers in Texas during this same

period are presented in Figure 8.52. The trends in Texas are generally similar to those for the country and Alabama. In all three, numbers generally increased from the early 1870s until the early 1880s when they began to decline.

There are no data on the distribution of cattle in Alabama counties in 1895; however, 1890 Census data indicate that the largest share (3.2%) of "other" cattle (all cattle excepting milk cows and working oxen) were on farms in Clarke County. Other counties with large share of the total herd were: Wilcox (2.5%), Jackson (2.4), Marengo (2.3), and Madison (2.3). These five counties accounted for only 12.7% of the state total, which is indicative of the widespread distribution of cattle in the state at that time. Some 34 counties reported 7,000 or more animals.

Hogs

The number of hogs in the United States had increased from 26.3 million in 1840 to 33.5 million in 1860. In 1870, after the end of the war, census data indicated that the number had fallen to 25.1 million. Much of this decrease was the result of large losses or high levels of consumption during the war, especially in the South. In 1860, the production of hogs was sufficient to provide 103 of the animals for each 100 persons in the country. By 1870, the ratio had declined to 65 per 100. In Alabama, the 1860 the ratio was 167 hogs per 100 persons, but by 1870 it had fallen to 70.

US Hogs

Census data for 1870 reveal that there were 25.1 million head on farms in the country. The largest share of the total herd in that year (11.4%) was on farms in Illinois. Other states with large shares of the total were: Missouri (10.4%), Indiana (8.5), Iowa (8.2), and Oklahoma (7.1). Approximately 45.6% of the total herd was located in these five states. Some 2.8% of the national total was on Alabama farms.

Note that the 1870 Census data for national hog numbers are much lower than those reported by NASS (25.1 versus 36.7 million). The USDA inventory was conducted on December 1, 1870. It is likely that the census data was taken at a different time of the year.

Value Per Head

Annual data published by USDA on the value per head of hogs in the United States during the period 1870-1895 are presented in Figure 8.53. The data indicate that although there were cycles in value during the period, it generally averaged $5.00 per head for the entire period. Economic factors (contractions and expansions) may have played some role in the elaboration of cyclic changes in value, but the most likely cause was cyclic changes in numbers of animals (Figure 8.54), or the supply-demand-price relationship. Apparently, hog farmers read price signals and responded to them throughout the period by either increasing or reducing herd size accordingly. Unfortunately, farmers were always behind the curve. For example, prices began to trend upward between 1873 and 1874, while numbers were still in the Liquidation Phase of a cycle (see below). Numbers did not begin to respond to improved prices until 1876, and in 1877 prices began to decline again. Then with prices declining sharply, numbers did not begin to decline until 1880.

Cycles of Abundance

Annual data obtained from a national inventory of hogs and pigs on December 1 of each year during the period 1870-1895 are presented in Figure 8.54. It is obvious from these data that there are also cycles in hog numbers. During the period, the data indicate that there were three fairly well-defined cycles (1875-1881, 1882-1887, and 1888-1892), as well as portions of two others (1870-1874 and 1893-1895).

The typical cattle cycle is 10 to 12 years in length. The cycle in hogs does not seem to be nearly so long. In this species, a cycle is generally completed in 3.5 to 4 years (Sterns and Petry, 1996). The shorter cycle in hogs is apparently the result of the difference in reproductive biology. As noted previously, cows typically produce a single calf each year. In swine, the sow can produce, on average, slightly more than two litters a year, with each litter consisting of an average of nine piglets.

Cycles of hog production consist of only two phases. When hog numbers are increasing, the cycle is said to be in its Accumulation Phase, and when decreasing, it is said to be in its Liquidation Phase. There is apparently no Consolidation Phase, as there is with cattle. Stearns and Petry (1996) noted that hog cycles are the result of factors present both inside and outside the industry. The primary factor is the time required to adjust herd size to changes in prices. Other factors include: variation in feed grain production; supplies and prices of competitive meats; and economic, social, and political factors that affect the demand for pork.

The role of the relationship between values and numbers in the elaboration of Cycles of Abundance was discussed in considerable detail in the preceding section and need not be repeated here. It is sufficient to note that number and value seem to be inversely related. When one increased, the other declined. Even though hog numbers in the country were cyclic from 1870 through 1895, they generally trended upward throughout the period.

Stearns and Petry (1996) suggested that cycles in hog production are related to corn prices. The primary use for corn in the United States in the second half of the nineteenth century was as a feed for hogs. Obviously, then, when hog numbers increased, the demand for corn also increased, and the increased demand for corn generally resulted in an increase in prices for the grain. At some point, however, the increasing price of the grain would begin to seriously erode profits for hog producers. At that point, numbers would begin to decline. A comparison of data on corn prices presented in Figure 8.28 and the data on hog numbers generally support the validity of the author's suggestion. For example, in the early 1870s with corn prices declining, hog numbers were increasing, however, after 1872 demand for the grain seemed to result in the beginning of a period with increasing prices, which continued through 1874. While there appears to be a relationship between the two, it is difficult to understand how numbers could respond so quickly to changes in prices. Between 1872 and 1873, prices began to increase, and in the same period numbers began to decline. It is obvious that other factors must have also been involved.

The presence of cycles in hog numbers and the complexity of farmer decisions and actions that they represent reveal a fascinating, yet altogether serious aspect of human behavior. It would be much better if there were no cycles. It would be better for both farmers and consumers if the process of supply and demand were more orderly. But no suitable system had yet evolved in the last half of the nineteenth century to replace this game played by hog farmers year after year in which they attempted to guess when to expand their herds and how much they could expand before there was an oversupply of pork on the market with the inevitable decline in prices. Year after year farmers carefully used changes in prices as a basis for making decisions as to how many pigs they should produce. Each year they used prices as signals to determine whether to sell their pigs or to keep more for brood sows. Unfortunately, under the best of conditions, these signals were always outdated when they were acted on. They inevitably were indicators of the relationship between supply and demand yesterday. Their problem was an extremely complex one. If they decided to reduce their herd size and prices remained high, they would forfeit good returns. Conversely, if they decided to increase herd size and prices declined their operation could sustain heavy losses. Unfortunately, even with their ever expanding experience in hog production, data presented in Figure 8.53 and 8.54 strongly suggest that, at least in the period 1870 to 1895, they never seemed to be able to guess correctly.

Data published by NASS indicate that in 1895 some 12.2% of all hogs in the country were in Iowa. Other states with large shares of the total were: Missouri (8.5%), Illinois (8.3), Indiana (5.2), and Texas (5.2). These five states accounted for 39.3% of the total.

Alabama Hogs

Census data indicate that there were 320 hogs available to each 100 Alabamians in 1850. For some reason, this ratio fell to 230 per hundred by 1860, but during the war years it declined even more sharply. In 1870, there were only 70 animals available for each 100 Alabamians.

Value Per Head

There are no annual data on prices received for Alabama hogs before 1924; however there are published annual data on farm value per head available beginning in 1867. Data for the period 1870-1895 are also presented in Figure 8.53. They indicate that there were also cycles in these values similar to those seen in the national data. In 1870, the average value was $4.45 a head. Then it declined rapidly to reach $2.60 a head in 1874. In the remaining years of the period, values went through three complete cycles and were into a fourth in 1895. Although values were cyclic, they generally remained in a range between $2.50 and $3.50 during the period, and there was little indication of an upward trend.

The cause(s) of these cycles must have been essentially the same as for hogs throughout the country. The periodicity of the two sets of data are quite similar, the primary difference being the smaller amplitude of the cycles for the values of Alabama hogs. In most years, values for Alabama hogs were about half of the corresponding national value.

Cycles of Abundance

Data from the annual December 1st inventories of Alabama hogs and pigs for the period 1870-1895 are presented in Figure 8.55. As was the case with national hog production (Figure 8.54), numbers also seemed to be cyclic over time in Alabama hogs. The data in Figure 8.55 suggest that there were three complete cycles (1876-1880, 1881-1886, and 1887-1890), as well as portions of two cycles (1870-1875 and 1891-1895). It is apparent that even at this early date hog production in Alabama was closely related to production in the other states (Figures 8.54 and 8.55). As detailed in a preceding paragraph, it is likely that this cyclic phenomena in both the Alabama and national data was the result of the relationship between value and number (Figures 8.53 and 8.55). That relationship was discussed in detail previously and need not be repeated here.

Even though cycles result in increases and decreases in number during the period 1870-1895, overall abundance increased substantially. It was at 1.1 million in 1870, and by 1895 it had increased to 1.6 million. However, as detailed in the preceding chapter, hog production declined sharply during the Civil War. The 1860 Census indicated that there were 1.7 million hogs in Alabama in 1860. Similar data indicate that hog production had not recovered in the state in 1870, and in fact, did not reach that level again for the remainder of the nineteenth century. Note that there are differences in data obtained by the censuses and that published by USDA (NASS), but these differences do not affect the conclusion that it took many years for hog production in the state to return to pre-Civil War levels. Alabama farms desperately need diversification, but they did not get it through hog production.

There are no county data on the distribution of hogs in Alabama in 1895; however census data for 1890 indicate that the largest share of the total number of hogs in the state at that time was on farms in Clarke County (2.9%). Other counties with large shares included: Jackson (2.8%), Henry (2.5), Marengo (2.2), and Madison (2.1). These five counties accounted for 12.5% of the total. This relatively low percentage is indicative of the widespread distribution of hogs in the state. A total of 36 counties reported 20,000 or more animals.

Milk

There are no data on the annual production of milk in the United States before 1924. The 1890 Census includes data on production for 1870, 1880, and 1890, but the data are apparently not comparable. That census also includes data on the production of butter and cheese in the country and the states for 1870, 1880, and 1890. The Statistics Division of USDA began to collect annual data on milk cow numbers in 1867.

US Milk

Census data published in 1890 indicate that in 1869 milk cows in the United States produced two billion pounds of milk (235.5 million gallons at 8.5 pounds per gallon). This quantity is equivalent to 53 pounds per person in the American population at that time. Some of this milk was converted (includes on-farm production only) into 514.1 million pounds of butter (13.6 pounds per capita) and 53.5 million pounds of cheese (1.4 pounds per capita).

Milk Cows

According to census data, the number of milk cows (cows that had calved-milk) in the North Atlantic and North Central regions increased at a moderate rate from 1850 to 1870. Numbers had also increased in the South Atlantic and South Central regions from 1850 to 1860, but farms in both of those regions suffered severe losses of animals between 1860 and 1870. In 1860, nationally there were 27 milk cows for each 100 persons. By 1870, this ratio had declined to 23. In 1860, there were about equal numbers of milk cows in both the North Atlantic and North Central regions; however, by 1870 there were substantially more in the Midwest (3.3 versus 2.8 million).

Annual data on the number of milk cows in the United States in the period 1870-1895 are presented in Figure 8.56. The inventory of December 1, 1870 estimated there were slightly less than 9.7 million in the country at that time. Although there were some year-to-year changes in the rate of increase, the total number trended upward for the period,1870-1892. In the period 1893-1895, it remained unchanged. From around 1876 through at least 1888, numbers increased at an exponential rate. The rate of growth slowed after 1888. Total herd size reached 15 million in 1890, and then increased very little through 1895 (15.2 million). There is little indications of cycles, in milk cow numbers during this period.

In 1895, the largest share of the national milk cow herd (9.7%) was on farms in New York. Other states reporting large shares included: Iowa (7.6%), Pennsylvania (6.0), Illinois (5.3), and Wisconsin (5.3). These five states accounted for 34.5% of the national total. The Alabama share was only 1.8%.

Milk Production

Data reported on American milk production in the 1890 Census were for all milk produced on farms in 1889. Data for milk produced in the United States, its regions, and Alabama are presented in Table 8.6. The data indicate that dairy cattle in the United States produced 44.3 billion pounds (5.2 billion gallons) that year. Farmers in the North Central Region reported the largest share of the total (52.1%). The highest production in an individual state was in New York where farmers reported producing 12.7% of the national total. Other states with large shares of the total included: Iowa (9.3%), Pennsylvania (7.1), Illinois (7.0), and Ohio (6.3). The large shares in New York and Pennsylvania, no doubt, are reflections of the growing urban areas in and around,New York City and Philadelphia. The large share in Iowa is somewhat surprising. Even in 1890, the population of the state was slightly over 1.9 million. Obviously, a large share of the milk they produced in 1889 was being shipped out of the state.

Dairy Products

The 1890 Census published data on the quantity of butter and cheese produced on American farms in 1869, 1879, and 1889. These data are presented in Table 8.7. The total quantity of butter produced was 514 million pounds in 1869. It increased by 51% to reach 777.2 million in 1879. The rate of increase between 1879 and 1889 was somewhat less (32%), but the level reached 1,024.2 billion pounds. Not nearly as much cheese was produced on farms, the larger share being produced in factories. Quantities produced were at levels of 53.5 million, 27.3 million, and 18.7 million in 1869, 1879, and 1889, respectively. It is likely that these data does not mean that American consumers had less cheese available; rather, more production was moving off-farm.

Butter. Data presented in Table 8.7 show that butter production increased in all regions between 1869 and 1889. In 1869, farms in the North Atlantic and North Central regions produced almost equal amounts, and together they produced 86% of the country's butter. In 1879, farms in the North Central Region produced about a third more than those in the North Atlantic Region. In 1889, production in the North Atlantic Region actually declined; while it increased by 42% in the Midwest. Obviously, by 1889 a considerable amount of butter was being shipped east by the farmers in the North Central Region. Farms in the other regions accounted for a rather small share of the country's butter production in any of the years.

Cheese. Data presented in Table 8.7 show that in 1869 most of the cheese produced on American farms (92%) was produced in the North Atlantic and North Central regions, and more than 65% was produced in the Northeast. However, in the two following decades, on-farm production fell sharply in both regions, as cheese-making moved off the farm into the factory. By 1890, those two regions accounted for 71% of the total, and the North Central Region had overtaken the North Atlantic Region in production. Farms in the South Atlantic and South Central regions did not produce much of the country's cheese in 1870, and the situation changed very little in the following two decades.

Alabama Milk

Data reported in the 1890 Census indicate that milk

cows on Alabama farms produced 471.7 million pounds of milk in 1889 (55.6 million gallons at 8.5 pounds per gallon). This quantity is equivalent to 321 pounds a person, considerably less than the national 726 pounds a person. Some of this milk was converted (only farm conversion included) into 6,100 pounds of cheese and 14.5 million pounds of butter (9.7 pounds a person).

Milk Cows

Annual inventory data, on the number of milk cows (cows that had calved-milk) in Alabama in the period 1870-1895 are presented in Figure 8.57. On December 1, 1870, there were 230,000 million milk cows in the state. Numbers trended downward between 1870 and 1876. Herd size was 200,000 animals in both 1876 and 1877. It began to trend upward in 1878, only to stall again in 1881 (225,000 animals). It remained at that same level through 1883, before beginning to increase again. After 1883, numbers generally increased at an exponential rate through 1891 (270,000). It remained at that level through 1893, before increasing to 275,000 in 1894 and to 280,000 in 1895.

There are no Alabama county data available on milk-cow numbers for 1895. However, 1890 Census data indicate that the largest share of the state herd that year was in Clarke County (2.5%). Other counties with large shares of the total included: Sumter (2.4%), Madison (2.4), Wilcox (2.3), and Dallas (2.3). These five counties only accounted for 11.9% of the state total. The average per county was 4,425 animals. A total of 57 of Alabama's 66 counties reported 3,000 or more head.

Milk Production

The Census of 1890 (Table 8.6) provides the only data on milk production during the period 1870-1895. In 1889, Alabama farmers produced 471.8 million pounds (55.5 million gallons). That census data revealed that Jefferson County farmers recorded the largest share of the state total (3.2%). Other counties with large shares included: Madison (2.9%), Jackson (2.5), Blount (2.4), and Dallas (2.4). These five counties only accounted for 13.4% of the state total, indicating that milk production was widespread in the state. In fact, 22 counties recorded productions of one million gallons or more.

Dairy Products

Data presented in Table 8.7 show that in 1869, 1879, and 1889 that Alabama was not a major player in the national production of on-farm butter and cheese. In those three years, Alabama farm-made butter accounted for 0.6, 1.0, and 1.4% of the national total, respectively. In 1859, it had been 1.3%, so by 1889 butter production had returned to its antebellum level. The state's share of cheese production did not reach 0.1% of the national total in any of the years.

Chickens

There were no data collected on American poultry before 1840. The census that year only reported the estimated value of all kinds of poultry in the states and counties. That year, the total value of all poultry in the United States was $9.3 million, and 31.6% of the total was reported from the North Atlantic Region. New York accounted for 39% of the total in that region.

US Chickens

There were no data collected on American poultry in 1850, 1860, or 1870; however, the 1880 and 1890 censuses reported the number of chickens in the states and counties. These data are presented in Table 8.8. In 1880; there were 102.3 million birds in the country when the national inventory was taken that year. Some 52.7% of the total was reported from the North Central Region. The high level of grain production in the region resulted in increased production of all kinds of domesticated animals. By 1890, the number of chickens in all of the regions, except the North Atlantic, had more than doubled, and the national total reached 258.9 million. The largest share (51.3%), was still on farms in the North Central Region. The share in the North Atlantic Region was down significantly from 1880.

In 1890, the largest share (8.8%) of the national flock of chickens was on farms in Missouri. Other states with large shares included: Illinois (8.3%), Iowa (7.8), Kansas (6.1), and Ohio (5.3). These five states accounted for 36.3% of the total. It is fascinating that in 1890 there was no indication of the twentieth century role that the Del-Mar-Va region and the Lower South would have on poultry production in the country and even the world.

Alabama Chickens

At the beginning of the twenty-first century, cash received from the sale of Alabama chickens (primarily broilers) is greater than $2 billion annually. The production of these animals was not nearly so important in the state at the end of the nineteenth century. There are no annual data available on the production of chickens in Alabama before 1924. Although chickens and eggs were extremely important on farms throughout the state before that time, they were not important enough in commerce to warrant the collection of annual statistics.

The 1870 Census did not collect data on chickens on Alabama farms. The 1890 Census include data indicating that there were 2.1 million domestic fowl (chickens) in the state in 1880 and 6.2 million in 1890 (Table 8.8). The share of the national total in 1880 and 1890 were 2% and 2.4%, respectively.

Census data indicate that the largest share of the Alabama flock of chickens in 1890 were on farms in Jackson County (3%). Other counties with large shares include: Blount (2.8%), Marshall (2.7), DeKalb (2.6), and Madison (2.6). These five counties accounted for only 13.7% of the state total. This small percentage is indicative of the widespread distribution of chickens on Alabama farms at the end of the nineteenth century. The state average per county was slightly less than 94,700 birds. That year, a total of 44 counties reported flocks of 80,000 or more birds. All of the top-five counties were located in north Alabama. As early as 1890, north Alabama was already becoming a center of poultry production in the state.

Eggs

Obviously, collecting data on egg production was difficult in the late nineteenth century when so many of them were consumed by the farm families. Further, unlike crops and animals, where accounting must be done only once a year, with eggs production it is almost continuous. Some accounting must be done each day. Because of the difficulty of obtaining data, there were no reports before the Census of 1880. The totals reported for the states and counties were from data collected in 1879.

US Eggs

Census data on eggs produced annually in the United States, its regions, and Alabama in 1879 and 1889 are presented in Table 8.9. The data indicate that there were 5.483 billion eggs produced in the country in 1879. This total was equivalent to 112 eggs per person. The larger share (52.3%) was reported from farms in the North Central Region. Most chickens were also reported from that region (Table 8.8). The next largest share (21.8%) was reported from farms in the North Atlantic Region. Production in the South was considerably lower than in either of those regions.

Between 1879 and 1889, egg production in the country increased by 79.4% to 9.836 billion. This 1889 total was equivalent to 159 eggs per person. During the same period, farmers in the North Central Region increased their share to 56.6%, while the share of the North Atlantic Region declined to 17%. Production in the South Atlantic and South Central regions increased by 60.6% and 83.8%, respectively; however, the combined production in those two regions still accounted for only 23% of national production.

In 1889, Ohio reported the largest share (8.6%) of national egg production. Other states reporting large shares included: Iowa (8.5%), Illinois (7.4), Missouri (6.5), and Pennsylvania (6.1). These five states accounted for 37.1% of total national production.

Alabama Eggs

Alabama egg production increased from 81.6 million in 1879 to 129.6 million in 1889; however, the state's share of the national total declined from 1.5% to 1.3% in the decade. In 1879, total egg production was equivalent to 66 eggs per person, a number considerably less than the national average of 112 eggs. In 1889, Alabama's production was equivalent to 87 per person.

In 1889, Jackson County farmers reported the largest share of egg production (3.2%). Other counties reporting large shares included: Madison (3.2%), Blount (2.7), Morgan (2.7), and Montgomery (2.4). These five counties accounted for only 14.2% of the state total. This relatively low percentage is indicative of the widespread production of eggs in the state. Some 16 counties reported egg production of 2.4 million (200,000 dozen) or more.

Agricultural Production

In 1867, USDA first began to publish annual data

on acres harvested of the various crops, but it would be a number of years before they would publish much data on the value of national agricultural production. In the period, 1870-1895, about the only data available was published in the censuses. These censuses published data indicating that the estimated value of farm products was $2.4, $2.2, and $2.5 billion in 1869, 1879, and 1889, respectively (Table 8.10). The data for 1869 are not comparable to those for 1879 and 1889. The 1869 data include values for betterments and improvements to stock. The data in Table 8.10 show that in both 1879 and 1889 the largest share of estimated value was reported from the North Central Region. In both years, just under half of the total for the country was reported from that region (45.7% and 45.2%, respectively). In both years, estimates for both the South Atlantic and South Central regions trailed far behind, likely as a result of the lingering effects of the Civil War. Of the two, estimates were considerably higher in the South Central Region. The difference between the two was likely the result of the value of much larger cotton crops in the South Central Region in both years. In 1879 and 1889, bales harvested in that region totaled 3.9 million and 5.1 million, respectively, but only 1.8 and 2.3 million, respectively, in the South Atlantic Region.

In reality, data from none of the years are strictly comparable; because the amount of improved farmland in use was changing. In 1869, 1879, and 1889, improved farmland in the country totaled 188.9, 284.8, and 357.6 million acres, respectively. This means that at least some of the increase in estimated value was not an increase in value itself, but an increase in the amount of farmland used to produce that value. To correct for this upward bias, total estimated value has been divided by acres of improved farmland. For example, in 1880 the quantity of improved farmland in the United States was estimated to be 284,771,000 million acres. Expressed as dollars/acre, estimated value would have been $7.77. The converted data, for the United States, its regions, and several states are presented in the right-hand portion of Table 8.10.

The per-acre values are not strictly what they appear to be. Data on estimated value were collected in 1869, 1879, and 1889, while data on acres of improved farmland were obtained in 1870, 1880, and 1890. It is likely that acres did not change very much in one year, but all of the converted data are incorrect to the extent that they did change.

In reviewing the data in Table 8.10, the effect of the inclusion of betterment and additions to stock are quite evident in 1869. The per-acre values for that year are uniformly much higher. Disregarding the 1869 data, national dollars/acre declined from $7.77 to $6.88 between 1879 and 1889 (a decrease of 11.4%). It is likely that the decline is a direct result of the decrease in prices of all the major crops during this period. Dollars/acre did not decline in the North Atlantic and Western regions between 1879 and 1889. It is likely that agriculture in those two regions was not as dependent on the production of the major crops as were the other two regions.

In both years, dollars/acre were higher in Alabama and Mississippi compared to Illinois and Indiana. This difference is primarily the result of the fact that the estimated value of an acre of cotton was considerably greater than the estimated value of an acre of wheat. For example, in 1885 the average yield of cotton in Alabama was 126 pounds of lint an acre, and the average price was 8.39 cents a pound; consequently, the value of an acre of cotton would have been $10.57. Conversely, yield and price for an acre of Illinois wheat were 8.5 bushels an acre and 77.2 cents a bushel. The value of this acre of wheat would have been $6.56.

The Bottom Line

At the end of the nineteenth century, American agriculture was in desperate straits. Farmers were harvesting unprecedented quantities of cotton, corn, wheat, and oats; yet when it became time to settle up and if they could pay their bills, they had little remaining with which to maintain their families until the next harvest.

USDA had begun to collect statistics on various aspects of farming soon after the end of the Civil War. Most of the data collected was related to some element of production: yields, acres, bales, or bushels harvested. Fortunately, they also collected some data on prices and even estimated gross income on individual crops, but there seemed to be little concern for the cost side of the all-important equation. If anyone in the department was overly concerned about net farm income, it was not obvious in the data that they were providing. USDA did not begin to publish annual estimates of this essential statistic for American agriculture until 1910, and estimates for the individual states did not appear until 1949. In 1910, the department published estimates indicating that total net farm income for the country was $4.1 billion, and that on a per farm basis it was $652. It is likely that in 1885 net farm income per farm values, especially for the South, would been considerably less than half of this amount.

Likely, by 1885 a large share of the country's 2.2 million farm population would have been overjoyed to get out of farming altogether. Unfortunately, while the American economy was growing rapidly at that time, it was not growing fast enough to absorb more than a tiny fraction of those desperate people. There were few opportunities for farmers to leave farming anywhere in the country, but in the South it was a virtual impossibility.

In all of the thousands of years of the history of agriculture, surplus production had never been a problem for very long. Following the Civil War, however, American agricultural surpluses quickly became a problem of epic, overwhelming proportions, and no one had the vaguest idea of what to do about it. In the following chapters, we will see that it was a problem that simply would not go away. Several generations have wrestled with it, and with apologies for getting way ahead of the game, well into the twenty-first century we will be no closer to a solution than our forebears at the end of the nineteenth century.

SERIOUSLY SEEKING SALVATION

Between 1885 and 1895 prices for cotton, corn, wheat, and oats, had fallen by 9.2%, 21.7%, 34.6%, and 30.8%, respectively. If the decline for cotton seems out of line, it is because by 1885 the price for the fiber was so low that it simply could not go much lower. While there was widespread agreement that the American agroecosystem at the end of the nineteenth century was an unmitigated disaster, few farmers anywhere seemed to know how they might solve their problems. Most of them saw outside forces as being responsible for their plight. They saw themselves as being manipulated and exploited by railroads, tariffs, government taxes, supply merchants, and greedy lenders. It would be years before many of them would understand that there were simply too many farmers producing too much cotton in the South and too much wheat and corn in the Midwest. Even if they had understood that they were at the center of their problems, there were few, if any, immediate options to remedy the situation. In the tenant farming South and the rapidly mechanizing North, there apparently were not many roads to salvation.

Meetings were held everywhere, and experts were consulted. Over time, these efforts led to the decision that salvation depended on farmers and their supporters joining together on a broad road of action that consisted of three parallel tracks: farmer organizations, improved information (research and extension), and diversification.

This so-called threefold path would provide marching orders for farmers throughout the country. How this path developed, how it functioned, and how well it served farmers will be considered in some detail in this section and in similar sections in the following chapters. By 1890, farmers already had a considerable amount of experience with the threefold path to salvation, and it was obvious to most of them, that organizations, better information, and diversification were not going to achieve the desired results so they began to discuss adding a fourth track to their battle plan. After widespread discussions throughout the country (Fite, 1984), farmers decided that to improve their lot they would have to bring the federal government into their agricultural marketplace as a direct participant. Getting the government involved, however, was somewhat difficult. Further, it was patently obvious that farmers only wanted the government involved to a point. They wanted government to help them market their crops, but not in making decisions on what or how much they should produce. With the value of hindsight, it probably would have served them well if they had stopped at this point to read the mythical story of Pandora's Box.

Salvation Through Better Information

Fite (1984) comments that if advice from experts concerning the merits of scientific farming and the benefits of diversified agriculture could have been somehow converted directly into cash, all American farmers would have been rich. After the end of the Civil War, exhortations by experts for the application of science and farmer education (extension) to farmers' problems began to increase rapidly, especially in the South. Although efforts to encourage the use of science in farming intensified in the latter half of the nineteenth century, they did not begin during this period. Gray (1958) notes there was almost no application of scientific principles to agriculture in the English colonies during the Colonial Period. After the American Revolution, however, some prominent farmers, such as George Washington and Thomas Jefferson, were experimenting with the scientifically-based agricultural methods that were evolving in Europe and especially in England. In turn, they freely shared the results of their field trials with their neighbors. After the War of 1812, interest in the application of science to agriculture and in crop diversification grew rapidly. Out of these early efforts came improved practices, such as plowing with the contour on hillsides, seed selection, production of Sea Island cotton, use of lime, and the use of clover as an adjunct to crop production.

During this period, the efforts of John Taylor of Virginia, who conducted various experiments on his farm and published the results of his work, were extremely important in improving agriculture in the Old Dominion. His book, *Arator* (1812), presented information on better methods of plowing, crop rotation, bedding, composting, handling manure, and restoration of wornout lands. Edmund Ruffin, also of Virginia, was arguably the most important early contributor to the improvement of agriculture in the South. His work on the role of soil acidity in determining the effectiveness of manure in crop production was seminal in its effect. For many years, he also experimented on the use of different kinds of lime to counter soil acidity. Because virtually all of the Ultisol soils of the South are highly acidic, efficient plant agriculture is difficult without the application of lime. In 1832, Ruffin published a book summarizing his work on the subject: *An Essay on Calcareous Manures*. He was not only a pioneer in the use of practical scientific research in agriculture, he was also closely identified with the establishment and function of agricultural societies and with the encouragement of the use of agricultural fairs to extend the results of experimentation to large numbers of farmers.

In the first half of the nineteenth century, efforts to encourage the use of science in agriculture and of the use of crop diversification was largely driven by an interest among knowledgeable and prominent planters to improve their farm operations and to make them more efficient, as well as more profitable. In the latter half of the century, those efforts were driven by the growing desperation among tobacco and cotton farmers, but with little avail. As the years passed, poverty was slowly inching its way up the social ladder, everywhere.

Farmers and planters who set the pace for agricultural change were scattered throughout the South. Agricultural societies were organized to encourage the use of science and technology in agriculture. Agricultural journals appeared and led the way in encouraging farmers to adopt improved practices. Unfortunately, much of the information provided by the societies and in the journals was too general to be of much value. As a result, it was soon realized that more specific, practical information had to be provided to the farmers through education. Further, it was also realized that practical information had to have a strong scientific base.

In 1862, Congress had passed and President Lincoln had signed, the Land-Grant College (Morrill) Act, whereby grants of federal land were given to the states as endowments to provide funds to establish college for the express purpose of teaching agriculture, mechanics, and military science. After the Civil War ended, the states of the former Confederacy were also allowed to take advantage of the provisions of the act. In 1872, Alabama accepted the gift of the East Alabama Male College at Auburn from the Methodist Church and chartered it as the Agricultural and Mechanical College of Alabama, the state's land-grant college.

Although there was no provision in the Morrill Act for using any of the funds for conducting agricultural research, the land-grant colleges were the best place to do it. In the 1870s, Southern states began to establish agricultural experiment stations associated with their land-grant colleges. North Carolina established an experimental farm in 1877, and Texas in 1878. The Alabama legislature provided the first funding for Auburn for agricultural research in 1883. Those funds were derived from a tax on the inspection of fertilizers. Auburn used its new funds effectively. By 1887, agricultural scientists at the college had published 32 bulletins based on their research.

Isaac Tichenor, first president of the Agricultural and Mechanical College of Alabama, set forth his philosophy for the operation of the new land-grant college. Tichenor was a firm believer in the emerging New South philosophy espoused by Henry Grady, long-time managing editor of the *Atlanta Constitution*. Tichenor believed that the future of Alabama and the South ultimately depended on the exploitation of its natural resources for industrialization; however, he realized that the lifeblood of the region's economy at that time was cotton and that the production of the crop was rapidly destroying the region. He also understood that he would never get the state's leaders to consider moving toward agricultural diversification and industrialization until the problems with cotton were solved. If he could solve the problems of the cotton farmer, through science and experimentation, maybe he could get the state's leaders to turn their attention to activities that would be more responsive to rapidly changing world conditions.

The deficiency of the Morrill Act in not providing funds for agricultural research was addressed in 1887 when Congress passed the Hatch Act, which authorized the use of federal funds for the development and operation of state agricultural experiment stations throughout the country. Then with both state and federal funds available, experiment stations across the South began to conduct a broad range of scientific and practical studies designed to solve farmers' problems. Much of the early work in all of the Southern stations was on the analysis and use of fertilizers and on the production of cotton. It was recognized from the outset that the proper use of high quality fertilizer would provide the most effective quick fix for one of the most severely limiting problems in cotton production—low crop productivity. Later, the focus of research on the stations was expanded to include more work on identifying and removing the barriers to increased diversification.

From the beginning of the 1890s, hundreds of bulletins resulting from research on the stations began to appear. It soon became apparent, however, that most of the bulletins were not getting to the mass of Southern farmers. Not many farmers wrote for experiment station bulletins. Further, widespread functional illiteracy made the use of the written word, as the medium for changing farming practices, generally ineffective at best. In an attempt to get useful information to the farmers, so-called institutes began to be widely used across the South. Institutes involved sending agricultural lecturers and scientists into farming communities to discuss crop rotation, animal and plant diseases, fertilizer recommendations, soil conservation, dairy and livestock farming, and other practical topics. Thousands of farmers attended these institutes, but the response was disappointing. For most farmers, hearing did not equal seeing.

The effort to provide improved information for all Alabama farmers is seen in the establishment of the Huntsville School for Negroes in 1875. In 1885, the name of the school was changed to State Normal and Industrial School of Huntsville, and annual state appropriations were increased to $4,000. By the latter part of the nineteenth century, it became obvious that the benefits of the first Morrill Act were not being shared equally by all of the people in the South. At that time, African Americans were not allowed to attend the large majority of colleges and universities in the region. As a result, the decision was made to provide federal funds to develop land-grant colleges for Southern blacks. Consequently, in 1890, a second Morrill Act enacted for that purpose became law, and it was decided that the Huntsville School would be designated as the 1890 land-grant college in Alabama. Under the provision of the two Morrill acts, the Alabama legislature was responsible for dividing the federal funds given to the state. As a result, in 1891 the Huntsville School received $2,727 in Morrill funds. These funds increased combined state and federal appropriations to $6,727 (Morrison, 1994).

That school's 1891 annual report noted that the Department of Agriculture was one of its academic departments, and they were making preparations to begin experiments in agriculture. According to Morrison (1994), as early as 1911 the school was conducting short courses for farmers during the winter months under direction of its agricultural instructors.

In 1881, Governor Cobb signed a bill establishing the Tuskegee Normal School for the training of black teachers (Mayberry, 1989). The bill also included a provision for a yearly appropriation of $2,000 for operations and maintenance. Dr. Booker T. Washington officially opened the school on July 4,1881. Although the school was established with public funds, gradually over the years Tuskegee would become known as a private institution. Then in 1882, Dr. Washington purchased a 100-acre abandoned plantation near the town as the permanent site of the institution (Mayberry, 1989).

The first annual Negro Farmers Conference was held at the school in February 1892. It was attended by upwards of 500 people from all over the South. It was judged to be so successful that Dr. Washington decided to make it an annual event. In 1893, the Alabama legislature amended the institution's founding charter to establish the Tuskegee Normal and Industrial Institute.

In the final analysis, there was never even the slightest indication that any of the tremendous effort (money and labor) expended in the last quarter of the nineteenth century in Alabama and the South to apply science to agriculture and to encourage diversification was having any beneficial effect on the lives of cotton farmers. As early as 1878, cotton prices fell to 10 cents, and for the next 20 years they continued to trend inexorably downward until they fell below five cents in 1889, the lowest price for cotton in history. For the entire period from the late 1880s to the end of the century, there was not a year when average cash receipts an acre totaled more than $15. And there were large areas in the South where returns were little more than half that amount. For the Southern landowner, this collapse of cotton prices was a disaster. Year after year, low returns were slowly eroding the value of the farmer's land, buildings, and equipment and his standard of living, which he had worked so tirelessly to create over the decades. Tenants, especially the share croppers, reached the bottom years earlier. They had little more to lose, as prices declined.

By the end of the nineteenth century, there was little doubt that quantum leaps had been made regarding the quantity and quality of information made available to Southern farmers. Unfortunately, as the amount of good information increased, the lives of farmers and their families were going in the opposite direction even more rapidly. Somehow, the experts had missed something. This disconnect should have given them pause. Tichenor's philosophy was getting nowhere fast. If solving the cotton farmers' problems was prerequisite for moving the state and the region toward industrialization, there was simply no way to get there from here.

These ever-worsening experiences for Southern farmers and their families were not unique. They were being shared by Midwestern corn, wheat, and oat farmers. Experts in that region also seemed to have missed something, and it was likely to have been something entirely different than in the South.

Salvation Through Organization

In the 1870s and 1880s, farmers throughout the country were slowly beginning to realize that many of their problems could not be solved by individual action and that they needed to join together or to organize so they could present a united front in seeking solutions. This sudden interest in organizing, no doubt, resulted from their realization that other parts of the American economy had already moved in that direction. Businessmen and industrialists were forming pools, trusts, and other combines of various kinds. Throughout much of the country, the working class was forming unions. So the farmers also began to organize.

The Grange

The first of a series of farmer organizations that would appear in the latter part of the nineteenth century was the National Grange of the Patrons of Husbandry. Simply known as the Grange, it was founded in 1867 by a USDA employee (Oliver Kelley) after a trip through the South. It was dedicated to promoting farmer unity, pride, and prosperity. The Grange spread rapidly through the Midwest, but was viewed with considerable suspicion in the South. However, it finally began to take hold in Alabama in 1872, and by 1877 it had grown to over 14,000 members. In Alabama, the Grange never became involved in politics. Rather it was a social organization that brought farmers together for picnics, barbecues, and singing conventions. While the national organization had no rules against black membership, few of this large group of Southern farmers ever became members. In Alabama, the Grange's primary thrust was farmer education. Through lectures, it avidly promoted diversification, application of science to agriculture, a wage system for farm laborers, and the formation of cooperatives to break merchant monopolies. Because of conflicts within the membership, there was little organized effort to accomplish any of these objectives. Nevertheless, a few short-lived, purchasing cooperatives were formed, and a Grange riverboat plied the Tombigbee and Black Warrior rivers hauling cotton for local farmers in an effort to get commercial shipping rates reduced.

In attempting to establish cooperatives, Grange members took as their model the Rochdale Society of Equitable Pioneers. This English society, established in 1844 by 28 workers who were dissatisfied with the treatment they were receiving from local merchants, established a supply cooperative. They also developed a list of 12 principles for the cooperative's operation. Three of the most important principles were: democratic control by members, payment of limited interest on capital, and net margins distributed to members according to level of patronage.

Grange members sought to establish supply cooperatives that would sell needed farm inputs at lower prices, but they also wanted a cooperative that was not part of the Rochdale movement. They also wanted to establish marketing cooperatives that would help members sell their crops for better prices. Unfortunately, in the establishment of cooperatives to sell their crops, they soon ran into an implacable foe—the Sherman Antitrust Act.

As industrialization grew in the United States in the latter part of the nineteenth century, it soon became apparent that citizens needed some protection from organizations and business arrangements (trusts, monopolies, etc.) established by large corporations to further their own, narrow interests. These concerns led to the passage of the Sherman Antitrust Act in 1890. The act had a salutary effect on business practices in the country, but it also made it illegal for farmers to organize themselves into marketing cooperatives. For some 30 years after passage of the act, courts consistently ruled that agricultural marketing cooperatives were illegal. So while it was relatively easy to establish the so-called supply cooperatives, it was difficult to find ways for farmers to cooperate in marketing their crops.

By the end of the century, the Grange could count few substantive accomplishments, and there was almost no continuing interest in the organization in Alabama. It did, however, show Alabama farmers that they could organize and that they could agree on the nature of common problems. It had also given ordinary farmers a better understanding of how the business sector of the agroecosystem functioned. In the final analysis, however, they could not agree as an organization on how it might solve their problems. Even had they agreed on an agenda, they simply did not have sufficient capital at their command to pursue it or any part of it. They were better informed than they had been, but to solve any of their problems, they still had to do so as individuals.

The Agricultural Wheel

The Wheel was the second farmer organization that took root in Alabama. First organized in Arkansas in 1882, by 1886 several local Wheel organizations had been established in North Alabama. In 1889, the Wheel claimed 75,000 Alabama members, who Rogers and Ward (1994) contend was more than a liberal estimate. The organization advocated cooperative manufacturers to reduce farmer costs and the designation of local cooperating storekeepers as Wheel Merchants. The Wheel also condemned the Louisville and Nashville Railroad, for in their view, the L&N charged excessive rates to ship their cotton. Although, the Alabama Wheel wanted to remain non-political, they did call for legislative action to deal with high freight rates. While they had good intentions, the Wheel accomplished very little for Alabama farmers. It ceased to exist when it merged, in 1889, with the better organized, rapidly growing, National Farmers Alliance.

The Farmers Alliance

The Alliance was probably the most influential farmer organization to take hold in Alabama. It was formally organized in Texas in 1878. Its membership was supposed to be restricted to genuine dirt farmers and others, such as small-town residents and rural ministers, who understood the importance of agriculture in local communities. Merchants, bankers, lawyers, and stockholders in corporations were excluded. The primary objectives of the Alliance centered on the formation of purchasing and marketing cooperatives. After early fearsome debate that almost destroyed the organization, members agreed not to pursue their objectives through political action. After a slow start, the Alliance gained momentum as it spread through adjoining states into the Upper South and finally into the Midwest. Along the way, its name was changed to the National Farmers Alliance. There was some early discussion of including black farmers in the organization; however, this was not to be. Instead, Alliance leadership encouraged the formation of a separate organization to represent this large group of farmers. As a result, the Colored Farmers Alliance was organized in 1888. There was never more than token cooperation between the two groups.

The Alliance spread rapidly throughout Alabama, and became popular in areas of the state where the Grange and Wheel had received scant attention, such as the Wiregrass and the piney woods in South Alabama. In 1889, the organization could boast that they had 125,000 Alabama members. When the Alliance arrived in Alabama, there was already a full-fledged agrarian revolt taking place across the South (Fite,1984). In 1889, there were some 645,000 Southern members of the Alliance. But Alliance men quickly learned what Grange members had learned earlier: while the formation of purchasing and marketing cooperatives would be good for Southern farmers, attempting to perform the functions of producer, merchant, and banker at the same time was fraught with problems.

The first thrust of the Alliance was an effort to get the railroads to lower rates on freight shipments. Also, in some counties, leaders following the Wheel practice of choosing official merchants who would reduce prices to the barest minimum in exchange for a guarantee of a much larger volume of business by Alliance members. The establishment of warehouses where members could store their cotton until they could sell it was the primary Alliance activity. Other ventures included a fertilizer company in Clayton, Barbour County; a cottonseed processing plant at Union Springs, Bullock County; and a large bank at Selma, Dallas County. The most ambitious of the Alliance venture was the establishment of a state exchange that would act as a central purchasing agent for all members. It would make huge cash purchases and resell to local Alliance organizations that would, in turn, resell to local members. After much jockeying for position, Montgomery finally became the site of the exchange, and it opened there in 1889. After all of the high expectations and fervent effort by Alliance members, by the end of 1891 it and most of their other ventures had either failed or were failing. Of all of the ventures that the Alliance attempted, only the effort to obtain a reduction of the prices for jute used in making bagging for cotton achieved any measure of success.

By the end of the nineteenth century, farmer efforts to pull themselves up by their bootstraps through organization had failed. The Alliance, like the Grange and the Wheel before it, had disappeared, and there was little remaining to show for their efforts. Farmers had simply been unable to solve any of their most pressing problems with these efforts. They had been unable to obtain any control over prices that they were paid for their crops and the prices they had to pay for goods they had to purchase. Finally, they had also been completely unsuccessful in solving the problem of obtaining sufficient credit at reasonable rates. Fite (1984) lists some of the reasons why efforts to organize and to engage in political action were unsuccessful:

1. Southern farmers included many kinds of producers (cotton, sugar, tobacco, rice), with varying economic interests (cash tenants, share tenants, and landowners). For example, many farmers were also merchants, and many bankers had large agricultural holdings. Tenants were too dependent on merchants and landlords to participate actively in some ventures. Similarly, markets for tobacco had a completely different base (addiction) in the

economy than did cotton (clothing) or rice (food). Farmers producing these different commodities had some of the same general problems, but their specific needs were considerably different, and as is always the case, the devil is in the details.

2. The well-known rugged individualism of Southern farmers often kept many of them from joining and working with their neighbors to present a united front to non-farm interests.
3. Tenant farmers, as well as many landowners, simply did not have sufficient capital to support ventures that they might agree to pursue in their organizations.

Salvation Through Diversification

Advocates of diversified farming argued that it was bad business for Southern farmers to concentrate on the production of cotton, while simultaneously purchasing his pork, meal, molasses, and other necessities at the local store. Most agricultural reformers agreed that the farmer should continue to produce cotton as his cash crop, because there was no suitable alternative. However, the reformers advised the reduction in the acreage of cotton, while increasing the production of livestock and other field crops. Furthermore, they argued, diversified farming could contribute to the rebuilding of the eroded and plant-nutrient-depleted soils. The endless production of cotton and corn, the so-called bare-ground-crops, had led to destructive soil erosion and mineral depletion for much of the land in the South, even before the Civil War.

Unfortunately, diversification made little, if any, progress. In 1870, farmers in the Old Cotton South harvested cotton from 23.2% of their improved farmland (9 million acres of cotton from 38.8 million acres of improved farmland), and in 1890, they harvested the crop from 26.9% (20.9 of 77.7 million acres). By 1900, the situation was even worse. That year, they harvested cotton from 28.6% of their improved acres (24 of 83.9 million acres). In Alabama, the situation was even more dismal. In 1870, farmers in the state harvested cotton from 30% of their improved farmland (1.5 million of 5 million acres), and by 1900, this percentage had increased to 36.8% (3.2 million from 8.7 million acres). This increase in the percentage of land used for cotton was discouraging enough, but even worse was the over twofold increase in acres harvested (9 to 24 million). There was simply no way that they could sell cotton from that much land at a reasonable price.

Fite (1984) has summarized some of the reasons why Southern farmers did not respond to repeated efforts to get them to diversify their farming operations and to apply the latest scientific advances. These include:

1. There was widespread resistance to book farming among Southern farmers. After all, a high percentage of these farmers were functionally illiterate, and they tended to be somewhat suspicious of the motives of those who were not. Like ignorance, poverty of any degree tended to make farmers highly conservative and unresponsive to trying new ways.
2. Landowners were reluctant to allow their tenants to grow anything but cotton on their land. It was the only crop, at that time, with any proven value.
3. Lenders would not generally loan money for the purchase of livestock; because, during that period, livestock throughout the South was subject to diseases that often resulted in heavy losses. Further, natural forages in the South were not very nutritious or productive. While native cattle could grow and reproduce on that nutritional base, improved breeds generally had a difficult time doing so.
4. Most tenants did not control enough land to support a mixed agriculture, including livestock production.
5. The tenant system, by its very nature, was difficult to reform. With cotton prices so low and the cost of purchased goods so high, virtually everything the tenant produced above bare subsistence was taken by the landowner to satisfy debts. Consequently, there was little incentive to trying to improve production and diversity, or increase efficiency.
6. Tenants tended to move so frequently that they seldom took time to make any improvements on a farm that they intended to leave at the end

of the year.

7. During the period, tenants and many landowners did not have sufficient capital to make the investments required for improving their farms and applying scientific methods. In their yearly cycles of production, they were never able to accumulate enough savings to afford capital improvements or to purchase land. Further, merchants and landlords lent money at very high interest rates. Much of what a tenant earned in a year had to be used to pay interest. Long-term credit for tenants was simply not available
8. The absence or uncertainty of satisfactory markets for crops other than cotton affected the tenant's capacity to diversify. Uncertainty exposed the farmers at or near the subsistence level to unacceptable risks.
9. The climate and climate-related problems in Alabama and much of the Lower South made crop production an extremely unpredictable undertaking, As a result, borrowing money at high rates of interest to diversify or to apply improved farming procedures was an extremely risky proposition.
10. Poor soils also affected the ability of tenants to produce enough of any crop to be profitable, even with improved practices and diversification. Fertilizers could be used to increase production on poor land, but this increased the cost of production and reduced any comparative advantage that an Alabama tenant farmer might otherwise have had.
11. Of all the factors affecting the ability of all farmers in Alabama and the South to improve their lot through the application of science-based methods and diversification, nothing was more important than the fact that in Alabama and the South there were simply too many people trying to farm at that time. There was simply no way that thousands of tenant farmers could have increased their production of any crop significantly without producing even larger amounts of the already unmanageable surpluses. Prices of everything they produced were already unacceptably low, and any additional increase in quantity was likely to bring even lower returns.

In the North Atlantic and North Central regions, it became apparent in the early years of the period that some diversification was needed there as well. The production of grains was increasing much faster than demand. As a result, prices, especially for wheat, were declining, and none of them were keeping pace with the costs of production. As a result, Northern farmers were encouraged to produce more livestock and dairy products. By doing so, they could convert grain that was being produced in excess into products of higher value. In effect, they could market their excess grain as beef, pork, milk, butter, eggs, and cheese. With the growing industrialization and urbanization in the Heartland, there was also a rapidly growing demand for all of these products.

Northern farmers did increase production of livestock somewhat; however, their hearts were still mostly with corn and wheat. For example, in 1879 farmers in the North Central Region,which includes the Corn and Wheat Belts, harvested wheat from 70.2 million acres. In 1899, acres harvested of the crop increased to 119.3 million.

Salvation Through Political Action

For many years after the Civil War, government at any level was largely ineffectual in doing anything. At that time, it was widely understood that government had little or no role to play in providing direct assistance or relief for individuals or groups. For this reason, farmers, in the beginning, did not include direct political action as one of their paths to salvation. In Alabama, farmers were effectively dragged into advocating political action.

As detailed in the preceding section, one of the most successful activities undertaken by the Farmers Alliance in Alabama was the effort to get prices reduced on the bagging that farmers had to use to bind their cotton into bales. The jute venture on the part of the Alliance was important for two reasons: It resulted in temporary lower prices on at least one item that the farmers had to purchase, and in its efforts to obtain this redress, the organization landed smack dab in the middle of Alabama politics, whether it wanted to be there or not. And it learned quickly that Alabama politics at the end of the nineteenth century was a bloodsport.

Beginning in 1889, the Alliance and its membership became directly involved in politics when they attempted to get then Commissioner of Agriculture Reuben F. Kolb elected to be the candidate for governor of the Democratic Party, which was tantamount to being elected governor of Alabama. This effort failed in 1890, and similar efforts failed again in 1892 and 1894. As Alabama Alliance members pushed their program to obtain redress to their problems through political action, their efforts became enmeshed with a much broader, more powerful National Populist movement. In the end, they all lost. There was no solution to the farmer's misery in direct political action, at least in the short-term (Rogers and Ward, 1994).

Out of these years of fruitless efforts, one idea did emerge that, while found unacceptable at that time, would play a defining role in the lives of farmers in the future. At its national meeting at Ocala, Florida, in 1890, the concept of the so-called Sub-Treasury Plan was introduced to Alliance delegates. Under this plan, the federal government would establish a Sub-Treasury with warehouses and/or grain storage facilities in every county in the country that produced agricultural products worth at least $500,000 a year. When farmers were ready to sell their crops, they could do so through normal commercial channels or they could take them to Sub-Treasury storage facilities in their counties. There, they would receive a certificate of deposit indicating what they had stored, and they would be allowed to borrow at an interest rate of 1% per annum, an amount of money equivalent to 80% of the current value of their stored crops. Farmers could redeem their crops within a year by repaying the loan, including interest. If commodity prices had increased sufficiently, during the year, beyond their level a year earlier, they could then sell their redeemed crops on the open market, realizing a worthwhile monetary gain. If, however, prices fell during the period, so that it would be of no benefit to the farmer to recover their deposited crop, to settle the loan, the Sub-Treasury would take possession and sell them for the best price they could obtain. In this situation, any loss in the transaction would be borne by the federal government. In essence, the Sub-Treasury Plan would have allowed farmers to borrow money directly from the federal government by using their crop as collateral. This proposed arrangement was much the same as the one available commercially at that time under the old crop-lien system. The difference, of course, was that with the Sub-Treasury Plan, We the People would have finally absorbed any loss that occurred. The plan would have directly involved the federal government in the marketing of agricultural products from private farms. In the long term, it would also probably have required printing money (inflating the currency), as large amounts of federal money would have been required to cover large losses. Further, when Sub-Treasury warehouses sold the excess crops to satisfy loans, excess-supply problems would have been exacerbated.

The Alliance's request for direct involvement of the federal government in the marketplace of American agriculture, as proposed by the Sub-Treasury Plan, was likely the result of the growing influence of European socialism and progressivism in the United States in the latter half of the nineteenth century. European socialism had its genesis in the realization that the benefits of the European Agricultural and Industrial revolutions were not being equally shared by all members of society and that it was grinding its workers into a wretched, poverty-stricken, amorphous mass, the urban working class. It had created a small number of extremely wealthy people and a modest middle class, but it had left behind an enormous working class that was probably worse off, in a relative sense, than it had been before the Agricultural and Industrial revolutions. While many solutions to this very real problem were proposed, most involved the direct intervention of government.

It is difficult to understand how Louis Blanc's call for government intervention in the lives of European industrial workers was translated into the request for the Sub-Treasury Plan for American farmers. In fact, it was not. They were inalterably opposed to government involvement in crop production. They just wanted government to buy their crops. They wanted to have their cake and eat it too.

The provisions of the Sub-Treasury Plan were hotly

debated at the national level, and finally effectively killed in Congress. However, it would rise like the mythical Phoenix from its ashes in less than four decades to become the foundation of a new National Agricultural Policy, which would carry and sometimes push and pull American farmers through the remainder of the twentieth century and into the twenty-first.

Direct political action on the part of American farmers would likely have involved the establishment of a national agricultural policy, and that would have been unacceptable to most farmers. From the beginning, British America was a land of farmers, and they remained at the heart of the United States for well over a century. Large numbers of industrial workers did not appear until after the Civil War. Yet long after the United States became the greatest industrial country in the world, most of its citizens clung to the Jeffersonian vision of a country where chivalrous and learned leaders would walk hand-in-hand in perfect harmony with God s finest creation—the farmer. Many saw farmers as working closely with the Creator producing food and fiber. They saw them as the salt of the earth. Farmers were thought to be virtuous, honest, democratic, hard-working, self-reliant, and independent. However, for well over a century with farmers, especially Southern tenant farmers, sinking deeper and deeper into abject poverty and misery, this Jeffersonian ideal among the broader citizenry was never mobilized so as to provide any relief for their suffering (self-reliant and independent) brothers through governmental action. We the People might admire farmer in their lives and struggles, but they could not see governmental intervention as being the solution to their problems. Lack of concern among the American people was not the only problem that resulted in the lack of governmental intervention. For example, there were politically powerful, commercial interests in the cotton economy who were suffering little, if at all, from the farmers' plight. They were in a position to get their share regardless of the price the farmer received for cotton or wheat. Many of these interests saw governmental involvement in supply and demand as a direct threat to their socioeconomic status and well being.

The powerful spirit of self-reliance and independence had its limits. Once the situation got bad enough, these independent, self-reliant souls embraced government patronage like a duck takes to water, and once they got wet almost none of them wanted to get out of the government pond.

ALABAMA AND HER SISTER STATES IN 1890

A primary thrust of this book is to document the establishment and evolution of the Alabama agroecosystem. An underlying theme, however, is to determine the extent that the rate of evolution of Alabama agriculture has contributed to its relatively low position among all the states in the production of all goods and services. I have posited that Alabama's status in this regard is much lower than it should be considering the wealth of its natural resources. To test this hypothesis, I aim to compare, over time, the rates at which Alabama and three sister states (Mississippi, Illinois, and Indiana) have moved through the different stages of societal evolution proposed by Lenski (1998): Hunters and Gatherers, Simple Agricultural, Advanced Agricultural, Industrial, and Post-Industrial stages.

Data indicate that as early as 1860 Alabama and Mississippi were already falling behind both Illinois and Indiana in the rate they were moving toward the Industrial stage. From the 1890 Census, I have chosen some additional data to be used in the further evaluation of the rate at which these states are evolving. They are presented in Table 8.11:

Farms per 1,000 persons: These data show that by 1890 farms were generally less important in the total populations of Illinois and Indiana than in those of Alabama and Mississippi.

Average Size of Farms. There was not much difference in the average size of farms in the four states, although they were a little smaller in Indiana.

Percentage of Improved Farmland. These data show the extent farmers in each state have converted total farmland into improved farmland. These differences probably are a reflection of differences in the geophysiography of the different states. A large share of the land on most Alabama and Mississippi farms is really poorly suited for farming, because of wetness, topography, etc. Generally, a higher percentage of the land in Illinois and

Indiana, deposited by extensive glaciation, is suitable for crop production.

Further, in 1889 a large share of the improved Alabama and Mississippi farmland was used for the production of cotton. With the poor soil conservation practices in use at that time, land used for the production of cotton quickly became non-productive or worn-out. In both states, worn-out land quickly reverted to unimproved status.

Value of Farm Real Estate. These estimates include values for land, buildings, fences, and other improvements, but the land itself contributes the greatest share of the total. In 1890, land values were still depressed from the effects of the Civil War, but much of the differences resulted from the lower economic productivity of Alabama and Mississippi soils, and the fact that a large share of the land on individual farms in those two states had little agricultural value.

Value of Farm Implements and Equipment. In 1890, implements and equipment on Alabama and Mississippi farms were little changed from 1820; whereas, there were a number of completely new items being manufactured specifically for use in the North Central Region in the last quarter of the nineteenth century. The higher values for Illinois and Indiana reflect their availability in those states.

Grain Crop Yields. In 1890, yields of corn, oats and "all" wheat were half as high in Alabama and Mississippi as in Illinois and Indiana. That relationship had changed very little since 1860. Yields of all of the grain crops were generally lower in 1890 than in 1860. Apparently, crop weather throughout the Eastern United States was less suitable for field crop production in 1890 than in 1860.

Valuation of Livestock. (dollars per farm, on July 1) In 1860, livestock production in Illinois and Indiana trailed far behind Alabama and Mississippi, but the data published in the 1890 Census indicate that those two states, had not only caught-up, but were also now far ahead. With all of the grain available, it had been just a matter of time before this happened.

Estimated Value of Farm Products. (dollars per farm) This comparison is somewhat misleading, because cotton was the primary crop in Alabama and Mississippi, while wheat was the primary crop in Illinois and Indiana. Further, while the price of cotton was somewhat higher than wheat (8.6 cents versus 1.6 cents a pound), yield was much higher for wheat than for cotton (571 versus 198 pounds an acre). The value of farm products was generally lower in Alabama than in the other three states. In 1890, the yield of cotton was lower than in Mississippi (169 versus 198 pounds an acre). Mississippi farmers harvested a much larger cotton crop.

Other Comparisons. The other comparisons presented in Table 8.11 are related to industrial development in the four states, and they generally show the degree to which Illinois and Indiana had moved into the Industrial stage of societal evolution. These differences were quite striking in 1890, but the more important consideration was whether Alabama and Mississippi could catch up in the future.

The combined data for all of these comparisons indicate that in the late nineteenth century the agricultural and industrial economies of Alabama and Mississippi were falling further and further behind those of Illinois and Indiana.

9

The Golden Age of American Agriculture

1895-1915

PRECIS

In the latter part of the nineteenth century and the beginning of the twentieth, Europe and America experienced a period of unrivaled material prosperity, the Age of Progress. There were new machines, new industries, new sources of energy, new things to sell and purchase. This industrial and commercial transformation was accompanied by transformations of the societies and economies of the several nation states. Generally, conservative governments, concerned about the advances of socialism, began to enact legislation that benefited the middle class. In this rapidly evolving world, the United States became the most economically powerful and successful country on earth.

In the period 1895-1910, the US Gross National Product increased from $228.8 to $423.3 billion, or by 85%. At the same time, the Consumer Price Index hardly increased at all (25 to 28). This exploding national prosperity lifted the agricultural economy as well. Prices for most agricultural commodities, even cotton, generally increased during the period. It was the Golden Age of American Agriculture.

Unfortunately, this increasing national prosperity barely reached the South. There could be relatively little revolution in industry because the region had so little of it to begin with. In agriculture, most of the region's farmers, and especially those in Alabama, just drifted deeper into poverty and misery. From 1890 to 1910, acres of cropland harvested per farm in Alabama fell from 34.5 to 27.4, and the tenancy rate increased from 48.6 to 60.2. The Golden Age of Agriculture in Alabama and the South was not very golden. Instead, it was decidedly grayish.

INTRODUCTION

As detailed in the preceding chapter, after the Civil War industrialization, commercialization, urbanization, and technological advances were savagely tearing American farmers in the Old Northwest out of their insular self-sufficiency. In a relatively short time, technology and industrialization provided them with a wide array of efficient, labor-saving equipment and machinery (gang plows, spring- and disk harrows, cultivators, hay loaders, silos, and cream separators). With this vast new arsenal, their work ethic, and one of the very best agricultural environments (soils and climate) in the world, they were producing enough food and fiber to meet the needs of a growing domestic population and to command the agricultural markets of Europe.

Unfortunately, farmers were their own worst enemy. They had reached a point where they could produce much more food and fiber than they could sell profitably, but they could not pay their bills. They had no control over the prices that they paid for farm inputs and had no practical way to limit production. The harder they worked, the "broker" they became. The experts advised them that the solution to their problems could be realized only through organization, crop diversification (more livestock), and scientific farming (research and extension). They quickly, to the extent practical, began to apply all of the experts' recommendations, especially organization and scientific farming. Unfortunately, at the end of the century, none of these silver bullets seemed to have had much positive effect. They were still in a land awash with food and fiber, a land of decreasing prices and increasing costs. They were caught securely and steadfastly on the horn of plenty. There seemed to be nothing to look forward to in the new century except "another day older and deeper in debt" (Johnstone, 1940).

Conditions on farms were desperate throughout the country, but they were worse, much worse, in the South. The world of the Iowa wheat farmer and the New York dairyman, even with all of their problems, was in a different universe from that of the cotton tenant in Alabama and the South. Most of them were not just desperate, they were destitute—in every sense of that word. Many were truly without means of subsistence.

At the end of the nineteenth century, industrialization, commercialization, and urbanization had reached few parts of the South and none had reached its corners. The economies of virtually all of the small towns and setttlements of the region depended almost entirely on agriculture—cotton agriculture. Farmers were trying to farm soils (Ultisols) that had been worn-out by thousands of years of endless hot summers and torrential rains, a most inhospitable climate. There were too many of them, and their large families did not help matters. Unfortunately, they had no place to go. Industrialization and commercialization in the South lagged far behind the North. An inordinately large number of Southern farmers lived out their lives in what seemed to be inescapable poverty. Their small farms, unpainted houses, shabby clothes, no education, little recreation, and poor general health, all testified to their perpetual hard times. They had too many problems making ends meet on a daily basis to worry much about what the new century might bring (Fite, 1984).

However a new century was approaching. There was not much to suggest that it would be any different for farmers, especially in the South, than the century they were leaving, but the country and the world were changing rapidly, in many ways. Just possibly, these changes might bring better days, or would they?

NATIONAL POLITICS

The Republican Party had ample opportunity to determine and control agricultural policy and legislation during the period 1895 to 1915. For most of this period, they were in control of the presidency, both houses of Congress, and the chairmanships of the agriculture committees of both houses.

The Presidency

Grover Cleveland (D, NY) was elected president in 1893. He had served an earlier term (1885-1889) in that office. Other persons serving as president during this period included: William McKinley (R, OH) (1897-1901); Theodore Roosevelt (R, NY) (1901-1909); William Howard Taft (R, OH) (1909-1913); and Woodrow Wilson (D, NJ) (1913-1921).

Secretaries of Agriculture

Sterling Morton (D, NE) served as secretary of agriculture from 1893-1897. Other persons serving in the position included: James Wilson (R, IA) (1897-1913) and David Houston (D, MO) (1913-1920).

The Senate

Democrats controlled the Senate in the 53rd Congress (1893-1895), but Republicans regained control in the 54th (1895-1897) and maintained it through the 63rd (1913-1915).

Senate Majority Leaders

Senate majority leaders were not identified by the Democrats until 1920, and by the Republicans until 1925.

Chairmen of the Senate Committee on Agriculture and Forestry

James Z. George (D, MS) was chairman of the Senate Agriculture Committee in the 53rd Congress (1893-1895). Other persons serving in that position included: Redfield Proctor (R, VT) (1895-1908); Henry C. Hansbrough (R, ND) (1908-1909); Jonathan P. Dolliver (R, IA) (1909-1910); Francis E. Warren (R, MA) (1911); Henry E. Burnham (R, NH) (1911-1913); and Thomas P. Gore (D, MS) (1913-1921).

The House of Representatives

The Democrats also controlled the House in the 53rd Congress, but in 1895 it shifted back to the Republicans who maintained control through the 61st (1909-1911). The Democrats gained control in the 62nd (1911-1913) and maintained it through the 64th (1915-1917).

Speakers of the House

Charles F. Crisp (D, GA) was speaker of the House in the 53rd Congress (1893-1895). Other persons serving in that position included: Thomas B. Reed (R, MA) (1895-1899); David B. Henderson (R, IA) (1899 and 1903); Joseph G. Cannon (R, IA) (1903-1911); and James Beauchamp (D, MO) (1911-1919).

Chairmen of the House Committee on Agriculture

William H. Hatch (D, MO) was chairman of the House Agriculture Committee in the 53rd Congress (1893-1895). Other persons serving in that office included: James W. Wadsworth (R, NY) (1895-1907); Charles F. Scott (R, KS) (1907-1909); John Lamb (D, VA) (1911-1913); and Asbury Lever (D, SC) (1913-1915).

The Political Situation

By 1895, the Populist Party was on its way to oblivion. Earlier, disgruntled farmers had united around some of the elements of European socialism concerning the use of government to address moral, social, and economic ills. The Populists had reached their zenith in the presidential election of 1892 when its candidate, James B. Weaver, received over a million popular votes (8.5% of the total), carried four states, and received electoral votes from two others. While the party never made more than a small splash in American politics, several of its more progressive ideas (the role of unions, government involvement in transportation and the conservation of government-owned, natural resources) would resurface as major thrusts in Theodore Roosevelt's administration.

The economic turmoil and labor unrest of Cleveland's second administration likely played an important role in his defeat in the election of 1896. As a result, William McKinley (R, OH) was elected president. The Spanish-American War (1898) and its aftermath occupied the country's politicians for most of McKinley's term in office. It was called by some a "splendid little war." It did not last long, and there was not a great loss of life or property, but the way it was prosecuted resulted in the recognition of the United States as a major military power in the world. As a result of the war, America entered the game of international imperialism by annexing the Philippines, the Hawaiian Islands, and Puerto Rico. These moves resulted in widespread political discussions and disagreement as to whether the Constitution provided for such acts.

The "splendid little war" also helped to restore economic prosperity to the United States. The growing world prestige and economic prosperity at home propelled McKinley into a second term in 1900, but he was assassinated before he completed the first year. His vice president, Theodore Roosevelt (R, NY) of Rough Rider and San Juan Hill fame, replaced him. Roosevelt completed McKinley's term, and in 1904 was elected to the office in his own right.

The United States was poorly prepared to deal with the explosive growth of industrialism and commercialism at the end of the nineteenth and the beginning of the twentieth century. This unfettered growth resulted in a wide range of activities (abusive trusts, adulterated foods, unclean meat, etc.) that were not very supportive of the common good of most American citizens. Most of Roosevelt's two terms were given to trying to put a human heart into capitalism without throwing the baby out with the bathwater. Roosevelt demanded a "square deal" for labor, but at the same time supported the open shop. The Federal Meat Inspection Service was strengthened.

He pushed for the enactment of the Pure Food and Drug Act and acts creating the Department of Commerce and Labor and the Interstate Commerce Commission. At the same time, he pushed for a more vigorous enforcement of the Sherman Antitrust Act, and he was unwavering in his efforts to clean up corruption that had come to characterize the federal bureaucracy since the Civil War. Almost single-handedly, he forced the country to begin to think about the conservation of its seemingly endless abundance of natural resources. Roosevelt decided not to seek the nomination of the Republican Party in 1908. Instead, he worked for the nomination and election of William Howard Taft of Ohio.

Roosevelt was a dynamic leader, but surprisingly he was the first president, and certainly the first Republican, to actively work to bring elements of European socialism into American government. Of course, several of his progressive ideas had actually originated in the Populist Movement in the nineteenth century. The so-called Gilded Age of the latter part of the nineteenth century was characterized by the reluctance of government to interfere in private enterprise, except in the areas of tariffs and railroads. After the depression of the late 1890s, however, a significant portion of America's emerging middle class began to demand that government become involved in a wide range of economic, political, social, and moral reforms. Although the Populist Movement was dead, many of their ideas lived on. The Progressives chose Roosevelt as their leader, and he did not disappoint them. After his term as president ended, progressivism came much easier for Woodrow Wilson.

Taft, at the time, was generally thought to be a weak president. However, more of Roosevelt's legislative agenda was actually enacted under Taft than under Roosevelt himself (Morison, 1965). His administration also vigorously enforced the provisions of the Sherman Antitrust Act by bringing suit against a large number of business trusts. He was able to get the Payne-Aldrich Tariff (1909) enacted; it reduced, to some degree, protection of the American economy from foreign competition. With Republicans generally in control of both the presidency and Congress from the Civil War to the end of the century, protectionism through the use of high tariffs had steadily grown. It finally reached its zenith in 1897 with the passage of the Dingley Tariff, the highest since the Civil War. Taft's efforts stopped this growing trend in protectionism, but it fell far short of what he wanted. One of the most significant of Taft's actions was the request in 1909 for legislation for amending the Constitution to allow the national collection of income taxes. In 1910, as a result of problems in the world supply of gold, the cost of living increased sharply in the United States. In fact, the cost of basic food items increased by 30%. This increase resulted in increased hardship for laborers. Likely as a result of this increase in the misery of the working class, the Republicans lost the House in the 1910 elections and had their margin of control severely reduced in the Senate. After this defeat, it did not appear that the Republicans could retain control of the presidency with Taft in 1912. Also, toward the end of his administration, ex-President Roosevelt began to express broad disagreement with Taft's performance. When he could not get wide support in the Republican Party for his complaints about Taft, he formed his own Bull Moose Party.

In the latter part of the nineteenth century and the beginning of the twentieth, some elements of European socialism had begun to seep into the political environment of the United States: first as part of the platform of the Populist Party, then as centerpieces of the Republican administration of Theodore Roosevelt, and finally into the leadership of the Democratic Party. At their 1912 convention, after considerable soul-searching, the Democrats nominated a completely different kind of a candidate for president of the party of Jefferson and Jackson. They nominated an academic—Woodrow Wilson, president of Princeton. Several reasons have been advanced explaining why the Republicans lost the election of 1912, and certainly the splitting of the party by Taft and Roosevelt was a major one. The fact remains, however, that the Democrats won with a Progressive candidate and neither the party nor the country would ever be the same.

Morison (1965) comments that "Wilson was a great leader because he sensed the aspirations of plain people." With Wilson's election, Democrats also took control of both houses for the 63rd Congress. They had not had control of both bodies since the 46th (1879-1891). With the Democrats firmly in control of the reins of government, it was plainly obvious that the course of America's business would not be business as usual. For years, both the presidency and Congress had been closely related to big business interests of the Republican Party, but Wilson and the 64th Congress were horses of a different color.

The Republicans had controlled both houses from the 54th (1895-1897) through the 61st (1909-1911). But throughout this long period of Republican control, in fact since the end of Reconstruction, the Solid South had been sending the same Democratic congressmen and senators to Washington. In the process, these individuals were amassing years and years of seniority. As a result, by the early part of the twentieth century, Southerners controlled many of the standing committees of Congress. With Southern agriculture still languishing when the Democratic Party took control of the presidency and Congress in 1913, Southern congressmen and senators began to push hard for legislation bringing the federal government into a direct support role of American agriculture.

In the interest of humanity, Wilson's legislative agenda included: reducing protectionism through the enactment of new tariff laws, increasing emphasis on natural resource conservation, improving regulation of banking, and improving regulation of larger business interests (Morison, 1965). Under Wilson, the 16th Amendment, which allowed for the collection of a national income tax, was finally ratified. The Underwood Tariff Act, approved by Congress in 1913. reduced protectionism to the lowest level since the Civil War. The act also included provisions for the actual collection of the income tax as authorized by the 16th Amendment. Continuing his efforts for getting his legislative agenda written into law, Wilson got Congress to pass the far-reaching Federal Reserve Act in 1913. Then in 1914, he was able to get legislation establishing the Federal Trade Commission, and in the same year Congress approved the Clayton Antitrust Act. Other legislative goals that Wilson might have pursued had to be shelved in July, 1914, when Austria-Hungary declared war on Serbia initiating World War I. Wilson immediately proclaimed that the United States would maintain strict neutrality in the conflict. Unfortunately, in May 1915, the American tanker, *Gulflight*, was torpedoed by a German submarine, and six days later, the *Lusitania* was sunk off the coast of Ireland, resulting in the loss of 128 American lives. Still Wilson refused to be forced into active intervention in the conflict. By late 1915, however, public pressure for the support of France and Great Britain was growing so rapidly that Wilson reluctantly had to recommend a plan of preparedness for America's entry into the conflict.

ALABAMA POLITICS

By 1896, the Populist Movement was essentially over in Alabama. The political revolt of small farmers, small businessmen, and laborers had accomplished very little. Rogers and Ward (1994) wrote their epitaph: "Populism failed, and because it failed, the same poverty continued and an 'elite' still ruled. . . . It is the Populist claim to our attention that, in spite of everything, a true and native crusade almost succeeded."

Governors

Governors of Alabama during the period 1895-1915 were as follows: William Oates (D, Henry County)(1894-1896); Joseph Johnston (D, Jefferson) (1896-1900); William Samford (D, Lee) (1900-1901); William Jelks (D, Barbour) (1901-1907); Russell Cunningham (D, Jefferson) (1904-1905) (served while Jelks was ill); Braxton Comer (D, Jefferson) (1907-1911); and Emmet O'Neal (D, Lauderdale) (1911-1915).

Commissioners of Agriculture

Alabama commissioners of agriculture during the period 1895-1915 were: Hector Lane, 1891-1896; Issac Culver; 1896-1900; Richard Poole, 1900-1907; Joseph Wilkinson, 1907-1911; and James Wade, 1915-1919.

The Political Situation

The Populist revolt of the early 1890s had failed, but it had thrown a real scare into the Democratic elites who controlled Alabama's political machinery that had enabled them to keep the rich rich and the poor poor. The Democratic political elite fully understood that with effective leadership African Americans, poor farmers, and organized labor potentially had the votes to put them out of business. So, beginning in the later years of the decade, they set about changing the Constitution so that such a revolt could never happen. The 1901 Constitution was written in such a manner that it could immediately disenfranchise virtually all of Alabama's black, male citizens without directly violating the 15th Amendment. Disenfranchising poor whites took a little longer, but the language of the Constitution finally eliminated most of them from the voter lists as well. The remainder of the Constitution simply reinforced the status quo.

Attempts to regulate the railroads kept the attention of the governors and legislators throughout this period. Alabama had few railroads before the Civil War, and most of those were destroyed during the conflict. When the war ended, the state made a concerted effort to rebuild and expand the railroad system. To do so, they granted all sorts of concession to the companies who built them. The railroads became extremely powerful politically. This power did not cause much of a problem until industrialization began to increase. Then the question of freight rates on the shipment of raw materials and finished products became extremely important. Governors and legislators generally agreed that some way had to be found to reduce rates, but the railroads, at least initially, had most of the big guns on their side. For almost this entire period (1895-1915), the battle raged over establishing an independent, elected, railroad commission and providing them with the power to deal effectively with the large, well-financed companies. Although some progress was made during this period, much of the progress was still tied up in the courts.

The question of the role of labor unions came violently to the surface of the political pot during the period. Just as before, governors, especially Governor Comer, used their police powers to virtually destroy the unions.

Two other perennial problems, child-labor and convict leasing, also held much of the attention of the politicians for the entire period. There was some, but very limited, progress on both of these problems. Both of them were just too important to Alabama's leading business interests to allow much progress.

At the national political level, Populism had also failed to win at the polls, but much of its philosophy finally became embodied in the Progressive Movement. President Theodore Roosevelt became its chief spokesman urging that the country should provide a Square Deal for all its citizens. Progressivism became widely popular throughout the country, and finally made its way to Alabama. Governor Comer was Progressivism's most ardent supporter. Unfortunately, with the governor's political baggage, the president'sSquare Deal never made it down to Alabama's tenant farmers and mill workers. Rogers and Ward (1994) concluded that the "circumscribed vision of the Democratic Comer Progressives" was just too limited to reach those who desperately needed their attention and support.

Alabama politicians did almost nothing in a positive sense for Alabama's farmers during the period 1895-1915. Truthfully, there was very little that they could do. There were simply too many farmers, and all of them were producing cotton—way too much cotton. Many of them would have gladly traded their mules and 20 acres for any job off the farm. Unfortunately, there were too few of those to go around, and it seemed that there was little that politicians could or would do about that.

THE NATIONAL ECONOMY

It was not until late in the nineteenth century that the portion of the Gross National Product derived from manufacturing exceeded that derived from farming. However, from the beginning of the twentieth century, the national economy was increasingly driven by industrialization, urbanization, and consumerism. The American political system was not well prepared for this rapid change.

The Economic Situation

The Progressive Era in America and around the world began at the end of the nineteenth century. Industrialization was roaring ahead in the North under the protection of powerful tariffs. The Dingley Tariff Act had received approval in 1897. The United States had become the foremost industrial power in the world. American captains of industry reigned supreme. Agriculture in much of the country, with the exception of the South, was once again moving ahead with the help of mechanization and the applications of science. As a result, fewer farmers were needed to provide the food and fiber required by the growing population. Fortunately, some displaced farmers were able to find employment in the rapidly expanding economy.

Europe was also in the middle of a dynamic period of material prosperity characterized by new industries, new sources of energy, and new goods. Everywhere there was economic boom. And there was no war. Across the entire region, the middle class was growing rapidly as real wages increased (Spielvogel, 1994). Unfortunately, Southern agriculture was only managing to pick up just a few of the crumbs falling off the table of growing national and international prosperity. Cash (1991) comments that Southern agriculture was changing at the beginning of the twentieth century in that its "descent into disaster" was slackening.

The period 1895-1915 witnessed some topsy-turvy conditions in the American economy. Within that 21-year period, it cycled through several periods of prosperity and a banking panic. During the early 1890s, the United States had suffered through the most serious depression since the administration of George Washington.

Fortunately, after 1895 the wave of unrest began to recede, and prosperity began to return. Then the prosperity was enhanced and accelerated by the "splendid little war" with Spain (April 1898-February 1899). In the fall of 1903, principally in October, the value of the stock market plunged 23% in the so-called Rich Man's Panic. Fortunately, this panic did not ripple through the economy, and the good times that began around 1895 continued through 1906.

In 1907, the period of prosperity that had begun before the turn of the century came to an abrupt halt when the financial community experienced a nerve-wracking panic. Failed speculation in the stock market by the owner of a large New York bank led to concerns on the part of depositors about the bank's soundness. As a result, they quickly began to withdraw such large amounts of their deposits that the bank was forced to close. There were similar runs on other banks as well. Soon the crisis spread to the stock exchange. Although the panic was essentially a financial one, it quickly spread to the general national economy. The GNP actually decreased 33 billion dollars from 1907 to 1908 (Figure 9.1). Fortunately, before the financial panic could spread further, the financier J.P. Morgan organized a syndicate to loan money to the threatened banks. When the depositors were assured that their funds were indeed available to them if they wished to withdraw them, the runs ceased, and the panic ended. Prosperity returned to the country in 1909, and good times continued through 1912. In 1913, the economy encountered another of those rough spots generated by advancing Progressivism. Growth slowed in 1913, and the country entered another recession in 1914, which continued through 1915.

It was during this period that the long-considered idea of a national tax on income came to fruition. It had been considered for a number of years as a means of providing a mild, leveling effect on society. Finally, in 1913 Congress passed the 16th Amendment that gave that body the power "to lay and collect taxes on incomes." It was ratified less than a year later, and soon the country was given its first graduated income tax.

Congress also passed other legislation in 1913 that would have far-reaching effects on all Americans. It had been obvious for some years that the old National Banking System, established in 1863, was no longer adequate to deal with national banking problems of the twentieth century. As a result, they created the Federal Reserve System (the Fed). It was created to serve several functions. One of the primary ones was to provide an elastic currency to help eliminate panics in the economic system and runs on bank deposits during periods of economic stress. Later, the Fed began to perform an entirely new, more modern function: controlling the money supply through the manipulation of interest rates. By manipulating interest rates that the Fed charged private banks for money loaned to them, it was envisioned that they could affect the business cycles and economic activity that would, in turn, have short run effects on rates of employment, inflation, and real economic growth.

The Gross National Product

Annual estimates of the US Gross National Product

(GNP) for the period 1890-1915 are shown in Figure 9.1. In 1895, when the severe economic problems of the early 1890s were coming to an end, the GNP was $228.8 billion (in constant, 1982 dollars). Unfortunately, some of the leftover effect of the severe recession and reduced economic activity remained. As a result, in 1896 the GNP declined 2.2% to $223.7 billion. After 1896, as prosperity spread through the economy, GNP began to increase, and by 1903 had reached $331.8 billion. In the fall of 1903, the so-called Rich Man's Panic staggered the stock market, and in the following year (1904) the GNP declined to $328.2 billion. It began to grow again in 1907, finally reaching $399.5 billion. That year, a severe financial panic again hit the American economy, and by the end of 1908 the GNP had lost 8.3% of its value ($399.5 to $366.4 billion). The economy began to recover in 1909 (GNP at $411.3) and continued to grow through 1913 when the GNP reached $460 billion. It grew steadily from 1909 through 1912 when the economy encountered one of those leftover rough spots. Growth began to slow in 1913, then entered a short recession in 1914. In that year, the GNP was down to $444.4 billion, and it continued to decline, reaching $439.6 billion in 1915.

The Consumer Price Index

Consumer Price Index (1967 = 100) data for the period 1870-1915 are presented in Figure 9.2. The CPI was 27 from the mid-1880s to the mid-1890s, but with the severe economic contraction from 1893 through 1894, it declined to 25 and remained there through 1901. Then, as the economy recovered, it returned to 27 and remained there until 1906. It increased slightly in 1907 as the result of the panic, but returned to 27 the following year. Following the panic, the economy began to grow rapidly, and the CPI also began to increase. It reached 30 in 1914 when World War I began; then increased slightly in 1915 as the United States gradually moved toward participation in the conflict.

The Inflation Rate

Data on the Rate of Inflation in the United States during the period 1870-1915 are presented in Figure 9.3. In several of the years during the period 1895-1915, there was no inflation of the currency. The rate was zero. However, after 1901 it began to increase, and after 1911 it began to increase more rapidly, but never reached an annual rate higher than 4%.

THE AGRICULTURAL ECONOMY

American farmers ended the nineteenth century caught in a slowly tightening noose of increasing surpluses, decreasing prices, and increasing costs. In the South, the agricultural economy was little changed from a half century earlier. Cotton production was still at the center of the economy with virtually every economic activity revolving around the culture of the white fiber. Other crops (corn, hogs, cattle) were produced only because the farmers and their families could not eat cotton. The vast majority of Southern farms were small, poorly capitalized, and operated by tenants. In 1900, in 11 Southern states the average farmer actually harvested crops from only 30 acres of land. Further, the total value of land and buildings on the average American farm was $1,048, but was only $603 in Alabama. Across the South, tenants operated 48% of the farms, and in those areas where slave holdings had been highest, tenancy rates were generally between 70 and 80%. Agriculture was still king, but it was an exceedingly poor kingdom. However, a new day was dawning with the new century. Progressivism was growing in the land. Business and industry were booming. Incomes and the demand for food and fiber were growing. In 1895, American farmers could not be faulted for not knowing that they were on the threshold of a Golden Age of Agriculture.

Federal Agricultural Legislation

At this point in the history of American agriculture (1895-1915), there was almost no interest in legislation dealing with any aspect of agricultural production or of agricultural markets. Farmers in the North and West were beginning to do extremely well financially. Congress began to pass some legislation for consumers. In 1906, they enacted the Food and Drug Act that, for the first time, provided for governmental oversight for the quality and safety of food. That same year, Congress further intervened in the area of food quality by passing the Meat Inspection Act.

Beginning in 1913 with Democrats in full control of both the executive and legislative branches of government and with Southerners in control of many of the standing committees, Congress began to devote a considerable amount of attention to agricultural matters. In the next three years, they would enact several pieces of legislation of considerable importance to farmers. In 1914, they passed the Cotton Futures Act. One of the major features of that act was to authorize USDA to establish standards (color grade, staple length, strength, etc.) for cotton. The act also included provisions designed to minimize the speculative manipulation of the cotton market. The act was later declared to be unconstitutional, because it had been developed in the Senate, rather than the House. It was reenacted in 1916.

In 1914, Congress also enacted the Agricultural Extension (Smith-Lever) Act that provided federal funding to establish federal, state and local cooperative extension work throughout the country.

Alabama Agricultural Legislation

The ratification of the reactionary 1901 Constitution resulted in two changes in the Alabama Department of Agriculture. "Industries" was added to the name of the department, and the commissioner's term in office was increase from two to four years. Later, in 1907, the legislature created two boards to assist the commissioner. One of these, the State Board of Horticulture, was given the responsibility to encourage horticultural pursuits in the state by licensing nurseries and working to control plant diseases and insect pests affecting the industry. The second board, the State Livestock Sanitary Board, had the responsibility to help protect the health of livestock. Then in 1911, the legislature charged the department with the responsibility to regulate the production of feedstuffs by creating a Division of Food, Drugs, and Feed. In the same year, the Board of Agriculture was established. It was created with the purpose to conduct farm demonstration work in cooperation with the USDA and the Extension Service at the Alabama Polytechnic Institute at Auburn. This board was the forerunner of the Alabama Cooperative Extension Service.

Texas tick fever is a disease of cattle caused by a bacterium living within the bodies of ticks. The disease is probably about as old as cattle, and it was endemic in all of the descendents of cattle that the Spanish introduced to the South in the sixteenth century. While the disease was not a major problem among herded cattle, it was a serious problem among the improved breeds being established in the state. In 1906, the US Congress empowered the secretary of agriculture to begin a cooperative program to eradicate the disease. The first of these control measures in Alabama was applied in Madison and Limestone counties. Control of the disease required that cattle be driven into and through large concrete vats containing a chemical that would kill the ticks. Obviously, with untold numbers of ticks living in Alabama's woodlands, "all" cattle had to be dipped several times.

The legislature created the State Livestock Sanitary Board to deal with such problems as tick fever. Unfortunately, efforts to control the disease quickly ran into violent opposition by cattle herders. Eliminating the open range was unacceptable to them, but rounding up their far-flung herds and driving them to dipping vats was even more unacceptable. As a result of this opposition, the legislature ruled that efforts to eradicate the disease would not be undertaken in counties where more than half of the land was in open range districts.

The long-term, often violent battle over free roaming livestock began before the Civil War (Blevins, 1998). As might be expected, the legislature had great difficulty dealing with it. In the early twentieth century, there were still substantial numbers of cattle herders operating in the backcountry of the Lower Coastal Plain and the hills of north Alabama. In 1903, the legislature shirked its duty in the matter by voting to give the problem to local county officials. This action set the stage for some really violent local battles in the individual counties. Cattle herding was important in "Alabama" long before it became a state. Its success depended almost entirely on allowing cattle open range, and it was not going to "go quietly into the night." The 1903 Act was passed on the heels of the approval of the 1901 Constitution. Poor whites and blacks, disenfranchised by the Constitution, looked upon the act as another attempt to drive them entirely from the land.

Agricultural Exports

The value of average annual agricultural exports for each decade from 1860 to 1899 is shown in Figure 9.4. Annual data for the period 1900-1915 are shown in Figure 9.5. Exports increased rapidly after the Civil War, and reached an annual average value of $703 million in the 1890s. In 1900, they totaled $941.1 million, the highest in history to that time. Unfortunately, this growing level of exports was wreaking havoc among the agricultural economies of Europe. As a result, many of the America's trading partners began to raise duties on imported foodstuffs. Some of them imposed embargoes, as in the case of hog products. Also, exports from Russia, Argentina, Australia, and Canada were beginning to crowd American products off the European market. By 1904, exports were down to $825.2 million. Fortunately, with the exports of American foodstuffs declining, exports of cotton began to grow. In 1905, total agricultural exports followed cotton exports upward to $974.6 million, and were over $1 billion for the first time in 1906. They remained essentially at the same level in 1907, but declined to $901.4 billion in 1908 and to $869.2 billion in 1909 when cotton exports fell to the lowest level in a decade. Production of the crop was much lower than expected in two of those years. In 1910, cotton exports began to grow again, and that year total exports were again above $1 billion. Cotton exports remained high in 1911, 1912, and 1913, and total exports continued to grow in 1911 ($1.05 billion) and in 1912 ($1.12 billion). Total exports were slightly lower in 1913 ($1.11 billion), but then surged upward in 1914 ($1.5 billion) and 1915 ($1.5 billion) as the war in Europe began and quickly spread.

Throughout the latter part of the nineteenth century, agricultural exports made up between 70 and 80% of all American exports. At the beginning of the twentieth century, the relative importance of agricultural exports began to decline. They represented 65% of the total in 1900, 51.1% in 1910, and 35.5% in 1915.

Agricultural Prices

In 1910, USDA began to publish annual data on average prices that farmers received from the sale of their crops and livestock, and on the prices they paid for all of the inputs they purchased to sustain their farming operations. The average of annual estimates of prices received for the period 1910-1914 was established as a base for making comparisons with future estimates. This average was given an index number of 100. In 1910, average prices received for all crops and livestock was 4% higher than the average for the four-year period, so the price received index was 104. Similar manipulations were made with data on prices paid. In 1910, average prices paid for all inputs purchased was 2% higher than the average for the four-year period; consequently, the prices paid index for that year was 102. Data for these two indices for the period 1910-191, are presented in Figure 9.6. The data indicate that throughout this period both indices remained near 100, or very close to the average prices received and prices paid for the base period (1910-1914). With the two indices remaining so close together (at parity) for these years, economists named this period the Golden Age of American Agriculture. In fact, later when farm leaders called for changes in agricultural policies to restore farm income to parity with non-farm income, they used the period 1910 to 1914 as their benchmark.

The two indices can also be converted into the parity ratio by dividing the prices received index by the prices paid index. For example, the ratio for 1910 was 104/102 = 1.02 and the parity ratio was 1.02. Ratios for the period 1910-1915 are shown in Figure 9.7. When the ratio is greater than 1.0, farmers are generally receiving more money for their crops and livestock than they are paying for inputs needed to produce those crops. Conversely, when the ratio is less than 1.0, they are paying more for inputs than they receive from sales. A straightforward application of the ratios suggests that if they remained below 1 for very long farmers would have to quit farming or find some source of off-farm income with which to subsidize their operations.

Farmers

There were 18.4 million American farmers in 1870; they constituted 47.6% of the total population. By 1890, the number had increased to 29.4 million, but their share of the population was down to 46.7%.

The Total Population

Population data for all persons in the United States for the period 1870-1915 are presented in Figure 9.8. Data presented in the figure indicate that the total American population increased from 69.7 million in 1895 to 100.8 million in 1915. Included in these population totals are relatively large numbers of immigrants. For example, in the period 1901 to 1910, some 8.8 million immigrants arrived (mostly from Europe).

Population data for Alabama for the same period are presented in Figure 9.9. These data indicate that the population of the state increased from 1.67 million in 1895 to 2.34 million in 1915. In 1895, some 2.4% of the national population lived in Alabama. By 1915, the percentage had fallen to 2.3%.

The Farm Population

Data on the number of American farmers in the period 1870-1915 are presented in Figure 9.10. These data show that numbers were increasing throughout this period; however the increase between 1910 and 1915 was relatively small. With the total US population expanding rapidly, the percentage of farmers in the population declined from 41.9% in 1900, to 34.9% in 1900, and to 32.4% in 1915.

From another perspective, 42 American farmers were producing enough food and fiber to meet the needs of 100 persons in 1900. During this period, the application of science and technology to agriculture was resulting in a rapid increase in the efficiency of food and fiber production, and by 1915 only 32 farmers were required to meet the needs of 100 persons.

Estimates of the number of farmers in Alabama are not generally available before 1920, but it is highly likely that the percentage of farmers in the general population would have been higher than in the country as a whole.

Farmers in the Work Force

Data on the percentage of farmers in the American labor force for the period 1870-1910 are presented in Figure 9.11. As might be expected, the percentage of farmers in the labor force was declining more rapidly than the percentage of farmers in the population.

Tenancy

Farm tenancy increased throughout the United States during the period 1880-1910. It was 25.6% in 1880, and 37% in 1910 (Figure 9.12). While these national percentages were distressing enough, the situation in Alabama was much worse. In 1880, 46.8% of all Alabama farms were operated by tenants. By 1910, that percentage had increased to 60.2%, with share-tenants (croppers) operating 43.9% of all Alabama farms that year.

Farms

After the Civil War and through the end of the nineteenth century, numbers of farms in the United States (and Alabama) generally increased. In general, numbers in the North Central and Western regions increased as a result of the establishment of new farms. However, in the North Atlantic, South Atlantic, and South Central regions, many of the added farms were established by subdividing old farms. Land in farms also increased nationally, but not as rapidly as farm numbers. Consequently, the average size of farms began to decrease.

Farm Numbers

Annual data on the number of American farms are not available before 1910; however census data indicate that by 1900 there were 5.7 million farms in the country (Figure 9.13). There had been 2.66 million in 1870. There were 6.36 million in 1910 and 6.46 million in 1915. The 1915 estimate, which is not census data, is taken from USDA annual data. There are no census data available for 1915. The USDA annual estimates tend to be slightly higher than those for the censuses in the same year.

There were also no annual data on Alabama farm numbers before 1910; however census data indicate that by 1900 there were 223,200 farms in Alabama (Figure 9.14). The 1870-1910 data are from ensuses; however, the 1915 estimate is from the National Agricultural Statistics Service (NASS). There had been 135,900 in 1870. In the first decade of the new century, the number of Alabama farms continued to increase. In 1910, there were 262,900. The data show that numbers seemed to decline slightly in 1915 to 261,000, but this estimate was taken from a different source. However, annual data from NASS also show a decline from 1910 to 1915 (265,000 to 261,000). Both estimates indicate that sometime after 1910, farm numbers began to decline in the state. This decline was a clear exception to what was happening in the counry as a whole.

In 1910, 3.1% of Alabama farms were in Dallas County. Other counties with substantial numbers were: Wilcox (2.5%), Montgomery (2.5), Lowndes (2.4), and Madison (2.2). Together these five counties accounted for only 12.7% of the total. There were 32 of 67 counties with 4,000 or more farms.

Alabama farmers had never been content with their economic situation. They just did not have any other options. Although industrial development came late to the state, it finally came (Rogers and Ward, 1914). The iron and steel industry grew rapidly after the Civil War, but it was localized. While it certainly must have been responsible in reducing the number of farms in those local area, it seemed to have little effect on the numbers of farms in the whole state. The textile industry was another matter, however. Textile mills were scattered across a large portion of the state, and with the growth of this industry, the number of farms began to decline. Flynt (1989) noted that there were 9,000 men, women, and children employed in the mills in 1900. By 1910, this total had increased to 13,000, and in 1920 there were 16,000 laborers converting Alabama cotton into textiles.

Farm Size

Census data for average American farm size for the period 1870-1910 are presented in Figure 9.15. After the Civil War, farm size declined sharply in the South, as large farms were divided into small ones. This decline in the acreage of Southern farms also resulted in a reduction for the country as a whole. However, by 1880 declines in the South were less dramatic, and from 1890 through 1910, the national average remained near the same (140 acres).

The number of farms in Alabama more than doubled between 1870 and 1900 (Figure 9.14), and while farmland increased somewhat during that period (Figure 9.17), it did not increase nearly as rapidly as the number of farms. Consequently, the average size of each farm declined. It was 222 acres in 1870, but by 1890 it decreased to 126 acres. Average size in the state continued to decline in 1900 and 1910, reaching levels of 93 and 79 acres, respectively. Of course, during this period tenancy was also increasing rapidly (Figure 9.12).

Total Farmland

Annual estimates of total American farmland are not available before 1850; however census data are available from 1850 onward. Some of these data are presented in Figure 9.16. The settlement of the lands between the Missouri River and California was in full swing in 1870. The census that year estimated that there were 407.7 millions acres of farmland in the United States. By 1900, however, the westward expansion had ended (Edwards, 1940). In that year, the census indicated that the quantity of farmland reached 838.6 million acres; then it increased relatively little by 1910 (878.8 million). It continued to increase relatively slowly until it reached 917.3 million acres in 1915. There are no census data for 1915; however Lerner (1975) has supervised the publication of a set of data that includes annual estimates of total American farmland from 1850 to 1970. Lerner's estimate for 1915 is reflected in the figure.

Total Alabama farmland fell sharply following the Civil War, and did not exceed the prewar level again until it reached a level of 19.8 million acres in 1890. Even with the severe economic contraction in the early 1890s, Alabama farmers added 900,000 acres of new farmland to reach a total of 20.7 million acres in 1900 (Figure 9.17). The total did not change during the following decade.

USDA has estimated that only 59% of the land in the state—19.4 million acres—is arable or suitable for crop production (Land Capability Classes I - IV). The amount of farmland in the state exceeded that level in 1890, 1900, and 1910. After 1880 and through 1910, there literally was no additional land suitable for crop production remaining in the state. In fact, some of the farmland in use was not really suitable.

Use of Farmland

Census-derived data on the uses of farmland in the United States in 1899 and 1909 are presented as percentages in Figure 9.18. As would be expected, the larger share was used for crops in 1889, but almost as much was used for pasture (38% versus 35%). A smaller share was used for woodland (23%). The "other" category (6%) includes: building sites, roadways, and waste places. The data indicate that use of farmland remained about the same in both years. Cropland increased from 38% to 39%, while pasture and woodland fell from 35% to 33% and 23% to 22%, respectively. The percentage of "other" increased from 6% to 7%. There are no comparable data available that could be used to compute the percentage of the different categories of land use in Alabama before 1924.

The percentages of US land in pasture underestimates the actual amount of land used for that purpose. The percentages shown in the figure represents only the land used for pasture and nothing else. It does not include cropland that might be grazed, or woodland used for grazing.

There are no direct estimates of total cropland in the country before 1924. Data for earlier years were apparently obtained from an extrapolation of data on acres harvested. Because estimates of total cropland are essential for calculating farmland uses, the percentages shown in Figure 9.18 are no more reliable than the extrapolations.

Cropland Harvested

Cropland harvested is a much better indicator of the status of general crop production than total farmland. Census data on cropland harvested in the United States in 1879, 1889, 1899, and 1909 are presented in Figure 9.19. The data show that it was 166.2 million acres in 1879 and 283.2 million acres in 1899, and that it kept trending upward in the following decade, reaching 311.3 million in 1910.

There are no USDA data available on total cropland harvested in Alabama before 1924. However, Fite (1984) published estimates of average acres of cropland harvested per farm for states of the Old Confederacy. His data show that in 1879, 1889, 1899, and 1909 acres harvested per Alabama farm was 37.2, 34.5, 30.1, and 27.4 acres, respectively. These data are indicative of the increasingly dismal condition of Alabama agriculture during this period. Unfortunately, Alabama was not alone in this situation. The average acres harvested for all of the Southern states in 1910 was 27.9 acres. In contrast to these dismal data, in 1910 farmers in Iowa and Kansas harvested crops from 111.9 and 93.9 acres, respectively.

The underlying theme of this book is to determine whether or not the rate of development of agriculture in Alabama over the years has played any role in the fact that the state is near the bottom of all the states in the production of goods and services at the beginning of the twenty-first century. Certainly, in looking at the possibility of such a relationship, the declining levels of acres of cropland harvested, noted in the preceding paragraph, would have to be an important indicator of just how slowly Alabama agriculture was evolving at the beginning of the twentieth century.

Value of Land and Buildings

In 1890, the value of land and buildings (farm real estate) on American farms totaled $13.3 billion (Figure 9.20). It had been $7.4 billion in 1870. Value increased to $16.6 billion in 1900. Then in 1910, at the beginning of the most prosperous portion of the Golden Age of Agriculture, it more than doubled to $34.8 billion. In 1915, it was $39.6 billion. These increases indicate a rapidly growing investment in land and buildings on American farms; however, farm numbers were also increasing during that period (Figure 9.13). Furthermore, the Consumer Price Index increased about 12% between 1909 and 1915 (Figure 9.2).

The values in Figure 9.20 are biased somewhat by changes in both numbers of farms and farm size. A more meaningful measure of the change in value is to convert the total values to value per acre. These data, which are shown in Figure 9.21, indicate that value per acre increased very little between 1870 and 1900. However, when the good times really came, around 1910, farmers rapidly invested more money in land and buildings. By 1910, value per acre had increased to $40. It doubled between 1900 and 1910, and only about 12% of the increase was the result of the growth in the Consumer Price Index.

However, those particular CPI values shown in Figure 9.2 represent a combination of all parts of the economy (food, clothing, etc.). Consequently, it is possible that prices of land actually increased more than 12% during this period.

The value of land and buildings on Alabama farms totaled $111 million in 1890 (Figure 9.22). It had been $54.2 million in 1870. There was a substantial increase in 1900 ($134.7 million). But between 1900 and 1910, it more than doubled to $288.2 million, a rate of increase much greater than in the country as a whole (Figure 9.20). There are no Alabama data available for 1915.

When these data are computed on a per acre basis (Figure 9.21), the same trend seen in the national data is also evident in the Alabama data. Value per acre increased very little from 1870 to 1900 ($4 to $6), but then more than doubled to reach $14 in 1910. Obviously, this increase in the value of land and buildings was a national phenomenon. The data in Figure 9.21 also show that total value per acre was only about a fourth as high in Alabama as in the country as a whole.

Value of Farm Implements and Machinery

Data on the value of implements and machinery on American farms during the period 1850 and 1915 are presented in Figure 9.23. The total in 1890 was $494 million; afterwards it increased to $750 million in 1900, and then by 1910 it almost doubled to $1,265 million. It continued its sharp upward trend to reach $1,849 million in 1915. Obviously, some of this was the result of an increase in farm numbers, as well as the increase in inflation (Figure 9.13).

When total values are converted to value per farm, a slightly different picture emerges. In 1890, 1900, and 1910, the converted values were: $108, $131, and $199, respectively. Between 1890 and 1910, total value increased 182%. In the same period, value per farm increased 103%. By 1910, most of the new machines and equipment developed for agriculture during the latter part of the nineteenth century were readily available, and many American farmers were actively mechanizing their farms with the higher-priced machines.

Data on the estimated total value of implements and machinery on Alabama farms in the period 1850-1910 are presented in Figure 9.25. I have included the earlier data to remind the reader of the long-term effect of the Civil War on Alabama farms. Note that values did not return to the 1860 level until sometime between 1890 and 1900. It was $4.5 million in 1890, and increased sharply to $8.7 million in 1900. Then from 1900 through 1910, it almost doubled to $16.3 million. Farm numbers were increasing rapidly in Alabama during this period (9.14). Consequently, a substantial portion of the increase was likely the result of purchases for these new farms.

In 1890, 1900, and 1910 value per farm in Alabama was $29, $39, and $63, respectively. During the period 1890 through 1910, total value had increased 261%. However, value per farm only increased 113%. While some of the increase in the total value of machinery and equipment on Alabama farms was obviously the result of the increase in farm numbers, the per farm data indicate that Alabama farmers seem not to have used the extremely good returns on their cotton crops after 1900 to modernize their operations even minimally.

These per farm data are indicative of the deplorable condition of Alabama farms during this period. In 1890, Alabama farmers, on average, operated their farms with implements and machinery valued at $29. The national average that year was $108. However, in 1900 and 1910 the average size of American farms was increasing, while it was decreasing on Alabama farms (Figure 9.15). An atmosphere of desperation surrounded those Alabama farmers trying to make a living in 1910—using $29 dollars worth of implements and machinery to grow and harvest crops from 27 acres on a 79-acre farm. These dismally low values at the end of the nineteenth and the beginning of the twentieth centuries are another indicator of just how slowly agriculture in the state was evolving.

Tractors on Farms

One of the most significant events in the history of American agriculture began to unfold in this period (1895-1915) when tractors began to appear on American farms. These machines resulted in several extremely important changes in farming. For example, as the number of tractors increased, the number of horses and mules declined. Then as the number of these animals declined, the amount of feed (oats, corn, etc.) required also declined. An even more important effect was that a single individual was able to effectively farm more land. The first data on the number of tractors in the country was reported by USDA in 1910. The department estimated that there were a thousand tractors in the entire country that year. Annual data on the estimated number of tractors in the country indicate that by 1915 the number had increased to 25,000. A more useful comparison is the average number of farms with tractors. In 1910, with 6.4 million farms in the country, there would have been about one tractor for each 6,400 farms. By 1915, this number increased to one tractor for each 258 farms. There are no data on the number of tractors on Alabama farms before 1920.

Animal Power on Farms

The number of horses, mules, and asses increased much faster than the country's population in the period 1870-1895. In 1900, census data reveal that there were 21.2 million horses and 3.5 million mules and asses in the country. Some 52% of the horses were in the North Central Region. Horses were the choice there to provide power to the larger and larger pieces of farm machinery. At the same time, 54% of the mules and asses were in the South Central Region. Mules were the choice there as the number of tenant farms increased and the acres of cropland harvested per farm declined.

Data on the number of horses and mules in the country during the period 1895-1915 are presented in Figure 9.26. As is indicated by the data, the number of horses continued to increase during the period 1895-1915. The rate of increase, however, was generally lower than it had been in the preceding period. The number of mules also continued to increase, seemingly at a somewhat higher rate than in the preceding period (1870-1895). During the period 1895-1915, the number of horses was increasing faster than the number of mules. The increases of both of these animals is probably related to the increase in the number of farms.

Data on the number of horses and mules in Alabama during the period 1895-1915 are presented in Figure 9.27. In 1877, there were about equal numbers of horses and mules in the state. Even as late as 1895 numbers were generally similar. Afterwards, however, the number of mules increased at a much faster rate. In 1915, there were over five times as many mules as horses.

In 1910, 3.6% of the horses in the state were on farms in Madison County. Other counties with large shares of the total were: Sumter (3%), Wilcox (2.9), Limestone (2.7), and Dallas (2.7). These five counties accounted for only 14.9% of the total. There were at least 2,000 horses in 25 of Alabama's 67 counties.

Data presented in the 1910 Census indicate that 3.1% of the mules in the state were on farms in Montgomery County. Other counties with substantial numbers were: Madison (2.9%), Dallas (2.6), Marengo (2.5), and Jackson (2.5). These five counties accounted for only 13.6% of the total. There were at least 4,000 mules in 27 of the state's 67 counties.

Because of the difference in the numbers of farms in the country and the state, it is difficult to get a clear picture of the relative use of horses and mules. This problem was alleviated to a degree by converting the total numbers data to number per 100 farms. These data for horses are presented in Figure 9.28. The converted data indicate that there were many more horses per 100 farms on American farms during the period 1890-1915 than on Alabama farms. They also show that numbers in the United States changed relatively little during the period. In contrast, numbers in Alabama generally trended downward from 1870 onward.

Data on the numbers of mules per 100 farms in the United States and Alabama during the period 1890-1915 are presented in Figure 9.29. The converted data generally show that there were more mules on Alabama farms than on American farms, and that numbers in Alabama generally increased after 1890. They also showed that the numbers of mules in the United States also increased from 1880 onward, and at a higher rate than in the state.

Use of Fertilizer

In 1879, USDA published the first estimate of acres of cropland harvested, but there are no annual estimates of fertilizer usage until 1899 (Lerner, 1975). In 1899, fertilizer usage in the country totaled 5,206 million pounds, and in that year crops were harvested from 283.2 million acres of cropland. By combining these two estimates, it can be seen that farmers applied about 18.5 pounds of commercial fertilizer an acre of harvested cropland. This estimate is likely to be more realistic than using acres of improved farmland.

By the beginning of the 1890s, agricultural experiment stations in most of the states were working on problems related to the quality and use of commercial fertilizers in crop production. These investigations quickly demonstrated that the use of these materials as recommended would increase yields. The research also demonstrated that the increase in yields was sufficiently large to pay the additional cost of the fertilizer. This was extremely welcome news to farmers throughout the country who were searching desperately for some way to counter the declining prices of their commodities during this period. At this juncture, the only possible path out of their economic misery seemed to be to increase production. In 1899, just a decade after the meeting in Washington to establish the National Cooperative Research Program on the use of commercial fertilizers in crop production, farmers used 18.5 pounds of commercial fertilizer for each acre of cropland harvested. Twenty years after the 1899 meeting, they applied 31 pounds, an increase of 68%.

Plant Agriculture

Nationally, in the period 1870-1895 total acres harvested of the major crops (cotton, corn, wheat, and oats) produced increased from 78.9 million to 180.2 million acres (an increase of 129%). Acres harvested of all these crops increased significantly. These increases helped to fuel the excess production during the period that resulted in much lower prices and financial problems for most American farmers. In Alabama, acres harvested for these crops increased from 3.1 million to 5.8 million acres during the period, an increase of 88%.

Alabama Crop Weather

In 1941, Secretary of Agriculture Claude Wickard (1941) wrote: "Next to crop prices, nothing is more important to the farmer's business than the weather, and in fact the weather has a strong influence on prices." In Alabama, with its highly variable, subtropical climate and generally droughty soils, Wickard's statement is probably more applicable than in most other areas of the country. Accordingly, some general information on crop weather in the state is included for the period 1895-1915. The information is based on maps depicting Palmer Drought Indices (PDSI) for Alabama for June, July, and August for each year during the period. These maps were prepared by Cook, et al. (1999) for the NOAA Paleo Climatology Project. Data from the maps indicate that the weather was normal over at least 50% of the state in seven of those 21 years: 1895, 1898, 1903, 1906, 1907, 1908, and 1915); drier than normal in nine of those years: 1896, 1897, 1899, 1901, 1902, 1904, 1911, 1913, and 1914; and wetter than normal in five years: 1900, 1905, 1909, 1910, and 1912.

Cotton

According to census data, American farmers harvested 2.4 billion pounds of cotton in 1859, but after the Civil War pounds harvested did not reach that antebellum level again until some time between 1879 and 1889. Afterwards, harvests continued to increase inexorably year by year until they reached 9.1 million bales in 1894. As harvests increased, prices decreased. Farmers received 4.6 cents a pound for that enormous 1894 crop. In 1895, harvest was slightly lower, and the price increased to 7.6 cents.

US Cotton

In 1895, American cotton farmers harvested 7.2 million bales of cotton. They had harvested 4.4 million in 1870. In 1895, Texas farmers harvested the highest percentage of the total crop (26.6%). Other leading cotton-producing states included: Georgia (14.9%), Mississippi (14.1), South Carolina (10.7), and Alabama (10.1). Together, these five states harvested 75.6% of the total American cotton crop that year. Production of upland cotton had its beginning in the South Atlantic Region in the late 1700s. In 1895, most American cotton was harvested southwest of that region.

Exports

The United States exported 4.7 million bales of

cotton in 1895 (Figure 9.30). This level of exports was equivalent to 47.6% of the entire 1894, national crop (9.9 million bales). Although there was considerable year-to-year variation, exports generally trended upward through 1908 to 9 million bales. Although the banking panic of 1907-1908 had ended in the United States in 1909, it had had enough of an effect on the economies of European countries that the demand for cotton was significantly reduced. In 1909, exports were down to 6.4 million bales, the lowest since 1903. Prosperity returned in 1909, and the following year exports began to increase again. In 1911, some 11.1 million bales were exported (the highest level in history). These exports were a part of the first 15.7 million-bale crop in American history. They returned to more normal levels of around nine million bales during the period 1912-1913.

On July 28, 1914, Austria declared war on Serbia, beginning World War I. Anticipating a wider war in Europe and fearing that export markets would disappear (the United States had exported over nine million bales in 1913), all cotton exchanges in the United States closed on July 31. With no place to sell cotton, panic quickly spread through those farmers, merchants, bankers, and brokers whose livelihood depended on the fiber. Fortunately, the Cotton Exchange Scare of 1914 soon abated, and exports began to flow once more; however, they were lower in 1914 (8.7 million bales) than they had been in the three preceding years. Then with Europe fully involved in the war in 1915 and with the loss of the large German market, exports plummeted to 6.1 million bales. They had not been that low since 1896.

Prices

In most years, the cotton harvest is not generally completed until mid-December. As a result, the size of the harvest in one year weighs heavily on prices well into the following year. A large share of the American cotton crop was exported in those years, so changes in exports, up or down, had a significant effect on prices. In 1895, the annual average price received by American cotton farmers was 7.6 cents a pound (Figure 9.31). It had been 4.6 cents the preceding year. Unusually large crops (Figure 9.34) kept prices low through 1889. At the same time, exports were beginning to increase slowly (Figure 9.30), and prices also began to trend slowly upwards. As a result of poorer crops in 1899 and 1900, they reached 9.2 cents in 1900. Prices had not been that high since 1881. Increasing production in the early 1900s, sent prices down again to near 7 cents. Bales harvested stabilized at 12 million in 1905, and prices remained around 10 cents through 1908. Bales harvested fell to 10 million in 1909, and prices increased sharply to 13 cents. After 1910, prices and bales harvested generally teeter-tottered to reach 11 cents in 1915.

Data on prices and bales harvested in the period 1903-1908 clearly shows the teeter-totter relationship between the two. In 1903, price and bales were 10.5 cents and 9.8 million bales, respectively. In 1904, bales increased to 13.4 million, and price declined to 9 cents. In 1905, bales declined to 10.6 million, and price increased to 10.8 cents. In 1906, bales increased to 13.3 million, and price fell to 9.6 cents. With the level of year-to-year variation in bales increasing after 1908, year-to-year variation in prices also increased, but even with the increased variation, the teeter-totter relationship generally persisted.

While there was obviously a fairly predictable relationship between supply and prices during this period, economic contractions also played an important role by affecting demand. For example, prices apparently were not seriously affected by the banking panic of 1907 and 1908, but when prosperity returned in 1909, it increased to 13.6 cents, the highest since NASS began to report price data in 1876. Of course, bales harvested that year was also down sharply; however, the increase in price seemed to be much greater than expected from that particular decline in supply.

Acres Harvested

As progressivism and economic prosperity spread across America in the first decade of the twentieth century (Figure 9.1), cotton mills and industry in the Cotton Belt were beginning to absorb some of the excess Southern labor, mostly tenant farmers. Approximately a quarter of a million of them would be removed from the production sector of the agroecosystem. The increase in cotton prices and the removal of a modest number of tenant farmers from agriculture were certainly welcome, but there was little resulting improvement in the lives of over a million cotton farmers living in dire poverty. There were still too many people growing too much cotton on farms that were too small to realize much benefit. Sadly, there was little indication that American cotton farmers were paying any more attention to the decades of cries by the reformers to plant less cotton. Unfortunately, even if they had heard the message, those poor farmers remaining on those small farms had little choice but to plant more of the crop. That was the only crop that had a dependable market in the South.

In 1895, American farmers harvested cotton from 19.8 million acres (Figure 9.32). Acres harvested had been 9.2 million in 1870. Afterwards, despite some year-to-year variation, it climbed inexorably upward to reach a level of 35.6 million in 1914. The 7.4 cents price that year really put the fear of God in them, and they harvested the crop from only 30 million acres in 1915.

Much of the increase in acres harvested, described in the preceding paragraph, came as a result of each farmer growing more cotton, but there was also a sharp increase in the total number of farmers growing cotton. Fite (1984) reported that between 1899 and 1909, the number of Southern farmers growing cotton increased from 1.4 to 1.7 million. Cotton was becoming even more important to Alabama's economy than it had been a decade earlier, while the production of cereals was decreasing. In 1899, cereal crops were produced on 35% of the improved farmland in the state. A decade later, the percentage had decreased to 29%.

The crisis of Southern cotton farmers had mostly deepened since Secession; although its roots were surely older. Furthermore, while there had been some level of national concern for the plight of "God's finest creation" (the old planters, their tenants, and former slaves), over the years there had never been a groundswell of public or political effort to try to solve their problems. Among much of the populace, there was some feeling that cotton farmers were their own worst enemy and that if they would only heed the advice of the reformers and truly seek economic salvation, their situation would quickly improve. Further, much of the continuing crisis and creeping poverty over all of those years had been confined primarily to the ranks of tenants. While, they had not completely escaped the effects of the continuing descent into disaster, landowners of all classes, merchants, bankers, cotton brokers, and exporters had largely escaped the deeper levels of the tragedy. Through those long, tragic years with only the little wheels squeaking through the voices of the Grange, the Agricultural Wheel, the Farmers' Alliance, and the Farmers' Union, there was no widespread, national public or political outcry to address the issues. The closing of the cotton exchanges at the beginning of World War I was another matter altogether. Here was finally the kind of crisis that could also affect the big wheels. In fact, it could destroy them. If the usual two-thirds of the 16 million bale crop harvested in 1914 could not have been exported, the whole system, from bottom to top, would have quickly collapsed.

Finally and for the first time in history, federal officials at the highest levels with the angry yells of the concerned big wheels in their ears became deeply concerned about cotton crops and prices, and they moved quickly to seek a remedy. In August 1914, the secretary of the treasury, members of the Federal Reserve Board, congressmen, senators, bankers, merchants, and farmers, convened a meeting in Washington. After several days of discussions, the best solution they could devise was based on elements of proposals put forth earlier by the Agricultural Wheel, the Farmers' Alliance, and the Farmers' Union that involved establishing warehouses, and a provision whereby federal loans could be made to farmers based on the value of their stored crop. President Wilson's administration rejected this approach, and although Congress approved the Cotton Warehouse Act later in the year, they appropriated no funds for loans. The cotton exchanges reopened in November with prices just above seven cents, and suddenly the pressure was off. Those on the upper end of the cotton ladder were safe, and the federal government quickly lost interest in intervening directly in the long-simmering cotton crisis. The politicians would not revisit the issues for several years.

Yields

During the period 1895-1915, there was considerable year-to-year variation in average cotton yields in the United States (Figure 9.33). They ranged from 156.5 (1909) to 216.4 (1914) pounds an acre. The average was

189.3 pounds, and the standard deviation was 18.3. There was no apparent trend in the annual data. It is interesting that there were few times when farmers got either good or bad yields back to back. In virtually every case, a poor average yield was followed by a better one. Much of this variation was likely the result of year-to-year variation in crop weather. This high level of variation is a reminder that the genome of the cotton plant was never well suited for the soils and climate east of the Mississippi River.

From the earliest days of upland cotton production in the South, farmers fought a slowly-losing battle for economic survival within a volatile global supply-and-demand environment, while trying to balance prices received with production costs. Early in the twentieth century, another even more devastating element was added to the equation that they were finding so difficult to keep in balance; the boll weevil arrived. Soon after the little weevil arrived in Texas from Mexico in the last decade of the nineteenth century, this doggerel appeared: "De first time I see de boll weevil, he was sitting on de square./ De next time I see de weevil, he got all his fambly dere."

The weevil was first found in the United States in 1892, and in the next 30 years, it had spread to all the cotton growing areas of the country. Smith (2007) comments that the South had not fully recovered from the effects of the Civil War when the most devastating agricultural pest in the history of the United States appeared in Texas, and as it moved northeastward across the South, it left a trail of unimaginable devastation in the cotton economy and to the people depending on it.

Bales Harvested

Bales harvested is a product of acres harvested and yield. Although there was considerable year-to-year variation in yields, there seemed to be no trend in the period 1895-1915. However, acres harvested generally trended upward. As a result, bales harvested also trended upward (Figure 9.34). It was eight million bales in 1895, and reached 16 million in 1914. Then for reasons detailed in the preceding sections, in 1915 it fell sharply to 14 million. Although bales harvested generally trended upward with acres harvested, the effect of the year-to-year variation was also quite obvious. The seesaw pattern in bales harvested between 1904 and 1910 is quite similar to the pattern in yields.

Alabama Cotton

According to census data, Alabama led the country in cotton bales harvested in 1859, with 22.9% of the country's crop; however by 1895, Alabama farmers harvested only 9.3% of the total crop, and ranked fifth in country.

There are no data available on the harvest of cotton in Alabama's counties in 1895. Data in the 1890 Census indicate that farmers in Montgomery County led the state in bales harvested with 5% of the total. The same five counties that lead the state in 1869, continued to do so in 1889, except that Barbour replaced Russell. Farmers in the five leading counties harvested 21.3% of the total crop. In 1889, 16 counties reported harvests of 20,000 or greater bales.

Prices

Cotton prices were not published separately for the states during the period 1895-1915. However, it is likely that those received by Alabama farmers were essentially the same as those shown in Figure 9.31.

ACRES HARVESTED

In 1895, Alabama farmers harvested cotton from 2.6 million acres (Figure 9.35). Acres harvested had been at a level of 1.5 million in 1870. Generally speaking, acres harvested in both the country and the state seemed to be in lockedstep during the period 1895-1915. Year-to-year trends were similar. This is not really surprising because both were responding to the same basic signals: prices and desperation. Further, acres harvested in Alabama represented a large share of the national total.

Acres harvested in the state generally increased throughout the period. It was 2.5 million in 1900, 2.8 million in 1905, and 3.2 million in 1910. Farmers finally blinked in 1915. With the war in Europe spreading, they reduced acres harvested to 30 million. It had not been that low since 1904.

Yields

As was the case with annual average yields per acre of cotton in the United States during the period 1895-1915, those in Alabama were also extremely variable (Figure 9.33). In the state, they varied from 135 pounds in 1902 to 220 pounds in 1914. In the early part of the period (1895-1904), yields in Alabama were clearly lower than in the country, but after 1904 Alabama yields began to catch up; although in most years, they were slightly lower, but not to the same degree as in the earlier part of the period.

Year-to-year changes in both state and national yields were quite similar in many of the years. For example, in 1904, 1911, and 1914, they were both quite high, while in 1909 they were almost equally low. The average yield for the state for the period was 173.2 pounds an acre, and the standard deviation was 29.6 pounds. The Alabama average was lower than the average for the United States (173.2 versus 189.3 pounds an acre) during the period, and the amount of year-to-year variation was greater. The standard deviation for the United States was lower than in Alabama (18.3 versus 29.6 pounds).

Bales Harvested

After 1904, there seemed to be no trend in the average yield of cotton in Alabama (Figure 9.33), although there was a considerable amount of year-to-year variation. Generally, acres harvested trended upward for the entire period (Figure 9.35). As a result, bales harvested (Figure 9.36) also trended upward, and within this general trend there was a great deal of year-to-year variation. Bales harvested was 1.1 million in 1897 and 1.7 million in 1914.

According to census data, in 1909 some 3.5% of the state's cotton crop was harvested on farms in Dallas County. Other counties with substantial harvests included: Montgomery (3.4%), Marengo (2.8), Wilcox (2.7), and Pike (2.6). These five counties accounted for 15.4% of the total crop. Average harvest per county was 16,859 bales. At least 12,000 bales were harvested in 45 of the state's 67 counties.

Figure 9.37 shows the statewide distribution of bales harvested in 1909. This dot-density map was drawn using a computer program developed by the Environmental Systems Research Institute of Redland, California. Their so-called ArcGIS program divided the array of bushels harvested in the different counties into five groups. Counties in the group with the highest levels of harvest were assigned the largest dots. Similar maps will be presented for other commodities for this period and for some other periods as well.

Corn

Maize, or Indian corn, was the most important plant crop grown in pre-Colombian North America and Mesoamerica. In 1895, corn continued to hold that position of prominence. That year, American farmers harvested the grain from 90.5 million acres, far exceeding acres harvested of wheat (39 million acres), oats (30 million acres), or cotton (19.8 million acres).

US Corn

In 1895, American farmers harvested 2.5 billion bushels of corn. They had produced 1.1 billion in 1870. Farmers in Illinois harvested the largest share of the 1895 crop (16.2%). Other leading corn-producing states that year were: Iowa (14%), Missouri (10.5), Kansas (8.1), and Indiana (6.1). These five states harvested 54.9% of the national crop. Alabama farmers harvested only 1.7% of that crop.

Exports

With the economies of Europe expanding after 1894, their demand for American feed grains exploded. In 1895, American corn exports totaled 100 million bushels (Figure 9.38). A year later (1896), they were at 180 million. Exports remained unusually high through the remaining years of the century, and in 1899 reached 213.1 million bushels (the highest level recorded to that time). After 1899, European countries began to seek alternatives to the importation of such large quantities of feed grains from the United States, and after 1900 exports began to decline. In 1901, as a result of unfavorable weather, American farmers harvested only 1.7 million bushels of corn (Figure 9.42), the poorest crop since 1894. Consequently exports fell sharply to only 28 million bushels. They had not been that low since 1887. Exports began to increase following that disastrous crop, and by 1905 were 119.9 million bushels. However, after that year various European initiatives to reduce imports were beginning to be effective, and by 1909 the United States only exported 38.1 million bushels of the grain. Exports remained at near that level through 1915, except in 1910 and 1913. In 1910, farmers harvested an extremely good crop (2.8 billion bushels) and exported 65.6 million bushels; then in 1913 when they harvested a poorer crop, they exported only 2.3 million bushels. The 1913 export total of 10.7

million bushels was the lowest since 1870 when the same quality was exported.

Prices

Annual American corn exports during the period 1895-1915 averaged 2% of annual production; consequently, it is not likely that they contributed very much to year-to-year changes in prices. It is more likely that the supply-demand-price relationship played the predominant role. In 1895, bushels of corn harvested was 2.5 billion (Figure 9.42), the highest in history; the average price was 25 cents a bushel, the lowest recorded by USDA to that time (Figure 9.39). Bushels harvested increased slightly in 1896, and the price fell to 21 cents. In 1897, bushels harvested began to decline and prices began to increase. In 1901, in a terrible crop year, bushels harvested fell all the way to 1.7 billion bushels, and the price exploded upward to 60 cents. Bushels harvested and prices generally teeter-tottered for the remainder of the period, until in 1915 the price reached 65 cents.

Unquestionably, changes in prices followed bushels harvested during the period, but at least from 1895 through 1908 prices trended upward, indicating that even though there was definite pattern from year to year, overall demand seemed to be increasing. [During much of this same period, American cattle numbers were also increasing rapidly (Figure 9.66)].

Inflation of the currency only played a minor role in the general increase in prices. The Consumer Price Index increased by about 20% between 1895 and 1914, but in the same period corn prices increased by 167%.

Acres Harvested

Although there was considerable year-to-year variation, acres harvested trended upward slowly for much of the period 1895-1909. Then from 1900 through 1908, it remained largely unchanged (Figure 9.40). Between 1906 and 1908, prices increased sharply, and in 1909 acres harvested followed. Unfortunately, in 1909 prices tumbled, and in 1911 acres harvested once again began to follow. In 1895, American farmers harvested corn from 90 million acres. By 1915, acres harvested had increased to100 million.

In the period 1895-1915, bushels harvested and prices were locked together in a teeter-totter relationship whereby estimates of, and actual bushels harvested, strongly affected prices in the same year. Acres harvested and prices also tended to be locked together, but in a different manner. Prices in a given year had a major effect on acres planted the following year, which in turn was a major determinant of bushels harvested. Of course, while acres planted and bushels harvested were related, the exact relationship was dependent on oft-times highly variable yields.

Yields

Although there was considerable year-to-year variation, average annual yields of corn generally remained relatively unchanged at 30 bushels an acre from 1895 through 1906 (Figure 9.41); thereafter, they tended to trend slightly downward to 25 bushels. The 1901 average yield (18.2 bushels an acre) was the lowest ever recorded in the United States. It was the result of severely adverse weather in the Corn Belt.

The national averages include a large number of states, and much of the corn in the country was produced in states where the average yields were much higher than the one reported here. For example, in 1895 Illinois, Iowa, Missouri, Kansas, and Indiana, in that order, were the leading corn-producing states. That year, these five states harvested 55% of all the corn in the country. The average yield for those states was 35 bushels an acre, or about 10 bushels above the national average.

Bushels Harvested

From 1895 through 1906, bushels harvested (Figure 9.42) generally followed acres harvested upward (9.40); then afterwards, they followed yields downward (Figure 9.41) through 1915. Bushels harvested was quite variable during this latter period, reflecting the high level of year-to-year variation in yields. The total American corn harvest, excepting the 1901 crop, averaged 2.64 billion bushels a year.

Alabama Corn

In 1839, when the first national data was published on corn harvested in the United States, Alabama farmers harvested 5.6% of the total crop. Over time, however, the state's share slowly declined, until in the last decades of the nineteenth century, it ranged between 1 and 2%.

Census data for 1890 indicate that farmers in Jackson

County harvested the largest share (3.9%) of the 1889 state corn crop. Other counties reporting large shares included: Madison (3.7%), Lowndes (3.5), Marengo (2.6), and Sumter (2.6). These five counties only accounted for 16.3% of the total crop that year, which is indicative of the widespread production of the crop.

Prices

Data on corn prices in both the United States and Alabama during the period 1895-1915 are presented in Figure 9.39. Generally, year-to-year changes over the entire period were essentially the same for both. Of course, with Alabama producing such a small proportion of the total crop, Alabama corn had relatively little effect on national prices. Local farmers were able to obtain better prices than the national average, but the basic national average was the determining factor for trends in the state. As was the situation in the country as a whole, prices trended upward rather sharply during the period. They were 36 cents a bushel in 1895, and in 1913 and 1914 reached 92 cents.

The spike in prices in 1901 (60 cents in the country and 77 cents in the state) was likely the result of the extremely poor national crop that year. The sharply lower price in 1910 (48 cents in the country and 74 cents in the state) was the result of bumper crops across the country. A large crop in the United States and in Alabama resulted in a much lower price in 1915 in the state (79 cents), but had little effect on the average price in the country.

Acres Harvested

Year-to-year changes in acres of corn harvested in Alabama during the period 1895-1915 generally followed the same year-to-year pattern as the country as a whole (Figures 9.40 and 9.43). Generally, acres harvested remained more or less unchanged at around 2.6 million throughout the period. Alabama corn farmers seemed to get a real burr under their saddle in response to slightly higher prices in 1913 and 1914. In 1915, they harvested corn from 3.2 million acres, the highest level in history to that time. Unfortunately, as expected when that crop was sold, prices were lower than they had been in four years.

Yields

Data on average annual yields of corn, both in the country and the state, during the period 1895-1915 are presented in Figure 9.41. As the data indicate, average Alabama yields ranged between 10 to 15 bushels an acre for the entire period. In only one of the years during the period (1902) did the average fall below 10 bushels an acre. In that year, Palmer Drought Severity Index (PDSI) data indicated that most of the state was excessively dry in June, July, and August. There seemed to be no trend in the data.

The most important aspect of these yield data is that the Alabama average was consistently around half of the national average; however, if it is compared to those states producing most of the corn in the country, it is only about a third of their average. In retrospect, this difference should have been a cause of great concern among Alabama agriculturalists. A difference this great was economically untenable. It is likely, however, that they expected to do better over the years. Unfortunately, they were apparently unwilling to admit that the basic causes(s) of the differences (soils and climate) were never going to go away. This difference was obvious as early as 1839.

Bushels Harvested

With acres harvested trending upward (Figure 9.44) and yields generally unchanged during the period, bushels harvested probably also trended upward, although the considerable year-to-year variation made it difficult to determine how much. In most of the years, bushels harvested ranged between 30 and 40 million. The unusually low value in 1902 was the result of lower than expected acres harvested and low yields, which resulted from poor crop weather.

In 1909, farmers in Madison County harvested 3.3% of the state's corn crop. Other counties reporting substantial harvests included: Lauderdale (3.2%), Jackson (3.1), Limestone (2.8), and Houston (2.7). These five counties accounted for 15.1% of the total crop. Farmers harvested at least 500,000 bushels in 23 of the state's 67 counties. Figure 9.45 shows the distribution of corn harvested in Alabama in 1909.

Peanuts

Annual data on American peanut production were not published before 1909. In that year, American farmers harvested 354.6 million pounds of the crop. North

Carolina was the leading state with 34.3%. Other states harvesting significant quantities were: Virginia (27.8%), Alabama (11.0), Georgia (8.9), and Texas (7.4).

At the beginning of the twentieth century, American peanut production was largely restricted to the area of the country where they were first introduced: Virginia and North Carolina. Production increased rapidly immediately after the Civil War, but censuses did not include information on the crop until 1890 (the 1889 crop). Census data indicate that American farmers planted 204,000 acres of the crop in 1889 and 517,000 in 1899. NASS data show that acres planted increased from 700,000 to a million from 1909 through 1915.

From the beginning, farmers harvested a large share of their peanut crop by hogging-them-off, or by turning their hogs into the fields. Farmers harvested only 67% of acres planted in 1909, but in 1913 they only harvested 56%. The percentage increased in 1914 (57%) and 1915 (58%), probably when the war years provided them with a somewhat better off-farm market for their crop.

There was considerable year-to-year variation in yield but no apparent trend. The average for the period was 774 pounds an acre. With the increase in acres harvested, pounds harvested increased from 354.6 to 480.6 million pounds, between 1909 and 1915. The average price that farmers obtained for their peanuts remained at four cents a pound throughout the period.

USDA data indicate that between 1909 and 1915 peanut production increased in Alabama at a greater rate than in the remainder of the country. Acres planted increased from 90,000 to 250,000. In the period, acres harvested increased from 50,000 to 110,000. According to census data, Alabama farmers harvested peanuts from 78,878 acres in 1889. Farmers in the state apparently "hogged-off" a slightly higher percentage of their crop than the average for the country. During the period 1909-1915, they only harvested between 34 and 60% of acres planted. During the period, the average yield for peanuts in the state was about 10% lower than in the country as a whole (706 versus 774 pounds an acre). Pounds harvested more than doubled: 35 million to 77 million.

In 1909, 15.6% of the peanuts harvested in the state came from farms in Houston County. Other counties reporting substantial harvests included : Geneva (12.6%), Coffee (11.0), Dale (9.7), and Covington (7.9). These five counties accounted for 56.8% of the total crop. Farmers harvested at least 50,000 bushels in just nine of Alabama's 67 counties.

Soybeans

There was little data collected on the production of soybeans for beans in the United States before 1924. Prior to that time, the crop was grown primarily for hay. Tables in the 1910 Census presenting data on the production of field crops in Alabama did not list soybeans.

Alabama Hay

Because of differences in the kinds of hay produced in the different parts on the country, there is little value in using national data as a basis for understanding changes in the production of the crop in Alabama. As a result, relatively little national data will be used in this discussion.

During the preceding period (1870-1895), hay had become a relatively important crop in Alabama. Tons harvested had increased from 10,000 to 86,000 in the 26-year period. This increase had been a direct response to the growing number of livestock in the state and efforts by farmers to provide a more dependable feed supply for them.

Prices

In 1895, Alabama farmers received $10.25 a ton, for their hay (Figure 9.46). The price had been much higher in 1870 ($17.95). It remained around $10 a ton through 1897, and then began to trend upward sharply. The Consumer Price Index was also increasing during this period. Its increase was probably sufficient to exert some upward pressure on prices regardless of what was happening to supply and demand. In 1913 and 1914, hay prices were near $14.50. It declined in 1915 ($13.10) as a result of the harvest of an unusually large crop. There is little indication in any of the Alabama data for the reason for the sharp increase in price in 1907 ($15.25 a ton). In that year, Alabama farmers harvested the largest crop in history.

Acres Harvested

In 1895, some 86,000 acres of hay were harvested. It

continued to trend slowly upward through 1902 (114,000 acres). It declined slightly in 1903, but afterwards began to trend sharply upward (following prices to reach 404,000 acres in 1915. Apparently, feeding hay to cattle in Alabama came of age during this period.

Yields

For much of the period, Alabama hay yields generally ranged between 0.9 and 1 ton an acre (Figure 9.48). They were much lower than expected in 1901, 1902, 1912, 1913, 1914, and 1915. Palmer Drought Indices (Cook, et al., 1999) indicate that crop weather during most of these years was much drier than normal.

Tons Harvested

With little or no change in average yields, the year-to-year trend in tons harvested (Figure 9.49) was similar to the year-to-year trend in acres harvested. Alabama farmers harvested 86,000 tons of hay in 1895. They had harvested 10,000 in 1870. Tons harvested increased slowly through 1903 (122,000 tons), and then began to trend rapidly upward to reach 347,000 tons in 1915. The off-trend harvests in 1902 and 1914 were likely the result of the much drier than normal crop weather mentioned in a preceding paragraph.

Farmers in Madison County harvested 5.2% of Alabama's hay crop in 1909. Other counties with substantial shares of the total crop included: Montgomery (5.2%), Dallas (3.9), Hale (3.6), and Jackson (3.5). These five counties accounted for 21.4% of the total crop. In that year, farmers harvested at least 4,000 tons in 23 of the state's 67 counties. Figure 9.50 shows the distribution of bales harvested in Alabama in 1909.

Alabama Wheat

In 1895, American farmers harvested 542.1 million bushels of wheat of all varieties. Minnesota reported the largest share of the total crop (14.2%). Other states with large shares included: North Dakota (13.5%), South Dakota (7.3), Ohio (6.4), and California (5.9). These five states accounted for 47.3% of the total national crop. Farmers in Alabama harvested 0.1% of the total.

In terms of acres harvested, wheat was a relatively important crop in Alabama in the 1870s; however, by the end of the century farmers were rapidly losing interest in the crop. For a number of reasons, they decided that it would be cheaper to purchase the wheat they needed than to try to produce it themselves. In consequence, with Alabama playing such a minor role in the production of the national crop, relatively little of the national data will be used in these discussions.

Most of the wheat grown in Alabama is the winter variety. The crop is planted at the end of one year, but not harvested until the middle of the following year. Unfortunately, USDA did not begin to report national data on winter wheat until 1909. Before that time, all of their national data was based on the production of all varieties.

US Exports

American wheat (all) exports were 130.1 million bushels in 1895 (Figure 9.51). These shipments were equivalent to 24% of the entire 1894 crop of 541.9 million bushels. Exports had been 52.6 and 188.3 million bushels in 1870 and 1880, respectively. After 1880, they began to decline as a result of increased competition from other exporting countries and increased duties imposed by the importing countries. In 1896, exports began to increase again, reaching 221.1 million in 1987. They remained at near this level through 1902. Afterwards, they began to decline, and by 1904 they were down to 46.3 million bushels (The lowest on record). The crop that year was the smallest in a decade. They began to recover somewhat in the period 1905-1907, as the world economy improved, only to lose those gains from 1908-1910. However after 1910, exports began to increase, and by 1914, the first year of World War I, they reached 335.7 million bushels (the highest on record at that time). They declined somewhat in 1915 (239.6 million), but were still near the record level.

Several factors affected wheat exports. One of these was bushels harvested. For example, in 1904 exports were 46 million bushels, the lowest since 1871. That year bushels harvested totaled 556 million bushels, the smallest crop in several years. In contrast, in 1914 exports reached 336 million bushels, the highest level on record at that time. In the same year, bushels harvested reached 897 million bushels, also the highest on record at that time.

During this period, most wheat exports went to Europe, where wheat had been produced for thousands of years. The levels of imports into Europe depended on a number of factors; consequently, it is difficult to explain year-to-year changes without a knowledge of these factors and their interaction. For example, the beginning of the war in 1914 boosted exports, but by 1915 access to the German market began to decline, and as a consequence, exports also declined.

Prices

Annual average prices for winter wheat received by Alabama farmers during the period 1895-1915 are presented in Figure 9.52. Data for American wheat (all) are shown for reference. The improving economic conditions in the country after 1895 resulted in increasing average wheat prices. Although there was considerable year-to-year variation, average Alabama prices generally increased from 80 cents a bushel in 1895 to $1.20 in 1915. The latter part of this period is included in the Golden Age of American Agriculture when prices of all farm commodities were increasing; the Consumer Price Index (Figure 9.2) and inflation (Figure 9.3) were also increasing. The high year-to-year variation was generally associated with sharp changes in the bushels harvested in the country and/or sharp changes in exports. For example, the 1897 price was somewhat higher than expected. That year exports had increased sharply over 1896. Also, the 1904 price was high. In that year, bushels harvested was sharply lower than in 1902 and 1903. Alabama was only a minor player in national wheat production, but changes in prices received by Alabama farmers closely paralleled those received by farmers all over the country; although the year-to-year changes were similar, Alabama farmers were able to sell their grain for slightly higher prices.

Acres Harvested in Alabama

Acres of winter wheat harvested in Alabama stabilized around 50,000 acres in the early1890s (Figure 9.53). It remained near that level through 1897. Prices had begun to trend upward again after 1894 (Figure 9.52), but Alabama farmers waited until 1897 to begin to respond by harvesting more acres of the grain. That fall and the following spring, they planted enough wheat to enable them to harvest 72,000 acres of winter wheat in 1898. It had not been that high since 1886. Acres harvested continued to follow prices upward until 1900 when they harvested wheat from 127,000 acres.

Surprisingly, with prices still increasing, acres harvested began to decline again, and by 1909, it had reached a plateau of 18,000 acres, which was the lowest recorded to that time. Farmers seemed to have a little renewed interest in the crop at the beginning of the war. In 1914, they harvested wheat from only 14,000 acres. In 1915, they doubled that level to 34,000 acres. The problem with wheat production in Alabama during this period was not prices; it was yields.

Yields

Data on annual average Alabama winter wheat yields for the period 1870-1915 are presented in Figure 9.54. Data for national yields are presented for reference. Yields had been around five to six bushels an acre, from 1870 through 1895. Afterwards, although there was considerable year-to-year variation, yields generally trended upward through 1915 to around 10 bushels an acre. During the same period, national yields were almost twice as high. The unusually low yields in 1899 (5.1 bushels) and 1902 (4.5 bushels) were likely the result of abnormally dry to severely dry weather in the state in those two years. In looking at these yield data, it is easy to understand why Alabama farmers were abandoning wheat. Even though yields had trended upward during the period, they were just too low compared to those in other states.

Bushels Harvested in Alabama

Even though Alabama wheat yields trended upward slightly during the period 1895 through 1915, trends in bushels harvested generally followed the trends of acres harvested more closely (Figures 9.53, 9.54, and 9.55). From 312,000 bushels in 1895, bushels harvested increased sharply along with acres harvested to 1,206 thousand in 1900. However, by 1909 it was back down to 116,000 bushels, which was the lowest ever recorded, at that time. It remained near that level through 1914. The slight up-tick in acres harvested in 1915 is reflected in the slight increase in bushels harvested that year.

Farmers in Madison County harvested 22.3% of Alabama's 1909 wheat crop. Other counties harvesting substantial shares of the total crop included: Lauderdale (12.2%), Randolph (9.5), Limestone (8.1), and Talladega (6.8). These five counties accounted for 58.9% of the total.

Farmers in 13 counties harvested at least 2,000 bushels of wheat. Figure 9.56, depicts the distribution of wheat harvested in Alabama in 1909.

Alabama Oats

The number of horses, mules, and asses in the United States increased from 8.2 million in 1870 to near 24.7 million in 1890. In the same period, the quantity of oats, a principal feed for these animals, harvested increased from 267.9 million to 609.1 million bushels. Almost 80% of that 1890 crop was harvested in the North Central Region.

American farmers harvested 924.9 million bushels of oats in 1895. They had harvested 267.9 million bushels in 1870 and 417.9 million in 1880. Farmers in Iowa harvested 23.1% of the 1895 national crop. Other states harvesting significant quantities included: Illinois (11.5%), Minnesota (8.8), Wisconsin (7.9), and Nebraska (5.2). Farmers in these five states harvested 56.5% of the entire American crop that year. Alabama farmers harvested only 0.4% of the total crop.

In 1895, Alabama farmers harvested major and minor crops from 5.3 million acres. Oats were harvested from 5.3% (280,00 acres) of that total. In the United States in 1895, farmers harvested 45 bushels of oats for each head of draft animal (horses plus mules). In Alabama, farmers harvested only 13.9 bushels per head.

US Exports

In 1895, The United States exported 15.1 million bushels of oats. (Figure 9.57). Only 148,000 bushels had been exported in 1870. The 1895 exports were equivalent to 2% of the 1894 crop (750 million bushels). During the period 1895-1915, exports never amounted to more than 10% (1914) of total bushels harvested, and for most years it was less than half of that amount. Exports probably had little effect on prices in most of the years.

After 1895, with the economies of most European countries booming, oat exports increased sharply, reaching 73.8 million bushels in 1897. However, they did not remain at that level for long. By 1903, they were back down to 1.9 million. Except for unusually high levels in 1905 and 1912, exports remained below 10 million bushels through 1913. In 1914 and 1915, the beginning of the World War I resulted in oat exports exploding upward to 100 million bushels. Horses, of course, played an important role in that war.

US and Alabama Prices

Data on annual average oat prices during the period 1895-1915, both for the United States and for Alabama, are depicted in Figure 9.58. Alabama farmers received 42 cents a bushel in 1895, near the lowest level ever recorded. In 1901, prices increased sharply to 64 cents when American farmers harvested a much smaller than expected crop. Afterwards, they drifted downward, reaching 51 cents in 1906. In 1907, national bushels harvested was again much lower than expected, and the average Alabama price increased to 67 cents. Prices generally remained near that level through 1915. Between 1895 and 1915, Alabama oat prices increased from 42 to 67 cents a bushel, or by 60%. In the same period, the Consumer Price Index increased by 21%.

As might be expected, the year-to-year changes in oat prices for both Alabama and the United States followed essentially the same pattern. However, throughout the period average prices received by Alabama farmers were approximately 20 cents a bushel higher.

Acres Harvested in Alabama

Alabama farmers harvested oats from some 280,000 acres in 1895 (Figure 9.59). It had been as high as 410,000 in 1883. Afterwards, acres harvested in the state remained generally unchanged at around 225,000 acres for much of the period 1895-1915. When prices increased rather sharply after 1906 (Figure 9.58), farmers began to increase acres harvested slightly; however in 1911 with prices down again, they harvested fewer acres of the crop. The slight uptick in prices in 1912, 1913, and 1914 also seemed to stimulate some additional harvest. Acres harvested reached 240,000 in 1914. Then in 1915, even with little support from the 1914 price, farmers anticipated a war dividend and sharply increased acres harvested to 300,000. Unfortunately, as had happened so often in the past, the average price was actually lower in 1915 than it had been in 1914.

The number of horses and mules increased in the state from 272,000 in 1895 to 431,000 in 1915, an increase of 58.4%. Because most of the oats produced in Alabama were used to feed those animals, one would expect a commensurate increase in acres harvested of the crop; however, as demonstrated by the data in Figure 9.59, that did not happen. Either Alabama farmers were beginning to use a different feed for their draft animals, or they were getting their oats from out-of-state sources.

US and Alabama Yields

Data on the annual average yields of oats in both the United States and in Alabama for the period 1870-1915 are presented in Figure 9.60. Yields in the United States probably tended to increase slightly during the period; however, there was so much year-to-year variation that it is difficult to be certain. Mostly they ranged around 30 bushels an acre for the entire period. There is little doubt that they did trend upward in Alabama. The average yield was 12 bushels an acre in 1895, and 19 in 1914; however, even with the upward trend, Alabama farmers still were harvesting 8-10 bushels fewer oats an acre than other farmers in the country. In 1915, farmers in Iowa, the state with the highest production that year, harvested an average of 39.5 bushels an acre. Alabama farmers harvested only 15 bushels an acre. It would not be surprising if Alabama farmers were beginning to purchase their oats rather than trying to produce them.

Average yield was generally lower than expected in 1899, 1902, and 1915. Excessively dry weather probably reduced yields in the first two of those years, and a combination of excessively wet weather in north Alabama and excessively dry weather in the south may have resulted in the lower yield in 1915.

Bushels Harvested in Alabama

Alabama farmers harvested 3.9 million bushels of oats in 1895 (Figure 9.61). They had harvested 5.3 million in 1882. Because acres harvested (Figure 9.59) remained largely unchanged during the period 1895-1915 and yields generally trended upward, the year-to-year changes in bushels harvested followed the yield trend upward to 4.5 million bushels in 1915. Bushels harvested was down sharply in 1899 and 1902 as a result of lower than expected yields. It was also lower than expected in 1912 and 1913, but acres harvested was also lower than expected in those same two years.

Farmers in Elmore County harvested 4.3% of the 1909 Alabama oat crop. Other counties harvesting substantial shares included: Tallapoosa (4.2%), Talladega (4.0), Montgomery (3.3), and Lee (3.1). These five counties accounted for 18.9% of the total crop. Farmers in 26 counties harvested at least 50,000 bushels. Figure 9.62 depicts the statewide distribution of harvested oats in Alabama in 1909.

Alabama Sweet Potatoes

In the preceding period (1870-1895), acres of sweet potatoes harvested in Alabama had trended upward slowly from 28,000 to 61,000 (Figure 9.63). In the period 1895-1915, acres harvested ranged between 51,000 (1899) and 67,000 (1907). It declined sharply after 1896 when prices began to decline. Prices began to trend upward again in 1897, and acres harvested began to follow after 1900. Even though prices remained good, after 1909 for some reason farmers seemed to lose interest in the crop, and acres harvested began to trend downward, finally reaching 57,000 acres in 1914. The war seemed to rekindle Alabama farmers' interest in sweet potatoes again in 1915, and acres harvested increased to 63,000. In the period 1895-1915, crop value ranged between $1.4 (1897) and $4.7 million (1911).

Farmers in Jefferson County harvested 5.6% of Alabama's 1909 sweet potato crop. Other counties reporting large shares included: Baldwin (3.2%), Mobile (2.9), Montgomery (2.8), and Covington (2.8). These five counties accounted for 17.1% of the total. Farmers in 13 of the 67 counties harvested at least 54,000 Cwt.

Alabama Irish Potatoes

Although Irish potatoes were not a very important crop in Alabama during the period 1895-1915, average yields were slightly higher than for sweet potatoes, and prices were much better. Acres harvested was 8,000 in 1895 (Figure 9.64). It trended slowly upward to 14,000 in 1907. It remained at that level through 1910, before falling back to 10,000 acres in 1913. It remained at that level through 1914. Like the sweet potato, Irish potato acreage received a wartime boost, up to 11,000 in 1915.

Alabama farmers harvested 677,000 Cwt of Irish potatoes in 1909. Farmers in Mobile County harvested the largest share (12.4%) of that crop. Other counties reporting large shares included: Jefferson (7.2%), Baldwin (5.6), Marshall (5.2), and Lauderdale (3.5). These five counties accounted for 33.9% of the total crop. Crop value ranged from $465,000 (1896) to $1.2 million (1909).

Alabama Sorghum and Sugar Cane

According to the 1900 Census, Alabama farmers produced 93,300 tons of sorghum cane in 1899 and 1.2 million gallons of molasses. The total value of cane and molasses was $371,000. In 1909, they produced 72,400 tons of the crop, and made 809,361 gallons of molasses. That year, the total value of cane and molasses was $450,000. According to 1910 Census data, a large share of the sorghum molasses was produced from Jefferson County northward. DeKalb County led the state in the production with 37,000 gallons (4.6% of the total). Farmers in Cullman and Lauderdale counties produced slightly smaller quantities.

Alabama farmers harvested 267,900 tons of sugarcane in 1899 and produced 2.7 million gallons of syrup with a value of $1 million. In 1909, cane harvested declined to 226,600 tons, but the yield of syrup was considerably greater—3.1 million gallons. The value of the 1909 syrup production was $1.5 million. In contrast to the production of sorghum molasses, a larger share of the production of sugarcane syrup was reported from the southern tier of counties. Houston County led the state with 185,300 gallons (6% of the total). Farmers in Geneva and Barbour counties harvested slightly smaller shares.

Alabama Vegetables

Data obtained by the 1910 Census indicate that there were 55,822 acres of vegetables (excepting potatoes, sweet potatoes, and yams) grown in Alabama in 1899. The comparable acreage for 1909 was 69,468. Total crop value for the two years was $2.6 million and $5.4 million, respectively.

Alabama Peaches and Nectarines

Alabama farmers produced a variety of fruits; peaches were the most important from a commercial standpoint. Data published in the 1900 Census indicate that in 1899 farmers harvested 948,000 bushels of orchard fruits (apples, peaches and nectarines, pears, plums and prunes, cherries, and apricots). Apples (719,000 bushels) accounted for the largest share. A decade later (1909), they harvested 2.5 million bushels of fruit. Peaches accounted for some 57% of that total. Data presented in the 1910 Census indicate that there were 80,762 peach and nectarine trees of bearing age in Alabama that year. The total value of the 1909 crop was $1.8 million.

Farmers in Tuscaloosa County harvested 3.9% of the 1909 crop of peaches. Other counties reporting large shares of the total crop included: Morgan (3.6%), Lee (3.6), Blount (3.6), and Jefferson (3.6). These five counties accounted for 18.3% of the total crop. Farmers in 23 of the counties reported harvests of at least 1.2 million pounds of the fruit.

Alabama Strawberries

At the beginning of the twentieth century, Alabama farmers also produced a variety of small fruits, with strawberries accounting for 97% of the total. According to the 1910 Census, in 1899 Alabama farmers harvested 954,000 quarts of strawberries (761 acres) valued at $54,097. In 1909, they harvested 1.9 million quarts (1,232 acres) valued at $165,386.

Alabama Pecans

The 1900 Census indicate that Alabama farmers harvested 60,670 pounds of pecans in 1899. In 1909, they harvested 228,341 pounds valued at $30,540. The 1910 Census found that in 1909 there were 44,683 pecan trees of bearing age on 2,569 farms.

Farmers in Mobile County harvested 11.2% of the 1909 pecan crop. Other counties reporting substantial shares of the total crop included: Baldwin (7.5%), Bullock (7.2), Lowndes (5.2), and Monroe (4.6). These five counties accounted for 35.7% of the total crop. Farmers in 15 counties reported harvests of 5,000 or more pounds.

Alabama Flowers and Nursery

For the period 1895-1915, only data on the value of production for flowers, plants, and nursery are available. Those data indicated that the total value of production was $175,100 and $427,300 in 1899 and 1909, respectively.

The census stated that 3,199 acres were devoted to the production of these crops in 1909.

Alabama Timber

Data on lumber production in the different regions of the United States during the period 1869-1915 are presented in Table 9.1. These data show that by the beginning of the twentieth century lumber harvesting in the country was rapidly shifting to the South and West. By 1915, most of the saw timber had been cut in the North Atlantic and North Central regions, and lumbering was centered in the Southern Yellow-pine forests of the South. In 1915, 49% of the country's lumber was being produced in the South, but the West's share was growing. Only a little more than a third was being produced in the North.

Stauffer (1960) noted that the 1900 Census estimated the amount of forestland in Alabama to be 24,512,000 acres, and that reliable data earlier than 1899 are not readily available. Stauffer also noted that Dr. Fernow, chief of the US Forest Service, had estimated that 53% of Alabama's total land area was probably covered in forest in 1902.

Data from the 1910 Census indicate the value of all Alabama forest products of farms totaled $2.5 million in 1899 and $6.3 million in 1909. Harper (1943) published data indicating that rough lumber production was 21,000 board feet per square mile in 1900 and 35,000 in 1910.

Animal Agriculture

Agriculture in the Northern states had benefited from the Civil War, and it continued to expand well into the period 1870-1895. The 1890 Census include data showing that the value of livestock on American farms totaled $1.5 billion in 1870. This estimate and the estimate for 1880 ($1.5 billion) for the country were biased downward, somewhat, by the much lower postwar values of the former Confederate States. By 1890, the value of the livestock industry in the South was recovering, and as a result the national estimate reached $2.2 billion. In 1870, the value of livestock on Alabama farms totaled $26.7 million. It had been $$43.4 million in 1860, just before the beginning of the war, and it did not return to its antebellum level until after 1890. It was $23.8 million in 1880 and $30.8 million in 1890.

Cattle

Available data indicate that cattle numbers in the country increased at a faster rate than population during the period 1870-1895, and that generally speaking, people in the country had more beef available per capita than at the beginning of that period.

US Cattle

On January 1, 1895, the inventory of "all" cattle (excepting milk cows that had calved) was 34.3 million head. In 1870, it had been 22.1 million. Cattle producers in Texas reported the highest percentage of the 1895 inventory (16.2%). Other leading states in the cattle inventory were: Iowa (6.5%), Kansas (4.7), Missouri (4.4), and Nebraska (4.1). These five states reported 35.8% of the total. By 1895, the American cattle industry had moved westward, well beyond the Mississippi.

Farm Value Per Head

Figure 9.65 presents data published by USDA on farm value per head of cattle for the period 1895-1915. Average value for "all" cattle was $16.56 in 1895. Afterwards, values went through one well-defined cycle and a portion of a second. The primary cycle began in 1896 and ended in 1905. A second one began in 1906. With the data available, we cannot know when it ended. These cyclic changes are almost certainly the result of a teeter-totter relationship between values shown in Figure 9.65 and cattle numbers shown Figure 9.66. In these cycles, values went from $17 in 1895 to $24 in 1899, before trending downward to $14 in 1905. From that low level, it began to increase again, finally reaching between $31 and $32 in 1915.

Real Net Cash Returns Per Cow

In 2001, USDA published a time-series of data, on the estimated average real net cash returns (dollars per cow) for the United States dating back to 1913 (Matthews, et al., 2001). These estimates are given in real 1982-1984 dollars. From data presented in the publication, it appears that real net cash returns per cow declined from $152 in 1913 to $147 in 1914. However, in the first complete year of World War I (1915), it declined to $135. This was the period in which the farm value per head was increasing rapidly (Figure 9.65).

Cycles of Abundance

In 1895, inventory data indicate that the American cattle herd ("all" cattle excepting cows that had calved-milk) was near the end of the Declining Phase of a cycle that had reached its Consolidation Phase around 1890 at 45 million animals (Figure 9.66). The Declining Phase of this particular cycle finally ended in 1896 (34 million head), and the Expansion Phase of a new cycle began the following year. The three phases of this new cycle were completed in 1912, with a level of 36,200 animals. In 1913, still another new cycle began. Its Expansion Phase was not completed at the end of the period in 1915. So the data presented in Figure 9.66 show that cattle numbers in the United States went through one full Cycle of Abundance and portions of two others in the period 1895-1915. It is likely that these cyclic changes were the result of the teeter-totter relationship between numbers and values during the period.

Alabama Cattle

Data presented in the preceding chapter indicate that cattle numbers in Alabama increased at a somewhat faster rate than population during the period 1870-1895, and that generally speaking people in the state had more beef available per capita in 1895 than in 1870. However, availability had not increased nearly as much in the state as it had in the country.

There are no data available on the distribution of cattle in Alabama counties in 1895; however, 1890 Census data indicate that the largest share (3.2%) of "other" cattle (all cattle excepting milk cows and working oxen) was on farms in Clarke County. Other counties with large shares, included: Wilcox (2.5%), Jackson (2.4), Marengo (2.3), and Madison (2.3). These five counties accounted for only 12.7% of the state total, which is indicative of the widespread distribution of cattle in the state at that time. Some 34 counties reported herds of 7,000 or more animals.

Farm Value Per Head

Data on farm value per head for "all" cattle on Alabama farms during the period 1895-1915 are presented in Figure 9.65. These data show that changes in values for Alabama cattle followed essentially the same pattern as in the country as a whole. They increased sharply between 1895 and 1900, only to trend downward until 1903. From 1903 onward, they trended sharply upward again. These cyclic changes were apparently a result of the same teeter-totter relationship between numbers and values described for national values in a preceding section. While the cyclic changes in values were similar in the country and the state throughout the period, values in the United States were substantially higher than in Alabama.

Cycles of Abundance

Data on numbers of cattle in Alabama (excepting cows that had calved-milk) indicate that year-to-year changes during the period 1895-1915 were similar to those in the country (Figures 9.66 and 9.67). During the period, numbers in the state also went through one complete cycle and portions of two others. There were 583,000 animals in 1895, before falling to 443,000 in 1901. The new cycle reached its Consolidation Phase in 1905, with 673,000 animals. At the end of that cycle in 1910, there were 490,000 cattle in the state. In the Expansion Phase of the new, incomplete cycle, the number reached 572,000 in 1915.

Data presented in Figure 9.67 indicate that cattle numbers in Alabama probably declined slightly during the period 1895-1915. The presence of Cycles of Abundance makes such a determination difficult, but a comparison of numbers in the Consolidation Phase of the preceding cycle in 1891 with the number in the Consolidation Phase of the next cycle in 1905 shows that between those two high-points, numbers declined by 27,000 animals (700,000 versus 673,000).

The 1910 inventory of cattle (all) in Alabama indicates that the largest share (3.2%) was on farms in Marengo County. Other counties reporting large numbers included: Wilcox (3%), Clarke (2.7), Dallas (2.6), and Monroe (2.4). These five counties accounted for only 13.9% of the total. These data indicate the widespread distribution of cattle within the state. Inventories indicate that 49 counties reported 10,000 animals or more. Figure 9.68 shows the statewide distribution of cattle in Alabama in 1910.

Hogs

In the early 1800s, the South Atlantic and South Central regions were home for hogs. In 1840, 59.9% of the total number in the country were in those regions. The South's share continued to increase to 1850 (63.3%), but by 1860, its relative position was beginning to decline, and by 1900, only 30.8% of the total national herd could call the South home. In that year, 63.8% of the herd was in the North Central Region. In the early years, the South was home for hogs, but by 1900 home for hogs was where the corn was.

US Hogs

In 1895, a December 1 inventory of hogs and pigs indicated that there were 49.2 million head in the United States. The highest percentage of these animals was on farms in Iowa (12.2%). Other states with significant numbers of the animals included: Missouri (8.5%), Illinois (8.3), Indiana (5.2), and Texas (5.2). At that time, these five states accounted for 39.3% of all the hogs and pigs in the country.

Farm Value Per Head

Data on the farm value per head for hogs in the United States are presented in Figure 9.69. In 1895, the estimated average value was $5.09, and although there had been considerable year-to-year variation, the trend line had remained near that level since 1870; however, values began to trend upward in 1896, and by 1915 it was between $10 and $11 a head. Obviously, year-to-year changes were cyclic throughout the period, 1895-1915. To a greater or lesser degree, these ups and downs were repeated throughout the remainder of the period. In 1895, values were declining as numbers increased. Value reached $4.36 in 1897, and the following year (1898) farmers had fewer hogs on their farms. Then, in the same year, values began to increase again. This relationship became much less predictable after 1905.

Cycles of Abundance

The USDA inventory in 1895 indicates that there were 49.2 million hogs (and pigs) in the United States (Figure 9.70). There had been 36.7 million in 1870. In 1895, the hog numbers were in the Accumulation Phase of a cycle that had begun in 1893 (46.5 million head). This phase ended in 1897 with 53.3 million animals. The Liquidation Phase began a year later, and ended in 1901 with 47.9 million. These cyclic changes generally persisted through the remainder of the period.

The largest share (15.5%) of the country's 1915 hog inventory was on farms in Iowa. Other states reporting large shares included: Illinois (7.7%), Nebraska (6.1), Missouri (6.0), and Indiana (5.8). These five states accounted for 41.1% of the total. Farms in Alabama reported 2%.

Alabama Hogs

In 1870, six of the Southern states reported more hogs than Alabama. Only 8.1% of all hogs in the region were on Alabama farms. In 1895, five of those states reported more hogs, and Alabama's share had increased to 10%. In 1870, 2.9% of the country's hogs was on Alabama farms. In 1895, this percentage had increased to 3.2%.

There are no county data on the distribution of hogs in Alabama in 1895. However 1890 Census data indicate that the largest percentages, at that time, was on farms in Clarke County (2.9%). Other counties with large shares included: Jackson (2.8%), Henry (2.5), Marengo (2.2) and Madison (2.1). These five counties accounted for 12.5% of the total. This relatively low percentage is indicative of the widespread distribution of hogs in the state. A total of 36 counties reported 20,000 or more animals.

Farm Value Per Head

Data on the Alabama farm value per head of hogs during the period 1870-1915 are presented in Figure 9.69—along with the national data. These data indicate that the pattern of yearly changes in the values in Alabama were similar to those in the country as a whole. Although there was considerable evidence of cycles in the Alabama data, the trend line generally remained around $3 a head from 1870 through 1902, when values began to increase rapidly. While there is considerable year-to-year variation after 1901, there is somewhat less indication of cycles of values than there was in the national data.

Cycles of Abundance

Data on the inventory of hogs on Alabama farms on December 1st of each year in the period 1895-1915 are shown in Figure 9.71. Data for the period 1870-1894 are also presented as a source of reference. In 1895, there were 1.6 million hogs (and pigs) on farms in the state.

There had been 1 million in 1870.

The data presented in Figure 9.71 provide a clear example of existence of cycles in hog numbers. In 1895, numbers were in the Accumulation Phase of a cycle. It ended in 1897 with 1.6 million animals. The Liquidation Phase began in 1898, and ended in 1902 with 1.1 million. There were apparently two complete cycles during the periods 1903-1909 and 1910-1914. At the end of the Liquidation Phase in 1914, the number had declined to 1.1 million animals, the lowest level since 1876.

In 1910, 4% of all of the hogs in Alabama were on farms in Houston County. Other counties with large shares of the state herd were: Covington (3.3%), Clarke (3.3), Greene (3.2), and Coffee (2.8). These five counties accounted for 16.6% of the total. Virtually every county had a substantial number of hogs. Farmers in 26 counties owned at least 20,000 of the animals. Figure 9.72 presents data on the distribution of hogs in the state in 1910.

Fite (1984) commented that at the beginning of the twentieth century there was an overall decline in hog production in the South. Swine production was highly dependent on corn production. When cotton prices were low, farmers planted more corn, which could support additional hogs; however, during most of this period, acreage in corn declined as farmers planted more cotton. While Alabama acres of corn harvested did not decline during this period (Figure 9.44), it certainly did not increase enough to support a significant increase in hog production.

Milk

Apparently, it was difficult to obtain good data on milk production in the United States until well after the end of the nineteenth century. Throughout most of that early period, a sizeable share of milk produced in the country was not obtained from cows on farms, but rather from animals held by families in and around cities and towns. In the censuses of 1870 and 1880, estimates were made of the quantity of milk sold in the country in 1869 and 1879. Then the 1890 Census published estimates on the quantity of milk produced on farms, but without any estimate of the quantity produced from cows owned by non-farmers. Another problem resulted from the fact that farmers in different regions of the country were not equally careful in reporting data on milk production. In all regions, they tended to report less milk produced than expected, when considering the numbers of dairy cows reported, but the deviation from the expected was greater in some regions than others. In some regions, farmers apparently milked some of their cows for only part of the year. When submitting data for the censuses, they counted these animals as milk cows, but did not report the quantity of milk they produced.

US Milk

Census data published in 1890 indicate that cows on American farms produced 44.8 billion pounds (5.2 billion gallons at 8.6 pounds a gallon) of milk in 1889. This quantity was equivalent to 726 pounds per person in the population that year. Some of this milk was converted (includes on-farm production only) into 1 million pounds of butter (16.6 pounds per capita) and 18.7 million pounds of cheese (0.3 pound per capita)

Milk Cows

During the period 1870-1895, the number of milk cows in the United States had increased 57.5% (9.7 to 15.2 million). In the same period, the population had increased 80.3%. As a result, in the period the number of milk cows per 100 persons in the country declined from 25 to 22.

An inventory of milk cows (cows that had calved) on January 1, 1895, indicates that there were 15.2 million head in the country at that time. The highest percentage was on farms in New York (9.7%). Other leading states included: Iowa (7.6%), Pennsylvania (6.0), Illinois (5.8), and Wisconsin (5.3). These five states accounted for 34.5% of all milk cows in the country.

Data derived from the annual USDA, January 1 inventory of milk cows (cows that had calved) for the period 1895-1915 are presented in Figure 9.73. In 1895, the inventory indicated that there were 15.2 million milk cows in the country; there had been 9.7 million in 1870. In 1895, numbers were on a plateau. It had changed very little since 1890. However, there was a 10-month economic contraction in the period 1890-1891, and widespread bankruptcies and a depression in the period 1893-1894. It is likely that these economic uncertainties led farmers to decide to reduce the size of their herds. Then in the period 1896-1910, numbers began to increase again, reaching 19.4 million head in 1910. The upward trend ended in 1911 for a year (1911), but continued afterwards to reach 20.3 million head in 1915. There was no indication of cycles in numbers during the period.

Milk Per Cow

There are no annual data on the production of milk per cow in the United States before 1925. Data in the 1910 Census indicate that average milk per cow in the country was 424 gallons in 1899, but only 362 gallons in 1909.

Milk Production

When the 1910 Census was being prepared, it was obvious that the quantity of milk being produced on American farms in 1909 was being seriously underestimated. Farmers that year reported that they had obtained 49.3 billion pounds (5.8 billion gallons) from their animals. However, using data on the number of milk cows in the country and estimates of milk per cow, it was estimated that the total should have been near 63.8 billion pounds (7.5 billion gallons). Based on these estimated data, farms in the North Central Region reported the largest share (49.7%) of the total quantity. The New England area reported 22.4%. It is obvious from these data that at the beginning of the twentieth century most of the milk (72.1%) was still being produced in the North.

Sale of Dairy Products

The 1910 Census includes data on the sale of dairy products in the United States in 1899 and 1909. The data indicate that in 1899 American farmers sold dairy products valued at $281.6 million. Farmers in the North Atlantic and North Central regions sold 42 and 43.5%, respectively, of this total. Then in 1909, they marketed products valued at $473.8 million, an increase of 55%. In that decade, the North Atlantic Region's share, of the total declined to 36%, while the shares of all of the other regions increased. In that year, farmers in the North Central Region sold 47% of the total. However, it is likely that a large share of these products were actually shipped to and finally sold to the rapidly growing urban population in the North Atlantic Region. Estimates of the sale of dairy products were not made for Alabama in those years.

Alabama Milk

Census data indicate that Alabama farmers reported the production of 477.4 million pounds (55.5 million gallons) of milk in 1889. This level of production is equivalent to 320.7 pounds (37.3 gallons) per person. Farms in Jefferson County reported the largest share (3.2%) of total production. Other counties reporting large shares included: Madison (2.9%), Jackson (2.5), Blount (2.4), and Dallas (2.4). These five counties accounted for only 13.3% of the total. Average production per county was 7.2 million pounds. A total of 43 of Alabama's counties reported productions of 6 million or more pounds.

Milk Cows

In Alabama, milk cow numbers increased only 21.7% between 1870 and 1895 (230,000 to 280,000), while the population increased 67.7%. In consequence, cows per 100 people declined from 23 to 17. An increase in milk per cow might have compensated for some of the reduction in cows per 100 people in the United States, but probably not in Alabama.

Data presented in Figure 9.74 were obtained from the inventory of milk cows (cows that had calved) on January 1st of each year during the period 1895-1915. Data from 1870-1894 are shown for comparison. There were 280,000 milk cows on Alabama farms in 1895. In 1895, milk cow numbers in the state had reached the end of an upward trend that had begun around 1878. Surprisingly, after 1895 it began to trend downward and reached 260,000 in 1898. It remained at around that level through 1904 before beginning to trend sharply upward. There is little information to explain what might have caused the decline between 1895 and 1904. Numbers increased rapidly after 1904 to reach 360,000 animals in 1910. Then it remained at that level through 1915. Although herd size declined somewhat during the period 1895-1900, this did not appear to be a cyclic change in abundance similar to those that were so clearly obvious in cattle during the period (Figure 9.67).

In 1910, the largest share (3%) of Alabama's milk cow herd was on farms in Wilcox County. Other counties with large shares of the total, included: Jefferson (2.9%), Dallas (2.8), Montgomery (2.6), and Marengo (2.4). These five counties accounted for only 13.7% of the state total.

Farmers in 41 counties had 5,000 or more milk cows.

Milk Per Cow

Alabama data in the 1910 Census indicate that average milk per cow was 250 pounds (29.4 gallons) in 1909. The author of the census report on milk production in the state commented that this estimate was probably somewhat deficient because of the way that farmers responded to the questions on the census forms.

Milk Production

According to the 1910 Census, farmers in Alabama reported the production of 677.1 million pounds (78.7 million gallons) of milk in 1909. This level of production was equivalent to 321.2 pounds per person. Production in 1899 was equivalent to 320.7 pounds per person.

Based on estimates, in 1909 cows on Alabama farms produced only 1% of the total quantity produced in the country. Farmers in Jefferson County reported the largest share (4.5%). Other counties reporting large shares of the total included: Madison (2.5%), Montgomery (2.5), DeKalb (2.4), and Jackson (2.2). These five counties accounted for only 14.1% of the total production. Farmers in 40 counties reported productions of at least 8,400,000 pounds.

Chickens

The 1890 Census reported that there were 258.9 million domestic fowls (chickens) on American farms. There had been 102.3 million in 1880. The 1890 data indicated that 51.3% of the total was on farms in the North Central Region. The next largest share (22.1%) was in the South Central Region. The Alabama share was 2.8%.

USDA did not begin to differentiate broilers from chickens until 1939; before that time, broilers would have been included in all data on chickens. There are no annual data on numbers of chickens in the United States and any of the states before 1909.

US Chickens

The 1900 Census indicated that there were 233.6 million chickens (including Guinea fowls) in the United States. Some 52.8% were on farms in the North Central Region. Most of the chickens were where the corn was. Iowa farmers reported the largest share (8.1%) of the 1900 inventory. Others states reporting large shares included: Illinois (7.1%), Missouri (6.4), Ohio (6.1), and Texas (5.8). These five states accounted for 33.5% of the total.

Value Per Head

Data on the average value per head of chickens on American farms each year during the period 1909-1915 are presented in Figure 9.75. Values ranged from 42.2 cents in 1912 to 49.1 cents in 1914. Yearly changes seemed to follow a cyclic pattern during the period. Surprisingly, value for chickens did not receive the same wartime bounce received by cattle and hogs (Figures 9.65 and 9.69). Numbers were up substantially in 1915, but value was lower (again, the teeter-totter effect).

Numbers

Data on the estimated annual number of chickens on American farms during the period 1909-1915 are presented in Figure 9.76. Numbers range from a low of 340.2 million in 1909 to a high of 381.5 million in 1911. Year-to-year changes in numbers also seemed to be cyclic (Figure 9.75). The two patterns (values and numbers) appear to be related in the usual teeter-totter relationship seen typically in all crops and other livestock.

Alabama Chickens

The 1890 Census reported that there were 6.2 million chickens on Alabama farms. There had been 2.1 million in 1880.

There are no annual data on numbers of chickens on Alabama farms before 1924, but the 1900 and 1910 censuses reported that numbers in those two years was 4,738,000 and 4,590,000, respectively. In 1900, some 2% of all the chickens in the country were in Alabama; however by 1910 this percentage had declined to 1.6%.

Census data for 1910 reported only the numbers of all kinds of poultry in the counties. There was no report for chickens alone; however, at least 90% of all fowls in the state likely would have been chickens. The leading counties with numbers of all fowls were: Madison (2.6%), Montgomery (2.6), Dallas (2.3), Marshall (2.2), and Wilcox (2.2). In 1910, poultry were widely distributed in all counties. The five leading counties only accounted for 11.9% of the total. A total of 38 counties reported chicken totals of 70,000 or more birds.

Eggs

From the days of the earliest European settlements in the North Atlantic and South Atlantic regions, eggs were an important dietary item. For many years, most of the eggs were consumed by the farm families responsible for their production; however, with the growing urbanization in the country, markets for eggs also increased. Sometime before 1909, the quantity of eggs sold became greater than the number being consumed by the households responsible for producing them.

Census data indicate that chickens on American farms laid 9.8 billion eggs in 1889. The total for 1879 had been 5.4 billion. In 1889, farms in the North Central Region accounted for 57.1% of the total. Farms in the North Atlantic and South Central regions accounted for 17.3 and 15.2%, respectively. The Alabama share was 1.3%.

US Eggs

In 1909, USDA began to publish data on annual egg production in the country, along some data on disposition. These data indicate that of 25.3 billion eggs were produced nationally in that year. The USDA estimate for egg production in 1910 was substantially higher (25.3 versus 19.1 billion) than the census estimate. Census data indicate that in 1909 the largest share (52.7%) of the country's egg production was reported by the North Central Region. Shares reported by the other regions were: South Central (18.5%), South Atlantic (8.6), and Western (6.6).

Prices

Farmers received 20 cents (17.5 to 20.9 cents) per dozen for their eggs during the period 1909-1915 (Figure 9.77). Prices seemed to be cyclic and related to numbers (Figure 9.78) in the typical teeter-totter relationship. Surprisingly, egg prices did not seem to receive the same wartime bounce as values for cattle and hogs. Number was higher in 1915, but the average price was lower than it had been in 1914.

Numbers

Data on the estimated annual production of eggs in the United States during the period 1909-1915 are presented in Figure 9.78. Numbers during this period also appeared to be cyclic, ranging from 25.3 to 29.9 billion.

Uses

According to the first annual USDA report on the use of eggs during the period 1909-1915, American farmers sold 71% of the eggs they produced, consumed 25%, and used 4% for hatching (Figure 9.79).

Annual per capita consumption of eggs varied between 209 and 326 during the period (Figure 9.80). Average annual consumption also seemed to be cyclic over the period, generally teeter tottering with prices.

Alabama Eggs

There are no annual data on egg production in Alabama before 1924; however, the 1910 Census reported that chickens on Alabama farms produced 225.3 million eggs in 1899 and 266.8 million in 1909. These levels of production were equivalent to 151 and 126 eggs per capita, respectively, and were considerably lower than estimates of national per capita consumption for those years.

Census data also indicate that farmers in Jackson County collected 3.5% of all the eggs produced in the state in 1909. Other counties reporting large shares included: Madison (3.4%), Marshall (3.2), Morgan (3.0), and DeKalb (2.8). These five counties accounted for only 15.9% of the total. Farmers in 21 of 67 counties collected at least 300,000 dozen eggs from their flocks in 1909.

Agricultural Production

Census data indicate that in 1900 (June 1) and 1910 (April 15) the total value of domestic animals, poultry, and bees on American farms was $3.1 and $4.9 billion, respectively, and that the value of all crops in those years was $3 and $5.5 billion, respectively. The totals for those two periods (1899-1900 and 1909-1910) were $8.1 and $8.5 billion, respectively.

In 1910, USDA's Economic Research Service began publishing annual data on agriculture's annual contribution to the national economic product. Net farm value added estimated the dollar contribution of crops and livestock to the national economy. These data for the period 1910-1915 are presented in Table 9.2; summaries are presented in Figure 9.81. The data indicate that total value added by agricultural production for this period generally ranged between $6.5 and $7.9 billion, and that the contributions of crops and livestock were roughly equal. However, in five of the six years values for livestock were slightly higher than for crops.

Estimates of values for inventory adjustment included in Table 10.2 relate to value added of crops and livestock produced in the current year that will not be sold before December 31 (positive value), or to crops and livestock produced in the preceding year that were not sold until the current year (negative value).

In five of the six years, value added for cotton was higher than for any of the other crops. Cotton prices were unusually low in 1914 (Figure 9.31). In four of the six years, value added for feed crops was higher than for food grains. Wheat prices were considerably higher in 1914 and 1915 than they had been in most of the preceding years (Figure 9.52).

USDA also estimated the share of total value added accounted for by home consumption. For example, in 1910, total value added for crops was $3.6 billion, and of that total, 11.7% ($424 million) of the crops it represents was consumed by the farm families producing them. Similar data were provided for the value added of livestock. In 1910, for example, home consumption for livestock amounted to near 22.9% ($846 million) of the total. In 1915, these values were 10% ($391 million) for crops and 20% ($801 million) for livestock.

A more meaningful measure of the value added for home consumption is obtained by relating it to the number of farms involved. These converted data are presented in Figure 9.82. They show that value added per farm for crops was around $800 and $400 per farm for livestock. Generally, values added for crops consumed were about half of that for livestock throughout the period. This difference is primarily the result of differences of the quantities consumed and to some extent to differences in prices.

There are no value added data available for the states before 1949. Census data show that the total values of all crops in Alabama in 1900 and 1910 were $73.2 and $144.3 million, respectively. Values for all livestock were $36.1 and $71 million, respectively; although it is not certain how comparable the livestock data are. Apparently, values for crops were about twice as high as for livestock.

The Bottom Line

As detailed in the preceding chapter, in the early 1890s American farmers everywhere, were suffering from low prices for all their commodities. Of course, cotton farmers in the South had been struggling with this problem for decades, but it was a relatively new experience for farmers in the North. Fortunately, just when most farmers were about to go under for the third time, the national economy began to turn around. After declining in 1893 and 1894, the Gross National Product began to increase, and farmers were caught up in the rising economic tide. They did not realize it at the time but they were about to enter the Golden Age of Agriculture.

Beginning in 1910, USDA began to publish several different kinds of data that made it easier to measure the economic status (the bottom line) for farmers. Probably the most important was Net Farm Income (NFI). USDA's Economic Research Service defines net farm income as: "the farm operator's share of the net value added to the national economy, within a year, independent of whether it is received in a cash or non-cash form." USDA did not publish data on net farm income for the United States before 1910 and for the states before 1949. NFI data for the period 1910-1915, are presented in Figure 9.83. The values are expressed in current dollars. The data indicate the NFI did not change very much in relative terms during the period ranging from $3.4 billion (1911) to $4.4 billion (1912). It was substantially lower in 1911 and 1914 and higher in 1912 and 1915. These values, both low and high, generally reflect the relative magnitude of value added presented in Table 9.2. For example, the lower NFI value in 1911 is largely the result of the lower value added for feed crops and cotton.

NFI data are made more meaningful by converting the total data to a per farm basis. These converted data are presented in Figure 9.84. On a per farm basis, the NFI ranged between $525 (1911) and $667 (1915). Year-to-year changes were similar to the total NFI data shown in Figure 9.83. The converted data are sobering, but probably not nearly as sobering as they would be if data for the South could be isolated. It is likely the values for those myriad of tiny farms in that region would be more sickening than sobering.

Farmers in Alabama and the South were truly suffering. For example, in Alabama in 1910, there were 265,000 farms. The largest number in history, and 60.2%

of them were operated by tenants—up from 57.7% in 1900. As the number of farms increased, the average size decreased. It was 92.7 acres in 1900, and 78.9 acres in 1910. Of even greater significance, however, was the decline in cropland harvested per farm. It was 30.1 acres in 1900, but declined to 27.4 acres in 1910. In contrast, Iowa farmers harvested crops from 96.2 and 93.9 acres in 1900 and 1910, respectively. Fite (1984), commented that "farms of such size usually could not produce enough to provide an adequate family living." Fite (1984) succinctly described the condition of the majority of Southern farmers in the early 1900s:

> The romantic ideal of a rural society dominated by self-sufficient, independent, land-owning farmers, was further from reality in 1910 than anytime in the history of the South. The Golden Age in American Agriculture had by-passed most Southern farmers. Both white and black and tenant farmers (60.2% of all farm operators) continued to live in dire poverty.

STILL SERIOUSLY SEEKING SALVATION

As detailed in the preceding chapter, agricultural experts during the latter half of the nineteenth century had called for the application of science and education to farming and for the organization of farmers and for crop diversification as portions of the threefold path to salvation for the American farmer. Later, another path—political action—was added. Unfortunately, for most of the latter part of that period, the experts' suggestions seemed to have little effect. The economic condition of American agriculture steadily worsened, but just when everything seemed hopeless, progressivism appeared and American agriculture began to recover. Soon, it would be in the middle of its so-called Golden Age of Agriculture. Fortunately or unfortunately, depending on how you view the situation, even though their lot was improving rapidly (for most of them at least), the process by which American farmers and their supporters had established, in the latter part of the nineteenth century, to find salvation was also moving ahead rapidly. In fact, it was quickly beginning to achieve a life of its own —a life that became increasingly important in terms of costs. Moreover, the process itself quickly became so intense that it seemed to become largely divorced from the results it achieved.

Salvation Through Better Information

Thus far, lack of success had not deterred the reformers from continuing their efforts to change Southern agriculture through scientific research and farmer education. The work of agricultural experiment stations across the region was increasing both in amount and in breadth. For example, in 1898 the Agricultural Experiment Station at Auburn (the main station) had only ten staff members, but its research staff was beginning to grow. By 1908 there were 19, and in 1913 it had increased to 32 (Kerr,1985). There were no substations during that period.

The Horticulture Department was created as part of the experiment station in 1903. The Department of Animal Industry (Animal and Dairy Sciences) was added in 1907. The departments of Botany and Entomology were created a year later.

Also, efforts at farmer education were being intensified. Nationally, both research and farmer education had achieved critical masses in public support, primarily a result of the perception (rightly or wrongly) that the increased amount and quality of information these workers had made available to farmers were a primary reasons for the growing success of agriculture production in the Midwest. It was assumed that it was just a matter of time before research and education would also save the Southern farmer.

At Tuskegee Institute under Booker T. Washington's leadership, a Department of Agriculture was established in 1896. In the same year, Dr. Washington was able to get George Washington Carver to leave Iowa State University to accept the leadership of the new department (Mayberry, 1989).

In 1897, Governor William Oats signed a bill establishing at Tuskegee a branch agricultural experiment station and agricultural school for African Americans. The bill also provided $1,500 annually to support it. The same year, Dr. Carver was named its first director. Also, in the same year, the institute dedicated the Armstrong-Slater Memorial Agricultural Building to house the new programs.

In 1899, Tuskegee was first considered to be part of the land-grant college system when the federal government allocated 25,000 acres in west Alabama for its use. In the same year, the school became involved in international agriculture when some of it staff and students visited Togo in West Africa to put on agricultural demonstrations for local people.

In 1904, Dr. Washington established a Department of Records and Research at Tuskegee. Its purpose was to collect, analyze, and disseminate data related to the daily lives of African Americans.

The first local short course in agriculture was offered by Tuskegee in January 1904. It provided participants with six weeks of intense exposure to all aspects of modern agriculture. Although the course was judged to be a success, Dr. Washington was concerned that the school was not able to reach enough farmers with only a single annual course; consequently he asked Dr. Carver to design a moveable delivery system that could be driven through rural areas putting on local field days. Dr. Carver and his assistants designed a wagon to be drawn by two horses. The wagon contained a wide range of equipment (cultivators, a revolving churn, plows, cream separators, cheesemaking equipment, modern canning equipment, a variety of seeds, and fertilizers). Once the design was finalized, Dr. Washington visited with Morris K. Jesup, a New York banker, with a request for funds to build it. The Jesup Wagon made its maiden voyage on May 24, 1906, and it was an immediate success. It was soon attracting widespread attention throughout the state and the South.

In the fall of 1906, Dr. Seaman Knapp, special agent in charge of farmers' demonstration work for the USDA, visited Tuskegee to discuss with Dr. Carver the possibility of establishing a cooperative demonstration program for African American farmers in the South. In December, Thomas Cambell was employed as the first black demonstration agent in the country. A month later, John Pierce was employed as the agent for the Upper South. He was stationed at Hampton University in Virginia. Soon, with the support of federal funds, African American agents were working in several counties in south Alabama, as well as in adjoining states. As a result of this expanded involvement in farmer education, in 1910 Dr. Washington established the Extension Department at Tuskegee Institute.

In 1914, Congress passed the Smith-Lever Act, which establishing the cooperative Extension Service nationwide. Under the provisions of the act, the government partnered with the land-grant colleges in the various states to conduct extension work. In Alabama, Auburn was chosen to administer the extension program. Later, the administration of extension work for blacks was assigned to Tuskegee through a memorandum of agreement.

President Buchannan of the State Agricultural and Agricultural College for Negroes made an impassioned plea for some of those extension funds, but Governor Emmett O'Neal decided that they would go to the Alabama Polytechnic Institute at Auburn. Later, black extension agents were employed by Auburn, but they worked under the direction of white agents. Morrison (1994) commented that although the school was denied direct access to the funds, it nevertheless continued to service the farm population.

Salvation Through Organization

As noted in the preceding paragraphs, the reformers had not given up their efforts begun in the latter part of the nineteenth century to improve the lot of Southern farmers through scientific research and farmer education; the organizers had not given up either. The Farmers' Union was organized in Texas in 1902. It spread rapidly across the South and into the Midwest, and a national organization was established in 1905. It had the same essential objectives as the Grange, the Agricultural Wheel, and the Farmer's Alliance, but its primary venture was to attempt to gain control of cotton marketing through the establishment of warehouses where farmers could store their cotton and borrow money against its value. This general approach represented a combination of proposals made earlier by members of both the Wheel and the Alliance. The Union did establish some warehouses and attempted to withhold cotton from the market to try to set a minimum price. After it became apparent around 1908 that the Union could never control enough cotton to set prices, membership in the organization began to decline. Soon, its warehouses were going broke along with its cooperatives, which had been established to help farmers obtain the best prices for fertilizer, groceries, and other commodities they had to purchase.

What the farmers needed were marketing cooperatives. They needed an organization to help them with the orderly marketing of their crops. Unfortunately, following the passage of the Sherman Antitrust Act in 1890, the courts were continuing to find that cooperatives for marketing were illegal.

Salvation Through Diversification

Diversification in agriculture was never much of a problem in the country generally, but diversification had always been extremely troublesome in the South. Cotton had long been king, and it appeared that at the end of the nineteenth century it would always reign supreme. Data presented by Fite (1984) indicate that in 1890, 1900, and 1910 Alabama farmers harvested crops from 5.4 million, 6.7 million, and 7.3 million acres, respectively. In those three years, they harvested cotton from 54.4%, 47.6%, and 47.2% respectively, of those totals of cropland harvested. Between 1890 and 1900, Alabama farmers seemed to be reducing their dependency on cotton somewhat, but there was little change from 1900 to 1910. Alabama farmers still had no dependable alternative cash crop.

During the period 1895-1915, there was considerable variation in the numbers of cattle and hogs on Alabama farms, but overall farmers did seem to be increasing their herds of these animals, slightly. There was a definite upward trend in the number of milk cows in the state.

Data from the 1910 Census indicate that the total value of all plant crops harvested in Alabama in 1909 was $144.3 million and that cotton constituted about 60.3% ($87 million) of the total. Data from the livestock census conducted in 1910 estimated the value of all livestock, including bees, to be $65.6 million. Although the data was collected in two different years, it is obvious that the value of all livestock was still considerably lower than that of cotton. Alabama was still primarily a cotton state.

Salvation Through Political Action

By 1895, the Populist revolution in Alabama and the country had about run its course, and the farmers had gotten very little out of some eight years of sometimes violent agitation. The Populists had tried, but in the end conservative Democrats remained firmly in control in state houses throughout the South. But while they had lost, the Populists' cause had left a lasting impression on Southern senators and representatives. They all realized that virtually nothing could be done at the state level because of the continuing problem of race, so they began to turn their attention to Washington and federal legislation to help farmers. Unfortunately, the Republicans with their business and industry orientation were in control of the presidency and Congress and would be until 1913. Finally, in that year—with Democratic President Wilson and a Democratic Congress—Southern lawmakers finally had their chance. In the 63rd, 64th, and 65th Congresses, Southern congressmen led the way in beginning to enact significant federal legislation to help their farmer constituents. It appeared that salvation was finally coming.

10

The War to End All Wars

1915-1920

PRECIS

In 1915, American farmers were still basking in the good times of the Golden Age of American Agriculture when they were pulled into the vortex of the storm of the "war to end all wars" that was expanding rapidly in Europe. At first, they only had to respond to the growing needs of America's friends in the conflict for food and fiber. Later, as the United States began to mobilize for war and finally entered it, the demand for the products of their farms began to increase even more sharply. The increased demand and production effected many changes in agriculture. After 1915, prices received increased sharply. Indices for prices paid also increased, but not as rapidly.

The war ended in 1918, and by 1919 farmers found themselves all dressed up but with no place to go. Suddenly, prices received began to decline while prices paid exploded upward. In 1920, the parity ratio was down to 0.88. It would go even lower the next year. Net farm income per farm peaked in 1919, but by 1921 it had skidded down to $517, considerably below where it had been in 1915.

INTRODUCTION

As detailed in a preceding chapter, the period from about 1897 until 1915 was probably the best in the history of American agriculture (Genung, 1940). Farm commodity prices had been increasing as demand increased, and prices that farmers had to pay for inputs remained relatively low compared to prices received. Farming was finally becoming a stable business, especially for farmers in the North. While the American market for food and fiber was growing, exports were the cornerstone of this period of good times. However this Golden Age of American Agriculture bypassed the South. Government support of agriculture was primarily restricted to education and research, as ways to increase productivity. Throughout the period, the country was committed to a protective tariff system. When World War I began American agriculture quickly became transformed forever.

NATIONAL POLITICS

World War I controlled the politics of agriculture during the period 1915-1920. There was no partisanship involved in encouraging American farmers to produce enough food and fiber to meet the needs of the country and its allies. Older farmers in both the North and South had some experience in producing food and fiber during war time, but they could not even imagine what would be involved in feeding a whole country and part of the world at war.

The Presidency

The Democratic Party controlled the Office of President for the entire period. Woodrow Wilson (D, NJ) was elected president in 1913, and he was re-elected in 1917. Although seriously ill, he served to the end of his second term in 1921.

Secretaries of Agriculture

David Houston (D, MO) was appointed to the position in 1913. He served until 1920. Edwin Meredith (D, IA) was appointed to replace him.

The Senate

The Democrats gained control of the Senate in the 63rd Congress (1913-1915) and remained in control through the 65th (1917-1919). The Republicans barely regained control in the 66th (1919-1921) with only a 49 to 47 majority.

Senate Majority Leaders

There was no Senate majority leader during this period.

Chairmen of the Senate Committee on Agriculture

Thomas Gore (D, MS) was chosen chairman in 1913 (63rd Congress), and he remained in that position until 1919. He was replaced by Asle Gronna (R, IA) in the 66th Congress (1919-1921).

The House of Representatives

The Democrats controlled the House in the period 1913-1917 (63rd and 64th), but relinquished it in 1917-1919 (65th). The Republicans continued to control it in 1919-1921 (66th).

Speakers of the House

James Beauchamp (D, MO) served as speaker of the House in 1913-1917 (63rd and 64th). He also remained in that position in 1917-1919 (65th), although the Republicans had a one vote majority (215 to 214) in the House. There were six independents in the House at that time. As a result, Beauchamp was allowed to retain his position.

Chairmen of the House Committee on Agriculture

Aubrey Lever (D, SC) was chairman of the House Committee on Agriculture in the period 1913-1919 (63rd, 64th, and 65th). He was replaced by Gilbert Haugen (R, IA) who served in 1919-1921 (66th).

The Political Situation

Austria-Hungary declared war on Serbia in July, 1914, and within months, virtually all of Europe was embroiled in the Great War/World War I. At first, the United States took a position of strict neutrality. In the country there were large pockets of support for the Allies and the Central Powers. Although the country proclaimed its neutrality, it was obvious from the beginning that we would strongly support the British in their war effort. In February, 1915, Germany announced that all waters around the British Isles were a war zone and that neutral/non-belligerent ships in the zone would be destroyed. On May 13, 1915, a German submarine sank the *Lusitania,* which resulted in the death of over 1,000 civilians, including 128 Americans. After this event, President Wilson began to work with the Congress to get the necessary legislation enacted that would allow the United States to prepare for war.

Even on the threshold of war, the country's business had to proceed. For example, in July, 1916, the Democratic Congress approved and Wilson signed the Rural Credit Act that provided cheap mortgages to farmers. Later, the Workmen's Compensation Act became law. Soon afterwards, the Jones Act granted autonomy to the Philippine Islands, and in September the Adamson Act was approved that provided for an eight-hour day and other benefits to the railroad union brotherhood. Liberalism was in full control of the national government.

Between the worry over the war in Europe and political battles over legislation to help the ordinary citizen at home, the country prepared for the election of 1916. The election pitted the Democrat Wilson against the Republican Charles Evans Hughes, an associate justice of the Supreme Court. Much of the campaign revolved around neutrality and foreign policy. Both candidates had sizable constituencies of war mongers and neutralists. As a result, the election was extremely close; Wilson finally prevailed. With his victory, even though it was hotly contested, he was finally free of the narrow political considerations that had kept him from devoting his energies to mediating the end of the war, or failing that effort, to begin to prepare the United States for war (Morison, 1965).

Even though Germany had continued to bring as much pressure as possible on the United States to curtail its support for the Allies, Wilson continued to hope against hope that American involvement in open warfare could be avoided. Unfortunately, in his efforts to avoid war in the face of continued German provocation, he appeared weak and uncertain about the future. However, his position did not differ from a majority of the American people. Increasing losses of American shipping to German submarines, plus an effort to establish a German-Mexican alliance against the United States, finally tilted the balance of public opinion toward war. During Easter week in April, 1917, Congress passed a declaration of war on the German Empire. Even though the country was officially at war, months passed before the United States made a significant contribution. Preparedness had begun late, and proceeded by fits and starts.

As the country entered the conflict, Wilson, with no doubt about its outcome, began to worry about the peace to follow. He was deeply concerned that the peace would leave the country with foreign entanglements and treaties that might involve territorial commitments. As his concern mounted, he began to push his concept of a peace agreement in his so-called Fourteen Points. The provisions of this proposal provided for the settling of territorial conflicts in Europe without involving the United States directly in any of the disputes. It also provided for a League of Nations to oversee the process.

The first major American contribution to the Allied war efforts was to put in-place tactics (convoying, destroyer escorts, etc.) that quickly reduced the loss of Allied shipping to German submarines. Other efforts were not as successful. It required eight months after the declaration of war before the first contingent of American troops reached France, and they were not involved in battle for some time after arriving.

During the period, one of the most difficult political problems that Wilson had to deal with was his efforts to get legislation enacted to provide for the conscription of men for military service. Soon after the war began, it became obvious that the number of troops required could not be provided by volunteers alone. There was considerable resistance in the Congress to conscription. Although few, if any, members remembered the conscription riots of the Civil War era, the institutional memory was vivid. Finally, however, wiser counsel prevailed, and the Selective Service Act was finally enacted in May, 1917. Of the 29 divisions that finally participated in action in France, only 11 were formed from draftees.

Over time, the United States gradually increased their commitment of troops to the war effort, but initially they did very little to end the stalemate in the trenches that was slowly bleeding the combatants white. Finally, in early 1918 the Germans decided to end trench warfare by going on the offensive to capture territory that they had lost earlier in the war. This change in tactics finally gave the Americans an opportunity to make their presence felt. On 18 July, the Allies began a counter-attack lead by two American divisions. It completely halted the German offensive, and convinced the German High Command that they were eventually going to lose the war. Of all American troops and units in France, Alabama troops were some of the very best fighters and Alabama units some of the most effect units. Even after that victory, isolated battles continued until the Armistice was signed on November 11, 1918.

With the war slowly grinding to a halt, Wilson had to begin to think about the congressional election of 1918. To accomplish his objectives, he desperately needed continued control of Congress by the Democrats. Unfortunately, for whatever reason(s) the American public did not agree with him. One could argue that the president's liberal policies were acceptable to the electorate in war-time, but not in peace-time. The election returned the Republicans to control. As a result, much of the country and the world saw the loss of the election as a repudiation of the political stature of the president, and when he went to Paris to participate in the peace conference, he went with diminished prestige.

In the peace conference, Wilson tirelessly pushed his League of Nations with its preamble "to promote international cooperation and to achieve international peace and security." Meanwhile, in the Senate, which had to ratify any treaty ending the war and creating the League, there was general consensus among both Republicans and Democrats that some form of a League of Nations was needed, but there were serious reservations with portions of the version that Wilson wanted them to ratify. Unfortunately, just at this point, Wilson's health failed, and he was physically unable to deal with the politics required to obtain Senate approval. Also, probably related to the condition of his health, he was unwilling to accept any changes that they proposed. In March 1920, after two efforts to obtain the necessary three-fourths vote, the Senate returned the proposed Treaty of Versailles to the president with the formal notice of its inability to ratify. Except for his disability, the Democrats would probably have nominated Wilson for a third term in the election of 1920. Instead, they nominated Governor James A. Cox of Ohio. The Republicans nominated Senator Warren Harding, also of Ohio. Harding went on to win in a landslide with 61% of the popular vote.

ALABAMA POLITICS

Alabama state government had spent much of its time in the early part of the 1895-1915 period, in efforts to secure Democratic control of the state's political system through the disenfranchisement of blacks and poor whites. This objective was finally accomplished in the ratification of the repressive 1901 Constitution. Unfortunately, positive action on the state's pressing social and economic issues was not prosecuted with the same diligence. There were five primary areas of need that had remained unresolved for decades: railroad regulation, child labor, convict leasing, public health, and public education. Then in 1911, a sixth was added—mine safety—when an explosion at the Banner Mine in Jefferson County left 128 miners dead, most of them prisoners working in the mine under the convict leasing system. Although these needs surfaced in every session of the legislature, little had been accomplished in any of these areas when Governor O'Neal left office in 1915.

Governors

Governors of Alabama during the period 1915-1920: Edmund O'Neal (D, Lauderdale County) (1911-1915); Charles Henderson (D, Pike) (1915-1919); and Thomas Kilby (D, Calhoun) (1919-1923).

Commissioners of Agriculture

Commissioners of Agriculture during the period 1915-1920: James Wade (1915-1919) and M.C. Allgood (1919-1923).

The Political Situation

Former Governor Comer and Charles Henderson, president of the Railroad Commission, survived the first primary election held by the Democrats in 1914. Then following a blistering campaign, much of it swirling around the question of Prohibition, Henderson was elected governor by a margin of slightly more than 10,000 votes (Rogers and Ward, 1994). Henderson's first year in office was chaotic. The state's financial condition bordered on disaster, and the cotton market was seriously disrupted at the beginning of the war. The governor had opposed Prohibition, but soon after he entered office the legislature enacted a law establishing it throughout the state. Later, the legislature passed a new primary election law that defined how [white male] voters would conduct some of their political affairs and a constitutional amendment allowing counties to levy a special school tax.

The war in Europe had begun in 1914, and although President Wilson had worked diligently to keep the United States out of it, over time the country was drawn ever closer to it. Finally, in April, 1917, the United States declared war on the German Empire. With this declaration, Alabama's politics had to be re-focused on matters of greater importance than child labor and convict leasing. Later, some 74,000 young Alabamians were inducted under the Selective Service Act, and many more would serve as part of the Alabama National Guard.

While Alabama's young men drilled, fought, and died, ordinary citizens in the state were experiencing some shortages while benefitting hugely from wartime prosperity. The state's iron and steel industries were special beneficiaries. During the war, the government decided to build a nitrate plant at Mussel Shoals in northwest Alabama to support the country's munitions industry. Millions of dollars poured into that area, but it was not completed before the end of the war.

A relatively small number of young Alabama men had been minimally involved in the Spanish-American War at the end of the nineteenth century, but in World War I thousands of them were rudely forced out of their largely rural lives onto a world stage, which they could not imagine really existed. Their experiences in the greatest war that the world had ever experienced changed all of them, and they were never the same when they returned to Alabama. The old song: "How you gonna' keep 'em down on the farm, after they have seen Paree" more than adequately describes the effect the war had on those young men who returned to the state in 1918. The war also changed the entire country in a few short years in ways never before experienced, and these changes reverberated into every nook and cranny of the largest cities and the smallest hamlets. Alabamians had thrown themselves wholeheartedly into support for the war effort, and when the Armistice was signed in November, 1918, wild demonstrations and celebrations exploded across the state.

Near the end of his administration, Governor Henderson had requested that the Russell Sage Foundation in New York conduct an in-depth study of the social and economic condition of the state (Flynt, 1994). The report was submitted to the governor in December, 1918, shortly after Thomas Kilby from Calhoun County was elected to replace Henderson. The report described in fine detail a grim picture of neglect and inequity.

Kilby had campaigned on a progressive platform that called for reformation of state business practices, improvement of education, revision of existing tax laws, enforcement of Prohibition, abolishing convict leasing, and enacting workers' compensation legislation. In the election, he was pitted against the deeply conservative William Brandon from Tuscaloosa County. Kilby won by less than 10,000 votes. In doing so, he won the eastern and northern counties, but lost in the Black Belt and the southwestern counties.

The Sage Report had provided Kilby with much of his campaign platform. Unfortunately, before he could get at the task of fulfilling his campaign promises, he and the legislature had to deal with the pressing problems of ratifying the 18th (Prohibition) and 19th (Woman's Suffrage) amendments to the US Constitution. Once these two matters were behind him, he began to push his legislative agenda. Kilby was an extremely effective administrator, and he had financed his own political campaign; consequently, he owed nothing to special interests in the state. He was, from the beginning, an extremely effective governor, but at the same time the legislature had changed. It seemed that the war had brought a sense of urgency to everyone to get busy, to solve Alabama's many pressing problems. The legislature enacted into law virtually everything that the governor proposed. His first initiatives dealt with the basic problem of state finances. The Sage Foundation had found that the lack of funding for state government was at the root of many of the most vexing problems. Kilby proposed that taxes be equalized, that a graduated state income tax be approved, and that a severance tax be levied on coal and iron ore mined in the state. Although the Alabama Supreme Court struck down his income tax legislation, the state's financial situation was in much better condition. With the basic problem of financing state government mostly solved, Governor Kilby turned his attention to his other promises. He quickly moved on to convict leasing, the twin problem of prison construction and institutions to care for the insane and feeble-minded, child welfare (he created the Child Welfare Department), education (he created the State Board of Education), funding for public health, workers' compensation, and road construction. Progressive legislation literally poured off Goat Hill in Montgomery. To someone who had watched the political machinations of state government in the early years of the twentieth century, this new spirit of progressiveness would have appeared to them like a wild dream.

Without a doubt, the Kilby administration was the most successful in the history of the state to that time; however, amid his and the legislature's many accomplishments, two failures stood out: They were unable to deal effectively and fairly with the problems of convict leasing and with growing unionism among Alabama's industrial workers.

While a considerable amount of progressive legislation was enacted during this period, there was no practical way that the political process could reach the thousands of Alabama tenant farmers mired in poverty. Better cotton prices during the war years helped some, but their farms were just too small. Even with better prices, their harvests were just too small to make much difference in their lives. Also, most of them had little voice in politics. The 1901 Constitution had effectively taken it away. The politicians really did not have to worry about them; there were few votes out there in the tenant shacks and cotton patches.

THE NATIONAL ECONOMY

After the banking panic in 1907 and 1908, general prosperity returned to the United States in 1909. Although there were continuing problems with the cost of living versus real wages. World War I began in Europe in the summer of 1914. The American economy did not immediately respond to the hostilities. It was still preoccupied with structural problems related to the recent panic. Finally, the demand-pull of the war effort began to exert enough force to overcome those problems, and the economy quickly expanded to supply materials, equipment, and food that the European combatants were no longer able to provide for themselves. What had been an economic slowdown in 1914 and 1915 quickly became a boom. In the spring of 1917, the United States formally entered the war. After the declaration of war, the American economy not only had to expand to meet the wartime needs of our own war efforts, but it also had to expand still further to help meet Allied needs. As a result, the economic boom became even stronger. Unfortunately, the American economy did not get a prolonged postwar bounce. As detailed above, there had been some structural problems in the economy before the beginning of hostilities, and they quickly slowed everything down afterwards. In 1920, the economy entered a sharp, but short recession.

The Gross National Product

The Gross National Product did not immediately responded to wartime economic conditions for the reasons noted in the preceding paragraph. It had been at a level of 462.4 in 1913, but by 1914 it had declined to 444.4 (Figure 10.1). It continued to decline in 1915 (439.6). The United States was not officially at war until 1917; however, in 1916, the GNP began to respond to the devastation in Europe and the efforts of the United States to come to the aid of Britain and France. In 1917 when the United States finally declared war, the GNP had recovered some of its losses and had reached 481.7. In 1918, the year of the Armistice, the GNP exploded upward to 570.0. The problems, however, that had been there in 1914 and 1915, quickly re-appeared, and the GNP began to decline. In 1919, as the economy headed toward recession, it was back down to 528.3, and in 1920, the first full year of the recession, it declined further to 487.1.

The Consumer Price Index

The US Bureau of Labor Statistics initiated publication of date on the Consumer Price Index (CPI) in 1913, but there are reliable estimates of the annual indices available from earlier years. These earlier estimates and the bureau's data are presented in Figure 10.2. The CPI was 29.7 (1967 = 100) in 1913, 30.1 in 1914, and 30.4 in 1915. However, when the American economy began to fully respond to the war in Europe, prices roared upward, and by 1918 the CPI was at 45.1. Unfortunately, the contraction of the economy following the end of the war (Figure 10.1) was not accompanied by stabilizing prices. The CPI continued to grow by leaps and bounds. It was at 51.8 in 1919 and at 60 in 1920, the first full year of the 1920-1921 recession.

The Inflation Rate

The inflation rate was slightly above 1% in the period 1913 and 1914 (Figure 10.3); however, it increased sharply to near 10% in 1915 and 1916, but then was up even more to around 15 to 17% in the period 1917 to 1920. In fact, in 1920, the first year of the 1920-1921 recession, the inflation rate remained high (15.8%). This high Inflation rate with a decreasing GNP was an early example of stagflation in the American economy.

Interest Rates

The government did not report interest rates related to the American economy until 1934.

THE AGRICULTURAL ECONOMY

Genung (1940) has provided an excellent account of the effects of World War I on American agriculture. He commented that at the beginning of the period 1914-1920, American agriculture was still in a long period of quiet adjustment. There was little additional virgin land to exploit, and the farm economy was attempting to learn to deal with an influx of new labor-saving machines and the increasing commercialization of agriculture. Then within a six-year period, hostilities in Europe and America's response brought about changes on American farms that normally would require generations.

The official policy of the American government during the war was to stimulate food and fiber production, while restricting price gouging to the extent practical. August 10, 1917, President Wilson created the US Food Administration by executive order, and Herbert Hoover was made administrator. The purpose of the organization was to assure the supply, distribution, and conservation of food during the war; to facilitate the movement of foods; and to prevent monopolies and hoarding. These objectives were to be attained by voluntary agreements and a licensing system. The Grain Corporation was established at the same time. The primary functions of this agency were to regulate grain trade by purchasing, storing, and selling grain and grain products, and to control grain exports and imports. The corporation was re-organized in 1919. In August 1917, Congress passed the Food and Fuel Control Act. This Act gave the president extensive powers to manage the production and distribution of foods and fuel vital to the war effort.

Unfortunately, while it was possible to restrict explosive price increases, at least on some commodities, it was not possible to stop inflation, and it quickly became rampant in the country and throughout the world. Land prices soared, and all production costs increased. Costs for hired labor, fertilizer, farm implements, livestock feeds, taxes, interest, and freight rates increased sharply. And to add to the farmer's problems, the cost of living also grew rapidly (Figure 10.2).

Shortly after the war began, production, exports, and prices of most American agricultural commodities quickly began to increase. Only cotton, among the major crops, did not fare as well, at least early in the conflict. Corn hardly responded at all.

The response on Alabama farms to the wartime emergency was somewhat different. Only hog production increased significantly. Cattle numbers increased as a result of the Expansion Phase of an on-going cattle cycle, but the increase was not greater than normally expected. Cotton production during the period was actually significantly lower than it had been in the early 1900s. There was little change, if any, in corn production.

Late in 1920, the air quickly escaped from the wartime agricultural economy, and commodity prices (prices received) fell even more rapidly than they had increased at the beginning of hostilities. Unfortunately, prices paid and the cost of living did not decline to the same extent. American farmers quickly became caught in the worst price-cost squeeze in history.

For decades, interest in bringing the federal government into the agricultural marketplace had been simmering in the United States. Farmers felt that since the government had actively urged them to get in the wartime mess they were in that it should play an active role in getting them out. The collapse of agricultural prices in 1920 lit the fuse that would result in the explosion of interventionist legislation a decade later.

Federal Agricultural Legislation

With Democrat Wilson in the White House and the same party in control of both houses of Congress, a Federal Farm Loan Act was enacted in 1916. Securing adequate credit had been a problem for Southern farmers since colonial times. There had never been a surplus of money actively seeking borrowers in Southern agriculture. Further, loans made to farmers were extremely high risk ventures, and as a result, when they were available, interest rates were notoriously high and short term. A commission appointed by President Theodore Roosevelt in 1908, identified credit as one of the more serious problems facing the predominantly rural population of the time. Their report led to numerous studies over the next several years that included studies of the agricultural credit systems used by other countries. The 1916 Act resulted from these studies and the serious debate associated with them. The basic aspects of the act were based on the German Landschaft system that had been in place since 1769. The act, as approved by President Woodrow Wilson (1913-1921), was designed to establish a cooperative credit structure using 12 Federal Land Banks (FLBs) located in key cities across the country. Congress appropriated $125 million seed money to establish the banks, but they were to be operated primarily on capital provided by private investors. Its primary purpose was to make long-term, low-interest loans to farmers, using land or crops as security.

Congress enacted the Food and Fuel Control Act (Lever Act) in 1917. Its purpose was to give the president power to manage the production and distribution of foods and fuels vital to the war effort.

In 1919, Congress enacted legislation authorizing the establishment of the US Census of Agriculture. It provided that the first census under this act would be conducted in 1925 and at five-year intervals thereafter..

Alabama Agricultural Legislation

In 1915, the Alabama legislature enacted legislation requiring the commissioner of agriculture to supervise the farm produce business in the state. It also required him to work to encourage the immigration of people from other states to Alabama. To handle these responsibilities, a Division of Immigration and Markets was established. To assist farmers with the orderly marketing of their produce and products, the newly-established division began to publish the Alabama Markets Journal in 1916.

The problem of the eradication of tick-fever in cattle was discussed in the preceding chapter. It was extremely difficult to eradicate the disease as long as there were free-ranging cattle. The need for increased beef production in the country during World War I gave the State Livestock Sanitary Board the support it needed to move forward on eradication. In 1917, the board ruled that no tick-infested cattle could be shipped within the state or out of the state. In 1919, the legislature provided legislative support for this declaration. The act required that all counties construct enough vats and provide enough chemicals to make dipping practical. It also required that "all" cattle be dipped at two-week intervals and provided for the collection of fines from farmers refusing to do so. At first, there was widespread resistance to dipping. Many refused to comply, and in a few cases, the vats were destroyed with dynamite. But those in opposition were holding a losing hand There was no reason to own cattle if they could not be sold, and they could not be sold if they could not be transported.

US Agricultural Exports

American agricultural exports began to increase slowly after 1910 (Figure 10.4), as economic conditions improved in the United States and Europe following the bank panic of 1907-1908. Exports were at a level of $1,029 million in 1910, and they had increased to $1,112 million in 1913. World War I began in Europe in 1914, and exports responded by growing to $1,474 million. Then responding to the growing need of the Allies for food and fiber, exports exploded upward to $3,579 million in 1918, the year the war ended. The growth continued in 1919 ($3,850 million), although at a much reduced rate. In 1920, the bounce from the war had ended, and with the short recession of 1920 and 1921 underway, exports declined sharply to $2,606 million.

Agricultural Prices

The indices of both prices received and prices paid by farmers remained relative unchanged at around 100 during the period 1910-1915 (the Golden Age of Agriculture) (Figure 10.5). Then in 1916, both increased sharply as a result of the growing effect of World War I on all aspects of the American economy. In 1917 when the United States entered the war, the indices were nearly the same (178 versus 182). Then in 1918 and 1919, farmers at last got the long-end of the stick. In those two years, the prices received indices were greater than prices paid indices (206 versus 188 in 1918, and 217 versus 199 in

1919). Unfortunately, farmers lost their hold on the stick in 1920— at the beginning of the short recession when the prices received and prices paid indices were 211 and 241, respectively. In 1920, inflation was continuing to push prices paid upward. Unfortunately, farmers were not able or willing to reduce production rapidly, so in 1920 supply began to exceed postwar demand, and the prices farmers received for all of their commodities began to decline.

The parity ratios of the prices received indices to the prices paid indices for the period 1910-1920 are shown in (Figure 10.6). For example, in 1915 the prices received and prices paid indices were 99 and 101, respectively; consequently the parity ratio was 0.98 (99/101). These ratios show the relationship between prices received and prices paid more clearly. The ratio was above 1 in 1914, but slightly below in 1915. Then in 1916, it fell to 0.86. Ratios were near or above 1 in the period 1917-1919, but in 1920 at the beginning of the short recession it was down to 0.88. Any time the ratio falls below 1.0, farmers are in a price-cost squeeze, and the farther it falls below 1.0, the more desperate their situation becomes.

Farmers

The number of American farmers was first estimated in 1840, and from that time onward—while the number of farmers continued to increase—it was not keeping pace with the population increase in the country. For example, in 1890 using an estimate of the farm population, it was determined that farmers constituted 39.4% of the total population. Then in 1900, 1910, and 1915, it was 39.3, 34.9, and 32.4%, respectively. Obviously, with the percentage of farmers decreasing, they must have been getting much more efficient in producing food and fiber.

The Total Population

Data on the total American population (all persons) are shown in Figure. 10.7 for the period 1895-1920. In the period from 1870 through 1910, it had increased from 38.6 to 92 million. From 1870 to 1880, the population increased 2.7% a year. Between 1910 and 1920, it increased from 92 to 105.7 million, or at a rate of 1.3% a year.

Data on the population of Alabama during the period 1895-1920 are presented in Figure 10.8. These data show increases in population throughout the period; however, the increase between 1915 and 1920 was very limited (2.34 to 2.36 million). It is likely that this slow rate of growth was the result of large numbers of people leaving the state for war-work in the North.

The Farm Population

Annual data on the number of farmers in the American population in the period 1910-1920 are presented in Figure 10.9. Numbers had been increasing from the earliest years of the United States, and at least for a portion of the period 1915-1920, it continued to increase. It was at a level of 32.1 million in 1910. It reached 32.4 million in 1915, and 35.3 million in 1916, but afterwards, it began to decline sharply. Farmer numbers in the country reached the highest level in history in 1916. The loss of farmers between 1916 and 1919 was not great (136,000), but the upward trend had been broken. Apparently, the number of off-farm jobs increased sharply during the war years, and large numbers of marginal operators accepted those opportunities. After the wartime emergency ended, many war-related jobs were terminated, and farmer numbers increased to 32 million in 1920.

With the total population continuing to grow (Figure 10.7), and with farmer numbers growing at a much slower rate, and finally beginning to decline after 1916, the percentage of farmers in the population also continued to decline (Figure 10.10). Percentages of farmers in the American population were at a levels of 34.9, 32.4, and 30.1% in 1910, 1915, and 1920, respectively.

Fite (1984) reported that there were 95,000 black farmers in Alabama in 1920, There had been 110,000 in 1910. He further reported that total farmers in the state numbered 1,355,000 in 1920. The outbreak of World War I in 1914 might have resulted in the Cotton Exchange shutdown in the South, but it brought about the further expansion of industry in the North. Along with a rapid growth in industrial output came a shortage of labor. Normally these industrial jobs would have been filled by European emigrants; however, because of the war emigrants were no longer available. As a result, Northern labor recruiters began to look at the surplus agricultural labor in the South as a source of workers. By 1916, a flood of black farmers—attracted by better wages, schools, and housing, as well as by less racial discrimination—began to immigrate to Northern industrial cities. Alarmed at the loss of so many laborers, Southerners first attempted to slow the flow with legal restrictions on recruiting agents, and in a few cases by arresting workers before they could board their trains. Finally, cooler heads prevailed, and efforts were made to treat their black workers better and to increase wages. None of these remedies worked very well, and thousands of black laborers left the South.

The immigration of black farmers did little to reduce the excessively large numbers of farmers in the South. There were still too many farmers for the amount of land available, and these poverty-ridden farmers lacked essentially all of the inputs (sufficient land, management skills, capital, and markets for crops other than cotton) required to become prosperous and self-sufficient. They were caught in a net from which there was no escape, short of death or some outside influence to completely alter the system.

Farmers in the Work Force

Data on the percentage of farmers in the national labor force during the period 1870-1920 are presented in Figure10.11. These data indicate that the percentage declined from 31% in 1910 to 27% in 1920. This decline had begun much earlier, but it was first detected in data collected in 1840 and 1850. Improving agricultural technology seemed to have a life of its own. It grew inexorably, and as it grew relatively fewer farmers were required to provide the food and fiber required by a growing population.

Census data for Alabama indicate that the average number of wage earners in manufacturing increased from 52,902 in 1900 to 107,159 in 1920, an increase of 103%. During the same period, the population of the state increased only 29%. These data indicate the general degree of movement of people off the farms, and even the abandonment of farms and farming, as more manufacturing jobs became available after 1900. The wartime boom hastened this process.

Tenancy

Census data indicate that tenancy had been increasing in the country since the first estimates were made in 1810. In 1900, 35.3% of American farms were operated by tenants (Figure 10.12). Obviously, this is a misleading statistic. The percentage for the country is biased upward by the extremely high percentages in the South. It is highly likely that the percentage in the North Central Region was much lower than 35%.

Data on tenancy in Alabama are also presented in Figure 10.12. They indicate that 60.2% of Alabama farms in 1910 were operated by tenants. A decade later, the percentage had declined to 57.9. In 1910 and 1920, only Georgia, Mississippi, and Texas had more tenants than Alabama. In 1920, 17.5% of all the Southern tenants were share croppers.

Farms

Numbers of American farms had been increasing from the years of the early colonial settlements, and this trend continued into the early years of the twentieth century. There were 1.4 million farms in the country in 1850, and by 1900 this number had increased to 5.7 million. Much of this increase was the result of people continuing to move onto the Central Lowlands and the Great Plains to establish farms. In 1850, the largest share (33.8%) of farms was in the North Atlantic Region, and by 1900 the largest share (38.3%) was in the North Central Region. In 1900, it seemed that there was no upper limit to the number of farms in the country.

In 1915, 49% of all farms in the country was in the South (South Atlantic and South Central regions), and 34% was in the North Central Region. In that year, 6.6%, of all farms was in Texas. Other states with large shares of the total (6.458 million) were: Georgia (4.7%), Mississippi (4.2), Missouri (4.2), and Ohio (4.2). These five states accounted for 23.8% of the total.

Farm Numbers

As detailed in the preceding paragraph, American farm numbers had been growing steadily for many years (Figure 10.13), and by 1910 they had reached 6.362 million. Census data indicate that numbers continued to increase through the war years, finally reaching 6.448 million in 1920. Annual data published by NASS provide estimates of numbers during the period 1915-1920. These data show that numbers increased steadily from 6.458 million in 1915 to 6.518 million in 1920.

Farm numbers in Alabama followed the national trend upward through 1910 to a level of 265,000 (Figure 10.14). Census data suggests, however, that by 1920 they had declined to 256,000. Annual data published by NASS provides estimates of numbers between 1915 and 1920. The NASS estimates also show that numbers in Alabama began to decline after 1910, and that in 1915 they reached a level of 261,000 farms. It is likely that this situation was repeated in most of the Southern states.

Farm Size

The average size of American farms had fallen sharply between 1870 (153 acres) and 1880 (134 acres), primarily as a result of even larger declines in the South; however, afterwards it ranged between 134 and 148 acres through 1920 (Figure 10.15). The continued growth in numbers of American farms during the same period (Figure 10.13) seemed to have little effect on size.

In contrast to the lack of a real trend in American farm size in the period 1870-1910, in Alabama it trended steadily downward from 222 to 79 acres (Figure 10.15). Then it changed relatively little by 1920 (76 acres). As numbers of farms increased in the state (Figure 10.14), average size decreased. It is likely that this reduction was still the result of dividing larger old farms into smaller new ones. In Alabama in the early part of the twentieth century, about the only way that a new farm could be developed was to subdivide an older one. At that time, there were already more acres of land in farms than there were acres suitable for farming.

Between 1910 and 1920, with the numbers of Alabama farms declining rapidly, size changed very little. Apparently when farmers abandoned their farms for wartime work, their land was not incorporated into other farms. Probably many of them expected to return after the war.

Total Farmland

With American farm size remaining relatively unchanged (Figure 10.15) during the period 1870-1900 and farm numbers trending upward (Figure 10.13), farmland also increased (Figure10.16). It was at a level of 407.7 million acres in 1870, and 878.8 million in 1910. With both farm numbers and size continuing to increase, in the next decade farmland increased sharply to 955.9 million acres in 1920. In 1910, farmland constituted 46.2% of the total land area of the United States. With the increase in the next decade, by 1920 it constituted 50.2%. It appeared at that time the entire country would soon be covered with farms.

The Alabama farmland situation was quite different. Farm numbers declined after 1910 (Figure 10.14), and farm size had been declining at least since 1870 (Figure 10.15). As a result, total farmland had stopped increasing sometime between 1890 and 1900 (Figure 10.17). It changed very little between 1900 and 1910, and then fell slightly over a million acres from 1910 to 1920 (20.7 to 19.6 million). In 1900, when it was at or near its highest level in history, Alabama farmland constituted 63.7% of the state's entire land area. With farmland declining afterwards, the percentage of the total land area in farms declined. In 1920, it reached 60.3%. Ironically, only 59% of the land in Alabama was suitable for crop production (Table 3.24), and a good share of that was marginal.

Use of Farmland

Data (percentages) on the uses of American farmland during the period 1899-1919 are presented in Figure 10.18. These data indicate that the percentage of farmland devoted to cropland slowly increased from 38% in 1900 to 42% in 1920. The sharp increase between 1909 and 1919 was probably the result of the response of farmers to the government's request that they produce more food and fiber for the war effort.

This increase in cropland apparently came from the decline in forestland, which declined from 23 to 18% during the period. The percentage used for pasture remained at around 34% throughout the period. The percentage of "other" land also remained unchanged. There are no data on the uses of Alabama farmland before 1924.

Cropland Harvested

Estimates from the 1910 Census indicate that nationally crops were harvested from 311.3 million acres in 1909, and 348.6 million in 1919 (Figure 10.19). These estimates translate into some 49 acres of cropland harvested per farm in 1909, and 54 acres in 1919. These national estimates are relatively low, because they included large numbers of small farms in the North Atlantic, South Atlantic, and South Central regions. Cropland harvested per farm was much higher in the North Central Region.

In 1899, 1909, and 1919, Alabama farmers harvested crops from an average of 30, 27, and 28 acres, respectively. These estimates were very close to the average of cropland harvested for all the Southern states in those years. In contrast, Iowa farmers harvested crops from 96.2, 93.9, and 95.7 acres per farm in those three years, and estimates for the three years were considerably higher in Missouri than in Iowa.

There is probably no better measure than cropland harvested per farm that describes the level of utter hopelessness of Alabama and Southern agriculture at the beginning of the twentieth century. At the end of World War I, there seemed to be no light at the end of the tunnel.

Value of Land and Buildings

Census data on the value of land and buildings in the United States during the period 1870-1920 are presented in Figure 10.20. The 1915 data were not obtained from a census. Rather, it was taken from Lerner (1975). The data show the sharp increase in values during the period 1915-1920 ($39.6 to $66.3 billion), resulting from economic inflation during the war years (Figure 10.3). Value increased 67% during the period, while the Consumer Price Index increased by 97% (30.4 to 60.0) (Figure10.2). Unfortunately, the increase in value was less than the inflation of the currency during that period.

Data on the annual average value of land and buildings in the United States on an acre basis for the period 1870-1920 are presented in Figure 10.21. The larger share of these values is the value of land. Censuses estimated that the average value for 1900 was $20 an acre, and $40 in 1910. In 1920, after the end of the war average values reached a level of $69 an acre.

Between 1915 and 1920, as American farmers increased their food and fiber production in response to wartime demand, they added to their landholdings (Figure 10.16) and probably to the number of buildings on their farms, as well. However, the large year-to-year changes were primarily the result of increased prices. Also, as shown in Figure 10.15, the average size of American farms increased from 142 to 149 acres from 1915 to 1920, so they did purchase more land, but not enough to cause values to increase the way that it did. Obviously, American farmers did purchase a small amount of additional land during the war years, but they apparently paid a dear price for it.

US census data for Alabama for 1900, 1910, and 1920 indicate that total values of farm real estate were $134.6, $288.2, and $543.7 million in 1900, 1910, and 1920, respectively (Figure 10.22). They generally show the same pattern of increase as seen in the national data (Figure 10.20). Alabama data computed on an acre basis are also presented in Figure 10.21. These data indicate that per acre values in Alabama were well less than half of those in the country as a whole. This difference is probably a reflection of the fact that land was much cheaper in Alabama and that farmers in the state had fewer and less costly buildings on their farms.

Comparative data also show that the average value per farm for Alabama for 1910 and 1920 was the lowest of any of the Southern states (Fite,1984). The average value per farm in Alabama in those two years was $1,096 and $2,123, respectively. In Georgia, they were $1,647 and $3,663, and in Mississippi, $1,218 and $2,903. In contrast, in Iowa they were $15,008 and $36,616. As should be expected, differences in these values for the different states are to some extent related to the average size of farms in those states.

Value of Implements and Machinery

Annual value of implements and machinery on American farms during the period 1870-1920 are presented in Figure 10.22. Value in 1910 was $1.3 billion. It had been $750 million in 1900. Data from Lerner (1975) indicate that it was $1.8 billion in 1915. Then by 1920, value increased almost 100% to $3.6 billion. However, the Consumer Price Index increased by 97%, during the period 1915-1920. This comparison suggests that most of the increase in values during the war years was the result of the inflation of the currency.

Data on value per farm provide another perspective on changes in the value of implements and machinery during this period. These data are presented in Figure 10.23. They indicate that it increased from $131 in 1900, to $199 in 1910, and to $557 in 1920.

Decennial census data indicate that value of implements and machinery in Alabama increased from $16.3 million in 1910 to $34.4 million in 1920. It had been $8.7 million in 1900. Between 1910 and 1920, value increased by 111%. In the same period, the Consumer Price Index increased by 114%. Obviously most of the increase in value was the result of inflation.

Data on value per farm for the period 1870-1920 are presented for both the United States and Alabama in Figure 10.24. They show that in Alabama value increased from $63 in 1910 to $134 in 1920. Value per farm was much lower in Alabama than in the country as a whole, but farm size was also much lower (Figure 10.15).

Animal Power on Farms

For decades during the latter part of the nineteenth century, numbers of horses and mules had been increasing in the country; as the number of farms increased, the demand for on-farm power increased. There had been 17.8 million horses in the United States in 1895. Then numbers increased slowly, reaching 21.4 million in 1915, the highest level in history. It slowly declined to 20.1 million in 1920 (Figure 10.26). As will be discussed in a following section, the number of tractors in the country increased from 25,000 in 1915 to 246,000 in 1920. Apparently, this sharp increase in the numbers of these machines was sufficient to bring about the beginning of a long downward trend in the number of horses, which had once been essential sources of farm power.

In 1900, there were around 312 horses for each 100 farms in the country (Figure 10.27). This ratio increased to 332 in 1915, before declining to 308 in 1920. It is interesting that horses in large numbers were one of the first of the products of American agriculture to be purchased by English and French agents after the beginning of World War I. These animals were required in large numbers to pull guns and supply wagons. Altogether, the United States exported 1.5 million of these animals during World War I. There is no indication that any effort was made to increase the production of these animals. Apparently, exports were furnished from the culling of existing stocks (Genung, 1940).

There had been 132,000 horses in Alabama in 1895 (Figure 10.28). Numbers trended upward to 150,000 in 1901, the highest level in history, then trended downward through 1910, before trending upward again until 1915. It began to decline again in 1916, and reached a level of 130,000 in 1920. Horses were never as popular in Alabama as in the country as a whole. Data on numbers of horses per 100 Alabama farms for the period 1870-1920 are also presented in Figure 10.27. The ratio generally declined from around 125 in 1870, to 50 in 1910, and it remained near that level through 1920. During the period 1910-1920, there were about fivefold as many horses per 100 acres on farms in the country as a whole than in Alabama.

Mules had never been as popular in the country as horses. In 1895, there were almost sevenfold (17.8 versus 2.7 million) as many horses as mules in the country (Figure 10.26). However, after 1895, numbers of mules began to trend slowly upward to reach a level of slightly more than 5.6 million animals in 1920. This growth was likely the result of the rapid increase in numbers of these animals on farms in the South. With numbers of mules slowly trending upward on American farms (Figure 10.26), numbers per 100 farms also increased (Figure 10.29). The ratio was 55, 66, 78, and 87 in 1900, 1910, 1915, and 1920, respectively.

Numbers of mules in Alabama had by-passed numbers of horses around 1878. From that time onward, numbers of mules increased rapidly, reaching levels of 190,000, 244,000, 281,000, and 296,000 in 1900, 1910, 1915, and 1920, respectively. With numbers trending rapidly upward, the ratio of mules per 100 farms also increased (Figure 10.29). It was 85 in 1900, but then increased to 92, 108, and 116 in 1910, 1915, and 1920, respectively. During most of this period (1910-1920), there were around one-third more mules per 100 farms in Alabama than elsewhere in the country.

Mules were the obvious choice of power on those tiny tenant farms in Alabama and the South, and as farm numbers increased (Figure 10.14), mule numbers did also. It is interesting that numbers of mule did not decline with the decline in farm numbers in the state between 1910 and 1920. However, while numbers did not actually decline, the rate of increase slowed considerably.

Tractors on Farms

The appearance of tractors on American farms early in the twentieth century was probably the most important technological change in the history of American agriculture. In the period 1915-1920, the number continued to increase rapidly. There had been 25,000 of the machines on American farms in 1915, and by 1920 the number had increased to 246,000. In 1915, there was one tractor for each 258 farms. By 1920, this ratio had changed to one tractor on each 26 farms. There are no data on the number of tractors in Alabama until the publication of the 1920 Census. Data in that census indicate that there were 811 machines at that time on 256,000 farms, or approximately three on each 1,000 farms.

Use of Fertilizers

After the results of research on the correct use of commercial fertilizer began to reach farmers in the early 1900s, they began to increase the amount of the materials that they applied to their crops. Between 1899 and 1909, fertilizer use in the United States increased from 18.4 pounds to 31 pounds an acre of harvested cropland, an increase of 68%. They continued to increase usage of the materials in the period 1915-1920. In 1915, they applied 10,836 million pounds, and 14,352 million pounds in 1920. The only estimate of harvested cropland made during the period was in 1919 when it was estimated that farmers harvested crops from some 348.6 million acres. That year, farmers used 13,502 million pounds of fertilizer on their harvested cropland, or 38.7 pounds an acre. This estimate represents an increase of 24.8% over the 1909 rate of application.

Plant Agriculture

At the end of the Golden Age of American Agriculture (1915), acres harvested of corn, wheat, oats, and cotton in the United States were: 100.6, 60.3, 38.8, and 30 million acres, respectively. Acres of combined corn and oats harvested totaled 139.6 million acres. It is interesting that more than twice as many acres of feed crops were harvested than of food grain (wheat). Also in 1915, value added for all crops accounted for 49% of value added for all crops and livestock combined.

The value of crops added $3.6 billion to the national economy in 1910. By 1915, with the war in Europe just beginning, value added for crops increased to around $3.9 billion,

In terms of the amount of land dedicated to the production of the major crops, corn was much more important. In fact, in both 1915 and 1920, acres of corn harvested (100.6 million and 85.5 million) was more that twice as great as its three closest competitors: cotton, wheat, and oats.

Alabama Crop Weather

Alabama weather in June, July, and August was relatively wet during the period 1915-1920. Palmer Drought Severity Indices (PDSI) (Cook, et al., 1999) for those months and years indicate that it was wetter than normal in more than half of the state, in five of the six years (1915, 1916, 1917, 1919. and 1920). It was drier than normal in more than half of the State in only one year (1918).

Cotton

Bales of cotton harvested annually in the United States increased steadily from around four million in 1870 to around 14 million in 1915, an increase of 250%. In the same period, the American population increased from 38.6 to 100.8 million, an increase of 161%. Obviously, cotton production was growing more rapidly than the population. Historically, American farmers had produced more of the fiber than could be used by American mills; consequently, a substantial amount of it had always been exported. However, even with the opportunity for selling some of the crop overseas, supply always seemed to exceed demand, and for decades prices received remained mired at around 10 cents a pound.

US Cotton

In the preceding period (1895-1915), bales harvested had trended upward from nine million to 14 million. During the same period, prices had increased from around six cents a pound to near 12 cents. In 1915, American cotton farmers harvested 11.2 million bales. Farmers in Texas harvested the largest share of the total (28.9%). Other leading producers were: Georgia (17.1%), South Carolina (10.1), Alabama (9.1), and Mississippi (8.5). Altogether, farmers in these five states harvested 73.8% of the total crop.

Exports

Exports of American Cotton in 1915 were equivalent to 37.9% of the 1914 crop (6.1 million of 16.1 million bales). Cotton exports declined sharply in 1915 following the Cotton Exchange Scare of 1914 (Figure 10.30). They had been at a level of 8.7 million bales in 1914, and in 1915 they were down to 6.1 million. The scare was not the only factor leading to the reduction. By 1915, the United States had lost the important German market. Also, World War I, although largely confined to western Europe, affected worldwide consumption of cotton. Exports continued to decline through 1917 (4.4 million bales). They had not been that low since 1892. They recovered somewhat in 1918 and 1919 (6.7 million), but declined again in 1920 (6 million) when the short recession began. At no time in this period (1915-1920) did exports reach the level of those of the prewar years. The war had provided a boost for exports of most American crops, but not for cotton.

Prices

American cotton farmers harvested 16.1 million bales in 1914 (Figure 10.34), the largest in history at that time. That harvest, coupled with the Cotton Exchange Scare and rapidly declining exports (Figure 10.31), sent cotton prices rapidly downward. Farmers received an average of 7.4 cents a pound for their cotton that year, the lowest since 1898. As a result of that low price, only about 30.5 million acres were planted in 1915 (Figure 10.32). Farmers had planted slightly over 36 million acres the year before. With a smaller crop in 1915 (12 million bales), prices recovered somewhat to an average of 11.2 cents. Then with war-driven inflation (Figure 10.3) and American industrial production roaring ahead, cotton prices joined the parade even though exports continued to lag. By 1919, with exports finally increasing, prices reached a level of 35.4 cents, the highest ever recorded by USDA at that time. From 1915 through 1919, prices had increased 216% (11.2 to 35.4 cents). During the same period, the CPI increased only 70% (30.4 to 51.9). Then in 1920, the combination of the beginning of the short recession of 1920 and 1921, a larger than average crop (13.4 million bales), and declining exports brought the golden age of cotton prices to an end. In that year, they returned to a level of 15.9 cents.

Acres Planted and Harvested

In 1909, USDA began to collect data on acres planted on most crops. The data provide a much clearer picture of farmer response to market signals. Data on both acres of corn planted and acres harvested are presented in Figure 10.32. Rapidly falling prices in 1914 (Figure 10.32) finally got the attention of American cotton farmers. In 1915, they reduced acres planted by slightly more than 5.6 million acres (36.2 to 30.5 million). Never in history, to that time, had cotton farmers responded so violently to low prices. However, prices recovered to just above 11.2 cents in 1915, and farmers planted another large crop (34 million acres) in 1916. As detailed in a preceding paragraph, prices continued to spiral upward through 1919 (35.4 cents), and acres planted generally followed.

Hostilities ceased in Europe in 1918, and apparently cotton farmers—anticipating a decline in demand—only planted 34.6 million acres in 1919. However, as was so often the case, they guessed wrong. In 1919, average cotton prices reached 35.3 cents a pound, the highest level ever recorded by USDA, at that time. In 1920, they guessed wrong again. The average price had reached a record level in 1919, and farmers increased acres planted by almost 1.3 million acres. Unfortunately, by the time they harvested that crop, the short recession had forced the average price down to 15.9 cents, almost 20 cents lower than in 1919.

Data presented in Figure 10.32 indicate that farmers harvested 98% of their acres planted from 1911 through 1915, but afterwards the percentage harvested began to decline as the boll weevil became an increasing problem.

In some areas during this period, damage from the pest was so great that entire fields were abandoned. From 1916 through 1920, only 97% of acres planted were actually harvested.

Yields

The effect of the boll weevil attack on American cotton production during the period 1915-1920 can be clearly seen in the data on average yields (pounds per acre) presented in Figure 10.33. During the previous four years (1911-1914), average national cotton yields had ranged between 192 and 216 pounds an acre. From 1915 through 1919, they ranged between 164 and 178, and in four of those years (1916-1919), they were near 165 pounds. By 1920, farmers were applying technology that countered the most serious effects of the pest, and the average increased to near 187 pounds an acre.

Problems with yields during this period (1915-1920) tended to counter problems with exports (Figure 10.30). Even though exports were down substantially from earlier years, lower yields helped keep prices up.

Bales Harvested

Changes in American cotton yields during the period 1915-1920 (Figure 10.33)) were relatively much larger than those in acres harvested (Figure 10.32). Consequently, year-to-year changes in bales harvested resemble changes in yields. Bales harvested ranged between 11.4 million and 12 million from 1915 through 1919 (Figure 10.34). Then with average yields increasing, bales harvested also increased, and were at a level of 13.4 million in 1920. The effect of the boll weevil is also clearly obvious in these data. As important as cotton was to the American and world economy, the war years had a less important impact on production of cotton than it did on other crops. The effect of the boll weevil was much greater.

In 1920, Texas farmers harvested the largest share (32.3%) of the national crop. Other states reporting large shares included: South Carolina (12.1%), Georgia (10.5), Oklahoma (9.9), and Arkansas (9.0). These five states accounted for 74% of the total crop.

Alabama Cotton

For many years, Alabama farmers had been harvesting about an equal number of acres of cotton and corn, but the value was always much higher for cotton. The price received for cotton on a per-pound basis was always higher. For example, in 1915 Alabama farmers received an average of 11.1 cents a pound for their cotton, but only 1.7 cents a pound for their corn. This price disadvantage was off-set somewhat by higher yields of grain than cotton per acre. In 1915, Alabama farmers harvested an average of 155 pounds of cotton per acre, compared to 686 pounds of corn. The result of these different variables was that between 1895 and 1913, the value of the Alabama cotton crop ranged between $26 and $96 million. For corn, it ranged between $16 and $33 million.

In 1910, farmers in Dallas County harvested the largest share (3.5%) of the Alabama crop. Other counties reporting large shares included: Montgomery (3.4%), Marengo (2.8), Wilcox (2.7), and Pike (2.6). These five counties accounted for only 15% of the total crop. The average harvest per county was 16,860 bales. Some 22 of the state's 67 counties reported harvests of 20,000 or more bales.

Prices

For all practical purposes, annual average prices received for cotton in Alabama during this period were the same as those received by farmers in the country. These are presented in Figure 10.31.

Acres Planted and Harvested

In 1915, Alabama farmers planted 3.2 million acres of cotton (Figure 10.35). Cotton prices were near 11 cents a pound in 1915 (up from seven cents in 1914) (Figure 10.31). Alabama farmers responded to the 1915 price by reducing planting to 2.8 million acres in 1916; however, by that time the boll weevil was taking an increasing toll. In 1916, the average price was up again (17.4 cents) as a result of wartime demand and inflation (Figure 10.3). Even with much better prices, farmers again responded to the threat of weevil depredation by reducing planting even further in 1917 to 2.1 million acres. The average price increased sharply in 1917 to 27.1 cents, and Alabama farmers apparently decided that even with the threat of crop loss from weevils the increase in price was just too good to pass up. In 1918, they planted 2.5 million acres. Also, by that time new technology was becoming available that reduced the effects of weevil damage to a limited extent. In 1919, they continued to respond to good prices and better weevil control by planting 2.6 million acres. The average price of cotton reached 35.3 cents a pound in 1919. It had never been that high, but based on past experience, farmers did not expect prices to continue to increase, and in 1920 they reduced acres planted by 37,000 (2.651 to 2.614 million). This time, their guess about prices was correct. As a result of the beginning of the short recession in 1920, the average price for the year declined to 15.9 cents. Even though the price was down in 1920, it was still higher than it had been for the entire period from 1900 to 1915. Unfortunately, while prices received had increased spectacularly from 1915 through 1919, prices paid had increased as well (Figure 10.5).

Boll weevil damage to Alabama cotton was severe in some cases, but it seemed to have only a limited effect on acres harvested (Figure 10.35). During the period 1909-1914, farmers had harvested 98.2% of planted acres. Then in the period 1915-1920, they harvested an average of 97.2%. They did not abandon very many acres once they had been planted. Reduced yields did seem to have some effect on acres harvested. In 1916, average yield for the state was 95 pounds an acre (Figure 10.33), and that year farmers harvested only 94% of acres planted (the lowest percentage during the period.

Yields

In 1914, Alabama farmers harvested an average of 220 pounds of cotton an acre, the highest in history (Figure 10.33). Then the boll weevil arrived. In 1915, average yield was down to 155 pounds. It continued to decline sharply in 1916 (95 pounds). This was the poorest ever recorded in the state. Yields improved somewhat in 1917 (121 pounds) and in 1918 (156 pounds); however, they were down again in 1919 to 133 pounds, and in 1920 to 124 pounds.

To a limited extent, farmers had learned to cope with the weevil, but the most promising technology (cultural practices) for combating the pest did not work very well in the heavy clay soils of the Black Prairie where most Alabama cotton was grown. Establishing the crop as early as practical in the spring, so that the most susceptible stage in plant development was reached before weevil numbers peaked, reduced damage significantly; however, in the Black Prairie it was very difficult to cultivate those soils early in the year because of their inherent wetness.

Bales Harvested

The level of year-to-year variation in yields (Figure 10.33) was significantly greater than in acres harvested (Figure 10.35); consequently annual changes in bales harvested were more heavily influenced by yields. Alabama farmers harvested 1.7 million bales of cotton in 1914 (Figure 10.36); however, as a result of declining yields and acres harvested, by 1916 bales harvested was at a level of 530,000. They had not harvested fewer bales since 1878. Bales harvested was somewhat higher in 1918 (799,000), 1919 (711,000), and 1920 (661,000), but generally, farmers had not harvested less cotton since the late 1800s. Virtually all of this loss was the result of weevil depredation. Finally, Alabama farmers, especially those in the Black Prairie, were convinced that they had to find some alternative to cotton production.

In 1920, the largest share of the Alabama cotton crop was reported from Madison County (4.8%). Other counties reporting substantial shares were: Marshall (3.7%), DeKalb (3.40), Limestone (3.4), and Cullman (3.3). These five counties accounted for only 18.6% of the total crop. The average per county was 10,700 bales. The top-five did not include any of the traditional Cotton-Counties in central Alabama. All of those counties were especially hard hit by the boll weevil.

Corn

Corn had been the king of crops in North America for hundreds of years before the Europeans arrived, and it held its position through the colonial period, the early period of country-building, and into the twentieth century. In 1915, American farmers planted almost as many acres of corn as the acreages of cotton, wheat, and oats combined (100.6 versus 108.4 million).

US Corn

In 1915, American farmers harvested 2.8 billion bushels of corn. They had harvested 2.5 billion in 1895. Illinois farmers harvested the largest share of the 1915 crop (12.8%). Other states harvesting significant shares included: Iowa (11.4%), Nebraska (8.0), Missouri (7.5), and Indiana (7.1). Together, these five states harvested 46.9% of the total crop. That year, Alabama farmers harvested 1.6%.

Exports

The war began in Europe in 1914, and that year the United States exported 50.7 million bushels of corn (Figure 10.37), or the equivalent of only 2% (50.7 of 2,272.5 million bushels) of the 1913 crop. Surprisingly, with Europe fully engaged in hostilities in 1915, exports declined to 39.9 million bushels. Exports to Germany and its allies had ended. Exports recovered somewhat in 1916 (66.8 million); but in 1917, they began to trend downward, as demand for feed grains declined. The downward trend ended in 1919 when 16.7 million bushels were exported. Apparently, after 1919 the combatants began to rebuild their shattered livestock industries, especially hogs, and the demand for American feed grains increased sharply. Exports that year totaled 70.9 million bushels.

Prices

In 1915, American corn farmers harvested a good crop (2.8 billion bushels), and prices declined slightly from their 1914 level (Figure 10.38). At that point, the war in Europe seemed to have little effect on price; however, a year later price exploded upward to $1.13 a bushel. This was the highest recorded price for the grain at that time. Early in the war period, the government placed price controls on wheat to prevent large increases in prices of that important food grain. But no controls were established on prices for feed grains. The average price of corn increased sharply again in 1917 ($1.39), the year the United States entered the war. A smaller increase followed in 1918 ($1.45). Then in 1919, with hostilities ended, the price declined slightly ($1.44). Finally, with the short recession beginning in 1920, it fell to 54 cents. Much of the extraordinary run-up in prices was likely associated with wartime hysteria. Certainly, it was not the result of increases in exports (Figure 10.37), and the magnitude of the increase was far greater than any increase in demand resulting from the war effort.

Acres Harvested

Annual data on the number of acres of corn planted in the United States or any of the states are not available before 1926. Data on acres harvested are available as far back as 1866. As is shown in Figure 10.39, acres harvested declined slightly during the period 1910-1914. However, with the beginning of the war in Europe in 1914, American farmers increased acres harvested slightly in 1915 to 100.6 million acres, although there had been no price signal to encourage such an increase. In 1915, the average price was at essentially the same level as in 1913 and 1914 (66-67 cents a bushel). Farmers again harvested grain from around 100.6 million acres in 1916. As detailed in a preceding paragraph, in 1916 the price spurted upward to $1.13 a bushel (Figure 10.38). The following year, farmers responded to that strong signal by harvesting grain from 110.9 million acres, the highest level in history. They harvested 2.9 billion bushels of corn that year, near the highest in history at that time. Corn farmers were concerned about that large crop, so even with prices continuing to increase, they began to reduce acres harvested. In 1918, it was down to 102.2 million acres, and in 1919 it fell to 87.5 million. It had not been that low since 1894. In 1920, farmers seem to have decided that they had over-reacted to fears of a price collapse with the planting and harvest of their 1919 crop, so in 1920 they harvested corn from 2.7 million, more acres than in 1919. Unfortunately, by the time they completed the harvest of that crop, their worst fears came true. The average price for that year was 54 cents a bushel.

Yields

There was little change in average corn yields in the country during the war years (1915-1920). Although there was considerable year-to-year variation during the period, they ranged around 26 bushels an acre (Figure 10.40). The promise of improved yields of corn through the application of research and extension was largely unrealized by 1920. The national average yield had been near 25 for many years.

Bushels Harvested

Although there was considerable year-to-year variation in acres harvested during this period (Figure 10.39), bushels harvested seemed more dependent on yields (Figure 10.40). Annual data on bushels of corn harvested in the United States during the period 1915-1920 (the war years) are presented in Figure 10.41. During the period, bushels harvested ranged between 2.3 (1919) and 2.9 (1917) billion bushels. The average for the period was 2.60 billion bushels. Although there has been considerable year-to-year variation, generally speaking annual bushels harvested had remained unchanged since 1895. This is primarily the result of the fact that yields changed relatively little during that period. Certainly, World War I seemed to have only a limited effect on corn production in the United States.

Alabama Corn

Although there was considerable annual variation, bushels of corn harvested in Alabama generally ranged between 30 and 40 million from 1895 through 1915. Yields were also largely unchanged during the period.

In 1909, farmers in Madison County harvested 3.3% of the state's corn crop. Other counties reporting substantial harvests included: Lauderdale (3.2%), Jackson (3.1), Limestone (2.8), and Houston (2.7). These five counties accounted for 15.1% of the total Alabama corn crop. Farmers harvested at least 500,000 bushels in 23 of the state's 67 counties.

Prices

Year-to-year changes in average annual prices of corn in Alabama during the period 1915-1920 (Figure 10.38) were generally similar to those for the country for the same period. In both cases, they increased rapidly after 1915, reached a maximum level about the same time; then they were sharply lower in 1920. Although the trends were similar, Alabama farmers received considerably better prices for their corn than the national average. In 1917, the difference was only 11 cents a bushel (65 cents versus 79 cents a bushel), but in 1920 it was much greater—56 cents (54 cents versus $1.10).

Acres Harvested

Unlike their national counterparts, Alabama corn farmers responded immediately to the beginning of the war in Europe. They had harvested 2.8 million acres in 1914, and then with a negative price signal in 1915, they harvested 3.2 million acres (Figure 10.42). They responded by harvesting more acres in 1915, even though the average price had been slightly lower in 1914 than in 1913 (Figure 10.38). In 1916, with the even stronger negative, 1915 price signal for guidance, they increased acres harvested only slightly (30,000). However, with price sharply higher in 1916, they increased their acreage significantly in 1917 and harvested corn from 3.8 million acres, the highest in history. Alabama farmers had received an extremely good price for their 1917 crop, but they were really concerned about the 59.4-million-bushel crop that year; consequently, in 1918, before the end of hostilities in Europe and with prices continuing to climb, they made the decision to begin to reduce their acreage of the crop. In 1919, acres harvested was down to 3.3 million, and in 1920 the crop was harvested from only 3 million acres.

Yields

Figure 10.40 includes data on annual average yields of corn in the United States and Alabama during the period 1895-1920. Alabama corn yields had ranged around 15 bushels an acre since the early 1900s, and except for 1916, they remained near the same level through 1920. Apparently, it was extremely wet throughout the state in June, July, and August, 1916. This wet weather probably contributed to the low average yield that year. Since at least 1840, Alabama corn yields had been around 10-15 bushels an acre, lower than in the country as a whole. In 1920, there was little indication of progress in improving yields through better information (research and extension). The old adage "You can't fool mother nature" seemed entirely appropriate in Alabama in 1920. The corn genome had not been adapted to Alabama, and research and extension had done very little to make it feel more at home. This lack of progress did not bode well for the future.

Bushels Harvested

With average yields remaining relatively unchanged during the period 1915-1920, trends in bushels harvested were largely determined by year-to-year changes in acres harvested (Figure 10.42). As a result of the rapid increase in acres harvested in 1915, bushels harvested increased to 46 million (Figure 10.43). It was sharply lower in 1916 (32 million bushels) when poor weather sharply reduced the yield (Figure 10.40). In 1918, bushels harvested reached a level of 59.4 million, the highest level in history at that time. With acres harvested declining in 1918, 1919, and 1920, bushels harvested also declined. In 1920, Alabama farmers only harvested 43.6 million bushels. It had been near that level in 1895.

Census data for 1920 indicated that farmers in Jackson County harvested the largest share (3.9%) of the state's corn crop in 1919. Other counties reporting substantial shares included: Madison (3%), Covington (2.7), Coffee (2.6), and Marshall (2.6). These five counties only accounted for 14.8% of the total crop. The average per county was 652,200 bushels. A total of 45 of Alabama's counties reported harvests of 500,000 or more bushels.

Alabama Peanuts

Peanuts, like cotton, were primarily a Southern crop. Production of the crop in relative terms had increased rapidly following the Civil War, but as late as the beginning of the twentieth century, peanuts were not a major player in American agriculture. USDA did not begin to publish annual statistics on the crop until 1910, and as late as 1910 only 703,000 of the country's 478.4 million acres of improved farmland (0.15%) was planted in peanuts.

In 1915, American farmers harvested 432.8 million pounds of peanuts. In 1909, they had harvested 354.6 million pounds. In 1915, North Carolina farmers harvested 29.2% of the total national crop. Other states harvesting significant quantities included: Virginia (24%), Alabama (16.0), Georgia (11.6), and Texas (9.2).

Alabama farmers only produced around 8% of the country's peanuts in 1889 and 1899. By 1909, the percentage had climbed to around 9%. As early as 1909, peanut production was already confined primarily to several counties in southeastern Alabama. Some 57.1 % of the total state crop was harvested from just five counties in that area: Houston (15.6%), Geneva (12.6), Coffee (11.0), Dale (9.7), and Covington (8.2)

US Prices

No data are available on the annual average price of peanuts in Alabama for the period 1915-1920. Data for the country are presented in Figure 10.44. In 1915, the annual average price of peanuts in the United States was four cents a pound. It had been near this level since 1909 (The first year that annual average prices are available). With the increased national demand for oil as a result of the growing European conflict in 1916, the price increased to 4.8 cents. Then it increased sharply in 1917 (7 cents), the year the United States entered the conflict. It was down slightly in 1918 (6.5 cents) when hostilities ceased, but rebounded to 9.3 cents in 1919. Then with the beginning of the short recession in 1920, prices of all agricultural commodities declined sharply (Figure 10.5). Peanut prices were no exception. In 1920, American farmers received an average of only 4.8 cents for their crop.

Acres Planted and Harvested In Alabama

After 1910, Alabama farmers began expanding acres of peanuts planted, and by 1915 it was at a level of near 250,000 acres (Figure 10.45); then they responded to somewhat better prices after 1915 (Figure 10.44) by increasing their rate of expansion even more. In 1918, price was substantially lower than in 1917. As a result, in 1919 acres planted declined to 450,000. The price was up again in 1919, and they increased acres planted in 1920 to 460,000. Unfortunately, when they harvested those increased acres, the price was down below 5 cents.

In 1915, Alabama farmers only harvested 44% of acres planted, a considerably lower percentage than in the country (58%). Remember that historically, peanuts were most commonly harvested by hogging-them-off. However, after 1915, with prices increasing sharply, Alabama farmers began to harvest a higher percentage. In 1917, they harvested 83% of their planted crop. In the country as a whole, only 73% was harvested that year. The percentage harvested in the state began to decline after 1917 when the average price declined, but it did not return to the 1915 level, as was the case in the country as a whole. In 1920, Alabama farmers harvested 69% of the acres they planted. These data suggest that experiences with peanuts during the war years convinced Alabama farmers that commercial production of peanuts had a future in the state.

Yields

In 1913, American peanut farmers harvested an average of 824 pounds an acre (Figure 10.46); however beginning in 1914 average yield began to trend downward. It was at a level of 779 pounds an acre in 1915, and by 1920 had declined to 699 pounds an acre. Although there was considerable year-to-year variation, average yields of peanuts in Alabama also declined during the period. It was at a level of 700 pounds an acre in 1915, but by 1920 it had declined to 550 pounds. The average yield was much lower than expected in 1916, probably as the result of wet weather in the Wiregrass during the growing season. Throughout the period, average yields in Alabama was much lower than in the country. The difference was especially large in 1918 (600 versus 779 pounds), 1919 (550 versus 719 pounds), and in 1920 (550 versus 699 pounds).

Pounds Harvested in Alabama

Although average yields of peanuts generally declined during the period 1915-1920 (Figure 10.46), they were not sufficiently large to off-set the large increase in acres harvested (Figure 10.45). As a result, pounds harvested generally followed acres harvested upward from 77 million pounds in 1915 to almost 272 million pounds in 1917. Although acres harvested increased from 1917 to 1918, the increase was not large enough to off-set the unusually large decline in yield, and as a result pounds harvested declined to 240 million. With acres harvested even lower in 1919, pounds harvested also declined (165 million). In 1920, Alabama farmers harvested almost 175 million pounds of the crop when acres harvested increased slightly.

Soybeans

Substantial acres of soybeans were planted in the South in the early 1900s; they were used as a forage crop. In 1918, E.F. Cauthen of the Alabama Agricultural Experiment Station commented that interest in the crop was due largely to destruction of cotton by the boll weevil. He further commented that he expected production of soybeans would become increasingly important in Alabama's cropping system. The soybean had several advantages as a crop. Being a legume, it required less fertilizer than cotton. Because of the wide range of uses for the beans, demand was generally strong. Also, they could be grown with equipment already available on most farms.

It is interesting that as late as 1920 there would be little data available on the production of soybeans in the United States. The crop was destined to become one of the more important ones in the country. Some beans were being imported in the early 1900s, and as a result USDA began to publish data on prices as early as 1913. However, the agency did not begin to publish data on production statistics until 1924.

In 1913, American farmers received $1.79 a bushel for their beans (Figure 10.48). The average price increased to $2.21 in 1914 and remained near that level through 1916. Then prices generally increased rapidly during the wartime boom to reach $3.53 in 1919. However, in the first year of the postwar recession (1920), prices fell to $2.67.

It is interesting that from the early 1900s, the unique characteristics of the soybean was recognized by the market and valued accordingly. For example, in 1913 the average price of a bushel of soybeans was $1.76 or three cents a pound. In the same year, prices per pound, of wheat, corn, and oats were 1.5, 1.4 and 0.7 cent(s), respectively.

Alabama Hay

The number of tons of Alabama hay harvested increased from around 100,000 in 1895 to near 347,000 in 1915. During the same period, yields had declined from 0.95 ton to 0.86 ton. Tons harvested in the state in 1915 amounted to only about 0.4% of the 91.4 million tons harvested in the entire country. Farmers in Nebraska harvested 7.2%, of the total. Other states harvesting significant quantities of this crop included: Iowa (6.9%), Minnesota (5.6), New York (5.5), and Kansas (5.2). Altogether, farmers in these five states harvested 30.5% of the total.

Prices

The average price obtained by Alabama farmers for their hay in 1915 was $13.10 a ton (Figure 10.49). Average prices had been near that level for several years. The average remained near $14.00 in 1916; however by 1917 wartime inflation (Figure 10.3) and increased demand began to influence the price of hay. That year, the average reached $17.70 a ton. It had not been that high since 1870, and it continued to increase sharply in 1918 ($20.80) and 1919 ($24.50). Farmers had never received that much for hay. In the period 1915-1919, the average price for hay increased 87%. In the same period, the Consumer Price Index increased from 30.4 to 51.8, an increase of 70.4%. In 1920 when the short recession began (Figure 10.1), and prices for virtually all agricultural commodities declined (Figure 10.5), the average price for hay in Alabama also declined ($21.60). The magnitude of the decline, however, was not as great as it was for most other crops. As we will see in a later section, the cycling of cattle numbers continued through this period, but the push of wartime inflation was so powerful that changes in numbers had little, if any, effect on hay prices.

Acres Harvested

In Alabama, acres of hay harvested had been increasing since the early 1900s (Figure 10.50) as farmers began to provide a more dependable feed supply for their animals. In 1915, it reached a level of 347,000 acres and continued to follow prices sharply upward (Figure 10.49) through 1919 (544,000). Finally, with prices declining in 1920, acres harvested also declined to 492,000.

Yields

In Alabama, annual average hay yields had been trending downward since 1905 (Figure 10.51)), and they reached a level of 0.86 ton an acre in 1915. It was much lower in 1916 (0.76 ton), when abnormally wet weather interfered with harvest. Annual yields continued to decline through 1920 when farmers harvested only 0.75 ton an acre. It is likely that the cause of this general downward trend was the result of getting better estimates over time, rather than a real decline.

Tons Harvested

Even though yields generally declined during the period 1915-1920, tons harvested (Figure 10.52,) followed acres harvested upward (Figure 10.50). In 1915, tons harvested was at a level of 347,000. It continued to increase through 1919 when it reached a level of 544,000. With both yield and acres harvested lower in 1920, tons harvested fell to 492,000 tons.

Alabama Wheat

Alabama farmers have always planted mostly winter wheat. Although all of the early statistics were reported as simply wheat, they really referred to the winter varieties—where the seeds are planted late in one year, but the crop is harvested about the middle of the following year. Until 1909, USDA data relating to the crop were also reported as wheat; however, the statistics were for all varieties. Then in 1909, the agency began to issue separate reports on winter wheat. Consequently, from this point onward, all USDA data, when used, will be for winter wheat only.

In 1900, Alabama farmers had harvested 1.2 million bushels of winter wheat, but by 1909 bushels harvested had fallen to 116,000, then by 1915 increased to 323,000. The problem, of course, was yields. Throughout the period 1909-1915, the average annual yield was never higher than 10.5 bushels an acre, and most years it was considerably lower.

In 1915, Kansas farmers harvested the largest share (16.5%) of the American crop of winter wheat (640.6 million bushels). Other states reporting large shares of the total crop were: Nebraska (10.2%), Illinois (8.2), Indiana (7.6), and Oklahoma (6.0). These five states accounted for 48.7% of the national total. Alabama farmers harvested less than 0.1%.

Farmers in Madison County harvested 22.3% of Alabama's 1909 wheat crop. Other counties harvesting substantial shares included: Lauderdale (12.2%), Randolph (9.5), Limestone (8.1), and Talladega (6.8). These five counties accounted for 58.9% of the total. Farmers in 13 counties harvested at least 2,000 bushels.

US Exports

Data on the export of American wheat (all varieties) for the period 1895-1920 are presented in Figure 10.53. They were at a level of 71 million bushels in 1910, but then in the first year of the war (1914), exports increased sharply to a level of 336 million bushels. Afterwards, the United States lost a substantial share of it markets in Europe, and exports began to decline. In 1917, they were down to 133 million bushels. With the war ending in 1918, they again began to increase, and in 1920 reached a level of 369 million bushels. The short recession that began in the United States in 1920 appeared to have little or no effect on wheat exports. The unexpected decline in exports in 1919 was likely the result of the extremely high price of the wheat that year ($2.16 a bushel) (Figure 10.54).

US and Alabama Prices

Data on the prices of American wheat (all) and Alabama winter wheat for the period 1895-1920 are presented in Figure 10.54. USDA did not begin to publish prices for winter wheat until 1919. Prices had generally been slowly trending upward since 1895, but when national production spurted upward after 1909, national prices began to trend downward. In 1913, they reached a level of around 80 cents a bushel. Then with the beginning of the war, they began to increase, reaching a level of $1.20 in 1915. With the end of the war in 1918, the upward movement ended, and they remained near $2.00 through 1919. When the short recession began in 1920, the average price fell back to slightly less than $1.50. Throughout the period, Alabama wheat prices were higher than for those in the country as a whole. Apparently Alabama farmers were able to sell locally the relatively small quantity they produced for higher prices.

Acres Planted and Harvested in Alabama

Data for acres planted of Alabama winter wheat were first reported in 1909. Data on acres planted and harvested for the period 1909-1920 are presented in Figure 10.55. From the data, it is evident that Alabama farmers had almost abandoned the crop by 1910. As late as 1914, acres planted was at a level of 18,000. However, with prices increasing rapidly at the beginning of the war, acres planted also began to increase. In 1915, Alabama farmers planted the crop on 36,000 acres, and they continued to increase it through 1917 (43,000 acres). Then with the war ending in 1918, acres planted fell to a level of 34,000, and it remained at near that point in 1919. Problems in the national economy began in 1919 (Figure 10.1), and Alabama farmers, anticipating the worst, reduced acres planted to only 12,000 acres.

Between 1914 and 1915, acres harvested increased along with acres planted, and between 1915 and 1916 Alabama farmers harvested only 70% of acres planted in 1917. It is likely that poor crop weather reduced acres harvested that year. In the remaining years of the period, farmers harvested around 98% of acres planted.

Yields

Data on the yields of winter wheat in the United States and in Alabama during the period 1909-1920 are presented in Figure 10.56. The data show that although there was some year-to-year variation, Alabama wheat yields remained in the range of 7 to 10 bushels an acre for the entire period. Throughout the period, yields were only about 40 to 50% of the average for the country. With these yields, it is easy to understand why Alabama farmers lost interest in the crop. It is surprising that they planted any, much-less 10,000 acres. The lower than expected yields in 1916, 1917, 1918, and 1919 were probably the result of unsuitable crop weather in those years.

Bushels Harvested

With yields remaining relatively constant over the period 1915-1920, changes in bushels harvested in Alabama were similar to changes in acres harvested. With acres harvested up sharply after 1914 (Figure 10.57), bushels harvested also increased. Farmers harvested only 178,000 bushels in 1914, but 323,000 in 1915. They harvested fewer bushels than expected in the period 1916-1919 (around 240,000 bushels), as a result of lower than expected yields. Then, responding to the beginning of the down-turn in the economy in 1919, they planted significantly less wheat in the fall of that year. In 1920, they harvested only115,000 bushels.

Farmers in Talladega County harvested the largest share (7.1%) of Alabama's 1919 crop of winter wheat. Other counties reporting substantial shares included: Lauderdale (5.9%), Chilton (5.5), Madison (5.4), and Limestone (5.1). These five counties accounted for only 29% of the total crop. The county average was slightly over 3,300 bushels. A total of 32 of Alabama's 67 counties reported harvests of 2,000 or more bushels.

Alabama Oats

By 1915, the number of tractors on American farms was increasing rapidly, but the total number had not reached the threshold where the number of horses and mules had begun to decrease. As a result, through the period 1895-1915 the production of oats for use as animal feed continued to increase. In 1915, American farmers harvested a total of 1.4 billion bushels of oats. Iowa farmers harvested 14.2% of that crop. Other states harvesting significant quantities included: Illinois (13.6%), Minnesota (9.5), North Dakota (7.4), and Wisconsin (7.2).

Bushels of Alabama oats harvested generally increased from 3 million bushels in 1896, to 4.5 million in 1915. In the same general period, yields increased from around 14 bushels an acre to 18. Farmers in Elmore County harvested 4.3% of the 1909 Alabama oat crop. Other counties harvesting substantial shares included: Tallapoosa (4.2%), Talladega (4.0), Montgomery (3.3), and Lee (3.1). These five counties accounted for 18.9% of the total. Farmers in 26 of the counties harvested at least 50,000 bushels.

US Exports

Horses and mules played extremely important roles in Europe during World War I. Widespread mechanization of war machines was a generation away, and horses and mules ate oats primarily. Fortunately, Europeans produced lots of oats; consequently, the United States had never exported much of the grain to the Continent. For years, exports remained equal to around 1% of bushels harvested in the country. In 1914, however, they exploded upward to near 100.6 million, and remained at around that level until 1919 when it fell to 43 million. By 1920, it was essentially where it had been in the early 1900s—below 10 million bushels.

US and Alabama Prices

American oat exports had increased sharply in 1914 (Figure 10.58), but prices did not immediately respond to this increased demand. American farmers received 37 cents a bushel for their oats in 1915 (Figure 10.59). It had been around 40 cents since 1907. In 1916 with exports unusually high, the price increased to 47 cents. In 1917, the increase was even larger (68 cents). It remained largely unchanged in 1918. With the hostilities in Europe ending, however, it increased modestly in 1919 to slightly less than 75 cents as a result of a much smaller than expected American crop. From 1915 through 1919, the average price of oats in the United States had grown from 37 to 75 cents, an increase of 102.7%. During the same period, the Consumer Price Index had increased from 30.4 to 51.8, an increase of 70.4%. As would be expected, much of the increase in price had been the result of wartime inflation.

Year-to-year changes in oat prices during the period 1915-1920 were essentially the same in the United States and Alabama. However, Alabama farmers received between 30 (1915) and 48 cents (1920) more a bushel for their oats.

Acres Harvested

Before 1926, there are no data on acres planted for oats in Alabama. The state's farmers harvested oats from 240,000 acres in 1914. Then in 1915, acres harvested exploded upward to 300,000. Even though prices continued to increase after 1915 (Figure 10.59), Alabama farmers were rapidly losing interest in oats as a crop. In 1920, they harvested the crop from only 163,000 acres. The problem, of course, as we will see in a following section was yields. Alabama farmers received much better prices than most of the farmers in the country, but their yields were atrociously low.

US and Alabama Yields

In 1915, American farmers harvested an average of 37 bushels of oats an acre, the highest in history at that time. Between 1915 and 1920, yields varied between 27.9 and 37 bushels. There was considerable variation between years, and there was no indication of a trend during the period. Generally, oat yields had been trending upward on Alabama farms from near 15 bushels in 1905 to 19 bushels in 1914. However, from 1915-1919, they plunged to between 13 and 15 bushels. Yield recovered somewhat in 1920 (18 bushels). The lower than expected yields during the period 1915-1919 was likely the result of five years of extremely wet weather in most of the state (Cook, et al., 1999). During this six-year period, annual yields of oats in Alabama ranged between 15 and 18 bushels, or around 50% of average national yields.

Bushels Harvested

Bushels of Alabama oats harvested had generally trended upward, along with acres harvested, between 1905 and 1914 (Figures 10.60 and 10.62). Farmers harvested 4.5 million bushels of oats in 1915. After 1915, as acres harvested declined, bushels harvested also declined. In 1919, only 2.4 million bushels of the crop was harvested in the state. It had not been that low since 1902. Alabama farmers must have finally decided that it would be cheaper to purchase oats to feed their horses and mules than to try to produce them on their farms.

In 1919, farmers in Houston County harvested 7.4% of the state's oat crop. Other counties with substantial harvests included: Butler (5.2%), Elmore (5.1), Russell (3.9), and Monroe (3.5). These five counties only accounted for 25.1% of the total. All of these counties were located in the southern portion of the state. The county average was near 16,700 bushels. A total of 43 of Alabama's 67 counties reported harvests of at least 10,000 bushels.

Alabama Sweet Potatoes

In 1915, acres of sweet potatoes harvested in Alabama was at a level of 63,000 acres (Figure 10.63). It had been trending downward for several years. Then with the beginning of the war years, it increased sharply to reach 91,000 acres in 1919. With prices also increasing sharply during the period, the value of the crop increased from $3.9 million in 1915 to $11.1 million in 1919. With the beginning of the short recession in 1919, both acres harvested and prices declined in 1920.

Alabama Irish Potatoes

Alabama farmers harvested Irish potatoes from 8,000 acres in 1895 (Figure 10.64). Acres harvested had slowly trended upward to 14,000 during the Golden Age of American Agriculture, but declined in 1910 even though prices remained good. Then during the war years farmers significantly increased acres harvested to 18,000 acres in 1918. After the end of the war, acres harvested immediately fell back to 15,000 acres. The value of the crop ranged between $748,000 and $3.8 million during the period 1915-1920.

Alabama Sorghum and Sugar Cane

There are no annual data on the production of sorghum in Alabama earlier than 1924. Census data indicated that in 1919 Alabama farmers harvested 52,000 acres of the crop, which was used to produce 2.4 million gallons of molasses with an estimated value of $2.6 million. In 1909, they had harvested 16,100 acres and produced 809,400 gallons of molasses valued at $399,000.

Alabama farmers harvested 27,200 and 25,300 acres of sugar cane in 1909 and 1919, respectively, and produced 3.1 and 3.2 million gallons of syrup from those two crops with values of $1.5 and $4.2 million, respectively.

Alabama Vegetables

In 1909, Alabama farmers harvested 69,500 acres of vegetables (excluding acreages of sweet potatoes and Irish potatoes), and 12,800 acres in 1919. Values of these two crops were $5.4 and $1.3 million, respectively. In 1919, watermelons, cabbages, and tomatoes accounted for 48%, 12%, and 9%, respectively, of all acres planted.

Alabama Peaches

There are no annual data on Alabama peach production earlier than 1919. Census data indicate that Alabama farmers harvested 52 and 46.8 million pounds of the fruit in 1919 and 1920, respectively. They received 3.8 and 5.2 cents a pound, respectively for those crops. Crop value for the two years was $1.9 and $2.4 million, respectively. According to the 1920 Census, there were 2.1 million peach trees of all ages in the state.

Alabama Strawberries

In 1909, Alabama farmers harvested 1.8 million quarts of strawberries valued at $160,000 from 1,167 acres. In 1919, they harvested 2 million quarts valued at $404,800 from 1,359 acres.

Alabama Pecans

As was the case with peaches, there are no annual data on the harvest of Alabama pecans before 1919. In that year, it was estimated that Alabama farmers harvested 1.4 million pounds of pecans. In 1920, they only harvested 600,000 pounds, a much lower than expected crop. Pecans are a masting crop, and consequently they generally do not produce large crops year after year. The 1920 harvest was much lower than expected in an off year. The weather was excessively wet throughout south Alabama in June, July, and August of 1920, and this excessive moisture might have affected nut production. There are no annual data available on pecan prices earlier than 1923. The US census estimated that there were 434,000 pecan trees of all ages in Alabama in 1920.

Alabama Greenhouse, Sod and Nursery

The 1920 Census reported that Alabama farmers sold a variety of nursery and greenhouse products valued at $499,300 in 1919. These products included plants, flowers, bulbs, and flower and vegetable seeds. They also sold trees, shrubs, vines, ornamentals, etc. valued at $234,700.

Alabama Timber

Data on lumber production in the different regions of the United States during the period 1869-1920 are presented in Table 10.1. These data show that by 1920 there was relatively little lumber production remaining in the North Atlantic Region (6.6%) and the share of the North Central Region was still declining rapidly (15.6%). Further, it is apparent that lumber production had reached its peak in the South. By that time, a large share of the most accessible timber had been cut in the region. As a result, lumbering was moving rapidly to the almost unlimited conifer forests of the West. Between 1915 and 1920, the share of the country's lumber supply manufactured there increased from 22 to 34.6%.

Census data indicated that cash receipts for Alabama forest products totaled $12.7 million in 1920.

Animal Agriculture

As detailed in the preceding chapter, in 1895 most of American agriculture was just beginning to emerge from the most disastrous period since the Civil War. According to census data, in 1890 the value of livestock on American farms was $2.2 billion, but with the coming of the Progressive Era in the United States and around the world, the American livestock situation quickly changed. USDA data indicate that the value of livestock added $3.7 billion to the national economy in 1910, while crops added $3.6 billion. By 1915, value added for livestock and crops had increased to $4 billion and $3.9 billion, respectively. In both 1909 and 1915, the value added by meat animals accounted for the largest share (44%) of the total. Dairy products and poultry and eggs accounted for lesser shares, 17% and 13%, respectively.

Census data indicate that the value of all livestock on Alabama farms totaled $30.7 million in 1890 and $36.1 million in 1900. Later census data indicate that by 1910 value had increased to $65.6 million. Surprisingly, the value of mules accounted for the larger share (50%) of that total. Horses and "all" cattle accounted for lesser shares: 21.5% and 21.1%, respectively.

Cattle

Before 1920, there are no annual data on the number of beef cows or bulls, heifers, or calves destined for beef herds in the country as a whole or in the states. Numbers cited in this section were obtained by subtracting the number of cows that calved-milk from cattle and calves-all.

In the Consolidation Phase of a cattle cycle in 1904, there were 49 million cattle (excepting milk cows) in the United States. In the Consolidation Phase of the following cycle in 1918, there were 51.5 million. Between 1904 and 1918, American cattle numbers increased by 5.2%.

US Cattle

On January 1, 1915, there were 43.6 million cattle being held primarily for the beef industry in the United States. The largest share of these animals (14.4%) were on farms in Texas. Other states with significant numbers included: Iowa (7.2%), Nebraska (5.7), Kansas (5.5), and Missouri (4.2). Altogether, these five states accounted for 36.9% of the national herd that year.

Farm Value Per Head

Data on farm value per head of American cattle (all) for the period 1895-1920 are presented in Figure 10.65. Nationally, cattle numbers entered the Declining Phase of a cycle in 1906 (Figure 10.67). At that point, value was at a level of $19.65. Then as numbers declined, values began to increase, finally reaching a level $32 a head in 1915. In 1913, the Expansion Phase of a new cycle had begun, and the rate of increase in values began to slow. There was no increase between 1915 and 1916. In 1916, wartime inflation began to take hold, and values completely forgot about numbers. Consequently, values continued to increase, reaching a level of $41.79 in 1919. With the short recession at full force in 1920, value retreated to $40 a head. However, changes in the CPI played an important role in these changes in values.

Cycles of Abundance

A new cattle cycle began in 1913 with numbers at 36 million (Figure 10.67). The cycle reached the Consolidation Phase in 1918 (51.5 million head). It was in the early stage of its Declining Phase in 1920. There were 48.9 million cattle and calves in the country on January 1 of that year. There is little indication that the wartime emergency had any effect on beef cattle numbers in the United States during the period 1915-1920. Cycles in those days were so long that there was little opportunity to increase numbers to any extent within such a short period of time. Generally, herd size had been increasing since 1867. At the Consolidation Phases of the two previous cycles, in 1890 and 1904, there were 45 million and 50 million cattle, respectively, in the country. At the Consolidation Phase of the next cycle (1918), the estimate was 51.5 million. That estimate was not any greater than was expected from the normal growth of herd size that had been underway since the middle of the nineteenth century.

Real Net Cash Returns Per Cow

Data on real net return per cow for cattle for the period 1913-1920 are presented in Figure 10.66. The data show that at the end (1915) of the Golden Age of American Agriculture net returns for American cattle reached $125 a head. Then, with wartime inflation roaring ahead, it reached $195 in 1917. Unfortunately, at the same time prices paid were also increasing, and in 1918 net returns began to decline. It was $190 a head in 1918. Then, with the economic downturn beginning in 1919, it fell to $159; finally, in the midst of the short recession of 1920, it fell to $77.

Alabama Cattle

The 1910 inventory of cattle (all) in Alabama indicates that the largest share (3.2%) was on farms in Marengo County. Other counties reporting large numbers included: Wilcox (3%), Clarke (2.7), Dallas (2.6), and Monroe (2.4). These five counties accounted for only 13.9% of the total. These data indicate the widespread distribution of cattle within the state. Inventories indicate that 49 counties reported 10,000 animals or more.

Farm Value Per Head

Data on farm value per head for Alabama cattle for the period 1895-1920 are shown in Figure 10.65, along with similar data for the entire country. The data are not exactly comparable. The national data are for "all" cattle, excepting milk cows; while the Alabama data include milk cows. As a result, the Alabama data is biased upward slightly, because of the higher value of milk cows. Trends in values were generally similar for both American and Alabama cattle, although they were several dollars lower for Alabama animals. Both had begun to trend sharply upward after the beginning of the war in Europe. Alabama values increased from $20 a head in 1915 to $39 in 1919. Then with the beginning of the postwar recession value declined to $35.20.

Cycles of Abundance

in 1913, a new cattle cycle also began in Alabama (492,000 animals) (Figure 10.68). It is obvious from the data that numbers were in the Expansion Phase in 1914 and 1915, but with the year-to-year variation, it is difficult to determine what is taking place during the remainder of the period. Numbers in the state certainly did not follow the national pattern. Although there was considerable year-to-year variation after 1915, herd size reached a level of 624,000 in 1920.

Census data indicated that farms in Marengo County held the largest share (6.2%) of Alabama's cattle herd in 1920. Other counties with substantial shares included: Baldwin (4.6%), Sumter (4.4), Escambia (3.6), and Hale (3.5). These five counties accounted for only 22.3% of the total. The county average was near 4,800. On average, there were 120 cattle on each 100 farms in the state. A total of 41 counties reported numbers of 3,000 or more animals.

Hogs

Hog numbers had been falling behind the growth in the American population since at least 1870. In that year, the inventory of hogs and pigs indicated that there were 94 of these animals for each 100 persons in the population. In 1890 and 1910, the ratio was 75 and 67 animals, respectively, per 100 persons. In the early years, hogs were literally everywhere. However, as the availability of beef increased, pork consumption declined in much of the country.

US Hogs

The December 1,1915, inventory of hogs and pigs in the United States indicates that there were 24.5 million of these animals in the country at that time. Some 15.5% of the herd was on farms in Iowa. Other states with significant numbers at that time were: Illinois (7.7%), Nebraska (6.1), Missouri (6.0), and Indiana (5.8). Some 41.1% of all the hogs in the country were in those five states. About 2% of the total were on farms in Alabama.

Farm Value Per Head

As a result of cycles in hog numbers and because of the supply-price relationship, there are also cycles in farm value per head. There seemed to be one well-defined cycle between 1913 and 1916. In 1915, value was at a level of $10.43. Afterwards, however, values became caught up in wartime inflation. It increased rapidly in 1917 and 1918, before finally reaching a level of $23.28 in 1919. In the midst of the short recession in 1920, it fell to $20.00.

Cycles of Abundance

A new hog Cycle of Abundance began in the United States in 1914 (56.6 million animals), and by 1915, there were 60.6 million hogs and pigs in the country (Figure 10.70). Assuming that the 1916, inventory number was an outlier for some reason, the Accumulation Phase of the cycle that began in 1914 continued until 1918 (64.3 million) when the Liquidation Phase began. In 1920, the inventory counted 58.9 million animals.

Alabama Hogs

Data presented in Figure 10.71 indicate that there were two distinct Cycles of Abundance in Alabama hogs and pigs between 1903 and 1914. In these two cycles' numbers ranged between 1,000,000 and 1,250,000 animals. In 1915, at the beginning of a new cycle, there were some 1,060,000 hogs in the state, or about 1.9% of the national total.

In 1910, 4% of all of the hogs in Alabama were on farms in Houston County. Other counties with large shares were: Covington (3.3%), Clarke (3.3), Greene (3.2), and Coffee (2.8). These five counties accounted for only about 16.6% of the total. Virtually every county had a substantial number of hogs. Farmers in 26 counties owned at least 20,000 of the animals.

Farm Value Per Head

Data on farm value per head for Alabama hogs for the period 1895-1920 are also presented in Figure 10.69. These data indicate that year-to-year changes in values were generally similar in both the country and state, although they were always a few dollars lower in the state. Values increased sharply with wartime inflation after 1916, from $7.30 an animal to $17 in 1919. With the short recession in 1920, it fell to $13.70. At that point, value for Alabama hogs was $7 a head, less than the national average.

Cycles of Abundance

In Alabama, a new hog Cycle of Abundance began in 1915 with a population of 1.2 million animals (Figure 10.71). The end of the Accumulation Phase of the cycle was reached in 1918 (1.6 million animals). As had been the case with the national cycle (Figure 10.71), the Liquidation Phase began in 1919, and by 1920 herd size was down to 1.3 million animals. It is clear from the data presented in Figure 10.71 that the wartime emergency significantly affected swine abundance in Alabama. Number at the end of the Expansion Phase of the 1915 cycle was much higher than at the end of the phase in the 1910-1914 cycle (1.6 versus 1.3 million).

In 1920, farms in Pike County held the largest share (3.8%) of the state's hog herd. Other counties reporting substantial shares included: Geneva (3.7%), Houston (3.7), Coffee (3.5), and Covington (3.4). These five counties accounted for only 18.1% of the state's total herd. In 1920, hog production was still widely distributed in the state. The county average was 22,300 animals, and on average there were some 580 hogs per 100 farms. A total of 31 of the state's 67 counties reported at least 20,000 animals.

Milk

Census data indicated the milk production on American farms totaled 44.3, 61.8, and 63.5 billion pounds, respectively, in 1889, 1889, and 1909—an increase of 43% over the 21-year period. Using average population data for those years, pounds of milk available annually per person was 718, 824, and 700, respectively, in those three years.

US Milk

Before 1924, there were no annual data available on national milk production. Census data indicate that cows on dairy farms in the country produced 63.5 billion pounds of milk in 1909. The largest share (49.8%) was produced in the North Central Region. Shares produced in the other regions were: North Atlantic (22.4%), South Central (13.3), West (7.6), and South Atlantic (7.0).

Prices

There are no annual data on national or state milk prices before 1924.

Milk Cows

The number of milk cows in the United States generally trended upward through the period 1895-1915 (Figure 10.72), although the rate of increase slowed significantly in 1911,1912, and 1913. In 1915, numbers reached 20.3 million animals. Numbers continued to increase rapidly through 1918 (21.54 million); however, in 1919 and 1920, after the end of hostilities in Europe, numbers changed very little (21.54 million in 1919 and 21.46 million in 1920). The short recession of 1920 actually began in 1919. The wartime emergency did seem to have a limited influence on numbers of milk cows in the United States. The rate of increase during the period 1915-1918 was considerably greater than in any other similar period.

In 1920, the largest share (8.5%) of milk cows in the country were on farms in Wisconsin. Other states reporting substantial shares included: New York (7%), Minnesota (6.3), Iowa (5.2), and Illinois (4.9). These five states accounted for 31.9% of the total. About 2% of the total was on Alabama farms.

Milk Per Cow

Before 1924, there are no annual data available on milk produced per cow in the country or any of the states.Census data for 1920 indicate that milk cows on American dairy farms produced 66.3 billion pounds of milk in 1919. They had reported a total production of 63.5 billion pounds in 1909. The largest share (71.8%) of the 1919 total was produced in the North Atlantic and North Central regions. Shares produced in other sections of the country were, in descending order: The South (South Atlantic and South Central regions) (18.4%) and the West (9.8). The value of all dairy products in the country in 1919 was $1.4 billion.

Alabama Milk

Census data indicate that Alabama milk production totaled 471.8 million pounds in 1899 and 669.2 million in 1909. Based on average populations in those years, annual milk production per person was 317 gallons in both years.

Milk Cows

The number of milk cows in Alabama increased rapidly during the period 1905-1910 (Figure 10.73). Then there was almost no change between 1910 and 1915. In 1915, the estimated herd size was 360,000. In 1916, numbers began to increase rapidly again, and in 1919 reached 420,000 animals. It remained at the same level in 1920. In 1915, some 1.9% of the milk cows in the United States were on Alabama farms. Five years later, the percentage had inched up to 2%.

Milk Production

Alabama farmers reported the production of 676 million pounds (79.5 million gallons) of milk in 1919. Farmers in DeKalb County reported the larger share (3.1%) of the total. Other counties reporting substantial shares included: Cullman (3%), Marshall (2.7), Chambers (2.4), and Jefferson (2.3). These five counties accounted for only 13.4% of the state total. This small share of the five leading counties is indicative of the widespread distribution of milk production in the state that year. Farmers in 24 0f Alabama's 67 counties reported productions of at least 10 million pounds. The value of all Alabama dairy products was $15.2 million. That year the average milk production per cow was 226 gallons (1,944 pounds).

Alabama Chickens

The 1910 Census reported that 87.8% of the country's 6.4 million farms had fowls of one or more classes (kinds). In 1900, the percentage had been slightly higher (88.8%). The census also noted that virtually every farm reporting fowls of any kind had chickens. In 1910, chickens were an important part of the American farm-scape. The 1910 Census also reported that on April 1, 1910 there were 295.9 million fowls in the country. Approximately 94.7% of the total was chickens, 1.5% geese, 1.2% turkeys, and 1% ducks. The largest share (54.3%) of the total number of fowls was in the North Central Region. Shares in the other regions were: South Central (19.7%), North Atlantic (11.2), South Atlantic (9.4), and West (5.3).

Value Per Head

Annual data on value per head of chickens in the country during the period 1910-1920 are presented in Figure 10.74. There are no similar data for the state until 1924. The data indicate that value responded to wartime inflation. It was 49 cents a bird when the war began in 1914, but increased to 77.5 cents in 1918 when the war ended, and to 95.5 cents in 1919. Surprisingly, value did not fall in 1920 as a result of the short recession. Rather, it remained at essentially the 1919 level.

Numbers

Although there are no annual data on numbers of chickens in Alabama before 1924; the 1920 Census published data indicating that there were some 5.9 millions on Alabama farms as of January 1, of that year—or about 231 birds for each 100 farms. Only 1.6% of the national total was on Alabama farms. Farms in Madison County reported the largest share (3.1%) of the inventory. Other counties reporting substantial shares included: Jackson (2.9%), DeKalb (2.9), Marshall (2.9), and Lauderdale (2.7). These five counties only accounted for 14.5 % of the state's total. Some 42 counties reported inventories of 75,000 or more birds.

Census data also indicate that in 1910 the value of all chickens in the state was $4.7 million and that the average value per head was 79 cents. In 1920, the comparable value in the country as a whole was 97 cents.

Alabama Eggs

The percentage of all eggs in the United States consumed by the farm families responsible for their production remained very close to 25.5 throughout the period 1909-1915, while the percentage used for hatching varied between 3.9 and 4.5%. Data presented in the 1910 Census indicated that chickens on farms in the North Central Region produced 52.7% of all American eggs in 1909; whereas farms in the South Central, North Atlantic, South Atlantic, and West regions produced 18.5, 13.6, 8.6, and 6.6%, respectively. Alabama farms accounted for only 1.2% of total production.

Prices

Data on the average annual price of eggs in the United States for the period 1909-1920 are presented in Figure 10.75. Similar data for Alabama are not available until 1924. National prices had varied relatively little at around 18-20 cents a dozen through most of the Golden Age of American Agriculture and to the beginning of the war in Europe, but as the United States drew closer to direct involvement in World War I, prices began to increase, and by 1920 it had more than doubled to 43.5 cents a dozen.

Production

Before 1924, there are no annual data on the production of eggs in Alabama. Data in the 1920 Census indicate that there were 265.2 million eggs produced on Alabama farms in 1919. Approximately 42.5% of these were sold. The remainder were either consumed by the families responsible for their production or were used for on-farm hatching. Census data reported that chickens in Jackson County led the state in egg production in 1919 with 3.2% of the total. Other counties reporting substantial shares included: Marshall (3.1%), Morgan (3.1), Cullman (2.9), and DeKalb (2.8). These five counties accounted for only 15.1% of Alabama's total production.

Uses

Uses of eggs in the country during the period 1915-1920 remained essentially unchanged from the preceding period (1895-1915). Farmers sold 70%, consumed 24%, and used 4% for hatching. Data on per capita egg consumption in the United States are presented in Figure 10.76. There was considerable year-to-year variation. Consumption increased between 1914 and 1915 (293-311 eggs), while the price was falling from 20.5 to 19.4 cents a dozen. Then in 1916 and 1917, consumption declined, as prices increased; however, in 1918 and 1919 consumption continued to increase even though prices were continuing to increase. Apparently, personal incomes were increasing rapidly enough in the wartime economy that people felt that they could afford eggs. The short recession in 1920 finally forced consumption down to 297.

Agricultural Production

Beginning in 1910, USDA began to collect and publish annual data on the value added to the national economy by the agriculture sector through production and services. As part of that data compilation, USDA estimated the value added through crop and livestock production each year. These data for the period 1910-1920 are presented in Figure 10.77, along with total value added. These data indicate that the total began to trend upward slightly after around 1913, and that it was at a level of 7.9 billion in 1915. It increased at a moderate rate in 1916 ($8.8 billion) and then sharply in 1917 ($13.7

billion). The rate of increase began to slow in 1918 and 1919 when it reached a maximum of $16.6 billion in 1919. It declined in 1920 when the short recession began to seriously impact prices received (Figure 10.5).

The data presented in Figure 10.77 also show that value added for both crops and livestock remained nearly equal from 1915 through 1919, but that it declined for livestock in 1920.

Total value added grew from $7.9 billion in 1915 to $16.6 billion in 1919, an increase of 110%, but a considerable share of this increase was the result of the inflation of the currency. During the same period, the Consumer Price Index grew from 30.4 to 51.8, an increase of 70%. In very general terms, the difference between 110% and 70% is indicative of the real increase in the production of all agricultural commodities during this period. It also indicates the level of excess production that American agriculture will have to deal with in the coming years.

With prices received beginning to escalate after 1915, the value added by home consumption of both crops and livestock also began to increase (Figure 10.78). In 1915, the value added for the home consumption of crops was $391 million, and it represented 10% of the total. For livestock it was $801 million, and represented 20% of the total. In 1920, these values were $943 million (11.3% of the total) and $1,566 million (21.1% of the total), respectively. For both crops and livestock, the increase in value added by home consumption was considerably greater than the rate of inflation, indicating that farm families significantly increased the consumption of commodities that they produced during the war years. It is likely that the rapid increase in the prices they had to pay for purchased food would have encouraged them to do so.

Data presented in Table 10.2 show some interesting aspects of American agricultural production during World War I. In 1910, value added for cotton alone was considerably greater than for either food grains or feed grains. By 1915, it was still the largest contributor, but not much greater than food grains. In 1920, primarily as a result of the dramatic increases in the production and prices of wheat, cotton was no longer the leading contributor to the American agricultural economy.

Values for inventory adjustment included in Table 10.2 relate to value added of crops and livestock produced in the current year that will not be sold before December 31 (positive value) or to crops and livestock produced in the preceding year that were not sold until the current year (negative value).

Data on value added did not become available for the individual states until 1949. Further, there was no annual data available on cash receipts of Alabama agricultural commodities until 1924. The only data available was that collected by the US Bureau of the Census in 1910 and 1920.

Census data obtained in 1910 indicate that the estimated total value of all Alabama crops in 1909 was $144.3 million (Table 10.3). The value of cotton alone accounted for almost two-thirds of the total. The value of cereals (mostly corn) contributed only 22%. By 1919, total value of all crops had increased to $304.3 million, but the importance of cotton had decreased relative to that for cereals. Surprisingly, the relative importance of vegetables increased significantly from 1909 to 1919.

The estimated total value of all Alabama livestock (including all poultry and bees) on April 1, 1910, was $65.6 million (Table 10.4). The value of horses and mules accounted for 69% of the total. On January 1, 1920, total value for all livestock was estimated to be $113.8 million, but the horse and mule share was only 53%, while the meat-animals share increased significantly.

Unfortunately, values for crops and livestock, were determined in two different years (1909 versus 1910 and 1919 versus 1920), which makes a comparison of the values of the two questionable; however, according to the census data, the total value of crops in 1909 was over two times greater than for livestock in 1910 ($144.3 million versus $65.3 million). In 1919 and 1920, the relative difference between the two was even greater ($304.3 million versus $112.2 million). In the United States, there was little difference between the values of crops and livestock in either 1910 or 1920 (Figure 10.77). The greater value for Alabama crops was primarily the result of the extremely high values for cotton. At this point, the state had made only limited progress in diversification. King Cotton continued to rule the roost.

Several factors were driving agricultural production

in the state and country in the period 1915-1920. The most important of these was the increased demand for food and fiber related to World War I. The war disrupted agricultural production in Great Britain, France and countries allied with them. As a result, those countries had to increase their imports rapidly. For example, before 1915, the United States exported 107 million bushels of wheat annually. In 1915, the country exported 243 million bushels. The increased demand for wheat was greater than for any other crop, but there was some increase in demand for just about everything American farmers produced (Figure 10.4). This increased demand persisted well beyond the end of hostilities in 1918. Much of the increased demand for food and fiber, especially crops, had to be met through a sharp increase in acres planted, because really large increases in yields were years in the future.

Increased demand related to the war was the most significant factor affecting agricultural production, during the period, but demand was also increasing as a result of population increases in the country (Figure 10.7). Each year there were more bodies to clothe and mouths to feed. In the United States, the population increased from 100.8 million in 1915 to 106 million in 1920. In Alabama, the change was not nearly so dramatic. In fact, the state's population barely changed, because large numbers of poor farmers, especially African Americans, responded to increased job opportunities in other states. Alabama's population only increased 3,174 (2,345,000 to 2,348,174) people from 1915 through 1920.

Another factor affecting demand was the general growth of the American economy during the early part of the twentieth century (Figure 10.1). Primarily as a result of growing industrialization and commercialization, people were growing more affluent and more able to purchase food and fiber.

STILL SEEKING STILL SERIOUSLY SEEKING SALVATION

The American farmer had faced severe problems and even financial ruin in the latter part of the nineteenth century, and it was critically important that they find some way to improve their lot. The result of their concerns had been the initiation of a so-called threefold plan (better information, farmer organizations, and diversification) to bring about their Salvation. Later, they would add another leg to the stool—political action. Farmers had hardly initiated their plan before—in the early twentieth century—they entered the Golden Age of American Agriculture, and went from those good times directly into the wartime boom. Their plan was working better than they ever imagined that it could. Then, starting in 1919, they began to encounter the back-side of the boom, and they were quickly up to their armpits in alligators once again. As it turned out, their plan had not accomplished much of anything.

Salvation Through Better Information

During the period 1915-1920 the Alabama Agricultural Experiment Station continued to expand its program to develop more and better information for Alabama's farmers. By 1918, there were 22 staff members working at the main station in Auburn. Two new subject-matter departments were created during the period: The Department of Agronomy and Soils in 1916 and Agricultural Engineering in 1917. With these two new departments, there were a total of nine in the experiment station.

In 1914, Congress passed the Smith-Lever Agricultural Extension Act. A primary purpose of the act was to separate research and extension activities at the land-grant colleges. The Alabama legislature did not accept the provisions of the new law until January, 1915. John Duggar became the director of the Alabama Extension Service, a department within the experiment station. In 1920, Luther N. Duncan, became director of the newly created, independent Alabama Cooperative Extension Service.

Nationally, the passage of the act and wartime needs for increased farm production had a significant effect on extension. In 1914, 928 counties in the United States had extension agents. By 1918, that number had increased to 2,435 (Genung, 1940). The actual increase in extension personnel was from 2,601 in 1915 to 6,728 in 1918. There was some reduction in extension effort, for a few years after the war, but it was evident from those wartime experiences that extension was here to stay.

By 1916, Alabama had been overrun by the boll weevil, and despite all the research that had been conducted to control the disastrous pest, it quickly became obvious that the insect and the devastation that it caused was not a temporary phenomenon. Consequently, research emphasis shifted somewhat to seek ways to diversify Alabama agriculture. In this new emphasis, research on cattle and dairy production was increased.

In 1917, Congress passed the Smith-Hughes Vocational Act. Its purpose was to promote vocational agriculture, to train people who have entered or who are planning to begin farming. Of course, the funds were assigned to the land-grant college at Auburn, but again Auburn signed another memorandum of agreement to give the responsibility for vocational agriculture for Negroes to nearby Tuskegee Institute (Mayberry, 1989).

Morrison (1994) reported that in 1918 the Department of Agriculture of the State Agricultural and Mechanical College for Negroes at Normal had land, livestock, farming, and teaching equipment, as well as its outreach program in cooperative extension that included boys' and girls' club work. The president of the institution noted that the school could not undertake planned research in agriculture because of the lack of funds. In 1919, the school was designated a junior college, and its name was changed to the State Agricultural and Mechanical Institute for Negroes.

Salvation Through Organization

At the national level, the Non-Partisan League was formed in 1917. Beginning in South Dakota, the league soon spread to Minnesota. Its purpose was to protest the poor market conditions for farm commodities and to encourage reforms. Considered by some to be a socialist organization, it was crushed in 1918. Between 1915 and 1917, the International Workers of the World (Wobblies) succeeded in organizing thousands of wheat-harvest workers across the country.

The Alabama Division of the American Farm Bureau Federation was established in 1920. Initially, there were organizations in ten counties with a total membership of 1,100. In the early years, C.W. Rittenour of Montgomery was president of the organization.

Salvation Through Diversification

Agricultural diversification had never been a major problem in the North; except that they needed to convert more of their feed grains into livestock, and fortunately livestock production had increased rapidly throughout that region. Unfortunately, production of feed grains was increasing just as rapidly.

In the South and Alabama the lack of diversification was still an extremely serious problem. There seemed to be no way to find other cash crops to replace cotton. In 1920, however, census statistics indicated that farmers were finally making some progress. In 1909, cotton and corn were harvested from 38.5% and 26.5%, respectively, of all improved farmland in the Alabama. In 1919, those crops were harvested from 26.8% and 34%, respectively. Acres of cotton harvested in the state were down 100,000 acres at the end of that decade. Unfortunately, as we will see later, Alabama farmers were not replacing cotton as their primary cash crop. Instead, they were trying desperately to learn how to live with the boll weevil.

Salvation Through Political Action

The Democrats had taken control of Congress in 1913 (63rd Congress), and they had quickly enacted some legislation designed to improve the lot of American farmers. Soon after they came to power, however, politics in Washington had to turn its attention to the war. Improving the lot of farmers through additional legislation would have to wait.

At the end of hostilities in 1919, there was every indication that Salvation for America's embattled farmers had finally arrived. The parity ratio (indices of prices received/ indices of prices paid) was near 1.10 Unfortunately, the following year (1920), the backside of the boom arrived, and the parity ratio fell below 0.9. As a result, it was quickly and painfully obvious that the millennium had not yet quite arrived, but most American farmers continued to believe that better information, organization, diversification, and political action would finally lead them out of the wilderness. With their belief in the plan still firmly in place, they concluded with considerable guidance from their leaders that its lack of efficacy was simply because they were not doing enough of it.

The Bottom Line

Net Farm Income (NFI) in the first third of the twentieth century (1910-1920) generally followed the same pattern of change as the broader-based Gross National Product (GNP) (Figures 10.1. and 10.79). Both generally increased over the 10-year period; however, the overall rate of change was somewhat greater for the measure of national productivity than for farm income.

Most American farmers benefited from the World War I boom. NFI had ranged around $4 billion in the early years of the twentieth century (Figure 10.79), and it was at around the same level in 1915 ($4.3 billion). By 1916, however, the wartime emergency was beginning to affect the agricultural economy, and NFI began to increase. It was at a level of $4.6 billion in that year, but a year later (1917), it had exploded upward to $8.3 billion, an increase of 81.7%. Growth of the NFI slowed in 1918 ($8.9 billion), the year hostilities ended, and increased only slightly in 1919 to $9.1 billion. It was down sharply in 1920 to $7.8 billion after commodity prices began to decline (Figure 10.5) in the first full year of the short recession.

Data on total NFI shown in Figure 10.79 were computed on a per-farm basis to correct for changes in farm numbers during the period. These data are shown in Figure 10.80. Both figures show the same general trends. Data presented in Figure 10.80 indicate that NFI per farm increased from $700 in 1916 to almost $1,400 in 1919, but then declined to below $1,200 in 1920.

Data in Figure 10.79 show that total NFI increased from $4.3 billion in 1915 to $9.1 billion in 1919, an increase of 111%. During the same period, the Consumer Price Index (CPI) increased by only 70% (30.4 to 51.8). These data indicate that a considerable share of the increase in NFI was the result of inflation in the currency.

Unfortunately, the wartime boom had little staying power. Farmers lost most of what little real benefit they had realized when NFI declined by 14% ($9.1 billion to $7.8 billion in 1920); while CPI increased by 16% (51.8 to 60.0). By 1920, the American agricultural economy began to suffer a serious postwar depression. Soon, farmers other than desperately poor Southern tenants, were receiving prices too low to pay their debts. Even Midwestern grain and dairy farmers were hurting, and for all hurting farmers there was another problem beyond the low commodity prices. All sectors of the American economy had benefited from the wartime boom, and prices for manufactured goods had risen sharply. When the boom ended and the depression began, prices for goods did not decline as much as the prices that the farmers received for their crops (Figure 10.6). Consequently, they were caught in a serious cost-price squeeze. For example, in 1920 with prices down and costs up, prices paid by American farmers increased during the war years (Figure 10.5). However, prices received and quantity of commodities that they were able to produce and sell were sufficiently large to give them three extremely profitable years (1917, 1918, and 1919), and those years were even better in the South. Fite (1984) commented that by the end of 1916 Southern farmers were benefiting from the greatest prosperity in their lives. Some producers had problems with poor weather and others with boll weevils, but the majority were better off than they had ever been. In 1919, Southern farmers harvested their first $2 billion cotton crop. Prices were through the roof. They increased from 11.2 cents a pound in 1915 to 35.3 cents in 1919. Corn and wheat prices had not increased nearly to that extent. Cotton was not only still King, it was a most beneficent monarch as well. Concerns about diversification were a distant memory.

But what a difference a year makes. In 1920, the average price of cotton fell to 16 cents, not much higher than it had been in 1915. Unfortunately, prices paid continued to roar ahead. Soon, it took two bales of cotton to purchase the same quantity of non-farm goods that could be purchased with a single bale before the war.

By 1920, both the Golden Age of American Agriculture and the wartime boom were distant memories. The need for Salvation was just as great—especially in the South—as it had been in 1894. American farmers had had some exceptional years during that period, but their basic problem remained the same. There were too many of them, and they were too productive, and nothing in their plan for salvation was really addressing that problem.

11

All That Glitters Is Not Gold

1920–1930

PRECIS

At the end of the nineteenth century, American agriculture, especially in the South, was seriously degraded. Fortunately, the plight of most of the country's farmers had improved considerably during the Golden Age of American Agriculture and during World War I. Unfortunately, following the end of hostilities, farmers found it extremely difficult to downsize their operations and to bring their wartime production capacity in line with peacetime demand. Consequently when the national economy accelerated into the Roaring Twenties, the country's agricultural economy skidded to a halt and quickly lurched into reverse. Hardly a segment of agriculture was spared.

Unquestionably, American agriculture had made considerable progress in the past 30 years. In general, farms were better equipped with more and better machines, and yields for many crops had improved. However, most farmers, especially those in the South, still could not make a decent living farming.

INTRODUCTION

As detailed in a preceding chapter, the tranquility and prosperity enjoyed by American farmers during the Golden Age of American Agriculture (1895-1915) was shaken when Austria-Hungary declared war on Serbia (July 28, 1914) to begin World War I. It was truly shattered on April 6, 1917 when the United States declared war on Germany. American agriculture had begun to increase production after 1914, because of the expectation of growing demand and higher prices, but after the United States formally entered the war, American agriculture was changed forever. Americans and American farmers, had never before been so deeply involved in such an enormous conflict on such a broad scale. As a result, farmers quickly increased total farmland, acres planted, equipment and machinery inventories, and livestock numbers. Prices received for their crops quickly more than doubled, but prices paid for inputs increased at about the same rate. Furthermore, farmers' cost-of-living, along with all other American families, also increased. Then, on 11 November, 1918, the greatest and most costly war the world had yet known came to an end. Unfortunately, the war had given birth to an industrial and agricultural behemoth. With the end of hostilities, the problem of quickly taming this awesome creature became an immediate and extremely difficult task. Mobilization itself had been an extremely difficult problem, but demobilization, especially for the American farmer, was equally if not more difficult.

NATIONAL POLITICS

In 1914, the nasty, partisan politics of taxes, tariffs, public works, banking, and currency of the early years of the twentieth century were quickly replaced with equally nasty but not so partisan, politics, when Austria-Hungary declared war on Serbia. Then, for the next five years, America's politics first revolved around efforts to remain neutral in the conflict while maintaining support for the Allies, then around military preparedness, and finally on the war itself. When it finally ended, nasty, partisan politics turned its attention to the peace conference in Paris, the League of Nations, and foreign policy.

The Presidency

The Democratic Party controlled the presidency for only a single year (1920) of this period (1920-1930). A Republican became president in 1921, and other members of the Republican Party held the presidency throughout the decade.

The Democrat, Woodrow Wilson (NJ) had been elected president in 1913. He served two terms (1913-1921) before being replaced by Republican Warren G. Harding (OH) in 1921. Harding died while in office in 1923, and was succeeded by his vice president. Calvin Coolidge (R, MA) (1923-1929) completed that term, and was then elected to serve a term himself. He was replaced in 1929 by Republican Herbert Hoover (CA) who served until 1933.

Secretaries of Agriculture

Edwin Meredith (D, IA) served as secretary of agriculture from 1920 to the end of Wilson's tenure in 1921. Other persons serving in that cabinet position were: Henry C. Wallace (R, IA) (1921-1924); Howard Gore (R, WVA) (1924-1925); William Jardine (R, KS) (1925-1929); and Arthur Hyde (R, MO) (1929-1933).

The Senate

The Republican senators barely controlled the 66th Congress (1919-1921) 49 to 47. They were much more firmly in control of the 67th (1921-1923), 59 to 37. They continued their control in the 68th (1923-1925), 53 to 42 and the 69th (1925-1927), 56 to 39. They barely retained control of the 70th (1927-1929), 48 to 47, but expanded it again in the 71st (1929-1931), 55 to 39.

Senate Majority Leaders

The Republicans did not officially designate a Senate majority leader until 1925. Henry Cabot Lodge Jr. (R, MA) was the first official Senate majority leader of the Republicans (68th Congress). Lodge died in 1925, and Charles Curtis (R, KS) was elected to the post. Curtis continued in that position until 1929 when he was elected vice president. James E. Watson (R, IN) was elected to succeed him.

Chairmen of Senate Committee on Agriculture

Asle Gronna (R, IA) served as chairman of the Senate Agriculture Committee from 1919 to 1921. Other persons serving in that capacity included: George Norris (R, NE) (1921-1925) and Charles McNary (R, OR) (1925-1933).

The House of Representatives

Republicans regained control of the House of Representatives in 1917 (64th), and were still in control in 1929 (70th).

Speakers of the House

Frederick H. Gillett (R, MA) was elected speaker of the House for the 66th Congress (1919-1921). He continued in that position through 1925 and was replaced by Nicholas Longworth (R, OH) in the 69th Congress. He continued in that office through the end of the 71st Congress (1929-1931).

Chairmen of the House Committee on Agriculture

Asbury F. Lever (D, SC) was chairman of the House Agriculture Committee in the 65th Congress (1917-1919), and in 1919 he was replaced by Gilbert N. Haugen (R, IA) who remained in the position through 1931.

The Political Situation

Wilson ended his second term (1917-1921) an extremely sick man, and when he vacated the Office of the President he left the country in a sick mood. The country could reach no agreement on the League of Nations. Serious race riots throughout the country left hundreds dead and wounded. Wilson's attorney general arranged for the arrest of thousands of alleged communists in 33 cities around the country. Morison (1965) comments that the people were plain "tired of being pushed around by rationing, wartime restrictions, drafts and such." Wilson was too ill to be nominated for a third term. In his place, the Democrats nominated James A. Cox of Ohio, with Franklin D. Roosevelt for vice president. The Republicans nominated Warren G. Harding, also of Ohio. Harding won the 1920 election with 61% of the popular vote.

Harding's primary effort in the early months of his administration was to attempt to slow the naval arms race that was beginning to capture the imagination of some of the world's major powers. The primary outcome of months of haggling left the Japanese with the most powerful navy in the western Pacific and with the loss of strong American military bases west of Pearl Harbor.

In the summer of 1921, Franklin D. Roosevelt who had been Wilson's secretary of the navy was stricken with polio. The illness and his efforts to cope with it would have far-reaching implications for the United States, especially its farmers.

World War I had left a pall of starvation, misery, and desperation over much of Europe. Harding's administration, along with private charity, made a determined effort to alleviate the worst of the suffering. The government worked to radically reduce the war debt and to reduce the excessively heavy burden of Germany's war reparations. Unfortunately, America's official generosity was unpopular, and efforts were quickly made by the Republican Congress to recapture some of the giveaways. In September, 1922, they passed the Fordney-McCumber Tariff, which was the most stringent tariff legislation ever enacted by the United States. The tariff made it virtually impossible for Europeans to trade themselves out of their economic misery by exporting goods to America.

By every measure, Harding's administration was the most corrupt since Grant's. It was beset by one scandal after another. Even though scoundrels seemed to be everywhere, their dishonesty seemed to have little effect on America's rapidly growing economy. Mass production in industry was producing a growing wealth of consumer products. This new consumerism quickly gave rise to the so-called Roaring Twenties. In the midst of the national furor over the misdeeds of his appointees and the rapidly growing economy, President Harding died. In August, 1923, Vice President Calvin Coolidge (R, MA) took the oath of office to replace him.

Coolidge commented that the "business of America was business," and the business of the Roaring Twenties roared ahead in his administration. He firmly believed that government should interfere as little as possible with the lives of American citizens. He vetoed congressional efforts to provide veterans of World War I with financial assistance, but his veto was over-ridden. Later, he vetoed the McNary-Haugen bills, which were designed to provide much-needed economic relief for farmers—who were not benefiting from the boom of the Roaring Twenties. But most of America was reveling in the Coolidge prosperity. The country quickly became the richest in the world, and its economic boom soon lifted economies throughout the world.

Coolidge might have wanted the Republican nomination for the 1928 election, but he made little or no effort to secure it. As a result, his secretary of commerce, Herbert C. Hoover of California, was selected. The Democrats nominated Alfred C. Smith. Hoover won with 58% of the popular vote. Hoover took the oath of office early in 1929, and in October the stock market crashed destroying the fortunes and lives of untold numbers of Americans. The exact causes of the sudden economic downturn were many and complex, but overly optimistic speculation in stocks was certainly a contributing cause. The crash seemed to initiate a downward spiral of prices, production, and employment. Then the Republican Congress probably made things worse in 1930 when they enacted the Hawley-Smoot Tariff, the most regressive ever enacted by the United States Congress. The economic contraction quickly became a depression, and the Hoover administration was unable to do anything to slow its progress. The administration was probably hampered, at least in the beginning, by Hoover's Coolidge-like philosophy that government action was likely to make the situation worse.

ALABAMA POLITICS

Flynt (1994) comments that the people living in Alabama in 1920 believed that they were witnessing the dawning of a new era when, in fact, they were living in the twilight of an old one. Alabamians continued to be plagued with problems that had been with them for decades. Some 16% of all Alabamians, 10 years of age or older, were illiterate. The state still had no place to house its convicts, so they were leased to owners of coal mines and timber products industries. As many as 1,200 children under the age of 16 were working in textile mills. Only 50% of the state's black elementary-age children attended school at all, and those that did only attended for about four months a year. Census estimates of assessments of property value for tax purposes averaged only 25% of true value, although the tax code required that that the rate should be 60%.

Governors

Thomas E. Kilby (D, Calhoun County) was elected governor in 1919 and served until 1923. In 1923, he was replaced by William W. Brandon (D, Tuscaloosa County), who served until 1927, and was replaced by Bibb Graves (D, Montgomery County). Graves served until 1931.

Commissioners of Agriculture

Miles C. Allgood (D, Blount County) was elected commissioner of agriculture in 1919; he served until 1923. James M. Moore was elected to succeed him in 1923. Samuel M. Dunwoody was elected in 1927, but died in 1929. Governor Graves appointed Seth P. Storrs (D, Talladega) to fill Dunwoody's unexpired term. Storrs was elected to the office in 1931.

Political Situation

In 1920, Governor Kilby began his term in office with a bang. The Sage Foundation in its report on conditions in Alabama had provided him with a clear roadmap of problems that desperately needed attention. He made the most critical of them the basis of his legislative agenda, and by 1922 they had enacted every bill he proposed: state finances, care for the insane and feeble-minded, treatment of children, education, public health, road construction, and state docks.

William W. Brandon won the race for governor in 1922, but Kilby had effectively dealt with so many of the state's most pressing problems that Brandon had little to do in his term. Brandon was fortunate in that the economic expansion of the Roaring Twenties was pouring new money into the state's coffers to fund the programs established under Kilby. Brandon attempted to deal with the convict leasing problem; although he did make some progress, in many ways it was worse at the end of his administration than at the beginning.

Bibb Graves had lost the 1922 election to Brandon, but when the election was decided he began to campaign for 1928. Graves was duly elected. A primary element of his campaign was the abolition of the convict leasing system. In June 1928, legislation removing the last convicts from the mines went into effect. With taxes flooding into the state treasury, Graves was able to convince the legislature of the need to expand funding for all levels of education. His education reform package called for the expenditure of $25 million, and was the largest amount ever appropriated for that purpose in Alabama history. Graves had been elected governor in 1927, and the stock market crashed in the fall of 1929. With the sudden end of the Roaring Twenties, taxes necessary to fund the expanded social and educational services put in place under Kilby, Brandon, and Graves began to dry-up.

Flynt (1994) commented that in the 1920s the Democratic Party had controlled Alabama politics since Reconstruction. This long period without any serious opposition had resulted in the evolution of a political structure with little or no capability to deal with disagreement within its ranks. More opposition may or may not have improved the effectiveness of state government, but one thing it did accomplish was to guarantee years of seniority for Alabama's congressional delegation. Once elected to Congress, most of them kept their positions as long as they were physically able to serve. Ever growing seniority virtually guaranteed, given the management structure of the Congress, that state's power in the national body would steadily increase.

In 1928, the national Democrats nominated a Roman Catholic, Alfred C. Smith of New York, as its candidate for president in the forthcoming national election. Almost over night, the Alabama Democratic Party came apart at the seams. The Republicans nominated Herbert Hoover of California. Suddenly, Yellow-Dog Democrats were forced to vote for an Irish Catholic or a Republican. The political bloodletting was a spectacle to behold, and it was to continue well after the election. When it all ended, Smith carried Alabama by around 7,000 votes. Most of his margin of victory came out of the Black Belt. Hoover actually won 27 of Alabama's 67 counties.

Alabama, Arkansas, Georgia, Louisiana, Mississippi, and South Carolina, along with Massachusetts and Rhode Island, were the only states carried by Smith. This Yellow-Dog response to the election would serve the state well after 1932.

THE NATIONAL ECONOMY

World War I hostilities ended in November 1918, and the economic boom generated by the war ended as well. The Gross National Product (GNP) began to decline in 1919. Then as Europe began to recover and the United States began to demobilize, the 1919 economic down-turn was quickly replaced by a sharp, short-lived depression (1920-1921). Fortunately, this depression did not last long, and it was soon replaced by the Roaring Twenties's speculative boom. Then near the end of 1929, the Roaring Twenties begat the beginning of the Great Depression. In October 1929, the stock market crashed wiping out 40% of the paper values of all common stock. Mobilization for World War I had resulted in the development of an industrial complex of almost unimaginable capacity. After the end of the war, however, it quickly began to over take the capacity of consumers to consume, and by 1929 the country was awash with unneeded industrial capacity— machines and people, especially farmers.

The Economic Situation

The speculative boom of the 1923-1929 period was really a continuation of the prosperity that had begun before World War I. The recession of 1920-1921 had only slowed it temporarily. It was called a speculative boom because Americans were beginning to take their savings out of banks and bonds to speculate in the stock market.

The widespread prosperity and the American role in World War I ushered in a period of enhanced self-confidence and general elation. Everyone expected the good times to last forever. In this period, installment buying first appeared in the American economy and resulted in a rapid increase in domestic consumption. Real estate and stock market speculation exploded on the national scene; however, not everyone shared equally in the benefits of the boom. Many industrial workers (textiles, coal, railroads) certainly were unaware that national wealth was growing rapidly. As a result, the country witnessed a series of dogged strikes.

The great speculative boom of the 1920s began to unravel at the seams in 1929. The severely over-priced stock market's crash began on October 23 during the last hour of trading. Over the following few days, there were spectacular losses. This downward spiral of stock prices continued almost unabated until 1932.

Several factors were probably responsible for the beginning of this truly enormous American disaster. These included: excessive public and governmental optimism about the future, agricultural and industrial over-expansion, labor-saving machines that led to a reduced need for workers, surplus capital that fueled speculation in real estate and stocks, and the decline of international trade (exports) as economic stagnation began to grow in Europe. Then the enactment of the Hawley-Smoot Tariff Act in 1930 probably made matters even worse than they should have been.

In Alabama, mobilization associated with World War I had resulted in rapid growth of the state's industrial economy (Flynt, 1994), and soon it became the most diversified industrial state in the South. In 1925, there were 66,800 industrial wage-earners working in some 30 different categories of industry. Of these workers, 23,400 were in textile mills; 24,200 in iron and steel mills; 8,500 in railroad shops; and 10,700 in a variety of other industries. While the increased industrialization was a God-send, it barely scratched the surface of Alabama's poverty-stricken population. Fite (1984) reported that the Alabama farm population was 1.3 million in 1925. The 66,800 wage-earners represented only 2.6% of the state's population of 2.5 million persons.

The Gross National Product

The GNP increased steadily after 1915 and was at a level of $570 billion (constant 1982 dollars), in 1918 (Figure 10.1), but then the bottom fell out. By the end of 1921, it had declined over $117 billion ($510 billion to $453 billion).

In 1922, the GNP at the beginning of the speculative boom was $520 billion. Just seven years later in 1928 it had increased to $668 billion. This increase represented real financial growth. The Consumer Price Index (CPI) remained relatively stable during this period (Figure 10.2). Probably the greatest single factor explaining this remarkable prosperity was the postwar development of new industries (automobiles, electrical appliances, radios, petroleum, road building, etc.) that began to employ more people at higher wages.

Because the 1929 crash came so late in the year, it did not affect GNP as much as it might have had it occurred earlier. It actually increased $47 billion from 1928 to 1929 ($668 to $710 billion), but then declined to $643 billion in 1930. The Great Depression had arrived.

The Roaring Twenties began in 1922 and came to a screeching halt in 1929. It required only eight years for Americans of all sorts to realize that "all that glitters is not gold."

The Consumer Price Index

Consumer prices (Consumer Price Index) did not immediately begin to decline when the short, postwar depression (1919-22) began (Figure 11.2). They continued to increase even after the GNP (Figure 11.2) began to decline in 1919. The CPI increased from 45.1 in 1918 to 51.8 in 1919 and then to 60 in 1920, before it began to fall back to 53.6 in 1921 and then to 50.2 in 1922. Surprisingly, consumer prices remained relatively stable during the speculative boom of the Roaring Twenties, but they certainly did not retreat very much either.

As early as 1927, the CPI began to warn of some unusual economic activity when, amidst the wave of speculation, it declined from 53 in 1926 to 52 a year later. It declined again in 1928 to 51.1; then remained stable at that level in 1929, before beginning its depression-driven, downward spiral in 1930.

The Inflation Rate

The inflation rate is another way of quantifying changes in consumer prices. The inflation rate (Figure 11.3) shows the effect of the short, but severe recession in the early 1920s. It was still at a war-driven level of 15.8% in 1920, but in the next two years, it was less than 0 (-10.7% in 1921 and - 6.3% in 1922). It climbed back above 0 at the beginning of the speculative boom in 1923, but as noted in a preceding paragraph, it did not increase as rapidly as might have been expected. In fact, it remained below 3% through 1925, and was 0.0, or less, from 1927 through 1930. The 1929 and 1930 values (0 and -2.5%), respectively, are indicative of the rapidly worsening economy leading toward the Great Depression.

THE AGRICULTURAL ECONOMY

Genung (1940) commented that World War I was an important turning point for American agriculture. American agricultural exports to Europe had begun to decline around the turn of the century. As a result of a trade imbalance with the United States, many of those countries were turning elsewhere for their imports of agricultural products. The war reversed this trend temporarily, when the United States made large loans to them that could be used for the purchase of food and fiber. After the war, when the loan program was terminated, exports began to decline once more. Another cause was growing efforts by the Europeans to increase their own production of agricultural commodities. Growing competition from Canada, Argentina, Australia, and New Zealand also contributed to the decline. These countries had also increased exports during the war, and they were determined to maintain as much of their share as possible after the hostilities ceased. The decline in exports took place over a relatively short period time—a period much too short to allow for an orderly readjustment of

American agricultural production. The problem was made more difficult by the growing productivity of the farm sector, partially as a result of the growing replacement of animal power on farms with mechanical power.

It had been obvious since the end of the nineteenth century that American farmers lacked the capability of saving themselves from themselves. With Southern cotton farmers, this fact became apparent much earlier. It was also becoming obvious that improved information, farmer organizations, and diversification were not going to bring Salvation. This left only the fourth part of the fourfold path—political activism.

Federal Agricultural Legislation

As a result of declining parity ratios after 1920, the Federal Land Bank System was no longer able to provide the amount of short-term credit that farmers needed; consequently, Congress approved the Agricultural Credits Act of 1923. This act added 12 intermediate credit banks (FICBs) to the Farm Credit System. Unfortunately, these institutions were flawed by geographical imbalance and a long and complicated loan approval process. As a result, the system became less and less responsive to farmer needs as the Great Depression approached.

In 1921, Congress enacted the Packers and Stockyards Act and the Grain Futures Trading Act. A year later, they passed the Capper-Volstead Act, which exempted farmer cooperatives from the antitrust laws. In 1923, they passed the Agricultural Credits Act, but in all of this effort the Republican Congress was steadfastly steering clear of enacting legislation that would involve the federal government directly and heavily in the marketplace of American agriculture.

As a result of the melt-down of American agriculture during the recession of 1919-1922, farmers began to apply so much pressure to Congress that they decided they had to do something even if it was against their better judgement. The result was the introduction of the McNary-Haugen Farm Relief Bill in 1924. This legislation was designed to encourage the establishment of member-owned cooperatives that would remove price-depressing surpluses of certain crops from the domestic market by selling the excess, above national needs, on the world market at going world prices. It would, in effect, establish a two-price system. And it was expected that by protecting the controlled domestic supply behind a strong tariff fence that prices would increase at home. Losses incurred in the sale of the excess on the world market would be recouped by charging a tax on domestic sales. The legislation would have been much more helpful to grain farmers, because they only exported 15 to 30% of their crop annually. It did not appeal to cotton farmers who regularly exported 50 to 60% of their annual production. As a result, Southerners did not wholeheartedly support the bill. Only after cotton prices plunged again in 1926 (to 10 cents) did cotton farmers join the effort to get the legislation approved. Finally, it passed in 1927, after tobacco and rice were added to the crops included in order to gain more support.

The bill, as passed, provided an appropriation of $250 million to help establish and stabilize the necessary cooperatives. President Coolidge (1923-1929) vetoed the bill, and a revised one passed in 1928. He was strongly opposed to any direct federal involvement in the market; however, he was a strong supporter of the establishment of member-owned cooperatives as a solution to the growing problems in agriculture. In 1925, Secretary of Agriculture William Jardine commented that "the most distinct and significant movement in American agriculture in this decade is the almost universal trend toward cooperation in marketing and distribution of farm products. . . . A movement of this magnitude with its tremendous economic and social significance, must be analyzed and guided so that its highest possibilities may be realized." With encouragement from the Coolidge Administration, Congress passed the Cooperative Marketing Act of 1926, which established the Cooperative Service within the Department of Agriculture. Coolidge's support of this legislation and of cooperatives, in general, did not change his basic reluctance to see the federal government involved directly in supporting them. As a result, he vetoed a revised McNary-Haugen Bill in 1928. He commented that this bill constituted "unwarranted federal interference in the economy." But the die had been cast. The barrier in Congress to federal intervention in the marketplace had been shattered. In 1929, President Hoover signed the Agricultural Marketing Act that, among other things, provided funds to member-owned cooperatives to hold crops off the market until prices increased. The old Sub-Treasury Plan of the Agricultural Alliance was about to see the light of day.

Alabama Agricultural Legislation

In 1923, the Alabama legislature enacted important legislation regarding the Alabama Department of Agriculture and Industries. It called for several changes in the department and its operations. It increased the department's capability to support the growing complexity of agriculture in the state. The primary legislation was known as the Agriculture Code of Alabama. It abolished the old Board of Agriculture, and established the new State Board of Agriculture that was designed to serve as the controlling authority for the re-organized department. The act also abolished the old Board of Horticulture and the Livestock Sanitary Board and transferred their responsibilities to the commissioner of agriculture and the department. The act reorganized the department into four divisions:

1. Division of Agricultural Chemistry, which replaced the old Division of Food, Drugs and Feed.
2. Division of Plant Industry, which assumed the activities of the old Board of Horticulture.
3. Division of Weights and Measures, which took responsibility for the inspection and regulation of measuring devices and weights used in commercial transactions.
4. Clerical and Records Division, which was organized to handle the department's administrative functions.

Under the act, the new state board was given oversight responsibility for the work of the state veterinarian at Auburn. The board also oversaw Alabama's share of the cooperative soil survey work with USDA and farmer education work being conducted by the Extension Service.

Operating under the provisions of the code, in 1924 the department entered into an agreement with the US Bureau of Agricultural Economics for the development of a Crop and Livestock Reporting Service. This new cooperative effort is reflected in much of the data on Alabama agriculture used in this book. During this same year, the department also became more active in promoting industrial development in the state.

The code was revised and re-written in 1927. The new version left departmental organization unchanged, but added more responsibilities and enforcement powers. It also provided for the establishment of a new state laboratory and an Industrial Development Board, which was to intensify efforts to attract new industry to Alabama.

In 1923, the Alabama legislature, following the lead of other Southern states, enacted legislation that established a yield tax on harvested timber (Nix, 1994). The tax was assessed as a percentage of the total value of timber harvested. It was considered to be a property (ad valorem) tax. The tax rate was set at 8%, and the tax was collected the year the timber was harvested.

Agricultural Exports

American agricultural exports had increased sharply during the war years, 1915-1919 (Figure 11.4). They were at $1.5 billion in 1915 and $3.8 billion in 1919. The primary reason for the increase was the quantity of food and fiber shipped to Europe during and immediately after World War I. Exports did not immediately decline after the war. In fact, the annual dollar amount remained at near the same level ($1.7 billion) for most of the period 1921-1929. A $200 million loan made to assist German recovery after World War I encouraged American businessmen, already involved in the speculative boom in the United States, to invest heavily in Europe, as well. These investments helped to create a new era of prosperity on the Continent between 1924 and 1929. This prosperity, in turn, helped to maintain the market for American agricultural products, which had been established during the war. The stock market crash in 1929 seriously affected the economies of both the United States and Europe, and as a result American exports began to fall rapidly. In 1930, they were back near the 1900-level of $1 billion.

Although significant quantities of food and fiber continued to be exported to Europe during the 1920s, the quantity was somewhat less than the dollar amount would indicate. Prices for food and fiber had increased substantially during the war (Figure 11.5), and they remained considerably higher than prewar levels until 1929.

Throughout the nineteenth century, American agricultural exports made up between 70 and 80% of all exports, and for much of that period, cotton was the most important crop exported. However, between 1865 and 1880 wheat exports increased sharply. At the beginning of the twentieth century, the relative importance of agricultural exports began to decline, until in the decade of the 1920s, they comprised only 42% of the total.

Agricultural Prices

Indices of prices received for agricultural commodities and prices paid for inputs required in farm production during the period 1915-1930 are presented in Figure 11.5. The indices were computed using prices for the period 1910-1914 as a base value of 100. By 1919, shortly after the end of World War I, the indices were at levels of 217 and 199, respectively. They had both been at levels of 100 in the period 1910-1914. Agricultural prices collapsed in the middle of 1920. Davis (1940) noted that the prices received index in July was 10 points below the June index. Then in August, it was down an additional 15 points, and it declined 15 more points before the end of September. Altogether, the index declined 40 points between early June and the end of September. The 1919-1921 depression had begun. By the end of 1921, the prices received index was down to 124. It had declined 41% in a year. With the beginning of the speculative boom in 1922, it began to increase and reached 156 by the end of 1925. Then ranged between 140 to 150 through 1929. In 1930 at the beginning of the Great Depression, it fell sharply to 125.

In 1920, the prices paid index was considerably higher than the prices received index (241 versus 211). It also declined sharply in 1921. In that year, the annual averages for the two indices were 167 and 124, respectively. The prices paid index remained at around 170 through 1924 when it began to drift downward to near 160. It remained near that level until 1929. Then in 1930, at the beginning of the Great Depression, it fell another 10 points to near 150.

The sudden implosive collapse of prices received in 1920 was totally unexpected, and it quickly created an alarming disparity with prices paid. Farmers everywhere began to protest vehemently. Membership in existing agricultural organizations began to increase, and new organizations were born. Protest was so vociferous that Congress quickly created a joint commission of inquiry to study the situation. This activity was followed by calling for a national agricultural conference in 1925. This sharp decline in prices received, in reality, was the opening salvo that ultimately forced the federal government to stick its nose under the tent of American agriculture.

Figure 11.6 presents data on the ratios (parity ratios) of the indices of prices received to the indices of prices paid for the period 1910-1930. In 1919, the ratio was 1.09 (217/199). Then, in 1920, at the beginning of the short recession, it fell to 0.88, and it continued to fall in 1921, reaching 0.74. After 1921, the ratio trended slowly upward to 0.9 and remained there until the beginning of the Great Depression. From 1920 through 1924, the ratio was considerably less than 0.9, which indicates the reason for the farmer's growing desperation. Anytime the ratio is below 1 farmers are being squeezed between prices received and prices paid, and the further the ratio falls below that level, the more painful the squeeze becomes.

Farmers

According to census data, the American farm population grew to 32 million between 1900 and 1910 and remained essentially at that level until 1917. It had been increasing steadily since at least 1840. Then, the year the war ended (1918) it reached 32 million when thousands of farmers were either called to service or left the farm for war work.

The Total Population

Data on the census of all persons in the American population indicate that it increased from around 106.8 million in 1920 to 123.5 million in 1930 (Figure 11.7), an increase of 15.6%. The population of Alabama was 2.359 million in 1920 and 2.647 million in 1930 (Figure 11.8), an increase of 12.2%. Alabamians made up 2.2% of the American population in 1920 and 2.1% in 1930.

The Farm Population

Data on the number of persons in the American farm

population during the period 1910-1930 are presented in Figure 11.9. These data show that numbers increased from 32 million in 1920 to 32.1 million in both 1921 and 1922, as servicemen and war workers returned to their farms. Then with the parity ratio (Figure 11.6) near 0.87 in 1923, numbers began to decline sharply finally reaching 30.5 million in 1927. It remained near that level through 1930. Hordes of farmers seem to have decided that they had no future in American agriculture. Fortunately, at that time, the economy was roaring ahead, creating thousands of new jobs in new industries.

As a result of the upward trend in the population (Figure 11.7) and the continued decline in the farm population, the percentage of farmers in the total population declined from 30.1% in 1920 to 27% in 1925, and then to 24.9% in 1930 (Figure 11.10).

Fite (1984) reported that total Alabama farmers numbered 1,355,000 in 1920; 1,343,000 in 1925; and 1,344,000 in 1930. The number of farmers did not fall nearly as fast in the period, as it did in the United States as a whole. Likely, there were relatively fewer jobs away from the farm in Alabama compared to the country as a whole.

The percentage of farmers in Alabama's population was around two-times as high during the period as in the country. It was 57.3% in 1920; 52.9% in 1925; and 50.8% in 1930 (Figure 11.10)).

The number of black farmers in Alabama continued to decline between 1920 and 1930, but at a much lower rate than between 1910 and 1920 (Figure 11.11). There were 110,000 in 1910; 95,000 in 1920; and 94,000 in 1930. The sharp decline between 1910 and 1920 was the result of large numbers of this group of farmers going to the North to take jobs during World War I.

Efficiency of Farmers

An important characteristic of American farmers has been their overall efficiency in providing the food and fiber required by the American public. One measure of this efficiency is the ratio of the total number of farmers to the total number of citizens. For example, in 1920 the ratio of the total population to the farm population was 3.3 (106.8 million/32 million). This ratio indicates that one farmer was able to provide for the food and fiber needs of 3.3 people. Other ratios for the period 1900 to 1930 are presented in Figure 11.12. The data indicate that a single farmer met the food and fiber needs of 2.5 persons in 1890, and that by 1930 one farmer could meet the requirements of 4.1 persons. Thus between 1900 and 1930, farmer efficiency increased by 64%. Of course, there are several problems with the validity of the ratio, but it does provide a rough measure of the changing efficiency of the American farmer over time. These ratios further emphasize the importance of creating ever increasing numbers of off-farm jobs for redundant farmers. Year after year the country required fewer farmers.

Farmers in the Work Force

Data on the percentage of farmers in the labor force during the period 1870-1930 are presented in Figure 11.13. During the period 1920-1930, the percentage continued the downward trend that began early in the nineteenth century. It was at 27% in 1920, and reached 21% in 1930. The downward trend consisted of two parts: an absolute decrease in numbers of farmers in the country, and an absolute increase in the numbers of persons in the labor force. Of the two factors involved, the increase in the labor force probably has more effect on the percentages.

Tenancy

Farm tenancy in Alabama was 57.9% in 1920 (Figure 11.14). It had been 60.2% in 1910. It had fallen slightly during the war years when so many tenant farmers had found off-farm employment; however, with demobilization underway, tenancy began to increase again and in 1925 reached 60.4%. Even after the speculative boom was well underway, tenancy continued to increase in Alabama, reaching 64.7% in 1930—near the highest level in history.

The percentage of Alabama farms operated by tenants increased steadily from 1900 through 1930 (Figure 11.44), but actual numbers generally followed the trends of farm numbers (11.15). The really surprising thing about these data is that when numbers of farms began to increase after 1925, the actual number of tenants also increased. There were almost 22,000 more tenants (144,200 versus 166,700) in the state in 1930 than in 1925. One can possibly understand why some landowners would come back to their farms when the price situation began to improve, but it is difficult to understand why tenants would do so unless they had no other choice.

According to census data, about 31% of Alabama's tenants in 1920 were cash tenants. These tenants personally provided all of the inputs (labor, tools, seed, animals, etc.) required to produce their crops and livestock. They simply rented land. Some 27% were share tenants. This group provided their labor and some of the inputs required; the remaining inputs were provided by the land-owners. Share tenants were paid with a share of the crops and livestock produced. The largest group (32%) were the so-called croppers. They brought nothing to the production enterprise, but their labor. The land-owner provided everything else, including living expenses. In this relationship, the land-owner received a much larger share of crops and livestock produced.

One of the reason there were so many tenants in Alabama stemmed from the fact that in 1920 Alabama ranked fourth in the country for the number of children under five years per native white mother (Odum 1936). It had likely been at that level for decades. The reproductive rate of Alabamians was 67% higher than necessary to sustain the farm population. Large families were an asset to farm operation, but they required more resources, and when the children grew to adulthood and established their own families, they had few opportunities to earn a living other than tenant farming. Reproductive rates in none of the Midwestern farming states were nearly this high.

Year-to-year trends of tenancy in the country as a whole were similar to those in Alabama (Figure 11.14), but the levels were considerably lower. It is likely, however, that the national data were biased upward by the high levels of tenancy throughout the South.

Farms

Numbers of American farms increased rapidly during the Golden Age of American Agriculture and during the years of World War I, but with hostilities concluded and with diminishing requirements for food and fiber by our European allies, it would be interesting and important to see whether or not farm numbers would or could continue to increase

There were 6.5 million American farms in 1920. Some 6.8% of them were in Texas. Other states with large shares included: Georgia (4.8%), Mississippi (4.2), North Carolina (4.2), and Kentucky (4.2). These five states accounted for 24.2% of the total, and all of them were in the South.

Farm Numbers

Data on the numbers of farms in the United States during the period 1910-1930 are presented in Figure 11.15. Numbers had trended rapidly upward from 1910 to 1920; then in mid-1920, prices received began to fall. Prices paid also declined, but not to the same extent (Figure 11.5). It was quickly obvious that farming was a losing proposition, and farm numbers began to trend sharply downward. By 1927, the country had lost 58 million farms. The Depression forced many farmers off their land, but with the speculative boom in full-swing in the mid-1920s, many of these eternal optimists decided to take another look at farming. In 1928, numbers began to increase. With all kinds of ominous clues indicating that the boom was ending and that it would likely be replaced by at least a serious recession, they continued to establish more farms. In 1930, on the eve of the greatest economic disaster in the history of the United States, there were 6,546,000 farms in the country, the highest in history.

There were 256,000 farms in Alabama in 1920 (Figure 11.16). There had been 265,000 in 1910. Farm numbers in Alabama followed essentially the same trend as those in the country, during the period 1920-1930. The number trended downward sharply through the early 1920s. This decline in Alabama had begun sometime between 1900 and 1910. Surprisingly, it continued right through the war years until 1924. In that year, numbers were down to 250,000, some 15,000 fewer than in 1910. Numbers remained at that level in 1925. Unfortunately, however, after 1925 a sizeable number of Alabamians, just like their counterparts throughout the country, became infected with the bug of the speculative boom, and on the eve of the Great Depression, numbers began to increase again, finally reaching 261,000 in 1930, not far below the 265,000 in 1910.

Farm Size

Estimates of the average size of American farms are not available on an annual basis before 1979; however, the US Censuses have recorded this data for well over a century. Data in the 1920 Census indicate that average size in the United States in that year was 148 acres (Figure 11.17). It had been near that level since the turn of the century. It remained near that level in 1925, but increased to 157 acres in 1930. Those new farms (Figure 11.15) established in the late 1920s were larger than existing farms.

Census data indicate that average size of Alabama farms was 76 acres in 1920 (Figure 11.17). It had been 93 acres in 1900. Average size was larger in Georgia (82 acres) and smaller in Mississippi (67 acres) than in Alabama in 1920. Average size was 157 acres in Iowa that year. In 1930, average size in Alabama fell to 68 acres. This decrease was probably related to the increase in tenancy detailed in a preceding section (Figure 11.14). In 1930, average size increased in Georgia to 86 acres, but declined further in Mississippi to 55 acres. It also continued to increase in Iowa to 158 acres.

Total Farmland

Census data indicate that in 1920 there were 955.9 million acres in farmland in the country (Figure11.18). There had been 838.6 million in 1900. With farm numbers declining (Figure 11.15), acres fell to 924.3 million in 1925. The speculative boom of the Roaring Twenties and the improved prices situation (Figure 11.6) pulled four million acres back into farming after 1925. It was up to 986.8 million in 1930, but by that year, it was quickly becoming obvious that the country had much more farmland than it needed.

Census data (Figure 11.19) indicate that acres of farmland in Alabama in 1920, 1925, and 1930, totaled 19.6, 16.7, and 17.6 million acres, respectively. The trends in the Alabama data were essentially the same as that in the country as a whole.

Uses of Farmland

In 1919, 42% of all American farmland was used for crops, and it remained near the same level through 1929 (Figure 11.20). A smaller percentage (35%) was used as pasture, and it increased to 38% in 1929. The small increase in pastureland came at the expense of forestland on farms, which declined from 18 to 15%. Other uses (house lots, roads, wasteland, etc.) remained around 6%.

There are no data published on the uses of farmland in Alabama before the 1925 Census of Agriculture; consequently it is not possible to estimate the percentage used for crops earlier than 1924. The percentages for 1924 and 1929 are shown in Figure 11.21. The share in cropland changed little between the two years (53 to 54%). Very little farmland was used for pastureland during this period. The percentage increased from 3 to 4%. Forestland was much more common, and it increased from 35 to 37% during the period. The percentage of other land decreased from 10 to 6%.

The percentage of Alabama farmland in cropland was much higher than in the country in 1924 and 1929 (Figure 11.20). There were a large number of small farms in the state in 1924. With such a large number of small farms, most of their acreage had to be used for crops. The data show that very little of Alabama's total farmland was used for pasture. Those hundreds of thousands of small farms did not produce enough cattle to need much pastureland.

Total Cropland

The 1920 Census reported that there were 955.9 million acres of improved land in farms in the United States. In 1925, the first US Census of Agriculture was published. That census no longer included data on improved land in farms. Those data were replaced with estimates of total cropland. Further, the estimates of total cropland obtained before 1954 were collected in the year before publication of each census. The census data for 1924 and 1929 indicate that there were 505 and 522.4 million acres, respectively, of cropland in the country. For the same two years, Alabama cropland totaled 8.8 and 9.4 million acres, respectively. In 1929, 1.8% of the country's cropland was in Alabama.

Cropland Harvested

Census data on cropland harvested in the United States were available as early as 1879 (1880 Census). Some of these national data are presented in Figure 11.22. Cropland harvested generally followed the same trends as total farmland (Figure 11.18) in 1919, 1924, and 1929. It trended upward from 283.2 million acres in 1899 to 348.6 million in 1919. Then with prices received for agricultural commodities falling rapidly after 1920 (Figure 11.5), cropland harvested fell to 344.6 million acres (68.2% of total cropland) in 1924. However, with the advent of the Roaring Twenties, farmers began to plant and harvest more crops and cropland harvested increased to 359.2 million acres in 1929 (68.8% of total cropland).

The trends in the Alabama cropland harvested data were similar to those in the country. Cropland harvested in the state increased from 6.7 million acres in 1899 to 7.3 million, in 1919. Then it declined to 6.6 million in 1924 (75.3% of total cropland) before increasing to 7.1 million (75.6% of total cropland) in 1929.

Fite (1984) reported that in 1919, farmers on individual Alabama farms harvested crops from an average of only 28.4 acres of cropland (Table 11.1). In 1899, it had been 30.1 acres. So, not only were farm numbers decreasing (Figure 11.16), but acres harvested were also diminishing. The average for 11 Southern states in 1919 was 29.4 acres. That same year, Iowa farmers harvested crops from 95.7 acres. In 1929, acres harvested in Alabama was even lower than it had been in 1919 (28.4 versus 27.6), and the average in all of the Southern states was also lower (29.4 versus 28.7). However, in Iowa acres harvested increased to 103.6 acres.

Value of Land and Buildings

USDA data on total value of land and buildings on American farms during the period 1910-1930 are presented in Figure 11.24. These data show that value was $34.8 billion in 1910, and that it increased to $66.4 billion in 1920. However, after the short recession in 1920 and 1921, it fell to $47.6 billion in 1927. It is not surprising that value would decline between 1920 and 1927. Farm numbers were also declining (Figure 11.15), but what is surprising is that value did not begin to increase afterwards when numbers began to increase sharply. For this to have happened, the price of land must have been decreasing. It is possible that land became severely overpriced during World War I, and that it was reverting to its normal level in the late 1920s. Unfortunately, there are no data available to evaluate this hypothesis.

This decline in value after 1920 supports the supposition that the Great Depression actually began on American farms in the early 1920s. Possibly, the farm economy was on a different track than the national economy.

During the period, inflation of the currency played only a very limited role in changes in value. The Consumer Price Index (Figure 11.2) remained essentially unchanged.

Value data when computed as dollars an acre (Figure 11.25) show essentially the same pattern as the total value data, except that the decline between 1925 and 1930 was more pronounced. It was $69 an acre in 1920, but then declined to $54 in 1925, and to $48 in 1930. It had not been that low since 1916.

Only census data are available on the total value of land and buildings on Alabama farms during the period 1900 to 1930. They are presented in Figure 11.26. The data indicate that value in Alabama followed the same trend as in the country as a whole, at least through 1925. It increased from $288.2 million in 1910 to $543.7 million in 1920, before falling to $414.9 in 1925. Then, surprisingly, in contrast with national data, it increased to $502.4 million in 1930. Farm numbers and land in farms increased substantially in Alabama between 1925 and 1930 (Figures 11.16 and 11.19).

Data on value in dollars an acre for Alabama farms are also presented in Figure 11.25. These data show that values in the state were much lower than in the country as a whole. In 1920, it was 41% of the national level, but in 1930 it had increased to 60%.

Value of Farm Machinery and Equipment

When the first (1925) Census of Agriculture was published, the editor chose to substitute the terminology "farm machinery and equipment" for "farm implements and machinery" that had been used since 1850. Further, in 1910 USDA began to publish data on the value of machinery and equipment in the country on an annual basis. These data for the period 1910-1930 are presented in Figure 11.27. Total value had been $1.8 billion in 1915. Then during the wartime boom, it increased to $3.6 billion in 1920. With farm numbers declining after 1920 (Figure 11.15), value also began to decline, finally reaching $2.8 billion in 1923. Although farm numbers did not began to increase until 1927, value of machinery and equipment increased after 1923, and finally reached $3.3 billion in 1930. It is difficult to explain why value increased while numbers were continuing to decline. As we will see in a later section, net farm income was increasing rapidly during this period. Possibly, farmers were using some of this new-found wealth to upgrade the complement of machinery and equipment on existing farms.

Obviously, these data are dependent on changes in farm numbers. As a result, values in Figure 11.27 have been computed on the basis of value per farm. These data, which are presented in Figure 11.28, also generally followed the same trend as farm numbers (Figure 11.15), which means that the short recession resulted in many farmers leaving their farms, but it also meant that farmers who kept their farms reduced the amount of money that they had been spending to maintain their machinery and equipment. In 1920, value per farm was $557. Five years later, with farm numbers down sharply, it had declined by over $100 ($557 versus $422). Surprisingly, when farmers began to move back to their farms after 1927, they did not increase the value of machinery and equipment as much as might have been expected. In 1930, value per farm was still about $32, below what it had been in 1920. Farmers were rapidly establishing new farms, but they could not maintain the pace of equipping them. This is another indication that the Great Depression arrived early on American farms.

US census data on the total value of machinery and equipment on Alabama farms during the period 1900-1930 are presented in Figure 11.29. There are no annual data available. They show the same general pattern of change as the national data. Value in the state increased from $16.3 million in 1910 to $34.4 million in 1920, but then declined to $23.8 million in 1925. This decline in value was also accompanied by a sharp reduction in farm numbers (Figure 11.16). But then as farm numbers began to increase afterwards, values also increased, reaching $33.5 million in 1930.

The total value of machinery and equipment data for Alabama, computed on a per farm basis, are also presented in Figure 11.28. They follow essentially the same pattern as the national data. Values in 1910, 1920, 1925, and 1930 were $63, $134, $100, and $130, respectively. The most glaring difference are in values in the same year. Values per farm in the state were only one-fourth as large as the average for the country. Some of this disparity, but not all of it, was the result of the difference in the average size of farms in the state and the country (Figure 11.17). They were slightly more than half as large in Alabama. Of course, the difference in average size was only part of the disparity. Those thousands upon thousands of poor tenant farmers in Alabama simply could not afford to own more than the bare necessities. Technologically, these farms had changed very little since 1840.

Animal Power on Farms

Mules powered the explosion of Southern cotton production in the nineteenth century, and horses had done the same for Midwestern grain production. The numbers of both continued to increase well into the twentieth century. However, by 1920, times were changing, especially for horses. Around 1915, farm tractors began to make in-roads in the horse population. Data on the numbers of horses in the country in the period 1900-1930 are presented in Figure 11.30. In 1915, there were 21.4 million horses in the country. By 1925, numbers were down to 16.7 million, and even with farm numbers increasing rapidly after 1927 (Figure 11.15), it continued to fall and reached 13.7 million in 1930.

In Alabama, numbers of horses had begun to decline at the beginning of the twentieth century (Figure 11.31). In 1915, there were 140,000 horses in the state. By 1920, numbers were down to 130,000. Then by 1930 there was only about half that number (66,000). Tenant farmers did not have much use for horses.

Horses had always outnumbered mules, and by 1915 there was only about one mule for every five horses in the country. While the popularity of horses began to wane, around 1915 numbers of mules continued to increase. In fact, numbers of mules did not begin to decline until around 1925. Then between 1925 and 1930, numbers declined from 5.9 to 5.4 million. Numbers did not decline rapidly in the country, because they remained so popular in the South.

In Alabama, numbers of mules continued to increase along with numbers of tenant farmers. In 1915, there were 281,000 mules in Alabama, and 296,000 in 1920. The small but growing number of tractors in the state seemed to have little effect on numbers of mules. Their number increased from 296,000 in 1920, to 303,000 in 1925, and to 322,000 in 1930. Numbers declined between 1921 and 1922, but otherwise the reduction in farm numbers (Figure11.16), did not seem to affect mule numbers. Tenancy (Figure 11.14) would not reach its highest level (64.7%) in Alabama until around 1935.

Total numbers of work animals are generally positively related to numbers of farms. For this reason, the total numbers in Figures 11.30 and 11.31 have been converted to number per 100 farms. These data are presented for horses for both the country and state in Figure 11.32, and for mules in Figure 11.33. The converted data for horses show the same general trend as the totals data, but they further emphasize the national preference for horses compared to mules. Similarly, the converted data show the overwhelming preference for mules in Alabama. It is likely that national data are biased upward by much larger numbers of mules in the South.

Tractors on Farms

In 1910, there was only one tractor for every 6,400 farms in the country, and by 1915 the ratio had increased to 1 per 258. Then just five years later, in 1920, this ratio had increased to an astounding 1 per 26 farms (approximately 4 tractors per 100 farms). In that year, 3.8% of all farms had tractors. By 1925, this percentage had increased to 8.5% (9 per 100 farms), and even with the growing economic uncertainty on farms in the late 1920s, the percentage increased to 14 (15 per 100 farms) in 1930.

As might be expected, the large majority of Alabama farmers did not have the resources to purchase. Further, on most of those farms there was not enough cropland being farmed to justify the purchase of a tractor. There are no data on tractors on Alabama farms until 1920. In that year, census data indicated that there were only 811 tractors in the entire state. Only 0.3% of Alabama farms had tractors. By 1925, 1% had them, and by 1930 the percentage had grown to 1.8%—about a seventh as high as the country as a whole.

Use of Fertilizer

About the time the first Census of Agriculture was taken in 1925, USDA began to publish more detailed data on the use of fertilizer in the United States. Some of that data indicates that in 1923 North Carolina farmers with its large number of small farms used the largest share of fertilizer (16.1%). Other states using large amounts included, in descending order: South Carolina (10.8%), Georgia (10.5), Alabama (7.0), and Virginia (6.6). These five states accounted for 51.4% of all the fertilizer used in the country that year. All of those five states had large numbers of farmers trying to grow crops (mostly cotton) on land worn-out decades earlier.

Table 11.2 presents data on the use of fertilizer in the various regions of the country. As expected, in all years the largest share was used by farmers in the South Atlantic and South Central regions. Farmers in the North Central Region used relatively lesser amounts. Soils in that region, deposited during the last ice age, contained such large amounts of residual plant nutrients that farmers had to use relatively less commercial fertilizers to get acceptable yields. Further, virtually none of the land in the region had been subjected to depredation by cotton production.

Data on fertilizer use per acre of harvested cropland on American farms during the period 1899 to 1929 are presented in Figure 11.34. These data indicate that use increased relatively little between 1919 and 1924 (38.7 to 39.7 pounds an acre). Prices of agricultural commodities fell sharply after 1920 (Figure 11.5). Farmers probably understood the value of using fertilizer; they just did not have the money required to purchase it, even if it were beneficial. Prices received began to slowly increase after 1921, and farmers apparently began to increase their purchases of fertilizer again. In 1929, they used 44.4

pounds of fertilizer for each acre of harvested cropland.

Data presented in Table 11.2 indicate that throughout that period (1923-1929) Alabama farmers used 7 to 8% of the total quantity of fertilizer used in the country. In 1924, they used 914 million pounds of fertilizer and harvested crops from slightly more than 6.6 million acres. They used 137.6 pounds of fertilizer for each acre of cropland harvested in 1924 and 189.9 pounds in 1929. Both estimates are three to four times as great as the national averages for those two years (39.7 and 44.4 pounds, respectively).

Plant Agriculture

In 1915, the value for all crops harvested added $3.9 billion to the national economy. By 1920, as a result of wartime inflation, value added had increased to $8.3 billion. It fell sharply during the short recession to $4.1 billion in 1921. In 1915, value added by crops accounted for 48.9% of the total, and by 1920 the percentage had increased to 52.9%.

In terms of value added, corn was the most important crop produced in the country in the period 1915-1920. In all six years of the period, except 1920, corn's value added was greater than those of wheat and cotton combined. Value added for cotton was greater than that for wheat in only two of the years (1916 and 1917). In terms of the amount of land dedicated to the production of the major crops, corn was also much more important. In fact, in the period 1915-1920, acres of corn harvested was greater than those for both wheat and cotton in all but one (1919) of the years.

Although there were no value added data available for Alabama crops for the period 1915-1920, census data for 1920 indicated that the value of all crops produced in Alabama in 1919 totaled $304.3 million. This value includes those for nursery and greenhouse and forest products. This value represents the end of the wartime economy in 1919. The value for cotton accounted for the larger share of the total in 1919 (59%). The value for corn accounted for the next larger share (19%). In 1919, Alabama farmers harvested more acres of corn than cotton (3 versus 2.5 million acres).

Alabama Crop Weather

Palmer Drought Severity Indices for Alabama (Cook, et al., 1999) for June-August in each year from 1920-1930 indicate that in these months at least half of Alabama was wetter than normal in five of the years (1920, 1923, 1927, 1928, and 1929), drier than normal in five of the years (1921, 1924, 1925, 1926, and 1930) and normal in only one year (1922). A total of four storms originating in the tropics passed through Alabama during the period (1922, 1923, 1926, and 1928). All were classified as tropical storms.

Cotton

Although there was considerable year-to-year variation in bales of cotton harvested in the United States, it had been trending upward for decades. It finally reached 16 million bales in 1914, the highest level in history. Then the boll weevil struck with a vengeance. The march of this pest across the South was much more destructive than the march of Union forces across the South half a century earlier. Between 1914 and 1915, bales harvested declined almost a third to 11.2 million. Fortunately, the average price responded positively to the reduction of the crop, and farmers actually made more money on smaller crops. Bales harvested remained at near the 1915 level until 1920 when measures were finally put in place in the fields to reduce the destructiveness of the insect to a limited degree.

US Cotton

In 1920, American cotton farmers harvested 13.4 million bales. In 1900, they had harvested 10.1 million. Texas farmers harvested 32.3% of the 1920 cotton crop. Other leading cotton-producing states that year included: South Carolina (12.1%), Georgia (10.5), Oklahoma (9.9), and Arkansas (9.0). Alabama farmers harvested 4.9%.

Exports

In 1920, the United States exported 6 million bales of cotton (Figure 11.35), which was equivalent to 53.6% of the 1919 crop of 11.1 million bales (Figure 11.39). Exports remained generally unchanged in 1921 (6.3 million bales), but declined in 1922 (5 million) following a sharply smaller American crop in 1921. Exports responded to the speculative boom of the mid-1920s.

In 1926, they reached 11.3 million bales (the highest in history, at that point). This extremely high level followed the harvest of the record (at that time) crop in 1925 (16.1 million bales). However, a much smaller share of the even larger 1926 crop (18 million bales) was exported in 1927 (7.9 million bales). Near the end of the 1920s, the speculative boom was ending in Europe, and the economic events leading to the Great Depression were beginning to appear. By 1930, exports were down to 7.1 million bales and represented the equivalent of only 48.1% of the 1929 crop.

Prices

American cotton prices during the period 1900-1930 are presented in Figure 11.36. Cotton prices have always been dependent to some limited extent on exports. However, the tendency of cotton farmers to produce more than the market could absorb often completely overwhelmed exports as a price-determining factor.

Cotton prices averaged 35 cents a pound in 1919, a year after World War I ended. Between 1919 and 1920, bales harvested increased from 11.1 million to 13.4 million, and in 1920 the average price fell to 15.9 cents. In 1921, bales harvested fell sharply as a result of a much lower average yield (Figure 11.38). In response, price increased slightly. Then from 1922 through 1929, bales harvested and prices teeter-tottered through the period. In 1930, they both declined as the Great Depression settled-in on American agriculture. Between 1928 and 1930, prices remained around 14.5 cents.

Acres Planted and Harvested

In 1919, the average price of American cotton was 35.3 cents, the highest level in the twentieth century, to that time. In response to that unusually high price, farmers planted 35.9 million acres, and when they harvested that crop, the price was down to 15.9 cents. In 1921, farmers responded by reducing acres planted to 29.7 million, and the price increased to 17 cents. This slight increase in price and the euphoria of the speculative boom sent farmers into a frenzy. Between 1922 and 1925, acres planted increased from 32.2 million to 46 million. They planted another huge crop in 1926 (45.8 million acres). Under the onslaught of these enormous crops, prices began to fall after 1923, and by 1926 they were down to 12.5 cents. In 1927, cotton farmers finally emerged from their dream-world and planted only 39.5 million acres. As expected, average price increased to 20.2 cents. Prices and acres planted teeter tottered in 1928 and 1929, with acres planted reaching 44.4 million in 1929. In 1930, both acres planted and prices declined, as the Great Depression began. Generally, acres harvested was considerably higher at the end of the period than at the beginning. This is somewhat surprising, since prices generally declined.

American farmers harvested a high percentage of the acres they planted (Figure 11.37) during the period 1920-1930. The average for the 11-year period was 96.8%, but in seven of the years, it was higher than 97%. In those years, virtually all of the cotton was harvested by hand, and pickers were paid by the pound, so there was almost no crop too poor to be picked.

Yields

American cotton yields were extremely variable during the period 1920-1930. Farmers harvested an average of 133 pounds an acre in 1922 and 193 pounds an acre in 1926. That 133-pound average yield in 1922 was the lowest recorded in the United States since 1866. By 1921, the boll weevil had just about completed its invasion of the South. It would finally reach the Carolinas in 1922. The Cotton Belt did not experience the full effect of the insect's "march to the sea" until around 1921, and the 133-pound yield was the result. Fortunately, by 1922 farmers were beginning to use chemicals and cultural methods to reduce the full impact of the invasion, and average yield began to increase. By 1925, it had increased to 174 pounds an acre, and to 193 in 1926. By 1926, average yield was back to the level of the pre-weevil period. Then, the insect mounted a counter-attack, and yields began to decline again. They were 162 pounds an acre in 1927, and they remained at near that level through 1930.

Bales Harvested

American cotton farmers harvested 13.4 million bales in 1920 (Figure 11.39). Then in 1921 with acres harvested at a 15-year low (Figure 11.37), and with yield at 133 pounds (Figure 11.38), bales harvested experienced a free-fall to 7.9 million. It had not been that low since 1895. Then with both acres harvested and yields increasing, bales harvested increased rapidly. The total crop reached 18 million in 1926, the highest in history to that time. After that record-setting crop with both acres harvested and yields lower, bales harvested declined to 13 million bales in 1927. The size of the crop increased slightly in 1928, and remained around 14 million bales through 1930.

The disastrously low parity ratios of the early 1920s (Figure 11.6) that followed the end of World War I again brought forth a chorus of calls for cotton farmers to control production. The president of the American Cotton Association, J.S. Wannamaker, made one of the most far-reaching and drastic proposals that had been made when he suggested, in 1920, that farmers be strongly encouraged to voluntarily enter into legally-binding contracts to reduce their 1920 acreage by 50% when they planted their 1921 crop. If a farmer violated his contract and planted more, all of his crop above the 50%-level could be destroyed. Further, it was recommended that fields be measured to prevent over-planting. As might be expected, there was virtually no interest in such a proposal. A decade later, in the midst of the Great Depression, Wannamaker's proposal would not seem that outlandish. There was widespread discussions in cutting production in 1921 to raise prices, but the result was the same as it had been for almost three-quarters of a century: No organized effort resulted.

In the middle years of the 1920s, other changes began to take place that would make it even more difficult to control production. Cotton farming began to spread rapidly into Oklahoma and Texas, and producers in those states quickly began to mechanize, which reduced their labor costs. This change gave them a distinct advantage over producers in the Deep South. They could produce unbelievably large quantities of cotton, and produce it much more cheaply than tenant farmers in the old Cotton Belt.

Following the 1926 calamity (18 million bales at 12.5 cents a pound), there again was widespread discussions among agricultural reformers concerning ways to get farmers to reduce cotton production. As a result, they mounted the most intensive campaign ever to get farmers to voluntarily reduce their 1927 crop acreage by 25%. Farmers responded by cutting acreage by 12%, but this reduction was largely the usual response to a bad crop year (1926), rather than a change in philosophy. Prices did recover somewhat in 1928 and 1929, and farmers immediately increased the amount of land devoted to the crop to over 45 million acres, which resulted in the production of between 14 and 15 million bales each year. Planned production in response to market demands was still in the future—maybe even after the Second Coming. A calamity would have to get much worse than anything that they had seen in the past before any kind of effective program could be implemented.

In retrospect, it might be somewhat difficult to understand why cotton farmers, through all of those years from the 1870s onward, could never agree to cut production; however, in the real world in which those farmers lived, it just never made sense to do so. After all, hundreds of thousands of them were living on farms of less than 50 acres, and they were cultivating only a fraction of that acreage. In Alabama in 1900, 154,000 of 257,000 farms (59%) were less than 50 acres, and well over half,were operated by share tenants. Further, at least a portion of their cultivated land had to be devoted to producing food. Additionally, cotton was the only crop they could produce that had a guaranteed market and that could be exchanged for at least some cash. Still another factor, were the low yields. For most of this period, production in much of the Southeast was considerably less than 200 pounds an acre. Then there was the problem of the declining quality of the soil base that had resulted from years of uncontrolled erosion and inadequate fertilization. Furthermore, the soil had not been very good to begin with. This deterioration made it necessary to continuously find new land to clear. Unfortunately, there was almost none to be found.

Controlling production also ran counter to 10,000 to 12,000 years of accumulated human experience and culture. Never in the years since the coming of agriculture had it been possible to obtain more by producing less. Throughout this long period of time, with the almost countless years of famine and starvation, there was never much room for error. Producing more was always better than producing less. The need to produce as much as you possibly could for as long as you could was etched indelibly in the minds of the species. It would take much more than the lectures from reformers to erase it.

When the individual tenant cotton farmer began to weigh all of those factors in his mind as he began to make preparations for the next crop, there was only a single reasonable alternative—plant more. In his analysis that was the only possible solution to earning more money to pay his debts, to purchase some decent clothes for his family, to fix his house, or to purchase his own place. He hoped, in fact in many cases he prayed, that he might grow a good crop in a year when prices were good. Unfortunately, the basic principle of the teeter totter made that combination (a big crop and high prices) extremely unlikely. Only when there was every indication that prices would be really bad when he was ready to sell his next crop, did he decide to plant less.

Alabama Cotton

During the period 1910-1914, Alabama farmers annually harvested 10.4% of the national cotton crop. However, in 1915 the boll weevil began to seriously reduce production in the state. That year, farmers harvested only 9.1% of the national total. Then in 1916, the Alabama cotton crop began to suffer from the full impact of the insect damage. In the period 1916-1920, Alabama cotton farmers averaged harvesting only 5.4% of the total national crop.

In 1919, farmers in Madison County harvested the largest share (4.8%) of the Alabama cotton crop. Other counties reporting substantial shares included: Marshall (3.7%), DeKalb (3.4), Limestone (3.4), and Cullman (3.3). These five counties only accounted for 18.6% of the total crop. All of them are north of Birmingham. The county average was 10,700 bales, and the average per farm was 2.8 bales. Some 53 counties reported harvests of 6,000 or more pounds.

Prices

There are no annual prices for Alabama cotton before 1979.

Acres Planted and Harvested

Year-to-year changes in acres planted in Alabama during the period 1920-1930 generally followed those in the country as a whole. Generally, cotton farmers everywhere responded to prices in the same way. As a result, prices and acres planted generally teeter tottered across the period, but the net result was that acres planted generally increased throughout the period—from 2.2 million in 1921 to 3.6 million in 1930 (Figure 11.40). Acres planted were off-trend in 1925 and 1927 when prices plunged the year before. This relatively long period of increase in acres planted is somewhat surprising since prices generally declined throughout the period (Figure 11.36).

Alabama cotton farmers harvested a slightly higher percentage of acres planted than did farmers in other states. In Alabama, the average percentage of acreage actually harvested was 98% for the 11-year period. The national average for the same period was 96.8%. In Alabama, in six of the 11 years, farmers harvested 98% of acres planted. With so many small cotton farms, Alabama farmers had to get all the cotton they could from every acre.

Yields

Alabama farmers suffered grievously from the ravages of the boll weevil. During the period 1910-1914, average yield was 200 pounds an acre, then in 1920 they harvested an average of 124 pounds an acre. In 1923, many farmers lost almost their entire crop, and they harvested only 93 pounds an acre, the lowest in history. Problems with the boll weevil, were intensified by wetter than normal weather from June to August, 1923. After that year, farmers were better able to cope with the pest, and yields began to increase. Farmers harvested 206 pounds an acre in 1926. Yield was lower in 1928 (154 pounds an acre), probably as a result of wetter than normal weather in late summer plus continuing problems with the weevil. In 1930, the average yield was 196 pounds an acre.

Generally speaking, for much of the period average yields in Alabama were essentially the same as in the country; however, it appeared that farmers in the state had more problems with the boll weevil than farmers in other states. One reason for the difference was that a significant portion of Alabama's cotton was produced in the Black Belt during that period. The structure of the soils in that region did not lend themselves to changes in cultural practices that were used to reduce the depredations of the insect, to some extent, in other regions.

Bales Harvested

Data on bales harvested in Alabama during the period 1900-1930 are presented in Figure 11.41. Bales harvested slowly trended upward from the beginning of the twentieth century until 1914 when it reached 1.7 million. Then the full force of the boll weevil invasion began to exert itself. In 1916, bales harvested fell to 530,00 bales, the lowest in the twentieth century. Although there were some ups-and-downs afterwards, bales harvested was still at a level of 583,000 in 1923. By 1924, control measures against the pest were beginning to pay off, and yields began to increase. In 1926, bales harvested totaled 1.5 million. It was lower in 1927, 1928, and 1929 when yields were lower. In 1930, the average yield was near the pre-weevil level , and bales harvested reached 1.5 million. At the end of the Roaring Twenties, bales harvested were not back to their Golden Age of American Agriculture level, but they were certainly better than they had been in the early 1920s.

In 1929, farmers in Madison County harvested the largest share (4.2%) of Alabama's cotton crop. Other counties reporting substantial shares included: Marshall (3.8%), Limestone (3.6), Cullman (3.5), and DeKalb (3.3). All of these counties were north of Birmingham. These five counties accounted for only 18.4% of the total crop. The county average was slightly less than 19,600 bales; the farm average was 5.1 bales. A total of 41 counties reported harvests of 15,000 or more pounds.

The boll weevil changed agriculture in Alabama and the South. By the 1920s, the insect was causing so much damage that tenants in some areas completely abandoned their farms. Flynt (1989) commented that in 1921 the pest reduced the cotton crop in Alabama by 32% compared to the year before. Its depredations would finally lead to the complete abandonment of cotton production in Alabama's Black Belt, which for years was one of the leading cotton-producing areas in the world. Afterwards, the boll weevil would become a permanent part of the cotton economy of the South. Farmers would finally learn to reduce losses caused by the insect to a tolerable level, but always at a relatively high cost. This higher cost added another nail to the coffin of Southern cotton production.

Corn

In 1900, American farmers harvested 35 bushels of corn, for each person in the country, and the ratio remained essentially the same in 1905. However, after 1910 the ratio began to decline. In 1910, 1915, and 1920 bushels harvested per person was 30, 28, and 25, respectively. While there are no data to support the hypothesis, I suspect that overall demand was beginning to decline in the country, as wheat replaced corn in the diets of consumers, especially in the South. Also, around 1915 numbers of work animals were beginning to decline.

US Corn

In 1920, American corn farmers harvested 2.7 billion bushels of the grain. Iowa harvested the largest share of the 1920 crop (15.8%). Other leading states contributing to the total harvest were: Illinois (10.9%), Nebraska (9.0), Missouri (7.0), and Indiana (6.8). These five states harvested 49.5% of the national corn crop. Alabama farmers harvested only 1.6% of the total.

Exports

In 1920, the United States exported 2.6% of the 1919 corn crop (70.9 million of 2.7 billion bushels). From 1907 to 1917, exports ranged around 45 million bushels (Figure 11.42). They were considerably lower in 1918 and 1919 when prices were unusually high (Figure 11.43). Prices were extremely low in 1920, 1921 and 1922, and exports exploded upward, reaching 180 million bushels in 1921. After 1922, prices increased to around 75 cents a bushel, and exports began to decline. In 1930, exports reached 3.3 million bushels, the lowest since 1869.

Prices

During the period 1920-1930, corn exports represented such a small proportion of total American production that it is not likely they affected prices very much. In fact, as suggested in the preceding paragraph, it is likely the reverse was true. In most of the years, exports seemed to be dependent on prices.

During the war years, American farmers had received extremely high prices for their corn ($1.44 a bushel) in 1919 (Figure 11.42). In 1920, they harvested 2.7 billion bushels, but by that time price was down to 55 cents. During this period, the teeter-totter relationship between supply and price was tenuous at best. Bushels harvested generally trended down for the entire period (1920-1930), while prices in most of the years remained relatively stable at around 70 to 80 cents. Both the highest price ($1.02) during the period and the lowest bushels harvested (1.9 billion) came in 1924. Otherwise, there seemed to be little relationship between the two. In 1930, with bushels harvested at its lowest level during the period (1.8 billion bushels), price was at 55 cents.

Then as the 1920-21 short recession began, corn prices declined sharply along with prices of all other agricultural commodities (Figure 11.5). The average price received for that year was 54 cents a bushel.

Acres Harvested

Because data on acres of corn planted each year in the United States are not available in NASS records before 1926, I have elected to use only acres harvested for the entire 1920 -1930 period. American farmers harvested corn from 90.1 million acres in 1920 (Figure 11.44). Neither the 1920-1921 short recession nor the speculative boom of the mid-1920s seemed to have any effect. It generally trended downward for the entire period; although there was some year-to-year variation. In 1920, acres harvested totaled 90.1 million, but by 1930 it had declined to only 85.5 million.

I suspect that the downward trend in acres harvested after 1920 could be the result of increasing competition from sorghum grain as an animal feed. There are no NASS annual data on the production of sorghum until 1929. In that year, American farmers harvested 50 million bushels from 3.5 million acres. Although these are the first published data on production of the grain, it is likely that acres planted had been increasing for some time. The sorghum harvest in 1929 represented only 2.3% of the 2.1 billion bushel corn crop. However, the growing presence of a competitive product may have been sufficient to cause a slow downward trend in acres of corn planted. Also remember that total numbers of work animals was beginning to decline, during the period (Figure 11.30).

Yields

During the period 1920-1930, American corn yields tended to trend slightly downward, from about 30 bushels an acre in 1920 to 27 in 1929. There was relatively little year-to-year variation in the data (Figure 11.45), except in 1924 and 1930. In both of those years yield was much lower than expected. The low yield in 1924 was likely the result of unusually wet weather throughout the corn belt in June, July, and August. The low yield in 1930 was likely the result of unusually dry weather in the same region

Bushels Harvested

With both yields and acres harvested generally trending downward during the period 1920-1930 (Figures 11.44 and 11.45), bushels harvested also generally trended downward from 2.7 billion bushels in 1920 to 2.1 billion in 1929. Prices (Figure 11.43) seemed to have little effect on bushels harvested. The teeter-totter phenomenon seems to have been put on-hold in corn production during the Roaring Twenties.

Alabama Corn

Alabama farmers harvested 1.1% of the national corn crop in 1900. The percentage increased slowly until it reached 1.6% in 1920. As might be expected, corn production was widely distributed in the state in 1920. Farmers in 48 of the 67 counties harvested at least 500,000 bushels of corn that year. Farmers in Jackson County harvested the largest share (3.8%). Other counties with large shares included: Madison (2.9%), Covington (2.6), Coffee (2.6), and Marshall (2.5). These five counties only accounted for 14.4% of the total state crop.

Prices

Average Alabama corn prices during the period 1920-1930 were generally 30 to 40 cents a bushel higher than average national prices (Figure 11.43). Year-to-year changes generally followed the same pattern.

Acres Harvested

There are no data available on acres of corn planted in Alabama before 1926; consequently, I have chosen to use only data on acres harvested in this section. In 1920, the average price of Alabama corn was $1.10 a bushel (Figure 11.43). The price had not been that low since 1915. The following year (1921), Alabama farmers apparently decided that they could compensate for low prices by planting and harvesting more acres. As a result, they harvested 3.4 million acres in 1921 (Figure 11.47). They had only harvested 3 million a year earlier (1920). Unfortunately, in 1921 the price declined to 73 cents, and farmers decided that they could not plant and harvest themselves out of economic difficulty. In 1922, acres harvested began to decline, and in 1928 reached a level of 2.6 million. It increased slightly in 1929, and in 1930 reached 2.8 million.

Yields

Figure 11.45 includes data on the average number of bushels of corn harvested an acre in Alabama and the country during the period 1900-1930. Throughout the period, yields in Alabama were 12 to 14 bushels an acre lower than the national average. Yields ranged from a low of 10.5 bushels an acre in 1930, to highs of 14.5 bushels in 1920 and 1926. Year-to-year variation was relatively low, and there did not seem to be a trend in the data over time.

With corn yields ranging from 12 to 14 bushels an acre below the national average, it is difficult to understand how Alabama farmers could continue to produce the crop. A comparison with corn belt states makes the case even more compelling. For example, in 1920 the average yield in Alabama was 14.4 bushels. For Illinois, Indiana, and Iowa, it was 35.0, 39.5, and 46.0, respectively. There was no real economic reason to continue to produce the crop, but they had to do it because corn was the staff-of-life for thousands of tenant farmers. It was a mainstay in their diets, they fed it to their yard chickens, they used it to finish their hogs in the fall before hog-killing time, and occasionally they would give a few ears to their mule. Those small farms ran on corn. It provided the energy to plant, cultivate, and harvest their few acres of cotton, their only source of cash. Southern farmers planted corn, because they had no other choice (Locher and Cox, 2004).

Bushels Harvested

With yields remaining relatively unchanged during the period 1920-1930, bushels harvested was primarily determined by acres harvested. Bushels harvested generally trended downward for the entire period, from around 43.6. million bushels in 1920 to 29.1 million in 1930. The downward trend was interrupted in 1921 when acres harvested was higher than expected and in 1926, 1927, and 1929 when average yields were better than expected.

Farmers in Lauderdale County harvested the largest share (2.9%) of Alabama's 1929 corn crop. Other counties reporting substantial shares included: Jackson (2.9%), Limestone (2.8), DeKalb (2.7), and Cullman (2.5). These five counties only accounted for 13.8% of the total crop. The county average was slightly less than 532,600 bushels, and the per-farm average was almost 139 bushels. A total of 43 counties reported harvests of 400,000 or more bushels of corn.

Alabama Peanuts

Some areas in Alabama, Georgia, North Carolina, Texas, and Virginia were producing substantial quantities of peanuts before World War I. During this period, peanuts were used primarily as feed for hogs. Presses for removing the oil were not widely available. As a result, market prices for peanuts were extremely low. From 1910 through 1915, the national price ranged between 4 and five cents a pound. Then after World War I began in Europe, prices began to increase rapidly, reaching better than nine cents a pound in 1919. They fell rapidly during the postwar bust, reaching 4.75 cents in 1920, near their prewar level. Nationally, farmers planted 700,000 acres of peanuts in 1910. There are no annual data on peanuts before 1910. Then, from 1910 through 1918 acreage increased rapidly to 1.9 million. The increase was especially high after World War I began in 1914. After 1918, there was considerable variation in acres devoted to the crop, but by 1920 acres planted reached 1.7 million.

In 1920, Alabama farmers harvested the largest share (25.1%) of the national peanut crop. Other states reporting substantial shares included: Georgia (23.4%), North Carolina (18.3), Virginia (15.9), and Texas (7.0). These five states accounted for 89.7% of the country's total.

US Prices

In 1920, the annual average price for peanuts in the country was five cents a pound (Figure 11.49), and it declined during the short recession to reach 3.8 cents in 1921. Prices began to increase after 1921, finally reaching 6.4 cents in 1923. Pounds harvested began to increase in the country in 1923, and in 1924 prices began to decline. Prices and pounds harvested continued their teeter-totter relationship until 1930, when both prices and pounds harvested declined. Prices generally declined after 1923 from 6.4 cents to 3.5 cents in 1930.

ACRES PLANTED AND HARVESTED

With peanut prices falling sharply between 1919 and 1920 (Figure 11.49), Alabama farmers began to reduce acres planted (Figure 11.50). Between 1920 and 1922, they declined from 460,000 to 320,000 acres. Prices had increased in 1922 and 1923, and farmers increased acres planted to 361,000 in 1924. Unfortunately, but predictably, in 1924 the average price fell to 5.8 cents, and in 1925 acres planted resumed their downward trend. Then in 1926, prices began to increase again, and in 1927, acres planted began to follow, reaching 406,000 acres in 1928. However, by that time prices had once again begun to decline, and acres planted followed them down to 334,000 acres in 1930.

Alabama farmers harvested an average of 58% of their acres planted during the period 1920-1930. Percentages ranged from 69% in 1920 to 42% in 1926. percentage harvested had been much higher (around 80%), during the war years of 1917 and 1918.

YIELDS

In 1920, the average yield of Alabama peanuts was 550 pounds an acre, and in most of the years in the period 1920-1930, it remained near that level. It was lower than expected, however, in 1922 and 1923, and higher than expected in 1927. Throughout the period, annual average yields of Alabama peanuts were around 150 pounds less than those in the country as a whole

POUNDS HARVESTED

With yields remaining generally unchanged (Figure 11.51) during the period 1920-1930, changes in pounds harvested (Figure 11.52) followed those in acres harvested (Figures 11.50). Pounds harvested was 144.6 million in 1920. Then it declined along with acres harvested to near 100 million pounds in 1922, and it remained generally at that level through 1930.

Farmers in Coffee County harvested the largest share (20%) of Alabama's 1929 peanut crop. Other counties reporting substantial shares included: Henry (15.4%), Dale (11.2), Pike (11.1), and Geneva (9.2). These five counties accounted for 66.9% of Alabama's total peanut crop.

Soybeans

Data on American soybean prices is available from 1913 onward; however comprehensive information (acreage, yield, and production) is generally not available for years before 1924. In 1913, farmers received $1.79 a bushel for their beans (Figure 11.53). Then prices increased rapidly during the wartime boom, reaching $3.53 in 1919. During the short recession, however, they trended downward to $2 in 1922. Prices began to recover during the early part of the speculative boom (1922-1924), only to begin falling again in 1925. By 1927, it was down to $1.80, and they remained near that level through 1929. During this period, prices were essentially where they were in 1913. Then in 1930, the average price fell to $1.34, the lowest on record. That low price was likely the result of the beginning of the Great Depression, but the enormous 1930 crop (13.9 million bushels) did not help the situation.

In 1924, American farmers planted 1.6 million acres of soybeans. Then acreage generally trended upward for the remainder of the decade, until they planted 3.1 million acres in 1930. In the early years, a high percentage of planted soybeans were, like peanuts, used as forage. The beans were not harvested. Throughout the period 1924-1930, farmers harvested beans from between 25% and 29% of their planted acres. With upward trends in both yields and acres harvested in the period 1924-1930, bushels harvested more than doubled (4.9 million to 13.9 million).

Alabama farmers planted 95,000 acres of soybeans in 1924, but harvested only 3,000 acres (3%). They harvested a much lower percentage of their planted acres than farmers nationally. In 1929, they planted 111,000 acres, and harvested only 7,000. Alabama soybean yields during the period 1924-1929 were only half of national yields. It remained at around six bushels an acre for the entire period. With relatively constant yields, bushels harvested generally reflected the increase in acres harvested during the period. In 1924, farmers harvested 20,000 bushels. In 1929, they harvested 42,000. The 1929 crop represented less than 1% of the national crop.

Alabama Hay

In 1920, American farmers harvested 91.7 million tons of hay. They had harvested 75.2 million tons in

1910. Texas farmers harvested the largest portion (6.4%) of the 1920 crop. Other leading hay-producing states were: Iowa (5.9%), New York (5.8), Minnesota (5.6), and Wisconsin (5.5). Those five states produced 29% of the country's total. This relatively low percentage is indicative of the widespread production of sizeable quantities of hay in states throughout the country. Because of differences in the kinds of hay produced from one part of the country to another, little of the national data will be used in this section.

Farmers in Hale County harvested the largest share (4.6%) of Alabama's 1920 hay crop. Other counties reporting significant shares included: Madison (4.6%), Dallas (3.9), Marengo (3.8), and Jackson (3.6). These five counties accounted for 20.5% of the state's total crop.

Prices

Alabama hay prices (Figure 11.54) and tons harvested (Figure 11.57) had generally followed the Expansion Phase of a cattle cycle upward to $24.50 a ton in 1919. Then with tons harvested at an all-time high, prices began to decline. By 1921 it was down to $15.40. Tons harvested declined sharply after 1921, and in 1922 prices began to increase again, finally reaching $18.90 in 1925. In 1926, tons harvested increased sharply, and the price fell just as sharply to $15.90 and remained near that level through 1930.

Acres Harvested

In 1920, the 1913-1927 cattle cycle in Alabama reached its Consolidation Phase, and in 1921 the Declining Phase began. A year after cattle numbers began to decline, farmers began to reduce acres of hay harvested (Figure 10.55). Acres harvested declined sharply in 1927, and then more slowly through 1929. However, it was up slightly in 1927. As noted in the preceding paragraph, a new cattle cycle began in 1928, and acres harvested increased slightly in 1930.

Yields

Generally, Alabama hay yields had been trending downward since around 1905 and 1906, and in 1919 yield was 0.80 ton an acre (Figure 11.56). There was considerably more year-to-year variation in the 1920s, but yields generally declined from 0.80 ton in 1921 to 0.70 ton in 1930. Yields were much lower than expected in 1924 and 1925. Palmer Drought Severity Indices (Cook, et al., 1999), indicate that the crop weather in June, July and August was drier than normal over much of the state in 1924 and extremely dry throughout the state in 1925.

Tons Harvested

With yields remaining relatively stable during the period 1920-1930, tons harvested were largely determined by trends in acres harvested (Figures 11.55 and 11.57). Alabama hay farmers harvested 492,000 tons in 1920. Tons harvested increased slightly in 1921, before beginning a downward trend that continued through 1925 (349,000 tons). In 1926, farmers harvested 0.87 ton of hay an acre, the highest yield since 1911, and as a result tons harvested increased to 472,000. Acres harvested were up slightly in 1927, and even though yield was lower, farmers still harvested 419,000 tons. In the period 1928-1930 yields remained around 0.70 ton an acre, and acres harvested remained around 500,000, so that farmers harvested around 360,000 tons of hay each year in those years.

Census data indicated that farmers in Montgomery County harvested the largest share (5.8%) of the Alabama hay crop in 1929. Other counties reporting substantial shares included: Dallas (5.2%), Madison (5.0), Jackson (4.0), and Hale (3.5). These five counties accounted for only 23.5% of the total crop. The county average was 4,980 tons. A total of 36 of the counties reported harvests of 3,000 or more tons.

Types of Hay Harvested

Volume 1 of *Alabama Agricultural Statistics* listed the acres of the different types of hay harvested in Alabama during the period 1924-1930. In Table 11.3, the totals for the different types for the entire period have been converted into percentages. As can be seen from the table, farmers harvested more peanut hay (35.1%) by a wide margin. Most of the hay harvested that was specifically identified was a by-product of production of crops for other purposes. Alfalfa and lespedeza were the exceptions. The "other hay" category was the largest, and there are no clues to what kinds were included. None of the grass hays are mentioned. These data were collected long before much of the research and development work was done on improving pasture grasses.

Alabama Winter Wheat

In 1920, American farmers harvested 613.2 million bushels of winter wheat. In 1910, they had harvested 429.9 million. Texas farmers harvested the largest share of the 1920 crop (23.6%). Other leading states in the production of winter wheat were: Nebraska (9.5%), Oklahoma (9.1), Illinois (6.8), and Missouri (6.5). These five states accounted for 55.5% of the total crop. Alabama farmers harvested less than 0.1% of the total 1920 crop.

Exports

In 1920, American wheat (all varieties) exports reached the highest level on record (369.3 million bushels) (Figure 11.58). That year, the United States exported the equivalent of 49.3% of the 1919 crop. Then, except for a sizeable increase in 1924 (260.8 million bushels), following a better than expected crop, exports trended downward to reach 108 million bushels in 1925. Between 1925 and 1926, the average price of wheat fell over 20 cents, and in 1926 exports increased sharply to 219.2 million bushels. Afterwards, prices continued to decline, and exports began to trend downward, finally reaching 131.5 million bushels in 1930. Likely, the Roaring Twenties were ending much sooner in Europe than in America.

Prices

The average price of Alabama's winter wheat crop in 1920 was $2.77 a bushel (Figure 11.59). It had been $1.21 in 1915. In 1921, prices began to decline, and except for 1925 and 1928, they continued to trend downward through 1930, finally reaching a level of $1.21. With exports declining rapidly after 1920 (Figure 11.58), there was just too much wheat being produced in the country to support higher prices. The higher than expected prices in 1925 and 1928 was likely the result of the extremely poor national crop in 1925 and the extremely poor Alabama crop in 1928.

Acres Planted and Harvested

Data on acres planted of Alabama winter wheat is not available before 1909. Data on acres planted and harvested during the period 1909-1930 are presented in Figure 11.60. The data indicate that Alabama farmers responded to the opportunity for increased income from wheat during the war years. They planted and harvested 18,000 acres of winter wheat in 1914, the year the war began, but a year later they increased it to 35,000 acres. Acres planted remained around that level through 1919. After that year, however, they seemed to quickly recall their earlier history with the crop and continued their long-term plans to eliminate it from their mix of crops. In 1920, they harvested only 12,000 acres of wheat. There is some indication that when the Roaring Twenties really began to roar that the old wheat farmers began to feel some faint stirring in their breasts for the crop. Acres planted was up slightly in 1922 (11,000 acres) and 1923 (11,000 acres). However, with prices continuing to decline, reality prevailed, and they resumed their efforts to eliminate winter wheat from the state. By 1930, they were only planting and harvesting 2,000 acres of the crop in the entire state. It is difficult to realize that 30 years earlier they had harvested 127,000 acres. In most years during the period 1920-1930, Alabama farmers harvested virtually all of their acres of wheat planted.

Yields

Throughout the period 1920-1930, yields of Alabama winter wheat ranged between 10 and 11 bushels an acre, or four to five bushels an acre less than across the country. Yields seemed to increase just slightly about the middle of the period, but then began to decrease again. Except for some unexpected variability around 1925, annual average yields for American winter wheat generally ranged between 14 and 16 bushels and acre (Figure 11.61). There appeared to be no trend involved.

Bushels Harvested

With yields of wheat remaining generally unchanged during the period 1920-1930 (Figure 11.61) and acres harvested declining (Figure 11.60), bushels harvested also declined. Alabama farmers harvested 115,000 bushels in 1920, but only 20,000 in 1930. Only 30 years earlier, Alabama farmers had harvested slightly over 1.2 million bushels of winter wheat.

Alabama Oats

In 1900, American farmers had harvested 945.5 million bushels of oats to feed 17.9 million horses and 3.1 million mules. In 1920, there were 20.1 million horses and 5.7 million mules in the United States, and American farmers harvested 1.4 billion bushels of oats to feed them. Iowa farmers harvested the largest share (16.2%) of the 1920 oat crop. Other leading oat-producing states were: Illinois (12%), Minnesota (9.7), Wisconsin (6.5), and Nebraska (5.5). These five states produced 49.9% of the total national crop. Alabama farmers produced 0.2%.

US Exports

From 1900 through 1913, the United States had generally exported two million bushels of oats; then when World War I began in 1914, exports totaled 100 million bushels. Afterwards, they remained above 100 million through 1918. The war ended in 1918, and by 1919 exports had declined to 43 million. In 1920, with the war ended but prices still abnormally high, exports fell to nine million bushels. In 1921 and 1922, prices were unusually low, and exports increased to 22 million in 1922. The price increased to 38.8 cents in 1923, and exports declined. Exports were unusually high in 1925 when American farmers harvested large crop. Afterwards, they mostly trended down to reach three million bushels in 1930.

US and alabama Prices

In 1919, Alabama farmers received $1.15 a bushel for their oats (the highest in history). Unfortunately, two years later at the end of the short recession, the price was down to 67 cents (Figure 11.64). Bushels harvested fell sharply after 1921, and prices began to increase in 1922, reaching 40.7 cents in 1923. They remained near that level through 1929, before falling to 32.2 cents at the beginning of the Great Depression. Figure 11.64 also includes data on American oat prices. Trends for Alabama and national oat prices were essentially the same. However, Alabama prices were 30 to 40 cents a bushel higher throughout the period.

Acres Planted and Harvested

There are no annual data on acres planted for Alabama oats before 1926. In 1920, Alabama farmers harvested oats from 163,000 acres to feed the state's 130,000 horses and 296,000 mules (Figure 11.65). They had harvested the crop from 300,000 acres in 1915. Farmers harvested more acres in 1921 (187,000) hoping that prices would rebound following the steep decline in 1920 and 1921 (Figure 11.64), but then quickly decided that that was a losing proposition. In 1922, acres harvested began a sharp decline that ended in 1924 at 100,000 acres. After 1924, the downward trend continued, but the rate of decline was lower. Acres harvested was down sharply in 1928 as a result of extremely wet weather. In 1930, farmers harvested oats from only 95,000 acres; it had not been that low since 1871. In 1930, there were 65,000 horses and 322,000 mules in the state. During the period 1920-1930, the total number of work animals (horses plus mules) had declined 9% (426,000 to 387,000), while acres harvested of oats declined 42% (163,000 to 95,000).

US and Alabama Yields

Data on the annual average yields of oats (bushels an acre) in the United States and Alabama for the period 1900-1930 are presented in Figure 11.66. Yields in the state generally varied between 15 and 20 bushels an acre over the 11-year period. There was no apparent trend. There was considerable year-to-year variation, and it is surprising how regular it was. It is difficult to imagine a natural phenomenon that could account for such a regular pattern. In most years, yields in the United States were 10 to 15 bushels an acre higher.

Bushels Harvested

There appeared to be no trend in Alabama oat yields during the period 1920-1930, (Figure 11.66). Consequently, trends in bushels harvested for the period (Figure 11.67) were largely determined by trends in acres harvested (Figure 11.65). Farmers harvested 3.4 million bushels of oats in 1921, and although there was considerable year-to-year variation (as there was in acres harvested), bushels harvested generally trended downward throughout the period, to reach 1.5 million bushels in 1930. The large amount of year-to-year variation from 1924 through 1930 was likely the result of the variation in yields during the same period.

Alabama Sweet Potatoes

Alabama farmers had significantly increased acres harvested for sweet potatoes during the war years, and they continued to harvest the crop from a substantial acreage (82,000-88,000) during the period 1920-1922 (Figure 11.68), even though prices were falling rapidly during those years. However, after 1922 they quickly made the decision to significantly reduce their acreage of the crop. In 1924, acres harvested was down to 60,000. Prices increased in 1924 and 1925, and acres harvested also increased. However, even though prices began to decline again in 1926, acres harvested generally trended slowly upward through 1930 to a level of 68,000. The average annual yield (Cwt per acre) varied from 37.8 in 1925 to 52.9 in 1926 and 1928. There was no apparent trend in yields during the period. The total value of the sweet potato crop in the different years ranged from $5.4 million in 1930 to $9.7 million in 1920.

Farmers in Baldwin County harvested the largest share (5.5%) of Alabama's 1929 sweet potato crop. Other counties reporting large shares included: Dallas (4.9%), Cullman (3.2), Covington (2.9), and Monroe (2.8). These five counties accounted for 19.3% of the total crop. The county average was 53,206 Cwt.

Alabama Irish Potatoes

Volume 1 of Alabama Agricultural Statistics included data on some of the characteristics of the production of Irish potatoes in Alabama during the period 1866-1946; Volume 3 introduced a new section entitled "Alabama Truck Crops." That section includes data on commercial early Irish potatoes. These potatoes represent a share of the total quantity of all Irish potatoes harvested in the state. The potato data discussed in this section combines data on the spring (early) and summer crops.

As detailed, in several preceding sections, all of Alabama's major crops (cotton, corn, wheat, oats) suffered grievously from the recession of 1919-1921, but for some reason, Irish potatoes seemed to prosper. Data on total acres of all Irish potatoes harvested during the period 1900-1930 are shown in Figure 11.69. Acres harvested was 15,000 acres in 1920, and it generally increased to a maximum of 31,000 acres in 1928. Prices were generally good throughout the period, and they were especially good in 1923, 1924, and 1925. There was generally less evidence for the effect of teeter totter for Irish potatoes during this period than for most of the other crops considered in the preceding sections of this chapter. Toward the end of the period, however, it seemed to make an appearance. In 1927, prices reached the unusually high level of $3.33 per Cwt ($2.00 a bushel). Then in 1928, farmers harvested Irish potatoes from 31,000 acres, the highest level in history. Unfortunately that huge crop forced the price down to $1.25 per Cwt. (75 cents a bushel). The price had not been that low since 1896. In response, in 1929 acres harvested fell to 22,000. The annual average yield (Cwt., an acre) varied from 38.4 (1925) to 42.9 (1922). The dollar-value of the crop during the period ranged from $1.6 million (1921) to $3.8 million (1927).

Baldwin County reported the largest share (27.4%) of the 1929 Alabama Irish potato crop. Other counties reporting substantial shares included Mobile (4.6%), Escambia (4.4), Madison (4.2), and Limestone (3.4). These five counties accounted for 44%, of the total crop.

Alabama Sorghum and Sugar Cane

According to census data in 1929, farms in the South Central Region produced 66.3% of the country's sorghum molasses. Alabama farms accounted for 16.3% of the regional total. The South Atlantic and North Central regions accounted for 19.3 and 13.5%, respectively.

There are no annual data on the production of sorghum in Alabama earlier than 1924. In the period 1924-1930, acres harvested of sorghum for molasses ranged between 22,000 acres (1928) and 31,000 (1930). Molasses produced ranged between 1.3 million gallons (1924) and 1.8 million (1929). Crop value ranged between $1.2 million (1929) and $1.4 million in 1921. The first annual data on the harvest of sorghum for forage was published in 1928. That year, farmers harvested a crop valued at $510,000 from 21,000 acres, and in 1929 they harvested a crop valued at $566,000 from 27,000 acres.

There are no annual data on Alabama sugar cane and syrup production until 1924. In the period 1924-1930, farmers harvested the crop from 20,000 acres. Average annual production of syrup for that seven-year period was 2.3 million gallons. The average annual value of syrup sold was $2.2 million.

Alabama Vegetables

The 1950 Census of Agriculture published data indicating that in 1919 Alabama farmers harvested a variety of vegetables for home use valued at $13.3 million. A decade later, home-use value had declined to $10.2

million. The data also indicate that Alabama farmers harvested vegetables for sale from 12,749 acres in 1919 and 36,782 acres in 1929. The value of vegetables sold in 1919 totaled $1.3 million and $2.5 million in 1929. In 1919, a large share of the total acreage in vegetables was in watermelons, cabbages, and tomatoes. In 1929, these same vegetables continued to account for much of the acreage, but the acreage of sweet corn was also beginning to increase.

Annual data on the harvest of vegetables in Alabama are not available before 1928, although Alabama farmers were growing significant quantities much earlier. Records published in the first volume of Alabama Agricultural Statistics (February, 1948) included the following crops: commercial early Irish potatoes, cabbage, cucumbers, strawberries, and watermelons. These data were published under the heading "truck crops."

Truck crops are a loosely defined group of crops (usually vegetables and some fruits) that are highly perishable that must be transported (trucked) to market soon after harvest. Unfortunately, production of most of the group is highly dependent on local climate. Because of Alabama's relatively mild and short winters, it would seem that the state would be an ideal location for the production of these crops. By planting and harvesting them early here, it should be possible to truck them north into large urban markets long before the climate is suitable for producing them there. While on the average Alabama's winters are short and mild, crops are not grown on the average. In too many years, the late winter period is too wet and cold for planting, and the date of the last killing frost in most of the state is notoriously unpredictable. These weather problems were discussed in detail in Part I of this book. Most years, weather conditions are suitable for the early production of truck crops. Unfortunately, in those bad-weather years virtually the entire investment in the crops can be lost.

One of the primary characteristics of truck crops is that generally not many acres are planted. Their production is usually highly labor intensive. Ordinary farmers would never have had the labor or machinery available to handle large acreages. Acres harvested of the several truck crops during the period 1928-1930 are listed in Table 11.4. During that period, farmers harvested, on average, more acres of commercial early Irish potatoes (12,300), watermelons (8,380), and strawberries (5,870) and fewer acres of cucumbers (2,030) and cabbage (1,930). Data on commercial early Irish potatoes represents a share of the total quantity of potatoes harvested in the state, as discussed in a preceding section.

Values of the different vegetables harvested in 1928, 1929, and 1930, and included in the truck crops group, are presented in Table 11.5. Average values for the three-year period of commercial early Irish potatoes ($1,270,000) and strawberries ($1,110,000) were considerably greater than for cabbage ($360,000), cucumbers ($310,000), and watermelons ($300,000).

Alabama Velvet Beans

The velvet beans is a tropical legume. They were first introduced to the South around 1875, and over the years they became an important regional crop. They were commonly planted with corn, because they are able to fix nitrogen; consequently, intercropping resulted in better production of corn. Further, the vines of the plant make a very good hay, and the seed pods are useful as a livestock feed. In the South in the 1920s, much of the corn was still being picked by hand. The presence of the vines of the velvet bean plant draped over the corn stalks interfered, to a degree, with the harvest of the corn, but after it was out of the field, cattle could be turned into the fields to graze on the vines and seed pods. Seed pods of the velvet bean plant could be harvested by hand and fed to livestock. However, the seed pods are covered with spiny hairs that could cause a rash on the hands of some people picking them. Velvet beans were certainly grown in Alabama in the late nineteenth century, but the first records on the yearly production of the crop are not available before 1924.

In 1924, Alabama farmers harvested velvet beans from 311,000 acres, primarily from land planted with corn (Figure 11.70). Acres harvested declined somewhat after 1924, but began to increase after 1926. It varied from 275,000 acres in 1926 to 364,000 in 1929. Yields were surprisingly high, varying from 600 pounds an acre in 1925 (a drought year) to 900 pounds an acre in 1928 (a wetter than normal year). With acres harvested remaining relatively unchanged during the period, tons harvested was largely determined by yield. Tons harvested ranged from 87,000 in 1925 (a drought year) to 148,000 tons in 1928 (a wetter than normal year). Except for 1925, all harvests were greater than 100,000 tons. Farmers probably also harvested a large quantity of velvet bean hay during the period, although it is not specifically identified in the available records on hay production. Data on prices received by farmers from the sale of their beans are not available before 1929. In 1929 and 1930, farmers received $14.00 and $13.50 a ton, respectively. The value of the crop was available only in 1929 ($2 million) and 1930 ($1.4 million).

Alabama Cowpeas

Cowpeas have been grown in the South since the Spanish first colonized portions of the region. There are many varieties of the crop. They are also called black-eye peas and crowder peas, along with many other local names. Because they are not cold tolerant, they are grown only in this region. They were widely cultivated in home gardens from the beginning of the colonial period. Because cowpeas are able to fix nitrogen, they have often been planted with corn and sorghum to increase the yield of those two grass crops. During the period 1924-1930, farmers averaged planting 81,000 acres annually, and an additional 111,000 acres were inter-planted with other crops.

During the period 1924-1930, farmers harvested cowpeas both for the peas produced and for hay. Acres harvested for peas ranged from 45,000 acres in 1924 to 73,000 acres in 1930, and generally trended upward during the period (Figure 11.71). Acres harvested for hay ranged from 56,000 acres in 1924 to 26,000 acres in 1929, and generally trended downward during the period. Cowpea yields did not vary very much during the period. They ranged from 4.5 bushels an acre in 1926 to 8 bushels in 1927. Bushels harvested ranged from 248,000 in 1924 to 536,000 in 1927. Prices ranged from $1.79 a bushel in 1930, at a time when virtually all agricultural prices, were declining, to $2.86 in 1924. The value of the crop was at its lowest point in 1926 ($499,000) when bushels harvested was also lowest, and it was highest in 1927 ($1.1 million) when bushels harvested was also highest.

Alabama Pecans

There are no records on the annual production of pecans before 1919. According to census data, 26.2 million pounds of pecans were harvested in the United States in 1929. Some 76.1% of the crop was harvested in the South Central Region, 64.5% of it in the western portion (Arkansas, Louisiana, Oklahoma, and Texas). Texas alone accounted for 36.7%; Alabama farmers harvested 5.9%.

Although pecan trees will produce good quantities of nuts throughout Alabama, they are most productive in the somewhat warmer, wetter climate, and sandy soils of south Alabama. In that region, most of the nuts are produced in the extreme southwest counties (Escambia, Baldwin, and Mobile). There are two primary factors in the year-to-year production of pecans. The most important one is the masting habit of the tree. This physiological characteristic was discussed in some detail in a preceding chapter. As result of this characteristic, it is very unusual to get good crops in two successive years. Another important factor is associated with powerful storms originating in the tropics that strike that area occasionally. Most of these storms, pass through that area in the late summer or early fall when the crop is maturing. The most severe of these storms, hurricanes, can down the trees themselves, or blow leaves or nuts off. Also, the heavy rains that often accompany these storms can result in the deterioration of the quality of the individual nuts. Drought on those sandy soils can also be a problem, but that weather phenomenon is not common in that area.

The US census reported that there were 434,000 pecan trees of all ages in Alabama in 1920. The US Census of Agriculture reported that the number of trees had increased to 710,000, in 1925 and to 743,000 in 1930. The alternate-year-bearing (masting habit) is quite obvious in the Alabama pecan harvest during the period 1919-1930 (Figure 11.72). The only period when the phenomenon was not apparent was in 1925 and 1926, when relatively moderate crops were harvested in both years. Tropical storms in 1922, 1926, and 1928 passed through Alabama's primary pecan-producing area during the period 1919-1930. A fourth storm (1923) passed northward through extreme eastern Mississippi, close enough to Mobile and Baldwin counties to have affected pecan production. However, given the dates when these storms were active in the state, it does not appear that they affected production any, if at all.

Excepting the off-years, pecan production seems to have trended upward from 1.4 million pounds in 1919 to between 3.1 and 3.5 million pounds in 1929 and 1930 (Figure 11.72). The number of trees was also increasing during this period; however, given the length of time required for a tree to grow to maturity and to begin bearing, it is not likely that those trees were the reason why production increased. The highest crop value ($908,000 in 1923) was a result of a moderately good crop and an unusually high price (36.3 cents a pound). In contrast the lowest crop value ($392,000 in 1927) was the result of the combination of the smallest crop harvested during the period and a price of 31.1 cents. In this period, the lowest crop value was 43% of the highest ($392,000 versus $908,000). This comparison represents only a relatively short period of time, but it fairly accurately characterizes the extreme volatility of commercial pecan production in Alabama.

Alabama Peaches

Census data indicate that American peach production totaled 2 billion pounds (42.8 million bushels) in 1929. Production was fairly evenly distributed in the different regions with the exception of the West. The largest share (36.4%) of the national crop was harvested in that region. California alone accounted for 28.7%. Harvests in the other regions were: South Central (19%), South Atlantic (18.5), North Central (15.3), and North Atlantic (10.7). Producers in Alabama harvested 1.2% of the total crop.

Certainly, fruit was grown on Alabama farms from the years of the first settlers. Apparently, however, they were not marketed in sufficiently large quantities before 1919 to warrant systematic record keeping. The US census estimated that there were 2.1 million peach trees of all ages in Alabama in 1920; by 1925 the number had decreased to 1.8 million, and in 1930 to 1.7 million.

Data on pounds of Alabama peaches harvested during the period 1919-1930 are presented in Figure 11.73. Generally speaking, considering only the good years, harvests increased from 52 million pounds in 1919 to 68.6 million in 1930. In the 12-year period (1919-1930), only seven of the years could be considered good. In the other years (1922, 1923, 1926, 1927, and 1929), harvests were considerably lower.

From the earliest years, Alabama peach production has been reduced by late freezes. This weather phenomenon was discussed in detail in Chapter 2. Late freezes were likely implicated in low production in both 1923 (29.5 million pounds) and 1927 (22.7 million pounds). In 1923, the last killing frost in Tuscaloosa occurred on April 1, and in 1927 on March 23. The low production in 1929 was likely related to the extremely wet weather that year. The last killing frost at Tuscaloosa was on February 25, but Tuscaloosa received 13.7 inches of rain in March of that year, and Cullman, 11.8 inches. These heavy rains would have made it almost impossible to control pests during the early and late bloom.

The value of the peach crop harvested in Alabama during the period 1919-1930 varied from $708,000 in 1927 to $2.4 million in 1920. A wide range of crop prices had always plagued Alabama peach production, and, in fact, in the country as a whole. Production of perennial crops is quite different from the production of annual crops. With annual crops, acres planted can be adjusted from year to year depending on the expected price, but that luxury is not available in the production of orchard crops. Tree numbers cannot be easily or quickly adjusted. Farmers take whatever crop they get, and hope for the best. In the case of peaches, the crop is produced so widely that competition for markets is really fierce. As a result of these two factors, prices vary widely.

Greenhouse, Sod and Nursery

Decennial census data indicate that American cash receipts from the sale of nursery products totaled $20.4 million in 1919, while receipts from the sale of greenhouse products totaled $77.4 million. In the same year, the sale of Alabama nursery and greenhouse products, flowers and vegetable seed and plants, etc., totaled $499,300. The sale of trees, shrubs, and ornamentals totaled $234,700.

Data in the 1930 Census indicate that receipts from the sale of nursery and greenhouse products in the country totaled $145.7 million. Some 69.5% of these sales took place in the North Atlantic and North Central regions. Only 14.8%, took place in the South Atlantic and South Central regions. About 0.5% of all sales ($783,400) were reported from Alabama.

Alabama Timber

Data on lumber production in the United States during the period 1869-1930 are presented in Table 11.6. These data show that lumber production in the East continued to decline during the period 1920-1930. In 1930, only 17.2% of the national total was manufactured in the combined North Atlantic and North Central regions. In 1869, 75% was produced in those two regions. Production in the South Atlantic and South Central regions was also continuing to decline, while it was increasing rapidly in the West.

Data on the value of forest products cut on American farms in 1909, 1919, and 1929 are presented in Table 11.7. Total values for 1909, 1919, and 1929 were: $195.3, $394.3, and $242 million, respectively. The higher value in 1919 was likely related to the wartime boom. In 1919, the largest share of total value was reported from the South Atlantic Region. In 1929, it was reported from the North Central Region.

Table 11.8 presents data on the relative values of the different kinds of products in 1929. The value of firewood was the most important product cut from farm forests that year. It was more than three times as valuable as sawlogs and veneer logs (62.5 versus 19.7%). In 1929, most of the farm homes in the country still used wood as a source of heat and for cooking.

There is almost no reliable information on the extent, characteristics, and use of Alabama's forest resources during the period 1920-1930. Much of what is available was collected and summarized by Stauffer (1960). Some of his findings and data obtained from Censuses are presented in the following paragraphs.

LAND IN FORESTS IN ALABAMA

In 1924, 1926, and 1929, the Alabama State Commission of Forestry estimated that forestland in the state totaled 20 million, 22 million, and 22 million acres, respectively. According to Stauffer, the 1926 value is considered to be quite reliable. It is generally used as the base estimate for comparisons of land use changes over time. This value indicates that 67% of all land in the state was in forest in 1929.

Volume of Timber

The commission estimated that the volume of Alabama timber in 1924, 1926, and 1929 to be 50 billion, 48 billion, and 44 billion board feet, respectively. The 1920 Census reported that there were 35.3 million acres of marketable timber on Alabama farms in 1919.

Timber Use

Stauffer (1960) also reported that Alabama sawmills produced 2.2 billion board feet of lumber in 1925 and 1.3 billion in 1930. The sharp decrease was likely the result of the beginning of the Great Depression.

Census data published in 1930 indicate that in 1929 Alabama forests provided the following quantities of products: saw and veneer logs (408.9 million board feet), firewood (1.6 million cords), pulpwood (33,975 cords), fence posts (2.6 million), railroad ties (580,303), and poles and piling (103,512). The estimate value of all of these products was $8.5 million (Table 11.7). The relative values of these products are shown in Table 11.8. As might be expected, the relative value of firewood was somewhat lower in Alabama than in the country as a whole (62.5 versus 57.7%). However, the relative value of sawlogs and veneer logs, was considerably greater (19.7 versus 32.4%).

Gulf States Paper Company established the state's first continuously operating paper mill in Tuscaloosa in 1927. International Paper Company established the second such mill in Mobile in 1929. No one in their wildest imagination could have foreseen the long-term implications of the establishment of these mills relative to the forest resources and general economy of the state.

Animal Agriculture

In 1915, value added to the American economy by the production and sale of livestock and livestock products was $4 billion, and value added for all crops totaled $3.9 billion. By 1920, value added for livestock and products had increased to $7.4 billion, but it had increased at a slightly higher rate for crops ($8.4 billion). In 1915, value added for meat animals comprised 43.8% of the total for all livestock and products; however, by 1920, it had fallen to 41.3%. The value of dairy products and poultry and eggs increased in importance during World War I.

Census data indicated that the value of all domestic animals, including poultry and bees, on Alabama farms in 1910 was $65.6 million. In 1909, the value of all crops harvested in the state totaled $144.3 million. As a result of the value of cotton, the value of livestock was less than half that of crops. These two estimates were made in two different years, 1909 and 1910, respectively. There are no comparable data for 1915. However, in 1920 the value of all domestic animals, including poultry and bees, reached $112.8 million, almost two times what it had been in 1910; however in 1919, the value of all crops more than doubled to $304.3 million.

Cattle

The American cattle herd completed a 16-year (1897-1912) Cycle of Abundance in 1912, and a new one began in 1913. The new cycle reached the Consolidation Phase in 1918 at a level of 51.5 million animals (less milk cows). The Declining Phase began in 1919, and by 1920 herd size was down to 48.9 million.

US Cattle

The national cattle and calves inventory (excepting milk cows) on January 1, 1920, counted 48.9 million animals. Texas farmers had the largest share (14%). Other leading cattle states were: Iowa (7%) Nebraska (5.3), Kansas (4.7), and Missouri (4.0). These five states accounted for 35% of the total. This relatively low percentage indicates the widespread distribution of cattle in the country. Approximately 1.3% of the national herd was on farms in Alabama.

Farm Value Per Head

Data on farm value per head for "all" cattle in the United States (less milk cows) in the period 1900-1930 are presented in Figure 11.74. The data indicate that value had increased rapidly during the inflationary war years to reach $41.79 in 1919. It declined slightly in 1920 with the short recession of 1919-1921; then fell sharply through 1922. Afterwards, it declined more slowly reaching a level of $22.52 in 1925. Then, with cattle numbers trending downward in the Declining Phase of a cycle (Figure 11.76), values began to increase in 1926, finally reaching a level of around $40 in 1929 and 1930.

Cycles of Abundance

Before 1920, NASS data included only two categories of cattle: cows that calved-milk and cattle and calves-all. Beginning in 1920, NASS began to publish data listing ten categories. However, to keep the data comparable, I continue the practice of using only the two categories. All the data used in this section and the same sections in other chapters will be for "all" cattle, excepting cows that had calved-milk.

As noted in a preceding paragraph, there were 48.9 million cattle and calves (less cows that had calved-milk) in 1920 (Figure 11.75). In 1915, cattle numbers in the Expansion Phase of a cycle had been following the wartime surge of farm values upward since 1913 (Figure 11.74). Normally with numbers increasing farm values should have been declining; however World War I had short-circuited the teeter-totter relationship. With the war ending in 1918 (50.2 million head) cattle producers began to worry about the unusually large number of animals on American farms, and even with farm value still increasing they began to reduce herd size in the Descending Phase of the cycle. This was one of the rare times when farm values followed numbers downward. After 1919, they both trended downward together. The decline in farm value ended in 1922, and then remained generally unchanged through 1925. Farm values began to rise sharply in 1926 under the influence of the growing economy of the Roaring Twenties; however, producers were not certain that farm values were going to continue to increase. Consequently, they did not begin to rebuild herd size in the Expansion Phase of a new cycle until 1929—on the eve of the Great Depression. By 1930, numbers had grown to 37.9 million. This cycle had lasted 16 years (1913-1928), and when it ended there were 2.3 million fewer cattle (37.3 versus 35 million) in the country than there had been at the cycle's beginning in 1913.

Real Net Cash Return Per Cow

Mathews, et al. (2001) published data on real net return per cow during the period 1913-1930. These authors estimated that American cattle farmers received a net return of $77 (1982-1984 dollars), per animal sold in 1920 (Figure 11.76). In 1916, it had been $140. Net return continued to decline through 1923 ($41)—along with agricultural prices (Figure 11.5)—then improved somewhat through 1927 as the speculative boom boosted prices. In 1927, farmers obtained $71 an animal. In 1929, however, agricultural prices began to decline again, and by the end of 1930 returns had fallen to $58 a head. Throughout most of this period, the American cattle inventory was declining rapidly (Figure 11.76). However, the decline in total herd size seemed to have had little direct effect on net returns during this period. General economic conditions in the country seemed to be more important. During this period, net returns generally tracked parity ratios (Figure 11.6).

Alabama Cattle

A new cattle cycle had also begun in Alabama around 1913 (Figure 11.77). Its Expansion Phase was somewhat bumpy during the war years, nevertheless trended upward to reach the Consolidation Phase in 1920 at a level of 624,000 animals. In 1920, 1.3% of the national cattle herd was in Alabama.

Census data for 1920 indicate that farms in Marengo County held the largest share (6.2%) of Alabama's beef cattle herd. Other counties with substantial shares included: Baldwin (4.6%), Sumter (4.4), Escambia (3.6), and Hale (3.5). These five counties accounted for only 22.3% of the total. The county average was near 4,800. On average, there were about 120 cattle on each 100 Alabama farms. A total of 41 counties reported herds of 3,000 or more cattle.

Farm Value Per Head

Data on Alabama farm value per head of cattle during the period 1900-1930 are presented in Figure 11.74, along with the national data. However, the data are not strictly comparable. The Alabama data are for all animals, while the national data are for "all" cattle, excepting milk cows. The inclusion of milk cows in the Alabama data would tend to bias the values upward, because values for milk cows were generally higher.

The data in Figure 11.74 show that farm value in both the country and the state generally followed the same trends. Both had increased sharply during World War I, only to fall through 1925. Afterwards, both increased through 1929. In 1930, the national value declined slightly, while the Alabama value increased. Although year-to-year trends were similar during the period, in each year they were $7 to $10 lower in Alabama. They would probably have been lower if the value of milk cows had not been included in the Alabama data.

Cycles of Abundance

Annual Alabama cattle inventories during the period 1920-1930 indicate that in 1920 numbers ("all" cattle, excepting cows that had calved-milk) had reached the Consolidation Phase of a cycle that had also begun around 1913 (Figure 11.77). In 1920, there were 624,000 cattle and calves in the state. In Alabama, the Declining Phase of the cycle began a year later than in the country. In Alabama, the Declining Phase ended in 1927, a year earlier than elsewhere in the country, with an inventory of 411,000 animals. By 1930 numbers had increased to 451,000. The total Alabama herd size was 81,000 animals less at the end of the 1913-1927 cycle than at the beginning (492,000 versus 411,000).

In the 1920s, most of Alabama's cattle and calves were marketed in the state, although a substantial number of animals were slaughtered on the farm for home use. In 1924, an estimated 10,000 cattle and 12,000 calves were slaughtered for this purpose. Also, in this period, there was a large beef slaughter and packing industry in the state. Alabama's cattle industry was years away from becoming a cow-calf operation in which slaughter and packing were all done out-of-state.

Hogs

Hogs and pig numbers in the United States went through several Cycles of Abundance between 1900 and 1920. In 1920, numbers reached the end of the Liquidation Phase of a cycle that had begun in 1916. In that year, the nation's hog inventory totaled 58.9 million animals. That cycle reached the end of its Accumulation Phase in 1918, at a level of 64.3 million.

US Hogs

On December 1, 1920, American farmers had 58.9 million hogs and pigs on their farms. Of the total, 14% were on farms in Iowa. Other leading hog states were: Illinois (8.2%), Missouri (7.0), Indiana (6.1), and Nebraska (6.0). These fives states accounted for 41.3% of the total inventory. Some 2.3% of the national total were on Alabama farms.

Farm Value Per Head

Data on American farm value per head for hogs during the period 1900-1930 are presented in Figure 11.78. Trends in farm values during the period 1920-1930 seem to have been largely determined by the topsy-turvey economy of the period: wartime inflation that carried over into the period of the short recession of 1920-1921, as well as the speculative boom of the mid-1920s. Coming out of wartime inflation in 1920, value was at a level of $20.00. At the end of the short recession in 1922, it had fallen to $10.58. Then in the middle of the speculative boom in 1927, it had increased to $17.19. Afterwards, it began to fall again, and in 1930 reached $13.45. Numbers of hogs seemed to have little effect on the elaboration of these changes in values.

Cycles of Abundance

As noted in the preceding paragraph, there were 58.9 million hogs and pigs on American farms on December 1, 1920 (Figure 11.79). At that time, numbers were at the end of the Liquidation Phase of a Cycle of Abundance that had begun in 1921. Surprisingly, in the midst of the short recession, a new cycle began. At the end of that Accumulation Phase, numbers were at a level of 69.3 million, but with values below $11. Consequently, the cycle lost its steam, and by 1925 there were only 52.1 million. Values increased in 1925, and a new cycle began in 1926. At the end of that Accumulation Phase in 1927, numbers were at a level of 61.9 million. Afterwards, both values and numbers declined. In 1930, the inventory revealed that there were 54.8 million hogs and pigs in the country. Changes in numbers during the period also seemed to be more dependent on changes in the economy than on changes in values.

Alabama Hogs

Alabama hog numbers went through two complete cycles between 1903 and 1914 (Figure 11.80). The second one reached the end of its Liquidation Phase in 1914 at a level 1.06 million animals. Then a new cycle began in 1915. The end of the Accumulation Phase of that cycle came in 1918. In 1920, the cycle was in its Liquidation Phase at a level of 1.35 million animals. In that year, 2.3% of the American hog herd was on farms in Alabama.

Census data indicate that farmers in Pike County reported the largest share (3.8%) of the 1920 Alabama hog inventory. Other counties reporting large shares included: Geneva (3.7%), Houston (3.7), Coffee (3.5), and Covington (3.4). These five counties accounted for 18.1% of the state's inventory, indicating the widespread distribution of hog production in the state. In 1920, 20 of Alabama's counties reported inventories of at least 25,000 animals.

Farm Value Per Head

Figure 11.79 also includes data on Alabama farm value per head of hogs during the period 1900-1930. These data indicate that year-to-year changes in values in Alabama were similar to those in the country as a whole, except that after 1925 when national values were increasing, values in Alabama remained largely unchanged and remained near the same level through 1930. At the end of the period of wartime inflation in 1920, Alabama values was at a level of $13.70. Then at the end of the short recession in 1922, it was $9.20. Afterwards, it slowly trended upward, reaching a level of $10.50 in 1930. Although trends in values were similar in both Alabama and the country, they were considerably lower in Alabama.

Cycles of Abundance

Alabama hog production cycles during the period 1920-1930 were somewhat different from those elsewhere in the country (Figures 11.78 and 11.79). In 1920, Alabama hog and pig numbers (1.3 million) were in the middle of the Liquidation Phase of a cycle that had begun in 1915. Elsewhere in the country, at the same time, numbers were at the end of a cycle. It would be expected that numbers in Alabama would have reached the end of their cycle shortly after 1920. The appearance of the data suggest that Alabama farmers had a notion about beginning to increase their herds in 1921, but with values at $10.70, they changed their minds and continued to liquidate. By 1925 numbers were down to 776,000. It had not been that low since 1909. Values in Alabama remained around $10 after 1921, but when national values increased to $13 in 1925, Alabama farmers just could not wait any longer. In 1926, they began to rebuild their herds. Unfortunately, their prices received hardly budged. In 1928 when they began to liquidate, they received $10.40 a head.

At the end of the cycle in 1902, there were 1,115,000 hogs and pigs in Alabama. At the end of the cycle in 1925, however, there were only 776,000. In that 24-year period, the state lost 339,000 (30%) of its hogs and pigs. These data suggest that Alabama farmers were quickly discovering another crop for which they had little, if any, comparative advantage.

Houston County farmers reported the largest share (4.9%) of the April 1, 1930 inventory of hogs and pigs. Other counties reporting substantial shares were: Geneva (4.5%), Coffee (4.4), Covington (3.9), and Dale (3.6). These five counties accounted for only 21.3% of the state's total. The county average was 12,400 animals. On average, there were 320 hogs per 100 Alabama farms. A total of 32 counties reported inventories of 10,000 or more head.

Milk

In the first two decades of the twentieth century, American milk production did not keep pace with population growth. Between 1909 and 1919, it increased 4.5% (64.2 to 67.1 billion pounds) while the population increased 15.5% (90.7 to 104.8 million). Production was widely dispersed. A substantial share was produced on farms with relatively few cows, primarily for home consumption. Dairy farms with 100 cows were considered to be large.

Census data indicate that in 1919 some 72.2% of the milk produced in the country was produced in the North Atlantic and North Central regions, 18.1% was produced in the South Atlantic and South Central regions combined, and 9.7% was produced in the Mountain and Pacific regions.

US Milk

The first annual data on American milk production was published in 1924. In that year, milk cows in the United States produced 89.2 billion pounds of milk. The largest share of the total was reported from Wisconsin (11.3%). Other states reporting large shares included: New York (7.8%), Minnesota (7.4), Iowa (5.6), and Pennsylvania (4.8). These five states accounted for 36.9% of the total.

Prices

Although American dairy farms were relatively small in the early 1920s, producers were already experiencing marketing and pricing problems. During this period, a classified pricing system was established whereby milk handlers paid for fluid milk according to its use (Blaney and Manchester, 2001). Data presented in Figure 11.81 gives the average wholesale price of whole milk in the United States during the period 1910-1930. By the 1920s, milk producers and milk handlers had generally been able to agree on the pricing of milk. As a result, during most of the period milk traded over a relatively narrow range of prices. It was $3.22 per Cwt in 1920 as a result of the run-up in prices during World War I; however, from 1921 through 1930, prices ranged between $2.11 (1922) and $2.53 (1929). The demand for milk declined so rapidly in 1930 at the beginning of the Great Depression that the price also declined to $2.21 per Cwt.

Milk Cow Numbers

In 1920 there were 21.4 million American milk cows and heifers, two years old or older, kept for milk. The largest share of the total was in Wisconsin (8.5%). Other states with large shares included: New York (7%), Minnesota (6.3), Iowa (5.2), and Illinois (4.9). These five states accounted for 36.4% of the national total.

In 1867, USDA began to publish annual data on cows that had calved-milk as part of their cattle inventory. These data have been used in all the chapter presentations subsequently. In 1924, the agency began to publish annual data on milk cows, and there is a small discrepancy in the data of the two series. For example, in 1924 the total for cows that calved-milk was 22,331, while the total for milk cows was 21,417. Beginning with the 1924 data, only the NASS milk cow data will be used. This creates a slight problem in Figure 11.82, which presents data on the number of American milk cows during the period 1900-1930. The bump in the line at 1924 represents the transition point between the two data sets.

The combined data in the figure indicate that milk cow numbers had trended sharply upward from 1900 through the end of World War I in 1918. Afterwards, with the fixing of milk prices, numbers were generally unchanged at a level of around 21.4 million through 1928. In 1930, the American herd size reached 22.2 million, the highest in history to that point.

Milk Per Cow

Annual estimates of the average quantity of milk produced per cow in the United States were not available before 1924. In that year, the average was 4,167 pounds (Figure 11.83). Quantity increased annually by around 82 pounds through 1929, when it reached 4,579 pounds. It decreased slightly, to 4,508 pounds, at the beginning of the Great Depression. It is likely that with a declining demand for milk farmers reduced the quantity and quality.

Pounds Produced

With milk per cow trending slightly upward during the period 1924-1930 (Figure 11.83), pounds produced also trended upward from 89.2 billion in 1924 to 100.2 billion in 1930 (Figure 11.84), or at a rate of 1.57 billion pounds a year. Even though milk per cow declined in 1930, the sharp increase in number that year was sufficient to keep pounds increasing.

In 1924, total American milk production was equivalent to 779 pounds (91.7 gallons) a person. By 1930, these values had increased to 811 pounds (95.4 gallons) a person.

Alabama Milk

In the period 1909-1919, Alabama milk production kept pace with population growth much better than the country as a whole. During this period, Alabama's population increased by 10.9% (2.108 to 2.337 million), and milk production increased by 10.2% (676 to 669.2 million pounds).

Census data indicated that farmers in DeKalb County reported the largest share (3.1%) of the total quantity of milk produced in 1920. Other counties reporting substantial shares included: Cullman (3%), Marshall (2.7), Chambers (2.4), and Jefferson (2.3). These five counties accounted for only 13.4% of the state total. This small share for the five leading counties is indicative of the widespread distribution of milk production in the state. Farmers in 24 of Alabama's 67 counties reported productions of at least 10 million pounds. The value of all dairy products on Alabama farms was $15.2 million. That year (1920) the average milk production per cow was 226 gallons or 1.944 pounds

Prices

There were no average annual data on prices for Alabama milk before 1924. Data for the period 1924-1930 are presented in Figure 11.85 along with corresponding data for the entire country. As a result of agreements between milk producers and handlers, prices for whole milk varied within a relatively narrow range during the period. They trended upward from $3.05 per Cwt in 1924 to $3.22 in 1929. Then with the beginning of the Great Depression in 1930, the price fell below $3.00 to $2.96. The agreement on prices in the country allowed for lower production per cow and higher production costs by allowing Alabama farmers to sell their milk for higher prices. Throughout the period, Alabama farmers received an average of 71 cents per Cwt. more than the national average.

Milk Cow Numbers

The same problem regarding the data sets on cows that calved-milk and milk cows described for American milk cow numbers is also present in the Alabama data. Figure 11.86 includes data on cows that had calved-milk for the period 1900 through 1923, and data on milk cows for the period 1924-1930. The combined data show that numbers increased sharply during the war years to reach 420,000 animals in 1919 and 1920, the highest level in history at that time. This number represented 2% of the national total.

With the short recession underway after 1919, Alabama numbers began to decline rapidly, finally reaching 335,000 head in 1924. With the new, highly advantageous pricing plan in place in 1924, numbers changed very little from year to year. Although annual increases were small, they did trend upward to reach 360,000 animals in 1930.

The largest share of Alabama milk cows in 1930 was on farms in Dallas County (2.9%). Other counties with substantial shares included: Jefferson (2.7%), Marengo (2.6), Montgomery (2.3), and Madison (2.2). These five counties accounted for only 12.7% of the total. This small percentage indicates the widespread distribution of milk production in the state. Some 45 of Alabama's 67 counties reported at least 5,000 head.

Milk Per Cow

There is no specific data on the average annual quantity of milk produced per cow in Alabama before 1924. Data for the period 1924-1930 were also presented in Figure

11.83, along with corresponding data for the whole country. These data indicate that milk per cow in Alabama generally increased from 2,790 pounds in 1924 to 3,230 pounds in 1929. As was the situation with milk per cow in the country as a whole, the average in Alabama was also down (3,150) in 1930. Milk per cow for Alabama milk cows averaged 30% less than the national average throughout this period.

Alabama farmers could not have know it at the time, or if they did knew it, they would not have accept it, but this difference in milk per cow indicated a level of comparative disadvantage that would plague Alabama's dairy industry for the remainder of the twentieth century and beyond, and, over time, without massive government intervention, it would mean the eventual end of dairy farming in the state.

Pounds Produced

There was no annual data published on Alabama milk production before 1924. Data on pounds produced during the period 1924-1930 are presented in Figure 11.87. With milk per cow slowly trending upward and with numbers increasing after 1926 (11.86), pounds also increased, especially in 1929 and 1930. Pounds increased from 935 million in 1924 to 1,134 million in 1930.

Pounds of milk produced in Alabama in 1924 was equivalent to 374 pounds (44 gallons) a person, or about half of the production-per-person in the United States in 1924. In 1930, pounds per person in Alabama increased to 428 pounds (50.4 gallons). Fortunately, during the period 1924-1930, milk production on a per-person basis increased at a faster rate in Alabama than in the country as a whole (14.6% versus 4.1%).

Alabama Chickens

In the early 1900s, American chickens were produced in millions of small back-yard flocks. In 1910, 88% of all farms had chickens. Until the 1920s, chicken meat was a luxury food that families only had for Sunday dinner. The primary reason for raising chickens was for the eggs they produced. Families only ate the excess birds in their flocks, as well as unproductive hens culled from their productive layers. Efforts to produce chickens for meat alone had generally not been successful economically. However, by the mid-1920s, the production of chickens for meat was beginning to reach significant levels.

Census data from 1920 indicated that 56.1% of the country's chicken flock was in the North Atlantic and North Central regions, and 36.7% was in the South Atlantic and South Central regions. Another 7.2% were in the West Region. These numbers more or less reflected the distribution of the population at that time. Approximately 1.6% of the country's chickens was on Alabama farms.

US Farm Value Per Head

Data on American farm value per head of chickens during the period 1909-1930 are presented in Figure 11.88. Value had increased sharply during the inflationary period associated with World War I to reach 97 cents in 1920; then it began to decline in the years of the short recession when the prices of most agricultural commodities collapsed. It reached 75 cents in 1923 before beginning to trend upward. It finally reached 91 cents in 1927, but when numbers reached 475,000 (the highest in history), value declined. Numbers were lower in 1929 and 1930, and value began to increase again. Surprisingly, the beginning of the Great Depression did not seem to affect value. It increased from 91.1 cents in 1929 to 92.8 cents in 1930.

Alabama Prices

The are no annual data on Alabama chicken prices before 1924. Figure 11.89 includes data on the annual average price that farmers received for their chickens in Alabama and the United States during the period 1924-1930. These data show that Alabama prices trended upward from 69 cents in 1924 to 74 cents in 1926. In 1927, Alabama chicken numbers reached 12.3 million, over 18% higher than in 1924 (Figure 11.90). That year, prices declined to 72 cents. Numbers were down sharply in 1928 (10.3 million), and prices increased to 74 cents. With numbers still down in 1929, prices continued to increase to 79 cents, but they fell sharply to 67 cents in 1930 with the beginning of the Great Depression. Nationally, prices continued to increase between 1929 and 1930. Prices of Alabama chickens were from seven cents to 18 cents lower than elsewhere in the country during most of the period, but at the beginning of the Great Depression in 1930, the difference was near 25 cents.

Alabama Chicken Numbers

The 1919 inventory indicated that there were slightly less than 8.1 million chickens on Alabama farms. The largest share (3%) was on farms in DeKalb County. Other counties with substantial shares included: Geneva (2.4%), Madison (2.4), Marshall (2.3), and Chambers (2.2). These five leading counties accounted for only 12.3% of the state total, which indicates how widespread chicken raising was in Alabama. In 1919, 43 of 67 counties reported at least 100,000 birds in the inventory.

Annual data on chicken numbers in Alabama are not available before 1924. Data on the annual, April 1, inventory of chickens during the period 1924-1930 are presented in Figure 11.90. In 1924, these data show that there were 10.4 million birds in the state. Afterwards, with numbers generally interacting with prices as detailed in the preceding section, inventories increased to 12.3 million in 1927, fell to 10.3 million in 1928, and then increased to 11.9 million in 1930.

Census data published in 1930 show that the largest share (3.1%) of the April 1 inventory of Alabama chickens was located on farms in DeKalb County. Other counties with large shares included: Marshall (3.1%), Cullman (3.0), Jackson (2.8), and Madison (2.6). All of these counties are north of Birmingham. These five counties accounted for only 14.6% of the inventory.

Use of Chickens

During the period 1924-1930, 71% of all chickens produced on Alabama farms were consumed by the families who produced them. Throughout the period, 7.5 to 8.7 million birds were consumed annually on the farms that produced them, and some 2.6 to 3.5 million were sold (Figure 11.91). As might be expected, numbers consumed generally followed the trend of the numbers of birds in the annual inventory. In 1930, chicken raising was not yet much of a commercial enterprise in Alabama.

Alabama Eggs

Egg production has a long and storied history in the United States. Chicken coops and hen houses in millions of backyards and on millions of farms were the source of the eggs that, when matched with bacon, provided the fuel that got the country moving every morning for 200 years. That's how Anglo-Saxons have greeted the day for far longer. Butter and egg money launched untold numbers of important enterprises. Endless hours were spent looking for the nests that were hidden so carefully by the hens, who were trying to accumulate enough eggs for a setting. From the earliest days of the colonial period, keeping foxes, skunks, chicken-snakes, and egg-sucking dogs away from the hens and their eggs was a continuing problem for farm women. But those times, along with brown eggs, are gone. Beginning around 1920, technology began to take hens and biddies out of the hands of American housewives, and put them in huge houses. Sadly, no one knew the names of their hens anymore.

The 1920 Census published data that indicate that chickens on farms in the North Atlantic and North Central regions produced 64.3% of all of the eggs in the United States. Approximately 26.6% were produced in the South Atlantic and South Central regions. The West Region accounted for only 9.1%. The data indicate that most of the eggs were being produced where most of the people were. Approximately 1.3% of the country's eggs were produced on Alabama farms.

US and Alabama Egg Prices

US egg prices trended sharply upward during the inflationary period associated with World War I to reach 44 cents a dozen in 1920 (Figure 11.92). Then with agricultural commodity prices declining at the beginning of the short recession of 1920-1921, prices fell sharply to reach 25 cents a dozen in 1922. Prices began to recover in 1923 and reached 30 cents in 1925; however with egg numbers increasing rapidly, prices began to decline again in 1926. From 1926 onward, prices generally responded to changes in numbers to reach 30 cents in 1929. Then with the depression beginning in 1930, the price fell to 24 cents. Surprisingly, egg numbers were up in 1930; however, it is likely that this increase came as a result of the higher prices in 1929, rather than being related to the 1930 price.

There were no annual data on Alabama egg prices earlier than 1924. Data on prices in the state and the country during the period 1924-1930 are shown in Figure 11.94. Generally, Alabama prices followed the same trends as those elsewhere in the country, but in all of the years they were around one cent a dozen lower.

US Egg Consumption

Annual egg consumption per capita in the United States had generally trended downward from 1911 through 1917 (Figure 11.94), but increased in 1918 and 1919, to reach 301 per person in 1919. During the short recession of 1920 and 1921, consumption changed very little; however at the beginning of the Roaring Twenties, it increased sharply in 1922 and 1923 to reach 324 eggs a year. Afterwards, however, with prices increasing (Figure 11.92), consumption began to decline, and in 1925 reached 316. Prices declined in 1926 and 1927, which resulted in an increase in consumption. In 1927, it reached 340, the highest level ever recorded. Unfortunately, prices increased in 1928 and 1929, and once again consumption began to decline. Even though the price declined sharply in 1930 at the beginning of the Great Depression, consumers were not inclined to increase consumption. In 1930, it was down to 329, but at that level, consumption was still 18% higher than it had been in 1917.

Uses of Eggs in the US

In the period 1920-1930, uses of eggs in the United States began to change slightly from those first recorded in 1909. Between 1920 and 1930, the percentage sold increased from 71 to 79%, while percentage consumed on farms declined from 24 to 19%, and percentage used for hatching declined from 4 to 2%.

Alabama Eggs

Annual estimates on the production of eggs in Alabama are not available before 1924. Census data indicate that Alabama farms produced 265.2 million eggs in 1919, or about 1.3% of the national total. The largest share (3.2%),of the Alabama total was reported from Jefferson County. Other counties with large shares included: Marshall (3.1%), Morgan (3.1), DeKalb (2.8), and Lauderdale (2.8). These five counties accounted for only 15% of the state total, indicating the widespread distribution of egg production. A total of 27 of Alabama's counties reported egg productions of 4 million or greater.

Annual Alabama egg numbers also generally followed prices upward in 1924 (423 million) and 1925 (450 million) (Figures 11.93 and 11.95). Egg producers, expecting prices to continue to increase, added to their production capacity for 1926 and 1927; however, with numbers increasing in 1926, prices began to decline. Farmers finally began to reduce production in 1928 (481 million), and prices began to increase. Farmers did not respond to increasing prices this time, and numbers continued to decline through 1930. In the period, numbers increased from 423 million in 1924 to 514 million in 1927, before declining to 457 million in 1930.

Uses of Eggs in Alabama

During the period 1924-1929, 43% of all eggs produced on Alabama farms were consumed by the families responsible for their production. (Figure 11.96). With egg prices sharply down in 1930 (Figure 11.93), the producers reduced sales and increased consumption. As a group, American farmers only consumed around 20% of the eggs they produced during the same period.

No estimates are available for the percentage of Alabama eggs used for hatching during the period 1924-1930; however by subtracting the annual total of eggs sold and eggs consumed from 100, a rough estimate can be obtained. These computations indicate that 6% of all eggs produced were used each year for hatching.

Agricultural Production

Annual estimates of the total value added (billions of dollars) to the national economy by agriculture during the period 1920-1930 are presented in Figure 11.97. As depicted in the figure, it exploded after the United States entered World War I and reached $15.7 billion in 1918. It had been slightly less than $14.4 billion in 1916. This increase was the result of two different forces: Farmers responding to the government's request to produce more food and fiber increased their production, and at the same time inflation was also extremely high (Figure 11.3). Value added continued to increase in 1919 ($17.4 billion) before beginning to decline in 1920, as the American economy, especially agriculture, entered a short postwar deflationary period. In 1921, it fell to near $11 billion, as the short repression gripped the country. Then the speculative boom of the Roaring Twenties pulled value added upward through 1925 ($14.1 billion), and it remained at near that level through 1929.

However, when the country began to stumble toward the Great Depression in 1930, it declined to $11.2 billion.

During the period 1920-1925, crops and livestock contributed about equal amounts to total value added (Figure 11.97), but afterwards the crops share began to decline as prices received also began to decline (Figure 11.5). The livestock share continued to increase slowly until 1930 when it also began to fall.

Data on the share of value added accounted for by home consumption during the period 1910-1930 are presented in Figure 11.98. The data have been converted to value added per farm to make it more meaningful. Annual trends in home consumption followed the same general pattern as total value added. It was $385 per farm in 1920 as a result of the high prices associated with wartime inflation. Then it fell sharply to $264 in 1921 during the short recession and remained near that level through 1929 before falling slightly in 1930. As was the case in the preceding period (1915-1920), value added contributed by livestock to home consumption was twice as great as that contributed by crops. In fact, value added for crops generally trended slowly downward from 1923 onward.

Data on value added by specific crops, groups of crops, groups of livestock, and livestock products for 1915, 1920, 1925, and 1930 are presented in Table 11.9. These data show that value added for cotton was probably greater than for any other specific crop, and except in one year, it was greater than for any group of crops. In three of the four years, value added for food grains was greater than for feed grains. The data also show the growing value added of oil crops, primarily soybeans.

As expected, value added for meat animals (primarily cattle and hogs) accounted for the major share of all livestock and products in all four of the years.

There were no data available on value added for Alabama agricultural commodities for the period 1920-1930; however, the Alabama Field Office of the National Agricultural Statistics Service (NASS) began to publish data on cash receipts for Alabama crops and livestock in 1924. Data for the period 1924-1930 are presented in Figure 11.99. Total receipts zig-zagged year after year throughout the period. There was no trend. The reason for this phenomenon can be seen in year-to-year changes in receipts for crops. Further, the zig-zagging in receipts for crops was the result of the same kind of changes in the values for cotton (Table 11.10). Except in 1930, total receipts for the crop varied between $159 (1928) and $192 million (1925). In 1930, at the beginning of the Great Depression total receipts fell to $120 million.

Except for 1930, cash receipts for crops ranged between $130 (1928) and $168 million (1925). In 1930, it was $97 million. The data in Figure 11.99 also show that receipts for livestock accounted for only 20% of total receipts, ranging from $23 to $29 million. There was much less year-to-year variation in the livestock data.

Generally, there are no Alabama data available on cash receipts for specific crops, livestock, and products during the period 1920-1930; however, in 1924 the Alabama Agricultural Statistics Service (AASS) began to publish data on the value of crops and livestock produced on Alabama farms. Table 11.10 includes some of these data for the period 1924-1930. As expected, the value of cotton was greater than the value of all other major crops combined. Year-to-year variation in cotton values was extremely high, varying from $66.5 million in 1930 to $132 million in 1925. Alabama was obviously still primarily a cotton state in the late 1920s, and its farmers were continuing to be whip-sawed by highly variable cash returns from the crop.

Data on values for livestock and products show that the value of horses and mules continued to be far greater than for any of the others. In the early part of the period, the value for horses accounted for about 20% of the total for horses and mules, but by 1930 the percentage was down to around 1%. The value of milk cows was second highest throughout the period, and it generally increased throughout the period, especially after 1927.

Data on the annual value (per-farm basis) of crops and livestock consumed by farm families during the period 1924-1930 are presented in Figure 11.100. Because of higher prices, the value of livestock and livestock products consumed were considerably higher than for crops. Per-farm values for livestock ranged between $161 and $187, and for crops, $104 to $113. Although the values are not strictly comparable, those of Alabama farm families tended to be somewhat higher than for the country as a whole, especially near the end of the period (Figures 11.98 and 11.100).

STILL SEEKING SALVATION

At the end of the 1920s, American agriculture had reached an impasse. Things were not quite as bad as they had been in 1895, but they were not much better either. O.V. Wells (1940), writing in the 1940 Yearbook of Agriculture, attempted to define the agriculture problem in the United States. He commented that one of the basic problems with agriculture was that it was composed of a large number of relatively small, individual operations, and that as a result, it would always seem to the individual farmers that it would be to his advantage to produce as much as he possibly could—because the individual farmer believed that the quantity he alone produced had little effect on prices. Wells also concluded that most farmers found it almost impossible to emulate industrial concerns and to reduce fixed costs when prices began to decline. He also pointed to the increasing efficiency of production as a major part of the larger problem. Efficiency was increasing so rapidly at the beginning of the twentieth century that there seemed no practical way to slow the increased production that accompanied it. Wells also acknowledged that the only way to effectively limit production was to drastically reduce the number of producing units. Unfortunately, before farmers could leave their farms, they had to be assured of finding some way to support their families. This meant that before the number of farms could be reduced a commensurate number of off-farm jobs had to be created; this solution was largely out of the farmers' hands.

All of Well's observations are correct, but they represented only part of an even larger problem. There was a wide range of efficiencies among farmers and farms. Some were so inefficient that no reasonable price was high enough to allow them to make a decent living for their families. Further, there was a wide range of farm efficiencies from region to region. Without question, farmers wished to receive the highest possible price for all the food and fiber they produced. In contrast, consumers wished to pay the lowest price possible for the food and fiber that they had to purchase. At the same time, there was also a broad range of consumers in terms of ability to pay. There were many consumers who were so poor that prices could not reasonably go low enough for them to be able to afford even the most basic necessities.

Another major problem derives from the fact that food is a necessity for life, and the demand for it is a constant, daily matter, and that producing it in the required quantity so that it was available when needed was still a highly unpredictable art. As a result, producing a surplus from year to year was a necessity. With these restrictions, it was obvious that finding a completely suitable balance between the needs of a broad range of producer efficiencies and consumer abilities to pay was practically impossible. While there were a number of strategies that, over time, might have improved the situation, most of them were too draconian to be acceptable in a liberal, democratic society. The only path to salvation still seemed to lie through better information, farmer organization, diversification, and political activism. Unfortunately, this path to Salvation virtually guaranteed that everyone would survive but that few would prosper.

In 1921, national farm income on a per-farm basis was $517. It had never been that low before. The ratio of the indices of prices received to prices paid was at a level of 0.74. It had never been that low before, either. At that point, there was little indication that the fourfold path to salvation for the American farmer would ever make a positive difference in their lives. However, by 1921 the bureaucracy established to promote salvation had taken on a life of its own. In this situation, any lack of progress was translated into a need for more time and more resources. Everyone, in the bureaucracy was still convinced that it was just a matter of time before the path would lead to the solution of all of agriculture's problems.

Salvation Through Better Information

In 1925, Congress passed the Purnell Act that expanded the scope of agricultural research to include investigations of the social and economic problems associated with agriculture. It also expanded federal funding to further the development of the agricultural extension system.

In 1927, the Alabama legislature provided funds to Auburn to establish an Alabama Agricultural Experiment Substation System in which substations were to be established in most of the state's physiographic areas. The initiative provided for the establishment of five substations (Table 11.11) and 10 experimental fields (Table 11.12). The substations were designed to conduct applied research on a wide range of crops and livestock. The experimental fields were designed for research, usually on field crops, on a more limited scale. Without question, this sudden demonstration of largesse on the part of the state was related to the additional funds flowing into the state's coffers during the speculative boom. This was a period of real economic growth, and inflation was almost nonexistent during most of the period.

As a result of the increased emphasis on agricultural research in the state, the number of staff of the main station of the Alabama Agricultural Experiment Station at Auburn was increased from 21 in 1923 to 48 in 1928. After the establishment of the substations in 1927, the number of staff at each of those units increased from none in 1923 to three in 1928 (Kerr, 1985).

There was one new department created in the School of Agriculture: the Department of Agricultural Economics (1928).

Salvation Through Organizations

Still another effort to organize farmers began in 1919 when the American Farm Bureau Federation was founded. It first appeared in Broome County, New York, in 1911, as the Farm Bureau. In the beginning, the federation planned to be a "non-partisan, non-political, business organization;" however, through the work of an Alabama planter, Edward A. O'Neal, who became president of the national organization, the federation established a powerful political coalition between Southern cotton farmers and Midwestern corn and wheat producers. Over the years, the American Farm Bureau Federation would become the most powerful farmer organization in the country and would play a major part in encouraging the intervention of the federal government in the marketplace of American agriculture.

Luther N. Duncan, director of the Alabama Agricultural Extension Service, presided at the organizational meeting of the Alabama Farm Bureau Federation in 1921. This pattern of Extension Service involvement in establishing state organizations was repeated across the country. Dr. Duncan's interest in the organization would continue for many years, as the Extension Service and the federation worked closely in advancing the cause of agriculture in the state. By the end of 1922, there were federation organizations in 36 of Alabama's 67 counties.

Virtually all of the early attempts by farmer organizations to establish cooperatives had failed; however, these experiences had lead many Southern agricultural leaders to believe that they could be successful in stabilizing cotton prices by establishing cooperatives if they had some federal financial support for them.

With encouragement from the Coolidge administration, Congress passed the Cooperative Marketing Act of 1926 which established the Cooperative Service within the Department of Agriculture. Coolidge's support of the legislation and of cooperatives, in general, did not change his basic reluctance to see the federal government involved directly in supporting them.

President Hoover (1929-1933) also strongly supported the formation of member-owned cooperatives, although he too was generally opposed to direct federal involvement. But as agricultural problems continued to mount, he signed the Agricultural Marketing Act of 1929 that established the Federal Farm Board and gave it two primary responsibilities: strengthening farmer cooperatives and engaging in direct efforts to stabilize plunging agricultural prices through purchase and storage of commodities, primarily through the cooperatives. Congress appropriated $500 million to support the intent of the act. While price-stabilization efforts failed, cooperatives appeared everywhere with the availability of federal funds to support their efforts. By the end of 1930, there were 12,000 cooperatives in the country.

Salvation Through Diversification

Diversification was never a serious matter in those regions outside the South, except that it would have been of considerable benefit to the North Central Region if its farmers could have converted more of their feed grains into livestock. Because of Cycles of Abundance in cattle and swine, the primary livestock animals, it is difficult to determine just how much progress they were making in meeting this goal. According to census data, in 1900 the value of all livestock on farms in the North Central Region was $1.577 billion. By 1920, the value had increased to $3.880 billion, but a considerable amount of this increase was the result of wartime inflation. By 1925, with agricultural prices declining, this value was down to $2.532 billion, and by 1930, the fist year of the Depression, it had increased to $3.049 billion. These data indicate that the region was increasing the production of livestock, but it is difficult to determine just how much.

In 1920, Alabama farmers appeared to be making an effort to diversify their agriculture. They were growing potatoes, vegetables, berries, nuts and tree fruits, and nursery and greenhouse crops. In addition, they were selling substantial quantities of forest products from their farms. They were also producing and selling a considerable number of cattle and calves, hogs and pigs, chickens, milk, butter and cheese, and eggs. Actually, it terms of variety, there was not much else that they could have produced. Unfortunately, cotton continued to be their primary cash crop. They could sell some of all of those crops listed, but to make something resembling a "living," they had to grow and sell large amounts of cotton. In Alabama in 1909, 1919, 1924, and 1929, acres harvested of cotton as a percentage of total acres harvested was: 48.20, 42.8, 44.5 and 50.1%, respectively. These percentages indicate that as late as the beginning of the Great Depression Alabama farmers had made little progress reducing the amount of cotton they grew.

Salvation Through Political Action

The Golden Years of American Agriculture and the World War I years had generally been among the best years in history for American farmers. The short-recession after the wartime boom, however, caught farmers outside the South in its net, just as it had in the late 1800s. Suddenly, stress in the agricultural system became a national problem rather than one involving only Southern cotton farmers. The national problem suddenly received national attention.

National farm organizations finally agreed on a potential solution to their problems, and the McNary-Haugen Farm Relief Bill was introduced in Congress in 1924. This legislation proposed the formation of a federal agency that would attempt to establish and maintain agricultural prices at Golden Age levels. It included the provision whereby the agency would purchase surplus agricultural crops and either store them or sell them on the world-market at world prices. Any losses to the agency would be recovered by equalization fees paid by producers. The legislation was warmly received by Midwestern grain farmers, but not by Southern cotton farmers. Cotton farmers were concerned, with good reason, over the prospects of trying to export their crop into a world awash with agency-dumped cotton.

The McNary-Haugen Farm Relief Bill differed from the old Sub-Treasury Plan in one extremely important way. Under one provision of that plan, accumulated surpluses would eventually be placed on the national market for the government to recover its investment. This, of course, would have simply compounded the accumulating surplus problem. With McNary-Haugen, surpluses would have been dumped overseas without any consideration of the effect that these surpluses would have on local agricultural economies.

The 1924 version of McNary-Haugen was not enacted. Then, in 1926, another version was introduced, but it suffered the same fate. Its supporters re-crafted the legislation and submitted a third version in 1927. This one included some special provisions for cotton and tobacco farmers. It easily passed both houses of Congress, but it was vetoed by President Coolidge who did not like its potential price-fixing or bloated government bureaucracy aspects. There was also concern that there was no provision to force farmers to limit production. Efforts to establish key elements of European socialism in American agriculture had failed temporarily. The plight of the American farmer was not yet severe enough to warrant it; however, the time when farmers' problems would change from severe to disastrous was little more than a moment away.

The Coolidge veto was upheld by a Republican Congress, but the problem did not go away. The parity ratio had remained around 0.9 for most of the Roaring Twenties. After much discussion about possible solutions, it finally became obvious that there was no other acceptable choice, but to finally bring the public treasury into American agriculture, both to help the inefficient producers and the poorest consumers. Unfortunately, this solution guaranteed that the most inefficient producers and the poorest consumers could remain in the game. They had little reason to do otherwise, and frankly, in a brother's-keeper society, there was little other choice. Unfortunately, that solution virtually guaranteed that politics would ultimately be deeply involved in deciding what role the public treasury would play. It also guaranteed that an ever increasing mass of middle-men and bureaucrats would develop in the nexus of these two adversarial group (farmers and consumers). The country, primarily as a result of Republican political ascendancy, had avoided this solution for many years. Finally, in 1929, with the agricultural problem growing worse daily, a Republican Congress and Republican President Hoover finally decided that something had to be done. In 1929, the president signed the Agricultural Marketing Act of 1929. It established a Federal Farm Board that would lend public treasury funds to farm organization cooperatives to be used to allow cooperatives to hold certain commodities off the market until prices improved. And so the die was finally cast. The act did not accomplish much, and it really did not work very well, but it did demonstrate that farmers finally had the ear of Congress.

The Bottom Line

Unfortunately, the boom in American agriculture associated with World War I became a bust after the war ended. Net Farm Income (NFI) had reached $9 billion in 1919, the highest in history (Figure 11.101), but by 1921 it was back to the point where it had been during the Golden Age of American Agriculture ($3.4 billion) when the American agricultural economy began to suffer a serious postwar recession. However, after 1921, it began to respond to the boom resulting from the Roaring Twenties. It finally reached $6.7 billion in 1925. It fell to around $6 billion in 1926, and remained at near that level until 1929. Then in 1930, it plunged to slightly less than $4.3 billion.

Figure 11.102 presents essentially the same data, but on a per-farm basis. Year-to-year trends were generally the same, but these data provide a more relevant picture of the changes. In 1920, NFI per farm was at a level of $1,200. A year later, it was down to $517, a decline of 131%. The whipsaw effect on farmers, their families, and their communities of a decline of this magnitude was staggering. (These data are in current dollars and are affected by changes in inflation.) Between 1921 and 1925, net farm income per farm increased from $517 to $1,041, an increase of 101%. Fortunately, the Consumer Price Index actually declined slightly during this five-year period. So this gain in farm income represented a real increase. Unfortunately, a large share of this gain would be lost in 1930 with the beginning of the Great Depression.

There were no NFI data for the individual states before 1949; however, there is ample evidence that through the 1920s Alabama's agriculture remained mired in a world of share cropping, one-mule farms, slavish reliance on cotton, and ruinous economic cycles, and that it would languish throughout the decade. Fite (1984) reported that in 1929 the "value of farm products per farm," in the state was only $856. This was the lowest value in 11 Southern states and half that in Virginia, and a quarter of Iowa's. Extravagant dreams were never a problem for the 78% of all Alabamian's, who lived on those pathetic farms in 1920. The prosperity of the decade of the Golden Age of American Agriculture had passed them by, and whatever they had gained in the World War I boom, was quickly lost in the short recession that followed the war.

The number of Alabama farms increased slightly during the 1920s. There were 256,000 in 1920 and 259,000 in 1929. Unfortunately, their value decreased by nearly 8%. In 1928, Cleburne County farm income dropped 37% in one year, and almost as much in nearby Cherokee County (35%) (Flynt, 1994). Flynt comments that "within this rural world, the Roaring 1920s sounded more like a desperate whimper." Alabama agriculture had been in a rolling recession for at least three decades before the coming of the Roaring Twenties, and it would not see better times for another two decades, at least.

ALABAMA AND HER SISTER STATES IN 1930

One of the basic reasons for writing this book was to collect data on the establishment and evolution of the Alabama agroecosystem so as to test the hypothesis that Alabama was unable to effectively use its wealth of natural resources in the early years because it progressed so slowly through the simple agriculture stage of societal development (Lenski, 1987). I proposed to test this hypothesis by following the agricultural and industrial development of Alabama and three sister states (Mississippi, Illinois, and Indiana) that were admitted to the Union at about the same time and while all four were still on the Western frontier. Table 11.13, includes data on the status of several of their agricultural and industrial characteristics in 1930. At that time, all four states had been recognized as states for about 110 years.

Population. The data in Table 11.13 show that the population of Illinois and Indiana had grown much more rapidly than in Alabama and Mississippi in the past century, and are probably indicative of the fact that the agricultural and industrial economies of the those Midwestern states had been more encouraging and supportive of population increases.

Farms per 1,000 Persons. These data generally show the relative importance of farming in the four states, and are also indicative of the level of industrialization and commercialization in each. They also generally indicate the degree to which each state's economy was dependent on farming. The data do indicate that Indiana was a little more dependent on agriculture than Illinois.

Average Size of Farms. In 1850, farms were considerably larger in Alabama and Mississippi than in Illinois and Indiana; 289, 309, 158, and 136 acres, respectively. Of course, these estimates in Alabama and Mississippi reflect extremely large plantations. By 1930, the farm size situation had changed dramatically. Size remained stable in Illinois and Indiana, but declined by 76% and 82% in Alabama and Mississippi. In the Deep South, there were just too many people with nothing to do but farm.

Acres of Cropland Harvested. These data clearly show the relative status of the evolution of agriculture in the four states. Harvesting crops from only 20 to 25 acres of cropland a year would seem to indicate a stunted state of agricultural development. Acres harvested this low would tend to indicate that Alabama and Mississippi, at that point, had hardly advanced through Lenski's (1987), simple agriculture stage. Certainly, with acres harvested this low, agriculture would be in no position to provide any of the capital required for moving on into the advanced agriculture stage. Further, there would be little possibility that the agriculture economies in those two Southern states would be able to provide any of the capital to move them into the industrial stage.

Value of Farm Buildings. These data are generally related to farm size, but these differences in values are much too great to be the result of differences in the average sizes of the farms involved.

Value of Implements and Machinery. These data are also somewhat dependent on farm size and especially on the amount of cropland farmed, but differences in cropland harvested are probably not great enough to explain the magnitude of the differences in the values of implements and machinery.

Farm Tractors. Probably tractor numbers are more indicative of the relative states of agricultural development in the four states than any of the other data. There are several reasons why Alabama and Mississippi farmers had significantly fewer tractors than Illinois and Indiana farmers, but most of them are largely the result of the fact that because of the status of the agricultural economies in those two Southern states, they could not afford them. If tractor numbers could be used as a measure of the status of evolution of agriculture in the two states, one would likely conclude that Alabama and Mississippi had barely entered the advanced agriculture stage as late as 1930.

Grain Yields. (bushels per acre) These data clearly show that in 1930 Alabama and Mississippi were continuing to fall further and further behind in yields of these important grain crops.

Manufacturing Establishments. These data indicate that the number of manufacturing establishments per million persons was twice as large in Illinois and Indiana, and would likely indicate that those two states were well into the industrial stage of societal evolution in 1930.

Total Value of Manufactured Products. These data

suggest that Illinois and Indiana not only produced a relatively larger quantity of manufactured products, but also that the products probably were of higher relative value. A large share of products produced in the two Deep South states were related to the forest products industry and were of relatively lower value.

Annual Wages. The wages data suggest that Alabama and Mississippi workers had fewer opportunities for off-farm work, and that the pay-rates were likely lower.

Little more than a century earlier, these four states were on the western frontier. The differences in their development, or lack of it, had been phenomenal.

12

From Purgatory to the Inferno

1930-1940

PRECIS

In the period 1915 to 1929, American farmers seemed to be forever confined to Dante's *Purgatory*. Throughout the period, the parity ratio (indices of prices received / indices of prices paid) generally ranged between 0.80 and 0.90. It was 1 or above in only two years. Unfortunately in 1930, it began to decline, and by 1932 it was down to 0.58. Instead of going to a far better land from *Purgatory* as Dante had, farmers had plunged directly into the *Inferno*. This was the rapidly deteriorating situation that Franklin Roosevelt and the Democratic Party faced when they took control of the federal government in 1933.

Quickly, Congress with its powerful Southern leadership enacted some of the most intrusive legislation ever in a democratic country. To reverse the downward spiral in agriculture, they passed and the president signed the Agricultural Adjustment Act of 1933. By paying farmers for plowing up cotton, not planting wheat, and slaughtering lambs, pigs, and pregnant cows, the federal agricultural reformers sought to increase prices received for farm commodities by removing surpluses from the market. Even Roosevelt and his secretary of agriculture were disturbed with what they were doing.

By 1939, the year World War II began in Europe, results of government efforts to improve the lives of its long-suffering farmers through direct intervention in their affairs was a mixed bag. There were both winners and losers, especially in the South. Those at the top of the pecking order were clear winners, while those at the bottom—those with little to lose—were generally losers. Farmers owning large acreages with no tenants were primary beneficiaries. In fact, many of those with large farms and tenants quickly discovered that under the provisions of the legislation their tenants were a distinct liability. They were much better off without them. At the end of the 1930s, it was not clear whether federal intervention in American agriculture had done more good that harm, but one thing was abundantly clear: it was here to stay.

INTRODUCTION

Near the end of his life, the Italian poet Dante Alighieri wrote his timeless classic, epic poem, *The Divine Comedy*. In it he described his imaginary travels through Hell and Purgatory to Heaven. Unfavorable political and economic events between the end of World War I and the beginning of World War II forced American farmers to travel this path in a much less favorable direction—from Purgatory to Hell.

After 1920, American agriculture began to experience hard times, and by the end of the decade, it was well on its way to disaster. In 1920, Net Farm Income (NFI) on a per-farm basis was $1,196. By 1926, it had fallen to $919, and in 1930 it was $650. In 1919, the Parity Ratio was 1.09; by 1925, it had declined to 0.95, and in 1930 it reached a level of 0.83. The Golden Age of American Agriculture and the boom accompanying World War I were distant memories. In 1921, farmers probably wondered if those years had been real, or were they only a dream.

As one viewed Alabama and Southern agriculture at the end of the third decade of the twentieth century, it was difficult to find many positive changes that had taken place in the past generation. Tenancy was increasing at a steady rate, while cropland harvested in all of the Southern states was declining. Diversification, scientific farming and improved efficiency— despite the expenditure of millions of dollars for agricultural research and extension—had hardly scratched the surface in meeting the needs of Southern farmers, especially the hordes of share tenants. Although no farmer in the country could know it at the time, before the situation could improve conditions would get considerably worse than they had ever been.

NATIONAL POLITICS

For much of the latter part of the nineteenth century and the first three decades of the twentieth century, both the executive and legislative branches of the federal government were mostly controlled by the Republican Party. The government's role in American agriculture for well over a half century had largely been determined by the politics of that party, a party dominated primarily by Eastern commercial interests. Times, however, were about to change. Politics in the United States was about to undergo a sea-change of immense proportions.

The Presidency

Except for the years 1913 to 1919 when Woodrow Wilson (D, NJ) was president, between 1897 and 1933 six Republican presidents held the reins of executive power. A Democrat was elected to the position in 1932, and that party controlled the office throughout the remainder of this period (1930-1940).

Calvin Coolidge (R, MA) served as president from 1923-1929. He was succeeded by Herbert Hoover (R, CA) who served from 1929-1933. In 1932, Franklin Roosevelt (D, NY) was elected, and would serve until his death in 1945.

Secretaries of Agriculture

William Jardine (R, KS) served as secretary of agriculture under President Coolidge, leaving that office in 1929. Others serving in that position included: Arthur Hyde (R, MO) (1929-1933) and Henry Wallace (D, IA) (1933-1940).

The Senate

The Senate was under control of the Republicans from 1930-1933, but the Democrats took control in 1933, and maintained it through 1940 (Table 12.1).

Senate Majority Leaders

James Watson (R, IN) was chosen Senate majority leader in 1929 when Charles Curtis (R, KS) was elected vice president. Watson served as Senate majority leader until 1933 when Joseph Robinson (D, AR) replaced him. Robinson served until 1937, and was replaced by Alben Barkley (D, KY) who served until 1947.

Chairmen of the Senate Committee on Agriculture

Charles McNary (R, OR) was first chosen as chairman in 1925, and he held the office until replaced in 1933 by Ellison Smith (D, SC). He served until 1944.

The House of Representatives

The Republican Party controlled the House in the 71st Congress (1929-1931), but lost control in the 72nd (1931-1933). The Democrats maintained control throughout the remainder of the decade (Table 12.1).

Speakers of the House

Nicholas Longworth (R, OH) was elected speaker in 1925 (69th Congress), and he served in that position until 1931 when he was replaced by John Garner (D,

TX). Garner served as speaker until 1933, when he was replaced by Henry Rainey (D, IL). Rainey died in office in 1934 and was replaced by Joseph Burns (D, TN). Burns also died in office (1936) and was succeeded as speaker by William Bankhead (D, AL). Then Bankhead died in 1940 and was replaced by Sam Rayburn (D, TX) who served the remainder of that year.

Chairmen of the House Committee On Agriculture

Gilbert Haugen (R, IA) was first chosen as chairman of the Committee on Agriculture in 1919. He served in the position until 1931 when he was replaced by Marvin Jones (D, TX). Jones served as chairman until 1941.

The Political Situation

Herbert Hoover, a Republican and a mining engineer from California, was elected president in 1928, and a few months after he took office the stock market crashed. American prosperity had fueled an economic boom in the Roaring Twenties that quickly spread around the world. Unfortunately, it was built on a base of precarious European currencies, wild speculation in stocks, a chaotic American banking system, and industrial uncertainty. The stock market crash resulted in a downward spiral in prices, production, consumer buying, employment, and foreign trade. Furthermore, the highly protective Hawley-Smoot Tariff, enacted by the Republican Congress in 1930, added fuel to the growing economic and political inferno.

In 1931, President Hoover was convinced that the worst was over, but it was to continue to worsen until mid-1932. Although the president was deeply troubled by the pain and suffering throughout the country, he strongly believed that the economy, given time, would right itself. He also felt that whatever the federal government might do would cause more harm than good. Instead of asking Congress to become involved, he became an out-spoken supporter for a nation-wide charity drive.

In the mid-term elections of 1930, amid the worsening Depression, the Democrats won control of the House and came within two votes of winning control of the Senate (Table 12.1). This sea-change in Congress encouraged both parties to get government involved in the economic situation. Despite some misgivings, in 1932, Hoover signed an act creating the Reconstruction Finance Corporation. This act was to provide government money to be lent to railroads, banks, agricultural agencies, and commerce.

Hoover, carrying the dead weight of the on-going Depression, was soundly defeated by Democrat Franklin Roosevelt in the election of 1932. Roosevelt had served as the secretary of the Navy in both Wilson administrations. He had been stricken with polio in 1921, and made his first of many re-habilitation trips to Warm Springs, Georgia, in 1924. On these train trips through the countryside of Virginia, the Carolinas, and Georgia, he was able to observe, first-hand, the widespread desolation and poverty of the people of the South.

Roosevelt was extremely popular in the South where the effects of the Depression were most severe. In fact, Democrats (any Democrat) had been extremely popular in the South since the end of Reconstruction. In the 1932 election, the South gave Roosevelt 32.6% of his 472 electoral votes. He never forgot that the Solid South was the heart of his and the Democratic Party's national coalition: blacks, blue collar workers and union members, state and city political machines, people on relief, the elderly, and intellectuals, as well as Irish, Italian, and Jewish white ethnics. Alabama, Arkansas, Georgia, Louisiana, Massachusetts, Mississippi, Rhode Island, and South Carolina had been the only states to give their electoral votes to Al Smith, the Catholic Democrat, in the 1928 election.

In the 1932 election, the country gave the Democrats a 193-member majority in the House and a 25-member majority in the Senate. From every indication, the old party of Jefferson and Jackson had been given a powerful mandate to vastly increase the level of government intervention in the everyday welfare of the governed.

In the first 100 days of the 73rd Congress, at the urging of the Roosevelt administration, 123 pieces of precedent-shattering legislation that would change the country for the forseable future were enacted (Morison, 1965). This legislation included:

1. Emergency Banking Act (March 9, 1933)
2. Economy Act (March 29)

3. Established the Civilian Conservation Corps (March 31)
4. Act abandoning the gold standard (Aril 19)
5. Federal Emergency Relief Act (May 12)
6. Agricultural Adjustment Act (May 12)
7. Emergency Farm Mortgage Act (May 18)
8. Tennessee Valley Authority Act (May 18)
9. Truth in Securities Act (May 27)
10. Home Owners Act (June 13)
11. National Industrial Reovery Act (June 16)
12. Glass-Segall Banking Act (June 16)
13. Farm Credit Act (June 16)

One of the most far-reaching of the acts approved during that period was the National Industrial Recovery Act (NRA). Its purpose was to use the government's power (intervention) to support the recovery of the country's industrial base. Depending on how one views it, the NRA was a unique experiment in the management of a large segment of the country's economy by the government, or it was an outright power grab by socialists. The NRA sanctioned, supported, and in some cases, enforced the alliance of industries. Companies were required to write industry-wide codes of fair competition that effectively fixed prices and wages, established production quotas, and imposed restrictions on entry of other companies into the alliances. Employees were given the right to organize and bargain collectively and could not be required, as a condition of employment, to join or refrain from joining a labor organization. The NRA was the embodiment of the Saint Simon-Blanc socialist philosophy.

From the beginning, the NRA suffered widespread criticism. Some of its provisions effectively fixed prices and wages, and as a result, it actually made the economic situation worse. It was widely agreed that the NRA was a failure as public policy. It was also an extremely good example of over-reach in governmental intervention. In 1935, the Supreme Court invalidated the compulsory codes provision, effectively emasculating the entire act. The court further refused to endorse Roosevelt's argument that the national crisis was sufficient reason to demand such radical intervention. This court decision was one of the events that led to Roosevelt's failed efforts at court packing. NRA supporters apparently decided that they would never get the votes to pass it again. As a result, they decided not to rewrite it to make it acceptable to the court. While this reversal of fortune for Roosevelt freed Americans from the threat of immediate government domination of the country's industrial economy, organized labor emerged much stronger that it had been before.

In several preceding chapters, I have commented on the genesis of European socialism. It generally came out of European liberalism at the beginning of the nineteenth century, but it did not have much political effect on the Continent until later. From the beginning the movement was quite diverse, but there was one basic belief shared by all of its followers: economic justice and equality can only be advanced by the increased involvement of the government in the affairs of the governed. Aspects of this basic belief appeared in America when the Farmers Alliance in 1889 proposed that legislation be enacted to establish the Sub-Treasury Plan to assist farmers with the marketing of their crops. Later elements of it would appear in the platform of the Populist movement at the end of the century. Further, some elements would appear in every administration from that of Theodore Roosevelt (1901-1909) to Herbert Hoover (1929-1933). The political philosophies of the movers and shakers of the first Franklin Roosevelt Administration were obviously quite diverse, but no group in the history of freely elected democracies had ever done so much to advance the involvement of government in the affairs of the governed as in its first 100 days.

Roosevelt easily won the election of 1936 against the Republican candidate, Alfred Landon of Kansas. Moreover, Democrats increased their margins in both the House and Senate (Table 12.1). Shortly after he was re-elected, Roosevelt received his first legislative defeat. He attempted to pack the Supreme Court by adding justices more supportive of his legislative agenda. His efforts were rebuffed by Congress, but he soon accomplished the same objective after several of the members of the court retired.

In 1937, the American economy entered a 13-month period of economic contraction. This recession, plus Roosevelt's continuing raids on the US Treasury, resulted in the formation of a conservative coalition in the Senate. This coalition, lead by Senator Robert Taft (R, OH) and Richard Russell (D, GA), worked across party lines to end the expansion of the president's programs. If the election of 1936 had been held in 1937 during the recession, it is likely that the outcome would have been somewhat different.

In the late 1930s, the United States was still muddling through the final stages of the Great Depression and recession, and it was ensconced in a warm-cocoon of pacifism, almost unparalleled in its history. Unfortunately, the Japanese were not equally interested in peace when they began a campaign to bring China, Indochina, Indonesia, the Philippines, Burma, and India under their hegemony. Then in 1938, Germany annexed Austria and Czechoslovakia, and in 1939 invaded Poland. Britain and France quickly declared war on Germany, and World War II had begun. By 1940, Germany had conquered virtually all of Western Europe, and Great Britain and the British Empire stood alone with little hope of withstanding the onslaught of the Nazi and their Axis allies. Within these threatening circumstances, Roosevelt proposed a scheme that would allow the United States to provided direct assistance to Great Britain, halt Japanese aggression, and begin to strengthen American capacity to wage war—all without openly entering the conflict.

It was at this complex juncture in American history when the country had not completely emerged from a catastrophic period of economic distress and when the world seemed bent on devouring itself that Roosevelt decided to seek an unprecedented third term as president. In the election of 1940, with unparalleled support for both his national and foreign policies, Roosevelt easily defeated the Republican candidate, Wendell Willkie of Indiana.

Morison (1965) comments that by 1938 the New Deal was firmly entrenched in American life. He further noted that the Roosevelt administration "saved twentieth-century capitalism by purging it of gross abuses and forcing an accommodation to the larger public interest." Roosevelt himself commented "The only sure bulwark of continuing liberty is a government strong enough to protect the interests of the people, and a people strong enough and well enough informed to maintain its sovereign control over its government" (Morison, 1965). Surprisingly, Roosevelt seemed to be completely unaware of what his politics would do to the balance between these two competing elements.

In the preceding paragraphs, some of the more important political events of the period 1930-1940 were recounted. However, because of our specific interest in the agricultural economy during this period, some of the political events related to New Deal efforts to provide relief for farmers also need some attention.

The change of control of Congress in 1931 and 1933 brought many Southerners into nationally-prominent leadership positions. Democrats had taken control of the political machinery in virtually all of the states of the Old South following the demise of Radical Reconstruction, which followed the end of the Civil War. With virtually no Republican opposition, these states tended to return the same Demoncrats to Congress year after year. As a result, these long-serving individuals accumulated many years of seniority. When the Democrats finally took control of Congress, these individuals assumed leadership positions on many of the important committees. The leadership changes were especially significant on the all-important agricultural committees and subcommittees of both houses (Table 12.2). Marvin Jones, a Texan who had been born on a cotton farm, became chairman of the House Agriculture Committee. Ellison D. Smith from South Carolina, became chairman of the Senate Agriculture Committee; although he had never been involved in cotton farming, Smith had been such a strong supporter of the crop that he had been given the nickname, "Cotton Ed." Another Southerner on the Senate Agriculture Committee, Alabama lawyer and planter John Bankhead, was vitally interested in farmer problems. He was also a member of the important Banking and Currency Committee. These Southerners, plus a number of others, serving as chairmen of important sub-committees, played a critical role in forming the legislation that would forever change American agriculture. The elements of European socialism that had attracted agricultural interests in the Populist Movement in the latter part of the nineteenth century emerged full blown in the Congress, guided to fruition by this Southern band-of-brothers. Ernstes, et al. (1997) have written extensively on this phenomenon.

At the beginning of the 1930s, Southern agriculture was still hopelessly mired in the nineteenth century and forever saddled with poor soils and a generally inhospitable agricultural climate. To attempt to solve the South's long-standing agricultural problems by political means was pure wishful thinking—with just a touch of ignorance thrown in.

ALABAMA POLITICS

Throughout the early 1920s, Alabama politics continued to grapple with state finances, education, child welfare, labor relations, prisons, and convict leasing. Then toward the end of the decade, under the leadership of Governor Bibb Graves, the state began to make real progress in solving problems related to all of these areas, in fact, in almost every facet of public life (Flynt, 1994). Then the Great Depression came.

Governors

Bibb Graves (D, Montgomery County) served as governor from 1927 to 1931. He was succeeded by Benjamin Miller (D, Wilcox), who served until 1935. Miller was replaced by ex-Governor Graves, who served until 1939. Frank Dixon (D, Jefferson) succeeded Graves and served until 1943.

Commissioners of Agriculture

Seth Storrs (D, Elmore County) was elected commissioner of agriculture in 1929 and served until 1935. He was replaced by Robert Goode (D, Wilcox), who served until 1939. Haygood Patterson (D, Montgomery) became commissioner in 1939 and served until 1943.

The Political Situation

Flynt (1994) commented "Given the dominating presence of President Roosevelt and New Deal liberalism, it was nearly inevitable that state politics would divide basically along the lines of FDR's supporters and opponents." Those opposing Roosevelt and his New Deal in the state were generally associated with planters and industrialists. Those supporting the president generally drew their strength from organized labor and farmers with limited resources. Also involved in this division of support was Roosevelt's insistence that the benefits of the New Deal be distributed without consideration of race or creed.

The gubernatorial election of 1930 quickly underscored the New Deal/anti-New Deal divisions in the state. The election pitted Governor Graves's choice, Lieutenant Governor William Davis, against Benjamin Miller, a Camden (Wilcox County) attorney. In his campaign, Miller attacked virtually all of the reforms of the Graves administration as being far too extravagant for the state. He also attacked Graves and by association, Miller's relationship with the Ku Klux Klan (Graves had once been Grand Dragon). Miller won the primary and the general election where he was opposed by Hugo Locke, the candidate of the Jeffersonian Democrats. The new governor, inaugurated in January 1931, inherited a state government in dire economic straits. State debt associated with Graves's reforms was massive, and with the beginning of the Depression, revenues were at an all-time low. Miller was finally able to get the legislature to approve an income tax and an inheritance tax, but these measures did not provide much relief. He had to reduce the salaries of state employees drastically. There were no funds to provide relief for the beleaguered schools, the jobless, or the hordes of desperate farmers. Whatever aid they received came from New Deal programs. By January, 1934, some 200 schools had to close their doors because of insufficient funds.

Early in Miller's administration, the Scottsboro Boys situation erupted. The governor was besieged with requests to pardon the boys, but there is little evidence that he ever became seriously involved.

In the 1934 gubernatorial election, ex-Governor Graves faced Frank Dixon, a Birmingham (Jefferson County) attorney. Most of the pro-New Deal people supported Graves, and he finally bested Dixon in a run-off. Graves had always supported the labor movement in the state, in principle if not in practice. But from the beginning of his second administration, with working people in desperate straits, he put principle into practice. In 1935, the Department of Labor was established to administer federal wage-and-hour laws, mediate labor disputes, and regulate child labor.

Graves's second administration did not result in the enactment of nearly as much progressive legislation as his first. The state was still mired in the Depression. Advances in public programs could only be made through cooperation with federal authorities. He was able to provide some support to education by getting legislation enacted that required schools to remain in session for at least seven months and to provide free textbooks for grades 1 to 3. In addition, he abolished the Child Welfare Department, and transferred its functions to the newly-created Department of Public Welfare. The Alabama State Employment Office was also established to provide assistance to the unemployed.

An inordinate amount of Graves's time was taken by continuing problems related to the Scottsboro Boys situation. He was beset from every direction with advice, and at one point promised to pardon all but one of them. However, after conducting private, personal interviews with each of them, he reversed his decision.

In the 1938 election, Frank Dixon continued the pattern—seemingly firmly established in Alabama politics—that defeated candidates for governor would win the second time around. Dixon had lost to Bibb Graves in the 1934 election, but he defeated Chauncey Sparks, a Eufaula (Barbour County) banker in 1938. Organized labor backed Dixon in the election, while business interests supported Sparks. Once in office, Dixon quickly abandoned his New Deal supporters by abolishing the Department of Labor and transferring its functions to the newly created Department of Industrial Relations. Dixon had campaigned on a platform of reforming state government, and he quickly set about doing it. He adopted a merit system for state employees, replacing the old well-worn spoils system. He also replaced the existing pardon, parole, and probation system with a new Pardon and Parole Board. Dixon established the Alabama Teachers Retirement System and several new departments including: Finance, Conservation, Personnel, Commerce, Revenue, and State Docks and Terminals.

Dixon was an effective, visionary administrator, but much of his success was related to the rapid improvement in the state's economy that began in 1938. Alabama's economy improved even more after 1939 when the federal government began preparing for war. In 1940, much of Dixon's time was given to helping Alabama prepare to carry its share of the load if and when war actually came.

The Alabama Department of Archives and History has written that "Alabama's congressional delegation, during the 1930s, was one of the most liberal and influential of any Southern state." They further noted that senators John Bankhead Jr, Hugo Black, and Lister Hill and congressmen Henry Stegall, Bob Jones, George Huddleston Sr., and Luther Patrick provided firm and active support for Roosevelt's New Deal legislation.

THE NATIONAL ECONOMY

The speculative boom of 1926-1929 resulted in world-wide overproduction of basic commodities, such as wheat, rubber, coffee, cotton, copper, silver, and zinc. The over-abundance of agricultural commodities was worsened by bumper crops in Europe in 1929. During the same period, serious weaknesses and stresses were developing in the economies of the United States and Europe. These included the tremendous amount of money tied up in the stock market, much of it borrowed; the large volume of mortgage and installment-buying debt; the chaotic American banking system; and the precarious position of European currencies. The first failure in this enormously complex house of cards was the spectacular stock-market crash, which began in October of 1929. This event set in motion a downward spiral in prices, production, employment, and foreign trade. The rapid decrease in international trade was pushed along by the highly restrictive Hawley-Smoot Tariff of 1930.

The Gross National Product

The Great Depression actually began in 1929, and it would not end until 43 months later in 1933, less than a year after Roosevelt's first 100 days. Maybe Hoover was correct, after all. While economic activity began to grow, after that time much of the misery resulting from the Depression would last until the beginning of World War II. The US Gross National Product (GNP) began to fall in 1930. It had been at a level $709 billion in 1929 (Figure 12.1), but fell to $643 billion in 1930, and it continued this downward trend through 1933, when it reached $499 billion. It had not been that low since the short recession of 1921. The GNP began to recover after 1933 as a result of, or in spite of, some of the various New Deal programs that began to take effect in the economy. The New Deal, however, did not solve all the economic woes of the country. Another economic contraction began in 1937, and continued for 13 months. This short downturn is evident in the data presented in Figure 12.1. The economy began to recover in 1938 and in 1940 the GNP was considerably higher ($772.9 billion) than it had been in 1929 ($709.6 billion).

The Consumer Price Index

The Consumer Price Index had begun to trend sharply downward after 1925 (Figure 12.2). It was 51.3 in 1929, and continued to fall until it reached 38.8 in 1933. It had not been that low since 1917. It recovered somewhat in 1934 to 40.1, and remained near that level until 1940.

The Inflation Rate

The inflation rate was 0 in 1929 (Figure 12.3), but with the beginning of the Depression in 1930, it began to trend downward, reaching -10.3 in 1932. Afterwards, it began to increase, reaching 3.3 in 1934. It remained near that level until the beginning of the recession of 1937-1938 when it fell below 0 again. After the beginning of World War II in Europe, it increased sharply to 9.6 in 1940.

Federal Discount Rates

The Federal Discount Rate was 1.5% in February, 1934, and it remained at that level until August, 1937 when it was reduced to 1.0.

The Economic Situation

In 1929, an economic dislocation of unprecedented proportions hit the industrialized world. It began in 1929 in all countries, but ended at different times, in different countries. The most heavily industrialized countries (United States, England, France, Germany, Canada, and Japan) were most affected, and within those countries the Depression was most brutal in the largest cities. In the United States, all construction virtually ended. The unemployment rate was at 3.1% in 1929 and 16.1% in 1931. It continued to increase to 25.2% in 1933. The rate began to fall after 1933, but was still 16.5% in 1938. The Index of Industrial Production was 109 in 1929 and 75 in 1931. It continued to decline, reaching a level of 69 in 1933. It also began to improve sometime after 1933, but was still 89 as late as 1938. A massive series of runs on the country's banks forced 4,004 of them to close in 1933. In rural areas, farmers suffered through price declines of 40-60%. The mining and lumbering sectors of the economy were probably the most seriously affected. Demand almost ceased to exist, and workers in those enterprises had few other opportunities available to them. Segments of the economy began to improve after 1933, but most did not fully recover until the beginning of World War II.

Just when the economy seemed to be recovering from the Depression, it was beset with still another economic contraction. This one began in 1937 and lasted for most of 1938. Industrial production once again declined sharply, and unemployment increased from 14.3% in 1937 to 19% in 1938. Obviously, there were some things that the New Deal, with all of its nation-changing legislation, could not do.

THE AGRICULTURAL ECONOMY

America's agricultural economy expanded rapidly during the years of World War I, much of it the result of inflation of the currency. In 1920, the short recession wrung the inflation out of the system, and by 1921 the agricultural economy had returned to about where it had been in 1916. It grew at a moderate rate through the Roaring Twenties, only to smash into the brick-wall of a world-wide depression in 1930.

Federal Agricultural Legislation

Needs of the mostly-impoverished American agricultural economy occupied center-stage when the Democratic Congress began its 100 days of frenetic legislative activity in 1933. Because of the numerical superiority of Democrats in Congress, they must be given the major share of the credit for enacting the plethora of agricultural legislation; however the Republicans also contributed. After the Democratic clean-sweep in the national election, the only areas in the country where the Republicans still held a majority were in the Midwestern farm belt.

Ernstes, et al., wrote that "through a convergence of forces in the early 1930s, they [Southern Democrats] were able to guide and implement sweeping agricultural legislation which affected the lives of 35 million people. Through the Depression and the war years, conventional wisdom suggests that southern Democrats controlled the agricultural agenda (Table 12.2)." . . . "Yet southern domination of agricultural legislation is not apparent and data indicates that the committees under southern leadership, were generally evenhanded in the distribution of government largess." Actually, the Southerners had no choice but to be evenhanded. They had the power and position on the committees to shape and control the flow of legislation, but they did not have the votes in Congress to pass any of it. To get what they wanted for the South, they had to give the other regions virtually anything they wanted. Once in place, such an arrangement was a recipe for political back-scratching that would undergird American agricultural policy and legislation in perpetuity.

When the 73rd Congress was organized in 1933, it was not clear what, if any, farm-relief legislation might be passed, but it was apparent that whatever they approved would have to include provisions to deal with the ever growing problems of the Southern cotton farmer. There was no question but that many congressmen felt for him, but there was considerable concern as to whether or not they could reach him.

Southern cotton farmers, especially share tenants, had known little but depression since the Civil War. Then general agriculture, nationwide, had shared the malaise in the late nineteenth century and again for a short period after the end of World War I. Even during the Roaring Twenties, all of agriculture was depressed because of the disastrous parity ratio situation described in the preceding chapter. After 1929, depression began to spread through the entire American economy. As a result, the 73rd Congress had many nasty problems on their plate, including: bank failures, growing unemployment, closing businesses, severely reduced industrial output, and the longer-term, but increasingly severe, farm depression. Most everyone agreed that something must be done to help farmers, but there was considerable discussion as to whether their problems were the cause or the result of the broader national recession.

One group held that the primary cause of the broader national economic contraction was that the narrower depression in the farm sector after World War I had seriously eroded the purchasing power of the country's enormous farm population for goods and services. In 1929, the American farm population comprised 25% of the total population. Through their representatives, the farmers insisted that the federal government had a responsibility to move quickly to restore prosperity on the farms. Another group countered that agriculture was suffering from the general population's eroded purchasing power for farm products, and that only by increasing the incomes of that broader group could there be any improvement in the farm situation.

With these concerns in mind, Democrats, with considerable Republican assistance, quickly began to develop legislation to provide a New Deal for the counry's beleaguered farmers, but there was considerable concern over how quickly Congress could act. A Republican-led Congress had discussed elements of the McNary-Haugen Farm Relief Bill from 1924 until 1927, but when they finally passed it, President Coolidge vetoed the bill. However, things were different in 1933. There was considerably more pressure to do something, even if it was wrong. Roosevelt was inaugurated in March of 1933, and on May 12 Congress passed the revolutionary Agricultural Adjustment Act (AAA). It was so revolutionary that Roosevelt himself commented that "I tell you frankly that it is a new and untrod path, but unprecedented conditions calls for the trial of new means to rescue agriculture." By enacting this legislation, socialists Saint-Simon, Blanc and Company were given a seat at the head of the table in the resurrection of American agriculture.

The primary thrust of the Agricultural Adjustment Act was to bring supply and demand into balance, by controlling production and storing surpluses in government warehouses when they accumulated. The preamble to the act included the following comment:

> it is hereby declared to be the policy of Congress... To establish and maintain such balance between the production and consumption of agricultural commodities ... as will reestablish prices to farmers at a level that will give agricultural commodities a purchasing power ... equivalent to the purchasing power of agricultural commodities ... in the period, August 1908-July 1914.

This language in the preamble was extremely important, because it was now the law of the land that the federal government would use the parity ratio as the basis of its agricultural policy. The immense economic power of the government would be used to keep the ratio near 1.0, or that the indices for prices received would be kept very close to those of prices paid.

Agricultural leaders had been attempting to get farmers to reduce production for well over a half century, but to no avail. Now, the federal government was going to try its hand, but they were going to take a completely different approach to the problem of overproduction. In all of the earlier efforts, farmers had been encouraged to voluntarily reduce the amount they planted and hope that by planting less they would receive more in return. Of course, farmers had never agreed to this approach, so the government did something that had never been tried before. In fact, it had never even been proposed before. They were going to use public funds to pay farmers not to produce. Unfortunately, very few farmers agreed with this approach. Most of them felt that they could make more money by not taking the government's money and planting what they wanted to plant; consequently, by 1937 quantities harvested for cotton, corn, wheat, and oats were larger than in 1933. As the old adage goes: "The road to hell is paved with good intentions." An unusually interventionist Congress had the best of intentions, but unfortunately they did not seem to understand farmers. Surplus production could have been reduced or eliminated only by allowing the marketplace to force large numbers of farmers out of farming. Unfortunately, given the nature of the national political situation, this was an unacceptable solution.

The activist political intervention that led to the enactment of the Agricultural Adjustment Act also led to the informal establishment of Dobson's (1985) so-called Triangle of Beneficiaries (politicians, their constituents, and bureaucrats). Dobson wrote concerning the Triangle: "... lawmakers following natural inclinations to acquire influence and obtain reelection have strong incentives to perpetuate government programs that aid constituents. Bureaucrats administering the programs also have incentives to see that such programs are continued." Instead of Triangle of Beneficiaries, I will use the term "Triangle" numerous times in the following sections and chapters. It is never used in a derogatory sense, but rather as shorthand term to remind the reader that the actions of the Triangle is the way our system of politics and economics, with all of its warts and idiocies, worked after 1932.

The typical Southern congressman during this period had a rather limited constituency in 1930. Over the years, virtually all African Americans and a large share of the poor whites in the region had been disenfranchised. The Populist (agrarian) up-rising of the late nineteenth century had scared witless the entrenched Southern political power structure, and they had responded by taking the vote away from those unreliable constituents.

Further, the 1901 Alabama Constitution was written in a manner that virtually all of the state's business was the product of the legislature. County and municipal home-rule did not exist. As a result, the voting constituents of most Southern congressmen tended to be landed farmers, trying to make a living producing 10-cent cotton with a horde of destitute tenants.

It is also important to note that the AAA identified (mandated) six crops for immediate governmental attention: corn, cotton (upland), peanuts, rice, wheat, and tobacco. It probably is not surprising that four of the six were Southern belles. A fifth, sugar cane, would be added to the list in 1934. The Triangle identified the five Southern belles as mandated crops even though they were of very limited national importance. For example, in 1933 American farmers harvested 1.2 million acres of peanuts, 35%, of them in Georgia alone; 798,000 acres of rice, 49% in Louisiana; 1.7 million acres of tobacco, 26% in Kentucky and 39% in North Carolina; and 211,000 acres of sugar cane, 93% in Louisiana.

In 1936, the Supreme Court decided that a key provision of the 1933 Agricultural Adjustment Åct was unconstitutional, but with Triangle support it was quickly rewritten to meet court objections. Even with cotton, corn, and wheat harvests higher by an average of 31% in 1937 compared with 1932 and with the country in another recession, the Triangle pressed forward. Apparently, they never once considered that food and fiber production and consumption were simply too complex to be managed by any governmental entity. From every perspective, food and fiber production and consumption are several orders of magnitude more complex than industrial production and utilization. While the industrial Triangle seems to have understood the situation and refused to try to rewrite the National Industrial Recovery Act, the agricultural Triangle with its Southern band-of-brothers did not. There is much wisdom in Alexander Pope's: "fools rush in where angels fear to tread."

Once the Democratic Congress and the Triangle determined that the sky did not fall when the federal government stuck it nose under the tent of American agriculture with the AAA, they opened Pandora's Box even wider, and a number of additional programs to assist farmers came pouring out. By 1938, most of the basic New Deal legislation for agriculture was in-place. It included the following:

1. In the latter part of 1933, Congress moved to address credit problems of farmers. It passed the Farm Credit Administration (FCA) Act that essentially restructured the farm credit system established under the Farm Loan Act of 1916. Then in October 1933, Congress established the Commodity Credit Corporation (CCC) to operate under the auspices of the FCA. The CCC was a government-owned and operated corporation within USDA. It was created to stabilize, support, and protect farm prices and income. For years, farmers had been borrowing money from landlords, merchants, and banks, usually at relatively high rates, using their crops as security for the loan. Under CCC provisions, they could borrow on very attractive terms from the federal government instead. In the first year, cotton farmers could borrow ten cents a pound on their expected crop. If the price of cotton advanced above that amount, they could sell their crop on the open market, pay off their loan, including the interest, and pocket the difference. If the price went below the loan rate, they could forfeit their crop to the CCC, and the federal government would absorb the loss. This is essentially what the members of the Agricultural Alliance had proposed with their Sub-Treasury Plan near the end of the nineteenth century. Later, Production Credit Associations (PCAs) were established throughout the country (200 were established in the South), under the provisions of the Farm Credit Administration Act. These organizations provided farmers with low-cost loans for crop production expenses.
2. Agricultural Adjustment Amendment of 1935 gave the president authority to impose quotas on imports when they interfered with AAA programs.
3. Soil Conservation and Domestic Allotment Act of 1936 provided for soil conservation and soil building payments to participating farmers. It also introduced the concept of parity in setting goals for farm incomes.
4. Agricultural Marketing Agreement Act of 1937 provided authority for the development of federal marketing orders for farm commodities. Because the Supreme Court had found provisions of the original AAA unconstitutional, this 1937 act re-enacted and amended certain provisions of the 1933 act.
5. Agricultural Adjustment Act of 1938 continued most of the provisions of the AAA of 1933 that had been declared unconstitutional. It continued to authorize mandatory supply controls through acreage allotments and marketing quotas, and permitted support for a large number of additional crop and livestock products. In its amended form, it would serve as the foundation of commodity program policy for the next five decades.
6. In 1939, an all-risk crop insurance program was

initiated for interested farmers. It was designed to prevent economic distress in case of crop failure or as a result of drought, flood, hail, and other natural disasters.

From the beginning, New Deal programs began to accumulate enormous surpluses of various food commodities. Obviously these could not be sold on the open market, because they would compete with new crops farmers were trying to sell. Fortunately, Congress had included provisions in the AAA of 1933 to give surpluses to needy families and to schools. By 1939, the federal government was distributing food to some 12.7 million unemployed and poor persons. Food donations also served to encourage the development and expansion of locally-operated, school lunch programs. While the disposition of the surpluses for these purpose was extremely beneficial to really needy people, it served to enormously broaden the base of persons dependent on government largesse. The role of government in taking care of needy people was a central tenet in the belief system of many of the Progressives (Socialists) in Roosevelt's administration. Using surpluses for this purpose allowed them to kill two birds with one stone: eliminate surpluses and build political coalitions.

The decision to give surplus commodities to needy people represents a watershed event in the evolution of the American agroecosystem. Dealing with surpluses had been at the heart of the farm problem from the beginning. Further, there could never be a solution until some way could be found to have excess production units (farms) taken out of production. Reducing surpluses by reducing the number of production units was not as straightforward as it seemed, because of rapidly increasing farmer efficiency. Between 1910 and 1930, the farm population as a percentage of the entire population had declined from 34.9 to 24.9%, but even with that reduction there were still enough farmers with their ever-increasing level of efficiency to produce surpluses of virtually everything.

The marketplace would likely have reduced farm numbers substantially in the early 1930s if the government had not intervened, but the prospect of hundreds of thousands of starving men, women, and children from abandoned farms was unacceptable. The marketplace eliminated thousands of excess units of industrial production capacity during the period, but of course food production was a different matter altogether. Purchases of industrial goods, even clothes, could often be postponed indefinitely, but not food. The marketplace, left unregulated, would likely have eliminated tens of thousands of farms, and it is highly likely that commodity prices would have escalated rapidly. This eventuality would have also been unacceptable, especially to politicians.

Reducing surpluses by giving food to poor and needy people did not solve the politicians' problem either. The more they gave away, the more farmers produced. The succeeding chapters will explore how the Triangle tries to manage this problem over the next six decades.

Alabama Agricultural Legislation

In 1940, the State of Alabama rewrote much of its legal code. As part of this effort, the Alabama Agricultural Code was rewritten to increase the authority of the commissioner of agriculture, including the authority to reorganize the department as he wished, as long as its basic foundation was maintained.

Agricultural Exports

American agricultural exports had increased rapidly in the 1920s with the annual average for the decade reaching $1.9 billion. Unfortunately, the world-wide economic depression that quickly reduced demand for American farm products began to take hold in Europe in 1928, a year before the stock market crash in the United States. After 1928, agricultural exports began to decline; in 1930 they reached $1 billion. By 1935, they had fallen to $750 million (Figure 12.4). They recovered slightly in 1937 and 1938, only to continue to drift downward afterwards, finally reaching $520 million in 1940. Agricultural exports had not been that low since the 1820s.

Agricultural Prices

From 1930 through 1937, year-to-year changes in the indices of agricultural prices, especially the indices of prices received (Figure 12.5) generally tracked annual changes in the Gross National Product (Figure 12.1). Afterwards, while annual estimates for the GNP continued to increase, indices for prices paid and especially for prices received began to decline.

Indices of both prices received and prices paid declined sharply after 1920 (Figure 12.5), but the prices received index suffered the greatest decline. Then after 1929 and as the Depression deepened, both indices began to decline. Unfortunately, even though both were declining, the prices received indices declined at a much faster rate. For example, in 1930 the two indices were 125 and 151, respectively; however, by 1932 they were 65 and 112, respectively. In that short period, the prices received indices declined by 48%, while those of prices paid declined by only 26%.

After 1933, the prices paid indices increased to 120 in 1943, and remained at near that level through 1940, as prices paid for farm inputs remained relatively stable.

The prices received indices also increased rapidly after 1933 to 120 in 1937. Unfortunately, after 1937 those programs lost their hold as prices received indices began to decline once more, and by 1938 they were below 100, and remained near that level through 1940.

At first, AAA programs seemed to help farm prices, but after 1937 they seemed to have lost their effectiveness to do so. In 1937, productions of cotton, corn, wheat, and oats were about twice what they had been in 1934, and those levels of productions were maintained in 1938, 1939, and 1940. For example, in 1934 American farmers harvested 526.2 million bushels of wheat (all). In 1940, they harvested 814.6 million. In the same years, they harvested 1.1 and 2.2 billion bushels of corn and 9.6 and 12.6 million bales of cotton, respectively. There was simply no way that AAA programs in place at that time could support prices under those conditions. By 1937, farmers throughout the country had learned to game the system very effectively. They simply produced everything that they could, sold it to the Commodity Credit Corporation for the support price and let the government worry about surpluses.

The relationship between these price indices over the period 1930-1940 are shown as parity ratios in Figure 12.6. Ratios fell rapidly in the early years of the Depression, reaching 0.58 in 1932. The ratio had been about 1 during the Golden Age of American Agriculture, 1910-1914. (For farmers, a ratio of 1 would represent parity between prices received and prices paid.) Ratios increased rapidly to 0.94 in 1937, but were back down to 0.8 in the period 1938-1939. As late as 1938, AAA had done little to solve the problem of the unacceptably low parity ratios. With the beginning of World War II in Europe in 1939, the ratio increased to 0.81 in 1940.

Ahearn (2008) reported data showing that American consumers spent an estimated 24.2% of their disposable income on food in 1930. Then with food production increasing rapidly after the passage of the Agricultural Adjustment Act, prices began to decline relative to disposable income, and in 1935 only 23.4% was required for the purchase of food. By 1940, the percentage had declined to 20.7. During these terrible economic times, it was a blessing for consumers to have food prices declining, but it made the plight of the farmer increasingly bleak.

Farmers

American farm population had begun to trend sharply downward during the short recession in 1922. In 1927, it reached 30.5 million, and it remained near that level through 1930. It had not been that low since 1900.

The Total Population

There were 123.5 million persons in the United States in 1930 (Figure 12.7). By 1940, that number had increased to 132.4 million, an increase of 7.2%. In the same period, Alabama's population increased from 2.647 to 2.845 million (Figure 12.8), an increase of 7.5%.

The Farm Population

USDA estimated that in 1930 there were 30.5 million people in the American farm population (Figure 12.9). The number had been generally trending downward since 1916. When people—who could not find non-farm employment during the depths of the Depression—became farmers again, the farm population began to increase, reaching 32.4 million in 1933. It had not been that high since 1917. By 1934, the US economy was beginning to grow again (Figure 12.1), and farm population numbers began to decline. By 1940, the number had declined to 30.5 million, or the same level as 1930. Fite (1984) reported that the farm population in Alabama remained largely unchanged between 1930 and 1940, around 1.3 million (Figure 12.10). It had not been near that level since 1920.

The data presented in Figures 12.8 and 12.9 indicate that the American farm population was about the same in 1940 as it had been in 1930, while the total population was increasing. Consequently, the percentage of the farm population in the general population declined from 24.9 to 23.2% between 1930 and 1940 (Figure 12.11). Data presented in Figures 12.8 and 12.10 show that between 1930 and 1940, the percentage of the farm population in the total population in Alabama declined from 50.8 to 47.2% during the period. The percentage of the farm population in Alabama's total population continued to be about twice as high as it was in the country as a whole.

While data presented in Figure 12.10 show that there was little change in the number of persons in Alabama's farm population during the period 1930-1940, the number of black farmers declined sharply. Alabama lost 22% (93,795 to 73,338) of its black farmers during the period. The Depression was an unusually difficult time for all Alabama farmers, but it was even more difficult for blacks.

Farmers in the Work Force

The USDA has published estimates of the percentage of the American labor force made up by farmers back to 1790 (90%). Of course, through the years numbers of farmers increased at a much lower rate than the labor force in general. In 1930, the percentage was down to 21%, and by 1940 it had declined to 18% (Figure 12.13). Machinery and other improved agricultural technology were rapidly taking farmers out of the fields.

Tenancy

Following the Civil War, especially in the South, the economics of farming had made it inceasingly difficult for an individual to own and operate a farm. By 1900, for example, 57.7% of Alabama's farms were operated by tenants, and by 1930, the first year of the Great Depression, tenancy had increased to 64.7%, the highest recorded level in history. The situation improved just slightly by 1935 (64.5%) when the economic situation had begun to improve. By 1940, when many new off-farm jobs were being created as the country prepared for war, the tenancy situation in the state improved dramatically when the rate fell to 58.8%. The year-to-year trend in national rates of tenancy were similar to those in Alabama, but were about 20% lower in all of the years (Figure 12.14).

According to census data, tenant numbers in Alabama in 1920, 1925, 1930, 1935, and 1940 were: 70,395, 70,539, 88,545, 100,705, and 78,573, respectively. The percentages of share croppers in those same years were: 29.4, 35.9, 42.4, 34.5, and 28%, respectively. The decline in the percentages between 1935 and 1940 (100,705 to 78,573) reflect the increasing availability of off-farm employment, resulting from the growing economic activity associated with the general mobilization for war. It is likely that a large share of this reduction was in black tenants. After all, between 1930 and 1940 Alabama lost 20,000 of its black farmers (Figure 12.12).

Farms

The number of farms in the United States trended upward from 1910 through the war years until the prices of agricultural commodities began their decline after 1920. Numbers then declined through 1927. With conditions looking better as a result of the Roaring Twenties, new farms began to be established at a remarkably high rate. When the Depression began, numbers were at the highest level in history to that time (6.546 million).

The trends in farm numbers were somewhat different in Alabama. They declined steadily from 1910 to 1924 and 1925, but began to increase during the last years of the Roaring Twenties. When the Depression began, numbers were 261,000, or near the 1910 level of 265,000.

Farm Numbers

In 1929, the year the Depression began, American farm numbers were at the highest level in history to that time (6.512 million) (Figure 12.15). Surprisingly, with the Depression deepening, farm numbers continued to increase at the highest rate in history. In 1935, there were 6.814 million. Apparently, numbers increased in 1928 and 1929 because potential farmers seemed to feel that prospects for farming were improving (Figure 12.6), but afterwards it continued to increase because non-farming prospects were increasingly limited. However, in 1936, with the economy beginning to grow again (Figure 12.1) and off-farm employment growing (non-farm employment increased from 27 million to 29 million between 1935 and 1936), farm numbers began to trend sharply downward, and finally reached 6.350 million in 1940, a level even lower than in 1910.

The pattern of change for the number of Alabama farms during the 1930s was similar to the national pattern (Figures 12.15 and 12.16), except that numbers in the state began to increase in 1925, instead of 1928. There were 261,000 farms in the state in 1930, and numbers increased as the Depression deepened. The increase continued until 1935 (274,000), also the highest in history. Afterwards, as off-farm job opportunities increased, numbers began to decline rapidly, reaching 241,000 in 1940, some 24,000 less than in 1910.

Farm Size

Data on average American farm size for the period 1910-1940 are presented in Figure 12.17. Average size was 157 acres in 1930; then with farm numbers increasing rapidly through 1935 (Figure 12.15), average size declined to 153 acres. By the 1930s, much of the land suitable for farming was already in farms. New farms had to be created by dividing old ones; then with farm numbers falling sharply after 1935, average size also increased sharply, reaching 174 acres. Farms were not generally abandoned. They were added to other farms to make larger ones.

Farm numbers in Alabama were trending upward in 1930 (Figure 12.16), and as a result average size declined to 68 acres. It had been 70 acres in 1925; then, surprisingly, with farm numbers continuing to increase, average size also increased to 72 acres. It is not immediately obvious where this additional land was obtained. In 1940, with farm numbers declining, average size increased to 83 acres.

Data on the percentage of American farms by size group for 1935 and 1940 are presented in Table 12.3. These data show that the highest percentage of farms in the country in both 1935 and 1940 was in the 50–179 acre group, 42.3 and 42.6%, respectively. The data also show that the percentages of farms in the 1-9 and 10–49 acre groups declined slightly between the two years; while percentages in the larger-sized groups all increased. The data simply reflect the shift toward larger-sized farms noted in the preceding paragraph.

Data on the percentage of Alabama farms by size groups for 1935 and 1940 are also presented in Table 12.3. The data show that in both years more than 50% of all Alabama farms were smaller than 50 acres. The data also show that, between the two years, percentages of all groups above 10–49 acres increased, while those below declined. Surprisingly, there were relative fewer farms in the 1–9 acre group in Alabama than in the country as a whole.

Total Farmland

Data on the amount of American farmland in the period 1910-1940 are presented in Figure 12.18. The data show that total farmland in the country in 1930 was 986.8 million acres, up from 924.3 million in 1925. Even though average size was down slightly in 1935 (Figure 12.17), farm numbers increased enough (Figure 12.15) to raise total farmland to 1,054.5 million acres; then numbers declined sharply from 1935 to 1940, but the increase in average size during that period was large enough to raise total farmland to 1,060.8 million acres.

Data on the amount of Alabama farmland in the period 1910-1940 are presented in Figure 12.19. These data indicate that total acres in 1930 was 17.6 million acres. It had been 16.7 million in 1925. With both average size (Figure 12.17) and farm numbers (Figure 12.16) increasing from 1930 to 1935, total farmland also increased, reaching 19.7 million acres. Between 1935 and 1940, average size increased from 72 to 83 acres, but it was not sufficient to overcome the loss of 33,000 farms. Consequently, total farmland fell to 19.1 million acres.

Use of Farmland

Figure 12.20 includes data on uses of American farmland in the period 1919-1939 expressed as percentages. These data indicate that the use of farmland for cropland decreased slightly during the period 1929-1939 (42 to

38%), while use for pasture increased slightly (34 to 43%). Uses for farm woodland and "other" remained about the same, 17 and 5%, respectively. It is interesting to note that in 1939 more farmland was used for pasture than for crops.

There are no data that can be used as a basis for determining the uses of farmland in Alabama before 1924, and no data were collected in 1939. Between 1929 and 1934, shares of farmland used for crops declined from 54 to 49%, while shares used for woodland increased from 37 to 42%. It is still amazing that such small shares of farmland (4%) would have been used for pasture.

These data indicate that in this period, cropland and woodland occupied larger shares of Alabama farmland than in the country as a whole, but that pasture was much more important in the country than in the state. They also show that relatively more Alabama farmland was used for crops than in the country as a whole.

Total Cropland

The Census of Agriculture published data on total cropland for 1924 and 1929, but not for 1934 and 1939. Instead, they included data on the land available for crops for those latter two years. These two categories of farmland use are quite different.

Cropland Harvested

American cropland harvested had generally trended upward from 311.3 million acres in 1909 to 359.2 million in 1929 (Figure 12.22). Then it declined by almost 18% to 295.6 million acres in 1934, before increasing to 321.2 million acres in 1939.

Between 1929 and 1934, American cropland harvested declined by 63.6 million acres. Although, it is likely that AAA efforts to reduce cropland harvested played some role in this reduction, poor crop weather also played a role. The Palmer Drought Severity Indices data (Cook, et al., 1999) indicate that in June, July, and August of 1934, the entire country from Pennsylvania to parts of California and Oregon and even into the upper South was in the throes of a severe drought. Acres harvested of cotton, corn, wheat, and oats were 16.4, 22.0, 6.5, and 8.7 million acres, respectively, lower in 1934 than in 1929. Declines in these four crops alone account for 53.6 million acres of the total 63.6 million reduction.

By 1939, harvested cropland was up to 321.2 million acres, but although it increased, it was still considerably lower than the 1929, pre-Depression level. Acres harvested of cotton, corn, wheat, and oats combined, were still 32.4 million acres below 1929 levels. In 1939, good weather returned to the eastern corn belt, but it was still unusually dry in the western portion.

Figure 12.23 presents data on the acres of cropland harvested per farm in the United States during the period 1924-1939. It generally followed the same year-to-year trends as total cropland harvested. It increased in 1929 (48.6 to 53.6 acres) before falling to 53.2 acres in 1934. Then it increased to 55.2 acres in 1939.

Between 1924 and 1929, Alabama farmers increased cropland harvested from 6.6 to 7.1 million acres (Figure 12.24), apparently in an attempt to counter the worsening parity ratios. Then between 1929 and 1934, it increased to 7.2 million. Cropland harvested declined in the country during this interval (Figure 12.22). The exceptionally dry weather that plagued the mid-section of the country in 1934 was not nearly as severe in Alabama. Between 1934 and 1939, cropland harvested fell to 7.1 million, primarily as a result of the sharp decline in acres harvested of cotton. In 1938, the average price for cotton was 8.6 cents a pound. In 1939, Alabama farmers planted 335,000 fewer acres of cotton than they had in 1938.

Data on the average number of acres of Alabama cropland harvested per farm during the period 1924-1940 are also presented in Table 12. 23. Year-to-year trends in the Alabama data were similar to those in the country as a whole, except that the slight decline between 1929 and 1934 was unexpected. Total cropland harvested actually increased in that interval. While the year-to-year trends were similar in both the country and the state, acres harvested per farm in Alabama was only about half as high as in the country. Farmers and their families on those dirt-poor, one-mule farms could hardly plant, cultivate, and harvest crops from more that 25-30 acres.

Table 12.4 contains data on acres of cropland harvested per farm in Alabama and in several surrounding states. Data for Iowa and the United States as a whole are presented for comparison. These data clearly show that farmers in Alabama and in the other Southern states harvested crops from much smaller amounts of land than elsewhere in the country and especially in important farm states like Iowa. All of the Deep South states were suffering equally from the large numbers of tenant farmers and their small farms.

Value of Land and Buildings

A large share of value of land and buildings on farms is the value of the land itself. Census data for 1920 indicate that the value of land alone (including fences, drains, etc.) accounted for 82.6% of the total value of land and buildings. Trivelli (1997) has suggested that there are many factors that act and interact to determine the value of farm land at any one time and that their importance ebbs and flows over time. The value of food and fiber produced on the land is only one of them. For example, during the period 1920-1930, wartime inflation had pushed land values sharply higher; then the short recession of 1920-1921 forced them down again, and they continued to decline even during the Roaring Twenties. In this period (1923-1930), the value of the land for producing food and fiber likely had little to do with its total value.

Census data on the value of land and buildings on American farms for the period 1910-1940 are presented in Figure 12.25. The data indicate that values had been trending slowly downward since 1922, before finally reaching $39.6 billion in 1930. Afterward, values declined rapidly to $30.7 billion in 1933. Then with farm numbers increasing rapidly (Figure 12.15), values began to increase slowly. However, with farm numbers trending rapidly downward after 1935, they did not get higher than $34.8 million through the remainder of the period.

Average value of land and buildings on a per-acre basis for American farms trended upward throughout the first two decades of the twentieth century to reach $69 an acre in 1920 (Figure 12.26). Thereafter, they trended steadily downward to reach $48 in 1930 and $31 in 1935. With the worst part of the Depression over, values an acre increased to $32 in 1940.

During the 1920s, value of land and buildings per acre fell from $69 to $48. Provisions of AAA apparently stopped the decline, but they did very little to reverse the trend. In fact, with the Consumer Price Index declining 16% during the period, the real values in 1935 and 1940 were likely higher than they appear.

There are no annual data on the value of land and buildings on Alabama farms during this period; however, census data indicate that values were $502.4 million in 1930 and $368.8 million in 1935. The quantity of Alabama farmland was also increasing during this period (Figure 12.19). This is essentially the same pattern seen in the national data (Figure 12.25), except that the rate of decline was not quite as great in Alabama (26.6 versus 31.3%). Then, as the Depression was replaced by preparations for war in 1940, values increased to $408.8 million. The Consumer Price Index increased only by 2.1%.

In 1930, the average value of land and buildings per acre on farms in Alabama was $29. It fell to $19 in 1935 before increasing to $21 in 1940. These values per acre are much lower than those in the country as a whole, and they are just one more indication of the deplorable status of Alabama farms in this period. If anyone ever desperately needed a New Deal, it was farmers in Alabama. The New Deal did not seem to have done much for them by 1940, but maybe times would be better in the future. After all, much of the push to get the government involved was to get help to those poor souls on those small Southern farms.

Fite (1984) reported that value per farm in 1940 in Mississippi, Alabama, Georgia, Louisiana, and Iowa was: $1,632, $1,746, $2,223, $2,359, and $12,614, respectively.

Value of Machinery and Equipment

Data on the total value of farm machinery and equipment on American farms during the period 1910-1940 are presented in Figure 12.28. The data show that value climbed slowly during the 1920s to reach $3,302 million in 1930. Afterwards, it plunged to $2,168 million in 1934. AAA programs seem to have been very beneficial to the value of farm machinery and equipment nationally. With farm numbers declining rapidly (Figure 12.15), values began to increase, finally reaching $2,998 million in 1938. Apparently, the recession of 1937 and 1938 slowed the increase. Values increased very little through 1940.

Data on the value of machinery and equipment per farm are presented in Figure 12.29. These data show the same general kind of changes as total value during the period 1930-1940. Value was $422 following the 1920-1921 short recession, but increased to $525 at the beginning of the Depression in 1930. It fell sharply after 1930 to reach $325 in 1935. Then after the infusion of money from New Deal programs, it climbed back to $502 in 1940. In 1940, value per farm was only 4.4% below where it had been in 1930, and the Consumer Price Index was still 16% lower (50 versus 42.0).

There are no annual data on the total value of machinery and equipment on Alabama farms in the 1930s and no census data for 1935. Census data for the period 1900-1940, except for 1935, are presented in Figure 12.30. The trends in these data are essentially the same as for the country (Figure 12.29). Alabama values were $23.8, $33.5, and $29.6 million in 1925, 1930, and 1940, respectively. It is unfortunate that the 1935 data are not available. The value for that year would probably have been below $29.6 million.

In 1930, value of machinery and equipment per Alabama farm was $130 (Figure 12.29). The comparable value for the average American farm was $525. By 1940, it had declined to $123. The comparable national value was $478. Some of the difference in values per farm in the country and the state, shown in Figure 12.29, was likely the result in differences in the average size of farms (Figure 12.17), but those differences were not great enough to account for the much larger differences in value.

As might be expected, tenant farms had lower investments in machinery and equipment than farms whose owners operated them. For example, on Alabama farms in 1930 values of machinery and equipment per farm for owner-operated and tenant-operated farms were $230 and $72, respectively. In 1940, they were down to $191 and $66. It is difficult to comprehend how a farmer could have operated a farm with machinery and equipment valued at $66. Tenant farmers in Alabama at the beginning of the Depression were already destitute, and it would be extremely important to see whether or not the New Deal could reach them.

Value of farm machinery and equipment per farm are but another indicator of the deplorable state of agriculture in Alabama in the period 1930-1940; however this situation was really nothing new. It had been in a deplorable state for decades. As early as 1880, the Alabama value was considerably less than one-third as high as the country ($28 versus $101). Alabama agriculture was not alone in this deplorable condition. In 1930, the average value for 12 Southern states was $184 or only about a third of the national average. From the very beginning, the Atlantic Coastal Plain (Figure 1.9) was no place for crop production. There was simply no comparative advantage in trying to grow crops commercially in its Ultisol soils and its Humid Mesothermal (Cfa) climate. Moreover, the New Deal could not solve those problems, but it did encourage farmers and politicians to keep trying.

Tractors on Farms

There was considerable speculation whether some American farmers might receive enough funds under the provisions of New Deal programs to purchase more tractors. The number of tractors on American farms did increase substantially between 1930 and 1940 (Figure 12.31), but it is not apparent that New Deal programs were responsible. USDA data indicate that tractor numbers increased by almost 24% between 1929 and 1932, but only 10% between 1933 and 1936. In 1930, there were 15 tractors per 100 farms in the United States, and numbers remained at essentially the same level in 1935; then by 1940, it increased to 25 per 100.

The 1930 Census of Agriculture estimated that there were 4,664 tractors (almost two per 100 farms) in Alabama that year. There were no census data for 1935, but by 1940 there were an estimated 7,638 tractors on 241,000 farms (three per 100 farms). Farm mechanization had begun in the state, but it had a long way to go.

Animal Power on Farms

Data on the total number of horses and mules on American farms during the period 1915-1940 are presented in Figure 12.32. The data show that between 1930 and 1940 the number of horses in the country declined from 13.7 to 10.4 million, respectively. During the same period, farm numbers first increased in the early part of the Depression, but after 1935 they began to decline rapidly. When corrected for changes in farm numbers, horses declined from 210 per 100 farms in 1930, to 174 and 164 in 1935 and 1940, respectively (Figure 12.33). During the same period, numbers of tractors per 100 farms increased from 14 in 1930, to 15 and 24 in 1935 and 1940, respectively (Figure 12.31). Clearly, as early as 1930-1940, even during this severe Depression, American farmers seemed to be able to find the funds to replace horses with tractors.

During this same period (1930-1940), numbers of mules in the country declined from 5.4 million in 1930, to 4.8 and 4 million in 1935 and 1940, respectively (Figure 12.32). On a per-100-farm-basis, numbers declined from 82 in 1930, to 71 and 64 in 1935 and 1940, respectively (Figure 12.34). Obviously, American farmers were also replacing their mules with tractors.

In the 1930s, Alabama farmers continued to depend primarily on animals for power. In 1930, there were only around two tractors per 100 farms in the state (Figure 12.31). Annual data on numbers of mules and horses in Alabama during the period 1915-1940 are presented in Figure 12.35. These data show that there were 66,000 horses in the state in 1930, and that in 1935 and 1940 numbers first declined to 50,000, but then increased to 63,000. During this same period, numbers of farms increased sharply during the first half of the Depression, but after 1935 began to decline rapidly (Figure 12.16). On a per-100-farms-basis, horse numbers also declined during the heart of the Depression (Figure 12.33). Numbers were at a level of 25 per 100 farms in 1930, 18 in 1935, and 26 in 1940. Most Alabama farmers did not own any horses, and those who did probably did not use them for crop production.

Mule numbers in the state remained steady as farm numbers increased in the early part of the Depression (Figures 12.35 and 12.16), but after 1935 both mule and farm numbers began to decline. In 1930 and 1935, mule numbers held at 322,000, but by 1940 it had declined to 306,000. On a per-100-farms-basis, mule numbers also declined during the heart of the Depression. There were 123 per 100 farms in 1930, but by 1935 it was down to 118, before increasing to 127 in 1940 (Figure 12.34). When out-of-work people began to move back to farms in the early part of the Depression, apparently many of them did not purchase mules. These data also indicate that during the Depression few Alabama farmers were replacing mules with tractors. There were only two tractors per 100 farms in the state in 1930 and three in 1940 (Figure 12.31).

Use of Fertilizer

In 1929, American farmers used 44.4 pounds of commercial fertilizer for every acre of harvested cropland (Figure 12.36). In the depths of the Depression, in 1934, use declined to 37.5 pounds, but it began to increase afterwards, reaching 48 pounds an acre in 1939.

Farm Modernization

Farm modernization in the North Atlantic and North Central regions had begun in the latter part of the nineteenth century, but there was not much change on Southern farms until the New Deal legislation went into effect in the mid-1930s, and there was precious little of it even then. Not everyone, however, was happy with any farm modernization. Efforts to modernize farms drew some criticism from agricultural fundamentalists who envisioned Thomas Jefferson's farmscape being replaced by industrial monstrosities devoid of all rural elements. The criticisms were especially strident from a group of Nashville writers and scholars (the Nashville Agrarians) who deplored the move toward industrial farming. They begged and warned farmers not to be a part of it. These criticisms were largely led by people who had never tried to make a living farming. However, Jeffersonian views were deeply imbedded in American culture, and probably even more so in the South.

Plant Agriculture

Data on acres harvested of the major crops grown in Alabama and the United States in 1919 and 1929 are presented in Table 12.5. In both 1919 and 1929, corn was harvested from more acres of American cropland than any other crop; although acres of hay harvested was not far behind. In both years, acres of corn harvested was greater than for winter wheat and oats combined. Cotton was also an important crop. Acres of cotton harvested was generally similar to acres of winter wheat and oats harvested. National data for 1940 records the appearance of two new cash crops: sorghum and soybeans. They were insignificant in 1940, but they would not remain that way very long.

In 1920 and 1930, acres of cotton and corn harvested were about equal in Alabama; they accounted for most of the acres harvested in the state in those years. The relative importance of hay, winter wheat, and oats was not nearly as great in Alabama as in the country as a whole.

Alabama Crop Weather

During the period 1930-1940, farmers in the state and the country not only had to contend with one of the worst economic depressions in history, but they also had to contend with the most severe drought in the twentieth century. This was the decade of the disastrous Dust Bowl of the Southwest and the Southern Great Plains (Figure 1.9). In July 1934, severe drought was present over 61% of the United States. In July, 1936, severe drought affected over 44% of the country. In the Dust Bowl itself, the drought continued for roughly seven years (1933-1940).

While Alabama never became a Dust Bowl, the state did suffer from drought during the period, especially in the early years. Palmer Drought Severity Indices (PDSI) (Cook, et al., 1999) for June, July and August during the period 1930-1940 indicate that crop weather in more than half the state was much drier than normal in six of the 11 years (1930, 1931, 1933, 1934, 1936, and 1937), much wetter than normal in three years (1938, 1939, and 1940), and normal in two years (1932 and 1935). Drought conditions were especially bad in Alabama in those years (1934 and 1936) when it was so severe across the country. A total of six storms originating in the tropics passed through the state during this period. Five were classified as tropical storms (1932, 1934, 1937, 1939, and 1940). In 1936, a minimal hurricane also passed through.

Cotton

American cotton farmers had responded to the speculative boom of the 1920s by increasing their acreage of the crop. Acres planted had generally trended upward from 30 million in 1921 to 44 million acres in 1929. This large increase of acres planted resulted in a concomitant increase in bales harvested. Prices suffered, falling from a high of 29 cents a pound in 1923 to 17 cents in 1929. Then, in 1930, they dropped to 9 cents.

American cotton yields in 8 of 10 years of the 1920s were below 175 pounds an acre; then just at the time when farmers seemed to be somewhat interested in reducing bales harvested, yields began to edge upward, further compounding their surplus production problems.

US Cotton

American farmers harvested 13.9 million bales of cotton in 1930. The largest share (29%) of the total was reported by Texas farmers. Other states reporting large shares included: Georgia (11.4%), Alabama (10.5), Mississippi (10.5), and South Carolina (7.2). These five states accounted for 68.7% of the total.

Exports

Data on American cotton exports during the period 1915-1940 are presented in Figure 12.37. The data indicate that exports had increased sharply during the 1920s when the economies throughout the world were booming. They had reached 11.3 million bales in 1926 before falling to 7.1 million in 1930. With prices below 10 cents a pound in 1931 and 1932, even at the beginning of the Depression, American cotton was a bargain. Exports increased to 9 million bales in 1931 and 1932. After 1932, even though prices were still hovering around 10 cents, world demand was finally responding to a world-wide depression, and exports began to decline. The general downward trend continued through 1938 when they reached 3.5 million bales. They returned to 7 million bales in 1939, the first year of the war in Europe before falling to 1.2 million as the war intensified. Cotton exports had not been that low since the Civil War.

Prices

Data on average annual cotton market prices in the United States during the period 1915-1940 are presented in Figure 12.38. These data indicate that prices had responded positively to the generally good economic times of the mid-1920s. They were at 29 cents a pound in 1923, but afterwards, with 8 million bales harvested

(Figure 12.42), they began to decline, finally reaching 9.5 cents in 1930. After 1929, year-to-year changes in cotton prices generally tracked those changes taking place in the larger national agricultural economy during this period of severe economic malaise. Year-to-year changes in the agricultural prices for all crops in the country are depicted in Figure 12.5. Of course, after 1932, cotton prices, as well as the prices of all other crops, were affected to greater or lesser extent by various New Deal programs.

With cotton prices at seven cents a pound in 1932 when the Southern congressmen took over the seats of power in the 73rd Congress, they had all the power they needed to use the US Treasury to help Southern cotton farming. Congress quickly put a floor, in the form of a support price set at a level of 10 cents a pound, under cotton prices (Figure 12.39). Under the support-price program, if farmers could not sell their cotton for an amount equal to that price, they could borrow money from the federal Commodity Credit Corporation (CCC). Loans were based on the quantity of their crop that they wished to be covered by the program and the support price. During the following year, if the market price increased beyond the support price, they could sell their crop on the open market and pay-off their loan plus interest. If, however, the market price remained below the support price, the CCC would take ownership of their crop as payment for the loan. The old Sub-Treasury Plan had finally emerged from oblivion.

In 1933, with support prices in place, the average market price went back to 10 cents. The support price level was set at 12 cents for 1934, and the average market price increased to 12.4 cents. Unfortunately, with the Depression in full-cry, the government could not justify keeping the support price at 12 cents, so it was lowered to 10 cents for 1935, to nine cents for 1937, and to eight cents for 1938. By 1938, market prices had followed support prices back down to 8.6 cents. The support price was increased to nine cents for 1939 and 1940, and average market prices followed, to 9.1 cents in 1939, and 9.9 cents in 1940.

If this cotton price situation gave pause to the New Deal cause, there is little indication of it, nor is there any record of how the Southern congressmen explained it when they faced their cotton-farming constituents back home. But, at least, someone was finally trying to do something. Unfortunately, the time to do something was long past. The die was cast when the first slave picked the first leaf of tobacco in Virginia in the early 1600s.

Acres Planted and Harvested

Data on the acres of cotton planted and harvested in the United States during the period 1915-1940 are presented in Figure 12.40. Data in Figure 12.38 show that market prices had begun to trend downward after 1927. Acres planted increased relatively little in 1928 and 1929, before reaching 43.3 million in 1930; then in 1931, it began to follow prices down. Prices finally reached 5.7 cents in 1931, but began to increase in 1932. At that point, acres harvested had fallen to 36.5 million. The price increased to 7.2 cents in 1932, and farmers increased acres planted to 40.2 million in 1933.

It quickly became evident after the passage of the Agricultural Adjustment Act in the middle of the 1933 cotton-growing season that if the legislation was going to benefit cotton farmers in that year some way had to be found to reduce production. The only practical way of doing it was to destroy between 25 and 50% of the plants already in the field; however, the willful destruction of a growing crop was opposed by some 10,000 years of cultural resistance. Secretary of Agriculture Henry Wallace remarked that "to destroy a growing crop is a shocking commentary on our civilization. I could tolerate it only as a cleaning up of the wreckage from the old days of unbalanced production." There was widespread opposition. Some contended that the "destruction of a maturing crop was contrary to God's will." The farmers were equally reluctant to destroy even a single stalk until the federal government offered to pay them between $7 and $20 for each acre they plowed under depending on their expected yield. In return for the payments, farmers had to sign a binding contract that they would destroy a specific amount of growing cotton on their land. When the destruction had been verified by field observation, payments were sent to them.

Most cotton farmers throughout the country put their qualms aside, and were soon participating enthusiastically. Receiving $7 to $20 an acre for cotton they did not have to cultivate or pick was just too good to pass up. Before harvest season, over one million farmers had signed contracts for the removal of almost 10.5 million acres from production. For their efforts, farmers received $110 million in direct payments from the US Treasury.

In 1932, American cotton farmers had harvested 98% of the acres they planted. A year later, after plowing under a substantial portion of acres planted, they harvested 73%. The federal program to "sign-up, plow-up, check-up, and pay up" seemed to be a success. Nothing like this had ever happened in American history. Soon, the market price of cotton increased to 10 cents (Figure 12.38), almost double what it had been a year earlier. But while the program provided a temporary solution to a desperate solution, it involved the federal government in everyday management decisions of private enterprise to an extent that had never been imagined. Socialism in its fullest glory had finally come to the cotton fields.

With the New Deal cotton program firmly in place by the end of 1933, farmers responded by planting only 27.9 million acres in the spring of 1934. Unfortunately, it quickly became obvious that not all farmers had agreed to sign contracts to destroy a part of their 1933 crop. If as expected, prices increased with the removal of over 10 million acres of production, those non-participating farmers would be able to sell the production from their entire unrestricted acreage for the better price. In effect, they could have their cake and eat it too. It seemed to all the program participants that it was unfair that farmers outside the program would benefit. By late 1933, there was strong pressure to replace voluntary control with compulsory control. As a result of these concerns, Congress passed the Bankhead Cotton Control Act in April 1934. This act limited the amount of cotton that individual farmers could sell without suffering stiff penalties. Marketing quotas were based on each farmer's production history over recent years.

By plowing under millions of acres of cotton in 1933, the AAA had an immediate impact on acres harvested (Figure 12.40). Farmers harvested some 6.5 million fewer acres in 1933 than they had the year before (1932). Acreage reduction also carried over into the following year (1935) when farmers actually planted fewer acres than they had harvested the year before (27.9 million versus 29.4 million). Unfortunately, AAA was not powerful enough to keep the farmers out of the field for very long; consequently, in 1936, acreage began to creep upward. Farmers planted 30.6 million acres that year and then 34.1 million in 1937. Acreage might have continued to climb, except for the fact that prices were going in the opposite direction (Figure 12.38). Consequently, in 1938, with the average price below 10 cents, farmers sharply reduced acres planted to 25 million, and it remained near that level through 1940. In 10 of the 11 years in the period 1930-1940, American cotton farmers harvested 97.5% of all acres planted.

As is shown in Figure 12.40, American cotton farmers were harvesting cotton from fewer acres in 1939 than they had in 1932, some 34% less (23.8 million acres versus 35.9 million); however, it is not clear that these reductions were the result of New Deal programs. Cotton acreage was beginning to decline fairly rapidly after 1929 and continued to do so in 1930, 1931, and 1932. The 1938 and 1939 planting levels of 25 and 24.6 million acres were likely at the level they would have been without the AAA.

Yields

Data on national, annual cotton yields during the period 1915-1940 are presented in Figure 12.41. As had always been the situation with cotton production in the Lower South, year-to-year variation in yields was extremely high. As a crop, cotton is at risk for an extremely long time, and given the erratic climate in the region, considerable variation had to be expected. Further, the genome of the cotton plant has never seemed well adapted in the South east of the Mississippi River.

Acreage reductions encouraged by AAA rules might have finally reduced supply and increased prices to more acceptable levels, except for the fact that yields began to increase rapidly about the time that the legislation was put into effect. Apparently, farmers chose to rent their poorer land to the government and to apply money saved on production costs on those rented acres to increase yields on their better lands.

American cotton yields during the period 1927-1932 averaged 172 pounds of ginned cotton an acre (Figure

12.41); then in the six years (1934-1939) after 1933, they averaged 217 pounds an acre, an increase of 26%. The 1937 yield was especially troublesome (270 pounds). Assuming that the yield might remain that high, in 1938 acres planted would have had to be reduced at least 26% just to keep bales harvested at the pre-New Deal level. Further, to have achieved truly meaningful reductions, acreage set-asides—considerably beyond those required to compensate for increased yields—were needed. In fact, to have reduced supply enough to really move prices upward, it probably would have been necessary to have reduced acreage at least 50%, and probably more, below the 1933 level of 40.2 million acres. Unfortunately, there was no way that such a large amount of land could be taken out of production in such a short period. Further, at the time the federal government was not prepared to provide the amount of money required to do so. Also, as we shall see, there were other valid reasons why acreage reductions of that magnitude were unacceptable

Bales Harvested

Data on bales of cotton harvested nationally for the period 1915-1940 are presented in Figure 12.42. These data show that bales harvested had increased significantly during the 1920s and were at a level of near 14 million in 1930. With yields generally increasing (Figure 12.41) and acres harvested generally decreasing (Figure 12.40), bales harvested declined slightly in the period 1930-1940. Bales harvested were higher than expected in 1931 and 1937 as a result of much higher than expected yields, and lower than expected in 1934 and 1935 as a result of lower than expected yields. Except for those unexpected levels of bales harvested, they declined from 13.9 million in 1930 to 11.8 million in 1939. In 1940, they increased slightly to 12.6 million.

New Deal efforts to manage the Southern cotton disaster at first looked promising, but it was soon apparent that its efforts to adjust cotton production (supply) to meet demand—so as to achieve satisfactory market prices—was not going to work, at least in the short-term. Average market prices did increase smartly from 6.5 cents in 1932 to 10.2 cents in 1933. Afterwards, they were at levels of 12.4, 11.1, and 12.4 cents in 1934, 1935, and 1936. Unfortunately, that was as good as it got. For the remainder of the decade, they never got higher than 9.9 cents.

As late as 1940, government warehouses were still bulging with cotton that the government had been forced to accept as payment for farm loans. For the period 1933-1940, production was near or above 12 million bales for six of the eight years, and in 1937 farmers harvested the largest crop ever recorded in the United States, 18.9 million bales. Given the low world-wide demand for cotton during the Depression, under AAA provisions the federal government had to store and finally purchase much of this huge crop. The New Deal had succeeded in removing some of the burden of too much cotton from the backs of farmers, but in doing so they simply shifted it onto the backs of the general population.

Alabama Cotton

In 1930, there were 207,000 Alabama cotton farms, 30% of them operated by owners and 70% by tenants. The response of Alabama cotton farmers to the economic conditions of the 1920s mirrored that of the country as a whole. The state's farmers had also responded to the speculative boom of the Roaring Twenties by increasing the acreage devoted to the crop. From 1916 through 1922, they had planted substantially less than three million acres a year, but in 1922 they began to increase acreage, and in 1930 they planted 3.6 million acres.

In 1929, farmers in Madison County harvested the largest share (4.2%) of Alabama's cotton crop. Other counties reporting substantial shares included: Marshall (3.8%), Limestone (3.6), Cullman (3.5), and DeKalb (3.3). These five counties accounted for only 18.4% of Alabama's total crop. The county average was 19,600 bales. The farm average was 5.1 bales. A total of 41 counties reported harvests of 15,000 or more pounds.

Prices

The Alabama Agricultural Statistics Service has published data on the annual average market prices for cotton sold during the period 1930-1940, but they are so similar to national prices to be largely indistinguishable from the data presented in Figure 12.38.

Acres Planted and Harvested

Data on acres of cotton planted and harvested in Alabama in the period 1915-1940 are presented in Figure

12.43. As was the case elsewhere in the country, acres planted had trended up sharply during the 1920s and had reached 3.6 million acres in 1930. Yearly changes in cotton acreage in Alabama after the passage of AAA fairly closely followed the pattern for the country from 1933 through 1939 (Figures 12.40 and 12.43). The large difference in acres planted and acres harvested in 1933 are the result of plowing-up cotton in the state. In 1934, 1935, and 1936, annual average cotton prices were 12.4, 11.1, and 12.4 cents, respectively. Alabama farmers, desperately seeking some way to dig themselves out of the deep hole they were in, accepted these slightly higher market prices as an indication that the government programs would ultimately be their salvation. With this assurance, in 1937 they went into their fields and planted 18.9 million acres of the crop, and they harvested cotton from almost every acre that they had planted. That year the average price fell back to 8.4 cents. Finally, convinced that they could not beat the system, they sharply reduced their acreage to around two million acres in 1938, 1939, and 1940. Except for 1933, Alabama farmers harvested cotton from an average of 98.5% of acres planted.

Yields

Data on Alabama's annual average yields of cotton during the period 1915-1940 are shown, along with national yields, in Figure 12.41. These data show that although there was considerable year-to-year variation, yields in the state were generally similar to those in the country as a whole. Both were near 210 pounds in 1930, and they generally increased through 1936 when the state average reached 290 pounds an acre, the highest in history. Alabama yields were down sharply (180 pounds) in 1939 and 1940 as a result of extremely poor crop weather.

Bales Harvested

Data on bales of cotton harvested in Alabama during the period 1915-1940 are shown in Figure 12.44. The data show that bales harvested were at a level of 1.5 million in 1930. Data presented in Figure 12.43 indicate that acres harvested generally trended downward during the period, while data shown in Figure 12.41 indicate that yields generally trended upward. As a result, bales harvested was sort of a composite of the two. Bales harvested declined along with acres harvested from 1.5 million in 1930 to around 950 thousand in 1932, 1933, and 1934. In Alabama, the enactment of AAA had little, if any, effect on bales harvested. They increased sharply after 1934, reaching 1.6 million bales in 1937, the highest level since 1914. Afterwards, bales harvested declined sharply, along with acres harvested, to reach near 780,000 in 1939 and 1940. Lower than expected yields in those two years also contributed to the relatively low levels of bales harvested.

Census data indicate that farmers in Madison County harvested the largest share (5.9%) of the 1939 Alabama cotton crop. Other counties reporting substantial shares included: Marshall (4.8%), Limestone (4.7), DeKalb (4.2), and Cullman (3.8). These five counties accounted for 23.5% of the state's total crop. Not a single county from the old Cotton Arc was included in the top five counties.

Corn

American corn acreage had reached its highest level during the Golden Age of American Agriculture and the years around World War I, but afterwards it began to decline. For over a century, bread made from corn had been a staple in the diets of a substantial segment of the Southern population, but that use had probably been trending downward since early in the twentieth century. Also, the number of horses and mules on American farms peaked around 1920. Farmers did not use much corn for their work animals, but they did use some, especially in the South where oat production was not very practical. Both of these factors worked together to begin to reduce demand for the grain.

US Corn

In 1930, American farmers harvested 1.8 billion bushels of corn. The largest share (19%) of the total was harvested in Iowa. Other states with large shares included: Nebraska (12.5%), Illinois (12.5), Indiana (6.1), and Minnesota (5.6). These five states accounted for 55.7% of the national crop that year. Alabama farmers harvested only 1.7%.

Exports

Data on American corn exports for the period 1915-1940 are presented in Figure 12.45. The data indicate that they were 3.3 million bushels in 1930. This quantity represents 0.5% of bushels harvested in 1929. Exports had been steadily trending downward since 1925. Historically, the United States had always exported a small share of its corn. In Europe, the demand was limited for Indian corn, or maize. Europeans did not use corn as food, and they usually fed their animals oats and fodder. Exports generally remained well below 10 million bushels through 1936, but then increased sharply in 1937 to almost 140 million. These years with unusually high levels of corn exports coincided with years with unusually large crops and with relatively low market prices. By 1940, exports were back down to 14.8 million, near where they had been in 1930.

Prices

The national market price of corn had been around 80 cents a bushel in 1927 and 1928, but with the Depression beginning in 1929, corn prices began to decline. Bushels harvested increased sharply in 1931 and 1932 (Figure 12.50), and market prices continued their decline, finally reaching 29 cents in 1932 (the teeter-totter phenomenon). They had not been that low since the late nineteenth century. After 1929, year-to-year changes in corn prices generally tracked those changes taking place in the larger national agricultural economy during this period of severe economic malaise. Year-to-year changes in the agricultural prices for all crops in the country are depicted in Figure 12.5. Of course, after 1932 corn prices as well as the prices of all other crops were affected to a greater or lesser extent by various New Deal programs.

Market prices began to recover after the AAA legislation was enacted in 1933. As part of the federal program, a support price of 45 cents was established for the grain in 1933 (Figure 12.47). The market price in 1933 reached 49 cents as a result of the support price and declining bushels harvested. In 1934, American farmers harvested the smallest corn crop of the twentieth century, to that time. As a result, market price reached 80 cents. With the market price at 80 cents in 1934, the support price was reduced to 45 cents for 1935, and the market price fell back to 63 cents when bushels harvested rebounded. Then in 1936, American farmers brought in another extremely small crop, and the market price increased sharply to $1.03, the highest level since 1924. With market prices that high, the government reduced the support price to 50 cents for 1937. In that year, with bushels harvested increasing, the market price fell sharply to 49 cents, or to the support price level. Support prices were increased to 57 cents in 1938, and kept at that level in 1939, but increased to 61 cents in 1940. With support prices up and bales harvested down in 1938, 1939, and 1940, market prices slowly increased to reach 60 cents in 1940.

Acres Planted and Harvested

Data on acres of corn (for grain) harvested in the United States are presented in Figure 12.48. There are no annual data on acres planted until 1926; consequently, to show trends over a longer period of time, I have chosen not to include any of the acres planted data in this chapter. American farmers had been harvesting 85 million acres of corn in the late 1920s, and in 1930 they harvested the crop from 85.5 million acres. In 1931 and 1932, acres harvested began to increase even though market prices were declining (Figure 12.46). However, in 1932 the market price finally reached 29 cents, and in 1933 corn farmers finally began to reduce acres harvested. In that year, 92.1 million acres were harvested. Because of poor crop weather, acres harvested were extremely low in 1934 (61.2 million) and in 1936 (67.8 million). Even though market prices increased slowly in 1938, 1939, and 1940, acres harvested drifted downward, finally reaching 76.4 million in 1940. It had not been that low since the late 1800s.

Except for the two poor crop years of 1934 and 1936, in the period 1930-1940 American farmers harvested 84.5% of their acres planted. In 1934 and 1936, they harvested 60.9 and 66.5%, respectively. Midwestern farmers, especially dairy farmers, harvested substantial amounts of their corn for silage, rather than for grain.

Yields

American corn yields had been trending slowly downward for a number of years before 1930 (Figure 12.49). In that year, average yield was even lower than expected (20 bushels an acre) as the result of a nation-wide drought. Yields were also much lower than expected in 1934 and 1936 as a result of poor crop weather. Hybrid corn was first introduced to farmers in the early 1930s, but by 1935 still only 1% of the total acreage of corn was planted with these improved seeds. Yields were lower than expected in 1934 and 1936 as a result of poor crop weather. After 1936, however, either as a result of the use of the improved genetic material or in spite of it, yields seemed to trend slightly upward. By 1940, 30% of corn acreage in the country was being planted with the improved genetic material. In that year, average yield reached 30 bushels an acre.

Bushels Harvested

Data on bushels of American corn harvested are presented in Figure 12.50. With yields increasing slightly during the period (Figure 12.49) and with acres harvested trending downward (Figure 12.48), bushels harvested generally remained at near the same level, although year-to-year variation made that determination difficult. Bushels harvested were lower than expected in 1930, 1934, and 1936, because of the low yields resulting from the extreme drought in the Corn Belt. A sharp increase in acres harvested in 1932, coupled with a higher than usual yield, resulted in a harvest of 2.6 billion bushels. With both acres harvested and yields remaining relatively stable from 1937 through 1940, bushels harvested also remained relatively stable at around 2.4 billion bushels.

Alabama Corn

Farmers in Lauderdale County harvested the largest share (2.9%) of Alabama's 1929 corn crop. Other counties reporting substantial shares included: Jackson (2.9%), Limestone (2.8), DeKalb (2.7), and Cullman (2.5). These five counties only accounted for 13.8% of the total crop. Corn production was widely distributed in the state. The county average was 532,600 bushels, and the per-farm average was 139 bushels. A total of 43 counties reported harvests of 400,000 or more bushels.

Prices

Figure 12.46 also included data on average corn market prices in the United States, as well as in Alabama. The data indicate that the year-to-year trends of both were essentially the same. However, throughout the entire period (1930-1940), Alabama's market prices were as much as 20 cents a bushel higher. Alabama's market prices remained well above support prices in most of the years during the period.

Acres Harvested

Data presented in figure 12.51 show that acres of corn (grain) harvested in Alabama generally trended downward from around 1917 through 1929, when it began to increase. Apparently, returns from cotton became so poor that Alabama farmers turned to corn production to try to provide some additional income for their farms. Even though market prices (Figure 12.46) were falling sharply after 1928, Alabama farmers increased acres harvested. With the market price at $1.08 cents in 1929, farmers in the state harvested the grain from 2.8 million acres in 1930. Then with market prices down to 43 cents in 1931, acres harvested increased to 3.2 million the following year (1932). The 1932-1933 disaster gave them pause about corn production in Alabama, and in 1933 they reduced acres harvested slightly. However, in 1933 market prices with AAA support were up sharply to 78 cents, and in 1934 and 1935 they began to increase acres harvested again. In 1935, it reached 3.7 million acres. In 1935, the market price declined slightly, and Alabama farmers apparently interpreted that slight decline as an indication that they were about to begin a downward trend. As a result, in 1936 they reduced acres harvested to 3.4 million. Unfortunately, market price did not decline in 1936; rather it increased sharply to $1.06. They continued to reduce acres harvested in 1937 (3.3 million). Some unknown signal resulted in an increase to 3.6 million in 1938, although market price had fallen to 59 cents in 1937. Unfortunately, when they harvested those additional acres, the average market price was still 59 cents. They slightly reduced acres harvested in 1939, but the price increased. Then in 1940, they increased acres harvested one more time to 3.5 million, and as expected the price declined.

Of course, Alabama farmers knew about the teeter-totter relationship between supply and prices. But they had no other choice but to play the game, and it seemed altogether reasonable to them that an increase in the price in a given year strongly suggested that they should plant more the following year, but what they did not seem to appreciate was the fact that virtually every other corn farmer in the country responded to the same signal in exactly the same way. While the response to the signal might have been an understandable one for the individual, it was a disaster for the group.

Yields

Data on corn yields for both the United States and Alabama for the period 1915-1940 are included in Figure 12.49. The data indicate that Alabama's corn yields throughout the 1920s had ranged between 12 and 14.5 bushels an acre, and they remained in this same general range throughout the 1930s, although the data for the period 1932 through 1938 seemed to indicates a slight upward trend. Throughout this period, Alabama yields were 20 to 30 bushels an acre lower than those in Iowa. It is not known to what extent Alabama farmers were using hybrid corn during this period. Yields were lower than expected in 1930, 1939, and 1940 as a result of poor crop weather.

In 1938, the average yield of corn in several Southern states, including Alabama, was 14.3 bushels an acre (Fergus, et al., 1944). The net cost of production an acre, including rent, was only $15.42. Unfortunately, with such a low yield, the net cost of production for a bushel was $1.08, the highest of any region in the country. In contrast, in Illinois and Iowa, with a substantially greater net cost an acre ($21.15), net cost a bushel was considerably less ($0.47 versus $1.08). From the beginning, the Southern states had almost no comparative advantage in the production of corn, but they kept producing it because they had nothing else except cotton that had much of a market value in the 1930s.

Bushels Harvested

With corn yields in Alabama remaining more or less stable during the period 1930-1940 (Figure 12.49), yearly trends in bushels harvested (Figure 12.51) generally followed those of acres harvested (Figure 12.50). Bushels harvested were lower than expected in 1930, 1936, 1939, and 1940 as a result of lower than expected yields . Except for those off-years, bushels harvested trended upward from 36.4 million in 1932 to 49.9 million in 1938.

Census data showed that farmers in DeKalb County harvested the largest share (4.9%) of the 1939 Alabama corn crop. Other counties harvesting substantial shares included: Marshall (4.1%), Jackson (4.0), Madison (3.9), and Cullman (3.6).

Alabama Peanuts

Because of the biology of the plant, peanut production is restricted primarily to the Lower South and parts of the Southwest. In 1930, the entire American crop totaled 697.4 million pounds. The largest share (30%) of the total crop was harvested in Georgia. Other states reporting large shares included: North Carolina (25.6%), Virginia (14.2), Alabama (13.3), and Texas (7.1). These five states accounted for 90.2% of the entire national crop.

Prices

Only national peanut market prices data are available for the period 1915-1940. They had been as high as 9.3 cents a pound in 1919 at the end of the period of wartime inflation, but they slowly trended downward to reach 3.5 cents in 1930 (Figure 12.53). After 1929, year-to-year changes in peanut prices generally tracked those changes taking place in the larger, national agricultural economy during this period of severe economic malaise. Year-to-year changes in the agricultural prices for all crops were depicted in Figure 12.5. Of course, after 1932 peanut prices, as well as the prices of all other crops, were affected to a greater of lesser extent by various New Deal programs. As the Depression intensified, peanut prices fell to below two centsover the next two years. With the support of New Deal programs, it increased to 2.8 cents in 1933, then, and for the remaining years of the decade, they remained relatively stable at between 3 and 4 cents.

Acres Planted and Harvested

Data on acres planted and harvested for Alabama peanuts during the period 1915-1940 are shown in Figure 12.54. In 1930, Alabama farmers planted 334,000 acres of peanuts, but harvested only 161,000 (48%). These data indicate that many Alabama farmers still allowed their hogs to harvest about half of their crop for them. Market prices had begun to trend downward in 1928, and reached 3.7 cents a pound in 1929. Alabama farmers responded to that price signal by planting 92,600 acres of peanuts in 1930.

Surprisingly, when market prices fell from 3.7 cents in 1929 to 3.5 cents in 1930, farmers interpreted this decline as a signal to increase acres planted, and they really responded with a bang, planting 411,000 acres in 1931. Market price declined again in 1931, to 1.6 cents. In 1932, acres planted totaled 521,000, the highest level in history at that time. In 1932, market price remained at 1.6 cents. In 1933, Alabama peanut farmers finally roused out of their dream-world, and in 1933 they reduced acres planted to 430,000. In the year AAA was enacted (1933), the market price increased for the first time since 1927. It increased to 2.9 cents, and farmers became excited again. In 1934, acres planted reached 475,000. Market prices increased relatively little between 1934 and 1940, but farmers steadily increased acres planted until they reached 553,000 in 1929. It was slightly lower in 1940 (520,000 acres). During the period 1930-1940, farmers harvested an average of 54% of acres planted; percentages tended to be somewhat higher in the latter part of the decade.

Yields

Alabama peanut yields ranged between 500 and 600 pounds an acre during the period 1930-1933 (Figure 12.55), then increased to almost 800 pounds in 1936. They remained near that level through the remainder of the decade, except for a disastrous, drought-induced crop failure in 1939. For much of the period 1920-1933, average Alabama peanut yields were 150-200 pounds lower than those of the country as a whole. However, after 1933 yields in the state increased sharply, and after 1934 there was little difference between the state and the country. Apparently, federal intervention in peanut production encouraged Alabama farmers to increase the level of inputs for their peanuts.

Pounds Harvested

With acres harvested (Figure 12.54) generally trending upward for the entire period and yields (Figure 12.55) trending upward after 1933, pounds harvested remained around 100 million pounds through 1933. Afterwards, it increased rapidly to reach 203 million in 1935. Although there was considerable year-to-year variation, pounds harvested remained around the same level through 1940.

Soybeans

Soybeans provide an interesting insight into the historical development of agriculture. Soybeans had been grown as a crop in East Asia for thousands of years before being introduced to North America. Rice made the trip much earlier, probably arriving in the Southeast around 1694, but it took soybeans almost 200 years longer. Even after the plant arrived, its characteristics were not adequately appreciated until much later. USDA did not publish any production statistics until 1924. Rice made the trip as part of the slave trade, but there was no similar commerce between East Asia and North America.

US Soybeans

In 1930, American farmers harvested 13.9 million bushels of soybeans. Farmers in Illinois harvested the largest share (50%). Other states with significant shares included: Indiana (15.2%), North Carolina (9.6), Iowa (7.3), and Missouri (6.9). These five states accounted for 89% of the total harvest. Alabama farmers harvested only 0.3%.

Even at this early date, it was becoming increasingly obvious that soybeans would be another major crop for the North Central Region. The soils and climate there were near ideal to provide all the requirements of the genome of the plant. In hindsight, it is obvious that conditions were so ideal there that it made little sense for other regions to become involved in soybean production. Unfortunately, such a wise decision could not be made in the state agricultural experiment stations across the country. Virtually all of them invested large sums of scarce research dollars to join the soybean bonanza.

Exports

The United States exported few soybeans until 1932 when they shipped 2.2 million bushels. Exports remained at around that level until 1938 when 11 million tons were shipped overseas. However, the next year, there were 300,000 bushels. Apparently, there were no exports in 1933, 1934 and 1936.

Prices

The market prices for soybeans was the same ($2.50 a bushel) for both the United States and Alabama in 1924 (Figure 12.57); however by 1930 it was 66 cents higher in Alabama ($2.29 versus $1.34). Market price for the crop fell to 53 cents in 1932, but then began to trend upward, reaching 80 cents in 1939 and 89 cents in 1940. At this point in the history of the production of soybeans in the United States, price was probably not a very good indicator of the balance between supply and demand. Markets for the crop were not very well organized, but this was about to change, and change rapidly.

Acres Planted and Harvested

Soybean acres planted and harvested in the United States climbed steadily through the 1930s (Figure 12.58). In 1930, farmers planted and harvested 3.1 and 1.1 million acres of the crop, respectively. In 1940, they had more than tripled those totals to 9.6 and 4.8 million acres, respectively. Throughout the decade, farmers used a significant portion of their crop for forage, rather than harvesting the beans. In 1930, they only harvested a third of their crop (1.1 million acres). By the end of the decade in 1940, farmers planted 10.5 million acres of soybeans, and they harvested 4.8 million acres, or 46% of acres planted.

Yields

Although there was considerable year-to-year variation, American soybean yields generally trended upward throughout the decade. It was 13 bushels an acre in 1930 and almost 21 bushels in 1939. Yields were down sharply in 1940 to 16.2 bushels. Palmer Drought Severity Indices (PDSI) (Cook, et al.,1999) indicate that the weather was unusually dry throughout the Soybean Belt in June, July, and August of 1940.

Bushels Harvested

With both acres harvested and yields (Figures 12.57 and 12.58) trending upward through the decade, bushels harvested followed. Harvest was 13.9 million bushels in 1930, and reached 90.1 million in 1939. The unusually good average yield and the unusually poor yield in 1935 and 1940, respectively, are obvious in the harvest data.

Alabama Soybeans

No data were published on the production of soybeans in Alabama before 1924. Between 1924 and 1930, acres of soybeans planted increased from 95,000 to 150,000 (Figure 12.61); however, Alabama farmers harvested only a small share of beans. During that period, acres harvested ranged between 3,000 to 8,000 acres. No data were collected on bushels harvested in the counties during that period.

Prices

Data on market prices of soybeans sold in Alabama during the period 1930-1940 are also presented in Figure 12.57. Throughout the period, market prices for Alabama soybeans were considerably higher than for the country as a whole; although there was considerable year-to-year variation, those in Alabama tended to trend upward during the period. It was $2.29 a bushel in 1930 and $2.43 in 1939.

Acres Planted and Harvested

Data on acres planted and harvested of soybeans in Alabama during the period 1924-1940 are presented in Figure 12.61. Acres planted generally increased during the period 1930-1940, from 150,000 in 1930 to 271,000 in 1938. There was little change in 1939 and 1940. However, while acres planted generally increased, acres harvested increased very little. Alabama farmers harvested the crop for beans from only 8,000 acres in 1930 and from 11,000 acres in 1940, or 5% and 4% of acres planted, respectively. At that time, there were few processing plants for the crop in the Lower South. Most of the crop had to be harvested for hay. However, since soybean is a legume, planting it increases the amount of nitrogen in the soil.

Yields

Data on soybean yields in Alabama during the period 1924-1940 are also presented in Figure 12.59. Generally, yields of the crop in Alabama changed very little during the period 1930-1940. They ranged between 5 to 7 bushels an acre in most of the years, and were 8 to 15 bushels an acre below those in the country as a whole. With no real market for their beans, Alabama farmers were unlikely to spend very much money on the crop,

Bushels Harvested

With yields of Alabama soybeans generally remaining stable during the period 1930-1940 (Figure 12.59), and acres harvested increasing steadily for most of the period (Figure 12.61), bushels harvested also increased steadily. They increased from 40,000 bushels in 1930 to 72,000 in 1940. It was much higher than expected in 1936 as a result of greater than expected yields. After 1935, year-to-year variation was quite high as the relationship between yields and acres harvested became more variable.

Alabama Hay

American farmers harvested 91.7 tons of hay in 1920. They harvested 74.5 million tons in 1930, with Nebraska farmers harvesting 6.8% of the national crop that year.

Other states harvesting significant shares were: Wisconsin (6.7%), Minnesota (6.6), California (6.4), and New York (6.2). Together, these five states harvested 33% of the total national crop. In 1930, Alabama farmers harvested 363,000 tons of hay, or 0.5% of the total national crop.

Data published in the Census of Agriculture showed that Alabama farmers harvested 333,700 tons of hay of all kinds in 1929. Montgomery County reported the largest share (5.8%). Other counties reporting substantial shares included: Dallas (5.2%), Madison (5.0), Jackson (4.0), and Hale (3.5). These five counties accounted for only 23.5% of the state's total. Twenty-three counties reported harvests of 5,000 or more tons.

Prices

Alabama hay prices remained relatively stable at around $15 a ton from 1926 through 1930 (Figure 12.63), but when the Depression struck in earnest after 1930 they fell sharply, reaching $7 a ton in 1932. After 1929, year-to-year changes in hay prices generally tracked those changes taking place in the larger, national agricultural economy during this period. Year-to-year changes in the agricultural prices for all crops in the country are depicted in Figure 12.5. Of course, after 1932 hay prices, as well as the prices of all other crops, were affected to a greater or lesser extent by various New Deal programs.

After the enactment of New Deal farm legislation in 1933, the average price climbed back to $10 a ton and remained near that level through 1940.

Acres Harvested

As we will see in a following section, a new cattle Cycle of Abundance began in Alabama in 1928, as farmers began to increase their herd numbers (Figure 12.88). As a result of increasing herd size, Alabama cattle farmers began to increase the harvest of hay on their farms. In 1930, they harvested the crop from 516,000 acres (Figure 12.64). Acres harvested increased again in 1931 (681,000). They harvested fewer acres in 1933 as a result of sharply lower prices (Figure 12.62). After 1933, the upward trend in acres harvested resumed. The 1928-1938 cattle cycle entered its Declining Phase in 1935 and ended in 1938. However, Alabama farmers continued to increase acres harvested through 1937 (988,000 acres). Acres harvested remained near that level through 1940.

Yields

Figure 12.65 presents data on the average annual yield of hay on Alabama farms during the period 1930-1940. Unfortunately, the scale used in the figure tends to exaggerate the year-to-year variation. During most of the period, yields varied between 0.70 and 0.75 ton an acre. Obviously, 1937 and 1938 were exceptional years during that period for the growth and harvest of hay.

Tons Harvested

There was relatively little change in the average yield of hay in Alabama during the period 1930-1940 (Figure 12.66). Tons harvested were more dependent on year-to-year changes in acres harvested (Figure 12.64). Total harvest in the state in 1930 was 363,000 tons. Afterwards, tons harvested generally trended upward to 988,000 tons in 1937; then, with both acres harvested and yields lower, tons harvested also declined. In 1940, Alabama farmers harvested 715,000 tons. Tons harvested were higher than expected in 1931 and 1932 when acres harvested were higher.

Alabama Wheat

In 1920, American farmers harvested 613.2 million bushels of winter wheat. In 1930, they harvested 633.9 million, with Kansas harvesting the largest portion of the crop (29.3%). Other top-five winter wheat harvesting states were: Nebraska (11.3%), Texas (6.0), Oklahoma (5.9). and Illinois (5.4). Altogether, these five states harvested 58% of the country's crop that year. Alabama farmers harvested less than 0.01% of the national crop.

Census data indicated that farmers in four counties north of Birmingham harvested 60.5% of the state's wheat crop in 1929. Farmers in Madison County harvested the largest share (24.9%). Other counties with large shares included: Madison (15.5%), Hale (12.3), Jackson (7.8), and Limestone (5.1). A total of 34 counties did not report any wheat harvest.

US Exports

In 1930, the United States exported 131.5 million bushels of wheat (Figure 12.67). Data presented in the figure represents the export of (all) wheat. However, winter wheat has traditionally been the primary wheat produced and exported. The 1930, exports were equivalent to 24%

of the 1929 national crop.

Exports were as high as 219.2 million bushels in 1926; however, world economic conditions began to deteriorate about that time. The Consumer Price Index began to decline after 1926 (Figure 12.2), and exports began to decline. The Hawley-Smoot Tariff Act was enacted in 1930, and immediately foreign countries began to retaliate. This situation further intensified the worsening economic conditions in the United States and abroad. Wheat exports fell to 41.2 million bushels in 1932 and remained near that level through 1936. Economic conditions began to improve slightly in 1934 (Figure 12.1), but exports did not begin to increase until 1936. They increased sharply to about 100 million bushels in 1937 and 1938. Surprisingly, they quickly fell to 54 million bushels in 1939 and to 40.6 million bushels in 1940.

US and Alabama Prices

After 1929, year-to-year changes in winter wheat market prices generally tracked those changes taking place in the larger, national agricultural economy during this period. Year-to-year changes in the agricultural prices for all crops in the country are depicted in Figure 12.5. Of course, after 1932 winter wheat prices, as well as the prices of all other crops, were affected to a greater or lesser extent by various New Deal programs.

American winter wheat market prices had been as high as $1.48 a bushel when economic conditions began to deteriorate in the late 1920s (Figure 12.1). By 1930, they were down to 69 cents a bushel (Figure 12.68), and they continued to decline in 1931 and 1932 to around 39 cents. Severe drought (beginning of the Dust Bowl) began to reduce winter wheat production in the Lower Great Plains in 1932, and with the resulting reduced production, market prices began to increase. They were between 78 and 84 cents a bushel from 1933 through 1935. Provisions of the Agricultural Adjustment Act of 1933 probably also provided some boost. Bushels harvested remained lower than normal through 1936, and market prices continued to trend upward through 1936 to around $1 a bushel. Winter wheat harvests were improved in 1937 and again in 1938, and market prices began to decline, reaching 57 cents in 1938. Bushels harvested fell again in 1939 and 1940, and market prices increased to 69 cents.

When the market price for wheat plunged below 60 cents (57 cents) in 1938, USDA immediately established a support proce for the crop at 60 cents. Bushels harvested were down somewhat in both 1939 and 1940, and market prices increased to around 69 cents. Support prices were 63 and 64 cents, respectively, in those two years.

Market prices for both US and Alabama winter wheat are shown in Figure 12.68. These data show that year-to-year trends were generally the same for both. Alabama farmers generally received slightly higher prices for their wheat during the period 1930-1940 than did their counterparts throughout the country. Of course, they had very little to sell, and it is likely that none of it ever left the state. The difference was somewhat greater before the enactment of AAA.

Acres Planted and Harvested

Alabama wheat farmers had begun to get out of the wheat production after World War I (Figure 12.69). There were just too many problems with the crop, and yields were never very good. Prices never were good enough to make the misery worthwhile. From 1926 through 1930, they planted less than 5,000 acres of the crop each year. Surprisingly, Alabama farmers began to increase their acreage of the crop in the early 1930s, although prices were declining and the droughts were becoming increasingly intense. The reason, of course, was that cotton prices had fallen from 16.8 cents in 1929 to 9.5 cents in 1930; and in 1931 it went down to 5.7 cents. Alabama farmers desperately needed a crop to replace cotton. In 1932, they planted 7,000 acres of winter wheat, the most since 1924. After 1932, they apparently began to get the message and reduced acreage. In 1933 (the 1934 crop), they planted only 5,000 acres. The enactment of AAA in 1933 rekindled their interest somewhat, and that fall they planted 10,000 acres (the 1934 crop). Even with the effects of AAA, prices did not return to the World War I level. As a result, Alabama farmers apparently decided, once again, that winter wheat was not a profitable crop for them. Also, they were sure that AAA was going to raise the price for cotton. After 1934, acres planted of winter wheat generally trended downward to 6,000 acres in 1939. The beginning of World War II in Europe, in

1939, apparently again rekindled their interest, and they planted 7,000 acres that fall.

Figure 12.69 also contains data on acres of winter wheat actually harvested in Alabama during the period 1930-1940. Farmers harvested virtually all of their planted acres throughout the period.

US and Alabama Yields

Annual average yields of winter wheat in the United States ranged around 15 bushels an acre throughout the 1920s (Figure 12.70). In 1930, the average was 15.4 bushels; however as a result of the widespread drought (Dust Bowl) of the early 1930s, it declined sharply. In 1933 and 1934, it was between 12 and 13 bushels. Although yields began to increase after 1934, the national average did not return to the 15-bushel level until 1939. In 1940, it was 16.4 bushels.

Year-to-year trends in wheat yields in Alabama during the period 1930-1940 were generally similar to winter wheat yields in the country as a whole (Figure 12.70). Throughout the 1920s, yields in the state had ranged around 10 bushels an acre. They were near that level in 1930 and 1932. In 1933 and 1934, the average was 8 bushels. It had not been that low since 1919. Yields began to recover in 1935 and increased to 13 bushels in 1938. They remained near that level through 1940. Yields in the state were only two-thirds of the average for the country during the period. With yields this far below the national average, it is understandable why Alabama farmers had lost interest in this crop.

Bushels Harvested

Alabama farmers only harvested 20,000 bushels of winter wheat in 1930 (Figure 12.71). Then, with yields varying relatively little during the period 1930-1940, bushels harvested was more dependent on acres harvested. Acres harvested increased slightly in the early 1930s, and bushels harvested followed, reaching 70,000 bushels in 1932. They remained near that level through 1940.

Alabama Oats

American farmers harvested 1.3 billion bushels of oats in 1930. They had harvested 1.4 billion in 1920. Iowa farmers harvested 18.3% of the 1930 crop. Other states harvesting significant quantities were: Minnesota (13.1%), Illinois (11.3), Wisconsin (7.6), and Nebraska (5.6). Together, these five states harvested 56% of the total national crop. Alabama farmers harvested 1.5 million bushels, or 0.1% of the total 1930 crop.

Census data indicate that farmers in five counties in central Alabama threshed 37.1% of the 1929 crop of oats. Farmers in Hale County threshed the largest share (10.8%). Other counties reporting large shares included: Montgomery (8.7%), Talladega (6.5), Marengo (5.7), and Macon (5.4). Farmers in only 16 counties reported threshing more than 5,000 bushels.

US Exports

The United States had exported 100 million bushels of oats during the years of World War I (1914-1918), but by the early 1930s, total exports were less than 5 million (Figure 12.72). In 1930, only 0.2% of the 1929 national crop was exported. Exports remained low throughout the period 1930-1940. In only one year (1937) during that period were exports greater than 10 million bushels in a year. Oats continued to be a major crop in Europe, and as a result, under normal conditions there was little market for American oats there.

US And Alabama Prices

After 1929, year-to-year changes in oat prices generally tracked those changes taking place in the larger national agricultural economy during the Great Depression. Year-to-year changes in the agricultural prices for all crops in the country were depicted in Figure 12.5. Of course, after 1932 oat prices as well as the prices of all other crops were affected to a greater or lesser extent by various New Deal programs.

In 1927, the annual average market price of American oats was near 50 cents a bushel (Figure 12.72). Between 1929 and 1930, bushels harvested increased from 1.113 billion to 1.274 billion, and market prices declined from 40.3 to 31.1 cents. Market prices continued to decline in the first years of the Depression to reach 14.8 cents in 1932. Bushels harvested were down sharply in 1933 and 1934, and market prices began to increase, reaching 46.4 cents in 1934. Farmers harvested a much larger crop in 1935, and the market price fell to 25.7 cents. Supply and market prices teeter-tottered through the remaining years of the period. In 1940, with farmers harvesting

1.246 billion bushels of the grain, the market price fell to 29.8 cents.

Figure 12.73 includes data on both national and state oat market prices. Year-to-year changes were essentially the same in both Alabama and the country as a whole; however, those received by Alabama farmers were usually 10 to 30 cents a bushel higher. These data suggest that, as might be expected, the same factors determining oat prices for the entire country were at work in Alabama as well. The differential probably indicates that most of the relatively small harvests of Alabama oats were marketed locally where they received somewhat better prices.

Acres Harvested

There are no data on acres of oats planted in Alabama before 1926; consequently, I have chosen to use only acres harvested for the discussion in this section. In 1920, there were 130,000 horses and 296,000 mules in Alabama. To help feed these animals, farmers harvested oats from 163,000 acres in that year (Figure 12.74). In 1930, there were 66,000 horses and 322,000 mules in the state, and farmers harvested oats from 95,000 acres to feed them. Apparently, Alabama farmers received some kind of signal before planting time in 1931 that encouraged them to increase acres planted and harvested substantially. In that year, they harvested oats from 172,000 acres. Unfortunately, the average price fell to 31 cents by the time they harvested that crop (Figure 12.73). The following year (1932), acres harvested was back down to near 90,000, and they received 34 cents for that crop. Whoever started that rumor certainly did not do Alabama farmers any favors. With the 1932 market price at 34 cents, they planted fewer acres in 1933, harvesting the crop from only 61,000 acres. As a result, their average price increased to 58 cents. In 1934, the market price had increased to 69 cents, and in 1935 acres harvested was increased to 97,000. When farmers harvested that crop, it was down to 61 cents. After 1935, farmers seemed to completely disregard market price signals as they steadily increased acres harvested to 112,000 in 1939. In the spring of 1940, likely responding to the beginning of the war in Europe, they sharply increased acres harvested to 146,000.

Generally, Alabama farmers harvested the grain from between 76 and 82% of their planted acres in those years (1931, 1934, 1935, 1937, 1938, and 1939) when yields (Figure 12.75) were not severely reduced by drought. When yields were 16 or less bushels an acre (1930, 1932, 1933, and 1936), farmers harvested 75% or less of their planted acres. Oats were used as a grazing crop on many farms in the state, and, of course, the grain would not have been harvested from those grazed acres.

US and Alabama Yields

Figure 12.75 presents data on annual average oat yields (bushels an acre) in the United States and Alabama during the period 1930-1940. Yields in the country ranged around 30 bushels an acre during the period. In most years, oat yields in Alabama were 10 to 15, or more, bushels an acre lower than the national average. Alabama yields were especially low in 1930 (drought), 1932 (excessive moisture), 1933 (drought), 1936 (drought), and 1940 (excessive moisture). In 1931, 1934, and 1938—when environmental factors seemed to have little effect on yields—farmers harvested slightly better than 20 bushels an acre. Year-to-year variation was extremely high, but there seemed to be little evidence that yields improved during this period.

Bushels Harvested

Acres harvested (Figure 12.74) generally had the larger effect on bushels harvested of oats (Figure 12.76) in Alabama than yield (Figure 12.75). Consequently, except for an extremely large harvest in 1931 (3.8 million bushels) and a low one in 1930 (1.5 million bushels), 1932 (1.4 million bushels) and 1933 (1 million bushels) bushels harvested tended to trend upward slightly, reflecting a slight upward trend in acres harvested, especially during the latter part of the period. Bushels harvested were considerably higher in 1940 (1.9 million bushels) when acres harvested received a boost, apparently from the beginning of the war in Europe.

Alabama Sweet Potatoes

In 1930, Alabama farmers harvested 2.9 million Cwt of sweet potatoes, with a market value of $5.5 million. That crop accounted for 5.6% of cash receipts ($96.9 million) obtained from the sale of all Alabama plant crops, except timber. They had harvested 4 million Cwt in 1920.

The Census of Agriculture published data indicating

that Alabama farmers harvested 3.6 million Cwt in 1929. Baldwin County reported the largest share (5.5%) of the state's total. Other counties reporting substantial shares included: Dallas (4.9%), Cullman (3.2), Covington (2.9), and Jefferson (2.8). These five counties accounted for only 19.3% of the total state crop. Some 31 counties reported harvests of at least 54,000 Cwt.

Data on acres of sweet potatoes harvested in Alabama during the period 1930-1940 are presented in Figure 12.77. Farmers harvested 68,000 acres of the crop in 1930. In 1931, acres harvested began to increase rapidly, finally reaching 110,000 in 1935. Hard times became even more difficult in Alabama around 1925, and many Alabama families had to return to the farm. Farm numbers began to increase in 1926 and reached a peak in 1935 (Figure 12.16). Afterwards, numbers began to decline rapidly. It is likely that sweet potatoes were an extremely important component of the food supply for many of those families forced back onto farms during the early 1930s.

Except years 1931, 1933, 1936, 1939, and 1940, when unfavorable weather affected production, yields ranged between 43 and 54 Cwt. As a result of the combined effects of acres harvested and yield, farmers harvested a maximum of 5 million Cwt in 1934 and a minimum of 1.7 million in 1940 when excess soil moisture significantly reduced yield (27 Cwt). With the wide range in Cwt harvested crop value also varied considerably. It was as high as $6.5 million in 1934 and as low as $2.9 million in 1940.

Alabama Irish Potatoes

The Census of Agriculture reported that 965,700 Cwt of Irish potatoes were harvested from Alabama farms in 1929, whereas a decade earlier Alabama farmers had harvested 657,000 Cwt. Baldwin County reported the largest share (27.4%) of the 1929 crop. Other counties reporting large shares included: Mobile (4.6%), Escambia (4.4), Lauderdale (4.3), and Madison (4.2). These five counties accounted for 44.9% of the total crop. A total of 31 counties reported harvests of at least 6,000 Cwt.

In 1930, Alabama farmers harvested 1.2 million Cwt of Irish potatoes with a value of $3 million. In that year, the sale of Irish potatoes accounted for 3.1% of cash receipts obtained from the sale of all Alabama farm plant crops except timber.

Data on acres of Irish potatoes harvested annually during the period 1930-1940 are presented in Figure 12.78. Alabama farmers harvested the crop from some 27,000 acres in 1930. Acres harvested had been trending upward since around 1915. Farmers responded to declining prices of Irish potatoes in the early 1930s with moderate increases in acres harvested. They had generally harvested 20,000 to 25,000 acres of the crop in the mid-to-late-1920s. In the early 1930s, acres harvested ranged from 33,000 to 35,000. Surprisingly, even though market prices remained well below pre-Depression levels throughout the 1930s, farmers began to significantly increase their acreage after 1936. Again, the likely reason was the decline in cotton market prices in the early 1930s. By 1940, they were harvesting potatoes from 51,000 acres. When national market conditions were right, Irish potatoes were a good crop for a relatively small number of Alabama farmers.

Except in 1932 and 1933 when unfavorable weather reduced production, yields ranged between 51 and 56 Cwt an acre during the period. As a result of the relationship between acres harvested and yields, Cwt harvested ranged from 1.2 million to 2.9 million Cwt.

Except for 1931, prices remained relatively stable at around $1 per Cwt throughout the period 1930-1940; however as a result of the large changes in acres harvested, crop value varied from $1.7 million in 1931 to $2.9 million in 1939 and 1940. Crop value was actually highest in 1936 ($4 million) when for some reason, farmers received an unusually high price ($2.17 per Cwt).

Alabama Sorghum

There are no annual data on the production of sorghum in Alabama before 1924. In the period 1930-1940, acres of sorghum harvested in Alabama averaged 79,000, but that was much higher than harvests from 1931-1934 (Figure 12.79). An average of 53% of the crop during the period was harvested for making molasses. The sharp increase in acres harvested after 1930 was largely attributable to an increase in molasses-making. The number of farms in Alabama increased sharply during this period (Figure 12.16). Molasses was likely an important part of the diet of many of those families who were forced back to the farm during the Great Depression.

In the period 1930-1940, sorghum harvested from some 41% of the acreage was used for forage and 6% for silage. Molasses production averaged 2.4 million gallons a year (1.5 to 3.9 million) during the period, and cash receipts from its sale averaged $1 million a year ($775,000 to $1.4 million). There are no estimates of acres of sorghum harvested for grain in Alabama before 1944.

Alabama Sugar Cane

Alabama farmers harvested an average of 25,000 acres (20,000 to 30,000) of sugar cane a year in the period 1930-1940 (Figure 12.80). The increase in acres harvested after 1929 is probably also related to the increase in farm numbers during that period (Figure 12.16). The crop was used to produce an average of 2.9 million gallons (1.5 to 3.9 million) of syrup annually. Average cash receipts were $1.4 million ($900,000 to $1.8 million).

Alabama Vegetables

The 1930 Census of Agriculture published data indicating that the total value of vegetables harvested in Alabama in 1929 for sale and home use totaled $11.3 million. Farmers in Baldwin County harvested the largest share (13.4%). Other counties reporting large shares included: Mobile (5.6%), Jefferson (5.3), Dallas (3.1), and Cullman (2.5). These five counties accounted for 29.9% of the state's total.

The Alabama Agricultural Statistics Service began collecting annual data on vegetable production in 1928. In the first Bulletin issued (1948), they provided some production data on seven vegetable crops under the heading truck crops. The relevance of these crops to the agricultural economy of the state was discussed in the preceding chapter. Data on the acres harvested of these crops during the period 1928-1940 are presented in Table 12.6. Although year-to-year variation was high, acres harvested for commercial early Irish potatoes was considerably higher than for any of the other crops (7,500 to 27,800). Acres harvested of this crop tended to increase toward the end of the period. Data on the production of these potatoes is not included in the data on Irish potatoes discussed in a preceding section. Acres harvested of those potatoes also increased during the period (Figure 12.78). Acres harvested for cabbage, cucumbers, tomatoes, and snap beans were generally the lowest, and intermediate for strawberries and watermelons.

Table 12.7 includes data on the value of the various truck crops in the different years. Values for the commercial early Irish potatoes were highest, but then there were considerably more acres harvested involved. Values for cabbage, cucumbers, and watermelons were generally lowest. Values were intermediate for strawberries.

Alabama Velvet Beans

Average annual acres harvested for Alabama velvet beans ranged from 306,000 to 510,000 during the period 1930-1940 (Figure 12.81). Most velvet beans were inter-cropped with corn. Acres harvested were 340,000 in 1930, before declining to 306,000 in 1931. Afterwards, however, acreage began to increase rapidly, reaching 510,000 in 1934. There was relatively little change after 1934. The increase in acres harvested seemed to be related to the increase in farm number, but it did not decline when numbers began to fall after 1935 (Figure 12.16).

Average prices were as low as $4.50 a ton in 1932, but generally ranged around $12 for much of the period. Annual value of the crop averaged $2.4 million, but was as low as $778,000 in 1932.

Alabama Cowpeas

Cowpeas were planted on an average of 499,000 acres in Alabama each year from 1930 through 1940. On 36% of the acreage they were planted alone. On 64%, they were inter-planted with other crops. Acres harvested averaged 322,000 a year during the period (Figure 12.82). They were 73,000 acres in 1930, but by 1932 had increased to 181,000. Except for 1935, they were above 130,000 for the remainder of the period. Apparently, acres planted and harvested followed farm numbers upward in the early 1930s, but remained at a relatively high level even when numbers declined later.

Approximately 45% of acres harvested during the period were harvested for peas, 34% for hay, and 21% for other purposes. The annual average price was as low as 68 cents a bushel in 1931, but afterwards it trended upward reaching $1.65 in 1940. The value of the pea crop varied between $571,000 in 1930 and $1.6 million in 1937. In most years, it ranged around $1 million.

Alabama Pecans

Alabama farmers, primarily in south Alabama, harvested 3.1 million pounds of pecans in 1930. The value of this crop was equivalent to only 0.7% ($723,000 of $96.9 million) of total cash receipts for all Alabama plant crops (excepting farm forest products).

Data on the quantity of pecans harvested annually in Alabama during the period 1919-1940 are presented in Figure 12.83. There are no annual data on pounds harvested before 1919. According to the Census of Agriculture, Alabama farmers established 276,000 new pecan trees, increasing their growing stock from 434,000 to 710,000 between 1920 and 1925, an increase of 64%. Pounds harvested in the good years increased from four million pounds in 1931 to 10 million pounds in 1939. The effects of masting were quite evident in 1930, 1931, 1933, 1938, and 1940. Surprisingly, it did not seem to be present in 1936. The upward trend in pounds during the period was probably the result of some of those trees, which had been planted in the early 1920s, beginning to produce nuts. Harvests in 1938 and 1940 were much lower than expected. It is likely that a combination of masting and drought in late summer contributed to the greater than expected reduction in 1938, and masting plus extremely heavy rainfall likely reduced production in 1940.

Pecan prices had generally declined throughout the late 1920s. Then between 1930 to 1931, they declined from 23.1 cents a pound to 13.3 cents. Unfortunately there was little improvement during the remainder of the period. They were 11.2 cents in 1940. It is probable that those rapidly increasing harvests were creating problems in the market. With harvests generally increasing and prices stagnated, value increased during the period from around $723,000 in 1930 to $996,000 in 1939.

Alabama Tung Nuts

The tung nut tree, a native of China, is a member of the Family Euphorbiaceae. This is an extremely large and diverse family of plants. It includes the Christmas poinsettia. Tung nuts are a source of tung oil, which is rich in unsaturated fat. It is used in paint, printer's ink, and water-proof coatings. The oil dries quickly and polymerizes into a glossy, water-proof coating on wood.

A few Alabama farmers began to plant tung oil trees on their farms in the late 1920s. The first data on the harvest of the nuts is found in the 1930 Census of Agriculture. There are no annual data before 1945. In 1929, they harvested 5,500 pounds from some 715 trees. The value of the nuts was $278. Pounds harvested increased sharply by 1939. In that year, farmers harvested near 22,300 pounds, worth $332.

Alabama Peaches

Data on the quantity of peaches harvested annually during the period 1930-1940 are presented in Figure 12.84. Alabama farmers harvested 68.6 million pounds in 1930. The value of that crop was $1.7 million, which was equivalent to 1.8% of the cash receipts ($96.9 million) for all other crops, except forest products.

According to the Census of Agriculture, Alabama farmers established an additional 550,000 peach trees, increasing their growing stock from 1,691,000 to 2,241,000 during the period 1930-1940. Pounds harvested, however, did not seem to respond to the increased number of trees. There was a significant amount of year-to-year variation in the data. Yields were reduced significantly in 1933, 1936, and 1940 as a result of unfavorable weather, primarily early-spring freezes. However even with the variation, there seems to be little evidence of an upward trend in pounds harvested. In four of the 11 years of the period, pounds harvested was around 80 million.

Prices declined from 2.5 cents a pound in 1930 to 1.3 cents in 1931. They rebounded somewhat in 1933, then remained around two cents in most years through 1940. Annual crop values varied considerably as a result of the variation in pounds harvested. They were lowest ($269,000) in 1932 when only 15.2 million pounds of the fruit was harvested, and highest in 1931 ($1.7 million) as a result of a moderately good crop (68.6 million pounds) and a good price (2.5 cents).

Alabama Greenhouse and Nursery

There are no annual data on cash receipts for Alabama greenhouse and nursery production until 1950. The 1930 Census of Agriculture published data indicating that the sale of greenhouse and nursery products in the United States in 1929 was $145.7 million. Together, the North Atlantic and North Central regions accounted for 69.5% of the total. Together the South Atlantic and South Central regions accounted for 14.8%, and the West Region accounted for 15.7%. Alabama sales accounted for 0.5%.

Sales of horticultural crops in Alabama in 1929 totaled $783,445. The largest share ($601,447) came from the sale of nursery products (trees, shrubs, vines, ornamentals, etc.). The remainder ($181,988) came from the sale of flowers and flowering plants. By 1939, the total sales of horticultural crops had increased to $860,994. Sales of nursery products declined to $515,890, while the sale of flowers and flowering plants increased to $196,239. In 1940, the census included data on a third category of horticultural crops—vegetables and plants grown under glass. These included flower seeds, vegetable seeds, vegetable plants, bulbs, etc. The total for this category was $148,865.

Alabama Timber

Data on American lumber production during the period 1869-1940 are presented in Table 12.8 These data show that the westward movement of lumbering was continuing and that the gains for the West Region were accompanied by almost equal losses in the North and South.

The 1950 Census of Agriculture published data indicating that the sale of all Alabama forest products in 1929 totaled $3.2 million. Table 12.9 includes data on the quantity of the various products cut on Alabama farms in that year. Data for 1939 indicated that the sale of all products totaled $1.4 million. Apparently the Depression hit Alabama's forest products industry especially hard.

Stauffer (1960) commented that although a number of estimates had been made of the volume of standing timber in Alabama, none of them were well founded. The first statewide, comprehensive, systematic study of Alabama's forest resources (the so-called First Survey) was undertaken in 1934 by the Forest Survey Group of the Southern Forest Experiment Station. The study was authorized by provisions of the McSweeney-McNary Forest Research Act of 1928. It was completed in 1936. The results of the survey was published in Lawson, et al. (1943). Some of the results of this survey are detailed in the following section.

Land in Forests

Data obtained by the 1934-1936 survey indicate that 58% (18.9 million acres) of Alabama's total land area was in forests in 1936.

Volume of Pine and Hardwood Growing Stock

The 1934-1936 survey also reports that total timber growing stock in Alabama, as of January 1, 1936, included 25.1 billion board feet of pine and 13.4 billion board feet of hardwood (including cypress).

Removal of Pine and Hardwood

Lawson, et al., 1943, also reported data from the first survey regarding the lumber cut derived from timber removed from Alabama woodlands. According to the survey, lumber cut in 1936 totaled 1.8 billion board feet of pine and 476 million board feet of hardwood.

Uses of Timber Removed from Alabama Woodlands

Table 12.10 presents data obtained by the first survey on the uses of timber removed from Alabama woodlands, during that period. Today, it is difficult to imagine that 23% of all the timber removed in the mid-1930s was used as fire wood or as a source of heat. In fact, almost half as much wood was used for heating, as was used for lumber (23 versus 55.4%). (As detailed in the preceding chapter, 57.7% of the 1929 withdrawal was used for firewood.) It is also difficult to imagine that during this period almost as much wood was used for railroad cross ties as for pulpwood (3.6 versus 3.9%). Data presented in Table 12.10 show that 55.4% of the timber removed in 1935 and 1936 was used for lumber. These data represent a major change in the use of the timber harvested in the state.

Animal Agriculture

The American livestock industry had benefited immensely from the prosperity generated by World War I. In 1920, value added by livestock marketed in the country had reached $7.4 billion. Then the recession of 1921 and 1922 turned the industry upside down. In 1921, value added declined to $5.7 billion, and the industry had not completely recovered in 1930 ($6.3 billion). Then the depths of the Great Depression hit the industry.

Cash receipts for Alabama livestock and livestock products were $24.5 million in 1925. There are no annual data available earlier than that, so it is not possible to determine exactly how much the 1921-1922 short recession had hurt Alabama's livestock industry. It did recover somewhat during the Roaring Twenties, but by 1930 it was getting a preview of the coming Depression. In 1930, receipts were down to $22.6 million.

Cattle

Data presented in the preceding chapter indicate that an extended cattle Cycle of Abundance ended in 1928 with a herd size of 35 million. It had begun in 1913 with a herd-size of 37 million. It had reached the Consolidation Phase in 1918 with a herd size of 51.5 million.

US Cattle

On January 1, 1930 there were 61 million cattle and calves in the United States. This number includes "all" cattle and calves and cows that had calved-milk. The largest share (10.6%) of the national herd was on farms in Texas. Other states with large shares included: Iowa (6.5%), Kansas (5.1), Nebraska (5.0), and Wisconsin (5.0). These five states accounted for 32.2% of the national total. About 1.4% of the national herd was on farms in Alabama. AAA had relatively little direct effect on the cattle industry. American cattle producers had opted-out of those AAA programs when they were being established. There were no support prices for cattle.

Farm Value Per Head

Farm value per head for cattle (excepting milk cows) in the country fell sharply during the period 1930-1940 (Figure 12.85), but the decline seemed to be the result of the cyclic relationship between values and numbers, rather than economic problems. The minimum farm value was not reached until 1934 when the economy was beginning to recover. Further, in 1934, the Cycle of Abundance was at the very end of an Expansion Phase that had begun in 1929.

Data on farm value per head for cattle (excepting milk cows) for the period 1915-1940 are presented in Figure 12.85. After 1925, with cattle numbers declining (Figure 12.86) farm value per head began to increase, finally reaching $42.77 in 1929. By that time, a new cycle had begun, and in 1930 ($40.38) values began to decline. That cycle reached its Consolidation Phase in 1934, and by that time values had fallen to $12.54. When the Liquidation Phase of the cycle began in 1935, values were $14.12. Afterwards, they generally trended upward and reached $30.90 in 1940. Surprisingly, the beginning of World War II in 1939 did not seem to change the trajectory of the upward trend at all.

Cycles of Abundance

American cattle numbers (excepting cows that had calved-milk) began to increase in 1929 (Figure 12.86) as part of the Expansion Phase of a new Cycle of Abundance. In 1930, there were 38 million of these animals in the country. At the end of the Expansion Phase in 1933, the cattle inventory counted 44.4 million head. Apparently the Consolidation Phase was completed in a single year (1934) with an inventory of 47.5 million. The Declining Phase was abnormal; there was little change in numbers until 1939 and 1940 when the war completely altered the dynamics of the relationship. The 1940 inventory counted 43.4 million cattle in the country.

Real Net Cash Returns Per Cow

Data on real net returns per cow provide a completely different perspective on the economics of cattle production than farm value per head. The return data reflects the actual relationship between prices received and prices paid in cattle production. As a result, year-to-year changes in returns generally tracked those changes taking place in the larger, national agricultural economy during this period of economic depression. Year-to-year changes in the agricultural prices in the country are depicted in Figure 12.5.

Real net returns (dollars) per cow had fallen during the latter part of the 1920s, and in 1930 it had reached $58 per animal (Figure 12.86). In 1932 (-$14) and 1933 (-$13) in the depth of the Depression, returns were actually negative. They did improve somewhat during the latter part of the decade, reaching $50 in 1935. Afterwards, returns slowly declined to reach $26 in 1940.

Alabama Cattle

Census data in 1930 indicate that 4.3% of the cattle (including milk cows) in Alabama were in Montgomery County. Other counties with large shares included: Wilcox (4%), Dallas (3.6), Lowndes (3.6), and Sumter (3.0). These five counties accounted for only 18.5% of the state's total, indicating the widespread distribution of cattle. A total of 28 counties reported herd numbers of 10,000 or more animals.

Farm Value Per Head

Data on farm value per head for Alabama cattle (including milk cows) are also presented in Figure 12.85. These data indicate that values for cattle in the state were $33.10 in 1930, but fell to $11.50 in 1934. Fortunately, after 1934 values began to increase and reached $23.60 in 1940. Year-to-year trends in values were similar for cattle in the country and the state. These values are biased upward somewhat by the inclusion of milk cows in the Alabama data. Values for milk cows were generally around 50% higher. Values for Alabama cattle were several dollars less than those for the country as a whole for most of the period (1930-1940), except during the depths of the Depression when there was little difference. It is likely that if the value of milk cows had not been included the difference between national and state values would have been greater.

Cycles of Abundance

Year-to-year trends in Alabama cattle numbers (excepting cows that had calved –milk) generally followed those of the country during the period 1930-1940 (Figure 12.88). In 1930, there were 451,000 animals in the state, and numbers were increasing in the Expansion Phase of a cycle that had begun in 1927. It ended in 1934 with an inventory of 648,000 head. The cycle reached the Consolidation Phase around 1935 with 666,000 animals. This phase was abnormal, as it was nationally. Then after 1935, cattle producers began to reduce herd size in the Declining Phase of the cycle. This phase ended in 1938 with an inventory of 537,000 head. Afterwards, a new cycle began, and in 1940, the inventory counted 627,000 animals in the state. In both the United States and Alabama, trends in numbers after 1934 were abnormal. It is likely that some aspects of the latter part of the Depression, the recession of 1937-1938, and the beginning of the war in 1939 interfered in some way with the normal teeter-totter relationship between prices and supply.

Hogs

On April 1,1930, the total value of hogs on American farms was $641.1 million. Farms in the North Central Region accounted for 76.7% of the total. Shares in the other regions were: South Central (10.7%), South Atlantic (5.6), West (4.3), and North Atlantic (2.6).

US Hogs

According to the USDA inventory, in 1930 there were 54.8 million hogs and pigs in the country. The largest share (19.2%) was in Iowa. Other states with large shares included: Nebraska (8.8%), Illinois (8.0), Minnesota (6.7), and Missouri (6.4). These fives states accounted for 49.1% of the total. Approximately 1.6% of the national herd was on Alabama farms.

Farm Value Per Head

Farm value per head for hogs and pigs in the country also fell sharply during the period 1930-1940 (Figure 12.85), but the decline seemed to be the result of the cyclic relationship between values and numbers, rather than the economic cycle. The minimum farm value was not reached until 1934 when the economy was beginning to recover. Further, the low value seems to have been associated with the Accumulation Phase of a Cycle of Abundance that had ended in 1932.

Farm value per head for hogs in the United States had increased sharply during the 1920s, but by 1930 it had drifted down to $14 (Figure 12.89). Between 1929 and 1930, hog numbers (Figure 12.90) that had been declining changed relatively little (55.70 to 54.84 million). In 1930, values were $13.45, and in 1931 they began to fall sharply. The Accumulation Phase of a new Cycle of Abundance began in 1931, but with values continuing to decline, it only lasted two years. It ended in 1932 with an inventory of 62.13 million. The Liquidation Phase of the cycle ended in 1934. At that point, values had declined to $4.21. Surprisingly, both numbers and values began to increase in 1935, but this situation did not last long. In 1936, values reached $12.71, and with numbers continuing to increase, trended sharply downward to reach $7.78 in 1940.

Cycles of Abundance

The teeter-totter relationship between numbers of hogs and farm value per head was discussed in the preceding paragraph. Data on hog numbers in the country during the period 1915-1940 are presented in Figure 12.90. The data indicate that there were parts of four Cycles of Abundance (Stearns and Petry, 1996) in hogs in the country between 1915 and 1930, and that a cycle, which began in 1926, was ending in 1930 at a level of 54.8 million animals. The Accumulation Phase of a new cycle began in 1931. At the end of this phase in 1932, the inventory was 62.1 million. The Liquidation Phase began in 1933 in the depths of the Depression, and when the cycle ended in 1934, there were only 39.1 million hogs (and pigs) in the country. Approximately a third of the country's hog herd was lost in a single year (1934). A new cycle began in 1935, and the Accumulation Phase ended in 1939 at 61.1 million animals. However, in the following year (1940) the Liquidation Phase of the cycle began with an inventory of 54.353 million head.

The cycle(s) described for this period are quite different than those observed before the beginning of the Depression. During the 1930s, the changes in economic conditions were so rapid and so severe that the normal relationship between supply and demand became quite turbulent and unpredictable.

Alabama Hogs

USDA data indicate that there were 831,200 hogs on Alabama farms on April 1, 1930, or 1.6% of the national total. The largest share of Alabama's herd was in Houston County (4.9%). Other counties with large shares included: Geneva (4.5%), Coffee (4.4), Covington (3.9), and Dale (3.6). All of these counties are southeast of Montgomery. These five counties accounted for only 21.3% of the state total. Some 20 counties reported totals of 15,000 or more head.

Farm Value Per Head

Figure 12.89 also presented data on farm value per head for hogs in Alabama for the period 1915-1940. Year-to-year trends in values were essentially the same in the country and the state during the period 1930-1940; although the trends were similar, values in the state were several dollars lower, except in the depths of the Depression when there was little or no difference. The teeter-totter relationship between hog numbers and farm value per head seen in the US data appears to be about the same for the Alabama data.

Average farm value per head was $10.50 in 1930. It had been near that level for several years. However, beginning in 1931 values began to fall sharply, and by 1934 reached $3.95. It had not been that low since 1902. Fortunately, values began to increase somewhat after 1934, and by 1937 they were back to near $9, just slightly better than in 1915. Unfortunately, when values began to increase in 1935, farmers began to hold more sows off the market, and number began to increase (Figure 12.91). When numbers reached 11 million animals, values began to falter, and in 1938 they began to fall, finally reaching $5.40 in 1940.

Cycles of Abundance

In 1925, hog numbers in Alabama were down to 776,000 (Figure 12.91). They had not been that low since USDA began publishing annual data in 1876. It appeared that Alabama was going to abandon the hog business. Then, after completing a new cycle, numbers reached 845,000 in 1929. Then in 1930, the Accumulation Phase of a new cycle began with an inventory of 870,000 head. That phase ended in 1932 with an inventory of 1.053 million. Its Liquidation Phase ended in 1934 with 889,000 head. The Accumulation Phase of the new cycle began in 1935, and it continued until 1939 when numbers reached 1.3 million. Unfortunately, values had begun to decline after 1937. As a result, between 1939 and 1940 Alabama farmers liquidated 11% of their herds (61,165 to 54,353). As noted in a preceding paragraph, characteristics of the Cycles of Abundance for hogs were quite different during the Depression years and the years before.

Milk

America's dairies provided 790 pounds (91.8 gallons of milk) per capita in 1925. In 1930, the amount had increased to 812 pounds (94.4 gallons). In 1930, 55.5%

of the whole milk produced on American farms was used for manufactured products. Included in these products were: 2.1 billion pounds of butter, 1.8 billion pounds of evaporated and condensed milk, 510 million pounds of cheese, and 255 million pounds of ice cream.

US Milk

American dairy farmers obtained 100.2 billion pounds of milk from their herds in 1930. Wisconsin farmers reported the largest share (11.2%). Other states contributing significantly to that total included: Minnesota (7.6%), New York (7.0), Iowa (5.9), and Illinois (4.6). Together these five states produced 36.4% of the total. Alabama dairies accounted for 1.1%.

Prices

After 1929, year-to-year changes in milk prices generally tracked those changes taking place in the larger, national agricultural economy of the Depression years. Year-to-year changes in the agricultural prices for all agricultural commodities in the country are depicted in Figure 12.5.

There are no data on annual prices of milk in the United States before 1924. The annual average price for milk was $2.07 per Cwt in 1930 (Figure 12.92). After 1930, with milk production increasing rapidly (Figure 12.95) prices began to decline, finally reaching $1.28 in 1932. They changed very little in 1933 ($1.30). In 1933, the Agricultural Adjustment Act (AAA) was enacted. It authorized the secretary of agriculture to enter into marketing agreements with handlers, processors and others, and to issue licenses to handlers and processors, to raise the prices of agricultural commodities, including milk (Blaney and Manchester, 2001). Production declined after 1933, and prices began to increase in 1934, reaching $1.99 in 1937. Production remained at near the same level from 1934 to 1937 before beginning to trend upward in 1938 and downward to $1.69 in 1940.

In 1935, the AAA was revised, and one of the revisions replaced marketing agreements for milk with marketing orders. The revision also included more specific standards for milk than the original act. Then, the Agricultural Marketing Act of 1937 replaced the 1935 act, and became the foundation of the modern federal milk marketing order system. Unfortunately, the provisions of the new act were inadequate in the short-term to keep prices from responding to the increase in production. However, after 1937, they seemed to be working better. Prices did not change very much between 1938 and 1940.

Milk Cow Numbers

The number of milk cows in the United States remained relatively constant, at around 21 to 22 million during the 1920s. As milk prices began to decline after 1929 (Figure 12.92), American dairy farmers, apparently began to increase herd size to maintain their income. The American herd size was 22.2 million animals in 1930, but by 1934 it had increased to 25.2 million (Figure 12.93). Prices of milk did not begin to increase until 1934. After 1934, milk cow numbers began to decline, reaching 22.3 million in 1939. With prices down after 1937, numbers increased to 23.7 million in 1940.

Milk Per Cow

In 1930, average milk per cow in the United States was 4,508 pounds. In that year, individual milk cows on California farms produced more milk per year (6,550 pounds) than those in any other state. Other leading states, with respect to the production of milk, per cow were: New Jersey (6,400 pounds) Rhode Island (6,350), Maine (5,990), and Connecticut (5,770).

Average milk per cow in the United States increased regularly in the late 1920s, and in 1929 it reached 4,579 pounds (Figure 12.94). However for the next four years, it trended sharply downward from 4,508 pounds in 1930, to 4,033 pounds in 1934. Apparently, with milk prices declining in the early 1930s (Figure 12.92), dairy farmers began to try to save money by reducing the care they provided for their animals. Likely, they either reduced the amount of feed given to each animal, or they reduced the quality of the feed. Possibly they did both. Prices began to increase in 1934, and the following year milk per cow began to increase. Milk prices declined again in 1938, 1939, and 1940, and while milk per cow did not decline, it increased very little in those three years. In 1940, it reached 4,622 pounds of milk per cow.

Pounds Produced

Total American milk production increased regularly in the mid- and late-1920s. It was 100.2 billion pounds in 1930 (Figure 12.95). Even though milk per cow declined through the early 1930s (Figure 12.94), the increase in dairy cow numbers in the country (Figure 12.92) was sufficiently large to fuel the continued increase in pounds produced. It reached 104.8 billion pounds in 1933. With only a slight increase in numbers in 1934, and with milk per cow continuing to decline, pounds fell sharply to 101.6 billion. Milk per cow began to increase again in 1935, but this increase was off-set by the continued decline in herd size. As a result, pounds continued downward to 101.2 billion. In the period 1937-1939, herd size remained fairly stable while milk per cow increased. These conditions resulted in a rapid increase in pounds. In 1940, with both herd size and milk per cow increasing, pounds reached 109.4 billion, the highest in history.

Alabama Milk

Alabama's dairies provided its citizens with 387 pounds (45 gallons) of milk per capita in 1925. In 1930, the amount increased to 430 pounds (50 gallons). In both years, per capita production in Alabama was only about half of that of the country as a whole. In 1930, Alabama dairymen and their families consumed 77.8% of the milk produced on their farms.

Prices

Figure 12.92 also includes data on annual average prices of milk in Alabama during the period 1930-1940. Year-to-year trends in prices obtained by Alabama dairy farmers were similar to those of farmers elsewhere in the country; however, throughout the period, prices in Alabama were at least 50 cents per Cwt higher. In 1930, they were 89 cents higher ($2.96 versus $2.07). This differential generally represents the increased cost of producing milk in Alabama. This difference in cost is primarily related to the lower milk per cow in the state (Figure 12.94). For example, in 1935 milk per cow in the United States was 39% higher than in Alabama (3,020 versus 4,184 pounds). That year, the average price of milk was 50% higher in the state than in the country ($2.22 versus $1.69 per Cwt).

Annual data on milk prices (dollars per Cwt) for the country and the state during the period 1924-1940 are presented in Figure 12.92. Obviously the quantity of milk produced must have had some influence on those prices; however, it appears that the most important influence was the changes taking place in the economy during the period. Data on the indices of prices received for all agricultural commodities during the Depression years are presented in Figure 12.5. These data show that the indices began to decline in 1930, finally reaching their lowest point of the period in 1932. Afterwards, they increased through 1937 when the recession of 1937 and 1938 forced them down again. Between 1938 and 1940, they remained generally unchanged. The year-to-year changes in milk prices, followed those of the indices of prices received throughout the period. Changes in Alabama milk prices followed almost the same pattern.

Alabama milk prices were $2.96 per Cwt in 1930. They declined sharply to reach $1.84 in 1932. Afterwards, they slowly trended upward to reach $2.42 in 1937. They fell to $2.26 in 1939, only to increase slightly in 1940 to $2.36.

Milk Cow Numbers

The Census of Agriculture reported that as of April 1, 1930 there were on Alabama farms 332,000 cows and heifers born before 1928 and kept mainly for milking. The largest share (3.5%) of these animals was on farms in Montgomery County. Other counties reporting large shares included: Dallas (3.4%), Marengo (3.2), Jefferson (3.1), and Lowndes (3.1). These five counties accounted for only 16.3% of the total. A total of 43 counties reported inventories of at least 4,000 animals.

There were 360,000 dairy cows in Alabama in 1930 (Figure 12.96), and herd size was growing rapidly even though milk prices were declining (Figures 12.92 and 12.96). Alabama dairy farmers were also increasing herd size to try to compensate for declining income. Numbers increased rapidly through 1934 when they reached 420,000 animals, near the highest in history. In 1935, after the AAA was revised, farmers were convinced that prices were going to get even better, and they began to sell-off their cows. By 1938 and 1939, herd size was down to 361,000 animals, essentially where it had been in 1930. In 1940, it inched-up slightly to 363,000 in response to the beginning of the war in Europe, and their expectations that both demand and prices would increase.

Milk Per Cow

In 1930, Alabama dairy farmers obtained an average of 3,150 pounds of milk from each cow, or about 70% of the national average (Figure 12.94). Then, as was the case elsewhere in the country, with prices declining in the early 1930s and with farmers in increasing economic distress, they began to cut corners in the management of their herds. Consequently, milk per cow began to decline. In 1934, it was down to 2,920 pounds. After 1934, with milk prices increasing, milk per cow also began to increase. The upward trend slowed after 1938, and declined in 1940 to 3,150 pounds.

By 1930, there were many ominous signs available to those interested in the lack of comparative advantage of Alabama agriculture. One of the best signs of this serious problem was in milk production. Throughout the period 1930-1940, milk per cow in Alabama was around 30% less than elsewhere, and if the comparison had been made with the major milk-producing states, the difference would have been much greater. Alabama's summer temperatures are simply too hot for maximum milk production. It is too hot for the genetic potential of the cow for producing milk to fully express itself. It should have been obvious well before 1924 that Alabama dairy farmers could not compete in milk production, and if they remained in the dairy business, it would be possible only through the application of substantial federal assistance.

Pounds Produced

In the early 1930s, the increase in the Alabama milk cow herd size (Figure 12.96) more than compensated for the decrease in milk per cow (Figure 12.94). Pounds produced increased from 1,134 million in 1930 to 1,224 million in 1932. After 1932, however, the magnitude of the decrease in milk per cow was great enough that it generally neutralized the continued increase in herd size with the result that pounds remained stable at around 1,230 million through 1935. Afterwards, the different combination of numbers and milk per cow resulted in a slight decline in production through 1937, followed by a slight increase in 1928 and 1929. The sharp decline in milk per cow resulted in a sharp decline in pounds in 1940 to 1,143 million. Generally, Alabama milk production remained relatively stable at around 1,200 million pounds from 1933 through 1939.

Alabama Chickens

Data published in the Census of Agriculture indicate that on April 1, 1930 the total value of chickens on farms in the United States was $321.6 million. The North Central Region accounted for the largest share (50.3%). The South Central Region accounted for the second largest share (16.4%), and Alabama accounted for 1.2% of the national total and 7.6% of the total for the South Central Region.

The Census of Agriculture reported that Alabama farmers raised 10.7 million chickens in 1929. The largest share (3.5%) was reported from Cullman County. Other counties reporting substantial shares included: DeKalb (3%), Marshall (2.7), Madison (2.6), and Blount (2.5). These five counties accounted for only 14.3% of the state's total. Some 57 counties reported productions of at least 2,100,000 birds. These data indicate the widespread production of chickens in the state in 1929.

US and Alabama Value Per Head

After 1929, year-to-year changes in value per head for chickens generally tracked those changes taking place in the larger, national agricultural economy during the Great Depression. Year-to-year changes in the agricultural prices for all agricultural commodities in the country are depicted in Figure 12.5.

Value per head for all chickens in the country was 92 cents in 1930 (Figure 12.98). Afterwards, they began to decline, finally reaching 42 cents in 1934. They began to increase in 1935, reaching 60 cents in 1940.

Value data for Alabama chickens are also presented in Figure 12.98. Generally, the year-to-year trends are the same for both, except that values for chickens in the state reached their lowest level (33 cents) in 1933, rather than 1934. Afterwards, they also followed the indices of prices received upward, reaching 50 cents in 1936, and remaining at that level in 1937. After 1938, they generally trended downward again to 42 cents in 1940. For most of the period, values for American chickens were between five cents and 25 cents higher than birds in Alabama, but were generally closer during the heart of the Depression.

Numbers

Annual data on numbers of chickens produced on Alabama farms during the period 1924-1940 are presented in Figure 12.99. These data indicate that there was considerable year-to-year variation in numbers, and there seemed to be a slight downward trend over time. Further, the Depression seemed to have little effect on numbers. For most of the period, numbers ranged between 10.5 and 12.3 million.

The numbers data presented in Figure 12.99 suggest that the production of chickens in Alabama during the period 1930-1940 was somewhat cyclic. Generally, numbers seemed to fall sharply in one year, then recover over a two-year period. The teeter-totter relationship between supply and price seemed to have little or nothing to do with this phenomenon.

Broilers

The national broiler industry had been growing rapidly in some areas of the South for several years before the beginning of World War II. In 1934, farmers in Delaware, eastern Maryland, and northeastern Virginia (the Delmarva Region) produced seven million birds for urban markets in the East. The production of poultry that had been a housewife business from colonial times became a rapidly growing commercial enterprise when feed dealers decided to provide chicks and feed on credit to farmers who wanted to enter the business. From the beginning, only a relatively few farmers in any single state became involved in this production-credit arrangement. However, in some limited areas, such as the Delmarva Region, the industry quickly began to contribute significantly to total farm income. USDA did not begin to collect information on broilers separate from chickens until 1934. That year, total production in the country was 34 million birds. The first data on broiler production in Alabama was reported by the Alabama Cooperative Crop Reporting Service in 1939. They reported that Alabama farmers produced 1.3 million broilers weighting 3.1 million pounds. Farmers received an average of 20.7 cents a pound for their birds. In 1940, production increased to 1.7 million birds, weighing 4.1 million pounds. The average price increased to 21 cents.

Eggs

The 1930 Census of Agriculture published data indicating that in 1929, the total value of all eggs produced in the United States was $799.3 million. The North Central Region accounted for the largest share (45.9%). Shares of the other regions were: South Central (15.7%), North Atlantic (15.5), West (13.9), and South Atlantic (8.9). Among the states, the largest share (6.4%) was received by farmers in California. Other states obtaining large shares of the total included: Iowa (6.3%), Missouri (6.0), Ohio (5.4), and Pennsylvania (5.4). These five states accounted for only 29.5% of the total. Farmers in Alabama received 1.3%.

US Eggs

Eggs produced in the United States in 1925, 1930, 1935, and 1940 were sufficient to provide consumers with 302, 313, 262, and 301 eggs, respectively, per capita.

Prices

After 1929, year-to-year changes in egg prices generally tracked those changes taking place in the larger, national agricultural economy during the Depression. Year-to-year changes in the agricultural prices for all agricultural commodities in the country are depicted in Figure 12.5. There are no annual data on egg prices before 1924. Data presented in Figure 12.100 indicate that national egg prices closely tracked year-to-year changes in the indices of prices received (Figure 12.5) for all agricultural commodities during the period 1930-1940. Both fell sharply after 1929, reaching a low-point in 1931 and 1932 before beginning to increase. Indices continued to increase through 1937, but after 1935 egg prices began to decline. In 1934, prices and numbers (Figure 12.101) seemed to re-establish their teeter-totter relationship. In that year, with numbers declining, prices began to increase. Then, in 1936 numbers began to increase, and prices began to decline.

Numbers

Nationally, egg numbers trended downward from 39.1 million in the early years of the Depression, finally reaching between 33 and 34 billion in 1935 (Figure 12.101). Afterwards, numbers generally trended upward to 39.7 million in 1940. Numbers had fallen slightly between 1937 and 1938 as result of the recession. During the first half of the decade, numbers seemed to be only roughly related to prices. They did trend downward as the

Depression deepened, but did not begin to increase until the recovery was well underway (Figure 12.1). Further, producers did not seem to notice that prices were falling rapidly in 1936.

Uses

As detailed in the preceding paragraph, egg numbers declined by 14% between 1930 and 1935 (Figure 12.101); however, the percentage of total numbers of eggs produced that were sold declined only slightly during the period. Eggs sold were 79% in 1930, but the percentage declined to only 76% in 1932, 1933, and 1934. By 1940, the percentage had climbed back to 79%. The increase in eggs sold during the economic recovery came at the expense at both eggs consumed and hatching eggs. In 1940, they were 20% and 1%, respectively.

Egg consumption per capita had begun to decline after 1927, before reaching 329 in 1930 (Figure 12.102). Afterwards, consumption generally followed numbers downward, and by 1935 it was down to 278. Many people simply could not afford to purchase eggs during that period. With the worst of the bad times over after 1935, consumption began to increase, reaching a level of 317 in 1940. The increase in consumption slowed, during the recession of 1937 and 1938, but it did not decline.

Alabama Eggs

According to the 1930 Census of Agriculture, farmers in Baldwin County reported the largest share (4.8%) of eggs produced in Alabama in 1929. Other counties with large shares included: Marshall (3.6%), DeKalb (3.5), Cullman (3.5). and Jackson (3.1). These five counties accounted or only 18.5% of the state's total. A total of 25 counties reported productions of 6 million or more eggs.

Prices

Data on average annual prices of American and Alabama eggs during the period 1924-1940 are presented in Figure 12.100. The data show that year-to-year trends were quite similar for the country and the state; however, in most years Alabama prices were a few cents per dozen lower.

Numbers

There are no annual data on the number of eggs produced in Alabama before 1924. The Great Depression had little effect on egg production in Alabama (Figure 12.103). Numbers were 457 million in 1930, and increased to 483 million in 1931. Thereafter, there was little change until the recovery was well under way. Between 1937 and 1939, it increased from 490 to 529 million; then unexpectedly fell to 486 million in 1940. Alabama egg producers seemed to be dancing to a completely different tune during the period. The estimate for 1935 (433 million) seems to be an error, but it was reported like this in Bulletin 1 of Alabama Agricultural Statistics, published in 1947.

Uses

The Depression had a significant effect on the way Alabama farmers used the eggs produced on their farms. From 1930 through 1937, eggs sold generally closely tracked the year-to-year changes in the Gross National Product (Figure 12.1). When percentages sold returned to near their pre-Depression level of around 52% after 1936, they remained near that level through 1940. During the period, percentages declined from 52% in 1929 to 44% in 1932. Then by 1937, it had returned to 52%.

In the late 1920s, farmers and their families consumed about 42% of the number produced. However, in 1930 the pattern of use began to change. The demand for eggs began to decline. Eggs sold declined from around 53% to around 45% during the heart of the Depression. At the same time, eggs consumed increased to around 50%. By 1935, economic conditions began to improve somewhat, and egg use returned to its pre-Depression pattern.

Apparently there were no data collected on the number of eggs used for hatching on Alabama farms during this period. In the United States, during the 1930s farmers used only 2% of total egg numbers for hatching.

Agricultural Production

Annual data on total value added to the US economy for crops and livestock provide a useful measure of changes in agricultural production over time. These values for the period 1915-1940 are presented in Figure 12.105. They clearly show the effects of changes in prices received for agricultural commodities (Figure 12.5) during this period of extreme economic contraction (Figure 12.1). Total value added was $12.9 billion in 1929 at the end of the Roaring Twenties and before the stock market crash had time to affect it very much. Afterwards, it closely tracked the indices of prices received, first downward to 1932 ($5.8 billion) then upward to 1937 ($11.1 billion). After the recession of 1937 and 1938 slowed the economic recovery (Figure 12.1), in 1938 the Gross National Product began to increase rapidly, once more. Unfortunately, prices received and total value added had hit a brick wall. Total value added remained near its 1935 level through 1940 ($9.9 billion).

Data on value added by crops and livestock separately during the period 1930-1940 are also presented in Figure 12.105. Year-to-year trends for both were similar, and generally also followed the trends of the indices of prices received throughout the period. Values for crops generally ranged between $2.4 and $4 billion. The unexpectedly high value in 1935 was the result of an unusually large value for cotton that year. The even greater deviation from expected in 1937 was the result of larger than expected values for both cotton and food grains. Value added for livestock ranged between $3.5 and $6.3 billion. In most years values for livestock were $1 to $2 billion greater than those for crops.

Data on the value added on a per farm basis of crops and livestock used (consumed) by American farm families during the period 1915-1940 are presented in Figure 12.106. Trends in value added per farm for livestock from year to year were very similar to those of indices for prices received during the period 1930-1940. The Depression obviously affected values for crops as well, but not nearly to the same extent. These data indicate that over the years home consumption for American farm families is slowly declining. Near the end of World War I, they consumed about $260 worth of livestock and livestock products that they had produced on their farms. By 1940, it was down to around to $120. Those values for crops were $130 and $70, respectively.

Data on value added by individual crops and groups of crops for 1925, 1930, 1935, and 1940 are presented in Table 12.11. In all of the four years included, cotton remained the most valuable crop produced in the United States. Although there were significant, year-to-year changes, food grains, feed grains, fruits and tree nuts, and vegetables generally contributed around the same value in all of the four years. Value added for all of the crops show the effect of the status of all commodity prices in the heart of the Depression. Values for 1933 and 1934 likely would have been even lower.

Table 12.11 also include data on the value added by groups of livestock and by livestock products. Of course, meat animals, which includes both cattle and hogs, contributed the largest share of total value added. Dairy products and poultry and eggs, in that order, contributed slightly lesser amounts.

Data on the total cash receipts from the sale of selected crops and livestock by Alabama farmers during the period 1924-1940 are presented in Figure 12.107. These data also reflect the roller-coaster changes in prices received (Figure 12.5). Certainly, there must have also been changes in the quantities marketed, especially in the heart of the Depression when demand was lower. However, it is likely that the largest portion of the changes were the result of lower prices. Total receipts were around $60 million during the period 1931-1933 as the Depression deepened. With the economic recovery underway, total cash receipts increased to $110-$120 million in the period 1934-1937. After 1937, they generally followed the indices of prices received downward to near $90 million in 1940.

Figure 12.107 also includes data on cash receipts for crops and livestock. The data also show that most of total receipts were from the sale of crops. Obviously, cash receipts for livestock were much lower than for crops throughout the period. However, the difference was nearly as great during the heart of the Depression and in the late 1930s. This large difference reflects the continued overwhelming importance of cotton and cotton prices to Alabama's economy.

Data on the cash receipts (on a per-acre basis) of crops and livestock and products consumed by Alabama farm families during the period 1924-1940 are presented in Figure 12.108. As expected, because of higher prices per unit consumed, receipts for livestock consumed were higher than for crops. The value for a meal of sausage was greater than the value of a meal of corn-pone. Year-to-year changes were quite similar to those now familiar changes in the indices of prices received. The effect of the Depression on the consumption of crops was not nearly as dramatic as for livestock. During the 1920s, the values for home consumption in Alabama were considerably lower than in the country as a whole, but by the end of the Depression the difference had declined considerably.

Table 12.12 presents data on the cash receipts of some specific crops, livestock, and livestock products on Alabama farms in several years during the period 1925-1940. These data show that in all of the years cash receipts for cotton were greater than for any of the other crops. The dominance of cotton, however, generally declined over the years, and in 1940 cotton receipts were not much greater that corn receipts. Except for cotton and sweet potatoes, receipts for all of the crops remained relatively stable in all of the years.

Data on the cash receipts for horses and mules showed that as a group these animals were still the most valuable of all the livestock on Alabama farms. Further, because the receipts for horses were relatively low, those for mules alone were greater than for any other animal. It is also noteworthy that the receipts for horses and mules tended to increase in all of the years, except in the heart of the Depression. It is obvious from these data that animals still powered Alabama farms as late as 1940. The increase in the receipts of "all" cattle is also significant. Alabama farmers finally seemed to have found an animal crop that might compete with cotton in its agricultural economy.

STILL SERIOUSLY SEEKING SALVATION

In 1930, the farm population in Alabama accounted for slightly more then 50% of the total population. The national average was less than half of that percentage. There were 261,000 farms in the state; 65% of them were operated by tenants. In 1929, farmers had harvested crops from only 27.6 acres of land. The average price of cotton was 9.5 cents a pound. The average value of farm products raised on Alabama farms was $856, the lowest among 11 Southern states (Table 12.13).

By 1940, some farm conditions in Alabama had improved; some had not. There were 20,000 fewer farms in the state, and tenancy was down to 59%. In 1939, farmers in Alabama harvested crops from 7.3 million acres (averaging 30.7 acres per farm). Included was two million acres of cotton, 25.9% of total acres harvested. The average annual price of cotton was 9.2 cents. Finally, in 1940, the average value of all farm products on Alabama farms was $349, still the lowest in the South. The average for 11 Southern states was $711 (Table 12.13).

In 1933, the Southern band-of-brothers had begun to push mightily to save the American farmer, but in 1939 and 1940, it was obvious that either they had not pushed hard enough or that they were pushing in the wrong direction. Farmers, especially in Alabama, were still seriously in need of salvation.

Salvation Through Better Information

As noted in a preceding chapter, even George Washington was convinced that better information was the key to better farming in America. In the latter part of the nineteenth century, the federal government and many of the states had begun to move forcefully to provide it. By the mid-1930s, hundreds of millions of dollars were being expended nationally for that purpose—a substantial amount of it in Alabama.

Budget information for the Alabama Agricultural Experiment Station could not be located earlier than for fiscal year 1939-1940. In that year, it budgeted revenues for agricultural research totaling $443,500. Of this total, $162,300 was provided by the Alabama legislature and $172,700 was provided by USDA. An additional, $25,400 came from grants and gifts, and $73,000 from the sale of commodities produced through field research.

In 1933, there were 52 full- and part-time employees working for the Alabama Agricultural Experiment Station on the Auburn campus (Kerr, 1985). An additional seven were employed on the various substations across the state. As a result of the federal farm legislation enacted in 1933, the experiment station was able to employ several additional staff members. In 1938, there were 62 full- and part-time employees on campus and nine on the substations. There were eight departments in the College of Agriculture in 1933. That number remained unchanged in 1938.

Data on the percentages of funds from all sources budgeted by the Alabama Cooperative Extension Service for extension projects in fiscal years 1928-1929, 1934-1935, and 1939-1940 are presented in Table 12.14. Extension work in the states was to be conducted on a cooperative basis; this meant that the individual counties were expected to provide substantial funds to support it. Total funds budgeted for extension declined from $621,5000 to $511,400 between 1928-1929 and 1934-1935. Federal funds actually increased slightly during the period, but as a result of the Depression, state and county funds declined and had not returned to the 1928-1929 level even in 1939-1940. Fortunately, federal funds more than doubled between 1934-1935 and 1939-1940.

As might be expected, the largest share of extension funds was allocated to work in the counties (county agents, home demonstration agents, and club work agents). After all, this is what extension work was about. In the three periods, percentages of total funds allocated to these three groups of agents declined from 80.4% in 1928-1929 to 73.4% in 1939-1940 (Table 12.15). In the three periods, work of the county agents received about 61% of the funds, while home demonstration and club work received around 36% and 3%, respectively. Individual projects of the subject matter specialists generally received from 1% to 3% of the funds, respectively.

Morrison (1996) commented that in 1932 Walter Gravitt (a county extension agent) and Lucy Upshaw (a county home demonstration agent) were the two employees of the Agricultural Extension Department of the State Agricultural and Mechanical Institute for Negroes at Normal. In 1938, a brooder and incubation house was added to the agricultural facilities available to the faculty and students of the Department of Agriculture.

According to Mayberry (1989), in 1940 Tuskegee dedicated the Agricultural Extension Building. It was constructed with Works Progress Administration (WPA) funds to provide a headquarters building for farm and home demonstration work in seven Southern states.

Salvation Through Organization

Only a single organization working in the area of food and fiber production in Alabama was established during the period 1930-1940. The State Commission of Forestry had been established in 1923, but it was abolished by the Department of Conservation Act of 1939. Under that act the work of the commission was transferred to the Forestry Division within the Department of Conservation.

Salvation Through Diversification

Progress on diversification was still painfully slow in Alabama. In 1925, Alabama farmers harvested cotton from 44.4% of total cropland harvested. In 1930, the percentage had increased to 50.1%. Then in 1935, as a result of AAA programs, the cotton share was down to 29.4%, and in 1940 it was still lower (27.1%). Alabama farmers finally seemed to accept the idea of diversification, but in reality they finally realized that they had no other choice. Most of them had embraced the New Deal with high-hopes. Their expectations seemed to be realized in 1934 when the price for cotton increased to 12.1 cents, and it was 12.2 cents in 1936. Unfortunately, it fell to 8.5 cents in 1938.

No one realized it, at the time, but 1938 represented a new-day in cotton production and diversification in Alabama. In 1937, Alabama farmers had harvested the crop from 2.7 million acres. In 1938, acres harvested was down to 2.1 million. A decade later, it would be down to 1.5 million. King Cotton had finally been taken with a severe case of the dwindles.

Salvation Through Political Action

Landed farmers, especially in the South, had been politicking since the late nineteenth century to get the government to provide financial support for their operations (the Sub-Treasury Plan). Finally, in 1933 their efforts came to fruition. As noted by Ernstes, et al. (1997), a Southern juggernaught was firmly established in Congress in the election of 1932, and when it took control of the agriculture committees and subcommittees in the 73rd Congress, it lead the way in getting the government involved as never before.

Franklin D. Roosevelt, a patrician urbanite, first visited Warm Springs, Georgia, in 1926, and he returned to the small town periodically until his death there in 1945. In his trips between, first New York and then Washington, DC, he had numerous opportunities to view first-hand the conditions of much of Southern agriculture and its people as they approached and entered the period of the disastrous Great Depression. Doubtlessly, this exposure to the ever-worsening economic conditions and human misery that he observed along the route through rural Virginia, North and South Carolina, and Georgia, likely shaped the proposals put forth by his first administration to try to alleviate the appalling conditions he had witnessed.

The other part of the equation that would play a leading role in the government's response to the ever-worsening conditions in American agriculture was the so-called Southern band-of-brothers (Ernstes's Southern juggernaught). These congressmen had grown-up with the ever-worsening problem in the South, and in a sense they were part of it. Many of them were descendants of those people who had first brought slaves to the region, who had pushed for Secession, who had lost billions of dollars in the Lost Cause and in the freeing of their chattels, who had taken over the reins of state government after the end of Reconstruction, who had been party to the disenfranchisement of poor blacks and whites in the early twentieth century, and who either would not or could not provide opportunities that would have moved hundreds of thousands of redundant farmers off the land. These congressmen, who would play a leading role in putting government into the marketplace of American agriculture and in determining the role that it would play there, were part of a unique political environment. A large share of those who were suffering most in the environment, either could not or would not vote. The primary political constituents of those who would help write and enact Roosevelt's farm program were landed cotton farmers who had never been able to find a suitable solution to their farm labor needs. Slavery had never worked very well, and share tenancy, had not either. While those landed politicians did not have to endure the depth of suffering, deprivation, and out-right misery of their share tenants, they did suffer in their own way. With cotton below 10 cents in 1930, 1931, and 1932, their share of cash receipts was for all practical purposes nothing. They were slowly drowning in their own increasing indebtedness.

The crux of the country's agricultural problem at the beginning of the Depression was that there were hundreds of thousands of farms and farmers producing food and fiber for which there were no buyers. The problem was extremely serious all over the country, but in the South it was catastrophic. The problem of supply exceeding demand had been growing in the South for over 300 years. The Depression simply high-lighted it. The problem had originated in British mercantilism and European nicotine addiction in the early seventeenth century when the Virginia Colony began the production of tobacco with slave labor. Thereafter, for about 200 years virtually all colonial investment went into creating more and more agricultural jobs (slaves), and nothing was invested in diversifying agriculture or in industrialization. This uneven pattern of investment persisted widely in the region until well after the end of the Civil War.

The problem of excess agricultural labor and its resultant excess production was already a problem in the late 1920s, but it began to get even worse as economic conditions in the country deteriorated, and more people were forced back onto farms. The American farm population was estimated to be near 30.6 million in 1929, but by 1933 it had increased to almost 32.4 million.

It was amid this extremely complex array of problems in the country's agriculture that the Roosevelt administration, the Southern band-of-brothers, and the 73rd Congress began their work. The response of the federal government to the agricultural crisis of the Great Depression was the enactment of the Agricultural Adjustment Act of 1933. In effect, the act provided federal funds to pay farmers not to produce corn, wheat, cotton, rice, peanuts, and tobacco; as well as milk, butter, pigs, and lambs. Note that four of the crops were mainstays of Southern agriculture.

The act was passed in May after the normal farming season was well under way. Consequently, there was no opportunity to make sweeping changes. However, many farmers were paid to plow-up around a quarter of their 1933 cotton acreage, and six million pigs and 220,000 pregnant cows were slaughtered. Farmers who participated in the AAA programs received their first checks in late 1933. They averaged just $19 each.

After the provisions of the act were implemented, indices of prices received did begin to increase. The index had been 65 in 1932, but increased to 70 in 1933, and by 1937 it had reached 123 (Figure 12.5). Indices for prices paid also began to increase, but not as rapidly. As a result, the parity ratio reached 0.94 in 1937 (Figure12.6). It had not been that high in several years. Further, net farm income per farm increased from $300 in 1932 to $900 in 1937.

It cannot be assumed that all of these seemingly beneficial changes in American agriculture after 1933 were the result of the implementation of the provisions of AAA. By 1934, the American consumer was returning to the market. The Gross National Product began to increase rapidly in 1934 (Figure 12.1). Between 1934 and 1937, non-farm employment increased from 24 million to 31 million. This improving employment situation served to pull some of those redundant farmers out of agriculture. The estimated American farm population was near 33.4 million in 1933, but by 1937 it was down to 31.3 million. It has been suggested that by the time that AAA went into effect the Depression was ending, and that the benefits were the result of the expanding economy. Supporters of this position cite the fact that while cotton and corn prices did improve after the enactment of AAA they both declined in 1937 at the beginning of the recession, and that they both remained near 1933 levels until the economy began to recover after the war in Europe began. After 1937, AAA seemed powerless to keep those prices up. Similarly, as noted in the preceding paragraph, net farm income per farm also increased after the enactment of AAA. However, when the recession began in 1937, it fell back to about where it had been in 1930.

When the first AAA checks were distributed in the fall of 1933, thousands of share tenants received little or nothing. Many of the landowners simply kept the tenant's share as payment against cash advances that the tenants had been given. The payments had been a godsend to landowners who had extended credit, but share tenants received almost nothing. By 1934, there was widespread criticism by social reformers and some government officials on the distribution of AAA benefits.

There was really nothing new about this. Landowners had always received the largest share of returns, because they provided the larger share of total production costs. The enactment of AAA did not change this age-old relationship or its consequences, which dated back, at least, to the manorial system of medieval Europe when the lord of the manor took a large share of what his serfs produced.

As might be expected in a free enterprise system, landowners acting in self-interest soon understood that if they ridded themselves of their share tenants, increased income would be available for them. Without share tenants, they would receive all of the land diversion payments without having the misery of dealing with tenants. Throughout the winter of 1933-1934, increasing numbers of share tenants were forced off their farms, mainly in the cotton regions. In Alabama, the first noticeable decrease in the number of farms (274,000 to 273,000) took place between 1935 and 1936. The future of this huge throng of miserable folk became increasingly grim as the federal-intervention "camel" thrust its nose further and further into the "tent" of Southern cotton farming.

AAA administrators made a half-hearted attempt to prevent share tenants from being forced off the land and to see that benefit payments were shared equally, but the agency's timid action failed completely. By 1935, after three years of AAA, it was clear that the program was essentially run by and for landowners. The benefits were programmed, probably unintentionally, to go to farmers with larger amounts of land and relatively large amounts of production. There was immediate concern among some agricultural reformers that so much federal money was going to farmers who needed it least. It seemed to them that while most of the debate surrounding the need for passage of the AAA legislation had centered on the desperate plight of the share tenants that were receiving almost no help from it. As a result, there were discussions about the need to cap the amount that any farm could receive. This restriction was never adopted, because of the need to obtain the full cooperation of the farmers with the largest holdings. The primary purposes of the AAA legislation was to reduce surpluses in order to increase market prices and farm income. While the largest farms were relatively few in number, their production was far out of proportion to their number. Surpluses could not be reduced without their participation, and they were not likely to take any acreage out of their highly productive farms without receiving a proportionate share of the payments.

The federal government tried several other approaches during the depth of the Depression to solve the problem of the abject poverty among share tenants especially African Americans in the South. Of all the classes of farmers, it was the blacks who felt the winnowing process most intimately, and they responded by leaving in large numbers. In 1930, there were 840,000 black farmers in the South. By 1940, that number had been reduced to 652,000. Decades would pass before they would begin to return, and then they would not return as farmers.

The federal government might have done more to assist those being forcibly driven from agriculture and to have reduced the terrible privation associated with it; however, the dominant economic, social, and political views of the time precluded any such assistance. Congress did pass the National Industrial Recovery Act of 1933 and the Federal Emergency Relief Act of 1935, and Congress created the Resettlement Administration and the Farm Security Administration, both efforts to provide some help to the share tenants being forced off their farms. However, the extremely limited successes of all of these efforts increasingly began to convince even the most die-hard advocates of federal assistance that there was no practical way to turn large numbers of share tenants into independent farmers with acceptable incomes.

The problem that AAA could not solve had been with Southern share tenants for generations—a 100% increase of nothing is nothing. The farms of the share tenants were, and always had been, too small. Even if they plowed up all of their acreage of cotton or held all of their land out of production, they received so little money that it made very little difference in their economic situation or to the economic condition of the country. Their farms were too small, and there were simply too many of them. The New Deal was in over its head in trying to solve those two related problems.

The New Deal's early efforts did little to solve agriculture's twin problems of over-supply and low prices, and they did virtually nothing for hundreds of thousands of tenant farmers. New Deal efforts, however, did have a significant effect on American agriculture. Federal payments to the big landowners were sufficient to allow them to continue or to begin the modernization of their farms. Also, the various government programs significantly reduced the landowners' costs of credit and their financial risks. The changes did not occur immediately but over the years. Because they had more money, they began to use more tractors, apply better fertilizers to their crops, use more chemicals to control pests, and improve their breeds of livestock—while at the same time taking better care of them. All of these changes increased their efficiency, along with their production. Fewer farmers could now produce more food and fiber. Farm modernization also meant changing the relationship between land and people. It meant the organization of larger farms, the increased use of capital, a different way of using labor, and above all improved business and management skills.

By 1939, the year World War II began in Europe, landed farmers in the United States had probably gotten more than they expected from political action, but less than what they wanted. They were being paid rather well for producing less cotton, corn, wheat, rice, peanuts, tobacco, milk and butter, and fewer pigs and lambs, but they never wanted to produce less. What they really wanted and what they had wanted since they had failed to get the Sub-Treasury Plan at the end of the nineteenth century, was the opportunity to produce as much as they wanted of everything at an acceptable, government-guaranteed price.

While the landed did not get everything that they wanted in the early years of AAA, they got something even better, or so they thought. They got an alliance (the Triangle) that would stand the test of time. The Triangle would give them everything that they wanted, and more, over the next 75 years. Unfortunately, as we shall see in the following chapters, while the Triangle gave them virtually any legislation that they wanted, the federal government could never give the landed farmers the agriculture they really wanted: to be able to realize a return from the investment of capital and labor on their farms equal to the return they could realize if they applied equal resources to an off-farm business.

On balance, with all of its problems, direct governmental involvement in the marketplace of American agriculture might have died an early death except for the fact that the politically powerful landed farmers, especially in the South, loved it. In all of history, Southern cotton farmers had never had it so good—guaranteed reasonable prices, with no slaves and fewer tenants with whom to share their income. With their close and direct involvement and with the Southern band-of-brothers in the Triangle, anything they wanted was what they got. Unfortunately, the primary thing that the Triangle accomplished was to prop-up a regional agriculture system with almost no comparative advantage and to slow the regions desperately needed move toward increased industrialization.

The Bottom Line

Net farm income represents the bottom line for American farmers. It represents the farm operator's share of the net value added to the national economy by agriculture. As in preceding chapters, total annual national net farm income data have been converted into dollars per farm. This conversion comes close to measuring the net income of the individual farm operator. These converted data for the period 1915-1940 are presented in Figure 12.109. Net farm income per farm was $65 in 1930. It was trending downward from the higher levels of the latter part of the 1920s. Afterwards, it closely tracked trends in the indices of prices received for the remainder of the period (Figure 12.5). The indices for prices paid remained relatively stable during the period. Net income per farm continued to decline through 1932 when it reached $304. It is difficult imagine from the vantage point at the beginning of the twenty-first century that the annual farm operator's share of agriculture's gross income could have ever fallen to near $300 per farm.

Fortunately, net farm income per farm began to increase after 1932 and reached $774 in 1935. As a result of a severe drought throughout much of the eastern portion of the country, it fell to $640 in 1936. In 1937, it reached $900. In 1938, indices of prices received fell sharply and remained at much lower levels in 1939 and 1940. In 1938, net farm income per farm fell to $668 and remained near that level through 1940.

There are no data available on net farm income for the individual states before 1949; however data published by Fite (1984) provides some insight into the income situation of Alabama and Southern farmers during the Depression. Data in Table 12.13 show the approximate value (dollars per farm) of farm products in several of the states of the South Central Region and of Iowa in 1929 and 1939. These values were likely to have been considerably larger than net farm income. In 1929, the value for Alabama was the lowest ($856) of the seven states listed. Even Mississippi ($910) and Arkansas ($988) were higher. Unfortunately, the situation in 1939 near the end of the Depression was even worse. In the 1930s, value in Alabama declined by 59% to $349. Values for all the states listed declined during the period, but the rate of decline was greater for Alabama than for any of the other states. The $349 is an average. One can only guess what the values were for farms that were below average. Some of those must have been well below $200.

New Deal agricultural programs added a new dimension to American agriculture—government payments. Many Americans probably agreed that payments to destitute farmers were entirely justified in the depths of the Depression. Unfortunately, once Pandora's Box was opened, it could never be closed again. The Triangle could always make a case for their continuation. In the beginning, they were fairly innocuous. In 1933, the average payment was only $19 per farm (Table 12.110). In the first full year (1934) of AAA payments, farmers received an average of $66 per farm. They increased to $84 in 1935. Then they were reduced ($41), as the economy began to recover (Figure 12.1). Unfortunately, the recovery of the economy was interrupted by the 13-month recession of 1937-1938. As a result of the recession, payments were increased to $51 in 1937 and to $68 in 1938. By 1938, AAA had completely lost control of agricultural production in the country, and supplies of all commodities exploded. As a result, the parity ratio fell sharply. Part of the increase in payments in 1938 was designed to compensate farmers for this decline in their income. The ratio continued to decline in 1939, and payments were increased to $118. The ratio improved slightly in 1940, and payments were reduced to $114.

Alabama farmers began to receive checks (payments) from the federal government in 1933 under the provisions of the Agricultural Adjustment Act, but there is relatively little information available on the amounts they received. Brackeen (1943) published information on the number and value of checks received by farmers in the state in 1940. His data are presented in Table 12.16. Over one-third of the checks were for less than $20.01, and almost two-thirds were for less than $40.01. These were pitiful amounts of money for farmers in such dire straights, but for a farmer with farm products valued at $300, a $40 check probably would seem like a godsend.

These statistics, as depressing as they are, do not convey in any sense the misery and deprivation that they represent among the people in Alabama, especially the thousands of tenant farmers, both black and white, during the Depression. Flynt—in his 1989 *Poor but Proud* and the chapter he prepared in the 1994 *Alabama-the History of a Deep South State*—clearly and sympathetically describes the conditions that his family and most Alabama families had to live through in those years. Fite, in his 1984 *Cotton Fields No More*, wrote of the deplorable conditions of families throughout the South. He commented that surveys revealed on over 600 plantations in the South, the average annual net cash income of share-cropper families was $122 in 1934, and in some areas it was as low as $10 to $20. He wrote that untold numbers of farmers and their families eked out an existence in their constantly deteriorating, weather-beaten and leaky shacks eating mainly fat meat, cornmeal, and molasses (the three Ms) and wearing patched and ragged clothes. "Lethargic because of dietary diseases and ground down by current hopelessness, millions of southern families, did not have a living but a mere subsistence." Fite recounted comments made by Renwick C. Kennedy, who had lived in Camden, Alabama, all of his life. Kennedy commented, "homes without a match or a cake of soap, men too weak from hunger to work, naked children, people taking their meals from blackberry bushes and plum thickets, tattered cotton rags for winter clothing." Of course, conditions were not this bad for all farmers, but they were the way-of-life for altogether too many. Wherever salvation was, it was a long way from Alabama.

13

Salvation in the Tents of Mars, the God of War

1940-1950

PRECIS

World War II enabled American agriculture to pull itself out of the Great Depression. It was not the Triangle of farmers, politicians, and bureaucrats and the New Deal programs that resolved American agriculture's long-standing problem of supply exceeding demand— although the federal system of subsidies and parity would become an enduring characteristic of American agriculture. Rather it was the unprecedented demands of war that saved the American farmer. Efforts to limit supply were essentially forgotten, and American farmers produced for the Home Front, War Front, and allies. Unfortunately, Southern farmers were only able to increase cotton production; with other agricultural products, they failed to produce even enough to meet internal needs of the region. This was due, in part, to the South not having solved, or even seriously addressed, the labor supply issue. Not only did Southerners rally to the flag and military service, but a significant part of the labor force, including African Americans, left the land to work in defense industries, often outside the South. (Many, especially blacks, never returned to the land; yet even at the end of the decade, 41% of Alabama farms were operated by tenants.) Mechanization of Southern agriculture, which could have helped alleviate the labor problem, lagged far behind the rest of the country. Of course, all those natural adversities (Ultisols, erratic rainfall, less than optimal weather, etc.) that had always plagued agriculture in much of the South, and especially in Alabama, persisted. Although Alabama farmers still lagged behind those elsewhere in the South in agricultural productivity when the war ended—and, of course, Southern farmers lagged behind those elsewhere in the country—few lived in abject poverty.

INTRODUCTION

When World War II erupted in Europe in September of 1939, the Great Depression was mostly over. The Gross National Product was well above where it had been in 1930, and it was growing rapidly. Non-farm employment was also higher, but all was not well in American agriculture. Prices of cotton, corn, wheat, oats, and soybeans were far below where they had been in 1929. Net farm income per farm and the parity ratio were about where they had been just after the short recession of 1921.

The great majority of Southern farmers were still deeply mired in the Depression. There had been a slight decline in the level of tenancy, but acres harvested per farm were continuing to decline. Cotton was still king, but his majesty was very, very sick. Even with the millions of dollars that the government had spent to improve the plight of cotton farmers, carry-over surpluses were enormous. At the end of 1939, government warehouses were filled with 12 million bales of unsold cotton, which amounted to almost a full year's production. The price was hovering below 10 cents. Further, New Deal price supports had made American cotton less and less competitive in the international market.

American agriculture had suffered mightily in the Depression, and it had not completely recovered. It was, however, about to be drenched with a large, powerful dose of tonic. An event powerful enough to pick it up and shake it thoroughly was about to begin.

NATIONAL POLITICS

Throughout the 1930s, American politics had been deeply enmeshed in dealing with an event unprecedented in the country's history—the Great Depression. Before the United States, its government, and its economic sector could fully comprehend the completely new world of federal government involvement in the private sector, the country found itself even more deeply enmeshed in an event unparalleled in world history—World War II.

The Presidency

The Democratic Party controlled the Office of President throughout the period 1940-1950. Franklin D. Roosevelt (D, NY) (1933-1945) remained president until his death in 1945. He was replaced by Vice President Harry Truman (D, MO) (1945-1953).

Secretaries of Agriculture

Henry Wallace (D, IA) was serving as secretary of agriculture at the beginning of 1940, but later in the year he was replaced by Claude Wickland (D, IN) who held the office until 1945. When Truman became president, he appointed Clinton Anderson (D, NM) to the office. Then in 1948, Charles Brannan (D, CO) was appointed to the position. He served until 1953.

The Senate

The Democrats controlled the Senate from 1940 until 1947 when they were replaced by the Republicans for two years (1947-1949). Then in 1949, the Democrats regained control of that body for the remainder of the decade (Table 13.1).

Senate Majority Leaders

Alben Barkley (D, KY) was elected to the position of Senate majority leader in 1939, and he continued in the office until 1947, when Lewis White (R, ME) replaced him. White remained in the position through the 80th Congress (1947-1949). With the Democratic victory in 1949, White was replaced by Scott Lucas (D, IL). He remained Senate majority leader until 1951.

Chairmen of the Senate Committee on Agriculture

Ellison Smith (D, SC) became chairman in 1933, and he served in the position until 1944, when he was replaced by another Southerner, Elmer Thomas (D, OK). Thomas remained chairman until 1947. Arthur Cooper (R, KS) became chairman in 1947 after the Republican victory, but he served only a single term. Elmer Thomas was elected to the position again in 1949, and he served until 1951.

The House of Representatives

Control of the House mirrored that of the Senate. The Democrats controlled it from 1940 until 1947, the Republicans from 1947 to 1949, and the Democrats for the remainder of the decade (Table 13.1).

Speakers of the House

William Bankhead (D, AL), served as speaker of the House in the 75th Congress (1937-1939), and he remained in the position through the 76th. He died in 1940, and was replaced by Sam Raburn (D, TX). Raburn remained speaker until 1947, when Joseph Martin Jr. (R, MA) was elected to the office. He only served a single term, and in 1949, Raburn again became speaker. He served in the office until 1963.

Chairmen of the House Committee on Agriculture

Marvin Jones (D, TX) was elected chairman in 1931, and he held the office until 1941, when he was replaced by

Hampton Fullmer (D, SC). Fullmer remained chairman until 1945. He was replaced by John Flannagan Jr. (D, VA). Flannagan only served a single term before being replaced by Clifford Hope (R, KS). Hope also served only a single term, and in 1949 he was replaced by Harold Cooley (D, NC). Coolley remained chairman until 1955.

The Political Situation

Roosevelt was overwhelmingly re-elected in 1936 in his contest with Alfred Landon (R, KS). By 1938, the essentials of all the New Deal programs in agriculture, industrial policy, and organized labor were in place. Through the day-to-day effects of these programs, a majority of the American public seemed to slowly accept the idea that the federal government was ultimately responsible for the people's welfare, employment, and security (Morison, 1965). European socialism was slowly settling in. It was in this general environment that American politics had to face the beginning of the war between Japan and China in 1937 and the annexation of Austria and invasion of Czechoslovakia by Germany in 1938. Then, in September of 1939, German forces invaded Poland, and two days later Great Britain and France declared war on Nazi Germany.

Roosevelt responded to the beginning of these hostilities by adopting his short-of-war policy. It consisted of three objectives: (1) provide England and France with enough war materiel to allow them to keep fighting, (2) gain time to allow the United States to re-arm itself, and (3) restrain Japan through diplomacy and naval deterrence. Then with the Germans advancing rapidly toward Paris and the British Expeditionary Force being evacuated from Dunkerque, Roosevelt announced that the United States was no longer neutral, but merely non-belligerent. With this changing attitude in the presidency, the country began to prepare for war. Then in September 1940, the future became alarmingly clearer when Japan joined the European Axis powers. This event meant that if the United States went to war, it would have to fight in both European and Pacific theaters.

In the middle of the rapidly worsening world situation, Roosevelt had to make a decision about the election of 1940. If he chose to seek the nomination, he was likely to get it, and if nominated, he was likely to win. However, no one had ever served three terms as president, and he was somewhat reluctant to be the first. With the uncertain prospect of changing presidents at such a dangerous juncture, Roosevelt decided to seek the nomination. In November, he was elected in a contest with Wendell Wilkie (R, NY). Roosevelt interpreted his re-election as an endorsement of both his domestic and foreign policies. With solid support from Congress, he increased materiel support for the Allies and accelerated preparations for war.

With enemies on the move in Europe and in the Pacific, every day brought the United States a step closer to war. Then on December 7, 1941, the Japanese attacked Pearl Harbor. By destroying so much of the US Pacific Fleet in their surprise attack, the Japanese gained considerable short-term tactical advantage, but it only served to silence all objections to the war in Congress and to unite the country as it had never been united before. On December 8, Congress declared war on Japan. Three days later, Germany and Italy declared war on the United States; the die had been cast. Roosevelt's government agreed to a Europe First war strategy.

By mid-1942, the Axis powers had conquered all of Western Europe, and were in control of territory as far west as French North Africa. In October, the Allies began the long journey to Berlin by attacking Nazi German forces, and its allies in North Africa. After an Anglo-American victory in North Africa, the Allies began to push through Sicily and northward up the Italian peninsula to Rome. A few days after Rome fell, on June 4,1944, the so-called Second Front was established on hostile beaches in Normandy. The struggle to break-out from the beachhead was extremely ferocious and bloody, but by mid-July, Allied armies were beginning the final lap on the journey to Berlin. Paris was liberated on August 25, and in March 1945, American forces crossed the Rhine at Remagen. Then a little over two months later, with the Russians moving into Berlin from the east and Allied forces from the west, the Nazis surrendered. The instrument of unconditional surrender was signed on May 7, 1945.

In the Pacific Theater, the Japanese forces literally

exploded across the Pacific and to Australia's doorstep. Even British India was threatened by Japanese forces pushing west from Indochina. In June 1942, US naval forces won a decisive battle with the Japanese at Midway. After this battle, American forces began a savage and bloody journey to the Japanese Home Islands, a journey that led American forces island hopping through the Western Pacific. With the Allied forces, overwhelmingly American, poised to begin the invasion of the Home Islands, Roosevelt and Churchill meeting in Potsdam presented the Japanese with an ultimatum to surrender or be subjected to complete and utter destruction. When the Japanese government did not respond, the order was given to drop atomic bombs on Hiroshima and Nagasaki. On August 14, 1945, the Japanese emperor agreed to the terms of the Potsdam ultimatum.

In 1944, with hundreds of thousands of young Americans in harm's way throughout the world, and the country involved in the unprecedented events of the world at war, there was still the matter of politics to deal with. A president and a Congress had to be elected. Of course, Roosevelt was nominated for a fourth term. In November, he won an easy victory over Thomas Dewey (R, NY). The Democrats also returned sizeable majorities to both houses of Congress. Unfortunately, President Roosevelt would not live to see either war concluded. With his health deteriorating rapidly, he traveled to his beloved warm springs in Georgia in the early spring. On April 12, 1945, he suffered a massive cerebral hemorrage, and died instantly. He was deeply mourned by the American people and by much of the world. He was replaced in office by Vice President Harry Truman of Missouri. Truman served as president for less than a year before both wars ended, and he was immediately faced with the huge, multi-faceted problem of supervising the demobilization of an enormous American war machine. In a very short time, President Truman was faced with another foreign power determined to conquer the world. Before the Axis powers had surrendered, the Soviet Union had begun what would eventually be called the Cold War.

Truman had served in World War I, and he vividly remembered the recession that accompanied the demobilization that occurred when it was ended. He was determined to prevent that from happening again. Try as he might, he was unable to get the smooth transition that he wanted. There was simply too much pent-up consumer demand and too much money left over from the high wages and salaries earned during World War II. He had wanted Congress to continue price controls for a short period, but Congress refused. Quickly, take-home pay for many industrial workers fell by as much as 50%, while the cost-of-living began to increase rapidly. The price of beefsteak was quickly increased from 50 cents a pound to $1.00. Soon, there were plenty of attractive new automobiles available—at double the price of the 1941 models. Prices of eight basic commodities increased by 25%. Amid all of this economic turmoil, Truman had to deal with an intransigent organized labor, flush with strength gained during the war years.

All of these problems overwhelmed the president and the Democratic Congress, and as a result, an unsympathetic public returned a Republican majority to both the House and Senate for the 1947-1949 term (80th Congress). The Republican Congress quickly solved one of the president's postwar problems, by enacting the Taft-Hartley Act that revolutionized labor relations in the country. Truman vetoed the act, but the Republicans passed it over his veto.

In 1947, the economies of most of the adversaries in World War II were in shambles. The economies of both France and Italy were on the verge of collapse—with strong prospects of communist takeovers. To remedy this situation, Secretary of State George Marshall proposed that the United States underwrite the economic recovery of all of Europe. Such a far-reaching proposal was unprecedented in world history. The Marshall Plan was introduced to a Republican Congress by the Democratic president in December 1947. As might be expected, such an unusual commitment of American taxpayer's money precipitated a considerable amount of discussion, but the act was finally approved in April 1948. Truman vetoed it, but the Republican Congress overrode his veto. Morison (1965) comments that this was the best thing that United States could have done for Europe. It was also the best thing the United States could have done for itself.

In 1948, Truman faced his first election as the incumbent president. In its nominating convention, the Democratic Party adopted a strong position on civil rights. Quickly, many Southern Democrats withdrew from the convention and formed a splinter party, the so-called Dixiecrats. Before the November election, the Dixiecrats, in convention in Montgomery, Alabama, nominated Strom Thurmond of South Carolina as their candidate. The Republicans nominated Dewey again. With a Republican Congress in place and with Truman's many domestic and foreign problems, the Republicans expected an easy win. The president, however, waged a spirited campaign, soundly denouncing the Republican Congress at every opportunity. When the votes were counted, he had won a clear majority in the Electoral College, although he won only 50% of the popular vote. Apparently, the public took Truman's criticism of the Congress to heart. The Democrats regained control of both houses of Congress. In early 1949, Truman began his first full term as president, but he would not have long to appreciate his victory.

In 1950, Truman was still trying to complete the return of the United States to peacetime, while overseeing the rebuilding of the economies of Europe and Japan. In June, a communist North Korean army crossed the 38th parallel, and moved rapidly through South Korean territory. United Nations' forces began a counter-offensive in September, and by early November they had driven the communist enemy back to near North Korea's border with Communist China. At this point, China decided to intervene. By December, UN forces had retreated south of the 38th parallel; however, in the ensuing months, UN forces recovered the momentum, and by mid-March they had fought their way north to recapture Seoul and reach the 38th parallel.

ALABAMA POLITICS

The New Deal as a political movement came to Alabama in 1933; in the depth of the Depression, the Democratic Party captured the presidency and Congress. It was well received by Alabama's organized labor and small farmers, which included most Alabamaians. But from the beginning, planters and industrialists were firmly opposed to the New Deal. There is ample reason to question how much difference it actually made in turning the economy of the state around, but its relief efforts played an extremely important role in alleviating some of the worst cases of hunger and deprivation among the state's neediest people. Unfortunately, it carried within its various programs a seed guaranteeing its destruction. If the New Deal was going to help the neediest people in the South, it had to help poor blacks, and that was simply unacceptable to a large number of Alabama voters at that time.

Governors

Frank Dixon (D, Jefferson County) was elected governor in 1939. He served until 1943 when he was replaced by Chauncey Sparks (D, Barbour County). James Folsom (D, Cullman County) was elected in 1947 and remained in office until 1951.

Commissioners of Agriculture

Haygood Patterson (D, Montgomery County) was elected commissioner of agriculture in 1939. He was replaced by Joseph Poole (D, Butler County) in 1947, who served until 1947 when Patterson was elected to the office once more. Patterson's second term in the office ended in 1951.

The Political Situation

The second Graves administration (1935-1939) began in the middle years of the Great Depression, and with Alabamians just trying to keep body and soul together, it was not a time to get much progressive legislation enacted. The majority of its accomplishments were the result of the state's cooperation with the federal government in the federal attempt to alleviate some of the most pressing problems among the neediest Americans.

Frank Dixon was elected governor in 1938 in a contest with Chauncey Sparks. Dixon had been narrowly defeated by Bibb Graves in 1934. He began his term in 1939 (near the end of the Depression and the beginning of America's preparation for war), and he finished his term in the middle years of World War II. These events had a major effect on Dixon's tenure. Several military installations were established in the state, as were manufacturing plants for the production of ammunition, ships, etc. Iron, steel, and coal production expanded rapidly. Agriculture flourished. The Selective Service System was established, along with the Office of Price Administration. Everyone who was not in service and wanted a job, had one. Wages, when overtime was added on, were extremely good. Most everyone had lots of money. With wartime rationing, however, there was not much to buy.

In 1942, Chauncey Sparks was elected governor in a contest with James Folsom and Chris Sherlock. Bibb Graves would have been a much more formidable opponent in his bid for an unprecedented third term, but he died in March 1942. Sparks took office in 1943 at about the time the Allied forces were beginning their offensive against the Axis powers in North Africa and the Japanese were evacuating what remained of their badly beaten forces on Guadalcanal. Most of Spark's early years were involved with continuing the wartime activities put in place during the Dixon administration. Then about halfway through his tenure, he had to contend with dismantling the huge military complex in the state, and the loss of the dynamic economic stimulation associated with the war. With the downsizing of the economy, he also had to deal with numerous labor problems. Organized labor had become extremely powerful during the war. As a result of these continuing problems, he re-established the State Labor Department. Sparks was a strong supporter of education at all levels. He was able to get state funding for education doubled and the school year expanded from seven to eight months. He was able to obtain funding to establish the University Medical College in Birmingham, and he took special interest in providing additional funding for the expansion of agricultural teaching, research, and extension at Auburn. One of his most noteworthy accomplishments was to use some of the extra money coming to state government during the war to reduce state indebtedness by 25%.

In 1942, James Folsom ran for governor again, but was defeated by Chauncey Sparks. Folsom ran again in 1946, and defeated Handy Ellis in a run-off election. Around the time that he was taking office in 1947, President Truman was endorsing a strong civil rights package that included repeal of the poll tax, enforcement of legislation forbidding job discrimination because of race, and making lynching a federal offense (Flynt, 1994). With the president taking this position, the second civil war for states rights began, and Folsom was right in the middle of it. Early in his tenure, he was able to increase funding for education and road construction. He failed in his attempt to get legislation enacted that would have improved the lives of Alabama's blacks, poor whites, and women. He attempted to get the poll tax law repealed, but the Senate blocked the effort; however, he used his power to appoint county registrars who would register qualified blacks. As a result, the percentage of blacks registered to vote increased from 1% in 1947 to 5% in 1952. Folsom also appointed the first woman to the Alabama Court of Appeals—at a time when women could not sit on juries in Alabama.

THE NATIONAL ECONOMY

The national economy became mired in a deep depression in the early 1930s, and various New Deal programs seemed to halt a downward spiral at the very brink of a complete economic meltdown. After the middle of the decade, the American economy seemed to be on the way to recovery, but when Nazi Germany invaded Poland in September 1939, and Britain and France declared war, the United States and its economy were dramatically transformed.

The Economic Situation

Conte and Karr (2001) comment that the hurried beginning of preparations for World War II resulted in the creation of a huge military-industrial complex in the United States. This complex quickly began to pump billions of dollars into the economy. It pulled millions of Depression-ridden farmers off farms, especially in the South, and for the first time, large numbers of women also entered the workforce. Fortunately for the state's economy, Alabama's military-industiral complex was not completely dismantled at the end of the war in 1945. Almost before the hot war ended, the Cold War began. With the Iron Curtain descending over Eastern Europe and the threat of a communist invasion of Western Europe, the United States was forced to keep a large military force in the field.

The complex had to keep busy providing equipment, weapons, and food for this large force. Also, a considerable investment was made in developing new weapon systems, including the hydrogen bomb. Economic aid flowed to a war-ravaged Europe. These funds, in turn, created vast markets for American goods of all kinds. The Marshall Plan only began American involvement in providing financial aid and related services to needy countries. In 1944, the United States led the way in establishing the World Bank and the International Monetary Fund.

Globilization became a widely-used buzz word in the last few years of the twentieth century, as corporations, goods, money, etc. began to move around the globe with ever increasing velocity.

American farmers began to participate in a global economy soon after they achieved self-sufficiency in food production in the seventeenth century and began to produce surpluses that could be sold to European countries and to European settlers in the Caribbean. Fish and naval stores were marketed in Europe even before there was sufficient surplus food to sell. Later, American tobacco, indigo, and rice entered international markets in large quantities. However, the sale of all of these crops and materials was dwarfed by the sale of Southern cotton abroad. Until well into the twentieth century, the economies of several Southern states depended heavily on the sale of their cotton crop on the international market. This relationship provided many Southerners with an early global perspective not shared by farmers in other regions of the country.

World War II, however, speeded up the process of globalization by several orders of magnitude. Because it was truly the first worldwide war, World War II resulted in a revolutionary turnover for the world's peoples, their cultures, and their economies. Soon after December 1941, millions of Americans were uprooted from their homes to go to places both in the United States and overseas that many of them had never even heard of. And wherever they went, unprecedented quantities of American war materiel, food, clothing, and medicine went with them. After the war, they returned to a different country than the one they had left. Within less than a decade, America had become the leader of the Free World.

World War II ushered in a period of global conflict, and the uneasy peace that followed did the same for supply and demand. Although globilization opened the world's markets, it also transformed competition for markets and comparative advantage into international matters. In time, American automobile makers were competing in a world market with Japan and Germany. Midwestern grain farmers had to compete with their counterparts elsewhere in the world. Alabama corn and soybean producers were competing with those in Iowa, Illinois, and Indiana.

The Gross National Product

The Gross National Product had fallen at the beginning of the Great Depression, but by 1939 it had recovered those losses to stand at $717 billion (constant 1982 dollars) (Figure 13.1). Then with World War II, the GNP almost doubled to $1.4 trillion in 1943. With the buildup of the arsenal of war largely complete, it declined slightly in 1944 when the growth in the Consumer Price Index slowed. With hostilities ending in September 1945, it fell to $1.1 trillion. It declined slightly in 1946, but then trended upward to reach $1.2 trillion in 1949. It remained at that level in 1950.

The Consumer Price Index

The Consumer Price Index declined from 50 in 1930 to 40 in 1933 (Figure 13.2). It remained near that level through 1940 and increased relatively rapidly in 1941, a trend that continued through 1943. Its rate of growth slowed afterwards, as price controls became effective, and it reached 53 in 1944. When the war ended in 1945, CPI was 54. In 1946, money saved during the war began to chase temporarily scarce goods, and the CPI increased sharply to reach 72 in 1948. It remained near that level (71.4) in 1949 but increased slightly to 72.1 in 1950. In June 1950, North Korea invaded South Korea, and in August the CPI began to inch up again. Between 1940 and 1950, the value of the US dollar declined by 72%.

Inflation Rate

The American inflation rate fell as low as -10% during the Depression (1932), but it had rebounded after 1933, and reached 1% in 1940 (Figure 13.3). With the beginning of the war, inflation began to increase sharply, and in 1942 the rate reached 10%, but with price controls becoming increasingly effective, it was down to 2% in 1944 and 1945. Congress refused to continue the controls after the end of the war, and with consumers buying everything in sight, inflation again increased sharply. By 1947, it was 15%. Afterwards, production of consumer goods began to catch-up with demand, and by 1949 inflation was -1%, and in 1950 it climbed back to 1%.

Federal Discount Rates

The Federal Discount Rate (interest rate) was 1% in 1940 (Figure 13.4). It remained at that level until the end of October 1942 when the Federal Reserve lowered it to 0.5%. It was unchanged until April 1946 when the threat of increasing inflation led to raising it back to 1%. Then it was raised again to 1.25% in January 1948, and to 1.5% in August 1948. Two years later, in August 1950, it was increased to 1.75%.

THE AGRICULTURAL ECONOMY

Some of the disastrous effects of the Great Depression on American agriculture were detailed in the preceding chapter. American farmers, especially tenant farmers in the South, probably were more severely affected by the events of that period than any other group in the United States. Sporadically, American farming had experienced severe ups and downs from 1783 onward, but for all practical purposes, it had remained little changed. For a number of reasons, in 1933 President Roosevelt and the 73rd Congress changed American agriculture forever. From that time forward, the federal government (the Triangle) would be deeply involved in every decision made daily on every farm in the country. The results of direct governmental participation for the first eight years (1933-1940) were not very promising. In fact, it would not have been at all surprising if direct governmental involvement in American food and fiber production had just died aborning. At that critical time, however, the machinations of Hideki Tojo, Benito Mussolini, and Adolf Hitler intervened, and Americans forgot about paying farmers not to farm.

Federal Agricultural Legislation

The coming of the World War II years allowed Congress a period of respite in its complicated efforts to manage the supply and prices of agricultural commodities. As wartime demands increased, accumulated stocks of agricultural commodities were quickly depleted. In response, production controls were gradually removed, and price supports were raised in successive steps to encourage increased production. Direct subsidies were used to encourage production of some crops. However, Congress did not have much time to rest on its laurels. Shortly after the end of hostilities, Congress had to put its shoulder to the New Deal wheel once again, and the Southern, Congressional agricultural phalanx was still there to lead the way (Table 13.2). Four pieces of important agricultural legislation were passed during the period 1940-1950. These included:

1. The Stegall Amendment of 1941 gave the secretary of agriculture the power to support prices on some non-basic commodities (hogs, eggs, certain dry peas and beans, potatoes, sweet potatoes, etc.) at 85% of parity or higher. In 1942, the act was altered to increase the minimum rate of support to 90% and to require that these supports be continued until two years after the end of the war.

2. The Commodity Credit Corporation Act (CCC) of 1948 re-established the CCC that had first been established in 1933 and specified that its administrators use certain federal funds to support farm prices and incomes.

3. The Agricultural Act of 1948 made price supports mandatory at 90% parity for certain basic crops including: corn, cotton, peanuts, rice, tobacco, and wheat. Four of these favored crops were extremely important to the South. It also required that in 1950 parity determinations take into consideration average prices for the past 10 years, as well as the 1910-1914 base period.

4. The Agriculture Act of 1949 (as amended), along with the Agricultural Adjustment Act of 1938 that constituted the central core of federal legislation, was formulated to deal with farm income and with commodity surpluses and prices.

Alabama Agricultural Legislation

In 1945, the Alabama legislature replaced the 1923 yield tax on timber with a severance tax (The Alabama Forest Products Severance Tax Act) (Nix, 1994). This act placed a tax on severed timber and overturned all other timber taxes. The tax was to be levied by the state on those harvesting the timber. Severers must report all harvests of timber on a quarterly basis. In addition to the severance tax, the act also stated that it was unlawful for local governments to tax forest products. The act provided for the imposition of a privilege tax on manufacturers who refine or process timber harvested from lands in the state. This tax rate was levied at one-half of the severance tax. All proceeds from these taxes were to be used by the Alabama Forestry Commission for forest protection.

Agricultural Exports

American agricultural exports had remained relatively constant at around $700 million during the latter part of the 1930s (Figure 13.5). As late as 1941, they were still slightly less than $670 million. However, as might be expected, in 1942 they began to increase rapidly, and in that year passed $1 billion. The rate of increase slowed slightly in 1943 and 1944 before sprinting upward to $4 billion in 1947. Exports were generally lower in the period 1948-1950, falling back to $2.9 billion in 1950. During the period 1940-1949, agricultural exports accounted for 22% of all American exports. During the World War I period, they had averaged 45%.

Agricultural Prices

At first, indices for both prices received and prices paid had responded positively to the New Deal initiatives (Figure 13.6). After 1937, however, their effectiveness began to wane, especially for prices received. By 1939, they were 95 and 123, respectively. The index for prices received increased slightly after the beginning of the World War II in Europe. It was up to 100 in 1940, or back to where it had been in the 1910-1914 period. In 1941, both indices began to trend sharply upward. The Stegall Amendment of 1941 provided the necessary support to boost prices received. At the same time, the Office of Price Administration, also established in 1941, restricted increases in prices for all items, except agricultural commodities. As a result, after 1940 the index for prices received began to increase at a faster rate than the index for prices paid. By 1948, the two were 287 and 260, respectively. Both fell after 1948, especially the index for prices received, and ended at near equality in 1950 (258 versus 256). Both indices increased from 1949 to 1950. In June 1950, North Korea invaded South Korea, and in August prices began to increase (Figure 13.6).

Between 1940 and 1948, the prices received index increased by 187% (100 to 287). During the same period the Consumer Price Index increased 72% (42 to 72.1).

Parity ratios had plunged below 0.60 in the early days (1932) of the Depression (Figure 13.7) but began to increase after 1939 and reached 1.05 in 1942. It had not been that high since 1918. It continued to increase until 1943 when it reached 113. It had never been that high. Afterwards, it remained near that level through 1948. Then with the indices for prices received falling rapidly, the parity ratio fell to 1.00 and remained there through 1950.

The World War II period was truly a watershed period for American agriculture. Farmers had been in a cost-price squeeze since 1920; then in 1942 the situation quickly changed. Through 1948, farmers found salvation in the tents of the god of war. They were making extraordinarily good profits from their farming operations, and with wartime rationing in place, opportunities for purchasing consumer goods were limited. Even after the war ended in 1945, agriculture continued to prosper in 1946, 1947, and 1948. They used savings accumulated to purchase all sorts of things, but a large share was used to buy new machinery and equipment for their farms and to improve infrastructure (Fite,1984). In this unique situation, the spoils of war fueled the modernization of America agriculture, but as we will see in the forthcoming chapters, this was not a completely unmixed blessing.

Agricultural Productivity

Beginning with 1948, USDA's Economic Research Service began to provide annual estimates of indices of agricultural productivity. These indices (estimates) were ratios of indices of agricultural outputs to agricultural inputs. Indices for 1947 were used as the base year. The ratio of these indices was set at 1.00 for 1947. In 1948, the indices of outputs and inputs were estimated to be 0.507 and 1.035, respectively (Figure 13.8). From these indices, the index of productivity was computed to be 0.490 (0.507/1.035). In other words, the index in 1948 was only 46% of what it was in 1947. Data presented in the figure indicate that the index of productivity remained at near the same level in 1948, 1949, and 1950.

Farmers

By 1940 and in less than a century, American farmers had been asked to respond to national wartime emergencies on several occasions. The Spanish-American War in 1898 had not lasted long enough to require much change in the activities of American farmers, although a number of farm boys was drawn away from farms, especially in the South. For the United States, World War I had not lasted very long either, but American agriculture had gotten deeply involved in that war. For the first time in American history, the country's farmers were asked to produce food and fiber for our own population under wartime conditions, as well as to provide large quantities of agricultural commodities for our allies. This arrangement required that farmers quickly upsize during World War I, then downsize just as quickly after the end of hostilities. This was a tall order, and farmers did not fare very well trying to do it. The short recession of 1921 resulted in a sharp decline in farm commodity prices; however, this short period of extreme difficulty was quickly followed by the Roaring Twenties. Unfortunately, this period of rapid economic expansion was quickly followed by the early warning signs of the Great Depression. Then, when the disastrous economic conditions down on the farm in the mid-1930s were slowly beginning to improve, farmers were once again requested to report for duty.

The Total Population

Data on the total population of the United States (Coulson and Joyce, 1999) in the period 1910-1950 are presented in Figure 13.9. The population was 132.4 million in 1940, but it was only slightly larger in 1945 (133 million). Population estimates do not include Americans living outside the United States; consequently, men and women in uniform in 1945 were not included in the estimate. Annual data show that the population actually declined during the period 1943-1945. The sharp increase to 151.9 million in 1950 includes the return of servicemen and women to the United States, plus the normal increase expected during the period. In this case, however, the increase was considerably greater than normal. In 1946, with the war over and prosperity returning to the country, married couples produced the largest numbers of babies in the history of the country, the Baby Boom.

Data on the total population in Alabama during the period 1910-1950 are presented in Figure 13.10. The data show that the total population in the state in 1940 was 2.84 million. Alabama also lost a share of its population to military service, but in addition to this loss, large numbers of Alabamians also left to take jobs in defense plants in other states. The loss is much more obvious in the annual data. The population did not begin to decline until after 1942, but in 1945 it was 2.78 million. After the end of hostilities, Alabama's population was up in 1946 (2.91 million), but not back to the 1942-level (2.94 million). It appears that the population recovered at least a portion of its losses to military service and benefited some from the Baby Boom. It is also obvious that a substantial number of Alabama's losses never returned. The state's population was not much higher in 1950 than it had been in 1942.

The Farm Population

During the depths of the Depression, hundreds of thousands of non-farmers had to return to farms to survive economically. As a result, the American farm population reached 33.4 million in 1933 (Figure 13.11). It had not been that high since 1917. After 1933, it began to drift downwards to reach 30.5 million in 1940. With many farmers being conscripted and many others leaving their farms to work in defense plants, the number began to trend rapidly downward, reaching 24.4 million in 1945, the year the war ended. It had not been that low since sometime between 1880 and 1890. After World War II, many of those who had left the farm began to return, and in 1947 the farm population reached 25.8 million. However, many of those who had returned seemed to have determined that the war really had not improved the agricultural economy all that much, and they left to find jobs in the postwar economy. In 1950, the number was back at a late nineteenth century level (23 million).

The percentage of farmers in the total American population had remained relatively stable (near 24%) between 1930 and 1940 (Figure 13.12); however, once World War II began the percentage began to decline. It was down to 17.5% in 1945, and continued downward to 15.3% in 1950.

With many, less efficient farmers leaving their farms during the war, those remaining had sharply increased their efficiency. In 1945, 24.4 million farmers (near the 1890 level) managed to provide food and fiber for all of the country's citizens, its armed forces, and millions of our allies, plus their armed forces. This was truly a miracle of agricultural production, and it presaged a view of other miracles to come. With the reduced demand for agricultural commodities after the war, it was quickly obvious that the country no longer needed 24.4 million farmers. Fortunately, the postwar economy was able to absorb much of the excess labor force.

The exodus of farmers from Southern farms actually began in the middle of the Depression, as New Deal agricultural programs resulted in large numbers of share tenants being pushed off the land. Then beginning in the early 1940s, war-related industries and businesses and the armed forces pulled hundreds of thousands more away. According to Fite (1984), between 1940 and 1945 the Southern farm population was reduced by 3 million people, a decrease of 22%. In some individual states, the rate of loss was even greater. During the war, many of the eligible men on Southern farms were drafted into the armed forces. Alabama provided 250,000, many of them from its rural areas. Even larger numbers, who for one reason or another were not eligible for the draft, took jobs in war-related industries and businesses. For example, in 1940 the total population of Mobile was 79,000, and its total workforce was 17,000. By early 1943, the city's population had grown to 125,000. The Census Bureau estimated that altogether some 89,000 people moved to Mobile County during the war years. Most of the increase represented poor farmers from southern Mississippi and Alabama who had moved there to work.

Relatively fewer owner-operator farmers left the land for industrial jobs during the war. The wartime economy made farming much more attractive for most of them than it had ever been. Also, the federal government made an effort to get farmers to remain on the land to help meet the increased need for food and fiber.

A higher proportion of white farmers left their farms compared to blacks. In 1945, there were 81,000 fewer farms in the South operated by whites than there had been in 1940. While there was also a decrease in the number operated by blacks, the relative change was much lower. Racial discrimination was still a powerful force in the South at that time. The majority of the better-paying jobs created during the war were not available to blacks; consequently, they had less incentive to leave their farms.

Fite (1984) published information indicating that the Alabama farm population remained around 1.34 million from 1920 through 1940 (Figure 13.13). During the war, 275,000 (1.34 to 1.07 million) farmers left their farms. Because there are no annual estimates, it is not possible to know what happened to the population after 1945; however, by 1950 Alabama farms had lost another 100,000 farmers (1.07 to 0.96 million).

The percentage of farmers in Alabama's total population also remained relatively stable during the Depression years (Figure 13.12). It was at a level of 47.2% in 1940, but began to decline afterwards. It was down to 38.5% in 1945 and 31.4% in 1950. Throughout the period, the percentage was about twice as high in Alabama as in the country as a whole. Alabama had always had too many farmers, and most of them would have readily agreed, but the state's political and commercial leadership had done very little to provide them with other opportunities.

Fite (1984) presented data indicating that Alabama's black farmer population had begun to decline sometime after 1910 (Figure 13.14), and in 1940 it had reached 73,300. Some black farmers were conscripted for military service, but many more found jobs in defense industries, especially outside the South. By 1950, there were 16,000 (22%) fewer black farmers in the state than there had been in 1940; however the loss of black farmers from 1940 to 1950 was not as great as the loss of whites. Some 28% of white farmers were gone from Alabama's farms in 1950.

The American farm population had been generally declining since the beginning of the twentieth century. This decline, with the ever-growing national labor force, meant that the percentage of farmers in the labor force would decline fairly rapidly. Data indicating the magnitude of the decline are presented in Figure 13.15. In 1940, farmers constituted 18% of the American labor force, but by 1950, the percentage was down to 12.2%.

Tenancy

Rates of tenancy were extremely high in all the Southern states for most of the first 35 years of the twentieth century. In fact, it was so high that it biased tenancy rates in the entire country upward into a range of 38 to 42% during that period (Figure 13.16). In Alabama, tenancy rates remained 60%, or above, through 1935 (Figures 13.16). However, the rate began to decline afterwards. It was 58.8% in 1940, but declined to 49.1% by 1945—when tenant farmers in the state were provided with opportunities to leave those largely uneconomical farms. The erosion of tenant farming continued through the remainder of the decade to reach 41.4% in 1950. This reduction during the decade was certainly helpful to Alabama's general agricultural economy, but even at 41.4% there were still far too many farmers unable to own and operate their farms without considerable assistance.

Farms

The spurt in the increase in US farm numbers associated with the Depression ended after 1935 and was quickly replaced with a decline amounting to 1% a year. It is likely that a substantial share of this decline was the result of the loss of tenant farms in the South. If so, it represented the squeezing-out of a large number of uneconomical farming units in the country's agricultural economy.

Farm Numbers

In 1940, the largest share of farms in the United States was in Texas (6.6%). Other states with large shares included: Mississippi (4.8%), North Carolina (4.7), Kentucky (4.2), and Missouri (4.1). These five states accounted for 24.4% of all American farms. Some 3.8% of the country's farms were in Alabama. In the ten states with the largest share of farms, nine of them had had substantial numbers of slaves before the Civil War and tenant farmers afterwards.

The decline in farm numbers that began after 1935 continued unabated through 1940, when it reached 6.3 million (Figure 13.17). The war years seemed to have little effect. Throughout the period 1940-1950, numbers trended downward at about the same rate; although the rate of decline did seem to slow slightly between 1944 and 1945. By 1950, there were 5.6 million. During the period 1940-1950, the country had lost 702,000 farms, or about 11% of the 1940 number.

In general, the decline in US farm numbers resulted from creating one farm out of two. This was certainly the situation in the South in the 1940s where there were so many small tenant farms. When a tenant farmer left his farm, for whatever reason, numbers declined by one, but the landowner's farm was increased in size. Even where the farmer owned his farm when he left it, he most likely sold it to a neighbor. In that situation, numbers decreased by one, but the amount of land in farms remained the same.

Changes in Alabama farm numbers followed the same trends as the national numbers during the period 1940-1950. Numbers in the state also increased at the beginning of the Depression, but then began a long-term, downward trend after 1935 (Figure 13.18). There were 241,000 farms in the state in 1940 and by 1945, the number had fallen to 225,000. At this point, the Alabama trend diverged from the national trend. From 1945 to 1950, there was relatively little change (225,000 to 220,000). During the period 1940-1950, Alabama lost 9%, or 21,000 (241,000 to 220,000), of its farms. This rate of loss was lower than in the country as a whole (9 versus 11%). Alabama desperately needed to lose more of its farms, but there were simply not enough off-farm jobs available.

Farm Size

The average size of American farms remained relatively stable at around 153 acres from 1920 to 1935 (Figure 13.19). Events related to the New Deal forced farmers off many smaller farms, especially in the South; consequently, average farm size began to increase, and by 1940 it was up to 174 acres. The continued decline in numbers of small farms (Figure 13.17) resulted in further increases in size, and by1945 and 1950 it reached levels of 195 and 216 acres, respectively. Size increased 10% during the period, or near the same rate as the decline in numbers (11%).

Data presented in Table 13.3 give the percentage of American farms in different size groups in 1935, 1940, 1945, and 1950. These data show that in 1940 some 37.5% of American farms were less than 50 acres. They also show that the relative share of smaller (10 to 179 acres) farms decreased between 1940 and 1950 from 71.8% of the total to 67.4%. At the same time, both the share of the smallest farms (1 to 9 acres) and the largest farms (180 acres and larger) increased. These data provide some additional detail to support the conclusion that the average size of American farms was increasing during the period.

Changes in the average size of farms in Alabama during the period 1920-1950 were also shown in Figure 13.19. Generally, the trends in size in the state were similar to those in the country as a whole; except that between 1940 and 1945, size of Alabama farms increased very little (83 to 85 acres). In 1940, 1945, and 1950 size in Alabama was 83, 85, and 99 acres, respectively, an increase of 19% during the decade. Throughout the period 1940-1950, the average size of Alabama farms was less than half the acreage of those in the counry as a whole. Further, the difference increased during the decade.

Data presented in Table 13.3 give the percentage of Alabama's farms in different size groups in 1935, 1940, 1945, and 1950. These data show that in 1940 some 53.3% of all farms in the state were smaller than 50 acres, but that by 1950, the percentage had declined to 50.8%. As was the situation nationally, the percentage of smaller farms (10 to 179 acres) decreased during this period, but both the percentage of the smallest farms (1 to 9 acres) and the largest farms (180 acres and larger) increased.

Total Farmland

The combination of American farm size and farm numbers during the period 1920 to 1950 (Figures 13.17 and 13.19) resulted in a farmland total of 1.061 billion acres in 1940 (Figure 13.20). During the 1940s, size increased at a slightly higher rate than the decline in numbers. As a result, farmland grew to 1.142 billion acres in 1945 and to 1.163 billion acres in 1950. The 1950 total was the highest level of farmland ever reported in the United States. This increase had been in progress for well over a century and was probably the result of the conversion of forestland and/or grazing land into farmland. Available data indicate that the quantity of forestland declined from 900 million acres in 1850 to 600 million in 1920. However, after 1920 the amount of land in forests began to increase again. USDA data indicate that the amount of American grazing land declined from 661 million acres in 1920 to 400 million acres in 1950.

During the period 1940-1950, total acres of farmland increased by 9%. The loss of large numbers of smaller farms during the 1940s was a welcome development, but an increase in total farmland was not. If 1.142 billion acres was sufficient to feed and clothe the United States at war—plus millions of persons in foreign lands—1.116 billion acres was likely too much farmland for the country at peace after 1950—especially if most of the farmland was being used as cropland.

There were 19.1 million acres of farmland in Alabama in 1940 (Figure 13.21). There had been 19.7 million acres in the state in the middle of the Depression in 1935. As a result of the combination of changes of size and number (Figures 13.18 and 13.19), the total in 1945 remained unchanged at 19.1 million, but increased to 20.9 million acres in 1950. It is likely that most of this additional farmland came from the conversion of forestland.

Use of Farmland

In 1934, 39% of American farmland was used for the production of crops (cropland), 43% was used for pasture, 18% was in forestland, and 4% was used for other purposes (Figure 13.22). After 1934, when many farms were being combined, numbers declined (Figure 13.17), and farm size began to increase (Figure 13.19). As these changes continued, some cropland and forestland were converted into pasture. This trend continued to at least 1944 when cropland and forestland were down to 35%

and 14%, respectively, and pasture was up to 46%. In 1949, cropland remained at 35%, but some of the pasture (42%) had been returned to forests (19%).

In Alabama in the 1930s, a much larger share of farmland was used for crops than for the country as a whole (Figures 13.22 and 13.23). In 1934, some 49% of all farmland was devoted to crop production, compared to 39% in the country. Also, much larger shares were devoted to farm forests (42 versus 18%). At the same time, smaller shares were devoted to pasture (4 versus 39%). There are no data available on land use in the state in 1939, but by 1944 land used for crops was declining (43%), and in 1949 it reached 42%. Most of this lost cropland was converted into forestland, although a small amount was put in pasture.

Alabama's cattle industry began to expand rapidly during this period, and the use of farmland for pasture was also beginning to increase. Pasture percentages represent only land used for nothing else but pasture. Before 1934, Alabama's cattle farmers met the grazing needs of most of their cattle from grazed cropland and forestland.

Total Cropland

As American farmers attempted to compensate for declining parity ratios, total cropland in the United States had generally trended upward following total farmland (Figures 13.24 and 13.20) from the mid-1920s through the 1930s. It was down to 513.9 million acres in the middle of the Depression. Although New Deal programs had attempted to reduce cropland during the 1930s, by 1939 it was up to 530.6 million acres. It first declined in the mid-1940s, as large numbers of farmers left their farms, but returned to 478.3 million acres in 1949 as farm size surged upward.

Trends in year-to-year changes in total Alabama cropland during the period 1929-1949 were similar to those in the country as a whole (Figures 13.25 and 13.24), except that during the heart of the Depression, Alabama did not lose cropland. In 1939, there were 10.4 million acres in the state, 8.3 million in 1944, and 8.7 million in 1949. In 1949, 1.8% of the country's cropland was in Alabama.

Cropland Harvested

Data on farm numbers, size, and uses of farmland are interesting and informative, but the most important data on farm characteristics is the quantity (acres) of cropland harvested. This characteristic has a more direct effect on the agricultural economy than any of the others. In 1929, American farmers harvested crops from 359.2 million acres (Figure 13.26). In 1934, this total slipped to 295.6 million acres, as a result of New Deal efforts to take cropland out of production; however, regardless of these efforts, acres rebounded to 321.2 million in 1939. Then, government efforts to encourage greater food and fiber production to support the war effort resulted in an increase in cropland harvested in 1944 (352.8 million acres). Fortunately, after the war ended farmers began to reduce acres, and by 1949 the total was down to 344.6 million. Unfortunately, this total acreage was essentially the same as it had been in 1925 when the country's population was 23% smaller. Only time would tell whether or not farmers could market the crops harvested from this quantity of farmland, and earn a suitable return on their investment.

During the Great Depression, Alabama farmers seem not to have paid much attention to AAA efforts to reduce cropland harvested. They harvested crops from more acres in 1934 than in 1929, and even in 1939 acreage was about where it had been in 1929 (Figure 13.27). Further, Alabama farmers did not respond to federal government efforts to increase agricultural production during the war. In 1944, they only harvested crops from 6.2 million acres of land. However, farm numbers were declining rapidly during this period. Farmers continued the trend to reduce cropland harvested through 1949 when they harvested crops from only 5.7 million acres.

As stated repeatedly in previous chapters, cropland harvested per farm was a continuing problem for Southern farmers. With generally lower yields of most crops—compared to those in the states of the North Central Region—Southern farmers could not expect to make an acceptable living while harvesting crops from small acreages. Data (Fite, 1984) presented in Table 13.4 show the crux of the problem. In 1940, Alabama farmers and those in adjoining states harvested crops from an average of only 29.4 acres, only 2.7 more than in 1909. In the same year, farmers in Iowa harvested crops from an average of 94 acres, or 64.6 more acres—and on much more productive land. In the same year (1939), American farmers harvested crops from 48.5 acres, and this estimate includes all of those low values on small Southern farms. In 1949, average cropland harvested per farm by the six Deep South states was down to 24.1 acres. In Alabama, it fell to 27.1 acres, even lower than in 1910 (27.1 versus 27.4 acres). These poor unfortunate souls, attempting to make a living on farms in Alabama and in the Lower South, were obviously headed in the wrong direction, and it would take a miracle to save them. It was obvious, in 1949, that the god of war had not been powerful enough to do it.

Percentage of Cropland Harvested

Data on the percentages of cropland harvested in the United States and Alabama during the period 1924-1949 are presented in Figure 13.28. These data show that percentage harvested in the country as a whole was 62.4% in 1939. It had been lower (57.6%) in the heart of the Depression. It increased sharply to 78.3% in the middle of the war years, but declined to 72% in the immediate postwar period (1949). In Alabama in 1924, 1929 and 1934, percentage harvested had ranged around 75% and was slightly higher than in the counry as a whole. In 1939, percentage harvested did decline slightly in the state, but returned to pre-Depression levels in 1944. Then in 1949, it was down to 65.9%.

Value of Land and Buildings

In 1940, the total value of land and buildings on American farms was $33.6 billion (Figure 13.29). Most of the total value represents the value of land. Total value had increased slightly between 1935 and 1940, but did not return to the pre-Depression level of $47.9 billion until around 1945 when it reached $54.6 billion. In 1941, value increased slightly to $34.5 billion. Then in the war years, it increased rapidly to reach $54.6 billion in 1945. After the war ended, it continued to increase sharply to reach $77.1 billion in 1949. It declined slightly to $75.2 billion in 1950. Altogether, values increased by 123.8% during the 11-year period (1940-1950). A substantial portion of this increase was the result of inflation of the currency. Between 1940 and 1950, the Consumer Price Index increased by 71.7% (42 to 72.1) (Figure 13.2). It is interesting that total value was not affected by the violent economic contraction of 1945 (Figure 13.1).

The sharp increase in value between 1940 and 1950 took place while farm acreage was increasing by 9.2% (Figure 13.20). Some of the increase must have been the result of the purchase of additional farmland.

The total value of land and buildings, shown in Figure 13.29, has been converted to dollars per acre. This conversion corrects the data for changes in farmland acreage. These data are presented in Figure 13.30. The converted data show essentially the same trend. However, the rate of increase between 1940 and 1950 was not as great. Converted value was $32 in 1940, and in 1945 and 1950 they increased to $41 and $65 an acre, respectively. Between 1940 and 1950, they increased by 103.1%. These data indicate that the increase in the quantity of farmland could have accounted for as much as 9.2% of the $41.6 billion increase in total value depicted in Figure 13.29, and inflation could have accounted for another 71.7%. However, even when corrected for inflation, the data in Figure 13.30 indicate that there was some real increase in the value of land and buildings in the period 1940-1950. Even inflated values would have been of little concern unless farmers added to their landholdings during the period when prices received were increasing, but had to pay for it when they were declining.

Data included in the 1950 *Census of Agriculture* for total values of land and buildings in Alabama for the period 1920-1950 are presented in Figure 13.31. Trends of the data for Alabama farms during the period 1940-1950 were similar to those in the country as a whole. In 1940, 1945, and 1950 they were $408.8, $559.8, and $1,023.6 million, respectively. However, the rate of increase in Alabama's total values was somewhat greater for the period (150.4 versus 123.8%).

Alabama total value data converted to value per acre for the period 1920-1950 are also presented in Figure 13.30. In 1940, 1945, and 1950, they were $21, $29, and $49, respectively. These converted values for Alabama farms were considerably lower than those for the country. In 1950, the converted values in Alabama and Iowa were $49 and $161 an acre, respectively. It is likely that farms in Iowa had more and better buildings and improvements than farms in Alabama, and it is also likely that this better infrastructure accounted for some of the difference in value. However, much of it reflects differences in the basic agricultural productivity of the geophysical environments (climate and soils, primarily) of the two states. They are likely to correctly reflect the levels of return that could be expected from investment in agricultural land and infrastructure.

Value of Machinery and Equipment

The total value of farm machinery and equipment in the United States declined through the years of the Depression, but began to recover in 1935. By 1940, total value had reached $3.1 billion, close to the 1930 total (Figure 13.32). Value began to increase sharply after 1940, aided by inflation (Figure 13.2), and reached $6.3 billion in 1945. After the end of World War II in 1945, the upward trend skidded to a halt when the national economy experienced a short but violent contraction (Figure 13.1). Value dropped in 1946 and 1947 to $5 billion. Afterwards, values began to trend sharply upward once more. From 1947 to 1950, values increased from $5.1 to $12.2 billion (139.2%). Between 1940 and 1950, total value increased from $3.1 to $12.2 billion, an increase of 297.6%.

Two factors make it difficult to determine how much farmers were increasing the real value of farm machinery and equipment during the period 1940-1950: First, there was a significant decline in farm numbers (Figure 13.17), and second, inflation was rampant (Figure 13.2). It is not likely that the decline in farm numbers had much effect on the total value of machinery and equipment. Data on farm characteristics for the period indicate that while numbers declined (Figure 13.18), acres of harvested cropland (Figure 13.26) did not decline at the same rate. In this process, smaller farms became part of larger farms, and the machinery and equipment on the smaller farms simply remained as part of the larger farms. Further, the Consumer Price Index only increased by 71.7% during the period. Apparently, inflation was not a major contributor to the increase in total value either. These data suggest that American farmers used the profits from the good prices they had received from the sale of their commodities during and immediately following the war (Figure 13.6) to add significantly to their inventory of machinery and equipment. Farm machinery and equipment had not been widely available during the war years.

This large increase in the investment in machinery and equipment after the war, plus the relatively large acerage of cropland on farms, did not bode well for the Triangle's efforts to effectively manage future agricultural supply and prices.

Bias introduced as a result of declining US farm numbers could be partly corrected by converting the total value data into value per farm. Part of these converted data are presented in Figure 13.33. Converted values for 1940, 1945, and 1950 were $482, $1,085, and $2,154, respectively. The converted data show the same general trend as total values. However, dollars per farm shows a much larger increase between 1940 and 1950 (347% versus 298%).

For some reason, the 1950 *Census of Agriculture* did not include any data on the value of machinery and equipment on Alabama farms in 1935 or 1950. Census data for the period 1890-1945, except for 1935, are presented in Figure 13.34. These data show that value on Alabama farms increased from $29.6 to $57.1 million between 1940 and 1945—thus following the same general trend as the national data. However, the rate of increase between the two years was not as great as for the country as a whole (100 versus 125%).

The converted Alabama data (Figure 13.33) show that value per farm increased from $128 in 1940 to $256 in 1945, or at about the same rate of increase as for the country. Unfortunately, the data also indicate how far the state had to go to catch up with the rest of the country. In 1940, value per farm was only about a fourth as large in the state as in the country. In 1945, the relative value for the state was even lower. However in both years, average farm size nationally was about twice that of the state (Figure 13.19).

The wartime economy provided the funds for American farmers to significantly up-grade their farm machinery and equipment, but the war's effects on agriculture were much broader. The war spawned a vast array of new technology that was soon adapted for use in the country's food and fiber production system. Included were vastly improved international transportation networks, improved communications systems, more efficient and durable machines, improved manufacturing processes, electronics, widespread production and use of plastics, and an explosion in the development of new chemicals. All these hold-over effects of the war would completely transform the productivity of American agriculture in the next two decades.

Tractors on Farms

The rapid increase in the number of tractor (number per 100 American farms) from 1925 to 1935 is surprising (Figure 13.35). During that period, tractors increased from 9 to 15 per 100 farms; however, that increase was not nearly as surprising as the increase in the latter part of the Depression. In that period, numbers increased from 15 to 25. Considering the number of American farms, that increase represented an awful large number of new tractors. Apparently, lots of American farmers converted most of the value of those AAA checks into new tractors, even in those hard times.

Data in the 1940 Census indicate that there were 25 tractors per 100 American farms in that year. Then by 1945, this ratio increased by almost 71% to 41 per 100. This rate increase was truly astounding. The rate of increase slowed somewhat between 1945 and 1950, but sometime during this period, tractors appeared on more than half of the farms in the country. In 1950, the ratio reached 64 per 100.

Government payments related to New Deal programs were sufficient to allow some Southern farmers to purchase tractors for their farms in the 1930s. For example, Fite (1984) notes that between 1930 and 1940 the percentage of farms with tractors in 11 Southern states more than doubled (3 to 6.3%). Census data indicate that Alabama farmers did not use much of their payment money to buy tractors for their farms. There are no data available on the number of tractors in Alabama in 1935. In 1930, there were only two tractors per 100 Alabama farms, and in the subsequent decade (1940) the ratio increased to only three per 100. The ratio reached eight per 100 in 1945, and then more than doubled to 21 per 100 in 1950. Most of the farms in the state were just too small to justify buying a tractor, even if the money had been available. Further, government checks received by most Alabama farmers were not large enough to purchase a tractor wheel, much less the whole tractor.

Data presented in Figure 13.35 show that in 1940 there were about eight times more tractors per 100 farms in the country than in the state. In 1945, there were only five times more, and in 1950 three times more. Alabama was catching-up quickly, but they had miles and miles to go.

Animal Power on Farms

The use of animal power on American farms, especially the use of horses, began to decline around the end of World War I and at the same time that farm numbers began a sharp downward trend. By 1940, there were only 10.4 million horses in the country, and by 1945 and 1950 those numbers had declined to 8.7 and 5.3 million, respectively (Figure 13.36).

Consideration of the trends in the numbers of horses in the country is more meaningful if they are related to numbers of farms. Figure 13.37 presents data on the number of horses on each 100 American farms during the period 1925-1950. The data show that there were 164 horses on each 100 farms in 1940. These numbers primarily reflect the animal power requirements in the North Central Region to operate the huge machines used in the planting, cultivating, and harvesting of corn and wheat. Tractors were an ideal source of power on the large, level fields found in the Midwest. By 1945, the number had declined to 146, and in 1950 it was down to 94. During the same period, numbers of tractors per 100 farms increased from 24 in 1940, 41 in 1945, and 64 in 1950. It was obvious by 1950 that the use of horses to supply power on American farms would soon be a thing of the past.

The decline in the total number of mules on American farms did not begin until around 1925 (Figure 13.36), and by 1940 it had declined to 4 million, or less than about half the number of horses in the country. By 1945 and 1950, numbers had declined to 3.2 and 2.2 million, respectively. This rate of decline closely paralleled that of the numbers of American farms during the same period (Figure 13.17). Converting total numbers to numbers per 100 farms corrects the data for changes in farm numbers and provides a better opportunity to consider changes taking place on individual farms. Figure 13.38 presents data indicating that there were 64 mules per 100 American farms in 1940, and that afterwards these numbers declined to 54 in 1945 and 38 in 1950. This decline is obviously also related to the increase in tractor ownership in the country during the same period (Figure 13.36).

Figure 13.39 presents data on annual estimates of the number of horses and mules in Alabama during the period 1925-1950. Horses had never been as important on Alabama farms as mules, and as the data show in 1925, there were 200,000 fewer horses than mules in the state. Horse numbers declined slightly during the heart of the Depression, but by 1940 they had increased to 63,000. Horse numbers remained around that level throughout the decade, even with farm numbers declining rapidly in the state (Figure 13.18).

Converted Alabama data reported as horses per 100 farms are presented in Figure 13.37. Throughout the decade, this number remained relatively stable between 25 and 30 per 100 farms, indicating that most Alabama farms did not have horses and suggesting that few of them were being used as a primary source of power in crop production.

The number of mules on Alabama farms during the period 1925-1950 are also shown in Figure 13.39. The total number of mules began to trend downward, along with farm numbers (Figure 13.18), around the middle of the Depression. In 1940, there were 306,000 mules on Alabama farms, but by 1950 number had declined by 38% to 190,000.

When mule numbers are related to farm numbers (Figure 13.38), a slightly different picture emerges. Numbers per 100 farms remained between 121 to 127 from 1925 to 1945, and only after 1945 did they begin to decline sharply. In 1950, there were 86 mules per 100 farms. In that year, the number of tractors in Alabama reached 20 per 100 farms. Tractors were finally beginning to replace mule-power. That was truly was a sign of an agricultural revolution in the state.

Use of Fertilizer

When related to changes in the number of acres of cropland harvested in the United States, fertilizer use changed relatively little between 1919 and 1939 (38.7 to 48.0 pounds an acre), increasing an average 1% a year (Figure 13.40). However, in 1934 it fell to 37.5 pounds, a level lower than in 1919. Then in the first few years of World War II (1939 to 1944), it increased at a rate of 9% a year (48.0 to 107.6 pounds). Prices received were so good during this period that farmers could afford to purchase just about any amount of fertilizer they thought was needed (Figure 13.6). After the end of the war, fertilizer use continued to increase rapidly. Between 1944 and 1949, it increased from 73.9 pounds to 107.6 pounds, an average increase of 6.7 pounds an acre per year. After almost a century, American farmers were finally beginning to listen to the experts.

Plant Agriculture

Throughout the period 1930-1940, corn was the most important crop in the country in terms of acres harvested (Table 13.5). Acres harvested was about two times greater than for either cotton, wheat, or oats. However, while more acres of corn were harvested, cotton was the primary target of the New Deal legislation. Acres of cotton harvested seemed to be directly linked to the unacceptable share-tenant destitution in the South. With the enactment of the AAA legislation, acres of cotton harvested declined from 44.4 million in 1930 to 23.9 million by 1940, a reduction of 44%. At the same time, the new programs resulted in reductions in acres harvested of all of the major crops. By 1940, acres harvested for corn, wheat, and oats declined by 10.6, 12.2, and 8.8%, respectively. At least in terms of acres harvested, the new AAA programs seemed to be accomplishing their purpose. Agriculturalists could not know at the time, but the increase in acres of soybeans harvested between 1930 and 1940 (1.1 to 1.4 million) foretold an astounding change coming in American agriculture.

Data also presented in Table 13.5 indicate that in Alabama, cotton and corn were the primary crops planted and harvested in 1930 (3.6 and 2.8 million acres harvested, respectively). In Alabama, they far overshadowed acres harvested of wheat, oats, and peanuts. Acres harvested for cotton was 29% greater than corn. Following the initiation of the New Deal programs, the order of importance was reversed. In 1940, acres harvested for corn was 81% greater in Alabama. The increases in acres of soybeans harvested (8,000 to 11,000 acres) did not seem to indicate much promise for the crop at that time, but the increase (161,000 to 310,000) in peanuts did. The increase in acres of hay harvested (521,000 to 971,000) seemed to indicate that an increase in cattle production was taking place.

Between 1935 and 1940, New Deal programs certainly changed crop production nationally, but an even more powerful world-wide event was just beginning that would change it even more drastically.

Alabama Crop Weather

Alabama data on Palmer Drought Severity Indices for the months of June through August during the period 1940-1950 are presented in Table 13.6. These data indicate that more than half the state was abnormally dry to severely dry in four of the 11 years of this period (1941, 1942, 1943, and 1944); abnormally moist to wet in four years (1945, 1946, 1949, and 1950); and normal in two years (1947 and 1948). A total of six tropical storms passed through the state during the period (1944, 1945, 1947, 1948, 1949, and 1950).

Cotton

Primarily as a result of Southern dominance on key agricultural committees and sub-commitees in both houses of Congress beginning in 1933, the plight of the cotton farmer took center stage in much of the agricultural legislation enacted during the Depression. As a result, the year New Deal cotton legislation went into effect (1933), the annual average price climbed back above 10 cents, and went on to 12.4 cents in 1934. It was 11.1 cents in 1935 and 12.4 in 1936. Then in 1937, 1938, and 1939 the props collapsed under the cotton basket. Prices fell below 10 cents again. While low prices were a major problem, the amount of government-owned cotton stored in government warehouses was an even greater one.

US Cotton

American farmers harvested 12.6 million bales in 1940. Texas farmers harvested the largest share of the total (25.7%). Other states reporting large harvests included: Arkansas (11.9%), Mississippi (9.9), Georgia (8.0), and South Carolina (7.7). These five states accounted for 63.4% of the total. Alabama farmers harvested 6.3%.

Exports

The beginning of World War II sharply reduced the export of American cotton (Figure 13.41). In 1940, exports were only 1.2 million bales, whereas in 1939 they had been over six million. Exports remained well below two million bales through 1943. They grew to near two million in 1944 and to 3.7 million in 1945, the year the war ended. They remained near the same level in 1946 before falling back to two million in 1947. The 1946 crop was the smallest harvested since 1921. It is likely that the size of this crop, together with the higher price, affected exports in 1947. By 1949, exports were back near six million bales, or near where they had been in the early years of the Depression. In 1950, farmers harvested another small crop, and with prices sharply higher, exports declined to 4.3 million bales. The 1950 exports were equivalent to 27% of the 16.1 million bale crop in 1949.

Prices

AAA legislation in 1933 placed a floor under cotton prices. These support prices were set by law at around nine cents a pound in the years between 1933 and 1940 (Figure 13.42). There are no data available for 1944 and 1945. As a result of the enactment of the Steagall Amendment, beginning in 1941 the government began to increase support prices to encourage increased production needed in the war effort. As a result, support prices increased from 14 cents in 1941 to 23 cents in 1946. However, because the amendment required that the higher level of support prices be continued for two years after the end of the war, they reached 29 cents in 1948. They remained near that level through 1950.

Average annual cotton market prices hovered around 10 cents a pound during most of the Depression years

(Figure 13.43). Even though exports were reduced severely at the beginning of the war, demands by American textile mills were so great that in 1943 acreage controls on cotton were removed. The increased demand also resulted in an increase in market prices to 20 cents in 1942, double those in 1939 and 1940. Cotton prices continued to increase slowly throughout the war, finally reaching 22 cents in 1945. The following year, farmers harvested an extremely poor cotton crop (Figure 13.46), and the market price increased to 33 cents. The teeter-totter relationship between supply and market price seemed to be re-established after 1945. With bales harvested increasing after 1946, market prices began to decline, reaching 29 cents in 1949. Then with the harvest of another poor crop in 1950, market price increased to 40 cents. (The Korean Conflict began in June 1950.)

During the period 1940-1950, market prices increased from 10 cents a pound to 40 cents, an increase of 300%. The Consumer Price Index increased 72% during the period. Throughout this period, market prices were never as low as support prices.

Acres Planted and Harvested

Data on acres of cotton planted and harvested during the period 1925-1950 are presented in Figure 13.44. Surprisingly, cotton farmers did not respond to government exhortations to produce more cotton at the beginning of the war. In fact, war-induced labor shortages and more profitable, alternative crops resulted in reduced acres planted. In 1940, they planted 25 million acres, but only 18 million in 1945, a decrease of 29%. However, in 1946 the market price went to 33 cents, and in 1947 acres planted increased to 21.8 million. After 1946, market prices began to decline, but with support prices increasing, acres planted followed them upward to 28.3 million in 1949. With both support price and market price lower in 1949, farmers reduced acres planted to 18.9 million in 1950.

In seven years of the 11-year period (1940-1950), American cotton farmers harvested the crop from 97% of acres planted. In three of those years (1945, 1946, and 1950) crop weather was unusually wet in the heart of the Coton Belt.

Yields

Year-to-year variation in annual average yields of cotton was extremely high during the period 1940-1950 (Figure 13.45). They ranged from a low of 232 pounds an acre in 1941 to a high of 311 pounds in 1948. However, even with the large amount of variation during the period, there did seem to be an upward trend from about 240 pounds to 290 pounds. Fite (1984) suggested that one of the reasons for increasing yields was the continued movement of cotton production onto irrigated acreages in Texas, Arizona, and California.

Bales Harvested

Year-to-year trends in bales harvested tended to follow trends in acres harvested fairly closely (Figures 13.46 and 13.44). With acres harvested generally trending downward from 1940 through 1945, but with yields increasing, bales harvested generally remained between 9 to 12 million. However, once acres harvested began to trend upward, bales harvested followed, reaching 16.1 million in 1949. With both acres harvested and yields down sharply in 1950, bales harvested fell to 10 million.

So at the end of the decade, despite heroic Triangle efforts, American cotton farmers found themselves in a familiar predicament. Their 1950 cotton crop sold for 40 cents a pound, the highest in history. Unfortunately, they only had 10 million bales to sell.

Alabama Cotton

Alabama farmers harvested 6.3% of the counry's cotton crop in 1940. Farmers in Madison County harvested the largest share (5.9%). Other counties with large shares included: Marshall (4.8%), Limestone (4.7), DeKalb (4.2), and Cullman (3.8). These five counties accounted for only 23.5% of the state's total crop, which indicates the widespread distribution of cotton production. None of these counties were in the old Cotton Arc of south-central Alabama where so much of the cotton was harvested before and immediately after the Civil War.

Prices

Market prices for cotton in Alabama during the period 1940-1950 were essentially the same as for the country as a whole, as shown in Figure 13.43.

Acres Planted and Harvested

Trends of acres of cotton planted and harvested in Alabama during the period 1940-1950 were generally the same as those for the country as a whole (Figures 13.47 and 13.44), except that the rate of decline between 1940 and 1944 was much higher in the state (31.1 versus 18.7%). Alabama farmers planted two million acres of cotton in 1940, but only 1.4 million in 1944. They had not planted fewer acres since 1871. Acres planted continued to decline slightly in 1945, but began to increase afterwards. The decline between 1946 and 1947 is interesting: market price had increased sharply from 1945 to 1946 (Figure 13.43), but rather than planting more in 1947, as expected, they planted less, apparently expecting that market price would likely fall just as sharply. It did decline, but not very much. In 1949, 1.9 million acres were planted, but then fell sharply to 1.3 million in 1950.

In 1925, Alabama farmers harvested cotton from 37.8% of their total cropland (Figure 13.48). In 1930, they increased the percentage to 38.1%. New Deal programs provided them with the opportunity to reduce cotton production in the state, and they accepted it. In 1935, they harvested the crop from only 23.1% of their total cropland. They continued to reduce the amount of land devoted to the crop in 1940 and 1945, and in 1950 acreage planted in cotton was down to 15.3% of Alabama farmers' total cropland.

Yields

Data on average annual cotton yields in the US and Alabama during the period 1925-1950 are shown in Figure 13.45. As expected, year-to-year variation in yields were much higher in Alabama (190 to 353 pounds an acre). The range was about twice as large in the state data (163 versus 79) as in the country as a whole. The national average is likely to be less variable, because it represents the average of all cotton producing states. Even with this high level of variation, yields in Alabama did seem to improve somewhat. The increase in yields between 1940 and 1945 is interesting. In that six-year period, yields increased from 190 pounds to 324 pounds, an increase of 70.5%. During the same period, acres planted fell from two million to 1.4 million, a decline of 31.3%. The extremely rapid increase in yields was probably the result of marginal operators with their low yields, who were leaving farming for jobs in defense industries. It is difficult to explain such a rapid increase any other way. Yields were much lower than expected in 1946, 1949, and 1950. All of these low yields were probably the result of extremely wet weather during those years. In seven of the 11 years during the period, Alabama yields were higher than the national average.

Bales Harvested

Annual trends in bales of cotton harvested in Alabama during the period 1940-1950 are shown by the data presented in Figure 13.49. Acres planted and yields, operating in opposition, resulted in controlling the year-to-year variation in bales harvested, somewhat. Alabama farmers harvested between 800,000 and one million bales in nine of the 11 years during the period. Bales harvested were out of this range in 1948 (1.2 million) when unexpectedly high levels of acres harvested and yields acted in concert and in 1950 when low levels of both had the opposite effect. The 575,000-bale crop in 1950 was the smallest harvested since 1917.

Data presented in the 1950 *Census of Agriculture* indicates that farmers in Limestone County harvested the largest share (6.3%) of the state's 1949 crop. Other counties with substantial shares included: Madison (6.2%), Lawrence (5.4), Marshall (4.6), and DeKalb (4.5). These five counties accounted for only 27% of the total. Most counties reported far smaller crops. Only17 counties reported 15,000 or more bales.

Corn

By 1940, it is surprising that any of the states outside of the North Central Region would bother to produce corn for grain. In that year, American production was 2,206.8 million bushels, and the farmers in that region harvested 70% of it. Shares harvested in the other regions included: South Central (18.3%), South Atlantic (8.6), North Atlantic (2.3), and West (0.7).

US Corn

In 1940, Iowa farmers harvested 20.5% of the country's corn crop. Other states reporting large shares included: Illinois (14.3%), Indiana (6.3), Minnesota (6.1), and Missouri (5.5). These five states accounted for 52.6% of the total crop. Farmers in Alabama harvested only 1.8%.

Exports

When World War II began in 1939, the United States exported 44 million bushels of corn (Figure 13.50). The following year, exports fell to 14.8 million. They declined through 1942, but began to increase in 1944, and in 1944 reached 17.1 million bushels. Beginning in 1945, the export data become very confusing because of shipments abroad under the Army Civilian Supply Programs and similar programs. In 1947, exports totaled only 9.5 million bushels. Bushels harvested that year was the lowest since 1936, a severe drought year. In 1948, 1949, and 1950 exports were near 110 million bushels in all three years.

Prices

New Deal legislation had provided support prices for corn beginning in 1933 (Figure 13.51). They had trended slowly upward to reach 60 cents a bushel in 1940. Afterwards, to comply with the Steagall Amendment, they began to increase more rapidly, reaching $1.10 at the end of the war in 1945. However, the amendment required that they were to be continued for two years after the end of hostilities. As a result, in 1947, it was $1.40. They were kept near that level through 1950.

American average annual corn market prices had been as low as 30 cents a bushel in 1931 and 1932, but it was 60 cents in 1940, just slightly below the support price level—60 cents versus 61 cents (Figure 13.52). In 1941, the market price was still slightly lower—74 cents versus 75 cents. During the war years of 1942-1945 and the postwar years of 1946 and 1947, the demand for corn forced market prices to levels well above support prices. In 1947, when bushels harvested fell to 2.1 billion, the market price rose to $2.16 a bushel, the highest in history to that time. In that year, the support price was set at $1.37. Between 1947 and 1948, American farmers increased bushels harvested from 2.1 to 3.3 billion (Figure 13.55), and in 1948 the market price fell to $1.28, well below the support price of $1.44. They harvested another large crop in 1949 (29.5 billion bushels), and the market price continued to decline to $1.24. The support price was $1.40. A smaller crop in 1950 (27.6 billion bushels) resulted in the slightly better market price of $1.52. The support price was set at $1.47 that year.

Acres Planted and Harvested

American farmers had begun to restrict acres of corn planted when the New Deal programs were initiated in 1933 (Figure 13.53), and by 1940 they were only planting 88.7 million acres of the crop. Acres harvested generally increased in the war years, reaching 8.5 million acres in 1944. Afterwards, they trended downward to reach 7.2 million acres in 1950. It is surprising that the farmers did not respond to the sky-high prices after 1945 (Figure 13.52). In most years, farmers harvested corn from 85 to 87% of acres planted. Considerable quantities of the crop were harvested green for silage, or farm animals were turned into the fields to harvest the grain.

Yields

American corn yields had been trending upward since the mid-1930s (Figure 13.54), and by 1940 they had reached 30 bushels an acre. In that year, hybrid seed were being planted on 30% of the total corn acreage in the country. From 1940 through 1950, yields trended upward to 38 bushels an acre, as the acreage planted with the improved seeds increased. Yields were lower than expected in 1947, as a result of unusually wet weather in the heart of the Corn Belt in June, July, and August.

Bushels Harvested

Bushels harvested generally followed the upward trend in yields during the period 1940-1950 (Figures 13.54 and 13.55). Farmers harvested 2.2 billion bushels of corn in 1940, and 3.3 billion in 1948. Bushels harvested were lower than expected in 1947, as a result of poor crop weather, and higher than expected in 1948 when yield was above trend. Acres harvested declined in 1950, and bushels harvested also declined to 2.8 billion.

Alabama Corn

Alabama farmers harvested only 1.8% of the national crop in 1939. Census data showed that the state's five leading corn-producing counties were: DeKalb (4.5%), Madison (4.2), Jackson (3.8), Marshall (3.2), and Limestone (2.9). These five counties harvested only 18.6% of the state's corn crop that year. All of these counties were north of Birmingham. Farmers in 35 counties reported harvests of 500,000 or more bushels.

Prices

Data on support prices for American corn are shown in Figure 13.51, and the average annual market prices for corn in Alabama during the period 1925-1950 are presented in Figure 13.52. Trends in state market prices for the period 1940-1950 were similar to those for the country as a whole; however in most years, market prices in Alabama were several cents higher. In the latter war years, state market prices were as much as 40 cents a bushel higher than national prices.

Acres Planted and Harvested

Acres of corn planted and harvested in Alabama increased dramatically from around 1925 until 1935, even continuing to increase during the heart of the Depression (Figure 13.56). In 1940, both acres planted and harvested were around 3.6 million. Surprisingly, after 1940 they began to decline rapidly, and by 1949 were 2.6 million. Within 10 years, Alabama lost over a quarter of its corn acreage. There is little indication that Alabama farmers responded at all to the high market prices for corn (Figure 13.52). A considerable share of this loss could be attributed to a reduction in cropland during the period. Census data indicate that there were 10.40 million acres in the state in 1940. By 1950, the total had declined to 8.72 million acres. Farm numbers also declined sharply during this period (Figure 13.18). However, some of the loss of corn acreage was definitely the result of disenchantment with the crop. In 1940, farmers harvested corn from 33.2% of their cropland. In 1950, they harvested the crop from only 28%.

Throughout the period, in the different years acres planted and harvested were generally closer in Alabama than in the country as a whole (Figure 13.53). In the period 1940-1946, Alabama farmers harvested 97% of acres planted; afterwards, however, the percentage began to drift downward to 91% in both 1949 and 1950.

Yields

Data on yields of corn in the United States and Alabama during the period 1940-1950 are presented in Figure 13.54. Yields also trended upward on Alabama farms, but at a lesser rate than in the country as a whole. During the period, yields increased from 12 bushels an acre to around 20. Throughout the period, Alabama's yields were almost 20 bushels lower than the national average. Part of this difference was likely the result of differences in the intensity of the application of cultural practices. For example, during this period, the South lagged far behind the Midwest in the use of hybrid seed. Nevertheless, most of the difference in Southern and Midwestern yields was likely the result of differences in soils and climate. Even as early as 1850, the average Alabama yield was slightly over 10 bushels an acre lower than for the country as a whole.

Bushels Harvested

In 1940, Alabama farmers harvested 39.7 million bushels of corn and 46.6 million in 1941. For the next several years, however, with acres harvested declining rapidly (Figure 13.56), bushels harvested also declined (Figure 13.57). In 1947, Alabama farmers harvested 38 million bushels. At the end of the period (1948-1950), bushels harvested increased, when both acres harvested and yields increased (Figures 13.54 and 13.56). In all three years, bushels harvested ranged around 50 million.

According to census data, farmers in DeKalb County harvested the largest share (5.6%) of the 1949 Alabama corn crop. Other counties reporting large shares included: Marshall (4%), Cullman (3.8), Jackson (3.7), and Blount (2.7). These five counties accounted for only 19.8%, of the total. Farmers in 38 counties harvested 500,000 or more bushels.

Alabama Peanuts

From the beginning, peanuts were primarily a Southern crop. With the passage of time, the crop became even more localized in the region. In 1940, just five Southern states accounted for 91% of the total national crop. Farmers in Georgia harvested 34.1%. Other states with large shares included: North Carolina (20.8%), Alabama (12.9), Virginia (12.2), and Texas (11.0). With the concentration of this crop growing in the South, Southern congressmen took an avid interest in it, and as production increased their interest in providing protection for it also increased. Peanuts was one crop where the Old South had a comparative advantage, at least compared to the remainder of the country.

Prices

Congress first authorized support prices for peanuts in 1941 (Figure 13.58). They were set at 4 cents. Then under the provisions of the Stegall Amendment, they were increased to 10 cents in 1947, and remained near that level through 1950. There are no annual data on market prices for Alabama peanuts before 1982.

Market prices for peanuts had remained below four cents a pound throughout the 1930s (Figure 13.59), and they were only slightly greater than three cents in 1940. Then, during the war years, they began to increase rapidly, and by 1947 they were 10 cents. In the following three years, they remained near that same level. A comparison of Figures 13.58 and 13.59 indicates that price supports played an important role in determining market prices throughout the period. Market prices and support prices were essentially the same, or only slight higher, throughout the period.

Acres Planted and Harvested

Alabama peanut acres planted grew slowly during the 1930s, and by 1940 had reached 520,000 acres (Figure 13.60). With the national demand for oil increasing by 1943, it almost doubled to 820,000. Acres planted generally trended downward through the following six years, and in 1950 it was 405,000. At that point, acres planted was lower than in 1940. This decline was surprising given the relatively higher prices of the crop during the period. It is likely that most of this apparent decline in acres planted was the result of the decline in farm numbers and cropland during the period. Alabama's harvested cropland fell from 7.1 million acres in 1939 to 5.7 million acres in 1949 (Figure 13.27).

Throughout the early part of the period (1940-1950), acres planted were around 200,000 acres higher than acres harvested; however, after 1943 the difference began to diminish, and by 1950 it was only around 70,000 acres higher. Historically, peanut producers had always allowed their hogs to harvest most of the nuts that they produced, and apparently this practice continued well into the 1940s (Figure 13.60). During World War II, the number of peanut processing facilities was increased. As a result, farmers began to harvest more of their peanut crop for processing. In 1950, farmers harvested 83% of acres planted. In 1945, only 77% were harvested.

Yields

Data on annual average yields of US and Alabama peanuts during the period 1925-1950 are presented in Figure 13.61. From 1925 to 1933, yields of peanuts in Alabama were 100 to 150 pounds an acre lower than for the country, but afterwards there was relatively little year-to-year difference. The average yield in the state was 800 pounds in 1941, but it declined to near 630 pounds in 1947. Afterwards, it began to increase rapidly, finally reaching 970 pounds in 1950. It is difficult to explain why yields increased so rapidly after 1945, but it is likely that with market prices increasing so rapidly (Figure 13.69), farmers simply took better care of their peanut crops. The average yield was lower than expected in Alabama in 1946 (550 pounds). Plots of Palmer Drought Severity Indices (Cook, 1999) indicate that southeast Alabama was much wetter than normal in June, July, and August of 1946.

Pounds Harvested

Alabama farmers harvested 227.8 million pounds of peanuts in 1940 (Figure 13.67), whereas they had harvested 92.6 million in 1930. With both acres harvested and yields increasing (Figures 13.60 and 13.61) after 1940, pounds harvested also trended sharply upward, reaching 416.2 million pounds in 1943. Afterwards, with acres harvested trending downward and yields trending upward, pounds harvested remained relatively stable at around 300 million pounds.

Census data indicated that farmers in Henry County harvested the largest share (16.5%) of the 1949 Alabama peanut crop. Other counties reporting substantial shares included: Houston (15.3%), Coffee (13.2), Pike (10.4), and Geneva (8.8). These five counties accounted for 64.2% of Alabama's total crop.

Alabama Soybeans

Soybeans had been planted in the United States for many years, primarily as a hay and fodder crop. National production, however, was so low that the USDA did not begin to publish annual statistics on acres planted and harvested until 1924. By 1929, significant quantities of beans were being crushed for oil. Then, because of the wartime need for additional vegetable oils, authorities urged farmers to increase their plantings of the crop; support price levels were increased to encourage them to do so.

In 1940, American farmers harvested 78 million bushels of soybeans. Illinois farmers harvested the largest share (44.7%). Other states with large shares of the crop included: Iowa (18.2%), Indiana (12.0), Ohio (11.5), and North Carolina (2.9). These five states accounted for 86.4% of the national total. Alabama farmers harvested slightly less than 0.1% of the national crop that year. Because the state played such a minor role in American soybean production, only a limited amount of national data will be used in this presentation.

The four North Central Region states accounted for 86.4% of the total 1940 national crop. In hindsight, about 1940 the USDA should probably have refused to fund any more soybean research and extension projects in states out of the North Central Region.

Alabama farmers had been interested in soybeans for a number of years, but mainly because, as legumes, they added nitrogen to the soil. The crop, with beans attached, also made an excellent forage. Acres planted had increased rapidly through the 1930s, but Alabama farmers never had much interest in harvesting the beans; however, with prices increasing rapidly after 1940, the beans in the bin began to look awfully good.

US Exports

In 1940, the United States exported only 300,000 bushels of soybeans (Figure 13.63) and only 900,000 in 1943, well into the war years. They increased somewhat, to around three million bushels in the period 1944-1947. After the Marshall Plan was approved in 1948, overseas shipments of soybeans increased dramatically to as much as 27.8 million bushels in 1950.

Prices

Congress first authorized support prices for soybeans in 1941 (Figure 13.64). The support price for 1941 was set at $1.05 a bushel. The government increased it to $1.60 in 1942, then to $1.80 in 1942, and to $2.04 in 1943. It remained near $2.04 through 1950. Between 1947 and 1948, market prices fell from $3.33 to $2.27, and in 1948 the support price was increased to $2.18. Throughout the period, market prices remained so high that there was little need to tinker with support prices.

National soybean market prices generally remained below $1 a bushel throughout the 1930s, and in 1940 they were 89 cents (Figure 13.65). With the increased demand resulting from the war effort, they increased rapidly to $2.08 a bushel by the end of the war in 1945, and continued upward to $3.33 in 1947. Market prices generally trended down after 1947, and in 1950 they were $2.47. It is surprising that market prices did not respond to the increase in exports (Figure 13.63) after the passage of the legislation establishing the Marshall Plan in 1948; however, the sharp increase in bushels harvested (Figure 13.67) was large enough to supply all the needs for the crop. In fact, bushels harvested was so large that market prices declined sharply in 1948 and 1949. Support prices seemed to provide a useful floor for market prices through 1945, but afterwards demand was apparently so great that market prices remained well above support prices through 1950.

Data on market prices for Alabama soybeans during the period 1940-1950 are also presented in Figure 13.65. From 1940 to 1945, Alabama farmers harvested only 0.1 to 0.2% of the total national soybean crop. During those years, market prices received by Alabama farmers were as much as $1.24 a bushel (1944) higher than those received by American farmers; however as the state's share of the total crop increased (1.3% in 1947), market prices became essentially equal. After 1947, with demand declining, market prices moved downward toward support-price levels, but never reached them.

Acres Planted and Harvested

Acres of soybeans planted in Alabama reached 275,000 in 1940 and were somewhat higher in 1941 (350,000). It remained near that level through 1943, before beginning to trend sharply downward, finally reaching 137,000 acres in 1950. Part of this decline resulted from the decline (16%) of harvested cropland in the state. but the larger cause was that it was no longer economically practical to produce soybeans for hay. As shown in figure 13.66, Alabama farmers had never harvested very many of their planted acres. After 1943, with indices for prices paid increasing so rapidly (Figure 13.6), it was no longer practical to plant a crop that would not be harvested. Of course, this situation was about to change, and change dramatically.

In 1940, Alabama farmers harvested soybeans from only 4% of acres planted, and the percentage remained near that level through 1943. Then with acres planted declining and acres harvested increasing, the percentage began to increase, and by 1950 they harvested 49.6% of acres planted. There had never been much interest in commercially extracting the oil from soybeans. Consequently, there were few processing plants anywhere. This situation changed during World War II, and by 1950 markets for soybeans were expanding rapidly. As markets expanded, acres harvested also increased.

US and Alabama Yields

American soybean yields generally trended upward throughout the 1930s, and were at 16.2 bushels an acre in 1940 (Figure 13.67). Although there was considerable year-to-year variation, yields remained at around 18 bushels an acre through 1947. They increased sharply to 21 bushels in 1948 and remained near that level through 1950. Average yield was lower than expected in 1947. Palmer Drought Severity Indices data (Cook, 1999) indicate that the entire soybean belt was much wetter than normal in 1947.

Throughout the 1930s, Alabama soybean yields had been 10 bushels an acre lower than the national average (Figure 13.67). In 1940, the average yield in the state was 6.5 bushels an acre, while the national average was 16.2 bushels. In 1941, however, state yields began to increase rapidly, and by1950 they were only slightly lower (19 versus 21.7 bushels) than the national average. In some of those years, the state average actually exceeded the national one. This sharp increase represents the transition from soybeans as a forage crop to soybeans as a cash crop. After Alabama farmers entered the soybean business, they began to improve their cultural practices. This increase in yields, beginning around 1941, coincided with the upward trend in acres harvested (Figure 13.66).

Bushels Harvested

With Alabama soybean acres harvested slowly increasing (Figure 13.66) and yields increasing rapidly (Figure 13.67), it is not surprising that bushels harvested also increased sharply during the period 1940-1950 (Figure 13.68). It was 72,0000 bushels in 1940, but increased to 1.3 million in 1950. Alabama's 1950 crop represented 0.4% of that year's 299.2 million-bushel national crop. Alabama farmers were not really in the soybean business yet, but they were making progress.

Census data indicated that farmers in Baldwin County harvested the largest share (74.5%) of Alabama's 1949 soybean crop. Escambia and Mobile counties accounted for 11.7 and 2.8%, respectively. These three southwestern counties accounted for 89% of the state's total crop. Jackson County reported an additional 3.7%.

Alabama Hay

American farmers harvested 96 million tons of all kinds of hay (dry) in 1940. They had harvested 74.5 million in 1930. Leading hay producing states in 1940 were: Wisconsin with 7.5% of the total, Iowa (6.8), Minnesota (6.5), New York (5.9), and California (5.3). Alabama farmers harvested 0.7% of the national total that year.

Prices

The average price received for hay in Alabama in 1940 was $10 a ton (Figure 13.69). A new cattle cycle began in Alabama in 1939. Cattle numbers reached their highest level in that cycle in 1945, although there was little increase above 1944. Numbers began to decline in 1946, and reached the end of that cycle in 1949. The rapid increase in the numbers of cattle at the beginning of the cycle resulted in a concomitant increase in hay demand and hay prices. Prices increased rapidly during this period, finally reaching $28 a ton in 1944. They remained relatively unchanged near that level in 1946 and 1947, even though cattle numbers had begun to decline rapidly. In 1948 and 1949, prices also declined sharply, as numbers continued to decline. They were $22 in 1949. Cattle numbers in the state began to increase again in 1950, and the average hay price followed them upward to reach $24 that year. Between 1940 and 1944, prices increased from $10.10 to $28.10, or by 178%. During the same period, the Consumer Price Index increased by 25% (42.0 to 52.7). While inflation of the currency played a role in the increase in prices, it was only a small one.

Acres Harvested

In 1940, acres harvested of all kinds of hay in Alabama totaled 971,000 tons (Figure 13.70). Then as prices increased (Figure 13.69), acres harvested also began to increase. In 1943, farmers harvested 1.2 million acres of hay, the highest in history to that time. Although cattle numbers and hay prices continued to increase sharply through 1944, acres harvested began to decline after 1943. In 1950, Alabama farmers harvested hay from only 801,000 acres, less than in 1940.

Yields

In 1940, the average annual hay yield in Alabama was 0.74 ton an acre (Figure 13.71). It had been around that level since the late 1920s. During the period 1943-1945, they fell to 0.7 ton an acre. They had not been that low since 1933. Yields were unusually low in those years as a result of abnormally dry weather. Yields were up again in 1945, 1946, and 1947 to around 0.72 ton, but then increased rapidly in 1949 and 1950 to reach 0.85 ton. Acres harvested were declining rapidly (Figure 13.70). In 1949 and 1950, apparently farmers were only harvesting hay from their better pastures.

Tons Harvested

Data on hay harvested annually in Alabama during the period 1940-1950 are presented in Figure 13.72. With acres harvested decreasing, and yields increasing (Figures 13.70 and 13.71), tons harvested remained at about the same level (700,000 tons) in most of the years of the period. Generally, World War II seemed to have very little effect on the quantity of hay harvested. Tons harvested did increase during the early 1940s, but the increase seemed to be the result of the 1939-1949 cattle cycle, rather than the effects of the war itself.

Alabama Wheat

American farmers harvested 592.8 million bushels of winter wheat in 1940. They had harvested 633.8 million bushels in 1930. At the beginning of the war years (1940-1945), Kansas produced 21.3% of the total national crop. Other states with substantial harvests included: Oklahoma (9.8%), Ohio (7.1), Illinois (6.6), and Nebraska (5.7). These five states accounted for 50.5% of the total. Alabama farmers harvested only 75,000 bushels, or 0.01% of the total. In the 1940s, Kansas further solidified its position as the wheat capital of the United States.

At mid-century, the wheat plant genome seemed to be increasingly happy in the Temperate Oceanic climate (Dc) and on the Mollisol soils of Kansas and Oklahoma (Nuttonson, 1955). At the same time, Alabama farmers were hardly in the wheat business at all, but they had been before and would be again.

US Exports

Data on wheat exports includes all kinds of wheat produced in commercial quantities. Generally, winter wheat comprised the larger share. The United States exported 40.6 million bushels in 1940 (Figure 13.73). Exports had been near that level since the early 1930s. The 1940 exports amounted to 7.2% of the 1939 crop of 565.7 million bushels of winter wheat.

Surprisingly, wheat exports did not increase as World War II spread around the globe. They remained near the 1940-level of 40 million bushels through 1944. At the end of 1944, the Allies had liberated much of Western Europe and were poised to enter Germany. In the Pacific, American forces had returned to the Philippines. In early January, the Soviets began their broad attack from the east. By the middle of 1945, hundreds of millions of people had been liberated around the world, but many of them were facing starvation. At that time, the export of American wheat literally exploded upward, reaching 319 million bushels that year. They continued to increase afterwards, reaching 505 million bushels in 1948. Exports declined sharply in 1949 to 300 million bushels, as local agricultural production (especially in Europe) began to come back on line.

US and Alabama Prices

Congress approved support prices for wheat in 1938 (Figure 13.74). It was set at 60 cents a bushel for that year and increased to 63 cents in 1939. With the enactment of the Stegall Amendment, it was raised sharply to 98 cents in 1941. It was raised more slowly during the period 1942-1944, then set at $1.38 in 1945. Surprisingly, it was again raised sharply in 1946 and 1947, and in 1948 increased to $2.00. It remained near that level in 1949 and 1950.

American winter wheat market prices were 70 cents a bushel in 1940 (Figure 13.75). They had been very near the same level in 1930. Exports had remained generally unchanged in the early years of World War II (Figure 13.73). However, market prices began to increase sharply after 1940 when Germany was sweeping across Europe. They continued to trend upward throughout the war, reaching $1.50 a bushel in 1945, and continued upward rapidly through 1946, finally reaching $2.24 in 1947. However, the demand for wheat began to slip after 1947; market prices retreated to $1.86 in 1949, but were up moderately in 1950 ($2.01). During the period, the year-to-year changes in market prices usually paralleled those of the indices of prices received for all crops (Figure 13.6).

Market prices were below support prices in 1940 and 1949. In those years, bushels harvested were so great that they could not be completely accommodated by the support prices. From 1943 through 1947, demand and market prices were so high that support prices had little or no effect, but in 1948 the market price fell to the support-price level of $2.00 a bushel, and was only slightly higher in 1950.

Market prices for Alabama wheat are also shown in Figure 13.75. In the late 1930s and early 1940s, state market prices were around 25 cents a bushel higher than elsewhere in the country, but afterwards the differential decreased, and by 1949 and 1950 the state prices were only a few cents a bushel higher. Throughout the period, year-to-year changes in Alabama market prices were very similar to those in the country as a whole.

Acres Planted and Harvested

Alabama farmers planted 7,000 acres of winter wheat in the fall of 1939 (the 1940 crop) (Figure 13.76). They had only planted 2,000 acres in 1929 (the 1930 crop). At the end of the 1930s, Alabama farmers, unlike those in the country as a whole, had responded to the increase in market prices by planting more of the crop (Figure 13.75). They increased acres planted significantly each year from 1940 through 1944, except in 1943 when there was little or no change from the preceding year. In 1944, acres planted reached 25,000. Even though market prices continued to increase through 1947, and exports (Figure 13.73) through 1948, Alabama farmers began to reduce acres planted in 1945. With the Allies advancing rapidly toward the German borders in the fall of 1944, state farmers only planted 24,000 acres of wheat. Then with the war over in Europe (May 1945) and the Pacific (September 1945), farmers sharply reduced acres planted that fall (the 1946 crop) to only 15,000 acres. From 1947 through 1950, acres planted generally declined, until in 1950 only 13,000 acres of the grain were planted. It appeared that Alabama's short flirtation with wheat was little more than a nod to the god of war.

Figure 13.76 also contains data on acres harvested for Alabama winter wheat during the period 1940-1950. In general, year-to-year trends in acres planted and acres harvested were similar. Throughout the period, Alabama farmers harvested around 88% of acres planted. Alabama farmers had traditionally used a considerable amount of their wheat crop for grazing.

US and Alabama Yields

Data on the average annual yields (bushels per acre) of winter wheat in the United States and in Alabama during the period 1940-1950 are presented in Figure 13.77. In 1940, the average yield in the country was 16.4 bushels an acre. It had been 15.4 in 1930. Although there was considerable year-to-year variation, they trended upward for most of the period from 16 bushels in 1940 to around 19 bushels in 1947 and 1948. In 1943, 1949, and 1950, average yields were much lower than expected, as a result of excessively wet weather in the heart of the Wheat Belt.

Alabama farmers harvested 12.5 bushels an acre in 1940 (Figure 13.77). They had harvested only 10 bushels in 1930. Throughout the period, yields in Alabama also trended upward. State farmers harvested 12 bushels an acre in 1940 and between 15 and 16 bushels an acre in 1950. In fact, year-to-year trends in yields in both the country and the state were remarkably similar. The average yield in 1943 was much lower than expected, as a result of severely dry weather during June, July, and August. Throughout most of the period, average yields in Alabama were considerably lower than those elsewhere in the country. They were somewhat closer in 1949 and 1950, when national yields were much lower than expected. However, in the period 1940-1948, state yields were 20% below national yields. This difference in yields is probably the primary reason that Alabama farmers got out of the wheat business.

Bushels Harvested

Alabama farmers harvested 75,000 bushels of winter wheat in 1940 (Figure 13.78). A decade earlier they had harvested a mere 20,000 bushels. During the 1940s, the trend in bushels harvested was more strongly affected by acres harvested than by yields. In fact, the graphs of bushels harvested and acres harvested are remarkably similar (Figures 13.78 and 13.76). Bushels harvested increased sharply from 75,000 in 1940 to 216,000 in 1942. It was down slightly in 1943 as a result of poor yields. It reached 359,000 bushels in 1944. Then, as acres harvested declined, so did bushels harvested. It reached 174,000 in 1946, and remained slightly below 200,000 for the remainder of the period.

Alabama Oats

In 1940, American farmers produced 1.5 billion bushels of oats, just slightly more than the 1.3 billion bushels produced in 1930. Iowa was the leading oat-producing state in 1940, with 13.6% of the total national crop. Other states reporting substantial harvests included: Minnesota (12.3%), Illinois (10.1), Wisconsin (6.7), and Michigan (4.2). These five states accounted for 46.9% of the national total. Alabama farmers produced 2.6 million bushels, or 0.2% of the national total. They had produced 1.5 million bushels in 1930, or 0.1% of the total. Alabama was still marginally in the oat business in 1940. With 369,000 horses and mules in the state (2.5% of the national total), Alabama farmers really needed to stay in the business of growing oats, if they could afford to do so.

US Exports

American oat exports were one million bushels in 1940 (Figure 13.79). Exports did not increase very much in the early years of World War II. They increased to five million bushels in 1941, and remained at near that level through 1944. However, after France and the Low Countries were mostly liberated by the fall of 1944, exports literally exploded in 1945 to 21 million bushels, and remained in the range of 21 to 26 million bushels through 1948. After 1948, production of the crop began to recover throughout Western Europe, and by 1950 American exports were down to near seven million bushels.

US and Alabama Prices

Congress first approved support prices for oats in 1945 (Figure 13.80). It was set at 48 cents for that year, then raised annually to reach 70 cents in 1948. It remained around that level through 1950.

American farmers received 30 cents a bushel (market prices) for their oats in 1940, essentially the same amount that they had received in 1929 (Figure 13.81). Oat market prices in the United States did not receive an immediate boost from the beginning of the war in Europe in 1939. America was not directly involved in hostilities until late 1941; market prices increased sharply that year to slightly more than 40 cents a bushel. They continued to climb through 1947, although they were down slightly in 1944 and 1945. In 1947, they reached $1.04, the highest level since 1919. After 1947, market prices were down sharply, and they remained in a range of 66 to 79 cents through 1950. During the period, yearly changes in market prices usually paralleled the annual changes in the indices of prices received (Figure 13.6).

Support prices had little effect on market prices between 1945 and 1947, when demand was so high. However, in 1948, the market price fell almost to the support-price level (72 versus 70 cents). Subsequently, the market price was slightly lower in 1949, and slightly higher in 1950.

Although market prices and exports were both relatively high during some of the war years, it is not likely that sales abroad, alone, had much of an effect on market prices. The highest level of exports during the period (26 million bushels in 1946) was equivalent to only1.7% of the 1945 crop of 1.5 billion bushels. Usually, only a small fraction of the American oat crop ever entered the international market.

American oat prices increased 247% from 1940 through 1947 (30 cents to $1.04 a bushel); however, only a quarter of the increase can be attributed directly to inflation of the currency. During that period, the Consumer Price Index increased from 42 to 66.9 (59%).

Figure 13.81 also includes data on average oat market prices in Alabama during the period. Trends in the country and the state were usually similar during the period, although Alabama prices were slightly higher. In 1940, the state market price was 16 cents higher, but in the latter half of the decade, the difference was as much as 45 cents. Throughout this period (1940-1950), bushels of Alabama oats harvested was such a small share of the national crop that it is likely that most of it was used locally.

Acres Planted and Harvested

Data on acres planted of oats in Alabama are not

available before 1926. State farmers, like their national counterparts, also responded to the beginning of the war in Europe in 1939 and to increasing prices (Figure 13.81) by planting more oats. They had planted only 145,000 acres of the crop in 1939, but increased their acreage to 203,000 in 1940 (Figure 13.82). They continued to increase acres planted in 1941, and in 1942 they planted 398,000. However, even though prices were continuing to increase, Alabama farmers became somewhat disenchanted with oats after 1942, and acres planted generally trended downward through 1950 to a level of 277,000 acres. Both farm numbers and harvested cropland were trending downward during this period. Between 1945 and 1949, acres planted declined by almost 29% (318,000 to 239,000). From 1945 through 1950, harvested cropland fell by 7%.

Figure 13.82 also includes data on acres harvested each year during the period 1940-1950. Alabama farmers generally harvested a much lower percentage of acres planted than did their counterparts in most of the other states. During that period, Alabama farmers never harvested more than 78% (228,000 of 294,000 acres in 1944). After that year, they began to harvest smaller and smaller portions of their acres planted, until in 1950 they harvested only 72,000 of 277,000 acres (26%). Alabama cropland was excessively wet during the spring and summer of 1949 and 1950; consequently, it is likely that both years the crops deteriorated so badly that most of the acres were never harvested.

US and Alabama Yields

Although there was considerable year-to-year variation, American oat yields trended slightly upward during the period 1940-1950 (Figure 13.83). If there was a real upward trend, however, it was not greater than one to three bushels an acre. During the entire period, average yields were in the range of 30 to 35 bushels.

Figure 13.83 also includes data on average oat yields in both the country and the state. Yields in the state definitely trended upward during the period; however they were at least 10 bushels an acre less than the national average. In the early 1940s, the difference was even greater. Although Alabama yields trended upward through the early 1940s, after 1944 they tended to remain at around the same level (23 to 26 bushels an acre).

Bushels Harvested

With annual average yields changing relatively little during the period 1940-1950 (Figure 13.83), the trends in the data on bushels harvested (Figure 13.84) were quite similar to the trends in the data on acres harvested (Figure 13.82). Farmers harvested 2.6 million bushels in 1940 and 6.3 million in 1945, but in 1950 they harvested even less of the grain than in 1940 (1.9 million versus 2.6 million bushels). In 1940, Alabama farmers harvested 6.9 bushels of oats for each head of work animals (horses plus mules). In 1950, they only harvested 6.4 bushels per animal. Obviously, farmers were feeding their animals something other than oats, or they were purchasing oats somewhere else.

Alabama Sweet Potatoes

Data on acres of sweet potatoes harvested in Alabama during the period 1940-1950 are presented in Figure 13.85. Alabama farmers had been planting and harvesting fewer acres of the crop since the mid-1930s. Acres harvested were down sharply in 1940 (64,000), when Baldwin County received extremely heavy rains in June and July. Cullman County received 13.3 inches in June. After prices increased 25% ($1.72 to $2.15 per Cwt), between 1941 and 1942 farmers significantly increased acres harvested in 1943 to 91,000. However, even though prices continued to increase through the war years, Alabama farmers seem to have lost their confidence in the crop. Acres harvested trended downward through the late 1940s to reach 35,000 in 1950. Between 1945 and 1950, acres harvested declined by 50%. They had not harvested fewer acres since 1871. The decline in farm numbers and harvested cropland can account for part of the decline, but no more than about a third of it. The decline in farm numbers may have played a larger role than suspected. The farms that were lost during the period probably accounted for a larger share of acres harvested, especially for sweet potatoes.

Alabama sweet potato harvest in 1940 totaled 1.7 million Cwt. Cash receipts from the sale of that crop were $2.9 million. This represented 4.7% of total cash receipts ($60.8 million) for all plant crops, excluding farm forest products, originating on Alabama farms that year. In 1950, sweet potato production totaled 1.7 million Cwt, with a market value of $6.7 million. Cash receipts for this crop represented 2.9% of the value ($231.8 million) of all farm products, except timber. In 1950, Alabama farmers harvested 6.4% of the national crop. Louisiana reported the largest share (22%). Alabama sweet potato production ranked seventh in the country that year. All of these data indicate that Alabama farmers had found still another crop for which they had little or no comparative advantage.

Alabama Irish Potatoes

Data on acres of Irish potatoes harvested in Alabama during the period 1925-1950 are presented in Figure 13.86. These data show acres harvested generally trending upward after the mid-1920s, and reaching 51,000 acres in 1940. It remained around that level through 1944 when, with market prices still extremely high by historical standards, farmers began to reduce acres harvested. By 1950, it was down to 32,000. It had not been that low since the early 1930s.

Cwt of Irish potatoes harvested in Alabama in 1940 was 2.6 million. With the market price at $1.08 per Cwt, cash receipts were $2.8 million, which represented 4.7% of cash receipts for all Alabama plant crops ($60.8 million), excluding farm forest products. In 1950, Cwt harvested totaled 2.2 million with an annual market price of $2.18 per Cwt. Total cash receipts for that crop were $4.9 million. These cash receipts represented 2.1% of the value ($231.8 million) of all Alabama farm products, except timber. In 1950, the Irish potato harvest in the state ranked twenty-first in the country.

Alabama Sorghum

Data on total acres of sorghum harvested in Alabama during the period 1925-1946 are presented in Figure 13.87. There are no annual data on total acres harvested after 1946. The data show that acres harvested ranged around 64,000 through much of the period, but increased sharply in 1946 when farmers begin to plant and harvest sorghum for grain in substantial quantities.

In 1940, almost equal amounts of sorghum was harvested for the production of molasses (31,000 acres) and for forage (28,000 acres). A much smaller share was used for silage (5,000 acres). Usage and total acres harvested remained around the same through 1944, when farmers began to increase acres planted for grain.

Between 1940 and 1946, around 30,000 acres of sorghum were harvested annually for making molasses, but afterwards acres harvested for that purpose declined rapidly, and by 1950 only 8,000 acres were harvested. During the period 1940-1946, between 23,000 and 30,000 acres were harvested for forage, but no data was collected on that use afterwards. The share harvested for silage remained around 4,000 to 6,000 acres throughout the decade.

In the 1940s, sorghum contributed relatively little to the overall state economy. For much of the period, farmers harvested only about 60,000 acres of the crop. In the early part of the period, sorghum molasses contributed around $2 million dollars a year to the economy, and forage a lesser amount. The crop would have been of considerable importance in several of the counties. In Talladega, Madison, and Montgomery counties, the value of sorghum products, except molasses, ranged between $51,000 and $88,000. In a number of counties, value was greater than $40,000.

Alabama Sugarcane

Between 1940 and 1945, Alabama farmers harvested an average of 22,000 acres of sugarcane for making syrup (Figure 13.88). Subsequently, acres harvested declined sharply, and in 1950 reached 10,000 acres. Throughout the decade, yields of cane harvested averaged around 118 gallons of syrup an acre. Prices increased from 60 cents a gallon in the 1940s to $2.10 in 1946. Afterwards, it declined to $1.30 in 1950. Value of the syrup ranged from $900,000 in 1940 to $4.9 million in 1946.

Alabama Vegetables

Data published in the 1940 *Census of Agriculture* indicate that in 1939 Alabama farmers produced vegetables valued at $10.5 million. Only 12.8% of the total crop was produced for sale. The remaining 87.2% was produced for home use. Farmers in Blount County produced the largest share of the total (3.7%). Other counties reporting large shares included: DeKalb (2.9%), Tuscaloosa (2.9), Cullman (2.8), and Marshall (2.5). These five counties accounted for only 15% of the total, indicating the widespread production of vegetables in the state in 1939.

The Alabama Agricultural Statistics Service has provided data on the harvest of the most important vegetables (truck crops) in the state for the period 1940-1950. The relevance of this group of plant crops to the state's agricultural economy was discussed in the preceding chapter. Data in Table 13.7 lists the acres harvested of each crop for the period. In most of the years, acres harvested for most of the crops (cabbage, cucumbers, strawberries, and snap beans) varied between 1,000 and 5,000. The quantity harvested per acre was quite high. There was not likely to have been a market for the produce from larger acreages. Acres of watermelons and tomatoes harvested were somewhat higher (6,000 to 8,000 acres) in most years. Commercial early Irish potatoes had the greatest number of acres of any other truck crop. The Statistics Service provided data on two groups of Irish potatoes. The commercial early group is included under truck crops. The other group has already been discussed in a preceding section. The Satistics Service also published data on acres harvested for two different crops of tomatoes (late spring and late summer) and snap beans (mid-spring and late summer). These have been combined into a single total for the table entry of each crop. Acres harvested for some of the crops increased during the war and declined afterwards. For others, the war seemed to have little effect.

Table 13.8 includes data on the total value of the various crops harvested in the different years. The value of the commercial early Irish potatoes was highest, but of course, acres harvested was considerably higher for that crop as well. Among the other crops, values for cabbage and cucumbers were the lowest. Values for strawberries and tomatoes were part of a third group, and watermelons and snap beans seemed to be part of a fourth group. Changes in acres harvested during the war years were not very obvious with most of the crops. For all the crops, with the possible exception of strawberries, values increased during the early- to mid-1940s, and declined afterwards.

Alabama Velvet Beans

Acres of velvet beans harvested averaged around 460,000 a year in the period 1940-1943, but afterwards began to decline rapidly. By 1949, farmers harvested the crop from only 70,000 acres (Figure 13.89). Yields remained near 850 pounds an acre for the entire period. Prices increased throughout the period from $23.30 a ton in 1940 to $36.00 in 1950. Farmers must have been getting tired of working with those stinging beans, but of even greater concern was the inadequate gross returns from producing the crop. Even with the 1950 price of $36.00 a ton, the gross return on 850 pounds (0.42 ton) of beans was only $15.12 an acre, and many of those stinging pods had to be picked by hand. Alabama farmers were still desperately trying to diversify their agriculture, but they were rapidly coming to the conclusion that velvet beans would not help very much.

Alabama Cowpeas

Data on acres of cowpeas harvested on Alabama farms during the period 1925-1950 are presented in Figure 13.90. Cowpeas were either planted alone or inter-planted with other crops. Data for both (circles and diamonds, respectively) are presented in the figure. Further, the data for the inter-planted acreage has been converted to its equivalent planted-alone acreage. Finally, the planted-alone and converted data are combined and are represented by the triangles in the figure.

Alabama farmers also began to loose interest in cowpeas during the period 1940-1950. In 1941, they harvested the crop from 368,000 acres. They used 47% of the harvested crop for peas, 32% for hay, and either grazed or plowed-under the remaining 21%. However, in 1950 they harvested the crop from only 61,000 acres, and they used 51% for peas, 7% for hay, and 42% was grazed or plowed under. Yields remained nearly the same at 5.8 bushels an acre throughout the period. Prices increased from $1.65 to $4.08 a bushel between 1940 and 1950. The problem was the same as with velvet beans. Gross returns were just too low. Even at $4.08 a pound, the gross return per acre was only $23.60, and most of those peas had to be hand-picked.

Alabama Pecans

In 1940, farmers in south Alabama harvested 5.7 million pounds of pecans from a relatively poor (masting-year) crop (Figure 13.91). Cash receipts from the sale of that crop ($638,000) were equivalent to 1% of the amount obtained from the sale of all plant crops ($60.8 million), excepting forest products, produced on Alabama farms that year. They had harvested 3.1 million pounds in 1930. Pecan harvest, in 1950, totaled 11.7 million pounds, in another masting-year, and the crop sold for slightly less than $3.5 million. In 1950, the sale of pecans constituted 1.5% of the sale of all products from Alabama farms, excepting forest products. Alabama pecan production ranked third in the country that year.

Although there was considerable year-to-year variation, pounds harvested generally trended upward during the period, a trend that had begun in the mid-1930s (Figure 13.91). According to the *Census of Agriculture*, Alabama farmers increased the number of pecan trees significantly during the early 1940s. In 1940, there were 785,000 trees, of all ages, on Alabama farms, but by 1945, farmers had planted an additional 186,000 trees. However, it is not likely that these new trees would have contributed very much to the upward trend in production during the period 1940-1950. More likely, this upward trend was the result of another large increase in tree planting in the late 1920s.

The high level of year-to-year variation in pounds harvested was probably the result of the interaction of the masting characteristic of pecans and local weather, as described in preceding chapters. Pounds harvested, was higher than expected in 1944, which should have been a masting-year. Similarly, in 1947 and 1950 it was lower than expected. South Alabama was moderately wet in the late summers of 1946 and 1947. Precipitation at Fairhope was 62% and 119% above normal, respectively, in those two years. Also, tropical storms were active in that area in September and October of 1940, 1941, 1945, and 1950.

Pecan prices increased sharply during the war years. As a result, value of production also increased. Farmers had received $638,000 (11.2 cents a pound) for their crop in 1940 and $3.2 million (36.3 cents a pound) for the 1946 crop. In 1948, prices were back down to 13.5 cents, and they received $2.7 million for a 20-million-pound crop.

Alabama Tung Nuts

There are no annual data on tons of tung nuts harvested in Alabama before 1945, but the *Census of Agriculture* published data showing that Alabama farmers harvested 11.1 tons of the nuts in 1939. Cash receipts were $332. Annual data show that tons harvested averaged 1,200 a year during the period 1945-1950.

Alabama Peaches

In 1940, Alabama farmers harvested 33 million pounds (688,000 bushels) of peaches from a freeze-damaged crop (Figure 13.92). Cash receipts from the sale of that crop were $688,000, the equivalent of 1.1% of the cash receipts ($60.8 million) for all plant crops produced on Alabama farms, excepting the value of forest products. Farmers had harvested 68.6 million pounds (1.4 million bushels) in 1930. In 1950, pounds harvested totaled only 10.6 million pounds, the lowest ever recorded. With that relatively small 1950 crop, Alabama ranked eighteenth in the country in peach production. Farmers received $759,000 from the sale of that crop, or 0.3% of the cash receipts from the sale of all plant crops ($231.8 million), excepting forest products, produced on Alabama farms.

Data on pounds of peaches harvested from Alabama farms during the period 1940-1950 indicate that peach crops were likely reduced by late freezes three times (1940, 1943, and 1946) during this 11-year period. After the third damaging freeze (1946), farmers seemed to lose some interest in the crop. Pounds harvested trended downward for the remainder of the period from 60 million pounds in 1946 to 10 million in 1950—even though prices were increasing (5 cents a pound in 1947 to 7.2 cents in 1950).

Alabama Greenhouse and Nursery

Census of Agriculture data indicate that the cash receipts for various horticultural crops sold by American farms in 1929, 1939, and 1949 totaled $145.7, $129.6, and $392.1 million, respectively (Table 13.9). Well over half of these receipts in each year were from the sale of flowers and flowering plants. The return of military personnel to civilian life, plus the home mortgage provisions of the

GI-Bill, set-off a boom in home construction after the end of the war. It is likely that the landscaping of these new homes with purchased horticultural crops largely explains the surge in value between 1939 and 1949.

Data on cash receipts from the sale of Alabama horticultural crops in 1929, 1939, and 1949 are also presented in Table 13.9. There was over an 800% increase ($861,000 versus $5.2 million) in the cash receipts from those crops between 1939 and 1949. Receipts from nursery products (trees, shrubs, vines, ornamentals, etc.) increased from $516,000 to $3 million). Receipts for flowers and flowering plants increased from $96,200 to $2.1 million. Receipts for vegetables and plants grown under glass (flower seeds, vegetable seeds, vegetable plants, bulbs, etc.) decreased from $149,000 to $87,000. These data suggest that Alabama farmers might have finally found crops for which they had a comparative advantage. Production of crops will be watched with a great deal of interest in the coming years.

Alabama Timber

Data on lumber production in the United States during the period 1869-1950 are presented in Table 13.10. These data show that the westward movement in lumber manufacture was continuing during this period. In 1950, the share of production in the West Region (48.7%) was almost equal to the combined shares of all of the other regions. With the exception of the South Atlantic Region, shares of all of the other regions were continuing to decline. In 1869, only 2.8% of all lumber was manufactured in the South Atlantic Region. At the time, the share there was the lowest in the country; however, afterward it had generally trended upward, finally reaching 11.8% in 1950.

There is very little information available on timber resources and harvest in Alabama during the years of World War II. Stauffer (1960) commented that a survey of timber resources in Alabama was conducted in 1945, but that the data obtained was not considered to be as accurate as that obtained in the 1934-1936 survey.

Lawson, et al. (1943) reported that in 1941 approximately 58% of the land area of Alabama was in forests. Leading counties, in percentage of land in forests were: Washington (89%), Baldwin (83), Clarke (82), Choctaw (79), and Escambia and Mobile counties with 78% each. Limestone County had the lowest percentage of land in forests (26%).

The 1950 *Census of Agriculture* included data on the quantity of some forest products sold from Alabama farms in 1929 and 1949. A summary of these data are presented in Table 13.11. These data show that the quantity of all of those forest products increased between 1929 and 1949, except sawlogs and veneer.

Census data on the value of some forest products sold from Alabama farms in 1949 are shown in Table 13.12. As might be expected, the larger share (72%) of the $10 million total value came from the sale of standing timber. The sale of forest products (firewood, fence posts, pulpwood, etc.) accounted for only 28% of the total.

Animal Agriculture

The Great Depression had wreaked havoc on the American livestock industry. The value of livestock sold in the United States decreased from $6.30 to $5.87 billion in the period 1930-1940, a decline of $434 million. The situation had been even worse in 1933 when the value of marketed livestock was only $3.4 billion.

In 1940, the livestock situation in Alabama was not nearly as dire as it was elsewhere in the country. In fact, in 1940 the value of livestock marketed was actually $39.5 million higher than in 1930 ($26.5 million versus $22.6 million), but it had been as low as $13.3 million in 1932. The livestock situation was definitely improving in Alabama, but it still had a long way to go.

Cattle

The Depression had a devastating effect on cattle numbers in the country. A new cycle began in 1929, but it was quickly aborted after 1934. At the apex of the Accumulation Phase of this cycle, there were four million fewer cattle (excepting milk cows) in the country compared to the same position (1918) in the preceding cycle. Through the early Depression years, people simply could not afford to buy beef.

US Cattle

There were 68.3 million cattle and calves (including

milk cows) in the United States in 1940. The largest share (10.2%) was on Texas farms. Other states with large shares included: Iowa (6.9%), Minnesota (5.0), Nebraska (4.2), and Kansas (4.0). These five states accounted for 30.3% of the national total. Farms in Alabama only accounted for 1.5%.

Farm Value Per Head

Farm value per head for American cattle (all) had fallen as low as $12.54 in 1934 (Figure 13.93), but then with the heart of the Depression behind, values began to increase, and the annual average reached $33.50 in 1940. With the beginning of World War II, farm value generally followed indices of prices received for all agricultural commodities upward (Figure 13.6), finally reaching $108.30 in 1949. Its upward trend was interrupted at the end of the war. It declined slightly to $99.20 in 1950. Values had never been that high.

From 1940 through 1950, values increased by 221%. A little less than a third of this increase could be attributed to inflation of the currency. The Depression had been a disaster for American cattle farmers, but the war years were an unparalleled bonanza.

Cycles of Abundance

American cattle numbers (excluding cattle that had calved – milk) completed a relatively short and mild 10-year cycle in 1938, and a new one began in 1939 (Figure 13.94). The January 1, 1940 inventory indicated that numbers had increased to 43.4 million in its Expansion Phase. This phase ended in 1943. The Consolidation Phase began in 1944 and ended in 1945 with numbers at 57.8 million. The Declining Phase began after 1945 and ended in 1948. At that point, the number was estimated to be 52.6 million. At the completion of the 1940s cycle in 1948, there were 11.8 million more cattle in the country than at the end of the 1930s cycle in 1938 (52.6 million versus 40.8 million). A new cycle began in 1949, and by 1950 numbers had increased to 54.1 million in the Expansion Phase of that cycle. During this period, the teeter-totter relationship between supply and price was tenuous, at best. Farmers appeared to be using the status of the general economy (Figure 13.1) to make herd management decisions.

Real Net Cash Flow Per Cow

Data on real net cash returns per cow for American cattle during the period 1925-1950 are presented in Figure 13.95. Returns had been below zero in 1932 and 1933, but then began to trend upward until reaching $59 (real dollars, 1982-1984 =100) in 1936. It declined again in the economic turmoil of the late 1930s, finally reaching $26 in 1940. With the beginning of World War II, returns increased sharply. It was $82 in 1941. Then in the period 1941-1950, it fell below $100 in only two years: $75 in 1944 and $86 in 1949. It was $102 in 1950.

Alabama Cattle

Fite (1984) commented that between 1940 and 1945 there was a sizeable increase in the numbers of cattle and calves in all of the Southern states. During this period, the acreage in cotton in the country (primarily the South) was reduced over six million acres. A considerable portion was converted into pasture and forage. In some areas the conversion was substantial. For example, cotton farmers in the Black Prairie of Alabama, which had been one of the most important cotton-producing areas in the country, used the push of the AAA and the pull of wartime demand for livestock to convert large acreages of level cotton land to pastures and to hay production. The boll weevil had been particularly devastating in that area, and the farmers there desperately needed an opportunity to reduce the level of their dependence on cotton. In 1944, Montgomery County, which was once one of the leaders in cotton production in the Black Belt, was receiving 57% of its farm income from livestock (primarily cattle) and livestock products and only 22% from cotton. A decade earlier, those percentages would probably have been reversed.

Census data indicated that only 1.5% of American cattle (all) were on Alabama farms in 1940. Some 5.5% of the Alabama total was on farms in Montgomery County. Other counties reporting significant shares were: Dallas (4%), Wilcox (3.8), Marengo (3.7), and Hale (2.5). These five counties accounted for 19.6% of the state's total. All of these counties had once been part of the old Cotton Arc across central Alabama.

Farm Value Per Head

During the period 1940-1950, trends in farm value per head for Alabama cattle were similar to those of cattle in the country (Figure 13.93). While the trends were similar, values for cattle in the state were lower throughout the period. In 1940, the difference was $9 a head, but in 1949 it was $27 less.

Cycles of Abundance

Trends in cattle numbers (excepting milk cows) in Alabama during the 1940s generally mirrored those of the country (Figures13.96 and 13.94). In the state, a 1930s cycle that had begun in the late 1920s ended in 1938, and a new 1940s cycle began. In the second year of its Expansion Phase in 1940, there were 627,000 head of cattle (and calves) in the state. This cycle was in its Consolidation Phase from 1944 to 1946, and by that time numbers had increased to 870,000. When this cycle ended in 1949, numbers had declined to 750,000. A new 1950s cycle began in 1950, and numbers increased to 844,000. While numbers were increasing and decreasing during these cycles, the abundance of cattle was increasing significantly. At the end of the Declining Phase of the 1930s cycle in 1938, there were 537,000 cattle in the state. At the end of the Declining Phase of the next cycle in 1949, there were 750,000, an increase of almost 40%.

Hogs

The December 1, 1940 inventory of hogs and pigs in the United States indicated that there were 54.9 million of the animals on its farms. By far the largest share (65.8%) was in the North Central Region. Shares in other regions were: South Central (17.3%), South Atlantic (9.6), West (4.9), and North Atlantic (2.4). The conversion rate of feed to body mass in these animals is quite high. So high, in fact, that it makes little sense to try to produce the animals very far away from the center of grain production.

US Hogs

The largest share of hogs and pigs in the country on December 1, 1940 was in Iowa (16.6%). Shares of some other major hog-producing states were: Illinois (9.6%), Indiana (7.2), Missouri (6.6), and Minnesota (6.2). These five states accounted for 46.2% of the total. The Alabama inventory indicated that the state's share was 1.9%.

Farm Value Per Head

Year-to-year changes in farm value per head for American hogs and pigs generally followed those of the indices of prices received throughout the period (Figures 13.6 and 13.97). Both increased from 1940 through 1948 and began to decline afterwards. Values were $7.78 in 1940 and increased to $45 in 1948. By 1949, wartime pressure on agricultural prices had abated somewhat, and the indices for prices received began to decline. As a result, in 1949 values for hogs had declined to $38.20, and in 1950 it was down to $27.10.

Values declined sharply in 1944 after hog numbers had reached the unprecedented level of 83.7 million the preceding year. (Figure 13.98). Even wartime demand was not powerful enough to keep values moving upward with that many hogs on American farms.

Altogether, between 1940 and 1950 values increased by 246%, but a third of the increase was the result of inflation (Figure 13.2). During most of this period parity ratios remained above 1 (Figure 13.7). American hog farmers had never had it so good.

Cycles of Abundance

Between 1940 and 1941, values per head had increased from $7.78 to $8.34, and the December 1, 1941 and 1942 inventories indicated that hog numbers had increased from 54.4 to 60.6 million (Figure 13.98). Numbers followed values upward until 1943 when they reached 83.7 million. With so many hogs on American farms, value fell sharply in 1944 (Figure 13.97). As a result, by December 1, 1944, American hog numbers had declined by 24.5 million head (83.9 to 59.4 million). Those completely unrealistic numbers in 1942 and 1943 had gotten the attention of hog producers. Afterwards, even though values continued to increase rapidly through 1948, farmers did not respond very forcefully. Numbers increased slightly (61.3 million in 1945), but then began to trend downward through 1947 (54.6 million). In 1948, values reached the highest level in history ($42.80). That extraordinarily high value was just too good to ignore. In December 1948 numbers increased to 56.2 million, and even though values declined sharply in 1949 and 1950, numbers continued to increase, finally reaching 62.3 million in 1950.

There was no clearly-defined cycle in American hog numbers during the period 1940-1950 (Figure 13.98). An Accumulation Phase of a cycle began in 1941 at 60.6

million animals, but there was no real Liquidation Phase. Numbers declined by 24.5 million (83.9 to 59.4 million) in a single year. During the war, the changes in economic conditions were so rapid and so severe that the normal relationship between supply and demand became quite turbulent and unpredicable. Except for the portion of the Accumulation Phase, from 1942 through 1943 numbers changed relatively little. They remained at 50 to 60 million animals throughout the period. In fact, except for that short period, numbers had not really changed very much since the late 1920s and early 1930s.

Alabama Hogs

Alabama Agricultural Statistics Service data indicate that Dallas County reported the largest share (2.4%) of the state's hog herd in 1940. Other counties with large shares included: Wilcox (2.2%), Madison (2.2), Montgomery (2.1), and Jackson (2.0). These five counties accounted for only 10.9% of the herd. There were hogs-hogs-everywhere. The old Anglo-Saxon love of hogs and pork was still quiete evident.

Farm Value Per Head

Trends in Alabama hog farm value per head, over the period 1940-1950, were similar to those for the country (Figure 13.97). Throughout the period, however, those for the state were lower, and the difference grew over the period. In 1940, it was $2.40 a head; in 1950 it was $8.40. Between 1940 and 1950, Alabama hog values went up from $5.40 to $18.70, an increase of 246%, about the same as for the country as a whole.

Cycles of Abundance

Year-to-year changes in Alabama hog and pig numbers generally mirrored those of the country during the period 1940-1950 (Figures 13.99 and 13.98). Although numbers reached an all-time high in 1943 (1.6 million), it was much lower for most of the period. Except for that unusually high level, numbers increased from 1 million in 1940 to 1.2 million in 1950.

Census data indicated that the largest share (4.6%) of Ålabama's hog herd in 1950 was on farms in Houston County. Other counties reporting substantial shares included: Geneva (4.5%), Covington (4.5), Coffee (4.0), and Pike (4.0). These five counties accounted for only 21.6% of the state's herd that year. A total of 26 counties reported hog numbers of 15,000 or more animals. It is somewhat surprising that such a large share of the state's hog herd would have been located on farms in those five southeastern counties, when the center of corn production was in north Alabama. In 1949, only 7.8% of the state's corn harvest was reported from those five southeastern counties.

Milk

American dairies produced 109.4 million pounds of milk in 1940. The North Central Region accounted for the largest share (57.6%). Shares reported by other regions were: South Central (13.3%), North Atlantic (11.6), West (11.6), and South Atlantic (6.0).

Milk available per capita had declined in the United States from 1930 to 1935 (811 to 792 pounds). People could not afford to purchase it, so production declined. However, after the worst part of the Depression had passed, demand and production began to increase. Between 1935 and 1940, per capita availability increased from 792 to 826 pounds.

US Milk

American dairy farms produced 109.4 billion pounds of milk in 1940. Wisconsin was the leading producer with 11.6% of the total. The four next highest producers were: Minnesota (7.7%), New York (7.0), Iowa (6.0), and Illinois (4.7). These five states accounted for 37% of total production. Cows on Alabama dairy farms produced 1.1 billion pounds, or 1% of the national total.

Prices

Federal milk marketing orders were first authorized by the Agricultural Marketing Act of 1937 (Blaney and Manchester, 2001). These marketing orders are the basis for the milk prices shown in Figure 13.100. The national data presented in the figure are average values from all the orders in force throughout the country during a particular year. For example, the national annual average price of milk under this system in 1940 was $1.73 per Cwt. During World War II, demand for milk and milk products increased sharply as a result of increased income from wartime factory work and large purchases by the armed forces. To meet this increased demand, the federal government encouraged increased production through guaranteed higher prices to producers. Under wartime price controls, prices to consumers were kept down by paying subsidies to processors to offset the higher prices they were having to pay producers for their milk. These efforts are reflected in the rapid increase in prices. By 1948, prices had been increased to $4.66 per Cwt, even though the war had ended three years earlier. In 1949, Congress enacted the Agriculture Act of 1949 that established a permanent price support program; however, milk marketing orders that applied only to fluid milk were continued under the 1937 act. Prices were reduced somewhat in 1949 and 1950 to near $3.80 per Cwt. Generally, the year-to-year changes for milk were very similar to those of the indices of prices received for all agricultural commodities (Figure 13.6).

Milk Cow Numbers

In 1940, there were 22.2 million milk cows in the United States (Figure 13.101). In the late 1930s, American dairy farmers must have foreseen an increased demand for milk. Milk cow numbers had declined sharply after 1934, but by 1939, farmers were beginning to rebuild their herds, although prices were near historic low levels (Figure 13.100). With the government encouraging increased milk production and with prices increasing, numbers of milk cows increased sharply. By 1944, there were 25.6 million animals, the highest in history. Milk prices changed very little from 1944 through 1945, and apparently farmers interpreted that lack of change as a signal that they were producing too much milk. In 1945, numbers begin to decline, and the downward trend continued through the end of the period. In 1950, numbers reached 21.9 million animals, or about where it had been in the early 1930s.

Milk Per Cow

The average annual amount of milk produced per cow in the United States was 4,622 pounds in 1940 (Figure 13.102). It remained at about that level through 1945. After 1945, however, milk per cow began to trend upward, and by 1950 it was 5,314 pounds, almost 700 pounds per animal higher (15%) than it had been in 1940. The milk price situation (Figure 13.100) was encouraging dairy farmers to improve their management practices. The lower than expected values of milk per cow in 1943 and 1944 seemed to be related in some way to the slowing of the increase in prices in 1944 and 1945.

Pounds Produced

Pounds of milk produced in the United States had been trending upward since around 1938, and in 1940 it reached 109.4 billion pounds (Figure 13.103). With milk per cow generally unchanged (Figure 13.102) during the early part of the period (1940-1945), but numbers of milk cows (Figure 13.101) increasing rapidly, pounds increased sharply in 1941 and 1942. However when milk per cow declined in 1943 and 1944, pounds followed. In 1945, milk per cow began to increase again. Unfortunately, by that time numbers were beginning to decline. As a result, pounds did not change very much through 1950. The overall effect of changes in these primary factors (numbers and pounds per cow) was that pounds increased from 109 billion in 1940 to 118 billion in 1943, but then remained around that level through 1950.

Alabama Milk

Alabama dairy farms produced only enough milk in 1930 to provide each person in the state with 428 pounds. Consumption and use above this level had to be met by purchasing it from other states. By 1935, pounds per capita had increased to 453 pounds, but then declined to 402 pounds in 1940.

In 1940, Montgomery County had the highest percentage (4.2) of Alabama's 363,000 milk cow herd. Other counties reporting substantial numbers included: Dallas (3%), Hale (2.8), Marengo (2.7), and Jefferson (2.6). These five counties accounted for only 15.3% of the state's herd. In 1940, there were hundreds of small dairies scattered across Alabama, most of them with relatively few cows.

Prices

Data on average prices (dollars per Cwt) for milk in Alabama each year during the period 1940-1950 are presented in Figure 13.100—along with corresponding prices for the country as a whole. Although Alabama prices were always higher, the year-to-year trends were remarkably similar. This similarity reflects the effect of the federal milk marketing order process. Annual price changes were similar from region to region, except for allowances for differential production costs. Generally, production costs were relatively high in Alabama, primarily as a result of lower milk per cow values (Figure 13.102). Consequently, prices determined by the marketing orders affecting the state were higher.

Milk Cow Numbers

The general unity within American agriculture was again demonstrated by the similarity of year-to-year changes in milk cow numbers in Alabama and the United States during the period 1940-1950 (Figures 13.104 and 13.101). This similarity indicates that Alabama farmers generally responded in the same way as did farmers across the country to changing economic signals. Alabama farmers increased their herd size from 363,000 animals in 1940 steadily through 1944 (416,000 animals) in response to increasing prices. Like their fellow dairy farmers throughout the country, they guessed that the small decline in prices from 1943 to 1944 presaged the beginning of a general reduction in marketing order prices that would continue for some time. So they began to reduce the number of animals in their herds. Herd reduction continued in the state through 1949 (348,000 animals). Alabama dairy farmers did break with farmers in other states by increasing herd size in 1950 (362,000 animals), while average numbers continued to decline in the country.

Approximately 4.6% of all milk cows on Alabama farms in 1950 were in Montgomery County. Other counties reporting large numbers included: Hale (3.4%), Dallas (3.1), Perry (3.0), and Marengo (3.0).

Milk Per Cow

Data on average annual milk per cow in Alabama in each year from 1940 through 1950 are presented in Figure 13.102, along with similar data from the country as a whole. Average milk per cow on Alabama dairy farms was 3,150 pounds in 1940, some 47% (1,472 pounds) below the national average (4,622 pounds) that year. Although milk per cow declined slightly in 1943, it generally trended up slowly during the entire period, and was at a level of 3,570 pounds in 1950. In that year, the average in Alabama was 1,744 pounds lower (3,570 versus 5,314 pounds, respectively; almost 49%) than the national average. In terms of milk per cow, Alabama dairy farmers lost ground during World War II.

From the beginning, significantly lower milk per cow values in Alabama have made dairy farming in the state a marginal proposition at best. Fortunately, federal milk marketing orders that provided better prices for milk produced in the state (Figure 13.100) during the war years helped the situation somewhat. Unfortunately, even with higher prices, the differences in milk per cow were probably too great to overcome. For example, in 1950, the average milk cow in Alabama produced 35.70 Cwt of milk per year. With an average price of $5.01 per Cwt, annual gross income per cow was $178.80. Using national averages, the gross income per cow (53.14 Cwt @ $3.75 per Cwt) in the country was $199.30. Alabama dairy farmers likely compensated for this difference by accepting a lower return on investment. This might have been an acceptable solution for the short term, but it was really no solution at all for the long term. Looking at the milk per cow situation in 1950, one would have had to conclude that the prognosis for the long-term health of commercial milk production in Alabama was extremely dismal.

Pounds Produced

As was the case with American milk production, year-to-year changes in pounds produced were generally similar to year-to-year changes in milk cow numbers. Pounds was at a level of 1.1 billion in 1940 (Figure 13.105). Then, as numbers rapidly increased (Figure 13.104), pounds followed, reaching 1.4 billion Cwt in 1944 and 1945, before beginning to decline. Small increases in milk per cow resulted in increases in pounds in 1949 and 1950 to 1.3 billion.

Census data indicated that dairy farmers in Montgomery County lead the state (12.3%) in the quantity of whole milk sold in 1949. Other counties reporting substantial sales included: Hale (7.3%), Mobile (6.5), Jefferson (5.4), and Marengo (5.1). These five counties accounted for 36.6% of all sales.

Alabama Chickens

In 1920, chickens could be found on 90.5% of all American farms (Perry, et al., 1999); afterwards, however, the percentage began to decline. It was down to 85.4% in 1930, and to 84.5% in 1940. According to these authors, agricultural research in the 1940s brought important new technologies to the poultry industry. Better breeds became available. Nutrition research provided the basis for much better commercial feeds. Disease treatment and control were improved significantly. Procedures for managing poultry in confinement, candling eggs, and sex determination of chicks were developed and/or improved. These new technologies and the improved cultural practices resulting from them made it possible to raise chickens (broilers) on a commercial scale. Soon, chicken meat began to compete with beef and pork in consumer's market baskets. Before the improvement in production technology and its use for producing more chicken meat at competitive prices, Americans ate about 20 pounds (boneless weight) of poultry each year, and most of it came from surplus and culled birds from egg-producing flocks. As a result, both production and consumption was seasonal. Broilers were first placed in a different category from chickens in 1939.

Value Per Head in the US and Alabama

Figure 13.106 includes data on the annual average value per head of chickens in Alabama and in the United States during the period 1925-1950. The value of chickens in the United States was 60.5 cents each in 1930. It had been 42 cents in 1934. After 1940, values generally trended upward to $1.44 in 1947. It remained at that level in 1948 before increasing to $1.66 in 1949. Then value declined sharply to $1.36 in 1950. Generally, year-to-year changes in value per head were similar to those of the indices of prices received for all agricultural commodities (Figure 13.6) during the period 1940-1947. In 1948 and 1949, values did not decline as the indices for prices received had. In those years, national chicken numbers were the lowest that they had been since 1942. Apparently, these lower numbers had kept values from declining. In 1950, numbers increased substantially, and values declined.

Year-to-year trends in the value of chickens in Alabama were similar to those in the country as a whole; however, throughout the period the annual average value of Alabama chickens was from 10 cents to 30 cents lower than the national average. In 1940, the Alabama value was 42 cents. It had been 33 cents in 1933. From 1941 through 1945, it trended rapidly upward to reach $1.11. Values continued to increase through 1949 ($1.34), but the annual rate of increase was much lower than during the war. Value declined sharply in 1950 to $1.15.

Numbers

Numbers of chickens on Alabama farms were 9.7 million on January 1, 1940, and most of them were being held for the production of eggs. Numbers had not changed very much since 1925 (Figure 13.107). Then with World War II raging in Europe and the Pacific, and values increasing rapidly (Figure 13.106), numbers exploded upward to 18 million in 1943. By the end of hostilities, however, numbers had fallen to 14.6 million, and they continued to trend downward, reaching 11 million in 1950. With values continuing to increase and with egg prices up and holding steady, it is surprising that numbers began to decline after 1943.

Census data indicated that the largest share (5%) of chickens, four months old and older, on Alabama farms on January 1,1950 were in Cullman County. Other counties reporting substantial shares included: DeKalb (4.5%), Blount (3.7), Marshall (3.6), and Jackson (3.5). These five counties accounted for only 20.1% of the state's total. There were chickens on just about every farm in Alabama. Inventories in 29 counties reported 75,000 or more birds.

Home Consumption

In 1940, Alabama farm families consumed 67% (based on numbers) of the chickens produced on their farms; then as values continued to increase, percentage consumed declined, reaching 41% in 1943. It was not that families were consuming fewer chickens. Consumption remained relatively constant. They were producing more birds, which forced the percentage downward. After 1944, the Alabama Agricultural Statistics Service began to report home consumption only on the value of the birds consumed. For example, in 1945 estimated gross income from the production of chickens was $12.7 million, and the value of home consumption was $5.9 million, or 46% of the total. The share of home consumption based on value ranged between 42% and 47% in the period 1945-1950.

Alabama Broilers

As detailed in the preceding section, there was a significant increase in chicken production in Alabama and elsewhere in the country during the war years, but broiler production literally exploded. Even before the beginning of the war, that industry had been growing rapidly for several years in some areas of the South. In 1934, farmers in Delaware, eastern Maryland, and northeastern Virginia (the Delmarva Region) produced seven million birds for urban markets of the East. The production of poultry had been a housewife's business from colonial times, but in the 1930s chicken processors began to provide farmers, who wished to enter commercial production, with broiler chicks and feed on credit. By World War II, this so-called vertical integration of poultry (broiler) production and processing was well under way. Wartime demand accelerated its development. The US Army quickly became the industry's largest customer. From the beginning, only a relatively few farmers in any single state became involved in this unique production-credit arrangement. However, in some limited areas, such as the Delmarva Region, the vertically-integrated industry contributed significantly to total farm income. In 1940, there were 106 million broilers produced in the United States. Alabama farms produced 1.7% of the national total.

US and Alabama Prices

In vertically-integrated broiler production, farm-gate prices are not available. Instead, the prices reported in this section are based on wholesale prices of processed, whole birds. In essence, it is a live-weight-equivalent price. Annual average prices for broilers in the country and in the state for the period 1940-1950 are presented in Figure 13.108. Prices for Alabama broilers are not available before 1939. Year-to-year changes in these prices also are similar to those of the indices of prices received for all agricultural commodities (Figure 13.6). In 1940, the average national price was 17.3 cents a pound. It did not increase very much in 1941, but afterwards trended rapidly upward to reach 28.8 cents a pound in 1944. They continued to increase through 1948 (36 cents), but at a much lower annual rate. It declined to 28.2 cents in 1949, and remained near that level in 1950.

Alabama broiler prices generally followed the same year-to-year trends as national prices during the period 1940-1950 (Figure 13.108), except that Alabama prices began to decline after 1947, rather than after 1948. During the period 1940-1947, Alabama prices were around 4 to six cents a pound higher than the national average. Afterwards, there was very little difference.

Numbers

In 1939, Alabama farmers produced 1.3 million broilers, and in 1940 they produced 1.7 million (Figure 13.109). Wartime demand hastened the development of Alabama's broiler industry. By 1945 when the war ended, numbers had almost tripled to 5.1 million. That was only the beginning. Broiler production had found a home in Alabama, especially on Sand Mountain, part of the Cumberland Plateau in northeastern Alabama (Figure 1.9). The industry contracted slightly following the end of the war. However, it began to grow again after 1947, and through 1949 numbers increased at an exponential rate to reach 10.5 million. The rate of increase slowed after 1949, and number reached 13.1 million in 1950. In the 11 years (1940 through 1950), numbers increased by 670%. Seemingly, some Alabama farmers had finally found an agricultural commodity that they could love.

Market Weight

In 1940, broilers' average market weight marketed by Alabama producers was 2.4 pounds. It remained at that level through 1947. Subsequently, as a result of selective breeding, improved nutrition, and better production practices average weight began to increase. It reached 2.5 pounds in 1948, 2.6 pounds in 1949, and 2.7 pounds in 1950.

Feed Efficiency

It was obvious from the beginning that if the American broiler industry were to grow and be successful, feed utilization efficiency had to be improved. In 1938, American broiler producers had to feed their birds about four pounds of feed to obtain one pound of gain. Fortunately, agricultural research quickly began to find ways to solve the problem. Soon, as a result of selective breeding, better information on the nutritional requirements of rapidly growing broilers, and improved feed formulation and manufacture efficiency quickly began to improve. By 1948, the amount of feed required for a pound of weight gain was down to 3.6 pounds, a gain of some 10% in a decade.

Eggs

By the 1940s, egg production had been slowly changing for a number of years. In 1910, 87.7% of American farms kept chickens primarily to produce eggs for home consumption. In 1930, 85.4% of farms had chickens, but between 1930 and 1950 farms with chickens declined to 78.3%. At the same time, specialized egg-producing farms were beginning to appear throughout the country.

US Eggs

In 1939, Iowa, Ohio, Texas, Pennsylvania, and Illinois (in descending order) were the top egg producing states. California ranked tenth, producing less than half of the number produced by Iowa (Madison and Harvey, 1997). According to USDA data, American egg production totaled almost 39.7 billion in 1940. Alabama chickens produced 1.2% of those eggs.

Prices

American egg prices were dismal during the heart of the Depression (Figure 13.110), falling to a low of 13.8 cents a dozen in 1933. They began to recover afterwards. Unfortunately, as prices improved, they pulled numbers (Figure 13.111) up behind them, and as numbers increased, prices declined, to reach 18 cents a dozen in 1940. Generally, year-to-year changes in egg prices were similar to those of the indices of prices received for all agricultural commodities.

With preparedness for war in full-swing in 1941, prices began to increase, and by 1943 they were 37 cents a dozen. The upward trend in prices had signaled farmers to increase numbers, and by 1944 they reached the highest level in history—58.5 billion (Figure 13.111). Prices immediately fell to 35.2 cents. In 1945, numbers began to decline, and prices resumed their upward trend, finally reaching 47.2 cents in 1948. Afterwards, with the indices of prices received declining, egg prices also began to decline. In 1950, it was down to 36.3 cents a dozen.

Numbers

After 1935, price signals had encouraged farmers to increase egg numbers, and they had responded with gusto. Numbers were 39.7 billion in 1940 and 58.5 billion in 1944 (Figure 13.111). However, the sharp decline in prices between 1943 and 1944 (Figure 13.110) gave them pause, and between 1944 and 1945 numbers began a slow decline, finally reaching 54.9 billion in 1948. In that year, the average price reached the highest level in history to that time (47.2 cents). The following year, farmers added more hens to their flocks, and numbers began to increase. In 1950, they reached 59 billion, the highest in history to that time. Unfortunately, prices declined in 1949 and then fell sharply in 1950 (teeter-totter phenonomon).

Uses

Data on per capita consumption of eggs in the United States during the period 1925-1950 are presented in Figure 13.112. There was a slight decline between 1940 and 1941. This decline was related to the fact that USDA did not collect consumption data on military personnel overseas. The data in the figure only represent consumption in the United States itself. After 1942, people had enough money to purchase all the eggs that they wanted, and by 1945 consumption reached 402 eggs, the highest in history to that point. In 1946, consumption declined sharply to 377, but afterwards trended upward slowly to reach a level 386 in 1950.

Throughout the 1930s, farm families had consumed around 22% of the eggs produced on their farms; they sold around 77%. During World War II, however, with both prices and numbers increasing, the percentage sold began to increase. It was 79% in 1940 and 88% in 1950. Even with per capita consumption increasing, numbers were increasing so much more rapidly that home consumption declined from 20% in 1940 to 12% in 1950. A substantial number of eggs was still being used for hatching on farms, but with numbers increasing so rapidly, by 1946 less than 1% were still being used for that purpose.

Alabama Eggs

Alabama poultry farms produced enough eggs in 1925, 1930, 1935, and 1940 to provide annually 174, 172, 159, and 150 eggs, respectively, for each person in the state. Assuming that Alabama's annual average per capita egg consumption was near the national average, Alabama consumers had to look to producers in other states for more than half of their eggs.

Prices

Annual average prices of American and Alabama eggs generally followed the same year-to-year trends during the period 1940-1950, and in most years they were essentially at the same level (Figure 13.110). Both trended sharply upward throughout most of the period.

Numbers

Alabama's laying flocks produced 486 million eggs in 1940 (Figure 13.113), which represented 1.2% of the national total that year. Egg numbers in the state had not changed very much in the preceding 15 years. With price increasing in 1941, numbers began to increase sharply, and by 1943, reached 777 million. The decline in the average price of eggs in 1944, however, signaled that there were too many eggs coming onto the market. In that year, numbers began to decline, reaching 622 million in 1948. In 1949 and 1950, Alabama egg producers joined their counterparts elsewhere in the country by increasing numbers. They reached 700 million in 1950. Unfortunately, by that time, prices had fallen to 42.9 cents a dozen.

According to census data, Cullman County's egg production was the highest (8.1%) in the state in 1949. Other counties reporting large shares included: Baldwin (8.1%), Blount (7.0), DeKalb (5.6), and Marshall (4.3). These five counties accounted for 33.1% of Alabama's total production.

Uses

In the late 1920s, Alabama farm families consumed around 43% of the eggs produced on their farms. When the Depression began, the percentage began to increase slowly, reaching 51% in 1932 and 1933. Afterwards, it began a slow decline, and by 1940 it reached 45% (218 million). With numbers increasing through the war years home consumption began to decline again, and reached 30% (211 million) in 1950.

Agricultural Production

As might be expected, the advent of World War II brought about a major change in farm policy and agricultural production in the United States. From World War I onward, surpluses of agricultural commodities and prices had been primary concerns. After December 1941, however, the emphasis quickly shifted to maximum production to meet the needs of the country at war. Further, because of wartime interruptions of agriculture worldwide, the international demand for American farm products literally soared. Consequently, the federal government began to exhort farmers to plant, plant, plant.

All agricultural production contributed (value added) a total of $9.9 billion dollars to the American economy in 1940 (Figure 13.114), just before the United States entered World War II. (A decade earlier, just before the beginning of the Great Depression, the total value added by agricultural production had been $10.3 billion.) Of this 1940 total, crops contributed $4 billion, and livestock contributed $5.9 billion. The difference between crop and livestock contributions was even greater in 1950 ($13.2 versus $18.9 billion). Total value added trended rapidly upward from 1940 ($9.9 billion) through 1943. In 1944, the growth in the Consumer Price Index (Figure 13.2) slowed, and the growth in value added also slowed ($22.2 billion). Although the rate of increase was not as great after 1943, it reached $34.7 billion in 1948. Between 1940 and 1948, total value added grew from $9.9 to $34.7 billion, an increase of 250%. During the same period, the Consumer Price Index increased by only 71.7% (42 to 72.1) (Figure 13.2). Farmers had truly benefited from their sojourn in the tents of the god of war. After 1948, indices of prices received (Figure 13.6) and total value added declined sharply to $29.2 billion. The index was up slightly in 1950, and value added increased to $31.3 billion.

Not all crops and livestock produced on farms actually reached the market. Of course, commodities consumed by families producing them are also a contribution (value added) to the national economy. In earlier years, a significant portion of both crops and livestock was consumed by the families who produced them. In 1940, just as the country was still emerging from the Depression, American farm families consumed the equivalent of 12.1% of total value added of American agriculture ($1.2 of $9.9 billion). By way of contrast, in 1930 just before the beginning of the Depression, home consumption accounted for 14.6% ($1.5 of $10.6 billion) of total value added. In 1950, home consumption, accounted for only 6.5% ($2.1 of $31.5 billion) of the total.

Because farm numbers were falling so rapidly during this period (Figure 13.17), it is instructive to correct the home consumption data for these changes. Figure 13.115 presents data showing the value added by home consumption of crops and livestock on a per-farm basis. In 1940, individual farm families consumed crops, livestock, and livestock products valued at $73 and $117, respectively. In 1950, the corresponding values, were $111 and $254, respectively. The value of livestock and livestock products in a family meal were generally much higher than the value of the crops.

As prices increased, the per farm value of home consumption of crops increased during the early years of World War II. Prices for all livestock and livestock products increased sharply during the war years; consequently, the value of home consumption per farm increased at a much higher rate than for crops. At the end of the war, the value of livestock home consumption was much higher than for crops, and well above prewar levels.

USDA did not begin to publish annual value added data for the individual states until 1949. Before that time, they only published data on cash receipts. Figure 13.116 includes cash receipts for crops, livestock, and livestock products marketed by Alabama farms for the period 1925-1950. These data indicate that total cash receipts were near $87.3 million in 1940. They increased sharply during the early war years, reaching a level of $277.1 million in 1944. The year the war ended, there was very little increase, but in 1946, with indices of prices received continuing to explode upward (Figure 13.6), receipts followed, reaching $431.6 million in 1948. Prices received were lower in 1949 and 1950, and receipts followed them down. Generally, these year-to-year changes were similar to those of the indices of prices received for all agricultural commodities (Figure 13.6).

In contrast to the relationship between value added by crops and livestock in the United States (Figure 13.114), total cash receipts for crops in Alabama were considerably higher than those for livestock and livestock products. In 1940, receipts for crops was $34.3 million greater ($60.8 versus $26.5 million). In 1950, the difference was even greater— $100.4 million ($231.8 versus $131.4 million). Of course, this reversed relationship was entirely the result of the production of cotton in the state.

In Alabama in 1940, total value of home consumption accounted for 38% ($53.4 of $140.7 million) of total cash receipts. By 1950, this percentage had declined to 22.1 ($103.1 of $466.4 million). In the United States, these pencentages in 1940 and 1950 were 12.2 and 6.6, respectively. These data on home consumption clearly demonstrate that as late as 1950, Alabama farms were significantly less commercialized than farms in much of the country. Alabama farm families were much more dependent on the crops and livestock that they produced than farm families elsewhere in the country. They were making progress, but subsistence farming was still extremely important in Alabama, as it had been since 1819.

Figure 13.114 presented data on annual total value added for crops, livestock, and livestock products in the United States during the period 1925-1950. Table 13.13 includes data on value added for some individual crops and for some kinds of livestock and livestock products in 1940, 1945, and 1950. In two of the three years (1940 and 1950), the production of feed crops contributed more to the American economy than the production of food grains. Also, in two of the three years (1940 and 1950) the production of cotton contributed more than any of the other crops or groups of crops. Value added for oil crops (primarily soybeans) increased in all three years, but the rate of increase was not greater than for the other crops. By 1950, soybeans had not made the contribution that the crop would make later. Surprisingly, in 1945 value added for vegetables was the highest of any of the crops or groups of crops.

Not surprisingly, the production of meat animals contributed more to the economy than any other kind or group of livestock or livestock product. In 1950, it was greater than all of the others combined. The contribution of dairy products, was the second most important, but not nearly as important as meat animals. Poultry and eggs were the third most important contributor, and it is obvious that broilers had not yet made much of a contribution.

Data on values of some crops and livestock produced in Alabama in 1940, 1945, and 1950 are presented in

Table 13.14. These data show that cotton continued to be the most important crop produced in Alabama. Peanuts and soybeans were beginning to contribute to the state's economy, but cotton remained the primary crop. Corn remained the second most important crop; although with its consistently relatively lower yields, it is difficult to understand why production remained so high.

By 1950, the value of horses and mules was finally beginning to decline. The decline in cropland harvested and the increase in tractor numbers were finally taking their toll on these important old sources of farm power. The increase in the values for other livestock and livestock products generally was the result of the increase in prices during the period. Broilers, however, were another matter. Poultry farmers could not even imagine how important this animal would become in the future. The increase in the value of cattle (all) also provided another glimpse into the future. In 1950, their value was approaching that of cotton. With cattle and broilers leading the way, Alabama was on the threshold of becoming a livestock state.

STILL SERIOUSLY SEEKING SALVATION

In the early 1930s, American farmers had been seriously seeking salvation for well over a quarter of a century, and at that point nothing that they had done had seemed to have made much difference. They were still in desperate straits. About the only thing that had seemed to help was World War I. Improved information, farmer organization, diversification, and political action had accomplished very little. Farmers found out that even the good-times resulting from World War I did not last very long either. Then things got so bad by 1932 that it soon became obvious that cataclysmic political action was the only course of action that offered any hope of providing timely relief from their desperate situation. The New Deal, for many American farmers, was exactly what the doctor ordered, and it did quickly lift the spirits for most all of them, and the incomes for some. But by 1940, net farm income in dollars per farm was only $56 higher than it had been in 1930 ($706 versus $650). Cataclysmic political action (The Triangle) did not seem to work very well either. They needed another war, but in the meantime those activities (better information, etc.) that they had begun to put in place in 1895 continued to motor along under their own power.

Salvation Through Better Information

Data on all funds available to the Alabama Agricultural Experiment Station for agricultural research in fiscal years 1939-1940 and 1949-1950 are presented in Table 13.15. In 1939-1940, with the state still recovering from the Depression the legislature provided slightly less money ($162,300 versus $172,700) than USDA for research. In 1949-1950, however, the state, flush with public funds harvested from the strong postwar economy, provided almost twice as much ($533,500 versus $331,500) as the federal government. With appropriated funds, grants, and gifts and sales all increasing sharply, total funds available for research increased from $443,500 to $1,383,100 between 1939-1940 and 1949-1950, an increase of 212%. With these increases, the public sector was sending a strong message to Alabama farmers concerning the state's commitment to providing farmers with better research information.

The most immediate effect of the World War II years on agricultural research at Auburn was the reduction of the research staff (Kerr, 1985). A large share of the research staff had been involved in the Reserve Officer Training Corps (ROTC) as students, and they held reserve commissions. By late 1941, five staff members had already reported for military duty. By the end of 1942, 16 were on active duty. The total increased to 19 and 23 in 1943 and 1944, respectively. At the end of 1945, 21 were still on active military duty.

World War II brought a considerable amount of additional money to the state treasury, and the election of 1942 brought Chauncey Sparks into the Office of Governor (1943-1947). Sparks was deeply concerned about education in Alabama and especially about education as it related to agriculture. As a result, several changes were made in the Experiment Station on Auburn's main campus. A new Department of Forestry was established in 1947. In the same year, departments of Research Data Analysis and Research Information were also established, and the Department of Poultry Husbandry became the Department of Poultry Science. Finally, in 1950, the experiment station added the Department of Home Economics Research. With these changes in the structure of the main station came changes in the physical plant. Additional land was purchased near Auburn for expansion of fisheries research and for a Swine Production Unit. By 1948, the staff on active military duty had returned, and the number working at the main station had increased to 92.

Probably the most significant changes in the experiment station during the period 1940-1950 were made off-campus. In 1943, using additional funds provided by the legislature, a new Forestry Research Unit was established in Autauga County. Similar units had been established in Coosa and Barbour counties in 1940. Then in 1944, another unit was established in Fayette County. In 1944 and 1945, respectively, the Upper Coastal Plain and the Piedmont substations were established at Winfield in Marion County and Camp Hill in Tallapoosa County. In 1946, a Plant Breeding Unit was established at Tallassee in Elmore County. In 1947, the Lower Coastal Plain Substation was established near Camden in Wilcox County. In 1948, the experiment station added two horticulture substations, one at Cullman in Cullman County and another at Clanton in Chilton County. Between 1943 and 1948, the number of personnel assigned to off-campus units increased from nine to 11.

Even though the war was bringing unprecedented prosperity to American farmers, Triangle bureaucrats were well aware that there would be serious troubles ahead for American agriculture once wartime demand ended. With this concern in mind, they encouraged Congress to enact legislation (the Agricultural Research and Marketing Act of 1946) that would provide federal funds for marketing research. It was probably evident to the bureaucrats by that time, that New Deal legislation was not going to solve the basic problem in American agriculture—overproduction. The experiment station at Auburn received the first installment of these funds in 1948, and they resulted in a significant increases in personnel working on marketing research in the Department of Agricultural Economics.

Data on funds budgeted for extension work in Alabama by the Cooperative Extension Service in fiscal years 1934-1935, 1939-1940, 1944-1945, and 1949-1950 are presented in Table 13.16. In 1939-1940, the Alabama economy was still in the process of recovering from the Depression. Consequently, state and county funds provided for extension work did not increase very much above mid-Depression levels. However, with the war ended, public support and public funds for extension began to increase rapidly. Between 1944-1945 and 1949-1950, funds provided by the state ($338,900 to $593,300), the counties ($350,000 to $503,700), and the federal government ($718,000 to $1,255,60) increased by 59, 44 and 72%, respectively. Between 1939-1940 and 1949-1950, the total of all funds available for extension work in the state increased from $1.05 to $2.38 million, an increase of 126%. During the same period, inflation increased by 72%. Obviously, the relatively large increase in funds was not as large as it seemed when inflation is considered.

Data on the percentages of all funds used for different extension projects in Alabama in 1939-1940, 1944-1945, and 1949-1950 are presented in Table 13.17. As was the case in the preceding period (1939-1940), the larger share was used to fund projects related to county work (county agents, home demonstration agents, and club agents). In 1939-1940, 1944-1945, and 1949-1950, the total percentage committed to county work was: 73.4, 78.2 and 81.2%, respectively. Obviously a significant share of the new funds received for extension work (Table 13.16) were put into county projects. The shares of the total allocated to the three groups of agents were essentially the same in this period, as in 1930-1940. Shares of funds allocated to work by subject matter specialists also remained about the same. Allocations to work in agronomy and home economics were somewhat higher than for most of the other specialties. During the period, there seems to have been declining support for animal industries. The allocation to marketing was surprisingly high in 1949-1950 when there had been none in 1939-1940 and 1944-1945. This increase is obviously the result of funds provided under the Agricultural Research and Marketing Act, which was passed by Congress in 1946.

Between 1939-1940 and 1949-1950, the total amount of money available for research and extension in Alabama increased from $1.5 to $3.8 million, an increase of 152%. Most of these funds were provided directly or indirectly by the public, and while almost half of the increase was probably related to inflation, it did represent a significant commitment on the part of the public to purchase improved information for their farmers.

In 1948, the State Agricultural and Mechanical Institute for Negroes at Normal became the Alabama Agricultural and Mechanical College. According to Morrison (1994), several new buildings relating to the mission of the Department of Agriculture were constructed in the period 1940-1950: potato house, canning plant, silo, farm shop, blacksmith shop, feed house, dairy barn, loafing house, and milking house. Construction of several of these buildings were indicative of the school's growing interest in dairying.

Ex-President Jimmy Carter wrote in his book *An Hour Before Daylight* (2002) that the farming techniques being used in the South at the beginning of World War II were little changed from colonial times. With the war over and with the fruits of wartime research purchased with public funds cascading down around them in the form of plastics, new chemicals, better machines, improved fuels, etc., Southern taxpayers were ready to ante-up to improve their food and fiber production system. Now, only time would tell whether or not they might receive a worthwhile return on their investment.

Salvation Through Organizations

During the period 1940-1950, the American Farm Bureau continued to be the primary national farmer organization in the country. There would not be a new umbrella national organization for farmers for several decades to come. Instead, farm commodity groups began to establish organizations to represent their interests at the national level. In Alabama during this period, several groups of farmers with interests in specific crops or areas of agriculture formed state organizations. These included:

1. Alabama Seedsmen Association, formed in 1941.
2. Alabama Crop Improvement Association, formed in 1942. The primary purpose of the organization was to encourage the production, protection, and use of improved varieties of crops.
3. Alabama Cattlemens Association, formed in 1944 to protect, promote, and advance the beef cattle industry in Alabama.
4. Alabama Farm and Power Equipment Dealers Association, Inc., established in 1944 with the goal of promoting the general welfare of the retail farm equipment dealers in Alabama and Florida.
5. Alabama Poultry Association, established in 1947 to support the activities of operators of hatcheries in the state.
6. Alabama Forest Products Association, formed in 1949 to deal with increased governmental regulation of the forest products industry and to assist the industry with problems resulting from increased competition from other states and countries.
7. American Dairy Association of Alabama, incorporated in 1950.

Salvation Through Diversification

For years, the personnel of the Alabama Agricultural Extension Service and the Alabama Agricultural Experiment Station had been preaching diversification throughout the state, and finally, during the war years, those efforts seemed to be paying off. In 1940, Alabama farmers harvested cotton from 27.6% of the state's total cropland. A decade earlier, they had harvested the crop from 50%. Then by 1950, the percentage had declined to 22.8%. Surprisingly, percentage of corn harvested declined by about the same amount (48.6 to 42.6%). The percentage of soybeans harvested, as well as cattle numbers increased by around 30%. Production of chickens also increased, and broiler numbers literally exploded.

While there were significant changes in the make-up of crops and livestock on Alabama farms during the 1940s, the most significant change was not farmers seeking alternative crops, but farmers seeking alternative occupations. Between 1940 and 1950, Alabama lost 9% (21,000) of its farms. During World War II, farmers left their farms, some for military service and others for work in defense plants. When the war was over, they just never came home. Of even greater significance, Alabama lost 1.4 million acres (19.4%) of its harvested cropland.

Salvation Through Political Action

In 1933, political action on the part of farmers resulted in legislation that forced the federal government directly into all aspects of American agriculture—to an extent that could have only been imagined a few years earlier. While the evidence indicates that government involvement did not work very well, it did result in the establishment of a powerful Triangle relationship between farmers, politicians, and bureaucrats that quickly became self-perpetuating. From that point onward, perpetuating the relationship seemed to become at least as important as achieving better conditions for farmers.

In 1940, the Triangle determined that government payments would be set so that each farm would receive nearly $120. (Figure 13.118). Then with parity ratios beginning to increase after 1940 (Figure 13.7), payments were reduced. In 1941, average payments (dollars per farm) were $85. In 1941, Congress enacted the Steagall Amendment, which added price supports to a number of commodities that had not been included in the original New Deal legislation. As a result of the provisions of this act, payments began to increase again, reaching almost $130 a farm in 1944. After 1946, farmer incomes (net farm income) were increasing so rapidly that there was little need for payments. By 1949, they were down to slightly more than $30 a farm, not much higher than in 1933. Unfortunately, after 1948 the parity ratios began to decline. Congress passed the Agricultural Acts of 1948 and 1949, respectively, and in 1950 payments authorized by those acts were increased to $52 a farm.

Political action on the part of farmers was relatively limited during the war years. There was little need for it. During the 1940s, the politics of agriculture were driven by Lend-Lease, wartime demand for more food and fiber from fewer farmers, and the Marshall Plan. As a result, increases in prices received were strongly encouraged, while controls were placed on prices paid. At the same time, acreage limitations were largely eliminated. Farmers could not have asked for more. Between 1940 and 1950, American farmers spent 11 years in and around the tents of the god of war, and they profited immensely from it. For much of the period, real farm income had never been so high, and price supports and government purchases of excess commodities were still available if needed. They had finally reached their Nirvana—a place without uncertainty and deprivation and without the endless cycles of boom-and-bust.

During the 1940s, the Triangle did relatively little tinkering with the agricultural economy. The cost of food as a share of disposable income remained relatively constant at around 20%, although it was a little lower during the years when the country was actually at war (Clauson, 2008). Elitzak (2008) reported that in 1950 farmers receive about 41 cents of each dollar American consumers paid for food. The parity ratio increased substantially during the early part of the period, and remained unusually high during the remaining years (Figure 13.7).

The year 1946 was a fateful year for American agriculture. World War II had ended, and the country was beginning the difficult process of demobilization. The New Deal's National Industrial Recovery Act of 1933 had been declared unconstitutional in 1935, and it had been so unpopular that it was not re-written. The Agricultural Adjustment Act of 1933 had also been declared unconstitutional in 1935, but by that time the Triangle was so well entrenched it was re-enacted in 1938. In 1946, American industry began the difficult process of down-sizing, but almost everyone seemed to agree that the marketplace should be allowed to allocate resources in the transition and in the future. Of course, the enactment of the Marshall Plan helped to smooth the process to some extent.

In 1946, the parity ratio was 1.13. It had been that high only one other time, in 1942. The ratio increased to 1.15 in 1947. Even with the ratio that high, it was obvious to everyone that the surplus problem would quickly reappear and that the ratio would begin to fall rapidly. In 1946, the Triangle could also have opted to allow the marketplace to begin to allocate resources in the production of food and fiber, but they made the fateful decision that they could control surpluses by legislating demand through the School Lunch Program. However, as detailed in a preceding section, once they got-stuck hitting the tar-baby, they seemed to think that they had no other choice but to keep hitting. As a result, they had to enact additional corrective legislation at almost yearly intervals.

In the meantime, American industry—although beset by government-mandated labor problems—went on to become a twentieth century miracle. American agriculture also became a twentieth century miracle—but in reality, it rapidly became a virtual house-of-cards that required ever increasing amounts of regulation and tax money to keep it standing.

The Bottom Line

As noted in preceding chapters, Net Farm Income (NFI) data converted to a per-farm basis is being used as a measure of the bottom line for American farmers. These converted data provide a useful indication of how farms fared over time. Annual converted NFI national data for the period 1925-1950 are presented in Figure 13.119. Dollars (per farm) for American farmers had been as low as $30 in 1932, but with New Deal programs, it slowly trended upward to $706 in 1940. Then with indices of prices received increasing sharply at the beginning of the war (Figure 13.6), dollars per farm followed. The rate of increase slowed somewhat in 1944, but regained its strength after 1945 and reached a level of $3,044 in 1948. Afterwards, indices of prices received began to decline, and dollars per farm followed. It was down to $2,218 in 1949, but increased to $2,414 after the beginning of the Korean Conflict in June 1950.

Living in the tents of the god of war during World War II resulted in a period of almost unparalleled prosperity for American farmers. Net Farm Income on a dollars per farm basis increased by 331%. During the same period, core inflation increased by 72%. The government had allowed prices received to inflate almost without restriction, while tightly controlling prices paid. This policy of encouragement and restraint on the part of the government resulted in the parity ratios growing to unprecedented levels (Figure 13.7). In 1943, 1946, and 1947, the ratio was above 1.1. It had never been that high before, even in the Golden Age of American Agriculture. In fact, the ratio remained near 1.1 from 1943 through 1948.

The war did result in unparalleled prosperity for all of American agriculture, because a rising tide raises all boats. Nevertheless, there was still a wide gulf between the fortunes of the individual states. Data on Net Farm Income (dollars per farm) for several states in 1950 are presented in Table 13.18. Dollars per farm was highest in California ($5,864). Other states with high values included, in descending order: Iowa ($5,123); Florida ($4,342); Illinois ($3,614); and Texas ($2,593.) Of the 14 states included in the table, Alabama ranked thirteenth, with $1,207. Only North Carolina ($1,005) was lower. Although the list is not exhaustive, it is obvious that the fortunes of the Southern states were substantially inferior compared to the remainder of the country. As late as 1950, the inability and/or unwillingness of the South to intelligently and effectively solve the problem of agricultural labor still adversely affected the Southern economy, even after three centuries.

Fite (1984) commented that Southern farmers did not respond very positively to wartime requests to increase their production. Except for cotton, they never more than barely met their own needs for farm commodities, and at times they did not even do that. Then with the turmoil of the exodus of a substantial portion of their labor supply, Southern farmers simply were not in a position to make a much greater contribution within that five-year period. There were just too many adjustments that had to be made. While their contribution was less than what they might have wished, they did try, and in trying, they began the slow process of joining the mainstream of American agriculture. Although Southern farmers were unable to carry their share of meeting wartime food and fiber demands, the war did have a dramatic effect on the income of most of them. Compared to farmers in other parts of the country, most Southern farmers were still poor when the war ended in 1945. Nevertheless, they were also enjoying unprecedented prosperity by their own standards. On a per capita basis, average annual net incomes of Southern farmers increased from $150 in 1940 to $454 in 1944. At the same time, per capita income for all American farmers, including Southern ones, was $530 in 1944.

USDA did not begin to publish Net Farm Income data for the individual states until 1949. Data presented in Table 13.19 reflects value per farm of farm products from several Southern states, plus Iowa, in 1929, 1939, and 1949. These data provide a general view of how well farmers were faring in those states. Increases for all of those states were similar to the increase in national values. The 1949, Alabama value ($1,419 per farm) was again the lowest among these neighboring states, as it had been in 1929 and 1939.

The first published USDA data on Net Farm Income for Alabama indicated that in 1949 and 1950 it was $1,154 and $1,207 dollars per farm, respectively, or about half the NFI for the country as a whole ($2,218 and $2,414, respectively). In 1950, Alabama's agriculture still had a tremendous gap to close if it were to eventuslly catch up with American agriculture elsewhere. Some 41% of its farms were still operated by tenants. Crops were being harvested from only an average of 27 acres per farm. There were only 20 tractors per 100 farms. Farmers still had no viable alternative to cotton. In 1950, cotton was harvested from 23% of total harvested cropland. Yields for most crops were much lower than the national average. In 1950, the bottom line for Alabama agriculture was that there was little indication that it could ever become truly competitive in an increasingly global economy. It seemed that it might keep up, but it could never catch up.

14

Agriculture Finds Its Groove

1950–1970

PRECIS

Both World War II and the Korean Conflict created a degree of economic hope in agricultural America, including Alabama. Then came the country's age of agricultural modernization. In the state, as in the country as a whole, modern equipment and fertilizers, and an increasing knowledge of scientific farming (the results of public funds spent on agricultural research and dissemination of knowledge) increased productivity and farm incomes. Progress in Alabama agriculture, however, did not grow at the rates experienced outside the South; Alabama agriculture and Alabama farmers were barely keeping up, but not catching up. Furthermore, investment in modernization did not pay off in net income per farm. America's century-old agricultural economy problem continued: over production and the resulting low prices and no effective mechanism to manage either. Coupled to this national phenomenon, Alabama continued to be overly dependent on cotton, for which the state had no comparative economic advantage. In fact, with only a few exceptions (chicken broilers, stocker cattle, and perhaps forest products), Alabama had no comparative advantage in agriculture at all—because of the ultimate worn out soils, erratic rainfall and violent weather, unpredictable last frost, and subtropical summers. It was not just the unseen hand of nature that adversely affected Alabama agriculture, so too did attempts by man (specifically the Triangle of farmers, politicians, and bureaucrats who tried to artificially manipulate the marketplace). By 1970, it was obvious that government attempts to manage American agriculture from the top (with price supports, acreage allotments, direct subsidy payments, acreage diversions, government-sponsored export programs, new parity formulas, and direct government purchase and storage of commodities that had been established by the Triangle) simply had not worked as intended. In fact, they have been counterproductive. Only off-farm work enabled the Alabama farmer to survive.

INTRODUCTION

In 1950, in just over three decades, American farmers had gone from the Golden Age of American Agriculture, to the Great Depression, to the prosperity of World War II; no one seemed sure where they would be going next. If history were any guide, they would be headed back to economic privation. The world in 1950, however, was different than it had been previously. The United States had emerged from World War II as the most powerful country on earth, as well as the most powerful in history. All of the country's infrastructure remained intact, and its industry, strengthened by new technology developed during the war years, was poised to deliver a rapidly growing peacetime economy and a rapidly growing middle class. Foreign trade seemed destined to expand far into the future. With prospects for a growing number of non-farmers with more money to spend and willing consumers waiting around the world, the future of American agriculture had never looked better. It seemed likely that all of the new technology (better machines, improved transportation, new chemicals, electronics, plastics, etc.) that had made American industry a wonder of the world could now be used to transform agriculture. However, American agriculture had not really changed much, especially in the South, since the beginning of the twentieth century. If agriculture did encounter minor obstacles, there were always New Deal legislation and programs to smooth the way. The Korean Conflict, as inconclusive as it was, did not dim the future in any way. However, if history had ever taught anything, it is that providing food and fiber has always been a very complex business. But all those problems of past feast and famine seemed just details for history books. After all, had not the United States, in the war, just taken up history and given it a good shaking. This was a new day for the country and the world, and American farmers were there, peering down from atop their success in the 1940s, waiting for it to come to them.

NATIONAL POLITICS

During World War II, American politics had been united and focused on saving the world from totalitarianism. There had been a considerable amount of partisan bickering leading up to the war, but after Pearl Harbor everyone united behind the war effort. Unfortunately, World War II had hardly ended before the Cold War began. It was a completely different kind of war, and it presented the American political system, attempting to return the country and the world to peacetime, with some very difficult problems.

The Presidency

The two parties more or less divided the presidency during the period 1950-1970. Harry Truman (D, MO) held the office from 1950 to 1953. He was succeeded by Dwight D. Eisenhower (R, KS) from 1953-1961. John F. Kennedy (D, MA) followed as president from 1961-1963. Then Lyndon B. Johnson (D, TX) served from 1963-1969. Finally, Richard M. Nixon (R, CA) served from 1969-1974.

Secretaries of Agriculture

Charles Brannan (D, CO) was Truman's secretary of agriculture; he served in that position from 1948-1953. Other persons holding the office included: Ezra Benson (R, UT) (1953-1961); Orville Freeman (D, MN) (1961-1969); and Clifford Hardin (R, IN) (1969-1974).

The Senate

The Senate was controlled by the Democratic Party for most of this 21-year period (1950-1953 and 1955-1970). The Republicans were in power for only two years, 1953-1955 (Table 14.1).

Senate Majority Leaders

Scott Lucas (D, IL) was elected to the office of Senate majority leader at the beginning of the 81st Congress in 1949. He held the office from 1949 to 1951. Other persons holding the office included: Earnest McFarland (D, AZ) (1951-1953); Robert Taft (R, OH) (1953); William Knowland (R, CA) (1953-1955); Lyndon Johnson (D, TX) (1955-1961); Mike Mansfield (D, MT) (1961-1977).

Chairmen of the Senate Committee on Agriculture

Elmer Thomas (D, OK) served as chairman of the Senate Committee on Agriculture from 1949-1951. Other persons serving in that position included: Allan Ellender (D, LA) (1951-1953); George Aiken (R, VT) (1953-1955); and Allan Ellender (D, LA) (1955-1971).

The House of Representatives

The situation in the House with respect to control mirrored that of the Senate. The Democrats were in control from 1950-1953 and from 1955-1970. The Republicans controlled the House only from 1953-1955 (Table 14.1).

Speakers of the House

Sam Raburn (D, TX) served as speaker of the House in the 81st and 82nd congresses (1949-1953). Other persons holding that office during this period were: Joseph Martin (R, MA) (1953-1955); Sam Raburn (D, TX) (1955-1961); and John McCormack (D, MA) (1961-1971).

Chairmen of the House Committee on Agriculture

Harold Cooley (D, NC) served as chairman of the House Committee on Agriculture in the 81st and 82nd congresses (1949-1953). Other persons holding that office during this period were: Clifford Hope (R, KS) (1947-1949); Harold Cooley (D, NC) (1949-1967); and William Poage (D, TX) (1967-1975).

The Political Situation

As detailed in the preceding chapter, Truman completed most of Roosevelt's fourth term as president. He was elected as the result of his own candidacy when he unexpectedly defeated Thomas Dewey in 1948. While completing Roosevelt's term, he promised that his administration would give American citizens a Fair Deal. In his second term, he was not able to get much of his legislation enacted, although Democrats had majorities in both the 81st (1949-1951) and 82nd (1951-1953) congresses (Table 14.1). Domestic affairs were increasingly dominated by concerns over the possibility of communist subversion of the government. Truman's administration took most of the blame for the communist victory in China in 1949. All of these events came together in 1950 when Congress enacted the McCarran Internal Security Act that required the registration of communists and communist-front organizations. The act was passed over the president's veto. Truman's efforts to get legislation enacted on his Fair Deal initiatives had to be put on hold in late 1950 when the Korean Conflict began. For most of the remainder of his second term, Truman was involved in ending the hostilities on the Korean peninsula. Truman was always a strong supporter of organized labor, but he opposed them in 1952 by seizing the steel industry to prevent a strike. He claimed that his action was justified by presidential powers during times of national emergency. The Supreme Court disagreed with him. Truman chose not to seek the nomination for another term in the election of 1952.

In the election of 1952, World War II Army General Dwight Eisenhower overwhelmingly defeated the Democratic Party candidate for president—Adlai Stevenson of Illinois. Eisenhower had campaigned on a platform of Dynamic Conservatism. His administration pursued policies of rigorous economic conservatism and moderately liberal social initiatives. He tried to reduce the size of government and contain postwar inflation, and to achieve some of his social objectives he promoted the establishment of the Department of Health, Education and Welfare. He also joined Congress in raising the minimum wage from 75 cents to $1 an hour and in extending Social Security benefits to 10 million Americans who had not been included in the original legislation.

Eisenhower had barely gotten his cabinet in place when he had to make a decision on an event that would lead to one of the most divisive periods in American history—the Vietnam War. In 1954, the French begged the president to use the US Navy to rescue Vietnam from communism. Eisenhower declined to do so and agreed for the partitioning of the country into a communist-controlled North and a South informally allied with the United States. He decided against providing the French with any direct military assistance, but he did agree to send a few hundred advisors to help the South counter communist guerrilla incursions from the North.

In 1956, the Eisenhower Administration became deeply involved with the Suez Crisis. After Egypt's President Nasser nationalized the Suez Canal Company, the United Kingdom, France, and Israel took control of the canal. They feared that the nationalized waterway would be operated in such a way as to jeopardize their economic interests. The United States, using financial leverage against the United Kingdom, forced a ceasefire and in 1957 pressured the three countries to withdraw their forces. After this short-lived conflict, the United States became the primary protector of Western interests in the Middle East. Eisenhower had to take a more forceful action to protect these interests in 1958 when he sent 14,000 Marines to Lebanon to help to quell a rebellion against the pro-Western government.

One of Eisenhower's major contributions during his first term was to encourage the United States to develop an interstate highway system modeled on Germany's autobahn system. During the war in Europe, he had been deeply impressed with the role that it played in Nazi Germany's war effort. As a result of his strong support, Congress enacted the Federal-Aid Highway Act of 1956 that authorized transfer of federal funds to the states to begin the construction of the system.

The role and place of African Americans in American society came into question long before the United States became a country. Some progress was made during the Civil War and Reconstruction, but in the twentieth century much remained to be done. Truman had made some progress, but no one had done very much on one of the most vexing of all the problems—segregated schools. In 1957, two black girls applied for enrollment in Central High School in Little Rock, Arkansas. This action lead to widespread public disorder and threatened mob violence in the area. The threat of outright anarchy prompted Eisenhower to intervene by sending federal troops to maintain order. On another racial-equality issue, Eisenhower completed the desegregation of the armed forces begun by Franklin Roosevelt, and with administration support in 1957, Congress enacted the first new civil rights law since 1890. Under its provisions, it was supposed to protect the right of African Americans to vote by removing some of the obstacles imposed by state law and state officials.

Eisenhower's Dynamic Conservatism was immensely popular, and he was easily nominated for a second term. Then in the election of 1962, he easily defeated the Democrat Stevenson for a second time. Eisenhower received 57.6% of the popular vote and 457 of 530 electoral votes.

John Kennedy was elected president in 1960 in an extremely close election with Richard Nixon, Eisenhower's vice president. Kennedy had only been in office for a short time (April 1961) when he had to deal with the disastrous Bay of Pigs situation in Cuba. Later, in August 1961, the young president and the country had to accept the Berlin Wall and the physical separation of East and West Berlin, and a year later, Kennedy had to manage the dangerous Cuban Missile Crisis. Communist China chose this time when American foreign policy was focused on Cuba to invade territory in northern India. India pleaded for help, and Kennedy quickly ordered the shipment of military equipment and supplies to the region.

No other American president had ever had to endure such a tempestuous series of events so early in his tenure. In the beginning, Kennedy followed Eisenhower's lead of just providing military advisors to help South Vietnam in its on-going war with the communist North; however as pressure from the North increased, American involvement increased.

The dangerous involvement of the Soviets in Castro's Cuba was only part of a determined effort to establish communism throughout Latin America. To counter this effort, the Kennedy Administration established the Alliance for Progress, which provided economic aid to the most seriously threatened countries in the region. Early in his presidency, the Peace Corps was established as an American, non-military force to help improve agriculture, education. and health in developing countries around the world.

Early in his tenure, Kennedy became concerned with the proliferation of nuclear weapons testing. Neither Truman nor Eisenhower had accomplished very much in trying to reach agreement with the Soviets on any matter, but Kennedy was determined to get them to agree to limit testing to underground sites. As a result of his determined efforts, he was able to sign a Partial Test Ban Treaty in August 1963. He believed that the treaty was his most significant accomplishment.

In his inaugural address, Kennedy asked the country to follow him across a New Frontier where the federal government would fund improved education and medical care for the elderly, intervene to end the recession, and actively work to end racial discrimination. Early in the 87th Congress, Kennedy sent the Congress over 20 recommendations for new legislation on economic recovery, increased emphasis on national defense and foreign aid, conservation of natural resources, and federal aid for housing and schools. However, little of this proposed legislation was enacted. The country was not really ready for such a large increase in federal involvement in their lives. Although the Democrats held sizable majorities in both houses of Congress, Southern Democrats worked with Northern Republicans to block most of Kennedy's legislative initiatives. While he had great difficulty in getting Congressional support of his domestic agenda, he was able to get them to appropriate the funds to bring the country into the Space Age. In 1957, the American people had been surprised and shocked by the Soviet Sputnik satellite, and they demanded that Congress provide Kennedy with the money he needed to catch up.

Kennedy's most discouraging personal failure as president was his inability to get meaningful civil rights legislation through Congress. During his fight to get the legislation he wanted, he had to send 400 federal marshals and 3,000 troops to Mississippi to maintain public order when James Meridith attempted to enroll at the University of Mississippi. Later, he had to send marshals and troops to Alabama when Governor Wallace personally blocked a University of Alabama doorway to prevent the enrollment of two black students. These uses of federal power, while badly needed for the good of the country, made it even more unlikely that Southern Congressmen would support any of his legislation, especially if it was remotely related to civil rights. In November 1963, Kennedy decided to go to the home state of Vice President Johnson to attempt to improve his public image in the South. While in Dallas, he was assassinated.

Lyndon Johnson (D, TX) was sworn in as president on the day Kennedy died. Probably no person in the history of the country took the office with more political experience. Yet his early days in the office were extremely difficult for him. It seemed that everyone in the country wanted to honor President Kennedy, who had been killed in his prime. Many chafed at the idea that this Texas politician, in every way different from John Kennedy, would now get credit for all of Kennedy's great initiatives.

A central theme of Kennedy's New Frontier was supposed to have been an all-out war on poverty in the country. Unfortunately, the press of world events, a difficult relationship with Congress and his untimely death prevented Kennedy from making any progress on this agenda. When Johnson became president, he made the War on Poverty a central theme of his administration. the consummate politician began to use arm-twisting tactics, which he had developed to an art-form while in Congress, to good effect. Where Kennedy was unable to get the time of day out of the Congress, Johnson was soon getting just about everything he wanted. Then with his landslide election in 1964, Johnson's long coattails swept enough conservative Republican Congressmen out of office to destroy the coalition that had made life so miserable for Kennedy. The 89th Congress was much easier to work with. Out of that Congress came a new Office of Economic Opportunity, the Equal Opportunities Act, The Revenue Act of 1964 (tax reduction), the Civil Rights Act of 1964, Head Start, the Food Stamp Act of 1964, Adult Basic Education, an increase in the minimum wage ($1.25 to $1.60 an hour), the school breakfast program, Job Corps, the Manpower Act, and the Voting Rights Act of 1965. In 1933, in an attempt to alleviate the effects of the disastrous Depression on the country's citizens, especially its farmers, Congress asked the federal government to intervene in their private affairs in ways that no one could have even imagined just two decades earlier. There is an old saying that originated among the Bedouins of North Africa to the effect that it is unwise "to allow a camel to put its nose under the edge of your tent, for soon you will have a camel in your tent." In 1933, the camel got its nose under the edge of the tent, and in 1964 and 1965, he ate the whole thing. For better or worse, after 1964 and 1965, the lives of Americans and many people throughout the world would never be the same—socially, politically, or economically.

Johnson followed Kennedy's lead in the War on Poverty, civil rights, Vietnam, and the space program. Unfortunately, he found that not everyone was as enamored of these new initiatives as Kennedy's think tanks had been. Implementation of parts of the legislation resulted in widespread unrest in many of the country's large cities. Rioting left hundreds of homes and buildings burning. When Kennedy was assassinated in 1963, there were 16,000 American so-called advisors in Vietnam. Then in August 1964, after the Gulf of Tonkin Incident, Johnson increased the number and vastly expanded their role. Apparently, Johnson deplored the war, but he did not feel that he could do anything that would make the country appear to be weak in the eyes of the world. With this general frame of reference after late 1964, events there led him to continuously escalate the level of the American involvement. As the level of intervention increased, the war required more and more of the president's time, and after the Tet Offensive in early 1968, his administration was dominated by the war. The Great Society became a distant memory. As American intervention increased, so did resistance to the war. It finally became so intense that Johnson decided that he would not seek the Democratic nomination for president for the 1968 election. One of Kennedy's legacies, the space program, became extremely productive with Johnson's strong commitment. It is unfortunate that with all of the support that he provided, he was just an ex-president when the program achieved Kennedy's and the American people's goal of landing men on the moon. It is ironic that Kennedy's adversary in the heated 1960 election had become president by 1969, and it was he who spoke to the astronauts on the lunar surface.

Republican Richard Nixon was narrowly defeated by the Democrat John Kennedy in 1960. Then during the 1964 campaign, Nixon worked ceaselessly to get Goldwater elected. Although Goldwater lost badly, the growing numbers of conservatives deeply appreciated Nixon's efforts, and in 1968 they were instrumental in giving him the nomination for the presidency. Nixon won a narrow victory over Hubert Humphrey of Minnesota. It is likely that the margin would have been larger, except for the candidacy of George Wallace of Alabama, a third party nominee.

Nixon had campaigned on a platform of ending America's involvement in Vietnam "with honor." Once in office, it seemed that "with honor" just meant more of the same when he ordered the bombing and invasion of Cambodia. Soon, however, he seemed to have a change of heart, and began the process of turning the actual fighting over to the South Vietnamese. Accordingly, he went to Vietnam in July 1969 to meet with President Nyugen Van Thieu. However, there was no quick way for the United States to disengage and leave Vietnam—thus ending American participation in the war. For the first two years, the Nixon Administration was still deeply involved in war matters. From the beginning of his first term, it was apparent that his primary interest would be in international affairs. In 1969, he began talks with the Soviets on the limitation of strategic nuclear arms. Further, it is likely that he was beginning to formulate the plans for improving the country's relationship with Communist China.

Although the Democrats still had strong majorities in both houses of Congress, continued efforts to remake America along the lines established by Kennedy's Brain Trust had rapidly diminished after 1965. There were few new domestic initiatives in the early years of Nixon's first term. In 1970, he did combine a number of existing agencies in the government to establish the National Oceanic and Atmospheric Administration within the Department of Commerce. About the same time, Congress decided to act on long-simmering concerns about the continued loss of passenger rail service. As a result, in 1970 the for-profit National Railroad Passenger Corporation (the forerunner of Amtrak) was established.

ALABAMA POLITICS

For the first half of the 1940s, Alabama politics had been primarily concerned with activities related to World War II; however after the war, when national Democrats installed a strong civil rights plank in its platform for the 1948 election, all hell broke loose. The descendants of Old Confederates from across the South met in Montgomery to "secede" again. Alabama politics would never be the same.

Governors

James Folsom (D, Coffee County) served as governor from 1947 until 1953. Other persons holding the office during the period 1950-1970 included: Gordon Persons (D, Montgomery) (1951-1955); James Folsom (D, Coffee) (1955-1959); John Patterson (D, Russell) (1959-1963); George Wallace (D, Barbour) (1963-1967); Lurleen Wallace (D, Barbour) (1967-1968); and Albert Brewer (D, Morgan) (1968-1971).

Commissioners of Agriculture

Haygood Patterson (D, Montgomery County) served

as commissioner of agriculture from 1947 until 1951. Other persons elected to the office during the period 1950-1970 included: Frank Stewart (D, Montgomery) (1951-1955); A.W. Todd (D, Franklin) (1955-1959); R.C. Bamberg (D, Perry) (1959-1963); A.W. Todd (D, Franklin) (1963-1967); Richard Beard (D, Jefferson) (1967-1972).

The Political Situation

Folsom's first term in office was marred by personal problems and accusations of scandal, misconduct, and corruption. As a result, he was not able to accomplish very much that would have improved the lot of most Alabamians. He was able to get the legislature to appropriate additional funds to improve education, increase support for the elderly, and improve roads. Folsom was relatively liberal on racial matters, but with Alabama's political environment at that time, he was unable to eliminate any of the gross inequalities deeply imbedded in Alabama's "way of life."

Gordon Persons had been defeated by James Folson in the election of 1946, but ran again in 1950 and defeated Chauncey Sparks. Persons became Alabama's 46th governor. Although a considerable amount of the legislation that he got the legislature to enact was designed to help at least some of the citizens, some of the legislation was counterproductive. He was able to get legislation enacted that reformed the state's welfare system, its pardon-parole system, and its merit system. The legislature also provided additional funds for education and roads. Persons supported the establishment of the Alabama Educational Television System, making Alabama one of the first states in the country to have public television. In contrast to these positive accomplishments, he signed right-to-work legislation that made it difficult for unions to organize workers in the state. He signed a bill eliminating the cumulative feature of the poll tax, but the bill did nothing to eliminate the tax. He also supported legislation that became known as the Little-Boswell Amendment, which established voting qualification designed to disqualify African Americans at the ballot box. During his administration, Attorney General Albert Patterson was murdered in Phenix City while trying to reduce organized crime in the town. The public safety situation became so unstable that Persons had to place the town and county under martial law. In 1954, also during the Persons Administration, the US Supreme Court declared separate-but-equal public schools unconstitutional.

Despite all the unfavorable publicity surrounding the first Folsom administration (1947-1951), Alabamians had not had enough of "Big Jim" Folsom. He ran for governor again in 1954. Much of his second administration was like his first, plagued with scandal and corruption. He did get some legislation enacted to provide additional services to the public. Probably his most important accomplishments were helping to get the Tennessee-Tombigbee Waterway Bill enacted and establishing the inland docks program. Folsom was probably the first governor to fully appreciate the potential of encouraging outside industry to locate in the state. In retrospect, it is surprising that Folsom accomplished as much as he did. The flames of the civil rights movement, ignited during World War II, were beginning to grow into an inferno in the state. In 1955, Alabamians were just beginning to realize that the Supreme Court decision regarding segregated public schools applied to them as well as the citizens of Topeka, Kansas. Then in February 1956, Autherine Lucy was admitted to the University of Alabama, and 10 months later the Montgomery Bus Boycott began.

The administration of Governor John Patterson (1959-1963) was able to take credit for some solid advances that improved the lives of Alabamians, as well as improving the state's economic infrastructure. However, Patterson will probably be best remembered for his support of a sharply, racially-segregated Alabama. He established his reputation while attorney general. In that office, he successfully banned the NAACP from operating in the state, and developed the legal framework necessary to combat the African American boycott of white businesses in Tuskegee and the bus boycott in Montgomery. With his well-established position on civil rights and with support from the Ku Klux Klan, Patterson was able to defeat George Wallace in the race for governor in 1958.

Patterson carried his position on civil rights with him from the office of attorney general to the governorship.

In early 1960, he engineered the expulsion of black students holding a sit-in at Alabama State University. He also refused to cooperate with federal officials attempting to implement the provisions of the Civil Rights Act regarding voter registration. He did very little to help protect out-of-state Freedom Riders who came to the state. He would leave the task of "standing in the school house door" to Governor Wallace, but in his own way Patterson did everything he could to preserve white supremacy in Alabama.

As suggested in a preceding paragraph, the Patterson Administration can be credited with several positive accomplishments that benefited all of the state's citizens: The legislature approved bond issues for highway and school construction and for facilities for the mentally ill, old age pensions were supplemented, and free hospitalization approved for elderly patients. A small-loan law was enacted with the intent of curtailing the predatory practices of loan sharks on the poor. Efforts begun by Folsom to improve Alabama's waterways and state docks were strengthened. Construction was begun on two new state buildings (Highways and Industrial Relations). However, the most significant contribution to Alabama's future was probably the decision of the federal government to establish the George C. Marshall Space Flight Center in Huntsville.

In 1958, George Wallace had been defeated by John Patterson in the election for governor. Then in 1962, Wallace was elected with the largest number of votes ever given to a candidate for that office up to that time. Between 1958 and 1962, Wallace had transformed himself from a Folsom liberal on race relations to a Patterson white supremacist. [In 1958, the NAACP had supported Wallace's candidacy.] Patterson had governed in a storm of civil rights activity, but under Wallace, the storm grew into a maelstrom of events from which there seemed to be no escape. In his inaugural address, Governor Wallace unequivocally established his position on race relations with his famous declaration "Segregation now. Segregation tomorrow. Segregation forever." In his first year in office, he stood in the schoolhouse door, and the Birmingham race riots occurred. The bombing of Birmingham's 16th Street Baptist Church resulted in the death of four young black girls. Encouraged by the reception of his segregationist message by people in Alabama and other Southern states, Wallace decided to enter the presidential primaries in Wisconsin, Maryland, and Indiana that led up to the 1964 election. He received surprising support in all three states. Buoyed by all of the media attention and his reception in the primaries, he called the legislature into special session in September 1965 to get them to pass an amendment to allow a sitting governor to run for a second term. In a close vote, his request was denied. With the legislature working full-time to enact Wallace's legislation to perpetuate segregation, they accomplished little else.

With no hope of succeeding himself as governor and the 1966 election approaching, Wallace prevailed on his wife, Lurleen, to seek the office as his stand-in. She finally agreed and won the May Democratic primary with 54% of the vote. This vote assured her of election in November. Initially, Lurleen worked to push her husband's segregation agenda, but she eventually escaped from his influence to get the legislature to increase funding for the state's mental hospitals and to pass a $160-million road bond bill. She also sought and obtained legislation requiring banks to pay interest on state deposits and to establish a program for the development of Alabama's parks and historic sites.

In March 1967, the federal government ordered Alabama to begin desegregating its public schools that fall. Shortly thereafter, she asked the legislature to seize all state public schools and put them under police power. In June 1967, it was discovered that the cancer for which she had been treated in 1966 had returned. She died on May 7, 1968, and Lieutenant Governor Albert Brewer succeeded her as governor.

Governor Brewer was a racial moderate with no interest in continuing the fight to preserve segregation; he was able to turn his attention to the state of the state. During his administration, appropriations for public schools received the largest increase in history, and state funding allocations to local school systems were equalized. Medicaid benefits were increased; an anti-air pollution bill was enacted. Brewer created the Alabama Development Office from several existing organizations. He made numerous out-of-state trips to encourage industry to relocate to Alabama. The Court of Appeals was divided in a manner so as to provide relief from overcrowded dockets. He also created the first Ethics Commission to promote honesty and integrity in state government. In 1970, Brewer ran for governor, and was defeated by George Wallace.

THE NATIONAL ECONOMY

Between 1930 and 1943, the American economy had taken a roller-coaster ride of unprecedented proportions. In 1930, the Gross National Product was near $700 billion. Four years later in 1933, it had fallen to $500 billion. Then a decade later in 1943, it was almost $1,400 billion. Further, in 1930 the Consumer Price Index was near 50; by 1933, it was below 40. Then with the war years and postwar years boosting it upward, the CPI reached 72.1 in 1948.

The Economic Situation

The Eisenhower years (1953-1961) are often described as a time of complacency. In the mid-1950s, federal expenditures resulting from foreign aid programs and overseas military commitments were continuing to grow. Unfortunately, the modest growth of the economy during this period was interrupted by the recession of 1957-1958. In 1957, following the formation of the European Common Market, American foreign trade began to decline. Soon, the income from exports was no longer sufficient to offset the increasing costs of imports. As a result, the country began to experience a balance of payments situation that resulted in a worrisome outflow of gold. The Federal Reserve Board decided to combat this problem by raising interest rates to make foreign investment in the American economy more attractive. Unfortunately, the higher interest rates slowed business borrowing and expansion, which quickly slowed the economy and ultimately forcing it into a recession. Although the actual recession did not last very long, unemployment increased rapidly to 7.8% (four million unemployed) before the economy began to recover.

By contrast, the 1960s and 1970s were a time of great change. President Kennedy (1961-1963) brought with him a much more activist administration and the so-called new economics. One of the members of his Council of Economic Advisors set the tone of his administration by commenting that the country should "accept the fact that the federal government has an overarching responsibility for the country's economic stability and growth."

Kennedy asked the country to respond to the challenges of a New Frontier. He sought to accelerate economic growth by increasing government spending and cutting taxes. He also pressed for medical assistance for the elderly and for inner cities. Relatively few of his initiatives became law before his assassination, but Congress and President Johnson (1963-1969) quickly moved to enact much of Kennedy's agenda and considerably more, under the banners of the Great Society and the War on Poverty. Through new programs such as Medicare, food stamps, and various federal education initiatives, the government sought to spread the benefits of America's rapidly growing economy to more of its citizens. As these programs grew, federal deficits began to increase. At the same time, the costs of the Vietnamese War were growing rapidly, necessitating even greater expenditures. Instead of increasing taxes to pay for these two "wars," the government increased its borrowing. Putting increasing amounts of borrowed money into the economy fueled a growing prosperity as well as inflation.

The American economy went through several cycles and parts of cycles during the 20-year period, 1950-1970. It was still in the middle of the post-World War II boom when the Korean Conflict began in 1950. Then that war, and the postwar adjustment that followed, continued until 1956. A short, shallow recession (1957-1958) followed the postwar readjustment period. However, a period of business expansion began in the latter part of 1958 and continued until 1970.

The Gross National Product

In 1950 when the Korean Conflict began, the Gross National Product was $1.2 trillion in constant 1982 dollars (Figure 14.1). Then, as a result of the increased business activity associated with the war, the GNP increased 10.3% to $1.3 trillion in 1951. It continued to grow moderately in 1952 and 1953 and then, as a result of the postwar readjustment, decreased 1.3% to $1.4 trillion in 1954. It resumed its upward trend at rates of 5.6, 2.1, and 1.7% in 1955, 1956, and 1957, respectively. As the low rate of growth in 1957 indicated, the economy was entering another period of economic slowdown. In fact, it had not really completely recovered from the post-Korean Conflict readjustment of 1953-1954. Unemployment had continued to increase throughout the period, and there were chronic problems in several sectors of the economy, including agriculture. Then, late in 1957, there were massive layoffs in the auto and steel industries, and the country plunged into recession as Eisenhower Prosperity ended. As a result, the GNP decreased by 0.8% in 1958 to $1.54 trillion. Fortunately, the economy began to recover the following year, and the GNP began to grow again. However this recovery was short-lived, and in 1960 and 1961, the economy experienced an 11-month period of contraction that slowed the growth of the GNP slightly. Afterwards, it experienced an extended period of growth that lasted until 1968 when it reached $2.36 trillion. In 1968, the Federal Reserve became concerned about inflation, and in April 1969 it increased the discount rate from 5.5 to 6%. This increase resulted in a decrease in business activity, and an 11-month economic contraction began. In 1970, the GNP declined by 0.3% to reach $2.42 trillion.

The Consumer Price Index

The Consumer Price Index had increased 80% from 1940 to 1948 (42 to 72.1), then remained relatively stable during the remainder of the decade before reaching 72.1 in 1950 (Figure 14.2). The CPI increased 8% in 1951 to 77.8 as a result of economic conditions related to the Korean Conflict. The CPI edged up slowly during the period 1952-1956, generally remaining around 80 during that period. Then around 1957 and 1958, it began to grow at an exponential rate to reach 116.3 in 1970. During the period 1950-1970, the CPI increased 61% (72.1 to 116.3).

The Inflation Rate

The inflation rate varied considerably during the period of readjustment following World War II and the Korean Conflict. During the period 1950-1960, it ranged between -3.7 (1955) and 8 (Figure 14.3). Then it entered a period of stability until 1960, at a level of between 1 and 2%; however in 1961, during the early days of the Kennedy Administration (1961-1963), it began to increase rapidly, and it continued to increase during the years of the Johnson Administration. It finally reached 5.9 in the second year (1970) of the Nixon Administration (1969-1974).

Federal Discount Rates

Interest rates charged by the Federal Reserve Board for their best customers had been 1.75% and lower from 1934-1950 (Figure 14.4), and they remained in that range until early 1956 when the Federal Reserve began to increase rates so as to make foreign investment in the American economy more attractive. In 1956 (April 13), they raised the rate to 2.75%, and the increases continued until 1957 (August 23) when they reached 3.5%, an unheard of rate for that time. Interest rates were lowered again in 1958, but as inflation slowly began to increase in 1961, the Federal Reserve began to increase interest rates again, and they generally followed the rate of inflation upward. In April 1969 interest rates reached 6%.

THE ALABAMA ECONOMY

In 1963, the Bureau of Economic Analysis of the US Department of Commerce began to publish estimates of Gross State Product for the individual states. The GSP is a measure of the total economic output of the state. In 1963, 1965, and 1970, Alabama's GSP totaled $7,328 million, $8,978 million, and $12,455 million, respectively. During the same years, similar values for Illinois were $36,619 million, $46,124 million, and $62,931 million, respectively. These data indicate that the values for Illinois were roughly five times as large as those for Alabama in 1963. It is likely that the Illinois economy was considerably larger; the Illinois population, however, was also around three times as large. Assuming that GSP is positively correlated with population, a valid comparison of values could only be made after correcting for differences in population. The values for population differences for Alabama, Illinois, and a number of other states have been corrected by dividing them by the numbers of people in each state. These corrected values are presented in Table 14.2. as dollars per person. In 1963, values for Alabama and Mississippi were lower than for the other states, even the Dakotas. The Alabama value declined between 1963 and 1965 ($2,182 to $2,008 a person). In that year, it was the lowest of any of the states listed. During this period, Alabama was in an extremely difficult time of social upheaval; however, it is not clear that the declining GSP was the result of the social unrest. By 1970, the value for Alabama had increased to $3,616, and was higher than Mississippi and about equal to North Dakota.

Table 14.3 contains data on the contribution (percentages) of farming to the GSP of the same states listed in Table 14.2. For example, in 1963, farming contributed 13.84% of the dollars in the Alabama GSP. These data indicate that the contribution of farming was lowest in those highly urbanized, commercialized, and industrialized states of Pennsylvania and Massachusetts, and greatest in the largely rural states of North Dakota and South Dakota. Alabama and Mississippi percentages were intermediate between these two extremes.

While there were large differences in the contribution of farming to the economies of the different groups of states, agriculture's contribution to all of their economies declined with time. Between 1963 and 1970, farming's contribution to the Alabama economy declined from 13.84 to 2.35%. The rate of decline in both Alabama and Mississippi was greater than for any of the other states.

THE AGRICULTURAL ECONOMY

As detailed in the preceding chapter, the sojourn of American farmers in the tents of the god of war, during the period 1940-1950, was extremely beneficial for them. In nine of the 11 years during the period, the parity ratio was 1.0, or above, and in four of those years, it was 1.1 or greater The ratio had never been so high for so long. Further, national Net Farm Income, expressed as dollars per farm, increased from $706 in 1940 to $3,044 in 1948, an increase of 331%. During the same period, the Consumer Price Index increased from 42 to 72.1, an increase of only 72.6%. Farmers had never had it so good. Unfortunately, farmers in Alabama and the South, did not fare as well as those in many of the other states, but they were waiting on Triangle promises with great expectations.

Shortly after the end of the Korean Conflict, American farmers found themselves in a completely new world. Between 1951 and 1952, the parity ratio fell from 1.07 to 1.00, and in 1953 it fell again to 0.92. Quickly farmers were caught in a slowly tightening cost-price squeeze. They responded to this deteriorating situation in four ways:

1. By increasing productivity through increased mechanization and adopting scientific agriculture, thereby reducing the cost per unit of crop produced.
2. By clamoring for more governmental assistance.
3. By abandoning farming altogether.
4. By simply hunkering down and hoping for the best.

As events evolved, clamoring for more federal assistance was the only solution that offered very much promise.

Federal Farm Legislation

Government efforts to provide direct assistance to farmers through legislated supply management had begun in 1929. That effort failed, but in 1933 the Triangle continued that failed effort by legislating a major acreage reduction program. In the early 1950s, it was difficult to assess the effectiveness of those efforts. Shortly after the passage of the court-approved Agricultural Adjustment Act of 1938, American agriculture became involved in a 14-year period of prewar, wartime, and postwar efforts to provide food and fiber for rapidly changing groups of consumers. Farmers would usually have fared well under those conditions without any legislation, but after World War II they entered a new world. Then the real testing time for the Triangle finally began.

In 1933, Congress had decided to become actively involved in controlling the supply of agricultural commodities and their prices with the passage of the Agricultural Adjustment Act. The original 1933 AAA was found unconstitutional, so they enacted a new version in 1938. When legislation accomplishes few of its objectives, legislatures are prone to pass additional laws to deal with the continuing difficulties, rather than refraining from government efforts to manage from above. So it has been in the national agricultural sector and particularly in cotton production. Congress continued to try to get it right throughout the era of modernization of American agriculture (1950-1970).

The Marshall Plan for the economic recovery of Europe following World War II was formally initiated in June 1947. Under this plan, large quantities of American agricultural products were purchased by the federal government and sent to Europe. These exports served to prevent the buildup of farm surpluses in the United States, although marketing quotas did have to be re-established on some crops.

Both the Korean Conflict (1950-1951) and government authorities encouraged farmers to again increase production to meet wartime needs. At the same time, all acreage allotments and marketing quotas were suspended. Farmers responded by increasing the production of most crops. Cotton farmers harvested 5.1 million more bales of cotton in 1951 than 1950. Farmers also planted more acres of corn (380,000), soybeans (128,000), and wheat (7.1 million) in 1951 than in 1950. Unfortunately, once the system was stimulated, it was difficult to retrench, so surpluses began to accumulate after 1951.

By the beginning of the Eisenhower Administration (1953-1961), Congress began to be concerned about the ever-increasing costs of federal farm programs. Politicians from non-farm states were beginning to question the wisdom of using tax dollars to encourage farmers to continue current levels of production when existing surpluses were already depressing prices, which necessitated even more government assistance. The legislation that required that price supports be set at 90% of parity was of special concern. As a result, in 1954 Congress passed a new law that established a sliding scale of payments whereby prices could be supported at between 82.5 and 90%.

Writing for the Economic Research Service of the Department of Agriculture, Bowers, et al. (1984) concluded that neither manipulating price supports nor paying farmers not to plant had "stemmed the tide of ever-increasing output." By 1960, "the surplus problem had reached crisis proportions. Total carryover of corn stocks had climbed to an all-time high of 1.8 billion bushels . . . . The wheat carryover stood at 1.4 billion bushels, nearly all of which was held by the government. Stored supplies of barley, grain sorghum, and some other crops were also near historically high levels . . . . Meanwhile, prices and farm income had been steadily falling from the high levels of the Korean war period."

According to Bowers, et al. (1984) when President Kennedy took office in 1961, he sent a proposal to Congress for a program to phase out existing regulations. He further proposed that committees of farmers develop plans for mandatory production controls be approved by a two-thirds vote of producers. Unfortunately, the proposal impinged on prerogatives assumed by the Triangle, and the proposal never made it past the first congressional hurdle. With no hope for mandatory production controls, Kennedy turned his attention to disposing of surpluses through donations of food to the poor.

In general, the administrations of Kennedy (1961-1963) and Johnson (1963-1969) made the control of overproduction the primary goal of their farm policy efforts. Farmers were offered what were, in effect, rental payments for portions of their land that would be taken out of production the following year. At the same time, the federal government made a determined effort to expand the export market of American agricultural products. However, its most serious effort to deal with the problem of overproduction was to give more of the surpluses to poor and disadvantaged people. During this period, the ratio of the average farmer's income to the average non-farmer income increased from about 50 to 75%.

Southerners continued to play a major role in farm legislation during the period 1950-1977 (Table 14.4). Allan Ellender (D, LA) served as chairman of the Senate Agriculture Committee from 1951-1953. He was replaced by George Akin (R, VT) in 1953, but returned to the office in 1955, and remained in that position until 1971. Harold Cooley (D, NC) was chairman of the House Agriculture Committee at the beginning of the period in 1950. He served in that office until 1955 when he was replaced by Clifford Hope (R, KS). Cooley returned to the office in 1957, and remained in it until 1967 when he was replaced by William Poage (D, TX). Poage was in the position at the end of the period in 1970. Emstes, et al. (1997) noted that in 1950 (81st Congress) seven of the nine congressional subcommittees dealing with agricultural legislation were chaired by Southern Democrats, and in the 82nd Congress (1951-1953), all six committees were chaired by Southern Democrats. From 1955-1974, from 50 to 70% of majority members of the House Agriculture Committee were Southern Democrats, and they held at least two-thirds of all subcommittee chairs. The dominance of Southern Democrats during this period (1955-1974) was not as obvious on the Senate Agriculture Committee, although 30 to 47% of majority members were Southern Democrats. They controlled at least half of all subcommittee chairs. In the 84th (1955-1957) and 87th to 91st (1961-1971), Southern Democrats were chairs of all five subcommittees. One of the most significant events affecting the role of Southerners in working on agricultural legislation during this period (1950-1970) was the election of Jamie Whitten (D, MS) as chairman of the powerful Agricultural Appropriations Subcommittee, a position that he would hold for a total of 44 years.

During the period 1950-1970, Congress intensified its efforts to honor its promise regarding parity ratios with price supports and supply management by enacting several significant pieces of legislation. The more significant of these are listed and described in the following section. These are exceedingly complex pieces of legislation. Space limitations allow only a cursory consideration of each. Bowers, et al (1984) contains much more information.

1. The Agricultural Trade Development and Assistance Act of 1954 (Public Law 480) provided the basic authority for the disposal of surplus agricultural commodities. It provided for the payment of export commodities in foreign currencies that could then be used to fund economic development projects in those countries making the purchases.
2. The Agricultural Act of 1954 established a flexible price support system for basic crops (excluding tobacco) at 82.5 to 90% of parity and authorized a Commodity Credit Corporation (CCC) reserve for foreign and domestic relief.
3. The Agricultural Act of 1956 became known as the Soil Bank Act. It allowed for the short- and long-term removal of land from production with annual payments to farmers who took their cropland out of production. It consisted of two parts: the Acreage Reserve Program and the Conservation Reserve Program. Under the former, farmers growing wheat, corn, rice, cotton, peanuts, and several types of tobacco retired land producing those crops on an annual basis, during the period 1956 through 1959, in return for annual rental payments from the federal government. Under the latter program, farmers retired cropland under contract for three, five, or 10 years in return for annual payments. Under this program, the government and participating farmers shared the cost of planting trees and grasses on the retired acreage.
4. The Feed Grain Act of 1961 established a level for support prices and required that producers divert a percentage of their base feed grain acreage to be eligible for program payments.
5. The Food and Agricultural Act of 1962 continued the feed grain support program, and authorized an emergency wheat program with voluntary diversion acreage.
6. The Food and Agricultural Act of 1965 was the first of the multi-year farm legislation programs. It authorized four-year commodity support programs for wheat, feed grains, and upland cotton. It further authorized the long-term diversion of land under a cropland adjustment program for feed grains and cotton. Finally, it repealed the Food and Agriculture Act of 1956 (the Soil Bank Act).

Alabama Agricultural Legislation

Free-roaming livestock was becoming a problem even before the Civil War. However, because of the politics involved, the legislature could never agree to deal with the problem on a statewide basis. Rather, in 1903, the legislature had put the problem in the hands of officials in the individual counties. Surprisingly, it was not addressed on a statewide basis until almost a half-century later. For all practical purposes, however, with the exception of some areas in the heavily forested region of southwest Alabama, there was little free-range for livestock anywhere in the state. In 1951, an act closing the range throughout the state was finally passed. This act officially brought to an end an agricultural tradition dating back to the time when sixteenth century Spanish explorers herded cattle and hogs through the wilds of the Lower South.

By 1950, with cotton production in the state declining rapidly, cattle production was increasing. Most of these animals were being marketed through local stockyards, and new stockyards were being built throughout the state. Consequently, farmers and stockyard owners were having to deal with a variety of problems that were new to both parties. In 1951, the legislature passed an act to prevent unfair stockyard practices and to standardize procedures for the marketing of cattle throughout the state.

Agricultural Exports

Agricultural exports had increased rapidly during World War II, before reaching $3 billion by 1950 (Figure 14.5). They grew to $4 billion in 1951 during the last year of the Korean Conflict, but declined below $3 billion in 1952 and 1953. However, after 1953, exports generally trended upward at a relatively high rate to reach $5 billion in 1961. Exports in 1962 of $2 billion were a clear exception to the general trend of the 1960s. This was the year of the Cuban Missile Crisis and increased tensions in Asia as China invaded Indian territory in the Himalayas. Exports increased again in 1963, and in 1964 reached $6.3 billion. They remained at this general level through 1969, probably as a result of unsettled world conditions (The Vietnam War, the war between India and Pakistan, the Cultural Revolution in China, the Six Day War in the Middle East, and the Warsaw Pact invasion of Czechoslovakia). The growth of exports resumed in 1970 to reach $7.2 billion.

In the period 1950-1969, American agricultural exports accounted for 22% of total exports. At the turn-of-the-century, they had accounted for 70% of the total.

Agricultural Prices

Thompson (1962) writing in the 1962 USDA Yearbook of Agriculture, *After One Hundred Years,* commented that USDA-led programs had been greatly expanded since 1933. Successive congressional legislation (the Agricultural Adjustment Act of 1933, the Agricultural Adjustment Act of 1938, the Agricultural Act of 1948, and the Agricultural Act of 1949) had sought to stabilize and/or increase farm prices and income by a combination of programs, primarily by support prices and supply management to support and strengthen prices received by American farmers.

Provisions for support payments were included in the Agricultural Adjustment Act of 1933, but through the World War II years they played only a minor role in the establishment of agricultural prices. After 1950, however, they began to play an increasingly important role. Support prices for the major commodities were increased during the Truman Administration (1945-1953), but at the beginning of the Eisenhower Administration (1953-1961), Congress began to be concerned by the increasing cost of government agricultural programs. As a result, support prices were lowered. Unfortunately, by 1959 market prices for many commodities were declining. With decreasing commodity prices, farm income began to decline and became a subject of debate in the 1960 presidential campaign. Shortly after Kennedy took office, USDA began to increase support prices again, and they continued to increase during the Johnson Administration (1963-1969).

In 1949 and 1950, indices of prices received and prices paid were nearly equal at around 250 (Figure 14.6). They remained relatively close together through 1952. Afterwards, however, they began to diverge rapidly. Indices for prices received declined to 233 in 1955; then slowly increased to 274 in 1970. Indices for prices paid remained around 277 through 1957, but then began to trend upward, reaching 382 in 1970. These data clearly show the growing imbalance between the indices after 1952.

By 1970, it should have been obvious that the price supports, acreage allotments, direct payments, acreage diversions, government-sponsored export programs, new parity formulas, and direct government purchase and storage of commodities that had been established by the Triangle simply had not worked as intended. Indices for prices received had trended slowly upward after 1955, but not nearly enough to bring them in-line with the indices for prices paid. It was evident to many that tinkering around the margin of the problem simply was never going to work. The obvious solution was to take a large share of the less efficient producers completely out of the system. Unfortunately, there was simply no politically acceptable way of doing this.

Parity ratios remained above 1 for most of the 1940s (Figure 14.7), and although the parity ratio began of decline toward the end of the decade, it increased sharply in response to the Korean Conflict. Unfortunately, after those hostilities ceased and demobilization was completed, the ratio began to decline. It was down to near 1 in 1953. The decline continued through 1970 when it reached 0.72. In two decades, American farmers had returned to the dire economic straits that usually characterize their condition. In fact, in 1970 it appeared that they were going to be ruined economically. However, with each downward move of the parity ratio, the Triangle seemed to grow stronger.

The preamble to the first Agricultural Adjustment Act (1933) stated that it was government policy to keep parity ratios at the 1908-1914 level of 1.0. Obviously, this promise had not been kept. The Triangle, with its political limitations, seemed powerless to do so.

Agricultural Productivity

Figure 14.8 presents data showing changes in the indices for agricultural inputs, outputs, and productivity for the period 1950-1970 (Indices in 1987 = 1.00). The input index remained close to 1.10 from 1950 until 1964; then it declined slightly through 1970. It was somewhat lower than expected in 1954, as the country's economy experienced an 11-month economic contraction (Figure 14.1) in 1953 and 1954—as the economy turned downward during the period of postwar adjustment. Then in 1959 and 1960, the index moved upward slightly, as the 8-month economic contraction of 1957-1958 came to an end. It declined slightly during the 10-month economic contraction of 1960-1961. After 1963, it was generally lower, as the Consumer Price Index began to increase rapidly (Figure 14.2). Then the inputs index remained slightly below 1.1 through 1970.

Agricultural technology on most American farms began to improve rapidly following the end of World War II and the Korean Conflict. As a result, indices of outputs usually trended upward for the entire period 1950-1970. It was 0.503 in 1950, and 0.719 in 1970. Those years when the indices were lower than expected were generally associated with periods of economic contraction (Figure 14.1).

The index of productivity is the ratio of the index of outputs to the index of inputs. In 1950, it was: 0.503/1.094, or 0.460. This means that the 1950 index of productivity was 46% of the index (100) in 1987. Between 1950 and 1970, the indices of productivity trended upward, along with the indices of outputs, while the indices of inputs remained at essentially the same level. The index of productivity reached 0.67 in 1970. This increase in agricultural productivity was a major factor in the ability of farmers to continue to operate even though the parity ratios declined (Figure 14.7). While this relationship would seem to work in theory, as we will see in the coming years, it does not work very well in practice.

Farmers

The 1940s were a period of unparalleled prosperity for a large percentage of American farmers; however, a sizable number used wartime realities to leave farming altogether. Between 1940 and 1950, the country lost 24.5% of its farmers, whereas Alabama lost only 9%.

The Total Population

Data published by Coulson and Joyce (1999) on decennial estimates of the American population for the period 1920-1950 are presented in Figure 14.9. The data indicate that the national population increased from 151.9 million in 1950 to 203.3 million in 1970. The data also indicate that the rate of increase was considerably greater in the period 1950-1970 than in the period 1920-1945. In the earlier period, the population increased at a rate of

0.94% a year, but in the latter period the rate of increase was 1.61% a year.

Data on the Alabama population are presented in Figure 14.10. These data indicate that Alabama's population increased from 3.06 million in 1950 to 3.44 million in 1970. They also indicate that it increased very little between 1965 and 1970. Annual data show that the population reached 3.46 million in 1966, but declined afterwards to 3.44 million in 1969. By 1970, it had begun to increase again. This was a period of intense civil/racial unrest in the state.

The Farm Population

Most American farmers returned to their farms after World War II (Figure 14.11). However, it did not take long to remind many of them that there was very little left for them down on the farm; consequently they began to leave in droves. By 1950, the farmer population numbers were down to 23 million. Numbers continued to decline until 1954. In 1953 and 1954, a 10-month economic contraction (Figure 14.2) temporarily reduced the availability of off-farm jobs. As a result, numbers actually increased between 1954 and 1955 (19 to 19.1 million). Afterwards, however, numbers began to decline again, and by 1970 there were only 9.7 million persons remaining in the national farm population. Between 1950 and 1970, the country lost 57.9% of its farm population (13.3 million). Unfortunately for the farmers remaining on their land, this exodus had very little effect on prices received for their commodities (Figure 14.6). Those remaining just increased their production. During the period 1950-1970, the indices of farm outputs increased by 42.9% (Figure 14.8).

With the total national population increasing (Figure 14.9) and the farm population decreasing (Figure 14.11), the percentage of farmers in the population also began to decline (Figure 14.12). In 1950, 15.3% of the total population was in the farm population, but by 1970 the percentage had declined to 4.8%. From another perspective, in 1950 some 15 farmers were producing enough food and fiber for 100 persons, and by 1970 five farmers were meeting the needs of 100 Americans.

Fite (1984) published data on the Alabama farm population for the period 1920-1970. They are presented in Figure 14.13. As the data show, numbers in the farm population remained relatively static from 1920-1940, but sometime between 1940 and 1950, they began to decline. In 1950, there were only 960,000 on farms in the state. Numbers continued to decline—to 519,000 in 1960 and 216,000 in 1970. Between 1950 and 1970, it declined by 77.5%. The percentages of the Alabama farm population in the total population are also shown in Figure 14.12. As the data indicate, in 1950 the percentage for the state was twice that of the country (15.3 versus 31.4). Afterwards, it declined much more rapidly. In 1970, the percentage for the state was down to 6.3%, or just slightly higher than in the country (4.8 versus 6.3%).

Data published by Fite (1984) on the numbers of black farmers in Alabama during the period 1920-1970 are presented in Figure 14.14. These data indicate that numbers declined from 57,200 in 1950 to 9,900 in 1970, a decrease of 83%. During the same period, Alabama's total farm population declined by only 77.5% (Figure 14.13). In 1950, the number of black farmers in the state was equal to 6% of Alabama's total farm population. By 1970, the percentage had declined to 4.6%.

Farmers in the Work Force

USDA data (Figure 14.15) show that in 1950 about 12.2% of the national labor force were farmers. In 1920, the percentage had been 27%. With the total labor force increasing rapidly between 1950 and 1970, and numbers of farmers declining, their contribution fell sharply. By 1970, farmers made-up only about 4.6% of the total national labor force.

Tenancy

Farm tenancy in the United States had begun to decline after 1930. Percentages in 1930, 1940, and 1950 were 42.4, 38.7 and 26.8%, respectively. Afterwards, tenancy continued to decline at about the same rate, finally reaching 12.9% in 1969 (Figure 14.16).

Tenancy rates in Alabama also began to decline after 1930 (Figure 14.16), reaching 41.4% in 1950. Between 1950 and 1969, rates in Alabama declined more rapidly than in the country, and by 1969 only 10% of the state's farms were operated by tenants. Tenancy in Alabama was finally lower than in the country as a whole. From the beginning, tenant farming—with the farmer and his family cultivating small patches with their one mule—was not an economical or efficient way of producing food and fiber. Nevertheless, for thousands farmers across the South, there was no alternative. Tenancy in the Southern states had begun to decline when New Deal programs made it profitable for the landowners to rid themselves of tenants. The decline continued when new jobs created by the wartime economy drew thousands of tenants away from the land. Then after the early 1950s, declining parity ratios made it increasingly difficult for farmers to remain on those small farms in a tenant relationship. Fortunately, the American economy was expanding fast enough to absorb many of them into the non-farm labor force.

Farms

Numbers of farms in the United States reached an all-time high of 6.8 million in the middle of the Great Depression, but afterwards, they began to trend downward at a fairly regular rate, reaching 5.6 million in 1950. During this period, the country lost 17.6% of its farms, at a rate of 75,000 each year.

During the same period (1935-1950), Alabama lost 19.7% of its farms, at a rate of 3,375 each year.

Farm Numbers

In 1950, national farm numbers had reached 5.7 million. Numbers continued to decline afterwards, finally reaching 2.9 million in 1970 (Figure 14.17). In that 21-year period, the country lost 2.7 million (47.8.1%) of its farms, at a rate of 128,500 each year. Declining parity ratios (Figure 14.7), and the resulting cost-price-squeeze that began in the early 1950s, provided farmers an increasingly powerful incentive to get out of the farming business.

During this period 1950-1970, farming was rapidly becoming more capital intensive. Investment required to remain competitive was becoming almost prohibitively expensive. While Net Farm Income increased at a higher rate than the Consumer Price Index (102 versus 61%), average investment in land and buildings increased from $65 to $196 per acre (201%), investment in machinery and equipment increased from $2,258 to $9,770 per farm (333%), capital consumption increased from $469 to $2,341 per farm (400%), and purchased inputs increased from $1,962 to $8,645 per farm (341%). With the rapidly escalating costs (investments) of these inputs compared to returns, it was difficult to remain in farming.

In 1950, 6.1% of the country's farms were in Texas. Other states with substantial shares were: North Carolina (5.3%), Mississippi (4.6), Tennessee (4.3), and Missouri (4.3). These five states accounted for 24.6% of all American farms. All of these states had large number of slaves before the Civil War, and large numbers of tenant farms afterwards.

During the period 1940-1970, trends in Alabama farm numbers were essentially the same as for the country as a whole (Figures 14.17 and 14.18). Numbers in 1935, 1940, and 1950 were 274,000; 241,000; and 220,000, respectively. Then between 1950 and 1970, numbers declined from 220,000 to 80,000. In that 21-year period, Alabama lost 138,000, or 62.7% of it farms— 6,600 a year. The national cost-price squeeze was even more vicious on the thousands of inefficient farms in the state. As was the situation with the national data, the rate of loss was much greater in the period 1950-1970 than it had been from 1940 to 1950 (6,600 versus 1,900 a year). The lack of comparative advantage of Alabama agriculture was rapidly changing the face of farming in the state.

In 1950, 3.9% of American farms were in Alabama The largest share of the total number was in Cullman County (3.7%). Other counties with substantial shares included: DeKalb (3.3%), Marshall (2.9), Madison (2.4), and Jackson (2.3). All of these counties are just south of the Tennessee border. These five counties accounted for only 14.6% of all Alabama farms. A total of 27 counties reported 3,000 or more farms.

Farm Size

Average farm size in the United States did not increase very much in the early years of the twentieth century. In 1930, the average size of an American farm was 157 acres; however, after 1935, farm numbers began to decline (Figure 14.17). As numbers declined, farm size increased as a result of consolidation. By 1950, farm size was up to 216 acres (Figure 14.19). Then in the early 1950s, parity ratios began to decline (Figure 14.7), resulting in an increasingly vicious, cost-price squeeze. At that point, numbers began to decline more rapidly, increasing the rate of farmland consolidation. In 1964, size reached 352 acres, and by 1969 it had increased even more to 389 acres.

Data presented in Table 14.5 show the percentages of American farms, in different size groups, in 1950, 1954, 1959, 1964, and 1969. These data show that between 1950 and 1969 the percentage of smaller farms (1 to 179 acres) declined from 72.6% of the total to 59.9%. Obviously, the percentage of the larger farms (180 and larger acres) increased by an equal amount. The national decline in the smaller farms was most obvious in the 10- to 49-acre group (27.5 to 17.3%). In the larger farms, the increase was most obvious in the 180- to 499-acre group (17.9 to 26.6%).

During the period 1920-1950, average farm size in Alabama generally followed the national trends, and in 1950 it reached 99 acres (Figure 14.19). In the early 1950s, size began to increase rapidly, as the rate of decline in farm numbers increased. In Alabama, average farm size reached 116 acres in 1954. Size continued its rapid rate of growth in the late 1950s and throughout the 1960s, reaching 143, 165, and 188 acres in 1959, 1964, and 1969, respectively. Even with this rapid rate of growth through 1970, farms in the state were still less than half the size of those in the country as a whole. If size had anything to do with efficiency and productivity, Alabama still had a long way to go.

Data presented in Table 14.5 show the percentage of Alabama farms in different size groups in 1950, 1954, 1959, 1964, and 1969. In Alabama, percentages declined only in those farms smaller than 50 acres (50.8% to 35%). Percentages in all other size groups increased. The increase in the 180- to 499-acre group was especially large (7.7% to 17.1%).

Total Farmland

In 1950, USDA began to publish annual estimates of the quantity of farmland in the individual states. Before 1950, only census data were available. To maintain continuity, this chapter reflects census data, even though the two estimates differ somewhat. Census data estimates of total farmland in the country for the period 1920-1969 are presented in Figure 14.20.

Acres of total farmland in the country reached the highest level ever recorded in 1950 (1,163 million acres). Subsequently, farm numbers began to decline rapidly (Figure 14.20). Although many farms were being consolidated as indicated by the increase in farm size (Figure 14.19), consolidation was insufficiently high to offset the rate of decline in numbers. By 1969, farmland had declined to 1,063 million acres.

Census data on the total acres of Alabama farmland during the period 1920-1969 are presented in Figure 14.21. During the period, acres in Alabama usually followed the same trend as farm numbers (Figure 14.18). It was 20.9 million in 1950 and 1954, before falling sharply in 1959 and 1964 to reach 13.6 million in 1969.

Use of Farmland

Census data indicated that total farmland in the United States declined throughout the period 1949-1969 (Figure 14.20); however the percentage of cropland increased from 35 to 43% (Figure 14.22). In 1949, the percentage of cropland was lower than that of pastureland (35 versus 42%). Percentages of cropland and pasture were the same in 1954 and 1959, at levels of 40 and 41%, respectively. In 1964, the percentage in pasture was again higher (44 versus 40%). In 1969, the percentage in cropland was up to 43%, the highest level in the period. For some reason, the census did not include comparable data on pasture or other land in 1969. The percentage of total farmland in forestland declined from 19% in 1949 to 10% in 1969. The land used to increase cropland and pastureland came from forestland.

Census data on the use of Alabama farmland in 1949, 1954, 1959, 1964, and 1969 are presented in Figure 14.23. The percentage of farmland used as cropland was 42% in 1949, but was generally lower in 1954, 1959, and 1964. In 1969, cropland returned to its 1949 level. In contrast to the national data, the share of total farmland in pasture was much lower in Alabama throughout the period. Pasture did increase from 8% in 1949 to 17% in

1964, reflecting the continued growth of the Alabama cattle industry. There are no comparable data available for pasture or other land in 1969. The percentage in forestland was considerably higher than cropland in all the years except 1969; it range from 50% in 1949 to 39% in 1969.

Total Cropland

Total cropland in the United States (Figure 14.24) trended downward, along with total farmland (Figure 14.20) from 1949 (478.3 million acres) to 1964 (434.2 million). In 1969, cropland increased to 459 million. It had not been that high since 1944. Indices for prices received for agricultural commodities began to trend upward after 1964 (Figure 14.6), although parity ratios were continuing to trend downward (Figure 14.7). This increase in prices they received for their crops seemed to have encouraged farmers, and they apparently increased their total cropland in response. They were further encouraged by the continued expansion of the economy during the period. Its growth was especially fast between 1960 and 1966 (Figure 14.1). In fact, the years between 1958 and 1970 have been characterized as a period of extended business expansion.

The 1939-1969 trend of changes in total Alabama cropland was essentially the same as for the country as a whole (Figure 14.25). It trended down from 8.7 million acres in 1949 to 5.1 million in 1964. The rate of decline in Alabama was somewhat greater than in the country. Between 1964 and 1959, total cropland in Alabama increased by about 14%, to reach 5.8 million acres in 1969. Alabama farmers seem to have been getting excited about their prospects.

Cropland Harvested

As might be expected, nationally cropland harvested generally followed the same trend as total farmland during the period 1939-1964 (Figures 14.26 and 14.20). Total cropland had trended downward along with total farmland from 1949 through 1964 (Figure 14.24), but then increased in 1969. Cropland harvested had followed essentially the same trend, but instead of increasing in 1969, the decline continued.

As expected, percentage of cropland harvested in the United States (Figure 14.27) also trended slowly downward during the period 1949-1969, from 72% in 1949 to 66.1% in 1964. Then in 1969, it plunged to 59.5%. Palmer Drought Severity Indices data (Cook, et al., 1999) indicate that the weather in June, July, and August 1969 was abnormally wet to excessively wet from New England to eastern Nebraska. It is possible that this wet weather resulted in fewer acres of cropland being harvested than expected.

Cropland harvested in Alabama also trended downward throughout the entire period 1949-1969. It was 5.7 million acres in 1949 and 2.7 million in 1969 (Figure 14.28). The percentage of cropland harvested had also trended downward throughout the period from 74.6% in 1944 to 46.7% in 1969 (Figure 14.27).

According to Fite (1984) farmers in the South were harvesting crops from 32.2 acres of cropland in 1950. Then in 1960 and 1970, the average increased to 52.6 and 84.2 acres, respectively. In the same years, Alabama farmers harvested crops from 27.1, 37.4, and 51.6 acres, respectively, an increase of 90%. From 1950 to 1970, Southern farmers more than doubled the amount of cropland harvested. They were making progress, but they had a long way to go. Farmers in Iowa, harvested crops from 111.0, 141.0, and 149 acres of cropland in those same years.

It is not surprising that Alabama lost so many farms during this period (1950-1970). The percentages cited above were averages. Many farmers harvested crops from even fewer acres. Most Alabama farmers had never realized decent returns from such small harvests, but for all those years they never had an alternative. Fortunately, the expanded national economy that grew out of World War II and the Korean Conflict finally provided many of them with opportunities to leave their small farms.

Value of Land and Buildings

Doherty (1962) described some of the far-reaching changes taking place in American agriculture following World War II and the Korean Conflict. He noted: (1) that by 1962 three of five Americans living in rural areas did not live on farms, (2) that 24% of the 3.7 million farms in the country at that time were operated by individuals who depended almost entirely on other than farm income for their living, (3) that the trend to larger farms and heavily capitalized, least-cost production was putting intense pressure on small farms, and (4) that thousands of farmers had abandoned farming for off-farm work. All of these changes ultimately affected the price of farm real estate and the value of land and buildings.

Census data indicated that the value of land and buildings in the United States had fallen to $32 billion during the heart of the Depression (Figure 14.29), and had remained near that level until World War II began in 1941. By 1950, value had increased to $75 billion. As might have been expected, value did not decline after the wars. Rather, it continued to grow rapidly, reaching $207 billion in 1969. During the 21-year period, value increased by 174.8%. During the same period, the Consumer Price Index increased by 52.2% (Figure 14.2), and the quantity of farmland decreased by 8.6%.

Because of the changes in farm numbers and farm size during the period, the value of land and buildings data expressed in dollars-per-acre are more representative of changes that took place during the period. Values for American farms were around $30 an acre in the mid-Depression years and remained near that level until 1941 (Figure 14.30). Then at the beginning of World War II, they began to increase and finally reached $65 an acre by 1950. During the period 1950-1969, they continued to increase, reaching $196 an acre in 1969. These data indicate that values increased 201% ($65 to $196) during the period. The Consumer Price Index only increased 52% (72.1 to 109.8) (Figure 14.2). These comparisons indicate that only a third of the increase in national farm real estate values during that period was related to inflation. Two-thirds of the change represented a real increase in value.

Census data on the total value of land and buildings on Alabama farms during the period 1920-1969 are presented in Figure 14.31. These data indicate the trend in values in the state was similar to the national value trend (Figures 14.30). In Alabama, they declined to around $368 million in the heart of the Depression, but began to increase thereafter, reaching $1,023.6 million in 1950. Subsequently, the rate of increase became exponential, and values reached $2,730.8 million in 1969.

Census values for Alabama during the period 1920-1969 expressed as a per-acre basis are also presented in Figure 14.30. The data show the same trends as the total values. They were $19 an acre in 1935, $58 in 1950, and $200 in 1969. The rate of increase was a little smaller than expected between 1950 and 1954. This lower rate of increase was likely related to the period of economic contraction of 1953 and 1954. The decline in the rate of increase during this period was not as severe in national values (Figure 14.30).

Throughout the period 1935-1950, Alabama values on a per-acre basis were $10 an acre lower than those in the country (Figures 14.30), and the difference increased between 1950 and 1954. After 1954, however, state values began to close the gap, and in 1964 state values were slightly higher ($200 versus $196 an acre).

Prevatt (2000) published data indicating that in 1950 the value of land on Alabama farms accounted for 64.7% of the total value of land and buildings. By 1969, the land share had increased to 67.7%. According to Prevatt, the average value of farmland increased from $31 an acre in 1950 to $126 in 1969, an increase of 306%. Obviously, during the period values for land were increasing at a higher rate than for buildings.

The value of farmland reflects both its inherent capacity for the production of agricultural crops, as well as non-farm influences (suburban sprawl, for example). In contrast, cash rents for farmland are a measure of current economic returns realized from farming the land. By comparing the rent obtained from a piece of land to its value, it is possible to determine whether productive capacity or non-farm influences are having the greatest effect. If the ratio of cash rent to market value is high, agricultural productivity is having the greatest influence. Conversely, a low ratio indicates that non-farm influences are more important. For example, in 1960 the ratio in North Dakota, primarily an agricultural state, was 9% (Barnard, 2002). This ratio suggests that the average annual cash rent for cropland in North Dakota in 1960 was equal to 9% of the market value of the land. In the same year, the ratio in New Jersey, primarily an urban state, was near 3.7%. A comparison of these two ratios suggests that farmland values in New Jersey in 1960 were being influenced by non-farm factors. A comparison of the ratios in 1968 is even more telling. In that year, the ratio for North Dakota was 9%, or essentially the same level as in 1960; however, the New Jersey ratio had fallen to 2.8%, which indicates an increasing non-farm influence on farmland values.

Data on rent-to-value ratios for Alabama land are not available before 1967. Prevatt (2000) reported that these ratios for Alabama cropland for 1967, 1968, and 1970 were 8.8, 8.2, and 7.5%, respectively. These ratios indicate that compared with North Dakota, Alabama non-farm influences were affecting land values to some extent, but not nearly so severely as in New Jersey. The fact that they were decreasing indicates that in the late 1960s these influences were becoming increasingly important in Alabama. This declining ratio may partially explain the fact that in the 1960s average values of Alabama land were increasing faster than those in the country as a whole.

Value of Machinery and Equipment

Mechanization of grain production in the North Central Region began in the late nineteenth century, and that region began to exploit tractors for farm power around 1915. Unfortunately, this was not the case in Alabama and the South. As former President Jimmy Carter (2001) noted, at the beginning of World War II farms in the South resembled those in colonial times with respect to farm mechanization. Modernization required that Alabama use improved machinery and that they substitute machine power for human and animal power on their farms. The higher farm income associated with the wartime economy during World War II had allowed farmers in Alabama and the South Central Region to begin the extremely expensive process of mechanizing their farms, but it was not until after the end of the Korean Conflict that the process began to move-forward rapidly.

Lerner (1975) supervised the publication of a set of data on the value of machinery and equipment on American farms. Some of these data are presented in Figure 14.32. The data indicate that value was at a level of $12,166 million in 1950. Afterwards, it trended upward to reach $34,052 million in 1970. Changes in the annual rate of increase during the period were influenced by the economic contractions (1953-1954, 1957-1958, 1960-1961, and 1969-1970), described in a preceding section and shown in Figure 14.1. Between 1950 and 1970, value increased by 180%. During the same period, the Consumer Price Index increased only 61% (72.1 to 116.3). Obviously, during the period, American farmers added a substantial amount of real value to their stock of machinery and equipment.

National data on the value of machinery and equipment converted to a per-farm basis for the period 1950-1970 are presented in Figure 14.33. These data show the same sharp increase in value as the unconverted data. Value per farm increased from $2,154 in 1950 to $11,547 in 1970. The converted data correct for the large loss of farms between the two dates (Figure 14.17). They show that value per farm increased by 436% during the 21-year period.

There are no annual data available for the value of machinery and equipment on Alabama farms during the period 1950-1970, and for some reason, the census did not collect any data in 1950, 1954, 1959, and 1964. Available data indicate the total value per Alabama farm in 1969 was $395.4 million. Converted data for Alabama farms indicated that value per farm was $5,909 in 1969. In comparison, it was $8,054; $12,293; $13,504; and $9,287 in Mississippi, Iowa, Illinois, and the United States, respectively.

Tractors on Farms

Between 1940 and 1945, tractor numbers in the country increased by an astounding 64% (Figure 14.34). Numbers continued to increase afterwards, but not at the same rate. In 1950, there were 67 tractors per 100 farms; then between 1954 and 1959, numbers finally reached 100 per 100 farms for the first time, and by 1969 numbers per 100 farms reached 162.

In 1940, Alabama lagged far behind the other regions of the country with respect to the number of farms with tractors (Figures 14.34 and 14.35). That year, there were only three tractors per 100 farms in Alabama compared to an average of 25 in the country. Fortunately, after

World War II, this situation began to change rapidly, and by 1950 there were 21 tractors per 100 Alabama farms. Numbers continued to increase rapidly during the remainder of the period, and by 1969 reached 103. Numbers in Alabama in 1969 were still below the average for the country (103 versus 162), but Alabama farmers were rapidly closing the gap.

Fite (1984) comments that the increased use of tractors did not simply substitute machines for horses and mules. Tractors were much more versatile, and a tractor with essentially the same size as a large farm-power animal was several times more powerful. Further, mules did not have power takeoffs, which could be used to meet a wide variety of farm power needs. Tractors could be used to power grain, and soybean combines, hay balers, corn pickers, peanut harvesters, post-hole drills, spraying equipment, and pecan tree shakers. In addition, tractors could be used to pull stumps and to level land.

In the South, the increasing availability of this increased versatility and power had far-reaching effects on farm practices. Studies on the relationship between tractor use and farm operation, demonstrated that farmers with the machines tended to operate larger farms and plant fewer acres of row crops, such as cotton and corn. At the same time, they planted more acres in small grains, forages, and pastures. Farmers with tractors also tended to have more hogs, milk cows, and beef cattle. Tractors had another important effect on Southern agriculture. Because they were expensive to own and operate, it was not practical to use these machines in conjunction with old, outmoded production methods. For example, farmers could not afford to use tractors to produce 15 bushels of corn an acre; consequently, tractor ownership and operation forced them to rapidly upgrade their practices. They had to use the best seeds available, use more and better fertilizers, and be more efficient in their application. They also had to be more aware of the costs of pests to their operations and to be ready to control them when necessary (Fite, 1984).

Animal Power on Farms

The increase in the number of tractors on American farms during the period 1950-1970 was a clear indication of the increasing modernization of agriculture. Another indication was the decrease in the number of horses and mules (Figure 14.36). By 1955, numbers of these animals had reached such a relatively low level that USDA no longer enumerated them separately. The data show that numbers of horses continued the same downward trend that had begun around 1915. In 1950, there were 5.3 million horses on American farms, and by 1955 numbers had fallen to 3.1 million. There were always more horses than mules in the country. The larger share of mules was found on Southern farms, and their numbers did not begin to decline until around 1925. In 1950, there were 2.2 million mules in the country, but by 1955 there were only 1.4 million.

Converting the total numbers of horses and mules in the country to numbers per 100 farms adjusts the data for changes in the numbers of farms. These converted data for horses for the period 1935-1955 are presented in Figure 14.37. These data indicate that the number of horses per 100 American farms declined from 94 to 67 between 1950 and 1955. During the same period, the number of mules per 100 farms declined from 38 to 31 (Figure 14.38).

The increasing numbers of tractors on Alabama farms also lead to sharp reductions in the number of farm-power animals in the state. There had always been more mules than horses on Alabama farms. The ratio was as high as about six to one in the mid-1930s (Figure 14.39). In the following years, mule numbers declined more rapidly than number of horses, and by 1950 the ratio was down to around three to one (190,000 versus 56,000). In 1955, there was apparently some change in the way that these animals were counted in the state. Data in Figure 14.39 indicate increases in the numbers of both animals between 1950 and 1955. Even with the apparent change in the method of enumeration, numbers of both continued to decline, reaching 151,000 and 50,000, respectively in 1950 and 1955.

On average, in 1940 there were 127 mules on every 100 Alabama farms (Figure 14.38). By 1950, there were only 86. With the change in the method of enumeration, the number of mules per 100 farms actually increased between 1950 and 1955 (86 to 94). Records were not maintained on the number of mules in the state after 1955. The situation with regard to the number of horses in the state was generally similar to that of mules. There were 30 horses per 100 farms in 1945 and 25 in 1950 (Figure 14.37). Then with the change in the method of enumeration, numbers of horses per 100 farms also increased from 1950 to 1955 (25 to 31). The relative decline in the number of horses during the period 1945-1955 was not as great as for mules, probably as a result of the fact that horses were increasingly being kept for recreational purposes by Alabama farm families.

Use of Fertilizers

Fertilizer use in the United States changed relatively little during the Depression (Figure 14.40), but with the beginning of World War II, use began to increase rapidly. Between 1939 and 1949, it increased from 48 pounds to 107.6 pounds an acre of harvested cropland. Prices received (Figure 14.6) during the war were so good that farmers bought all the fertilizer they wanted, and the results were so positive that they really became convinced of its efficacy. Consequently, after the end of the Korean Conflict they continued to increase use of fertilizer. In 1954, they used 136.8 pounds an acre, and by 1969 that total was up to 285.2 pounds, an increase averaging almost 8.6% a year. At this rate, the quantity used annually would double in a decade.

Fertilizer use began to increase rapidly in Alabama during the World War II years (Figure 14.41). In 1949, Alabama farmers used 356 pounds for each acre of cropland harvested. Then in the next 21 years (1949-1969), it increased to 764 pounds, an increase of 109%. Throughout the period, use in Alabama was well over twice as high as in the country. Crop production in much of the country, especially on the Alfisol and Mollisol soils found in the country's primary corn, wheat, oats, and soybean producing areas (Figure 3.2), requires significantly less fertilizer than crop production on the Ultisols of the Atlantic Coastal Plain (Figure 1.9). Unfortunately, even with the significantly higher use of fertilizer, yields of most Southern crops were much lower than those in other areas of the country.

Plant Agriculture

Data on acres harvested of major crops in the United States and Alabama in 1940 and 1950 are presented in Table 14.6. There are several points of interest in the national data. First was the declining importance of cotton (23.9 to 11.2 million acres) over the 11-year period. Second was the overwhelming importance of feed grains (corn, oats, sorghum, and especially corn) compared to the food grains (wheat) in American agriculture. Third was the rapidly growing importance of soybeans (4.8 to 42.2 million acres) over the period.

Harvested cropland in Alabama declined by 19.4% (7.11 to 5.73 million acres) between 1940 and 1950. This loss is evident in the Alabama data. Acres harvested of all the major crops, with the exception of soybeans, declined during the period. Even the old standby (sweet potatoes) rapidly lost favor with farmers. These data clearly demonstrate the passing of an era in Alabama agriculture. Farmers in the state were finally beginning to understand that they had little comparative advantage in crop production.

Alabama Crop Weather

Data on Palmer Drought Severity Indices (Cook, et, al, 1999) for June through August during the period 1950-1970 in Alabama are presented in Table 14.7. These data indicate that more than half of the state was abnormally dry to in drought in nine of the 21 years of the period (1951, 1952, 1953, 1954, 1955, 1956, 1960, 1962, and 1966); abnormally moist to wet in six years (1955, 1957, 1958, 1961, 1967, and 1970); and normal in six years (1959, 1963, 1964, 1965, 1968, and 1969). Tropical storms passed through the state in 11 of the 21 years. Tropical storms hit the state in 1950, 1953, 1954, 1955, 1956, 1957, 1960, 1965, and 1970. Hurricanes Easy and Hilda passed through the state in 1950 and 1964, respectively.

Cotton

Under pressure from New Deal agriculture programs to bring supply in line with demand, the acreage devoted to cotton production in the United States had begun a slow downward trend during the Depression years.

Total acres planted had been 44.4 million at the beginning of the Depression in 1929. In 1950, less than half that amount (18.9 million acres) was actually planted. However, even with this level of reduction, the economic condition of cotton farming was little better than it had been two decades earlier. Federal government support prices were higher, but so were costs. Unfortunately, prices were still much too high for American cotton to compete effectively in the domestic market with the ever-increasing supply of low-cost, man-made fibers or in the international market with producers from India, Brazil, China, and the Soviet Union. The international market was especially important, because the United States, even with all of the government programs, continued to produce far more cotton than could be used by the national economy.

The problems of most Southern cotton farmers were further exacerbated by the continued reduction in their acreage allotments. In 1933, Congress, under New Deal farm programs, established a cotton allotment for each farm; the federal government was still tinkering with the allotments in 1950. For example, in 1955, 80% of the cotton allotments in North Carolina were six acres or less. Unfortunately, there was no practical government price-support policy that could meet the needs of a cotton farmer with a six-acre planting allotment. Over time, many of these operators with small allotments simply sold their farms, along with their allotments. Some leased their allotments to other farmers, who in this manner accumulated enough acreage to allow them to reach a critical mass for their overall operation.

US Cotton

In 1950, American farms harvested 10 million bales of cotton. Texas farmers harvested 29.4% of the total crop. Other states reporting large shares were: Mississippi (13.3%), Arkansas (10.9), California (9.8), and Alabama (5.7). These five states accounted for 69.1% of the total.

Exports

Export data for cotton are reported on a fiscal-year basis. For cotton, both the fiscal year and the foreign trade year begin August 1 of one year and end July 31 of the following year. With the beginning of World War II in Europe in 1939, American cotton exports had declined sharply, and through 1943 remained around one million bales (Figure 14.42). There had been no need for cotton support prices during the war, but after it ended, they were put in place again in an attempt to help farmers with a rapidly growing cost-price squeeze. Exports began to increase afterwards, and reached 5.7 million bales in 1951. Exports were extremely important to American farmers during the period 1950-1970. Unfortunately, stocks of many commodities began to accumulate in the early 1950s; consequently, the Agriculture Act of 1954 established marketing quotas for a number of crops and changed the system of mandatory support prices to one with flexible supports. However, even with the marketing quotas, stocks continued to accumulate. As a result, USDA began to slowly reduce support prices. By 1955, cotton exports were down to 3.3 million bales. In 1956, Congress (Triangle) directed the Commodity Credit Corporation to sell cotton that it owned at the world price and absorb losses that would result. With plenty of cotton available at the lower price, exports immediately skyrocketed to 7.9 million bales, the highest level since 1933. Between August 1960 and July 1961, exports totaled 6.9 million bales. Those exports were equal to 48.6% of the 1960 crop. However, with decreasing commodity prices, farm income began to decline and became a subject of debate in the 1960 presidential election. Shortly after Kennedy took office, USDA began to increase support prices again. Unfortunately, these support levels were well above the world price for cotton, and exports began to drift downward again, reaching 2.4 million bales in 1968. They recovered somewhat in 1969 and 1970.

Several international events (the Cuban Missile Crisis, the Chinese invasion of Indian territory in the Himalayas, the Vietnam War, etc.) affected, to a lesser or greater extent, exports of all agricultural commodities during this period. It is difficult to determine what specific effects these international events had on cotton exports. For example, the Berlin Crisis began in November 1958, and in the 1958 foreign trade year, exports fell to 2.9 million bales of cotton.

Prices

Support prices for cotton for the period 1950-1970 are shown in Figure 14.43. They were 28 cents a pound in 1950. Then with market prices beginning to decline (Figure 14.44), the government began to increase support prices. They reached 32 cents in 1955. By 1954, the Eisenhower Administration was showing concern for the cost of New Deal farm programs. Between 1955 and 1956, the support price for cotton was reduced from 32 cents a pound to 29 cents. In 1960, the market price fell to near 30 cents. That year, Kennedy was elected president, and in 1961 the support price was increased to 32 cents. Although there were year-to-year changes subsequently, they generally ranged between 29 and 34 cents through 1968. After 1963, market prices began to trend downward, but the government waited until 1969 to boost support prices. They were raised to 35 cents in 1969 and to 37 cents in 1970. Between 1950 and 1970, support prices increased from 28 to 37 cents.

Cotton market prices increased rapidly during World War II and through the postwar period, reaching 40 cents a pound in 1950 (Figure 14.44). Between 1949 and 1950, bales harvested (Figure 14.47) fell from 16.1 million to 10 million, and in the same period, market price increased from 29 to 40 cents a pound (teeter-totter phenomenon). In 1951, bales harvested recovered (15.1 million) and remained near that level through 1953. As a result, market prices began to decline, reaching 32 cents in 1953. Teeter totter generally continued through 1965, with market prices reaching 29 cents that year. In 1966 and 1967, adverse crop weather (Cook, et al., 1999) resulted in unusually low cotton harvests, but market prices did not seem to follow. Bales harvested recovered in 1968, but market prices generally continued to drift downward to 22 cents in 1969 and 1970.

Clearly, throughout this entire period, cotton market prices were headed in the wrong direction. All the scaffolding erected around cotton production by the Triangle had come to naught. In relative terms, cotton farmers were worse off in 1970 than they had been in 1933. In that year, the average market price for cotton was 10 cents a pound. In 1970, it was 22 cents, an increase of 120%. Unfortunately, the Consumer Price Increase grew 200% during the same period. The only thing saving cotton farmers was the growing efficiency of production (Figure 14.8).

A comparison of Figures 14.43 and 14.44 shows that support prices had little effect in 1950, 1951, and 1952. During this period, market prices remained comfortably above the support prices. From 1953 through 1963, however, support prices provided an important floor for the prices that farmers received for their cotton. After 1963, the support-price program was unable to handle the quantity of cotton harvested, and average market prices fell below support-price levels and remained there through 1970. In that year, the average market price was 15 cents below the support price (22 cents versus 37 cents).

Acres Planted and Harvested

American cotton farmers had not responded very positively to increasing prices that were the result of the beginning of World War II (Figures 14.44 and 14.45). Acres planted trended downward from 1940 through 1944, apparently responding to concerns about the loss of markets overseas (Figures 14.42) rather than better prices. They finally began to increase acres planted in 1946, and by 1951 it reached 29.3 million acres. Acreage had been much lower in 1950, apparently as a result of unusually wet conditions throughout the heart of the Cotton Belt. After 1950, acres planted generally tracked prices downward, as farmers used the average price received the preceding year to make decisions regarding current-year planting. For example, in 1950, as a result of a weather-limited crop, market price increased sharply to 40 cents. Then in 1951, farmers planted 29.4 million acres. As a result of the unusually large harvest that year, the average market price declined to 38 cents (teeter totter). The following year (1952), farmers planted 28.1 million acres. With acres planted usually tracking prices and with prices generally trending downward for the remainder of the period, acres planted also generally trended downward. Between 1951 and 1970, acres planted declined from 29.3 million to 11.9 million acres. Except for a few years in the early 1950s, farmers annually harvested around 95% of acres planted (Figure 14.45).

Acres planted and harvested were considerably lower than expected in 1957 and 1958. This unexpected decline was probably related to the eight-month economic contraction during that period (Figure 14.1). Acres planted and harvested were also lower than expected in 1966 and

1967. These deviations were probably related to adverse crop weather in both years (Cook, et al., 1999).

At least some of the reduction in acres of cotton planted was probably the result of the decline in farm numbers (Figure 14.17) and cropland harvested (Figure 14.26) during the period. The country had lost 48% of its farms between 1950 and 1970 and 21% of its harvested cropland between 1950 and 1969. The cotton situation was more complicated, however. In 1950, acres of cotton harvested accounted for 5.2% of all cropland harvested. By 1969, this percentage had declined to 4%.

Yields

American cotton farmers responded to the continuing problems with their operations in the 1950s and 1960s by modernizing them. With the help of agricultural scientists, they thoroughly modernized all aspects of the production process, including mechanized planting, fertilization, pest control, and harvesting. Vastly improved genetic material (seeds) became available. All of these elements together resulted in a significant increase in yields. National yield data, presented in Figure 14.46, show that although these yield estimates are actually averages, there was considerable year-to-year variation. But even with the variation, it is obvious that average yields did increase. They were around 270 pounds an acre in the early 1950s and 520 pounds in the mid-1960s. Yields generally declined after 1965. With prices hovering around 21 cents a pound during this period (Figure 14.44), it is likely that farmers skimped on their cultural practices to save money.

Bales Harvested

Annual national data on bales of cotton harvested during the period 1940-1970 are presented in Figure 14.47. With acres harvested generally decreasing (Figure 14.45) and yields generally increasing (Figure 14.46), bales harvested generally ranged between 14 and 16 million bales between 1951 and 1965. Lower than expected bales harvested in 1956, 1957, and 1958 were the result of rapidly declining levels of acres harvested during those years. The lower than expected harvests from 1966 through 1970 were largely the result of low yields.

Unfortunately, while modernization made cotton farmers much more productive, increased yields outpaced reductions in total production expected from New Deal-mandated acreage limitations. For example, cotton farmers planted 14.2 million acres in 1965, some 25% fewer than the 18.9 million they planted in 1950; however,with the increase in yields, they actually produced more cotton on 4.7 million fewer acres in 1965 (14.9 versus 10 million bales), than they had in 1950. Ten million bales had been too much for the market in 1950, and 14.2 million were much more than was needed in 1965. As a consequence, the 1965 market price was 28% lower (40 versus 29 cents a pound) than it had been in 1950. Cotton farmers were clearly winning the battle of modernization and efficiency, but, regardless of the improvements and the continuation of New Deal programs, many of them were still losing the war for economic survival.

The year-to-year variation seen in the bales harvested data in Figure 14.47 emphasizes the major difficulty in arriving at a governmental policy for stabilizing cotton production. The average bales harvested for the 21-year period was 12.9 million, but the standard deviation was 2.5 million bales. With a standard deviation of 2.9 million bales, bales harvested could easily change as much as 5.8 million or more from year to year. Between 1950 and 1951, bales harvested increased by 5.1 million (10 to 15.1 million). Then between 1965 and 1966, it declined by 5.3 million (14.9 to 9.6 million). Under such conditions, setting acreage allotments was to little avail.

Triangle efforts at supply management seems to have worked fairly well for cotton production during the period. At least, bales harvested was at near the same level in 1951 and 1965. There was a significant reduction in acres harvested, but it was generally canceled out by an increase in yields. Unfortunately, 15 million bales was still considerably more cotton than the market could absorb.

Alabama Cotton

Alabama farmers harvested 5.7% of the total national cotton crop in 1950. Within the state, farmers in Madison County harvested the largest share (7.9%). Other counties reporting large shares included: DeKalb (7%), Limestone (6.5), Cullman (5.7), and Lawrence (4.9). These five north Alabama countries accounted for 32% of the state's total crop. Some cotton was harvested in every county of the state, but most of it was harvested in a relatively small number of counties. Census data indicate that only 10 of the state's 67 counties reported harvests of 20,000 or more bales in 1949, and that most of those counties were also in the northern half of the state. The old Cotton Arc was no more.

Prices

Average annual market prices for cotton in Alabama during the period 1950-1970 were essentially the same as the national prices shown in Figure 14.44.

Acres Planted and Harvested

Trends in acres of cotton planted and harvested in Alabama during the period 1950-1970 were essentially the same as national trends (Figures 14.48 and 14.45). Although acres planted were lower than expected in a few of the years, they trended down from 1.6 million acres in 1953 to 565,000 acres in 1970. Alabama farmers responded the same way to price signals as farmers in other states. Except for 1967, Alabama farmers harvested 97% of their planted acres throughout the period. The lower than expected values in 1958 and 1959 and in 1966 and 1967 were probably the result of the same factors affecting cotton planted and harvested in the country during those years.

Acres of cotton planted in Alabama declined by 57.6% between 1950 and 1970. At least a portion of this reduction was the result of the decline in total farmland. NASS data indicate that total Alabama farmland declined by 30.5% in the same period. Also, the number of farms (Figure 14.18) and acreage of cropland harvested (Figure 14.28) declined during the period. Alabama had lost 63% of its farms between 1950 and 1970 and 53% of its harvested cropland between 1950 and 1969, but the cotton situation was more complicated. In 1950, acres of cotton harvested accounted for 22.8% of all Alabama cropland harvested. By 1969, this percentage had declined to 20%.

Yields

As expected, there was considerably more year-to-year variation in the average annual Alabama cotton yields during the period 1950-1970 than there was in the national data (Figure 14.46). The national data represents an average of yields from a number of states. Although there was considerably more variation, Alabama yields generally trended upward during most of the period from aound 280 pounds an acre in the early 1950s to 500 pounds in the mid-1960s. As was the case with yields in the country, low prices (Figure 14.44) likely led to Alabama farmers skimping on cultural practices that also resulted in lower yields from 1965 through 1970.

Bales Harvested

Annual bales of Alabama cotton harvested, which represents the combination of yield and acres harvested, generally trended downward from around 900,000 bales in the early 1950s to between 400,000 and 500,000 in the late 1960s (Figure 14.49). Then in the final years of the period, bales harvested fell to between 400,000 to 500,000. From 1951 and 1970, bales harvested declined by 44%.

Efforts at supply management seem to have worked about as well in Alabama as elsewhere in the country. Bales harvested was at essentially the same level in 1965 as it had been in 1951.

NASS data indicates that farmers in Madison Country harvested the largest share (11.3%) of Alabama's 1970 cotton crop. Other counties reporting substantial shares included: Limestone (8.9%), Lawrence (6.4), Cherokee (4.8), and Colbert (3.6). These five counties accounted for 35% of the state's entire cotton crop.

Corn

The first annual data on acres harvested of crops in the United States was collected in 1866. In every year since, in terms of acres harvested, corn has been the most important crop in the country. In most years, total acres harvested of corn has been more or less equal to the sum of acres harvested for all of the major grains (wheat, oats, and sorghum) grown in the country. In 1929, 1934, 1939, 1944, and 1949, acres of corn harvested as a percentage of all cropland harvested were: 23.2, 20.7, 24.4, 24.1, and 22.4%, respectively.

US Corn

In 1950, American farmers harvested 1,480 million bushels of corn. Farmers in Iowa harvested the largest share (16.5%). Other states reporting large shares included: Illinois (14.8%), Nebraska (8.7), Indiana (7.4), and Missouri (6.1). These five states, all in the North Central Region, accounted for 53.5% of the total national crop.

Farmers in Alabama harvested only 1.9%.

Exports

Corn exports are reported on a fiscal year basis (July 1-June 30). Corn exports increased at an exponential rate (Figure 14.50) from 1950 (117 million bushels) through 1961 (416 million), as the demand for feed grains, especially in Europe, skyrocketed. The upward trend was interrupted in 1962, the year of the Cuban Missile Crisis. Growth resumed in 1963, and in 1965 corn exports reached 687 million bushels. Exports did not increase for the remainder of the decade, as sales of the grain abroad experienced the same unsettled international conditions as other agricultural exports.

Prices

The government regularly increased support prices for corn after 1940, and in 1950 they reached $1.47 a bushel (Figure 14.51). They slowly increased through 1954 ($1.62), the year Eisenhower became president. In 1955, they began to decline, finally reaching $1.06 in 1960, the year that Kennedy was elected. In 1961, support prices began to increase again, and reached $1.35 in 1967. They remained at that level through 1970.

American corn market prices increased rapidly during World War II. By 1947, they had reached $2.16 a bushel (Figure 14.52), but by 1950 they had retreated to $1.52. After 1948, bushels harvested began to decline, and in 1949 market prices began to increase. Afterwards, bushels harvested generally trended upward through 1960, and market prices declined, reaching $1.13 in the same year. Afterwards, market prices increased slowly, reaching $1.16 in 1966. After 1967, bushels harvested began to decline, and market prices increased to $1.33 in 1970.

A comparison of Figures 14.51 and 14.52 shows that from 1952 through 1970 market prices were consistently lower than support prices. In 1967, the difference was as much as 32 cents ($1.03 versus $1.35). Apparently American farmers were marketing much more corn than the support-price program could accommodate.

Acres Planted and Harvested

During the 1950s, acres of corn planted in the United States continued the downward trend that had begun in the 1930s; however, after 1960 it changed very little. American farmers had planted 80 million acres in 1950, and by 1958 acres of corn planted had declined to 64 million, as acres planted followed market prices downward (Figure 14.52). In 1958, Congress enacted another version of the Agriculture Act of 1938. The new version allowed farmers to hold a referendum on the future of government-sponsored corn programs. They chose to have acreage allotments eliminated and support prices reduced. With allotments eliminated, farmers sharply increased acres planted to 83 million in 1959 and 81 million in 1960. As a result, bushels harvested increased sharply, and prices declined. By 1961, the bold new move toward unrestricted production was over, and acres planted fell sharply to 66 million and remained near that level through 1970. Between 1961 and 1970, acres planted reached the lowest levels recorded since USDA began collecting data in 1926. Throughout the period 1950-1970, American farmers harvested corn from 86% of their planted acres. In the dairy regions, substantial amounts of corn were cut for silage, rather than harvested for grain.

Yields

Annual average American corn yields increased from 28.9 bushels an acre in 1940 to 39.4 bushels in 1954 (Figure 14.54), or at a rate of increase of 2.4% a year. Then from 1955 through 1970, they increased from 42 bushels an acre to 72.4 bushels, or at a rate of 4.5% a year. In the late 1940s and early 1950s, the use of hybrid corn was still spreading across the United States, especially the South. By 1955, it was in widespread use throughout the Corn Belt. From that time onward, annual yields began to increase more rapidly, as farmers learned how to manage the new genetic potential more effectively.

The average yield was much lower than expected in 1964, primarily as a result of unusually hot, dry weather throughout the Corn Belt in June, July, and August. The lower than expected yield in 1970 seems to have been the result of much wetter crop weather than usual in June, July, and August throughout the Mississippi Valley.

Bushels Harvested

During the period 1950-1970, average yields trended sharply upward (Figure 14.54), while acres harvested trended downward (Figure 14.53). The rate of increase in yields was considerably greater than the rate of decline in acres harvested. As a result, bushels harvested also trended upward from 2.8 billion bushels in 1950 to 4.7 billion in 1969 (Figure 14.55). The greater than expected levels of acres harvested in 1959 and 1960, as well as adverse crop weather in 1964 and 1970 increased the year-to-year variability.

It does not appear that Triangle efforts at supply management of American corn production worked very well during this period. Despite their efforts, bushels harvested increased by 1.9 billion bushels, or by 70%. Acres harvested did decline during the period, but it was more than offset by an increase in yields. If there was any consolation, it is that without those additional Triangle efforts, it could have been worse.

Alabama Corn

In 1950, Alabama farmers harvested 1.9% (52.5 million bushels) of the national crop. Census data indicate that farmers in DeKalb County harvested the largest share (6.7%) of the state's 1969 crop. Other counties reporting large shares included: Marshall (4.4%), Madison (3.9), Cullman (3.8), and Lawrence (2.4). These five North Alabama counties accounted for 21.2% of the state's total crop. Census data also indicated that in 1949 farmers in 39 of Alabama's 67 counties harvested 500,000 or more bushels of corn.

Prices

Data on the average annual market prices of corn received by Alabama farmers during the period 1940-1970 are also shown in Figure 14.52. Those data indicated that trends in market prices for Alabama farmers were essentially the same as for the country as a whole, but that in most years, as a result in the differential in support prices, Alabama market prices were five cents to 10 cents a bushel higher. State market prices increased sharply from $1.48 to $1.89 a bushel from 1950 through 1952, but then quickly declined to $1.15. From 1955 through 1968, with the exception of 1966, prices remained in a relatively narrow range ($1.10 to $1.20). Then in 1969 and 1970, they were up sharply to $1.36 and $1.58, respectively, after bushels harvested declined (Figure 14.58).

Acres Planted and Harvested

Alabama's acres of corn planted trended steadily downward during the period 1950-1970 (Figures 14.56 and 14.53), as it had in the country (Figure 14.56). However, in the state, the rate of reduction was much greater. Alabama farmers planted 2.5 million acres in corn in 1950, but in 1970 they planted only 700,000 acres, a reduction of 72%. This rate of reduction in acres planted was four times greater than for the country as a whole (18% versus 72%). The Agriculture Act of 1958 seemed to have had a very limited effect. Acres planted were only slightly higher than expected in 1959 and 1960. Alabama farmers harvested corn from a larger portion of their acres planted than farmers in the country as a whole. During the period 1950-1970, they harvested from 79 to 91% of their acres.

Acres of corn planted declined by 73.9% in Alabama during the period 1950-1970, while NASS data indicated that total farmland declined only 30.5%. In 1950, Alabama farmers planted over 11% of their farmland in corn. By 1970, that percentage was a little less than 5%.

Yields

Alabama corn yields generally increased during the period 1950-1967 from around 20 to around 48 bushels an acre. In 1968, 1969, and 1970, yields decreased precipitously, almost back to 1950 levels (Figure 14.54). Yields were somewhat lower than expected in 1952, 1954, and 1962 as a result of abnormally dry to severely dry weather in June, July, and August of those years. It is not clear what caused the lower average yields in 1968, 1969, and 1970. The weather was mostly normal in 1968 and 1969, but wetter than normal in 1970.

While average yields in Alabama increased during the period, yields in the country increased at a much faster rate. In 1950, the difference was only 16.7 bushels, but by 1967, it was more than twice as large (33.1 bushels). Year by year, it was becoming increasingly obvious that Alabama had virtually no comparative advantage in corn production. The situation is even more dire when making the comparison between Alabama and a state in the Corn Belt, such as Iowa. This comparison is made in Figure 14.57. Yields in Iowa were over twice as high in the early 1950s, and it became even larger over time. The magnitude of the difference grew larger, especially in the last 10 years. Based on these data alone, there would have been little justification for growing a single acre of corn in Alabama.

Bushels Harvested

During the period 1950-1970, acres of corn harvested in Alabama declined (Figure 14.56), while yields increased (Figure 14.54). Because the rate of decline in acres planted was greater than the rate of increase in yields for most of the period, bushels harvested also trended downward (Figure 14.58). Bushels harvested were at their highest level in 1955 (58.9 million bushels), but by 1970 they were 14.7 million, the lowest level of the twentieth century. The much lower than expected yields in 1952, 1954, 1966, 1968, 1969, and 1970 resulted in much lower than expected levels of bushels harvested.

Bushels of corn harvested in Alabama declined 29.9 million bushels (54%) during the period 1950-1970. It is difficult to attribute this decline to Triangle efforts at supply management. With those abysmally low yields, it is surprising that any corn was produced in the state.

Alabama Agricultural Statistics Service data show that farmers in DeKalb County harvested the largest share (7.1%) of Alabama's 1970 corn crop. Other counties reporting substantial shares included: Houston (5.8%), Geneva (5.0), Jackson (5.5), and Coffee (3.3). These five counties accounted for 26.7% of the state's entire crop. It is surprising how much corn was harvested south of Montgomery in 1970. These data also showed that 18 of Alabama's 67 counties harvested at least 250,000 bushels in 1970.

Alabama Peanuts

In 1950, American farmers harvested 2,035 million pounds of peanuts. Farmers in Georgia harvested the largest share (34.4%) of the country's total crop. Other states reporting large shares included: Texas (16.2%), Alabama (16.0), North Carolina (12.4), and Virginia (11.2). These fives states located on the Atlantic Plain accounted for 90.2% of the total national crop.

US Prices

The Agriculture Act of 1949 established support prices for peanuts at between 75 and 90% of the then current level. The act also guaranteed that all peanuts harvested from USDA-approved allotments would receive the support price, which varied between 10 and 12 cents throughout the entire period, but was increased to 13 cents in 1970 (Figure 14.59). Support prices for all commodities generally increased during the Truman Administration (1945-1953), declined during the Eisenhower Administration (1953-1961), and increased during the Kennedy and Johnson administrations (1961-1969).

Peanut market prices increased rapidly throughout the 1940s, and finally reached 10.9 cents a pound in 1950. Afterwards, they continued to increase, reaching 12 cents in 1954 (Figure 14.60). Unfortunately, stocks of many commodities began to accumulate in the early 1950s. Consequently, the Agriculture Act of 1954 established marketing quotas for a number of crops, including peanuts, and changed the system of mandatory support prices to one with flexible supports. However, even with marketing quotas, stocks of peanuts continued to accumulate. As a result, USDA began to slowly reduce support prices, and by 1959 peanut market prices were down to 9.5 cents. In 1960, support prices began to increase again, reaching 12.8 cents in 1970. The market price quickly increased to the same level. This sequence of events, related to the market price of peanuts, indicates that this narrowly regional crop had acquired some exceedingly powerful friends in the Triangle.

Acres Planted and Harvested

Acres of peanuts planted in Alabama began to decline even before the World War II ended, and by 1950 it was down to 405,000 acres (Figure 14.61). In 1941, Congress, concerned about the increasing stocks of unneeded peanuts, decided to place a limit on acres planted in the country at about 1.9 million. Alabama's share of that total would have been about 260,000 acres, but before the allotment regulations could be established, the Japanese attacked Pearl Harbor. During World War II, with the demand for vegetable oil growing rapidly, there was no allotment enforcement. By 1951, however, it was obvious that something had to be done about the peanut situation, and the allotment regulations, which were still on the books, were enforced for the first time. In 1952, Alabama farmers planted 260,000 acres. In subsequent years, some Alabama farmers began to give up on the crop, and acres planted trended slowly downward, reaching 234,000 acres in 1958. In 1959, USDA lowered the national allotment further to 1.5 million acres, and acres planted in Alabama continued to decline to 183,000 in 1967. In the final three years of the period, farmers slowly began to increase their acreage, and in 1970 they planted 195,000 acres.

In the early 1950s, Alabama farmers harvested peanuts from only 80% of acres planted. The remainder were apparently hogged-off. However, by 1955, the percentage harvested had increased to 90%, and by 1965 farmers were harvesting 96%.

Yields

Yields generally increased in the state from 900 pounds an acre in the early 1950s to 1,650 pounds in 1970. Year-to-year variations were relatively high from 1950 through 1966, and especially so from 1950 through 1962. Yields were much lower than expected in 1951, 1954, 1957, 1959, 1962, and 1966. Crop weather during June, July, and August was abnormally dry to severely dry in four of those years, and abnormally wet in one (Table 14.7). Trends in the state data were similar to those for all the states combined, but in the national data the deviations from the expected were not as great. Yields in the state and the country were similar during the early 1950s, but began to diverge over time. By the mid-1960s, yields on Alabama farms were only 70 to 80% of those on farms in the other peanut-producing states.

Pounds Harvested

From the early 1950s through 1962, increases in peanut yields (Figure 14.62) in Alabama generally countered declines in acres harvested (Figure 14.61), so that total pounds harvested remained around 200 million (Figure 14.63), except in those years when yields were lower than expected. However, afterwards increasing yields began to control the relationship, and pounds harvested began to increase, reaching 315.4 million pounds in 1970.

Peanut production in the United States provides an interesting study in the socio-political machinations of Dobson's (1985) Triangle of beneficiaries, politicians, and bureaucrats in manipulating American farm policy. In 1933, Congress, led by the overwhelming presence and power of Southern Democrats (Ernestes, et al., 1997), enacted the New Deal legislation designed to ensure the survival of the cotton farms of the Southern planters. Of course, to accomplish this objective, they had to bring more important crops from other regions into the tent. While deeply involved in this urgent effort to save King Cotton, the peanut sneaked into the tent as well. The goober pea had a few good and powerful friends in high places.

In 1933, there were 109.8 million acres of corn, 44.8 million acres of wheat, 43.7 million acres of oats, and 430,000 acres of peanuts planted in the country. Peanuts were planted on sandy land in a relatively few congressional districts on the seaward margin of the Atlantic Coastal Plain, and at that time peanut farmers were actually harvesting only 39% of their planted acres. Farmers harvested peanuts on just 169,000 acres in 1933. Unfortunately, getting the peanut into the tent meant that lots of other minor agricultural products—including turpentine, rosin, dates, figs, and prunes—had to be admitted, as well. Many more equally deserving crops would be admitted later. This Depression-driven exercise in backscratching on the part of the Triangle virtually guaranteed that it would be impossible to ever develop a workable national food and agricultural policy.

Unfortunately, the machinations required to bring the peanut into the tent were just the beginning. Producing peanuts economically on the Atlantic Coastal Plain was a questionable proposition from the beginning, but to keep the peanut viable in the ensuing years would require a level of legislative and regulatory politicking that would be truly astonishing. Getting the peanut into the tent required a few good friends in high places and a lot of back-scratching; to keep it there required an ever-increasing infusion of the same.

The problems of the government's efforts to regulate peanut production were already obvious by 1970, but there is another point of view. Crop production in those sandy areas where the peanut was grown was marginal under the best of conditions. At least the peanut could be grown in generally predictable quantities. Given the alternatives that farmers in those areas had, the almost religious fervor of its congressional supporters is more understandable.

Soybeans

Most of the plants used in North American agriculture

were brought here soon after the Eastern portion of the continent was settled in the early seventeenth century. Their early importation was directly related to the ocean currents and trade winds rotating clockwise around a huge area of high pressure in the southern portion of the North Atlantic Ocean. For hundreds of years, the characteristics of the oceanic winds and currents in the Pacific Ocean precluded direct trade between most of Asia and North America. The soybean plant evolved in the temperate climate zone of Central Asia. It was domesticated in China sometime between the seventeenth and eleventh centuries B.C. The plant did not reach North America until the middle of the eighteenth century. Before being brought to America, it was first carried from China to England. The plant played a relatively minor role in American agriculture before the middle years of the Great Depression. In the late 1930s and early 1940s, American farmers were beginning to learn what Asian farmers had known for centuries: the soybean is one of the most useful plants ever domesticated. Its stems and leaves made very good forage. Its beans contained a relatively large quantity of high quality oil and protein; in addition, the plant produced its own nitrogen.

US Soybeans

By the middle of the twentieth century, the soybean plant had found a home in North America. In 1950, farmers in just five states in the North Central Region harvested 75.8% of the total national crop. Illinois farmers harvested the largest share (32%). Other states reporting large shares included Iowa (14.2%), Indiana (12.4), Missouri (9.0), and Ohio (8.2), In that year, Alabama farmers harvested only 0.4% of the national crop.

Exports

At the end of the Korean Conflict, the entire Western world was just beginning to learn how versatile soybeans were and that the United States was their best source of the crop. As a result, exports began to trend rapidly upward. In 1950, exports were 27.8 million bushels (Figure 14.64), whereas in 1940, they had totaled only 300,000 bushels. By 1960, they reached 135 million bushels, and in 1968, 287 million. At the end of the 1960s, national economies were expanding throughout most of the world, and in 1970 American soybean exports reached the unimaginable level of 434 million bushels.

Prices

Data on soybean support prices for the period 1941-1970 are shown in Figure 14.65. In 1950, the support price was $2.06 a bushel. Then, after 1953 (the year of Eisenhower's election)—with the support price at $2.56 and with the cost of the program growing—the government began to reduce support prices; by 1960, it was down to $1.85. In 1961, when the Kennedy Administration came into power, support prices began to increase. This increase continued in the Johnson Administration, finally reaching $2.50 in 1968. It was reduced to $2.25 in 1969, and it remained at that level in 1970.

American soybean market prices generally ranged between $2.00 and $3.00 a bushel during the period 1950-1970 (Figure 14.66). Year-to-year changes seemed to have little relationship to the runaway growth in bushels harvested (Figure 14.69). They were $2.47 in 1951, $1.96 in 1959, $2.75 in 1966, and $2.35 in 1969. For most of the period, support prices and market prices followed the same trends. In 16 of the 21 years during the period, market prices were higher than support prices.

Acres Planted and Harvested

Acres of soybeans planted in the United States increased 187% (15 to 43 million acres) during the period 1950 to 1970 (Figure 14.67), or at a rate of 9% a year. There seemed to be little relationship between acres planted and market prices. Farmers seemed to be responding to other signals when determining how much to plant from year to year—probably irrational exuberance.

This enormous increase in acres planted came at the same time when American farmland was being abandoned at an average rate of 1.5% a year. Between 1950 and 1970, American farmers increased the percentage of their total farmland devoted to soybeans from 1.2 to 3.9%. In 1950, American farmers harvested soybeans from 92% of acres planted (Figure 14.67). In 1970, they increased the rate to 98%.

Yields

Average annual soybean yields in the United States trended upward from 20 bushels an acre in 1950 to 27 bushels in 1970. There was considerable year-to-year variation, probably the result of adverse crop weather.

Yields of soybeans in Iowa during this period are shown for comparative purposes in Figure 14.71. They also trended upward during this period, but at a generally higher rate than for the country as a whole. In 1950, Iowa yields were 22 bushels an acre, and by 1970 they had increased to 32 bushels.

These soybean yield data are indicative of the astounding improvement in American agricultural technology during this period. They are a fitting tribute to the accomplishments of a small army of agricultural researchers and extension agents.

Bushels Harvested

Average annual bushels of American soybeans harvested remained relatively stable at around 300 million bushels (Figure 14.69) during the period 1950-1953 when yields were declining, and acres harvested were increasing (Figures 14.68 and 14.67). After 1953, with both yields and acres harvested trending upward, bushels harvested increased rapidly through 1958 to 580.2 million. Bushels harvested were lower in 1959 and 1960, but began to increase afterwards, reaching 1,107 billion in 1968. They remained generally unchanged in 1969 and 1970. Between 1950 and 1968, bushels harvested increased from 299.2 million to 1,107 million, an increase of 270%. The ancient Asian food crop had established a new kingdom on the Mollisols of the North Central United States.

Alabama Soybeans

In 1950, Alabama farmers harvested crops from 5.7 million acres of land. Soybeans were harvested from only 55,000 (0.96%) of those acres. Census data indicate that farmers in Baldwin (56.7%) and Escambia (19.2%) counties harvested the largest shares of Alabama's soybeans in 1950. Other counties reporting somewhat lower quantities were: Jackson (8.1%), Mobile (3.2), and Madison (1.8).

Prices

Data on Alabama soybean prices are also presented in Figure 14.66. Trends for state and national prices were essentially the same, except in the early part of the period. From 1950 through 1955, Alabama farmers received slightly lower prices for their soybeans than the national average; however, from 1956 through 1970, prices received were essentially the same.

Acres Planted and Harvested

From 1950 through 1970, Alabama farmers increased acres of soybeans planted from 137,000 to 625,000 (Figure 14.70). In fact, Alabama's average annual rate of increase was twice as great (18 versus 9%) as in the country as a whole (Figure 14.67); however, the actual yearly changes were quite different. There was essentially no increase in acres planted in the state from 1950 through 1960; from 1961 through 1966, they increased slowly. Then, between 1962 and 1969, Alabama farmers doubled their plantings from 300 thousand to over 600 thousand acres. This significant increase in acres planted is surprising at a time when there was a 63% loss of farms and a 30% reduction in farmland. In 1950, Alabama farmers devoted less than 1% of their farmland to soybeans, but in 1970 they devoted 4% to the crop. As a result of this suddenly increased emphasis on soybeans, in 1970 Alabama was actually planting over 10% more acres (625,000 versus 565,000) in soybeans than in cotton. In 1950, farmers had planted 0.6% of their total farmland in soybeans. By 1970 that percentage was up to 4.2%. Alabama farmers had finally found the crop of their dreams—or had they?

As noted previously, in earlier years, farmers, especially in the South, planted soybeans as a forage crop and harvested very few of the beans. By 1950, American farmers were harvesting 90% of their acres planted in soybeans, and by 1970 the percentage had increased to 99%. However, in 1950 Alabama farmers were harvesting only half of their acres planted in soybeans, but by 1970 the percentage was at about the same rate as for the country as a whole.

Yields

Although there was considerable year-to-year variation, Alabama soybean yields also increased during the 21-year period (1950-1970). Levels were around 18 bushels an acre in the early 1950s and around 26 bushels in 1967. The annual average yield was much lower than expected in 1954. According to Cook, et al. (1999), the entire state was extremely dry during June, July, and August of that year. Yields were also considerably lower than expected during the period 1959-1965. During this period, Alabama farmers began to increase acres of soybeans planted significantly (Figure 14.70). It is possible that as they increased acres planted rapidly they did not maintain their cultural practices as they should have.

Except for a few years during the period 1950-1970, Alabama yields were a little lower than in the country as a whole (Figure 14.68). They were also considerably lower than those in Iowa (Figure 14.71). Year-to-year variation was also somewhat greater in Alabama, as well as in Iowa, than for the country; however, this difference is not entirely unexpected. The national numbers are really averages of averages from many states, and would tend to be more stable that the average from a single state.

Alabama soybean farmers and supporters of that industry probably did not recognize it at the time, but differences in yields here and in the North Central Region, already apparent in this period, were handwriting on the wall for the future of the crop. Those differences were telling indicators of differences in comparative advantage that would only grow larger with time.

Bushels Harvested

In 1950, Alabama farmers were harvesting around 1.3 million bushels of soybeans annually (Figure 14.72). Then over the next 15-year period, bushels harvested increased at a rate of 19% a year, as both acres harvested (Figure 14.70) and yields (Figure 14.71) trended upward. The only break in this general upward trend came in 1954 when average yield fell to 12 bushels an acre as a result of unusually dry weather during June, July, and August. Bushels harvested fell from 1.8 million bushels in 1953 to 1.1 million in 1954. Then, in 1965 Alabama farmers harvested five million bushels of the crop. During the period 1965-1970, the increase in production was astounding. It was five million bushels in 1965, and just five years later in 1970, it had reached 14 million. Even with this relatively high level of production, however, Alabama remained only a minor player in national soybean production. In 1969, the large Alabama crop was only 1.2% of the 1.1 billion bushel national crop.

NASS data indicated that farmers in Baldwin County harvested the largest share (20.3%) of Alabama's 1970 soybean crop. Other counties reporting substantial shares included: Madison (6.5%), Limestone (5.3), Mobile (4.4), and Jackson (4.3). These five counties accounted for 40.7% of the state's entire soybean crop.

Alabama Hay

As cattle production increased in Alabama, tons of hay harvested also increased. In 1950, acres of hay harvested exceeded those of wheat, oats, sorghum, soybeans, peanuts, sweet potatoes, and Irish potatoes combined. Surprisingly, in 1950 a large share of Alabama's hay crop was being harvested in the Low Hills section of the Lower Coastal Plain. That year, farmers in Covington County harvested the largest share of the state's crop (4.5%). Other counties reporting large shares included: Geneva (4.5%), Houston (4.3), Coffee (4.2), and Pike (4.2). These five counties accounted for 21.6% of the state's total crop.

Prices

In 1950, a new Cycle of Abundance in cattle was just beginning in Alabama (Figure 14.106). Consequently, hay prices began to increase (Figure 14.73). In 1950, farmers received an average of $24 a ton for their hay, but by 1952 the average price was up to $32. However, even though cattle numbers continued to increase through 1954, prices began to decline after 1952, and in 1956 reached $22 a ton. Apparently, the increase in the size of the cattle herd in the state had triggered the harvest of more hay than could be sold at those higher prices.

Still another Cycle of Abundance in Alabama cattle began around 1960, and prices began to increase again. However, this cycle did not result in a large increase in prices. Farmers sold their hay for an average of $28 a ton in 1962. Again, even though cattle numbers continued to increase through 1965, hay prices began to decline in 1963. By 1965, they were down to $25 a ton.

Another Cycle of Abundance began in the late 1960s, and hay prices began to increase again, reaching $28 a ton in 1968 and remained at that level through 1970.

Acres Harvested

Data on acres of hay harvested in Alabama during the period 1950-1970 are presented in Figure 14.74. Generally, acres harvested trended downward during the entire period. Farmers harvested 700,000 to 800,000 acres in the early 1950s, but by the late 1960s, they were only harvesting around 500,000. The effects of teeter totter are apparent in Figure 14.74. For example, prices were up sharply in 1951 (Figure 14.73), and farmers began to increase acres harvested in 1952. With prices declining after 1952, acres harvested began to decline in 1956. Further, in 1962, prices increased to $28 a ton, and in 1963 acres harvested began to increase.

Yields

Although there was considerable year-to-year variation Alabama hay yields trended upward through the period 1950-1970 (Figure 14.75). Yields were relatively stable at between 0.8 and 0.9 ton an acre through the late 1940s and early 1950s, but began to increase after 1954. By 1970, they were over 1.7 tons. Throughout the period, yields responded positively to the establishment of improved grasses and generally improved management. Certainly, the increasing availability of tractors on Alabama farms also played a significant role in the increase.

Tons Harvested

Data on tons of hay harvested annually on Alabama farms during the period 1950-1970 are shown in Figure 14.76. The downward trend in acres harvested (Figure 14.74) coupled with the upward trend in yields (Figure 14.75) kept tons harvested in the range of 600,000 to 700,000 tons from 1950 through 1962. The sharp increase in 1955 to 870,000 tons was the result of the relatively high level of acres harvested and a greater than expected yield that year. After 1963, with only limited change in acres harvested, but continuing increases in yields, tons harvested also began to trend upward. In 1970, Alabama farmers harvested 930,000 tons of hay, the highest on record to that time.

I can remember well my grandfather's unrelenting efforts to control native Bermuda grass in his Butler County (Alabama) cotton patches in the 1940s. Then in 1943, Glen Burton, at the Coastal Plains Experiment Station in Tifton, Georgia, released his now-famous Coastal Bermuda grass for general farm use. Within a few years, thousands of acres of eroded cotton land were covered with a lush carpet of Bermuda or fescue grasses (Fite, 1984). Replacement of worn-out cotton land with grass land was certainly a positive change, but there were some unexpected costs involved. Farmers quickly learned that good pastures demanded almost as much attention as some of their other crops. Good production of high quality forage required the regular application of fertilizer and the control of soil acidity through liming. Farmers also learned that, even with the best of care, no grass would produce well during the short-term and sometimes long-term droughts and periods of prolonged cold, characteristic of the highly variable climate of the Lower South.

Much of the change from row crops to forage was the result of the Soil Bank Act of 1956 when the federal government began to pay farmers for taking cropland out of production and planting it with trees and grasses. They not only paid participating farmers for the estimated economic loss resulting from not growing crops, but they also cost-shared the establishment of these conservation areas. Fite (1984) commented that "no single thing did so much to change the farmscape in the South as the Soil Bank."

The Soil Bank program was established in 1956. Three years later, 3.7 million acres of cropland was harvested in Alabama; in 1969, it had declined about 1 million acres (27.2%) to 2.1 million. At the same time, cropland used only for pasture or grazing had increased from 1.4 to 2.1 million acres (48.6%).

Winter Wheat

American and Alabama winter wheat farmers had good reason to be optimistic at the beginning of the 1950s. Exports had surged upward after 1944, and while they had declined somewhat in 1949, they were still three times greater than they had been in the early 1940s. Of even greater importance, market prices had increased along with exports. For the first time in recent history, they were above $2 a bushel. At the beginning of that decade, winter wheat production was centered in the lower Great Plains. In 1950, Kansas farmers produced almost a fourth of the 740.6 million-bushel wheat crop. Together, farmers in surrounding states (Nebraska, Missouri, Oklahoma, and Colorado) produced another 31% of the total. The five leading wheat-producing states were: Kansas (24%), Nebraska (11.8), Washington (8.0), Ohio (6.3), and Oklahoma (5.7).

Alabama farmers had lost interest in wheat production after the end of World War I, and it remained dormant until the early years of World War II, when acres planted trended upward to reach 25,000 acres in 1945. However, the enthusiasm of the American wheat farmers, as they approached the 1950s, was not shared by their Alabama counterparts. In 1950, farmers in the state only planted 11,000 acres of winter wheat. They planted almost three times as many acres of sweet potatoes and Irish potatoes. That year they harvested only 0.02% of the 740 million-bushel national crop of winter wheat.

US Exports

American wheat (all) exports were 374 million bushels in 1950 (Figure 14.77). In that year, exports were equivalent to 34% of the 1949 national crop. Exports grew to near 480 million bushels in 1951; however, as a result of the postwar adjustment following the end of hostilities in Korea, exports declined sharply in 1952 and 1953 to 220 million bushels. After 1954, exports trended upward through 1961 to 685 million bushels. As a result of unsettled world conditions in 1962 (the Cuban Missile Crisis and the Chinese invasion of India), exports slipped again to 604 million bushels. In 1965, the United States exported 852 million bushels, the highest in history at that time. As a result of the Six Day War (the third Arab-Israeli war) and the Warsaw Pact invasion of Czechoslovakia, exports declined in 1966, 1967, and 1968 to reach 544 million bushels. They recovered at the end of the decade and reached 741 million bushels in 1970.

US and Alabama Prices

Data on support prices for wheat during the period 1938-1970 are presented in Figure 14.78. They had been increased during World War II to reach $1.99 a bushel in 1950, and they had been increased to $2.24 in 1954. After the Eisenhower Administration came to power, support prices began to decline, reaching $1.78 in 1960, the year of Kennedy's election. Afterwards, they began to increase, reaching $2.82 in 1970.

American winter wheat market prices were $2.01 a bushel in 1950 and $2.12 in 1954 (Figure 14.79). Afterwards, with support prices declining, they began to trend downward, finally reaching $1.75 in 1959 and 1960. The increase in the support price seemed to boost market prices in 1961, and low bushels harvested boosted the market price even higher in 1962. Bushels harvested increased in 1963, 1964, and 1965, and market prices declined to $1.35. Once again, an increase in the support price (to $2.57 in 1966) seemed to boost the market price temporarily. But in 1967, 1968, and 1969 bushels harvested exceeded 1.4 billion, and market prices began to decline once more, reaching $1.25 in 1969. A slight decline in bushels harvested in 1970 seemed to result in an equally slight increase in the market price to $1.30.

The support-price program seemed to have only limited effect on market prices during the period. The market price was slightly higher than the support price in 1950 ($2.01 versus $1.99) and equal to it in 1962 ($2.00 versus $2.00), but in the remaining 19 years of the period, market prices were lower. Farmers, in an effort to maintain income from wheat production, simply marketed more of the crop than the support-price program could accommodate.

It is obvious that year-to-year trends in market prices for winter wheat in Alabama were essentially the same as in the country as a whole. In some of the years, it was slightly higher in the state, but in other years it was lower.

Acres Planted and Harvested

In 1950, Alabama farmers planted 14,000 acres of winter wheat (Figure 14.80). In response to the upward surge of market prices, American farmers increased their acreage of the crop significantly in the early 1950s, (Figure 14.79). Alabama farmers did not respond in the same manner. In 1950, 1951, and 1952, acres planted in the state remained essentially unchanged at near 13,000, but began to increase slightly in 1953 and 1954, and then rapidly in 1955, 1956, and 1957. In 1957, Alabama farmers planted 160,000 acres of wheat, the highest on record at that time. It is difficult to determine what motivated them to plant so much wheat at that particular time. Market prices had been relatively good in 1950, 1951, and 1952, but by 1953, they were beginning to decline. However, between 1953 and 1958, Alabama lost over a million acres of cotton (Figure 14.48). It is likely that the increase in acres of wheat planted simply represented an effort to find something productive to do with all of that land taken out of cotton production. Unfortunately, with prices declining, this was a terrible time to plant that much wheat.

After 1957, with market prices declining, acres planted in Alabama fell sharply to 64,000 in 1960. Acres planted was much lower than expected in 1962, with weather throughout the state drier than normal in the early fall. Market prices were up sharply in 1966, and farmers responded by planting 130,000 acres in 1967 and 140,000 in 1968. Unfortunately, when they harvested those crops, market prices were down again. In 1969 and 1970, acres planted fell to 121,000.

In 1950, Alabama farmers planted less than 0.1% of their total farmland in winter wheat. By 1970, that percentage had increased to 0.8%.

Figure 14.80 also contains data on acres of wheat actually harvested in the different years in Alabama. During the period 1950-1970, Alabama farmers did not harvest more than 81% (1953) of acres planted, and in most years they harvested significantly less. They only harvested 61% in 1967. A considerable amount of winter wheat was planted in the state to provide early season grazing, and in some instances, yields were simply not good enough to pay for the cost of harvesting.

US and Alabama Yields

Average yields of winter wheat in the United Sates trended upward during the entire period (1950-1970), although there was considerable year-to-year variation (Figure 14.81). The average was 17 bushels an acre in 1950 and 33 bushels in 1970. Data on the annual average yields of winter wheat in Alabama during this period are also presented in Figure 14.81. There was considerable year-to-year variation in both sets of data. State yields were 16 bushels an acre in 1950, but trended upward to 27 bushels in 1970. These data indicate the yields obtained by Alabama farmers during the period 1950-1962 were generally similar to those obtained by farmers in other states; however afterwards, with the exception of 1966, yields on Alabama farms were lower by about one to three bushels an acre.

Bushels Harvested

Data on the bushels of winter wheat harvested in Alabama during the period 1950-1970 are presented in Figure 14.82. With changes in acres harvested being of greater magnitude than changes in yield, the graph presenting bushels harvested resembles the graph presenting the acres harvested data, rather than the one presenting yield data. Although increasing yields had a more significant effect after 1965. Alabama farmers harvested only 170,000 bushels of winter wheat in 1950. As a result of the sharp increase in acres harvested and some increase in yields, bushels harvested trended rapidly upward to near 2.3 million bushels in 1957 and 1958. Even with yields continuing to increase, reductions in acres harvested were sufficiently large to reduce bushels harvested to 840,000 bushels in 1962. Toward the end of the period, the combination of both increased acres harvested and higher yields resulted in the harvest of 2.8 million bushels in 1968, the highest in history at that time. Afterwards, it declined to 2.4 million bushels in 1970.

In 1950, Alabama farmers produced 0.02% of the country's winter wheat crop. In 1970, regardless of the fact that Alabama farmers had increased their acreage from 10,000 to 140,000 acres, they actually produced a slightly smaller percentage of the national crop than they had in 1950. Winter wheat production was much more important in Alabama agriculture in 1970 than it had been in 1950, but not in relationship to national production.

Oats

The Census of Agriculture published data indicating that in 1949 American farmers harvested crops from 344.6 million acres of land. USDA has published data indicating that farmers harvested oats from 11% of those acres. They harvested corn, wheat, cotton, and soybeans from 22.4, 15.8, 8.0, and 3% of cropland, respectively.

During the period 1950-1970, a large share of the oats produced in the United States was used as feed for horses and mules. During this period, however, numbers of these animals were rapidly declining. Between 1950 and 1955, the number of the two combined declined by 40% (7.5 to 4.5 million). It is likely that numbers continued to decline equally rapidly through the remainder of the period. As a result, the demand for oats declined at about the same rate.

For all practical purposes, all of the oats produced in the country in 1949 were harvested in the North Central Region (84.6%). The oat genome was so successful in that environment that there was little need for anyone else to produce oats elsewhere. The other four regions generally shared equally in the production of the remaining 15.4% of the total national crop.

US Oats

At the beginning of the 1950s, oat production was centered, in states along the eastern edge of the central Great Plains (Figure 1.9). In 1950, Iowa farmers harvested 19.8% of the 1,369 million-bushel oat crop. Other states reporting large harvests of the crop included: Minnesota (13.8%), Illinois (11.4), Wisconsin (10.2), and South Dakota (6.3). Together these states harvested 61.4% of the total national crop. In 1950, Alabama harvested only 1.9 million bushels (1.4%).

Exports

In fiscal year 1951, the United States exported the equivalent of only a tiny fraction of the 1.4 billion-bushel, 1950 national oat crop. The country never exported very many of its oats, because oats were produced in abundance in Europe where many of America's major trading partners were located. Except in wartime or other emergencies, European countries produced about all of the oats that they needed.

Oat exports began to increase sharply just as World War II was ending in 1945. Production had been seriously reduced in Europe during the war years, and livestock producers there were anxious to begin rebuilding their herds. Also, large quantities of the grain were exported under the Marshall Plan. In 1946, oat exports reached 26 million bushels (Figure 14.83), the highest level since World War I. With oat production largely recovered in Europe after 1948, US oat exports began to decline, reaching seven million bushels in 1950. Then, as a result of the postwar adjustment following the end of hostilities in Korea, exports continued to decline, finally reaching three million bushels in 1953. The late 1950s was a period of rapid economic expansion throughout Europe; at the same time, oat market prices (Figure 14.85) in the United States were declining. As a result, exports trended sharply upward, reaching 29 million bushels in 1955. A year later, after Britain, France, and Israel attacked Egypt, which had nationalized the Suez Canal and after Britain and the United States refused to finance the Aswan High Dam, exports declined slightly, but began to increase slowly in 1957 and then more rapidly in succeeding years, reaching 49 million bushels in 1959. The European Common Market was established in 1957, and agricultural exports to those countries soon began to decline. Also, the Federal Reserve Board had begun to increase interest rates in 1955; this move soon began to reduce exports to some extent. As a result, exports fell sharply after 1959, and in 1964 reached five million bushels. This sharp decline was also probably abetted by the Cuban Missile Crisis and the Chinese invasion of India in 1962. Exports increased sharply between 1964 and 1965, only to fall sharply afterwards. It is likely that both the Six Day War in the Middle East and the Warsaw Pact invasion of Czechoslovakia,contributed to the decline in the late 1960s.

Prices

Support prices for oats were also increased by the federal government in the late 1940s and early 1950s (Figure 14.84). Federal support was 71 cents a bushel in 1950, and finally increased to 80 cents in 1953. With the Eisenhower Administration in power in 1954, support prices began to decline, finally reaching 50 cents in 1959 and 1960. After Kennedy's election (1960), they were increased again. Federal support was raised to 62 cents in 1961, and remained near that level through 1970. Throughout the period from 1955 through 1970, with the exception of 1959 and 1960, support prices varied in the narrow range of 61 to 63 cents a bushel. In general, from 1950 through 1970, bushels of oats harvested in the United States (Figure 14.88) trended downward. As a result, market prices were largely determined by support prices. The annual average national oat market price (Figure 14.85) was 79 cents a bushel in 1950 (six cents above the support price), and by 1955 it was down to 60 cents. Afterwards, they generally remained near the support price level, through 1970 when it was 62 cents.

Acres Planted and Harvested

In 1950, 45 million acres of oats were planted in the United States (Figure 14.86). It followed market prices (Figure 14.85) upward in the early 1950s, reaching 47.5 million in 1955, but afterwards began a long-term downward trend that reached 20.7 million acres in 1967.

It increased slightly after 1967 to 24.4 million in 1970. There seemed to be no relationship between acres planted and market prices for most of the period.

Many farmers used oats for grazing, and as a result, they actually harvested grain from only a portion of their acres planted. During this period, American farmers never harvested more than 90% of acres planted, and in many years, the percentage was considerably lower (Figure 14.86). In 1965 and 1970, only 77.1 and 76.2%, respectively, were harvested.

Yields

Year-to-year variation in annual average oat yields was relatively high in the United States during the period 1950-1970. The oat plant is at risk for a relatively long period. From the time the seed is planted until the grain is harvested, there are numerous environmental factors (weather, insect pests, and diseases) that can adversely affect yields. The sum of these effects appear as year-to-year variation. Even with this variation, national average annual yields of oats trended upward during the period. They were 35 bushels an acre in 1950 and between 50 and 60 bushels in 1968 and 1969.

Bushels Harvested

The magnitude of the downward trend in acres of American oats harvested during the period 1950-1970 (Figure 14.86) was sufficiently great to mask the general upward trend in yields (Figure 14.87). As a result, bushels harvested also trended downward (Figure 14.88). In 1950, farmers harvested 1,369 million bushels of oats, but by 1967 it was down to 793.8 million. Then bushels harvested was up slightly to around 940 million in the late 1960s. The high level of variation between 1953 and 1959 was the result of unique combinations of yields and acres harvested in some of those years.

Alabama Oats

In 1949, Alabama farmers harvested crops from 5.7 million acres of land. Oats were harvested from 80,000 of those acres (1.4%). NASS data indicated that Alabama farmers harvested 1.4% of the national crop of oats in 1950. Census data indicated that farmers in Talladega County harvested the largest share (6.1%) of the state's oat crop that year. Other counties reporting large shares included: Dallas (4.9%), Hale (4.0), Montgomery (3.8), and Madison (3.7). These five counties, three of them south of the fall line (Figure 1.17), accounted for 22.5% of Alabama's oat crop that year.

Prices

Data on Alabama oat market prices during the period 1950-1970 are also presented in Figure 14.85. They generally followed the same trends as national prices. Apparently, for most of the period and especially after 1955, support prices (Figure 14.84) were the primary determinants of market prices. However, Alabama market prices were generally around 20 cents a bushel higher during the entire period 1950-1970, and the differential was as high as 35 cents in the early part of the period. Apparently, this variation was the results of differences in regional support prices.

Acres Planted and Harvested

Alabama farmers responded to the relatively high market prices that they were receiving for oats in 1950, 1951, and 1952 (Figure 14.85) by sharply increasing acres planted (Figure 14.89). They planted 277,000 acres in 1950, and five years later in 1955, they planted 561,000 acres. Afterward, however, even with market prices remaining relatively stable at around 62 cents a bushel, acres planted began to trend downward, finally reaching 119,000 in 1970. Alabama farmers appeared to pay scant attention to market prices. Rather they seemed determined to get out of the oat business as quickly as practical.

Unfortunately, market prices began to decline after 1952, and reached 94 cents a bushel in 1954, the year before they planted that extremely large crop. Market prices continued to decline through 1955, and acres planted quickly followed and continued to trend downward through 1970, although the rate of decline was reduced somewhat after 1966. In 1970, farmers planted 119,000 acres of the crop. Another factor, other than market prices, was probably important in farmer annual decisions on acres planted during this period: Alabama farmers took 770,000 acres of land out of cotton between 1950 and 1970, and some of that lost cotton acreage likely became acres planted in oats.

In 1950, Alabama farmers planted 1.3% of their total farmland in oats. By 1970 that percentage had fallen to

0.8%.

Figure 14.89 also includes data on acres harvested. Alabama farmers have always grazed a large portion of their oat crops. Throughout this period (1950-1970), farmers harvested a relatively constant 26% of acres planted, a percentage much lower than the national average. In 1955, they harvested 148,000 of 561,000 (26%). In 1970, they harvested 31,000 of the 119,000 acres (26%). Oats were rapidly becoming a less important plant crop in Alabama.

Yields

Data on Alabama oat yields are also included in Figure 14.87. Trends in yields in the state during the period 1950-1970 were similar to those in the country as a whole. Average oat yields on Alabama farms increased from 25 bushes an acre in 1950 to 40 bushels in 1970. There was a considerable amount of year-to-year variation throughout the period. In every year except 1953 and 1956, average yields in Alabama were significantly lower than those in the country as a whole. In some years, the differential was as great as 15 bushels an acre. The oat genome certainly was not well suited to the Alabama agro-environment during this period. Yields were considerably lower than expected in 1955, 1963, and 1967, likely as a result of poor crop weather, plant diseases, or pests.

Bushels Harvested

Bushels of oats harvested (Figure 14.90) by Alabama farmers during the period 1950-1970 followed the trends of acres harvested during that period (Figure 14.89). They harvested between one and two million bushels in 1950 and 1951, but with acres harvested increasing sharply, bushels harvested also increased, and by 1954 reached 5.5 million. After 1954, bushels harvested generally followed acres harvested downward; even though yields were trending upward at the time. In 1965, bushels harvested reached 1.4 million, and it remained near that level through 1970. In 1955, 1957, 1958, and 1963, bushels harvested were somewhat lower than expected as a result of the lower than expected yields in those years.

Alabama Sweet Potatoes

Alabama farmers harvested 1.37 million Cwt of sweet potatoes in 1949. Some 23.3% of the crop was harvested in just two counties: Cullman (16.8%) and Baldwin (6.5). Other counties reporting large shares included: Tuscaloosa (2.6%), Perry (2.4), and Sumter (2.2). These five counties accounted for 30.5% of the state's entire crop. But sweet potato production was spread throughout the state with 29 counties reported harvests of 16,200 Cwt or more.

Data on acres of sweet potatoes harvested annually in Alabama during the period 1940-1970 are presented in Figure 14.91. Alabama farmers harvested sweet potatoes from 35,000 acres in 1950. They had harvested the crop from 64,000 acres in 1940. Although there was some year-to-year variation, acres harvested trended downward throughout the entire period, and was 4,800 acres in 1970, a decline of 87.1% in the two decades. This rate of loss is considerably greater than the rate of loss (62.7%) of farms during the same period (Figure 14.18). It is likely that a large share of the farms that were lost during the period would have been those where the production of sweet potatoes, for use by the families involved, would have been extremely important.

Farmers in the state harvested 1.7 million Cwt of sweet potatoes in 1950, but with the general decrease in acres harvested during the remainder of the period, they only harvested 398,000 in 1970. They received $6.7 million and $1.7 million, respectively, for the 1950 and 1970 crops.

Alabama Irish Potatoes

Data on the acres of Irish potatoes harvested annually during the period 1940-1970 are presented in Figure 14.92. Alabama farmers harvested the crop from 32,000 acres in 1950, whereas they had harvested 51,000 acres in 1940. Although there was some year-to-year variation, acres harvested usually trended downward throughout the entire period. Acres harvested were 16,700 in 1970, a decline of 47.8% during the two decades. Total cropland harvested in the state declined by 52.8% during the same period.

The sharp increase in acres harvested in 1953 was the result of an unusually large increase in the average market price received by farmers in 1952 ($3.75 per Cwt). This price was the highest received by Alabama farmers since 1920. It had been $2.00 in 1951. With this increase, farmers increased acres harvested by 31% (29,000 to 38,000). Unfortunately, when they harvested that large crop, the average market price was down to $1.57.

Alabama farmers harvested 3.9 million Cwt of Irish potatoes in 1950, but with the general decrease in acres harvested during the remainder of the period, they only harvested 2.1 million Cwt in 1970. They received $10.4 million and $8.1 million, respectively, for their 1950 and 1970 crops.

Census data indicate that Baldwin (45.8%), Escambia (9.8), DeKalb (5.3), Jackson (4.8), and Cullman (4.6) counties harvested the largest shares of the 1949 crop of Irish potatoes. These five counties accounted for 70.3% of the state's crop that year.

Alabama Sorghum

The demand for grain-fed beef increased substantially through the 1940s and 1950s, as consumers developed a taste for the tender, heavily-marbled beef that could be produced in feedlots through the feeding of various combinations of grains. Grain sorghum quickly became an essential and growing part of this process.

In 1950, American farmers produced 233 million bushels of sorghum grain. Texas (61%) and Kansas (19.1%) produced 81% of the total. Altogether, five states (Texas, Kansas, New Mexico (3.6%), Nebraska (2.1), and California (2.0) accounted for 89% of the national total. Alabama farmers contributed only 595,000 bushels (0.25%) of the crop.

US Exports

In 1950, the United States exported a significant share of its sorghum grain. That year, 75 million bushels of the 1949 crop of 148.5 million bushels was exported. There was considerable year-to-year variation in exports during the period 1950-57, but they usually remained below 75 million bushels. They were much lower than expected in 1952, 1953, and 1956 as a result of unusually low levels of bushels harvested in the country. After 1957, exports began to trend upward, finally reaching 266 million bushels. Afterwards, however, they declined to 106 million in 1968. They increased in 1969 and 1970 to 144 million.

The extremely high export levels during the period 1964-1967 were a result of unusually large harvests. In 1965, American farmers harvested 672.7 million bushels of the grain, the highest level on record to that time.

US and Alabama Prices

As had been the situation with all of the grains, the government increased support prices for sorghum sharply after 1940. In 1950, it was set at $1.87 a bushel (Figure 14.94). It was increased to $2.43 in 1953. After Eisenhower was inaugurated, the support prices were lowered, reaching $1.52 in 1959 and 1960; then after Kennedy's inauguration, the government began to increase them again. It reached $2.14 in 1967, and remained at that level through 1970.

Data on the annual average market prices that Alabama and American farmers received for their sorghum grain during the period 1950-1970 are presented in Figure 14.95. There are no market price data available for Alabama before 1949. Bushels of American sorghum grain harvested declined sharply from 1950 through 1952, and in the same period market prices increased from $1.88 a bushel to $2.82. Afterwards, bushels harvested trended upward through 1960, and market prices declined reaching $1.49 that year. Market price was much lower than expected in 1955 ($1.74). That year, the support price was reduced from $2.28 to $1.78. Bushels harvested fell by 140 million bushels between 1960 and 1961, and market price increased to $1.80. Also, in 1961, the support price was raised to $1.93. Bushels harvested generally increased from 1961 through 1968, and in 1967 the support price was increased to $2.14. In 1968, market prices fell to $1.69. Market prices increased in 1969 ($1.91) and 1970 ($2.04).

Generally, market prices received by Alabama farmers were somewhat higher than those received by farmers elsewhere in the country; trends, however, were essentially the same for both. Market prices for Alabama sorghum grain were generally higher than support prices from 1950 through 1960. Afterwards, it was either equal to it or lower.

Acres Planted and Harvested

Very little of the sorghum grown in Alabama was harvested for grain before 1946; consequently, there is little information on the subject before that time. Data published by the Alabama Agricultural Statistics Service in 1950 indicate that in 1944 and 1945 some 2,000 and 6,000 acres, respectively, were harvested for that purpose. Before that time, most of the crop was grown for silage and for the production of molasses.

With the Alabama market price at $2.30 in 1949 (Figure 14.95), Alabama farmers planted 72,000 acres of the crop in 1950. Surprisingly, with market prices increasing rapidly in 1951 and 1952, farmers chose to reduce acres planted in those years. In 1952, they planted only 40,000 acres. Surprisingly, afterwards, with market prices and support prices declining, they chose to sharply increase acres planted. In 1957, it was up to 82,000 acres. It declined rapidly thereafter, reaching 34,000 acres in 1962. With market prices relatively stable from 1962 through 1968, acres planted remained around 40,000. With the sharp increase in market prices after 1968, acres planted increased to 54,000 in 1970. In most of the years during the period 1950-1970, Alabama farmers harvested sorghum for grain from around 43% of acres planted (Figure 14.96). In 1969, Alabama farmers harvested the crop from 17,000 acres, or 0.6% of total harvested cropland.

US and Alabama Yields

Data on the comparative yields of sorghum grain in the country and the state for the period 1950-1970 are presented in Figure 14.97. In the early 1950s, yields in the state and the country were similar; although even then, they tended to be slightly lower in Alabama. However, after 1955, while they trended up in the state, the national rate of increase was much higher. Yields in the country and the state were around 20 bushels an acre in 1950. By 1970, they had increased to the low 50s in the country and to the low 30s in the state. Obviously, Alabama was quickly learning that there was very little comparative advantage in the production of sorghum grain either.

Bushels Harvested

With only a modest increase in yields during the period 1950-1970 (Figure 14.97), trends in bushels harvested for grain in Alabama (Figure 14.98) were more dependent on trends in acres harvested (Figures 14.96). Bushels harvested varied widely during the period. It was 189,000 in 1964 and 840,000 in 1955. In 13 of the 21 years, it was less than 500,000 bushels.

Acres Harvested for Silage

Data presented in Figure 14.99 indicate that between 1940 and 1953, Alabama farmers harvested about 7,000 acres of sorghum for silage. Afterwards, acres harvested for that purpose began to trend upward, reaching 26,000 acres in 1958. Then it declined, remaining between 12,000 and 21,000 through 1970.

Acres Harvested for Molasses

In the early 1930s, Alabama farmers had harvested 50,000 to 60,000 acres of sorghum each year for the production of molasses; subsequently, acres harvested for that purpose had slowly trended downward until it reached 8,000 acres in 1950 and 3,000 acres in 1959. Data on this use was not collected after 1959. In 1931, farmers produced 3.9 million gallons of molasses, but only 195,000 gallons in 1959.

Acres Harvested for Forage

Data on acres of sorghum harvested for forage was not collected after 1946.

Alabama Sugar Cane

Alabama farmers harvested around 30,000 acres of sugar cane during the middle of the Depression (1935-1937). Afterwards, acreage of sugar cane began to decline, reaching 10,000 acres in 1950 and 3,000 in 1961. No data were collected on the harvest of that crop after 1961. In 1937, Alabama farmers produced 3.9 million gallons of sugar cane syrup and 1.2 million in 1950, but only 285,000 gallons in 1957.

Alabama Vegetables

In 1949, USDA began to publish annual data on the value of vegetables harvested on Alabama farms. Data for the period 1950-1970 are presented in Figure 14.100. Value of the state's vegetable crop was $12.5 million in 1950. It increased sharply to $20.1 million in 1953; then it trended downward to $12.4 million in 1955. Afterwards, it trended upward to reach $21.5 million in 1964, and subsequently varied around that general level through 1970. Between 1950 and 1970, value increased from $12.5 to $23.5 million, an increase of 88%. During the same period, the Consumer Price Index increased by 61%.

Annual data were first collected on Alabama vegetables in 1928. These early data (1928-1946) were first published in Volume 1 (1948) of *Alabama Agricultural Statistics*, under the heading "Truck Crops." In a following volume, the same crops after 1945 were listed under the heading, "Commercial Vegetables for Fresh Market." The vegetable crop data are listed under this heading for the remainder of the period.

Census data for 1949 indicate that Alabama farmers harvested vegetables for sale from 46,300 acres. The largest share (19.5%) of this total was reported from Baldwin County. Other counties with significant shares included: Blount (13.4%), Chilton (7.7), Mobile (6.7), and Houston (5.6). These five counties accounted for 52.9% of the state's total crop.

Data on acres harvested for these vegetable crops for the period 1950-1970 are presented in Table 14.8. These data show that in terms of acres harvested, commercial early spring Irish potatoes was the most important crop during the early part of the period. These data represent only a portion of total acres harvested of Irish potatoes. From 1950-1954, farmers harvested an average of 22,400 acres of the crop. There are no data available for the period 1955-1965; however, after 1965 the Agricultural Statistics Service began to include data on the crop in their annual bulletins. The 1966-1970 data indicate that acres harvested of commercial early spring potatoes was trending rapidly downward during that period.

The sharp increase in acres harvested of the commercial early Irish potatoes between 1952 and 1953 (21,200 to 31,800) was the result of the sharp increase in the average price that farmers received for their crop between 1951 and 1952 ($2.00 to $3.75 per Cwt). As a result of that high price in 1953, farmers increased acres harvested by 50%. Unfortunately, when that large crop was harvested, the price was down to $1.57.

Watermelons and tomatoes were also important crops during the 21-year period. Farmers harvested an average of 15,000 acres of watermelons and 5,700 acres of tomatoes. Acres harvested of the remainder of the crops generally ranged from 1,000 to 4,000 acres.

Data on the value of production of the various crops for the years 1950, 1955, 1960, 1965, and 1970 are presented in Table 14.9. The value for all of the crops increased substantially from 1965 to 1970. During this period, the Consumer Price Index increased by 20%, and much of that increase came in 1969 and 1970. Because of the high levels of acres harvested, the value of production for commercial early Irish potatoes was highest, ranging from $3.57 to $4.20 million. Values for tomatoes and watermelons were somewhat lower for most of the period, but by 1970 both were higher than the value of the Irish potatoes.

Alabama Velvet Beans

In 1950, Alabama farmers harvested velvet beans from 100,00 acres. In 1940, they had harvested 500,000 acres of the crop. Acreage harvested declined rapidly through the 1950s, and by 1961 had sunk to 16,000 acres. Data collection on the production of this crop was discontinued, after that year. In 1940, the value of the crop was $2.5 million, but by 1950 and 1960 it was down to $1.5 million and $600,000 respectively.

Alabama Cowpeas

Near the end of the Depression in 1937, Alabama farmers planted "alone" 255,000 acres of cowpeas and "inter-planted" another 592,000. The acreage "inter-planted" was equivalent to 296,000 acres planted "alone." These data indicate that there were 551,000 acres of the crop planted that year. Some 193,000 acres of the total acreage were harvested for peas. The remainder was harvested for hay, grazed, or plowed-under. Surprisingly, acres harvested declined during World War II, and in 1950 only 31,000 acres were harvested for peas and 4,000 acres were harvested for hay. Afterwards, acres harvested for peas continued to decline, reaching levels of 23,000 acres in 1955; 9,000 in 1960; 6,000 in 1965; and 5,000 in 1967. Data collection was discontinued after 1967.

The large 1937 crop was valued at $8.52 an acre. At that time, it was obvious that cowpeas had a very limited future as a commercial crop in Alabama. The price-cost ratio was simply too low. In their rush toward diversification, Alabama farmers tried many crops. Unfortunately, few of them ever provided the alternative to cotton that they so desperately needed.

Alabama Pecans

Annual data on pounds of pecans harvested in Alabama during the period 1940-1970 are presented in Figure 14.101. In 1950, farmers harvested 11.7 million pounds, ranking third in the country. Thereafter, pounds harvested trended upward. Pounds harvested in the good (masting) years increased from 26 million pounds in 1951 to 61.3 million in 1963. Surprisingly, however, the harvests increased very little in the poor (non-masting) years when the crops varied between 10 and 20 million pounds. Between 1965 and 1969, the masting phenomenon seemed to be absent in the production of the pecans. Pounds harvested seemed to remain stable at around 30,000 pounds, an intermediate point between expected harvests in good and poor years.

Cash receipts (value) from the sale of pecans in 1950 were equivalent to 1.5% ($3.5 million of $231.8 million) of the sale of all plant crops (excluding the sale of farm forest products). As a result of the high year-to-year variation in pounds harvested, receipts also varied widely. It was at its lowest level ($1.1 million) in 1957 when only four million pounds were harvested. It was highest ($12.9 million) in 1968 when 31.5 million pounds were harvested. Also, pecan prices generally varied inversely with pounds harvested. For example, farmers received 28.7 cents a pound for the 4,000-pound crop in 1957, but only 16.8 cents for the 61,300-pound crop in 1967.

Alabama Tung Nuts

In the 1930s, Alabama farmers, primarily in the southern tier of counties, decided to begin the production of tung nuts as a means of diversifying their crop production. They harvested 2.8 tons of the nuts in 1930 and 11.2 tons in 1940. By 1950, tons harvested had grown to 1,000 tons a year in the good (masting) years. There were 252,000 trees of nut-bearing age in the state at that time. Tons harvested continued to increase in the good years, finally reaching 2,800 tons in 1954 and 3,800 tons in 1958. In the poor years, tung nut harvests generally ranged from 400 and 900 tons. In one of the better years (1954), farmers harvested 2,800 tons, and they received $60 a ton (three cents a pound) for the crop. Obviously, with prices in this range, it was not likely that farmers would continue to produce the crop in the state. After 1966, pounds harvested declined to the point that data were no longer collected on the crop.

Alabama Peaches

In 1950, Alabama farmers harvested 10.6 million pounds (220,000 bushels) of peaches, ranking eighteenth in the country in production of the crop. In 1940, they had produced 33 million pounds (688,000 bushels), and the largest harvest on record was in 1941 with 113.2 million pounds. In 1950, cash receipts from the sale of peaches was equivalent to only 0.2% ($759,000 of $231.8 million) of the sale of all plant crops (excluding the sale of farm forest products) in the state.

Census data indicated that Alabama farmers harvested 14.3 million pounds of peaches in 1949. Some 47.6% of the total crop was harvested in two counties: Chilton (34.6%) and Blount (11.2). Other counties reporting significant harvests included: Franklin (3.3%), Marion (2.6), and Morgan (2.5). These five counties, all located north of Montgomery, accounted for 56% of the state's total peach harvest.

Data on the pounds of peaches harvested in Alabama during the period 1950-1970 is presented in Figure 14.102. From 1950 through 1957, with the exception of 1955, pounds harvested ranged from 10 to 50 million. In 1955, however, it was less than 24,000 pounds, the result of a period of freezing weather in the late spring of that year. Of course, poor crop weather affected pounds harvested in other years as well. After 1957, pounds harvested began to trend upward, reaching 70 million pounds in 1964. Afterwards, it began to trend downward, reaching 12.5 million in 1964. Between 1964 and 1970, pounds harvested ranged from 10 to 30 million pounds, or near the 1950-1956 level, and not much greater than in the period 1949-1951.

Even though pounds harvested varied widely during the period, the value of peach production remained relatively stable. In most of the years, it varied between $2 and $3 million. The inverse relationship between prices and pounds harvested was extremely strong. For example, in 1961 farmers harvested 60 million pounds of peaches and received 4.4 cents a pound when they sold them. Then in 1970, they harvested a mere 15 million pounds and received 9.6 cents a pound.

Alabama Greenhouse and Nursery

Census data indicated that sales of American horticultural specialties in 1949 totaled $392.1 million dollars. This total was equivalent to 3.3% of all crops sold in the country that year ($0.392 billion of $11.958 billion). Nationally, flowers and flowering plants accounted for 60% of all horticultural sales.

Census data on sales of Alabama's horticultural specialty crops in 1939, 1949, 1954, 1959, 1964, and 1969 are presented in Table 14.10. In 1949, total sales of all Alabama's horticultural specialties were somewhat less important in the overall crop production situation than they were in the country as a whole. Sales that year were equivalent to 2.1% of the sales of all crops in the state ($5.2 of $250.8 million).

The total value of nursery crops increased from 1949 through 1964, reaching $3.00 million in 1949, $4.14 million in 1954, $5.27 million in 1959, and $5.60 million in 1964. The total was somewhat lower in 1964 ($5.52 million). The 1964 and 1969 data are not exactly comparable to the earlier data: the early data were obtained from all farms, whereas the 1964 and 1969 data were taken only from Class 1-5 farms (farms with sales of at least $2,500). It is not likely, however, that this would have made much difference in the nursery products data. Sales of flowers and flowering plants, etc. also increased during the period ($2.12 to 2.71 million) and did not decline between 1964 and 1969. In fact, it almost doubled ($2.71 to $4.94 million). Sales of flower seeds, etc. did not increase very much during the period; unfortunately there were no data available for 1964 and 1969.

Alabama Timber

Data on lumber production (percentages) in the country's different geographical regions during the period 1869-1970 are presented in Table 14.11. These data indicate that in 1970 the largest share (56.4%) of lumber produced in the country was still being manufactured in the West Region. Its share in 1950 was 48.7%; however, the data seemed to indicate that region's share was beginning to decline. In 1960, the share of the West Region was 59.7%. Then in 1965, the percentage declined to 58.7%, and it was down to 56.4% in 1970. With this decline in the percentage manufactured in the West Region, the percentages for all of the other regions increased slightly.

Between 1940 and 1950, Alabama lost 1.7 million acres of cropland. Taking that quantity of land out of cultivation provided a unique opportunity for the loblolly pine. Within a short time after fields were abandoned, windblown seeds from loblolly pines located in border woodlands quickly covered the abandoned fields. Within a few years, what had been fields of growing cotton became fields of rapidly growing pine trees. In most cases, this was a completely unplanned, unintended event with extremely fortuitous consequences. Almost no one abandoned a field in those days to create a stand of timber, but the enormous quantity of almost pure stands of loblolly pine created in this manner gave rise to the tree-farming industry in the South. Unfortunately, the establishment of succeeding stands would not come so easy. Because of the nature of their seeds, the pines colonized the old fields first; however, hardwoods, particularly sweetgum, soon also invaded the old abandoned fields and thrived in the shade of the pines, especially in the absence of fire. When the pines matured and were harvested, they were quickly replaced by almost pure stands of hardwood. Pine seeds from surrounding areas still fell into the harvested areas, but they did not grow well in the shade of the hardwoods. Forest managers soon found that this natural process of tree succession could be prevented only by expensive land treatment to remove the hardwoods followed by planting pine seedlings.

Published information on the early production of Alabama timber is very limited. According to Stauffer (1960), the US Forest Service conducted the first comprehensive field study of Alabama timber production during the period 1934-1936; however they did not follow up with a similar study in the following decade. The next systematic study was published in 1953 (Wheeler, 1953).

Census data indicated that in 1949 sales of forest products sold off Alabama farms totaled $10 million. The largest share (7.5%) was reported from Washington County. Other counties reporting large shares included:

Monroe (5.9%), Baldwin (3.0), Marengo (2.9), and Cullman (2.6). These five counties accounted for only 21. 9% of the state total. This relatively low percentage is indicative of the widespread distribution of merchantable forest products in Alabama. A total of 41 counties reported sales $100,000 or more.

Land In Forests

Stauffer (1960) reported that in 1954 forests constituted 63.9% (20,756,200 acres) of the land area of Alabama (Table 14.12). This was the second lowest estimate ever published. It had been as low as 57.8% (18,770,700 acres) in 1936. The number of farms in Alabama had reached a maximum level in 1935 (274,000), but by 1954 the number had declined to 168,000 (21.2 million acres of farmland). As farm numbers and acres of farmland decreased, the area in timber increased. By 1963, the number of farms in the state was down to 105,000 (16.4 million acres of farmland), and the acreage in forests had increased to 21,770,000 acres (67%). Data published by Murphy (1973) indicate that the amount of forestland had actually decreased slightly by 1972 to 21,350,100 acres (65.7%), even though the number of farms was down to 78,000 (14.6 million acres of farmland). Murphy commented that the decrease was likely the result of clearing forestland for pasture. Cattle numbers in the state were increasing rapidly during this period.

Forest Ownership

The percentages of public and private land in forests remained largely unchanged during the period 1953-1972—4.7% public and 95.3% private (Table 14.13). Data are not available for 1953, but by 1963 the forest products industry owned 18.7% of all forestlands, and by 1972 their share had increased to 19.7%, as the industry moved to guarantee the supply of timber required in their manufacturing operations. While industrial ownership of forestlands was increasing, the share owned by farmers was trending downward, along with the number of farms. In 1953, farmers owned 46.9% of all forestland in the state. By 1963, their share had declined to 35.1%, and it continued downward to 31.6% in 1972.

Forest Type

Data on changes in the percentages of six forest types in Alabama during the period 1953-1972 are presented in Table 14.14. Both of the pine forest types (longleaf-slash and loblolly-shortleaf) declined during the period. In 1953, the percentage of the loblolly-slash type was almost twice as large as the percentage of the oak-hickory type. By 1972, however, it was only slightly larger (29.9 versus 27.7%). Obviously, during this period, oaks and hickories were rapidly replacing loblolly and shortleaf pine in the state. When the increase in the oak-pine percentage (15.7 to 23.5%) is considered, the replacement of pines was even more significant. The loss of over half of the longleaf-slash type (14.4 to 6.9%) during the period, was even more significant.

Tree Plantations

There is relatively little published information on the establishment of artificially- established plantations before 1975, although a state forest tree nursery was established in Sumter County as early as 1926. In 1960, Stauffer (1960) reported that Alabama nurseries distributed 600 million seedlings to Alabama landowners during the period 1952-1960. From 1963-1972, some 600,000 acres of tree plantations were established with 40% being artificially regenerated.

Growth and Removal of Commercial Timber

Data on the growth and removal of commercial softwoods and hardwoods from Alabama woodlands during the period 1953-1971 is presented in Table 14.15. In all of the three years studied (1953, 1962, and 1971), the growth of both softwoods and hardwoods exceeded removals, although removals of softwoods almost doubled (373.3 to 739.6 million cubic feet) during the period.

Uses of Harvested Timber

Table 14.16 presents data on the various uses made of timber harvested from Alabama woodlands in 1951, 1962, and 1971. The most important change in the use of timber during this period was the rapidly growing importance of pulpwood (18.7 to 60%) and the concomitant relative reduction in the use of timber for sawlogs (54.8 to 31.4%). The percentage used for pulp almost tripled during the period, and in 1971 almost twice as much was used for that purpose as for sawlogs. The increase in the use of timber for veneer is also significant (2.4 to 5.4%) and represents the growth of the plywood industry in the state. The reduction in the use of timber for fuelwood represents the major shift in the source of energy for heating in the state during the period.

Animal Agriculture

During the period generally associated with World War II and the Korean Conflict (1941-1951), American livestock producers had responded very positively to the market opportunities generated by those conflicts. The value added of livestock production in the United States had increased from $7.9 to $22.3 billion, an increase of 182%. During the same period, the Consumer Price Index increased 76% (44.1 to 77.8%). At the beginning of this period (1941), value added of livestock production was 56% larger than for crop production ($7.9 versus $5 billion). At the end of the period (1951), the ratio had increased to 59% ($22.3 versus $14 billion).

In 1950, cash receipts from the sale of all livestock produced in Alabama was $209.9 million, but was only 79% as large as that of all crops ($209.9 versus $267 million). In 1940, total livestock cash receipts for the state had been $26.5 million, and equal to only 44% for crops (26.5 million versus 60.8 million). All of these cash receipts were related to animal production in a wartime economy. With World War II over, many farmers worried that livestock production would quickly return to the conditions it suffered through in the years leading up to the war. However, the United States in 1952 was a different place than it had been in the late 1920s and 1930s.

Cattle

In 1950, USDA estimated that there were 78 million cattle of all types (including cows that had calved-milk) in the country. The North Central Region, with a substantial number of feedlots and a large number of dairies, reported the largest share (42%). The South Central Region, with its large number of pastured cattle and a substantial number of small dairies, reported the second largest share (24.3%). The West Region, with its large number of rangeland cattle, ranked third (20.6%). The North Atlantic and South Atlantic regions, with little comparative advantage in cattle production, ranked fourth (7%) and fifth (6.6%), respectively. Nevertheless, the American dairy industry had it roots in the North Atlantic Region, and in 1950 both New York (2,116,0000) and Pennsylvania (1,704,000) still had substantial numbers of milk cows.

US Cattle

In 1950, Texas farmers reported the largest share (10.4%) of cattle ("all" cattle and calves, including cows that had calved-milk) in the United States. Other states with large shares included: Iowa (6.2%), Nebraska (4.9), Wisconsin (4.8), and Kansas (4.6). Yet these five states accounted for only 30.9% of the national total. This relatively low percentage is indicative of the widespread distribution of cattle in the country.

Prices

In preceding chapters, farm value per head has been used as a measure of the annual value of cattle; after 1949, however, USDA began to publish data on cattle prices per Cwt. As a result, from 1950 onward these newly available data will be used to express the relationship between supply and demand in the cattle market in Alabama and the United States. Data on the annual average price per Cwt of cattle (all) marketed in the United States during the period 1950-1970 are presented in Figure 14.103. In 1950 and 1951, average prices were $23.30 and $28.70, respectively. With prices at this high level, farmers began to build their herds of brood cows again, and a new Cycle of Abundance began in 1950 (Figure 14.104). The price data are for "all" cattle; however, the numbers data in Figure 14.104 are for "all" cattle, excepting cows that have calved–milk.

With the 1949-1958 cycle going through its Expansion and Consolidation phases between 1950 and 1956, prices began to decline, finally reaching $14.90 in 1956. With prices at this level, the cycle entered its Declining Phase. With the supply of cattle declining after 1955, prices began to trend upward in 1957. The Expansion Phase of a new cycle began in 1959. The following year, with cattle numbers increasing, prices responded by beginning to trend downward, reaching $18 in 1964. The Consolidation Phase of the cycle lasted from 1964 through 1966, and the Declining Phase began in 1967. Surprisingly, it only lasted one year (1968) before a new cycle began. After 1964, prices did not seem to respond to numbers at all. They began to increase in 1965 and were

$27.10 in 1970. After 1964, indices of prices received for all agricultural commodities began to increase (Figure 14.6), and continued to do so through 1970. Apparently, both cattle numbers and prices responded to this increase.

Cycles of Abundance

Nationally, cattle (excepting cows that had calved–milk) numbers increased from 80 million head in 1950 to 110 million head in 1970, an increase of 50% (Figure 14.104). The well-documented boom-and-bust cyclic nature of cattle numbers, which was apparent in the 1900-1930 and 1930-1950 periods, was also apparent in the 1950-1970 period. The teeter-totter relationship between prices and numbers in the elaboration of the Cycles of Abundance during the period was detailed in the preceding section and will not be repeated here. There was a classic Cycle of Abundance (1950s cycle) between 1949 and 1958, although its Declining Phase was certainly not typical. Then a 1960s cycle began in 1959. It seemed to reach a Consolidation Phase in the period 1964-1966, but instead of numbers declining afterwards in the typical Declining Phase, they began to increase again in what seemed to be the Expansion Phase of another cycle.

Real Net Return Per Cow

Real net returns per cow (real dollars, 1982-1984 = 100) generally increased during the war years of the 1940s and were $102 per animal in 1950 (Figure 14.105). After the end of the Korean Conflict in 1951, with cattle numbers extremely high (Figure 14.104), returns plummeted to $4 in 1956. Then with numbers declining, it trended upward to $70 an animal in 1958. Numbers generally increased after 1958, and returns trended downward to $11 in 1964. In 1965, the indices of prices received for all agricultural commodities began to increase (Figure 14.6), and returns followed, finally reaching $60 in 1969 and $55 in 1970. Obviously, there was little money to be made in the cattle industry during much of this period.

Alabama Cattle

The importance of cattle herding among Scotch-Irish immigrants and their descendants in the pioneer agriculture of Alabama and the South was discussed in an earlier chapter. For a century and a half, Southern consumers had to be content with the poor quality beef that woodland-foraging animals produced. In fact, this was the only beef available. With the coming of farm modernization, there was a widespread effort to improve cattle herds through a combination of improved pastures and improved animals. The result was a vastly improved slaughter animal. As a result, there was widespread optimism about the future of cattle production in the region. Consequently, cattle numbers increased rapidly.

The 1950 Census published data that indicate the largest share (6.7%) of Alabama's cattle (all) herd that year was in Montgomery County. Other counties reporting significant shares of the total included: Dallas (4.8%), Lowndes (3.9), Marengo (3.8), and Wilcox (3.2). These five counties accounted for only 22.4% of the state's total. This relatively small percentage is indicative of the widespread distribution of cattle production in the state. In that year, 32 counties reported cattle herds of 15,000 head or more. The five counties listed above had been in the center of the old Cotton Arc during the heyday of the production of cotton in the state.

Prices

Figure 14.103 also included data on cattle prices in Alabama during the period 1950-1970. The data indicate that the relationship between prices and supply was essentially the same in Alabama as it was in the country as a whole. Although the trends were essentially the same, Alabama farmers received $1 to $4 less per Cwt for their cattle.

Cycles of Abundance

In Alabama, trends in cattle numbers were similar to those in the country as a whole in the period 1940-1970. However, the overall increase was somewhat higher in the state. Numbers in the state increased from 1.1 million in 1950 to between 1.9 and 2 million in 1970, an increase of 75%. The cyclic nature of cattle production in the country is also apparent in the Alabama cattle numbers (Figures 14.104 and 14.106). The numbers data indicate that a new cycle also began in Alabama in 1950, but it ended a year later (1959) than in the country as a whole. The new cycle began in 1960, and ended in 1967. There was not a clearly defined end to this particular cycle in the country. During the period 1950-1970, state cattle numbers were highest in 1970—1.95 million head. That number represented 1.7% of the 112. 4 million national herd in that year.

The move toward increasing cattle herds in Alabama represents an interesting study in human adaptation to changing events. There are no natural grasslands in the South; however, as a result of information obtained from research in crop science, it is possible to create ,through proper management, truly exceptional pastures. As a result, the growth of cattle production in the state and region has been joined at the hip with advances in the development and management of pasture grasses. Unfortunately, it is difficult to produce a plant that is highly functional in both humid and dry conditions. Improved pasture grasses in the South do not function well under the drought conditions that often occur in the region. Further, these grasses, which function so well in our moist hot climate, produce very little new vegetation under extreme cold conditions, which may last several weeks in some winters. Also, optimum grass production required close attention to fertilization and lime requirements, which are relatively costly. In sum, the comparative advantage of the grass-cow production system of Alabama versus the grain-cow system of the Midwest was a very complex situation that had not completely worked itself out in 1970.

During the period 1950-1970, as Alabama farmers were expanding and improving their cattle herds, they encountered another perplexing problem with the use of pastures. With the globalization that followed World War II, American consumers began to be exposed to increasing quantities of a different kind of beef—beef produced in Midwestern feedlots using the inexpensive, abundant feed grains widely available in that region. This beef was tender and well marbled with white fat. Although it could be produced relatively cheaply, it soon became apparent that even the best grass-fed beef, with its yellowish fat, could not compete for the consumer's dollar with the feedlot product. At the same time, it also became apparent that Southern farmers could not produce enough feed grain locally, at competitive prices, to sustain their own feedlot operations. Further, they could not afford to have the grain shipped to the region to produce beef competitively. As a result, over time the South became a supplier of stocker cattle for the Midwestern feedlots. Animals grown to a specific size on grass were shipped to the feedlots for finishing.

Hogs

NASS data indicated that there were 62.3 million hogs and pigs in the United States on December 1, 1950. Some 70.8% of the total number were on farms in the North Central Region. The second and third largest shares were on Southern farms, 16.4% in the South Central Region and 6.9% in the South Atlantic Region. The West (3.6%) and North Atlantic (2.2%) regions had even less comparative advantage in hog production than did Southern farmers, and they were wise enough not to try to force the issue. The South had little comparative advantage either, but farmers and their leaders had not accepted the fact at that time. In the environment of the North Central Region, the genome of the hog found a near perfect home. Trying to compete in hog production with farmers in that region was an exercise in futility.

US Hogs

The North Central Region lead the country in numbers of hogs in 1950; as might be expected, the five states with the highest number of hogs were also found there. The largest share of the total was reported from Iowa (20.3%). Other states with large shares included: Illinois (10.5%), Indiana (7.5), Missouri (7.2), and Minnesota (6.0). These five states accounted for 51.5% of the national total. Only 2% of the total number of hogs was on farms in Alabama.

Prices

Dollars per Cwt, rather than farm value per head, are being used for the following discussion on hog prices. Average annual prices (dollars per Cwt) of American hogs during the period 1950-1970 are presented in Figure 14.107. The data indicate that prices were extremely cyclic throughout the period. During the period, prices bounced back and forth between $14 and $22, but most of the values were in a relatively narrow range of $17 to $20. This cyclic pattern is similar to those described in earlier chapters, and is the result of the teeter-totter relationship between prices and supply. Generally, however, the relationship was somewhat more ragged for hogs than for cattle. In 1952 and 1953, hog numbers declined, and in 1953 and 1954 prices increased. Similarly, in 1954 and 1955, numbers increased again, and in 1955 and 1956 prices declined. This teeter-totter relationship between prices and numbers generally persisted throughout the period, until finally in 1966, 1967, and 1968 numbers increased, and in 1967 and 1968 prices declined.

Cycles of Abundance

Data on numbers of hogs counted annually on December 1 in the United States during the period 1940-1970 are presented in Figure 14.108. The teeter-totter relationship between hog prices and numbers resulted in the elaboration of four complete Cycles of Abundance, as well as portions of two others, during the period 1950-1970. Most of the cycles were extremely short—three or four years. The longest one lasted from 1961 through 1965. Its Accumulation Phase lasted only two years (1961-1962), and its Liquidation Phase lasted from 1963 through 1965. While the cyclic nature of hog abundance was quite evident, it did not result in much change in total numbers. There were about 60 million animals in the country in 1950, and about the same number in 1968.

Alabama Hogs

The 1950 Census of Agriculture published data indicating that the largest share (4.6%) of the state's hog herd that year was in Houston County. Other counties reporting significant numbers included: Covington (4.5%), Geneva (4.5), Pike (4.0), and Barbour (3.3). These five counties accounted for only 20.9% of the state's total. Some 39 counties reported totals of 10,000 or more hogs.

Prices

Annual average prices (dollars per Cwt) received by Alabama farmers from the sale of hogs during the period 1940-1970 are also presented in Figure 14.107. These data were essentially the same as those for the country as a whole. Where there were slight differences, the national prices are slightly higher. Alabama prices were also quite cyclical, and for the same reasons as the national ones: The teeter-totter relationship between prices and supply and the lag time required for farmers to adjust to changes. Another important factor was the fact that the annual inventory was taken in December. As a result, farmer's responses could not be made until the following year.

Cycles of Abundance

During the period 1950-1970, Alabama hog numbers usually followed the same cyclic trends as the national numbers (Figure 14.109). Although the cycles were similar, the overall trend in the Alabama numbers was downward. At the end of the Accumulation Phase of the cycle in 1950, there were 1.2 million hogs in the state, but at the end of the same phase in the cycle in 1968, number had declined to 1 million. The decline in numbers was even more obvious in 1965 (0.7 million).

Milk

In the late 1940s and early 1950s, a long-term evolutionary process began to bring change in the American dairy industry (Blayney, 2002). During the period, the development and adoption of technological innovations, changes in milk production and marketing systems, and specialization began to bring about major structural changes. USDA reported that milk production in the United States totaled 116.6 billion pounds in 1950. The largest share (50.6%) of the total was reported by dairy farmers in the North Central Region. The American dairy industry had its beginnings in the North Atlantic Region, and milk producers there, especially in New York and Pennsylvania, showed they could still compete in the milk and milk products market. That region reported 16.9% of the total. The South Central (12.8%) and West (12.7%) regions reported similar shares of total productions. Only 6.9% of the total was reported from the South Atlantic Region.

US Milk

In 1950, milk cows on American dairy farms produced 116.6 billion pounds of milk. In 1940, they had produced 109.4 billion pounds. Wisconsin was the leading milk producing state in 1950 with 14.8 million pounds (12.7% of the national total). Wisconsin was followed by New York (7.5%), Minnesota (6.9), Iowa (5.3), and California (5.1). These five states accounted for 37% of the country's milk supply. Alabama dairy farms reported only 1.1% of the national total.

Prices

National milk prices during the period 1950-1970 were controlled under the provisions of the Agricultural

Marketing Agreement Act of 1937. The prices shown in Figure 14.110 represent the average price across the country, resulting from the application of a number of regional and sectional federal milk marketing orders. The annual average price in 1950 was $3.75 per Cwt. Prices increased rapidly during the Korean Conflict to $4.68 in 1952; however, during the years of the postwar adjustment and the first years of the Eisenhower Administration (1953-1956), they were adjusted downward. Prices changed relatively little from 1954 through 1965 ($3.86 to $4.26). Between 1966 and 1970, they were raised sharply ($4.84 to $5.78), as the indices of prices received for all agricultural commodities increased.

During the period 1950-1970, the Consumer Price Indices increased 68.2% (72.1-121.3%). Approximately a third of this increase came in the period 1966-1970 when milk prices were escalating rapidly. Altogether, milk prices increased 54.1%, during the period 1950-1970 ($3.75-$5.78 per Cwt), which was somewhat lower than the increase in the Consumer Price Index for the period.

Milk Cow Numbers

There were 23.7 million milk cows in the country on January 1, 1940 (Figure 14.111). The number increased sharply during the early years of World War II to reach 25.6 million in 1944. Thereafter, it began to trend downward. In 1950, there were 21.9 million in the country. Cow numbers increased slightly following the short-lived, upward spurt in prices in 1951 and 1952 (Figure 14.110). From 1955 through 1970, however, they declined as prices fell further and further behind increases in the Consumer Price Index (Figure 14.3). Altogether, the American dairy herd declined by 45% (21.9 to 12 million animals) during the period 1950-1970.

Milk Per Cow

In 1950, the national average annual milk production per cow was 5,314 pounds (Figure 14.112). In 1940, it had been 4,622 pounds. In 1950, it was highest in California with an annual average of 7,710 pounds. Production in California was followed by New Jersey (7,450 pounds), Rhode Island (7,200), Wisconsin (6,850), and New York (6,810). Average production in Alabama was 3,570 pounds. Milk per cow in the United States trended slowly upward through the 1940s and early 1950s. In 1954, it began to increase rapidly, as dairy farmers moved to adopt advances in animal nutrition and health, improved breeds, and improved cow and herd management. As a result of the adoption of these new practices, milk per cow increased to 9,751 pounds, in 1970. From 1954 through 1970, milk per cow in the country increased 4,094 pounds (72%), or at a rate of 241 pounds a year.

Pounds Produced

American milk production was 109.4 billion pounds in 1940 and 116.6 billion in 1950 (Figure 14.113). During the period 1950-1970, numbers of milk cows generally trended downward (Figure 14.111), while milk per cow trended upward (Figure 14.112). However, the rate of loss of milk cows during the period (83%) was greater than the rate of increase in milk per cow (72%). As a result, the change in cow numbers had the slightly greater effect on total milk production. The short-lived increase in numbers during the period 1953-1954, as well as the rapid growth of milk per cow after 1952, translated into a sharp increase in milk production (124.9 billion pounds in 1966). Similarly, the sharper decline in numbers in the period 1957-1959 resulted in a decrease in production. Finally, the increased rate in loss of animals after 1964 resulted in steep decline in production. As a result of the interplay of these two factors (cow numbers and milk per cow), milk production generally increased during the early part of the period (1950-1964) then decreased for the remainder of the period. In 1970, total production was 117 billion pounds, or only slightly higher than it had been in 1950.

Alabama Milk

NASS data showed that dairies in Alabama produced 1.1% of the country's milk in 1950. Census data indicate that dairy farms in Montgomery County reported the largest share (12.3%) of gallons of milk sold in the state in 1949. Other counties reporting large shares included: Hale (7.3%), Mobile (6.5), Dallas (4.8), and Perry (4.3). These five counties accounted for 35.2% of the state's total. Some 35 counties reported sales of 200,000 gallons or more.

Prices

Data on the average annual prices of milk in Alabama during the period 1940-1970 are also presented in Figure

14.110. In 1940, it was $2.36 per Cwt. Then in 1950, it reached $5.01. Except for some up-and-down variation in the early 1950s, prices generally trended upward throughout the entire period. Alabama milk prices increased from $5.01 to $6.83 per Cwt, an increase of 36%. During the same period, the Consumer Price Index increased 68%. As a result, Alabama farmers were receiving substantially fewer un-inflated dollars per Cwt for their milk in 1950 than in 1970.

Alabama prices were higher than in the US throughout the period. This reflects differences in federal milk marketing orders for different regions of the country. While Alabama prices were higher, year-to-year changes were similar. This similarity reflects the more or less uniform methods used to establish prices from region to region.

Milk Cows Numbers

There were 363,000 milk cows on Alabama farms on January 1, 1940, and in 1950 there were 362,000 (Figure 14.114). Year-to-year trends in numbers were generally similar to those in the country as a whole (Figure 14.111). They increased slightly in the early 1950s, along with prices (Figure 14.110), but after 1953 numbers trended steadily downward through 1970 to 122,000. Year-to-year trends in numbers in the state were similar to national changes (Figure 14.111). Altogether, Alabama milk cow numbers declined by 66% (362,000 to 122,000 animals) during the period 1950-1970. Alabama's rate of loss was substantially greater than in the country as a whole (66% versus 45%), during the same period. Dairy farmers in Alabama apparently were just beginning to see the futility of attempting to compete with those in California and Wisconsin.

Milk Per Cow

In 1940, average milk per cow for Alabama's dairy herd was 3,150 pounds annually (Figure 14.112). By 1950, the average had increased to 3,570 pounds. Milk per cow decreased slightly in the early 1950s, but after 1955 it generally trended upward to reach 6,689 pounds in 1970. In 1940, average production per cow was 1,472 pounds (4,622 versus 3,150 pounds) greater in the country as a whole than in Alabama. In 1950, the difference was 1,744 pounds (5,314 versus 3,570 pounds), and in 1970 the difference was even greater, 3,062 pounds (9,751 versus 6,689 pounds). The milk cow genome was definitely unhappy in Alabama, and dairy farmers were paying for it through the nose.

As a result of the federal marketing order program, Alabama dairy farmers received a higher price for their milk during the period 1950-1970 (Figure 14.110). However, the growing difference between milk per cow in the state and the country was rapidly eroding whatever comparative advantage the greater price might have provided. In fact, it is difficult to understand why any milk at all was being produced in the state. In 1950, the value (milk per cow multiplied by average price) of milk produced by an average cow in the country was $2,042 greater than for an average cow in Alabama. In 1970, the difference was $10,675. If the comparison were to be made between an average cow in Alabama and Wisconsin, the difference would have been even greater. It is difficult to comprehend the sacrifices (private and public) that had to be made during this period to keep any Alabama farmers in the dairy business.

Pounds Produced

Alabama dairy farms marketed 1.1 billion pounds of milk in 1940 and 1.3 billion in 1950 (Figure 14.115). Even though milk per cow declined slightly in the early 1950s (Figure 14.112), the increase in numbers during the period was sufficient to keep production between 1.25 and 1.30 billion pounds. After 1953, however, numbers declined so rapidly (Figure 14.114) that it completely overshadowed a slight upward trend in milk per cow (Figure 14.112). As a result, production also declined rapidly. By 1964, it was down to 814 million pounds. It had not been that low in the state since at least 1924, and probably earlier. After 1966, milk per cow was increasing at a much faster pace As a result, production remained relatively stable through 1970, although cow numbers continued to decline. In 1970, production in the state was 816 million pounds. In 1950, Alabama dairy farmers produced 1.1% of the milk in the country. In 1970, that percentage had fallen to 0.7%.

Alabama Chickens

In the preceding chapter, data on the total annual production of chickens in Alabama was used as the basis for the presentation. Those data included chickens produced and sold during an entire year. In this chapter, data on the number of chickens on farms on January 1 or December 1 will be used instead. The data represent estimates of all chickens (laying hens, pullets old enough to lay, pullets not old enough to lay, and other chickens) on all Alabama farms on either of those inventory dates. The data do not include numbers of broilers. About 70% of each estimate represents chickens (hens or pullets) old enough to lay, and most of the remainder represents pullets growing to the point where they could also be added to laying flocks.

US and Alabama Prices

Annual average value per head of all chickens sold in the country during the period 1940-1970 are shown in Figure 14.116. Value had increased sharply during the war years to $1.36 in 1950, but had fallen to $1.26 by 1956. It remained near that level through 1969, but increased to $1.35 in 1970, as the indices for prices received for all agricultural commodities increased (Figure 14.6).

Figure 14.116 also includes data on value per head for Alabama chickens during the period 1950-1970. Throughout the period, Alabama values were lower than for those in the country as a whole. The difference was as low as six cents in 1960, but as high as 21 cents in 1970. The general trend in values was similar for the state and the country, except that after 1960 they tended to decline in Alabama. In 1950, the annual average value was $1.15 a head, but by 1967 and 1968 it was down to 94 cents. With the rapid increase in prices received at the end of the period, it increased to $1.05.

Numbers

Annual data on numbers of all chickens (excluding broilers) on all Alabama farms during the period 1950-1970 are presented in Figure 14.117. From 1950 through 1968, the inventory was taken on January 1. In 1969 and 1970, it was taken on December 1. On January 1, 1950, the inventory indicated that there were 8.2 million chickens in the state. Afterwards, numbers slowly trended downward along with values (Figure 14.116) through 1956 (6.4 million). In 1957 (6.8 million), numbers began to increase rapidly, reaching 21 million in 1969, an increase of 212%. Between 1956 and 1957, the number of broilers produced in Alabama increased by 26% (82.5 to 103.6 million), and by 1969 Alabama farmers were producing over 376 million. The rapid increase in chickens was the direct result of the growing need for eggs for hatching broiler chicks. It is surprising that this need for additional eggs was not reflected in an increase in value.

Total Value

Data on the total value (millions of dollars) of all chickens on Alabama farms on either January 1 or December 1 in each year of the period 1950-1970 are presented in Figure 14.118. The data show that trends in total values were similar to those for value per head. In 1950, total value was $9.4 million. Then values declined through 1955 ($6 million). Afterwards, they trended upward to $22 million in 1969. In 1950, total value for chickens accounted for 4.5% of value added from the production of all livestock and livestock products in Alabama. In 1969, the percentage was down to 4%.

Alabama Broilers

As noted in a preceding chapter, commercial poultry production in the South began to expand in the early years of the Depression. Its development was most pronounced in the Delmarva area of Delaware, Maryland, and Virginia. In 1934, farmers in that area were "growing-out" about seven million broilers a year. After World War II, as self-service supermarkets became increasingly popular across the South, consumer interest in processed and frozen poultry increased rapidly. This convenience for consumers, plus its relatively low price compared to those of beef and pork, resulted in the explosion of the industry in parts of Alabama, Arkansas, Georgia, North Carolina, and Texas. In Arkansas, for example, the number of broilers marketed increased from 49.2 million in 1950 to 450.8 million in 1970, and the total weight marketed in the same period increased from 137.7 million to 1.5 billion pounds. Similar increases took place in the other Southern states as well. As Fite (1984) noted, no area of Southern agriculture expanded as rapidly after World War II as the broiler industry.

During the 1950s, vertical integration of the broiler industry developed. Large, farmer-owned cooperatives, such as Goldkist, and major feed companies, such as Ralston Purina, became involved in the industry. They were soon controlling the production of millions of birds annually throughout the region through their cooperative arrangements with local growers.

Under vertical integration, a company or corporation generally controlled all phases of producing, processing, marketing, and distributing the broilers. Actual day-to-day production activities was the responsibility of private farmers under pre-arranged contract with the companies. In this process, the company provided the chicks, feed, medications, supplies, and a serviceman's expertise. The company generally arranged for the harvest of the finished broilers and transportation to processing plants. The independent grower provided fully-equipped houses, labor, utilities, litter, and supporting farm equipment. When the birds were removed, the grower was paid for his management, labor, and investment based on a pre-arranged contract schedule.

Vertical integration of the broiler industry brought a measure of prosperity to isolated areas in several of the Southern states that had suffered greatly before and during the Depression. Critics of the industry, however, commented that formerly independent farmers had become little more than laborers or hired hands for large cooperatives or corporations. But that was only one side of the story. Broiler production before vertical integration had been an extremely risky operation for resource-limited farmers. Afterwards, much of the risk had been removed, and that which remained was shared by the producer and the cooperative or company involved. As vertical integration proceeded, the companies assumed more-and-more of the risks associated with market fluctuation.

It was only natural that vertical integration would eventually result in increased consolidation in the industry. In this process, the larger, more efficient companies bought out smaller ones. By the beginning of the 1960s, 20 companies controlled 32% of the entire industry: just four companies controlled 12%.

USDA reported that American farmers produced 631.4 million broilers in 1950. The largest share (47.2%) was produced in the South Atlantic Region. Farmers in Maryland produced 12.8% of the national crop. The second largest share (19.5%) was produced in the South Central Region where Arkansas accounted for 7.8% of the national total. The South Atlantic and South Central regions combined accounted for 66.7% of total national production. The North Atlantic and North Central regions accounted for almost equal shares (12.5 and 12.4%, respectively). Farmers in the West Region produced 8.3%.

Despite the importance of broiler production in the Delmarva region in the late 1940s, broiler production was still looking for a home. The genome for the chicken (the red jungle fowl) was probably established in the tropics of Southeast Asia, and it was likely that it would be much happier in an area considerably south of Maryland and Delaware. Although the red jungle fowl of the tropics never had the luxury of a diet rich in feed grains, it was determined through early research that the bird could grow rapidly and very efficiently when presented with such a ration. Unfortunately, the further the genomes, which were raised on feed grains, moved away from a tropical climate and toward the temperate zone, the unhappier they became. Unfortunately, it seemed that the genomes of chickens and feed grains appreciated different climates. Fortunately, it was determined that through selective breeding, the feed efficiency (pounds of feed per pound of gain) of the chicken could be improved to the point that it would be economical to grow chickens in the sub-tropics (Georgia, Alabama, Mississippi, and Arkansas) and ship the feed grains to them.

Data presented in Table 14.17 show the percentages of the American broiler flock inventoried in Delaware, Maryland, Georgia, Alabama, and Arkansas in 1950, 1955, 1960, 1965, and 1969. It shows that in 1950 some 12.8% of the total flock was in Delaware, but by 1969 only 4.8% was there. In contrast, in 1950 only 2.1% of the flock was in Alabama, but by 1969 the state was home to 12.6% of the total. Seemingly, as feed efficiency improved, the broiler industry would move southwestward. Considering the Georgia data in the table, it appears that the industry had moved toward the South, but data for Alabama and Arkansas also indicate that it was moving westward as well. The Georgia share of the national flock declined in 1965 and 1969, while shares in Alabama and Arkansas increased. These changes were not the result of declines in production in any of the states. Production in all of them increased. Rather, the changes were the result of increases being greater in Alabama and Arkansas.

Prices

Alabama broiler prices had increased rapidly during World War II reaching 38 cents a pound in 1946 and 1947 (Figure 14.118). Thereafter, they began to decline, and reached 29 cents in 1950. Throughout the 1950s, with numbers increasing rapidly (Figure 14.119), prices declined sharply, and by 1960 they were below 14 cents. Although there was some year-to-year variation, prices remained around 14 cents through 1970.

The broiler was a godsend for consumers in the country. By 1970, there were few anywhere who could not afford to buy a pound of broiler meat occasionally. Further, this low price effectively placed a limit on pork and beef prices.

Numbers

Data on numbers of broilers produced annually in Alabama during the period 1940-1970 are presented in Figure 14.119. Generally, declining prices are a clear signal for farmers to reduce production, but that general principle certainly did not seem to function in Alabama's broiler production industry during the period. As detailed in the preceding paragraph, prices trended sharply downward throughout most of the 1950s while broiler numbers soared. Farmers produced 13.1 million in 1950 and 376.1 million in 1970. Clearly there was something strange about the price-supply relationship during the period. It will be interesting to see how long this unusual situation could continue. In 1950, Alabama farmers accounted for 2.1% of total national production, but by 1970 the percentage had increased to 12.6%.

Market Weights

Market weights of Alabama broilers generally trended upward from the early 1950s through 1961 (Figure 14.120). It was 2.7 pounds in 1950 and 3.4 pounds in 1961. After 1961, market weight only increased another 0.1 pound to reach 3.5 pounds in 1966. It remained at that level through 1970. This increased market weight is a clear testimonial to the value of scientific research and extension. It is also an extremely positive indicator of the value of close cooperation between producers and processors.

Total Value

With broiler numbers increasing (Figure 14.119) and prices declining (Figure 14.118) during the period 1950-1970, it would be expected that there would be relatively little change in the total value of production, but this was clearly not the case (Figure 14.121). In 1950, it was $9.5 million, and in 1970 it reached $160.6 million. The difference, of course, was the sharp increase in market weights (Figure 14.120). This increase is the reason why the price-supply relationship did not seem to function. The increase in market weights offset the decline in prices.

In 1950, the total value of broiler production in Alabama was equivalent to 4.6% ($9.6 of $209.9 million) of the value added by the production of all livestock and livestock products in the state that year. By 1970, the percentage had increased almost sevenfold to 29.1% ($160.6 of 551.6 million). It certainly appears that in this period of modernization of Alabama agriculture that the old jungle fowl genome had found a home in Alabama.

Feed Efficiency

Of course, the increasing total value resulting from the increase in market weight is not necessarily the end of the story. The increase in market weights must have cost the farmers something. Certainly the increase should have required more feed; consequently, net returns might not have been as attractive as the total value indicated. Fortunately for farmers, however, feed efficiency also improved during the period. In 1948, it was 3.60 pounds of feed per pound of gain, and in 1968 it was 2.25 pounds. This decline (3.60 to 2.25 pounds of feed), or an improvement of 37.5%, essentially paid for the increase (2.5 to 3.5 pounds) in market weight of 40%.

Alabama Eggs

Until the mid-1950s, the Alabama egg industry was a small but mirror image of the American egg industry. Some eggs were consumed by the households on the farms where the eggs were produced. A small percentage was used for hatching eggs to maintain the laying flocks, but the larger share was sold to be used as table eggs or for the manufacture of foods containing eggs. With the explosion of the broiler industry in Alabama and the Lower South, the egg industry in this region changed radically. It no longer resembled the national industry.

Total egg production in the country reached 58.7 billion in 1950. Some 88% were sold, 11.7% were consumed by families on the farms where they were produced, and 0.3% were used for hatching.

In 1950, Alabama farms produced 718 million eggs, ranking the state 24th in the country in egg production. Census data indicate that in 1949 Cullman County reported the largest share (8.1%) of the state's total egg production. Other counties with large shares included: Baldwin (8.1%), Blount (7.0), DeKalb (5.6), and Marshall (4.3). These five counties reported 33.1% of the state's total. A total of 28 counties reported productions of two million or more eggs.

US Consumption

About the middle of the Great Depression, American egg consumption (eggs and egg products) began to trend upward (Figure 14.123). During World War II, it increased sharply, reaching 402 eggs per person in 1945—the highest level in history to that time. After 1945, per capita consumption declined to around 385, and remained at that level through 1950. It remained near that level through 1952, then began to decline rapidly, finally reaching 313 in 1965. It generally remained near that level through 1970.

US and Alabama Prices

Data on US and Alabama egg prices (cents per dozen) during the period 1940-1970 are presented in Figure 14.124. The data show that national prices increased rapidly during the years of World War II and the Korean Conflict and began a slow decline afterwards. During the early 1950s, national prices ranged around 45 cents, but after 1953 they began to trend slowly downward from 39 cents in 1955 to 31 cents in 1968. Prices were higher in 1969 and 1970 (34 and 40 cents, respectively) as prices received for all agriculture commodities increased (Figure 14.6). By 1967, it was down to 31 cents. With increasing consumer prices (Figure 14.2), egg prices increased to 40 cents in 1969, and remained near that level in 1970.

Year-to-year trends in Alabama egg prices during the period 1950-1970 were similar to those in the country as a whole (Figure 14.123). Throughout the period, however, state prices were from two to eight cents a dozen higher. They were as high as 50 cents in 1953 and as low as 39 cents in 1967 and 1968. In 1970, it was 44.2 cents. It is likely that Alabama egg prices were somewhat higher as a result of the fact that higher prices for hatching eggs would have been included.

Numbers

Egg production in Alabama was 718 million in 1950 (Figure 14.125). It increased slowly in the early 1950s, reaching 743 million in 1954. Afterwards, however, production began to increase rapidly. It reached 2.8 billion in 1969 and 1970. In the period 1955-1969, egg production increased by close to 300%. During the same period, Alabama broiler production increased from 57.8 million to 352.7 million, or 510%. Virtually all of the increase in egg production in Alabama during this period was the result of the increased demand for hatching egg sfor the production of broiler chicks.

Uses

In 1950, in Alabama, egg consumption by households on farms that produced them accounted for 30% of total production. Some 70% were sold; however by 1970, the percentage used for home consumption had declined to well below 1%. Obviously, home consumption had not really declined to that degree. The decline in the percentage was the result of the extremely large increase in eggs sold for hatching.

Agricultural Production

Data on total value added to the American economy by the production of crops and livestock during the period 1940-1970 are presented in Figure 14.126. These data show that total value increased from $31.3 billion in 1950 to $51.3 billion in 1970, an increase of 64%. The Consumer Price Index increased by 61.6% (72.1 to 116.2) during the same period—indicating that there was little real increase in total value added during the period. This lack of a real increase from 1950 through 1970 is in sharp contrast to the 1940-1950 period when the increase far exceeded inflation.

The larger share of the total increase in value added during the period 1950-1970 resulted from the production of livestock. It was higher than the value added from plant production throughout the period. The difference ranged from $2.9 billion in 1955 to $10.3 billion in 1970.

Data on the value added for American crops and livestock in 1950, 1955, 1960, 1965, and 1970 are presented in Table 14.18. In 1950 and 1955, cotton accounted for the larger share of total crop value, however, afterwards the value of that crop began to decline while the value of feed crops began to increase. The value for food grains did not change very much, whereas the value of oil crops did increase. This increase indicates the growing importance of soybeans in the country's agricultural economy.

The value of meat animals accounted for more than half of total value added for livestock in all years and its share increased throughout the period. Values for dairy products and poultry and eggs also increased.

Added value accounted for by home consumption was generally twice as large for livestock compared to crops, but both declined in real terms throughout the period. In 1970, the values of home consumption as a share of total value was 1.1% for crops and 1.8% for livestock. In 1910, these shares had been 11.6 and 23.1%, respectively.

Value added data for the contributions to the American economy by the production of crops and livestock in Alabama during the period 1950-1970 are presented in Figure 14.127. During the period, total value increased from $476.9 million to $789.3 million, an increase of 66%. Total value changed relatively little from 1950 through 1963. It increased slightly between 1962 and 1963, and then remained little changed through 1967. Most of the total increase for the entire period came after 1967 when the indices for prices received and the Consumer Price Index were both increasing rapidly.

At the beginning of the period, value added for crops in Alabama was slightly higher than for livestock; however, after 1956, the relationship was reversed. From 1956 onward, the value of livestock trended upward, while the value for crops remained at about the same level. In 1970, the value for crops was less than half of that of livestock; however these data are biased somewhat by the relatively poor crops of cotton in the late 1960s and in 1970 (Figure 14.49). The lack of an increase in the value of crops is very telling. Given the relatively large increase in the Consumer Price Increase, the real value added for crops declined at a very disappointing rate. Without the astounding increase in value added by poultry and eggs, agricultural production in Alabama during the period would have been really dismal.

Data on value added by different Alabama crops and livestock in 1950, 1955, 1960, 1965, and 1970 are presented in Table 14.19. These data indicate that cotton was the most important crop throughout the period, except in 1970 when bales harvested was much lower than expected as a result of adverse crop weather. The value of oil crops increased rapidly. In 1970, Alabama farmers were really getting excited about soybeans. It is obvious that they really needed some crop to get excited about.

In 1950 and 1955, value added by meat animals was considerably larger than that of poultry and eggs; in 1960, 1965, and 1970, the relationship was reversed. However, the value of meat animals in any given year is heavily dependent on the status of Cycles of Abundance in cattle and hogs in that year.

Probably the most significant change in value added during the period was the real decline in home consumption. In 1950, values of home consumption for crops and livestock as a share of their respective totals were 14.7 and 32.2%, respectively. By 1970, these percentages had declined to 4.2 and 2.3%. The sharp decline in the percentage for livestock was the result of the rapid growth of the broiler industry. In 1924, the value of crops consumed by families on farms that produced them was 19.1% of cash receipts for all crops. In the same year, the value of livestock consumed at home was actually larger than cash receipts.

STILL SERIOUSLY SEEKING SALVATION

In several preceding chapters, I have discussed some of the measures that farmers took beginning around 1895 to improve their lot; this effort, I have called Seeking Salvation. Unfortunately, none of the measures seems to have worked very well, if at all, except in war time. One of the measures of how well farmers did is the parity ratio. In the period 1910-1914, it was at a level of 1.00, and except when being buoyed-up by wartime economies, it has never been that high again. It was around 0.75 during the short recession in 1921 and 1922, 0.54 at the beginning of the Depression in 1932, and 0.72 when the great episode of agricultural modernization was winding down in 1970. Despite strenuous effort by the Triangle, no measures have been found that would get American farmers to produce only the amount of food and fiber that they could market for reasonable prices.

Salvation Through Better Information

Along with private efforts to modernize agriculture in the period 1950-1970, the public sector (both federal and states) significantly increased its support for providing more and better information to farmers through research and extension. In 1950, there was little argument that better information was the key to increased production of food and fiber.

Some data on the budgets of the Alabama Agricultural Experiment Station in fiscal years 1939-1940, 1949-1950, 1959-1960, and 1969-1970 are presented in Table 14.20. The data show that state funds for agricultural research at Auburn increased from $533,500 in 1950 to $2,644,700 in 1970, an increase of 396%. During the same period, federal funds increased from$331,500 to $1,522,100, an increase of 359%. The total of funds from both sources increased from $865,000 to $4,167,000. These data generally reflect the level of political pressure that farmers were able to bring to bear on elected officials to provide additional funds for new and better information through research. Considering all sources, funds for agricultural research increased from $1,383,100 to $7,083,400, an increase of 412%. A substantial share of this increase resulted for the sale of products produced in experiments. But even the funds derived from those sales are public funds.

With the additional funds, four new research departments were created at the main station at Auburn during the period 1950-1970. These included: Home Economics Research (1950), Animal Health Research (1954), Research Data Analysis (1957), and Fisheries and Allied Aquacultures (1970). Two new substation departments were also established during the period: the Ornamental Horticulture Field Station at Spring Hill in Mobile County (1951) and the Foundation Seed Stock Farm at Thorsby in Chilton County (1954).

Kerr (1985) commented that six new buildings related to the mission of the Alabama Agricultural Experiment Station were completed in 1961: Funchess Hall, the Biological Sciences Building; the Lambert Meats Laboratory; a new wing on the Animal Science Building; and two new buildings for the School of Veterinary Medicine (McAdory Hall and Sugg Laboratory).

During the period 1950-1970, the number of employees at the main station of the Alabama Agricultural Experiment Station at Auburn increased significantly. There were 129 in 1953, 164 in 1958, 176 in 1963, and 198 in 1968. Personnel employed on the substations increased from 20 in 1953, to 25 in 1958, to 22 in 1963, and to 25 in 1968.

Data on scientific man-years (SMYs) of research purchased by the Alabama Agricultural Experiment Station with appropriated state and federal funds in fiscal years 1950-1951, 1955-1956, 1959-1960, 1964-1965, and 1969-1970 in the different subject matter departments are presented in Table 14.21. The data were derived from budgets prepared by each of the departments and approved by the university before the beginning of each fiscal year. The data generally reflect the number of full-time equivalents of professorial-level positions in each department paid from research funds. For example, in 1950-1951 there were 4.40 full-time equivalents in the Department of Agricultural Economics paid from funds appropriated specifically for research in the various areas of agricultural economics. This total would have involved a number of assistant professors, associate professors, and professors in the department. For most of them, a substantial share of their salaries would have been paid from funds appropriated for teaching or funds derived from student tuition. The 4.40 total represents the accumulated shares of all of their salaries derived from appropriated research funds. None of these totals include any vacant positions. In many of the departments, there were vacant positions at the time the annual budgets were prepared, and it was not practical to determine when each of them might have been filled.

The data in Table 14.24 show that the total number of SMYs purchased for agricultural research increased sharply between 1950-1951 and 1955-1956 (43.86 to

70.20). This growth generally reflect the increased public support for agricultural research that followed the end of World War II. It also reflects the improved economy that persisted in the state following the war. There was also a marked increase between 1959-1960 and 1964-1965 (72.50 to 82.40). This increase is indicative of the overall modernization of agriculture in the state and country detailed in this chapter.

In 1950-1951, the larger share of research SMYs (10.68) was purchased in the Department of Agronomy and Soils. This emphasis was not unexpected. At that time, and for many preceding decades, the primary concern in Alabama agriculture was in crop production. In 1924, the share of cash receipts from the sale of crops accounted for 87% of sales total. In 1950, value added for crops produced in Alabama accounted for 55% of total value added. The share of SMYs purchased in that department continued to increase throughout the period, reaching 15.48 in 1969-1970, but the rate of increase was not nearly as great as in most of the other departments. In 1970, only 29% of total value added in Alabama was derived from the production of crops.

As expected, the growing interest in livestock production, especially in broiler production, resulted in sizable increases during the period in SMYs purchased in the Departments of Animal Science (3.66 to 10.48) and Poultry Science (2.80 to 5.96) between 1950-1951 and 1969-1970. Given the sharp increase in value added for poultry and eggs ($29.9 to $267.6 million), it is surprising that the increase in poultry science was not greater. The increase in research effort purchased during the same period in the Department of Zoology and Entomology was also substantial (4.43 to 12.46). This department was home to three relatively large programs other than zoology. It also included entomology, fish culture, and wildlife management. The increase in the department reflects growth in all of those areas. The increase in SMYs purchased in forestry (5.39 to 10.37) clearly indicate the growing importance in the forest industry in the state.

Budget data for the Alabama Cooperative Extension Service in fiscal years 1939-1940 through 1969-1970 are presented in Table 14.22. These data show that funds provided by the state ($539,300) and counties ($503,700) for agricultural extension work totaled $1,097,000 in fiscal year 1949-1950. In the same year, federal funds totaled $1.233,600. In fiscal year 1969-1970, the state ($3,021,600) and counties ($924,200) contributed a total of $3,945,800 for extension work, and the federal government contributed $3,301,400. Considering all sources, funds for extension increased from $2,383,200 in 1949-1950 to $7,547,000 in 1969-1970, an increase of 217%.

Data on the percentages of funds from all sources budgeted for different projects by the Alabama Cooperative Extension Service in fiscal years 1949-1950, 1959-1960, and 1969-1970 are presented in Table 14.23. The data for 1949-1950 and 1959-1960 are generally comparable. Unfortunately, by 1969-1970 the projects categories had changed considerably. It is difficult to determine how comparable these data are to the other two years. For example, the 1969-1970 budget did not separately list county agent work and home demonstration work. I have assumed that the funds for these two projects are included in county extension work. Further, it is difficult to identify all of the work done by subject matter specialists in the 1969-1970 budget. Further, there were two new projects: rural resource development and production and development. Assuming that the data are more or less comparable, it is obvious that most of the funds were expended for work in the counties. In the three fiscal years, the percentages ranged from around 70 to 80%. Club work remained at around 1% throughout the period.

Table 14.24 shows data on the number of different classes of professional personnel employed by the Cooperative Service in 1940, 1950, and 1971. There are no data available for 1960 and 1970. These data show that total personnel increased 58% between 1950 and 1971 (444 to 700). A large share of this increase was the result of employing more subject matter specialists (39 to 81). County agents only increased by eight persons (384 to 392).

Combining the totals in Tables 14.23 and 14.25, the Cooperative Extension Service and the Agricultural Experiment Station budgeted a total of $3.8 million for extension and research in 1949-1950 and $14.6 million in 1969-1970, an increase of 288%, or about 13.7% a year. The Consumer Price Index only increased 61% during this period. A large share of those extension and research funds were provided by the public sector. This show of public-sector commitment to purchasing more and better information for Alabama's farmers through extension and research is astonishing. It is obvious that the modernization that took place in Alabama agriculture during this period carried over into public support.

Morrison (1994) comments that in 1952 agricultural teaching and research were included in the Division of Agriculture of the Alabama Agricultural and Mechanical College (Alabama A&M). Included in the division were: agricultural education, animal husbandry, poultry, agronomy, horticulture and ornamental horticulture, and vegetables and fruits. Several new agricultural buildings were constructed during this period including: a greenhouse, silo, breeder house, six laying houses, and a brooder house. In 1969, Alabama A&M College's name was changed once more—to Alabama Agricultural and Mechanical University.

In 1967, the contribution of the country's 1890 institutions (Black colleges) to agricultural research was recognized when Congress, under PL 89-106, appropriated $283,000 for their use. Each of the colleges received $10,000, and the remaining funds appropriated under the act were allocated based on the formula used for the disbursement of Hatch Act Research Funds. Funding remained at the same level from 1967 through 1971.

In 1963, the Department of Records and Research established at Tuskegee in 1904 was reorganized as the Department of Social Science Research with the purpose of studying race relations and other regional problems. In 1967, that department became part of the Carver Foundation. In 1970, this area of work was reorganized, once again, as the Division of Behavioral Science (Mayberry, 1989).

As a result of a federal directive to eliminate racial discrimination in education, the State of Alabama was ordered to end efforts to maintain separate-but-equal extension programs. As a result, in October 1965, the entire extension staff at Tuskegee was relocated to Auburn.

A review of some of the improvements in agricultural production in Alabama during this period (1950-1970) indicates that the tax-paying public received a very good return on their investment in research and extension. For example:

1. The average market-weight of broilers increased from 2.7 to 3.5 pounds.
2. Broiler feed efficiency declined from 3.50 to 2.25.
3. Milk-per-cow increased from 3,700 pounds to 6,700 pounds a year.
4. Cotton yields increased from 300 to 500 pounds an acre.
5. Corn yields increased from 20 to 40 bushels an acre.
6. Oat yields increased from 25 to 40 bushels an acre.
7. Soybean yields increased from 22 to 27 bushels an acre.
8. Farmers increased their fertilizer use from 365 to 764 pounds an acre of cropland harvested.

While there is obviously no one-to-one correlation between funds expended for research and extension and improvements in agricultural production, these data indicate that extension and research personnel in Alabama were extremely effective in using public funds to accomplish their assigned responsibilities. Unfortunately, these accomplishments do not tell the entire story. A comparison of these improvements with those in the country as a whole indicate that while research and extension personnel were effective in helping Alabama farmers keep up with their national counterparts, they were not able to help them catch up. For example, milk per cow increased from 3,700 to 6,700 pounds a year in Alabama, but nationally it increased from 5,200 to 9,800. Alabama dairymen were keeping up, but they were not catching up. Similarly, Alabama's corn yields increased from 20 to 40 bushels an acre. Nationally, they increased from 39 to 80 bushels. If state data were to be compared with other individual states, such as Wisconsin's milk production and Iowa's corn production, Alabama deficiencies would be even more striking.

A more telling comparison is in net farm income. In the period 1950-1970, the NFI per Alabama farm increased from $1,207 to $2,751. The national average increased from $2,414 to $4,871. Alabama farmers were keeping up, but they were not catching up.

Helping Alabama farmers keep up, but not catch up, might have been marginally acceptable to extension and research personnel and their administrators, but it was unequivocally unacceptable to farmers, especially in the long-term. They were purchasing more farm machinery and equipment, using more fertilizer, purchasing better seed, applying more chemicals, and using better cultural practices. However, all the other farmers in the country were doing the same. This unfortunate situation was not lost on a majority of Alabama farmers. There were 220,000 of them on the land in 1950 and only 82,000 in 1970—a loss of 62.7%. During the same period, the state lost 30.5% of its farmland. In the country, comparative losses were: 47.8 and 8.3%, respectively.

Alabama research and extension personnel were caught between a rock and a hard place. No matter how smart they were or how hard they worked, there simply was no way that they could help their farmers overcome the resistance to catching up imposed by an inferior geophysical agricultural environment (erratic weather and poor soils) and by having to use plants and animals poorly adapted to the sub-tropics. Apparently in Alabama, Salvation could not be achieved by more and better information. The basic problem was, and always will be that it's nigh impossible to get blood out of a turnip.

Salvation Through Organization

Several organizations were established during the period 1950-1970 that would play important roles in advancing Alabama agriculture. These included:

1. The American Dairy Association of Alabama. Its purpose was to undertake activities designed to promote greater sales of milk and milk products.
2. The Alabama Nurserymen's Association, established in 1951. Its purpose was to establish a professional association for those involved in and allied to the ornamental horticulture industry.
3. The Alabama Poultry Industry Association, established in 1953. Its purpose was to provide an umbrella group for all other organizations serving the poultry industry. Under this umbrella, the Alabama Poultry Processors Association, the Alabama Feed Association, the Turkey Growers Association, and the Poultry Producers Association were established the same year. The Alabama Egg Association was organized in 1957-1958.
4. The Alabama Poultry Industry Association, reincorporated in 1962. It was an umbrella organization for the Alabama Hatchery Association, the Alabama Feed Association, the Alabama Poultry Processors Association, the Alabama Poultry Producers Association, the Alabama Egg Association, and the Alabama Turkey Association.
5. The Alabama Council of Cooperatives was incorporated in 1957, to promote the welfare of cooperatives in Alabama.
6. The Alabama Turfgrass Association, established in 1961. This organization united turfgrass management personnel of Alabama and northwest Florida.
7. The Alabama Soybean Association, was established in 1968. It was organized to provide coordination services for all persons involved in producing, marketing, distributing, and using soybeans.

Salvation Through Diversification

As detailed, in a preceding chapter, farm diversification was one of the primary goals of Alabama agricultural reformers, beginning in the late nineteenth century. It was never a matter of the variety of crops that Alabama farmers produced. From the beginning, farmers in the state produced a wide variety of plants and animals. The problem was that farmers were committing too many of their scarce resources to a single crop—cotton. Data in Table 14.19 indicate that in 1950 Alabama farmers were producing relatively large quantities of a variety of both plant and animal crops. Cotton, however, remained the stack pole of the state's agricultural economy, contributing 36.4% of total value added of all agriculture commodities, excepting farm forest products ($130.8 million/$359.4 million). However, by 1970, the situation had changed drastically. In that year, the value added by the cotton crop represented only 8.4% ($64.7 million/$765.3 million) of the total value of all commodities. If reducing the relative contribution of cotton to Alabama's agricultural economy is a suitable measure of diversification, the state's farmers finally achieved that goal during this period of the modernization of its agriculture (1950-1970).

Salvation Through Political Action

As discussed in a preceding chapter, the level of government payments (subsidies) to farmers in any one year provides a useful measure of the intensity, real or imagined, of the political pressure being applied on the Triangle to fix agriculture. For example, during World War II, demand and prices for agricultural commodities were so good that there was little political pressure to provided payments to farmers. They were faring extremely well without subsidies; however, the need was so great for additional commodities that the federal government provided farmers with subsidies anyway to obtain additional production.

As a result of the strong agricultural economy associated with World War II and the Korean Conflict, parity ratios remained around 1 in the early 1950s; consequently, demand for government subsidies in support of agriculture remained muted. As a result, payments to farmers through 1955 remained at postwar levels of around 50 dollars a farm annually. The payments that they did receive were primarily those authorized by the Soil Conservation and Domestic Allotment Act of 1935.

Unfortunately, the parity ratio fell below 1 in 1953, and farmers were quickly caught-up in a rapidly deteriorating cost-price squeeze. By 1954, the ratio was down to 0.88, and the clamor for federal intervention exploded. If the level of Triangle response is any measure of the intensity of political pressure applied, it must have been massive. The politicians responded by enacting the Agriculture Act of 1954 (flexible price supports), the Agriculture Act of 1956 (Soil Bank), the Food and Agriculture Act of 1962, the Agricultural Act of 1964, the Food and Agricultural Act of 1965 (the first multi-year farm legislation), and the Agriculture Act of 1970. Most of the provisions of these acts were designed to promote supply management in the agricultural economy (Outlaw and Klose, 2001). The nature of each of these various acts were discussed in a preceding section.

In 1950, payments per farm in the United States were $52, and in 1955 they were $49 (Figure 14.128). However, as a result of the provisions of the Soil Bank Act and the Wool Act, they increased to $256 per farm in 1956. Payments remained near that level until 1961 when payments for feed grains and wheat were added. Payments for cotton were added in 1964. By 1965 and 1970, payments were $734 and $1,260, respectively. Payments increased from $49 to $1,260 per farm in the period 1955-1970. Unfortunately, as payments increased, parity ratios declined. In 1970, with payments at $1,260 a farm, the ratio had fallen to 0.72, the lowest level ever recorded at that time. Seemingly, the Triangle had completely lost control of American agriculture. This should have given them pause, but it did not. Brer Rabbit just hit harder.

Figure 14.128 also presents data showing the extent of federal government payments to Alabama farmers during the period 1950-1970. Data on payments received under the provisions of the various acts, for the various states, are not available before 1949. As might be expected, trends in Alabama payments followed the same general pattern as for payments at the national level. From 1950 through 1956, Alabama payments also ranged between $25 and $49 a farm. They did not begin to increase for Alabama farmers until 1957. From 1960 through 1965, payments were much lower than for the country as a whole. Payments for feed grains and wheat, which were phased-in in 1961, had only limited effect on payments coming to Alabama. However, after payments for cotton were phased-in in 1964, the amount coming to the state also began to increase. Between 1965 and 1966, national payments for cotton increased by over tenfold. With the high level of cotton production in the state, payments increased sharply. They were $364 a farm in 1965 and $838 in 1966. Alabama payments reached $965 in 1966, the same level as the country. Afterward, however, they remained essentially unchanged through 1970 when acres of cotton harvested declined sharply.

The end of this period (1950-1970) is an appropriate time to consider what American farmers have obtained from their attempts to purchase Salvation Through Political Action. Between 1933 and 1951, food and fiber production in the country were heavily influenced by war. For much of that period, their Salvation was in the hands of international events. But afterwards, the country entered a time of peace, so it is finally time to look closely at what political action has accomplished in two decades of intense effort.

The modernization of virtually all of American agriculture was finally completed in the period 1950-1970. It also represented the most active period, thus-far, in Triangle efforts to manage American agriculture, especially supply. Several pieces of legislation were enacted to reduce commodity surpluses, but none of them seemed to work. All of their efforts were being swamped by a rising tide of agricultural productivity (Figure 14.8). Bowers, et al. (1984), writing for the Economic Research Service of USDA, commented that "by 1960 the surplus problem had reached crisis proportions."

Uncontrolled surpluses drove food prices downward relative to costs. In 1950, the cost of food for consumers accounted for 20.6% of average disposable income. By 1970, the percentage had declined to 13.9% (Ahearn, 2008). The farmer's share of the food-dollar was 41 cents in 1950, but by 1970 it was down to 31 cents (Elitzak, 2008). In 1960, 52.8% of total farm household income was obtained from off-farm sources. By 1970, the percentage had increased to 63.1 (Ahearn, 2008). It is indeed fortunate that the national economy was growing fast enough to absorb the uncompensated labor pouring off the country's farms; otherwise the farm crisis would probably have lead to civil unrest. This off-farm-income safety-valve provided the Triangle with the cover they needed to continue to attempt to manipulate the country's agriculture. Triangle policies, with their largely uncontrolled surpluses, were quickly resulting in production of the cheapest food in the world, but they were rapidly eliminating farming as a livelihood. Thomas Jefferson, the advocate of an agricultural country of small farmers, was probably spinning in his grave.

Unfortunately or fortunately, depending on point of perspective, American farmers in 1933 had purchased a largely unrevocable contract as a partner in Dobson's "Triangle of beneficiaries." At this point, about the only thing that they could do was to: hire a lawyer to interpret the fine print of Triangle rules and regulations, hope for the best, and put on their best suit to go look for a part-time job.

The Bottom Line

Data on American net farm income (on a per farm basis) for the period 1940-1970 are presented in Figure 14.129. It was $2,414 in 1950, but increased sharply in 1951 after prices received began to increase (Figure 14.6). Afterwards, however, the net farm income per farm followed prices received down through 1955, then upward again through 1970. After 1964, it began to increase at a faster rate than prices received as both the Gross National Product and the Consumer Price Index began to increase more rapidly (Figures 14.1 and 14.2). In 1970, it reached $4,871 per farm. It had increased by 101% in the 21-year period. The Consumer Price Index had increased by only 61% during the period.

After 1932, net farm income included government subsidy payments. In 1950, the net income per farm of $2,414 included $50 per farm in payments. Without payments, it would have been $2,364. That year, payments comprised 2.1% of net farm income per farm. Similarly, in 1970 net farm income per farm, with and without payments, was $4,871 and $3,611, respectively. In 1970, payments comprised 25.9% of net income per farm. Net income per farm without payments increased from $2,364 to $3,611, an increase of 52.7%. If these data had been used, the increase would have been less than the increase in the cost of living index during the period (61 versus 53). The most important portion of net farm income per farm was being provided by American tax payers under the supervision of the Triangle.

The increase in agricultural productivity (Figure 14.8) of 46% (0.460 to 0.672) would have also affected the bottom line during the period. This increase would have made the declining parity ratios (Figure 14.6) somewhat more tolerable.

No net farm income data was available for Alabama before 1949. Data for the period 1950-1970, calculated on per farm basis, are shown in Figure 14.130. As the figure shows, there was considerable year-to-year variation, but it generally trended upward throughout the entire period. It was $1,207 in 1950 and $2,751 in 1970. The large year-to-year variation was primarily the result of the year-to-year variation in total value added (Figure 14.127). For example, between 1953 and 1954, net farm income per farm declined from $1,712 to $1,179. During the same period, total value added declined from $564.7 million to $455.6 million. In turn, the decline in total value added was largely the result of a much smaller than expected cotton crop (963,000 versus 728,000 bales) in 1954.

Alabama's net farm income per farm increased by about 128% in the 1950-1970 period, but considering the change in the Consumer Price Index, only about half of that increase was real. Unfortunately, while Alabama's net farm income per farm increased sharply in relative terms, Alabama farmers were in about the same position in 1970 as they had been in 1950. Table 14.20 presents data for several other states in 1950 and 1970. Alabama had been at the bottom of the list in 1950, and except for Tennessee, it was still at the bottom of the list in 1970. Of even greater concern is the fact that the rate of increase for most of the states on the list was much higher than for Alabama.

Net income per farm for Alabama also included subsidy payments. Without payments, Alabama's net farm income per farm was $1,170 in 1950 and $1,782 in 1970. Using these data, the rate of increase between 1950 and 1970 would have been 52.3%, a value considerably lower than the increase in the Consumer Price Index.

It is amazing to contemplate the degree of modernization that took place on American farms during the period 1950-1970. Much of it was fueled by the growing economy that emerged from the war years. Data in Table 14.21 summarizes some of the many changes that took place. Some of these data are taken from the agricultural census, and some are taken from annual USDA reports. As a result, not all the periods involved were for 1950-1970, but they are for that period unless indicated by a superscript letter. The data show that agricultural productivity increased from 0.460 to 0.672, an increase of 46%; the number of tractors per 100 farms increased from 67 to 162, over a twofold increase; the value of farm machinery and equipment increased from $1,105 to $9,970 per farm, almost a ninefold increase; the value of land and buildings increased from $65 to $196 an acre, a threefold increase; and the use of commercial fertilizer increased from 108 to 285 pounds an acre of cropland harvested, a twofold increase.

These were truly amazing changes. Unfortunately, added investments in machinery and equipment and in land and buildings did not actually result in increased net farm income. Between 1945 and 1969, total investments increased from $2,064 to $4,764 per farm, just over a twofold increase; however, without government (tax-payer) subsidy payments, net farm income per farm actually decreased during the period when considering inflation of the currency. This is not really surprising. The increase in total value added for the period was also lower than the increase in the Consumer Price Index (55.3 versus 61.4%).

The problem, of course, was declining parity ratios (Figure 14.7). It was 1.09 in 1945, but by 1969 it had declined to 0.73. No amount of investment in farm infrastructure could completely counter this rapidly deteriorating situation. Apparently it did result in some improvement in productivity, but that increase was not large enough to counter rapidly rising prices paid indices (Figure 14.6). In fact, it is likely that the increase in productivity was increasing the magnitude of the problem. In 1940, American farmers had produced more food and fiber than they could market at reasonable prices, when the per farm value of machinery and equipment was $502, and the value of land and buildings was $5,298. There was simply no way that increased investment in infrastructure alone was going to solve the problem of oversupply. American agriculture was finally all dressed up, but with nowhere to go.

Data on some changes in Alabama agriculture during the period 1950-1970 are presented in Table 14.22. These data show that the changes in Alabama agriculture were even more striking than in the country as a whole: The number of tractors increased from 21 per 100 farms in 1950 to 103 in 1970, almost a fivefold increase; between 1945 and 1969, the value of machinery and equipment increased from $256 to $5,909 per farm, a 23-fold increase; between 1945 and 1969, the value of land and buildings increased from $2,506 to $37,569 per farm, respectively, almost a 15-fold-increase; and the use of commercial fertilizer increase from 267 pounds an acre of harvested cropland to 685 pounds, almost a threefold increase.

The return on investment in farm infrastructure was even more dismal in Alabama than in the country. Between 1945 and 1969, the combined value of machinery and equipment and land and buildings increased from $2,762 to $43,478 per farm, almost a 16-fold-increase. Unfortunately, between 1949 and 1969, net farm income for the state increased from $1,154 to $2,941 per farm. Alabama farmers had increased their investment in infrastructure by almost $40,000 per farm, but the net income had only increased around $1,500.

ALABAMA AND HER SISTER STATES IN 1970

The US Bureau of Economic Analysis reported that the Gross Domestic Product for Alabama in 1970 was $3.4 billion. On a per capita basis it was $3,616, the third lowest in the country. Only the GDPs for Arkansas ($3,461) and Mississippi ($3,279) were lower. It is difficult to comprehend how Alabama— with a world-class seaport, an unmatched navigable river system, plentiful timber, and abundant water supplies, as well as large deposits of coal, iron ore, and limestone—would have a GSD that was the third lowest in the country in 1970. At the very beginning of this book, I posited that Alabama's relatively lowly economic status had been the result of the extremely slow pace with which an economy progressed through Lenski and Nolan's (1998) Simple Agriculture and Advanced Agriculture stages of development and emerged into their Industrial Stage. To test this hypothesis, we have followed the progress of Alabama's agricultural development and those of three of its sister states at intervals from 1860 onward. As detailed in a preceding chapter, Mississippi, Illinois, and Indiana were chosen as sister states because they, along with Alabama, were admitted to the Union within a few years of 1819, and all of them were located on the frontier when admitted.

Table 14.28 includes data on some agricultural, industrial, and commercial characteristics of these four states in 1970. While lying with numbers is a well known argument tactic, it would appear that the data presented in the table would support the conclusion that by all those measures cited, agriculture, commerce, and manufacturing in Alabama and Mississippi have lagged behind those in Illinois and Indiana, and in some of those characteristics, they have lagged far behind.

The histories of all four of these states began at about the same time, and generally under the similar circumstances. In 1819, the populations of Alabama, Mississippi, Illinois, and Indiana were: 116,016; 71,034; 50,918; and 135,398, respectively. Alabama and Indiana likely had the larger populations because they were both located closer to Eastern population centers.

Even after only about 40 years (1860) of statehood, differences in agriculture, commerce, and industrialization in Alabama and Mississippi and in Illinois and Indiana were already appearing (Table 6.37). The highly successful mixed farming of corn, hay, hogs, and cattle in the Ohio Valley provided a solid base for the establishment of urbanization, commercialization, and industrialization for people immigrating from the East. Meanwhile, by 1860, the failed agricultural economies of the old Southern colonies had been carried westward into the valleys of the Tennessee, Alabama, Tombigbee, and lower Mississippi. In 1860, there were 435,000 slaves in Alabama and 437,000 in Mississippi, and cotton sold for 12 cents a pound. As Antonio commented in Shakespeare's *The Tempest,* "What's past is prologue." Certainly the relative situations in the four sister states had not changed very much by 1970.

15

It Was the Best of Times, It Was the Worst of Times

1970–1985

PRECIS

American farmers began to make a serious effort to modernize their farms during the period 1950-1970. They had accumulated a considerable amount of money during the wartime emergency, and beginning in the early 1950s, there were many new things to purchase to improve their operations. As a result, they went on a buying spree that would increase in intensity into the early 1970s. Unfortunately, they had little to show for their efforts. The parity ratio had fallen steadily throughout this period, but suddenly in 1972 the ratio began to increase, and in 1973 it rose to 0.91. All the investment in modernization seemed to be beginning to pay off. Now it would be just a matter of time, but just as quickly Demeter's (Greek goddess of agriculture) warm smile became an angry frown. In 1973, the Organization of Petroleum Exporting Countries (OPEC) began to restrict the sale of crude oil to the West, and in the fourth quarter of that year, the country entered a serious recession. In 1974, a pall began to cover the country's fields, pastures, and feedlots once more. The parity ratio began to decline, and it trended steadily downward to reach 0.5 in 1985. The worst of times had come again. It appeared at this point that the Triangle had, with good intentions, decreed that like Sisyphus, American farmers were fated to push the same stone up the same mountain forever.

The declining parity ratio was only one of the problems bedeviling the Triangle and American agriculture in the period 1970-1985. Surpluses of government-owned commodities were reaching unsustainable levels. Farm household income was increasingly being sustained by off-farm work. Fewer and fewer young people were becoming farmers; rural communities were being devastated. Problems related to Triangle management of agriculture had been mounting for years, but now the situation was becoming a conflagration. It was obvious at this point that they had tried, with no avail, about every solution allowable under the US Constitution, but they had one more ace—a black one—up their sleeve: stand back and throw more money at the problems. Between 1980 and 1983, government payments (dollars per farm) increased from $527 to $3,909, a sixfold increase. In 1983, they amounted to 65% of total national net farm income. At this point, the Triangle had, with the use of taxpayer dollars, taken almost full responsibility for the livlihood of American farmers, but alas all was not lost. From 1970 through 1985, the share of consumer income spent for food fell from 13.9 to 11.7%, and the amount of food that could be given to the ever-increasing numbers of needy people was increasing year by year.

In a relative sense, Alabama farmers fared no better in the period 1970-1986 than they had before. Annual yields of all major crops and milk per cow were still much lower than the national average. They continued to harvest crops from a much smaller share of total cropland. Alabama's net farm income (dollars per farm) was among the lowest in the country. Finally, without the growing contribution of pulpwood, broilers, and hatching eggs to the state's agricultural economy, Alabama would have been considerably more dismal than it was.

INTRODUCTION

As detailed in the preceding chapter, in the period 1950-1970 farmers throughout the United States made unprecedented progress in modernizing their farms. In that period, the value of farm machinery and equipment on a per-farm basis increased from $2,258 to $11,546, a fourfold increase. The increase in Alabama was even more amazing. In 1945, the value on a per-farm basis was only $256. In 1969, it reached $5,909, a 22-fold increase. About 15% of the national increase was the result of inflation, but only about 5% of the increase in Alabama was the result of rising prices. Farmers had finally found their groove.

During the period 1950-1970, indices of prices paid for all farm inputs increased from 256 to 382, or about 50%. Unfortunately, indices of prices received only increased from 256 to 258. Farm modernization allowed farmers to rapidly increase productivity. Unfortunately, it also allowed them to produce more food and fiber than they could sell at reasonable prices. In response to continuing problems with farm income, the Triangle continued to erect more-and-more scaffolding around agriculture. In 1950, federal subsidy payments for all programs totaled $52 a farm. By 1970, the total had increased to $1,260. In Alabama, the total increased from $37 to $969. Obviously, these payments helped. In 1970, 25% of the national net farm income came from federal payments. In Alabama, 39% was derived from that source. In 1970, it was obvious that American farmers were still seeking the Salvation that they began seriously to search for at the end of the nineteenth century.

NATIONAL POLITICS

In 1950, the North Koreans invaded South Korea, and the United States quickly became involved in the first hot war with a communist country. This event began a long period of often hostile confrontation with communism at home and around the world. At home, there was great concern with the possible subversion of the American government by communist agents. Then there was the Berlin Blockade and the Cuban Missile Crisis. The United States had avoided getting involved in the French war with the communists in Southeast Asia, but soon found itself deeply involved fighting them on the same battleground. These battles, both hot and cold, left the country sharply divided along political lines. Amid these efforts to contain the communists, some American politicians began to consider the possibilities of breathing new life into Roosevelt's New Deal. These efforts resulted in the establishment of a wide range of new government programs under the sponsorship of the Kennedy's New Frontier and Johnson's Great Society. While these new programs were of great benefit to many of the country's disadvantaged citizens, they added new fault lines to an already fragile political system.

The Presidency

Republicans held the presidency for most of the period, 1970-1985. Republican Richard Nixon of California, elected in 1968, held the office until resigning in August 1974. He was replaced by his vice president, Gerald Ford (R. MI), who served until 1977. Other persons elected to the office include: Jimmy Carter (D, GA) (1977- 1981) and Ronald Reagan (R, CA) (1981-1989).

Secretaries of Agriculture

Clifford Harding (R, IN) served as Nixon's secretary of agriculture from 1969 until 1974 when he was replaced by Earl Butz (R, IN). Other persons serving in that position during this period (1970-1985) include: John Knebel (R, VA) (1976-1977); Bob Bergland (D, MN) (1977-1981); and John Block (R, IL) (1981-1986).

The Senate

The Senate was controlled by the Democrats from

1970-1981. The Republicans took control of the body in 1981 and remained in control until 1987 (Table 15.1).

Senate Majority Leaders

Mike Mansfield (D, MT) was chosen as Senate majority leader in 1969 for the 91st Congress. He continued in that position until 1977. Others elected to the position include: Robert Byrd (D, WV) (1977-1981) and Howard Baker (R, TN) (1981-1985).

Chairmen of the Senate Committee on Agriculture and Forestry

The Senate Committee on Agriculture was re-named the Committee on Agriculture and Forestry in 1884. The name remained the same until 1977, when it became the Committee on Agriculture, Nutrition and Forestry.

Allen Ellender (D, LA) was chosen as chairman of the committee for the 91st Congress in 1969. He served until 1971. Other persons elected to the office include: Herman Talmadge (D, GA) (1971-1981) and Jesse Helms (R, NC) (1981-1987).

House of Representatives

Democrats controlled the House for the entire period (1970-1985) (Table 15.1).

Speakers of the House

John McCormack (D, MA) (1963-1971) was speaker when this period began in 1970. He had been chosen for that position for the 88th Congress in 1963. Other persons chosen for that office include: Carl Albert (D, OK) (1973-1977) and Thomas O'Neil (D, MA) (1977-1987).

Chairmen of the House Committee on Agriculture

The chairman of the House Committee on Agriculture in 1970 (91st Congress) was William Poage (D, TX). He served in that position until 1975. He was followed by Thomas Foley (D, WA) (1975-1981) and E. "Kika" de la Garza (D, TX) (1981-1995).

The Political Situation

Richard Nixon (R, CA) defeated Hubert Humphrey (D, MN) in the 1968 election for the presidency. Four years later, he defeated George McGovern (D, SD) in the election of 1972—by the largest margin in American political history. Once in office, he immediately began to develop a strategy for ending the country's involvement in the Vietnam War. After his visit to South Vietnam in 1969, American involvement began to decline, and by 1973 there were no American troops remaining there. Subsequently, when Nixon began to improve relations with China, it became much easier to work with the Soviet Union. As a result, in 1972 the United States and the Soviet Union signed the SALT I Treaty, which moved the two countries closer to nuclear peace. Nixon also strongly supported Pakistan in the Indo-Pakistan War of 1971—as a means of tempering the effects of Russian-Indian collaboration on the subcontinent. Nixon authorized an airlift of arms to Israel in the Yom Kippur War of 1973, saving Israel from possible defeat at the hands of the Arab coalition.

On the domestic front, Nixon's policies tended to be centrist or even liberal. He imposed wage and price controls in an attempt to slow the growth in inflation following the end of the Vietnam War. He was also responsible for indexing Social Security for inflation and creating the Supplemental Security Income (SSI) program. He was responsible for establishing the Environmental Protection Agency (EPA), the Occupational Safety and Health Administration (OSHA), the Drug Enforcement Administration, and the Office of Minority Business Enterprise. He initiated the first significant, federal affirmative action program, and worked diligently to achieve the first large-scale integration of public schools in the South.

After large numbers of American troops began to arrive in Vietnam in 1966, inflation in the United States had begun to increase rapidly. When Nixon took office in 1970, the inflation rate was 6%. Given his efforts to expand the size of the federal government during this period, he apparently gave little attention to this growing problem, and when he resigned from the presidency, the inflation rate had increased to 11%.

Nixon used his enormous victory in the 1972 election to enlarge the powers of the presidency at the expense of Congress. During this period, he impounded billions of dollars worth of federal spending by expanding the powers of the Office of Management and Budget. This expanded use of presidential prerogative probably added to his problems in trying to handle the Watergate situation, which began with the break-in in June 1972. The affair, while smoldering at first, finally broke into an open flame and finally into an inferno for President Nixon, forcing him to resign in August 1974.

Vice President Gerald Ford (R, MI) assumed the presidency in August 1974. Shortly thereafter, in the mid-term elections of 1974, the American electorate, sickened by the Watergate scandel, gave the Democrats a large majority in both houses of Congress (Table 15.1). As a result, few of Ford's initiatives received any attention, and when he vetoed legislation that they passed, they quickly over-rode his veto. Probably the most important contribution of his administration was the approval of the Helsinki Accords in August 1975. This agreement significantly reduced Cold War tensions in Europe. His most seriously debated act was his pardon of Nixon in September 1974. It was obvious to Ford, if it had not been to Nixon, that inflation was a rapidly growing problem for the country. He attempted to deal with it by jaw-boning the public, but little else. With inflation still near 6%, and the memory of Watergate still fresh in the electorate's mind, Jimmy Carter (D, Ga) defeated Ford in the 1976 presidential election.

Carter was viewed as an outsider in Washington, and his feuds with Democratic leaders in Congress made it difficult for him to get much of the legislation that he wanted. Nevertheless, he was able to get some important environmental legislation enacted and to get some improvements made in Social Security. He was also responsible for the formation of the Department of Energy and for the deregulation of the trucking, airline, rail, finance, communications, and oil industries.

Carter's major international accomplishments included: the Camp David Accords of September 1978 that provided some important steps toward Arab-Israeli peace in the Middle East, the Panama Canal Treaties that gave control of the canal to Panama, establishment of full diplomatic relations with the People's Republic of China, and agreement on a second version of the SALT Treaty. He had much more difficulty in dealing with the Soviet invasion of Afghanistan and the Iranian hostage crisis.

As noted in a preceding paragraph, inflation had begun to increase in the latter part of the second Johnson Administration, but neither Johnson, Nixon, nor Ford were able to devise effective plans to control it. Then with Carter's difficulty of dealing with Congress, the problem was allowed to grow even faster. When Carter took office in 1977, the inflation rate was 6% and rising. By 1979, it was 11%. With inflation, interest rates, and unemployment rising, the so-called misery index also began to increase rapidly. With the misery index rising and the Iranian hostage crisis unresolved, Carter lost the 1980 election to Ronald Reagan (R, CA).

In Reagan's 1980 landslide election, the Senate was also returned to Republican control, and the Democratic control of the House was reduced substantially. Reagan also benefited from some of the efforts begun by Carter. The Iranian hostages were finally released the day after Carter left office. Also, the increase in interest rates, which intensified during Carter's administration, finally broke the back of inflation. In May 1981, soon after Reagan took office, the interest or discount rate reached 14%. In 1980, the inflation rate had been between 13 and 14%, but in 1981 it fell to slightly above 10%, and it continued to decline rapidly in the following years. Carter's misery became Reagan's "morning in America."

With these increasingly strong winds of change at his back, Regan moved aggressively into his agenda. He quickly reduced the federal income tax by 25%, and with inflation declining, he engineered a decrease in interest rates. He also escalated the Cold War with the Soviet Union by significantly increasing military spending. He rejected *détente*, calling instead for the destruction of communism. These changes resulted in a dramatic increase in deficits and the national debt, but they also resulted in a positive change in the mood of the country. Consequently, the economy began to grow impressively.

Reagan was unable to get the changes in federal welfare programs and in abortion rights that he wanted, but he was able to push the federal courts in a more conservative direction by the judicial appointments that he made.

With positive changes in national affairs all around him, Reagan was re-elected in 1984 with a 49-state landslide in his contest with Walter Mondale (D, MN).

ALABAMA POLITICS

In the preceding period (1950-1970), Alabama was beset by the most tumultuous social and political upheaval in its history. It included the outbursts resulting from the Supreme Court decision on Brown v. Board of Education in 1954, the Rosa Parks arrest on a bus in Montgomery in 1955, the attack on the Freedom Riders in Montgomery in 1961, the bombing of Civil Rights targets in Birmingham, and the Selma to Montgomery March in 1965. The unparalleled period of racial turmoil finally came to a head in 1965, when Congress enacted the Voting Rights Act. By that action, the American people, through their elected representatives, said "enough already." There were still many bitter days to come, but Alabama had been forced to turn the corner.

Governors

Lieutenant Governor Albert Brewer (D, Morgan) became governor on May 7, 1968 when Governor Lurleen Wallace died. Brewer was defeated in the 1970 election by George Wallace (D, Barbour). Wallace was shot while campaigning for president in Maryland in May 1972. Lieutenant Governor Jere Beasley (D, Barbour) served as acting governor from June 5 through July 7, 1972 while Wallace was convalescing. Forrest "Fob" James (D, Lee) became governor in 1979 and served until 1983. He was succeeded in the office by George Wallace.

Commissioners of Agriculture

Richard Beard (D, Jefferson) served as commissioner of agriculture from 1967 until 1972. He was succeeded by M.D. "Pete" Gilmer who served until 1974. McMillian Lane (D, Montgomery) was commissioner from 1974 through 1972, and Albert McDonald (R, Madison) from1983-1991.

The Political Situation

In 1970, in an extremely heated election, former Governor George Wallace defeated the incumbent Governor Albert Brewer by a narrow margin. In 1972, Wallace again entered the presidential primaries, this time as a Democrat. During the campaign, he was shot by a would-be assassin resulting in complete paralysis in both legs. While he was recovering from the wound, Lieutenant Governor Jere Beasley, became acting governor. Beasley serve in this capacity for a month in mid-1972. With Wallace's quest for the presidency ended, he returned to state politics a changed man.With no more national windmills to fight, he threw himself into bettering Alabama with a vengeance, even though he was severely handicapped. He easily won the 1974 election. During these successive administrations, Wallace was able to get the legislature to provide funds for the largest expansion of highway construction in the state's history, record appropriations for education, vital improvements in the Alabama Law Enforcement Agency, improved health care, and old age pensions. He also was able to establish the Alabama Office of Consumer Protection. Albert Brewer had begun the task of lifting Alabama out of its obsessive parochialism, and a severely handicapped George Wallace worked incessantly to keep that movement on track.

In the 1978 election, Alabama's electorate seemed to be looking for an end to the cronyism in state government that had evolved over the years in the different Wallace Administrations. As a result, they turned to a candidate who had never held elective office. In the election, Forrest "Fob" James (D, Lee) defeated the experienced politician Bill Baxley (D, Houston) in the primary. James had been a Republican earlier, but changed parties to campaign as a born-again Democrat. He easily defeated the Republican candidate, Guy Hunt (R, Cullman) in the general election. He had the unfortunate experience of entering office just as the deep recession of 1979-1982 was getting under way. He was able to achieve some worthwhile goals during his term, but most of his efforts had to be given to making ends meet. Once in office, he had to cut state spending by 10%, institute a hiring freeze, and actually layoff a considerable number of state workers. Even with these financial problems, he was able to get approval for an education reform package, improve the State Mental Health System, relieve overcrowding in state prisons, re-establish the financially-strapped State Medicaid System, and give stiffer penalties for drug offenders. He was also instrumental in improving the state highway system by earmarking funds from the oil-windfall for the highway program. But even with these noteworthy accomplishments, James was unable to levy a fuel tax, eliminate income tax deductions for Social Security payments, or begin rewriting Alabama's outmoded 1901 constitution.

One would have assumed that James's defeat of Baxley in the 1978 election meant that the Alabama electorate was finally tired of the kind of state government that they had been getting with the traditional yellow dog Democrats, but that assumption would be wrong. When George Wallace offered himself as a candidate for governor in the 1982 election, he was received as a long-lost friend. He easily defeated George McMillian (D, Jefferson) and Joe McCorquodale (D, Clarke) in the first primary and McMillian in the second primary. Then, finally, he defeated the Republican candidate, Emory Folmar (R, Montgomery) in the general election. In these contests, Wallace received unusually strong support from African Americans. Wallace had defeated Albert Brewer in the 1970 election by outright race-baiting. Blacks were a liability in 1970, but by 1982, Wallace needed them badly. After being shot in 1972, Wallace became a changed politician, and blacks in the state seemed to sympathize with him and to accept this badly handicapped man as their champion. Unlike Governor James who took office in the middle of a roaring recession, Governor Wallace took office with the country's economy roaring ahead. Wallace's growing health problems limited what he could accomplish in his final administration. Probably his most important contribution was engineering the approval of a constitutional amendment that established an unspendable Trust Fund to receive monies from taxes and fees from the state's oil and gas industry. Interest from the funds was to be used only to maintain the stability of the General Fund, which supported all non-education parts of state government. Probably his second most important accomplishment was engineering enactment of legislation that brought much needed changes in job-injury laws. He also worked with the legislature to pass a $310 million education bond issue.

THE NATIONAL ECONOMY

At the beginning of the seventh decade of the twentieth century, the American economy was in the middle of a period of rapidly increasing prices that had begun around 1957. During the 1960s, costs related to Kennedy's new economics, the Vietnam War, and Johnson's War on Poverty and other Great Society programs escalated rapidly. Unfortunately, the government was unwilling to increase taxes to pay for them. As a result, prices increased rapidly and inflation became rampant, especially after 1964.

The Economic Situation

Between October 1973 and March 1974, an oil embargo by members of the Organization of Petroleum Exporting Countries (OPEC) pushed energy prices higher and created shortages. This embargo was the result of overwhelming support for Israel by the Western World during the Yom Kippur War of 1973. Even after the embargo ended, energy prices remained high, adding to the already high rate of inflation and eventually causing increased unemployment. There was another short-term reduction in oil supplies from November 1978 to June 1979—a result of the war between Iran and Iraq. This shortage resulted in another round of price increases.

Inflation continued to grow into the late 1970s, along with the federal expenditures and the federal deficit. The inflationary spiral seemed to feed on itself. People began to expect that prices were going to continue to increase, so they began to increase their buying, apparently assuming that if they delayed purchases they would simply have to pay more later. This increased demand pushed prices even higher. Increasing prices led to demands for higher wages. Labor contracts increasingly came to include automatic cost-of-living adjustments, and the government began to peg some payments, such as Social Security benefits, to the Consumer Price Index. These practices helped labor and retirees cope with inflation, but they also worked to perpetuate it.

As government increased its borrowing to meet its growing obligations, interest rates began to increase. With high energy costs and high interest rates, business investment languished, and unemployment increased to uncomfortable levels. The stagnating economy with continued inflation led to the coining of a new term to describe this phenomenon—stagflation.

When Carter (1977-1981) became president, he attempted to combat economic weakness and high unemployment by further increasing government spending while at the same time establishing voluntary wage and price guidelines to control inflation. Both of these efforts were largely unsuccessful. Then, in 1979, the Federal Reserve Board began to restrict the amount of money it had been supplying to the rapidly inflating economy. As a result, competition for money increased rapidly, pushing interest rates even higher. Soon consumer spending and borrowing by business began to slow dramatically, and the economy soon fell into a deep recession. The recession officially began early in 1980 and lasted until early 1983. During this period, employment rates went above 10%, and business bankruptcies increased dramatically. Farmers were hit especially hard, as agricultural exports declined, crop prices fell, and interest rates increased.

The Gross National Product

Between 1960 and 1965, the Gross National Product grew at an exponential rate (Figure 15.1). Afterwards the rate of growth began to slow, and in the fourth quarter of 1969 a short recession began. It ended in the fourth quarter of 1970 with the GNP at $2.4 trillion. In 1971, it began to grow rapidly once again. In 1973, OPEC placed an embargo on the shipment of oil to the United States, and in the fourth quarter of that year, another recession began. It lasted until the first quarter of 1975. Afterwards, it began to grow once more, but with consumer prices increasing rapidly (Figure 15.2), the growth in the GNP began to slow in 1979. That year, the Consumer Price Index increased by 13.3%, the largest increase in 33 years. In the first quarter of 1980, a short recession began. It only lasted until the third quarter of that year. Then with unemployment increasing, another recession began in the third quarter of 1981. It lasted until the fourth quarter of 1982. After the end of that recession, the GNP began to increase once again, finally reaching $3.6 trillion in 1985. Unfortunately, a large share of this almost unimaginable increase in the GNP was the result of inflation of the currency.

The Consumer Price Index

At the end of the 1960s, the Consumer Price Index had been growing at an exponential rate, and it generally continued to do so throughout the 1980s (Figure 15.2). The CPI was 116.3 (1967 = 100) in 1970 and 247.7 in 1980. The rate of increase slowed somewhat during the early 1980s, but the upward trend continued. The CPI reached 318.5 in 1985. Between 1970 and 1985, the CPI increased by 174%. With this rate of increase, the real value of a 1970 dollar would have been near 36 cents in 1985.

The Inflation Rate

In 1970, the inflation rate was 6% (Figure 15.3). It had been between 1% and 2% in 1960. Early Nixon Administration (1969-1974) efforts to contain its upward spiral were, at first, generally effective as the rate fell back to 3% in 1972. Unfortunately, these efforts worked for just so long before inflation began to rocket upward again. It reached 11% in 1974. The recession of 1974-1975, described previously, forced it back to 5.8% in 1976, but the array of forces pushing it upward were just too powerful. In 1980, inflation reached 13%. Then in 1980, the Federal Reserve Board decided to stop inflation regardless of the consequences. By 1983, the inflation rate fell back to 3%. Although this strong medicine cured the inflationary fever, the unfortunate side effects brought on the deep recession of 1980-1983. Fortunately, in 1985, the recession was over, and the inflation rate was 3.5%.

Federal Discount Rates

The Federal Reserve System (Fed) had generally held interest rates below 3.5% during the 1950s; however, as inflation began to increase in the early 1960s, the Fed began to increase rates, and in December 1969, it reached 6%. It held them at between 4.5% and 6% during the period 1970-1972 and into 1973 (Figure 15.4). As inflation began to increase dramatically in 1973, it raised rates in an attempt to slow inflation. In August of that year, the rate was raised to 7.5%, and the following April (1974), it was raised to 8%. In 1975 and 1976, inflation fell rapidly, and the Fed began to reduce rates. By the end of 1976, the rate was down to 5.25%.

Unfortunately, the interest rate increases of 1973 and 1974 were neither large enough nor maintained long enough to cool the embers of the inflation inferno. After 1976, it began to grow again, and reached 13.4% in 1980 (Figure 15.3). As the inflation rate grew, the Fed once again began to increase interest rates. This time however, they were determined that rates would be high enough and remain in place long enough to reduce the fuel supply (money) for the conflagration. In mid-1981, they raised the rate to 14%. Interest rates that banks charged their customers were much higher. In the latter part of 1981, the prime rate, the interest banks charge their best customers, edged past 20%. Although the Fed began to lower rates late in 1981, they kept them in the double-digit range until the latter part of 1982. The rate was still 8% at the end of 1984 when the inflation rate was back to 3.5%. Then they reduced it an additional half point to 7.5 in 1985.

THE ALABAMA ECONOMY

Data on the Alabama Gross State Product (GSP) in billions of current dollars for the period 1963-1985 are presented in Figure 15.5. The data indicate that the Alabama's GSP increased from $12.4 billion in 1970 to $53.7 billion in 1985, an increase of 333.1%. Almost half of this increase, however, was the result of inflation of the currency (Figure 15.2). Table 15.2 presents data on GSPs (dollars per person) for Alabama and several other states in 1965, 1970, 1975, 1980, and 1985. These data indicate that economies of Alabama and Mississippi were the poorest of all of the states listed in all of the years. In most of the years, the economy of Alabama was slightly better than that of Mississippi. The data also show that the economies of Alabama and Mississippi kept up with the economies of the other states during this period, but that they were not able to catch up.

Table 15.3 presents data on the contribution (percentages) of agriculture to the GSPs of Alabama, Mississippi, Illinois, Indiana, Pennsylvania, Massachusetts, North Dakota, and South Dakota in 1965, 1975, 1980, and 1985. These data clearly show the relative differences in the role of farming in the economies between clearly rural and highly industrialized, commercialized states (North Dakota and South Dakota versus Pennsylvania and Massachusetts). It also generally shows the effects of the transition from agriculture to industrialization, where two of the states are somewhat further along in this process (Alabama and Mississippi versus Illinois and Indiana).

The economies of all these states were seemingly less dependent on farming in 1985 than in 1970. Unfortunately, these data are somewhat misleading, because of the differential rates of inflation in the general economy and the agricultural economy during the period. For example, Alabama's total GSP increased by 152% while its GSP for farming increased by only 39%, and the Consumer Price Index increased by 98%.

THE AGRICULTURAL ECONOMY

Farmers in the country and the state spent billions of dollars modernizing their farms during the preceding period (1950-1970). This investment resulted in a significant increase in agricultural productivity. Unfortunately, with the increase in productivity, the parity ratio of the indices of prices received to prices paid continued to decline. As a result of the interaction of all of these relationships, modernization resulted in an increase in net farm income; however, the increase was considerably more modest than the magnitude of the investment.

The federal government (Triangle) had first jumped head first into the American agricultural economy with the passage of the Agricultural Adjustment acts of 1933 and 1938. Then for the next four decades, it made repeated efforts to fine tune its participation through additional legislation. Unfortunately, throughout this period, except for the war years (1941-1945 and 1951-1952), the plight of the farmer grew worse year by year, as prices received fell further and further behind prices paid. However, by 1970 the roots of federal involvement were so intertwined within the national and agricultural economies that the only alternative seemed to be to keep trying to find a formula that would improve the lot of those making a living on the land. Finding such a magic formula, however, would become increasingly difficult. In the immortal words of Bob Dylan, "The Times They Are A-Changin'."

At the beginning of this period (1970-1985), Southern Democrats were still firmly in control of national agricultural legislation (Ernstes, et al., 1997). Throughout the period, Southerners, either Democrat or Republican, served as chairmen of both House and Senate committees on agriculture (Table 15.4). Until the 97th Congress (1981-1983), Southerners made up more than half of all the members in the subcommittees in both houses. Throughout this period, Jamie Whitten (D, MS) served on the House Committee on Appropriations, and from 1979 through 1992, he served as chairman. But even with this level of legislative power, Southern lawmakers were still a minority in Congress, and they could not have enacted any legislation without support from farm interests and their legislators in other regions. Southerners might have controlled the flow of legislation through the agriculture committees, but that legislation had to have pork in it for everyone. Southern legislators were primarily concerned with protecting the interests of farmers producing cotton, peanuts, rice, and sugar in the South, but to enact the legislation they wanted, they had to provide protection for virtually every other crop produced anywhere in the country.

This level of control by Southerners was the result of the willingness of rural districts in the South to re-elect the same agriculturally-friendly Democratic congressmen year after year. In the early 1970s, this cozy relationship began to change. The Supreme Court had upheld the one-person-one-vote principle in 1964. This ruling resulted in major changes in congressional elections in the South where many rural districts were over-represented relative to urban districts. Further, about this time, many progressive Democrats began to be dissatisfied with the seniority system in the House, and especially with the degree of control exerted over legislation by elderly, conservative Southerners. The effect of these changes was compounded in the election of Ronald Reagan in 1980. Reagan's coattails brought a large number of Southern Republicans to Congress. By 1985, half of the senators and 43 of 116 House members representing Southern states were Republicans. Between 1973 and 1985, the number of Southern Democrats on the agricultural committees declined by a fourth in the House and by half in the Senate (Ernstes, et al., 1997). These were momentous changes, and after Reagan's election, they culminated in calls by his administration for a complete phaseout of all agricultural subsidies. Fortunately or unfortunately, depending on an individual's perspective regarding subsidies, the meldown in the country's agricultural economy, which occurred about that time, prevented any serious consideration of the Reagan Administration's proposal.

In 1970, American farmers could not know it, but they were entering a short period of almost unrivaled prosperity. Some observers commented that the Golden Age of American Agriculture was returning. That period (1909-1914) had been the time when the prices farmers received for their crops had been greater than the prices they had to pay to produce them. The early 1970s were the beginning of a period of the best of times. Observers predicted that farm surpluses would soon be a thing of the past. They predicted that the expanding world population would need more and more of American commodities and that the world's growing needs would provide stable and profitable markets well into the foreseeable future. Farmers were encouraged to get to work planting from fence row to fence row. In 1975, President Ford's (1974-1977) secretary of agriculture, Earl Butz, predicted that world markets for American farm products would continue to expand. He further suggested that farmers should disregard the prospects of increasing harvests and focus instead on the frightening challenge of keeping food supplies adequate to meet the needs of a rapidly growing population. Unfortunately, the best of times did not last very long. They were quickly replaced by the worst of times.

Soon, farmers would learn new lessons about the fickleness of international markets and about agro-politics. They would also learn that over the years American consumers had grown to like inexpensive food and that they were extremely reluctant to accept higher prices, even when the costs of production were increasing rapidly. The government was also beginning to face a growing taxpayer revolt over the ever-increasing cost of federal agricultural support programs. Finally, in this period, farmers would encounter a new force that had never before been encountered in the entire history of agriculture—the environmentalist, who would bring changes to the farmers' already complex world, a complexity they in their wildest dreams could never have imagined.

Federal Agricultural Legislation

In an attempt to keep food inexpensive, protect the livelihood of millions of American farmers, and cater to the wishes of environmentalists, Congress (Triangle) passed several significant pieces of legislation in the period 1970-1985:

1. The Agricultural Act of 1970. This act was expected to remain in effect through 1973. It established the cropland set-aside program and provided for potential payments on half of a farmer's production, based on differences between market and support prices. It also plowed new ground by setting a $55,000 limitation on the amount of money that one person could receive for removing land used for any one crop from production. This provision was included in response to growing concerns that a few farmers with large landholdings were receiving exorbitant payments from the government for not producing crops.
2. The Agriculture and Consumer Protection Act of 1973. This act was passed in an agricultural environment different from any since the 1930s. Demand for American farm products exceeded production when the legislation was being debated. It ended direct subsidies or payments for land held out of production This omnibus act created target prices and deficiency payments to replace former direct price support payments. Under the provisions of this part of the act, the government set an ideal or target price for a commodity. If the market price fell below the target price, the government paid the farmer the difference—that is, made a deficiency payment. In this manner, the government helped maintain the farmer's income when commodity prices were unusually low. In return for this guarantee of income stability, the farmer was required to make a mandatory acreage reduction of the crops covered under the program. It also further reduced to $20,000 the annual payment that an individual could receive for any crop, included in a federal farm program, and authorized government payments to farmers who lost crops because of natural disasters such as floods and droughts. The 1973 act was the first federal agricultural legislation that specifically included the recognition of the growing role of consumers in setting farm policy in its title, and it resulted from the growing concern for the increase in food prices during this inflationary period. The increase in the prices of farm commodities in the early 1970s were a godsend to the farmers, but it resulted in a different reaction among consumers, who became concerned with the rapid increase in food prices. This concern was translated into political action in August 1975 when President Ford (1974-1977) restricted grain sales to the Soviet Union until the effect of large-scale exports on domestic supplies and food prices could be determined.
3. Food and Agriculture Act of 1977. This omnibus act retained the basic provisions of the 1973 act, but introduced several important modifications. It increased price and income supports, established a farmer-owned reserve for grain, provided for farmer income protection if exports were suspended for domestic reasons, and created a new two-tiered peanut program that allowed peanut producers to sell specified amounts of their crop (quotas) at the supported price and the remainder at the lower, world market price.
4. Congress passed a new crop insurance program in 1980 that was designed to help farmers cope with the effect of adverse weather conditions on their operations. The new insurance program was designed to replace disaster loans as a means of dealing with these problems.
5. Agriculture and Food Act of 1981. This omnibus act, which focused on making American commodities competitive abroad, contained credit and subsidy provisions to expand exports. The act also set high minimum loan rates, high annual target prices, alternative acreage reduction pro-

cedures, as well as continued the farmer-owned grain reserve and offered protection when foreign embargoes on American exports resulted from US foreign policy decisions.

6. The Omnibus Budget Reconciliation Act of 1982. This act increased loan rates, provided for a voluntary acreage reduction and a paid diversion program, permitted advance program payments, increased funding for export promotion, and slightly modified some food stamp provisions.
7. In 1983, Congress passed legislation establishing the Payment-in-Kind (PIK) program. It was designed to reduce production, stocks, and government payments and to raise commodity prices and farm income. Under the program, instead of paying farmers in cash for their grain, they were issued certificates that could be exchanged for grain to be sold whenever prices improved, or they could be traded like stock certificates. Other legislation provided for acreage reduction and paid land diversion.
8. Agricultural Adjustment Act of 1984. This act froze target prices established in the 1981 act and authorized paid, land diversions for feed grains, upland cotton, and rice.
9. Food Security Act of 1985. This act was intended to govern the country's farm policies for the next five years. It retained target prices, but substantially reduced loan rates. It also established a dairy buyout program. Direct and PIK payments were used to encourage farmers to set aside acreage. Under this act, a Conservation Reserve Program (CRP) was introduced. The CRP established a long-term cropland retirement program. It provided participants (farm owners, operators, or tenants) with an annual per-acre rent, plus half the cost of establishing a permanent land cover (usually grass or trees). In return, participants agreed to retire highly erodable or environmentally-sensitive cropland from production for 10 to 15 years. An enrollment mandate was established at 40-45 million acres. The 1985 act also contained provisions for an Export Enhancement Program (EEP) by providing American exporters with bonuses that would, in effect, allow them to sell American agricultural products overseas at reduced prices—thus making them more attractive in international markets. This legislation recognized the important role that exports played in American agriculture. It also recognized that as a result of noncompetitive prices our exports were declining rapidly on world markets.
10. Farm Credit Legislation of 1985. This legislation established the Farm Credit System Capital Corporation, which was designed to rescue the Farm Credit System from insolvency. The system had loaned farmers enormous sums of money during the euphoric period of the late 1970s and early 1980s. Then, as the 1980-1983 recession deepened, they could not repay the large sums of money that they had borrowed. By early 1985, the old Farm Credit System was drowning in bad loans.

Alabama Agricultural Legislation

Finally, at the beginning of the last quarter of the twentieth century, Southern and Alabama agriculture entered the modern world. It has become thoroughly modernized, generally adequately capitalized, and broadly diversified. Unfortunately, life in this new world is not always a bed of roses. In fact, it can be downright unfriendly to people attempting to wrest their fortunes from the soil. Much had changed in their world, but the more it changed, the more it stayed the same. Sure, they owned larger farms, drove tractors, harvested crops from more acres that they did not share with many tenants, sowed better seeds, harvested everything mechanically, and had better sources of credit at reasonable rates. Nevertheless, after all this time and all these changes, they still could not sell what they could produce at prices necessary to compensate them for their lower yields or higher costs of production.

In the late 1970s, Alabama farmers were in the middle of probably the most difficult period in the history of agriculture in the state. At the national level, the ratio of the indices of prices received to prices paid was 0.70, and still falling. This national situation was probably even worse in Alabama. In casting about for some source of relief for farmers, agricultural leaders began to look at changes that might be made in the Alabama tax code. The 1901 Alabama Constitution dictated that all property in the state would be taxed according to its market value. However, in 1969, a suit was filed challenging the way taxes were assessed. This suit resulted in a court decision that the existing system was, indeed, unconstitutional (Hyman, 1999). This decision, in effect, would have resulted in the tripling of property taxes in the state. In response, in 1978, the legislature passed a constitutional amendment that divided all property into four classes, each with its own rate of assessment. Homes, farms, and forests were included in Class 3. Then, in 1982, the legislature passed the so-called current use law, which decreed that taxation of farm and forest property would be based on the current use of the property, rather than the value under any alternative use. At that time, similar laws were already in effect in many states throughout the country. This legislation had the immediate effect of lowering taxes on farm and forest property in the state, especially property close to urban areas. However, it also allowed farmers, already near economic collapse, to keep their farms rather than selling them to pay their taxes. In enacting the legislation, the legislature commented that it was "particularly concerned about the preservation of its agricultural and forest property and seeks through its property tax structure to preserve such property by providing additional preferential tax treatment for such property."

Agricultural Exports

In 1970, American agricultural exports totaled $7.2 billion, and they increased slowly to $9.4 billion in 1972 (Figure 15.6). In the early 1970s, weather in Europe suddenly became considerably cooler, resulting in reduced harvests of some crops. This reduction forced Europeans to purchase additional quantities of the affected crops from the United States. Also in the early 1970s, the Soviet and Japanese governments made decisions to provide more meat for their populations, but neither country produced enough grain to increase livestock production. As a result, both countries, especially the Soviets, began to buy large quantities of grain from the United States. In 1973, the first year of the Soviet purchases, exports exploded upward to $17.7 billion and continued upward to $22 billion in 1974. While grain exports were increased by the largest percentage, exports of other crops increased as well. Then weather improved in Europe after 1974, and the growth in exports slowed. In 1975, they were slightly lower than they had been in 1974 ($21.8 versus $21.9 billion).

Rapid growth in exports resumed in 1978, and in 1980 and 1981 they reached $41 billion and $43 billion, respectively. The Agriculture and Food Act of 1981 included provisions for credit and subsidies designed to increase exports. Unfortunately, in 1982 an overvalued dollar, high prices of agricultural crops produced in the United States relative to those on the world market, foreign subsidies of crop prices, and a global recession sharply reduced American exports. By 1985, they were down to $29 billion, and they went even lower ($26.2 billion) in 1986.

Soviet grain purchases rescued the American farmer in the early 1970s. Unfortunately, they put him right back in economic trouble in December 1979 when the Soviets invaded Afghanistan, and President Carter responded by announcing an embargo on additional sales of agricultural products to them. However, the United States did honor commitments to grain sales made earlier in the decade.

Agricultural Prices

As noted in the preceding paragraph, in the early 1970s Japan and most European countries, especially the Soviet Union, quietly entered the American grain market and began to make massive purchases from the US government-owned grain reserves. These reserves had been established when the government began to purchase surplus grain in an effort to increase prices. Wheat prices more than doubled ($1.76 to $3.95 a bushel) from 1972 to 1973. In the same period, corn prices increased from $1.57 to $2.55 a bushel, cotton prices increased from 27 cents to 44 cents a pound, and soybean prices increased from $4.37 to $5.58 a bushel. The United States was in the midddle of a period of unparalleled agricultural prosperity. Indices for prices received and prices paid had begun to diverge in the early 1950s (Figure 15.7), and as late as 1970, they were 274 and 382, respectively. However, after 1971 they began to converge again, and in 1973 reached 447 and 490, respectively. Unfortunately, after 1972, the Consumer Price Index also began to increase rapidly (Figure 15.2); consequently, prices paid for farm inputs also began to increase, and by 1980 the indices for prices received was trailing far behind (624 versus 978). The two continued to diverge, and in 1985 were 579 and 1131, respectively.

In 1970, the parity ratio of the indices of prices received to prices paid had declined to 0.7 (Figure 15.8), which means that for each dollar that farmers spent producing their crops, they received only 70 cents when they sold them. Then, when exports began to increase (Figure 17.5), the ratio also began to increase, and soon had climbed to 0.91. It was the best of times. Unfortunately, as detailed in the preceding paragraphs, conditions quickly changed, and the ratio fell rapidly. By 1980, it was 0.6, lower than it had been in 1970. The ratio continued to trend downward through the 1980s, finally reaching 0.51 in 1985. It was the worst of times.

Many economists contend that this decline in the parity ratio was not really as bad as it appeared. They contend that increases in productivity tended to offset declines in the ratio in farm operations.

As discussed in the following section (see Figure 15.9), indices of productivity were increasing rapidly during this period. However, the price situation was probably much worse than the economists were willing to admit. In 1950, off-farm income constituted only 52% of total farm-household income (Fernandez-Cornejo, 2007). By 1970, the percentage had increased to 65%, and by the early 1980s it reached 85%. Productivity was increasing rapidly, but farmers were being forced to earn more and more of their income from off the farm. Apparently, with the parity ratio declining, farmers were finding it increasingly difficult to make a living farming, regardless of the increase in the efficiency of their production.

Agricultural Productivity

Indices of agricultural inputs trended slowly upward during the early 1970s (Figure 15.9); then they increased rapidly after 1977 to the highest level of the twentieth century, 1.18 (the 1987 index = 1.00). In 1980, as the parity ratio trended sharply downward (Figure 15.8), inputs also began to decline, and by 1985 it was down to 1.05. Inputs had not been that low in years.

Indices of agricultural outputs generally trended upward during the period 1970-1985 (Figure 15.9), although there was considerable year-to-year variation. It was 0.72 in 1970 and 1.02 in 1985. Poor crop weather played an important role in the variation. For example, the decline in 1974 was the result of a severe drought throughout much of the Corn Belt that resulted in a significant reduction in yields. The decline in 1980 was the result of a drought/heat wave in the Central and Eastern United States, and the even sharper decline in 1983 was the result of widespread freezes, flooding, and storms over most of the country. The National Climate Data Center included both 1980 and 1983 in its list of Billion Dollar Weather Disasters during the period 1980-2003.

Indices of agricultural productivity also generally trended upward (Figure 15.9) throughout the period 1970-1985, as agricultural technology continued to evolve and improve. It was 0.67 in 1970 and reached 0.96 in 1985. Throughout the period, the changes in productivity seemed to be independent of changes in inputs; however, unfavorable weather events, as noted in the preceding paragraph, played a major role in year-to-year variation.

Farmers

As detailed in the preceding chapter, the number of farmers in Alabama, as well as in the United States as a whole, fell sharply during the period 1950-1970. Those who remained on farms invested heavily in modernizing them. As a result, their productivity increased, and their net farm income also increased, but not at the same rate as their increased investment.

The Total Population

Data published by Coulson and Joyce (1999) on estimates of the American population (all persons) for the period 1950-1985 are presented in Figure 15.10. These data show that the population increased from 203.3

million in 1970 to 237.9 million in 1985, or at a rate of 1.06% a year. This rate of increase is contrasted with that of the preceding period (1950-1970) of 1.16% a year.

Similar data on the Alabama population (all persons) are presented in Figure 15.11. These data indicate that Alabama's population increased from 3.44 million in 1970 to 3.97 million in 1985, or at a rate of 0.96% a year. This rate is contrasted to that for the preceding period (1950-1970) of 0.59% a year. Between 1966 and 1969, Alabama's population actually declined. In 1970, 18.7% of Alabama's population lived in Jefferson County. Other counties with large shares of the population included: Mobile (9.2%), Madison (5.4), Montgomery (4.8), and Tuscaloosa (3.4). These five counties accounted for 41.5% of the state's total population.

The Farm Population

Data presented in Figure 15.11 show that during the period 1970-1985 the American farm population (farmers) declined from 9.7 million to 5.4 million, or an average loss of 2.8% a year. It had trended downward steadily from 1950 through 1970; then the rate of decline slowed in 1971, 1972, and 1973, as the indices of prices received for farm commodities began to trend upward (Figure 15.7). In the period 1974-1978, the parity ratio (Figure 15.8) began to fall sharply, and the rate at which farmers were leaving their farms began to increase again. Apparently, when the national recession began in 1980 (Figure 15.1), there were fewer opportunities for farmers to find jobs in any occupation other than farming, and the rate at which they were leaving their farms began to slow once again.

With the American population increasing (Figure 15.10) and the farmer population declining (15.12), the percentage of farmers in the population declined rapidly during the period 1970-1985 (Figure 15.13). It had been 15.3% in 1950, but by 1970, it had declined to 4.8%. In 1985, it was down to 2.2%. These data indicate that by 1985 just 2.2% of the total population was providing food and fiber for the remaining 97.8%.

There are no annual data available on Alabama's farmer population for the period 1970-1985. Fite (1984) published data indicating that it was 216,000 in 1970. It had been 960,000 in 1950; then by 1980, it had declined to 88,000. Fite notes, however, that between 1970 and 1980, the definition of "farmers" changed. In 1980, the average number of farmers per state in 11 Southern states was 107,000.

Fite (1984) also published data on the number of black farmers in Alabama in 1970 and 1980. There were 9,900 in the state in 1970, whereas in 1950 there had been 57,200. In 1978 the number declined to 4,900. (This illustrates that the definition of what constitutes a "farmer" had been changed.) In 1978, the average number of black farmers per state in 11 Southern states was 5,090.

Farmers in the Work Force

Data has been published decennially on the percentage of farmers in the American labor force since 1840. Data for the period 1920-1980 are presented in Figure 15.14. These data show that the percentage of farmers in the labor force also trended sharply downward. It was 12.2% in 1950, and down to 3.4% by 1980.

Tenancy

Census data on the percentage of American farms operated by tenants during the period 1950-1982 are presented in Figure 15.15. These data show that as late as 1950 the national tenancy rate—buoyed upward by relatively high rates in the Southern states—was 26.8%. By 1969, the rate had declined to 12.9%, and it remained at near that level through 1982 (11.6%).

Data on the percentage of Alabama farms operated by tenants during the period 1950-1982 are also presented in Figure 15.15. These data show that as late as 1950, the percentage of tenants operating Alabama farms was considerably higher than the national percentage. (41.4 versus 26.8%). It is astonishing that as late as 1950 some 41.4% of Alabama's 220,000 farms were operated by tenants and that such a large number of Alabama farmers could not afford their own farms. After 1950, however, the rate of tenancy in the state fell sharply, and by 1982 it was actually lower in Alabama than in the country as a whole (6.8 versus 11.6%).

Farms

Census data indicated that between 1950 and 1970, the United States lost 49.3% of its farms, 8.5% of its land in farms, and 4% of its cropland. Similarly, Alabama lost 65.7% of its farms, 34.6% of its land in farms, and 33.6% of its cropland.

Farm Numbers

During this period (1970-1985), the cost of inputs in farming were continuing to escalate sharply. Investment required to remain competitive was becoming still more prohibitive. Net farm income increased at a lower rate than the Consumer Price Index, 155 versus 177%. At the same time, average investment in land and buildings increased from $196 to $735 per acre (235%), investment in machinery and equipment increased from $9,770 to $41,227 per farm (322%), capital consumption increased from $2,341 to $8,480 per farm (262%), and purchased inputs increased from $8,645 to $32,058 per farm (271%). With this dismal cost and return situation, it is surprising that farm numbers did not fall any further than they did.

USDA data indicate that in 1970 there were 2.9 million farms in the United States. Texas reported the largest share of the total (7.2%). Other states with large shares included: North Carolina (5.1%), Iowa (4.9), Missouri (4.8), and Illinois (4. 3). These five states accounted for 26.3% of the national total. Alabama accounted for only 2.8%.

In 1975, USDA began to use a new definition of a "farm" to differentiate those who simply lived in rural areas from those who actually farmed. From that time, farms were defined as those operations that earned or were capable of earning $1,000 or more from actual annual agricultural product sales. This change resulted in a large reduction in the number of farms in the country, as is shown in Figure 15.16. Then in 1978, the continuing four-decade-long decline in American farm numbers came to an end, and number remained around 2.44 million units through 1980. However after 1981, the downward trend resumed. This interruption of the long-term downward trend between 1975 and 1981 generally coincided with the period when the indices of prices received were increasing (Figure 15.7). By 1985, the total number of farms was down to 2.29 million.

The trend in farm number in Alabama during the period 1970-1985 was similar to the national trend (Figures 15.17 and 15.16). There were 82,000 farms in the state in 1970 and 52,000 in 1985. The rate of loss had slowed somewhat after 1960, but it continued through 1985. The loss of farms, as a result of the change in the definition of a farm, was slightly greater in the state than in the country as a whole (19.2 versus14.5%). The distribution of farms within the state in 1969 is shown in Figure 15.18.

Farm Size

As detailed in the preceding chapter, the average size of American farms trended sharply upward from 216 acres in 1950 to 389 acres in 1969, while numbers of farms were trending sharply downward (Figures 15.19 and 15.16). Obviously, this increase in size was the result of consolidation. Size continued to trend upward from 1969 through 1974, but changed very little thereafter, as the loss of farms slowed. In 1982, average size was 440 acres nationally. Clearly, by 1982 increases in average size were beyond the inflection point of the S-shaped curve. Such a change usually indicates that important changes are beginning to occur in the dynamics that have given rise to the variable (average size). Essentially, this change probably means that average size of American farms is beginning to approach an upper limit. The small increase in size between 1974 and 1978 may have been partially the result of the change in the definition of a farm in 1975. Changing the definition eliminated many extremely small farms, which should have resulted in an increase in average size.

Data on average farm size in the United States during the period 1970-1985 were discussed in the preceding paragraphs; however, those data do not provide any insight on the distribution of farms by size. Data in Table 15.5 show the percentage of farms in seven size classes in 1969, 1974, 1978, and 1982. For example, in 1969, 5.9% of American farms were from one to nine acres, and 2.2% were 2,000 acres or larger. These data also show that in 1969 over 63% of all farms were between 50 and 499 acres. Over the years, however, the percentage of farms in these size classes declined and the percentages in both smaller and larger size classes increased. By 1982, only 55% of all farms were between 50 and 499 acres. These data probably indicate that the percentage of hobby farms and commercial farms were increasing while the percentage of family farms was decreasing.

The trend in changes in average size of Alabama farms between 1969 and 1972 (Figure 15.19) was similar to the trend for the country as a whole. However, size of Alabama farms was much lower (188 acres in 1969 and 211 in 1982). As was the case with the national data, size increased substantially between 1969 and 1974, but very little afterwards. The increase between 1974 and 1978 was also likely the result of the change in the definition of a farm.

Data on the percentage of farms in seven different size classes for Alabama are also shown in Table 15.5. These data indicate that in 1969 some 70.4% of all farms in Alabama were between 10 and 179 acres and that the distribution of farms in the state did not change very much between 1969 and 1982. The most noticeable changes were the increases in the percentages in the one to nine-acre farms (5.1 to 6.3%) and the decreases in the percentages in the 50 to 179-acre farms (40.8 to 37.7%).

Total Farm Land

In 1950, USDA began to include estimates of total land in farms with their estimates of farm numbers. The data presented in Figure 15.20 are from that source. Total land in farms in the United States also trended downward during the period 1970-1985. There were 1.102 billion acres in farms in 1970 and 1.012 billion in 1985, which represented a total loss of 90 million acres or 8% of the 1970 total. The rate of loss was around 6,000 acres a year, except in 1975 when the definition of a farm was changed. A small share of this loss was, of course, the result of the changed definition.

USDA data indicated that changes in the total amount of farmland in the Alabama during the 1970s and early 1980s essentially mirrored changes in the number of farms (Figures 15.21 and 15.17). There had been 14.8 million acres of land in farms in the state in 1970, and by 1985 the total was down to 11.2 million, which represented a loss of 5.6 million acres or 38% of the total. Surprisingly, land in farms in the state stabilized in 1972, 1973, and 1974. During this period, indices for prices received spurted sharply upward (Figure 15.7). Apparently, Alabama farmers decided to hold onto their farmland during this period. Alabama may have "lost" as much as 300,000 acres of farmland as a result of the change in the definition of a farm.

Use of Farmland

Census data were used to compute the uses of American farmland (expressed as percentages) during the period 1964-1982 (Figure 15.22). For some reason, the data did not include estimates of the amount of land used for pasture or for "other" in 1969 and 1974. The uses of American farmland changed relatively little during the period. Also the amount used for cropland and pasture remained about equal (40-45%). It is likely that the large share used for pasture is the result of including Western rangeland in the estimates. The use of land for woodland remained between 9 and 13%, and the amount used for other remained 3 to 4%. It is somewhat surprising that the uses of farmland would remain relatively unchanged when the total amount was declining (Figure 15.20).

Data on the uses of farmland in Alabama during the period 1964 and 1982 are presented in Figure 15.23. These data indicate that the percentage used for cropland increased steadily from 1964 (34%) to 1978 (49%), while the percentage used for woodland generally trended downward. The percentage used for pasture ranged between 11 and 17%, but data for 1969 and 1974 were not available. Between 1978 and 1982, there was little change in any of the uses.

Total Cropland

Census data on total cropland in the United States during the period 1949-1982 are presented in Table 15.24. During the period 1969-1982, total land in farms was trending downward (Figure 15.20). Cropland also declined: 459 million acres in 1969 and 445 million in 1982. Total cropland was somewhat lower than expected in 1974. In that year, the entire eastern portion of the country (Wisconsin to east Texas, eastward) was wet to excessively wet in June, July, and August. It is likely that this wet weather reduced the amount of cropland in use that year. The change in the definition of a farm should have also resulted in a slight decrease. If it did, the decline cannot be detected in the census data. Further, because the new definition of a farm eliminated very small farms, it is unlikely that very much cropland was involved. Apparently, the sharp increase in the indices of prices received during the period 1972-1980 had only a limited effect on total cropland.

Table 15.6 presents data on the percentage of American farms of different size harvesting different quantities of cropland during the period 1950-1982. For example, in 1969, 16.2% of all American farms harvested crops from between one to nine acres. Further, in the same year, about the same percentage harvested crops from 50 to 99 acres and 100 to 199 acres. Generally, the percentage of farms harvesting from one to 199 acres declined during the period, and the percentage harvesting more than 199 acres increased.

Trends in Alabama total cropland in the period 1949-1982 were generally similar to those in the country as a whole (Figure 15.25). It trended sharply downward from 1949 (18.7 million acres) through 1964 (5.1 million), then changed relatively little from 1969 (5.8 million) through 1982 (5.1 million). From 1969 through 1982, Alabama "lost" 12.1% of its cropland.

Data on the percentage of Alabama farms harvesting crops from different amounts of land during the period 1956-1982 are presented in Table 15.6. For example, in 1969, 33% of Alabama's farms harvested crops from one to nine acres. Nationally, that percentage was 16.2%. Over 50% of Alabama farms harvested crops from one to 19 acres in that year. Generally, the percentage of Alabama harvesting crops from one to 19 acres declined during the period, while the percentage harvesting crops from larger acreages increased.

Percentage of Cropland Harvested

Data on the percentages of cropland harvested in the United States during the period 1959 and 1982 are presented in Figure 15.26. These data show that the percentages generally increased from 59.5% in 1969 to 73.2% in 1982. This increase indicates that as total cropland declined (Figure 15.24), the percentage harvested increased.

Data on the percentage of cropland harvested for Alabama during the period 1959-1982 are also presented in Figure 15.26. These data show that the year-to-year trends for the state were similar to those for the country as a whole. However, in all of the years, the percentages for Alabama were lower. In 1982, the percentage of cropland harvested in Alabama versus the United States as a whole was 63.9 versus 73.3. In 1969, the percentage had been 46.7 versus 59.5.

Value of Land and Buildings

Data presented in the preceding chapter indicated that the total value of land and buildings on American farms had increased from $75 billion in 1950 to $208 billion in 1970 (Figure 15.27). This increase was much larger than the rate of inflation during this period. However, that relatively steep overall increase was limited, compared to the rate of increase between 1970 and 1985. In 1970, total value was $208.2 billion. The lower rate of increase, recorded in the preceding period, continued through 1972 ($238.7 billion), However, afterwards, value exploded upward, reaching $843.7 billion in 1981, an increase of 250%. A large share of this increase was the result of inflation (Figure 15.3). Between 1973 and 1981, the Consumer Price Index increased by 104% (133.1 to 272.3) (Figure 15.2). However, some of this increase was likely fueled by the increased optimism by farmers based on the parity ratio reaching 0.9 in 1973 (Figure 15.8). After 1981, value began to decline. It reached $719.4 billion in 1985. The Federal Reserve had increased the discount rate to 14% in May 1981 (Figure 15.4). Afterwards, inflation began to decline rapidly (Figure 15.3).

Because of changes in the amount of farmland involved, a better measure of changes in the value of land and buildings is to convert the total value data to value an acre. These converted data for the period 1950-1985 are shown in Figure 15.28. The trend in the converted data is essentially the same. Value an acre was $246 in 1973 and $823 in 1981, an increase of 235%.

Data on the total value of land and buildings in

Alabama during the period 1950-1985 are presented in Figure 15.29. The trend for the state was essentially the same as for the country as a whole. In 1973, the total value of land and buildings was $3.9 billion, and by 1981 it had increased to $10.8 billion, an increase of 226%. The upward trend was interrupted between 1974 and 1975. After 1973, the parity ratio began to fall sharply (Figure 15.8), and by 1975 it was down to 0.75. Apparently, this decline in the ratio did not seem to affect the change in value in the country as much as it did in the state.

Total value data were also converted to value per acre for Alabama. These are also shown in Figure 15.28. Between 1970 and 1981, they increased from $200 to $910, an increase of 355%. Obviously, the trends of value per acre were similar for both the country and the state. In 1970, they were essentially the same, at around $200; however, between 1975 and 1976 those for Alabama were slightly higher. They were nearly the same from 1977 through 1979, but afterwards the Alabama values were clearly higher. Those in the state reached $910 per acre in 1981 before declining to $797 in 1985.

As might be expected, the value of farmland alone in Alabama followed the same trend between 1970 and 1985 as did value per acre (Figure 15.30 and 15.29). It was $141 in 1970 and $732 in 1981; however, the rate of increase in the value of the land alone was substantially greater than for land and buildings for that period (419 versus 355%). The value of buildings did not increase as rapidly as the value of land.

Average Cash Rents of Alabama Cropland and Pasture

Prevatt (2000) published data on cash rents for Alabama cropland and pasture during the period 1967-1985. Some of these data are presented in Figure 15.31. Rents for cropland increased exponentially from $15.40 an acre in 1967 to $35.00 in 1980. They continued to increase each year through 1983, but at a much lower rate. They reached $37.80 in 1983, before beginning to decline. Between 1970 and 1983, rents increased from $17.30 to $38.70, an increase of 124%. However, rents increased at a much lower rate than value. Of course, rents reflect the amount of money that a farmer might make from farming an acre of land during this particular period. With the ratios of the indices for prices received to prices paid falling rapidly (Figure 15.8), it is surprising that they could afford to pay as much as they did. While rents increased for most of this period, the increase did not keep up with inflation.

Rents for pasture on an acre basis were generally half, or less, as much as rents for cropland during the period 1967-1985 (Figure 15.31). Rents for pasture also increased during the period; however, the general rate of increase was not as great as for cropland over the same period. In contrast to the situation with cropland, rents for pasture changed relatively little after 1980, although they declined slightly after 1983. Rents for pasture increased from $7.60 an acre in 1970 to $17.40 in 1982, an increase of 129%. The average real net return per cow in the United States during this period was less than zero. Consequently, it is surprising that rents were as high as they were. During the same period, the Consumer Price Index increased from 116.3 to 288.6, or 148% (Figure 15.2). The increase in pasture rent was substantially lower than the increase in the CPI.

Rent-to-Value Ratios for Alabama Cropland and Pasture

Ratios of rent-to-value are useful indicators of the nature of the forces that are operating to determine the use of farmland at any one time. These ratios are obtained by dividing average annual cropland rents by average farmland values and expressing them as a percentage. If the ratio is high—that is, if the amount of rent per acre of land represents a relatively large proportion of the total value of that acre—then the primary pressure determining use is likely to be returns from agricultural production. However, if the ratio is low, then some use other than farming is responsible for the most pressure. For example, in 1970 USDA reported that the rent-to-value ratio for farmland in North Dakota (a major agricultural state) was near 9.0. In the same year, the ratio for farmland in New Jersey (a largely urbanized state) was between 2 and 3.0.

Annual rent–to-value ratios for Alabama cropland and pasture for the period 1967-1985 are presented in Figure 15.32. The ratios computed for cropland for the period 1970-1985 generally trended downward from 8.1 in 1970 to 4.4 in 1982. Afterwards, through 1985, ratios remained at near the same level. USDA data indicate that in 1985 the rent-to-value ratios were: Iowa (7.1), North Dakota (7.0), Illinois (6.5), Delaware (3.5), Maryland (2.5), and New Jersey (1.2).

The USDA data and the data presented in Figure 15.32 would seem to indicate that between 1967 and 1985 land values and rents in Alabama were in a period of rapid transition from a rural, agricultural state to an urban, industrialized one—to an economy more like Delaware, Maryland, and New Jersey than Iowa, North Dakota, and Illinois. Certainly, Alabama was in a period of transition, but another factor must have been operating that tended to lower Alabama's rent-to-value ratios. Alabama was certainly not becoming an urbanized state that rapidly. Returns on investment in crop production were declining in the state, making it extremely difficult for those renting farmland to pay for it.

The same downward trend in the rent-to-value ratios for pasture during the period 1970-1985 is also evident in Figure 15.32. Those ratios for pasture were lower than those for cropland.

Value of Machinery and Equipment

Data on the total value of farm machinery and equipment on American farms during the period 1950-1982 are presented in Figure 15.33. (There were no census data on value in 1950, 1954, 1959, and 1964.) Data in the figure from those years were taken from Lerner (1975). The remaining data are from the censuses for those years. Value in the United States had almost doubled during the period of agricultural modernization (1950-1970). However, as striking as that increase was, it was limited compared to the increases taking place after 1969. From 1969 through 1982, value increased from $25.3 to $93.7 billion, an increase of 270%. During the same period, the Consumer Price Index only increased 163% (109.8 to 288.6), so considerably less than half of the increase in value was real growth in investment. Most of the real growth in value took place between 1969 and 1974. During this period, the increase in value (90.3%) was considerably greater than the increase in the CPI (34.5%). It was during this period that the ratio of prices received to prices paid reached 0.91 (Figure 15.8). Between 1974 and 1978, increases in total value and the CPI were closer (32.2 versus 60.3%); however, between 1978 and 1982, the increase in value was less than half the increase in the CPI (20.6 versus 47.8%). The almost unbounded optimism that had characterized the farming community between 1969 and 1974 was replaced by realism in the period 1974 and 1978 and then by pessimism. It was the best of times, it was the worst of times.

Because of the change in the number of farms in the country during this period, converting the total value data into value per farm provides a more meaningful measure of the changes in the value of machinery and equipment in the country. These data for the period 1969-1982 are presented in Figure 15.34. The converted data show essentially the same trend. Value per farm increased from $9,770 in 1969 to $41,919 in 1982, an increase of 329%. While the upward trend was the same, the rate of the increase on the individual farms was considerably greater (329 versus 270%). These data indicate that investment in machinery and equipment on individual farms was considerably greater than the increase in the CPI (329 versus 179%). The converted data show an even higher rate of real growth. In this period, growth in value per acre was almost four times as great as the increase in the CPI (128.3 versus 34.5%).

Alabama farmers reacted to the optimistic outlook for agriculture in the early 1970s the same way that all American farms reacted. They quickly and significantly increased their investment in farm machinery and equipment (Figure 15.35). Census data indicate that the total value of investment (in millions of dollars) was $395.40 in 1969, $699.20 in 1974, $1,062.00 in 1978, and $1,167.20 in 1982. Unfortunately, these data show that in reality Alabama farmers increased their real investment relatively little. Between 1969 and 1983, total value increased by 195%; however, in the same period, the CPI increased by 162%.

Total value data for machinery and equipment on Alabama farms were also converted into value per farm (Figure 15.34). These data show that investment by individual Alabama farms was considerably lower than their counterparts in the country as a whole. The data also show that almost all of their real increase in investment took place between 1969 and 1974. During that period, value per farm increased by 123%, while the CPI only increased by 34%. Unfortunately, in the period 1978-1982, the rate of inflation (47.8%) was almost three times as great as the rate of their increase in investment (15%). Consideration needs to be given to the fact that the average size of Alabama farms was a little less than half that of farms in the country as a whole (Figure 15.19). It is likely, however that there is not a 1:1 relationship between farm size and the amount of machinery and equipment required to operate it.

Tractors on Farms

The increase in the number of tractors on American farms had been the primary reason for the phenomenal growth in the productivity and efficiency of American agriculture. For much of the period from 1910 onward, the number of tractors in the country increased rapidly. Data on the number of tractors per 100 farms for the period 1959-1982 are presented in Figure 15.36. The number continued to increase during this period, but no longer at such an extraordinary rate. In fact, as might be expected, the rate of increase was actually declining. Between 1959 and 1964, numbers increased by 20.7% (121 to 146); however, between 1964 and 1969 the rate declined to 11% (146 to 162). It increased to 14.8% (162 to 186) during the period 1969-1974 (the best of times). The rate of increase declined further to 10.2% (186 to 205) between 1974 and 1978, and then between 1978 and 1982, with the indices of prices paid skyrocketing (Figure 15.7), numbers actually declined (205 to 202).

Data on tractors in Alabama during this period are also presented in Figure 15.36. The trends were essentially the same in both the state and country. The data show that there were 103 tractors per 100 farms in the state in 1969. Then numbers increased to 117 in 1974 and to 137 in 1978. In 1982, numbers remained at the same level as 1978 (137). Throughout the period, the numbers in Alabama were consistently lower than in the country as a whole. In 1959, there were a little more than twice the number in the country; however by 1969, there were about 57% more, but by 1982 there were only 47% more. These differences are not as great as they might seem when differences in farm size are considered. From 1964 through 1982, the average size farm in the country was at least twice as large as the average size Alabama farm. When farm size is considered, the concentration of tractors was higher on farms in Alabama. Unfortunately, while farmers in Alabama had more tractors on their smaller farms, they also used less cropland to provide funds to pay for their tractors' maintenance and depreciation. Farmers in Alabama were likely almost tractor-poor.

Animal Power on Farms

In the period 1970-1985, animal power was still important on many of the smaller farms, especially in the South, but they were of such limited importance that the USDA no longer reported data on horses and mules separately. In fact, they had not done so since 1955.

Use of Fertilizer

Data on the pounds of fertilizer used per acre of harvested cropland in the United States and Alabama during the period 1959-1982 are presented in Figure 15.37. Farmers in other regions of the country did not have to use as much fertilizer to obtain good crops as did farmers in the South. The natural fertility of those old, wornout Ultisol soils was so low that Southern farmers had no choice. They generally were not competitive in crop production even with the use of fertilizers, but without it, farming was hopeless. Data in the figure show that farmers in the country increased their use of fertilizer from 162 pounds an acre of harvested cropland in 1959 to 285 pounds in 1969. Of course, this number is biased upward by the extremely high use in the South. Fertilizer use in the United States increased slightly from 1969 to 1974. After 1974, however, the ratio of the indices of prices received to prices paid (Figure 15.8) began to fall sharply; apparently American farmers began to reduce their use of fertilizer. It was down to 299 pounds in 1978 and slightly less than 299 in 1982.

Changes in the use of fertilizer were more striking in Alabama than in the country as a whole. In the state, use increased from 578 pounds an acre of cropland harvested in 1959 to 764 pounds in 1969, and it increased another 6.5% (814 pounds) in 1974. Between 1974 and 1978, use fell by 22.1% to 634 pounds. With the comparatively low yields that Alabama farmers obtained from their crops, most of them could hardly afford to use fertilizer even under the best of conditions. When prices paid began to rapidly outstrip prices received, something had to give, and they apparently chose to use less fertilizer. Use continued to decline after 1978, reaching 555 pounds an acre in 1982.

Plant Agriculture

Data on acres harvested of some major crops in the United States and Alabama in 1950, 1960, and 1970 are presented in Table 15.8. National data show that cotton was becoming a much less important crop than it had been in terms of acres harvested. Corn continued to be the most important crop grown in the country. National data also further documented the continued decline in oat production, as well as the growing importance of soybeans.

Alabama data in Table 15.8 show that cotton was also becoming less important in the state's agriculture. At the same time, corn production also seemed to be declining. The only obvious winner in crop production seemed to be soybeans. Taking all the data together, it appeared that crop production in the state declined substantially during the period 1950-1970. For years, comparative yield data—for virtually all of Alabama's major crops—suggested that for generations most of the state's farmers had been pouring water down a rat hole. In the period 1950-1970, it seemed that the chickens had finally come home to roost.

Alabama Crop Weather

Some data on the status of soil moisture on Alabama farms in the different years during the period 1970-1985 are presented in Table 15.9. This information was taken from the "Crops Review" section of several issues of *Alabama Agricultural Statistics.* Data in this table illustrate the inherent weather-related problems of trying to produce crops in Alabama. Soil moisture was excessive in the late winter and early spring in 12 of 16 years in the period. It was inadequate during portions of the growing season in nine of the 16 years. It was excessive enough to interfere with harvest and the planting of late fall crops in four of the 16 years. Obviously, it is virtually impossible to produce crops consistently for a global marketplace with this degree of weather variability. Certainly, the variability of crop weather is a major comparative disadvantage in Alabama agriculture.

Storms originating in the tropics affected Alabama crop production in several years of the period. Tropical storms produced heavy rainfall and flooding in 1970, 1971, 1972, 1975, and 1977. The two tropical storms passing through the state in 1985 were remnants of hurricanes Elena and Juan. Hurricanes Agnes (1972), Eloise (1975), and Frederick (1979) affected widespread damage to crops and farm infrastructure. The unusually wet weather in 1982 and 1983 was associated with an El Niño event.

Cotton

The first man-made fiber was manufactured in commercial quantities in France near the end of the nineteenth century. Its use slowly increased until World War II when production was increased rapidly. By the 1950s, the industry was supplying more than 20% of the total US fiber needs, and by 1965 the percentage had increased to 40%. The increasing availability of these textiles was beginning to exert considerable downward pressure on the production and use of cotton. During the preceding period (1950-1970), acres of cotton harvested in both the country and the state declined sharply. But because yields generally increased, the decline in the number of bales harvested was not as precipitous. Prices generally declined throughout the period.

US Cotton

In 1970, Texas farmers harvested the largest share (31.5%) of the national 10.2 million-bale cotton crop. Other states reporting large harvests included: Mississippi (16.1%), California (11.4), Arkansas (10.3), and Alabama (5.0). These five states accounted for 74.3% of the country's total harvest that year.

Exports

Although exports were highly variable from year-to-year, they generally increased during this period (Figure

15.38). Exports were four million bales in 1970, but increased rapidly to six million in 1973. Late that year OPEC placed an embargo on oil shipments to Europe and the United States. As a result, exports fell sharply in 1974 (3.9 million) and 1975 (3.3 million). Afterwards, however, they surged upward along with growth in the American and world economies, reaching nine million bales in 1984. However, after 1984—with high prices, an overvalued dollar, and a growing global recession—exports returned to six million bales. Then in 1985, with all agricultural exports trending sharply downward (Figure 15.6), cotton exports fell to two million bales. They had not been that low since 1968.

Prices

Data on support prices for cotton during the period 1950-1985 are presented in Figure 15.39. The federal government kept support (subsidy) prices at around 30 cents a pound from 1950 through 1966. Then with the parity ratio declining (Figure 15.8), the government began to increase the subsidy, and by 1984 and 1985 support prices were 81 cents a pound.

The annual national average market price for cotton was 22 cents a pound in 1970 (Figure 15.39), then it generally increased, along with inflation, through 1980. Inflation began to decline in 1981, and market price followed. Afterwards, it seemed to begin to respond to bales harvested (Figure 15.42) once again. In fact, the unusually high market price in 1980 (75 cents) was probably the result of the poor crop that year (11.1 million bales), and the sharp decline between 1980 and 1981 (75 to 54 cents) was likely the result of the sharp increase in bales harvested (11.1 to 15.6 million). Then, with bales harvested declining, market prices increased, but when they began to increase in 1984 and 1985, market prices began to decline once more, finally reaching 56 cents in 1985.

Market prices were below support prices in 1970, 1971, and 1972, then above them in the period 1973-1980. After 1980—with the recession deepening, exports declining, and acres planted increasing—market prices again fell below support prices. Under those conditions, there was simply too much cotton around for the support prices to have much effect.

Acres Planted and Harvested

Generally, acres planted followed market prices upward from 1970 through 1981 (Figures 15.40 and 15.39). Acres planted were lower than expected in 1975 after the market price had fallen between 1973 and 1974. With market prices increasing steadily between 1977 and 1980, acres planted followed. Unfortunately, farmers did not anticipate the immediate effect of the increasing federal discount rate on prices. In 1980, the rate was raised to 14%, and in 1981 inflation began to decline rapidly and the market price of cotton followed. Farmers planted another large crop in 1981 (14.3 million acres), and when they harvested it, market price had fallen to 54 cents. With the sharply lower market price in 1981, acres planted followed. Acres planted continued to decline through 1983 (7.9 million acres). However, with the market price at 67 cents in 1983, farmers went back to their fields, and acres planted increased in 1984. Of course, the market price immediately declined. Acres planted was down slightly in 1985 to 1.3 million, but the market was not impressed. Market price fell to 56 cents.

Data on acres harvested during the period 1950-1985 are also presented in Figure 15.40. These data show that during this period American farmers harvested around 93% of their planted acres.

Yields

Average yields of cotton varied widely in the country during the period 1950 to 1985 (Figure 15.41), ranging from a low of 402 pounds an acre in 1980, a year of a disastrous summer drought and heat wave, to a high of 628 pounds in 1985. Average yield, however, generally remained relatively unchanged from 1970 through 1980, at around 450 pounds an acre. However, they began to trend upwards afterwards, and in 1985 reached 628 pounds, the highest in history to that point. Adverse weather, other than in 1980, also affected yield during the period. In 1974, wet weather in the spring followed by drought in the summer reduced yields. Drought also reduced yields in 1978. Further, the El Niño weather pattern resulted in seriously reduced yields in 1983.

Bales Harvested

Annual bales harvested were extremely variable in the United States during the period 1970-1985, ranging from eight million in 1975 to 15.6 million in 1979 (Figure

15.42), the highest since 1962. Adverse weather affected bales harvested in five of the 16 years in the period. As detailed in a preceding section, wet weather and flooding resulted in reduced planting in 1975 and 1983 (Figure 15.40), and wet weather and/or drought affected yields in 1973, 1978, 1980, and 1983. Even with the degree of variation, yields and bales harvested generally trended upward during the period. The average production for the entire period was 11.8 million bales a year.

Alabama Cotton

Agricultural experts have long pleaded with Alabama farmers to diversify their agriculture and to plant less cotton. Farmers made some progress in planting and harvesting fewer acres of cotton in the 1920s and 1930s. However, as late as 1929 cotton was harvested from 50% of all of the cropland harvested in the state, but by 1944 they harvested cotton from only 22.2%. Unfortunately, the percentage would not go this low again, at least through 1969. In that year, the percentage was back up to 33.4%.

In 1970, Alabama farmers harvested 507,000 bales of cotton. Madison County reported the largest share (11.3%) of the total crop. Other counties reporting large shares included: Limestone (8.9%), Lawrence (6.4), Cherokee (4.8), and Colbert (3.6). These five counties accounted for 35% of the total. Some 14 counties reported harvests of 10,000 or more bales. Figure 15.43 shows the distribution of bales harvested in the state in 1970. By 1970, south Alabama was mostly out of the cotton business.

Prices

Annual average prices received by farmers in Alabama for their cotton were essentially the same as the national averages during the period 1970-1985; consequently, only the national data are presented in Figure 15.39.

Acres Planted and Harvested

In Alabama, cotton acreage had been trending downward since the 1930s. Farmers planted 1.3 million acres of the crop in 1950, but by 1967 acres planted were down to 513,000. The general decline finally ended in 1967, and acres planted began to increase during the last years of the 1960s.

In 1970, Alabama farmers planted 565,000 acres of cotton (Figure 15.44). They continued to increase their acreage through 1972 when they planted 601,000. This general increase in acres planted during the latter part of the 1960s and early 1970s was likely in response to the rapid increase in market prices (Figure 15.39) and the general feeling of euphoria concerning the future of agriculture. In 1973, acres planted declined sharply to 525,000 as the result of widespread flooding and generally wet weather during the spring. It rebounded to 600,000 acres in 1974 and remained near that level through 1974, before beginning a long-term decline that would take it down to 330,000 acres in 1985. Heavy rains and flooding in 1979 and again in 1983 (El Niño) also resulted in significantly reduced planting in the state.

Data on acres of cotton harvested in Alabama are also presented in Figure 175.44. The data show that in most years Alabama farmers harvested cotton from at least 96% of their planted acres. In 1969, 1974, 1978, and 1982, acres harvested represented 33.4, 33.2, 24.7, and 25.4%, respectively of total harvested cropland. These percentages were not much different from those in the 1930s and 1940s.

Yields

The overall trend in annual average Alabama cotton yields was similar to that of the country as a whole during the 1970s (Figure 15.41), although as might be expected, they were much more variable than the national average. During the period 1972-1977, state yields fell far below national yields. In 1977, as a result of drought conditions in the state, the yield was only 337 pounds an acre. It had not been that low in a decade. In the same year, the national yield was 520 pounds. Fortunately, in 1978 state yields increased to a level slightly above the national average. Yields also fell sharply in 1980 and again in 1983, as a result of adverse weather. In years without the poor weather conditions, Alabama cotton yields seemed to increase.

Bales Harvested

With yields (Figure 15.41) increasing relatively little in the period 1970-1985 and with acres harvested (Figure 15.44) generally trending downward, bales harvested also trended downward for most of the period (Figure 15.45). There were 640,000 bales harvested in 1971, the highest since 1965 (852,000 bales). From the 1971 high, they trended generally downward, reaching a low of 472,000 bales in 1981. Then, with the exception of 1983, the period ended with slightly better yields and a slight increase in acres harvested.

The 640,000-bale Alabama harvest in 1971 represented 6.3% of the 10.1 million-bale national crop harvested in the same year. In 1849, Alabama had harvested almost 23% of the country's cotton crop.

Corn

Historically, in terms of acres harvested, corn has been the most important crop produced in the country. In 1879, total harvested cropland was 166.2 million acres. This was the first year that cropland harvested data were available. In that year, American farmers harvested corn from 62.2 million acres of farmland, or 37.4% of the total acreage. By 1919, the percentage had declined to 25%. In that year, percentages for cotton, wheat, and oats were 9.9, 14.4, and 11.4%, respectively. From 1919 through 1969, the percentage for corn ranged from 20% (1969) to 24.4% (1924 and 1939).

US Corn

In 1970, American farmers harvested 4,152 million bushels of corn. Iowa farmers harvested the largest share (20.9%) of the total national crop. Other states reporting large shares included: Illinois (17.7%), Minnesota (9.2), Indiana (9.0), and Nebraska (8.8). These five states accounted for 65.6% of the total. Alabama farmers harvested 0.35% of the national total.

Exports

American corn exports increased rapidly from 1970 to 1979, along with exports of all agricultural commodities (Figures 15.6 and 15.46). Corn exports in both 1979 and 1980 were near 2.4 billion bushels. Then, as inflation soared in the early 1980s, corn market prices followed (Figure 17.46). At the same time, US currency became highly overvalued. As a result, corn exports, along with the export of other agricultural commodities, fell rapidly. They were down to only 1.2 billion bushels in 1985.

Prices

The government slowly increased support prices (subsidies) for corn from $1.06 a bushel in 1960 to $1.35 in 1970 (Figure 15.47). They remained near that level through 1975, but with inflation at 9%, they began to increase; by 1984 and 1985, they were $3.03.

American corn market prices generally remained below $1.20 a bushel from 1960 through 1965; then between 1965 and 1966, when there was little change in bushels harvested (Figure 15.50), they increased to $1.24. However in 1967, bushels harvested increased to 4.9 billion, the highest in history, and the market price promptly fell to $1.03. In 1968, bushels harvested began to decline, and market prices began to increase again. In 1970, with bushels harvested at a five-year low, the market price reached $1.33. Bushels harvested set a new record in 1971 (5.6 billion), and the market price fell to $1.08 (teeter totter).

With the Consumer Price Index (Figure 15.2) and corn exports (Figure 15.46) increasing rapidly, market prices for corn began to increase sharply in 1972, and by 1974 they reached $3.02. In 1975, 1976, and 1977, bushels harvested continued to set new records; in 1977 the market price fell to $2.02. In 1978, inflation was between 7 and 8%, and on it way up to between 13 and 14% in 1980; the market price for corn followed. It reached the unprecedented $3.12 a bushel in 1980. In that year, the federal discount rate was raised to 13%, and in 1981 both inflation and exports began to decline sharply. In 1981, the marker price of corn fell to $2.47. It recovered to $2.55 in 1982. In 1983, bushels harvested fell to 4.2 billion, as a result of the crop-weather disaster (El Niño), and market price quickly went to a new high of $3.21. In 1984 and 1985, bushels harvested began to recover, and market prices began to decline, finally reaching $2.23 in 1985.

Market prices for corn were below support prices for 1960 through 1971, but generally above them from 1972 through 1983, except for 1982. Then in 1984 and 1985, they were considerably lower. The overproduction of corn had returned with a vengeance.

Acres Planted and Harvested

American farmers planted 64.3 million acres of corn in 1969 (Figure 15.48), the lowest in recorded history at that time. Afterwards, acres planted followed market prices upward (Figure 15.47) through 1976 (84.6

million). Subsequently, market prices bounced rapidly back-and-forth between $2.00 and $3.20 a bushel, and acres planted generally followed. Market prices changed so rapidly and so sharply that it was difficult for farmers to make adjustments in planting. Acres planted were 60.2 million acres in 1983, a result of the El Niño crop-weather disaster. In 1983, the market price had responded to the small crop by going to $3.20 a bushel. Then in 1984 and 1985, farmers responded by planting over 80 million acres.

Data on acres harvested are also presented in Figure 15.48. These data indicate that American farmers harvested around 86% of acres planted in most years. Many farmers in the Corn Belt harvested large amounts of their corn for silage. The data shown in the figure pertain only to corn harvested for grain.

Yields

Annual average, national corn yields had begun to increase in the 1940s, and by the end of the 1960s had reached around 80 bushels an acre. Although there continued to be considerable year-to-year variation, they continued to trend upward during the 1970s, finally reaching 120 bushels an acre in 1985 (Figure 15.49). Adverse crop growing weather resulted in lower than expected yields in 1970, 1974, 1980, and 1983. The El Niño effect was especially damaging in 1983, resulting in extremely wet weather in the Western Corn Belt.

Bushels Harvested

As a result of the combination of the upward trends in acres planted (Figure 15.48) and average yields (Figure 15.49) bushels of corn harvested nationally increased from 4.1 billion in 1970 to 8.9 billion in 1985 (Figure 15.50). Although increases in acres planted and yields both affected total harvests, trends in yields seemed to be the more important factor. Bushels harvested in 1970, 1974, 1980, and especially 1983 were much lower than expected. In all four years, lower production was the result of both reduced acres harvested and lower yields. Clearly, corn farmers accepted the glowing predictions for American agriculture advanced by the federal government in the mid-1970s.

Alabama Corn

As late as 1959, corn was harvested from 48.6% of total harvested cropland in Alabama, but by1969 farmers in the state were apparently realizing the futility of living with those relatively low yields. Consequently, by 1969 the percentage was down to 24%.

In 1970, in a year when adverse weather severely limited the production of most crops, Alabama farmers harvested only 14.7 million bushels of corn. DeKalb County reported the largest share (6.9%) of the state's total. Other counties reporting large shares included: Houston (6.2%), Geneva (5.7), Jackson (5.4), and Coffee (3.5). These five counties accounted for only 27.8% of the total. Some 25 counties reported harvests of 200,000 or more bushels. Figure 15.51 shows the distribution of bushels harvested within the state in 1970.

Prices

Data on annual average prices that Alabama farmers received for their corn during the period 1970-1985 are shown in Figure 15.52. These data show that the year-to-year changes were essentially the same for both the state and the country. However, over the years, Alabama prices varied from seven to 48 cents a bushel higher than national prices.

Acres Planted and Harvested

By 1970, acres of corn planted in Alabama had declined to around 785,000 (Figure 15.53). Initially, state farmers did not respond to the rapidly increasing market prices at the beginning of the 1970s (Figure 15.52) by planting more of the crop. During the period 1970-1974, with the exception of 1973, Alabama farmers only planted around 700,000 acres a year. However, beginning in 1975, they finally caught the fever, and in 1976 planted 880,000 acres. Unfortunately, by the time they harvested that crop the market price had fallen to $2.15. The following year, they began to reduce acres planted, and by 1979 acreage was down to 502,000. Market price spiked again in 1980, and in 1981 Alabama farmers increased acres planted slightly. Once again, however, by the time they harvested that crop, the market price had fallen to $2.47. They again increased acres planted slightly in 1984, but with the same results. In 1985, acres of corn planted in Alabama were down to 370,000. In 1935, it had been almost 3.8 million acres of corn in the state.

Adverse weather conditions contributed to the year-to-year variation in acres planted during the period. Flooding and wet weather resulted in reduced planting in 1973, 1980, and 1983 (El Niño). In those years, planting was delayed so long that some farmers decided not to plant at all.

Data on acres of corn harvested in Alabama are also presented in Figure 15.53. These data indicate that Alabama farmers generally harvested from 80 to 88% of acres planted, a lower percentage than in the country (Figure 15.48). In 1969, corn was harvested from 24.2% of all harvested cropland in the state; however, by 1982 this percentage was down to 11.6%.

Yields

Data on annual average yields of corn in Alabama are also presented in Figure 15.49. These data show that although there was considerable year-to-year variation, average yields probably increased from around 40 to 60 bushels an acre during the period 1971-1984. They were lower than expected in 1974, 1977, 1980, and to some extent in 1983, as a result of adverse crop weather during the growing season. The data also indicated that yields on Alabama farms were increasing at a much slower rate than in the country as a whole. In the 1950s, Alabama yields were slightly less than 20 bushels lower than in the country; however, by 1985 it was more than 40 bushels lower.

Bushels Harvested

With acres planted (Figure 15.53) generally declining steadily and yields (Figure 15.49) generally trending upward, bushels harvested remained at about the same level during the period 1971-1981 (Figure 15.54). It was 28.2 million bushels in 1971 and 29.2 million in 1981. It was lower after 1981 as a result of rapidly declining levels of acres planted. In 1985, it was down to 24.4 million. Bushels harvested were lower than expected in 1970, 1977, 1980, and 1983 (El Niño) as a result of the low yields in those years. Bushels harvested were higher than expected in 1975 and 1976, as a result of the fortuitous combination of higher levels of acres planted and higher than usual yields. The 48 million bushels of corn harvested in Alabama in 1976, the largest crop harvested during the period 1970-1985, represented less than 1% of the 6.3-billion-bushel national crop that year.

Alabama Peanuts

In 1970, farmers in Georgia harvested the largest share (37.7%) of the 2.98-billion-pound American peanut crop. Other states reporting large shares included: North Carolina (14.9%), Texas (14.4), Alabama (10.6), and Virginia (10.5). These five states accounted for 88.1% of the entire national crop. In the same year, the Alabama crop totaled 315.4 million pounds. Farmers in Houston County harvested the largest share (21.6%). Other counties reporting large shares included: Henry (17%), Coffee (11.9), Geneva (10.4), and Barbour (10.2). These five counties accounted for 71.2% of the state's total crop.

US and Alabama Prices

In 1941, the government began to set support prices on peanuts at a level usually exceeding the cost of production. In 1970, the federal subsidy was set at 13 cents a pound (Figure 15.55). Afterwards, with inflation increasing rapidly (Figure 15.3), support prices were increased almost yearly until they reached 28 cents in 1982. Then with inflation declining, they were left at that level through 1985.

Generally, with acreage controls in place (Figure 15.56) throughout this period (1970-1985), market prices remained near support prices. However, they were a little higher in 1980 and 1981 and a little lower in 1982, 1983, and 1985. The greater than expected variations in market prices after 1977 are likely the result of the new two-tiered peanut program established under the Food and Agricultural Act of 1977. Under this program, farmers could produce a pre-determined quantity (quota) of peanuts to be sold on the American market for a guaranteed price (support price), or they could produce an unlimited quantity of the crop, but the quantity in excess of their quota had to be sold outside the United States. Alternatively, they could sell the excess in the country and also receive a guaranteed price for those "additional" peanuts, but it was set at less than 20% of the quota or support price. Those guaranteed high prices were generally much greater than those on the world market; consequently to prevent the importation of large quantities of the cheaper peanuts, the federal government (Triangle) established high tariffs, ranging from 104 to 160%, to prevent this from happening. Although the peanut was not a very important national crop, it had powerful friends in high places.

Acres Planted and Harvested

Federal agricultural programs had essentially frozen the country's peanut acreage at 1.5 million before the planting of the 1959 crop. As a result, acreage remained at near this level throughout the 1960s. This program was continued until 1977, and acres planted remained relatively constant at 1.5 million acres under the new two-tiered program. After 1978, farmers were required to own both a quota allotment and an acreage allotment.

The various acreage control programs kept acres of peanuts planted in Alabama near 200,000 acres from 1970 through 1980 (Figure 15.56). After 1982, the mandatory acreage-allotment requirement was abandoned. As a result of all of these changes in programs and market prices, Alabama farmers increased acres planted to 224,000 in 1981, but reduced it to near 180,000 in 1982 and 1983. They increased acreage to 221,000 in 1984, but in 1985 it was back at 201,000—near where it had been since 1970.

Data on acres harvested are also presented in Figure 15.56. They show that Alabama farmers harvested between 95 and 99% of acres planted during this period. Peanuts had become too valuable to be hogged-off as they had been in the past.

US and Alabama Yields

Data on annual average peanut yields in Alabama and Georgia are presented in Figure 15.57. Yields in Alabama were consistently several hundred pounds lower than in Georgia during the period 1970-1985. At least a portion of this difference was probably a result of the fact that Georgia was beginning to irrigate some of their peanuts during this period. Alabama yields increasing rapidly during the early 1970s, reaching 2,700 pounds an acre in 1977. Then in 1979, Alabama peanut farmers harvested 2,800 pounds an acre, the highest level on record at that time. In 1980, yields plummeted to 1,300 pounds an acre, mirroring the severe drought-induced decline experienced nationally. In the early 1980s, yields reached 2,900 pounds, and with the exception of 1983, remained near that level through 1985. Lower than expected yields as a result of adverse weather were most evident in 1980; however, dry weather during the growing season also resulted in lower yields in 1972, 1973, 1976, 1978, and 1983.

Pounds Harvested

Even with acreage controls, pounds of peanuts harvested in Alabama trended sharply upward throughout the 1970s, because yields were increasing (Figure 15.57). In 1970, Alabama peanut farmers harvested 315 million pounds of the crop, or 10.6% of the country's peanuts. Then, with acres harvested increasing only slightly, pounds harvested followed yields up and down through 1985. The 1979 crop was the third largest ever produced in the state (584.8 million pounds). That large harvest represented 15% of the national crop of near four billion pounds. That large 1979 crop was eclipsed by crops harvested in 1981 (602.7 million) and 1984 (648.6 million).

Soybeans

American soybean production grew at an unparalleled rate after 1953. Between 1953 and 1968, bushels harvested increased by over threefold. Equally telling is the fact that in 1954 the harvest of soybeans accounted for only 5.1% of total cropland harvested, but by 1969 that percentage had almost tripled to 15.1%. In the same year, the corn harvest only accounted for 20%. The old Asian native, after sleeping for awhile in the New World, was quickly making a place for itself.

US Soybeans

In 1970, the country's farmers harvested 1.1 billion bushels of soybeans. Illinois farmers harvested the largest share (18.7%) of the total crop. Other states reporting large shares included: Iowa (16.4%), Indiana (9.0), Arkansas (8.8), and Missouri (7.8). These states accounted for 60.7% of the total crop. Alabama farmers harvested 1.2% of the national crop.

Exports

Data on American soybean exports for the period 1950-1985 are presented in Figure 15.59. The trend was essentially the same as for all agricultural exports, except that the rate of increase was slightly higher for all exports (15.6). Soybean exports were 434 million bushels in 1970, and they generally trended upward to 875 million bushels in 1979. They were lower than expected in 1974 because adverse weather in the Midwest resulted in significantly reduced production. They also declined in 1980 as a result of adverse crop weather (El Niño), but then reached the maximum for the period in 1981 (929 million.) The Federal Reserve raised the discount rate to 14% in 1981. Afterwards, soybean exports generally trended downward with all agricultural exports.

Prices

Soybean support prices were kept relatively stable at around $2.25 a bushel throughout the 1960s and early 1970s, but when the parity ratio began to fall rapidly (Figure 15.8) after 1974, the government began to increase support prices (Figure 15.60). They were $2.25 in 1975, $2.50 in 1976, $3.50 in 1977, $4.50 in 1978 and 1979, and $5.50 in 1980; then they remained at that level through 1985.

Market prices were $2.85 a bushel in 1970 (Figure 15.60), but with the Consumer Price Index increasing rapidly, they also began to increase rapidly. By 1974, market prices reached $6.64, an unheard of price for soybeans. At that level, market price was almost $2 a bushel above the federal support price. Then market price hit a wall of resistance, and it bounced around—generally depending on annual bushels harvested (Figure 15.63)— between $5.00 and $7.83 for the remainder of the period. For example, in 1980, because of poor crop-weather, bushels harvested fell to 1.8 billion, and market prices responded by increasing to $7.60. By contrast, in 1982 bushels harvested increased to 2.2 billion, and the market price fell to $5.71 (teeter totter).

Acres Planted and Harvested

With high prices (Figure 15.60), increasing exports (Figures 15.59), and expectations of long-term market price stability, American soybean farmers responded by increasing acres planted. In 1970, they planted 43 million acres, and by 1973 acres planted had increased to 56.5 million. But when market prices hit-the-wall after 1974 (Figure 15.60), farmers quickly got the message, and acres planted fell to 52.5 million. It remained near that level through 1976. In that year, market prices moved upward to $6.81 a bushel, and farmers quickly lost their caution. Acres planted increased in 1977 (59 million), 1978 (64.7 million), and 1979 (71.4 million). Unfortunately, when they harvested the 1979 crop, the market price was down to $6.29 a bushel. At that price, acres planted began to trend downward. With the index of prices paid above 1,100 (Figure 15.7), they had little encouragement to increase their acreage. It simply cost too much to put in more soybeans. In 1985, acres planted was down to 63.1 million acres.

In addition to the ups and downs of market prices, adverse weather conditions played a role in the acreage of soybeans that farmers planted. This was especially true in 1983 when the 1982-1983 El Niño weather pattern resulted in widespread flooding during the planting season. The wet conditions lasted so long that many farmers decided not to plant at all.

Data on acres harvested are also presented in Figure 15.61. These data indicate that farmers generally harvested 97 to 98% of their planted acres. In 1969, acres of soybeans harvested accounted for 15.1% of all harvested cropland. By 1982, the percentage had grown to 21.3%. In the same year, acres of corn harvested accounted for 22.3% of all harvested cropland.

Yields

Although there was considerable year-to-year variation, American soybean yields generally trended upward during the period 1970-1985 (Figure 15.62). Data taken from a trend line would probably indicate that annual average yield increased from 27 bushels an acre in 1970 to 32 bushels in 1985. Somewhat reduced yields in 1970, 1974, 1976, 1978, 1980, and 1983 reflected the adverse weather (primarily lack of soil moisture) during the planting and growing season in the heart of Soybean Belt. The lower than expected yield in 1983 was the result of the El Niño effect.

Bushels Harvested

With both yields (Figure 15.62) and acres harvested (Figure15.61) generally increasing during the period 1970-1985, bushels harvested increased from 1.1 billion bushels in 1970 to 2.3 billion in 1979 (Figure 15.63). Afterwards, with year-to-year variation increasing, bushels harvested varied between 1.6 billion and 2.2 billion bushels. The adverse-weather-related, lower-than-expected yields in 1974, 1976, 1978, 1980, and 1983 translated into lower-than-expected bushels harvested in those years.

Alabama Soybeans

Soybean production also exploded in Alabama in the 1950s and 1960s. Apparently, farmers had finally found a crop to replace cotton, and they climbed on the soybean bandwagon in a rush. They harvested almost tenfold more bushels of the crop in 1970 than in 1950. In 1949, the harvest of soybeans accounted for less than 1% of total cropland harvested. but by 1969 the percentage had increased to 23.3%. In the same year, cotton accounted for 20%. In 1970, Alabama farmers harvested 13.8 million bushels of soybeans.

Farmers in Baldwin County harvested the largest share (20.3%) of the 1970 crop. Other counties reporting large shares included: Madison (6.5%), Limestone (5.3), Escambia (4.8), and Mobile (4.4). These five counties accounted for 40.7% of the state's crop. A total of 20 other counties reported harvests of 200,000 or more bushels. Figure 15.64 shows the distribution of bushels harvested in Alabama in 1970.

Prices

Data on the annual average market price of soybeans for the state and the country for the period 1970-1985 are presented in Figure 15.65. The year-to-year changes were almost identical in both the state and the country as a whole. It increased rapidly from $2.31 a bushel in 1970 to $7.01 in 1974. Thereafter, it see-sawed between $4.88 and $7.80 through 1985.

Acres Planted and Harvested

Data on acres of soybeans planted and harvested in Alabama during the period 1950-1985 are presented in Figure 15.66. The trends in state and national data were similar from 1970 through 1977. Acres of soybeans planted were 625,000 in 1970. Then they followed market prices up to 2.2 million in 1979. Acres planted in 1974 and 1976 were slightly lower than expected because of adverse weather in the planting season. Like national data, acres harvested in Alabama began to trend downward after 1979; however, the decline in the state was much more precipitous. In 1985, it reached 1.1 million acres. Alabama farmers seemed to be quickly losing interest in the crop. Data on acres harvested are also presented in Figure 15.66. The data indicate that farmers generally harvested from 95 to 97% of their planted acres during the period.

Yields

Data on annual average yields of soybeans in Alabama during the period 1950-1985 are also presented in Figure 15.62. Although there was considerable year-to-year variation, it was not great enough to effect the conclusion that there seemed to be no trend. Alabama soybean yields generally remained unchanged during the period, while it generally trended upward in the country as a whole. This is likely the reason that Alabama farmers began to lose interest in soybeans. By 1985, the average yield in the counry was seven to 10 bushels an acre higher. Average yields in Alabama were lower than expected in 1972, 1973, 1976, 1978, 1980, and 1983 as a result of adverse weather (generally dry weather) conditions. The 1980 yield of 15 bushels an acre, the lowest in years, resulted from a prolonged drought.

Bushels Harvested

The effects of the threefold increase in acres harvested (Figure 15.66) and the generally unchanged yields (Figure 15.62) combined to increase significantly the bushels of soybeans harvested in the state from 1970 through 1979 (Figure 15.67). The 2.2 million acres harvested in 1979, plus the higher than usual yield of 25 bushels an acre, combined to push bushels harvested to almost 54 million, the largest soybean crop ever produced in the state. It was four times greater than the 1970 crop of 13.8 million bushels. After 1979, bushels harvested generally followed acres harvested downward and reached 28.7 million in 1985. Bushels harvested was lower than expected in 1972, 1974, 1976, 1980, and 1983 as a result of lower than expected yields. Although Alabama farmers harvested a record crop of soybeans in 1979, it represented only 2.4% of the 2.3 billion-bushel national crop that year.

Alabama Hay

In 1970, American farmers harvested 127 million tons (dry) of all kinds of hay. The largest share (8.4%) was harvested on Wisconsin farms. Other states reporting large shares included: Minnesota (6.3%), California (5.8), Iowa (5.4), and Nebraska (4.8). These five states accounted for 30.7% of the total. Alabama farmers harvested 0.7% of the national hay crop.

Tons of Alabama hay harvested increased from

678,000 in 1950 to 930,000 in 1970; however, during those 21 years, the crop went through several cycles that were probably related to Cycles of Abundance of Alabama cattle during the same period. In 1970, Alabama farmers harvested 930,000 ton of hay. Based on census reports, farmers in Montgomery County harvested 5.6% (the largest share) of all the acres of hay (excepting sorghum hay) harvested in Alabama that year. Other counties with large shares included: Hale (3.7%), Marengo (3.3), Morgan (3.1), and Dallas (3.0). These five counties accounted for only18.7% of all the acres harvested in the state. A total of 12 counties reported harvests of 10,000 or more acres.

Prices

Alabama hay prices increased rapidly from 1970 ($28 a ton) through 1977 ($56.50 a ton), an increase of 102% (Figure 15.68). However, the rapid increase in the Consumer Price Index (116.3 to 181.5) (Figure 15.2) probably was responsible for over half of that increase. This was also a period of rapid expansion of the Alabama cattle herd. Although the CPI continued to increase at a rapid rate from 1977 through 1985 (195.3 to 318.5), the increase in hay prices moderated considerably. This moderation was likely the result of declining cattle numbers after 1976. Although prices increased more slowly, they finally reached $63 a ton in 1983, 1984, and 1985.

Acres Harvested

Acres of hay harvested in Alabama during the period 1970-1975 generally followed prices upward (Figures 15.68 and 15.69). Some 536,000 acres were harvested in 1970, and 630,000 in 1975. In 1986, acres harvested were reduced by adverse weather. Then in 1976, cattle numbers began to decline rapidly in the state. Afterwards, the rate of increase in acres harvested slowed, and only 700,000 acres were harvested in 1985. Acres harvested were also reduced in 1980 and 1983 by adverse weather.

Yields

Alabama hay yields had increased sharply between 1955 and 1970 (Figure 15.70), but increased very little through 1983. In most of these years, yields varied between 1.6 and 1.8 tons an acre. However, they increased sharply in 1984 and 1985 to reach 2 and 2.2 tons an acre, respectively. The low yields in 1972, 1977, 1980, and 1983 were the result of poor growing conditions.

Tons Harvested

With acres harvested generally increasing throughout the period 1970-1985 (Figure 15.69) and yields increasing after 1982 (Figure 15.70), tons of hay harvested trended upward for the entire period. It was 930,000 tons in 1970 and 1.2 million tons in 1982. Afterwards, it exploded upwards to 1.4 million in 1984 and to 1.5 million in 1985. Tons harvested were somewhat lower than expected in 1972, 1976, 1980, and 1983 as a result of adverse weather.

Alabama Winter Wheat

American farmers harvested 1.1 billion bushels of winter wheat in 1970. Kansas farmers harvested the largest share (27.4%) of the total crop. Other states reporting large shares included: Oklahoma (9.3%), Nebraska (8.4), Washington (8.3), and Colorado (5.4). These five states accounted for 58.8% of the total. Alabama farmers harvested 0.2% of the total national crop. Baldwin County reported the largest share (20.8%) of the total. Other counties with large shares included: Colbert (8.9%), Escambia (7.7), Lauderdale (4.4), and Madison (4.0). Some 22 other counties reported harvests of 30,000 or more bushels. Figure 15.72 shows the distribution of bushels of winter wheat harvested in Alabama in 1970.

US Exports

Data on the quantity of wheat (all) exported by the United States in the period 1950-1985 are presented in Figure 15.73. Wheat exports were 741 million bushels in 1970. Then in the 1970s, they increased sharply—with all agricultural exports (Figure 15.6)—as a result of sharply increased shipments to Russia, China, and North and West Africa. However, between 1972 and 1974, wheat market prices in the United States increased from $1.71 to $3.90 (Figure 15.74), and that increase effectively curtailed increasing exports. While exports did not decline, they did remain near the same level (1.1 billion bushels) until after 1978. Then with market prices falling after 1974, exports began to increase again, reaching 1.8 billion bushels in 1981. When the federal discount rate went to 14% in 1981 (Figure15.4), all agricultural exports, including wheat, began to decline. Wheat exports finally reached 909 million in 1985.

US and Alabama Prices

In the late 1960s, when winter wheat market prices declined to $1.20 a bushel (Figure 15.74), the lowest since 1942, the government immediately began to increase support prices (Figure 15.73). In 1970, the subsidies were up to $2.82, and in 1973 to $3.39. When market prices went to $3.72 in 1973 (Figure 15.74), support prices were quickly reduced, and by 1974 and 1975, they were down to $2.05. But when market prices plunged to $2.28 in 1977, the federal government began to increase support prices once again. It continued to increase them through 1984 and 1985 ($4.38).

In 1968, American farmers harvested the largest wheat crop on record at that time (1.2 billion bushels), and they received $1.20 a bushel when they sold it. Afterwards, bushels harvested declined, and with rapidly increasing inflation and increased exports (Figure 15.3), market prices spurted upward, reaching $3.90 in 1974 (Figure 15.74). That astronomically high market price really excited farmers, and in 1975 they harvested 1.6 billion bushels, another record crop. That record crop sent market prices spiraling downward to $2.28 in 1977. As a result, acres harvested began to decline again, and in 1978 bushels harvested were down to 1.2 billion. Of course, with acres harvested declining, market prices began to increase, reaching $3.88 in 1980. This time, farmers were determined to finally catch up. In both 1981 and 1982, they harvested 2.1 billion bushels. In 1983, 1984, and 1985, they harvested 2.0, 2.1, and 1.8 billion bushels, respectively. In every one of those years, market prices continued to decline. In 1985, farmers received $2.98 for their crop. In that year, the parity ratio reached 0.51 (Figure 15.8). It was the best of times, it was the worst of times. Even with high government support prices, farmers just could not beat the teeter-totter phenomenon, and it seemed that they would never learn that lesson. Actually, they had little choice but to keep trying.

Market prices were far below support prices for much of the period (Figure 15.74). They were higher in 1973, 1974, 1975, 1976, and again in 1979 and 1980. American farmers were just growing too much wheat.

Figure 15.75 presents data on average annual market prices of winter wheat in the United States and in Alabama during the period 1950-1985. The overall trends for both were essentially the same. In a few of the years, state market prices were considerably lower than national prices. When they were higher for Alabama, the difference was not very large.

Acres Planted and Harvested

Alabama farmers planted 100,000 acres of winter wheat in 1970 (Figure 15.76). They did not initially respond very positively to the increase in market prices in the early 1970s (Figure 15.75). Acres planted did increase in 1971 and 1972, but the increase was not nearly so dramatic in the state as it was in the country as a whole. After declines in 1973 and 1974, acres planted trended slowly upward at a rate of five to 10 thousand acres a year through 1978, when it reached 180,000 acres. However, after American wheat market prices began to increase sharply in 1978, Alabama farmers decided that this time they would join the parade. In the following four years (1979-1982), acres planted in the state increased from 220,000 to 850,000. Unfortunately, in 1981 market prices began to trend sharply downward. As a consequence, Alabama farmers left the parade as quickly as they had joined it. By 1985, wheat acres planted was down to 500,000.

Data on acres of wheat harvested are also presented in Figure 15.76. These data indicate that Alabama farmers harvested 71% of their planted acres in 1970, 71% in 1975, 80% in 1980, and 80% in 1985. Previously, Alabama farmers had planted a substantial amount of wheat as a cover crop and for grazing; however, as acres planted increased, they generally began to harvest a larger share of those acres. In 1949, acres of wheat harvested accounted for only 0.2% of all cropland harvested. By 1969, the percentage had increased to 1.2%; then with the run-up in acres harvested in the early 1980s, the percentage was 22.2% in 1982. Imagine farmers in the subtropics being in the wheat business.

US and Alabama Yields

Annual average yields of American winter wheat generally trended upward between 1970 and 1985 (15.77). It was 33 bushels an acre in 1970 and near 42 bushels in 1983. Yields were lower in the period 1971-1977 as a result of several years with adverse crop weather. The El

Niño effect in 1983 was especially damaging. In 1970, the annual average yield in Alabama was 28 bushels an acre (Figure 15.77), but it declined to 20 bushels as a result of a widespread outbreak of the so-called rust disease, which devastated the crop. Adverse weather also reduced yields to some extent from 1972 through 1979. Regardless of the adverse weather, however, yields slowly trended upward through 1978 to 26 bushels, before beginning to trend downward again, finally reaching 25.5 bushels in 1980, another year with adverse weather. Finally, in 1981 all of the signs were aligned, and farmers harvested 44 bushels of wheat an acre, the highest in history to that time. Afterwards, they declined to range between 32 and 39 bushels an acre through 1985. Throughout the entire period, except for 1981 and 1984, average yields of wheat in Alabama were from four to 12 bushels an acre lower than the national average.

Bushels Harvested

Annual trends in bushels of winter wheat harvested in Alabama during the period 1970-1985 were similar to those in acres harvested (Figures 15.78 and 15.76). Bushels harvested remained relatively stable at around two million from 1970 through 1979. It increased dramatically to 24.9 million in 1981, but declined to 12.8 million in 1985. The largest crop of winter wheat ever harvested in Alabama (24.9 million bushels, in 1981) represented only 1.2% of the national crop of 2.1 billion bushels that year.

Alabama Oats

In 1970, American farmers harvested 915.2 million bushels of oats. Minnesota farmers harvested the largest share (18.3%) of the total national crop. Other states reporting large shares included: North Dakota (13.1%), Wisconsin (11.6), South Dakota (11.2), and Iowa (10.3). These five states accounted for 64.5% of the total. Alabama farmers harvested 1.2 million bushels of the feed grain, or 0.13% of the national total.

US Exports

Large quantities of oats have historically been raised in Europe. American exports of oats to Europe have never been very high, except when war or adverse weather have reduced production there. In 1970, American oat exports totaled 19 million bushels (Figure 15.79). That year, they amounted to 2.1% of total bushels harvested in the United States. Exports remained at about the same level in 1971 and 1972 then rocketed upward to 57 million bushels in 1973. They had not been that high since 1918. In 1973, crop weather was bad in much of Europe, especially in the Soviet Union. Afterwards, exports generally trended downward to reach one million bushels in 1982. They remained at that level through 1985.

US and Alabama Prices

With market prices holding steady at around 60 cents a bushel throughout the 1960s, support prices were set at near the same level (Figure 15.80). Subsidies were 63 cents in 1970 but were lowered to 54 cents during the period 1971-1975. Then with inflation increasing (Figure 15.3), the government began to increase support prices. In 1977, they were $1.03. The rate of increase slowed afterwards, and in 1980 they were $1.08. Market prices had increased rapidly during that period. In 1980, with the inflation rate between 13 and 14%, support prices began to increase once more, and in 1983 they reached $1.60. They remained at that level through 1985.

The average annual, American market price for oats was 62 cents a bushel in 1970, but with the inflation rate at 6% (Figure 15.3), they began to increase rapidly, reaching $1.53 in 1974 (Figure 15.80). Between 1974 and 1975, bushels harvested increased from 600.7 million to 639 million, and as a result, the market price fell to $1.45. Bushels harvested declined in 1976, and the market price edged-up to $1.56. In 1977, bushels harvested increased to 752.8 million, and the market price immediately fell to $1.09. Bushels harvested generally declined through 1981; market prices responded by increasing. They were $1.88 in 1981. Subsequently, the teeter-totter phenomenon carried market prices down, up, and finally down to $1.23 in 1985.

Market prices were higher than, or equal to, support prices in 15 of the 16 years of the period.

Data on the annual average market prices for oats in the United States and Alabama during the period 1950-1985 are represented in Figure 15.81. Generally, year-to-year trends were the same for both. However, in most of the years, Alabama's market prices were around

20 cents a bushel higher.

Acres Planted and Harvested

Alabama farmers gave scant attention to the up's and down's of support prices and market prices for oats during the period 1970-1985. They had grown disenchanted with oat production after 1955. By 1954, there were so few horses and mules on Alabama farms that records on numbers were no longer kept separately, and according to census data, tractors per 100 farms had reached a level of 35. Alabama farmers no longer needed oats to fuel most of their farm power. After 1955, acres of oats planted began to decline, and were down to 119,000 in 1970 (Figure 15.82). The decline continued until acres planted reached 84,000 in 1973. It recovered just slightly, to around 90,000 to 95,000 acres between 1974 and 1982 when market prices were so high (Figure 15.81). However, when prices began to decline, Alabama farmers regained their determination to be done with oats, and acres planted began to decline again. In 1985, acreage of oats reached 80,000 acres. (As late as 1955, Alabama farmers had planted 561,000 acres of the crop.)

Data on acres of oats harvested are also presented in Figure 15.82. Historically, Alabama farmers had planted large acreages of oats as a cover crop and for grazing. In 1955, they only harvested the grain from 26% of their acres planted, and they continued to harvest about the same percentage through 1970. When prices began to increase after 1970, percentage harvested began to increase. They harvested 30% of planted acres in 1975, 33% in 1980, and 44% in 1985.

In 1954, acres of oats harvested in Alabama accounted for only 3.8% of total harvested cropland. By 1969, this percentage was down to 1.2%, and it remained at that level through 1982.

US and Alabama Yields

Data on annual average yields of oats on American and Alabama farms during the period 1950-1985 are presented in Figure 15.83. These data show why Alabama farmers became disenchanted with oats. Average yields in the country were 52 bushels an acre in 1970, and they generally trended upward to 58 bushels an acre in 1984. They were lower than expected in the period 1971-1976 as a result of the adverse crop weather described earlier. The effects of the severe drought in the Upper Midwest in 1976 and the El Niño in 1983 were especially limiting.

In 1970, the average yield in Alabama was almost 10 bushels an acre lower than the national average (40 versus 49.2). The effects of the adverse weather between 1971 and 1976 are also apparent in the Alabama data. In 1980, Alabama yields were still about 10 bushels an acre lower (42 versus 53); however, with the average price of $1.70 a bushel (it had never been that high before), Alabama farmers seemed to have pulled out all the stops on their cultural practices. For once, they seemed to do everything right, and in 1981 the average yield went up to 60 bushels an acre. In fact, it was slightly higher than the national average. Unfortunately, even with that high yield and the high prices, Alabama farmers apparently still could not seem to make any money on the crop, so yields drifted back down to 40 bushels in 1985. Even with the considerable year-to-year variation, Alabama oat yields did seem to drift upward during the period.

Bushels Harvested

With acres harvested in Alabama changing relatively little from 1970-1985 (Figure 15.82), changes in bushels harvested (Figure 15.84) were primarily determined by the slight increase in yields (Figure 15.83). Bushels harvested remained relatively stable at around 1.1 to 1.4 million from 1964 through 1970. Thereafter, they declined as a result of the lower yields. Then they followed yields upward to reach 2.4 million bushels in 1981; then followed them down to reach 1.4 million in 1985. The old European oat genome had never been suited for Alabama's subtropical environment. After over one-and-one-half centuries of trying to force it to fit, Alabama farmers finally seemed willing to agree with Mother Nature that it could not be done in a globalized, competitive world market. In 1985, only 0.3% of the country's oat crop was harvested in Alabama.

Alabama Sweet Potatoes

In 1970, Alabama ranked eighth in the country in the production of sweet potatoes. Acres of sweet potatoes harvested in Alabama during the period 1970-1985 are presented in Figure 15.85. Acres harvested remained relatively unchanged at around 5,000 throughout the period. It is difficult to believe the fact that Alabama farmers harvested 35,000 acres of sweet potatoes in 1950.

Although acres harvested remained relatively stable during the period, annual average yields increased from 85 Cwt an acre to 120 during the period 1970-1985. As a result, Cwt harvested also generally increased. It was 398,000 Cwt in 1970; 522,000 in 1975; 530,000 in 1980; and 679,000 in 1984. Severe drought reduced yields and Cwt harvested in 1977, 1980, and 1983.

Cash receipts for sweet potatoes were $1.7 million in 1970, $3.2 million in 1975, $8.7 million in 1980, and $11.4 million in 1985. This general upward trend in receipts reflects the upward trend in production and inflation.

In 1985, cash receipts for sweet potatoes accounted for 1.3% of cash receipts for all crops, including farm forest products, sold in the state that year.

Alabama Irish Potatoes

Alabama ranked 20th in the country in the harvest of Irish potatoes in 1970. Data on the acres harvested of the crop during the period 1970-1985 are presented in Figure 15.86. Acres harvested were 16,700 in 1970. Farmers had harvested 32,000 acres of the crop in 1950. Between 1972 and 1973, prices of Irish potatoes increased from $2.97 to $9.45 per Cwt. As a result, acres harvested also increased, and reached 23,000 in 1974. Prices declined somewhat after 1974, and acres harvested began to decline. Acreage harvested was 14,500 in 1980 and 13,000 in 1985. The decline in acres harvested over the period could not have been the result of a decline in prices alone. Average prices were $11.90 per Cwt in 1978 and $11.00 in 1981, and those unusually high levels seemed to have little or no effect on acres harvested. Years earlier, Alabama farmers had apparently lost confidence in their ability to compete in a national market.

Even though acres harvested generally trended downward during the period, the decline was not great enough to offset the relatively large year-to-year variation in yields (104 to 157 Cwt an acre). As a result, there was no real trend in Cwt harvested. During the period, it ranged from 1.1 (1980) to 3.3 (1974) million Cwt.

As a result of prices below $4 per Cwt in 1970, 1971, and 1972, crop values in those years were below $8.4 million. However, prices were substantially higher afterwards, and crop values increased. Value was $27.9 million in 1975 and $11.8 million in 1985.

Cash receipts for Irish potatoes accounted for 1.4% of crop receipts for all crops, including farm forest products, sold in the state in 1985.

Alabama Sorghum

In 1970, American farmers harvested 683.2 million bushels of sorghum grain. Texas reported the largest share (48.2%) of the total national crop. Other states reporting large shares included: Kansas (21.4%), Nebraska (11.0), Oklahoma (3.4), and California (3.4). These five states accounted for 87.1% of the total national crop. Reports of bushels harvested were available from only 24 states. Farmers in Alabama harvested only 0.1% of the national crop. According to the 1969 *Census of Agriculture* data, farmers in Montgomery County harvested the largest share (10.6%) of the state's total. Other counties reporting large shares included: Baldwin (8.3%), Perry (7.0), Pickens (6.8), and Houston (4.6). These five counties accounted for 37.7% of the state's total.

Data on acres of sorghum harvested in Alabama first appeared in the Census of 1890, although data on the gallons of sorghum molasses made in the state had been reported in the 1860 Census. The first annual data on acres harvested date to 1924. That data indicate that in 1924 farmers had harvested 54,000 acres of sorghum, with 26,000 acres used for making molasses and 28,000 used for forage. The first data on use of the sorghum for silage is from 1929. Then in 1944, data on the harvest of the crop for grain were first reported. However, by 1970 USDA was collecting and reporting data only on acres harvested for grain and for silage. In that year, only 5% of the sorghum harvested in the United States was used for silage. The great majority (95%) was harvested for grain. In Alabama, 52% was harvested for grain and 48% for silage.

US Exports

American sorghum grain exports had been trending upward since the early 1950s and were 144 million bushels in 1970 (Figure 15.87). That year, exports accounted for

21.1% of total bushels harvested. In the same year, wheat exports accounted for 54.8% of the national crop; while exports of corn accounted for only 12.4%. Sorghum exports began to increase after 1971, and reached 212 million bushels in 1972. Through 1978, they remained relatively unchanged at around 225 million. Sorghum exports did not benefit nearly as much from the adverse weather in Europe as the other grain crops. Then in 1973, market prices (Figure 15.88) began to rocket upward. They were $3.97 a bushel in 1973 and $5.25 in 1974. Sorghum market prices had never been that high, and those high prices quickly put an effective lid on exports. In the late 1970s, with inflation roaring ahead (Figure 15.3), all agricultural exports began to increase (Figure 17.6), and sorghum followed. But in 1979, a recession began, and by 1980 it was full-blown. In 1980, sorghum exports began to decline, and by 1982 they were down to 214,000 bushels. The recession ended in 1982, and exports began to increase again, only to be caught by the overvalued dollar in 1985. In that year, exports were 178 million bushels, not much higher than they had been in 1970.

US and Alabama Prices

American sorghum market prices began to decline after 1952, and during the period 1961-1968 generally ranged around $1.80 per Cwt. As a result, the government began to increase support prices in 1961, and by 1970 it was $2.25 per Cwt (Figure 15.88). With inflation increasing (Figure 15.3), support prices began to increase, reaching $2.61 in 1973. In the meantime, market prices shot up to $4.95 in 1974, and support prices were reduced slightly. However, market prices began to decline in 1985, and support prices began to increase once more. They continued to trend upward, finally reaching $5.14 per Cwt in 1985.

The average annual market price for sorghum was $2.04 per Cwt in 1970 (Figure 15.88). Then with inflation increasing (Figure 15.3), they began to increase rapidly, reaching $4.95 in 1974. Between 1974 and 1975, bushels of sorghum grain harvested increased from 622.7 million to 754.3 million bushels, and in 1975 the market price fell to $4.21. Market prices continued to fall until 1977 ($3.25). For some reason, bushels harvested seemed to have little effect on market prices in 1977, 1978, and 1979. Then in 1980, when farmers harvested only 579.3 million bushels of grain, it increased to $5.19, the highest on record at that time. Thereafter, the teeter-totter phenomenon carried them down, up, and down to $3.45 per Cwt in 1985. That much lower market price was the result of a 1,120.3-million bushel crop, the largest recorded to that time.

Market prices for sorghum grain were lower than support prices from 1961 through 1971 (Figure 15.88). In the remaining 14 years of the period 1972-1985, they were higher in six years, near the same in three years, and lower in five.

Year-to-year trends in market prices for sorghum grain in Alabama during the period 1970-1985 were essentially carbon copies of national trends (Figure 15.89). However, in slightly more than half of the years, those in Alabama were a few to several cents higher.

Acres Planted and Harvested

In 1970, acres planted of sorghum for grain in Alabama totaled 54,000 acres (Figure 15.90). The market price had increased from $1.75 in 1968 to $2.20 in 1980 (Figure 17.86), and Alabama farmers apparently thought they saw an opportunity to make a killing on sorghum. In 1969, they had planted 42,000 acres of the crop for all purposes, but with the opportunity before them, they increased acres planted to 112,000 in 1971. Unfortunately, when they harvested those increased acres, the market price was back down to $1.70. This experience apparently taught them a lesson, and when market prices began to rocket upward in 1972, they hardly budged. With market prices going to over $5, they only increased their acres planted by a few hundred, and they reduced them by a few hundred when market prices declined after 1974. Then a really unusual event captured sorghum production in Alabama. Agricultural administrators at Auburn had been concerned for some time that cattle production in Alabama had become primarily a cow-calf business. They decided that farmers could earn more from their cattle operations if they could put their calves in local feedlots, rather than sending them to the Midwest for finishing. Unfortunately, Alabama did not produce enough feed grain to support feedlot operations.

To remedy this situation, the Cooperative Extension Service was instructed to begin a concentrated effort to get Alabama farmers to produce more sorghum grain. This program was initiated in 1980 with 65,000 acres planted. Acreage increased to 90,000 in 1981; to 100,000 in 1982; to 125,000 in 1983; to 220,000 in 1984; and to 270,000 in 1985. Much of this sharp increase had taken place when market prices were declining. In 1983, when 125,000 acres were planted, the average market price in the state was $5.65 per Cwt. Unfortunately, afterwards, as acres planted increased, market prices declined. In 1985, when farmers planted 270,000 acres, the market price was down to $3.35. Farmers were planting five times as many acres of sorghum as they had in 1970, and considering inflation, they were realizing considerably less money for each bushel.

Data on acres harvested of sorghum for grain in Alabama is also presented in Figure 15.90. Throughout the 1950s, 1960s, and 1970s, Alabama farmers harvested 40 to 50% of their planted acres for grain. However, when the cattle feed program began to increase grain production, the percentage harvested for grain increased. It was 52% in 1980, 64% in 1981, 68% in 1982, and 85% in 1985.

Data on the acres harvested for silage are also presented in Figure 15.90. These data show that acres harvested for silage remained relatively stable at around 15,000 acres for most of the period, although it was somewhat higher in 1970 and 1971 and in 1984 and 1985.

US and Alabama Yields

Data on annual average yields of sorghum grain in Alabama and the United States during the period 1950-1985 are presented in Figure 15.91. These data indicate that national yields changed very little during the period 1970-1985. There was considerable year-to-year variation, but yields generally varied between 50 and 60 bushels an acre during the period. There seems to have been no trend. There also seems to have been little change in Alabama yields for most of the period. They generally varied between 30 and 40 bushels an acre; however, when the Alabama Cooperative Service began the campaign to get farmers to increase sorghum production, yields began to increase. It was 34 bushels an acre in 1970 and 37 bushels in 1981; however, in 1982 it began to increase, finally reaching 55 bushels an acre in 1985. It is not clear why yields increased so rapidly. Either the Extension Service was extremely effective in getting farmers to use better production technology, or only the better farmers responded to efforts to increase sorghum grain production in the state.

Except for the period 1983-1985, yields of sorghum grain averaged about 20 bushels an acre less in Alabama than in the country as a whole. In 1983, 1984, and 1985, the difference was much less. The sorghum genome was not any happier in Alabama than genomes of the other grains.

Bushels Harvested

With yields in Alabama remaining relatively unchanged during the period 1970-1981 (Figure 15.91), bushels harvested (Figure 15.92) generally closely followed the trends of acres harvested (Figure 15.90). Except for 1971, farmers generally harvested around a million bushels a year from 1970 through 1980. In 1981, with the effort underway to get increased production of the grain in the state, bushels harvested began to increase, and in 1985 farmers harvested 12.6 million bushels, by far the largest harvest in Alabama history.

Alabama Vegetables

Annual data on value added by Alabama vegetables to the national economy during the period 1950-1985 are presented in Figure 15.93. These data show that value added slowly trended upward through the 1950s and 1960s to reach $2.5 million in 1970. In the same year, national value added for vegetables totaled $2.8 billion. The Alabama contribution that year was 0.8% of the national total. Generally, value added in Alabama followed the Gross National Product upward (Figure 15.1.) Value added seemed to respond to the short recessions of 1973-1975, 1980, and 1981-1982, but reached the unprecedented level of $109.4 million in 1981. Between 1970 and 1985, value added by the production of vegetables in the state increased from $23.5 to $95.2 million, an increase of 305%. During the same period, the Consumer Price Index increased by 173%.

The Alabama Agricultural Statistics Service has published annual data on acres harvested of some vegetables since 1928. Initially, these vegetables were variously identified as truck crops and commercial vegetables for fresh markets; however, after 1970 they were just listed as vegetables. Also, after 1970 the number of vegetables included in the annual bulletins was reduced. In the period 1970-1985, only Irish potatoes, sweet potatoes, tomatoes, watermelons, sweet corn, and snap beans were included. Data on acres harvested of several of these are included in Table 15.10. Data on Irish potatoes and sweet potatoes have already been discussed. Data in the table show that more acres of watermelons (11,000 to 15,000 acres) were harvested than any of the other crops. They were followed by tomatoes (3,050 to 9,300 acres); sweet corn (2,300 to 5,000); and snap beans (1,100 to 1,500). However, acres harvested of tomatoes trended down during the period, and there were fewer acres of them harvested in 1985 than sweet corn. Acres harvested of sweet corn generally ranged between 2,300 and 4,000 during the period 1970-1980, but afterwards it was generally higher (4,200 to 5,000 acres).

Data on the value of these vegetables are included in Table 15.11. These data show that the value of tomatoes was much greater than any of the others during this period. Values of this crop ranged between $4.4 and $18.3 million. It was $4.4 million in 1970, and generally followed the rate of inflation (Figure 17.3) and prices up and down. It was $10.7 million in 1973, $18.3 million in 1979, but by 1985, value was back down to $4.8 million. The value of watermelons varied between $2.2 and $6.8 million, but in most years, the range was between $2 and $4 million. Values of sweet corn varied between $0.8 and $7 million, but in most years the range was much more restricted. Values for snap beans were the lowest of all the vegetables, ranging between $0.4 and $1.5 million.

Alabama Pecans

Data on the annual pounds of pecans harvested in Alabama during the period 1970-1985 are shown in Figure 15.94. The masting phenomenon continued to be quite evident throughout this period. Generally, but not always, a good crop was followed by a poor one. The largest crop (41 million pounds) was harvested in 1973. This was the highest level ever recorded at that time. The following year (1974), farmers harvested only 11 million pounds. While masting was an ever-present factor in pecan production, often hurricanes and drought also restricted pecan production during the period. Hurricanes Agnes in 1972, Eloise in 1975, and Frederic in 1979 restricted production to a degree. Hurricane Frederic was especially damaging. Farmers harvested only four million pounds of nuts that year. Drought affected production in 1978 and 1984. Even though there was considerable year-to-year variation, pounds harvested seemed to be a little higher from 1970-1973 than from 1974-1985.

Prices varied between 34.7 cents a pound (1971) to 83.5 cents a pound (1976) and were generally affected by pounds harvested in a given year. For example, in 1973 farmers harvested 41 million pounds of the nuts and received 36.5 cents a pound for them. In 1976, they harvested five million pounds, receiving 83.5 cents a pound.

As expected, with the variability in pounds harvested and prices, cash receipts for pecans was also extremely variable. The five-million-pound crop in 1976 and the four-million-pound crop in 1978 earned Alabama pecan farmers $4.1 million and $2.4 million, respectively. At the other extreme, cash receipts for the 33-million-pound crop in 1977 and the 34-million-pound crop in 1981 produced cash receipts of $19.8 and $17.4 million, respectively. Better prices compensated for lower harvests, to some extent, but never enough to completely make-up for the lost revenue. In 1976, farmers harvested only five million pounds of pecans, and they received 83.5 cents a pound for them. Unfortunately, the combination produced only $4.1 million in cash receipts. The higher price had compensated to a limited extent, but not nearly enough to save farmers from serious loss.

Alabama Peaches

Alabama peach harvests were variable from year to year throughout the period 1970-1985 (Figure 15.95), but there was little evidence of a tend. Farmers harvested 15 million pounds of the fruit in 1970, and pounds harvested generally remained around that level in most of the years of the period, except when adverse weather interfered. Throughout much of the period, farmers were plagued with cold, abnormally wet weather in the late winter and early spring (Table 15.9). In 1985, a late freeze severely damaged the developing crop, and only 1.5 million pounds were harvested.

As might be expected, with peach production so variable during the period, cash receipts for the crop were equally variable. Receipts varied from $500,000 in 1985 (freeze year) to $4 million in 1981. In 1980, Alabama produced only 0.5% of the country's peaches. As a result, the quantity of peaches produced in the state had only a limited effect on prices. Of course, receipts were directly related to pounds harvested, but production in other areas of the country was probably more important in determining how much money Alabama peach producers received for their crops in a given year. This relationship is the reason why Alabama peach producers received $3.2 million for a 22-million-pound crop in 1984 and $4 million for the same size crop in 1981.

Alabama Greenhouse, Sod and Nursery

Value of sales of nursery products, sod, bulbs, and florist plants by Class 1-5 farms (sales of at least $2,000) in the United States and its regions in 1969 are shown in Table 15.12. These data show that the total value of the sales of all of the products in the country in that year was $768.5 million and that the South led the country in sales with $198.3 million. The data also indicate that sales for florist plants ($429 million) was larger than for any of the others, and were followed by nursery products ($280.4 million), sod ($42.8 million), and bulbs ($10.9 million).

Since 1950, the Alabama Agricultural Statistics Service has published data on the total cash receipts for greenhouse, sod, and nursery products. Some of these data are presented in Figure 15.96. They show that sales increased from $13.5 million in 1970 to $114.2 million in 1985, an increase of 746%. The increase in the Consumer Price Index (Figure 15.2) accounted for slightly less than a fourth of the increase in cash receipts. In 1970, cash receipts of these horticultural products accounted for 6% of the cash receipts of all crops produced in the state, including farm forest products. By 1985, this percentage had increased to 15.2%. Cash receipts for nursery, sod, and greenhouse products were increasing at a much faster rate than for crops in general. Cash receipts for these products were also increasing as a share of all farm commodities, including farm forest products. In 1970, their share of the total was 1.8%. In 1985, it was 5.3%. Nursery, sod, and greenhouse products were rapidly growing in importance to Alabama's agricultural economy during this period.

Cash receipts for some specific horticultural products in the state in 1969, 1974, and 1982 are presented in Table 15.13. Data presented in the table show that receipts for nursery products and for flowers and flowering plants were $5.5 and $4.9 million, respectively, in 1969 and that they increased to $24.6 (347%) and $27.3 million (457%), respectively, by 1982. During the same period, the Consumer Product Index increased by 163%. Receipts for all of these horticultural crops were increasing at a rate faster than inflation.

In 1982, Mobile County reported the largest share (37.2%) of value for the sale of all nursery and greenhouse products in the state. Other counties reporting large shares included: Cherokee (7.7%), Madison (5.2), Montgomery (4.6), and Baldwin (4.5). These five counties accounted for 59.2% of total receipts.

Alabama Timber

Data on the production (percentages) of lumber in the geographical regions of the United States during the period 1869-1985 are presented in Table 15.14. These data show that lumber production began to decline in the West Region after 1960. That region had led the country in production since 1920, finally reaching 59.7% of the total in 1960. Then by 1985, the percentage was down to 52.5%. It had not been lower since 1950 (48.7%). Later, the environmental movement would force a major reduction in lumber production in the West Region, but at that time it is likely that the decline was because the most easily available timber had been cut. Moreover, Southern farmers had abandoned millions of acres of farmland after 1950. Much of this abandoned land was naturally seeded by surrounding forests. By 1985, some of the trees in these naturally-established plantations were beginning to reach sawtimber size.

Census data for 1969 indicate that the sale of forest products originating on the country's farms totaled $146.3

million (Table 15.15). Some 63% of all of these products originated on Southern farms, and 7.4% originated on Alabama farms. Further, sales of 70.8% of all standing timber, 37.3% of all firewood and fuelwood, 57% of all sawlogs and veneer logs, and 85.8% of all pulpwood originated on Southern farms. Alabama farms were the source of sales of 6.2% of the standing timber, 2.8% of the firewood and fuelwood, 9.2% of the sawlogs and veneer logs, and 13% of the pulpwood in the country.

Unfortunately, the 1969 *Census of Agriculture* did not have county-level data on the production of farm forest products. However, that data is available in the 1974 Census. Data presented in that census indicate that sales of standing timber or trees from Alabama farms totaled $10.5 million. Leading counties included: Baldwin with 5.2% of total sales, Sumter (4.6), Wilcox (3.9), Dallas (4.4), and Conecuh (4.0). These five counties accounted for 22.1% of total sales. Some 20 counties recorded sales of $200,000 or more.

By 1974, there was very little firewood and fuelwood being sold from farms in the state. Census data indicated that sales totaled only $58,000. Only Autauga ($9,000), Washington ($4,000), and Winston ($4,000) reported sales of more than $1,000. Alabama farms reported that sales of sawlogs and veneer logs totaled $2.6 million in 1974. Leading counties included: Butler with 9.3% of sales, Geneva (8.0), Washington (6.8), Coosa (6.4), and Marengo (5.4). Some 16 counties reported sales of $50,000 or more. According to census data, pulpwood valued at $2.8 million was sold from Alabama farms in 1974. Leading counties included: Chambers (5.9%), Marengo (5.5), Wilcox (5.6), Baldwin (4.6), and Coffee (3.8). Some 21 counties reported sales of $50,000 or more.

The US Forest Service Experiment Station in New Orleans has produced a series of extremely useful reports on the status and use of Alabama's forest resources. Two of these, Murphy (1973) and Rudis, et al. (1984), are particularly useful for this particular period 1970-1985. Some of their data are summarized in the following sections.

Land in Forests

Estimated acres of Alabama forestland increased from 20,756,200 in 1954 to 21,350,100 in 1972 (Table 15.16), an increase of 2.9%. They continued to increase slowly during the period 1972-1982, reaching 21,724,900 acres in 1982. As a result of this change, the percentage of land in forests increased from 65.7 to 66.9% (total land area = 32,476,160 acres) during the period 1972-1982.

Forest Ownership

The percentages of public and private forestlands remained generally unchanged at 4.7% and 95.3%, respectively, during the period 1953-1971 (Table 15.17). Farmer-owned forestland decreased a third (46.9 to 31.6%) in the same period. During the period 1972-1982, the percentage of all public land in forests increased slightly from 4.7 to 5.4%. With the share of public forests increasing, the percentage in private forestland decreased from 95.3 to 94.6%. Within the private-land segment of Alabama forestland, the downward trend in farmer ownership, noted during the period 1953-1972, continued. In 1972, farmers owned 31.6% of all Alabama forestland. In 1982, this percentage was down to 27.1%. The decrease in farmer-owned forestland was not the result of converting forestland into cropland. Rather, it was the result of an overall decrease in the amount of land owned by farmers in the state. While forestland owned by farmers was decreasing, ownership by the forest products industry was increasing. During the period 1972-1982, the percentage owned by the industry increased slightly from 19.7 to 20.6%. During this period, the forest products industry apparently continued to believe that they needed to own enough timberland to meet a significant portion of their requirements for forest products.

Forest Types

During the period 1953-1972, the percentage of the area of forestland in Alabama stocked with pine (longleaf-slash and loblolly-shortleaf) declined sharply (Table 15.18). During the same period, the percentage of area stocked with oak (oak-pine, oak-hickory, and oak-gum-cypress) increased. The trend of changes in the relative percentages of pine and oak forest types in Alabama, noted during the period 1953-1972, continued during the period 1972-1982. Of the six major forest types recognized in Alabama, only the percentage represented by the total area of the oak-hickory type increased (27.7% versus 33.6%). The percentages represented by the loblolly-shortleaf pine and longleaf-slash pine types decreased from 29.9 to 27% and from 6.9 to 6.8%, respectively. Even the percentage of oak-pine decreased (23.5 to 21.1%).

Tree Plantations

Alabama tree nurseries distributed almost 600 million seedlings to Alabama landowners during the period 1952-1960. From 1963-1972, some 600,000 acres of tree plantations were established with 40% being artificially regenerated.

During this period tree planting and establishment of tree plantations were still being driven to a degree by the Cropland Conservation Program of 1962 and the Cropland Adjustment Program of 1965. Also, demand for pulpwood for the pulp and paper industry was continuing to expand rapidly. Figure 15.97 contains data on the acres planted with pine seedlings during the period 1975-1985. At the beginning of this period, Alabama landowners were planting 140,000 acres of additional land with pine seedlings each year; however, the prospects for row-crop agriculture began to look so promising around the mid-1970s that farmers quickly reversed the trend of converting cropland to forest. In 1978, acres planted with pines was less than half (65,246 acres) what had been planted in 1977 (147,518 acres). Unfortunately, reality quickly returned to agriculture in Alabama, and farmers once again began to convert cropland into forest. By 1985, Alabama landowners were planting pine seedlings on well over 200,000 acres each year.

Volume of Growing Stock

As shown in Tables 15.19 and 15.20, the volume of growing stock of the major species of hardwoods in Alabama forests increased 47.5% (5.5 to 8.1 billion cubic feet) during the period 1953-1972. Growing stock of these major species continued to increase in the period 1971-1982 (8.1 to 9.2 billion cubic feet). Data on the total volume of growing stock of all hardwoods in the period 1971-1982 are also presented in Table 17.19. During the period, volume increased from 8.9 to 10 billion cubic feet, or by 13%. Much of this increase in the volume of growing stock of hardwoods resulted from significant increases in the volumes of red oaks, sweetgum, and yellow poplar.

The estimated total volume of growing stocks of the major species of softwoods increased from 5.3 to 7.3 billion cubic feet (37.7%) between 1953 and 1971, and by 9.6% (10.6 to 11.7 billion) from 1971 to 1982. Growing stock of all softwoods increased by 11.8% (11.3 to 12.6 billion) between 1972 and 1982

Growth and Removals

In the years 1953, 1962, and 1971, growth of both softwoods and hardwoods exceeded removals in Alabama's woodlands (Table 15.21). During the period 1971-1982, the volume of the growing stock of softwood in the state exceeded removals by 5.6 million cubic feet (641.8 versus 636.2 million). During the same period, the average annual increase in the growing stock of hardwood exceeded removals by 125.9 million cubic feet (344.5 versus 218.6 million). Obviously, the stock of hardwoods in Alabama's forests was increasing at a much faster rate than softwoods.

Stumpage Prices

Data presented at the 17th Annual Conference on International Forest Products Marketing, held in 2000 in Seattle, Washington, indicate that pine sawtimber stumpage prices in the South reached $100 per thousand board feet (MBF) around 1950, and generally remained near that level through 1970. Stumpage prices fell below that level during the period 1962-1966, when interest rates began to increase.

Stumpage prices for pine sawtimber in Alabama for the period 1977-1985 (Timber Mart-South) are shown in Figure 15.98. Prices for this product increased rapidly in the late 1970s as the American economy roared ahead (Figure 15.1). The price was $100 per MBF in 1977, and by 1979 it had almost doubled to $170. Then it fell sharply in 1980 when interest rates reached levels as high as 13%. Surprisingly, prices generally increased during the 1982-1983 recession (Figure 15.1). In 1983, it reached almost $180 per MBF, the highest in history, at that time. Prices trended downward in 1984 and 1985, even though discount rates were also declining (Figure 15.4) during this period. However, the Gross National Product had begun to trend downward again in 1984 (Figure 15.1). Between 1977 and 1985, stumpage prices for pine sawtimber in Alabama grew from $99.50 per MBF to $143.10, an increase of 43.8%. During the same period, the Consumer Price Index increased by 75.1%. As a result, Alabamians selling sawtimber in 1985 were getting considerably less for it in real dollars than they got in 1977. Stumpage rates for hardwood sawtimber were much lower than for pine (Figure 15.98) throughout the period, generally remaining unchanged at $50-60 per MBF.

Figure 15.99 includes data on Alabama pine and mixed hardwood pulpwood stumpage prices for the period 1977-1985. The data indicate that prices increased rapidly in the late 1970s and early 1980s as the pulp and paper industry expanded in the state. Also, the Consumer Price Index increased 50% (181.5 to 272) during the period 1977-1981. The price was $10 a standard cord in 1977, and by 1981 it had increased to $17. Prices remained generally unchanged in 1982 and 1983, largely as a result of the recession (Figure 15.1) during this period. However, as business conditions improved in 1984, prices began to increase again and by 1985 reached $19 a cord, the highest in history at that time. Stumpage prices for pine pulpwood in Alabama grew from $10.32 a cord in 1977 to $18.65 in 1985, an increase of 80.7%, or slightly more than the CPI (75.4%) during the same period. Stumpage prices for hardwood pulpwood remained much lower than for pine throughout this period (Figure 15.99). They increased somewhat after 1980, probably as a result of the rapidly growing demand for pulpwood. The price for this product was also slammed by the 1982-1983 recession.

Cash Receipts

The Alabama Agricultural Statistics Service has published annual data, beginning in 1950, giving the estimated annual cash receipts from the sale of farm forest products. Some of these data (1950-1985) are presented in Figure 15.100. Cash receipts generally trended downward from $24 million in 1950 to $14 million in the late 1960s. However, during the period 1955-1959, they were as high as $29.8 million. Receipts generally trended upward from 1970 through 1981 from around $14.4 million to $85.6 million, although they did decline in 1975, apparently as a result of the effects of the 1973-1975 recession (Figure 15.1), which followed the OPEC embargo on oil shipments to the West. Receipts also declined somewhat after 1980 as a result of the 1981-1982 recession. These trends are generally reflective of the stumpage prices for sawtimber and pulpwood and shown in figures 15.98 and 15.99. Over the entire 1970-1985 period, farm forest receipts increased from $14.4 million to $93.8 million, an increase of 551%. During the same period, the Consumer Price Index increased 343% (72.1 to 318.5). Throughout this period, the percentage of forestland owned by farmers was trending downward (Table 15.17). In 1970, receipts from the sale of all farm forest products was the equivalent of 6.2% of all agricultural crops and 1.9% of all farm commodities. By 1985, these percentages had grown to 11.1 and 4.4%, respectively. The sale of forest products from Alabama farms was growing increasingly important to the state's farmers during this period.

In 1972, the State Agricultural Statistics Service began to publish annual data on the cash receipts from the sale of non-farm commercial timber in Alabama. Data for the period 1972-1985 are presented in Figure 15.101. They show that receipts for these products grew from $56.8 million in 1972 to $283.9 million in 1985, an increase of 400%. In the same period, cash receipts from the sale of farm forest products increased only 258% ($26.2 to $93.8 million), and the CPI increased by 154%. These data indicate that receipts received from the sale of non-farm forest commercial timber generally increased at a higher rate than sales from all farm commodities during the period 1975-1985 (209% versus 60%). Cash receipts for non-farm commercial timber were also negatively affected by the recessions of the mid-1970s and early 1980s.

Forest Products Harvested

In 1951, almost three times as much Alabama timber was harvested for sawlogs as for pulpwood (54.8 versus 18.7%); however by 1971, the relationship was reversed. In that year, 60% was harvested for use as pulpwood and 31.4% for sawlogs.

Data published by Rudis, et al., 1984, indicate that in the 10-year period 1972-1981 some 7,473.3 million cubic feet of timber products were harvested from Alabama woodlands (Table 15.22). Approximately 57.2% was used for pulpwood, 40.8% for sawlogs and veneer logs, and 2% for poles. Some 74% of the total harvested was softwood and 24% hardwood.

Data published by McWilliams (1994) indicate that in 1982 some 666.7 million cubic feet of pulpwood was harvested from Alabama's woodlands and an additional 151.2 million cubic feet of wood for pulping were obtained by chipping slabs, edgings, miscuts, sawdust, shaving, veneer cores, etc. (plant residues) recovered from sawmills and manufacturing plants.

Murphy (1973) commented that 179 million cubic feet of plant residues were generated by the forest industries in Alabama in 1971. Two-thirds of this material was coarse enough that it could be chipped for use in the production of pulp. A total of 115 million cubic feet of residues were used in the manufacture of pulp and particle board, 29 million cubic feet were burned, and almost eight million cubic feet were used for other purposes such as the production of charcoal, animal bedding, or mulch. Two million tons of tree bark were also generated by forest industries in 1971. Approximately two-thirds of this material was also reclaimed and used.

All of these data show the continued growth in the importance of the pulp and paper industry to owners of forestland in Alabama. The data also demonstrate the growing efforts of the forest industries in the state to use to the fullest extent all the different parts of trees being harvested from the state's woodlands.

Animal Agriculture

During the period 1950-1970 when agriculture in the United States was being rapidly modernized, value added to the national economy by livestock production had trended upward from $18.1 to $30.8 billion, an increase of 70.2%. During the same period, the Consumer Price Index increased by 61% (Figure 15.2). The increase in the value added of livestock was generally similar to the increase in the value of plant crops through 1965. Thereafter, however, the value added for livestock surged upward, primarily as a result of the beginning, in 1962, of the Expansion Phase of a lengthy cattle cycle. Also, the national broiler industry was beginning to expand rapidly. In 1950, value added for livestock accounted for 58% of total value added (crops plus livestock), but by 1970 it had increased to 62% of the total.

Value added by livestock in Alabama increased from $209.9 to $557.9 million between 1950 and 1970, an increase of 166%. Almost two-thirds of that increase ($237.7 million) was the result of a tenfold increase in the value of poultry (primarily broilers) and eggs ($29.9 versus $267.6 million) during the period. In 1950, value added by livestock accounted for 45% of total value added (crops plus livestock) in the state; however, by 1970 it had increased to 71%.

Cattle

Between 1950 and 1970, national cattle numbers (excepting cows that had calved–milk) increased from 54.1 to 100.3 million. In this period, numbers went through two well-defined Cycles of Abundance. The first began in 1949 and ended in 1958, and the second began in 1962 and ended in 1969. There is evidence that there was a third one, 1959 through 1962, but if there was, it was not well developed.

US Cattle

In 1970, the largest share of cattle and calves (including milk cows) in the United States was in Texas (11.4%). Other states reporting large shares included: Iowa (6.6%), Nebraska (5.6), Kansas (5.3), and Oklahoma (4.4). These five states accounted for 33.3% of the total national herd.

Prices

Data on American cattle prices (dollars per Cwt) for the period 1950-1985 are presented in Figure 15.102. The price was $27.10 per Cwt in 1970. Then, with the national economy growing rapidly (Figure 15.1), cattle prices also increased, reaching $42.80 in 1973. At the same time, cattle numbers were also increasing in the Accumulation Phase of a cycle (Figure 15.104). With numbers increasing rapidly in 1974, prices began to decline, reaching $32.20 in 1975. After that year, numbers began to decline, and in 1976, prices began to increase again. In 1979, it reached $66.10. In 1980, the Expansion Phase of a new Cycle of Abundance began, and in the same year prices once again began to decline. In this new cycle, numbers did not increase very much, and even when the Declining Phase began in 1982, prices continued to decline. Price appeared to begin to respond to declining numbers in 1984, but fell to $53.70 in 1985.

Beef Consumption

In 1950, the annual per capita consumption of beef (Figure 15.103) was slightly lower than that of pork (63.4 versus 69.2 pounds), but considerably higher than for chicken (20.3 pounds); however, by the mid-1950s, beef had taken a commanding lead. Over the next 15 years, consumption of pork remained relatively stable, while the consumption of both beef and chicken trended steadily upward. By 1970, consumption of beef, pork, and chicken totaled 113.7, 66.4, and 40.5 pounds per capita, respectively. Beef consumption fell slightly at the beginning of the 1973-1975 recession (Figure 15.1), but soon recovered and reached 129.3 pounds in 1976 (the highest in history to that time). In 1970, the annual average retail price of choice beef was 99.9 cents a pound, but by 1976 it was up to $1.46. Retail prices remained near that same level in 1977, and beef consumption began to decline. It was down to 105.4 pounds in 1979, and remained near that level through 1985.

Cycles of Abundance

There were 100.3 million cattle (excepting cows that had calved–milk) in the United States on January 1,1970. At that point, numbers were in the Expansion Phase of a 1970s cattle cycle that probably had begun in 1969 (Figure 15.104). The cycle reached its Consolidation Phase in 1975 with herd size at 120.9 million animals. Prices trended sharply downward after 1973 (Figure 15.102), which likely triggered the end of the Expansion Phase. The Declining Phase ended in 1979 with the herd size at 100.1 million. The same year, average price reached $60, and as a result farmers began to rebuild their herds. A new cycle began in 1980 with the herd number near 100.5 million. This cycle was unusual. The Expansion Phase lasted only two years (1980 and 1981), and the Declining Phase began in 1983. It is likely that the 1980 and the 1981-1982 recessions, effectively killed this cycle before it could mature. With unemployment near 10% and consumer prices at an all-time high, American housewives were purchasing much less beef than they had just a decade earlier (Figure 15.103).

Real Net Return Per Cow

Through the 1960s and into the early 1970s, American cattle producers received a real return on their cattle of around $45 to $50 a head (Figure 15.105), but when the 1973 to 1975 recession began, returns fell sharply. As late as 1973, they received an average return per animal of $84. Then just three years later, in 1975, it was down to $79. Returns recovered in the late 1970s and early 1980s, as inflation reached 13% (Figure 15.3). Returns reached $167 in 1979, but when consumer prices increased 13% in 1979 (Figure 15.2), returns began a sharp decline. It was down to $67 in 1980, and when the 1981-1982 recession was over, it was below zero again, and it remained below that level through 1985.

American cattle producers could not know it at the time, but the end of the 1970s cycle in 1979 represented a watershed event. The industry would never be the same. Beef consumption per capita had peaked at 130 pounds around 1976, and it fell rapidly thereafter (Figure 15.103). By 1980, consumption was down to 115 pounds per capita. Obviously, there was a rapid and significant shift in the entire beef industry during that period. Reduced demand played a significant role in the rapid decline in cattle numbers, but severe weather in the late 1970s and early 1980s was also important, along with increased conversion of pasture to cropland and high interest rates that raised production costs.

Alabama Cattle

USDA data indicated that in 1970 the largest share (4.6%) of the Alabama cattle herd was on farms in Montgomery County. Other counties with large shares included: Lowndes (3.6%), Dallas (3.0), Marengo (2.8), and Cullman (2.7). These five counties accounted for only 16.7% of the state's total. This low percentage is indicative of the widespread distribution of cattle in the state. Some 25 counties reported cattle herds of 30,000 or more animals. The old Celtic love for cattle was still alive and well on farms throughout the state. Figure 15.106 shows the distribution of cattle (all) in Alabama in 1970.

Prices

Data on annual average cattle prices (dollars per Cwt) in Alabama for the period 1950-1985 are presented in Figure 15.102. Year-to-year changes in prices for cattle in Alabama were similar to those in the United States, except that in most years, those in the state were lower. Also, prices in Alabama were much more negatively affected by the recessions (Figure 15.1). Because there were no support prices involved, these differences represent the results of market forces.

Cycles of Abundance

Year-to-year trends in Alabama cattle numbers during the period 1970-1985 generally followed those of the national herd (Figures 15.107 and 15.104). In 1970, at the beginning of the 1970s cycle, there were 1.8 million cattle and calves (excepting cattle that had calved–milk) in the state. When the Expansion Phase ended in 1976, herd size totaled 2.8 million animals. This phase ended a year later than in the country, 1976 versus 1975, respectively. The Declining Phase of Alabama's 1970s cattle cycle ended in 1980 with a herd size of 1.7 million. This cycle also ended a year later than the national one (1980 versus1979). A new cycle began in 1981, but it was largely aborted by the period of economic malaise (Figures 15.1 and 15.3). When the 1970s cycle began in 1970, there were 1.8 million cattle in the state. At the end, in 1980, there were 1.7 million. The largest Alabama herd in the 1970s, 2.8 million in 1976, represented 2.4% of that year's 116.9-million national herd.

Hogs

In 1950, Americans on a per capita basis consumed more pork (69.2 pounds) than either beef (63.4 pounds) or chicken (20.3 pounds) (Figure 15.103); however over the next 20 years (to 1970), the consumption of pork (66.4 pounds) changed very little, while the consumption of beef (113.7 pounds) and chicken (40.5 pounds) increased sharply. Of course, with the increase in the population, the total consumption of pork also increased, but not to the same extent as the other two. Also, in those 20 years, the North Central Region further solidified its position as the country's primary pork-producing region. It had 70.8% of the hogs and pigs in 1950 and 73.3% in 1970. Most of this increase came at the expense of the South Central Region, where the percentage decreased from 16.4 to 10.6%. Apparently, it was finally dawning on farmers in that region that they had very little, if any, comparative advantage in hog production.

US Hogs

In 1970, the largest share (24%) of the national hog and pig herd was in Iowa. Other states with large shares included: Illinois (11.3%), Missouri (7.6), Indiana (5.6), and Minnesota (5.6). These five states accounted for 56.1% of the national total. Only 1.5% of the national herd was on Alabama farms.

Prices

In 1970, the annual, average price for hogs in the United States was $22.70 per Cwt (Figure 15.108). It declined to $17.50 in 1971, probably as a result of the enormous increase (Figure 15.109) in hog numbers between 1969 and 1970 (54 to 67.3 million). Thereafter, numbers declined sharply, and prices began to increase, reaching $38.40 in 1973. Then prices declined slightly between 1973 and 1974, as a result of the 1973-1975 recession (Figure 15.1). In 1975, numbers were down to 49.3 million, the lowest since 1953, and the price went to $46.10. Afterwards, combinations of changes in numbers, recessions (1980 and 1981-1982), high interest rates (20.4% in 1980), and high unemployment rates (10.8% in 1982), moved prices up-and-down between $38 and $47, through 1985. In 1985, it was $43.60 per Cwt.

Pork Consumption

As detailed in a preceding paragraph, pork consumption was 69.2 pounds per capita in 1950 (Figure 15.103). Afterwards, it declined to between 60 and 70 pounds, and remained in that range until 1974. It was a little more variable afterwards, but remained in that range through1985. In that year, it was 66 pounds per capita, or essentially where it had been in 1955 and 1969.

Cycles of Abundance

National hog numbers had generally varied between 50 and 60 million animals during the 1960s, but had increased sharply to just above 67 million in 1970 (Figure 15.109); however, the price fell to $17.50 per Cwt (Figure 15.108) in 1971. As a result of high hog numbers, the recession, and inflation, American farmers seemed to have decided that they owned too many hogs. By 1975, numbers had fallen to 49.3 million (it had not been that low since 1953). Then, in 1976, with prices between $40 and $50, they apparently decided to try their luck one more time. Consequently, between 1975 and 1979 numbers increased from 49.3 to 67.3 million. With numbers at 67.3 million, prices began to decline (teeter-totter phenomenon). It was down to $43.50 in

1979 and fell below $40 in 1980. In 1979, farmers began to reduce herd size, and by 1982 it was down to 54.5 million. Prices had begun to trend upward after 1980, and in 1982 reached $52.30 per Cwt, the highest level in history at that time. That unusually high price gave hog farmers the itch one more time, and in 1983 they increased their herd numbers just slightly—but right in the teeth of another price decline. They quickly began to reduce numbers, and in 1985 they were down to 52.3 million.

The data presented in Figure 15.109 indicate that there were portions of at least two cycles of abundance during the period 1970-1985. The Accumulation Phase of one of them began in 1965 and ended in 1970, and the Liquidation Phase ended in 1975. The Accumulation Phase of the second cycle began in 1976, and the Liquidation Phase probably ended in 1982. It appears that a new cycle began in 1983, but it was probably aborted by the national economic problems of that period. Without question, the herd size adjustment efforts by American hog farmers from 1970 through 1985 was the most hectic in the twentieth century.

Alabama Hogs

In 1970, the largest share (5%) of the Alabama hog herd was in Houston County. Other counties with large shares included: Marshall (4.4%), DeKalb (4.3), Geneva (3.6), and Jackson (3.8). These five counties accounted for only 21.1% of Alabama's total, indicating the widespread distribution of hogs in the state at that time. Some 20 counties reported herds of 20,000 or more animals. Figure 15.110 shows the distribution of hogs and pigs in Alabama in 1970.

Prices

Data on prices that farmers in Alabama received for their hogs during the period 1970-1985 are also presented in Figure 15.108. Year-to-year trends of both are almost identical to those in the country.

Cycles of Abundance

Hog numbers in Alabama had been generally trending downward in the state since the mid-1940s, and in 1970 it had reached one million (Figure 15.111). Afterwards, year-to-year trends in numbers during the period 1970-1985 were similar to those for the national herd (Figure 15.109), except that in Alabama the end of the Accumulation Phase of the 1970s cycle came in 1971, rather than 1970, and the end of the Liquidation Phase came in 1976, rather than 1975. A new cycle began in 1977, and in the Liquidation Phase numbers declined to 345,000 animals in 1985. Between the beginning of this cycle, in 1975 and 1985, numbers declined by 305,000 animals (650,000 to 345,000). Numbers had never been that low since the collecting and reporting of annual data began in 1866. Alabama farmers finally realized that they simply could not compete in the national market with Midwestern states. The surprising thing is that it took them so long.

Milk

As detailed in the preceding chapter, a long-term evolutionary process began to bring major changes in the American dairy industry in the early 1950s. During that period, the development and adoption of technological innovations, specialization, and changes in milk production and marketing systems began to bring about major structural changes. While many structural changes were taking place, the region where most of the country's milk was produced was changing very little. In 1970, milk cows in the North Central Region produced 48.6% of the country's milk supply. Some 18.4% was produced in the South (10.9 and 7.5% in the South Central and South Atlantic regions, respectively). The North Atlantic and West regions accounted for 17.9 and 15.1%, respectively. The relatively high percentage in the North Atlantic Region was primarily the result of high production in New York (9% of the national total) and Pennsylvania (6.2%). The high percentage in the West Region was the result of production in California (8.2%).

US Milk

Wisconsin milk cows produced the largest share (16%) of the country's milk supply in 1970. Other states reporting large shares included: New York (9%), California (8.2), Pennsylvania (6.2), and Minnesota (6). These five states accounted for 45.4% of the national total. Strong dairy industries had developed in New York and Pennsylvania in colonial times in response to the growing urban population centers there. Surprisingly, while the national dairy industry has changed in many ways since those early years, those two states have been able to sustain their competitive advantages.

Prices

After holding national milk prices relatively stable during the late 1950s and early 1960s (Figure 15.112), the federal government (Triangle) began to increase them slowly after 1965, when inflation began to increase (Figure 15.3). By 1970, price was $5.78 per Cwt. Then in 1973, with the Consumer Price Index increasing rapidly, the government began to increase prices more rapidly, and in 1981 price reached $14 dollars per Cwt.

Some provisions in the Omnibus Budget Control Act of 1982 were included in an effort to control the spiraling cost of the milk support program. These provisions froze dairy price supports at $13.10 per Cwt. Unfortunately, dairy herd size increased again in 1982 and in 1983 (Figure 15.113). In 1983, Congress enacted the Dairy and Tobacco Adjustment Act of 1983. The provisions of this act reduced the price to $12.60, beginning in 1984. The act also provided for a voluntary milk diversion program, whereby farmers who voluntarily reduced milk production between 5% and 30% of their base production level would receive $10 per Cwt for the amount of milk that they did not produce. As a result of the congressional intervention, milk prices began to decline, reaching $12.76 in 1985.

Milk Consumption

USDA data indicate that American per capita consumption of all dairy products was in 1970 (563.8 pounds), 1975 (539.1), 1980 (543.2), and 1985 (593.7), while consumption of beverage milk for those years was 269.1, 254, 237.4, and 229.7 pounds, respectively.

Milk Cow Numbers

As shown in Figure 15.113, the average number of milk cows in the country was 12 million animals in 1970. Afterwards, numbers declined slowly to reach 10.7 million in 1979. Herd numbers increased very slightly in 1980. Milk prices had been increasing rapidly since 1973, and apparently dairy farmers began to increase herd numbers in response. The price of milk continued to increase through 1981, and numbers continued to increase slightly through 1983, even though they had begun to decline after 1981. Farmers responded to the congressional acts of 1982 and 1983 by beginning to slowly reduce herd size. It was down slightly in 1983 and 1984, but increased to 11 million animals in 1985.

Milk Per Cow

In 1950, the average quantity of milk produced per cow in the United States was 5,314 pounds (Figure 15.114). By 1970, this average had increased to 9,751 pounds, an increase of 83%, or almost 4% a year. In 1950, milk production per cow was highest in California (7,710 pounds), followed by New Jersey (7,450 pounds), Rhode Island (7,200 pounds), Wisconsin (6,850), and New York (6,810). In 1970, production per cow was still highest in California (12,526 pounds); followed by Washington (11,881); Arizona (11,700 pounds); and Massachusetts (10,987 pounds).

Average annual milk per cow in 1970 was 9,751 pounds. Afterwards, it trended steadily upward throughout the 1970-1985 period, although there were some slight year-to-year deviations in the magnitude of annual increases, which were primarily the result of differences in weather and feed conditions (Miller, 2002). By 1985, production per cow had increased to 13,024 pounds, an increase of 34% (2% a year). Milk per cow increased very little during the period 1972-1975. It is likely that during the 1973-1975 recession (Figure 15.1) dairy farmers either did not provide their animals with as much feed or they used poorer quality feeds.

By 1985, Washington State had replaced California as the leading state in production of milk per cow (16,816 versus 16,102 pounds). Other leading states included: New Mexico (16,090 pounds); Arizona (15,674 pounds); and Oregon (14,380 pounds). Note that in 1985, there was not a single state east of the Mississippi River in the top five. In 1950, there had been four.

Pounds Produced

In 1950, total national milk production was 116.6 billion pounds (Figure 15.115), and by 1970 it had increased to 117 billion. In 1950, Wisconsin was the leading milk-producing state with a total production of 14.8 billion pounds. It was followed by: New York (8.8 billion), Minnesota (8.1 billion), Iowa (6.2 billion), and California (six billion). In 1970, Wisconsin continued to lead the country in total milk production with 18.4

billion pounds. Following Wisconsin were: New York (10.3 billion), Minnesota (9.6 billion), California (9.5 billion), and Pennsylvania (7.1 billion).

In 1970, total milk production in the United States was 117 billion pounds (Figure 15.115). It then trended upward in 1971 and 1972. Milk cow numbers (Figure 15.113) were declining during this period, but the increase in production per cow (Figure 15.114) was great enough to more than compensate for the loss of animals. Total production declined in the period 1973-1975, as milk per cow declined slightly. Production climbed again in 1976 and 1977, but slowed in 1978 and 1979 when milk per cow was slightly lower again. It spurted upward again in the period 1980-1983 as herd size increased; however. it declined in 1984, along with herd size and milk per cow. In 1985, production was 143 billion pounds, as herd size and milk per cow increased again.

Alabama Milk

In 1950, the availability of milk in the state and the country was 422 and 768 pounds per capita, respectively. In 1970, it declined in both state and country to 237 and 576 pounds, respectively. The rate of decline in Alabama, however, was considerably greater (43.8 versus 25%). Figure 15.116 shows the distribution of milk cows in Alabama in 1970.

Prices

Data on milk prices in Alabama for the period 1950-1985 are also presented in Figure 15.112. As might be expected, with the federal control of milk pricing in the country, the year-to-year trends in the period 1970-1985 were similar in both the country and the state. However, throughout the period, Alabama farmers received from $1.05 to $1.76 per Cwt more for their milk as a result of differences in regional milk marketing orders. The largest difference ($1.76) came in 1985 after Congress enacted legislation making significant changes in the national milk support program.

The federal government controlled Alabama milk prices throughout this period. They were $6.83 per Cwt in 1970 and $14.40 in 1980, an increase of 110.8%. During the same period, the Consumer Price Index (Figure 15.2) increased by 112.4%. However, if the entire period !970-1985 is considered, the Consumer Price Index outdistanced price increase by 173 versus112%.

Milk Cow Numbers

In 1950, there were 362,000 milk cows in Alabama, but by 1970 the number had declined to 122,000 animals (Figure 15.117). In 1950, Alabama had 1.6% of the milk cows in the country; however by 1970, the percentage had declined to 1%. Numbers generally declined from 122,000 in 1970 to 65,000 in 1985, a decline of 47%, a trend that had been established in the mid-1940s. It is surprising that the rapid increase in milk prices during this period (Figure 15.112) had little or no effect on the number of animals. By 1985, only 0.46% of the milk cows in the United States were on farms in Alabama. In 1970, the percentage had been 1%.

Milk Per Cow

In 1950, average annual milk production per cow in Alabama was 3,570 pounds (Figure 15.114). By 1970, the average had increased to 6,589 pounds. Afterwards, it trended steadily upward to 10,940 pounds in 1985. During this same period, average national milk production per cow increased from 9,751 to 13,024 pounds. In Alabama, milk production per cow, with the exception of 1978 and 1983, increased throughout the period 1970-1985. From 1970 through 1977, trends were similar in the state and the country; however, average yearly production in Alabama for that period was 2,969 pounds lower than in the country as a whole.

Beginning in 1979, average production per cow began to grow at a higher rate in the state, and by 1985 it was much closer to the national average than it had been in 1970 (10,940 versus 13,024 pounds). From 1970 through 1979, Alabama lost a large share of its dairy farms. It is likely that after 1979 that only the most efficient, best managed operations were still in business. Those remaining likely did a better job of getting the highest production out of their herds.

Pounds Produced

In 1950, total milk production in Alabama was 1,292 million pounds (Figure 15.118). In 1970, the total was down to 816 million. During the same period, national milk production increased from 116,602 million to 117,007 million pounds (Figure 15.115). In 1950, Alabama dairy farmers produced only 1.1% of all the milk in the country; however by the end of 1970 that percentage had declined to 0.7%.

During the period 1970-1985, milk cow numbers declined sharply in Alabama (Figure 15.117), while production of milk per cow increased (Figure 15.114). However, the relative magnitude of the reduction in the number of animals was considerably greater than the relative increase in milk per cow. As a result, pounds produced generally decreased throughout the period. It was 816 million pounds in 1970 and 547 million in 1985.

Production actually increased (relative to the preceding year or years) only in those years where the rate of increase in milk per cow was greater than the rate of decrease in herd size. For example, in the period 1971-1972, numbers declined 2.5% (120,000 to 117,000), while production per cow increased 2.7% (6,825 to 7,009 pounds). As a result, pounds produced increased from 819 to 820 million.

Alabama Chickens

In the early 1940s, Alabama's chicken business began to change drastically. Before that time, chickens had been produced in the state primarily as a source of table eggs. Only cockerels and culled laying hens were used for meat. However, with the coming of broiler production, the demand for eggs for hatching broiler chicks exploded along with the demand for more laying-hens. On January 1, 1960, there were eight million chickens (all chickens, except broilers) in the state, and by 1970 there were 20 million. Some 67% were hens and pullets of laying age.

The Alabama Agricultural Service published data indicating that on January 1, 1970 the largest share (11.2%) of the chicken inventory was on farms in Cullman County. Other counties with large shares included: DeKalb (10.6%), Blount (8.8), Walker (5.3), and Morgan (4.2). These five counties accounted for 40.1% of all the chickens in the state. Some 19 counties reported flocks of 300,000 or more birds. Figure 15.119 shows the distribution of chickens in Alabama in 1970.

US and Alabama Prices

Data on the annual, average price of chickens (dollars per head) in the country and the state for the period 1950-1985 are shown in Figure 15.120. Prices had remained relatively stable while numbers were growing rapidly in the period 1950-1970; however in the period 1970 through 1985, this was not the case. When the Consumer Price Index (Figure 15.2) really began to increase rapidly around 1972, chicken prices in Alabama followed. In 1972, the price was $1.15 a head; then it trended sharply upward to reach $2.20 in 1980. Prices in Alabama were also responsive to the recession of 1973-1975, falling sharply during that period. Rapid changes in broiler production probably contributed to the sharp increase in the price between 1979 and 1980. Between 1978 and 1979, broilers produced in the state increased by about 51 million. That increase triggered a massive demand for hatching eggs and for laying hens. After 1980, inflation began to fall rapidly (Figure 15.3), and prices declined to range around $1.90 from 1981 through 1985.

Numbers

The number of chickens on Alabama farms increased sharply from 1960 to 1970 in response to a similar increase in broiler numbers. In 1970, the January 1st inventory counted 19.8 million chickens (all, excepting broilers) in the state. Afterwards, the annual rate of increase in broiler production slowed, and numbers remained generally unchanged through 1973. Then with the recession of 1973-1975, broiler production remained generally unchanged and chicken numbers declined. In 1976, broiler production began to increase again, and chicken numbers followed, reaching 20.9 million in 1977. Broiler production remained essentially unchanged between 1976 and 1977, and the January 1st inventory of chickens was 300,000 lower than it had been on January 1, 1977. With broiler production continuing to increase, chicken numbers reached 22 million in 1979, the highest on record at that time. Numbers began to decline in 1980, and reached 16.6 million in 1983. They remained near that level through 1985.

Alabama Broilers

The trend of vertical integration of the broiler industry, which generally began in the preceding period (1950-1970), continued in the 1970-1985 period. Continued consolidation within the industry resulted in the rapid development of new information and technology and their rapid adoption throughout the industry. For example, between 1968 and 1978, feed efficiency (pounds of feed required to produce one pound of broiler) decreased from 2.25 to 2.13. It had been 3.60 pounds of feed in 1948. Further, as a result of the adoption of the new technology, the cost of production per bird fell rapidly. All of these advances led to an improved and less expensive consumer product, and the public responded by increasing their consumption. All of these factors worked together to bring about the rapid expansion of the entire broiler industry.

Broilers produced in the United States increased from 0.6 billion in 1950, to 1.8 billion in 1960, and to three billion in 1970. In that year, Georgia reported the largest share (15.2%) of total national production. Other states reporting large shares included: Arkansas (15.1%), Alabama (12.6), North Carolina (10.3), and Mississippi (8.3). Production in these five states accounted for 61.5% of the national total.

Chicken Consumption

A more plentiful, more attractive, and less expensive product (broilers) led to a growing consumption of chicken by American consumers, especially after 1975 (Figure 15.103). In 1970, the average American consumed 40.5 pounds of chicken annually. Probably as a result of the recession, consumption remained at about the same level in 1975. However, by 1985 it had increased to 58 pounds, an increase of 43%, or almost 3% a year.

Prices

Broiler prices are not prices that farmers receive when they sell them. Rather, those reported here are wholesale prices of processed whole birds. Price was 12.2 cents a pound in 1970 (Figure 15.122). Beginning in the early 1950s, price had trended sharply downward, reaching 12-14 cents in the early 1960s. It fell slowly through 1970, then increased sharply to 23.4 cents in 1973. This was the first year of the rapidly increasing inflation that would plague the country for a decade (Figure 15.3). It fell to 20 cents in 1974 in the midst of the recession (Figure 15.1). While inflation seemed to be the primary force driving price increases, the teeter-totter effect also seemed to play a role. For example, between 1975 and 1976, broiler numbers (Figure 15.123) increased by almost 35 million, and price fell from 25.3 to 22.3 cents. Inflation began to decline rapidly after 1980, and in 1982 price declined. However, the demand for broiler meat was too powerful for prices to be affected by falling inflation for very long. In 1984, it reached 32 cents, the highest level in history at that time. Finally, in 1985, with numbers continuing to explode upward, price blinked and fell to 28 cents a pound.

Numbers

Alabama farmers produced 13 million broilers in 1950. By 1970, they had increased their annual production to 376 million (Figure 15.123). The increase in numbers continued through 1972, then changed very little through 1975, probably as a result of the effect of the recession. Numbers reached 430.2 million in 1976 and generally continued to trend upward to reach 561.8 million in 1985. The 1981-1982 recession apparently resulted in a slight reduction in numbers. During the period 1970 through 1985, Alabama broiler numbers increased from 376.1 to 581.8 million, an increase of 49.4% or an average of 3.1% a year.

Growing demand for broiler meat—a relatively cheap, nutritious, attractive alternative to beef and pork—seemed to pull numbers along throughout the entire period. However, the teeter-totter phenomenon also seemed to play a role in some years. For example, price declined from 1975 to 1976, and remained near that lower level through 1977. This decline seemed to slow the rate of increase in numbers in both 1977 and 1978. Also, price declined from 1981 to 1982, and in 982 number was lower than in 1981. Then it seemed to follow prices upward and reached 536.6 million birds in 1984. It continued to increase in 1985 (561.8 million), even though price had declined sharply after 1984.

Market Weights

The average market weight of broilers sold in Alabama was 2.7 pounds in 1950 (Figure 15.124). Then between 1950 and 1970, it increased to 3.5 pounds. Afterwards, market weight continued to increase, reaching four pounds in 1985. Between 1940 and 1985, improved production technology being applied on the farms resulted in increasing market weight from 2.4 to four pounds. This increase was one of the reasons why broiler production was growing so rapidly. When coupled with improved feed conversions, production efficiency was increasing rapidly in the industry.

Cash Receipts

As a result of the increase in numbers and prices, cash receipts from the sale of broilers produced in Alabama increased from $160.4 million in 1970 to $629.2 million in 1985 (Figure 15.125). In 1970, cash receipts for broilers was equivalent to 29.6% of the cash receipts for all livestock and poultry in the state. By 1985, this percentage had increased to 48.4%. In 1970, cash receipts from the sale of broilers exceeded those from the sale of cattle and calves by some $4.2 million ($160.4 versus $156.2 million). At the end of 1985, however, this difference had increased to $282.9 million ($629.2 million versus $346.23 million). In 1970, cash receipts for broilers were equivalent to 21% of total cash receipts received from the sale of all agricultural commodities originating on Alabama farms, including farm forest products. By 1985, this percentage had increased to 29.3%. Chickens, mostly broilers, had finally come home to roost in Alabama.

Alabama Eggs

In 1970, poultry farms in the North Central Region produced the largest share (27.6%) of all eggs in the United States. Production in this region was followed by the South Atlantic Region (21%), the South Central Region (20.8), the West Region (16.6), and the North Atlantic Region (14.0). Among the states, California farmers gathered 11.9% of the total. Other states reporting large shares included: Georgia (7.5%), North Carolina (5.2), Arkansas (4.9), and Pennsylvania (4.7). These five states accounted for only 34.2% of the national total. These percentages are somewhat misleading. Eggs produced in the North Central Region were used primarily for human consumption, whereas in the South Atlantic and South Central regions, a large share was used for hatching eggs for broiler chicks.

Consumption

The consumption of table eggs in the United States had been declining for several years before 1970 (Figure 15.126). In that year, it reached 276 eggs per person, having been around 390 in 1950. Consumption continued to trend downward through 1977, apparently as a result of public concern about the effect of large quantities of eggs in their diets. Also, American lifestyles were changing rapidly during this period. People were spending less time preparing the typical bacon and eggs breakfast. It is also likely that the recessions of the early 1970s contributed somewhat to the decline. Consumption increased slightly in 1978 and 1979, but quickly resumed its downward trend. In 1985, it reached 216 eggs per capita.

In 1970, Americans consumed 33 eggs per capita in processed form (pasta, candy, baked goods, etc.). Consumption in this form remained in the range of 31 to 35 until 1984, when it began to trend upward. It was slightly over 38 in 1985.

US and ALabama Prices

In 1970, Alabama farmers received 44 cents a dozen for their eggs (table and hatching combined) (Figure 15.127). Prices declined in 1971 and 1972, probably as a result of the 1969-1970 recession (Figure 15.1). Between 1972 and 1973, prices increased from 34 to 53 cents as inflation began to increase rapidly in the country (Figure 15.3). As a result of the 1973-1975 recession, prices remained near the 53-cent-level through 1975. Afterwards, they followed inflation upwards, reaching 73 cents in 1981, the highest level in history to that time. However, prices fell sharply between 1981 and 1982 as the 1982 recession took its toll. Egg numbers (Figure 15.128) began to decline after 1980, and after the recession ended in 1983 prices began to increase, finally reaching 71 cents in 1984. After 1984, the Alabama Agricultural Statistics Service only published data on prices of table eggs. Prices of hatching eggs were considerably higher. The lower-than- expected price in 1985 reflects the removal of those higher-priced eggs from the estimates.

Figure 15.127 also includes data on average annual prices of eggs throughout the country. Trends for state and national prices were similar, because both were largely determined by national economics during the period. Nationally, a relatively higher percentage of all eggs was used for human consumption. In the South, and especially in Alabama, a large share was used for hatching (higher-priced eggs). Consequently, national prices tended to be lower, especially after 1975 when the demand for hatching eggs for broilers in the state increased sharply. Apparently, the 1982 recession (Figure 15.1) had a greater effect on prices in Alabama. In 1982 and 1983, prices in the state were lower. Of course, the lower price in 1985 was the result of the change in reporting, detailed in the preceding paragraph.

Numbers

Alabama's egg number situation during the period 1970-1985 was becoming increasingly complex, and the number produced for hatching increased steadily. The Alabama Agricultural Statistics Service did not begin to report data on the two separately before 1980. In that year, 25% of all eggs were used for hatching. The percentages used for hatching were: 27.7% (1983), 28.2% (1984), and 30.2% (1985).

Alabama poultry farmers gathered 700 million hatching and table eggs in 1950. By 1970, the total had increased to 2.8 billion. Egg numbers increased relatively little between 1970 and 1975 (2.9 million) as the growth in broiler numbers slowed (Figure 15.123). However, after 1975 broiler numbers did begin to increase rapidly, and as a result egg numbers also began to increase, reaching 3.3 billion in 1980. By that year, a large share of table eggs being consumed in the country and the state were being marketed through large grocery chains, and Alabama egg producers seem not to have been able to compete very well in this market. Therefore, the number of table eggs produced in the state fell rapidly while hatching eggs increased, but hatching eggs did not increase as rapidly as the loss of table egg production. For example, from 1983 through 1985 table egg production declined by 85 million, while hatching egg numbers increased by only 66 million. As a result of these interacting factors, total egg numbers were down to 2.9 billion in 1985.

Uses

The Alabama Agricultural Statistics Service did not begin to report the number of hatching eggs produced in Alabama until 1983. Production of hatching eggs in Alabama was 780 million (1983), 784 million (1984), and 846 million (1985). In Alabama, the percentage of total state production used for hatching was 27.75% (1983), 28.2% (1984), and 30.3% (1985). Nationally, where the requirement for eggs for hatching broiler chicks was much more limited, the share of national egg production used for hatching was lower: 8.3% (1983), 9.3% (1984), and 9.6% (1985).

Cash Receipts

Total cash receipts for all eggs sold in Alabama were \$101.6 million (1970), \$132.7 million (1975), and \$180.1 million (1980), an increase of 77.3% during that 11-year period. During the same period, the Consumer Price Index increased by 112.4%. It is not surprising then that Alabama poultry farmers were beginning to restrict their production of table eggs. In 1985, cash receipts for table eggs only totaled \$160 million.

In Alabama, cash receipts for all eggs accounted for 18.8% (1970), 15.9% (1975), and 15.3% (1980) of cash receipts from all livestock and poultry sold in the state in those years. The 1985 cash receipts for table eggs only accounted for 12.3%. Cash receipts for all eggs accounted for 13.3% (1970), 10.1% (1975), and 9.7% (1980) of all agricultural commodities sold from Alabama farms, including farm forest products. Cash receipts from 1985 only accounted for 7.8%.

Alabama Catfish

At the beginning of this period (1970), the list of animals used by Alabama's livestock produces had been generally fixed from well before the creation of the state. For all that long period of time, farmers produced cattle, hogs, a few sheep and some goats, and chickens; that was about it. After World War II, a few enterprising farmers began to produce bait minnows for use by sport fishermen. Also, at about that time, some farmers producing largemouth bass and bluegill sunfish in their ponds began to sell permits to the public that allowed them to catch a limit of fish. Although these small operations added to farm income in the state, the contribution was very limited. Channel catfish had been a commercial product in the state for decades. Large quantities were sold locally by fishermen who were taking them by hooks and traps from the state's creeks, rivers, and reservoirs. Long before 1970, there was a well-established market for channel catfish; however, they were not farm-raised in Alabama until the late 1950s. Although, this animal came late to Alabama's farmscape, it has certainly created a splash, especially in some portions of the state.

In the early years, it is unlikely that those responsible for collecting agricultural statistics in the country or the state felt that is was necessary to collect data on the production of farmed channel catfish. After all, this was not a traditional agricultural animal, although fish had been farmed in monastery and manor ponds during the European Middle Ages. There is relatively little information available on catfish farming in the United States for the period 1970-1985. The first USDA survey of the industry was not conducted until 1982, and the second one in 1989. Harvey (1988) tabulated the pounds of catfish processed, but little else. The Alabama Agricultural Statistics Service included catfish with other livestock and poultry until after 1985.

Harvey (1988) presented data (Figure 15.129) showing pounds of catfish processed in the United States. They increased relatively slowly from 5.7 million pounds in 1970 to 18.7 million in 1978; thereafter, the quantity processed surged upward to 191.6 million pounds in 1985. It is doubtful whether the production and processing of any other agricultural commodity in American history ever increased at such a rapid rate.

The USDA's National Agricultural Statistics Service "Catfish Growers Survey" for 1982 reported that there were 250 catfish operations in Alabama comprising a total of 8,200 acres, and that the state contained 25.3% of all catfish operations and 11.1% of the catfish pond acreage in the United States. The average catfish operation in the state had 33 acres of ponds. Cash receipts for 1982 totaled $14.2 million.

Data published by the national Marine Fisheries Service in *Fishery Statistics of the United States* indicate that the annual average price received by American catfish farmers was 61 cents a pound in 1983, 69 cents in 1984, and 72 cents in 1985.

Agricultural Production

Data on total value added to the American economy from the production of crops and livestock during the period 1950-1970 show that the total increased from $31.3 billion in 1950 to $51.3 billion in 1970, an increase of 64%. The Consumer Price Index increased by 61.6% (72.1 to 116.2) during the period—indicating that there was little real increase in total value. This lack of a real increase from 1950 through 1970 is in sharp contrast to the 1940-1950 period, when the increase far exceeded inflation.

Data presented in Figure 15.130 indicate that total value added for crops and livestock in the United States increased from $51.3 billion in 1970 to $142.8 billion in 1985, an increase of 178.4%. During the same period, the CPI increased by 173.9%. Here again, this comparison shows that in real dollars value added increased relatively little during this period. Trends in value added for both crops and livestock were similar during the period, and both contributed about the same amount to the total. The lower-than-expected value for crops in 1983 (an El Niño year) was the result of lower-than-expected values for food grains, feed crops, and cotton that year.

Data on value added for some specific crops in the United States in 1965, 1970, 1975, 1980, and 1985 are presented in Table 15.23. These data show that in terms of value added feed crops (primarily corn) and oil crops (primarily soybeans) were the most important (value-wise) produced in the country during this period. Values for food grains, feed crops, oil crops, fruits and tree nuts, and vegetables increased more during the period than did the CPI. The data also show that the values for home consumption of crops remained relatively constant ($0.23 to $0.28 billion) during the period; however, with the increase in total value, the relative importance of crops consumed by farm families producing them declined.

The data also show that the value for meat animals was highest for all livestock (Table 15.23). The value for dairy products was less than half of that of the meat animals. None of the increases in the values of any of the livestock and poultry were greater than the increase in the CPI. Values for home consumption of livestock and poultry varied between $0.53 and $0.90 billion during the period, indicating that the relative importance of consumption of these products by farm families did not keep pace with the increases in the values of the commodities themselves.

Data presented in Figure 15.131 show that total value added for crops and livestock produced on Alabama farms increased from $789.3 million in 1970 to $2,062.2 million in 1985, an increase of 161.2%. (The CPI increased by 173.8% during the same period.) Throughout the period, the value for livestock contributed more to the total than the value for crops. Values for both crops and livestock increased during the period, but the value for livestock seemed to increase at a slightly faster rate, probably as a result of the increasing production of broilers.

Data on value added from the production of some specific crops and livestock produced on Alabama farms in 1965, 1970, 1975, 1980, and 1985 are presented in Table 15.24. Increases in values for food grains (primarily wheat), feed crops (primarily corn), oil crops (primarily soybeans), and fruits and tree nuts during the period were all greater than the increase in the CPI during the same period. Through the entire period, the value added for oil crops accounted for the largest share of the total; however after 1980 Alabama farmers were beginning to lose interest in soybeans, because it was quickly becoming evident that they could not produce the crop competitively.

Values for home consumption of crops in Alabama declined from $9.7 million in 1970 to $5.6 million in 1985. In 1970, home consumption accounted for 4.2% of total value added; however by 1985 the percentage had fallen to 0.7%. The relative importance of value added for home consumption of crops in the state were considerably greater than in the country as a whole. In 1970, some 4.2% of the value added of all crops were consumed by Alabama farm families who had produced them. In the country as a whole, home consumption accounted for only 1.1% in the same year. By 1985, the comparative percentages were 1.6% and 1.1%, respectively.

The value for poultry and eggs (primarily broilers), was roughly equal to the value for meat animals in 1970 (Figure 15.24); however, by 1985 it was almost twice as large. Only increases in values for poultry and eggs, among the livestock and poultry, were greater than the increase in the CPI during the period 1970-1985. Values for home consumption of livestock and poultry increased from $12.6 to $23.1 million between 1970 and 1975, but declined thereafter, reaching $10.1 million in 1985. Home consumption of poultry and livestock was less important than for crops during the period. In 1970, home consumption only accounted for 1.6% of the total, and by 1985 it was down to 0.5%. The importance of home consumption of livestock and poultry was generally similar in the state and the country. The comparative percentages were 1.6% and 1.1% in 1970 and 0.5% and 0.5% in 1985.

Table 15.25 presents data on the cash receipts obtained from the sale of some crops and of livestock and poultry in Alabama in 1970, 1975, 1978, and 1985. These data show that in terms of cash receipts that cotton, peanuts, and soybeans, in that order, were the most important crops in the state in 1970, but that by 1975, soybeans had taken first place, moving cotton to third. By 1985, however, with farmers losing confidence in soybeans, cotton returned to its rightful place as king of the field crops in Alabama's farm economy. However, while cotton might have returned to its rightful place, other plant crops were the jewels of the future. Cash receipts for greenhouse, sod, and nursery increased from $13.5 million in 1970 to $114.2 million in 1985, an eightfold increase. Also, receipts for farm forest products increased from $14.4 million to $93.8 million in the same period, a better than sixfold increase.

Cash receipts for cattle and calves increased substantially between 1970 and 1975 (Table 15.25), but not as much as the increase in the CPI. Receipts for broilers increased from $101.6 million in 1970 to $629.2 million in 1985. Broilers had been the first-among-equals among livestock and poultry in 1970, but by 1985 broiler production was the gorilla in the room. Cash receipts for other livestock and poultry increased from $5.6 million in 1970 to $36.1 million in 1985. This increase was probably the result of the increasing importance of channel catfish in Alabama's agricultural economy.

STILL SERIOUSLY SEEKING SALVATION

Between 1960 and 1970 the parity ratio declined from 0.80 to 0.72. It had not been as low as 0.72 since 1933. Farmers were expected to cope with this increasingly severe cost-price-squeeze by increasing productivity. But increased efficiency alone was not sufficient to compensate for their deteriorating price-cost situation.

They had to make-up the difference by earning more and more of their income off the farm. Between 1933 and 1970, federal and state agencies had spent hundreds of millions of tax dollars trying to improve the economic life of farmers, but it was obvious in 1970 that farmers were Still Seriously Seeking Salvation. But, seemingly, in all of those efforts, the Triangle had learned very little about what might be done to improve the situation. In 1970, their consensus. seemed to be that "we are on the right track; we just need to spend more money."

Salvation Through Better Information

As detailed in several preceding chapters, it was decided many years ago that one of the solutions to the problem of low farm income was to provide farmers with better information, and this was to be accomplished by purchasing agricultural research and extension with public funds.

Data presented in Table 15.26 show funds available for the operation of the Alabama Agricultural Experiment Station during the period 1939-1940 through 1985-1986. These data show that funds available for research increased significantly between 1969-1970 and 1985-1986. During this period, total funds available increased from $7.1 million to $25.3 million, an increase of $18.2 million. Funds from the federal government, grants/gifts, and sales increased during the period, but the largest share of the increase ($10.9 million) came from funds appropriated by the Alabama legislature. These data do not show it, but the national recessions of 1980 and 1981-1982 (Figure 15.1) severely restricted the state's economy, which in turn seriously affected appropriated funds available for agricultural research (Kerr, 1985). High inflation (Figure 15.3) further decreased the effectiveness of the funds that were available. The country's economic situation improved somewhat after 1983, and state appropriations for research increased rapidly.

Kerr (1985) describes a number of significant changes in the administrative structure for agricultural research and extension at Auburn in the early 1980s. Unfortunately, in 1984 some of the more important changes were eliminated when a new university president was installed.

Data on the number of scientific man-years (SMYs) of agricultural research purchased by the Alabama Agricultural Experiment Station in Auburn's agricultural subject matter departments are presented in Table 15.27. The data indicate that the total number of SMYs increased relatively little between the 1969-1970 and 1984-1985 fiscal years (83.58 to 88.25 SMYs). This was a period of extreme economic malaise in the state and the country as a whole. As a result, there was little opportunity to receive very much additional funding for agricultural research. SMYs purchased in the departments remained at about the same level throughout the period. The only significant increase was in the Department of Botany and Plant Pathology (7.30 to 12.55). However, much of this increase was probably the result of adding research in microbiology in this department. In contrast, there was a sizeable reduction in the Department of Agronomy and Soils (15.48 to 12.71 SMYs).

Data on total extension work funds budgeted by the Alabama Cooperative Extension Service in the years 1964-1965 to 1984-1985 are presented in Table 15.28. The table also includes data on the amount received from federal, state, county, and other sources. The data show that total funds budgeted increased from $7.5 million in 1969-1970 to $24.5 million in 1984-1985, an increase of 227%, or about 14.2% a year. Unfortunately, this was a period of high inflation. During the period 1970-1985, the Consumer Price Index increased by 177% (38.8 to 107.6). As a result, the large increase in funds budgeted probably did not purchase as much extension work as the increase indicates.

Much of the total increase in extension funds budgeted came as a result of a large increase in state funds, $3 million in 1969-1970 to $13.3 million in 1984-1985. As a result, the share of state funds of the total increased from 40% in 1969-1970 to 54.2% in 1984-1985. During the period, county contributions increased by 125.6%.

Data on the number of professional personnel in different categories employed by the Extension Service in 1971, 1975, 1980, and 1985 are presented in Table 15.29. These data show that total personnel declined from 1971 to 1975 (497 to 468, then increased to 528 in 1980, only to fall to 446 in 1985). Most of this variation was the result of large changes in the number of county agents. The number of subject matter specialists increased significantly during the period (81 to 117).

In the period 1967-1971, USDA provided funds for agricultural research to Alabama A&M University and other 1890 Universities. (The Morrill Act of 1890 established African American land-grant universities.) Then in 1972, the USDA budget contained substantial funds for the use of these institutions for both agricultural research and extension. From 1971 through 1977, these funds were appropriated on an annual basis, but the 1977 Agriculture Act passed by the 95th Congress made them permanent. These funds could not be used for facilities, but in December 1981 PL 97-88 was enacted that appropriated funds for facilities on a non-matching basis (Morrison, 1994).

Under provisions of that 1972 legislation, funds became available that allowed Tuskegee to re-establish its extension program on its own campus (Mayberry, 1989). Also under this legislation, Tuskegee received its first funds from the Cooperative Research Service of USDA for agricultural research. Before that time, the presidents of the 1890 Universities had voted unanimously to provide an equal share of the total allocation to the privately-operated university.

Alabama A&M University became involved in international development work in the late 1960s with funds provided to them by PL 89-106 and later under Section 1445 of PL 95-113. With these funds, AAMU was able to increase its competence in agricultural research and technology transfer. With this experience, it was only natural that AAMU would receive a USAID Strengthening Grant. The grant to increase AAMU's competence in international technology transfer was approved in 1979, and provided AAMU with $100,000 a year for five years,

In 1981, the US Office of Civil Rights informed Alabama that it was in violation of Title III of the Civil Right Act of 1964, and asked the state to submit a plan for remediation. Because the state was unable to present an acceptable plan, in 1983 the Office of Civil Rights brought suit against the state in federal court. However, because of several critical problems with the trial, the suit was not settled until well after 1985.

Table 15.30 presents data on the number of professional employees working in extension for AAMU in 1980 and 1985. In both years, the total was around 20, with most being either specialists or urban/county agents. Further, in both years there were about half as many specialists as urban/count agents.

As noted in Chapter 14, available data suggested that no matter how diligent and dedicated Alabama agricultural research and extension personnel might be in the production and dissemination of new and better agricultural information, farmers using that information were able to keep up with farmers in other regions, but they could not catch up. This hypothesis was suggested by comparing the yields over time of most field crops grown in the state and in milk produced per cow with national average yields or with yields in some individual states (Figures 16.55, 16.62, 16.71, 16.81, 16.87, 16.97, and 16.112). In every case, yields of Alabama crops increased, but never enough to overtake national average yields or yields in other states.

The keep-up-but-not-catch-up phenomenon in Alabama agriculture persisted through the current period (1970-1985). The physical and biological impediments to agricultural production described in preceding chapters were just too great to overcome with research and extension. With their persistence, at least from 1950-1985, it is not likely that better research information will ever enable Alabama farmers to compete on a level playing field. If they are to continue to farm, they must find some way to compensate for persistently, significantly lower yields.

The results of the effect of this phenomenon are obvious in the decline in farms in Alabama versus the decline in the United States. Between 1970 and 1985, Alabama lost 36.6% of its farms, while the country only lost 22.2% (Figures 15.17 and 15.16). It is also partially the cause of differences in the net farm income per farm data presented in Table 15.31. Only the values for Mississippi and Tennessee were lower. Yet, differences in farm size likely played some role.

These data pose an important question for the future of agricultural research and extension in Alabama. If better information is not sufficient to allow Alabama farmers to compete in crop production, how long will the public sector be willing to invest its money in agricultural research and extension?

Salvation Through Organization

The American Farm Bureau was established in 1919. No other national farm organization was formed in the United States after that time until the American Agricultural Movement (AAM) was established in the mid-1970s. Farmers in the West were especially hurt in 1976 and 1977 by declining prices for cattle and wheat and increasing prices for production inputs. In the late summer of 1977, farmers in southeastern Colorado, western Kansas, Oklahoma, and northwestern Texas were threatening to stop planting crops and purchasing any supplies until Congress passed legislation that would virtually guarantee that their incomes would equal incomes of other workers in the national economy (100% of parity). Out of this farmer unrest came the formation of the American Agriculture Movement. By October 1977, the AAM had spread into the Southeast. Later that year, Georgia AAM supporters pushed their demands with a series of highly publicized tractor marches in Alma, Atlanta, and Plains (President Carter's hometown). Then in January 1978, they joined farmers from all over the country by staging a giant tractorcade in Washington. Congress responded to their concerns by providing more generous land diversion payments, extending payment plans on their loans with the Farmers Home Administration, and establishing a large emergency loan program. However, the AAM demand for price supports that would guarantee them 100% parity was rejected. The AAM was generally dissatisfied with these results, and they sponsored a similar visit to Washington in early 1979. But by that time the movement was rapidly loosing steam, and the congressional response was even less positive. As a result, the AAM more or less ceased to function as an organization.

Several agricultural organizations were established in Alabama, or others changed their status during the period 1970-1985. These included:

1. The Alabama Soybean Growers Association was established in 1968 for the purpose of bringing together all persons interested in the production, marketing, distribution, and use of soybeans. In the early 1970s, members of the organization voted to change its name to the Alabama Soybean Association. This Alabama organization is affiliated with the American Soybean Association.
2. The Alabama Forestry Commission was established by an act of the legislature in 1969. The commission replaced the Division of Forestry, which had been a part of the Alabama Department of Conservation since 1939. The purpose of the commission was to protect, conserve, and increase Alabama's timber and forest resources.
3. In 1972, members of the Alabama Poultry Industry Association voted to change the name of the organization to the Alabama Poultry and Egg Association.

Salvation Through Diversification

If the decline in the amount of cotton harvested in the state is used as an indicator of the level of diversification, Alabama farmers made significant progress in increasing it between 1969 and 1982. Census data indicated that in 1949, of total acres of cropland harvested, acres of cotton harvested accounted for 33.6%. It had been 50% in 1929. By 1969, the percentage had declined to 20%. Between 1974 and 1978, it declined from 20.9 to 9.4%, and finally in 1982 it reached 8.7%. The decline in cotton acres harvested was accompanied by an increase in acres of soybeans harvested. In 1969, that crop accounted for 23.3% of all acres of cropland harvested. It had been less than 1% in 1929. Then by 1982, acres of soybeans harvested reached 61.2% of the total. Alabama farmers had finally rid themselves of the white fiber albatross. Unfortunately, comparative data on net farm income for the period 1970-1985 (Table 15.26) seem to indicate that it did not make very much economic difference to Alabama farmers.

Salvation Through Political Action

The use of government payments as an indicator of the intensity of political pressure being applied by farm interests has been discussed in preceding chapters, along with the peculiar role played in this process by the Southern band of brothers in Congress. The effect of these phenomena is obvious in the period 1970-1985.

Government payments in the United States associated with the various legislative mandates supporting agriculture remained relatively high—about $1,250 a farm during the period 1970-1972 (Figure 15.132). However, they declined rapidly in the years 1974-1976 (Figure 15.8) to levels not seen since the 1933-1956 period. In 1974, payments reached $190 a farm. Unfortunately, the parity ratio had begun to decline after 1974, and was down to 0.66 in 1977. That year payments exploded upward to $1,244. The ratio increased slightly in 1978 and 1979, and payments declined to $564 and $527 a farm, respectively. After 1974, the ratio began to fall sharply again, finally reaching 0.53 in 1982. Payments were increased to $1,451 a farm that year and to $3,908 in 1983, the highest levels in the history of government payments. They remained near that level through 1985.

Government payments to Alabama farmers usually mirrored those for the country as a whole, except that in all years during the period 1970-1985 payments to farmers in the state were lower. From 1981 through 1985, they were much lower (Figure 15.132). The differential between the state and the national average payments after 1981 was the result of large increases in US payments for crops (feed grains, wheat, and cotton). As stated above, crop production was of lesser importance in the state at that time.

These data indicate that by the beginning of this period (1970), the actions of Dobson's (1985) Triangle (beneficiaries, politicians, and bureaucrats) had become even more firmly institutionalized. The immediate, seemingly automatic response to annual changes in the parity ratio in the form of government payments demonstrates the degree to which these three groups were locked together. They also support Dobson's conclusion that with this degree of institutionalization, the Triangle relationship would be extremely resistant to change.

Unfortunately, there is little indication that political action was having the desired effect. Even with the rapid growth of government payments after 1980 (Figure 15.132), the parity ratio continued to decline. It was 0.64 in 1980, but by 1985 it was down to 0.51. Similarly, surpluses continued to grow despite Triangle efforts to take land out of production, increase exports, and give as much away to "needy" people as possible. The cost of food as a share of disposable income continued to decline. It had been 20.6% in 1950, but by 1970 it was down to 13.9%. The decline continued in the period 1970-1985, reaching 13.2% in 1980 and 11.7% in 1985 (Clauson, 2008). The farmer's share of the food dollar also continued to decline. It had been 41% in 1950, but by 1970 it was down to 32%, and it continued to decline in the following years, reaching 31% in 1980 and 25% in 1985 (Elitzak, 2008). It seemed apparent that after a half-century of Triangle activism, political action was not going to provide farmers with the Salvation they so ardently pursued.

The Bottom Line

At the end of the year when farmers total all of their costs (cash and non-cash) and all their income (all forms), the difference between the two totals is net farm income (NFI). This value represents their total "pay" for all of the time they spent (physical labor, planning, running errands, etc.) on their farming operations for the year. That is the bottom line.

In the preceding 21-year period (1950–1970), when American farmers were investing enormous sums of money to modernize their farms, NFI (per-farm basis) increased by 101.8% in the country and 127.9% in the state. Increases in both the country and the state were considerably greater than the increase in the Consumer Price Index (61.3%).

Figure 15.133 presents data showing average annual NFI values (on a per-farm basis) for all American farms for the period 1950-1985. These data show that NFI began to respond to the increase in inflation(Figure 15.3) as early as 1965, but that after 1971 it literally exploded upward, increasing from $5,173 to $12,170 per farm in 1973. Subsequently, NFI declined in the following four years to reach $8,095 in 1977, as the national economy muddled through the energy-shock recession of 1973-1975 (Figure 15.1). NFI began to increase again after 1977, and by 1979 it reached $11,249; then it was savaged again by the 1980 and 1981-1983 recessions. In 1980, the NFI was also affected by a prolonged drought and heat wave that reduced crop yields significantly throughout the Central and Eastern United States. Unfortunately, in 1983 the agricultural economy was once again devastated by an El Niño weather disaster. In 1984 and 1985, the GNP increased sharply, and the NFI followed, reaching $11,129 in 1984 and $12,438 in 1985. With two recessions and two crop weather disasters, American farmers really needed a break.

When government subsidy payments are excluded, the average NFI actually declined between 1981 and 1985. (Federal government payments are included in NFI statistics.) When payments are excluded from NFI in Alabama, it declined by almost $1,000 a farm between 1982 and 1985. Apparently, American farmers could not find Salvation in Political Action either.

Regardless of ups and downs of the period 1970-1985, the national NFI grew from $4,871 a farm in 1970 to $12,438 in 1985, an increase of 155.4%. Unfortunately, the Consumer Price Index grew at an even faster rate. It was 116.3 in 1970 and 318.5 in 1985, an increase of 174%. In real dollars, American farmers were receiving less income in 1985 than they had received in 1970. It was the best of times, it was the worst of times.

Data on the annual NFI for Alabama farms (per farm basis) for the period 1970-1985 are also presented in Figure 15.133. These data show that the year-to-year trends were essentially the same for both the country and the state. However, in all except one year, national values were higher, but the magnitude of the difference varied considerably from year to year. For example, in 1973 the national value was $5,305 higher ($12,170 versus $6,865), but in 1982 the difference was only $903 ($9,905 versus $9,002), and in 1983, just after the end of the recession, the state value was $631 higher. The increase in the CPI during the period actually left Alabama farmers in better financial positions than their national counterparts. In the state, the NFI increased by 213.3%, considerably higher than the increase in the CPI (173.9%). The rapid increase in broiler production was probably the reason.

Data presented in Figure 15.133 indicate that the average NFI (per farm) in the United States was $4,871 in 1970 and $12,438 in 1985; however, there was considerable variation among the values in various states. Data presented in Table 15.31 show the NFIs for several states in 1950, 1970, and 1985. In 1970, the values ranged from a low of $1,824 in Tennessee to a high of $16,019 in California. The national average was $4,871. Of the states listed, only Tennessee had a lower NFI than Alabama. In 1985, Tennessee and California still had the lowest and highest NFIs, ($3,334 versus $45,661), and the national average had increased to $12,438. Only NFIs for Tennessee and Mississippi were lower than Alabama.

There are probably several reasons for the differences in NFIs between states. For example, farmers in California and Florida tended to produce crops (citrus, vegetables, etc.) with higher value, compared to those in Tennessee and Mississippi. In 1985, dairy farms in Pennsylvania produced 7% of the country's milk. However, one of the important reasons for the range in NFIs was differences in farm size. For example, in 1985 average farm size in Tennessee was 141 acres, whereas in California it was 391 acres; however, farm size alone was not sufficient to account for all of the large difference in NFIs of the states. The average size of farms in Pennsylvania was not much greater than in Tennessee (150 versus 141 acres); yet the NFI was almost four times greater ($12,276 versus $3,334). Average farm size in Iowa and Florida were relative close, 302 versus 310 acres, respectively; however the NFI for Florida was almost three times greater ($15,663 versus $45,395). Farm size in Georgia was 25% larger than in Alabama (270 versus 215 acres); however, Georgia's NFI was 67% higher.

Differences in government payments from state to state probably also influenced differences in NFIs to some extent. States that obtained a larger share of their NFI from the production of livestock were at a disadvantage because there were no direct support programs for livestock and poultry. For example, in 1985 in Alabama the sale of livestock and poultry provided 57% of all of the cash receipts received from all commodities, including farm forest products. Alabama farmers received no direct payments of any kind for production of those commodities.

American farmers were responding to the growing crisis in farm income in two ways: They were leaving farming, or those who remained were increasingly supporting themselves with non-farm income. As shown in Figure 15.16, numbers of American farms declined by 22.4% (2.95 million to 2.29 million) between 1970 and 1985. Some of this decline was the result of the new definition of a "farm" in 1973. In Alabama (Figure 15.17) farm numbers declined by 36.6% (82,000 to 52,000).

In contrast, the share of total farm household income derived from off-farm sources continued to increase. It had been 52.8% in 1960, but by 1970 it had reached 63.1%. Then in 1985, it reached to 67.8% (Cornejo, 2007). America's first farmers, the Native Americans, were unable to make a living farming. They had to continue to hunt, fish, and gather to make ends meet. Four thousand years later, farmers in the country are still unable to make a living farming.

16

Sharecropping with the Guvmint

1985–2002

PRECIS

In summarizing the information presented in this last chapter of recounting my journey to trace the evolution of the Alabama agroecosystem, I am reminded of the comment made by the great Yankee catcher Yogi Berra after watching Mickey Mantle and Roger Maris hit back-to-back home runs: "It's *deja vu* all over again." Alabama farmers have been going into their fields and pastures at first light and staying until half dark for almost 200 years—fighting poor soils, sheet erosion, droughts, floods, high humidity, late freezes, hurricanes, boll weevils, corn ear worms, and Bermuda grass. At the beginning of the twenty-first century, they are still about in the same position relatively speaking that they were in at the beginning of the nineteenth—still sucking hind tit! In 1867, farmers in Alabama and Illinois harvested 13 and 41 bushels of corn per acre, respectively. In 2001, farmers in the same states harvested 107 and 152 bushels. In addition, they harvested 35 and 45 bushels of soybeans per acre. Further, cropland rents were $36 and $119 per acre. Finally, milk cows in the two states produced 14,286 and 17,414 pounds of milk each. Like Sisyphus, Alabama farmers had been pushing agriculture up the mountain of environmental resistance for 200 years, and apparently they still have a ways to go.

While there was little new during this period in Alabama agriculture except broilers, hatching eggs, and catfish, at the national level the Triangle reached a new plateau in audacity and mendacity. After the Republicans won control of the House and Senate in 1994, a somewhat subdued Triangle passed the Federal Agricultural Improvement and Reform Act (FAIR). The act did not solve all of the problems resulting from six decades of Triangle ineptitude, but it did introduce a small dollop of sanity. Unfortunately, in 1997 the teeter totter intervened as it has been wont to do, and the parity ratio fell to 0.43 from 0.47. This sudden change provided the Triangle with a good opportunity to get back in the game. In 1997, federal payments per farm increased to $5,650 from $3,421 the previous year. Then in 1998, widespread adverse crop weather provided the Triangle with the opportunity to take over the entire ballpark. In 1999, payments increased to $9,814 and then to $10,541 in 2000. In this two-year period, they threw disaster payments into every nook and cranny in the land. Unfortunately, strewing this money around helped very little, if at all. In 2000, the parity ratio was down to 38, and 95% of total farm household income was obtained from non-farm sources.

INTRODUCTION

During the years 1973 and 1974, American farmers thought that the Golden Age of American Agriculture (1910-1914) had returned. Between 1972 and 1973, parity ratios increased from 0.74 to 0.91. The ratio had not been that high since the years of the economic expansion associated with the Korean Conflict. Unfortunately, for a number of reasons, the ratio declined to 0.86 in 1974, and afterwards began to decline rapidly, finally reaching 0.51 in 1985. In all the years since it was first computed (1910), the ratio had never been that low. The growing problem with farm income was a result of the inflationary increase in prices paid for farm inputs compared to the much slower growth of prices received for commodities at the farm gate. Farmers were still producing more food and fiber than they could market at reasonable prices, and as their financial situation steadily deteriorated, they tried to compensate by producing even more. Government payments to farmers had increased from the early 1950s through 1972, but after the ratio of the indices of prices received to prices paid reached 0.91 in 1973, annual payments were quickly reduced. By the mid-1970s, they were even lower than those of the early 1950s when considering inflation of the currency in the intervening years. After 1977, with the ratio back down to 0.66, growing political pressure quickly resulted in increasing levels of payments to farmers. In 1976, average payments per farm in the country was $294. In 1985, it was $3,362. In 1985, there was relatively little sharecropping remaining on American farms, but farmers sharecroppingwith the guvmint had entered a new stage.

USDA established the Economic Research Service in 1961, giving it an extremely broad mission that included research of economic development, as well as river basin and watershed development. In 1977, ERS was combined with USDA's statistics service to establish the Economics, Statistics and Cooperatives Service. Then in 1981, ERS was reestablished as a separate agency within the Department of Agriculture.

Since the early 1980s, ERS has published the results of a large number of studies on many aspects of the economics of agriculture. Results of ERS studies are published on a regular schedule—within years, between years, and over longer periods of time. These reports are extremely valuable in following changes in virtually all aspects of agriculture. This chapter will include infomation taken from some of these reports. As a result, it will contain some new sections that have not been included in the preceding chapters.

NATIONAL POLITICS

From 1914 through 1985, there were few years when the United States was not involved in one capacity or another in a hot or cold war. However, by 1985 even the Cold War was finally winding down. The Berlin Wall was demolished in 1987. The country was finally at peace. The American political system had not been in that position since the early 1900s. The country had generally prospered through all of those troubled times, and by 1985 the country was the most powerful and prosperous on earth. Its political system had served it fairly well during a period of almost endless conflict. Now it would have to function in this new environment, one that it had seldom encountered in its 200-year history.

The Presidency

The two parties essentially evenly shared the presidency during the 17-year period, 1985-2002. Republicans held the office for nine years and Democrats for eight.

Republican Ronald Reagan (1981-1989) was president at the beginning of the period 1985-1989. He was replaced by Republican George H.W. Bush (1989-1993) in 1989. In 1993, Democrats regained the presidency with Clinton (1993-2001); then relinquished it to Republicans in 2001 when G.W. Bush (2001-2008) took office.

Secretaries of Agriculture

John Block (R, IL) served as Reagan's secretary of agriculture during the period 1981-1986. Other persons holding that office during the 1985-2002 period included: Richard Lyng (R, CA) (1986-1989), Clayton Yeutter (R, NE) (1989-1991), Edward Madigan (R, IL) (1991-1993), Mike Espy (D, MS) (1993-1995), Dan Glickman (D, KS) (1995-2001), and Ann Veneman (D, CA) (2001-2002).

The Senate

The Senate was under the control of Republicans at the beginning of this period (1985-2002), and they remained in power until 1987 when the majority shifted to Democrats (Table 16.1), who remained in control until 1995. Republicans regained control in 1995, and retained it until 2001 when Senator Jeffords of Vermont, a Republican, decided to become an independent and to caucus with the Democrats. This change gave Democrats a one-vote majority in the upper house..

Senate Majority Leaders

Senate majority leaders during the period were: Robert Dole (R, KS) (1985-1987), Robert Byrd (D, WV) (1987-1989), George Mitchell (D, ME) (1989-1995), Robert Dole (R, KS) (1995-1997), Trent Lott (R, MS) (1997-2001), and Tom Daschle (D, SD) (2001-2002).

Chairmen of the Senate Committee on Agriculture, Nutrition, and Forestry

Jesse Helms (R, NC) served as chairman of the committee from 1981-1987. Other persons serving in that position included: Patrick Leahy (D, VT) (1987-1995), Richard Lugar (R, IN) (1995-2001), Thomas Harkin (D, IA) (2001), Richard Lugar (R, IN) (2001), and Thomas Harkin (D, IA) (2001-2002).

The House of Representatives

Democrats controlled the House for much of the period (1985-2002) (Table 16.1). They were in power in 1985, and retained it until 1995 when Republicans won the election of 1994. Republicans had not controlled that chamber since 1955. The 1994 Republican majority quickly began to erode, and in 2001 their majority was less than 10 votes.

Speakers of the House

Speakers of the House during this period were: Tip O'Neil (D, MA) (1985-1987), Jim Wright (D, TX) (1987-1991), Tom Foley (D, WA) (1991-1995), Newt Gingrich (R, GA), (1995-1999), and Dennis Hastert (R, IL) (1999-2002).

Chairmen of the House Committee on Agriculture

Kika de la Garza (D, TX) served as chairman of the House Committee on Agriculture from 1985 until 1995. Other persons holding that office included: Pat Roberts (R, KS) (1995-1997), Robert Smith (R, OR) (1997-1999), and Larry Combest (R, TX) (1999-2002).

The Political Situation

George H.W. Bush brought eight years' experience as Reagan's vice president to the presidential election of 1988, and although he initially trailed Michael Dukakis (D, MA) in the early polls, he pulled ahead in the final months. He won the office with 53.4% of all ballots cast.

When Bush took office in January 1989, Democrats held sizeable majorities in both houses of Congress. Foreign policy claimed much of Bush's attention in the early months of his administration. On December 20, 1989, the United States invaded Panama in Operation Just Cause with the specific purpose of removing General Manuel Noriega from power. The Panamanian dictator had once been a valuable ally of the United States, but he was increasingly using the power of his government to facilitate the shipment of drugs from South America to the United States. The military operation only lasted a few days until Noriega took refuge in the Vatican's embassy there. He surrendered to the US forces on January 3, 1990.

On August 2, 1990, Iraq invaded its smaller neighbor, Kuwait. On November 29, the United Nations passed a resolution urging all countries allied with Kuwait to use all necessary means to remove the Iraqi forces if they did not voluntarily leave by January 15, 1991. On January 17, 1991, US forces lead a coalition of 34 countries in an attack on Iraqi forces in the beginning of Operation Desert Storm. President Bush called a cease fire on February 27, 1991 after forcing all Iraqi forces from Kuwait. Coalition forces killed large numbers of Iraqi troops as they retreated toward Baghdad, but the president chose not to follow them to the Iraqi capital. He declared that the requirements of the UN resolution had been met when the Iraqi forces were driven out of Kuwait.

Although the Berlin Wall had been torn down in 1989, the Cold War did not officially end until Bush and

Gorbachev announced a strategic partnership to resolve bilateral and world problems in a meeting between the two in July 1991.

Because of Bush's popularity following the end of Operation Desert Storm, it was assumed that he would easily win the 1992 election. However, because of a short recession in late 1990 and early 1991, a reversal of his stated policy not to increase taxes, and concerns that he had not properly finished the Gulf War, he was easily defeated by Democrat Bill Clinton, who had previously served two terms as governor of Arkansas.

President Clinton took office on January 20, 1993. In the first two years of his first term, he had substantial majorities of Democrats in both houses of Congress. Where the Bush Administration had been forced to deal with numerous thorny foreign policy issues, most of Clinton's early concerns involved domestic issues. Early in his first term, he signed into law the Family and Medical Leave Act of 1993. This act mandated that employers grant unpaid leave to employees because of pregnancy or other medical conditions. He also quickly became involved in problems related to homosexuals in the armed forces. His leadership in this area lead Congress to implement the Don't Ask, Don't Tell policy. Clinton was also involved in another controversial policy matter when he provided leadership in getting Congress to approve the North American Free Trade Treaty. NAFTA had been negotiated by his Republican predecessor and the Canadian prime minister, but Clinton and other leading Democrats worked together to push it through Congress. Clinton signed, and the Senate ratified NAFDA in 1993. Other important accomplishments during his first term included passage of the Brady Bill, which provided for some governmental oversight on the purchase of handguns; expansion of the Earned Income Tax Credit, which provide additional financial aid for working class families with dependent children; and the Omnibus Budget Reconciliation Act of 1993, which provided the basis for the longest peacetime economic expansion in American history. Clinton failed in his efforts to get legislation enacted to provide universal health care coverage for American citizens. The public passed judgment on the considerable accomplishments in the first two years of his first term by electing Republican majorities in both houses of Congress. This was the first time that Democrats had lost both houses in the same year in 40 years.

In the 1996 presidential election, Clinton easily defeated his Republican opponent, Bob Dole (R, KS). In the Congressional elections, Republicans maintained their control of both houses. For the first time in decades, American citizens seemed to prefer divided government.

The second Clinton administration was beset with problems stemming from the scandal resulting from his affair with a young, female White House intern, and as a result the president was unable to get important domestic legislation enacted. But he was involved in two notable military events. In Operation Desert Fox (December 16-19, 1998), Clinton authorized the use of American airpower to reduce Saddam Hussein's persecution of minorities in Iraq, and in March 1999, Clinton authorized the use of airpower and troops to stop genocide of Albanians by Serbs. Near the end of his second term, Clinton made a determined effort to broker progress in the Arab-Israeli conflict, but after several days of intense negotiations by the two sides at Camp David, Clinton's efforts came to naught.

It was expected that with a strong national economy and relatively peaceful world conditions Albert Gore (D, TN), who had been Clinton's vice president, would be elected president in the 2000 election. Although he lost the popular vote, Republican George W. Bush received the majority of the electoral votes in a very controversial election. Bush was inaugurated on January 20, 2001. Republicans in the House and Senate narrowly won in the elections for the 107th Congress, but in June 2001 Senator James Jeffords (R, VT) became an Independent, and began to caucus with the Democrats. His action gave the Democrats a 50/49 majority in the Senate.

Bush had been in office only a short time when he began to hold townhall public meetings across the country to build support for a massive $1.35 billion tax relief program. The Economic Growth and Tax Reconciliation Act of 2001, although controversial, received considerable bi-partisan support. In the House, the vote was 240 to 154 with 39 members not voting. In the Senate, the vote was 58 to 33 with two members voting present and seven not voting. The act provided for the largest tax cut in American history.

The use of federal funds that would lead to destruction of human embryos for medical research had been forbidden by law since 1995. In August 2001, Bush signed an executive order lifting the ban on the use of federal funds for studies on 71 existing lines of stem cell derived from human embryos; however the ban on the use of federal funds for the development of new lines remained in place.

The terrorist attacks on the country on September 11, 2001 forced the Bush Administration to radically alter its agenda. The president immediately declared a Global War on Terrorism. One of the first acts under this announced policy was to order an invasion of Afghanistan. Bombing within the country began on October 7, 2001. One of the first pieces of legislation enacted after September 11, was the Patriot Act, which provided law enforcement agencies in the country with many new tools to increase their effectiveness in the fight against terrorism both at home and abroad.

Although most of the president's concerns in 2002 were related to the fight against terrorism, he did take time to provide leadership to strengthen education, efforts close to the heart of the First Lady. The results of these efforts was the passage of the Elementary and Secondary Education (No Child Left Behind) Act. Bush signed the legislation in January 2002.

ALABAMA POLITICS

At the beginning of the preceding period (1970-1985), Alabama was beset by racial tension and violence. Under those conditions, it was difficult for the state to make progress in solving any of its many problems. By 1980, however, as a result of the effects of wide-ranging federal legislation related to civil rights and a series of federal court decisions, there was little advantage left in the politics of race. Finally, Alabama was ready to move on to other things.

Governors

The seriously incapacitated George Wallace (D, Barbour) had been inaugurated as governor in January 1983. He served until January 1987. Other persons holding that office during the period 1985-2002 were: Guy Hunt (R, Cullman) (1987-1993), James Folsom Jr. (D, Cullman) (1993-1995), Fob James (R, Lee) (1995-1999), and Don Siegelman (D, Mobile) (1999-2003).

Commissioners of Agriculture

Albert McDonald (D, Marshall) was elected commissioner of agriculture in the 1982 election. He held the office until 1991. Other persons holding the office during this period (1985-2002) were: A.W. Todd (D, Franklin) (1991-1995), Jack Thomson (R, Colbert) (1995-1999), and Charles Bishop (D, Walker) (1999-2003).

The Political Situation

George Wallace began the last of his storied terms as Alabama's governor in 1983. His final administration should have been his most powerful one. The so-called Wallace Coalition included some of the most powerful political forces in the state (Alabama Education Association, organized labor, black political organizations, trial lawyer, etc.). Probably as a result of failing health, Wallace was unwilling or unable to marshal these forces in a period of good economic times to accomplish the many important things on the state's critical needs list. Consequently, his last hurrah was marked not by what he did, but by what he did not do that he could have done.

In the election of 1986, the Alabama Democratic Party was forced to name Bill Baxley as its candidate for governor; although he had not won the primary. This act seemed to infuriate the state's electorate, and they took out their frustration on Democrats by electing the little known Republican Guy Hunt (R, Cullman) to the office. Hunt received 56% of the vote. He narrowly won re-election in 1992, defeating Paul Hubbert, after trailing for most of the campaign. In 1993, he was convicted of using inaugural funds to pay personal debts. He was immediately removed from office. Later, the Alabama Board of Pardon and Parole issued him a full pardon, claiming that he was innocent of the charges.

The Hunt Administration faced almost impossible odds in trying to get legislation of any kind through the legislature with its overwhelming Democratic majority.

He was unable to get any important legislation enacted. He worked tirelessly to bring more industry to the state and to increase tourism, but virtually everything else he tried to do was stymied by the legislature. Hunt is generally credited with making Alabama a two-party state.

The day that Guy Hunt was convicted, Lieutenant Governor James E. Folsom Jr. (D, Cullman) was sworn in as governor. Shortly after taking office, he was approached by Mercedes-Benz concerning the possibility of locating a major auto manufacturing plant in the state. Folsom worked tirelessly to recruit the plant, and in October 1993 the company announced its intention of locating its first plant outside of Germany near Tuscaloosa. Apparently, Alabama's electorate gave Folsom little credit for his efforts to recruit the industry. In the 1994 election, they chose Fob James (R, Lee) as the state's 55th governor.

Fob James, who governed as a Democrat from 1979-1983, changed parties to become a Republican before the 1994 election. He was elected as part of the so-called Republican Revolution that swept the country in that election. In his first term (1979-1983), James actively pushed a broad economic agenda to improve the lives and future prospects of the state's citizens, but when he took office in 1995 his agenda had changed. He spent most of his second term avidly pushing a conservative social agenda on religion in schools, display of the Ten Commandments in public building, anti-abortion, and resistance to the excesses of the federal judiciary. Apparently, the state's electorate did not appreciate the intensity with which he pushed this agenda. He ran a lackluster campaign in the 1998, and was easily defeated by Don Siegelman (D, Mobile) in the general election.

From the beginning, Sielgelman was determined to change education in Alabama significantly. He signed an executive order to eliminate portable classrooms. He signed a bill to raise teachers' salaries to the national average. He implemented the nationally recognized Alabama Reading Initiative. He also created an Office of School Readiness. He worked tirelessly in support of a constitutional amendment for the creation of a state lottery that would have provided funds for free tuition at state universities for most high school graduates.

While improving education was his primary goal, he supported important initiatives on a wide range of other concerns of the state's citizens. He signed several bills designed to make Alabama's homes, schools, and streets safer. He continuously pushed for tougher DUI laws.

The national recession during the first three quarters of 2001 resulted in serious state budget problems in the middle of his administration. These budget problems limited many of the things that he could have accomplished, and as a result he worked tirelessly to broaden the state's tax base. He traveled widely, encouraging the movement of industries to Alabama. Mercedes-Benz had established it plant in Tuscaloosa before he became governor, but during the Sielgelman Administration, the company decided to double the size of its operation. Also, he was able to secure commitments from both Honda and Toyota to locate manufacturing plants in the sate.

THE NATIONAL ECONOMY

The economy of the United States during the period 1970-1985 can best be described as being in a period of deep malaise. In late 1969 and early 1970, the economy was in recession. Then in late 1971, inflation was roaring ahead so rapidly that President Nixon had to put wage and price controls in effect. OPEC's embargo on oil shipments to the West in November 1973 resulted in a recession that lasted through the first quarter of 1975, but by 1979 consumer prices were increasing rapidly again. In 1979, they increased by 13.3%. Unfortunately, in 1980 and 1981-1982, there were two more short recessions. Then in 1982, the national unemployment rate reached 10.8%.

The Economic Situation

By 1985, the economy was growing again. President Reagan (1981-1989) was rapidly implementing his program to reduce taxes, the size of the federal government, and its intrusiveness, as well as to dismantle some of the government-sponsored social programs. At the same time, Reagan with the support of a Democratic Congress was significantly increasing military spending. By 1986, the federal budget deficit was beginning to increase rapidly, and in 1986 reached $221,000 million. At the same time, the national economy was entering the longest period of sustained economic growth since World War II. During the period 1982-1987, the American economy created more than 13 million new jobs.

George H.W. Bush became president in 1989, and in the beginning of his administration worked to keep most of the Reagan program in place. He signed several important pieces of environmental and handicapped legislation. He also had to suffer through the debilitating savings and loan crisis. Toward the end of his administration, he was forced to renege of his promise of no new taxes. Unfortunately, the tax increases were not accompanied by reductions in spending, and as a result the government made little progress in reducing the federal budget deficit.

The Federal Reserve had increased discount rates to historically high levels during the early 1980s in an effort to cut off the supply of money fueling the inflationary spiral. In the succeeding years, the Fed very slowly released its firm grip on the money supply, lest the embers of inflation flare up again. In August 1986, the discount rate was down to 5.5%. It had not been that low since November 1976. Then in late 1987, the Fed began to be concerned about inflation and began to raise interest rates again. This move by the Fed probably played a role in the disastrous stock market crash in the fall of 1987. By February 1989, the discount rate had been increased to 7%. This increase resulted in a slowdown in aggregate spending, and by July 1990 the economy was in a short, shallow recession that lasted until March 1991.

The effects of the 1990-1991 recession continued long enough ("it's the economy stupid") to at least assist with the election of Democrat Bill Clinton (1993-2001) in 1993. He continued many of the policies of his predecessors. He worked diligently to reduce the size of the federal government. He also joined the Republican Congress in changing the welfare system. After the end of the 1990-1991 recession, the national economy began to expand rapidly. At the same time, corporate earnings increased, unemployment declined, and the stock market began to surge upward. The Dow Jones Industrial Average had been around 1,000 in the late 1970s. In 1999, it reached 11,000. As the economy continued to grow, tax revenues increased rapidly, and in 1998 there was a surplus in the federal budget for the first time in 30 years. Clinton, like his predecessors, pushed for the elimination of barriers that impeded trade among countries. As a result of his efforts and those of a Congress that was divided on the issue, the North American Free Trade Agreement (NAFTA) was put into effect in 1994. This agreement improved economic ties with Mexico and Canada, and generally moved the world forward toward increased global economic integration. The Clinton Administration presided over a portion of the longest peacetime economic expansion in history. It began in March 1991 and continued until March 2001. Surprisingly, throughout its course, while unemployment decreased steadily, there was no real indication of a return of inflation.

The Fed had been slow to reduce interest rates after the end of the 1990-1991 recession, but they were finally reduced to 3% in July 1992. However, in the succeeding months the watch-dog agency again began to become concerned about the possibility of inflation, and responded by increasing rates again. By May 2000, they were back up to 6%. In the meantime, capital spending and production by business was roaring ahead. By the middle of 2000, inventories were beginning to expand beyond demand. As a result of a reduction of the money supply and rapidly increasing inventories, the longest continuously growing economy in peacetime history came to an end, and a mild recession began in early 2001—just after Republican George W. Bush became president.

The Gross National Product

In 1991, the Bureau of Economic Analysis of the Department of Commerce replaced the Gross National Product (GNP) with the Gross Domestic Product (GDP) as the primary measure of American national production of goods and services. The GNP measured the total production of the American economy, but it did not include the production of those companies operating in the United States, but owned by foreign interests. It did, however, include returns obtained by American firms operating overseas. The GDP represents total productivity within the borders of the country without regard to ownership of the means of production, but it does not include returns of American companies operating beyond our borders.

Annual values for the GNP and the GDP are relatively

similar. Values for the GNP are slightly larger than those of the GDP when Americans receive greater returns for their overseas investments than foreign firms receive for their investments in America. When the opposite conditions apply, however, the GDP will be larger. Data in actual dollars for the GNP and GPD for the period 1970-2002 are presented in Figure 16.1. With the scale used in the figure, there are no observable differences in the two. However, the actual data show that in all the years of the period, the GDP was slightly more than 99% of the GNP. For example, in 1970, 1980, 1990, and 2000, the comparable values for the GNP and GDP were in billions of actual dollars: $1,045 versus $1,038.5; $2,823.7 versus $2,789.5; $5,837.9 versus $5,803.1; and $9.855.9 versus $9,817.0, respectively.

The data presented in the figure show that the GDP increased from $4,220 in 1985 to $10,500 in 2002. Because these are actual dollars, a substantial share of this increase is the result of the inflation of the currency. Both the GNP and the GDP show the effects of the recessions of 1990-1991 and 2001.

The Consumer Price Index

The Consumer Price Index (CPI) data used in the preceding chapters used 1967 as the base year (1967 = 100). Beginning in July 1998, the Bureau of Labor Statistics of the Department of Labor began to use the period 1982-1984 as the base period (1982-1984 = 100). Consumer price indices data for the period 1970-2002 using the new base period are presented in Figure 16.2. These data show that the CPI increased steadily throughout the period—from 38.8 in 1970 to 179.9 in 2002, an increase of 363.7%. It was 107.6 in 1985. The rate of growth slowed considerably between 1981 and 1986, as a result of increasing discount rates (Figure 16.4) and annual unemployment rates between 7% and 9.7%. It began to increase again after 1986, but the rate of increase slowed again between 1997 and 1999, and after 2000.

The Inflation Rate

The recession of 1980-1983 had effectively taken the air out of the inflationary balloon of the late 1970s. In 1980, the inflation rate had been above 13% (Figure 16.3). However, by 1985 it was down to just above 3%. It fell below 2% in 1986 before rising to above 5% in 1990. However, throughout the 1990s the inflation rate remained below 3% and actually reached 1.6% in 1998. It was a little higher in the period 1999-2001, but remained less than 3.5%. The base year for computing the CPI was changed in 1998, but this change had no effect on the computation of the inflation rate.

Federal Discount Rates

Some aspects of the trends in interest rates during the period 1985-2001 were detailed in a preceding chapter. In general terms, the Federal Reserve kept interest rates in double digits (14% in May of 1981) from mid-1979 until the latter half of 1982. Thereafter, the Fed slowly reduced rates until they reached 3% in July 1992 (Figure 16.4). With the economy expanding so rapidly and with concern for inflation, the Fed slowly increased interest rates. They reached 6% in May 2000 (Figure 16.5). By that time, however, the concern for the return of inflation was replaced by a concern for the short recession of 2001, and rates began to trend downward until they reached 1.25% in December 2001. At one point in 2002, the interest rate was as low as 0.75%.

THE ALABAMA ECONOMY

In 1998, when the Gross National Product was discarded in favor of the Gross Domestic Product (GDP), similar data on the economies of the individual state was changed from Gross State Product to Gross Domestic Product by State (GDPS). The two measures of goods and services produced by the individual states are essentially the same. Data on GDPS in millions of current dollars during the period 1970-2002 are presented in Figure 16.6. During this period (1985-2002), Alabama's GDPS increased from $53,688 to $123,805 million, an increase of 131%. During the same period, the Consumer Price Index (1982-1984 = 100) increased from 107.6 to 179.9, an increase of 67%. These data indicate that almost half of the increase in Alabama's GDP during this period was the result of inflation of the currency.

Data on GDPS in current dollars per person for Alabama and several other states in 1985, 1990, 1995, and 2000 are presented in Table 16.2. The data show that—when corrected for increases in the state's population growth—the GDPS for Alabama increased from $13,517 to $25,765 per person in the period 1985-2000. They also show that in 2000 the GDPSs for Alabama, Mississippi, and South Carolina were several thousand dollars lower than the remaining states and that increases from 1985 to 2000 were also lower for those same states.

Data in Table 16.3 show the contribution (percentages) of agriculture, forestry, fisheries, and hunting to the GDPS. (Hunting was included in the estimates after 1997.) These data indicate that the economies of Alabama and Mississippi became somewhat less dependent on the exploitation of natural resources during the period 1985-2000, but remained considerably more dependent than the economies of Georgia and Ohio. Among the states listed, the economy of Iowa made the most progress in reducing the role of agriculture, forestry, fisheries, and hunting.

Smith and Taylor (2000) have published data on the contribution of agribusiness to Alabama's economy. Some of their data on 1999 gross output (millions of dollars) of several types of these businesses in Alabama's six geographical regions and in the state as a whole are presented in Table 16.4. The authors define gross output as "the estimated value of total production before any deductions or adjustments." As might be expected, gross output of the different types of businesses (sectors) varied considerably. Gross output was greatest for the manufacturing of lumber, furniture, and pulp and paper ($14.5 billion), and least for the manufacture of machinery used in the various types of agribusinesses ($321 million). Similarly, there was considerable variation in the contribution that agribusiness made to the different regions of Alabama. For example, gross output was $12.2 billion for the counties of Alabama's Northeast Region, but only $4.4 billion for those counties in the state's West-Central Region.

Smith and Taylor (2000) also included employment in the different types of agribusinesses in Alabama in 1996. Some of these data are presented in Table 16.5. The data show that the number of jobs in agribusiness represented 21% (476,483/2,273,161) of jobs in all industries in Alabama. The data also show that the number of jobs in agribusiness was considerably higher in Alabama's Northeast Region (153,374). Except for the state's Northwest Region, agribusiness in the Northeast was more than twice as large as in any of the other regions, and represented 32% of Alabama's total. This large share is a result of the presence of the broiler production and processing industries in that region and of the presence of a large share of the state's population and retail businesses there also. Crop production in the western Tennessee Valley and the population centers in Madison and Morgan counties resulted in the second largest level of employment (92,324).

THE AGRICULTURAL ECONOMY

Data presented in the preceding chapter told the poignant story of the American agricultural economy in the preceding period (1970-1985). In 1973, the ratio of the indices of prices received to prices paid was 0.9. At that time, farmers were receiving almost as much money for the commodities they marketed, as the amount of money required to produce them. Unfortunately, just 12 years later, in 1985, the ratio was down to 0.51. Of course, during the period there was some increase in agricultural productivity, but not nearly enough to compensate for the decline in prices received. The political clamor was deafening. In 1933, the camel of federal intervention first thrust its nose under the wall of the tent of the country's agricultural marketplace. In the early 1980s, the whole camel moved in.

Federal Farm Legislation

In earlier chapters, we have followed the unique role, as described Ernstes, et al. (1997), that was played by Southern congressmen in producing the legislation that resulted in the expanding level of federal intervention in American agriculture. As early as the beginning of the twentieth century, the five belles of Southern agriculture (tobacco, rice, cotton, sugar, and peanuts) were experiencing great difficulty in competing in the ever-growing world marketplace. Until 1933, Southern congressmen were in no position to come to the aid of farmers growing those crops. Afterwards, Southern congressmen, as an essential part of the Triangle, led the way in producing an ever-increasing amount of supportive and protective legislation. Fortunately or unfortunately, depending on an individual's perspective, these Southern congressmen learned that once a lifeboat had been launched everyone around wanted to climb in.

The situation that quickly developed in producing a flood of legislation of greater and greater complexity is illustrated by an event that took place in the Senate several years ago. A well-known Alabama senator had just completed an impassioned speech about the value of sugar beets to the American economy when a visitor commented that he did not know that Alabama produced this crop. The senator responded that he was correct, "the state did not grow sugar beets, but they sure grew lots of peanuts."

Although there would be drastic changes in politics in Congress during this period (1985-2002), Southerners in the House—through the 103rd Congress (1993-1995)—continued to play a pivotal role in enacting legislation relating to agriculture and providing oversight for federal agricultural programs. While they did not hold majorities on the full House Committee on Agriculture, the chairman was a Southerner (Table 16.6). Also, through the 103rd Congress, Southerners held half of the majority of chairmanships of the subcommittees. Further, Jamie Whitten (D, MS), often called the real secretary of agriculture, served as chairman of the House Committee on Appropriations until 1993. The situation was somewhat different in the Senate. Republicans were in control of the Agriculture Committee only in the 98th and 99th Congresses (1983-1987), but during the period Southerners served as a majority of chairmen on several of the subcommittees.

For almost three-quarters of a century, the federal government had been attempting to solve the problems of the overproduction of agricultural crops, low commodity prices, and low farm income by a combination of the following (Hurt, 1994):

1. Expansion of exports.
2. Increasing food consumption through the school lunch program and the use of food stamps.
3. Making food available to developing countries through expanded foreign aid.
4. The use of federal funds to support the prices of agricultural commodities near world levels.
5. Direct payments to farmers who participated in production-control programs.
6. Acreage controls.
7. Voluntary production controls, linked to payments and specific loan rates.
8. The use of target prices and deficiency payments.

By the mid-1980s—after thousand of hours of Congressional hearings and debate, the expenditure of billions of dollars of public money, the publication of thousands of pages of often-confusing rules and regulations (legislation), and the employment of thousands of people to administer and conduct the hundreds of federal programs—the old bugaboo problems of overproduction, relatively low commodity prices, and low farm income were largely unresolved. Unfortunately, like Brer Rabbit and the Tar Baby, the harder Congress tried the "stucker they became." Now there was nothing else that they could do except to keep on trying and hope that sooner or later they could find a combination of legislative provisions that would work.

To keep on trying in 1985 was not as simple as it had been in the past. Federal farm programs were extremely expensive, and in a period of growing budget deficits, it was increasingly difficult to convince the public, especially in non-farm states, that farmers should be paid not to farm. The problem of securing continued public approval for federal agricultural programs was further compounded by the fact that farmers were increasingly becoming a smaller and smaller percentage of the American population. These programs were increasingly at the mercy of a growing population that had no connection with farms and farmers. As a result, farmer subsidy programs increasingly had to seek support from liberal urban legislators who wanted to guarantee abundant supplies of inexpensive food for poor people and from the growing numbers of environmentalists who welcomed land being taken out of production to increase wildlife habitat and wild space. Finally, there was a growing awareness that subsidies linked to production were distorting free marketing and trade in the United States and around the world, and that in the long term they would significantly affect the ability of the United Sates to export its surplus agricultural commodities. Further, the problem of keeping on would become even more complicated by a national realignment of the political landscape about the middle of the period (the 1994 elections). Until 1994, the lion's share of federal farm legislation had been enacted by large Democratic congressional majorities, but usually with considerable bi-partisan support. However, afterwards, Republican majorities weould play an increasingly important role.

Some of the more important pieces of legislation enacted by Congress during this period were:

1. The Food Security Act of 1985. This five-year omnibus law, described in the preceding chapter, had been written to try once again to solve problems in the agricultural economy that had emerged after World War I. Supposedly, this legislation would not be revisited by Congress until 1990. In the meantime, a number of unforeseen difficulties arose that required special legislation.
2. Farm Disaster Assistance Act of 1987. This act assisted farmers who experienced crop losses from natural disasters in 1986.
3. Disaster Assistance Act of 1988. This act provided assistance to farmers hurt by drought or other natural causes in 1988. Farmers with losses greater than 35% of production were eligible for financial assistance. Assistance purchasing feed was also available to livestock producers.
4. Disaster Assistance Act of 1989. This act provided financial assistance to crop and livestock producers suffering losses in 1988 or 1989 from natural disasters. To be eligible, producers with crop insurance had to have had losses of at least 35% of production.
5. Food, Agriculture, Conservation, and Trade Act of 1990. Among other things, this five-year omnibus farm bill froze minimum target prices and limited total acreage eligible for deficiency payments, but it allowed more planting flexibility. It also changed price support formulas for many commodities and altered rules of the operation of the national grain reserve.
6. Federal Agriculture Improvement and Reform Act of 1996 (FAIR). In 1994, Republicans won control of both the House and the Senate. They had not controlled the House, where most agricultural legislation originates, since the 83rd Congress (1953-1955) when Eisenhower was president. Although the majorities were not huge, especially in the Senate (52:48), the Republican Congress set about to undo over half a century of legislation that had increasingly placed farmers in the position of being sharecroppers with the guvmint. As a result, the Federal Agriculture Improvement and Reform Act of 1996 (FAIR) was enacted. It was expected to be in effect until 2002 and was designed to accelerate the process of making farm policy more market oriented. In effect, FAIR replaced grain and cotton target price payments with seven-year contracts providing fixed, but declining, annual market transition payments that were not tied to market prices. Farmers who were enrolled in grain and cotton programs in the past were eligible for participation in the new program. The act eliminated all acreage reduction programs and most planting restrictions. Further, it also eliminated the farmer-owned grain reserve. It lowered Commodity Credit Corporation loan rates and eliminated marketing loan repayment provisions. These two changes effectively removed USDA from playing a role in the management of commodity storage and management.

FAIR anticipated a major shift in American farm policy. For the first time in the 60-plus year history of government involvement in farmer assistance, farm-support payments were to be decoupled from production. The act represents the culmination of efforts by the United States, as detailed in a preceding section, to end the worldwide distortion of trade resulting from the use of production subsidies. Finally, with the passage of the act, it seemed possible that free markets would become a fair and equitable determinant of commodity prices and ultimately farm incomes.

While this revolutionary legislation was being debated and finally enacted, there were 29 Republicans and 21 Democrats on the House Committee on Agriculture. Pat Roberts (R, KS) was chairman. There were 23 Southerners (including members from Missouri and Oklahoma) on the committee: 13 Republicans and 10 Democrats. Southerners were chairman of four of seven of the subcommittees. Richard Lugar (R, IN) was chairman of the Senate Agriculture Committee. Seven of 18 of the committee members were Southerners, and Southerners served as chairmen of three of six of the subcommittees.

Revolutionary FAIR was enacted during a period of good economic times in the United States. The Gross Domestic Product (Figure 18.1) was increasing rapidly; yet inflation was generally below 3%. The prices received situation was still bleak. In fact, the index had changed very little since the late 1970s (Figure 16.9). By 1996, when FAIR was enacted, fortunately or unfortunately, depending on an individual's perspective, total farm household income was largely independent of income generated by their farms. By 1996, USDA estimated that from 85 to 95% of farm household incomes were generated by non-farm sources. So, in 1996, with the economy growing rapidly, with inflation relatively low, and with farm households increasingly dependent on growing off-farm incomes, FAIR seemed to be what farmers needed.

Unfortunately, while on-farm income was having a smaller and smaller effect on total farm household income, it still mattered. After 1996, the demand for American agricultural products began to decline. In the same year, American agricultural exports were $60.4 million. Then most of the national economies in Asia began to encounter serious difficulties, and by 1998 American agricultural exports were down to $51.8 million and still falling. With this decline in exports, demand for most American farm commodities also began to decline. With demand down, the indices of prices received also began to decline. Farmers and the Triangle quickly lost their interest in a market-driven agricultural economy, and just as quickly contacted their congressmen about returning to the good old days of taxpayer supported agriculture. This unfortunate situation also provided an excellent opportunity for the hundreds of thousands of Triangle people who had been employed to oversee and to operate the government (sharecropper system) to return to the table. The Republican Congress quickly and meekly acceded to the insistence of these stakeholders by enacting the Emergency Farm Financial Relief Act of 1998. This act provided for the early release of $5.5 billion in contract payments to farmers who requested them. In effect, it allowed farmers to receive an advance on the market transition payments due them the following year. As usual, this legislation did not solve the basic problem: too many farmers producing too much food and fiber. But the stakeholders quickly demanded more money for everyone.

7. Omnibus Consolidated and Emergency Appropriations Act of 1999. The act provided $5.9 billion in one-time emergency spending for USDA programs, including $2.9 billion for farmers who held grain and/or cotton contracts (market transition payments). It also included $2.345 billion for direct disaster payments to crop farmers with significant losses in 1998 or several previous years. In addition, the legislation provided $200 million in livestock feed assistance and $11 million for Georgia cotton indemnity payments. By early 2001, Congress had approved $25 billion in farm rescues since the 1999 act was passed, and they added an additional $5.5 billion later that year.

Farmers liked some provisions of FAIR. They liked the concept of finally being allowed to produce for the market. They especially liked the provisions whereby they could plant as much or as little of any crop they desired without jeopardizing eligibility for contract payments. They also liked the guaranteed annual payments (market transition payments). It is very likely, however, that they would not accept its declining payments provision. At the same time, they would like to have future legislation include marketing loans that were eliminated in FAIR. The 1985 legislation had allowed farmers to pocket differences in actual market prices and federal minimum prices for grains, cotton, and soybeans. What they really wanted was to have their cake and eat it too.

8. Agricultural Risk Protection Act of 2000. In 2000, the Triangle furthered their efforts to reduce risk to farmers and at the same time to continue to shift the cost of reducing it from farmers to taxpayers when Congress enacted the Agricultural Risk Protection Act of 2000. The preamble to this act states that its purpose is to "amend the Federal Crop Insurance Act to strengthen the safety net for agricultural producers by providing greater access to more affordable risk management tools and improved protection from production and income loss, to improve the efficiency of the Federal insurance program." Anyone who still honestly believes that the Triangle could effectively manage American agriculture should take the time to read this act. The primary thrust of the 100 pages of "ifs, ands, and buts" was to provide a substantial increase in premium subsidies.

   Government subsidies of crop insurance premiums had been part of the Federal Crop Insurance Program for several years. In 1999 and again in 2000, emergency premium discounts had been added to premium subsidies. The 2000 act revised subsidy rates and increased government funding of premium subsidies for the period 2001-2005. At the time of its passage, it was estimated that the act would increase government costs for premium subsidies by $8.2 billion during the 2001-2005 period. By 2000, aggregate premium subsidies (including discounts) had reached 60% of total premium costs (Dismukes and Vandiver, 2001). With the Triangle pushing, this percentage is likely to increase over time. It is not surprising that crop insurance and the larger safety net, which it provides, has quickly become extremely popular with both farmers and private insurance companies.

9. Farm Security and Rural Investment Act of 2002. A policy statement issued by the USDA in late 2001, regarding the enactment of the 2002 farm bill, commented that "commodity subsidies stimulate excess production, inflate land rents, and largely benefit only a small number of large farms." It is surprising that the president would have signed legislation so different from the USDA position. The flip-flop seems to have resulted from efforts of both political parties to garner favor with voters in several key farm states in the heart of the farm belt. With control of both houses of Congress up for grabs, neither party wanted to be caught in the position of voting against a large safety net for farmers just before mid-term elections.

   When members of the 107th Congress were discussing the 2002 farm bill, Democrats held a single vote majority in the Senate, 50:49. One member was listed as "other." Republicans held a nine vote majority in the House, 221:212. Two members were listed as "other." Thomas Harkin (D, IA) served as chairman of the Senate Committee on Agriculture, Nutrition, and Forestry. Six of the 21 members of the committee were Southerners, two Democrats and four Republicans (Table 14.6). Larry Combest (R, TX) was chairman of the House Agriculture Committee. Twenty-two of 51 committee members, including the chairman, were Southerners. Southerners served as chairmen of three of the five subcommittees. As a result of the makeup of the committees, especially in the House, the 2002 act seems to have had a distinctly Southern flavor. As an interesting aside, Combest was replaced as committee chairman in 2003—after he failed to vote for the 2002 act.

   The Farm Security and Rural Investment Act of 2002 authorized the expenditure of $180 billion over the next 10 years, a $73.5 billion increase over expenditures that would be incurred if the 1996 act were to remain in effect. This act replaced the production flexibility contracts of the 1996 act with fixed direct payments. Payments under the production flexibility contracts were supposed to end in seven years. There was no termination date for the fixed direct payments. The 2002 act also introduced counter-cyclical payments, expanded conservation land retirement programs, placed more emphasis on on-farm environmental practices, relaxed rules to make more borrowers eligible for federal farm credit assistance, and restored food stamp eligibility for legal aliens. The act also provided payments for some crops that had not been supported, at least recently. These included: peanuts, milk, lentils, honey, and wool.

Following the enactment of the 2002 farm bill, the Triangle began discussions on policy changes that might be incorporated into the next one (Young, et al., 2007). Acknowledging problems with past policies, proposals emphasized the need to provide a safety net that would protect farmers and ranchers against risks, such as lost income, limited access to credit, or devastation from natural disasters (USDA–Farm Bill Forums).

All of these events suggest—given the nature of the Triangle—that ultimately American farmers will become government employees with a defined set of benefits and with rights to form unions and engage in collective bargaining. Under those conditions, the government would control the production, processing, and distribution of food. If this seems preposterous, consider how close we were to that point with the peanut and tobacco programs and with the current fluid-milk program. The Socialists Henri de Saint-Simon and Louis Blanc really understood human nature much better than we could have ever imagined (Spielvogel, 1994). The Soviet communist politician Vladimir Ilyich Lenin commented that "Capitalists will sell us the rope with which we will hang them."

Agricultural Exports

American agricultural exports expanded rapidly from 1970 ($7.2 billion) through 1981 ($43.3 billion). Then, as a result of problems in the American economy, especially inflation, they trended downward through 1985 to reach a level of $29 billion in that year.

US Agricultral Exports

In 1986, US agricultural exports fell to $26.2 billion (Figure 16.7). They had not been that low since 1977. Finally inflation in the American economy began to abate. It declined rapidly after 1980, and by 1986 was down to 2% (Figure 16.3). As a result, in 1987 exports began to increase again, finally reaching $60.4 billion in 1996, the highest level in history. In 1997, turmoil developed in most of the economies in Asia, and the demand for American agricultural exports began to decline. In 1999, exports were back below $50 billion. Fortunately, they recovered slightly at the beginning of the new century, reaching $53.1 billion in 2002.

Agricultural exports continue to be extremely important to both American farmers and the national economy (Hanrahan, 2006). Roughly a third of the total production of harvested acreage is exported, including 32% of the wheat, 33% of the soybeans, 26% of the cotton, and 16% of the corn. In 1998, agricultural exports generated 800,000 jobs, including almost 500,000 in the non-farm sector.

Nearly all of the states exported some agricultural commodities during the period 1985-2002; however, the states with the greatest share of exports were California, Iowa, Nebraska, Kansas, Illinois, Texas, Minnesota, Washington, Indiana, and Wisconsin. These states accounted for 56% of the total of agricultural exports. Arkansas, North Carolina, Ohio, Florida, Missouri, Georgia, and South Dakota exported lesser amounts than the states listed above, but all of these exported at least one billion dollars worth of commodities a year.

The composition of agricultural exports has changed considerably over the years. Before 1991, bulk commodities such as grain oilseeds and cotton accounted for more than 50% of the total. Since that time, non-bulk such as wheat flour, feedstuffs, and vegetable oils have become more important, along with consumer-ready products such as fruits, nuts, meats, and processed foods. In 2000, high-value agricultural exports accounted for 63% of all exports.

For many years, countries with sizable farming sectors have adopted policies to protect them from foreign competition. These policies, usually involving export subsidies, generally resulted in the shrinking of international markets for agricultural commodities while reducing their prices. Further, these policies also resulted in increasing surpluses of farm commodities in exporting countries.

In the mid-1980s, the United States began to work with some 90 other countries to negotiate the gradual reduction of all policies that distort farm prices, production, and trade. This was a particularly difficult task for the United States, because virtually all of American agriculture was, at that time, precariously balanced on a fulcrum of subsidies, target prices, deficiency payments, land set-asides, marketing orders, quotas, farmer-owned grain reserves, etc. Progress in reaching agreements with other countries would require that most, if not all, of this rickety supporting structure would ultimately have to be dismantled, and there was no one around that knew exactly what would happen when this house of cards came tumbling down.

In these negotiations, the United States was particularly interested in the eventual elimination of European farm subsidies and the end to Japanese bans on rice imports. Finally, heads of the major industrial countries of the West agreed to move toward the goals of subsidy reductions and freerer markets. Over time, some progress was made in the European Union in achieving some of the goals, and by the mid-1990s trade tensions were reduced somewhat, although they were not eliminated. As the discussions were progressing, American imports began to increase once again, and generally continued to trend upward until 1996 when they reached $60 billion.

After 1996, financial difficulties, an economic slowdown in East and Southeast Asia, and a resurgence of competition in global corn, wheat, and soybean markets resulted in the beginning of a downward trend in American exports. These problems were exacerbated by complaints by European countries about the export from the United States of genetically-altered crops and animal products that had been produced with the use of synthetic hormones. These concerns resulted in a reduction in the quantity of American farm products that Europe purchased. By 2000, American exports were $50 billion, and were expected to increase an additional 2% in 2001.

The 2002 farm bill (Farm Security and Rural Investment Act of 2002), signed by President Bush, increased farm subsidies substantially above the 1996 farm bill (Federal Agriculture Improvement and Reform Act of 1996). These increases seemed to counter the earlier efforts of the United States to lower the level of agricultural subsidies worldwide. As might be expected, there was an immediate outcry concerning the level of subsidies in the new legislation.

Alabama Agricultural Exports

Before 1994, *Alabama Agricultural Statistics* published data on Alabama agricultural intermittently; however, beginning that year each bulletin has included information on the value of the most important commodities exported. Data for the period 1994-2002 are presented in Figure 16.8. These data indicate that yearly trends of Alabama exports were generally similar to those of the country as a whole (Figure 16.7). The state's exports were $340.8 million in 1994, but increased afterward, reaching $544.4 million in 1997, a year after American exports peaked. They declined in 1998 and 1999 before beginning to slowly increase again. In 2002, they were $418.1 million. Between 1994 and 2002, American exports increased from $340.8 to $418.1 million, an increase of 22.7%. During the same period, the Consumer Price Index (Figure 16.2) increased by 21.4%. So, although the total dollar value of the exports was higher, the real value increased very little.

When considering the CPI, the total value of Alabama exports was not greatly different in 1994 and 2002; however, there were considerable changes in the mix of products exported over time. Table 16.7 includes data on the value and percentage of the most important exports in 1980, 1984, and 2002. The major change during the period was the large increase in the value of poultry exported and the corresponding decline in the value of soybeans. The value of export of live animals and meat also declined, while the value of cotton and cotton products increased.

Agricultural Prices

As detailed in the preceding chapter, in 1973 the indices of prices received by farmers for the commodities that they marketed and the indices of prices paid for all the inputs required for producing those commodities were relatively close together (447 versus 490, respectively). Unfortunately, the two quickly began to diverge. The indices for prices received changed relatively little, while the indices for prices paid increased rapidly. By 1985, the two were 579 and 1131, respectively (Figure 16.9). In the current period (1985-2002), the situation only became worse. By 2002, the two indices were 620 and 1680, respectively. Obviously, as late as 2002, sharecropping with the guvmint was not helping the farmers' prices received at all.

There was some year-to-year variation in the indices. For example, the uptick in indices for prices received in 1983 and 1984 was associated with the increase in prices resulting from the El Niño weather that severely reduced agricultural production over a large part of the country. The increase in 1988 was the result of severe drought over much of the country, which restricted crop yields. Severe flooding in 1989 curtailed planting. Again in 1996, widespread drought resulted in reduced yields and increased prices. The sharp increase in the indices of prices paid in 1996 and 1997 was the result of the reduced demand for exports of American farm commodities associated with the turmoil in Asian economies.

With the indices of prices received and prices paid diverging sharply, the parity ratio trended sharply downward. It was 0.51 in 1985, and 0.37 in 2002 (Figure 16.10). It had never been that low. When Marvin Jones (D, TX), chairman of the House Committee on Agriculture and Ellison "Cotton Ed" Smith (D, SC), chairman of the Senate Committee on Agriculture and Forestry, led the charge in the 73rd Congress (1933-1935) to give the American farmer a New Deal in agriculture, the two indices were 70 and 109, respectively, and the parity ratio was 0.66. Except for the years of World War II and despite well-intentioned efforts of the Congress, the prices received situation had been getting gradually worse for almost three-quarters of a century. In 1785, Robert Burns, Scotland's Ploughman Bard, disturbed a mouse's nest while plowing his fields. The unhappy event lead to his writing one of his most famous poems, "To a Mouse." One of its verses includes the immortal lines: "The best laid schemes of mice and men/Go often askew,/And leaves us nothing but grief and pain . . . ."

Concentration in the Food Processing Industry

The sharp decline in parity ratios described in the preceding section was accompanied by an equally important phenomenon: the consolidation of large numbers of smaller firms involved in food production and processing into a smaller number of larger firms. Extensive consolidation results in concentration. Further, if consolidation and concentration continue beyond a certain point, the remaining large firms can significantly affect the price of food at the firm and/or the consumer level. When firms have this capability, they are said to possess market power. When four firms, together, have more than 40% of market share of a particular sector (Consolidation Ratio, CR4) that sector's competitiveness begins to decline.

Hefferman (1999) noted that in the late 1990s, the consolidation ratio for the beef packers was 79%. This CR4 percentage means that four consolidated beef packing firms controlled 79% of that sector. The CR4 values for several other sectors were: flour milling (62% CR4), pork packers (57%), dry corn milling (57%), broilers (49%), and turkeys (42%).

According to Azzam (2002), consolidation and concentration can also lead to the efficiency effect (either lower or higher production costs). Various combinations of market power and the efficiency effect can result in either increasing or decreasing food prices. For example, if concentration results in both increased market power and lower production costs and if market power dominates the efficiency effect, food prices increase and both farmers and consumers suffer. He further suggests that, most commonly, increased market power leads to higher food prices, because as fewer and fewer companies own larger and larger shares of the food business, tacit collusion rises and competition suffers.

Azzam further noted that among 33 food-processing industries, "rising concentrations had an upward effect on 'market power' in 28, a downward effect on costs in 14 and an upward effect on costs in 9. That led to higher food prices in 24 and to lower food prices in 3." He concluded that "although concentration led to lower costs in most cases, the 'market power' effect more than dominated the 'cost-efficiency' effect associated with rising concentration, resulting in higher food prices in most food industries."

Growing concentration means that food prices are largely freed from the rigors of competition. Highly concentrated industries are free to demand any price that they want. It is surprising that concentration among food industries would have occurred under the nose of the Triangle. It probably has been able to do so because the Triangle fervently wished that through concentration the efficiency effect would have been stronger than the market effect. As Azzam notes when this situation occurs, although food prices decline, costs decline even more, and both farmers and consumers benefit.

Agricultural Productivity

As noted in the preceding chapter, the index of agricultural productivity in the United States trended steadily upward during the period 1970-1985, from 0.64 to 0.96, an increase of 50%. It was much lower, however, than expected in 1974 and 1980, as a result of adverse crop weather in much of the Eastern portion of the country. It was also lower than expected in 1978 and 1979, as a result of the sharp increase in the index for inputs. The El Niño in 1983 also resulted in a sharp reduction in the index.

Data on the indices of American agricultural inputs, outputs, and productivity during the period 1970-2002 are presented in Figure 16.11. Note that the data for the period 1970-1985, in Figure 15.9, are different from the data for the same period in Figure 16.11. Around 2000, the government discontinued the collection of some of the data required for the calculation of the old series of indices. As a result, a new procedure was developed for making the calculations, but this method produced slightly different numbers. Figure 16.11 includes data developed using the revised procedure. The revised data show that the indices for inputs generally trended downward from 1.032 in 1985 to 0.993 in 1993. They increased, then declined, then increased and declined again in the period 1994-2002, but was 1.006 in 2002.

The index for outputs was 0.864 in 1985. Afterwards, in 1986, 1987, and 1988, they trended downward, as adverse crop weather, especially in the South, resulted in much lower than expected yields of major crops. Crop weather was much improved in the period 1989-1992, and indices increased from 0.867 to 0.950 in that period. It was again lower than expected in 1993 (0.910) and 1995 (0.959) as a result of early, cool, wet weather in the spring and excessively dry weather later. After the poor harvests in 1995, indices began to increase again, reaching 1.058 in 1999. Poor crop weather again plagued farmers in 2000, 2001, and 2002. As a result, indices declined slightly during those years. In 2002, the index was 1.006.

Data (new series) on the indices of agricultural productivity in the United States for the period 1970-2002 are also presented in Figure 16.11. Agricultural productivity is a measure of the efficiency with which inputs are transformed into outputs (Ball, 2007). In the data presented in the figure, the annual estimate of productivity is the ratio of the indices of outputs divided by the indices of inputs. As expected, the indices of productivity generally followed the year-to-year trends of the indices of outputs from 1985 through 1993. However, indices of productivity were lower in 1994 and 1995 and in 1997, 1998, and 1999, when indices for inputs were higher than expected. There was little change in indices in 2000, 2001, and 2002, as the indices for outputs declined slightly. If the indices for inputs had not also declined during that period, productivity would have declined sharply.

Bell (2007) published data showing that Alabama ranked 29th in the country in agricultural productivity in 1960, but that by 1999 the state had fallen to 33rd. Similar data showed that the average annual growth of agricultural productivity in the state was the lowest in the entire South during the period 1960-1999.

Farmers

In each of the preceding chapters (7 through 15), discussions address farmers as a specific group of people within the larger population; however, it is no longer practical to do so. In the 1950s when American agriculture was beginning its world-changing effort to modernize, everyone knew who farmers were and what they did. This situation has likely changed forever. The ever-worsening situation with regard to the relationship between prices received and prices paid, even when productivity was increasing rapidly, decreed that farmers would have to change. Over time, even with the massive intervention of the government, it became increasingly difficult to make a living on family farms. Although it has been increasingly difficult to make ends meet from the sale of farm commodities, there are still many people living in rural areas who were willing to accept losses in their operations to continue to farm or to be known as farmers. Obviously, they could not accept losses that would be incurred on large farms. As a result, most of these marginal operations are small—a substantial share of them no larger than nine acres.

Farmer Occupations

In 1974, some 61.7% of persons who operated America's farms claimed that farming was their principal occupation. In 1987, this percentage had fallen to 54.5%, and by 1997 it was down to 50.3%. Trends in the changes in the occupations of Alabama farmers after 1974 were similar to those in the country as a whole, except that in the state, the trend toward a lesser commitment to farming for a living was more advanced. For example, in 1974, only 45% of persons operating farms in Alabama indicated that farming was their primary occupation. In 1987, the percentage was down to 37.8%, and by 1997 it was 37.6%. Comparable percentages for the country were: 61.7% (1974), 54.5% (1987), and 50.3% (1997).

Types of Farmers

Banker and Hoppe (2005) divided farmers into two groups depending on whether or not their primary occupation was farming (Table 16.8).

The group of farmers who do not have farming as their primary occupation is further divided into three sub-groups:

1. Those who operate small, limited resource farms.
2. Those who operate farms, although they are retired.
3. Those who have a major occupation other than farming.

The group of farmers who give farming as their primary occupation is divided into four sub-groups:

1. Those whose farms have annual sales of less than $100,000.
2. Those whose farms have annual sales ranging from $100,000 to $249,999.
3. Those whose farms have sales ranging from $250,000 to $499,999.
4. Those whose farms have sales of $500,000 or more.

Table 16.9 provides some data on the characteristics of the seven types of farms that American farmers operate. It is quite obvious that the farmers and their families who do not have farming as their primary occupation are a major problem for those responsible for making agricultural policy in the country. These farmers own over a third of all assets of all family farms, yet they produce slightly less than a tenth of the total value of production. Unfortunately, they are quite numerous. Well over half of all of the family farms in the country are operated by farmers who do not consider farming their primary occupation. As a result of their numbers, they are politically potent, and they have wide-ranging, often highly divergent opinions of what they want to get out of farming.

Off-Farm Work

With agricultural productivity in America increasing at a greater rate than the population of the country for many years, something had to give, and it was the farmers that did the giving. A large percentage of them simply left commercial agriculture and have subsequently participated on its fringes. Virtually all who chose to remain had to seek off-farm work to supplement their on-farm incomes. The steadily worsening relationship between prices received and prices paid made it largely impossible to maintain a positive net income from their farm operations alone; even with the addition of substantial federal subsidies. As a result, by the beginning of the twenty-first century, up to 95% of farm household income came from off-farm sources.

Table 16.10 presents some data on the percentages of farm operators in the United States and in Alabama who worked off their farm for different amounts of time in 1982, 1987, 1992, 1997, and 2002. In the United States, 55% of all farmers worked off-farm in each of those years, and the larger share of those worked off-farm 200 days or more. A large share of this group must have worked close to full-time. The larger percentage not working off-farm in 2002 is likely the result of the USDA decision to include data on farms where the operators would not be expected to work away from the farm.

In Alabama, a much larger share of farmers worked off-farm in the different years than in the country as a whole. Except for 2002, slightly over two-thirds worked away for some of the time. Moreover, of those who worked away some, a much larger share worked 200 or more days. This relatively higher share working away more is likely a reflection of the general low profitability of farms in the state.

The Farmer's Plight Versus the Well-Being of the Country

The willingness of farmers to accept these momentous changes in their lives to continue producing food and fiber has resulted in far-ranging, unmitigated blessings for American consumers. The cost of food as a share of disposable personal income is at a historic low of 10%. No citizen, under any circumstances, need go hungry for even one day. Food and fiber production are being practiced in a more environmentally-friendly manner than ever before in the history of the country. Even the part-time farmers themselves are better off. Since the mid-1990s, total income of farm households has consistently exceeded total income of all American households. Even those farmers who operate on the fringes of commercial agriculture make important contributions to the rural economy of the country, and they provide a positive stewardship for a substantial share of its land resources.

While farmers might be technically better off than they have ever been, with their combination of off-farm and on-farm incomes and government subsides, they and their families pay an exorbitant price for opting to remain in commercial agriculture. Farming is a demanding occupation; it is hard work. Farming requires the combined skills of mechanic, engineer, accountant, economist, lawyer, computer expert, financial planner, meteorologist, hydrologist, technical writer, environmentalist, scientist, and veterinarian. In addition, farmers must balance the often rigorous and demanding schedules of land preparation, planting, pest control, cultivation, harvesting, and marketing with family obligations at PTA meetings, soccer matches, school plays, and special science projects with a full-time, off-farm job. The overpowering pressure of such a life explains why so few young families are choosing to enter or to remain in commercial agriculture, and why the average age of farmers in the country has increased from 50.3 years in 1978 to 55.3 years in 2002. One has to wonder whether our national agriculture policy is in the process of destroying our most valuable national asset—farmers, especially young farmers.

Tenancy

Data on tenancy rates in the country and the state during the period 1974-2002 are presented in Figure 16.12. In 1974, approximately 11.3% of all farms in the country were operated by tenants. It increased to 12.5% during the period of extreme national economic malaise around 1978. Afterwards, it drifted downward, reaching 11.3% in 1992. After 1992, NASS began to include some new kinds of farms in its counts. These farms were extremely unlikely to be operated by tenants. As a result, the tenancy rate was down to 10% in 1997 and further down to 7% in 2002.

The tenancy rate in Alabama had finally fallen below the national rate around 1969, and it was well below the national rate in 1974 (7.1 versus 11.3%). The agricultural situation had changed so much in Alabama that there was no place for tenant farming anymore. Alabama's tenancy rate also increased around 1978, as a result of the country's economic problems. It also declined afterwards, finally reaching a level of 6.1% in 1997. Then it fell sharply to 4.6% in 2002.

Looking at the 1997 and 2002 data, it is difficult to realize that a little over half a century earlier that well over 50% of the farms in the state were operated by tenants. The sharp decline in tenancy would seem to indicate that conditions in Alabama agriculture have improved significantly. That assumption is probably not correct. It was likely more difficult to make a living on farms in the state in 2002 than it was in the 1940s. The difference is that most of those excess farmers who were tenants in 1940 now have many other opportunities for employment.

Farms

In the preceding chapter, it was noted that in the period 1970-1985 trends in some of the characteristics of American farms, which had begun in the late 1940s and early 1950s, were continuing; while in others, the trends seemed to be reversed. Farm numbers were continuing to decline, although the rate of decline was not as high as it had been in earlier years. Sharp increases in farm size noted in earlier years had disappeared. Total cropland, which had fallen sharply through the early 1960s, generally stabilized during the period. However, the percentage of cropland harvested was beginning to slowly increase.

Types of Farms

Hoppe, et al. (2001) divided American farms into three major groups: small-scale family farms, large-scale family farms, and non-family farms.

Hoppe divided the major types into several sub-types. Definitions of the major types and sub-types are presented in Table 16.8. Data in Table 16.9 show that 98% of all farms in the country were family farms. The most surprising statistic is that 61.8% of all farms were operated by families (operators) whose primary occupation was not farming. In fact, 40.4% of this group farm because they like the lifestyle. This is really an amazing statistic. It is highly likely that there are few other businesses in the country whose owners operate them because they like the lifestyle. With almost 62% of all farms operated by families who do not farm for a living, it is not surprising that it is so difficult to develop an efficient, cost- effective agricultural policy for the country.

The farm dog is one of the difficult problems that managers of wildlife populations have to deal with. These animals are fed and maintained by the farm family, but allowed to hunt and kill at their leisure. In natural predator-prey situations, the populations of these two groups are tied together in a teeter-totter relationship. When the prey population increases, the size of the predator population follows, but when the predator population increases, the prey population begins to decline. Farm dogs are not a part of this relationship. The size of the prey population has little effect on them. They can continue to hunt and kill until the prey population is virtually eliminated, because they are fed whether they hunt or not. For farmers who earn most if not all of their living from off-farm occupations, the bottom line in farming is very different from those farmers who must recover their costs, as well as feed, shelter, and clothe their families with on-farm income from their operations. Farmers, like the farm dog, who farm for fun generally can operate outside the territory circumscribed by the law of supply and demand. Unfortunately, they can create serious problems for those farmers who do.

Table 16.9 contains some specific data on nine subgroups of farmers. For example, these data show that small-scale, family farms (limited-resource, retirement and lifestyle farms), in which farming was not the primary occupation of the operator, comprised 61.8% of all farms. They accounted for 9.1% of all production and owned 38.1% of all farm assets and 30.12% of all farmland. Farmers without a farming bottom line were responsible for the operation of a significant share of American production agriculture.

Farm Numbers

The number of American farms decreased 22.3% (2.949 to 2.292 million) during the period 1970-1985 (Figure 16.13). This downward trend continued from 1985 through 1993 when the number abruptly increased from 2.108 million to 2.202 million. This increase was a result of including farms in the count that had never been included before, when USDA (NASS) assumed responsibility for collecting data for the *US Census of Agriculture*. From 1994 through 1999, numbers remained around 2.19 million. Afterwards, however, they began to decline again, and in 2002 they fell to 2.135 million.

In this period, the hemorrhaging of farms in the country slowed considerably. Between 1950 and 1970, numbers declined by 48%. Then from 1970 through 1985, the decline amounted to only 22%, and from 1985 through 2003 it was much lower (7%). Further, the explosion in input costs, characteristic of the two preceding periods, had begun to calm down somewhat. They continued to increase, but at a much reduced rate. Spending on equipment, land and buildings, capital consumption, and purchased inputs increased. In all cases, however, the increases were either below or near the increase in the Consumer Price Index (67%) for the period. Surging investment requirements were no longer driving large numbers of farmers out of agriculture.

In 1985, Texas had the largest share (8.4%) of farms in the country. Other states with large shares were: Iowa (4.8%), Kentucky (4.4), Tennessee (4.1), and Illinois (4.0). These five states accounted for 25.7% of all farms in the country that year. Some 2.3% of all farms were in Alabama. The list of states with the most farms changed very little by 2002, except that Missouri replaced Illinois. The largest share (10.7%) continued to be in Texas. Other states with large shares included: Missouri (5%), Iowa (4.2), Tennessee (4.1), and Kentucky (4.1). In 2002, these five states accounted for 28.1% of all farms. Alabama had a slightly smaller share (2.1%).

During the period 1970-1985, Alabama lost 36.6% of its farms (82,000 to 52,000)—a higher rate of loss than the country as a whole. After 1985, the number of Alabama farms continued to decline through 1992, reaching 46,000 in that year (Figure 16.14). The number increased slightly in 1994 to 49,000 as a result of the changes in data collection. Numbers remained at the same level through 1998 when the downward trend resumed. In 2002, USDA (NASS) counted 45,000 farms in the state. During the period 1985-2002, the rate of loss of farms in Alabama continued to be considerably greater than in the country as a whole (13.5 versus 6.9%).

In 2002, Cullman Country had the largest share (5.1%) of farms in the state. Other counties with large shares included: DeKalb (4.8%), Marshall (3.7), Lauderdale (3.3), and Jackson (3.0). These five counties accounted for 19.9% of all Alabama farms. Some 15 counties reported having 750 or more farms. Figure 16.15 shows the distribution of farms within the state

Farm Ownership

According to census data, the overwhelming share of farms in the country are owned by families or individuals, rather than corporations (Table 16.11). It remained relatively constant at around 86% from 1982 through 1997, but increased in 2002 after USDA (NASS) began to include more smaller farms in the data. Family ownership in Alabama was more prevalent than in the country as a whole. It generally varied around 91% from 1982 to 1997, but also increased in 2002, with the addition in the data of more farms that would be expected to be owned by families.

A much smaller share of American farms are owned by corporations with boards of directors involved in management decisions (Table 16.11). In the years 1982, 1987, 1992, and 1997, the percentage of corporation-owned farms in the United States ranged between 2.7 and 4.4%. While they are relatively few in number, they are important because they produce food and fiber on a relatively large share of the country's farmland. For example, in 1987 corporations owned only 3.2% of the farms in the country, but 12.4% of the farmland. A decade later, they owned 4.4% of the farms and 13.9% of the farmland. Corporate farms are somewhat less common in Alabama than in the country as a whole (Table 16.11). In 1987, corporations owned only 1.3% of the farms and 4.2% of the farmland. A decade later, they owned 1.8% of the farms, and 5% of the farmland.

Farm Size

The average size of all American farms increased from 375 acres in 1970 to 447 acres in 1985 (Figure 16.16). Average size continued to trend upward until around 1992, when it reached 464 acres (the highest in history). After the early 1990s, it began to trend downward slowly, probably as a result of changes in data collection. Average size increased slightly in 2000 and 2001, and it was 436 acres in 2002.

Table 16.12 includes information on the percentages

of American farms in different size classes in 1982, 1987, 1992, 1997, and 2002. The data indicate that over 50% of all farms in the country ranged in size from 10-179 acres. It is difficult to see trends in the data. In 1992, USDA (NASS) took over the responsibility of the Bureau of the Census for collecting data for the *Census of Agriculture.* Afterwards, USDA began to include some types of smaller farms in the collection of data that had not been included before. These changes resulted in increasing the counts of smaller farms. These changes are especially evident in comparing data for 1992 and 1997. Percentages of all farms between one and 179 acres increased during the period. What is most surprising about these data is the persistence of farms of 49 acres or less. In the five years of data, at least 28% of all farms in the country were less than 50 acres. It is difficult to envision how the country can ever establish a farm policy that meets the needs equally well of farms nine acres or less and those with more than 999 acres. Both groups include about 8% of all farms.

Data presented in Figure 16.16 indicate that in the period 1970-1985 the average size of Alabama farms increased from 180 to 215 acres. This upward trend generally continued through 1992 when it reached 223 acres per farm—the highest in history. Average size declined after 1992, finally reaching 197 acres in 2002. This decline is likely the result of including more smaller farms in the count.

Table 16.12 also includes data on the percentages of Alabama farms in seven size classes in 1982, 1987, 1992, 1997, and 2002. The data show that the percentage of farms between 10 and 179 acres was much higher in Alabama than in the country as a whole, but that the percentage of really small farms (1-9 acres) in the state was lower. There were also relatively fewer farms larger than 499 acres in Alabama.

Farmland

During the period 1970-1985, total American farmland decreased from 1,102.4 million to 1,012.1 million acres. This downward trend continued through 2002 when the total was down to 940.3 million acres (Figure 16.17).

From 1970 to 1985, Alabama lost 24.3% of its farmland (14.8 to 11.2 million acres). In that period, its rate of loss greatly exceeded that in the country as a whole, and farms in Alabama also trended downward after 1985 as it had in the country as a whole (Figures 16.18 and 16.17). This trend was reversed temporarily in 1993 and 1994 when acreage actually increased; however, the downward trend resumed in 1995 and continued through 2002. There were 11.2 million acres in farms in Alabama in 1985. By 2002, the total was down to 8.9 million, a reduction of 2.2 million acres (19.6%). The relative loss of land in farms in Alabama was much greater (19.6% versus 6.8%) during this period than in the country as a whole.

There is currently widespread concern in Alabama about the loss of farmland to suburban sprawl and strip development. This loss is of special concern in the Tennessee Valley where each year the growth of towns and cities is consuming hundreds of acres of some of the state's best farmland. While the loss of farmland to non-farm use is of concern to some, it is somewhat misplaced. The real concern should be with the loss of farmland in farms. As detailed in a preceding paragraph, land in farms in Alabama fell 2.3 million acres (11.2 to 8.9 million) during the 16-year period 1985-2002. This decline translates into a loss of almost 144,000 acres a year.

There is no good reason to believe that the loss of farmland in farms will not continue well into the future in the state. Data on monetary gains versus losses on Alabama farms in 1997 suggest that the state still has a significant excess of land in agriculture. If present trends continue, it is not unrealistic to expect that land in farms in Alabama will fall well below five million acres or lower in the next quarter century, and if a reasonable definition of what constitutes a farm were adopted, the amount of farmland would be only a fraction of that amount.

The most unfortunate aspect of this long-term, downward trend in the amount of land in agriculture is that, in the long-term, Alabama farmers will have little voice in the ultimate fate of the millions of acres of rural land in the state that were once in agriculture. Technological change, economic cycles, public policy, and globalization now operate on such a broad scale that a state with very little comparative advantage in agriculture will have little opportunity to have its voice heard, much less heeded.

Uses of Farmland

Data on the uses (percentages) of farmland in the United States for 1982, 1987, 1992, 1997, and 2002 are presented in Figure 16.19. These data show that the uses of farmland in the country have not changed very much since 1982. In fact, the percentages have remained essentially unchanged. In all the years, slightly more has been used for cropland than for pasture (around 46% versus 42%). Both of these uses are far greater than uses as woodland and "other." Both of those have hovered around 8% and 4%, respectively.

Data on the uses of farmland in Alabama in some years between 1982 and 2002 are presented in Figure 16.20. These data show that relative uses for cropland (50%), woodland (33%), pasture (13%), and other (4%) remained generally unchanged through 1992. In 1997, use for cropland began to decline, while the use for woodland began to increase. In 2002, the use for cropland was down to 42%, while uses for woodland and pasture were up to 36% and 17%, respectively. These changes probably reflect the final realization that the returns on crop production in the state were not comparable to those from beef and tree production. It also may have meant that farmers were disillusioned with crop production and they simply put their cropland in storage.

Total Cropland

In the preceding period (1969-1982), total cropland in the United States ranged between 440 and 459 million acres, without a clear trend. After 1982, it generally trended downward to reach 431.1 million in 1997 and 434.2 million in 2002 (Figure 18.18). The slightly higher total in 2002 is likely the result of including some additional kinds of farms in the count that year. From 1982 to 2002, total cropland declined by only 2.5%. During the same period, total farmland declined by 8.5% (Figure 16.17).

In Alabama, total cropland declined from 5.8 to 5.1 million acres in the period 1969-1982, and the decline continued generally through 2002 (Figure 16.22) to reach 3.8 million acres. From 1982 to 2002, the state's cropland declined by 25.5%. The rate of loss was much greater in the state than in the country (25.5% versus 2.5%). In the same period, Alabama farmland declined by 24.6% (Figure 16.18). These data add further evidence that during this period Alabama farmers were rapidly losing interest in crop production.

Cropland Harvested

Cropland harvested in the United States generally trended upward during the preceding period, from 273 million acres in 1969 to 326.3 million in 1982, before falling sharply in 1987 to 282.2 million. In 1987, crop weather in June, July, and August was extremely dry east of the Mississippi and wet west of it. Drought conditions were extremely bad in much of the South. Further, in 1987 the parity ratio reached 0.49 (Figure 16.10). It had never been that low. In addition, exports had been trending downward for several years (Figure 16.7), resulting in reduced demand for agricultural commodities. Apparently, these conditions were sufficient to reduce the amount of cropland that farmers harvested. Cropland harvested began to increase afterwards, reaching 302.7 million acres in 2002, but it did not return to the 1982 level.

Data presented in Figure 16.24 show that the percentage of cropland harvested in the country increased from 68.9% in 1974 to 73.2% in 1982. Then it declined to 63.7% in 1987 (Figure 16.24) before it returned to range between 68 and 71.8% in 1992, 1997, and 2002.

Cropland harvested in Alabama increased from 2.7 million acres in 1969 to 3.3 million in 1982, before falling to 2.2 million in 1987 (Figure 16.25). It had not been that low in the twentieth century. The Alabama Agricultural Statistics Service reported that in 1987 drought conditions in the last half of the growing season were severe, and that a late freeze had damaged the peach crop in the state for the third year in succession. The pecan crop was also reduced. Apparently, as a result of this poor crop year, Alabama farmers decided to permanently reduce the amount of cropland that they harvested. In 1992, 1997, and 2002, they did not harvest more than 2.1 million acres, and in 2002 they harvested only two million.

Data presented in Figure 16.24 show that for many years Alabama farmers harvested crops from a much smaller share of their cropland than in the country as a whole. In 1982, they harvested 10% less (63.9 versus 73.2%). Afterwards, the difference was much greater. In 1997, American farmers harvested crops from 71.8% of their cropland, whereas Alabama farmers harvested only 49.5% of theirs.

Census data show that between 1930 and 2002 the average acres of cropland harvested declined by 35.8% in 12 Southern states, but by only 4.8% in 12 Midwestern states. The 72% decline in Alabama was the highest in the 24 states. These data probably reflect, in a general way, the comparative advantage of crop production in the two regions.

Conservation Reserve Program

As detailed in the preceding chapter, the Food Security Act of 1985 established the Conservation Reserve Program (CRP). The act mandated that USDA enroll 40 to 45 million acres of environmentally-sensitive cropland in the program. The Food, Agriculture, Conservation, and Trade Act of 1990 extended the CRP enrollment through 1995 and broadened the program's focus. The Federal Agriculture Improvement and Reform Act of 1996 continued the program through 2002 at a maximum enrollment of 36.4 million acres. Under the provisions of the various acts, enrollments in millions of acres were 9.9 in 1987, 22.8 in 1992, 31.9 in 1997, and 32.7 in 2002. These enrollments amounted to 2.2, 5.2, 7.2, and 7.5% of total cropland in the country in those years.

In those same years, farmers and landowners in Alabama enrolled 122,100; 270,200; 483,200; and 472,300 acres in the CRP. These enrollments amounted to 2.7, 6.4, 10.9, and 12.6% of total cropland in the state in those years. These data indicate that Alabama farmers and landowners responded more positively to efforts to take environmentally-sensitive cropland out of production than landowners in the country as a whole. It probably also indicates that these Alabama landowners were extremely pleased to have the government "rent" their less productive land. Returns from those rents were probably greater than returns they would have received from using them for crop production.

Value of Land and Buildings

Figure 16.26 presents data on the value per acre of land and buildings on American farms in the period 1970-2002. These data show that value declined from $801 in 1984 to $713 in 1985, a result of the economic problems described in a preceding chapter. It continued to decline through 1987 ($599). Afterwards, it began to trend upward. The trend was slowed slightly during the 1990-1991 recession. In 2002, value reached $1,210. During the period 1985-2002, value increased by 69%. During the same period, the Consumer Price Index increased by 67%.

As expected, the value of farm real estate varied considerably from region to region in the country. In 2001, value was highest in the Northeast ($2,640 an acre) and lowest in the Mountain States of the West ($486). Surprisingly, the average value for states in Appalachia (Virginia, West Virginia, North Carolina, Kentucky, and Tennessee) was higher than the average for the states of the Corn Belt ($2,250 versus $2,030). Obviously, there are some non-farm influences at play here. The average value for the states in the Southeast (South Carolina, Georgia. Florida, and Alabama) was also slightly higher ($2,140 versus $2,030) than for the Corn Belt.

Data also presented in Figure 16.26 indicate that the value of land and buildings on Alabama farms was $797 an acre in 1985. It generally remained unchanged through 1987, before beginning to trend upward. The 1990-1991 recession (Figure 16.1) had a greater effect on value in Alabama than it did in the country as a whole. But value began to increase again in 1992, and reached $1,700 in 2002. During the entire period (1985-2002), value increased 113%, or at a considerably higher rate than the increase in the Consumer Price Index during the same period (113 versus 67%).

USDA has published data showing that, with the exception of several states in Northeast, the value of land and buildings in Alabama increased more rapidly during the period 1980-1997 than any other state in the country. The increase between 1992 and 2002 was spectacular. During that period, it increased from $936 to $1,700 an acre, an increase of 81.6%, or by 7.4% a year. There are very few investments in the business that would have generated a higher return in such a short period of time. This is an interesting phenomenon, and it will be discussed at greater detail in a subsequent section.

Rents of Alabama Cropland and Pastureland

Cash rents are generally considered to be good indicators of current returns from the production of crops on an acre of land in a particular location. A large share of rented cropland has been used for the production of cotton; consequently, in the preceding period (1970-1985), cropland rents generally followed cotton prices upward through 1980. Afterwards, rents declined along with cotton prices.

By 1985, rents for cropland in Alabama were down to $29.50 an acre (Figure 16.27). They remained near that level through 1992. After 1991, prices for land increased sharply, and rents followed, reaching $39 in 1996. After the FAIR Act was passed in 1996, cotton prices fell sharply, and once again rents followed. In 2000, they were down near $30. Cotton prices were up and down for the remainder of the period, but rents went to $36 in 2001 and remained at that level through 2002.

The data presented in Figure 16.27 are averages for cash cropland rents for the entire state. Cash rents in different parts of the state were both higher and lower, depending on location (Prevatt, 2000). In 2000, the lowest annual rent was $29 an acre in north-central Alabama, generally east and west of Shelby County. The highest rents ($46) were paid in the Tennessee Valley counties in the northwestern corner of the state.

The availability of irrigation on cropland allows owners to attract significantly higher rents. For example, the average rent obtained for an acre of bare cropland in the northwestern corner of the state in 2000 was $46 an acre; however, if irrigation were available, rents were increased to $93. In the north-central region, the availability of irrigation resulted in increasing rents that could be obtained from $29 to $47 an acre.

Pastureland is rented only for the purpose of providing grazing for cattle; consequently pastureland rents are closely related to cattle prices. In the preceding period (1970-1985), pasture rents (Figure 16.27) generally followed cattle prices up and down through 1985. In that year, pasture rents averaged $16.60. They generally followed cattle prices upward through 1993, reaching $19.40 that year. After 1993, cattle prices were down sharply in Alabama, and rents followed. In 1995, they were down to $12.50 an acre. Afterwards, cattle prices recovered, and once again rents followed. They were $18 in 2002.

Across the state, cash rents for pasture was much less variable than rents for cropland. Farmers in the northeastern corner paid the highest rents ($24 an acre) for improved, permanent pasture. The lowest rents ($20) were paid in the counties east and west of Montgomery County.

Cash Rent to Value in Alabama

The ratio of the cash rent on an acre of cropland to its cash value provides a measure of the degree of non-farm influences on the value of that acre of land. Rents are indicators of the value of land for crop production; while values are indicators of the value of land for crop production plus an indication of its value in alternative uses. For example, in 2002, the average value of cropland in New Jersey was $9,000 an acre, but average rent was only $47. Obviously, there is some farm enterprise value in that land price, but much of it had to be a result of non-farm influence, in this case: rapid urbanization. The rent-to-value ratio of 0.52% ($47/$9,000 x 100) indicates that an acre of farmland in New Jersey would have been much more valuable for some urban use.

Data on the ratios of cash rents to land values in Alabama for the period 1970-2002 are presented in Figure 18.25. Data for the period 1970-1994 are from Prevatt (2000). Data for the remainder of the period (1995-2002) were computed from estimates of the average cash rents and average values of cropland per acre in the state published by the Alabama Agricultural Statistics Service for those years. The 1985-1994 data indicate that the ratio changed relatively little in that period. The 1995-2002 data indicate that the ratio generally trended downward during those years. The comparability of the two sets of data could not be determined. According to the latter set of data, the ratio declined from 3.1 in 1995 to 2.2 in 2002. Subsequent data have indicated that the ratio continued to decline after 2002.

Data on cropland rents, cropland values, and the ratio of the two expressed as a percentage for several states in 2002 are presented in Table 16.13. These data represent a range of situations in which there is virtually no non-farm influence on cropland value (North Dakota) and in which non-farm influence is likely the primary determinant of value (New Jersey). Declining ratios for states along the Atlantic coast from Georgia northward to New Jersey probably reflect the increasing non-farm influence of urbanization.

Alabama seems to be in an unusual situation. The ratio (2.2) indicates that there is likely some non-farm influence affecting the value of cropland ($1,600 an acre), but it is not likely to be urbanization. The population of Birmingham, the state's largest city was less than a million in 2002, and none of the remainder of its largest population centers even approach 500,000 persons. The decline in the ratio in the state is primarily the result of the sharp increase in the value of the land, and the increase in land value is likely to be the result of the relatively low level of property taxes levied on farm property. In 2001, USDA published data in *Agricultural Statistics—2001* showing that in 1995 the average tax levied on farms in the United States was $0.73 per $100 of full market value. Levies ranged from $0.14 in Alabama to $2.03 in Arizona. In Florida, Georgia, Mississippi, and Tennessee, levies were: $0.80, $0.52, $0.27, and $0.43, respectively. In the three major farm states of Iowa, Indiana, and Ohio, levies were: $0.94, $0.56, and $0.81, respectively. These data strongly suggest that low tax rates are likely to be the non-farm influence responsible for the declining ratio in Alabama.

The Alabama legislature enacted the Current Use Law in 1982. Between 1982 and 1995, the average value of farm real estate increased from $885 an acre to $1,260 (Figure 16.26), an increase of 42%. Between 1994 and 1995, the average value of Alabama farm real estate increased from $1,120 to $1,320 an acre, an increase of 12.5%. With land real estate values increasing at that rate, and with a tax rate of $0.14 per $100 of full market value, owning Alabama farmland was an extremely attractive investment opportunity.

While it seems likely that the low tax rate was an indirect cause of the sharp increase in the value of farm real estate, the important point is that value is unusually high relative to its capacity to provide a reasonable return on investment through farming. Further, it is so high that purchasing land for farming would be extremely impractical. Certainly, young farmers could never afford to purchase it for that purpose. While the high value is a formidable obstacle to the future of farming in the state, it does provide strong collateral for securing credit. Also, where an elderly farm owner has been unable to build a good estate from returns from the production of food and fiber, the increase in land value at a rate considerably above inflation has afforded him that opportunity.

Value of Machinery and Equipment

The value of machinery and equipment on American farms increased from $9,770 a farm in 1969 to $41,919 in 1982, an increase of 330%. In the same period, the value of machinery and equipment on Alabama farms increased from $5,909 an acre to $24,150, an increase of 309%. In both cases, the rate of increase in value greatly exceeded the rate of increase in the Consumer Price Index of 163%.

Data presented by the 2002 *Census of Agriculture* indicate that the average value of machinery and equipment on American farms increased from $41,919 a farm in 1982 to $66,570 in 2002 (Figure 16.29), an increase of 59%, which was considerably lower than the 86% increase in the CPI during this period.

On Alabama farms, the value of machinery and equipment increased from $24,151 a farm in 1982 to $42,705 in 2002 (Figure 16.29), an increase of 77%, also considerably lower than the increase in the CPI for the period. However, Alabama out-performed their counterparts in the remainder of the country on a percentage basis (77% versus 59%) in this all-important area of farm management. As a matter of comparison, in 2002 the value of machinery and equipment on the average Iowa farm was $100,442.

Data presented in Figure 16.29 show that the average value of machinery and equipment on American farms (per farm) was considerably greater than on Alabama farms. Of course, the average size of American farms was considerably larger. In 2002, the average size of each was 441 and 197 acres, respectively (Figure 16.16). When the 2002 value data is converted to a per-acre basis, it is greater on the Alabama farms ($217 versus $151). It is likely that a given complement of machinery and equipment is needed for operating a 197-acre farm, but that same complement might be adequate for the operation of much larger farms.

Value of Machinery and Equipment and Cropland Harvested

The average market value of machinery and equipment on an American farm in 2002 was $66,570 per farm. In the same year, crops were harvested from an average of 204 acres of land per farm. The ratio of these two values ($66,570/204 acres) indicates that American farmers maintained machinery and equipment valued at $326 for each acre of cropland harvested. In 1982, this ratio had been $232.

In 1997, the market value of machinery and equipment on Iowa farms was $100,442 per farm. In the same year, crops were harvested from 247 acres. These data indicate that the average Iowa farmer owned machinery and equipment valued at $406 for each acre of harvested cropland.

Only 44 acres of cropland were harvested on an average Alabama farm in 2002. So with an average of $42,705 in machinery and equipment on each farm, Alabama farmers had machinery and equipment valued at $971 ($42,705/44 acres) for each acre of cropland harvested. In 1982, this ratio had been $246 an acre harvested. The Alabama ratio for 2002 was much larger than for either the United States or Iowa ($971 an acre versus $232 and $406, respectively). Obviously, for Alabama farmers it would have been extremely difficult to maintain their level of investment in machinery and equipment. Unfortunately, this situation would be even more difficult than a comparison of these ratios indicate, because of the lower yields and higher production costs of most crops harvested on Alabama farms.

Tractors on Farms

Data presented in the preceding chapter indicated that the number of tractors on American farms (number per 100 farms) generally increased from 162 in 1969 to 205 in 1978, and that the number declined slightly in 1982 to 202. Numbers of tractors on Alabama farms kept pace with the country as a whole. They increased from 103 in 1969 to 137 in 1978, and remained unchanged in 1982.

Data on tractors on American farms (number per 100 farms) during the period 1978-2002 are presented in Figure 16.30. In the period, numbers increased from 202 in 1982 to 224 in 1992. It was lower in 1997 (206). This decline was probably the result of a reduction in the demand for food and fiber related to the turmoil in Asian markets during that period. The number increased to 216 in 2002, but did not return to the 1992 level. The 2001 recession possibly slowed the purchase of the machines, even into 2002.

Data on the numbers of tractors on Alabama farms during the period 1978-2002 are also presented in Figure 16.30. These data indicate the trend in numbers in the state generally paralleled that of the country, although they were much lower. The smaller numbers generally reflect the difference in the average size of the farms (Figure 16.16).

In 1982, in the United States the ratio of numbers of tractors to acres of cropland harvested was 1:66. Farms owned one tractor for each 66 acres of cropland harvested. In 2002, it had increased slightly to 1:72. In Iowa in 1997, the ratio was 1:80. Ratios in Alabama were much lower. In 1982, it was 1:49, and in 2002 it was even lower (1:27). The size and topography of many of Alabama's fields, especially on smaller farms, makes tractor work much slower. Alabama farmers also probably had lots of tractors that they did not need. This was likely to have been especially true on the so-called lifestyle farms.

Use of Fertilizer

Data presented in the preceding chapter showed that during the period 1959-1982 Alabama farmers used twice as many pounds of fertilizer per acre of harvested cropland than farmers in the country as a whole.

Publication of data used as a basis for the presentations on uses of fertilizers in preceding chapters was discontinued in 1983. As a result, data on US use of plant nutrients has been substituted. These data, in pounds of nutrient used per acre of harvested cropland for the period 1969-2002, are presented in Figure 16.31. These data show that the pounds of nitrogen applied generally increased throughout the period, from 52 pounds in 1969 to 80 pounds in 1997. There was little change between 1997 and 2002. Data was collected in 2002 from some additional types of farms that had not been included in earlier censuses. It is likely that these farms would have used considerably less nitrogen in their operations.

The data also shows that farmers applied much less phosphorus than nitrogen, and the use of phosphorus changed relatively little during the period. It tended to trend downward from 34 pounds in 1969 to 28 pounds in 1987, but began to increase slightly afterwards, finally reaching 31 pounds in 2002. They applied even less potassium than phosphorus. Amounts applied changed relatively little during the period. In most of the years, it varied around 34 pounds an acre.

Huang (2004) suggests that changes are taking place in production and marketing of fertilizer that may force American farmers to change the way they use it. He notes that the United States went from being a net exporter of nitrogen fertilizer in the 1980s to being a net importer in the 1990s. For years, American farmers benefited from the lower prices of imports, but recently, as a result of the increase in the price of natural gas, even imports are growing more expensive. Further, because of the increased competition, the American capacity for making nitrogen fertilizer has just about been eliminated. He further notes that the United States has long been a net importer of potash fertilizer. In 2002, American production of fertilizers containing potassium accounted for only 20% of domestic consumption. The country has long been a net exporter of phosphorus. In fact, it is the world's largest exporter of this fertilizer.

Plant Agriculture

Data on acres harvested of several major crops in the country and the state in 1970, 1980, and 1985 are presented in Table 16.14. These data show changes in acres harvested for these crops during the preceding period. The data in the table show that in terms of acres harvested corn continued to be the dominant American crop in that period (1970-1985); however, soybeans was replacing wheat as the second most important. Further, the data show that oats were becoming less important.

Data in Table 16.14 show that combined acres harvested of cotton and corn in Alabama remained at about the same level throughout the period, and that it declined from 500,000 acres in 1970 to 300,000 in 1985. The continued increase in acres of soybeans harvested was nothing short of amazing. It was obvious from the yield data that Alabama would likely never have a comparative advantage in the production of soybeans. Nevertheless, by 1985 there were three times as many acres of soybeans harvested as cotton. In fact, acres of soybeans harvested were considerably higher than for cotton and corn combined. The most surprising change, however, was the rapid increase in wheat production. Between 1970 and 1985, acres of wheat harvested increased from 85,000 to 400,000. The other amazing change was the sharp increase in acres of sorghum harvested, but as detailed in the preceding chapter, this was a unique event that did not continue long after 1985.

Production costs are at the very heart of production agriculture. Unfortunately, they are different for each crop, and for the same crop they change to some degree from year to year. However, even with these limitations, I have decided to include them for some of the major crops in this chapter. It is difficult to comprehend fully the complexity of producing and harvesting crops without an understanding of the ebb and flow of dollars through the processes involved.

Alabama Crop Weather

Some details of crop weather and the all-important soil moisture content in Alabama for the period 1985-2000 are presented in Table16.15. As the data in the table indicate, during this period Alabama farmers continued to be plagued by the same problems relating to soil moisture alluded to in several preceding sections: excessive soil moisture in the late winter and early spring, mini-droughts during the heart of the plant growing season, and excessive rain during the late-summer and fall harvest season.

As is shown in the table, soil moisture was generally excessive during January, February, and March in nine of the 18 years during the period 1985-2000. However, excessive soil moisture in late winter is primarily a concern in early land preparation. The condition usually does not result in significant decreases in either the acreage planted or yields of most field crops; although in some situations, excessive rain during this period can result in a deterioration of the winter wheat crop.

As the data in Table 16.15 indicate, soil moisture was less than adequate to support maximum plant growth during all or part of the period June-August in 10 of 18 years in the period. The effects of inadequate soil moisture will be clearly evident when examining the annual yield data for the individual crops in the following sections.

Late-winter and early-spring freezes plagued farmers in eight of the 18 years in the period 1985-2002 (1985, 1986, 1987, 1989, 1990, 1992, 1993, and 1996). The peach crop and early vegetable crops were severely damaged in most of those years. These conditions were almost yearly occurrences in the period 1985-1996, but did not occur during the remainder of the period, as an annual warming trend began.

Storms originating in the tropics (four tropical depressions, three tropical storms, and four hurricanes) directly affected farming in nine of the 18 years during the period 1985-2002. There was one storm in 1994, two in 1995, and three in 1998. Most of these storms were clustered in the period 1994-1998. Several of the storms brought much needed moisture to the state, but three of them—hurricanes Opal (1995), Danny (1997), and Georges (1998)–resulted in extensive damage to crops, especially in the southern counties. Hurricane Opal toppled large numbers of mature pecan trees throughout south Alabama, and large quantities of mature timber as far north as Birmingham.

Data presented in the preceding paragraphs, underscore once more the difficulty of farming in Alabama. In 17 years of this 18-year period, unfavorable weather interfered with some phase of crop production in some part of Alabama.

Cotton

Meyers and Macdonald (2002) commented that cotton, at the beginning of the twenty-first century, accounts for most (40%) of the fiber (man-made and natural) produced in the world. At least some cotton is produced in 80 countries, but the United States, China, and India together produce more than half of the world's total. US production is second only to China's, but leads the world in exports. American cotton exports account for 25 to 30% of all world trade in the fiber. Our cotton industry is the basis for some $25 billion in products each year and generates 400,000 jobs.

Although there was considerable year-to-year variation, American cotton acreage in 1985 was close to where it had been in 1970, 12 million acres. There was also considerable year-to-year variation in average yields, but they generally trended upward from 500 pounds an acre in 1970 to 600 pounds in 1985. Production was extremely variable during the period as a result of year-to-year changes in both acres harvested and yields. During the period, it was as low as 8.2 million bales in 1975 and as high as 15.6 million bales in 1981. The national yield in 1985 was 13.3 million bales. American mill use of cotton was eight million bales in 1970, and it generally trended downward to six million bales in 1985. Exports were highly variable during the entire period, but generally varied between three and six million bales. Prices increased rapidly from 20 cents a pound in 1970 to 60 cents in 1976, then remained near that general level until 1985.

US Cotton

Cotton is produced in substantial amounts in 17 states from Virginia to California, but the major concentrations of production is in the following areas: the high and rolling plains of Texas, the lower Mississippi River Delta, California's San Joaquin Valley, central Arizona, and southern Georgia.

Texas farmers harvested the largest share (29.3%) of the country's 1985 cotton crop of 13.4 million bales. Other states reporting large shares included: California (23.2%), Mississippi (12.3), Arizona (7.7), and Louisiana (5.5). These five states accounted for 78.1% of the total national crop that year. Alabama farmers accounted for 4% of the national crop.

Exports

Between 1984 and 1985, cotton exports fell sharply to around two million bales when relatively high prices for the American crop made it uncompetitive on the world market. They recovered in 1986, and remained between six and eight million bales until 1992 when uncompetitive prices again sent them downward to near five million bales (Figure 16.32). They returned to seven million in 1993, but then rocketed upward in 1994 to 9.4 million bales as a result of a rapidly-expanding world economy, an increasing demand for cotton textiles, and a relatively-low world crop, especially in China. The 1994 exports were the highest in nearly 70 years. Exports returned to more typical levels of around seven million bales in 1995, 1996, and 1997, but then plunged to 4.3 million bales in 1998 when American cotton production declined to slightly less than 14 million bales. The Asian financial crisis, which began in 1997, probably also affected 1998 exports. Exports recovered in 1999 and 2000 to levels of seven million bales. In 2001 and 2002, a rapidly improving Asian economy resulted in a growing demand for textiles, and with American cotton prices at their lowest level in years and with production at extremely high levels (17 million bales in 2000), American cotton exports reached a new record level of 11.9 million bales.

Mill Use

Mill use of cotton had trended downward in the early 1970s, when man-made fibers began to take a larger and larger share of the market; however by the mid-1980s, research on cotton fibers had developed ways to make even better fabrics by combining natural and man-made fibers. From that time, the use of cotton began to increase. In 1985, American mills used 6.4 million bales. Then, from 1985 until 1994 mill use trended upward to around 11 million (Figure 16.32), as the sale of American-made cotton textiles increased on the domestic and international markets. In the period 1995-1997, mill use held at around 11 million bales, but began to decline thereafter, and was down to near eight million in 2000. The rapid decline near the end of the century was the result of slowing of the growth of the American economy (Figure 16.1) and a rapid increase in the quantity of textiles imported from Asia.

Prices

Target prices and loan rates for cotton during the period 1970-1995 are presented in Figure 16.33. The target price for cotton was around 35 cents a pound in the early 1970s, but after 1975 trended sharply upward to reach 81 cents a pound in 1984. It remained at that level through 1986. The government began to reduce the target price for cotton in 1987, and by 1989 it was down to 73 cents. It remained at that level through 1995 when target prices were eliminated.

From 1975 through 1978, target prices and loan rates for cotton were similar, but afterwards, loan rates were much lower. In 1985, the loan rate was 24 cents lower (81 versus 57 cents). After 1985, the government began to lower the loan rate, and by 1989 it was down to 50 cents. It remained near that level through 2002.

Year-to-year changes in American cotton market prices were very complex. Market prices for corn are generally negatively related to bushels harvested; as bushels harvested increase, market prices tend to decline. Generally, market prices for cotton in the period 1985-2002 did not follow this pattern. In fact, in several of the years market prices seemed to be positively correlated with bales harvested. Although there was considerable year-to-year variation, from 1986 through 1995 market prices seemed to bounce back and forth between 50 cents and 70 cents. Further, there seemed to be a slight upward trend during that period. In 1993, mill use and exports totaled 17.3 million bales, and the 1993 market price was 58 cents. In 1994, the mill use plus exports total reached 20.6 million, and market price advanced to 72 cents. The total remained above 18 million bales in 1995, and the market price reached its highest level during the entire period (76 cents).

After the enactment of the FAIR legislation in 1996, market prices began to decline rapidly, and in 2001 they reached 32 cents. In that year, farmers harvested the largest crop in history (20.3 million bales). Bales harvested fell to 17.2 million in 2002, and market price rebounded to 46 cents. In the entire period (1985-1995), the market price was equal to or greater than the target price only in 1995; however, it exceeded the loan rate in nine of the 11 years.

The 2001 market price of cotton was 32 cents a pound. It had been 10.2 cents in 1933. From 1933 through 2001, the Consumer Price Index increased from 13 to 179.9, an increase of 1,283.8%. This means that the real market price of cotton in 2001 was 2.3 cents. This low, real market price must have been a kick in the teeth to Dobson's (1985) Triangle of Beneficiaries. It was probably the lighted fuse of low cotton market prices in 1932 (6.5 cents) that set off the explosion of federal legislation in 1933. So after almost 70 years and hundreds of millions of dollars spent on a bewildering array of programs for cotton farmers, they received a real 2.3 cents for their crop in 2001. In considering the reality of this situation, I am reminded of the remark made by Casey Stengel, the frustrated manager of the dismal New York Mets baseball team; after they had made some outrageous error, he remarked: "Does anybody here know how to play this game?"

Acres Planted and Harvested

Data on acres of cotton planted and harvested in the United States during the period 1970-2002 are presented in Figure 16.35. Cotton market prices fell from 67 cents in 1983 to 59 cents in 1984 (Figure 16.34), and between 1984 and 1985, acres planted declined from 11.1 to 10.7 million (Figure 16.35). The market price declined to 52 cents in 1986, but surprisingly farmers slightly increased acres planted in 1987. Even more surprisingly, farmers had guessed right: market price was up sharply in 1987. For once in their lives, cotton farmers harvested a good crop and sold it at a good price. Unfortunately, they attempted to outsmart the market in 1988 by increasing acres planted from 10.4 to 12.5 million. With that potentially large crop in the fields, market price fell to 57 cents. Chastened by their wrong guess, they reduced acres planted to 10.6 million in 1989, but with that relatively smaller crop in the fields, market price increased again. As a result of the higher 1989 market price, acres planted were increased in 1990. This time they were once again fortunate. Even though there was a larger crop in the fields, in 1990 the market price continued to increase. This teeter-totter relationship between market prices and acres planted generally continued for the remainder of the period. The relationship was especially interesting in 1993, 1994, and 1995. Both market prices and acres planted began to increase after 1992. Market price increased sharply between 1993 and 1994 (58 to 72 cents); then between 1994 and 1995, acres planted increased by 3.2 million acres (13.7 to 16.9 million). Acres planted in 1995 were the highest ever recorded at that time. Farmers had one more opportunity on the teeter totter before the end of the period. With the market price down to 32 cents in 2001, acres planted were reduced sharply in 2002 (15.8 to 14.0 million), but in 2002 the market price increased to 46 cents.

While market price the preceding year had a significant effect on acres planted in most years, adverse weather also affected it in at least one year during the period. In 1996, it was down sharply from 1995, even though the average market price had increased the previous year. In 1996, an early widespread drought, especially on the Southern plains, reduced planting significantly; later, widespread wet weather also negatively affected planting.

Data on acres of cotton harvested in the United States during the period 1985-2002 are also presented in Figure 16.35. From 1985 through 1997, farmers harvested around 95% of acres planted. Adverse crop weather affected acres harvested somewhat in 1986, 1992, 1996, 1998, 1999, 2000, and 2001. In 1998, a widespread drought and heat wave from Oklahoma eastward to the Carolinas resulted in a $6 to $9 billion loss to agriculture in that region. That year, cotton farmers planted 13.4 million acres, but only harvested 10.7 (79.8%).

The so-called Bt cotton has been available for planting for several years, and it has revolutionized efforts to control the European and southwestern corn borer, both highly destructive pests. Bt cotton is made by inserting genetic material obtained from the naturally-occurring soil bacterium, *Bacillus thuringiensis,* into the genetic material of the cells of the cotton plant. Once it is inserted, it becomes part of the genome of the plant, and when reproduction takes place, it is passed on to succeeding generations. The transferred material in a cell initiates the synthesis of an endotoxin, a protein that is toxic to larvae (worms) of the insect order Lepidoptera (butterflies and moths). When the larvae consume vegetative portions (cells) of the cotton plant, they also ingest the endotoxin that kills them. It generally does not harm the immature forms of other orders of insects. With the use of this genetically-altered plant, significantly less pesticide is released into the environment.

Another type of genetically-altered cotton also reduces the amount of pesticide required in the production of the crop. Weed control among the growing cotton plants is a major problem to farmers. It is difficult to control the weeds without killing the cotton. Fortunately, scientists have been able to insert a gene into the plant that makes it resistant to one of the most effective chemical weed killers–glyphosate. The highly effective and widely used herbicide, Roundup, contains glyphosate. Now, this effective pesticide can be used on weeds in cotton fields without damaging the cotton plants.

Boll Weevil Eradication

The National Cotton Council has commented that the National Boll Weevil Eradication Program ranks near Eli Whitney's invention of the cotton gin in terms of contributions to the advancement of the cotton industry. This program was initiated in 1970. As a result of the federal-state-grower partnership effort, the boll weevil has been essentially eradicated from 6.5 million acres. Most of the cotton growing regions, including the Southeastern states, are currently free of this devastating pest. Eradication is accomplished by a combination of insecticide applications; destruction, after harvest, of cotton stalks that provide shelter for over-wintering insects; and trapping with traps containing chemicals (pheromones) that attract weevils. Once the initial chemical applications are completed, further use of pesticides is generally not required. Growing cotton without pesticides significantly reduces the cost of production; while at the same time reducing the amount of toxic chemicals released into the environment.

Yields

In 1985, the average annual national cotton yield was 630 pounds of lint an acre. It had trended upward from 450 pounds in 1970 (Figure 16.36). Year-to-year variation was high through the entire period 1985- 2002. It generally ranged between 600 and 700 pounds an acre for the entire period. Usually, cotton yields are directly related to weather and/or damage by pests during the plant's critical growing and maturation periods. In 1985, the weather during the critical period was good, and the average yield was 628 pounds an acre, which was near the average for the period. However, in 1986, as a result of a widespread drought during the growing season, yield was only 547 pounds, far below the average. Then in 1987, good weather was again associated with improved yield (702 pounds). In 1993, 1995, 1997, 1998, 1999, and 2001 widespread adverse weather resulted in much lower yields than expected.

Bales Harvested

Acres harvested generally trended upward from 1985 through 1995 (Figure 16.35) while there seemed to be little, if any, real trend in yields during the entire period (Figure 16.36). Consequently, bales harvested usually followed the acres harvested trend. It was 13.3 million bales in 1985 and 19.7 million in 1994 (the highest in history) (Figure 16.37). Afterwards, both acres harvested and yields tended to be somewhat lower, and bales harvested were also generally lower. In 2001, both acres harvested and yields were higher than expected, and bales harvested reached 20.3 million (another record). Both controlling factors were down in 2002, and bales harvested were also lower (17.2 million). The much lower than expected value in 1998 was the result of the extremely low yield that year.

In 2002, Texas farmers harvested 29.5% of the country's 17.2 million-bale cotton crop. Other states reporting large shares included: California (12%), Mississippi (11.2), Arkansas (9.7), and Georgia (9.2). These five states accounted for 71.6% of the total national crop. Alabama farmers accounted for 3.3%.

Production Costs

Brooks (2001) published a detailed study on the costs and returns of producing cotton in 1997 in nine Agriculture Resource Management Survey (ARMS) regions of the United States (Figure 16.38). It is useful to compare data on costs and returns from several of these regions—including Prairie Gateway, Fruitful Rim, and Mississippi Portal with data from the Southern Seaboard, which includes most of Alabama, but unfortunately, it includes only the western end of the Tennessee Valley. In choosing 1997 for the study, Brooks chose a year with moderately high bales harvested (Figure 16.37).

Some characteristics of the cotton farms in these four regions are given in Table 16.16. The largest percentage of cotton farms in the country was located in Prairie Gateway (33%) in 1997; however, the largest percentage of cotton was produced in Fruitful Rim (26%). Cotton acreage per farm was considerably lower in Southern Seaboard (398 acres) than in Prairie Gateway (515 acres) or Fruitful Rim (543 acres).

Operating Costs. Operating (variable) costs are those expenses that are directly related to the production of cotton in a specific crop year. They must be paid each year. Generally, they are items purchased as the crop is being established, cultivated, harvested, and processed. They are purchased from local merchants who, in turn, have purchased them from specialized suppliers.

Operating (variable) costs per planted acre (seed, fertilizer, chemicals, ginning, etc) in this study were lowest in Prairie Gateway ($172.34 an acre) and highest in Fruitful Rim ($468.83) (Tables16.17). Costs in Southern Seaboard ($272.22) and Mississippi Portal ($291.06) were sandwiched between. Farmers in Southern Seaboard spent more on fertilizer ($50.75 an acre) than farmers in the other regions, but their costs were not much greater than those for farms in the Fruitful Rim ($45.39) or Mississippi Portal ($40.69). It is not surprising that fertilizer costs were higher in Southern Seaboard, because virtually all of the soils in the entire region belong to the Order Ultisol (Figure 3.2), which is characterized by low natural fertility, low organic matter, and high acidity. Because of low fertility, farmers in Southern Seaboard had to use more commercial fertilizer. With the exception of fertilizer, farmers in Fruitful Rim spent more on seed, chemicals, custom operations, fuel and electricity, purchased irrigation water, and ginning than farmers spent on those items in any of the other regions.

When operating costs are computed on a per-pound-of-lint rather than a per-acre basis, a somewhat different picture emerges (Table 16.18). Yields (pounds of lint per acre) were considerably higher in Southern Seaboard than in Prairie Gateway (714 versus 466 pounds); however, yields in both of these regions were considerably lower than in either Mississippi Portal (807 pounds) or Fruitful Rim (1,131 pounds). In 1997, the average cotton yield in Alabama was 597 pounds an acre, still greater than the average for Prairie Gateway (466 pounds), but much lower than the average in the other two regions. It was also much lower than the average for the Southern Seaboard (714 pounds an acre). On the basis of costs per pound of lint, operating costs are relatively consistent from region to region . Only six cents separate the lowest and highest regions (36 cents in Prairie Gateway versus 42 cents in Fruitful Rim).

Allocated Costs. Allocated costs are those expenses related to the production of a cotton crop, but which cannot be assigned in full measure to production. For example, all farms must pay taxes and purchase insurance, but they are not considered to be operating expenses. General farm overhead costs—which include such items as the cost of hand tools, building and maintaining farm buildings, and general farm transportation—are real costs, but in the strict sense not annual operating costs that must be paid to make a crop. Also, the cost associated with capital recovery of equipment is not a required annual cash outlay. It is the amount of money that should be set aside each year to replace equipment when it is worn out. It is a discretionary expense, and need not be paid each year, but must finally be paid if the farm is to remain viable. The opportunity cost for land item is also a special situation. If the farmers own their land, they do not have to pay this cost; however if the production and sale of a crop is not sufficient to pay them as much as they might receive in rent for the use of the land, their operation is not likely to be very profitable in the long term.

Total allocated costs in the study were significantly higher in Fruitful Rim ($367.85) than in any of the other regions (Table 16.17). Cotton farmers in that region used much more hired labor, set aside more money for replacement of equipment and machinery, and planted their crop on land with a much higher rental rate than in the other regions. Total allocated costs were lowest in Southern Seaboard ($212.99).

Total Production Costs. Total production costs per planted acre (operating costs + allocated costs) were more than twice as high in Fruitful Rim as in Prairie Gateway ($836.69 versus $386.96 an acre). Costs in Southern Seaboard ($485.22) and Mississippi Portal ($538.49) were between those two extremes (Table 16.18).

Again, when these costs are related to yield a different pattern emerges. The total cost of production per pound of lint produced was highest in Prairie Gateway (83 cents). It was 74 cents in Fruitful Rim, 68 cents in Southern Seaboard, and 67 cents in Mississippi Portal (Table 16.18).

A strong positive relationship is obvious between production costs and yields in these data. At one extreme, farmers in Prairie Gateway spent $387 an acre in production costs and obtained a yield of 466 pounds. At the other extreme, farmers in Fruitful Rim obtained a yield of 1,131 pounds from production costs of $837. Production costs and yields for the other two regions were between these two extremes. However, when relating yield per dollar of operating cost is considered, the results are somewhat different. As expected, the lowest yield per dollar of operating cost (1.20 pounds) was obtained by farmers in Prairie Gateway. The highest return (1.50 pounds) was obtained by farmers in Mississippi Portal. The return for farmers in Fruitful Rim was not much greater than in Prairie Gateway (1.35 pounds). These data suggest that some portion of the larger investment (production costs)—possibly custom operations, fuel and electricity, or ginning—in Fruitful Rim was not equally efficient in increasing yield.

Gross Value of Production. There was wide variation in the gross value of production among the regions (Table 16.18), primarily as a result of the differences in yield, but also to some local differences in prices received from the sale of the lint. Gross value tends to be quite variable from year-to-year, more so than production costs, because they are highly dependent on annual yields, which are in turn dependent to a large extent on weather and on prices that tend to be inversely related to production (teeter totter). Generally, investments in production are made assuming that the weather will be ideal and prices obtained will be high. Unfortunately, as has been so well established as to be almost a truism in agriculture, usually good crop weather and high prices for commodities tend to be mutually exclusive.

Returns to Management. Returns to management when only annual operating costs are considered (gross value of production minus operating costs) were positive in all four regions, ranging from $170.15 an acre in Prairie Gateway to $510.57 in Fruitful Rim (Table 16.18). Operating costs must be paid each year. As noted in a preceding paragraph, they generally are items that must be purchased annually from local merchants, who in turn must pay their suppliers on a timely basis. However, as long as farmers can realize enough return on their crops at the end of year to pay these costs, they can continue to farm for several years.

Returns when all costs (operating and allocated) are considered (gross value of production - total production costs), provide an entirely different perspective (Table 16.18). When using this measure, returns for farmers in Prairie Gateway were negative ($-44.46 an acre) ($342.50 - $386.96), although their total production costs were the lowest of the four regions ($386.96 an acre). Their yields were so low that they could not take advantage of their lower costs. Returns were highest in Fruitful Rim ($142.72 an acre). In Southern Seaboard ($79.21) and Mississippi Portal ($72.49), they were between those two extremes.

Alabama Cotton

Acres of Alabama cotton planted and harvested declined rapidly from 600,000 acres in 1970 to 300,000 in 1979, and remained near that level until 1985 . Yields generally mirrored those in the country as a whole. They were extremely variable. They were near 450 pounds an acre in 1970, and just above that level in 1985. Alabama bales harvested followed acres harvested downward from 500,000 in 1970 to 300,000 in 1980, but then began to increase to 400,000 in 1985.

In 1985, Alabama ranked seventh in the country in production with 4% of the total national cotton crop. Farmers in Limestone County harvested the largest share (14.3%) of the 545,000-bale state crop that year. Other counties reporting large shares were: Lawrence (11.9%), Madison (9.6), Colbert (8.9), and Lauderdale (6.2). These five counties accounted for 50.9% of the entire Alabama crop. A total of 11 counties reported harvests of 10,000 or more bales.

Receipts for cotton sales in 1985 represented 17.4% ($147.8 million) of the sale for all plant crops (including farm forestry receipts) and 6.9% of all farm commodities (including farm forestry receipts).

Exports

Data on the exports of cotton from Alabama for the period 1985-2002 is limited. Available data indicate that in the period 1996-2002 exports of the crop ranged from 14.2 to 22.6% of total exports. The average annual value of cotton exported was near $81.6 million.

Prices

Data on cotton market prices in Alabama during the period 1970-2002 are also presented in Figure 16.34. For all practical purposes, national and state market prices for the fiber were essentially the same. As a result, the comments made about year-to-year trends in national prices are appropriate for changes in prices in the state.

Acres Planted and Harvested

In 1985, Alabama cotton farmers planted 330,000 acres of the crop and continued to plant at near that level in 1986 and 1987 (Figure 16.39). Year-to-year changes in acres of cotton planted were less variable in Alabama than in the country as a whole (Figure 16.35). Except for 1986 and 1989, acres planted trended steadily upward to reach 590,000 acres in 1995. Afterwards, with market prices declining, acres planted followed, finally reaching 495,000 acres in 1998. Then, even with market prices continuing to decline, acres planted increased to 610,000 in 2001. That year's market price was sharply lower, and acres planted were reduced to 590,000 acres in 2002. When farmers harvested that smaller crop, market price had increased to 46 cents.

Alabama cotton farmers began to have their first serious problems making ends meet in the late 1860s. The situation had not improved very much in the early 1930s when Congress decided to intervene. Unfortunately, in 2001 the situation was even worse than it had been in 1932. In that year, Alabama cotton farmers received an average of 6.5 cents a pound for their crop. In 2001, when considering the prices they had to pay for the inputs required for making a crop and inflation of the currency, they received 2.3 cents a pound.

Data on acres of Alabama cotton harvested during the period 1970-2002 are also presented in Figure 16.39. In most years during the period, Alabama farmers harvested 98% of acres planted. Percentages were considerably lower in 1997 and 2002, when poor crop weather reduced yields to the point that it was impractical to harvest many of the fields.

Yields

Data on average yields of cotton on Alabama farms during the period 1970-2002 were also presented in Figure 16.36. As might be expected, yields for the state showed more year-to-year variation than average yields for all the states combined. Although there were several years with unusually high yields (1985,1992, 1994, 1996, and 2002), there seemed to be little change in the underlying trend during the period. In 10 years of the 18-year period, yields varied between 480 and 570 pounds an acre. Yields were below 500 pounds an acre in 1988, 1990, 1995, and 2000. Crop weather was extremely poor in all of these years (Table 16.15). In 12 of the 18 years, yields obtained by Alabama farmers were lower than those obtained by farmers in the country as a whole.

The years when yields were below 500 pounds an acre (1988, 1990, 1995, and 2000) were characterized by extremely long periods without sufficient rainfall to sustain adequate cotton plant growth and maturation. However, if the four years in which yields were highest (1985, 1992, 1994, and 1996) were considered to have had adequate moisture to support maximum production, then the remaining 12 years of the period would have to be considered to have had less than adequate rainfall or some other type of adverse weather. Four out of 16 years are pretty bad odds for farmers trying to make a living with any kind of food or fiber production.

Further, if disinterested observers were to consider the year-to-year variation in cotton yields depicted in Figure16.36 for the period 1985-2002, they probably would be forced to conclude that Alabama was not a very good place to grow cotton, at least during those years. Actually the variability is even greater than is indicated by the data presented in the figure. Those individual data points are averages representing estimates of yields from all the farms in the state. Unfortunately, the data on yields of individual farms are not available, but it is safe to say that the variability among the individual farms would have been considerably greater than the variability of the state averages. In the period 1985-2002, annual average yields in the state ranged from 409 to 795 pounds an acre, or 386 pounds. In 1997, the average yield for the state was 597 pounds, but the range was from 347 pounds in Coffee County to 864 pounds in Monroe, or 517 pounds. If data on individual farms in Coffee County were available, it is likely that a few yields would have been below 50 pounds an acre. Further, yields from individual farms are averages representing estimates from all the acres on the farm. It is likely that yields of a few acre-sized plots on an individual farm would have been as low as five pounds an acre.

The point of these observations is that acre-to-acre, farm-to-farm, county-to-county, and year-to-year variation in cotton yields were and are extremely high in Alabama. The magnitude of the variability indicates that the genome or genetics of the cotton plant is/are not well adapted to the Alabama agricultural environment. This basic genome-environment incompatibility and the difficulty that Southern farmers had coping with it were at the heart of the enactment of the Agricultural Adjustment Act of 1933. Unfortunately, this basic incompatibility could not be legislated away.

Between 1992 and 1993, the average yield of cotton in Alabama declined by almost 47% (766 to 409 pounds an acre). While most year-to-year changes were not so great, declines of 20 to 30% were not unusual. It is difficult to imagine any business surviving with such large swings in annual production. From these observations, one has to assume that someone, somewhere was paying an exorbitant price to keep Alabama farmers in the cotton business in the period 1985-2002.

Bales Harvested

Acres harvested of Alabama cotton trended upward during the period 1985-2002 (Figure 16.39), while there was no apparent trend in yields (Figure 16.36). As a result of these relationships, bales harvested also generally trended upward (Figure 16.40). There was considerable year-to-year variation in bales harvested, primarily as a result of the year-to-year variation in yields. In the good years, bales harvested generally increased from around 550 to 900 bales, and in the poor years they increased from around 380 to 600 bales. Unfortunately, the poor years were much more numerous than the good ones (12 versus 6).

In 2002, farmers in 35 of Alabama's 67 counties planted less than 500 acres of cotton. In the remaining 32 counties, bales harvested ranged from 2,100 (Morgan) to 80,000 (Limestone). Some 53% of the counties reported less than 11,000 bales harvested, and 75% reported less than 19,500. Farmers in Limestone County harvested 14% of the state's cotton crop in 2002. Other counties reporting large shares included: Madison (12.3%), Lawrence (9.0), Colbert (6.3), and Lauderdale (4.7). These five counties accounted for 46.3% of the state's total crop. In 1985, these same five counties had accounted for 50.9%. Figure 16.4, shows the distribution of bales of cotton harvested in Alabama in 2001. Production was so poor in 2002 that the data did not represent a very accurate picture of the distribution of bales harvested in the state that year.

Production Costs

Goodman, et al. (2001) prepared enterprise budgets for the production of the conventional tillage of cotton for north, central, and south Alabama, in which they estimated the variable (operating) and fixed (allocated) costs for producing the crop in those regions of the state in 2001 (Table16.19). The authors note that these are estimates and are to be use for planning purposes only; however, they do provide the reader with a general idea of the various costs of producing cotton in Alabama. These authors use somewhat different terminology from Brooks (2001).

Gross Receipts. The authors estimated gross receipts by using yields of 750, 700, and 750 pounds of lint and 1200, 1120, and 1200 pounds of cottonseed an acre, respectively, in north, central, and south Alabama and by using prices of 58 cents and four cents, respectively, for lint and cottonseed. While their estimated price was realistic (Figure 16.34), the estimate for average yield was not. The average yield for cotton in Alabama during the period 1985-2002 was 547 pounds of lint an acre. Using their estimated yields and prices, the authors determined that gross receipts would have been $483.00, $450.80, and $483.00 an acre, respectively, for the state's three regions.

Variable Costs. The estimated variable costs of production per planted acre for the three regions (Table16.19) were similar, as might be expected, $352.19 in north Alabama, $361.24 in central Alabama, and $368.41 in south Alabama. These small differences were primarily the result of the higher cost for fertilizer ($46.67, $59.83, and $60.65, respectively) and for chemicals ($87.06, $88.56, and $91.56, respectively).

Fixed and Other Costs. The combined total for estimated fixed and other costs were nearly identical in the state's three regions (Table 16.19), ranging from $137.27 per planted acre in north Alabama to $138.40 in south Alabama.

Total of All Specified Expenses. Combining variable costs, fixed costs, and other costs resulted in estimates of $489.45, $499.15, and $506.80 per planted acre, respectively, in the three regions (Table 16.19).

Net Returns Above all Specified Expenses

When the total of all specified expenses per planted acre in Table 16.19 are used in comparison with gross receipts, estimates are negative (-$6.45, -$48.35, and -$23.80, respectively). If the authors had used more realistic estimates of yields, the net returns would have been much more dismal.

Effects of Yield and Price on Net Returns

Enterprise budget data, prepared by Goodman, et al. (2001), for the production of conventional tillage cotton in north Alabama (Table 16.19) were used as a basis for determining the effect of various combinations of yield and price on net returns. Data from that region were used for this exercise, because so much of the state's cotton is produced there. Those estimates are presented in Table 16.20. These data show that with the specified expenses shown in Table 16.19 farmers could expect a positive return only when yields and prices were unrealistically high. In the period 1985-2002, average cotton yields in Alabama were 600 pounds an acre, or above, in only seven of the 18 years (Figure 16.36). Average market prices were $0.64 a pound, or above, in only seven years (Figure 16.34).

Corn

In terms of acres harvested, corn has been the most important crop in the United States from the beginning years; however, in 1999 corn fell to second-place behind soybeans. By 2002, acres of soybeans harvested exceeded that of corn by three million (72.5 versus 69.3).

US Corn

Acres of American corn harvested increased rapidly from 57 million in 1970 to 70 million in 1976; by 1985, it had reached 75 million. In 1985, Iowa farmers harvested 19.2% of the national 8.9-billion-bushel corn crop. Other states reporting large shares included: Illinois (17.3%), Nebraska (10.7), Indiana (8.5), and Minnesota (8.2). These five states accounted for 64% of the total national crop. These same five states accounted for 65.7% of the total in 1970. Alabama farmers accounted for only 0.3% of the national corn crop.

Exports

Slow economic growth in the early 1980s and government policies that increased the price of corn—making it less competitive on the world market—resulted in a rapid decline in American corn exports. They were 1.2 billion bushels in 1985 (Figure 16.42), the lowest level since the early 1970s. After the passage of the 1985 farm bill, which reduced loan rates, corn prices became more competitive, and in the following four years exports increased rapidly to reach 2.4 billion bushels in 1989. In the early 1990s, as a result of the break-up of the Soviet Union and rising Chinese exports, American corn exports declined, and in the next four years fell back to 1.3 billion bushels in 1993. They recovered again in 1994 and 1995 when the Chinese again became net importers rather than exporters. Unfortunately, the Chinese did not remain out of the export market very long. They began exporting large quantities of corn again in 1996, and their exports, along with a downturn in economic activity in Asia, pushed American corn exports down to 1.5 billion bushels in 1997. After that year, competitive export pressure subsided somewhat, and American corn exports rebounded to between 1.9 and 2 billion bushels. They remained near that level through 2002, because the factors that determine export levels (bushels harvested, prices, competitors' production, prices, world demand for animal feeds, etc.) remained more or less unchanged.

PRICES

The government increased support prices (subsidies) rapidly from $1.35 in 1976 to $3.03 in 1984 (Figure 16.43), and kept them at that level through 1987. Afterwards, they were slowly reduced until they reached $2.75 in 1990. They were kept at that level until the program was eliminated after 1995. Loan rates were also increased after the mid-1970s, and in 1983 they reached $2.65 (Figure 16.43). Afterwards, they were slowly reduced, finally reaching $1.57 in 1990; then they were slowly increased again to $1.89 in 1994. They were kept at that level through 2001, when they were increased to $1.98.

The American market price for corn reached the unusually high level of $3.21 in the El Niño year, 1983, and then fell rapidly to $2.23 in 1985 (Figure 16.44) as bushels harvested increased (Figure 16.47). It continued to decline in 1986, reaching $1.50 a bushel when farmers harvested yet another large crop. The price had not been that low since the early 1970s. The market price rebounded to $1.94 in 1987 and $2.54 in 1988, as bushels harvested declined. It remained relatively constant at around $2.30 a bushel during the period 1989-1994, although it was somewhat lower than expected in 1992 ($2.07) when farmers harvested 9.5 billion bushels of the crop. It increased sharply in 1995 to $3.24 a bushel, in response to the smaller, wet-weather-stressed crop (7.4 billion bushels) that year. Then, beginning in 1996, prices began to fall rapidly as farmers harvested a series of better than 9.2-billion-bushel crops. The decline continued until 1999, finally reaching $1.82 in 1999. Declining production after 2000 resulted in slightly better prices in 2001 ($1.97) and 2002 ($2.32).

Market prices were below the support (target) price in 10 of 11 years during the period 1985-1995, but only four of 11 years for loan rates.

Acres Planted and Harvested

Acres of American corn planted were 83 million in 1985 (Figure 16.45), but in the next two years (1986 and 1987) it plummeted to 66 million, as acres planted followed prices downward (Figure 16.44). Acreage began to recover somewhat in 1988, but did not return to the trend of 79 million acres until 1991. Then in 1993, the disastrous Midwestern flood resulted in another significant reduction in acreage. Planting returned to normal in 1994, only to be kicked in the teeth again by too much spring rainfall in 1995. Acreage was again on trend in 1996, 1997, and 1998. Although 1999 was a good crop year in the Corn Belt, spring planting followed the 1998 price downward to 77 million acres. Market prices began to increase after 2000, but apparently farmers were content to keep their acreage below 80 million.

Data on acres of corn (grain) harvested are also presented in Figure 16.45. They indicate that American corn producers generally harvested around 91% of acres planted during the period 1985-2002. A considerable share of corn acreage is harvested each year for silage, rather than for grain.

Bt corn has been available for planting for several years, and it has revolutionized efforts to control the European and southwestern corn borer, both highly destructive pests. It is produced using the same genetic engineering process described in a preceding section for the production of Bt cotton. As a result of its effectiveness and low relative cost, Bt corn is used widely in American corn production; however, its use is causing a furor on both the national and international scenes. There is considerable concern, especially in Europe, about the safety of genetically-modified crops such as Bt corn—both in the environment and as ingredients of foods, especially those eaten by children and the elderly.

Another concern stems from the fact that the pollen from a Bt corn plant also contains the toxin, and when carried by the wind onto the leaves of other kinds of plants within and on the borders of fields of corn, can be toxic to moth or butterfly larvae feeding on those leaves. While this toxicity has been produced in the laboratory with larger amounts of pollen, it seems to pose little threat to these animals under normal field conditions.

The same gene that gives the cotton plant resistance to the herbicide glyphosate (described in a preceding section) has also been incorporated into the genome of the corn plant. It is marketed as Roundup Ready Corn, and is being used widely throughout the Corn Belt and other corn-growing regions of the United States.

Yields

During the period 1985-2000, American corn yields

continued to trend upward (Figure 16.46). It was 120 bushels an acre in 1985, and reached between 135 and 140 bushels at the end of the century. Yields were down significantly in 1988, 1991, 1993, 1995, and 2002, as a result of adverse weather (usually excessive rainfall). The 2002 decline was the result of an extensive drought over much of the United States from spring through fall.

Bushels Harvested

Although there was considerable year-to-year variation, acres of American corn harvested did not change very much during the period 1985-2000 (Figure 16.45). However, with the steady increase in yields throughout the period (Figure 16.46), bushels harvested trended upward from eight billion in 1985 to 10 billion in 2000 (Figure 16.47). It declined sharply during the period 1986-1988, and in 1991, 1993, 1995, 1999, 2001, and 2002, as a result of reduced acres planted and/or reduced yields resulting from adverse weather.

Iowa farmers harvested 21.5% of the nine billion-bushel national crop of corn in 2002. Other states reporting large shares included: Illinois (16.4%), Minnesota (11.7), Nebraska (10.5), and Indiana (7.0). These five states accounted for 67.1% of the national crop that year. The same five states had accounted for 65.7% in 1970 and 64% in 1985. In the early nineteenth century, the genome of the corn crop domesticated by Native Americans, probably in Central America, truly found a home on the old hunting grounds of the Cahokia, Ioway, Illiniwek, Kaskkaskia, Kickapoo, and Mesquakie peoples in the Northcentral states. The whole country and, indeed, much of the world has benefited immensely.

Production Costs

Foreman (2001) has published a detailed study on the cost of corn production in 1996, in different Agriculture Resource Management Survey (ARMS) regions of the United States (Figure 16.38). Some of the data from her study are shown in Tables 16.21 and 16.22. Additional data prepared by the Economic Research Service of the USDA were also used in preparing the table. Data for only two of the regions, Heartland and Southern Seaboard, are included in this section. These two regions were chosen because the Southern Seaboard region includes Alabama, and because the Heartland region includes the primary corn-producing areas of the country. The Southern Seaboard, however, does not include the eastern portion of the Tennessee Valley in Alabama.

Operating Costs. Itemized operating costs in the Heartland and Southern Seaboard regions on a per-acre basis are presented in Table 16.21. The summary of the data in Table 16.22 indicates that these costs were similar in both ($157.70 versus $159.59). Heartland corn farmers spent more for seed, chemicals, and fuel. Southern Seaboard farmers spent more for fertilizer, soil conditioners, repairs, and hired labor.

Comparing operating costs in the two regions on the basis of per-bushel costs presents a somewhat different picture. Average yields were 93 bushels an acre in Southern Seaboard and 138 bushels an acre in Heartland. When yields are included in the equation, Heartland operating costs are considerably lower ($1.14 versus $1.68 a bushel).

The difference in fertilizer costs in the two regions ($49.90 in Heartland versus $60.11 in Southern Seaboard) is important in that it highlights a key difference in the two regions' comparative advantage in corn production. The soils in the Heartland region belong primarily to two orders, Alfisols and Mollisols (Figure 3.2). Soils in both of these orders, in their natural condition, are relatively high in nutrients and contain large amounts of calcium and organic matter. They require limited amounts of commercial fertilizer for corn production. In contrast, most of the soils in the Southern Seaboard region belong to the Order Ultisol. These soils are highly weathered and leached. As a result, they are nutrient-poor and acidic. These soils require relatively large amounts of commercial fertilizer for corn production.

Allocated Costs. Data on specific items included in allocated costs are included in Table 16.21; a summary is presented in Table 16.22. These data indicate that there was somewhat more difference, relatively speaking, in allocated costs ($199.14 in Heartland versus $179.18 in Southern Seaboard). The primary source of this higher cost was the opportunity cost of land ($91.43 in Heartland versus $31.11 in Southern Seaboard). Costs to Southern Seaboard farmers were greater for capital recovery for machinery and equipment ($60.50 versus $72.57, respectively) and for taxes and insurance ($6.32

versus $8.94, respectively).

Total Production Costs. The difference in the total production costs (operating costs + allocated costs) between the two regions was somewhat greater than the difference in operating costs ($356.84 versus $335.77). This greater difference was primarily the result of the higher allocated costs, especially the greater opportunity cost for land, as detailed in the preceding paragraph. When total production costs are related to yield, the cost of producing a bushel of corn in Southern Seaboard was 39% greater than in Heartland ($3.61 versus $2.59).

Foreman (2001) concluded that differences in production costs of corn in the two regions were primarily the result of smaller farms and lower yields in Southern Seaboard. Of the two, the difference in yields is certainly the most important. It is more important because, first, it represents the greatest difference in the two regions, and, second, it is the difference that is the most difficult to remedy.

Returns to Management

Table 16.22 also contains data on the returns to management above operating costs (gross value of production – operating costs). Primarily as a result of the difference in yields, returns were $50 an acre greater in Heartland ($225.95 in Heartland versus $175.10 in Southern Seaboard). When total production costs (operating + allocated) were considered, returns were much smaller. In fact, they were negative in Southern Seaboard and less than $30 an acre in Heartland ($26.81 in Heartland and $-4.08 in Southern Seaboard).

In a longer-term study, the USDA's Economic Research Service published data giving the returns to management for corn production in the Southeast ERS Farm Resource Region during the period 1975-1995. Return estimates were obtained by comparing gross value of production (yield x price) with cash costs (generally operating expenses). During the period 1985-1995, returns varied between $-28.95 and $104.03. Returns were negative for only two of the 11 years. However, when gross value of production was compared to total production costs (operation + allocated), returns were negative for all 11 years.

Alabama Corn

Acres of corn planted in Alabama were 700,000 in 1970 and 880,000 in 1976. Thereafter, it generally trended down rapidly until it reached 370,000 acres in 1985. Yields generally increased throughout the period 1970-1985, from 40 to 70 bushels an acre. Although there was considerable year-to-year variation, bushels harvested for much of the period (10 of 16 years) varied from 24 million to 30 million bushels a year.

Farmers in Baldwin County harvested the largest share (10.2%) of the state's 1985, 24.4 million-bushel corn crop. Other counties reporting large shares included: DeKalb (6.7%), Jackson (5.7), Madison (5.4), and Escambia (4.4). These five counties accounted for 32.4% of the total crop. These relatively low percentages are indicative of the widespread distribution of corn production. A total of 11 of Alabama's 67 counties reported harvests of 750,000 or more bushels. Data on the distribution of bushels harvested within the state in 1985 are presented in Figure 16.48.

Prices

Alabama corn market prices usually followed the same trends as national prices during the period 1985-2002 (Figure 16.44). This is to be expected because the same factors that determine prices in the country also generally determine prices in the states. This is particularly true for a state like Alabama that produces such a small share of the national crop. Although the trends were similar except in a very few of the years, Alabama prices were higher, sometimes substantially so. These differences are primarily the result of federal regulations that adjusts the posted county price, which is based on the distance farmers in a county have to travel to deliver their grain to one of the designated terminals.

ACRES PLANTED AND HARVESTED

Acres of corn planted in Alabama followed declining market prices downward from 370,000 in 1985 to 260,000 in 1989 (Figure 16.49). Market prices increased in 1987 and 1988, to reach $2.75 cents a bushel (Figure 16.44), but began to decline slowly afterwards. Surprisingly, with market prices declining in 1989, acres planted increased in 1990. With market prices continuing to decline in 1990, farmers planted a smaller crop in 1991 (260,000 acres). But with market prices still declining in

1991, they planted 330,000 acres in 1992. Afterwards, however, Alabama corn farmers seemed to tire of chasing market prices, and they hardly seemed to notice the sharp increase in 1995 and 1996. Acres planted remained near the 300,000-acre level through 1998; then with prices falling, the farmers bailed out once more, and by 2001 acres planted were 180,000. It had never been that low since annual data were first reported in 1866.

Data on acres of corn harvested in Alabama during the period 1970-2002 are also presented in Figure 16.49. Throughout the period 1985-2002, farmers harvested around 88% of their planted acres, slightly less than the national average. Percentages harvested were considerably lower in those years when poor crop weather damaged crops in some fields so severely that it was impractical to harvest the grain.

Yields

Data on average annual corn yields in Alabama during the period 1970-2002 are also presented in Figure 16.46. Although there was considerable year-to-year variation, yields in the state during the period 1985-2002 generally trended up from 75 to 105 bushels an acre in the good years, and 45 to 85 bushels in the poor years. Unfortunately, during this period, the poor years were almost as common as the good ones.

Figure 16.50 presents data on yields of corn (bushels per acre) in Alabama and Iowa during the period 1985-2002. The data show that yields in Iowa were considerably higher in all the years of the period. In 1993, average yield in Iowa was 25 bushels higher; in 1998, the difference was 82 bushels. In 11 of the 18 years of the period, yields in Iowa were 50 bushels or more higher. With the comparative advantage that Iowa farmers have in producing corn, one is continuously amazed that any farmer would put a grain of seed corn in the ground in Alabama.

Bushels Harvested

Even though Alabama corn yields generally trended upward (Figure 16.51) during the period 1985-2002, they did not increase enough to offset the overall downward trend in acres harvested (Figure 16.49). As a result, bushels harvested trended downward during the period from 25 million bushels in 1985 to 15 million in 2002 (Figure 16.51). It was actually much lower (10.7 million) in the drought-year of 2000. Alabama's highest corn production during the period came in 1992 when 27.7 million bushels were harvested. That harvest represented 0.3% of the 9.5 billion-bushel national crop that year.

Farmers in Jackson County harvested the largest share (10.7%) of the 15.8- million-bushel Alabama corn crop in 2002. Other counties reporting large shares of the total crop included: Madison (8%), Lawrence (7.4), DeKalb (6.9), and Colbert (5.0). These five counties accounted for some 38% of the total crop that year. This relatively low percentage is indicative of the widespread distribution of corn production in the state. A total of 10 counties reported harvests of 500,000 or more bushels. Figure 16.52 shows the distribution of bushels harvested within the state in 2002.

Production Costs

Novak, et al. (2001) prepared enterprise budgets for the production of corn for grain (dry land production) in north, central, and south Alabama for 2001 (Table 16.23). In the publication, the authors remind us that these data are estimates and are to be used only for planning purposes.

Gross Receipts. The authors used average yields of 105 bushels an acre (north Alabama), 85 bushels an acre (central Alabama), and 100 bushels an acre (south Alabama) in preparing their enterprise budgets. These were unusually optimistic estimates. The average yield for Alabama during the period 1985-2002 was 76 bushels (Figure 16.50). The average corn yield for Alabama in 2001 was the highest on record (107 bushels an acre), but within the state, they varied between a low of 68 bushels an acre in Dale County (south Alabama) to a high of 147 bushels an acre in Cullman County (north Alabama). They also used $2.35 a bushel for price in their computations. This was a realistic estimate as shown in Figure 16.44. Using these data estimates for gross receipts (dollars an acre) in the state's three regions were: $246.75, $199.75, and $235.00, respectively.

Variable Costs. The total variable (operating) costs were relatively close in the three regions ($152.90 an acre in north Alabama, $150.99 in central Alabama, and $160.29 in south Alabama). Costs were slightly higher in south Alabama as a result of small differences in all categories, except seed.

Fixed and Other Costs. Fixed and other costs included the annual share of the cost of replacing tractors and machinery used in the production of the crop, general farm overhead, and labor (Table 16.23). The totals of these costs per acre were $88.80 (north Alabama), $88.67 (central Alabama), and $96.04 (south Alabama). The higher total for south Alabama was the result of the higher cost for replacing tractors and machinery, as well as the higher cost of labor in that region.

Total Specified Expenses. Total specified expenses per acre for the three regions were: $241.70 (north Alabama), $239.66 (central Alabama), and $256.33 (south Alabama) (Table 16.23). Total specified costs per bushel harvested were $3.77 ($256.33/68) for Dale County and $1.64 ($241.70/147) in Cullman County. Obviously, these costs of production estimates are unusually low because of the record yields in 2001. Alabama's average statewide corn yield during the period 1985-2002 was slightly less than 76 bushels an acre. With an average yield of this magnitude, costs of production for the entire period would have been nearer those of Dale rather than Cullman County.

Net Returns Above all Specified Expenses

Using data cited in the preceding paragraphs, Novak et al., (2001) estimated net returns (gross receipts – all specified expenses) in the state's three regions as follows: $5.03, -$39.90, and-$21.33 an acre, respectively. If the authors had used a more realistic yield in their computations, net returns would have been much more dismal.

Effect of Yields and Prices on Net Returns

Enterprise budget data in Tables 16.22 and 16.23 for north Alabama were used as a basis for determining the effect of changes in yield and price on net returns. Data from only that region were used for this exercise, because so much of the state's corn crop is produced there. The estimates are presented in Table16.24. These estimates emphasize the problems of producing corn in the state. According to the tabular data, a yield of 95 bushels an acre would result in a positive return only when the price was between $2.35 and $2.60. In the period 1985-2002, the average yield was equal to, or greater than, 95 bushels an acre in only three of the 18 years. Fortunately, the average price was $2.35 or greater in 13 years of the period.

It is obvious from the data presented in Table 16.24 that cost of production is not the primary problem in the production of corn in Alabama. The basic problem is low yields. In a global market that sets the price of the outputs—corn in this case, as well as the prices of inputs—low yields should result in money moving out of corn production. It is difficult to understand why this has not happened in Alabama. Of course, federal subsidies help some, but they supposedly help everyone producing corn.

Alabama Peanuts

At the end of the twentieth century, peanuts continued to be a minor crop nationally. The farm-level value was less than 5% of the value of corn production. However, the crop was a key contributor to local economies in nine states. A total of 55% of the crop is produced in the Southeast (Alabama, Florida, Georgia, and South Carolina). Approximately 70% of total domestic production is used for direct consumption (peanut butter, peanut candy, roasted peanuts, and snack peanuts). Some 18% is crushed for meal and oil, and 12% is used for seed and other uses (Dohlman, 2002).

As noted in the preceding chapter, acres of peanuts planted in Alabama remained generally unchanged in a range of 195,000 to 210,00 acres during the period 1970-1985. There was more year-to-year variation for the period 1981-1985, as farmers adjusted their planting to the new quota rules. They planted 200,00 acres in 1985.

US Exports

Dohlman (2002) commented that exports had long been an important market for American peanut producers. These exported peanuts were the "additionals" produced under the two-tiered program. After 1980, exports accounted for 14 to 25% of total production; however, during the period 1985-2002, China became an increasingly competitive exporter, primarily as a result of their lower production costs. Data presented in Figure 16.53 show that although there was considerable year-to-year variation, in the good years exports trended downward from one billion pounds in 1985 to 720 million in 2002. In the poor years, they declined from 950 million to near 500 million. The especially sharp declines in the 1993-1994, 1998-1999, and 2000-2001 marketing years were the result of severe drought that left fewer peanuts available for export.

Alabama Exports

Data on the exports of peanuts from Alabama for the period 1985-2002 is limited. Available data indicate that in the period 1996-2002 exports of the crop ranged from 5.4 to 8.3% of the export of all agricultural products from the state. The average for the period was 6.5% annually. During the period 1994-2002, the value of the export of peanuts and peanut products ranged from $21.2 million to $41.7 million, with an average of $29.1 million.

US Imports

Before the 1994 Uruguay Round Agreement on Agriculture (URRA) and the North American Free Trade Agreement (NAFTA), which became effective the same year, importing peanuts into the United States was limited by federal regulation to 1.7 million pounds (shelled basis) a year. This level, established by the Agricultural Adjustment Act (AAA) of 1933, was intended to prevent lower-priced, imported peanuts from destroying the shield protecting domestic peanut production located primarily in the Southeast. The import quota was increased sharply to 27 million pounds in the 1990-1991 market year as a result of a severe drought in the Southeast that reduced peanut yields to the lowest levels since 1980, another year with extremely low rainfall. A total of 27 million pounds was imported that year (Figure 16.54). Under the URRA and NAFTA, the United States opened its domestic markets to increasing quantities of imported peanuts and peanut butter. By the end of the century, exports reached 200 million pounds a year, a level equal to 5% of domestic production.

US and Alabama Prices

Data presented in Figure 16.55 show support prices for quota peanuts and loan rates for "additionals" for the period 1978-2001. Data presented in Figure 16.56 show the annual average market price that Alabama farmers received for all of their peanuts (quota and "additionals") during the period 1970-2002. Data for American farmers are shown in the same figure.

In 1984 and 1985, American farmers harvested more than four billion pounds of peanuts, the two largest crops in history. A large share of these enormous crops had to be marketed as "additionals" at a lower price. As a result, the average market price received for all peanuts by Alabama farmers fell sharply to between 22 and 23 cents a pound. With the market price of peanuts down in 1985, and the ratio of the indices for prices received to prices paid continuing to trend downward (Figure 16.10), the government began to slowly increase the support price for quota peanuts until it finally reached 34 cents in 1995. Generally, Alabama market prices for all peanuts followed support prices upward, reaching 32 cents in 1994. After 1993, peanut exports began to decline sharply (Figure 16.53), and after 1994 imports began to increase. Then in 1994, American market prices began to decline, and in 1995 Alabama market prices also began to fall. In 1995, the government also reduced the support price for quota peanuts. After 1998, with exports below 700 million pounds and imports above 140 million pounds, market prices in both Alabama and the country as a whole went into free fall. By 2001, both were below 19 cents a pound, but in Alabama it was actually just slightly above 16 cents. The peanut—one of the five Southern belles that had been assiduously protected by congressmen from the region for almost 70 years—had finally been dragged into the real world, and unfortunately there no longer seemed to be anything they could do about it.

The Farm Security and Rural Investment Act of 2002 effectively ended the quota system for peanuts. It also provided a mechanism whereby owners of quotas would be reimbursed (the Peanut Buyout Program). The end of the quota system will certainly have an effect on peanut prices in the future. Under that act, the quota system has been eliminated. Peanuts will be treated the same way as other program crops, such as grains and cotton. Peanut farmers will participate in a system with marketing-assistance loan provisions, with a loan rate of $350 a ton; fixed, decoupled payments, and counter-cyclical payments. How this new world will affect the slightly wizened, older Southern lady is not certain; but with her old suitors now older and wiser, but still with considerable power in the halls of Congress, she is not likely to "go gentle into that good night."

Acres Planted and Harvested

Alabama farmers planted 201,000 acres of peanuts in 1985 (Figure 16.57). Then with market prices for all peanuts increasing rapidly (Figure 16.56), acres planted also increased rapidly, reaching 278,000 acres in 1991. Unfortunately, market prices had begun to decline after 1988. Alabama farmers did not immediately respond, but after 1991 they did so with a vengeance. By 1996, acres planted were down to 192,000. Surprisingly, after 1996, with market prices still falling, acres planted in the state began to stabilize, and from 1997 though 2002, they remained around 190,000 acres. Throughout this period 1970-2002, acres planted by Georgia farmers totaled well over twice the level in Alabama.

Data on acres of peanuts harvested in Alabama during the period 1970-2002 are also presented in Figure 16.57. The data show that in most years of the period Alabama farmers harvested around 99% of their acres planted. Exceptions were in 2000 and 2002. In 2000, a severe drought damaged a substantial share of the crop to the point that it was impractical to harvest it. In 2002, heavy rains around harvest time also resulted in severe damage to some of the crop.

Alabama and Georgia Yields

The average peanut yield in Alabama in 1985 was 2,950 pounds an acre (Figure 16.58); however, in 1986 it fell to 2,260, and in 1987 it was down to 2,115 pounds. Then for the next 16 years, with the exception of 1992 and 2001, yields were not above 2,400 pounds as farmers experienced increasing problems with diseases. Also, adverse weather wreaked havoc on Alabama peanut yields during some years of the period. Poor crop weather, generally drought, seriously reduced yields in six years of the 18-year period (1985-2002) and reduced them to a lesser degree in three additional years. There was little indication of an upward trend in average yields during the period. In fact, there is more indication of a slight downward trend.

Data on peanut yields in Georgia are also presented in Figure 16.58. Generally, year-to-year trends in yields were similar in both states, but in every year, especially after 1993, yields in Georgia were higher. The difference ranged from five pounds in 1993 to 1,210 pounds in 2000. In the 18-year period, yields in Georgia were 400 or more pounds higher. The primary reason for the differences between the two states in recent years has been the increased use of irrigation for peanut production in Georgia. For example, in 1998 Georgia farmers irrigated 68.4% of their peanuts. In the same year, Alabama farmers irrigated only 46.3%.

Pounds Harvested

In 1985, there were 590 million pounds of peanuts harvested in Alabama, among the highest in history (Figure 16.59). However, for the remainder of the period 1986-2000, with the exception of 1991 and 1992 when acres harvested were unusually high, pounds of peanuts harvested trended downward, as both acres harvested (Figure 16.57) and yields (Figure 16.58) declined. It finally declined to 380 million pounds in 2002. The especially poor yields in 1990, 1997, and 2000 are quite noticeable in the pounds harvested data. Alabama pounds of peanuts harvested reached its highest level during the period 1985-2002 in 1991 (638 million pounds). This large crop represented 13% of the 4.9 billion-pound national crop that year.

Farmers in Houston County harvested the largest share (19.7%) of Alabama's 379.2-million-pound 2002 peanut crop. Other counties reporting large shares included: Geneva (15%), Henry (13.2), Coffee(12.0), and Baldwin (11.1). These five counties accounted for 71% of the state's total crop that year.

Production Costs

Crews, et al. (2001) published information on estimated costs and returns from the production of peanuts (dry land) in Alabama in 2001. Some of their data are shown in Table 16.25.

Gross Receipts. In the enterprise budget prepared by Crews, et al. (2001), they used estimates of pounds harvested on a typical Alabama peanut farm of 1,880 pounds of quota peanuts and 620 pounds of additionals. They also used 30.5 cents and 16.2 cents a pound, respectively, as prices for the two classes. With this combination of yields and prices, gross receipts totaled $673.44 an acre.

Variable Costs. The per planted acre costs associated

with the use of chemicals accounted for the major share (38.8%) of estimated variable costs for producing peanuts ($183.61 of $472.60). The costs related to the control of pests and pathogens in Alabama peanuts are excessive. Because of its growth and reproduction habit, the peanut plant is at risk for a long period of time. Further, the hot, humid weather in southeast Alabama provides ideal environmental conditions for the development of harmful organisms. Budget costs of producing peanuts in Texas in 2002 estimated the costs of chemicals at $27.91 a planted acre, or 15.2% of the comparative Alabama cost. Crop insurance for peanuts ($35.00) and drying and cleaning also contribute significantly to these costs. Altogether, the total variable costs for producing an acre of peanuts were $472.60. These costs were over $100 an acre greater than for cotton ($472.60 versus $360.61—the Alabama average), which is Alabama's second most expensive crop to produce (Table 16.19).

Fixed Costs. Fixed costs associated with the depreciation of tractors and machinery ($122.92) and labor were also relatively high compared to the costs of those items production of other crops (Table 16.25). It was 50% higher than the depreciation of tractors and equipment required for the production of cotton, which indicates that the production of peanuts requires more equipment than the production of cotton. General farm overhead ($33.08) was also higher than for all other major crops produced in the state. The total for all fixed costs was $156.00.

Other Costs. Labor ($61.45) is the only item included in this budget category. It was also 50% higher than the comparable costs for producing cotton.

All Specified Expenses

The total of all specified expenses was $690.05 per planted acre. This total for Alabama was more than twice as large as the same costs for producing peanuts in Texas ($690.05 versus $313.32). It is also $200 an acre greater than the costs of producing cotton ($690.05 versus $498.47).

Net Returns Above all Specified Expenses

With gross receipts of $673.44, this hypothetical farmer could have easily met his variable costs of $472.99, but not all specified costs of $690.06. Considering all costs, the net returns would have been -$16.61. With this combination of yields, prices, and the ratio of quota peanuts to additionals (3:1), net returns would have been barely positive only if the price for additionals was at least 19.1 cents a pound.

The federal program to provide a safety net for Southern peanut farmers is probably the best example of sharecropping with the guvmint and of the Triangle gone berserk. It also lends credence to the old adage that "The road to Hell is paved with good intentions." It was put in place in the 1930s when Southern farmers desperately needed help, but like many efforts to do good, its unintended consequences became an albatross around the neck of American agriculture. Everyone knew that in the long term it was indefensible. It finally became the poster-child of Congressional efforts to protect a specific group of farmers against the inevitable effects of comparative advantage. In the end, only 16% of the people who owned peanut quotas actually grew peanuts. They owned the land that held the quota, but they did not farm. Instead they leased the quota to someone who wanted to grow the crop. In addition to the increasingly indefensible quota system, the program also guaranteed the farmers a price for their quota peanuts that was far above the world price. In the end, globalization forced Dobson's (1985) Triangle of beneficiaries to accept change, if not reality.

Soybeans

Soybeans are the poster child for the introduction of non-native crops. Introduced relatively recently, acres of soybeans harvested finally exceeded that of corn in the United States in the late 1990s. It is the almost perfect crop. Yields are extremely high, the protein in its seed is of excellent quality, and its above-ground structure, when dried, makes an excellent hay. Further, it produces its own nitrogen. If there is a downside. it is soybean's genome, which was largely determined in the temperate regions of Asia and is not really adaptable beyond the temperate regions of the United States.

US Soybeans

American soybean acreage increased rapidly from

4.3 million acres in 1970 to 70 million in 1979. However, from that peak it generally trended downward to 63 million acres in 1985. Changes in yields were not so spectacular as were the changes in acreage. Farmers harvested 27 bushels an acre in 1970 and 32 in 1985. With the relatively small increase in yields and the up-and-down cycles in acreage, bushels harvested increased by one billion (1.1 to 2.1 billion) in the period 1970-1885. Soybean market prices were $2.80 a bushel in 1970. They increased rapidly thereafter, reaching around $6.50 in the mid-1970s. They remained near that level until 1985.

In 1985, Illinois farmers harvested 18.2% of the 2.1-billion-bushel national soybean crop. Other states reporting large shares included: Iowa (14.8%), Indiana (8.8), Missouri (8.6), and Ohio (7.6). These five states accounted for 58% of the total crop. Alabama farmers harvested only 1.3%.

Exports

The high value of the American dollar compared to other world currencies resulted in the volume of soybean exports falling sharply to 600 million bushels in 1984 (Figure 16.60). The 1985 farm bill contained provisions to improve the competitiveness of American agricultural exports. As a result of some of its provisions and a generally more expansive world economy, soybean exports increased to 740 million bushels in 1985. They increased slowly to 757 million in 1986, and on upward to 802 million in 1987. In 1988, severe flooding in the Midwest resulted in a significant reduction in acres planted. The reduced planting also resulted in a sharp increase in the domestic price for the crop that year. As a result, exports fell to 590 million bushels, the lowest level since the mid-1970s, and they remained near that level through 1990.

Exports began to recover in 1991, and in 1992 reached 770 million bushels. Exports plummeted to below 600 million bushels in 1993 when American production fell again, as a result of more flooding in the Midwest and drought in the South. It recovered in 1994 (800 million) and continued to increase in 1995 (851 million) and 1996 (882 million), as a result of surging world demand and reduced global competition for export markets. In 1997, exports turned downward again, when the combination of tight American supplies, a strong American dollar, and economic slowdown in East and Southeast Asia resulted in lower demand. It was sharply lower in 1998, as a result of the almost total loss of the crop in the South to drought. Exports recovered quickly to set new records in 1999 (973 million bushels), 2000 (996 million), and 2001 (1.4 billion). Adverse crop weather resulted in the decline of bushels harvested by 100 million bushels in 2002, and exports declined slightly.

Prices

Federal support prices for soybeans for the period 1970-2001 are presented in Figure 16.61. It was set at $5.02 a bushel in 1985, but then slowly reduced to $4.50 in 1990, as market prices (Figure 16.62) increased rapidly. With market prices lower after 1988, the support price was raised back to $5.02 in 1991, and it remained near that level through 1996. After that year, market prices began to fall, and in 1997 the support price was increased to $5.26. It remained at that level through 2001. In 2002, the Farm Security and Investment Act re-established a target price for soybeans at $5.80 a bushel, and set a loan rate at $5.00.

In 1985, American farmers harvested 2.1 billion bushels of soybeans, the third largest crop in history to that time (Figure 16.65). With that large crop, the market price fell to $4.78 a bushel. It had not been that low since 1975. After such a low price, bushels harvested declined slightly in 1986. Then from 1987 through 1994, both bushels harvested and market prices remained relatively stable at around 1.9 billion and $5.75, respectively, except in 1988 when bushels harvested was unusually low and in 1993 when it was slightly lower than expected. After 1995, demand for soybeans began to increase, and by 1995 the market price was up to $6.72; bushels harvested quickly followed, reaching 2.4 billion in 1996. Unfortunately, 2.4 billion bushels of soybeans was a bit too much for the market to absorb at a market price of $7.35. Consequently, it began to fall rapidly, and by 2001 it was down to $4.38. Bushels harvested finally declined in 2002, and the market price increased to $5.53.

Except in 1986 and four years at the end of the period, market prices were considerably higher than the support prices.

Acres Planted and Harvested

As soybean acreage continued to trend downward from the highs of the early 1980s, American farmers planted 63.1 million acres of the crop in 1985 (Figure 16.63). Afterwards, acres planted generally followed market prices downward to 57.8 million acres in 1990. Then with market prices increasing, acres planted also increased. In 1996, the market price peaked at $7.35 a bushel before beginning to decline. In 1996, acres planted reached 62.2 million acres. Then, surprisingly, instead of following market prices downward, acres planted continued to increase and reached almost 74.5 million acres in 2000. It remained near that level through 2002.

Figure 16.63 also presents data on acres of soybeans harvested in the United States during the period 1985-2002. These data show that American farmers harvested 97 to 98% of their acres planted throughout the period, except in 1993, 1998, 1999, and 2000—when adverse crop weather reduced yields so much in some areas that it was impractical to harvest some of the fields.

The same herbicide-resistant gene that has been added to the genomes of cotton and corn plants (described in preceding sections) has been added to the genome of the soybean plant. Weed control among soybean plants with this gene can be managed so that fewer herbicide treatments are required, saving farmers both time and money. As a result, less pesticide is released into the environment. Approximately 65% of all soybeans planted in 2000 contained the gene.

Yields

The annual average national soybean yield was 34 bushels an acre in 1985 (Figure 16.64), and it remained around that level through 1991, as a series of years with somewhat adverse weather resulted in relatively poor growing conditions. Afterwards, however, yields seemed to trend upward slightly, finally reaching 40 bushels an acre in 2001. Yields deviated from this general trend in 1988 when a drought and heat wave in the Central and Eastern United States seriously reduced bushels harvested. Adverse weather also resulted in somewhat reduced yields in 1993, 1995, and 1999.

Bushels Harvested

With both yields (Figure 16.64) and acres harvested (Figure 16.63) trending upward after 1990, bushels harvested followed essentially the same pattern. Yields remained near 1.9 billion bushels from 1986 through 1991. Afterwards, they climbed steadily to reach 2.9 billion bushels in 2001. Exceptions to this general pattern were in 1988, 1993, 1995, 1999, 2000, and 2002 when yields were somewhat lower than expected as a result of adverse crop weather.

In 2002, Iowa farmers harvested the largest share (18.1%) of the national 3.7-billion-bushel crop of soybeans. Other states reporting large shares included: Illinois (16.4%), Minnesota (11.2), Indiana (8.7), and Nebraska (6.4). These five states accounted for 60.8% of the total crop. Alabama farmers harvested 0.13% of the 2002 national crop.

Production Costs

For the period 1997-2000, the USDA'S Economic Research Service developed data on the cost of soybean production in different Agriculture Resource Management Survey (ARMS) regions of the country. Data from this publication are presented for the Heartland and the Southern Seaboard in Table 16.26. These two regions were chosen because the Southern Seaboard includes most of Alabama, and because the Heartland includes the country's primary soybean-producing areas. The Southern Seaboard does not include the eastern portion of the Tennessee Valley. All of the data from the four different years have been averaged for each of the two regions.

Operating Costs. The data indicate that—except for fertilizer ($7.23 in Heartland versus $22.10 in Southern Seaboard)—there was relatively little difference in the operating costs required to produce an acre of soybeans in the two regions.The effect of the higher cost of fertilizer is apparent when total operating costs are computed on a per-acre basis (Table 16.27) ($75.89 versus $91.19, respectively). This difference becomes relatively larger when operating costs are computed on a per-bushel, rather than a per-acre basis. Average soybean yields for the period 1997-2000 were 44 bushels an acre in Heartland and 34 bushels an acre in Southern Seaboard. As a result of the difference in yields, per-bushel operating costs were over 55% higher in Southern Seaboard ($2.68 versus $1.72).

Generally, fertilizer costs for soybean production are lower than for either corn or cotton. A source of nitrogen is not required in commercial fertilizers for soybeans, because microorganisms within specialized structures on the roots of the plants are able to fix atmospheric nitrogen.

Allocated Costs. In the USDA (ARMS) study, several items within the allocated costs category were higher in Heartland than in Southern Seaboard (Table 16.26), but the most striking difference was in the much higher opportunity cost of land (rental rate). It was $84.49 per planted acre in Heartland and $34.70 in Southern Seaboard. As a result of this difference, when allocated costs are computed on a per-acre basis, they are 34% higher for Heartland ($177.52 versus $132.40). However, when total allocated costs are computed on a per-bushel basis, much of the difference disappears ($4.03 versus $3.89).

Total Production Costs. Primarily as a result of the higher opportunity cost for land (Table 16.27), total production costs were higher in Heartland ($253.42 versus $223.82 an acre); however, when total production costs are calculated on a per-bushel basis, the higher yield in Heartland reverses the ranking of the regions. On a per-bushel basis, total production costs were 82 cents higher in Southern Seaboard ($6.58 versus $5.76 a bushel).

Returns to Management

Table 16.27 also contains data on the returns to management above total operating costs. Primarily as a result of the difference in yields, returns were $63 an acre greater in Heartland ($152.93 in Heartland versus $90.10 in Southern Seaboard). However, when all production costs (operating + allocated) were considered, average returns were negative ($-26.60 an acre in Heartland versus $-42.53 an acre in Southern Seaboard).

Alabama Soybeans

Acres of soybeans planted in Alabama increased sharply from 625,000 acres in 1970 to 2.2 million in 1979 and 1980. Afterwards, it declined just as sharply to reach 1.1 million in 1985. Average annual yields did not increase during the period 1970-1985. With no increase in yields, bushels harvested followed acres harvested up and down from 13.8 million in 1970, up to 53.8 million in 1979, then down to 27.8 million in 1985.

In 1985, Alabama farmers harvested 1.3% of the 2.1-million-bushel national soybean crop. That year farmers in Baldwin County harvested the largest share (8%) of the 27.8-million-bushel Alabama crop. Other counties reporting large shares included: Madison (7%), Perry (4.6), Dallas (4.4), and Jackson (3.8). These five counties accounted for only 27.8% of Alabama's total crop. This relatively low percentage is indicative of the widespread production of soybeans in the state that year. A total of 19 counties reported harvests of 500,000 or more bushels. Figure 16.66 shows the distribution of bushels harvested within the state in 1985.

Exports

Data on the exports of soybeans from Alabama for the period 1985-2002 are limited. Available data indicate that for the period 1996-2002 exports of the oilseed ranged from 1.7 to 7.7% of total agricultural exports from the state. The average for the period was 4.1%. The value of the export of soybeans and soybean products ranged from $7.1 million to $41.6 million during the period 1994-2002. The average was $20.1 million. Values tended to decline during the period.

Prices

Prices for soybeans in Alabama closely followed national trends during the period 1985-2002 (Figure16.62). The state produced such a small portion of the national crop during those years that even extremely large swings in bushels harvested in Alabama, for whatever reason, had little effect on national or even Alabama prices.

Acres Planted and Harvested

In 1980 acres of soybeans planted in Alabama totaled 2.2 million, but acreage plummeted as the cold reality of the state's lack of comparative advantage became apparent. By 1985, acreage of soybeans was down to 1.1 million. Acres planted continued in free fall until it reached 650,000 acres in 1986. Afterwards, the decline continued, but at a lower rate, before finally reaching 240,000 acres in 1995. As a result of the increase in market prices in the mid-1990s (Figure 16.62), acres planted increased slightly in 1996 and 1997. With market prices declining in the late 1990s, acres planted began to fall again, and finally reached 140,000 acres in 2001. Between 1980 and 2001, Alabama lost over two million acres of soybeans.

Figure 16.67 also includes data on acres of soybeans harvested. These data indicated that in good years Alabama farmers harvested 94% of acres planted, a slightly lower percentage than in the country as a whole. In poor years, when adverse crop weather destroyed a portion of the crop, the percentage was considerably lower. For example, in the severe drought year of 2000, Alabama farmers harvested only 84.2% of acres planted.

Yields

Soybean yields for Alabama probably trended upward slightly in the period 1985-2002, but the year-to-year variation was so great that it is difficult to be certain (Figure 16.64). During the period, annual average yields ranged from 16 (1999) to 34 bushels (1996). If only yields in the years with good weather for soybean production (1985, 1992, 1994, and 1996) are considered separately, they did trend upward from near 27 bushels an acre to 35 bushels an acre in 2002. Unfortunately, in the other 13 years of the period, there is no indication of an upward trend. Yields obtained in these poor years explain why acres planted in Alabama plunged from 2.2 million acres in 1980 to 170,000 in 2002. Yields were much lower than expected in 1998, 1999, and 2000, primarily as a result of low soil moisture during critical periods in the production cycle (Table 16.15).

Data in Figure 18.58 also show that Alabama soybean yields were considerably lower than those in the country as a whole during the period 1985-2002. If Alabama yields are compared to those in Iowa, Illinois, or Indiana, the contrast was even more dismal.

It is obvious from these data that the soybean genome is not well-adapted for Alabama. For example, the 18-year (1985-2002) average annual soybean yield for Iowa was almost twice as high as for Alabama (42.4 versus 24.2 bushels an acre), but the variation associated with the Alabama average was almost twice as large as the variation associated with the Iowa average. The coefficient of variation for the Alabama and Iowa averages was 22.5 versus 12.2%, respectively.

Bushels Harvested

With little or no change in soybean yields in the state during the period 1985-2002 (Figure 16.64), bushels harvested (16.65) closely followed the negative exponential trends in acres harvested (Figure 16.67). As a result, bushels harvested fell sharply from 27.8 million in 1985 to 14 million in 1986. Afterwards, they continued to decline, but the rate was much lower. In 2002, Alabama farmers harvested only 3.7 million bushels; it had not been that low since 1962. The 2002 crop represented only 0.1% of the national crop of 3.7 billion bushels.

Farmers in Jackson County harvested 15.9% of the state's 3.7-million bushel soybean crop in 2002. Other counties reporting large shares included: Limestone (12.6%), Madison (12.3), Baldwin (7.5), and Lauderdale (7.2). A total of 52 counties reported less than 500 acres planted. Figure 16.69 shows the distribution of bushels harvested within the state in 2001. Data from 2001 was chosen for this figure because crop weather was so poor in 2002.

Production Costs

Goodman, et al. (2001A) published information on estimated costs and returns for the production of soybeans (dry land) in Alabama in 2001. Some of their data is shown in Tables 16.28 and 16.29.

Gross Receipts. Goodman, et al. used a yield estimate of 34 bushels an acre and a price of $5.50 a bushel for their 2001 costs and returns estimates for soybeans in Alabama. Consequently, gross receipts were $187.00 (34 bushels x $5.50 a bushel) (Table16.29). The author's estimate for average yield was unusually optimistic. Only in one year (1996) in the entire period 1985-2002 did the average yield in Alabama reach 34 bushels an acre (Figure 16.64). However, the estimate of market price was probably somewhat low (Figure 16.62).

Variable Costs. Data on estimated variable costs for producing soybeans in Alabama show that the total was $107.58 an acre and that the costs associated with seed, fertilizer and lime, and chemicals were similar.

Fixed and Other Costs. Totals for these two categories were $73.04 and $19.31, respectively. Costs associated with the replacement of tractors and machinery accounted for most of the fixed costs, while the cost of labor accounted for most of the "other" costs.

Net Returns above all specified expenses

Using this gross receipts estimate, net returns above total variable costs would have been $79.52; however, when all specified expenses ($199.93) were considered, return was -$12.93.

The Goodson, et al. (2001A) data presented in Table 16.29 demonstrate the basic problem with the production of soybeans in Alabama. In the period 1985-2002, the annual average yield in the state was 24 bushels an acre. With all specified expenses at $199.93 an acre and with a yield of 24 bushels an acre, Alabama farmers could not have earned a positive net return from the production of soybeans unless the average price was at least $8.33 a bushel. Market prices never reached this level in the entire period. Net returns from other combinations of yields and all specified expenses are presented in Table 16.30.

Alabama Hay

In 2000, American farmers harvested 152 million tons of hay from 59.8 million acres of land. Hay production trailed only wheat and corn in acres harvested. In 2000, Texas with 8.9 million tons harvested, California with 8.6, South Dakota with 7.4, Minnesota with 6.8, and Kentucky with 6.2 were the leading hay-growing states in the country. The total value of production of the national crop that year was $11.4 billion. About 70% of total production was used on the farms where it was grown. Hay production varies so widely across the country that it did not seem useful to compare it with production of the crop in Alabama, as has been the practice with most of the other crops.

Acres harvested of hay in Alabama increased from 535,000 acres in 1970 to near 700,000 acres in 1985, reversing a long period of decline during the 1950s and 1960s. Hay yields remained around 1.7 tons an acre during the period 1970-1983, although there was considerable year-to-year variation. After 1983, they began to increase and reached two tons an acre for the first time in history. The average was 2.2 tons an acre in 1985. With acres harvested increasing, tons harvested also increased, growing from 900,000 tons in 1970 to 1.5 million tons in 1985. Alabama hay prices increased rapidly from $28 a ton in 1970 to $56 in 1977; thereafter, though increases were more moderate, it still continued to grow. The price in 1985 was $63 a ton.

In 1985, farmers in Montgomery County harvested the largest share (5.4%) of Alabama's 1.5-million-bale hay crop. Other counties reporting large shares included: Cullman (4.7%), Perry (2.9), Lawrence (2.8), and Lowndes (2.7). These five counties accounted for only 18.5% of the total crop. This relative low percentage is indicative of the widespread production of hay in the state in 1985. A total of 23 counties reported harvests of at least 10,000 bales. Figure 16.70 shows the distribution of tons harvested in Alabama in 1985.

Prices

Alabama hay prices had trended upward to $63 a ton in 1985 (Figure 16.71). Then from 1985 through 1997, except for the unusual case in 1993 when it reached $79, they trended downward, as tons harvested increased (Figure 16.74). Tons harvested declined modestly in 1998, and prices began to increase, reaching $53 that year and $56 in 1999. As a result of the severe drought of 2000, Alabama's hay yield fell to 1.8 tons, but it did not seem to affect the price, which remained near $56 in 2000, 2001, and 2002.

Acres Harvested

Alabama farmers harvested 700,000 acres of hay in 1985 (Figure 16.72), and even though prices remained stagnant through 1997, acres harvested generally trended slightly upward. They reached 770 million acres in 1997. After prices began to increase in 1998, acres harvested also increased, reaching 920,000 in 2001, before falling to 825,000 in a modestly severe crop year in 2002. It was also much lower than expected in 2000, a severe drought year.

Yields

Alabama's hay yields increased rapidly in the early 1980s to reach 2.2 tons an acre in 1985 (Figure 16.73). Although there was considerable year-to-year variation, except for several years with adverse crop weather, yields generally remained near that level through 2002. Extremely low yields in 1986, 1990, and 2000 were the result of severe drought throughout the state. At the other extreme, 1994 was an exceptionally good year for crop production. As a result, the yield that year (2.7 tons an acre) was the highest ever recorded. The year 2001 was also a good crop year, with an average yield of 2.6 tons an acre, the second highest ever recorded.

Tons Harvested

As a result of the upward trend in acres harvested

(Figure 16.72), tons harvested also trended upward during the period 1985-2002, from 1.5 million to 1.8 million tons (Figure 16.74). The poor yields resulting from extreme drought in 1986, 1990, and 2000 are evident in the production data, as is the record yield in 1994. This general increase in hay production during this period is interesting: Alabama cattle numbers decreased some 20%, and farmers may have been holding their animals longer on hay, attempting to make their individual animals more valuable in a generally negative cattle production environment.

In 2002, farmers in Cullman County harvested the largest share (6%) of the state's almost 1.8-million-ton hay crop. Other counties reporting large shares included: DeKalb (4.6%), Marshall (3.9), Jackson (3.3), and Blount (3.1). These five counties accounted for only 20.9% of the total. This relatively low percentage is indicative of the widespread production of hay in the state. A total of 40 counties reported harvests of 20,000 or more tons. Figure 16.76 shows the distribution of tons harvested within the state in 2001. Crop weather was extremely poor in 2002 and skewed the distribution of production to a considerable degree.

PRODUCTION Costs

In 2002, Prevatt and Ball (2002) published data on the estimated costs and returns associated with the production of fescue for hay in Alabama in the crop year 2001-2002. The authors remind us that these data are estimates only and should be used only for planning purposes. Some of their data are presented in Tables 16.31, 16.32, and 16.33.

Gross Receipts. Prevatt and Ball (2002) used four tons for the estimate of tons harvested and $55 for the estimate of price in their publication on the costs and returns of fescue hay production in Alabama. With this combination of yield and market price, gross receipts totaled $220 an acre. The yield value in this estimate seems unusually optimistic. The highest yield recorded in the period 1985-2002 was only 2.7 tons an acre (Figure 16.73). The estimate of $55 a ton seemed to be very close to the actual data (Figure 16.71).

Variable Costs. The larger share of variable costs required for the production of an acre of hay is represented by the cost of fertilizer and lime ($75.93 of $105.43) (Table 16.31). Generally, the production of hay in Alabama requires greater expenditures for fertilizer and lime than for other crops. The relatively high cost of fertilizer is not surprising. Hay is usually cut several times a year. Generally, cutting results in rapid regrowth of the plant, which, in turn, requires extra plant nutrients. The cost of the operation of tractors and machinery is also relatively high ($27.70 of $105.43). Production requires the use of several items of specialized equipment including: mowers, windrowers, balers, and bale wagons. Further, these specialized machines plus tractors and trucks are used several times during the season. While the cost of tractor and machinery operation per acre is relatively high compared to that of the other grass crops (wheat - $10.59, corn - $16.50, and even soybeans -$18.75), it is relatively modest compared to the cost incurred for their use in the production of cotton ($46.49) and peanuts ($59.01).

Fixed and Other Costs. The amount of money that must be set aside to replace worn-out tractors and machinery constitutes the larger share ($56.50 of $79.10) of fixed costs (Table 16.31). Because of the amount of these types of equipment required, this is one of the larger of the variable costs. A single cutting of hay is a relatively labor-intensive operation, and when it is repeated several times a year, the cost of labor and total fixed costs increase proportionately ($58.88).

NET RETURNS OVER ALL SPECIFIED EXPENSES

With estimates—gross receipts of $220 (4 tons x $55 a ton) (Table 16.32) and of all specified expenses of $243.40—net returns would have been -$23.41. Gross receipts were more than adequate to cover all variable costs ($220 versus $105.43), but not all specified expenses ($220 versus $243.11).

Using the costs data presented in Table 16.32, Prevatt and Ball (2002) estimated net returns from several combinations of yields and prices. In these computations, all production costs are held constant. The results are presented in Table 16.33. Although these results are based on estimates, they are probably realistic enough to show that it is difficult to realize a positive net return producing fescue hay in Alabama. In the 18 years of the period 1970-2002, the annual average yield of hay was above 2.5 tons only two years, and it never reached three tons. With yields of two and three tons, there was no price between $35 and $75 a ton that would result in a positive net return. In good crop years, a lot of hay can be produced in Alabama, but it certainly is not inexpensive.

Alabama Winter Wheat

In 1985, Kansas farmers harvested 23.7% of the country's 1.8-billion-bushel winter wheat crop. Other states reporting large shares included: Texas (10.2%), Oklahoma (9.0), Colorado (7.4), and Washington (6.4). These five states accounted for 56.7% of the country's total crop. Alabama farmers harvested 0.7% of the 1985 total national winter wheat crop. In 1970, they had harvested only 0.2%.

According to the 1987 *US Census of Agriculture*, Alabama farmers harvested winter wheat from 7% of its harvested cropland that year. Percentages for soybeans, cotton, and corn were 25.8%, 15.5, and 10.5, respectively. Farmers in Baldwin County harvested the largest share (14.2%) of the state's 5.3-million-bushel crop that year. These five counties accounted for 36.7% of the total crop. This relatively low percentage is indicative of the widespread production of the crop in the state. A total of 20 counties reported harvests of 100,000 or more bushels. Figure 16.76 shows the distribution of bushels of winter wheat harvested in Alabama in 1985.

US EXPORTS

American wheat (all) exports were at the bottom of a steep decline in 1985 (Figure 16.77). Exports had not been lower since 1971. In 1986, with national bushels harvested at its lowest level in seven years, with world stocks decreasing, and with interest rates in the United States at 5.5%, exports began to increase, but only until 1987 when they reached 1.6 billion bushels. In 1988, they began another downward slide that continued until 1990 (1.1 billion bushels), as world supply exceeded demand. In 1991 and 1992, exports rallied somewhat as world and American production continued to decline and as world consumption began to eliminate the surplus. Then in 1992, world production literally exploded, and exports began to decline once more, to reach one billion bushels (the lowest level since 1986) in 1996. In the final years of the twentieth century, wheat exports increased minimally to 1.1 billion bushels; but afterwards they gave up those small gains, and in 2002 reached between 800 and 900 million bushels. Wheat exports had not been that low since the early 1970s.

World supply and demand were at the heart of the American wheat export cycles throughout the period 1985-2002, but there was another factor also involved, especially in the 1990s. The world wheat market was becoming increasing competitive. Canada, Argentina, Australia, and especially the European Union were increasing market share at the expense of the United States. Experts believe that this is a long-term situation and unlikely to be reversed for many years, if ever. Further, the 2002 farm bill, which increased subsidies for grain farmers, may encourage governments throughout the world to do the same. This increased protection from the rigors of the marketplace would likely reduce world trade in wheat.

US and Alabama Prices

Data on support prices and loan rates for American wheat are presented in Figure 16.78. The government had rapidly increased the support price for wheat in the latter part of the 1970s and the beginning of the 1980s. It was $4.38 a bushel in 1985. After 1987, market prices increased sharply (Figure 16.79), and in 1988 the support price was slowly lowered and finally set at $4 in 1990. It remained at that level through 1995, when the support price program was eliminated. The loan rate was also increased after the mid-1970s, and it reached $3.65 in 1983. Afterwards, it was lowered to $3.30 in 1985, and finally to $1.95 in 1990.

Data on market prices for wheat in both the country and the state for the period 1970-1985 are presented in Figure 16.79. In 1985, the annual, average market price of a bushel of American winter wheat was $2.98. Bushels of winter wheat harvested in the United States began to increase rapidly after 1979, and in 1981 market prices began to decline, reaching $2.33 in 1986. Bushels harvested declined sharply between 1985 and 1986, and in 1987 market prices began to increase again. In 1989, it reached $3.78. In 1990, American farmers harvested two billion bushels of the crop, and market price fell sharply to $2.62. In 1991, as a result of that decline in the market price, American farmers harvested only 1.4 billion bushels, and market price responded by increasing to $2.92. Market prices generally teeter tottered with bales harvested for the remainder of the period. In 1995, it was $4.41, but four years later, in 1999, it was down to $2.29. That low price really got the attention of the farmers, and they began to reduce production. By 2002, it was down to 1.1 billion bushels. Of course, as bushels harvested declined, market prices increased and in 2002, reached $3.41. You just simply cannot outsmart the teeter totter.

Market prices for the state's and the country's winter wheat generally followed the same trends throughout the period, but in most years they were slightly lower in the state (Figure 16.79).

Generally, the American market price for winter wheat remained well below the support price for the entire period 1985-2002, except in 1995 and 1996 when farmers harvested two relatively small crops, and market prices increased sharply as a result. Market prices remained well above the loan rate throughout the period, except in 1998 and 1999 when two unusually large crops were harvested.

Acres Planted and Harvested

Data on acres planted and acres harvested of winter wheat in Alabama during the period 1970-2002 are presented in Figure 16.80. In 1985, Alabama farmers planted 500,000 acres of winter wheat. They had planted 850,000 acres in 1982. Acres planted in 1985 would be the largest for the remainder of the twentieth century and the beginning of the twenty-first, as Alabama farmers seemed to lose some interest in the crop. Throughout most of the 1970s, acres harvested had ranged between 100,000 and 200,000 acres. After 1990, it returned to that level, and generally remained there through 2002. The downward trend was interrupted, temporarily, during the period 1988-1990, when American wheat market prices were up sharply (Figure 16.79). Surprisingly, the upsurge in market prices in 1995 and 1996 had little, if any, effect on acres planted. Most Alabama winter wheat farmers had already made the decision to get out of the business of growing the crop, and those high prices were not going to lure them back into it. They did respond to the unusually high market prices in 1995 and 1996 by slightly increasing acres planted, but acres harvested remained at the same level.

Data on acres harvested are also included in Figure 16.80. These data indicate that in most years, Alabama farmers harvested around two-thirds of their acres planted. Alabama farmers have always used a considerable share of their acres planted for grazing or as a cover crop.

The slight decline in acres planted in 1996 has an interesting history. In 1996, USDA became concerned about the potential for an outbreak of the Karnal bunt disease in Alabama wheat. As a result, they threatened to quarantine the sale of wheat outside the state. Because of their concern about the marketing of their 1996-1997 crop, many Alabama farmers did not plant in the fall of 1996. In the following years, fears of effects of the disease were allayed, and acreage returned to more normal levels in subsequent years.

US and Alabama Yields

Figure 16.81 includes data on the annual average yields of winter wheat in the country and the state during the period 1970-2002. The most obvious characteristic of the data is the high degree of year-to-year variation. Yet, even with this level of variation, average yield seems to have generally trended upward. Considering Alabama yields in the good years (1988, 1992, 1994, and 2000), yields seemed to have increased from 40 bushels an acre to 54 bushels; however, using the poor years (1986, 1989,1993, 1995, and 1998), the increase would have been only for around 26 to 42 bushels. The good years were well above the national average. Unfortunately, there were altogether too many of the poor years, and those poor years drove most Alabama wheat farmers out of the business after 1980. In the 18-year period (1985-2002), Alabama yields were lower than national yields in 10 of those years.

The wheat yield for 2000 demonstrates the unique climate-genome relationship of the crop in Alabama. The yield of 54 bushels in 2000 was the highest ever recorded in the state. Yet, yields of virtually all of the other field crops were severely reduced by the extreme drought that plagued the entire state for most of the year. Winter wheat benefits from the crop's growth and seed development taking place during the part of the year with the highest soil moisture. Even in a severe drought year, there was sufficient moisture available during the late winter and early spring months for the plants to germinate, grow, and mature.

Bushels Harvested

Bushels of winter wheat harvested in Alabama (Figure 16.82) during the period 1985-2002 usually followed the year-to-year changes in acres harvested (Figure 16.80). The slight upward trend in yields (Figure 16.81) seemed to have had little effect. Bushels harvested was highest in 1985 (12.8 million). Thereafter, it generally trended downward to 3.6 million in 1996. It remained around that level through 2002, although increasing yields toward the end of the 1990s did seem to boost it slightly.

Farmers in Madison County harvested the largest share (12.5%) of the 2.4-million-bushel wheat crop in 2002. Other counties reporting significant shares included: Limestone (11.2%), Baldwin (10.5), Talladega (5.5), and Lauderdale (5.3). These five counties accounted for 45% of the state's total crop. Figure 16.83 shows the distribution of bushels of wheat harvested in Alabama in 2001. Crop weather was extremely poor in 2002, skewing to a considerable degree the distribution of production. The 2001 data present a more realistic picture.

Alabama Oats

In 1985, Alabama farmers harvested oats from 35,000 acres (Figure 16.84). It is difficult to imagine that in 1954 they harvested the crop from 185,000 acres in a desperate attempt to find a substitute for cotton. The oat plant genome was poorly suited for the Alabama agricultural environment, but it took progressive globilization within the country, plus the demise of animal power on farms, to emphasize the comparative disadvantage of raising oats in Alabama.

Alabama's acres harvested have trended downward from 40,000 in the mid-1980s to 20,000 acres in 1999 (Figure 16.84). The downward trend nationally had been equally discouraging. During the same period, acres harvested in the country had trended down from nine million to between two and three million. In the meantime, imports of oats has increased from 30 million bushels in the mid-1980s to 100 million bushels in the late 1990s. As a result of the continuing decline in acres harvested, the Alabama Agricultural Statistics Service no longer reported statewide data on oat production after 1994, and no data at all after 1999.

The primary problem with the crop from the beginning was the relatively low and highly erratic yields (Figure 16.85). Between 1985 and 1999, the average yield in Alabama exceeded the national average in only three of 18 years, and since 1970 in only four of 30 years. Between 1985 and 1999, yields ranged between 35 and 60 bushels an acre. The other major problem was the result of relatively high variable costs of production. The amount of fertilizer required is high compared to areas where the majority of the country's oats are produced. Further, in Alabama's hot and humid climate there are a number of devastating diseases that must be dealt with; thus the cost of chemicals in the production cycle is also relatively high.

Alabama Sweet Potatoes

The period 1985-2002 saw a mainstay of the diets of early Alabama families fall on hard times. In the early years of the Great Depression, Alabama farmers harvested sweet potatoes from 105,000 acres. Afterwards, acres harvested began to trend downward. In the late 1970s and early 1980s, the crop was being harvested from only around 5,500 acres of cropland (Figure 16.86). Then in the current period (1985-2002), it fell to 2,700 acres in 2002. Alabama families continue to have plenty of excellent sweet potatoes available to them in their local supermarkets, but very few of them are harvested in Alabama. More than likely, the sweet potatoes purchased in Alabama supermarkets were harvested either in Louisiana or North Carolina. For the last several years, farmers in those two states have harvested almost two-thirds of the country's total crop. It would be interesting to know what gives those two states such a powerful comparative advantage in the production of sweet potatoes.

Alabama Irish Potatoes

Evidence has continued to accumulate that this period, which I have titled "Sharecropping with the Guvmint," has been an unmitigated disaster for crop production in Alabama. Unfortunately, the Irish potato is another victim. As noted in preceding chapters, Irish potatoes were never as important in the diets of Alabama families as sweet potatoes. In 1890, when Alabama farmers harvested 58,000 acres of sweet potatoes, they only harvested 6,000 acres of Irish potatoes.

In 1985, Alabama farmers harvested Irish potatoes from 13,000 acres, and acres harvested remained around that level until 1990, when it began to decline sharply (Figure 16.87). It continued to decline at a rate of about 6.7% a year to reach 3,300 acres in 2002. Although Irish potatoes are produced throughout the United States, for many years Idaho farmers have harvested the largest share of the crop, generally around 30%. Acres harvested in other states are much smaller.

Alabama Sorghum

In 1985, American farmers harvested 1.1 billion bushels of sorghum grain. Kansas farmers harvested the largest share (26.5%). Farmers in Texas (21.6%), Nebraska (13.8), Missouri (10.4), and Arizona (5.9) also harvested large shares. These five states accounted for 78.2% of the country's total crop. Alabama farmers harvested 1.13% of the national crop that year. In 2002, bushels harvested in the United States had fallen to 360 million, and Kansas and Texas still accounted for most of it: 34.7 and 33.9%, respectively. By 2002, the Alabama share had fallen to 0.1% of the national crop.

Sorghum harvested for grain, once looked on ostensibly with great promise, is another of Alabama's field crops that has fallen on hard times in the last 18 years (1985-2002). It was harvested for the production of molasses by some of the earliest English settlers in the state, but it was not harvested for grain until the early 1940s. Before 1981, Alabama farmers had never harvested more than 42,000 acres (1955) for grain. In 1981, the ill-fated Alabama Cooperative Service program began to encourage farmers to significantly increase their production of sorghum grain. As a result, in 1985 state farmers harvested 240,000 acres of the crop (Figure 16.88), but they quickly found that there was no one to purchase the grain from that acreage. By 1987, acres harvested was back to around 30,000, about where it had been since 1970. Afterwards, unfortunately, it began to trend downward, reaching 19,000 acres in 1993. After that year, the Alabama Agricultural Statistics Service no longer reported statistics on the crop on a statewide basis. In 1996, acres harvested fell to 10,000 acres, and by 1998 it was down to 6,000. After 1999, the Statistics Service no longer published any information on the crop. The data in Figure 16.88, after 1999, was obtained from NASS.

The reason for the demise of sorghum harvested for grain in Alaama is obvious in the data presented in Figure 16.89, which gives data on the annual average yields in the United States and Alabama for the period 1970-2002. In 17 of 18 years in the period 1985-2002, the average yield in Alabama was lower than the national average. For the entire period, the average American yields were slightly over 17 bushels an acre higher than the corresponding Alabama yields.

Obviously, low yields are not a recent problem with this crop. As early as 1957, it should have been obvious that Alabama farmers would never be able to competitively grow sorghum for grain. In that year, national and state yields were 29 and 19 bushels an acre, respectively. With this data available to them, it is difficult to understand why the farmers ever thought that they could compete.

Alabama Vegetables

In 1985, the value added to the national economy by the production of vegetables was almost $8.6 billion. The Alabama share was 1.1% of the total. By 2002, national value added had increased to $17.2 billion nationally, but the Alabama share had fallen to 0.2%. Data presented in Figure 16.90 shows the annual contribution of Alabama vegetable production to the national economy during the period 1970-2002. Throughout most of the 1980s and into the early 1990s, value added by Alabama vegetables ranged from $80 to $95 million, but after 1995 it began to fall sharply. By 2002, it was down to $40.1 million. Data presented in several preceding sections of this chapter showed that the production of most of Alabama's field crops had declined substantially during the period 1985-2002. The data presented in Figure 16.90 indicate that the production of vegetables did not escape the disease either.

The Alabama Agricultural Statistics Service has published data on the production of tomatoes and sweet corn for the period 1985-2002, and for watermelons for the period 1992-2002. Data on the acres harvested of those vegetable, for those periods, are shown in Table 16.34. The data indicate that farmers harvested considerably more acres of watermelons, and that acres harvested generally trended downward from around 8,000 acres to 4,000 acres during the period. Farmers also harvested slightly fewer acres of tomatoes compared to sweet corn, and acres harvested of both of those crops trended downward as well.

Data on the value of tomatoes, sweet corn, and watermelons harvested during the period 1985-2002 are presented in Table 16.35. These data show that fluctuation in prices for the different vegetables resulted in a different picture for value than for acres harvested. There was considerable year-to-year variation in the value of watermelons during the period, but they did not appear to trend downward to the same degree as acres harvested. The value of tomatoes increased substantially during the middle of the period and remained relatively high through the end of it. The value of sweet corn did appear to trend downward, although it recovered somewhat in 2001 and 2002.

Alabama Pecans

According to the Alabama Agricultural Statistics Service, in 1985 Alabama ranked fourth nationally in pecan production, with 7.6% of total national production, but by 2002 Alabama production had fallen to seventh, with only 2.9% of national production. In 2002, pecans were harvested in commercial quantities in 16 of Alabama's 67 counties, but virtually all of them were produced south of a line from Sumter County to Russell County; within that region, Baldwin County (58%) and Mobile County (20%) produced 78% of all the pecans grown in the state.

Data on the annual production of pecans in Alabama during the period 1970-2002 are presented in Figure 16.91. Generally, pounds harvested ranged between five million and 26 million pounds during the period 1985-2002; but in nine of the 18 years, it ranged between 10 million and 20 million. The sharp declines in pounds harvested in 1988, 1990, 1992, and 1994 were the result of alternate-year bearing (masting). The lower than expected harvest in 1995 was the result of the depredation of Hurricane Opal, which toppled large numbers of mature pecan trees. Two years later, in 1997, Hurricane Danny blew large numbers of almost mature nuts off the trees. Then the following year (1998), Hurricane Georges severely damaged the crop. The 2002 crop was probably low as a result of masting, but the heavy rains after September certainly did not help yield. It seems that if Alabama pecan producers did not have bad luck, they would not have any luck at all.

As might be expected with such high variation in production, cash receipts would be equally variable. In the period 1985-2001, they varied from a low of $3.2 million (1994) to a high of $15.4 million (1991).

Alabama Peaches

According to the Alabama Agriculture Statistics Service, in 2002 Alabama ranked ninth among all states in peach production, with 0.8% of the national crop. Data on annual pounds of peaches harvested in Alabama for the period 1970-2002 are shown in Figure 16.92. Data in the figure demonstrate the susceptibility of this crop to late winter and early spring freezes. Pounds harvested were reduced by the effects of this weather phenomenon in 1985, 1986, 1987, 1990, 1992, 1993, and 1996. The freezes were much more common in the first half (1985-1993) of the period. Reductions in the crop were especially severe in 1985, 1986, and 1996. In 2000, pounds harvested were reduced by a severe drought during the fruiting period. Altogether, pounds harvested were reduced significantly in nine of 18 years during the 1985-2002 period. Excepting those poor crops, normal production seemed to vary between 15 million and 22 million pounds a year. In 2002, farmers in Chilton County harvested 65% of Alabama's 19-million-pound peach crop. Limestone and Blount counties were a distant second and third with 7.4 and 6%, respectively.

Cash receipts from the sale of peaches were highly variable as a result of the effect of the freezes on production. During the period 1985-2002, receipts varied from a low of $200,000 (1996) to a high of $9.2 million in

2001. Receipts in 2001, were the highest in history.

Alabama Blueberries

The Alabama Agricultural Statistics Service did not begin publishing specific data on blueberries until 1992. Previously, they were included with other crops. In 2002, the service reported that Alabama ranked 11th among all states in blueberry production, with 0.3% of the total national crop. Cash receipts for blueberries harvested (used portion) on Alabama farms comprised 0.1% ($377,000) of total cash receipts for all plant crops originating on Alabama farms, including farm forest products. Obviously, blueberries are not very important in Alabama on a statewide basis, but they are of considerable importance on some of the sandy soils in the Southern Pine Hills (Figure 1.9).

Pounds harvested of blueberries (used portion) varied considerably during the period 1992-2001 (Figure 16.93). This crop is similar to peaches with regard to susceptibility to late-winter and early-spring freezes and to early-season drought. The effect of inclement weather on pounds harvested is obvious in the figure. Pounds harvested were much lower than expected in 1992 (360,000 pounds), 1993 (48,000 pounds), 1996 (390,000 pounds), 1998 (500,000 pounds), and 2000 (450,000 pounds). The first three poor years were the result of freezes. The severe winter storm of 1993 was especially damaging to the blueberry crop. The last two poor years were the result of early droughts.

Cash receipts for blueberries was $377,000 in 2002. They had been slightly higher in 1995 ($457,000) and lower in 1992 ($308,000); 1992 was the first year that blueberry data were included in the annual bulletins of the Alabama Agricultural Statistics Service. Lowest cash receipts ($47,000) were received in 1993, the year of the severe winter storm.

Alabama Floriculture

According to *Census of Agriculture* data, in 2002 there were 21,728 farms in the United States that produced bedding and garden plants, cut flowers and cut florist greens, or foliage plants and potted flowering plants under glass. A total of 363 (1.7%) of those farms were in Alabama. The largest share (16.5%) of these farms was in Mobile County. Other counties reporting large shares included; Baldwin (7.4%), Cullman (6.1), Calhoun (5.2), and Madison and Jefferson (3.6% each). These six counties accounted for 42.4% of the total. Of Alabama's 67 counties, 51 reported at least one farm producing the plants and plant products.

In 2002, there were 14,579 American farms producing these plants in the open. Some 406(2.8%) were in Alabama. The 2002 Census did not include any information of the value of products in the country or in the state. The most recent financial data for the country is the 1997 Census.

The Alabama floriculture industry has grown impressively in the last decade. The Alabama Agricultural Statistics Service has published data indicating that in 1987 there were 125 growers of floriculture crops in the state (Figure 16.94). By 1993, the number had increased to 205; however, in the following four years, the number declined to 161 (1997). It recovered somewhat in 1998, 1999, and 2000, reaching 191 growers in 2000. Afterwards, it began to decline again, finally reaching 152 growers in 2002.

The wholesale value of the sale of Alabama's floriculture crops was $22.6 million in 1987 (Figure 16.95). Afterwards, it generally trended upward to $82.2 million in 1999. Since then, it has been somewhat lower: $80 million in 2000, $59.1 million in 2001, and $66.2 million in 2002. Wholesale value increased impressively between 1987 and 1999 ($22.6 million to $82.2 million, an increase of 264%). During the same period, the Consumer Price Index increased by only 58%.

Although no single floriculture product has contributed a large share of the total value of crop production in the state, the combined value of all floriculture products is important to Alabama's agricultural economy, and the increase in the past decade has been especially impressive. In 1987, the value of the sale of floriculture crops accounted for 2.8% of the value of all crops, excepting the value of farm forest products. By 2002, this share had increased to 12.3%.

Table 16.36 presents data on the number of containers (pots or flats) of six important floricultural products sold by Alabama growers in 1987, 1990, 1995, and 2002. These data show that in terms of the number of containers sold, poinsettias were most important in all of the years. Sales of hardy/garden mums were about half as great. In three of the products (poinsettia, hardy/garden mums, and geranium cuttings), where all four years of data are presented, number sold increased. Numbers of Easter lilies sold were variable, but did not seem to increase. As would be expected, wholesale value of sales increased for poinsettia, hardy/garden mums, and geranium cuttings in the four years (Table 16.37). In 2002, the value for poinsettia was $5.4 million, somewhat less than values for sweet potatoes ($7.8 million), wheat ($7.4 million), and Irish potatoes ($6 million).

Alabama Nursery Crops

The Alabama Agricultural Statistics Service has not included specific information regarding the production and sale of Alabama's nursery plants in its annual bulletins. Fortunately, the *US Census of Agriculture* began to report some of this information for the state in 1978. Unfortunately, the 2002 Census contains very little specific information. Table 16.38 includes information on the sale of nursery plants in Alabama in the census years of 1978, 1982, 1987, 1992, 1997, and 2002, along with the number of Alabama producers in those years.

As is also shown in Table 16.38, the number of nurseries in the state increased almost 97% (189 to 372) from 1978 to 1997. Much of that increase came in the period 1978-1992. There was almost no change in number from 1992 to 1997 (297 to 298), but in 2002 the number increased to 372. Table 16.39 includes the names of the counties with the largest number of nursery crop producers in the different years. Of course, Mobile County is the primary source of nursery crops in the state, but by 2002 grower numbers in some of the north Alabama counties were beginning to increase.

The sale of nursery plants increased from $12.230 million in 1978 to $63.543 million in 1997 (Table16.38), a better than fivefold increase. The increase in the Consumer Price Index accounted for only about a third of the total increase in sales. The nursery business grew rapidly in Alabama from 1978 through 1997. Surprisingly, however, it expanded relatively little from 1992 to 1997. Sales increased only $1.8 million ($61.7 million in 1992 to $63.5 million in 1997). In the previous five years, sales had increased $9.3 million ($53.4 million in 1987 to $61.7 million in 1992).

Apparently, little data is being collected systematically on the relative quantities of the different nursery crops being produced in Alabama; however Dr. Jeff Sibley (personal communication), assistant professor of horticulture at Auburn University, has identified the six leading woody nursery crops produced in the state. In order of importance, they are: azaleas, hollies (*Ilex sp.*), junipers (*Junipererus sp.*), maples (*Acer sp.*), Indian hawthorne (*Rapliolepis sp.*), and cleyera (*Cleyera japonica*).

Alabama Sod

According to data presented in the *2002 Census of Agriculture*, farmers on 4,956 American farms harvested sod from 463,631 acres. Acres of sod harvested by Alabama farmers on 96 farms accounted for 5.6% of the national total.

The Alabama Agricultural Statistics Service has not included specific information regarding the production and sale of sod in the state in its bulletins. Fortunately, the *US Census of Agriculture* began to report this information for the state in 1978. Table 18.41 includes information on the sale of sod in Alabama in census years 1978, 1982, 1987, 1992, and 1997, along with the number of Alabama producers in those years.

The number of Alabama sod producers doubled (48 to 96) from 1978 to 2002 (Table 16.40). This increase is even more remarkable considering that Alabama farm numbers decreased substantially during that period. Sod production was one of the few bright spots in an otherwise dismal plant crop production picture during that period.

The value of the sale of sod in the state increased from $3.2 million in 1978 to $37.9 million in 1997, an elevenfold increase (Table 16.40). The increase in the Consumer Price Index only accounted for 10% of the increase in sales. The rate of increase was much greater ($3.2 million in 1978 versus $37.9 million in 1997) than the increase in the number of producers over that period. This is a phenomenal rate of increase. By 1997, sod was one of the more important crops in Alabama. Cash receipts for sod ($37.9 million) were greater than those for hay ($19.8 million) or for wheat ($12.2 million).

Table 16.41 lists the most important sod-producing counties in the state in census years 1982, 1987, 1992, 1997, and 2002. There were no county data listed in 1978. The data in the table indicate something of the spread of sod production both numerically and geographically in the state during this period. In 1982, only two counties had three or more producers. In 2002, ten counties had four or more producers, and 24 had from one to three producers. In 1982, the major producing counties were located in the central portion of the state. By 2002, they were scattered widely across the southern portion, with a large concentration of 21 producers in Baldwin County.

Alabama Timber

US census data indicated that lumber production in the country increased from 34.7 billion board feet in 1970 to 36.4 billion in 1985. In the period 1985-2002, production continued to trend upward, reaching 46.5 billion board feet in 2002 (Table 16.42). During this period, production increased by 28%.

Data presented in other chapters show that from around 1925 onward (Table 16.42) the largest share of lumber produced in the United States was produced in the West Region. The share from that region continued to increase until around 1960. In that year, mills in the region produced 59.7% of the total. Afterwards, its share began to decline, and by 2002 only 38.1% of the national total was produced there. Even with the decline, however, the West continued to produce the country's largest share. With the decline in production in the West, the total shares in the other regions began to increase.

Data published by the Alabama Agricultural Statistics Service indicate that in 1980 cash receipts from the sale of Alabama farm forest products ($82.1 million) and non-farm commercial timber ($197.6 million) totaled $279.7 million. By 1985, these totals had increased to $93.8 million and $283.9 million, respectively, ($377.7 million total). In 1980, cash receipts from the sale of all timber accounted for 13.7% of the sale of all agricultural commodities in the state, plus all timber. By 1985, this percentage had increased to 16.2%.

In 2002, the cash receipts received from the sale of farm forest products ($148.5 million) and non-farm commercial timber ($587.3 million) totaled $735.8 million. In that year, cash receipts from the sale of all timber accounted for 19.9% of the sale of all agricultural commodities plus the sale of all timber.

Land In Forests

There were 21.7 million acres of forestland in Alabama in 1982. In 1990, the US Forest Service completed the *Sixth Survey of Alabama Forest Resources* (McWilliams, 1992). This survey estimated that the area of forestland had increased 240,000 acres to 22 million acres, or 67.6% of the entire area of the state (Table 16.43).

The Seventh Survey of Alabama Forest Resources was conducted in 2000 (Hartsell and Brown, 2002). This survey indicated that the area of forested land in the state had increased since the 1990 survey to 71% of its total area. The 2002, Alabama Forest Resources Report included data showing that the area in commercial forests in the state accounted for 70.6% of the total land area. This estimate is slightly smaller than the 2000 estimate (70.8%); nevertheless, the discrepancies are not particularly significant.

The Alabama Forestry Commission (2002) published information on the estimate of commercial forestland in each of Alabama's 67 counties. Table 16.44 presents data on the total amount of forestland in the counties in each of the state's six agriculture reporting districts (Figure 16.96). Because there are different numbers of counties in each district, the average acreage of commercial forestland per county in each district provides the best picture of the distribution of forestland in the state. As would be expected, the lowest average (228.7 thousand acres per county) was in District 10. It includes much of the land in the Tennessee Valley, where a significant share of Alabama's cropland is located. The second lowest county average (264.9 thousand acres) was in District 20 in northeast Alabama. The eastern end of the Tennessee Valley also contains a substantial amount of farmland.

Counties in District 50 had the highest average (540.4 thousand acres). Most of that district is located in the Southern Pine Hills Physiographic District (Figure 1.9).

In 2002, this district contained only a small share of the state's cropland. The second highest county average (366.1 thousand acres) was in District 30—which includes large shares of the Fall Line Hills Physiographic District (Figure 1.9), the Piedmont Uplands Physiographic Section (Figure 1.13), and the Alabama Ridge and Valley Physiographic Section (Figure 1.9). There is relatively little farmland remaining in the district, except in the Coosa Valley Physiographic District (Figure 1.9).

Forest Ownership

In 1982, some 94.6% of Alabama's forestland was privately owned (Table 16.45). The 1990 survey (McWilliams, 1992) indicated that there had been only a slight change in that total, and the 2000 survey found private forest ownership at essentially the same level (near 95%). However, while there was little change in the totals, there were some important changes within the private-ownership category. Farmer-owned forestland declined from 1982 to 1990 (27.1 to 22.7%), as land in farms declined 1.7 million acres. The 2000 survey did not report a percentage for farmer-owned forestland. However, the downward trend certainly continued because of the continued loss of land in farms. The percentage of forestland owned by the forest industry declined from 20.6 to 16.3% during the period 1982-2000. Forestland owned by individuals, other than farmers, increased from 1982 to 1990 (38.5 to 41.9 %); it certainly must have continued to increase through 2000, although the 2000 survey did not include data on that class of owners. The increase in ownership by non-farming individuals is not unexpected. When farmers leave farming, if they retain the land, they become individual owners in the non-industrial private class.

Forest Types

The *1982 Survey of Alabama Forest Resources* (Rudis, et al., 1984) reported that oaks in combination with other species (oak-pine, oak-hickory, and oak-gum-cypress) comprised 66% of the forestland found in Alabama woodlands (Table 16.46). Of these three forest-type groups, the oak-hickory combination was dominant (33.6%). Combinations of pines (longleaf-slash and loblolly-shortleaf) comprised 34%. The loblolly-shortleaf (27%) forest-type was the dominant group in the pines combinations. Figure 16.97 shows the general distribution of the different types of forests within the state.

Data reported in the 2000 survey (Hartsell and Brown, 2002) indicate that there had been a clear and consistent increase of the amount of forestland with stocks of the loblolly-shortleaf pine forest-type during the period 1982-2000 (27 to 30.5%) (Table 16.46). These data also indicate that this increase came at the expense of declines in the areas stocked with the longleaf-slash pine (6.8 to 4.7%) and the oak-pine (21.1 to 18.3%) forest-types. Most of these changes came as a result of converting the areas with longleaf-slash pine and oak-pine forest types to loblolly pine plantations by planting seedlings.

Tree Plantations

Planting trees on Alabama's non-industrial private land increased from 223,608 acres in 1985 to 403,365 in 1988 (Figure 16.98). The especially large increase in 1986 was the result of the passage of the Food Security Act of 1985. Title XII of this act established the Conservation Reserve Program (CRP). Under the provisions of the CRP, farm owners, operators, and tenants were provided with an annual, per-acre rent, plus half the cost of establishing permanent grass or tree cover on highly-erodeable or environmentally-sensitive cropland, and maintaining the established cover for 10 to 15 years. The enrollment mandate of the act was for 40-45 million acres nationwide. Of course, not all CRP land was planted with trees. McWilliams (1992) suggested that the provisions of the 1985 CRP probably accounted for a third of the increase in tree planting.

McWilliams commented further that in 1990 there were 3.4 million acres of planted pine in Alabama, an increase of 81% since 1982. In that year, planted stands comprised 46% of the state's pine-type timberland. Between 1972 and 1982, an average of 61,700 acres were planted each year; however, in the period 1982-1990 the annual rate of planting increased threefold to 184,700 acres a year.

Pine planting fell sharply after 1988, and remained at around 250 thousand acres a year during the period 1990-1996 (Figure 16.98). In 1997, planting literally exploded upward to reach 437,500 acres, the highest level in history at that time. A primary reason for this large increase was the passage of the Federal Agriculture Improvement Act of 1996. The new act continued the CRP at a maximum national enrollment of 36.4 million acres at any one time through 2002.

McWilliams (1992) further commented that while plantations were playing a significant role in the management of the state's timber resources, naturally-established pine stands "have been the mainstay of Alabama's forest resources throughout history and continue to be so today, despite the recent 25-percent decrease in area."

Volume of Softwood and Hardwood Growing Stock

Table 16.47 presents information on the volume (millions of cubic feet) of various softwood species in Alabama woodlands in 1982, 1990, and 2000. As indicated by data in the table, softwood stocks in the state increased moderately from 1982 to 2000 (11.6 to 12.6 million cubic feet), probably as a result of the widespread planting of softwood seedlings in the state. The data also indicate that softwood stock were dominated by loblolly pine. In each of the three years, growing stocks of this species comprised more than 50% of the total for all species. Further, growing stocks of the loblolly increased dramatically (34%) during the period 1982-2000. Most of the seedling planting programs used this species. Growing stocks of longleaf and shortleaf pine decreased during this period. The decrease in shortleaf stocks was especially large (two to 1.2 million cubic feet). The widespread distribution of the little-leaf disease in this species probably contributed to the reduction. Surprisingly, stocks of cypress increased.

Table 16.48 includes information on the volume of growing stocks of various species of commercial hardwood tree species. As the data in the table indicate, the increase of the growing stocks of hardwood during the period 1982-2000 was even more dramatic (9.7 to 15.2 million cubic feet) than the increase in the volume of softwood detailed in the preceding paragraph. No single species of hardwood dominated this group in the manner that the loblolly pine dominated the softwoods. Red oaks represented a relatively large share of the total; however, there are a number of species included in this group. The most dominant single species of hardwood was the sweetgum, followed closely by the yellow poplar. Both species are aggressive competitors in Alabama woodlands. Both have winged seeds, which strongly support their dispersal. Volumes of both of these species increased dramatically during the period—the sweet gum 46% (1.6 to 2.4 million cubic feet) and the yellow poplar 95% (0.7 to 1.4 million cubic feet).

In 1982, the total volume of softwood in Alabama was significantly higher than the total volume of hardwood (11.7 versus 9.7 million cubic meters); however, by 2000 the relationship was reversed. The volume of hardwood was considerably greater (15.2 versus 12.7 million cubic feet). Removals of softwood were especially high during much of this period.

Growth and Removal of Softwood and Hardwood

Table 16.49 includes information on the average annual growth and removals (millions of cubic feet) of growing stock of all softwood and all hardwood in Alabama during the periods 1982-1989 and 1990-1999. From 1982-1989, average annual growth of softwoods was 657.8 million cubic feet. Then in the period 1990-1999, it increased to 844.1 million cubic feet, an increase of 31%. Average annual hardwood growth also increased within this time interval (565.9 to 596.2 million cubic feet); however, the rate of increase was much lower than that of softwood (31 versus 5%).

Average annual softwood removals also increased significantly from the 1982-1989 period to the 1990-1999 period (719.5 to 890.2 million cubic feet), or by 24%. Annual hardwood removals also increased between the two periods (370.1 to 406.8 million cubic feet), or by 10%. Obviously, softwood removals were much higher than those for hardwoods. In the 1990-1999 period, they were more than twice as high (890.2 versus 406.8 million cubic feet).

In both periods (1982-1989 and 1990-1999), softwood removals were significantly higher than growth—61.7 million cubic feet (9%) in the first period and 46.1 million cubic feet (5%) in the second. Hardwood growth was higher than removals in both periods.

Hartsell and Brown (2002) reported that final harvest was completed on 451,400 acres of Alabama timberland annually after 1990. Table 16.50 indicates in general terms the sources of those removals. Almost equal percentages (24 to 29%) came from oak-pine, upland hardwood, or natural pine stands. Altogether, 86% of all removals came from natural rather than plantation stands.

Rudis, et al. (1984) commented that during the period 1972-1981, some 7,473.3 million cubic feet of forest products were harvested in Alabama. Of this total, 5,530.6 million cubic feet (74%) were softwood. In the data provided by McWilliams (1992), of the 1,230.7 million cubic feet harvested in 1990, 68% of the total volume was softwood.

Stumpage Prices For Alabama Sawtimber

Lynn (2000) commented that new residential construction accounts for more than 35% of American lumber demand, that repair and remodeling account for 33%, and that non-residential construction, materials handling, etc., accounts for the remainder. He further comments that demand for lumber for repair and remodeling is relatively stable from year to year, and that demand for non-residential construction tends to track changes in the general economy. The most volatile of the three uses is housing starts. The volatility of this factor is obvious in Figure 16.99. In turn, housing starts are closely related to interest rates. In the early 1980s, the Federal Reserve Board increased discount rates to 14% to combat rampant inflation (Figures 16.4 and 16.5). Then, when the inflationary pressures eased somewhat, interest rates began to come down. They were down to 5.5% in August 1986; however, in 1987 the Federal Reserve began to be concerned about inflation again, and as a result they began to increase the discount rates once more. By February 1989, the rate was back up to 7%. Then in the third quarter of 1990, the 1990-1991 recession began. Throughout this period, from mid-1986 through early 1991, housing starts trended downward. In the latter part of 1991, the recession had ended, and the Federal Reserve began to reduce discount rates; housing starts began to trend upward again. The sensitivity of housing starts to slight increases in interest rates was demonstrated in 1995. As interest rates declined, new, single-family construction had been trending upward since mid-1991; however, in February 1995, the Fed increased the prime rate to 5.25%. It had been 3.5% in May 1994. In the first quarter of 1995, housing starts declined sharply. Yo-yoing in housing starts have a whipsaw effect on timber prices, especially for pine sawtimber.

Average stumpage prices for pine sawtimber in Alabama for the period 1985-2001 are shown in Figure 16.100. In 1985, prices were at $143 per thousand board feet (MBF-Scribner scale). They trended downward through 1987 to $129—as a result of a flood of low-priced log imports from Canada, and a reduction in housing starts in the South. Housing starts had been trending downward sharply since 1985. Pine sawtimber prices essentially remained flat until the end of the 1990-1991 recession. They spurted upward from 1993 through 1994, to $294, as home construction increased (Figure 16.99). In 1996, prices were down again to $259, as demand for lumber throughout the world slowed. Fortunately, the market began to recover in 1997, as the demand for lumber increased, fueled by growing home construction. In 1998, prices were $376, the highest in history. Then they generally trended downward through 2001—as a result of a global recession, slowing home construction, weak exports, and the increasing value of the dollar—to a level of $307 per MBF. However, as the economic conditions improved, in 2002 the price increased to $330.

Hardwood sawtimber stumpage prices during the period 1985-2001 are also shown in Figure 16.100. They generally followed the trends of prices for pine sawtimber during most of the period. They trended upward from $58 per MBF (Doyle) in 1985 to $170 in 1995. After being down in 1996 and 1997, they climbed upward again, to reach $196 in 1998, also the highest in history to that time. They remained at around that general level through 2002. Hardwood prices did not fall quite as sharply as pine from 2000 to 2001. Relatively less hardwood is used in the construction of single-family housing.

Stumpage Prices For Alabama Pulpwood

Stumpage prices for pine pulpwood during the period 1985-2002 are shown in Figure 16.101. The price was $19 per standard cord in 1985. They were generally down during the period 1987-1989, as a result of the flood of less expensive imports. Then they trended upward steadily from 1990 through 1995 as the economy grew. They were much lower in 1995, as a result of restructuring within the pulp and paper industry; however, by 1998 they had climbed to $33 per cord, also the highest in history. By the end of 1998, the pulp and paper industry was beginning to have serious problems with overcapacity and decreasing demand, as Asia began to enter a recession. Prices were sharply lower in 1999 and 2000. In 2001, they reached $16 a cord, and it remained the same in 2002 They had not been that low since 1989, and when considering the change in the Consumer Price Index, they had not been that low since the 1970s. Throughout the late 1990s and into the new century, weakness in all aspects of the pulp and paper industry intensified (Muehlenfeld, 1999).

Hardwood pulpwood stumpage prices generally followed the same trends as pine, although hardwood prices for most of the period were much lower (Figure 16.101). They generally trended upward from $4 a cord in 1985 to $27 in 1997. Prices of hardwood pulpwood seem not to have been affected as much by imports during the late 1980s as pine had been. Prices fell from 1995 to 1996, probably in response to the increase in interest rates. Price recovered somewhat in 1997 and 1998, only to be victimized by the growing malaise in pulp and paper in the late 1990s; however after 2001, price began to recover somewhat, and in 2002 price reached $21.

Cash Receipts from the Sale of Forest Products

Data on cash receipts from the sale of products from Alabama woodlands during the period 1970-2002 are presented in Figure 16.102. These data, published by the Alabama Agricultural Statistics Service, do not differentiate among the sales of softwood and hardwood, sawtimber, and pulpwood. In 1985, cash receipts from the sale of farm forest products totaled $93.8 million, but by 2002 they were $148.5 million. In this 18-year period, cash receipts increased by 58.3%. In the same period, the Consumer Price Index increased by 67.2%. In 1982-1984 dollars, cash receipts were lower in 2002 than in 1985.

In 1985, cash receipts from the sale of farm forest products accounted for 9% of cash receipts from all crops, including farm forest products ($93.8 of $846.9 million). In 2002, with cash receipts for crops declining, cash receipts for farm forest products accounted for 21.6%.

In 1985, cash receipts (estimated from severance taxes paid) from the sale of forest products originating on Alabama's farms were $93.8 million. They changed relatively little through 1991, but began to increase rapidly during the period 1992-1995, as all segments of the national economy roared ahead (Figure 16.1). In 1995, the average price for pine sawtimber and pulpwood stumpage reached $293 per MBF and $31 per standard cord, respectively, the highest prices in over 20 years (Figures 16.100 and 16.101). Then in 1995, growth in sales of paper slowed and finally declined in 1996, as over-supply and poor demand for the product forced a re-structuring of the pulp and paper industry. As a result, cash receipts fell to $150.8 million in 1996, but began to recover in 1997, and reached $200.8 million in 1998, the highest level in history. In 1999, they began to decline again, as a result of decreased demand for pulp and paper products, especially in Asia, and finally reached $148.5 million in 2002.

In 2002, farmers in Wilcox County received the largest share (9.3%) of total cash receipts from the sale of farm forest products in the state. Other counties reporting large shares included: Monroe (6.6%), Conecuh (6.2), Hale (4.8), and Marengo (4.5). These five counties accounted for 31.4% of the state's total. This relatively low percentage is indicative of the widespread distribution of marketable farm forest products in the state. In 2002, a total of 21 counties reported cash receipts of $2 million or more.

In 1985, cash receipts from the sale of forest products from all non-farm commercial timber sources (government land, the forest industry, etc.) were three times greater ($283.9 million versus $93.8 million) than from farms (Figure 16.102). Afterwards, year-to-year trends were similar, because the same economic factors affected both. However, cash receipts from non-farm wood grew at a much faster rate. By 1998, they were four time larger ($810 versus $220 million); however, by 2002 they were down to $587.3 million.

In 1989, the Alabama Agricultural Statistics Service began to divide non-farm forest wood cash receipts into its components: private non-farm, forest industry, and government. These data for the period 1989-2002 are shown in Figure 16.103 along with cash receipts from the sales of wood from farms, for comparison. As might be expected, trends for sales from all sources were generally the same. They generally trended upward through 1998; then downward through 2002. Also, the data indicate that sales from private, non-farm sources were considerably greater than for any of the other sources.

Kinds of Forest Products Marketed

Table 16.51 contains information on the quantity of different kinds of timber products that were harvested from Alabama's woodlands in 1989 and 2001. The data indicate that the total quantity of sawlogs increased 7.1% (342.7 to 367.3 million cubic feet), with most of the increase coming from the increase in hardwood logs (62.2 to 82.2 million cubic feet).

The total quantity of all pulpwood harvested increased 3.9% (680.3 to 706.5 million cubic feet) in the 12-year period. During the period, the quantity of softwood increased 16.2%, while hardwood decreased by 14.1%.

The processing of timber into green chips is significantly altering the way in which pulping material is removed from the forest. Traditionally, trees to be used in pulp production were cut into short sections and transported to the mill in that form. At the mill, the logs were de-barked and chipped before being placed in the so-called impregnation vessels. In the last 20 years, however, chipping technology has changed the way harvested trees are processed. Now substantial quantities of timber are chipped on-site, and the chips rather than the logs are transported to the mill. Hauling chips is much simpler and less expensive than hauling whole logs. On-site chipping also makes it possible to harvest wood from areas where it would not be cost-effective with traditional practices. For example, in mountainous areas of the state, it is now much more practical to harvest hardwood for pulp by chipping it on-location. Further, chipping on-site gives new meaning to the term clear-cutting. With this technology, literally all the wood on the site, regardless of size, can be removed.

Green chip production increased only 5.3% during the 1989-2001 period (193.9 to 204.2 million cubic feet). The amount of softwood chips decreased by 2.7%, while hardwood chips increased 29.3%.

Data in Table 16.51 also demonstrate the relative importance of the paper versus the lumber industry in Alabama. In 1989, some 71.8% of all wood removed from Alabama woodlands, excepting poles and piles, was used for manufacturing paper. The percentage (71.5%) remained unchanged in 2001, as a result of long-term capacity versus demand problems in the paper industry.

The Future

At the beginning of the twenty-first century, the future of some portions of the forest products market in Alabama are uncertain. *Timber Mart—South*, in their third quarter 2002 press release, commented that stumpage prices for pine sawtimber, chip-n-saw timber, pine pulpwood, and hardwood pulpwood had generally trended downward since the third quarter of 1997. The future of pine pulpwood sales is of special concern, because we have so much of it waiting in the wings. Stumpage prices for this product were at a five-year low in the third quarter of 2002. These discouraging trends were the result of a number of converging factors: reduced demand for paper associated with the 2001-2002 recession, the strength of the US dollar that hurt exports, substitution of other materials and technology (plastics and electronic media) for timber products, large quantities of conservation reserve pulpwood coming to the market, overcapacity in wood pulp production, and increased international competition.

International competition is the most difficult of those contravening factors to address. In their March 12, 2002 news release, the US International Trade Commission noted that American imports of wood pulp are duty-free and that, generally, internationally there is little or no duty imposed on the trade of this product. Consequently, it moves freely around the world in response to free market opportunities. The commission further noted that the American wood pulp international trade balance remained positive in 2000, but that it had declined $571 million ($696 million to $125 million) from 1996—as a result of declining exports and increased imports.

While imports of wood pulp (mostly from Canada and Brazil) are low, relative to consumption, when there is already an over-supply of native product on the American market, even a small quantity of imports have a significant impact. Further, there is every indication that imports will increase. The International Trade Commission's report commented that South America production is expected to continue to increase. Also, there are enormous quantities of old-growth hardwood in Eastern Europe that are beginning to find their way into world wood pulp markets at very attractive prices.

Muehlenfeld (1999) stated that pulp and paper manufacturers in the American South have lost their low-cost producer status in pulp and some grades of paper. Recently, the comparative advantage has shifted to such countries as Brazil and Indonesia. These countries have emerged as the world's low-cost producers as a result of newer mills, lower wood costs, and lower labor costs.

Another concern for Alabama pine pulpwood farmers is the increasing use of re-cycled paper in production (Skog, et al. (1998). They stated that products "such as tissue, newsprint and corrugating medium will be made with a greater percentage of recycled fiber." They also suggested that changes in paper-making technology would result in the use of increasing amounts of hardwood pulp.

Animal Agriculture

During the period 1970-1985, the value added to the American economy by livestock production grew from \$30.8 billion to \$68.9 billion, an increase of 123.2%, while the Consumer Price Index increased by 173.3%. All of the different components of the livestock sector increased during the period, but value added by poultry and eggs and by dairying more than doubled. In 1970, the value added by livestock was considerably higher than that added by crop production (59.3 versus 39.5%), but by 1985 value added by crops was larger (51.8 versus 48.2%). In 1970, value added of home consumption of livestock and products was equal to only 1.8% of the total value added of the sector, but by 1985 the home consumption share was down to 1%.

The value added by Alabama livestock production increased from \$557.8 million in 1970 to \$1,295 million, and in 1985 it increased 132.3%. The rate of increase was higher in Alabama than in the country as a whole (132.3 versus 123.2%), but both were lower than the rate of increase in the CPI of 173.3%. Much of the increase in the state was fueled by an almost threefold increase in the value added by poultry and eggs. Value added by livestock production was more than twice the value added by crop production in 1970 (70.6 versus 29.4%). The difference was somewhat smaller in 1985 (62.8 versus 38.2%) in 1985. The home consumption of livestock and products share of value added was 2.3% of total value added in 1970, but only 0.8% of it in 1985.

Cattle

Cash (2002) commented that at the beginning of the twenty-first century that the United States continues to be the world's leading producer of high-quality, grain-fed beef, and that most of the production is consumed within the country. In 1999, only 9% of total production was exported. In value, American exports exceed imports, but in quantity, imports are greater. Most of the cattle used by the beef processors are produced on small farms (total sales less than \$249,999). Cash further commented that small operations control 74% of all land dedicated to beef cattle production, and that 75% of all of the animals processed into beef spend some portion of their life on a small farm.

US Cattle

During the period 1970-1985, American cattle numbers ("all" cattle and calves excepting cows that had calved-milk) went through two complete cycles of abundance. A comparison of the Consolidation Phases of the two cycles indicates that there was a net loss of 16.4 million animals (120.8 to 104.4 million) between 1975 and 1982, or 13.6% of the 1975 number. In the two corresponding cycles in Alabama, the net loss was 874,000 animals (2.76 to 1.89 million), or 31.6% of the 1986 number.

In 2001, the five leading states in the numbers of "all" cattle were Texas (13.70 million), Kansas (6.70 million), Nebraska (6.60 million), California (5.15 million), and Oklahoma (5.05 million). These five states accounted for 38% of "all" cattle in the United States that year.

BEEF EXPORTS

Data on beef exports is presented in Figure 16.104. Exports trended upward steadily from 300 million pounds in

1985 to 2,447 million pounds in 2002. They had declined slightly in 2001 (2,269 million pounds) as a result of a strong US dollar, high American beef prices, and concern in Japan over bovine spongiform encephalopathy (BSE). Japan has been the largest importer of American beef for some time. In 2001, that country accounted for 44% of all American beef exported. In 2000, Japan, Mexico, Canada, and South Korea (in that order) accounted for 82% (by weight) of all beef exports. Beef exports account for 10% of total American exports.

BEEF IMPORTS

Data presented in Figure 16.104 show that beef and veal imports were 2.1 billion pounds (carcass weight) in 1983. Afterwards, they slowly trended upward to reach 2.4 billion pounds in 1988. They were down in 1989, probably as a result of the slowing of the growth in the American economy (Figure 16.1). They remained fairly constant at around 2.4 billion pounds through 1994. In 1995 and 1996, they were down to around 2.1 billion pounds, probably as a result of the concern over mad cow disease. Also, American cattle numbers were increasing rapidly during this period (Figure 16.107) After 1995, American cattle numbers began to trend sharply downward, and with that decline imports began to increase. By 2002, they were 3.2 billion pounds, or 31.5% higher than exports.

BEEF CONSUMPTION

Data on American beef consumption (pounds per capita) during the period 1985-2002 are presented in Figure 16.105. These data represent the equivalent weight of dressed beef. As a result, these quantities are higher than the consumption of choice retail beef. Generally, 1.43 pounds of dressed beef are required to produce one pound of choice retail beef. As is shown in the figure, beef consumption trended sharply downward from 108 pounds per capita in the mid-1980s to 93 pounds in 1993, probably as a result of a general concern about including too much red meat in diets. Consumption began to recover somewhat in 1994, but then dipped again in 1997, probably as a result of the mad cow disease furor, before continuing to move upward. Consumption reached 99.3 pounds in 2000, and remained near that level through 2002.

Prices

Data on annual average prices (dollars per Cwt) of American cattle during the period 1970-2002 are presented in Figure 16.106. These data show that the price was $53.70 in 1985. At that time, American cattle numbers were in the Declining Phase of a cycle (Figure 16.107). The cycle had reached its Consolidation Phase in 1982. The price continued to decline in 1986 ($52.60), but with numbers continuing to decline, prices began to increase in 1987, and by 1990 reached $74.60. As might be expected when prices began to increase, cattle producers began to add to their herds, and the Expansion Phase of a new cycle began in 1991. Then as sure as night follows day (teeter totter), with numbers increasing, prices began to decline, finally reaching $58.70 in 1996 (at the Consolidation Phase in the cycle). With numbers declining, prices once again began to increase, and the increase generally continued through 2001 ($71.30). The price declined in 2002 to $66.50 even though numbers were continuing to decline. Apparently, the price decline was the result of an increase in the supply of beef. Some reports suggest that the average slaughter weight of animals coming out of the feedlots had increased.

CHARACTERISTICS OF US CATTLE

Farms

In 1996, USDA (Short, 2001) surveyed 80% of US farms with cattle as part of a larger Agricultural Resource Management Study (ARMS). Some of the data are presented in Table 16.52. Obviously, some of these data are affected by the differences in the size of the regions and in the size of the operations within the regions, but the ratio of calves per cow and pounds weaned per bred cow are independent of those two variables. In both of those characteristics, operations in the Southeast were lower.

Cash (2002) published a paper describing some characteristics of cattle production on small family farms (farms with sales of $249,999 or less). According to the author, these farms control 74% of all land dedicated to American beef cattle production. A large share of the beef cattle operation in Alabama would be included in this group. Cash divided these farms into five categories, depending on annual sales and the primary occupation of the operator. Some of his data are presented in Table

16.53. Data in the table show that small farms with full-time operators control more acreage and have larger numbers of cattle. Generally, cattle only provide these operators with a portion of their farm income. Their cattle provide them with a supplemental income in their traditional mixed-output agricultural enterprise. These operators generated 29% of the total value of American cattle production in 1997.

The remaining five groups of small farms are in reality part-time operators. They are not really in the business of cattle farming. They likely derive most of their income from other sources; however, they produce 34% of all beef cattle and calves in the country. Retired and residential, these lifestyle farmers operate farms because they enjoy a rural lifestyle and they love cows.

Cycles of Abundance

The relationship (teeter totter) between prices and cattle numbers detailed in the preceding section on prices (Figure 16.106) resulted in the elaboration of portions of two Cycles of Abundance in cattle numbers in the country in the period 1985-2002. In 1985,the number was 98.8 million. At that time, the cycle was in its Declining Phase. It had begun in 1980, and it ended in 1990 with 85.8 million animals. Then the Expansion Phase of another cycle began in 1991. It reached the Consolidation Phase in 1996 with 94.1 million animals. The Declining Phase began in 1997 and continued through 2002. At that point, there were 87.6 million head in the country. Between the Consolidation Phases in 1982 and 1996, cattle numbers in the country declined from 104.4 to 94.1 million, or by 9.9%.

In 2002, the largest share (14.1%) of the country's 96.7-million cattle and calf herd ("all" cattle and calves, including cows that had calved-milk) was located in Texas. Other states reporting large shares included: Kansas (6.8%), Nebraska (6.6), California (5.4), and Missouri (5.4). All of these states were home to a large number of feedlot operations.

PRODUCTION Costs

Short (2001) published data on the cost of producing cattle in cow-calf operations in five regions of the United States as part of the 1996 Agricultural Resource Management Study (ARMS). Table 16.54 provides a summary of some of those data. Some characteristics of these farms were described in a preceding section (Table 18.52).

Operating Costs. Total operating costs per bred cow in the different regions were: North Central ($269.45), Northern Plains ($267.99), West ($232.64), Southern Plains ($210.61), and Southeast ($190.68) (Table16.54). Operating costs were lower in Southeast ($190.68) than in the other regions, primarily as a result of much lower total feed costs. Operators in this region likely made use of more grazing than in other regions. Operating costs were highest in North Central ($269.45), because farmers there had to depend more heavily on supplemental feed.

Ownership Costs. Total ownership costs per bred cow were also much higher in North Central ($227.07) than the other regions, primarily as a result of the much higher costs of capital recovery for tractors, vehicles, equipment, and machinery (Table 16.54). These costs were also relatively high in Southeast ($190.68). In both cases, capital recovery had to be covered by a much smaller number of bred cows, 50 and 57, respectively. These costs were less than half that amount in West ($98.70), which had the highest number (146).

Total Production Costs. Primarily as a result of the large differences in total ownership costs, total production costs per bred cow were also highest in North Central and lowest in West ($498.52 versus $331.34) (Table 16.54).

VALUE OF PRODUCTION

In Short's (2001) study, the value of production per bred cow (pounds produced x average regional price) was: Northern Plains ($304.12), West ($291.28), North Central ($245.69), Southern Plains ($233.29), and Southeast ($191.91) (Table 16.55).

NET RETURNS

Comparing these values of production to operating costs and total production costs (operating + ownership costs) provided the results shown in Table 16.55. When considering only operating costs, net returns were negative only in North Central (-$23.75) and ranged between $1.24 in Southeast to $58.64 in West. When considering total production costs, net returns were negative in all regions and ranged from -$40.06 in West to -$252.82 in North Central. The primary problem for farmers, as shown in these data, was the extremely low prices that farmers received for their calves that year.

NET RETURNS PER BRED COW

Data presented in Table 16.55 provided an estimate of net returns in five regions of the country in 1996; however, USDA has been providing these estimates for the entire country for many years. Data in Figure 16.108 provide these estimates on an annual basis for the period 1985-2002. These estimates are equal to the difference between the gross value of production and total operating costs. These data show that net returns per bred cow were positive from 1986 through 1993, generally while numbers were declining (Figure 16.107). Then from 1994, net returns went through a long period of negativity, finally reaching a low of -$159.63 in 1996. Returns became positive again in 2000 ($29.37) with numbers declining, but began to fall in 2001, probably as a result of the recession. In 2002, it fell into negative territory again (-$23.59). Altogether, returns were negative in six of the 17 years of the period 1985-2002. The trends in these data are similar to trends in American cattle prices shown in Figure 18.94.

The situation on net returns was really even more dismal than shown by the data in Figure 16.108. USDA also published data on the difference between gross value of production and all specified (operating plus ownership) costs. In all 17 years of the period, they were negative (Taylor, 2002).

Alabama Cattle

In 1985, the Alabama Agricultural Statistics Service reported that numbers of Alabama cattle and calves ("all" cattle and calves, including cows that had calved–milk) ranked 22nd in the country with 1.7% of the total. In 2002, state numbers ranked 23rd, with 1.5% of the national total. In 1985, cash receipts from the sale of cattle and calves in Alabama accounted for 16.1% of all farm commodities, including farm forest products. By 2002, that percentage had declined to 9.8%.

In 1985, a large share of Alabama's cattle was in the southern part of the state. The top five counties were: Montgomery (3.6%), Lowndes (3.5), Dale (3.4), Geneva (3.4), and Cullman (2.6). These five counties accounted for only 16.5% of "all" cattle in the state. A total of 29 counties reported at least 25,000 head. Four of these counties are in south Alabama. The lower portion of this region was a center of open range cattle herding in the early days of statehood. Figure 16.109 shows the distribution of cattle (all) within the state in 1985.

Prices

Data on the annual average price that Alabama farmers received for their cattle during the period 1970-2002 are also presented in Figure 16.106. Alabama prices were generally a few dollars per Cwt below those in the country as a whole. Prices were considerably lower in 1995 and 1996. Cattle numbers were much higher than expected in those years (Figure 16.110) and may have affected prices unduly.

Cycles of Abundance

Trends in Alabama cattle and calf numbers ("all" cattle and calves, excepting cows that had calved-milk) generally mirrored those of the country as a whole during the period 1985-2002 (Figures 16.110 and 16.107). However, the cycles in the state are somewhat more erratic than those in the country. By combining data from all of the states, the US data and the cycles appear to be much smoother. In 1985, Alabama's cattle numbers were also in the middle of a Declining Phase of a cycle that had begun in 1981. In 1985, near the end of that phase, numbers reached 1.8 million animals. A new cycle began in 1991 and reached the Consolidation Phase in 1995 (1.7 million animals). Then in 1996, the Declining Phase began; it continued through 2001. At that point, there were 1.2 million animals in the state. A new cycle began in 2002 (1.35 million animals).

A comparison of the number of cattle during the Consolidation Phases of the two cycles in 1982 (1,950,000) and 1995 (1,780,000) indicates that the state lost 170,000 animals between those years, a decline of 8.7%.

By 2002, the larger share of the state's cattle herd was in north Alabama. Four of the five counties with the most cattle were in that region: Cullman (4.7%), DeKalb (3.8), Montgomery (3.1), Marshall (3.0), and Lauderdale (2.9). These five counties accounted for only 17.5% of the state's total. Twenty-five counties reported at least 20,000 head. Figure 16.111 shows the distribution of cattle (all) within Alabama in 2002.

The Alabama Agricultural Statistics Service has published data indicating that in 1985 of the state's 54,000 farms, 42,000 (77.8%) included cattle in their operations. In 2002, the percentage of farms with cattle had declined to 55.3%. Using these data plus data on numbers, it appears that while the share of farms with cattle declined, the average number per operation increased from 42 to 55 head. Although diminished slightly with time, the old Scotch-Irish love for cattle seems to be alive and well in Alabama at the beginning of the twenty-first century.

PRODUCTION Costs

Prevatt and Miller (2001) have published estimated costs of producing weaned calves in Alabama in 2001, using a raised replacement system. Rather than using per bred-cow costs as Short (2001) did, they estimated the costs of producing calves using 85 cows, plus bulls, on 160 acres of land. These estimates are presented in Table 16.56.

Variable, Fixed, and Total ProductionCosts. According to the author's estimates, total variable costs for producing a crop of 58 calves was $38,722; total fixed costs was $13,496; and total production (all estimated costs) costs was $52,218 (Table 16.56).

Net Returns

Prevatt and Miller (2001) further estimated that their hypothetical cow-calf operation, using the raised replacement system, would produce for sale per year: 38 steer calves, 20 heifer calves, 13 cull cows, two cull heifers and 0.71 cull bull. Using average cattle prices for the last 10-year period, they estimated that these animals would bring their owner $38,009 in gross receipts (Table 16.57). When gross receipts are compared to total variable costs of $38,722, their net return would be -$713, and when compared with total specified costs, net return would be -$14,209.

Some 62.6% of cattle farms in the Alabama sold less than 20 calves in 2002 (Table 16.58), and 37.8% sold less than nine. Only 14.5% of all farms sold more than 50. It is likely that most of the Alabama cattle operations suffer some loss, especially when gross receipts are compared to all specified expenses. It is likely that most of these smaller operations either forego the payment of some of the expenses or pay them from some other source of income.

Hogs and Pigs

As detailed in preceding chapters, when the European hog was introduced to North America, it immediately went hog wild. It quickly found a home in the widespread hardwood forests. With the ferocious, semi-wild sow taking good care of her offspring, the population exploded, and within a few years, the old Anglo-Saxon meat source became a nuisance throughout the colonies. With no fencing available, farmers found that it was almost impossible to keep the semi-wild animals out of their small fields.

As noted in an earlier chapter, contracting and vertical integration in the broiler industry occurred several decades ago (1950s). Now, nearly all broilers are produced under pre-arranged contracts between growers and processors. Contracting in the pork industry lagged far behind its use in the broiler industry, but has increased rapidly in recent years, especially in the Eastern United States where farmers and food processors in broiler-producing areas have greater familiarity with this process.

In a study conducted by the USDA's Economic Research Service on hog production costs and returns in seven different regions of the country, in 1998 an estimated 40% of the hogs grown in the country were produced under contract between growers and processors. The study also determined that the proportion under contract increased to 41% in 1999. Contracting was much more prevalent in some regions than in others. In Mississippi Portal (Figure 16.38), only 7% of the hogs were produced under contract in 1999. In the same year, 60% of the hogs produced in Eastern Uplands, and 87% produced in Southern Seaboard were grown under contract.

While vertical integration has several benefits for producers and consumers, it results in two serious problems that are difficult to manage. Vertical integration requires vastly increasing the number of animals confined in an area. Further, because of the nature of their digestive system, swine require a relatively large amount of high quality, concentrated feed for maximum growth. As a result, these animals produce the largest amount of nutrient-concentrated manure per pound of gain of any of the meat animals. They produce a relatively large amount of manure containing a considerable amount of protein and fat. The bacterial decomposition of this manure results in the production of large quantities of really foul-smelling gases that are released into the air. It is easy to understand why these large swine operations do not make good neighbors.

Hog manure does not smell any worse now than it did 60 years ago when my father gave me a beautiful Poland China sow to use in a 4-H project. But two things have changed since that time. Now instead of having eight to 10 hogs roaming free in a fenced yard, thousands are confined in a single large building. Probably the most important difference is that city people, with no love for hogs or the smell of decomposing hog manure, are swarming into the countryside, especially in the Eastern United States. They spend a relatively large amount of money on a small plot of land, but would like to control the smells and sights of a much larger area. As might be expected, these changes have led to some conflicts that ultimately the hog producer is going to lose.

It is generally difficult to get very many people interested in zoning rural land, but the far-ranging smell of decomposing hog manure may rapidly change that perspective in many rural areas. As a result, in several states, all kinds of regulations are being developed and implemented to separate people and the smell of hog manure.

US Hogs and Pigs

As detailed in the preceding chapter, American hog and pig numbers went through portions of two major cycles in the period 1970-1985. Numbers in 1970 were at the end of a major Accumulation Phase (67.3 million animals), which probably began in 1965. The Liquidation Phase of that cycle began in 1971 and ended in 1975 with 49.3 million animals. The Accumulation Phase of another cycle began in 1976 and ended in 1979 (67.3 million). The Liquidation Phase began in 1980, and was continuing in 1985 at a level of 52.3 million. In these two cycles there was no net loss of animals. The Accumulation Phases of both ended at about the same level of animal abundance (67.3 million).

In 1985, Iowa farms accounted for the largest share (25.8%) of the country's 52.3 million hog herd. Other states reporting large shares included: Illinois (10.3%), Indiana (7.9), Minnesota (7.8), and Nebraska (7.5). These five states accounted for 59.3% of the national total. Alabama farms accounted for only 0.7%.

PORK CONSUMPTION

At the beginning of the twenty-first century, pork was unpopular in the American meat industry. Throughout the 1990s, the consumption of pork trailed far behind the consumption of beef and broilers (Figure 16.105). In 2002, consumption of the three most popular kinds of meat were: poultry (96 pounds per capita), beef (96.8 pounds), and pork (66.5 pounds). In 2002, pork consumption was very near where it had been in 1970 and 1985. As was noted in an earlier chapter, pork was not always the low man on the totem pole. Until the 1940s, Americans ate more pork than beef or poultry. However, since that time, hamburgers and chicken fingers have literally forced the other white meat far away from the trough. It seems that no amount of money spent on advertising or on research, can undo this quirk in American dietary behavior. Somehow the broiler and beef have forged ahead in the race for public acceptance, and now they apparently cannot be displaced.

Prices

American hog numbers followed prices down in the early 1980s (Figures 16.113 and 16.112) to reach $44.00 per Cwt in 1985. Then in 1986, with numbers at an eight-year low, prices began to increase again, reaching $51.20 in 1987. In 1987, with the prices increasing, numbers followed. This teeter-totter relationship generally continued throughout the period. (Hogs are inventoried on December 1 of each year, so that numbers would tend to have a limited effect on prices until the following year.) Increasing numbers during the period 1990-1994 led to much lower prices, around $45 per Cwt from 1992 through 1995. Numbers began to decline in 1995, and prices increased to $51.90 in 1996 and to $52.90 in 1997. Teeter totter resulted in prices reaching $42.30 in 2000 and $44.40 in 2001. In response, numbers increased slightly in 2001, but in 2002 the price fell to $33.40.

Cycles of Abundance

The teeter-totter relationship between prices and numbers described in the preceding paragraph resulted in the elaboration of at least four Cycles of Abundance in numbers of hog in the United States in the period 1985-2002. On December 1, 1985, there were 52.3 million hogs and pigs in the country, and in response to declining prices, hog farmers were in the process of reducing the sizes of their herds in the Liquidation Phase of a cycle (Figures 16.113 and 16.112). The Liquidation Phase of that cycle ended in 1986 at 51 million head, and the Accumulation Phase of a new cycle began in 1987. The Liquidation Phase of that cycle ended in 1989. The data in Figure 16.113 seem to indicate that there were at least three additional cycles in the remaining years of the period: 1990-1993, 1994-1996, and 1997-2000. Another cycle began in 2001. From the end of the Liquidation Phase of the 1983-1986 cycle to the end of that phase in the 1997-2000 cycle, hog numbers in the country increased by 8.1 million animals (51 to 59.1 million).

PRODUCTION Costs

The USDA's Economic Research Service has published data on the cost of hog production in several different regions of the country in 1999. Data on some characteristics of the operations and on the costs and returns of hog production in three of these regions (Heartland, Eastern Upland, and Southern Seaboard) are presented in Tables 16.59 and 16.60. The location of these regions is shown in Figure 16.38. Heartland was chosen because a large share of our hog production has traditionally taken place there. Eastern Highlands and Southern Seaboard were chosen because these two regions contain all portions of Alabama. Southern Seaboard also includes North Carolina, which produces a large share of the country's hogs.

There was considerable difference in some of the characteristics of the hog production operations in the three regions. For example, the average size of the market hog operations in Southern Seaboard (5,074 animals) was much larger than in either Heartland (1,522) or Eastern Upland (873) (Table 16.59). This large difference is the result of the unusually large increase in the number of hogs and pigs being grown in North Carolina in 1998 and 1999. In 1985, that state ranked fifth in the country in hog and pig numbers with 4.5% of the country's herd. By 1998, however, it had charged into second place with15.6% of the total.

Feeder pig operations were also much larger in Southern Seaboard (6,652 versus 845 in Heartland and 1,464 in Eastern Upland) (Table 16.59). In Heartland, only 29% of the operators produced hogs and pigs under contract. The remainder (71%) produced them as independent operators. In Eastern Upland, 60% of the hogs and pigs were produced under contract. In Southern Seaboard, relatively few animals were produced by independent operators; 87% were grown under contract.

Operating Costs. Feed costs (dollars per Cwt of gain) represented the largest share (around half) of total operating costs in all three regions and were generally similar in all three (Table 16.60). The cost of feeder pigs was another matter. Apparently the competition for the purchase of these animals was much greater in Southern Seaboard. Farmers there paid almost twice as much for their pigs as those in the other regions: ($16.53 in Southern Seaboard, $9.59 in Heartland, and $7.39 in Eastern Upland). Farmers in Southern Seaboard also spent more on marketing. Most of the other operating costs were similar; however, because of the higher cost for pigs, total operating costs were considerably higher in Southern Seaboard ($41.37), than in either Heartland ($34.49) or Eastern Upland ($33.06).

Allocated Costs. Total allocated costs (dollars per Cwt of gain) were considerably lower in Southern Seaboard ($13.06) than in either Heartland ($20.23) or Eastern Upland ($26.07) (Table 16.60). All categories of these costs were lower in Southern Seaboard than in the other two regions, but the largest differences were in opportunity cost of unpaid labor and capital recovery of machinery and equipment. These costs were considerably higher in Eastern Upland than in the other two regions.

GROSS VALUE OF PRODUCTION

The gross value of production (dollars per Cwt of gain) in the three regions were $39.19 in Heartland, $45.08 in Eastern Upland, and $51.23 in Southern Seaboard (Table 16.61). The primary difference in these values was the higher price that producers in Southern Seaboard received for their feeder pigs.

NET RETURNS

Hog producers in all three regions recovered their operating costs (Table16.61). The returns above total operating costs in those three regions were: $4.80 in Heartland, $12.02 in Eastern Upland, and $9.86, in Southern Seaboard. In contrast, producers in none of the regions were able to cover all of their total production costs. These values were -$15.43 in Heartland, -$14.05 in Eastern Upland, and -$3.20 in Southern Seaboard.

Alabama Hogs and Pigs

Hog and pig numbers in Alabama also went through portions of two Cycles of Abundance in the period 1970-1985. However, the elaboration of the cycles in the state resulted in a significant net loss of animals. At the end of the Accumulation Phase in 1971, numbers were one million, but at the end of the Accumulation Phase in 1979, it was down to 880,000 animals. Then by 1985, the Liquidation Phase had carried it down to 345,000. Alabama hog farmers were getting out of the business in droves. Between 1971 and 1985, the state lost two-thirds of its hogs and pigs.

In 1985, the Alabama Agricultural Statistics Service reported that hogs were produced in all of Alabama's 67 counties. In the state's six agricultural reporting districts (Figure 16.96), the largest share (30.6%) of its 345,000-head herd was on farms in District 60 in the southeastern Alabama. Surprisingly, among the counties, Autauga County reported the largest share (4.4%) of the state total. Other counties with large shares included DeKalb (5.1%), Houston (4.7), Henry (3.9), and Lauderdale (3.6). These five counties accounted for 21.7% of the state total. Twenty-four counties reported counts of 5,000 or more animals. Figure 16.114 presents data on the distribution of hogs and pigs in Alabama in 1985.

Prices

Data on prices of Alabama hogs (dollars per Cwt) during the period 1970-2002 are also presented in Figure 16.112. The data indicate that year-to-year trends in prices were similar in both the country and the state. From about 1985 onward through 2002, however, Alabama prices were three to four cents a pound lower than average prices in the country as a whole. Throughout the period 1985-2002, Alabama hog farmers were in the process of slowly liquidating their herds (Figure 16.115). It appears that under these conditions, they were operating in a buyer's market and seem to have been willing to accept lower prices to get rid of their hogs. The differences tended to be greater in the early 1990s. It was at this time that numbers were declining at a fairly rapid rate. There was no teeter totter here; it was mostly all teeter.

Cycles of Abundance

There were no apparent cycles in Alabama hog and pig numbers from 1985 through 2002 (Figure 16.115). There is little indication that hog farmers responded to any of the price signals shown in Figure 16.112. Slight upticks in numbers in 1986, 1990, 1993, and 2001 indicate that in those years they might have added a few brood sows to their herds.

In 1985, there were 345,000 hogs and pigs in the state. Numbers generally remained near that level until 1990, when they began to decline. They trended down slowly through 1990, then more rapidly through 1995. Afterwards, the decline slowed. By 2002, there were only 165,000 hogs and pigs on Alabama farms.

In 2002, the Alabama Agricultural Statistics Service reported hog and pig production for only six counties. There were 60,000 in DeKalb County, 30,000 in Pickens, and 13,000 in Sumter. No other counties reported more than 1,400.

In 1897, there were 1,640,000, hogs and pigs in Alabama, the highest number ever recorded. There have been many ups and downs in numbers since, but there have been many more downs than ups. After 1980, common sense began to replace Alabamian's love affair with hogs. Alabama has never had any comparative advantage in producing them, but given the farming systems that the yeomen and their descendents operated, it was much easier to produce hogs than cows. Hogs could pretty much take care of themselves. Cattle could not, primarily because cows do not have a rooter and consequently, they cannot root. Hogs are wonderful scroungers; cows are not. Until Alabama's open range was closed, hog production had a chance, but once hogs had to be maintained on high-quality rations, there was little future in their production.

Milk

By 1988, reduced fat dairy products were gaining wide acceptance among consumers, and for the first time sales of those products exceeded sales of whole milk products.

In 1995, the hormone bovine somatotropin was approved for use in the American dairy industry. The use of this hormone provided a safe and effective means of increasing milk per cow. By the turn of the century, it was being used widely throughout the American dairy industry.

At the beginning of the twenty-first century, farm receipts for milk in the United States were second only to beef in receipts obtained from animal agriculture, and were generally equal to receipts for corn (Miller, 2002). Approximately a third of milk produced in the country is used as fluid milk and cream products. The remainder is used in manufactured dairy products such as cheese (mostly cheddar), ice cream, butter, non-fat dry milk, etc. The per capita consumption of fluid milk slowly declined during the period 1985-2002 as a result of intense competition from other beverages (Figure 16.116).

US Milk

The number of milk cows in the United States declined from 12 million in 1970 to 11 million in 1985. During the same period, the average quantity of milk produced by each cow increased from 9,751 to 13,024 pounds a year. As a result, total milk production increased from 117 to 143 billion pounds. While quantity was increasing, per capita consumption was declining. Between 1970 and 1985, consumption of total fluid milk and cream products declined from 275.1 pounds to 240.8 pounds.

In 1985, American dairy cows produced 143 billion pounds of milk. The largest share (17.3%) was reported by Wisconsin farmers. Other states reporting large shares included: California (11.7%), New York (8.2), Minnesota (7.6), and Pennsylvania (7.0). Dairy cows in these five states accounted for 51.8% of total national production. Alabama's share was a mere 0.4%.

Prices

Milk pricing has depended heavily on two programs for over half a century: The Federal Price Support Program and Federal Milk Marketing Orders. In the 1990 Farm Act, Congress added a third program: the Dairy Export Incentive Program. The 1996 Farm Bill (FAIR) called for the elimination of the Price Support Program by the end of 1999. Instead, Congress later extended it to 2000 and then to 2001. The 2002 Farm Act (Farm Security and Rural Investment Act) extended the program through 2007. The 2002 Act also added a new milk-related program that pays farmers directly to compensate them for any counter-cyclic effects in the economy (Miller, 2002).

Figure 16.117 presents data on the annual average price (dollars per Cwt) of milk in the country during the period 1970-2002. Federal support prices and marketing orders generally kept market prices between $12 and $13 per Cwt from 1985 through 1995; although they were somewhat higher around 1989 and 1990 when Congress enacted the Dairy Export Incentive Program. The 1996 Act was supposed to simplify the American milk pricing system, but while it might have simplified the system, it created havoc in annual market prices, resulting in unusually high year-to-year variation that continued through 2002. This variation did not seem to be related to milk production (Figure 16.120).

Milk Cow Numbers

In 1985, there were 11 million milk cows in the country (Figure 16.118). Numbers had been generally trending downward for much of the century. This trend continued through the period 1985-2002. In 2002, numbers were down to 9.1 million. They declined sharply in 1986 and 1987, as milk prices began to decline (Figure 16.117). After 1987, however, the decline in numbers slowed somewhat. It continued at about the same rate until the enactment of the reform legislation in 1996. Between 1997 and 2002, there was relatively little change.

Milk Per Cow

Average milk production per individual cow in the United States increased steadily from 13,024 pounds in 1985 to 18,608 pounds in 2002 (Figure 16.119); however, there was considerable variation in average production per cow in the different regions. For example, in 2002 average pounds per cow in several states in the Upper Midwest (Iowa, Michigan, Minnesota, and Wisconsin) was 18,317 pounds. In the Lower South (Alabama, Arkansas, Florida, Georgia, Louisiana, Mississippi, Tennessee, and Texas), it was 14,574 pounds, and in the Pacific states (California, Oregon, and Washington), it was 20,792 pounds.

Pounds Produced

Although milk cow numbers declined significantly in the country during the period 1985-2002 (Figure 16.118), the increase in production per cow (Figure

16.119) was sufficient to keep the total production of milk increasing—143 billion pounds in 1985 to 170 billion in 2002 (Figure 16.120).

In 2002, American milk cows produced 170.6 billion pounds of milk. California farmers reported the largest share (20.6%) of the national total. Other states reporting large shares included: Wisconsin (13%), New York (7.2), Pennsylvania (6.3), and Minnesota (5.0). These five states accounted for 52.1% of the national total. These same five states were also the top five in 1985. That year, they reported 51.8% of the total. Also in 2002, California replaced Wisconsin as the top state, and Pennsylvania, which had been fifth in 1985, replaced Minnesota in that position.

Without question, the production and marketing of milk are the most heavily regulated of all of the food and fiber producing systems in the country. Between 1985 and 2002, milk production on American dairy farms increased by 19.3%—from 16.8% fewer milk cows. Unfortunately, in 1985 farmers received an average of $12.76 per Cwt for their milk. In 2002, they received an average of $12.19, and if inflation of the currency is taken into consideration, they only received $7.29. For all practical purposes, American dairy farmers were regulated out of business in that 18-year period.

Alabama Milk

The Alabama Agricultural Statistics Service reported that in 1985 milk production in the state ranked 40th in the country with 0.4% of the national total. They also reported that there were 1,600 dairy farms in the state. In 2002, they reported that milk production in the state ranked 43rd in the country with 0.2% of the total and that there were only 200 farms with milk cows.

In 1985, the state's statistics service reported the production of milk from 41 of Alabama's 67 counties. There was so little milk produced in the remaining counties that they were not included in the report in order not to disclose data on the production from individual operators; their production was included in the total for the state. The statistics service further reported that production was at least 10 million pounds in 23 counties. In that year, farmers in Hale County reported the largest share (7.9%) of the state's production of 547 of million pounds. Other counties reporting large shares were: Morgan (5.2%), Montgomery (5.0), Perry (4.6), and Winston (4.4). These five counties accounted for 27.1% of total production. Average production per county was highest in District 40 (Figure 16.96), which includes most of Black Prairie Physiographic District (Figure 1.9). (Until the coming of the boll weevil, this was the heart of Alabama's old Cotton Arc.) The lowest average production per county was reported from District 60, in the southeastern corner of the state.

Prices

Data on annual average prices received by Alabama dairy farmers for their milk during the period 1970-2002 are presented in Figure 16.117. Prices for milk in Alabama followed the same year-to-year trends as those in the country as a whole. This is not surprising given that they were closely regulated by the federal government. Year-to-year variation was a result of the same factors influencing milk prices throughout the country. Prices in Alabama were generally higher primarily as a result of regional differences in Federal Marketing Orders.

In 14 of the 18 years of the period 1985-2002, Alabama milk prices were between $14 and $15 per Cwt. Considering that during this period, the Consumer Price Index increased by 67.2% (Figure 16.2), Alabama dairy farmers were actually receiving only a little more than $9 per Cwt for their milk at the end of the period.

Milk Cow Numbers

In Alabama, milk cow numbers declined from 122,000 in 1970 to 50,000 in 1985. During the same period, milk per cow increased from 6,689 pounds to 10,940 pounds, but because of the large decline in numbers, quantity produced declined from 816 million to 547 million pounds.

In 1985, there were 50,000 milk cows in the state (Figure 16.121). From 1985 through 2002, numbers continued the downward trend that began in the mid-1940s. By the end of that period, the number was down to 20,000 head. In 1985, some 2.9% of the state's farms were considered to be dairy farms (1,500 of 52,000). By 2002, the percentage had declined to 0.4% (200 of 47,000).

Milk Per Cow

Average annual production of milk per cow in Alabama was 11,163 pounds in 1985, some 1,861 pounds below the national average of 13,024 pounds (Figure 16.119). Data presented in the figure indicate that annual average milk production per cow in the period 1978–1988 increased more rapidly in Alabama than in the country as a whole. However, as detailed in the preceding chapter, it is likely that these data are somewhat misleading. This period coincided with the sharp decline in milk cow numbers (Figure 16.121). It is likely that the increase in milk per cow during the period was really the result of less efficient operators leaving the dairying business.

After 1984, annual average milk prices began to drift slowly downward, finally reaching $14 in 1988 (Figure 16.117). In 1989, milk per cow declined sharply. It would appear from these data that when prices began to decline, the financially-strapped, Alabama dairy farmers began to spend less money on the care and nutrition of their animals.

After 1992, milk cow numbers (Figure 18.106) began to trend sharply downward once again, and milk per cow began to increase. But in 1995, the price fell to $14. Afterwards, milk per cow once again began to decline, and from that point onward, the national and state averages began to diverge significantly.

In retrospect, it is difficult to understand why there were any dairy farmers remaining in Alabama after 1950. By that time, with the inter-regional transportation of milk becoming so much easier, the country had little need for Alabama's relatively high-priced product. Further, the differential between national and state prices was never sufficient to accommodate the differential in milk per cow. They fairly quickly realized this fact of life, and by 1970 there were 243,000 fewer milk cows in the state than in 1952. Of course, getting out of the business was not a quick, simple matter, and there were few opportunities to use their land and infrastructure for anything else.

Once again, a segment of Alabama agriculture had been caught in the ever-present, genome-environment trap. The genome of the milk cow simply does not respond very well to the state's subtropical climate. Without suitable housing, milk production is generally reduced in the temperate climate of February, and there is no practical means of countering the effects of the tropical climate of June, July, and August (Figure 16.122). In the late 1970s, the Alabama Agricultural Experiment Station attempted to stanch the flow of dairy farms out of the business. A new modern dairy farm and dairy was developed on the E.V. Smith Research Center, near Milstead. Unfortunately, data presented in Figure 16.119 indicate that the expenditure of those funds for that purpose had little, if any, positive effect. Problems with genome-environment interactions generally cannot be readily solved with new facilities or even with new research programs.

Pounds Produced

In 1985, Alabama dairy cows produced 547 million pounds of milk (Figure 16.123). Then production trended slowly downward to 500 million pounds in 1994 as the decline in numbers slowed (Figure 16.121). After 1994, however, with numbers declining more rapidly and with very little increase in milk per cow (Figure 16.119), production trended sharply downward to reach 277 million pounds in 2002.

In 2002, the Alabama Agricultural Service reported milk production data from only nine of Alabama's 67 counties. No data were reported from counties with 500 or less milk cows, or where publication of the data would disclose information on production from individual operations. However, these data are included in the state total. Milk cows in Cullman and Morgan counties produced 11.2 and 10.1% of the state total of 277 million pounds. None of the other counties reported more than 5% of the total.

Alabama Chickens

By the mid-1950s, Alabama farmers had found another commodity for which they had little comparative advantage—production of table eggs. As a result, chicken numbers declined. Fortunately, at about the same time, they began to find another use for their chickens, a use in which they had little competition—production of hatching eggs for the production of broiler chicks. Consequently, through the 1960s, numbers increased rapidly. In 1985, there were 16.9 million head in the state, excluding broilers, (Figure 16.125). Alabama's share of the national flock of 368.5 million head was 4.6%.

In 1985, the largest share (18.3%) of the state's chicken flock was on farms in Cullman County. Other counties reporting large shares were: DeKalb (10.5%), Blount (10.2), Marshall (7.7), and Lawrence (5.6). These five counties accounted for 52.3% of the state's total flock. All of these counties, with the exception of Lawrence (District 10), are located in District 20 (Figure 16.96). As might be expected, 53.7% of the state's broiler production that year was also located in District 20.

US and Alabama Prices

Most of the chickens in the United States are kept for the production of table eggs, and changes in chicken prices probably reflect changes in numbers that in turn respond to the supply and demand for eggs. Data on the annual average price per head (dollars) of all chickens in the state and in the country, excepting broilers, during the period 1970-2002 are presented in Figure 16.124. As shown by the data, annual average national prices generally trended upward from $1.90 a head in 1985 to $2.72 in 1997. They remained at that level through 1999, before declining to $2.37 in 2002. The decline after 1999 was probably the result of rapidly falling egg prices (Figure 16.132).

In 1985, the annual average Alabama chicken price per head was $1.90 (Figure 16.124). The Alabama prices shown in the figure include prices for those chickens held for producing table eggs, as well as those held for producing hatching eggs. After 1998, the Alabama Agricultural Statistics Service no longer reported the numbers held for each purpose. In 1985, the numbers held for producing hatching eggs accounted for 34% of the total; however, by 1998, this percentage had increased to 57%. This increasing percentage is primarily the cause of the sharp increase in prices shown in the figure ($1.90 in 1985 to $5.40 in 1999). Prices of chickens used for producing hatching eggs are much higher.

It is difficult to explain the rapid decline in prices after 1999 when chicken numbers were lower. Data on Alabama table egg production (Figure 16.136) suggest that in the period 1999-2002 chickens held for that purpose might have declined by almost 42%. If this estimate is correct, it suggests that numbers of chickens held for the production of hatching eggs might have increased by almost the same amount. An increase in numbers of this magnitude could have triggered a sharp decline in prices.

Numbers

Data on the estimated number of chickens on Alabama farms on December 1 of each year in the period 1970-2002 are presented in Figure 16.125. These numbers include chickens being held for the production of table eggs, as well as for the production of hatching eggs. Percentage of chickens being held for the production of table eggs was declining rapidly during the period (66% in 1985 to 43% in 1998). The data in the figure show that in eight of the 18 years in the period total numbers remained close to 16 million head. These data suggest that the increase in numbers being held for the production of hatching eggs was about equal to the decline in the numbers being held for table eggs.

Broilers

The American broiler industry has evolved from millions of small, backyard flocks on farms scattered across the country's rural landscape to 50 highly-specialized, vertically- integrated, agribusiness firms primarily located in the South (Krutchen, 2002). At one time, chickens were produced on family farms, primarily as a continuing source of eggs. Roosters obtained in this process and old hens were fare for the table. Today, the United States is the world's largest producer of poultry (chicken and turkey) meat. The free market system responsible for producing 32 billion pounds of ready-to-cook, inexpensive food each year is truly a modern miracle, and it should be noted that the industry has developed with no direct government subsidy. It is the most completely, vertically-integrated of all of our food- and fiber-producing industries. It involves a unique corporate- and individual-farmer relationship. Competition within the industry has lead to almost unbelievable increases in productivity, especially in the on-farm segment of the system. As a result, when considering the Consumer Price Index, prices of broiler meat are considerably lower now than 20 years ago. This price situation has been a godsend for people with limited incomes. In addition to this obvious societal benefit, however, these low prices tend to serve as a regulator on prices of competing meat products.

All of these consumer-friendly factors related to broilers have not been lost on the country's chicken eaters. Since around 1975, when American broiler production really took off (Figure 16.127), the consumption of poultry has trended steadily upward from around 40 pounds per capita to almost 100 pounds (Figure 16.105). During the same period, pork consumption has remained largely unchanged, while beef consumption has generally declined. In 2002, poultry and beef consumption were essentially equal.

Contracts between growers and processors for the production of broilers were first used in the 1950s. Since that time, the use of this procedure has increased rapidly. Now, virtually all of the broilers grown in the country are produced under contract. Perry, et al. (1999) noted that over time there have been substantial changes in some of the details of contracting. For example, when contracting began, the agreements included a per-bird payment or a simple per-pound fee. Today, contracts usually include three factors that determine total compensation for grower services: (1) a base payment or fixed-fee per pound of live meat produced, (2) an incentive or performance payment that is a percentage of the difference between all settlement costs of all contractor flocks during a specific period and costs associated with individual farmer performance, and (3) a disaster payment.

The settlement costs are determined by combining chick, feed, medication, and other costs usually associated with the production of a crop or cycle of broilers. This total is then divided by the total pounds of live poultry produced that provides a cost-per-pound measure of producer performance. In this relationship, lower values or lower costs per pound are indicative of better managerial performance. For below-average settlement costs, growers are usually awarded a bonus. In contrast, growers are penalized when their settlement costs are higher than the average of a pool of producers.

Over the years, contracting and vertical integration in the broiler industry has become concentrated in fewer and fewer larger companies. As noted previously, in 1960 the largest four broiler processing and marketing companies in the United States handled 12% of all contracting. In 2001, this proportion had increased to 50%.

US Broilers

In 1985, broilers produced on American farms totaled 4.5 billion. Production in Arkansas accounted for 17% of the total. Other states reporting large shares included: Georgia (15.1%), Alabama (12.5), North Carolina (10.0), and Mississippi (7.4). These five states accounted for 62% of the national total.

Prices

Data on American annual average broiler prices for the period 1970-2002 are presented in Figure 16.126. By 1985, most of the broilers produced in the country were grown under contract. As a result, relatively few of them were sold on the open market. Subsequently, the price data shown in the figure are not true farmgate prices. Rather, they are a live-weight-equivalent price based on wholesale prices of processed whole birds. In the period 1985-2000, there was considerable year-to-year variation in prices, but for most of the period, they were between 30 and 40 cents a pound. Price was 30.1 cents a pound in 1985. Afterwards, there was considerable annual variation through 1991, when it reached 30.8 cents. Between 1991 and 1998, prices trended upward to 39.3 cents, but by 2002, it was down to 30.5 cents, or about where it had been in 1985. This year-to-year variation did not seem to be related to variation in numbers (Figure 16.127).

Numbers

Data on the annual numbers of broilers produced in the United States during the period 1970-2002 are presented in Figure 16.127. Over the 18-year period (1985-2002), the graphed data have the characteristics of a typical sigmoid curve or tilted S-shaped curve. Such a curve is common in biology and is often associated with phenomena in which the increase becomes self-limiting. In this situation, broiler numbers began a growth phase around 1986 with 4.6 billion birds and continued until reaching an inflection point around 1990 (5.9 billion). The inflection point should have served to alert the industry that there were factors operating that would likely limit increases in the future. Afterwards, numbers entered a declining phase that was continuing through 2002 (8.6 billion). In situations such as this, often the factors that led to the declining phase are removed, and another growth phase begins. The slight increase in numbers between 2001 and 2002 indicate that a new growth phase might have been beginning.

In 2002, Georgia reported the largest share (15%) of the total number of broilers produced in the country. Other states reporting large shares included: Arkansas (13.8%), Alabama (12.3), Mississippi (9.0), and North Carolina (8.6). These five states accounted for 58.7% of the national total. With neither Virginia nor Maryland in the top five states and with North Carolina in fifth place, it appears that the industry was continuing to move southwestward. These data indicate that the old Red Jungle-fowl genome had found a happier home in the Lower South.

MARKET AGE

There have been many changes in the American broiler industry since its inception, but none have been more amazing than the decline in market age (Figure 16.128). In 1950, broilers were ready for market at about 70 days. By 1970, market age had been reduced by 20% to 56 days. Then in 1985, it was down to 49 days. Not surprisingly, it has not changed very much since. Further, it is difficult to imagine it being reduced very much in the future. The importance of the reduction in market age is that it allows the farmer to increase the number of production cycles that can be completed within a year, which, in turn, increases the efficiency of the use of production facilities.

MARKET WEIGHT

Increase in market weight of broilers in the United States is another amazing story. In 1970, broilers were marketed at an age of 56 days and at a weight of 3.76 pounds. In 1985, they were marketed at 49 days, at a weight of 4.37 pounds. While market age did not change much between 1985 and 2000, market weight increased by 26% to 5.53 pounds.

FEED CONVERSION

In dollars corrected for inflation, broiler meat is much cheaper at the beginning of the twenty-first century than it was in the middle of the twentieth. A major reason for this phenomenon has been the decline in the quantity of feed required to produce a pound of broiler (feed conversion rate). In 1950, the rate was near 3.0 (Figure 16.130). Between 1950 and 1970, it declined from 3.0 to 2.25. Then by 1985, it was down to 2.0. It has not changed very much since. Without some miraculous breakthrough in nutrition and physiology, the industry average is not likely to decline much further. However, some good operators are already obtaining rates below 1.95, and they might be able to see some further improvement, but not much.

Alabama Broilers

In 1985, Alabama ranked third in the country in the production of broilers, with 12.5% of the national total. The largest share (23%) of the 561.8 million-bird production was reported from Cullman County. Other counties reporting large shares were: Marshall (9.4%), DeKalb, (8.5), Coffee (6.0), and Winston (4.5). Broiler production reports were obtained from 45 of Alabama's 67 counties. The largest share (54.7%) of birds was produced in District 20 (Figure 16.96), which encompasses much of the Cumberland Plateau Physiographic Section (Figure 1.09). The smallest share (1.4%) was reported from District 40, which includes most of the Black Prairie Physiographic District (Figure 1.09). A surprisingly large share (17.7%) was reported from District 60, in the southeastern corner of the state.

In 1985, cash receipts from the sale of broilers in Alabama accounted for 48.4% of total cash receipts of all livestock and 29.3% of the cash receipts for all farm commodities, including farm forest products.

Prices

Data on prices for Alabama broilers during the period 1970-2002 are also presented in Figure 16.126. From 1985 through 1987, Alabama prices were a few cents lower than average national prices, but afterwards there was little difference.

Numbers

Data on Alabama broiler numbers during the period 1970-2002 are presented in Figure 16.131. Production in 1985 was 561.8 million birds, and numbers were in an exponential growth phase similar to the one seen in national numbers in the same period (Figure 16.127). The sigmoid curve depicting the Alabama data also reached its inflection point at about the same time as the national data, but the declining phase was quite different. From 1992 through 1996, there was little change in numbers, as they ranged around 900 million birds. This phenomenon seemed to be associated with somewhat lower prices (Figure 16.126). After 1996, numbers began to grow exponentially again, but the growth phase was aborted in 2000 with the number at 1,038.7 million. This event seemed to be related to a sharp decline in prices in 1999 and 2000. In 2002, production was 1,051.3 million, or near where it had been in 2000.

In 2002, Cullman County reported the largest share of one billion birds produced that year. Other counties with large shares included: DeKalb (10.4%), Marshall (6.7), Blount (5.3), and Coffee (4.9). These five counties accounted for 43.5% of total production. A total of 19 counties reported production of at least 20 million birds. The Alabama Agricultural Statistics Service reported broiler production from 41 counties. Farmers in other counties produced small numbers of broilers, but these data were not published to protect the data from individual operations. Those numbers, however, are included in the state total.

PRODUCTION Costs

Strawn, et al. (2000) have published estimates on the costs of producing broilers in Alabama in 2000 and 2001. Table 16.62 includes information on some of the characteristics of the production procedures used to estimate the costs. The estimates are based on two production plans: (1) production of a large number of smaller birds (4.30 pounds) and (2) production of a smaller number of larger birds (6.60 pounds).

Total Cash Receipts. Total cash receipts from the sale of small and large broilers were $37,445 and $35,529, respectively (Table 16.62). The larger amount received for the heavier birds was not sufficient to overcome the smaller number of birds produced (107,666 versus 174,162).

Variable Costs. The estimated average total variable costs (Table 16.63) (dollars per "house") for producing small or large birds were $10,805 and $8,975, respectively. These numbers represent only a fraction of the real variable costs. These are only the costs that are the responsibility of the producer. The integrator provides the chicks, feed, medications, and veterinary needs. These integrator-supplied inputs are generally equal to well over 50% of the variable costs. Over 50% of the total variable costs provided by the production systems was for electricity and propane. These costs, related to keeping the birds warm in the winter and cool in the summer, are relatively high. The typical poultry house contains almost half an acre of floor space (20,000 square feet). It is extremely expensive to keep this much space either warm or not overly hot throughout the year. Electricity and propane costs were slightly lower for producing large birds because of the lower number of cycles (flocks) (5.10 versus 6.60) passed through the house during the year.

Fixed Costs. As expected, fixed costs were much higher than variable costs for both production systems (Table 16.63). Interest on buildings and equipment was a major portion of the fixed costs. Simpson, et al. (2000) commented that it costs $135,000 to build and equip a typical broiler house. Approximately 45% of the initial investment relates to construction and 55% to equipment purchase and installation. These houses are usually financed for 15 years. Depreciation of buildings and equipment is even higher than interest payments. This budget sets aside $7,575 a year for this purpose. These costs involve setting aside enough money each year to replace worn out machinery or buildings that become so dilapidated from constant use that they must be replaced. These costs do not have to be paid each year, but in the long-term they must be covered if the operator expects to remain in business.

ANNUAL RETURNS

Estimated annual returns above all specified expenses (dollars per "house") for producing small and large birds are presented in Table 16.64. In both production systems, receipts were sufficient to cover all variable costs, as well as all specified costs. Unfortunately, net returns (cash receipts – all specified costs) for both systems ($10,765 for small birds and $10,680 for large birds) would be relatively small to pay operators for their labor and risk. Simpson, et al. (2000) comment that most broiler producers operate four houses, usually with 20,000 square feet each. With all things being equal, the four houses would improve the bottom line substantially. Based on the estimates by Strawn, et al., with four houses a producer should clear between $40,000 and $50,000 a year.

Brown, et al. (2006) analyzed costs and returns from

a number of broiler operations in Alabama in 2001 and 2002 as part of the work of the Alabama Farm Analysis Association. Although broiler production was the primary business of the operations, most of the operators kept beef cattle to use waste litter. Also, most of the operators did some custom work for other farms. Data in Table 16.65 are based on average costs and returns for the entire operation on 15 farms in 2001 and 17 farms in 2002. Actual data have been converted to costs and returns per 1,000 square feet of broiler housing space on each farm. The data show that in 2001 the average return to management on the 15 farms was a negative $60.18 per 1,000 square feet of housing. In 2001, the average farm had 75,819 square feet of housing. As a result, return per farm was a negative $4,563 ($-60.18 x 75.82). In 2002, the average return to management per 1,000 square feet was $46.03; with an average of 100,781 square feet of housing on each farm, return per farm was $4,639 ($46.03 x 100.781). Given the amount of investment required to produce broilers, these are relatively low returns.

The estimates of returns to management are actually somewhat better than they appear in Table 16.65; unpaid labor is charged against gross farm returns. It is the value placed on uncompensated labor contributed by the operator(s) and family to the farm. It is based on estimates of what the operator(s) and their families might earn working for another farm. If these charges are added back to gross returns, the bottom line looks somewhat better. In 2000, the corrected estimate of return to management would be $133.33 rather than-$60.18, and in 2001 it would have been $194.42 rather than $46.03.

Taylor (2002) has published the results of a study on the economics of the contract broiler production system in Alabama. He concluded that returns to management and risk, based on production in four 20,000 square feet houses, averaged a negative $7,006 (adjusted for inflation and expressed in 2002 dollars) over the six-year period (1995-2000). Based on these data, he stated that "records from other impartial sources of information on profitability of contract poultry production in other states also show that profitability has decreased to the point where many contract producers have a poverty level of income."

In 2002, cash receipts for broilers in Alabama amounted to 51.7% of cash receipts of all farm commodities, including farm forest products. Obviously, broiler production is an extremely important industry in Alabama, but the studies by Brown, et al. (2006) and Taylor (2002) indicate that it, like most other commodities produced in the state, is not in a very healthy condition.

Eggs

The United States at the beginning of the twenty-first century is the second leading egg-producing country in the world. Of the total of 85.7 billion eggs produced in the country in 2001, approximately 72.8 billion (85%) were marketed for human consumption and the remainder as hatching eggs. Of those eggs sold for human consumption, 58.3% were sold at retail outlets as whole or shell eggs, primarily for home consumption; 22.8% were sold for use in food service businesses and institutions; 30% were further processed and marketed in liquid or dried form; and 0.8% were exported.

US Eggs

According to Madison and Harvey (1997), the number of American commercial farms with 3,000 or more laying birds declined from 2,000 in the mid-1980s to below 1,000 in the mid-1990s. Fewer farms and increasing production indicate that farm technology has changed dramatically. Individual nests in small laying houses with free-ranging birds have been replaced by enormous houses holding upwards of 100,000 birds each. Laying birds are confined in stacked cages with six or seven birds per cage. Cages are provided with automated feeding, egg collection, and manure removal. It is not uncommon to have well over a million laying birds on a single property. In 2000, the average layer laid 256 eggs a year.

CONSUMPTION

Surprisingly, with egg production remaining relatively constant from 1985 through 1990 (Figure 16.134) and a steadily-increasing American consumer population, per capita consumption continued the downward trend (Figure 16.132) that began in the 1950s (Madison and Harvey, 1997). In 1985, total annual per capita consumption was 255 eggs. It continued to decline through 1991 (234 eggs) before beginning to slowly trend upward to 180 in 2002. Shell egg consumption generally paralleled total egg consumption during the period 1980-2002. Consumption was 236 eggs in 1980, before trending downward to 172 eggs in the mid-1990s. It also began to increase afterwards, but at a lower rate than total eggs. In 2002, per capita consumption of shell eggs was 180. In 1980, Americans consumed a relatively small share (12.9%) of their eggs in processed form; however, afterwards the share began to increase, and by 2000 the percentage had more than doubled to 29.2%. It remained at near that level through 2002.

Madison and Harvey (1997) commented that the decline in per capita consumption through the early 1990s was a result of a change in lifestyles beginning in the mid-1980s. As American work habits changed, they had less time for breakfast preparation, which included the preparation of eggs. Also, it declined as a result of the perceived ill effects of cholesterol intake associated with egg consumption. The increase in consumption afterwards seemed to be largely the result of people eating more eggs in convenience and snack foods and apparently with less concern with the supposed negative attributes of eggs when they are eaten in that form.

Prices

American egg prices slowly trended upward through the late 1970s and early 1980s to reach 57.2 cents a dozen in 1985 (Figure 16.133). Afterwards, however, with numbers increasing (Figure 16.134), prices generally declined, reaching 52.8 cents in 1988. Numbers began to decline after 1987, and prices responded by increasing sharply. In 1990, they reached 70.9 cents. By the early 1990s, the production of eggs for hatching broiler chicks was beginning to force numbers sharply upward. At first, prices responded by declining, but then increased sharply. In 1996, the annual average price reached 75 cents. It had never been that high before. Afterwards, however, with numbers soaring, prices began a long decline, finally reaching 59 cents in 2002—about where it had been in 1976.

It is difficult to fully understand the changes in egg prices during this period, because they reflect prices for both market and hatching eggs. The problem is compounded by the increasing importance of hatching eggs over time. For example, in 1970 only 7% of all eggs produced in the United States was used for hatching. By 1985, this percentage had increased to 9.6%, and in 1997 it was 13.8%. Then by 2002, it had fallen back to 13.2%.

Numbers

Data shown in Figure 16.134 are combined production numbers for table (market) eggs and hatching eggs. As these data show, the total number was 68.2 billion in 1985. It had been near that level since the late 1970s. Numbers increased slightly from 1985 through 1987 (70.4 billion), then followed prices (Figure 18.116) down to 67.2 billion. Afterwards, they began to trend sharply upward, primarily as a result of a strong upward trend in the production of broiler hatching eggs. Further, beginning in 1990 increased exports also helped fuel growing production. In 2002, total egg production reached 87 billion.

Alabama Eggs

In 1985, the Alabama Agricultural Statistics Service reported that the state ranked ninth in the country, with 4.1% of total national egg production. Layers in Cullman Country produced the largest share (15%) of the state total of 2.7 billon eggs that year. Other counties reporting large shares included: DeKalb (12%), Marshall (11.0), Blount (10.7), and Lawrence (6.0). These five counties accounted for 54.7% of the state total. District 20 (Figure 16.96), primarily on the Cumberland Plateau (Figure 1.09), reported the largest share (57.1%) of the total. Districts 40 and 50 reported the smallest shares: 3% and 2.7% respectively. A total of 24 counties reported productions of 20 million or more eggs.

Prices

Data on the annual average prices of market eggs in Alabama during the period 1985-2002 are presented in Figure 16.135. The Alabama Agricultural Statistics Service did not report this type of data before 1985. Year-to-year changes in Alabama prices show a striking similarity to national prices for that period (Figure 16.133). However, while year-to-year changes were similar, average national prices were considerably higher, but those prices included some hatching egg data.

Alabama egg prices were 50.1 cents a dozen in 1985. With egg numbers relatively high in the early 1980s (Figure 16.136), prices trended downward, reaching 38.2 cents in 1988. During this period, numbers followed prices downward. Then with declining numbers, prices began to increase, finally reaching 60.9 cent in 1990. This teeter-totter relationship continued through 2002, with prices rising to 62 cents in 1996, before falling to 42 cents in 1999.

Numbers

The Alabama Agricultural Statistics Service began to report numbers of market (table) eggs and hatching eggs separately in 1983. These data are presented in Figure 16.136. They show that layers in the state produced 2 billion market eggs in 1983. Then, responding to the teeter-totter relationship between prices and numbers from 1984 through 1990, numbers sharply trended downward to one billion in 1990 before trending upward to 1.3 billion in 1994. Afterwards, with prices generally declining, numbers began to trend downward, finally reaching 635 million in 2002. Between 1985 and 2002, Alabama market egg numbers fell by 67.4% (1.9 billion to 635 million).

As Alabama's broiler industry expanded, numbers of hatching eggs produced in Alabama generally trended upward throughout the entire period 1985-2002—from 846 million in 1985 to 1.6 billion in 2002 (Figure 16.136). Numbers were down somewhat in the period 1994-1997. As shown in Figure 16.131, there was very little increase in broiler production in the state during that period; consequently, demand for hatching eggs increased very little. In 1983, only 27% of all eggs produced in the state was used for hatching; however, by 2002 about 72% was used for that purpose.

The Alabama Agricultural Statistics Service reported that in 2002 layers in the state produced 2.3 billion eggs (market and hatching). Cullman County reported the largest share (17%). Other counties reporting large shares included: DeKalb (15.1%), Marshall (9.9), Baldwin (7.4), and Blount (6.2). These five counties accounted for 55.6% of the total. Production data was reported for only 20 counties. Other counties reported limited production, but the data were not published to protect the identity of individual operations.

Production COSTs for BROILER HATCHING EGGS

Strawn, et al. (2001) estimated the costs and returns of producing broiler hatching eggs under contract in Alabama for the period 2000-2001. Table 16.66 lists variable and fixed costs (dollars per 1,000 laying hens) of egg production for one year. As was the case with contract broiler production, the integrator provided the birds, feed, and veterinary services. As a result, no costs related to these items appear in variable costs.

Gross Receipts. Gross receipts from the sale of eggs and for reimbursing the farmer for growing-out the laying hens was $3,900.15 per 1,000 hens (Table 16.66).

Variable Costs. The major variable cost items were electricity ($472.50), equipment repair ($241.15), and miscellaneous ($111.11). Total variable costs per 1,000 hens was $907.47 (Table 16.66).

Fixed Costs. As was the case with broiler production, fixed costs were much larger than variable costs ($2,223.51 versus $907.47) (Table 16.66). This difference is the result of the high cost of constructing buildings and of purchasing the required equipment. Strawn, et al. (2000) estimated that the total investment in the house, water well, pump, generator, incinerator, etc. would total $171,700. As a result, interest and depreciation related to these facilities were relatively high.

RETURNS TO MANAGEMENT

As shown in Table 16.67, gross receipts exceeded variable costs by $2,993 ($3,900 - $907) and all specified costs by $769 ($3,900 - $3,131). If the return per 1,000 laying hens is expanded to return per house of 9,000 layers, it becomes $6,922. This is a relatively small sum to pay the operator for his labor and his risk associated with investing in a $171,700 facility. These data represent a hypothetical situation. While actual data might deviate somewhat, the estimates provided by Strawn, et al. (2001) are useful guidelines for those interested in entering the business of producing broiler hatching eggs in Alabama.

Catfish

In 2001, American per capita consumption of beef, pork, poultry, and mutton combined was 219 pounds (Figure 16.5). Fish and seafood added only another 15 pounds to this total. Obviously, fish and seafood play a relatively minor role in the American food industry, and channel catfish play an even smaller role. Catfish constituted only 0.6% of total meat consumption. Americans consume most of their fish and seafood away from home. Consequently, the amount consumed per capita at any one time is a good measure of general economic conditions at that time. When the economy is good, employment is high and personal income increases. Under these conditions, Americans eat more of their meals away from home. Further, even when people decide to eat out choosing fish from a menu is seldom a matter of economics. Fish and seafood items are always among the most expensive items available. Another complicating factor is that much of the fish and seafood consumed in the United States is captured in the wild; therefore its availability is not nearly as predictable as farmed foods. Further, most of the important stocks of those aquatic animals, which constitute the bulk of the fish and seafood that we consume, are already being seriously over-exploited, forcing commercial fishermen to go further afield to locate new species. All of these factors that affect the consumption of fish and seafood, weigh heavily on the consumption of channel catfish.

Data published by the National Marine Fisheries Service in the 2004 edition of *Fisheries of the United States* on American per capita consumption of commercial fish and shellfish for the period 1985-2002 are shown in Figure 16.137. Consumption zig-zagged around an average of 15.5 pounds per capita during the entire period. The high for the period was 16.2 pounds in 1987, and the low was 14.6 pounds in 1997. There is little evidence in these data that fish and shellfish are becoming more important in the American diet.

US Catfish

The quantity of catfish processed in the United States increased from 5.7 million pounds in 1970 to 191.6 million pounds in 1985. Data on the cash receipts from the marketing of catfish is not available before 1983. In that year, American farmers received $83.9 million from their sale. In 1985, that value had increased to $139 million. Although the production of channel catfish had grown rapidly in the past 15 years, the industry continued to be highly regionalized. It was still largely confined to four contiguous states in the Deep South (Alabama, Arkansas, Louisiana, and Mississippi). A majority of production is also consumed in the South.

CATFISH CONSUMPTION

In 1990, annual American per capita consumption of dressed catfish was 0.6 pound (Figure 16.138). From that point, it trended steadily upward to 0.9 pound in 1993 before declining to 0.75 pound in 1994. After 1994, consumption began to increase again, finally reaching 1.16 pounds in 1999. It remained near that level through 2002. In that year, Americans consumed 1.1 pounds of farm-raised channel catfish; consumption of this species was exceeded only by shrimp (3.7 pounds per capita), canned tuna (3.1 pounds), salmon (2.02 pounds), and Alaskan pollock (1.12 pounds) (Table 16.68).

CATFISH IMPORTED

Data on the quantity of catfish imported into the United States during the period 1970-2002 are presented in Figure 16.139. These data show that as American production of channel catfish surged after 1980 (Figure 16.141), imports declined from 15 million pounds in 1980 to less than one million pounds in 1997. In the late 1990s, however, imports from Southeast Asia began to increase rapidly. By 2001, imports totaled 18 million pounds. This surge in imports resulted in extreme hardship in the domestic catfish industry.

Prices

American catfish prices paid to producers at the processing plant ranged between 71 and 76 cents a pound in nine of the 17 years in the period 1985-2002 (Figure 16.140). They were below that range in 1986 and 1987 when production outpaced marketing efforts—in 1991 and 1992 as a result of the effects of the 1990-1991 recession, and in 2001 as a result of the slowdown in the economy that began late in 2000 and the 9/11 terrorist attack. Prices paid producers and fisherman of all seafood items plunged after September 11, 2001, after eating-out also declined sharply. Prices above the 71 to 76-cent range were related to the rapidly expanding economy during the mid-1990s. Prices were also lower in 2001 and 2002. In 2002, the average price was 56.8 cents; it had not been that low since 1982. It is likely that the lower prices in 2001 and 2002 were related to the high level of imports in those years (Figure 16.139); however, in both years imports amounted to less than 3% of the total weight of catfish processed in the United States (Figure 16.141).

CATFISH PROCESSED

The quantity of channel catfish processed in the United States generally trended upward from 192 million pounds in 1985 to 604 million in 2002 (Figure 16.141). The trend was interrupted after 1992, probably as a result of the lingering effects of the 1990-1991 recession. Prices were sharply lower in 1991 and 1992 (Figure 16.140). Per capita consumption of all fresh and frozen seafood in the United States was down significantly during the same period, 1990-1993 (Figure 16.137). The upward trend in catfish processing was re-established in 1995 and continued through 1999 when it reached 597 million pounds. Processing remained near that level through 2002, probably as a result of lower prices (Figure 16.140), increased imports (Figure 16.139), and declining consumption (Figure 16.138).

Alabama Catfish

The Alabama Agricultural Statistics Service did not report cash receipts for channel catfish separately until 1987. In that year, receipts for the species accounted for 0.9% of total receipts for all livestock and poultry and 0.6% of all commodities produced on Alabama farms, including farm forest products.

Data on cash receipts for the sale of catfish for individual counties were not published before 1988. In that year, farmers in Hale County received the largest share (42.5%) of the state total of $23.2 million. Other counties with large shares included: Dallas (12.8%), Greene (7.6), Perry (7.2), and Sumter (3.6). These five counties accounted for 73.7% of the total. Farmers in District 40 (Figure 16.96) received 78% of all receipts.

These data show that as early as 1988 the industry was highly localized on the western extremity of the Black Prairie Physiographic District of the East Gulf Coastal Plain Section. Soils in that portion of the Black Prairie are well suited for pond construction. Also, the surface relief of land in that area is nearly ideal for capturing rainfall to fill the ponds. Finally, there is a substantial amount of ground water in the Eutaw Formation (see Figure 1.14) that lies several hundred feet below the surface of the land in that area; water found in that formation is pumped to the surface to fill ponds in that area.

CATFISH OPERATIONS AND ACRES IN PONDS

The National Agriculture Statistical Service reported that in 1989 there were 352 catfish operations (farms) in Alabama (Figure 16.142). The number remained at about the same level through 1992, when it began to decline. By 1997, it was down to 245. American catfish prices declined from 75.8 cents a pound in 1990 to 59.8 cents in 1992 (Figure 16.140). This decline seems to have been sufficiently large to eliminate a substantial number of marginal operators. Even though prices began to increase again in 1993, numbers continued to decline. Prices reached their highest level in history (78 cents a pound) in the period 1994-1996. Even though prices declined about five cents after 1996, a few new operations were established each year after 1998, and by 2000 there were 270. Unfortunately, just as farmers were beginning to gain more confidence in the future of the industry, prices declined sharply in 2001; 30 operations closed between 2001 and 2002.

The average number of acres per catfish operation trended upward from the upper 30s in 1989 to near 90 acres in 1998 (Figure 16.143). They generally remained near the same level through 2001. In 2002, the average increased sharply to 108 acres. It is likely that this sharp increase was the result of smaller operations leaving the industry after 2001 (Figure 16.142).

Total acres in Alabama catfish ponds remained at around 17,000 to 18,000 from 1990 to 1995 (Figure 16.144), but began to trend upward in 1996. In 2002, total acreage reached 25,900. In 1999, total acres in Mississippi and Alabama were 105,000 and 21,300, respectively, and increased to 111,500 and 25,900, respectively, by 2002.

Prices

Trends in prices paid for catfish in Alabama during the period 1985-2002 were similar to those of prices for catfish elsewhere in the country (Figures 16.140).

QUANTITY OF FISH SOLD PER ACRE

Alabama catfish farmers increased their stocking rates and production inputs during the period 1993-1999. As a result, pounds marketed per acre of water in production increased from 3,700 in 1993 to 5,500 in 1999 (Figure

16.145). Then in 2000 and 2001, catfish farmers began to reduce inputs as concern over prices increased and as pounds sold began to decline, reaching between 5,000 and 5,100 pounds in 2001. Pounds sold per acre in Alabama was much higher than in the United States as a whole.

TOTAL QUANTITY MARKETED

USDA data on the quantity of catfish sold by Alabama farmers during the period 1993-2001 are presented in Figure 16.146. As expected—with both total acres (Figure 16.144) and pounds sold per acre (Figure 16.145) trending upward during the period 1993-2001—total quantity sold annually also trended upward (Figure 16.146). Farmers sold 50 million pounds in 1993 and 115 million pounds in 2001.

CASH RECEIPTS

Data on cash receipts from the sale of Alabama farm-raised catfish during the period 1987-2002, published by the Alabama Agricultural Statistical Service, are presented in Figure 16.147. Although there was considerable year-to-year variation in catfish prices during the period (Figure 16.140), there was no general trend. However, with pounds marketed annually trending upward during the period 1993-2001 (Figure 16.146), cash receipts followed the same trend (Figure 16.147). During the period, they increased from $14.2 million in 1993 to $76 million in 2002. During the 1987-2002 period, cash receipts increased by 435.2% ($14.2 to 76 million). During the same period, the Consumer Price Index only increased by 58.4%. The sale of Alabama catfish made a significant, real contribution to the economy. In the period 1987-2002, the sale of farm-raised catfish contributed slightly less than $751 million to the economy of the most economically disadvantaged area of Alabama.

Cash receipts from the sale of catfish in 1987 accounted for 0.9% of total receipts for all livestock and poultry sold from Alabama farms and 0.6% of all commodities produced on Alabama farms, including farm forest products. In 2002, catfish receipts accounted for 3.2% of receipts for livestock and poultry and 2.8% of all commodities, including farm forest products. In 2002, catfish receipts exceeded those of both hog and dairy.

PRODUCTION COSTS

Crews, et al. (2001) published estimates of the cost of producing catfish in watershed ponds in west Alabama in 2001. Characteristics of the farming enterprise used to estimate costs are shown in Table 16.69.

Gross Receipts. The authors assumed a selling price of 70 cents a pound and with production at 5,480 pounds an acre (4,000 fish at 1.37 pounds each), gross receipts would have been $3,836.

Variable Costs. Variable costs in dollars per acre resulting from the production of channel catfish in watershed ponds in Alabama are presented in Table 16.70. The most significant of all the variable costs were those resulting from the use of pelleted floating feed ($1,226). This single cost accounted for 53.5% of total variable costs ($2,289). This high percentage underscores the central role of feed prices and feeding practices in catfish production. The second most significant cost was the total price for the 6,000 fingerlings stocked per acre ($500). This cost accounted for 21.8% of the total. Interest on operating capital ($195) was also another substantial variable cost.

Financing Costs. The authors estimated that annual principal and interest payments associated with the operation would have been $803 an acre (Figure 16.20).

Total Specified Expenses. The total of all specified expenses was $3,092. Of this total, 74% was associated with the variable costs of production (Figure 16.70).

Net Returns to Management

Net returns above variable costs were $1,546 an acre ($3,836 - $2,290). Net returns above all specified costs were $744 an acre ($3,836 - $3,092).

Hunting

Hunting has been important to Alabamians for thousands of years, and it remains an extremely important activity in Alabama's rural landscape at the beginning of the twenty-first century. Because there is so little public land in the state, most of the benefits of hunting accrue to private landowners in rural areas. Further, hunting brings large numbers of people into rural communities who would not go there otherwise. Large quantities of supplies needed by hunters are purchased locally. Hunters are concerned about the quality of rural environments and interested in maintaining those environments as habitat for game animals.

The essential role that hunting played in the lives of the earliest Native Americans living in Alabama was described in a preceding section. Even after these early inhabitants began to farm, hunting remained an essential part of their everyday lives. They never produced enough food with their primitive agriculture to meet all of their needs. Native Americans had their first experiences with commercial hunting after the English colonized the Carolinas and Georgia. The English fell in love with deerskin breeches, and the Lower South was the only place where these skins could be obtained in large quantities. In 1755, some 50,000 deerskins were exported from the region to England. By 1772, that number had increased to over 200,000 a year. The Indians exchanged deerskins (hence the term "bucks" for money) for English trade goods (pots, pans, axes, hoes, beads, blankets, guns, gun powder, etc.).

Hunting was also extremely important to the lives of the early pioneer settlers in the South. They also had difficulty producing enough food to meet their needs. Although settlers along the Atlantic Seaboard became self-sufficient in food production, pioneers pushing inland hunted to supplement the meager amounts of food that they were able to carry with them. Hunting also allowed them to survive until they could make their first crop.

Since time immemorial, hunting has been such an important human activity that it became an integral and permanent part of human behavior. Even after it was no longer necessary to hunt to eat, humans continued to do so. Little (1971) in his *History of Butler County, Alabama* noted that many of the early (Scotch Irish) settlers in that area spent considerably more time hunting than was necessary, even neglecting their crops and the care of their families and farmsteads.

Hunting remained critical to segments of the population in the Lower South long after the region became able to produce enough food to provide for its people's needs. In *Poor But Proud* (1989), Wayne Flynt writes about the Odoms, a cotton mill family who lived in Cordova in Walker County between the two world wars. "They maintained the antebellum folk tradition of markmanship . . . . Nor were hunting and fishing merely recreational. Game added precious protein to their diet." In *An Hour Before Daylight* (2001), Jimmy Carter (2001) comments on the importance of hunting and fishing to tenant farmers during his childhood in Plains, Georgia. "Hunting, fishing, and trapping were important sources of additional meat used to supplement the cheap pork fatback that was used for grease, for seasoning, and as a source of protein."

At the beginning of the twenty-first century, only a small proportion of the American population hunts for food anymore; although very little of the game that is taken by hunters is wasted. Most hunters take great pride in eating what they kill. And even when they do not intend to eat it themselves, most go to great lengths to see that people who can use the food, get it. For example, the Buckmaster American Deer Foundation collects donations from it members to fund Project Venison. Under this program, hunters take their excess deer to a local processing plant where it is prepared, packaged, and given to a local food bank. Local chapters of the foundation pay for the cost of processing.

THE NATIONWIDE ECONOMIC IMPACT OF HUNTING

The International Association of Fish and Game Commissioners recently employed Southwest Associates of Fernandina Beach, Florida, to study the contribution of hunting to the American economy during the 2001-2002 hunting season (Animal Use Issues Committee, 2002). The report entitled, "Economic Importance of Hunting in America," indicated that 13 million Americans (4.6% of the population age 16 and older) hunted. Some 11.1 million (85%) hunted deer. Hunting in the United States resulted in the expenditure of $25 billion. On average, each American hunter spents $1,896 (5.5% of their annual income) on the sport. Table 16.71 includes data on expenditures for some of the more important items. In addition, hunters paid $17 billion to those who helped provide them with the opportunity to hunt. Altogether, hunting in the United States supported the employment of 575,000 people.

Texas led the country in retail sales related to hunting with expenditures of $1,761.3 million. Pennsylvania ($1,165.1 million), Wisconsin ($960.1 million), New York ($891 million), and Alabama ($799.3 million) followed.

THE ECONOMIC IMPACT OF HUNTING IN ALABAMA

Hunting in Alabama supported 16,870 jobs in 2001, resulting in the payment of $364 million in wages (Table 16.72). In turn, these wage earners paid $37 million in federal income taxes. Hunters themselves spent $31 million on gasoline sales and fuel taxes. Obviously, most of the expenditures resulted from hunting by native Alabamians, but out-of-state or non-resident hunters were responsible for a substantial portion of the total, possibly as much as 28%.

Alabama hunters ranked fifth in the country in the retail purchase of hunting related items. Each Alabama hunter spent an average of $1,900 for these items. Altogether their expenditures totaled $799 million. Applying the economic multiplier to all of these hunting expenditures results in a $1.5 billion economic impact on the state's economy that year.

NUMBER OF HUNTERS

Data obtained in the survey estimated 422,780 hunters in Alabama in 2001. A total of 117,450 of these were non-residents. Some 89.6% of the total (378,750) indicated that they were primarily deer hunters.

HUNTING DAYS

Data obtained from hunter interviews were used to estimate that all hunters spent 7.6 million days afield 2001. Some 80.8% (6.3 million) of those days were spent hunting deer.

Agricultural Production

As detailed in the preceding chapter, value added to the American economy by the production of crops and livestock, combined, increased from $51.3 billion in 1970 to $142.8 billion in 1985, an increase of 178.4%. During the same period, the Consumer Price Index increased by 177.3%. In 1970, value added by the production of livestock contributed the larger share of the total (60%), but by 1985 the crop share was slightly larger than it had been (51.9%).

Annual data on value added by crop and livestock production in the United States during the period 1970-2002 are presented in Figure 16.148. Total value added increased from $142.8 billion in 1985 to $191.9 billion in 2002 (Figure 16.148), an increase of 34.4%. During the same period, the Consumer Price Index increased by 67.2%. Although there was considerable year-to-year variation, total value added generally trended slowly upward from 1986 through 1995, before increasing sharply in 1996 and 1997 as a result of a sharp increase in the production of feed crops in those years.

Value added for American crop production increased from $74.1 billion in 1985 to $98.4 billion in 2002 (Figure 16.148), an increase of 32.8%. Year-to-year variation was considerable. Value added in 1986, 1987, and 1988 was somewhat lower than expected, as a result of a decline in agricultural prices. Then in 1996 and 1997, it was somewhat higher than expected, as a result of the higher price situation. In 1985, value added by crops accounted for 51.9% of the total, but by 2002 it had declined slightly to 51.2%. It is likely that the temporary decline in value in the South's rapidly growing broiler industry contributed to this slight reduction.

During the period 1985-2002, value added by American livestock production increased by 36.1% ($68.7 to 93.5 billion). Year-to-year variation was much lower for livestock than for crops (Figure 16.148). The nature of livestock production contributes to greater stability, and it is much less susceptible to adverse weather.

Data on value added to the American economy by the production of some specific crops and groups of crops in 1980, 1985, 1990, 1995, and 2002 are presented in Table 16.73. These data show the importance of the production of feed grains (primarily corn) and oil crops (primarily soybeans) to the American economy. These two groups of crops accounted for 40-50% of total value added by crop production in all of those years.

National value added data for specific groups of livestock are also presented in Table 16.73. These data show the overriding importance of the production of meat animals (primarily cattle) to the economy. In all the years, value added by the production of meat animals accounted for at least half of the total. Value added for meat animals increased during the period, but this was largely the result of the inflation of the currency during the period. During the period, value added for poultry (broilers) and eggs increased at a much greater rate than inflation.

Data on the annual total value added from the production of crops and livestock in Alabama during the period 1970-2002 are presented in Figure 16.149. The data indicate that it increased from $2,062.2 million in 1985 to $2,954.5 million in 2002, an increase of 43.3%. The value added situation was quite different in the state than it was in the country as a whole. In Alabama, value added for crops actually declined during the period, while it increased sharply for livestock. Year-to-year changes in value added for livestock were largely determined by the annual changes in broiler numbers and prices (Figures 16.126 and 16.131). For example, the sharp decline in 2002 was the result of an unusually large decline in broiler prices.

As noted in the preceding paragraph, value added for crops in Alabama during the period 1985-2002 actually declined (Figure 16.149). It was $766.4 million in 1985 and $532.4 million,in 2002, a decline of 30.5%. When this decline is coupled with the 67.2% increase in the Consumer Price Index, it is obvious that crop production in the state did not contribute very much to its economy during the period. It is now clear that for at least 3,000 years crop production on the Atlantic Coastal Plain has been poor at best, but it took growing globilization to establish the fact firmly.

Data on the annual value added to the state's economy by livestock production during the period 1970-2002 are also presented in Figure 16.149. These data show that it increased from $1,295.8 million in 1985 to $2,422.1 million in 2002. It is obvious that these year-to-year changes were largely determined by the year-to-year changes in Alabama's rapidly growing broiler industry.

Data on value added to the American economy by production of specific crops and groups in Alabama in 1980, 1985, 1990, 1995, and 2002 are presented in Table 16.74. The data show that the production of feed grains (primarily corn) are much less important in Alabama than in the country as a whole. Cotton continues to be much more important. The sharp decline in the value for cotton in 2002 was the result of adverse weather that was especially damaging to the crop. Value added for oil crops (peanuts and soybeans) trended down during the period, primarily as a result of declining interest in both peanuts and soybeans by Alabama farmers (Figures 16.57 and 16.67). The decline in the values for food grains (primarily winter wheat) is also important. Alabama farmers have been trying for years to identify a crop that would replace cotton, but as these data indicate, it does not appear that winter wheat is a suitable candidate. Alabama's agro-ecosystem apparently does not lend itself to producing wheat competitively in a global market.

Similar data on the value added for livestock produced in Alabama are also presented in Table 16.74. These data show that in all the years, poultry and eggs (primarily broilers) are the tail that wags the dog in livestock production, and, in fact, a truly positive force in Alabama's agricultural production. In 1980, poultry and eggs accounted for 59% of the total value added for livestock and poultry. By 2002, that share had increased to 79%. Value added for meat animals (primarily cattle and hogs) has trended down in the last decade, as a result of declining interest in the production of both animals (Figure 16.110 and 16.115). The decline in the value added for dairy products is also important to the state's agricultural economy.

Data on cash receipts obtained from the sale of various crops produced in Alabama in 1980, 1985, 1990, 1995, and 2002 are presented in Table 16.75. These data generally show the same trends as the data on value added presented in Table 16.74; however, the data on cash receipts provide a more specific insight into changes taking place in agricultural production in those years. These data show the same declines in the production of wheat, peanuts, and soybeans, but they also identify the emergence of greenhouse, sod, and nursery as a growing contributor to the state's agricultural economy. They also show that while hay is not yet a major contributor, it is nonetheless increasing in importance. Additionally, the data show that other than cotton, the loblolly pine was likely the single most important plant produced on Alabama farms in those years.

Similar data on cash receipts from the sale of livestock in those same years are presented in Table 16.76. These data confirm the decline of the production of cattle, hogs, and dairy products, but show more clearly the amazing growth of broiler production in the state. The data also show that broiler production carried egg production (primarily broiler hatching eggs) along with it. They also show the growing importance of the production of channel catfish. By 2002, cash receipts from their sale was greater than those for hogs and dairy products combined.

STILL SERIOUSLY SEEKING SALVATION

As discussed in preceding chapters, when the Europeans began to arrive in North America in the sixteenth century, Native American farmers were unable to make a living farming. They had to supplement their farm income with off-farm work (gathering, hunting, and fishing). Now, after some 300 years, the situation remains essentially the same. Few American farmers can make a living by farming. In 2003, 88.8% of the average household incomes of operators of all family farms in the country had to be obtained by off-farm work (Hoppe and Banker, 2006). Farm earnings were negative for 60.3% for all households operating family farms. Some 35.8% of all farm households had negative net farm incomes. By 2002, the ratio of indices of prices received by farmers from the sale of their commodities to prices paid for the inputs required to produce those commodities had reached 0.37. After thousands of years, American farmers were still Seriously Seeking Salvation.

As detailed in the preceding chapters, beginning at the end of the nineteenth century, public agencies, private groups, and individuals began to search for ways to solve the problem of inadequate farmer income. Over time, the efforts of those groups have doubled, redoubled, and redoubled again. While these efforts have accomplished little of the goal of making it possible for farmers to earn a living on their farms, they have played an essential role in making our food and fiber supply the success story that it is today.

Salvation Through Better Information

Data on the funds budgeted by the Alabama Agricultural Experiment Station for research from the 1985-1986 through the 2001-2002 fiscal years are presented in Table 16.77. During the period, total funds budgeted increased from $25.3 to $43.1 million, an increase of 70%. During the same period, the US dollar was inflated by 64%. State funds increased by about 73%, but the increase in federal funds was much lower (18%). Funds received from sales actually declined. Research funds obtained from grants and gifts increased by almost threefold ($4.3 to $12.1 million). In 2001-2002, funds derived from these sources equaled slightly more than half of total state funds that year. These funds allowed the experiment station to actually do considerably more research than it could have done on appropriated funds alone.

Between 1994-1995 and 2001-2002, appropriated funds (state plus federal) for agricultural research at Auburn increased by 12% ($24.9 to $27.9 million). During the same period, the Consumer Price Index increased by 18% (Figure 16.2). These data indicate that the experiment station actually had less real dollars for research in 2001-2002 than in 1994-1995. The short recession in 2001 probably affected appropriations somewhat, but whatever the reason, taxpayers did not seem inclined to increase the money that they were paying for agricultural research enough to allow it to keep up with inflation.

Data on the number of scientific-man-year equivalents purchased by the Alabama Agricultural Experiment Station (AES) in its various administrative units during the period 1984-1985 to 2001-2002 are presented in Table 16.78. These data are indicative of the amount of reorganization that has taken place in AES in the last 15 years. In 1981, the School of Agriculture (it had had that name since 1928) became the School of Agriculture, Forestry, and Biological Sciences. Then in 1985, its name was changed to College of Agriculture.

From 1950-1951 through 1969-1970, the organization of the School of Agriculture remained relatively stable. Then between 1969-1970 and 1974-1975, the Department of Dairy Science was eliminated, and its teaching and research programs were included in the Department of Animal and Dairy Sciences. Additionally, between 1969-1970 and 1974-1975, the teaching and research programs in fisheries and aquaculture were removed from the Department of Zoology and Entomology, and the new Department of Fisheries and Allied Aquacultures was created (1970).

Between 1979-1980 and 1984-1985, microbiology was added to what had been the Department of Botany and Plant Pathology. The reconstituted department took the name: Department of Botany and Microbiology, but it still included plant pathology. Also, the Department of Forestry was removed from the School of Agriculture, Forestry, and Biological Sciences and made a school of its own. It was at this point that the School of Agriculture, Forestry, and Biological Sciences became the College of Agriculture.

Then after 1984-1985, all hell broke loose. Between 1984-1985 and 1989-1990, plant pathology was removed from the Department of Botany and Microbiology and placed in its own department. During the same period entomology was removed from the Department of Zoology and Entomology and placed in its own department. Afterwards, the Department of Zoology and Entomology would include only zoology and wildlife sciences.

Between 1989-1990 and 1994-1995, some personnel were given joint appointments with the College of Agriculture and the Alabama Cooperative Extension. Then between 1994-1995 and 2001-2002, the Department of Agricultural Engineering became the Department of Biosystems Engineering, Wildlife Sciences was removed from the Department of Zoology and Wildlife Sciences and placed in the School of Forestry. The personnel and programs of the Department of Zoology and the Department of Botany and Microbiology were removed from the College of Agriculture and placed in the Department of Biological Sciences within the College of Science and Mathematics.

Data on the number of scientific-man-year-equivalents (SMYs) of research effort purchased by the Agricultural Experiment Station in the various departments for the period 1984-1985 to 2001-2002 (presented in Table 16.78) were obtained in the same manner as described in Chapter 15. The totals include only tenured and tenure-track personnel, and do not include positions that were vacant at the time the budgets were prepared. Data in the table show that total SMYs purchased increased along with budgeted funds (Table 16.79) from 88.32 in 1984-1985 to 99.55 in 1994-1995. This increase of 11.23 SMYs was largely the result of increased purchases in the Departments of Agricultural Engineering (1.94), Agronomy and Soils (1.61), Animal and Dairy Sciences (1.61), Fisheries and Allied Aquacultures (5.39), and Forestry (3.74).

Budgeted funds of the experiment station declined in real dollars between 1994-1995 and 2001-2002. As a result, SMYs purchased also declined (99.55 to 91.34). This reduction generally was shared by all of the departments.

Bell (2007) reported that Alabama's national ranking in agricultural productivity declined from 29th in 1960 to 33rd in 1999 and that the increase in Alabama was the lowest in the South and among the lowest in the Eastern United States. Many factors act and interact to bring about increased agricultural productivity. Better information for farmers through research and extension is just one of them.

Budgeted funds of the Alabama Cooperative Extension Service, obtained from several different sources from 1979-1980 through 2001-2002, are reported in Table 16.79. The data show that funds increased from $18.3 million in 1979-1980 to $47.3 million in 2001-2002, an increase of 157%. During the same period the Consumer Price Index increased by 118%. Most of the increase in funding came from the state. Funds from that source increased from $8.9 million in 1979-1980 to $26.2 million in 2001-2002. Federal funds increased from $7 to $8.8 million. Federal funding reached a maximum level in 1994-1995 ($10 million), but declined afterwards. In relative terms funds from other sources increased the most ($0.7 to $9.9 million). County funds increased by less than $1 million ($1.6 to $2.3 million) over the entire period.

Data on the number of professional extension personnel employed by Auburn as part of the Alabama Cooperative Extension System, in the period 1980-2002, are presented in Table 16.80. The data show that county agents accounted for well over half of the positions throughout the period, but that the number of budgeted positions in that category generally declined from 382 in 1980 to 203, in 2000. Probably as a result of the large increase in budgeted funds between 1999-2000 and 2001-2002, ACES was able to add 37 additional positions in this category.

The number of positions in the district agent category also trended downward, and after 1995 the organization of ACES was changed so that the number of district agents was sharply reduced.

The number of positions in the specialist category remained relatively constant at between 116 and 119 through 1990, but was increased afterwards. There were 147 budgeted positions in this category in 2002, the highest number during the entire period.

During this period (1985-2002), Alabama Agricultural and Mechanical University (AAMU) continued its efforts to obtain equitable funding for its 1890 land-grant mission. In 1991, after several years of contentious adjudication, Judge Harold Murphy of the Federal District Court ordered the State of Alabama to give $10,000,000 to AAMU over a 10-year period. He also ruled that AAMU had no claim on the Federal Cooperative Extension Funds received by Auburn University.

Although the court denied the AAMU's request to share state extension funds, the extension program did make considerable progress when the James I. Dawson Extension Building was dedicated on the AAMU campus. It was constructed with funding provided by USDA. Dr. Dawson, who retired in 1993 as associate dean for the Cooperative Extension Program at the university had given almost half a century of exemplary service to agricultural education and extension throughout the state.

In 1995, the Federal District Court, in a remedial decree, unified the Alabama Cooperative Extension at Auburn University with the 1890 Cooperative Extension Education program of AAMU. It resulted in the formation of the Alabama Cooperative Extension System. Under the decree, the two universities were to focus on serving different segments of the state's population. It required that AAMU would serve urban and non-traditional clients and that Auburn University would serve traditional and rural clients. The court order creating the new system was to be vacated in 2005, but by that time the two universities would have the necessary agreements in place to maintain the system as a legally-mandated entity.

As a result of the federal decree, working groups from Auburn University, Alabama A & M University and Tuskegee University, operating under the leadership of Dr. Gaines Smith for AU and Dr. Virginia Caples for AAMU planned the first unified extension program in the country in which extension resources of 1862 and 1890 land-grant universities were combined. The unique program was expanded when the resources of a private university were included. Tuskegee University was made part of the system as a cooperative partner.

Table 16.81 presents data on the funds available for extension work at Alabama A & M University. These data clearly show the improvement in funding at AAMU following the implementation of the decree. In 1994, total funds available was $1.6 million. In 1996, the amount had almost doubled to $3.1 million. In succeeding years, the total increased, and in 2002 reached $5 million. Funds accounting for the larger share of this increase were provided by the state.

Data on the number of budgeted positions for professional personnel employed in extension work by AAMU are shown in Table 16.82 in several years during the 1980-2002 period. These data also clearly show the improvement in university capability following implementation of the decree. As late as 1995, there were only 23 professionals employed in extension work by the university. By 2002, the total had almost doubled to 43. In all the years, specialists and urban/county personnel accounted for most of the employees, and there were about half as many specialists as urban/county agents.

In 1997, the Federal District Court issued another order dealing specifically with the funding of agricultural research at AAMU. The court ordered that in fiscal year 1996-1997 the Alabama Agricultural Experiment Station at Auburn (AAES) transfer to AAMU the sum of $400,000 to offset the costs associated with the maintenance and operation of its Agricultural Research Station. Further, the court ordered that for fiscal year 1997-1998 and beyond that AAES provide AAMU with $200,000 annually for this purpose. In addition, the state was ordered to provide an additional $300,000 annually for the AAMU research station.

The Agricultural Research, Extension, and Education Reform Act of 1998 required that each state receiving Hatch funds submit a plan for research. Responding to this requirement, Auburn University, Alabama A and M

University, and Tuskegee University submitted Alabama's first unified plan in 1999. As a result of this requirement for unified research planning, in 2000 the three universities agreed to form the Alabama Agricultural Land-Grant Alliance. Although this is not a court-mandated system, it strongly encourages the cooperation of the three universities in conducting research on national agricultural research priorities.

Salvation Through Organization

In the period 1985-2002, efforts continued to improve farmer income through the activities of farmer organizations. Several new groups organized for the purpose of providing assistance to agriculture and forestry in Alabama were formed. Also, some organizations, formed earlier, either changed their names or made other changes in their structure or mission. These included:

1. The Alabama Agribusiness Council was established in 1989 to work with groups and organizations to effectively promote and enhance the business of agriculture and forestry in Alabama.
2. The Alabama Equipment Dealers Association, Inc., is the new name chosen, in 1997, for the Alabama Farm and Power Equipment Dealers Association, Inc.
3. The Alabama Farm Bureau ceased to exist in 1986 when members of the organization agreed that as a result of a dispute with the American Farm Bureau over their insurance business they would no longer use that name for their organization. Later, some Alabama farmers began efforts to realign themselves with the parent organization.
4. The Alabama Farmers Federation (ALFA) was the name chosen, in 1986, by the members of the Alabama Farm Bureau when they agreed that they would no longer use the original name for their organization. With the change in name, ALFA assumed control of all of the assets of the Alabama Farm Bureau.
5. The Alabama Horse Council was established in 1985 to promote the development of the horse industry in the state and to "establish one voice to speak for all horse interests in the state of Alabama."
6. The Alabama Loggers Council was established in 1992 as an adjunct organization of the Alabama Forestry Association. The purpose of the organization is to "represent the interests of its members and to enhance the public image of logging and the forest industry as a whole in Alabama."
7. The Alabama Pulp and Paper Council was established in 1990 as an adjunct organization of the Alabama Forestry Association. The purpose of the organization is to "represent the Pulp and Paper Industry in Alabama and to serve as a liaison to government agencies and the State Legislature and acts as an information source for the media and general public."

Salvation Through Crop Diversification

In the late nineteenth century, American agricultural experts suggested that crop diversification would help to alleviate some of the chronic problems farmers were having. This suggestion was aimed primarily at Southern cotton farmers, but it applied equally well to grain farmers in the country's Heartland. Experts suggested that grain farmers produce more livestock as a way of using their surplus grain. Increasing the production of hogs would have been especially helpful, because those animals require such high quality feeds. Further, hogs require a relatively large amount of feed for each pound of gain. Because of these two characteristics, large quantities of corn could have been marketed through the production and sale of hogs. Figure 16.150 presents data on the percentage of the country's hogs and pigs produced in the principal states of the Heartland Farm Resource Region (Illinois, Indiana, Iowa, Missouri, and Ohio) in selected years during the period 1900-2002. As the data indicate, from around 1900 through around 1925, farmers in that region produced about 40% of the country's hogs; afterwards, however, farmers began to heed the advice of the experts, and hog numbers began to increase, reaching 60% in 1975. After that year, numbers of hogs produced in the Heartland began to decline, and by 2002 its share was down to 46%. This decline in production and the loss of market share was likely the result of increasing severity of environmental regulations in the region, the increase in vertical integration in the industry, and the increased numbers of large pork producing factories in other regions. Family grain farmers had heeded the advice of the experts by increasing their production of hogs, but by the end of the twentieth and beginning of the twenty-first centuries, those family farms were rapidly losing their competitive edge. The presence of large quantities of inexpensive corn was no longer the advantage that it had been.

It was assumed that the increase of livestock production in the Heartland would also mean increased production of cattle. However, because of the nature of the digestive system of cows, they are able to grow extremely well on poorer quality feeds. In fact, they grow extremely well on grass alone, if it is of good quality. As a result, the presence of large quantities of corn in that region provided relatively little comparative advantage for cattle production on family farms. Figure 16.150 presents data on the percentage of the country's cattle (all) herd on farms in the region in selected years during the period 1900-2002. These data show that in 1900 numbers in the region accounted for only 23% of the national cattle herd. Except for the period 1925 through 1965, numbers declined in the region, and by 2002 reached 12%. For several reasons, after the mid-1960s the Heartland began to slowly lose its comparative advantage in cattle production and especially in feedlot operation.

In 1949, cotton was harvested from 33.6% of the total acres of harvested cropland in Alabama that year (Table 16.83). Two decades earlier, it had been 50%. As data in the table show, Alabama farmers were making a determined effort to reduce their dependence on cotton. By 1959, the percentage was down to 22.5%. During the period 1949-1959, the decline in the percentage of cotton harvested was matched by increases in the percentage of corn harvested. Unfortunately, by 1964 Alabama farmers were beginning to understand that the production of corn was not the solution to their problems either, and the percentage of that crop harvested also began to decline. Fortunately, acres of soybeans harvested was beginning to increase rapidly. Alabama farmers were also beginning to harvest more acres of winter wheat. As acres of soybeans harvested continued to increase, acres of both cotton and corn harvested declined. In 1982, acres of soybeans harvested reached an all-time high of 61.2% of all acres harvested. In the same year, percentages of acres of cotton and corn harvested reached all-time lows (8.7 and 11.6%, respectively). Unfortunately, by 1987 Alabama farmers were beginning to understand that they had virtually no comparative advantage in producing soybeans either. As a result, the percentage of acres harvested of that crop began to decline dramatically. In the same year, the percentage of acres of cotton harvested began to increase. In 2002—with the percentages of acres harvested of corn, soybeans, and wheat at 9.0, 7.8 and 3%, respectively—the percentage of acres of cotton harvested was 27.1%, almost back to where it had been 1949. The data show that after 1949 Alabama farmers had made a determined effort to reduce their dependence on cotton, but when efforts to produce more corn, soybeans, and winter wheat failed, they had little choice but to go back to cotton. Data presented by Brown, et al. (2006) show that at the beginning of the twenty-first century it is still difficult to make money, consistently, in Alabama producing cotton, especially in the central and southern portions of the state. Even with that inconsistency, returns from cotton are more acceptable than for the other major crops.

Salvation Through Political Action

The level of government payments to the farm sector is probably a good indicator of the extent and intensity of efforts on the part of Dobson's Triangle of Beneficiaries to solve the problem of farm income by political means. Data on levels of annual government payments (per farm) to American farmers during the period 1970-2002 are presented in Figure 16.151. The passage of the Agriculture and Food Act of 1981 resulted in a sharp increase in government payments. Between 1970 and 1980, payments generally remained below $1,000 per farm. In 1980, they were $527 per farm ($1.3 billion total). After the passage of the act, payments spurted upward to $1,451 ($3.5 billion total) in 1982 and to $3,907 ($9.3 billion total) in 1983. The Food Security Act of 1985, with its increased emphasis on conservation, added more fuel to the fire, and by 1987 the average payment per farm reached $7,567 ($16.7 billion total). Congress became concerned with the rapid increase in payments following the passage of the 1985 act, and decided to let some air out of the balloon with the passage of the Food, Agriculture, Conservation, and Trade Act of 1990. As a result, payments began to trend downward, and by 1997 they were down to $3,421 per farm ($7.3 billion total). However, it was sharply higher in 1993. This sharp increase was the result of the increase in disaster payments for feed grains, wheat, and cotton. That year, Midwest flooding and the drought and heat wave in the South caused widespread losses of those crops. Payments also increased sharply after 1997, as the result of funds allocated by Congress under their ad hoc and emergency programs for disaster and emergency assistance to farmers. Those funds were reduced significantly by the Farm Security and Investment Act of 2002.

Data on total payments to farmers, as a result of their participation in different programs for selected years in the period 1970-1995, are presented in Table 16.84. In 1985, 1990, and 1995, a large share of all payments went to farmers producing feed grains (corn) and wheat. However, in 1985 Congress passed the Conservation Reserve Program Act (CRP) as part of the Food Security Act of 1985; it provided payments to farmers for planting grass, shrubs, and/or trees on environmentally-sensitive farmland. This program replaced the old Soil Bank Act of 1956, which had limited payments. The CRP provided payment for taking a much larger amount of land out of production. In 1985, conservation payments totaled only $189 million. With the enactment of the CRP, they increased to $254 million in 1986, to $1,771 million in 1988, and to $1,898 million in 1990. Conservation had quickly become a major source of subsidies. Then in the 1990 farm bill, Congress added legislation (Wetland Reserve Program) offering farmers payments for protecting, restoring, and enhancing wetlands on their farms. Three years later (1993) conservation payments reached $1,967 million.

Following the sea change in the political landscape after the 1994 election, Congress (Triangle) passed the Freedom to Farm Act of 1996. The act eliminated most subsidies based on price and supply in an effort to make farm policy more market oriented. However, its most far-reaching provision was the Production Flexibility Contract Payments program. These payments were based on historical production of particular commodities and were largely independent (decoupled) of current market prices and most farm-level production decisions. With this act and these payments, Congress had finally decided that effecting positive changes in farm income through price supports and supply management alone were simply never gong to work. Finally, they had decided to confront the problem head-on. They were going to make payments to farmers as a direct means of increasing income (providing farmers with a safety net). Taxpayers were, for the first time, going to pay farmers directly for farming. For all intents and purposes, with the passage of the 1996 act farmers went on the federal payroll. Henceforth, they would receive some level of guaranteed income support. In the year the act was passed (1996), payments were $3,350 per farm ($7.3 billion total), and in 1997 they inched up to $3,421 ($7.4 billion total).

In 1996 and 1997, payments were near the same level they had been before the passage of the act; however, the nature of the programs providing them was quite different. In 1996, there were no payments resulting from the production of feed grain, wheat, cotton, etc. Instead, payments were based on production flexibility contracts, and the amounts farmers received were based on the quantities of those crops that they had produced in the past.

Needless to say, most farmers were ecstatic with the passage of the 1996 act. Unfortunately, in 1967 the mechanics of the teeter totter drove the index of prices received down 4.8% (712 to 678) and the index of prices paid up 2.8% (1531 to 1574). As a result, the parity ratio fell from 0.47 to 0.43 (Figures 16.9 and 16.10). Quickly it became obvious that the Production Flexibility Contracts Payments authorized under the FAIR act would not be adequate when farmers were losing so much money in their operations. For several reasons, the Triangle seized this opportunity to get back in the game, and in 1998 government payments increased to $5,650 per farm. In 1997, they had totaled only $3,421. Then in 1998, the Triangle stumbled into its best opportunity in years when the counry's agriculture was beset by widespread adverse crop weather. There was widspread agricultural disaster, but very quickly there was disaster everywhere. There was a rolling bandwagon of disasters all over the land. Quickly, the Triangle pulled the drawstring on the bag holding taxpayer funds, and they flooded out across the landscape (Table 16.85). In 1998, payments almost doubled to $9,814 per farm; then in 2000 they increased to $10,541. In 2002, federal payments for disaster relief totaled $2.264 billion.

The 2000 payments were the highest ever made to that time. Even the most calloused congressman had to admit that the 2000 payments were both excessive and obscene. As a result, payments began to decline in 2001 ($9,654 per farm), and the decline continued in 2002 ($5,134).

In 2000, total payments reached $22.9 billion (the highest in history), and the parity ratio reached the lowest level in history–0.38 (Figure 16.10). Further, in the same year, 88% of the total annual income of households responsible for operating all of America's family farms came from off-farm sources. This set of circumstances was enough to make even the most seasoned politician blanch, and it was under this unfortunate and complex set of circumstances that the politicians wrote the 2002 farm bill. The Farm Security and Rural Investment Act of 2002 continued marketing assistance loans, loan deficiency payments, price support programs for dairy and sugar, subsidized crop and revenue insurance, ad hoc emergency programs, market loss assistance programs, and all of the agro-environmental programs. The concept of decoupled direct payments. which was written into the 1996 act, was also kept, but with some changes. They replaced the Production Flexibility Contract Payments program with the Fixed Direct Payments program. These payments continue to provide income support (paid) farmers who have historically produced wheat, feed grains, rice, upland cotton, oilseeds, and peanuts (Young, et al., 2007). Payments were not related to current market prices or to most farm-level production decisions; consequently, they did not have an effect on a producer's cropping decisions.

The sharp decline in prices of agricultural commodities in 1996 had shown that direct income support alone was not sufficient to protect farmer income when prices decline sharply; therefore the politicians had to add another important element to the 2002 bill to completely insulate their new "employees" from the marketplace. As a result, they added the Counter-Cyclical Payments Program. This program also provides direct income support based on historical levels of production of those same crops. However, payments from this program are made only when market prices for the commodities fall below trigger prices set in the 2002 farm act. Producers can receive payments when prices are low for commodities they historically produced, even if they are not currently producing them. Congress set limits on payments that an individual could receive from programs of the 2002 act—$40,000 for Fixed Direct Payments, $65,000 for Counter-Cyclical Payments, and $75,000 for Marketing Loan Benefits.

Even though the data presented in Figure 16.152 is based on average payments per farm, not all farms received them. Hoppe and Banker (2006) reported that in 2003 (similar data are not available for 2002) only 39% of all American farms received payments (Table 16.86). Further, they reported that the percentage receiving payments was dependent on farm type. For example, considerably less than half of small family farms (limited/resource, retirement, residential and low-sales) received payments. Conversely, considerably more than half of all farms with at least $100,000 in annual sales (small family farms with medium sales, large family farms, and very large family farms) received them.

Government payments to American farmers have become an ever increasing part of their net farm incomes. Through the long period from 1933 through 1983, government payments did not make up more than 15% of NFI in most years; however with the financial difficulties faced by farmers in the early 1980s, payments began to increase. As a result, their contribution to NFI also began to increase (Figure 16.152). Payments comprised as much as 65.2% of total NFI in 1983, but the percentage generally trended downward afterwards to 12% in 1995. The following year, the government began to increase payments again, and their contribution to NFI also began to increase, reaching 50% in 2000. Payments began to decline once more in 2001; then they fell sharply in 2002. In that year, they constituted 30% of NFI. The data on payments to farmers do not include the funds they receive in subsidies and discounts for crop insurance premiums.

While government payments contributed significantly to growth of Americn net farm income, especially in the 1980s, inflation of the currency was also a factor in its growth. For example, between 1985 and 1990, the Consumer Price Index increased by 21%.

Other than farmers receiving direct payments based on their farming history, conservation was the most obvious winner in the 2002 act. For a number of years, farmers have received payments for their participation in the Conservation Reserve, Wetlands Reserve, and Environmental Quality Incentives programs. The 2002 act added an important new program—the Conservation Security Program. Under this program, farmers receive payments to reward good conservation performance. The increased emphasis on conservation is also apparent in the increase in payments. Between 2001 and 2002, payments for conservation programs increased from $1.966 billion to $2.167 billion (Table 16.85), and it is very likely that payments will increase substantially as more and more farmers improve their conservation performance. These changes do not bode well for future political funding for the production of food and fiber in the country. With the continued decline of the rural/farmer voting block and the rapid increase in the urban environmental voting block, it is highly likely that in future farm bills, funds will be increasingly shifted from food and fiber to environmental programs and feeding the needy.

Hoppe and Banker (2006) reported that in 2003 (there are no comparable data for 2002) that 14% of all American farms received payments for their participation in at least one of the various conservation programs (Table 16.86). Nine to 12% of limited/resource, residential, and low-sales farms received payments from this source. Surprisingly, 19% of retirement farms also received them, along with from 18 to 21% of medium-sales, large, and very large family farms. More non-family farms received payments (27%) than any other type of farm. While the percentage of small-family farms receiving conservation payments was relatively low, a larger share of their total payments from all sources came from that source.

Data on annual government payments to Alabama farmers (per farm) during the period 1970-2002 are also presented in Figure 16.152. Year-to-year changes were similar to those in the country as a whole. During most of the period 1985-2002, payments to Alabama farmers ranged between $1,000 and $3,000 per farm. These values were considerably lower than the national average. As noted in the preceding chapter, most federal payments are related to crop production, Furthermore, crop production in Alabama contributes a smaller share to NFI than in the country as a whole (Figures 16.148 and 16.149). In 1999, payments reached $3,740 per farm, the highest level in history at that time. They declined slightly in 2000, but when federal payments declined afterwards, those provided to Alabama began to increase. In 2001, they were up to $5,016, and in 2002 they reached $5,896, a level slightly higher than the national average ($5,134). The increase in 2001 was fueled by a large increase under the Ad Hoc and Emergency Payments program, and the increase in 2002 was the result of a payment of $135 million under the Peanut Buy-Out Program.

The contribution (percentage) of government payments to Net Farm Income in Alabama during the period 1970-2002 is also shown in Figure 16.152. Throughout the period 1985-2002, year-to-year changes in the percentages were generally similar in the country and the state. However, as a result of the nature of agricultural production in Alabama and the nature of government payments, the percentages were considerably lower for the state. In most years, they ranged between 5 and 15% of the state's NFI.

When farmers went into their fields in coastal Virginia in the early seventeenth century, they carried their weapons with them for protection against the risks of marauding Native Americans. They could have never imagined that four centuries later their descendents would be petitioning their elected representatives for legislation that would, in effect, protect them from all risks. Throughout history, evolution has created some strange things. Most of those strange creations could not compete in the give and take of the everyday world, and they have disappeared. It will be interesting to see whether farming-without-risk will be one of the winners or one of the losers. My bet would be that with the Triangle leading the way farming-without-risk, including the requirement that tax payers pay for it, will be a winner. The wildest hog in the deepest Southern swamp could be quickly and easily tamed for slaughter by giving it a few nubbins each day.

THE BOTTOM LINE

Annual average American net farm income (NFI) on a per farm basis trended slowly upward from 1950 through 1971, but then increased sharply afterwards. From 1972 through 1985, there was considerable year-to-year variation, but there were no apparent trends. During the period, it varied from a low of $6,615 in 1980 to a high of $12,438 in 1985. Data for Alabama showed a generally similar pattern, but in the state NFI varied from a low of $4,098 per farm in 1980 to a high of $9,860 in 1981.

Among the states of Alabama, California, Florida, Georgia, Iowa, Mississippi, Pennsylvania, and Tennessee in 1970 and 1985; NFI per farm was lower in Alabama, Mississippi, and Tennessee than in any of the other states, and in 1985 it was several times larger in California and Florida than in those three states.

US Net Farm Income

Data on the annual average American NFI on a per farm basis for the period 1970-2002 are presented in Figure 16.153. The data show that NFI per farm ranged between $8,000 and $12,000 in most of the years from 1974 through 1984. However, in 1985 it began to trend upward, reaching $21,557 in 1990. Thereafter, through 2001 it remained near that same level in most of the years before falling to $18,779 in 2002.

Several factors determine NFI in any given year. Among these are prices received, prices paid, government payments, exports, and crop weather. Weather plays an especially important role. For example, weather was extremely unfavorable for crop production from Texas eastward to the Carolinas in 1998. Crop growing weather was satisfactory over most of the country in 1999; although drought in the Lower South did cause some reduction in yields. Good weather and good crop yields, however, were sufficient to allow NFI to increase slightly in 1999. The following year, crop weather was generally good across these country, although dry weather again resulted in some reduction in cotton yields in the South. Overall, good harvests were sufficient to fuel a small increase in NFI. Crop production weather was generally good again in 2001 allowing NFI to increase again. In 2002, adverse crop weather throughout the country and a large reduction in government payments resulted in one of the largest, one-year NFI declines in history. In 2002, the enactment of the Farm Security and Rural Investment Act resulted in payments almost 50% lower than in 2001.

As shown in Table 16.87, the annual average NFI on a per farm basis was $12,438 in 1985; $21,558 in 1990; $18,110 in 1995; and $18,779 in 2002. Of course, these values represent the average of all of the states. Data presented in the table also give the annual average NFI values for some of the individual states in the country and the South. As would be expected, California's NFI—with its irrigated vegetables, fruits, nuts, etc.—was highest in all four years, although it was not much higher than Florida's with its vegetable and citrus industries. Tennessee's, with its large number of small farms, was lowest.

Data on Illinois, Indiana, Iowa and Ohio for 1995 and 2002 show the effects of poor crop weather. In all of these states NFIs were much lower than expected. States where animal production, especially hogs and broilers, contributed a large share to NFIs were not as seriously affected.

Data presented in Figure 16.153 simply show the annual estimates of national NFI in each of the years in the period 1970-2002. Unfortunately, not all NFIs result in positive gains for individual farms. In many cases, they are negative. Hoppe and Banker (2006) provide some data on this matter for 2003. There are no similar data available for 2002. They report estimates of the percentage of farms in each of eight different farms-types with negative NFIs (Table 16.88). A description of the different types of farms can be found in Table 16.8. Their data show that NFIs of 35.8% of all American farms were negative in 2003. Limited-resource family farms (household incomes of less than $20,000 annually) had the highest percentage (40.1%). Very large family farms (sales of $500,000 or more annually) had the lowest (18.3%). These percentages are another indicator of the general malaise of American agriculture at the beginning of the twenty-first century.

Alabama Net Farm Income

Data on net farm income on a per farm basis in Alabama during the period 1970-2002 are also presented in Figure 16.153. These data show that year-to-year trends in NFI during the period were similar for both the country and the state for much of the period. It was essentially the same level in 1983, at around $10,000 a farm. The two estimates were also generally similar during the period 1989-1997, at $20,000 a farm. Afterwards, however, the Alabama NFI began to trend upward, reaching $30,000 a farm in 1999. In 2000, it fell to $23,913—as a result of hot, dry weather over most of the state during the growing season and sharply lower broiler prices (Figure 16.126). It reached its highest level in history ($35,137) in 2001—as a result of favorable crop weather and recovering broiler prices. Then in 2002, the state's NFI fell sharply to $25,498 per farm—a result of adverse crop weather, especially for cotton, as well as the lowest broiler prices in over a decade.

Because value added by crops represents a relatively smaller share of the total value added, adverse crop weather does not affect NFI in Alabama as much as in the country as a whole. However, it probably played an important role in reductions in 2000 and 2002. In addition to those three years, it is likely that it also resulted in lower than expected increases in 1986, 1987, 1990, 1993, 1995, 1997, and 1998.

Data presented in Table 16.87 show that Alabama's NFI more than doubled between 1985 and 1990 ($8,619 to $18,136 per farm), probably as a result of the rapid growth in the broiler industry in the state (Figure 16.127). The rate of increase in Mississippi's NFI was equally large; broiler production also increased sharply in that state during the same period. The increase of 41% in the Alabama NFI between 1990 and 2002 ($18,136 to $25,498) was also impressive, but a large share of the increase ($3,071 per farm) was the result of the Peanut Buy-Out Program.

Among the states listed in Table 16.87 in 2002, the NFIs of Alabama and Iowa were in an intermediate group, between the low values of Mississippi. Pennsylvania, and Tennessee and the high values of California, Florida, and Georgia. Nevertheless, they were considerably higher than the national average.

US censuses of agriculture in 1992, 1997, and 2002 published estimates of the percentages of Alabama farms with net gains and losses in net cash returns. (Net cash return for a farm is not the same as its NFI.) The net cash return represents the balance of cash receipts and cash production expenses in a given calendar year. It represents the amount of funds that the farm operator has to meet family living expenses and to pay debts. Census data show that the percentage of Alabama farms with net losses in their net cash income accounts were 49.4% in 1992, 57.1% in 1997, and 54.4% in 2002.

ALABAMA AND HER SISTER STATES IN 2002

At the very beginning of this book, I posited the hypothesis that in relative terms Alabama's economy lags behind most of the other states at the beginning of the twenty-first century, and I suggested that given the abundance of natural resources in the state the long-term ineffectiveness of its agriculture has been the primary cause of this phenomenon. I further suggested that the relatively slow pace of advancement of Alabama's economy, through Lenski's stages of societal development, was the more specific cause. Because of the ineffectiveness of its agriculture, Alabama's economy moved slowly through the Simple Agriculture and Advanced Agriculture stages, and as a result, it was unable to move quickly and effectively into the Industrial stage. To test this hypothesis, we have followed the progress of certain aspects of the economies of Alabama and three of its Sister States (Illinois, and Indiana, and Mississippi) from the granting of statehood in the early nineteenth century through the twentieth century. In this concluding section, we consider some of the agricultural and economic characteristics of the states as they enter the twenty-first century, after almost two centuries of development.

Table 16.89 includes data on a number of characteristics of the agricultural economies of Alabama,

Mississippi, Illinois, and Indiana. All of the data are for 2002, unless noted otherwise. By that year, it was becoming increasingly difficult to make comparisons between agricultural production in the Sister States. For most of the period between their joining the Union and the middle of the twentieth century, the largest share of the total values of their respective economies was derived from the production of crops. However, the coming of the broiler industry, coupled with the growing lack of competitiveness in crop production, the situation began to change in Alabama and Mississippi, and by 2002 the share of total value added, by the value added from crop production in those two states, was considerably less than 50%, while the crop share in Illinois and Indiana remained well above 50%. Crop production was relatively less important than livestock production in Alabama and Mississippi in 2002, and the data in the table show why. Alabama and Mississippi corn yields have increased since 1849-1850, but are still far below those in Illinois and Indiana. Alabama and Mississippi soybean yields were also too low to be competitive under normal conditions. The most telling comparison, however, are the data on rents. In Alabama and Mississippi, they were well below those of Illinois and Indiana. Rent is clearly indicative of the level of return that the farmer expects to receive from farming the land. These data clearly show that expectations were considerably lower in Alabama and Mississippi in 2002.

The lower levels of milk-per-cow and lower indices of agricultural productivity further show the lack of competitiveness in the agricultural economies of Alabama and Mississippi compared to those in Illinois and Indiana. This conclusion has been obvious since the first census of agricultural production was published in 1840. The agricultural economies of Alabama and Mississippi have generally kept up, but there was little likelihood that they could ever catch up.

Data on some characteristics of the general economies of Alabama, Mississippi, Illinois, and Indiana are presented in Table 16.90. All of these states were located on the outer margins of the frontier when they were admitted to the Union around 1820. Although they began their journeys at about the same time and place, their progress has been quite different. In 2002, the Gross Domestic Products on a per-person basis were considerably lower in Alabama and Mississippi than in Illinois and Indiana, while the shares (percentages) of the GDPs derived from farming was about twice as large in Alabama and Mississippi. The GDPs on a per-person basis derived from manufacturing in Illinois and Indiana, were considerably higher. Alabama and Mississippi received a larger share (percentage) of their GDPs from the manufacture of wood products.

While the general economies of Illinois and Indiana were clearly more industrialized than those of Alabama and Mississippi, some of the data in the table suggest that none of the Sister States had advanced very far into the Post-Industrial stage of societal development. In all four of the states, the percentages of their GDPs derived from the manufacturing of computers and electronics were similar. All four states obtained less than 0.75% from this source. In the same year, Massachusetts and California obtained 2.89 and 2.62%, respectively, of their GDPs from that source.

One of the more interesting comparisons of the general economies of the four Sister States is ratios of federal dollars received to federal taxes paid. Citizens in both Illinois and Indiana paid more in federal taxes than they received in federal benefits. In both Alabama and Mississippi, the ratios were above 1.50. These are compelling indicators of the quite different outcomes achieved by the people of Alabama and Mississippi and Illinois and Indiana over their almost 200 years of history. Taking these data together with those presented in Tables 16.89 and 16.90 clearly indicate these differences are the result of continuously less effective agricultural economies in the two Sister States in the Deep South. This conclusion leads to accepting the hypothesis that the economies of Alabama and Mississippi developed more slowly than those of Illinois and Indiana because their agricultures moved more slowly through the Simple Agriculture and Advanced Agriculture stages of societal development. As a result of this general ineffectiveness, there was never enough capital accumulated to power early advancement of their economies into the Industrial stage.

17

What's It All about, Alfie?

PRECIS

From its colonial beginnings in the seventeenth century, the American South (and Alabama from its territorial beginnings in the early nineteenth century) has been inextricably tied to agriculture, especially staple crop agriculture (tobacco, rice, indigo, sugar, and cotton). This agriculture depended on inexpensive labor, first slavery, then the crop-lien system, and today's off-farm work. The Civil War made it abundantly clear that the South's economy must be diversified, but efforts to do so in the postwar period foundered on the lack of capital in the South. Yet by the 1880s, some industrialization had taken place—founded on non-Southern capital and inexpensive local labor. The larger profits, of course, left the South. Cotton, nevertheless, remained the foundation of its economy. That commodity, like other regional agricultural commodities, suffered from the economic market realities of supply and demand (surpluses, declining prices, and the teeter-totter relationship). Alabama's agriculture also suffered from the invisible hand of nature: Ultisols (ultimate, worn-out soils), erratic rainfall, violent storms, subtropical summers and temperate winters, and unpredictable last frost, as well as crops and animals whose genomes are not well-matched with Alabama's environment. The combination of these disadvantages has rendered Alabama agriculture uncompetitive, especially with the American Midwest. To combat this lack of competitive advantage, the Triangle (especially Southern politicians whose long tenure gave them disproportionate power in Congress, Southern cotton farmers, and federal bureaucrats) began to devise a system of federal intervention in American agriculture, which in time linked agriculture, food, and environmental policy. In effect, this system circumvented the natural, capitalistic workings of the marketplace by establishing a federal system of price supports and supply management. Without federal intervention, most Alabama farmers would likely have been forced to abandon agriculture. Although American food appears to be inexpensive, it is not. The actual cost of food in the United States is not merely what is evident at the checkout counter of the supermarket; the actual cost is also borne indirectly by the American taxpayer. This federal agriculture policy, which flies in the face of market capitalism, is likely unsustainable in the long term.

INTRODUCTION

In 1966, the movie *Alfie* was released. Based on a novel and play by Bill Naughton, *Alfie* is the story of a young Englishman who in his early life paid scant attention to the effect that his actions were having on the lives of those around him. However, when he is older, several reversals in fortune made him rethink the nature of his relationships with others. Unfortunately, he finds that many of the things that he has done were not easily undone. Burt Bacharach wrote the song "Alfie" for the movie. The first line of the song included these words: "What's it all about Alfie?" As we reflect on the evolution of agriculture in the world, the United States, and Alabama, over a 500 million-year period, this line in Bacharach's song, is clearly appropriate.

RETROSPECTION

As detailed previously, geophysical processes have blessed North America with one of the most beneficent agroenvironments on the planet. However, this blessing is not uniformly granted. Much of the continent is deserts, semi-deserts, mountains, and vast areas too cold for agriculture, and in Alabama and the South, farmers have been encumbered with a highly variable subtropical climate and poor agricultural soils (Ultisols). However, the region has been provided with a wealth of potentially valuable natural resources, such as navigable rivers, a world-class seaports, timber, and minerals. With this particular combination of resources, the Southern colonies should have quickly followed Great Britain into the Industrial Revolution in the eighteenth century. But this was not to be. With its mercantilist economic system, Great Britain never seriously considered allowing its colonies to compete in the production of manufactured goods. Consequently, the Southern colonies were forced, first into the plantation-based, commercial production of tobacco, then into the production of rice, indigo and sugar, and finally, after independence, into the production of cotton—all based on slave labor that had developed in the tropics for the commercial production of sugar. No one could have known it at the time, but by the 1660s, much of the political, social, and economic future of the South had largely been determined. By the 1780s, the plantation model had reached ceded Indian lands in the Broad River Valley of northeast Georgia, and by the early 1800s it had reached deep into Mississippi Territory.

From the beginning, production of cotton had been the source of major problems for Southern agriculture and for the country as well. Cotton farmers had always been able to produce more of the crop than could be sold in the country for reasonable prices. As a result, the continuing health of the industry was largely determined by exports. Cotton had been a relatively unimportant crop in the South before the American Revolution, but afterwards, demand soared in Europe as textile manufacturing was mechanized. Then, as cotton growing accompanied settlers westward into the Old Southwest, production of the fiber literally exploded. Southern farmers harvested two million pounds of cotton in 1790, 40 million in 1800, 80 million in 1810, and 180 million in 1820, the year after Alabama became a state. Cotton production, like tobacco production, required enormous quantities of hand labor, and also like tobacco, cotton quickly depleted the soil of its nutrients. Thus, as cotton production marched across the South, slave labor and an insatiable hunger for new land marched with it. In the latter part of the eighteenth century, tobacco production in the country, virtually all of it in the Upper South, reached 100 million pounds. By 1820, cotton was being grown throughout the Southeastern United States, and production in pounds was almost twice as great (180 million pounds) as tobacco. In 1790, there were 496,000 slaves in the old tobacco-producing states of the Upper South. In 1820, there were over 1.4 million in the cotton-producing states.

Colonial farmers in British North America first experienced the problem of surpluses and the resulting low prices when Virginia tobacco producers were able to grow more of the crop than could be used in the colonies or exported. In 1660, the English tobacco market became so glutted with the crop that prices fell so sharply that farmers were hardly able to survive. Surplus production and declining prices began to plague Southern cotton farmers soon after the Civil War. During the war, exports of cotton had fallen sharply. Because of the war effort, the Confederacy had to encourage production of food crops rather than cotton, and in an effort to encourage European governments to support them in the war, the Confederacy reduced exports. Consequently, Europeans began to search for new sources of the fiber. Cotton's annual average price was 29 cents a pound in 1869, but by 1876 it was slightly below 10 cents a pound. Supply continued to exceed demand throughout the 1880s, and in 1894 the average annual price had fallen to 4.5 cents.

After 1782, with no English restrictions to concern them, the states of the North Atlantic Region began to industrialize rapidly. In contrast, there was little indication that there was any concern in the South for the lack of industrial development. The future of the Cotton Kingdom seemed to be limitless. The overriding philosophy of the Southern cotton planters in those heady years was to produce more cotton, to purchase more slaves, to produce more cotton, ad infinitum. This was not an entirely new philosophy; it was little different from that of the sugar planters in late sixteenth century Brazil and that of tobacco planters in early seventeenth century Virginia.

In hindsight, it is easy to say that after the American Revolution the South should have freed its slaves and completely revamped its agricultural labor arrangement, but for all practical purposes this was virtually impossible. By the end of the eighteenth century, Southern planters owned thousands of slaves; freeing them would have meant the immediately loss of hundreds of millions of dollars. English colonies in the South had started this carousel, and had jumped aboard in Virginia in the early seventeenth century. By the beginning of nineteenth century, it was turning too fast for them to get off.

By 1820, 4.9% of the people in the states of the North Atlantic Region were involved in manufacturing, compared to only 2.6% in the major tobacco- and cotton-producing states of the South (Table 17.1). By 1840, 8.2% of the people of those same Northern states were employed in manufacturing and the trades, compared to 1.9% in the tobacco- and cotton-producing states.

Even if the South had wanted desperately to industrialize, after the Civil War it did not have the capital to do so. All that antebellum Southerners had amassed before that time had been invested in ever increasing numbers of slaves and acres of land. Data presented in Table 17.2 indicate the magnitude of this phenomenon. Entries in the Table show 1860 census data on the true value of personal property on a per person basis. Values in the cotton - producing states were considerably higher than in states in the Northeast and Midwest. At that time, of course, slaves were considered to be personal property. Values for Mississippi ($641.62), South Carolina ($629.20), and Alabama ($577.40) were almost three times greater than those in Ohio ($132.88), Illinois ($136.37), and Indiana ($119.27). When the slaves were freed, their owners lost millions of dollars of their [slave] personal property. Further, many Southerners had used a substantial share of their disposable wealth to support the Confederacy. When the war ended in 1865, Confederate bonds were worthless. Many Southerners literally had nothing left to maintain their families, much less to invest in industry. The only solution was to import capital. Not surprisingly, however, there was not much Northern or European interest in investing in the development of industries in the South that would produce goods in direct competition with their own.

Over time, industrialization in the North grew much faster than in the South, and by the beginning of the Civil War their economies were vastly different. Data in Table 17.3 show manufacturing investment (dollars per person) in 1840 and 1860 in the North Atlantic, South Atlantic, North Central, and South Central regions. In both 1840 and 1860, investment in manufacturing was considerably higher in states in the North Atlantic Region. This difference is also apparent in the data presented in Table 17.4. In 1860, the number of manufacturing establishments per one million persons was much lower in the South than in the North. Data presented in the two tables also show that these differences persisted at least until the beginning of the twentieth century.

The Civil War provided an impetus for Alabama to begin to use some of its abundant supplies of coal, limestone, and iron ore for the production of iron and steel for the war effort. Exploitation of these resources, however, increased slowly in the years following the war. In 1880, according to census data, there were only 1,626 Alabamians employed in the manufacture of iron and steel. In the same year, employment in those industries totaled 4,121 and 57,952 in West Virginia and Pennsylvania, respectively. Further, in 1880 there were only 14 iron and steel manufacturing establishments in Alabama, compared to 134 in Ohio and 366 in Pennsylvania.

Alabama's textile manufacture began before the Civil War. According to Rogers and Ward (1994), in 1860 there were 13 Alabama textile mills employing 1,300 workers. Alabama's textile industry grew slowly after the war, and as late as 1890, there were only 2,100 Alabamians employed in textiles. In 1890, there were 157,800 farms in the state,and a large share of their operators desperately needed to find other employment. Unfortunately, the number of jobs in the manufacture of iron, steel, and textiles was minuscule compared to the number of Alabamians desperately seeking employment.

In 1889, Henry Grady, managing editor of the *Atlanta Constitution* and a leading proponent of developing a

New South—based on industry rather than agriculture, succinctly described the South's problem in a speech that he delivered to the Bay State Club of Boston. He commented:

> I attended a funeral once in Pickens County (about 50 miles north of Atlanta) in my State. This funeral was peculiarly sad. It was a poor "one gallus" fellow, whose breeches struck him under his armpits and hit him at the other end about the knee—he didn't believe in *décolleté* clothes. They buried him in the midst of a marble quarry: they cut through solid marble to make his grave; and yet a little tombstone they put above him was from Vermont. They buried him in the heart of a pine forest, and yet the coffin was imported from Cincinnati. They buried him within touch of an iron mine, and yet the nails in his coffin and the iron in the shovel that dug his grave were imported from Pittsburg. They buried him by the side of the best sheep-grazing country on the face of the earth, and yet the wool in the coffin bands and the coffin bands themselves were brought from the North. The South didn't furnish a thing on earth for that funeral but the corpse and the hole in the ground. There they put him away and the clods rattled down on his coffin, and they buried him in a New York coat and a Boston pair of shoes and a pair of breeches from Chicago and a shirt from Cincinnati, leaving him nothing to carry him into the next world with him to remind him of the country in which he lived, and for which he fought for four years, but the chill of blood in his veins and the marrow in his bones.

At the end of nineteenth century, Alabama agriculture (crops) was too unproductive to provide the capital required to completely move the state from Lenski and Nolan's (1998), Simple Agriculture and Advanced Agriculture stages into the Industrial stage. As late as 1920, average yields of the major crops being grown in Alabama had changed very little since 1840. Fortunately, Northern capital finally began flowing into the South. By the 1880s, Northern factory owners were beginning to realize that they could produce industrial goods much more profitably in the South, because of the availability of almost unlimited quantity of cheap labor begging to work at almost any wage. Although industrialization was advancing rapidly in Alabama and the South in general at the end of the nineteenth century and the beginning of the twentieth, progress was woefully inadequate to provide alternative employment for very many farmers, and the situation was growing worse year by year. In 1900, there were 223,000 farms in Alabama. A decade later, there were 262,000, and as late as 1930 there were still 261,000. Most of their operators desperately needed to be permanently removed from farming.

During the Civil War, industrialization in the North increased sharply as a result of the rapid growth in the production of war materiel. Following the end of the conflict, some of this expanded manufacturing capability was directed to increased production and improvement of the equipment used in agriculture. As a result of better agricultural machinery, corn and wheat producers in the Midwest also began to experience the problems of surplus production and the resulting falling prices. In 1870, the annual average price of corn in the United States was 80 cents a bushel, but by 1894, it had declined to 50 cents. Changes in wheat prices followed a similar pattern.

In the late nineteenth century, as production increased and prices declined, American farmers began to seek direct governmental intervention in the agricultural marketplace. However, just when the government was beginning to show concern for the farmers' plight, the American economy and the demand for food and fiber began to grow rapidly (the Age of Progress). Soon, farmers were in the Golden Age of American Agriculture (1895-1915), and neither surplus production nor prices were of concern to them. During this period, the [parity] ratio of the indices of prices received to prices paid remained around 1.0 for several years.

The good times for farmers generally continued until after the end of World War I. After wartime demand slowed, however, surplus production and declining prices again became an economic issue, and in 1919 the parity ratio fell below 1.0. Once again, farmers began to clamor for government help. By 1930, the price for cotton was 9.5 cents. Prices for corn and wheat were 55 cents and

69 cents a bushel, respectively. With production costs higher, farmers were in much worse economic straits than they had been in 1894.

By the late 1920s and early 1930s, the states of the Old Confederacy had essentially become the economic, political, and social wasteland of the country (Odom, 1936). This was not a new phenomenon. It had been slowly developing since the mid-1660s with the establishment of Virginia plantation agriculture using slave labor. However, in the years after World War I and the Roaring Twenties, all the chickens finally came home to roost. In 1930, employment in manufacturing in the Southern states, as a percentage of the total population, still trailed that of the Northern states by a considerable margin (Table 17.5). Average employment in the six states of the North Atlantic Region was over three times greater than for the seven states in the South Central Region (11.3 versus 3.4%). It was about two times greater in the North Atlantic Region than in the six states of the South Atlantic Region (11.3 versus 5.9%). There was less difference in the percentages in five states of the North Central Region (11.3 versus 9.3%).

At least 50% of the South's population lived in rural areas. Almost 21% of the country's population lived in the South (South Atlantic and South Central regions combined), but only 12% of the country's wealth was found in the South. In 1930, per capita personal income for the non-farm, Southern population was only 56% ($946 versus $535) of that in the Northeast and 64% ($854 versus $535) of that of the Middle States (Table 17.6). Per capita income for the farm population in the Southeast was only 50% ($366 versus $183) of that of the Northeast, and 70% ($262 versus $183) of the Middle States. Among the Southeastern states, per capita annual income ranged from $129 in South Carolina to $186 in Louisiana.

Farms in the Southeast trailed far behind the rest of the country in the percentage of modern tools and conveniences, such as tractors, trucks, and electric motors (Table 17.7). In all of the categories listed in the table, Alabama and Mississippi ranked within the bottom three in the country.

Except for Arizona and New Mexico, in 1930 illiteracy for individuals above ten years of age in the states of the Old Confederacy ranked highest in the country, ranging from 6.8% in Arkansas to 14.9% in South Carolina.

Unfortunately, with all of its warts, cotton was still king in the South of the 1930s. Farmers in the region had no other choice. They were not competitive in the production of any other field crops, and for all practical purposes, there was no livestock industry. Further, because industrial development was so retarded in the South, there were few non-farm job opportunities. According to Odum (1936), in 1930 there were around two million cotton-farm families in the Southeast, or nearly one-third of all farm families in the country. Across the region, about half of these cotton-farm families were tenants, and in some subregions the rate reached 90%. At least half of Southerners obtained all of their cash income from cotton and its related economy. Approximately half of all the cropland in the region was dedicated to the crop, and in Alabama the percentage was increasing. In 1920, cotton was harvested from 36.2% of all harvested cropland in the state. By 1925, the percentage had increased to 43.4%, and in 1930 it was 50.1%.

Although there was considerable concern among politicians about the plight of American agriculture and especially Southern agriculture after the mid-1890s, little direct governmental intervention was forthcoming until the political realignment that accompanied the election of 1932. Beginning in 1933, the federal government—with the philosophies of early European socialists Henri Saint-Simon (1760-1825) and Louis Blanc (1811-1882) as its guide and with especially strong leadership from Southern congressmen—decided to take responsibility for the management of virtually all aspects of American agriculture by passing the Agricultural Adjustment Act (AAA) in May 1933. A month later, they attempted to do the same for American industry by passing the National Industrial Recovery Act (NRA). AAA legislation set in place a combination of price supports and supply management that would guide American agricultural policy for the next 60 years (Effland, 2000).

At the same time, this activist political intervention led to the establishment of Dobson's (1985) *Triangle of Beneficiaries* (politicians and their constituents and bureaucrats). The Triangle quickly set to work fixing American agriculture. The Supreme Court ruled that its first effort (Agricultural Adjustment Act, AAA) was unconstitutional. The National Industrial Recovery Act (NRA) had suffered a similar fate. However, unlike their brothers in the industrial Triangle, the agricultural Triangle chose to rewrite its legislation to make it acceptable to the court.

The Triangle attempted to reduce production of farm commodities with the 1933 legislation, but to no avail. As shown in Table 17.8, the supply situation was worse for the major crops in 1937 than it had been in 1933. Much of the motivation behind the passage of the Agricultural Adjustment Act in 1933 was to provide relief for the Southern cotton farmer. However, with no experience whatsoever in governmental management of any industry in the country, no one in Roosevelt's administration or in Congress had any idea of whether or not the AAA would help the Southern cotton situation. In fact, it did not. The price of cotton was slightly higher in 1933 (10.2 cents) as well as in the following three years, but when the national economy worsened in 1937, cotton fell to 8.4 cents. None of the programs established by AAA improved the plight of the large majority of Southern cotton farmers in any way. In fact, until the beginning of World War II, most of them were worse off than they had been in 1933. This failure should have given the Triangle pause, but it did not. They had the bit in their teeth and fire in the belly.

Apparently, it was the intent of the Agricultural Adjustment Act to adjust supply management and support prices, so that henceforth the parity ratio would always remain near 1.0. In 1934, the year after the AAA was passed, the ratio was 0.75, and in 1939 it was 0.77. The ratio began to increase in 1940, and it remained above 1.0 throughout the years of World War II and the Korean Conflict. However, in 1954, it fell below 1.0 again and generally trended downward afterwards. By 2002, it had fallen to 0.37 *(Agricultural Statistics)*. It should have been obvious as early as the mid-1950s that no combination of supply management and price supports was going to return American agriculture to its Golden Age level of 1.0. It also should have been patently apparent that the Triangle simply could not manage anything as complex as American agriculture.

It is not clear at this point whether or not the Triangle ever really intended to develop an agricultural policy for the country that would result in parity for farmers. Apparently, even as early as the end of World War II in 1945, they began to understand that the provisions of the AAA could never effectively deal with the problem of surplus production. Consequently, the Triangle began to hedge its bets. Rather than letting market forces deal with the problem of excess production, it began to look for ways to use progressive governmental policies to get the surpluses out of the marketplace. In 1946, the Triangle enacted the legislation providing for the School Lunch Program. From that point onward, agricultural policy and food policy were slowly merged.

With the merging of agricultural policy and food policy, the Triangle seems to have realized that it could have its cake and eat it too. The Triangle could keep both its progressive agricultural programs, with their evergrowing surpluses and cheap food, by giving much of the excess to supposedly deserving people or by shipping it overseas. Apparently, with this realization came the decision to throw parity under the wheels of the bus. By 1964, the parity ratio was down to 0.76, and in the same year, Congress established the Food Stamp Program. Later, they established the School Breakfast Program (1966), the Child and Adult Food Care Program and the Summer Food Service Program (1968), and the Women, Infants and Children Nutrition Program (1972). Someplace Henri Saint-Simon and Louis Blanc must have been exceedingly pleased.

While the Triangle seemed to have lost much of its concern for parity, it nevertheless seemed bothered when the ratio fell too rapidly. For example, between 1979 and 1982, the ratio fell from 0.70 to 0.53. Quickly, in 1981, the Triangle began to increase payments to farmers. In 1980, they reached $1.3 billion, then by 1983 they had been increased to $9.2 billion. Similarly, after 1996, declining ratios resulted in a perfect storm of payments. They increased from $7.4 billion in 1997 to $23.4 billion in 2000.

In the 1950s, farmers began to understand that the

Triangle was going to do little to help them very much, and they began to leave their farms to find additional income to supplement their ever-decreasing farm income. Fortunately, in the 1950s, the American economy was growing so rapidly that it could easily accommodate this influx of additional labor. In 1960, only 52.8% of farm household income was obtained from off-farm sources. However, by 1980 the percentage had increased to 76.8%, and in 2000 it reached 95.4% (Ahearn, 2008). Without this safety valve, the Triangle would have had an unmanageable farm crisis to contend with—a crisis that would have either torpedoed its cheap-food policy or required complete nationalization of the country's agriculture.

As a result of persistent problems with surplus production, prices received did not keep pace with the retail price of food; consequently, the farmer's share of the retail food dollar declined steadily. It was 41% in 1950, 32% in 1970, 24% in 1990, and 19% in 2000 (Eltizak, 2008). Further, with such relatively low costs for the basic ingredients, food processors, distributors, and retailers have had a field day preparing and marketing products in a variety of different ways, with different packaging, and with free toys and games. Competition for the consumer's dollar has been fierce. In 2000, packaging accounted for 8.5% of the food dollar and advertising for 4%.

While inexpensive food was a primary result of Triangle legislation, they understood that, while that policy was extremely rewarding politically, it would have to guarantee somehow that sufficient quantities were being produced to meet the requirement for an ever-expanding population. They also understood that a relatively small number of farms produced most of the country's food. As a result, year after year it enacted legislation to subsidize those farms producing most of the food by providing them with payments of one sort or another. In 2003, about 39% of all farms shared $16.4 billion in government subsidies for participation in various conservation, disaster, and commodity farm programs. The total grew to $23.4 million in 2000. Farms with sales of $200,000 or more shared 77% of the total amount of payments made available that year. Farms with sales of $500,000 or more shared 32% of those payments ($9.6 billion) linked specifically to commodity production (Hoppe and Banker, 2006).

It is likely that the cheap-food policy has also lead to poor use of our food resources. Kantor, et al. (1997) have estimated that 2% of the country's edible food supply was lost at the retail level and 26% at the combined food service and consumer levels. Considering all levels of food production and consumption, the estimated total waste is in excess of 40% (Bloom, 2007). Some losses occur at all stages of the food chain, from harvesting and processing, through preparation, through the hated leftovers in our home refrigerators, to the discarded broccoli, cauliflower, and spinach in the highly nutritious, "free" and "low-cost" school lunches. Almost all experts agree that this is a serious problem for several reasons. For example, an estimated 17% of the volume of the country's landfills consists of wasted food. Further, decaying food releases a considerable quantity of methane directly into the atmosphere. While the elimination of food waste is a lofty goal, a significant reduction would likely result in further increases in surpluses and even more problems for the Triangle.

Regardless of all of the perennial supply problems with Triangle oversight, America's food and fiber supply has become the envy of the world. It is unequalled in quantity, quality, and variety. In 1960, a single American farmer could feed 25 people. At the beginning of the twenty-first century, he can feed 130. In 1933, expenditures for food required 25.2% of the typical family's annual income. By 1970, 1980, 1990, and 2002, it required 13.9, 13.2, 11.1, and 9.8%, respectively (Clauson, 2008). Much of this success story is a direct result of the willingness of the Triangle to invest large sums of public money in agricultural research and extension. However, the ever-increasing levels of agricultural productivity, driven by this constant stream of research and extension funding, has been a primary factor contributing to the perennial problem of overproduction.

One of the primary results of Triangle policies has been inexpensive food for everyone. The percentage of income used to purchase food in the United States is probably the lowest in the world. However, food in America is not really all that inexpensive. For example, in 2002, as detailed in a preceding paragraph, American farmers received $12.9 billion for participation in the various commodity support programs, and this total did not include the billions of dollars that government spent annually to subsidize crop insurance premiums. If these payments were somehow added to the costs of producing, processing, and marketing, the cost of food would be considerably higher. If the billions of dollars spent by the government, through USDA—for the administration of its cheap-food program and for research and technology transfer—were added to food cost at the supermarket, the cost would be even higher still. Probably, the high cost of forcing our surplus commodities onto the world market should also be counted. Further, both federal and state tax revenues lost in the effort to keep a surplus of inefficient farmers on the land should definitely be included. Finally, the hundreds of billions of dollars of farm assets held on the hundreds of thousands of inefficient farms is also a part of the cost of food to the country. Most of these assets are inefficiently used. Food in the United States only seems to be inexpensive because most of the true costs of producing it are kept off the books.

Bringing order out of chaos to American cotton production has probably been the Triangle's most intractable problem. The inability to solve it underscores the limits that exist in the effectiveness of government intervention. The plight of hundreds of thousands of desperate cotton farmers in the South was a primary driving force behind the passage of the Agricultural Adjustment Act in 1933 and the first appearance of the Triangle. In 1931, cotton farmers harvested 17.1 million bales of the fiber, and sold it for 5.7 cents a pound. In 2002, after over 70 years of government intervention, 17.2 million bales were harvested and sold for 45.7 cents a pound. Between 1931 and 2002, the price of cotton increased by 701.9%. Unfortunately, during the same period the Consumer Price Index increased 1,083.6%. In real dollars, the average annual price in 2002 was only about 65% of what it was in 1931.

In the mid-1980s, the Triangle began to reach out to a completely new group of beneficiaries—the environmentalists. Environmentalism began to coalesce as a political movement after the publication of Rachel Carson's *Silent Spring* in 1962. It quickly morphed into a quasi-religious group in which its adherents replaced Mother Mary with Mother Gaia, the Greek Goddess of Earth, in their personal pantheons. Later it became politically active, and within a short time converted much of the federal government, especially the executive branch into an environmental theocracy. Largely urban in their orientation, this group, which had only a very limited interest in whether or not farmers could make a living farming, was vocally and politically active in determining how farmers farmed. With strong support from this politically-active group, Congress established the Conservation Reserve Program in 1985. This basic legislation also contained language that would later be extrapolated into the Wetland Reserve and Swamp-Buster programs. The Triangle was further strengthened with the inclusion of language establishing the Environmental Quality Incentives Program in the 1996 Farm Bill. Then the 2002 Farm Bill added still another activity: the Conservation Security Program. Under this program, farmers receive payments to reward good conservation performance. By 2002, the Triangle had effectively merged environmental policy with agricultural and food policy, creating a significantly more powerful Triangle than had ever existed before, and one that would accept few limits on the amount of taxpayer money they would spend and on the number government-funded programs that they will demand. The ethanol fuel mandate is an excellent example of what this new alliance is able to unleash on the country.

Probably, the Triangle first expected that the implementation of the Agricultural Adjustment Act would require only limited additional work for the Department of Agriculture; however by 2002, the mission of the department had exploded to include providing direct assistance in agricultural production, promoting exports of American farm commodities, assuring the safety of the country's food supply, protecting its natural resources, fostering strong rural communities, and fighting hunger in America and around the world. That year, the work of the Department of Agriculture required 131,400 employees, the expenditure of $76.6 billion, and almost 7,400 offices outside Washington to fulfill the total mission. Astonishingly, departmental expenditures were almost twice as large as the country's total net farm income for that year.

It has been argued in the preceding chapters that—because of relatively poor soils, highly unpredictable weather, and the use of plants with poorly adapted genomes—crop yields in Alabama and the South have been lower and production costs higher from the very beginnings. Consequently, the South has had very little comparative advantage in crop production. Globalization has added to the disparity. The combination of low yields and high costs has forced Southern farmers to accept lower returns on investment year after year since early in the nineteenth century. Unfortunately, the Triangle's policies have tended to obscure this inefficiency to the detriment of the South and the country. Had it not been for its intervention, higher per-unit production costs when coupled with low farm commodity prices would likely have driven a much larger number of Southerners out of farming. Between 1930 and 2002, average acres of cropland harvested in 12 Southern states declined by 35.8%, but only by 4.8% in 12 Midwestern states. The decline in Alabama (72%) was the highest of all 24 states. Without Triangle intervention, the loss would likely have been much greater, and the reallocation of resources out of agriculture into industrial and post-industrial development would have begun earlier and would have been much more substantial. While the transition would have been brutal to many of the individuals involved, it likely would have been a godsend for the region as a whole. A sizeable amount of this transition had taken place following the end of World War II. Unfortunately, it did not proceed nearly far enough at that time.

Data presented in Table 17.9 show the long-term effects of the poorly competitive agriculture on the economy of Alabama and the South. Data in the table present estimates of 1970 and 2002 Gross Domestic Product (GDP) on a per capita basis for some states in the North Atlantic, South Atlantic, North Central, and South Central regions. In both years, the average GDPs of states in the South (South Atlantic and South Central regions) were significantly lower than those for the states in the North Atlantic Region whose economies had reached the Post-Industrial Stage. Of the 21 states included in the comparison, both the 1970 and 2002 GDPs of Mississippi, Arkansas, and Alabama, in that order, were the lowest in the country. Obviously, these comparisons are much more complex than this analysis indicates, but the fact remains that the productivity of the economies of these three states have lagged behind for years, and it is likely that dependence on a noncompetitive agriculture for so many of those years is a primary cause for those poor performances. Obviously, Triangle policies had done little to remedy this situation.

As he neared the end of his life, Alfie found that there were decisions that he had made in his youth that could not be unmade. Only time will tell whether poor decisions made many years earlier regarding the use of Alabama's resources can even be unmade. Alabama's antebellum history was not necessarily predestined, but given the sequence of events, especially before it became a state, it would have been difficult for it to have evolved any other way. Then after the Civil War, the effects of the precipitious destruction of the plantation system were truly catastrophic; even under the best of conditions, it would have been difficult to cope with them. Unfortunately, for many years a majority of Alabamians refused to accept the fact that the old system was gone forever. This attempt to walk backwards into the future made the situation much more difficult than it should have been. It took another war (World War II) to finally eliminate most of the vestiges of the old system. During and after that war, Alabama's agriculture finally moved well into the Advanced Agriculture stage, and the pace of industrialization began to increase. Unfortunately, Alabamians had walked backward for so long that while the state appears to be capable of keeping up economically with other states in our growing country, it is not all that certain that it will ever be able to catch up.

So, "What's it all about Alfie?" Generally speaking, what it is all about is that the Triangle's progressive efforts in attempting to manage agricultural production in the country over a period of almost three-quarters of a century have not worked very well. The evidence presented in preceding chapters is overwhelming. Government, with the best of intentions, is generally incapable of effectively managing anything involving people, because it is virtually impossible to manage them even at the point-of-death.

The more intense the efforts to manage them, the more active the Law of Unintended Consequences becomes. The Soviets under Stalin learned this bitter lesson after efforts in the forced collectivization of Soviet agriculture, beginning in 1927, led to massive starvation. Chairman Mao's Great Leap Forward in China resulted in the same outcome.

Even if the Triangle's interventionist policies have not worked very well, it is not at all apparent that Adam Smith's "invisible hand" (1776) would have worked any better. It is highly likely, however, that without it we certainly would have had a completely different food production and distribution system than the one that we have today. For example, if consumers had to pay all the costs for the food they consume and there were no subsidies, prices would be much higher and probably more variable from year to year. Further, it is likely that considerably less food would be wasted, and there probably would be considerably less so-called free food available.

The development of the country's livestock industry provides some indication as to what might have happened with limited governmental intervention, but while Triangle policies have not generally directly affected that industry, they have had an extremely positive indirect effect. For example, the livestock industry would likely have been completely different without inexpensive feed grains made possible by Triangle programs. Essentially free markets have resulted in Cycles of Abundance in cattle and hogs that whipsaw producers periodically and that require the continuous, inefficient reallocation of resources. Further, more or less free market policies have given us a situation in which broiler meat is unbelievably inexpensive, but one in which profit margins for producers are increasingly dismal. However, even with Triangle policies, there were dramatic year-to-year changes in acres planted, acres harvested, yields, and prices for most of the major crops. Likely, year-to-year variations have been greater for Triangle-controlled crops than for those essentially free of control.

Providing an adequate supply of food and fiber for an ever-increasing population is a daunting problem for any country even under the best of conditions. While the agriculture that would have developed in the United States under free market conditions likely would have been quite messy, in the long term it would have been less costly to taxpayers. Further, without question, the ever-increasing costs associated with Triangle efforts to manage it will be unstainable, and efforts to reduce costs will ultimately lead ever closer to collectivisation, which the Soviets under Stalin showed also to be unsustainable.

After reviewing a half-century of Triangle activities, Dobson (1985) concluded:

> . . . lawmakers following natural inclinations to acquire influence and obtain re-election have strong incentives to perpetuate government programs that aid constituents. Bureaucrats administering the programs also have incentives to see that such programs are continued. On balance, the above comments suggest that USDA farm programs will remain highly resistant to change.

Given these realities, it is likely that the 75-year move toward more Saint-Simon, Blanc, and company involvement in agriculture will continue in the future. The ever-increasing effort by the Triangle to provide a larger, stronger, more permanent safety net for farmers is simply another step in this sequence of events. In this respect, the Triangle seems to be much like Hydra, the mythical Greek god-monster. Hydra had many writhing serpent heads attached to its body. When one was cut away, it was replaced by two.

Dobson, however, did not consider one important characteristic of the future of Triangle-managed agriculture—the unrelenting increase in the costs of management efforts. It is likely that they will finally become so expensive that the country simply cannot afford them. Saint-Simon, Blanc, and company never considered the nature of the end game of their politics. They never seemed to have any concern with counting the costs. The total FY 2002 USDA budget totaled around $63.4 billion, or $220 per capita. The FY 2003 budget was about $11 billion larger, or $256 per capita. The rate of increase in per capita outlays was 16%. The Triangle with its intertwined food production, food safety, food aid, and environmental policies has no mechanism for saying "no." True, it is impossible to remove all Hydra's serpent heads, but it is likely that some of them could be starved to death. It is fascinating to contemplate which of the heads would die first if all were deprived of ever-increasing quantities of taxpayer nourishment. After all, starvation finally ended Stalin's collectivization.

Darwinian evolution generally holds that unless animal or plant species evolve as their environments change they will finally disappear, although threatened genomes can be protected for a time by slowing environmental changes. This is the philosophy underlying the Endangered Species Act. If this policy fails, as it ultimately will, endangered species can be kept in special facilities, such as zoos. Ultimately, this effort will also fail, because over time the protected genomes will respond to the selection pressure of their new environment—the zoo. The organisms may be kept alive for some years for the public to enjoy, but over time the wild genomes will disappear.

For some 10,000 to 12,000 years, the genome of wild agriculture slowly evolved through fits and starts with its changing social, political, intellectual, climatic, and physiographic environments, to arrive at the beginning of the twentieth century relatively hale and hearty. However, in 1933, it was decided that the American agricultural genome was in great danger and that it should be permanently placed in a "zoo" to protect it. Quickly, its wildness began to disappear, and at the beginning of the twenty-first century it is increasingly dependent on its keepers—the Triangle and teh American taxpayer. In the meantime, for better or worse, its genome has changed so much that it can never be returned to its wild state, but this is exactly what Saint-Simon, Blanc, and company wanted.

In the final analysis, what Alfie had at the end of his life was what he had, and there was little that he could do about it. His misdeeds and their effects on the lives of others simply joined the swiftly flowing maelstrom of human experience.

Epilogue

I am well aware that it is long past time for a final period on what I have written. However, I request author's prerogative to make a few additional comments to bring my long personal voyage to closure.

Working on this project for lo these many years has left me with a mixture of joy, sadness, and some uncertainty. It has been a source of joy to be reminded of some of the many pleasures that a youngster could experience growing up on and around yeoman farms in south Alabama in the 40s—the pleasure of just being associated with a real old mule—watching and listening to a small flock of hens exchanging gossip while scratching and pecking amid dead leaves in the chicken yard—watching a litter of recently-born, pink piglets experience the miracle of their mother's milk for the first time—inhaling the sweet fragrance of bundles of recently pulled and dried corn fodder, neatly stacked in the barn—literally swimming in the fragrance of fresh sawdust resulting from cutting firewood with a crosscut saw from a recently felled pine log, and mixed with that of a small bunch of green pine needles saturated with kerosene that had been forced into the neck of a Royal Crown Cola bottle full of the liquid and used to remove accumulated turpentine from the saw blade from time to time.

And then working on the project has also saddened me— saddened that descendants of those fiercely independent pioneers who established thousands of small farms throughout the state, who still farm, must literally wait hat in hand while someone far away decides whether their dawn to dusk labor will result in profit or loss or in leaving farming forever.

I am also saddened that my wonderful state with its brilliant white beaches, Dougherty Plain sinkholes, Black Prairie, towering Talladega Mountains, massive sandstone ridges, limestone valleys, level Decatur Soils, and with its nodding sea oats, yellow jessamines, redbuds, Cahaba lilies, evening primroses, Joe-Pye weeds , and pure stands of majestic longleaf pines has so freely given of its amazing storehouse of natural resources for hundreds of years—only to limited avail.

Finally, I am intrigued, but uncertain, concerning the role that the genomes of my beloved Highland Scots (my maternal grandmother was a McClain), when removed to Alabama, has played in the unfolding of the panoply of events that I have described. For generations they had lived by their firm belief in natural liberty; been content to literally live in holes in the ground; regularly "reived" cattle from the Borderlands, but steadfastly refusing to store feed for their winter keep; and often wounded the cattle so as to collect blood to be mixed with oatmeal for making haggis. Then, in 1745, they willingly and even joyously followed Bonnie Prince Charlie to Culloden and oblivion.

Bibliography

Abernethy, Thomas Perkins. 1990. *The Formative Period in Alabama, 1815-1828* (with introduction by David T. Morgan). University of Alabama Press. Tuscaloosa, AL.

Ahearn, Mary. 2008. *Historic Data on Farm Household Income. USDA/ERS Briefing Room: Farm Household Economics and Well-Being.* Economic Research Service. US Department of Agriculture. Washington.

Akin, Wallace E. 1991. *Global Patterns: Climate, Vegetation and Soils.* University of Oklahoma Press. Norman, OK.

Alabama Agricultural Statistics Service. 1983. *Crop Weather.* May 31.

Alabama Agricultural Statistics Service. 1987. *Crop Weather.* April 6.

Alabama Agricultural Statistics Service. 1989. *Crop Weather.* June 12.

Alabama Agricultural Statistics service. 1989. *Crop Weather.* June 19.

Alabama Agricultural Statistics Service. 1989. *Crop Weather.* June 26.

Alabama Agricultural Statistics Service. 1989. *Crop Weather.* July 10.

Alabama Agricultural Statistics Service. 1989. *Crop Weather.* July 17.

Alabama Agricultural Statistics Service. 1990. *Crop Weather.* March 26.

Alabama Agricultural Statistics Service. 1990. *Crop Weather.* September 4.

Alabama Agricultural Statistics Service. 1991. *Crop Weather.* May 6.

Alabama Forestry Commission. 2003. *Forest Resource Report, 2002.* Montgomery, AL.

Aldrich, Samuel R., Walter O. Scott and Robert G. Hoeft. 1986. *Modern Corn Production.* A and L Publications. Champaign, IL.

Ali, Mir and Gary Vocke. 2002. "How Wheat Production Costs Vary. " *Wheat Yearbook.* Economic Research Service. US Department of Agriculture. Washington.

Allen, E. J. 1978. "Plant Density." Pgs 276-326 in P. M. Harris, editor. *The Potato Crop: The Scientific Basis for Improvement.* Chapman and Hall. London.

Animal Use Issues Committee. 2002. *Economic Importance of Hunting in America.* International Association of Fish and Wildlife Agencies. Washington.

Atkins, Leah Rawls. 1994. "From Early Times to the End of the Civil War." Pgs 1-222 in William Warren Rogers, Robert David Ward, Leah Rawls Atkins and Wayne Flynt. *Alabama: The History of a Deep South State.* University of Alabama Press. Tuscaloosa, AL.

Atwater, W. O. 1889. "Report of the Director of the Office of Experiment Stations." Pgs 485-544 in J. M. Rusk, 1889, *Report of the Secretary of Agriculture.* US Department of Agriculture. Washington.

Azzam, Azzeddine. 2002. *The Effect of Concentration in the Food Processing Industry on Food Prices. Cornhusker Economics.* Cooperative Extension. Institute of Agriculture & Natural Resources. Department of Agricultural Economics. University of Nebraska. Lincoln, NE.

Bailyn, Bernard and Philip D. Morgan. 1991. "Introduction." Pgs 1-31 in Bernard Bailyn and Philip D. Morgan, editors. *Strangers within the Realm: Cultural Margins of the British Empire.* University of North Carolina Press. Chapel Hill, NC.

Ball, Eldon. 2011. *Agricultural Productivity in the United States.* Economic Research Service. US Department of Agriculture. Washington.

Barnard, Charles. 2002. *Agricultural Land Values. USDA/ERS Briefing Room: Land Use, Value and Management. Economic Research Service.* US Department of Agriculture. Washington.

Barnard, Charles, Richard Nehring, James Ryan, Robert Collender and Bill Quimby. 2001. "Higher Cropland Value from Farm Program Payments: Who Gains?" *Agricultural Outlook.* November: 26-30. Economic Research Service. US Department of Agriculture. Washington.

Barr, Jewel T. and S. Radley Edwards. 1989. *Alabama Agricultural Statistics. Bulletin 31.* Alabama Agricultural Statistics Service. Alabama Department of Agriculture and Industries. Montgomery, AL.

Barr, Jewel T. and S. Radley Edwards. 1990. *Alabama Agricultural Statistics. Bulletin 32.* Alabama Agricultural Statistics Service. Alabama Department of Agriculture and Industries. Montgomery, AL.

Becker, Geoffrey S., editor. 1999. *Farm Commodity Legislation: Chronology, 1933-98.* Congressional Research Service. Washington.

Benson, Ezra Taft. 1957. "Foreword." Pgs v and vi in Alfred Stefferud, editor. *The Yearbook of Agriculture, 1957.* US Department of Agriculture. Washington.

Bergamini, David. 1969. *The Universe.* Time-Life Books. New York.

Bergerson, Paul H., Stephen V. Ash and Jeanette Keith. 1999. *Tennesseans and Their History.* University of Tennessee Press. Knoxville, TN

Berry, Wendell. 1977. *The Unsettling of America.* Sierra Club. San Francisco, CA.

Bidwell, Percy Wells and John I. Falconer. 1941. *History of Agriculture in the Northern United States, 1620-1860.* Peter Smith. New York.

Birdwell, Bobby T. 1987. "Some of the Best Farmland Is Here." Pgs 91-95 in William Whyte, editor. *Yearbook of Agriculture, 1987.* US Department of Agriculture. Washington.

Blakely, B. D., J. J. Coyle and J. G. Steele. 1957. Erosion on Cultivated Land. Pgs 290-307 in Alfred Stefferud, editor. *Yearbook of Agriculture, 1957.* US Department of Agriculture. Washington.

Blayney, Don P. 2002. *The Changing Landscape of US Milk Production.* Statistical Bulletin No. (SB-978). Economic Research Service. US Department of Agriculture. Washington.

Blevins, Brooks. 1998. *Cattle in the Cotton Fields.* University of Alabama Press. Tuscaloosa, AL.

Bloom, Jonathan. 2007. *The Food Not Eaten.* November 19. Culinate. Portland, OR.

Bolt, Don. 1983. "White Water, Proud People." *National Geographic.* Vol. 163 (4):458-472).

Bolton, Mike. 2003. "State Gets Lots of Help from Hunters." *Birmingham News.* January 19.

Bond, Carl E. 1996. *Biology of Fishes.* Saunders College. Fort Worth, TX.

Boorstin, Daniel J. 1958. *The Americans: The Colonial Experience.* Vintage. New York.

Booth, Arthur John. 1871. "Saint-Simon and Saint-Simonism." In *The History of Socialism in France.* Longmans, Green, Reader and Dyer. London.

Boschung, H. T. 1957. *The Fishes of Mobile Bay and the Gulf Coast of Alabama.* Ph. D. Dissertation. University of Alabama. Tuscaloosa, AL.

Bowen, Charles D., Bobby C. Fox, David Lewis, E. H. McBride and H. C. Buckelew. 1979. *Soil Survey of Blount County, Alabama.* Soil Conservation Service. US Department of Agriculture. Washington.

Bowers, Clement Gray. 1960. *Rhodendrons and Azaleas: Their Origins, Cultivation and Development.* Macmillan. New York.

Bowers, Douglas E., Wayne E. Rasmussen and Gladys L. Baker. 1984. *History of Agricultural Price-Support and Adjustment Programs, 1933-84.* Agriculture Information Bulletin No. (AIB485). Economic Research Service. US Department of Agriculture. Washington.

Bradshaw, Michael and Ruth Weaver. 1993. *Physical Geography: An Introduction to Earth Environments.* Mosby. Philadelphia, PA.

Brison, Fred R. 1974. *Pecan Culture.* Capital Printing. Austin, TX.

Broderson, William D. 1991. *From the Surface Down: An Introduction to Soil Surveys for Agricultural Use.* Soil Conservation Service, US Department of Agriculture. Washington.

Brooks, Nora. 2001. *Characteristics and Production Costs of US Cotton Farms.* Statistical Bulletin No. (974-2). Economic Research Service. United States Department of Agriculture. Washington.

Brooks, Robert P. 1913. *History of Georgia.* Library of Congress. Washington.

Brown, Harry Bates and Jacob Osborne Ware. 1958. *Cotton.* McGraw-Hill. New York.

Brown, Steve, Holt Hardin, Bob Lisec, Hal Pepper, Jerry Pierce and Jamie Yeager. 2006. *Alabama Farm Analysis Association: Summary Report* (2001-2005 Data) (AGEC 0801-06). Alabama Cooperative Extension System. Auburn University. Auburn, AL.

Budiansky, Stephen. 1998. "Taking the Measure of El Niño." *US News and World Report.* Vol. 124 (9):7.

Burne, Jerome. 1989. *Chronicle of the World.* Prentice Hall. New York.

Burns, Edward M., Richard E. Lerner and Standish Meachem. 1988. *Western Civilizations: Their History and Their Culture.* WW Norton. New York.

Burton, W. G. 1989. *The Potato.* Longman Scientific and Technical. Essex, England.

Byers, H. G., Charles E. Kellogg, M. S. Anderson and James Thorp. 1938. "Formation of Soil." Pgs 948-978 in Gove Hambidge, editor. *Yearbook of Agriculture, 1938.* US Department of Agriculture. Washington.

Byers, Horace G., M. S. Anderson and Richard Bradfield. 1938. "General Chemistry of Soil." Pgs 911-928 in Gove Hambidge, editor. *Yearbook of Agriculture, 1938.* US Department of Agriculture. Washington.

Cadwallader, Chris L. 1981. "Poultry Review." Pgs 54-60 in Denis S. Findley and Hugh E. Bynum. *Alabama Agricultural Statistics.* Alabama Department of Agriculture and Industries. Montgomery, AL.

Caldwell, Billy E. 1973. "Preface." Pgs ix-x in B. E. Caldwell, editor. *Soybeans: Improvement, Production and Uses.* American Society of Agronomy. Madison, WI.

Cardozier, V. R. 1957. *Growing Cotton.* McGraw-Hill. New

York.

Carlson, John B. 1973. "Morphology." Pgs 17-96 in B. E. Caldwell, editor. *Soybeans: Improvement, Production and Uses*. American Society of Agronomy. Madison, WI.

Carns, H. R. and Jack R. Mauney. 1968. "Physiology of the Cotton Plant." Pgs 41-73 in Fred C. Elliot, Marvin Hoover and Walter K. Porter Jr, editors. *Advances in Production and Utilization of Quality Cotton: Principles and Practices.* Iowa State University Press. Ames, IA.

Carter, E. A. and V. G, Seaquist. 1984. *Extreme Weather History and Climate. Atlas for Alabama.* Strode. Huntsville, AL.

Carter, Jimmy. 2001. *An Hour Before Daylight: Memories of a Rural Boyhood.* Simon and Schuster (A Touchstone Book). New York.

Cash, A. James. 2002. "Where's the Beef? Small Farms Produce More Cattle." *Agricultural Outlook.* December: 21-24. Agricultural Research Service. US Department of Agriculture. Washington.

Cash, W. J. 1991. *The Mind of the South.* Vintage Books. New York.

Chermock, Ralph L. 1976. *Hurricanes and Tornadoes in Alabama.* Geological Survey of Alabama. Information Series 46. University of Alabama, Tuscaloosa, AL.

Childe, V. Gordon. 1987. *The Aryans: A Study of Indo-European Origins.* Dorset. New York.

Childe, V. Gordon. 1950. *Prehistoric Migrations in Europe*, A. Aschenhoug. Oslo. Norway.

Childe. V. Gordon. 1958. *The Prehistory of European Society.* City of Westminster. London.

Chronic, Halka. 1983. *Roadside Geology of Arizona.* Mountain Press. Missoula, MT.

Clark, Ross Carlton. 1970. *A Distributional Study of the Woody Plants of Alabama.* PhD dissertation of the University of North Carolina at Chapel Hill. Chapel Hill, NC.

Clauson, Annette. 2008. *Food Expenditures by Families and Individuals as a Share of Disposable Income. USDA/ERS Briefing Room: Food CPI, Prices and Expenditures.* Economic Research Service. US Department of Agriculture. Washington.

Clements, John. 1991. *Flying the Colors: Alabama Facts.* Clements Research II. Dallas, TX.

Clewell, Andre F. 1985. *Guide to the Vascular Plants of the Florida Panhandle.* University of Florida Presses. Gainesville, FL.

Cline, Wayne. 1999. *Alabama Railroads.* University of Alabama Press. Tuscaloosa, AL.

Cochrane, Willard W. 1993. *The Development of American Agriculture: A Historical Analysis.* University of Minnesota Press. Minneapolis, MN.

Coleman, Kenneth, general editor. 1991. *A History of Georgia.* University of Georgia Press. Athens, GA.

Coleman, N. T. and A. Melich. 1957. "The Chemistry of Soil pH." Pgs 72-79 in Alfred Stefferrud, editor. *The Yearbook of Agriculture, 1957.* US Department of Agriculture. Washington.

Coleridge, Samuel Taylor. 1798. "The Rime of the Ancient Marinere." In William Wordsworth and Samuel Taylor Coleridge. *Lyrical Ballads.* Joseph Cottle, Bristol, England.

Conniff, Michael L. and Thomas J. Davis. 1994. *Africans in the Americas.* St. Martins. New York.

Conte, Christopher and Albert R. Karr. 2001. "The US Economy: A Brief History." Chapter 3 in *An Outline of the American Economy.* Bureau of International Information Programs. US Department of State. Washington.

Cook, Edward, David Meko, David Stahle and Malcolm Cleaveland. 1999. *Reconstruction of Past Drought Across the Coterminous United States from a Network of Climatically Sensitive Tree-Ring Data.* Paleoclimatology Program of the NOAA Climate and Global Change Program. National Geophysical Center. Boulder, CO.

Coolbear, Peter. 1944. "Reproductive Biology and Development." Pgs 138-172 in J. Smart, editor. *The Groundnut Crop: A Scientific Basis for Improvement.* Chapman-Hall. New York.

Cooper, H. P., Oswald Schreiner and R. E. Brown. 1938. Soil Potassium in Relation to Soil Fertility. Pgs 397-405 in Gove Hambidge, editor. *Yearbook of Agriculture, 1938.* US Department of Agriculture. Washington.

Coulson, David P. and Linda Joyce. 1999. *United States, State Level, Population Estimates: Colonization to 1999.* Publication, RMRS-GTR-111WWW. Rocky Mountain Research Station. US Forest Service. US Department of Agriculture. Fort Collins, CO.

Cope J. T., B. F. Alvord and A. E. Drake. 1962. *Rainfall Distribution in Alabama.* Alabama Agricultural Experiment Station. Progress Report Series 84. Auburn, AL.

Cotton, James A. 1989. *Soil Survey of Covington County, Alabama.* Soil Conservation Service. US Department of Agriculture, Washington.

Coulter, E. Merton. 1933. *A Short History of Georgia.* University of North Carolina Press. Chapel Hill, NC.

Crews, Jerry R., W. Robert Goodman, James L. Novak and Max W. Runge. 2001. *2001 Budgets for Major Row Crops in Alabama.* Alabama Cooperative Extension System. Auburn University. Auburn, AL.

Cutter, Elisabeth G. 1978. "Structure and Development of the Potato." Pgs 70-152 in P. M. Harris, editor. *The Potato Crop: The Scientific Basis for Improvement.* Chapman and Hall. London.

D'Andrea, Nicholas. 1973. "Early White Population In-Migration." Pgs 22-23 in Neal G. Linebck and Charles Taylor. *Atlas of Alabama.* University of Alabama Press. Tuscaloosa, AL.

Darlington, Philip J., Jr. 1957. *Zoogeography: The Geographical Distribution of Animals.* John Wiley and Sons. New York.

Darnay, Arsen J., editor. 1992. *Economic Indicators Handbook* (Time Series, Conversions, Documentation). Gale Research. Detroit, MI.

Davidian, H. H. 1995. *The Rhodendron Species*, Volume IV, *Azaleas*. Timber Press. Portland, OR.
Davidson, Keay and A. R. Williams. 1996. "Under Our Skin." *National Geographic*. 189(12):100-111.
Davis, Chester C. 1940. "The Develop." Pgs 297-326 in Gove Hambidge and Marion Julia Drown, editors. *Farmers in a Changing World. The Yearbook of Agriculture, 1940*. US Department of Agriculture. Washington.
Davis, Donald D. and Norman D. Davis. 1975. *Guide and Key to Trees of Alabama*. Kendall/Hunt Publishing. Dubuque, IA.
Dean, Blanche E., Amy Mason, and Joab Thomas. 1973. *Wildflowers of Alabama and Adjoining States*. University of Alabama Press. Tuscaloosa, AL.
Dean, L. A. 1957. "Plant Nutrition and Soil Fertility." Pgs 80-84 in Alfred Stefferud, editor. *The Yearbook of Agriculture, 1957*. US Department of Agriculture. Washington.
Democratic Staff. 2004. *Economic Concentration and Structural Change in the Food and Agriculture Sector: Trends, Consequences and Policy Options*. Committee on Agriculture, Nutrition and Forestry of the US Senate. Washington.
Dismukes, Robert and Monte Vandeveer. 2001. "US Crop Insurance: Premiums, Subsidies and Participation." *Agricultural Outlook*. December: 21-24. Economic Research Service. US Department of Agriculture. Washington.
Dismukes, Robert and Ron Durst. 2006. *Whole-Farm Approaches to a Safety Net*. Economic Information Bulletin No. (EIB-15). Agricultural Research Service. US Department of Agriculture. Washington.
Dobson, W.D. 1985. "Will USDA Farm Programs Remain Resistant to Change?" *American Journal of Agricultural Economics*. 67(2): 331-334.
Doherty, Joseph C. 1962. "Rural America in Transition." Pgs 585-589 in Alfred Steffenrude, editor. *After A Hundred Years. The Yearbook of Agriculture,1962*. US Department of Agriculture. Washington.
Dohlman, Erik. 2002. "Peanut Consumption Rebounding Amidst Market Uncertainties." *Agricultural Outlook*. March: 2-5. Economic Research Service. US Department of Agriculture. Washington.
Dolt, Robert H., Jr. and Roger L Batten. 1988. *Evolution of the Earth*. McGraw-Hill. New York.
Donahue, Roy L., Raymond W. Miller and John C. Schickluna. 1983. *Soil: An Introduction to Soils and Plant Growth*. Prentice-Hall., Englewood Cliffs, NJ.
Doyle, C. B. 1941. "Climate and Cotton." Pgs 348-363 in Gove Hambidge, editor. *Climate and Man. 1941 Yearbook of Agriculture*. US Department of Agriculture. Washington.
Dozier, William Alfred Jr. 1965. *Growth and Development of the Stuart Pecan*. Thesis, Auburn University. Auburn, AL.
Drahovzal, James A. 1968. "Physiography of Alabama." Pgs 7-14 in Charles W. Copeland, editor, *Geology of the Coastal Plain: A Guidebook*. Circular 47. Geological Survey of Alabama. Tuscaloosa, AL.
Dungan, Lewis A. 1986. *Soil Survey of Monroe County, Alabama*. Soil Conservation Service. US Department of Agriculture. Washington.
Durham, Walter T. 1990. *Before Tennessee: The Southwest Territory, 1790-1796*. Overmountain Press. Johnson City, TN.
Durnant, Will. 1950. *The Age of Faith*. Simon and Schuster. New York.
Durst, Ron. 2007. *Federal Income and Social Security Taxes. USDA/ERS Briefing Room: Farm Household Economics and Well-Being*. Economic Research Service. US Department of Agriculture. Washington.
Economic Research Service. 2000. "Farmers and the Land." In *A History of American Agriculture*. Economic Research Service, US Department of Agriculture. Washington.
Edminster, T. W. 1957. "Drainage Problems and Methods." Pgs 378-385 in Alfred Stefferud, editor. *Yearbook of Agriculture, 1957*. US Department of Agriculture. Washington.
Edwards, Everett E. 1940. "American Agriculture: The First 300 Years." Pgs 171-276 in Gove Hambidge and Marion Julia Drown, editors. *Farmers in a Changing World. Yearbook of Agriculture, 1940*. US Department of Agriculture. Washington.
Effland, Anne B. W. 2000. "US Farm Policy: The First 200 Years." *Agricultural Outlook*. March: 21-25. Economic Research Service. US Department of Agriculture. Washington.
Eicher, Don L. and A. Lee McAlester. 1980. *History of the Earth*. Prentice-Hall. Eastwood Cliffs, NJ.
Eisenhower, Milton S., editor and Arthur P. Chew, associate editor. 1935. *Yearbook of Agriculture, 1935*. US Department of Agriculture. Washington.
Ekelund, Robert B, Jr. and Robert D. Tollison. 1988. *Economics*. Scott, Foresman. Boston, MA.
Elizak, Howard. 2008. *Price Spreads from Farmer to Consumer*. ERS Data Sets. Economic Research Service. US Department of Agriculture. Washington.
Emigh, Eugene D. 1941. "Supplementary Climate Notes for Alabama." Pgs 759 – 760 in Gove Hambidge, editor. *Yearbook of Agriculture, 1941*. US Department of Agriculture. Washington.
Emmerson, Frederic V. 1920. *Agricultural Geology*. John Wiley and Sons. New York.
Engle, Carole R. and Gregory N. Whitis. 2000. *Cost and Returns of Catfish Production in Watershed Ponds*. Publication FSA9084-3M-4-00N. Division of Agriculture. Cooperative Extension Service. University of Arkansas. Little Rock, AR.
Ensminger, M. E. and R. C. Perry. 1997. *Beef Cattle Science*. Prentice Hall. Upper Saddle River, NJ.
Ensminger, M. E. and Richard Parker. 1984. *Swine Science*. Interstate Publishers, Kansas City. MO.
Ernstes, David P., Joe L. Outlaw and Ronald D. Knutson. 1997. *Southern Representation in Congress and US Agricultural Legislation*. AFPC Policy Issues 97-3. Agricultural and Food Policy Center. Department of Agricultural Economics.

Texas A&M University. College Station, TX.

Fanning, Delvin Seymour and Mary Christine Balluff Fanning. 1989. *Soils: Morphology, Genesis and Classification.* John Wiley and Sons. New York.

Farijon, Aljos. 1984. *Pines: Drawings and Descriptions of the Genus Pinus.* E.J. Brill/Dr. W. Backhugs. Leiden.

Faulkner, Ronnie W. 2008. *Progressive Farmer.* North Carolina History Project. John Locke Foundation. Raleigh, NC.

Fein, Jay S. and Pamela L. Stephens. 1987. *Monsoons.* John Wiley and Sons. New York.

Fenneman, N. M. 1938. *Physiography of the Eastern United States.* McGraw-Hill. New York.

Fenton, Thomas E. and Gerald Miller. 1982. *Iowa Soils.* Iowa Physical Environment Series. Iowa State University, Cooperative Extension Service. Ames, IA.

Fergus, E. N., Carsie Hammonds and Hayden Rogers. 1944. *Southern Field Crops Management.* J. B. Lippincott. Chicago, IL.

Fernandez-Cornejo, Jorge. 2007. *Off-Farm Income, Technology Adoption, and Farm Economic Performance.* Economic Research Report No. (ERR-36). Economic Research Service. US Department of Agriculture. Washington.

Findley, Dennis S. 1982. *Alabama Crop Weather Bulletin 1972-81.* Crop and Livestock Reporting Service. Montgomery, AL.

Fischer, David Hackett. 1989. *Albion's Seed: Four British Folkways in America.* Oxford University Press. New York.

Fischer, David Hackett and James C. Kelly. 2000. *Bound Away: Virginia and the Westward Movement.* University of Virginia Press. Charlottesville, VA.

Fite, Gilbert C. 1984. *Cotton Fields No More: Southern Agriculture, 1865-1980.* University of Kentucky Press. Lexington, KY.

FitzPatrick, E. A. 1983. *Soils: Their Formation, Classification and Distribution.* Longman. London.

Flynt Wayne. 1994. "From the 1920s to the 1990s." Pgs 411-630 in William Warren Rogers, Robert David Ward, Leah Rawls Atkins and Wayne Flynt. *Alabama: The History of a Deep South State.* University of Alabama Press. Tuscaloosa, AL.

Flynt, Wayne. 1989. *Poor but Proud: Alabama's Poor Whites.* University of Alabama Press. Tuscaloosa, AL.

Folmsbee, Stanley J., Robert E. Corlew and Enoch L Mitchell. 1960. *History of Tennessee.* Lewis Historical Publishing. New York.

Ford, Lacy K. 1986. "Yeoman Farmers in the South Carolina Upcountry: Changing Production Patterns in the Late Antebellum Era." *Agricultural History.* Vol. 60(4): 17-37.

Foreman, Linda. F. 2001. *Characteristics and Production Costs of US Corn Farmers.* Statistical Bulletin 974. Economic Research Service. US Department of Agriculture. Washington.

Fowler, Stewart H. 1969. *Beef Production in the South.* Interstate Printers and Publishers. Danville, IL.

Fox, Bobby C. 1989. *Soil Survey of Conecuh County, Alabama.* Soil Conservation Service, US Department of Agriculture. Washington.

Galle, Fred C. 1987. *Azaleas.* Timber Press. Portland, OR.

Gallup, Jere R. 1980. *Climatic Features and Length of Growing Season in Alabama.* NOAA/National Weather Service and Alabama Agricultural Experiment Station. Bulletin 517. Auburn, AL.

Garrett, J. C., Thomas L. Stuart and Roy D. Bass. 1948. *Alabama Agricultural Statistics.* Alabama Cooperative Crop Reporting Service. Montgomery, AL.

Genung, A. B. 1940. "Agriculture in the World War Period." Pgs 277-296 in Gove Hambidge and Marion Julia Down, editors. *Farmers in a Changing World. Yearbook of Agriculture, 1940.* US Department of Agriculture. Washington.

Geological Survey of Alabama. 1996. *Alabama Waterways Guide.* Geological Survey of Alabama. Tuscaloosa, AL.

Getz, Rodger R. and Larry M. Owsley. 1979. *Precipitation Probabilities and Statistics for Alabama.* NOAA/National Weather Service Environmental Studies Center and Alabama Agricultural Experiment Station. Weather Series 18. Auburn, AL.

Giraud, Marcel (translated by Brian Pearce). 1958. *A History of Louisiana* (Volume II) *Years of Transition, 1715-1717.* Louisiana State University Press. Baton Rouge, LA.

Gladwell, Malcolm. 2008. *Outliers.* Little, Brown. New York.

Godfrey, Robert K. 1988. *Trees, Shrubs and Woody Vines of Northern Florida and Adjacent Georgia and Alabama.* University of Georgia Press. Athens, GA.

Godfrey, Robert K. and Jean W. Wooten. 1979. *Aquatic and Wetland Plants of Southeastern United States (Monocotyledons).* University of Georgia Press. Athens, GA.

Godfrey, Robert K. and Jean W. Wooten. 1981. *Aquatic and Wetland Plants of Southeastern United States.* University of Georgia Press. Athens, GA.

Goodman, Bob, Dale Monks, John Everest, Ron Smith and Max Runge. 2001A. *Enterprise Budget for the Production of Soybeans (Dryland) in North Alabama.* Alabama Cooperative Extension System. Auburn University. Auburn, AL.

Goodman, Bob, Dale Monks, Mike Patterson, Ron Smith and Max Runge. 2001. *Enterprise Budget for the Production of Conventional Tillage Cotton in North Alabama.* Alabama Cooperative Extension System. Auburn University. Auburn, AL.

Gras, Norman Scott Brien. 1946. *A History of Agriculture in Europe and America.* F. S. Crofts. New York.

Gray, Lewis Cecil. 1958. *History of Agriculture in the Southern States* (Volume II). Peter Smith. Gloucester, MA.

Gregory, Walton C., Ben W. Smith and John A. Yarbrough. 1951. "Morphology, Genetics and Breeding." Pgs 28-88 in National Fertilizer Association. *The Peanut: The Unpredictable Legume.* William Byrd Press. Richmond, VA.

Gregory, Walton C., M. Pfluge Gregory, Antonio Krapovicas, Ben W. Smith and John W. Yarbrough. 1973. "Structure and Genetic Resources of Peanuts." Pgs 47-133 in American Peanut Research. *Peanuts: Culture and Uses, A Symposium.*

Griffith, Lucile. 1972. *Alabama: A Documentary History to 1900.* University of Alabama Press. Tuscaloosa, AL.

Grizzle, J. M. and W. A. Rogers.1976. *Anatomy and Histology of the Channel Catfish.* Alabama Agricultural Experiment Station. Auburn, AL.

Hammons, R. O. 1994. "The Origin and History of the Groundnut." Pgs 24-42 in J. Smart, editor. *The Groundnut Crop: A Scientific Basis for Improvement.* Chapman-Hall. New York.

Handley, J. E. 1953. *Scottish Farming in the 18th Century.* Faber and Faber. London.

Hanrahan, Charles E. 2006. *Agricultural Export and Food Aid Programs.* Order Code IB98006. Congressional Research Service. Library of Congress. Washington.

Harlan, Jack R. 1992. "Indigenous African Agriculture." Pgs 59-70 in C. Wesley Cowan and Patty Jo Watson, editors. *The Origins of Agriculture: An International Perspective.* Smithsonian Institution Press. Washington.

Harlan, Jack R. 1995. *The Living Fields: Our Agricultural Heritage.* Cambridge University Press. Cambridge, England.

Harper, Roland M. 1928. *Economic Botany of Alabama* (Part 2). Monograph 9 of the Geological Survey of Alabama. Tuscaloosa, AL.

Harper, Roland M. 1943. *Forests of Alabama.* Monograph 10, Geological Survey of Alabama. Tuscaloosa, AL.

Harris, Dwight M., Jr. 1968. *Soil Survey of Houston County, Alabama.* Soil Conservation Service. US Department of Agriculture. Washington.

Harris, Macarthur C. 1989. *Soil Survey of Sumter County, Alabama.* Soil Conservation Service. US Department of Agriculture. Washington.

Hartmann, Hudson T., Dale E. Kester, Fred T. Davis, Jr. and Robert L. Geneveve. 1997. *Plant Propagation: Principles and Practices.* Prentice Hall. Upper Saddle River, NJ.

Hartsell, Andrew J. and Mark J. Brown. 2002. *Forest Statistics for Alabama, 2000.* Southern Research Station Resource Bulletin SRS - 67. US Forest Service. US Department of Agriculture. Washington.

Harvey, D. 1988. "Aquaculture: Meeting Fish and Seafood Demand." *National Food Review.* 2(4) 10-13.

Harwood, Joy, Richard Heifner, Janet Perry and Agapi Sumwaru. 1999. *Managing Risks in Farming: Concepts, Research and Analysis.* Agricultural Economics Report No. 774. Economic Research Service. US Department of Agriculture. Washington.

Hatcher, Robert D., Jr. 1989. "Tectonic Setting of the Southern Appalachians." Pgs 1-10 in Robert D. Hatcher, Jr. and William A. Thomas, leaders. *Southern Appalachian Windows: Comparison of Styles, Scales, Geometry and Detachment Levels of Thrust Faults in the Foreland and Internides of a Thrust-Dominated Orogen. Field Trip Guidebook T167.* American Geophysical Union. Washington.

Hawkes, J. G. 1978. "Biosystematics of the Potato." Pgs 15-69 in P. M. Harris, editor. *The Potato Crop: The Scientific Basis for Improvement.* Chapman and Hall. London.

Hefferman, William. 1999. *Consolidation in the Food and Agriculture System. Report to the Farmers Union.*

Henderson-Sellers, A., and P. J. Robinson. 1986. *Contemporary Climatology.* John Wiley and Sons, New York.

Hewett, Caspar J. M. 2008. "Henri de Saint-Simon: The Great Synthesist." *Workshop on Progress of the Human Mind: From Enlightenment to Postmodernism.* University of Northumbria. Newcastle, England. September 2008.

Hickman, Cleveland P., Jr., Larry S. Roberts and Frances M. Hickman. 1990. *Biology of Animals.* Times Mirror/Mosby College. St. Louis, MO.

Hickman, Glenn L. and Charles Owkes. 1980. *Soil Survey of Mobile County, Alabama.* Soil Conservation Service, US Department of Agriculture. Washington.

Hildreth, A. C., J. R. Magness and John W. Mitchell. 1941. "Effects of Climatic Factors on Growing Plants." Pgs 292-307 in Gove Hambidge, editor. *Yearbook of Agriculture, 1941.* US Department of Agriculture. Washington.

Hiller, Ilo. 1996. *The White-Tailed Deer.* Texas A & M University Press. College Station, TX.

Hilliard, Sam Bowers. 1984. *Atlas of Antebellum Southern Agriculture.* Louisiana State University Press. Baton Rouge, LA.

Holliman, D. C. 1963. *Vertebrates of Alabama.* PhD dissertation. University of Alabama. Tuscaloosa, AL.

Hooks, W. Gary. 1973. "Physiography and Topography." Pages 5-7 in Neal G. Lineback and Charles T. Traylor, editors. *Atlas of Alabama.* University of Alabama Press. Tuscaloosa, AL.

Hoover, Ethel D. 1960. *Prices in the 19th Century: Studies in Income and Wealth.* Vol. 24:212. National Bureau of Economic Research. Washington.

Hoppe, Robert A. and David Banker, 2006. *Structure and Finances of US Farms: 2005 Family Farm Report.* Economic Information Bulletin No. (EIB-12)). Economic Research Service. US Department of Agriculture. Washington.

Hoppe, Robert A., James Johnson, Janet E. Perry, Penni Korb, Judith E. Sommer, James T. Ryan, Robert C. Green, Rob Durst and James Monke. 2001. *Structural and Financial Characteristics of US Farms: 2001 Family Farm Report.* Agricultural Information Bulletin No. (AIB768). Economic Research Service. US Department of Agriculture, Washington.

Horton, Douglas E. 1987. *Potatoes: Production, Marketing and Programs in Developing Countries.* Westview. Boulder, CO.

Howard, H. W. 1978. The Production of New Varieties. Pgs 607-646 in P. M. Harris, editor. *The Potato Crop: The Scientific Basis for Improvement.* Chapman and Hall. London.

Howell, A. D. 1921. "Mammals of Alabama." *North American Fauna*. 45:1-88.

Hoyt, Robert S. and Stanley Chodorow. 1976. *Europe in the Middle Ages*. Harcourt Brace Jovanovich. New York.

Huang, Wen. 2004. "US Increasingly Imports Nitrogen and Potassium Fertilizer." *Amber Waves*. February:1-2. Economic Research Service. US Department of Agriculture. Washington.

Hudson, Charles. 1976. *The Southeastern Indians*. University of Tennessee Press. Knoxville, TN.

Hughes, Travis H. 1973. "Geology." Pgs 2-4 in Neal G. Lineback and Charles T. Traylor, editors. *Atlas of Alabama*. University of Alabama Press. Tuscaloosa, AL.

Humphrey, Harry B. 1941. "Climate and Plant Diseases." Pgs 499-502 in Gove Hambidge, editor. *Yearbook of Agriculture, 1941*. US Department of Agriculture. Washington.

Hurst, R. Douglas. 1994. *American Agriculture: A Brief History*. Iowa State University Press. Ames, IA.

Hyman, L. Louis. 1999. "Current Use Taxes in Alabama." *Alabama's Treasured Forests*. Spring: 12-14.

Hyslop, James A. 1941. Insects and the Weather. Pgs 503-507 in Gove Hambidge, editor. *Yearbook of Agriculture, 1941*. US Department of Agriculture. Washington.

Imhof, T. A. 1976. *Alabama Birds*. University of Alabama Press. Tuscaloosa, AL.

Jackson, Harvey H. III. 1995. *Rivers of History: Life on the Coosa, Tallapoosa, Cahaba and Alabama*. University of Alabama Press. Tuscaloosa, AL.

Janssen, Johannes (M. A Mitchell and A. M. Christie, translators). 1966. *History of the German Peoples at the Close of the Middle Ages*. Volume I. AMS Press. New York.

Jarrell, William G. 1998. *Country Profiles: Alabama Vital Events*. Volume IV. Center for Health Statistics. Alabama Department of Public Health. Montgomery, AL.

Jenkins, Merle T. 1941. "Influence of Climate and Weather on Corn." Pgs 308-320 in Gove Hambidge, editor. *Yearbook of Agriculture, 1941*. US Department of Agriculture. Washington.

Johnson, Bob. 1995. "Erin Picks Pecans Before Their Time." *Birmingham News*. 8:18(August 5): C10.

Johnson, Jerry L. 1992. "Woodland Management and Productivity." Pgs 50-51 in Robert W. Stevens. *Soil Survey of Walker County Alabama*. Soil Conservation Service, US Department of Agriculture. Washington.

Johnstone, Paul H. 1940. "Old Ideals Versus New Ideas in Farm Life." Pgs 111-170 in Gove Hambidge and Marion Julia Drown, editors. *Farmers in a Changing World. Yearbook of Agriculture, 1940*. US Department of Agriculture. Washington.

Jones, Philip W., F. Douglas Martin and Jerry D. Hardy, Jr. 1978. *Development of Fishes of the Mid-Atlantic Bight*. FWS/OBS-78/12. Fish and Wildlife Service, US Department of the Interior. Washington.

Kantor, Linda Scott, Kathryn Lipton, Alden Manchester and Victor Olivera. 1997. "Estimating and Addressing America's Food Losses." *Food Review* (January-April). Economic Research Service. US Department of Agriculture. Washington.

Keeton, William T. 1967. *Biological Science*. W. W. Norton. New York.

Kellog, Charles E. 1938. "Soil and Society." Pgs 863-886 in Gove Hambidge, editor. *Yearbook of Agriculture, 1938*. US Department of Agriculture. Washington.

Kellog, Charles E. 1957. "We Seek; We Learn." Pgs 1-11 in Alfred Stefferud, editor. *The Yearbook of Agriculture, 1957*. US Department of Agriculture. Washington.

Kephart, Horace. 1921. *Our Southern Highlands*. Macmillian. New York.

Kerr, Norwood Allen. 1985. *History of the Alabama Agricultural Experiment Station, 1883 to 1983*. Alabama Agricultural Experiment Station. Auburn University. Auburn, AL.

Kincer, J. B. 1941. "Climate and Weather Data of the United States." In Gove Hambridge, editor. *Climate and Man, Yearbook of Agriculture, 1941*. US Department of Agriculture. Washington.

King, Philip B. 1977. *The Evolution of North America*. Princeton University Press. Princeton, NJ.

Kishlansky, Mark, Patrick Geary and Patricia O'Brien. 2005. *Civilization in the West*. Longmans, New York.

Knowe, Steven Alan. 1980. *Comparison of Height-Over-Age Curves for Young Loblolly Pine (Pinus taeda L.) Plantations*. Thesis, Auburn University. Auburn, AL.

Lamb, H. H. 1982. *Culture, History and the Modern World*. Methven. London.

Lambert, David and the Diagram Group. 1988. *The Field Guide to Geology*. Facts on File. New York.

Langer, William L. 1952. *An Encyclopedia of World History*. Book World. Taipei, Taiwan.

Lawson, James L., Luther Fuller, Walter L. Randolph, T. C. Reid and B. P. Livingston. 1943. *Farm Production and Marketing in Alabama*. Extension Service Circular 241. Alabama Polytechnic Institute. Auburn, AL.

Le Comte, Douglas. 1993. "Massive Winter Storm Thrashes East." *Weekly Weather and Crop Bulletin*. 80(11): 4-5.

Lee, Jasper. 1991. *Commercial Catfish Farming*. Interstate. Danville, IL.

Lenski, Gerhard and Patrick Nolan. 1998. *Human Societies: An Introduction to Macrosociology*. McGraw-Hill College. New York.

Lerner, William. 1975. "Consumer Price Indexes, Series E135-166." In *Historical Statistics of the United States, Colonial Times to 1970* (Part I). Bureau of the Census. US Department of Commerce. Washington.

Lerner, William. 1975A. "Farm Population, Farms, Land in Farms, and Value of Farm Property and Real Estate: 1850 to 1970, Series K 1-16." In *Historical Statistics of the United*

States, Colonial Times to 1970 (Part I). Bureau of the Census. US Department of Commerce. Washington.

Lewis, C. F. and T. R. Richmond. 1968. "Cotton as a Crop." Pgs 1-21 in Fred C. Elliot, Marvin Hoover and Walter K. Porter Jr., editors. *Advances in Production and Utilization of Quality Cotton: Principles and Practices*. Iowa State University Press. Ames, IA.

Levin, Harold L. 1994. *The Earth Through Time*. Saunders College Publishing. New York.

Leyburn, James G. 1962. *The Scotch-Irish: A Social History*. University of North Carolina Press. Chapel Hill, NC.

Lincoln, Bruce. 1981. *Priests, Warriors, and Cattle: A Study in the Ecology of Religions*. University of California Press. Berkley, CA.

Little, John Buckner. 1971. *History of Butler County Alabama, From 1815 to 1865*. Kessinger. Whitefish, MT.

Liu, Ke Shun. 1997. *Soybeans: Chemistry, Technology and Utilization*. International Thomson. New York.

Locher, Julie and Donna L. Cox. 2004. "Clabber, Corn Pone and Cured Hog." *Alabama Heritage*. Fall 2004. University of Alabama and the Department of Archives and History. Tuscaloosa, AL.

Lockwood John G, 1985. *World Climate Systems*. Edward Arnold. London.

Long, Arthur, R. 1978. "The Climate of Alabama." Pgs 1-16 in James A. Ruffner, editor. *Climates of the States*. US National Oceanic and Atmospheric Adminstration.

Loomis, R. S. and D. J. Conner. 1992. *Crop Ecology: Production and Management in Agricultural Systems*. Cambridge University Press. Cambridge, MA.

Lotterman, Edward. 1999. "District Hog Prices Hit Historic Lows." *Fedgazette*. Federal Reserve Bank of Minnesota. Minneapolis, MN.

Lovell, Tom. 1989. *Nutrition and Feeding of Fish*. Van Nostrand Reinhold. New York.

Lydolph, Paul E. 1985. *The Climate of the Earth*. Rowman and Littlefield. Totowa, NJ.

Lynn, Evadna S. 2000. "Lumber Markets Under Siege?" Pgs 9-11 in *Market Newsletter*. Timber Mart-South. Fourth Quarter.

Madison, Milton and David Harvey. 1997. "US Egg Production on the Sunny Side in the 1990s." *Agricultural Outlook*. May:12-14. Agricultural Research Service. US Department of Agriculture. Washington.

Manaster, Jane. 1994. *The Pecan Tree*. University of Texas Press. Austin, TX.

Masek, James C. 1996. "It All Runs Downhill Through Mobile Bay." *Alabama Wildlife*. (Winter) 96:36.

Mathews, Kenneth H., Jr., William F. Hahn, Kenneth E. Nelson, Lawrence A. Duewer and Ronald A. Gustafson. 2001. *The US Beef Industry: Cattle Cycles, Price Spreads and Packer Concentration*. US Beef Industry/TB-1874. Economic Research Service. US Department of Agriculture. Washington.

Mauney, J. R. 1968. "Morphology of the Cotton Plant." Pgs 23-40 in Fred C. Elliot, Marvin Hoover and Walter K. Porter Jr., editors. *Advances in Production and Utilization of Quality Cotton: Principles and Practices*. Iowa State University Press. Ames, IA.

Mauseth, James D. 1995. *Botany: An Introduction to Plant Biology* Saunders College Publishing. Philadelphia, PA.

Mayberry, B. D. 1989. *The Role of Tuskegee University in the Growth and Development of the Negro Extension System, 1881-1990*. Tuskegee University Extension Program. Tuskegee, AL.

McNeill. 1973. *The Rise of the West: A History of the Human Community*. University of Chicago Press. Chicago. IL.

Mcnutt, Robert B. 1981. *Soil Survey of Lee County, Alabama*. Soil Conservation Service. US Department of Agriculture. Washington.

McWhiney, Grady. 1988. *Cracker Culture: Celtic Ways in the Old South*. University of Alabama Press. Tuscaloosa, AL.

McWilliams, William H. 1992. *Forest Resources of Alabama*. Southern Forest Experiment Station Resource Bulletin, SO 170. US Forest Service. US Department of Agriculture. Washington.

Mederski, H. J., D. L. Jeffers and D. B. Peters. 1973. "Water and Water Relations." Pgs 239-266 in B. E. Caldwell, editor. *Soybeans: Improvement, Production and Uses*. American Society of Agronomy. Madison, WI.

Mell, P. H. 1890. *Climatology of Alabama*. Bulletin No. 18 (New Series). Agricultural Experiment Station of the Agricultural and Mechanical College. Auburn, AL.

Mettee, Maurice F., Patrick E. O'Neil and J. Malcom Pierson. 1996. *Fishes of Alabama and the Mobile Basin*. Oxmoor House. Birmingham, AL.

Miller, Bobby J. 1986. "Soil Fertility Levels." Pgs 74-76 in Paul G. Martin, *Soil Survey of Avoyelles Parrish, Louisiana*. Soil Conservation Service. US Department of Agriculture. Washington.

Miller, James J. 2002. *Background. USDA/ERS Briefing Room: Dairy* (July 25). Economic Research Service. US Department of Agriculture. Washington.

Miller, James J. 2002. *Policy. USDA/ERS Briefing Room: Dairy* (August 5). Economic Research Service. US Department of Agriculture. Washington.

Mitchell, Charles C., Jr and John C. Meetze. 1994. *Soil Areas of Alabama*. Circular ANR-340. Alabama Cooperative Extension Service. Auburn, AL.

Mohr, Charles. 1901. *Plant Life of Alabama* (Monograph 5). Brown Printing. Montgomery, AL.

Monroe, Watson H. 1941. *Notes on Deposits of Selma and Ripley Age in Alabama*. Bulletin 48. Geological Survey of Alabama. Tuscaloosa, AL.

Moorby, J. 1978. "The Physiology of Growth and Tuber Yield." Pgs 153-194 in P. M. Harris, editor. *The Potato Crop: The Scientific Basis for Improvement*. Chapman and Hall. London.

Moore, Albert Burton. 1934. *History of Alabama*. Alabama Book Store. University of Alabama. Tuscaloosa, AL.

Moore, J. D. and M. W. Szabo. 1994. *Alabama's Water Resources*. Geological Survey of Alabama. Tuscaloosa, AL.

Moore, John Hebron. 1968. "Two Cotton Kingdoms." *Agricultural History*. Vol. 60(4): 1-16.

Moore, Harry L. 1988. *A Roadside Guide to the Geology of the Great Smoky Mountains National Park*. University of Tennessee Press. Knoxville, TN.

Moreng, R. E. and J. S. Avens. 1985. *Poultry Science and Production*. Reston Publishing. Reston, VA.

Morison, Samuel Eliot. 1965. *The Oxford History of the American People*. Oxford University Press. New York.

Morrison, Richard D. 1994. *History of Alabama A & M University, 1875-1992*. Golden Rule Printers. Huntsville, AL.

Mount, R. H. 1975. *The Reptiles and Amphibians of Alabama*. Alabama Agricultural Experiment Station. Auburn University. Auburn, AL.

Mount, Robert H. 1986. *Vertebrate Animals in Alabama Needing Special Attention*. Alabama Agricultural Experiment Station. Auburn University. Auburn, AL.

Muehlenfeld, Ken. 1999. "The Demand for Solid Wood vs. Viber." *Alabama's Treasured Forests*. Summer: 7.

Munro, John M. 1987. *Cotton*. Longman Scientific and Technical. Essex, England.

Murphy, Paul A. 1973. *Alabama Forests: Trends and Prospects*. Forest Service Resource Bulletin SO-42. US Forest Service. US Department of Agriculture. Washington.

Myers, David G. 1989. *Psychology*. Worth. New York.

Nason, Alvin. 1965. *Textbook of Modern Biology*. John Wiley and Sons. New York.

National Agricultural Statistical Service. 1997. *Agricultural Statistics*. US Department of Agriculture. Washington.

National Climate Data Center. 1972-1991. *Climatological Data, Alabama*. 78(1)-97(12).

National Cotton Council. 2007. *Cotton Production Costs and Returns*. National Cotton Council of America. Cordova, TN.

Neathery, Thornton L., Robert D. Bently and George C. Lines. 1979. Cr*yptoexplosive Structure near Wetumpka, Alabama*. Geological Society of America. 87:567-573.

Nelson, B. 1956. "Propagation of Channel Catfish in Arkansas." *Proceedings of the Tenth Annual Conference of the Southeastern Association of Game and Fish Commissioners*. 10:165-168.

Nesheim, Malden C., Richard E. Austic and Leslie E. Card. 1979. *Poultry Production*. Lea and Febiger, Philadelphia, PA.

Newlin, J. J., Edgar Anderson and Earl Bressman.1949. *Corn and Corn Growing*. John Wiley and Sons. New York.

Nix, Steve. 1994. "Alabama Severance Tax." *Alabama's Treasured Forests*. Summer: 12-15.

Novak, James, Paul Mask, John Everest, Kathy Flanders and Max Runge. 2001. *Enterprise Budget for the Production of Corn for Grain in North Alabama*. Alabama Cooperative Extension System. Auburn University. Auburn, AL.

Nuttonson, M. Y. 1955. *Wheat-Climate Relationships and the Use of Phenology in Ascertaining the Thermal and Photo-Thermal Requirements of Wheat*. American Institute of Crop Ecology. Washington.

Odum, Eugene P. 1983. *Basic Ecology*. Saunders College Publishing. Philadelphia, PA.

Odum, Howard W. 1936. *Southern Regions of the United States*. University of North Carolina Press. Chapel Hill, NC.

Olmstead, L. B. and W. O. Smith. 1938. "Water Relations of Soils." Pgs 897-910 in Gove Hambidge, editor. *Yearbook of Agriculture, 1938*. US Department of Agriculture. Washington.

Olsen, Sterling R. and Maurice Fried. 1957. "Soil Phosphorus and Fertility." Pgs 94-106 in Alfred Stefferud, editor. *Yearbook of Agriculture, 1957*. US Department of Agriculture. Washington.

Osborne, W. Edward, Michael W. Szabo, Charles W. Copeland, Jr. and Thornton L. Neathery. 1989. *Geologic Map of Alabama. Special Map 221*. Geological Survey of Alabama. Tuscaloosa, AL.

Osborne, W. Edward, Michael W. Szabo, Charles W. Copeland, Jr. and Thornton L. Neathery. 1992. *Geologic Map of Alabama. Special Map 232*. Geological Survey of Alabama. Tuscaloosa, AL.

Otto, John Solomon. 1994. *Southern Agriculture during the Civil War Era, 1860-1880*. Greenwood. Westport, CT.

Outlaw, J. L. and S. L. Klose. 2001. *Supply Management*. Issue Paper (01-1) of the Agriculture and Food Policy Center. Texas A&M University. College Station, TX.

Owsley, Frank Lawrence. 1982. *Plain Folks of the Old South*. Louisiana State University Press. Baton Rouge, LA.

Overton, Mark. 1996. *Agricultural Revolution in England: Transformation of the Agrarian Economy, 1500-1850*. Cambridge University Press, Cambridge, England.

Parry. M. C. 1978. *Climate Change and Agricultural Settlement*. Archon Books, the Shoe String Press. Hamden, CT.

Patterson, Nick. 1993. "State Is of African Descent." *Birmingham News-Birmingham Post Herald*. October 16:D1.

Pearman, J. L., F. C. Sedberry, V. E. Stricklin and R. W. Cole. 1993. "Water Resources Data: Alabama Water Year 1992." *US Geological Survey Water-Data Report AL-92-1*. US Geological Survey. Tuscaloosa, AL.

Pearson, R. W. and L. E. Ensminger. 1957. "Southeastern Uplands." Pgs 579-594 in Alfred Stefferud, editor. *Yearbook of Agriculture, 1957*. US Department of Agriculture. Washington

Pendergast, Lolo and Diane Freeman. 1979. "CD Issues Hurricane Loss Figures." *Mobile Register*. 166(164): C4.

Perry, Janet, David Banker and Robert Green. 1999. *Broiler Farms' Organization, Management, and Performance*. Agriculture Information Bulletin No. 748. Economic Research Service. US Department of Agriculture. Washington.

Pielke, Roger A. 1990. *The Hurricane*. Routeledge. London.

Pierre, W. H. 1938. "Phosphorus Deficiency and Soil Fertility." Pgs 377-396 in Gove Hambidge, editor. *Yearbook of Agriculture, 1938.* US Department of Agriculture. Washington.

Pierre, W. H. and F. F. Riecken. 1957. "The Midland Feed Region." Pgs 535-552 in Alfred Stefferud, editor. *Yearbook of Agriculture, 1957.* US Department of Agriculture. Washington.

Pond, W. G. and J. H. Maner. 1984. *Swine Production and Nutrition.* AVI Publishing. Westport, CT.

Ponting, Clive. 1991. *A Green History of the World.* Penguin. New York.

Prentice, Agnes. 2000. "US Sorghum Production Costs and Returns Per Planted Acre, 1995-1999." *Commodity Costs and Returns.* Economic Research Service. US Department of Agriculture. Washington.

Prentice, A. N. 1972. *Cotton: With Special Reference to Africa.* Longman. London.

Prevatt, J. Walter and Donald M. Ball. 2002. *Fescue for Hay: Estimated Annual Costs Following Recommended Management Practices.* Alabama Cooperative Extension System. Auburn University. Auburn, AL.

Prevatt, J. Walter and Kelley Miller. 2001. *2001 Alabama Beef Enterprise Budgets.* Alabama Cooperative Extension System. Auburn University. Auburn, AL.

Probst, A. H. and R. W. Judd. 1973. "Origins, US History and Development and World Distribution." Pgs 1-15 in B. E. Caldwell, editor. *Soybeans: Improvement, Production and Uses.* American Society of Agronomy. Madison, WI.

Putman, Rory. 1988. *The Natural History of Deer.* Cornell University Press. Ithaca, NY

Rao, V. Ramanatha and U.R. Murty. 1994. "Botany: Morphology and Anatomy." Pgs 43-95 in J. Smart, editor. *The Groundnut Crop: A Scientific Basis for Improvement.* Chapman-Hall. New York.

Raven, Peter H., Ray F. Evert and Susan E. Eichorn. 1992. *Biology of Plants.* Worth. New York.

Raymond, Dorothy E., W. Edward Osborne, Charles W. Copeland and Thomas L. Neathery. 1988. *Alabama Stratigraphy.* Circular 140. Geological Survey of Alabama. Tuscaloosa, AL.

Reeves, Williard J. 1979. *Soil Survey of Dallas County, Alabama.* Soil Conservation Service, US Department of Agriculture. Washington.

Rhoad, A. O. 1941. "Climate and Livestock Production." Pgs 508-516 in Gove Hambidge, editor. *Yearbook of Agriculture, 1941.* US Department of Agriculture. Washington.

Rice, T. D. and L. T. Alexander. 1938. "The Physical Nature of Soil." Pgs 887-896 in Gove Hambidge, editor. *Yearbook of Agriculture, 1938.* US Department of Agriculture. Washington.

Richards, Eric. 1991. "Scotland and the Atlantic Empire." Pgs 67-114 in Bernard Bailyn and Philip D. Morgan, editors. *Strangers within the Realm: Cultural Margins of the British Empire.* University of North Carolina Press. Chapel Hill, NC.

Richards, L. A. and S. J. Richards. 1957. "Soil Moisture." Pgs 49-60 in Alfred Stefferud, editor. *The Yearbook of Agriculture, 1957.* US Department of Agriculture. Washington.

Ricklefs, R. E. 2000. *The Economy of Nature.* Chiron. New York.

Riley, Benjamin Franklin. 1964. *History of Conecuh County.* Weekly Packet. Bluff Hill, ME.

Ritter, Michael. 2007. *The Physical Environment.* Department of Geography/Geology. University of Wisconsin, Stevens Point, WI

Robertson, Ian. 1987. *Sociology.* Worth. New York.

Rogers, William Warren and Robert David Ward. 1994. "1865 through 1920." Pgs 225-408 in William Warren Rogers, Robert David Ward, Leah Rawls Atkins and Wayne Flynt. *Alabama: The History of a Deep South State.* University of Alabama Press. Tuscaloosa, AL.

Romer, Alfred Sherwood. 1955. *The Vertebrate Body.* W. B. Saunders. Philadelphia, PA.

Rosengarten, Frederick Jr. 1984. *The Book of Edible Nuts.* Walker. New York.

Rudis, Victor A., James F. Rosson, Jr. and John F. Kelly. 1984. *Forest Resources of Alabama.* Southern Forest Experiment Station Resource Bulletin SO-98. US Forest Service. US Department of Agriculture. Washington.

Rue, Leonard Lee III. 1978. *The Deer of North America.* Crown. New York.

Sanoff, Alvin P. 1991. "Conservation: The Legacy of Our Genes." *US News and World Report.* 10(1)(January 14):53.

Sapp, Daniel C. and Jacques Emplaincourt. 1975. *Physiographic Regions of Alabama. Special Map 168.* Alabama Geological Society. Tuscaloosa, AL.

Sartwell, Alexander. 1995. *Publications of the Geological Survey of Alabama and State Oil and Gas Board.* Geological Survey of Alabama. Tuscaloosa, AL.

Sauer, Carl O. 1941. "Settlement of the Humid East." Pgs 157-166 in Gove Hambidge, editor. *Yearbook of Agriculture, 1941.* US Department of Agriculture. Washington.

Sawyer, John E. 2003. "Soil Calcium: Magnesium Ratios." University Extension. Iowa State University. *Integrated Crop Management.* IC-490 (5)(April 21, 2003):40-42.

Schreiner, Oswald and B.E. Brown. 1938. "Soil Nitrogen." Pgs 361-376 in Gove Hambidge, editor. *Yearbook of Agriculture, 1938.* US Department of Agriculture. Washington.

Schultz, Robert P. 1997. *Loblolly Pine: The Ecology and Culture of Loblolly Pine (Pinus taeda L.).* Forest Service Handbook 713. US Department of Agriculture. Washington.

Scofield, C. S. 1938. "Soil, Water Supply and Soil Solution in Irrigation Agriculture." Pgs 704-716 in Gove Hambidge, editor. *Yearbook of Agriculture, 1938.* US Department of Agriculture. Washington.

Scotese, Christofer R. 1994. *Continental Drift.* The Paleomap Project.

Scott, Bob. 1993. "What's Causing All the Floods?" *Storm.*

1(2): 22-24.
Scott, Walter O. and Samuel R. Aldrich. 1983. *Modern Soybean Production*. S and L Publications. Champaign, IL.
Seyfert, Carl K. and Leslie A. Sirkin. 1979. *Earth History and Plate Tectonics*. Harper and Row. New York.
Shantz, H. L. 1938. "Plants as Soil Indicators." Pgs 835-860 in Gove Hambidge, editor. *Yearbook of Agriculture, 1938*. US Department of Agriculture. Washington.
Sherard, Hoyt, 1977. *Soil Survey of Lauderdale County, Alabama*. Soil Conservation Service. US Department of Agriculture. Washington.
Short, Sara D. 2001. *Characteristics and Production Costs of US Cow-Calf Operations*. Statistical Bulletin 974-3. Economic Research Service. US Department of Agriculture. Washington.
Sikora, Frank. 1995. "Pesky Tomato Virus Forces Changes in Straight Mountain's Way of Life." *Birmingham News*. 108(67): B1.
Sikora, Frank. 1995A. "Heat Wave Fries State's Chickens Before Their Time." *Birmingham News*. 126(136, August 18): D1.
Simpson, Eugene H., Harry B. Strawn, Michael K. Eckman, John P. Blake and James O. Donald. 2000. *Entering the Contract Broiler Business*. AEC-PS 0001. Alabama Cooperative Extension System. Auburn University. Auburn, AL.
Singer, Michael J. and Donald N. Munns. 1987. *Soils: An Introduction*. Macmillian. New York.
Singh, A. K. and C. E. Simpson. 1994. "Biosystematics and Genetic Resources." Pgs 96-137 in J. Smart, editor. *The Groundnut Crop: A Scientific Basis for Improvement*. Chapman-Hall. New York.
Skog, Kenneth E., Peter J. Ince and Richard W. Haynes. 1998. "Wood Fiber Supply and Demand in the United States." Pgs 72-89 in *Proceedings of the Forest Products Study Group Workshop*. North American Forestry Commission. Mérida, Yucátan, México. June 23, 1998.
Small, John Kunkel. 1933. *Manual of the Southeastern Flora*. Science Press. Lancaster, PA.
Smith, Bruce D. 1995. *The Emergence of Agriculture*. Scientific American Library. New York.
Smith, H. Arlen and C. R. Taylor. 2000. *Alabama Agriculture and Agribusiness Factbook, 2000*. Department of Agricultural Economics and Rural Sociology. College of Agriculture. Auburn University. Auburn, AL.
Smith, Ora. 1977. *Potatoes: Production, Storing, Processing*. AVI Publishing. Westport, CT.
Smith, R. H. 2007. *History of the Boll Weevil in Alabama, 1910-2007*. Bulletin 670. Alabama Agricultural Experiment Station. Auburn University. Auburn, AL.
Smith, Robert Leo. 1977. *Elements of Ecology and Field Biology*. Harper and Row. New York.
Smitherman, R. O. and R. A. Dunham. "Genetics and Breeding." Pgs 283-321 in C. S. Tucker, editor. *Channel Catfish Culture*. Elsevier, NY.
Smith-Vaniz, W. F. 1968. *Freshwater Fishes of Alabama*. Alabama Agricultural Experiment Station. Auburn University. Auburn, AL.
Soil Conservation Service. 1990. *Baldwin County, Alabama: Soil Survey Tables*. US Department of Agriculture. Washington.
Soil Conservation Service. 1993. *State of Alabama Hydrologic Unit Map with Drainage Areas by Counties and Sub-Watersheds*. I-B. 21. US Department of Agriculture. Washington.
Soil Conservation Service. 1970. "Soils of the United States." Pgs 85-88 in US Department of the Interior. *The National Atlas of the United States of America*. Washington.
Soil Survey Staff. 1975. *Soil Taxonomy: A Basic System of Soil Classification for Making and Interpreting Soil Surveys*. Handbook 436. Soil Conservation Service. US Department of Agriculture. Washington.
Solbrig, Otto T. and Dorothy J. Solbrig. 1994. *So Shall You Reap*. Shearwater. Washington.
Southerland, Henry DeLeon, Jr., and Jerry Elijah Brown. 1989. *The Federal Road through Georgia, the Creek Nation and Alabama, 1806-1836*. University of Alabama Press. Tuscaloosa, AL.
Spalding, Phinizy. 1991. *Colonial Period*. University of Georgia Press. Athens, GA.
Spielvogel, Jackson, J. 1991. *Western Civilization*. West Publishing. New York.
Stauffer, J. M. 1960. *A History of State Forestry in Alabama*. Alabama Division of Forestry. Alabama Department of Conservation. Montgomery, AL.
Stavrianos, L. S. 1971. *Man's Past and Present: A Global History*. Prentice Hall. Upper Saddle River, NJ.
Stearns, Larry D. and Timothy A. Petry. 1996. *Hog Market Cycles*. EC-1101. North Dakota State Extension Service. North Dakota State University. Fargo, ND.
Stefferud, Alfred. 1957. "Preface." Pgs vii-viii in Alfred Stefferud, editor. *Yearbook of Agriculture, 1957*. US Department of Agriculture. Washington.
Stein, Philip L. and Bruce N. Rowe. 1996. *Physical Anthropology*. McGraw-Hill. New York.
Steltenpohl, Mark G. and W. Bruce Moore. 1988. *Metamorphism in the Alabama Piedmont*. Circular 138. Geological Survey of Alabama. Tuscaloosa, AL.
Sternitzke, Herbert S. *Alabama Forests*. Southern Forest Experiment Station. US Forest Service. US Department of Agriculture. Washington.
Strawn, Harry B., Eugene Simpson and Jerry Crews. 2000. *2000-01 Budgets for Alabama Broiler Production*. AEC Bud 6, 1 June 2000. Alabama Cooperative Extension System. Auburn University. Auburn, AL.
Stevens, Robert W. 1984. *Soil Survey of Shelby County, Alabama*. Soil Conservation Service. US Department of Agriculture. Washington.
Stevens, Robert W. 1992. *Soil Survey of Walker County, Alabama*. Soil Conservation Service. US Department of Agriculture.

Washington.
Storer, Tracey. 1943. *General Zoology*. McGraw-Hill. New York.
Stribling, Lee. 1989. *White-Tailed Deer Management*. Circular ANR-521. Alabama Cooperative Extension System. Auburn University. Auburn, AL.
Sullivan, Charles L. *Hurricanes on the Mississippi Gulf Coast*. Gulf Publishing. Biloxi, MS.
Swingle, H. S. 1954. "Experiments on Commercial Fish Production in Ponds." *Proceedings of the Annual Conference of the Southeastern Association of Game and Fish Commissioners*. 8:69-74.
Swingle, H. S. 1958. "Experiments on Growing Channel Catfish to Marketable Size in Ponds." *Proceedings of the Annual Conference of the Southeastern Association of Game and Fish Commissioners*.12:63-72.
Szabo, Michael W. and Charles W. Copeland, Jr. 1988. *Geologic Map of Alabama. Special Map 220*. Geological Survey of Alabama. Tuscaloosa, AL.
Taylor, C. Robert. 2002 (March). *Restoring Health to Contract Poultry Production*. Auburn University College of Agriculture. Auburn, AL.
Terrell, John Upton. 1971. *American Indian Almanac*. World Publishing. New York.
Thomas, Joab. 1973. "Vegetation." Pgs 15-17 in Neal G. Lineback and Charles T. Taylor, editors. *Atlas of Alabama*. University of Alabama Press. Tuscaloosa, AL.
Thompson, Murray. 1962. "The Search for Parity." Pgs 543-556 in Alfred Stefferud, editor. *After One Hundred Years. Yearbook of Agriculture, 1962*. US Department of Agriculture. Washington.
Thompson, Philip D. and Robert O'Brien. 1965. *Weather*. Time Inc. New York.
Tondera, Bonnie Perkins, Laura French Clark and Mildred M. Gibson. 1987. *Wild Flowers of North Alabama*. University of Alabama Press. Tuscaloosa, AL.
Trewartha, Glenn Thomas and Lyle H. Horn. 1980. *An Introduction to Climate*. McGraw-Hill, New York.
Trewartha, Glenn Thomas. 1981. *The Earth's Problem Climates*. University of Wisconsin Press, Madison, WI.
Trimble, Donald E. 1980. *The Geologic Story of the Great Plains*. Bulletin 1493. US Geological Survey. US Department of the Interior. Washington.
Trivelli, Carolina. 1997. "Agricultural Land Prices." *SD Dimensions*. Sustainable Development. Food and Agricultural Organization of the United Nations. Rome.
Truog, Emil. 1938. "Soil Acidity and Liming." Pgs 563-580 in Gove Hambidge, editor. *Yearbook of Agriculture, 1938*. US Department of Agriculture. Washington.
Tucker, Craig S., editor. 1985. *Channel Catfish Culture*. Elsevier. New York.
Tucker, Craig S. and Edwin H. Robinson. 1990. *Catfish Farming Handbook*. Van Nostrand Reinhold. New York.
Tucker, Craig S. and John Hargreaves. 2004. *Biology and Culture of Channel Catfish*. Elsevier. London.
Vanderberry, Herb L. 1998. *Alabama Agricultural Statistics*. Bulletin 39 (1996-1997). Alabama Agricultural Statistics Service. Alabama Department of Agriculture and Industries, Montgomery, AL.
Vanderberry, Herb L. and William T. Placke. 1995. *Alabama Agricultural Statistics*. Bulletin 38. Alabama Agricultural Statistics Service. Alabama Department of Agriculture and Industries. Montgomery, AL.
Vanderberry, Herb L. and William T. Placke. 2001. *Alabama Agricultural Statistics*. Bulletin 43. Alabama Agricultural Statistics Service. Alabama Department of Agriculture and Industries. Montgomery, AL.
Vanderberry, Herb L. and William T. Placke . 2002. *Alabama Agricultural Statistics*. Bulletin 44. Alabama Agricultural Statistics Service. Alabama Department of Agriculture and Industries. Montgomery, AL.
Vanderberry, Herb L. and William T. Placke. 2003. *Alabama Agricultural Statistics*. Bulletin 45. Alabama Agricultural Statistics Service. Alabama Department of Agriculture and Industries. Montgomery, AL.
Van Bath, B. H. 1963. *The Agrarian History of Western Europe: AD 500-1850*. Edward Arnold. London.
Von Hagen, Victor W. 1961. *Realm of the Incas*. New American Library. New York.
Wahlenberg, W. G. 1960. *Loblolly Pine*. School of Forestry. Duke University. Durham, NC.
Wallace, David Duncan. 1951. *South Carolina: Short History*. University of South Carolina Press. Columbia, SC.
Walls, Jerry G. 1975. *Fishes of the Northern Gulf of Mexico*. T.F.H. Publications. Neptune City, NJ.
Ward, Henry S., C. H. M. Van Bavel, J. T. Cope, Jr., L. M. Ware and Herman Bouwer. 1959. *Agricultural Drought*. Bulletin 316. Alabama Agricultural Experiment Station. Auburn, AL.
Wattenberg, Ben J., editor. 1977. *The Statistical History of the United States from Colonial Times to the Present*. Basic Books. NY.
Weichert, Charles K. 1965. *Anatomy of the Chordates*. McGraw-Hill. New York.
Wellborn, Thomas L. 1997. *Channel Catfish: Life History and Biology*. Southern Regional Aquaculture Center Publication 180. Alabama Cooperative Extension Service. Auburn University. Auburn, AL.
Wells, H. G. 1956. *The Outline of History* (Volume I). Doubleday. New York.
Wells, O. V. 1940. "Agriculture Today: An Appraisal of the Agricultural Problem." Pgs 395-397 in Gove Hambidge, editor. *Yearbook of Agriculture, 1940*. US Department of Agriculture. Washington.
Wenke, Robert J. 1990. *Patterns in Prehistory: Humankind's First Three Million Years*. Oxford University Press. New York.
Went, Frits W. 1963. *The Plants*. Time-Life Books. New York.

Wheeler, Philip R. 1953. *Forest Survey for Alabama*. Forest Survey Release 73. Southern Forest Experiment Station. New Orleans, LA.

Whitaker, R. H. 1975. *Communities and Ecosystems*. Macmillan. New York.

Wikard, Claude R. 1941. "Foreword." Pg vii in Gove Hambidge, editor. *Yearbook of Agriculture, 1941*. US Department of Agriculture. Washington.

Williams, Harold and Robert D. Hatcher, Jr. 1983. "Appalachian Suspect Terranes." Pgs 33-53 in Robert D Hatcher, Jr, Harold Williams and Isidore Zietz, editors. *Contributions to the Tectonics and Geophysics of Mountain Chains*. Geological Society of America. Memoir 158. Boulder, CO.

Williams, Jack. 1992. *The Weather Book*. Vintage. New York.

Woodroof, Jasper Guy. 1966. *Peanuts: Production, Processing, Products*. AVI Publishing. Westport, CT.

Woodroof, Jasper Guy and Naomi Chapman Woodroof. 1927. "Development of the Pecan Nut (Hicoria pecan) from Flower to Maturity." *Journal of Agricultural Research*. 34:1049-1063.

Wright, G. C. and R. C. Nageswara. 1944. "Groundnut Water Relations." Pgs 281-335 in J. Smart, editor. *The Groundnut Crop: A Scientific Basis for Improvement*. Chapman-Hall. New York.

Wright, Lin H. 1979. "Shy Giant of the South." *Alumnus* (Spring, 1979). Mississippi State University. Starkville, MS.

Wright, Louis B. 1947. *The Atlantic Frontier: Colonial American Civilization*. Cornell University Press. Ithaca, NY.

Young, Edwin, Paul Westcott, Anne Effland and Erik Dohlman. 2007. *Basics of US Agricultural Policy. USDA/ERS Briefing Room: Farm and Commodity Policy*. Economic Research Service. US Department of Agriculture. Washington.

Selective Index of Alabama Agriculture Topics

D

E

F

G

H

I

M

N

O

P

S

T

www.ingramcontent.com/pod-product-compliance
Lightning Source LLC
Chambersburg PA
CBHW061321190326
41458CB00011B/3852
9781603062039